Advances in Experimental Medicine and Biology

Volume 1165

Advances in Experimental Medicine and Biology presents multidisciplinary and dynamic findings in the broad fields of experimental medicine and biology. The wide variety in topics it presents offers readers multiple perspectives on a variety of disciplines including neuroscience, microbiology, immunology, biochemistry, biomedical engineering and cancer research.

Advances in Experimental Medicine and Biology has been publishing exceptional works in the field for over 30 years and is indexed in Medline, Scopus, EMBASE, BIOSIS, Biological Abstracts, CSA, Biological Sciences and Living Resources (ASFA-1), and Biological Sciences. The series also provides scientists with up to date information on emerging topics and techniques. 2018 Impact Factor: 2.126

Content published in this book series is peer reviewed.

More information about this series at http://www.springer.com/series/5584

Bi-Cheng Liu · Hui-Yao Lan · Lin-Li Lv
Editors

Renal Fibrosis: Mechanisms and Therapies

Springer

Editors
Bi-Cheng Liu
Zhongda Hospital, Southeast University
School of Medicine
Southeast University
Nanjing, Jiangsu, China

Hui-Yao Lan
Department of Medicine and Therapeutics,
Li Ka Shing Institute of Health Sciences
Chinese University of Hong Kong
Hong Kong, Hong Kong

Lin-Li Lv
Zhongda Hospital, Southeast University
School of Medicine
Southeast University
Nanjing, Jiangsu, China

ISSN 0065-2598 ISSN 2214-8019 (electronic)
Advances in Experimental Medicine and Biology
ISBN 978-981-13-8870-5 ISBN 978-981-13-8871-2 (eBook)
https://doi.org/10.1007/978-981-13-8871-2

This Springer imprint is published by the registered company Springer Nature Singapore Pte Ltd.
The registered company address is: 152 Beach Road, #21-01/04 Gateway East, Singapore 189721, Singapore

Preface

The high prevalence of chronic kidney disease (CKD) is a tremendous global burden. In developed countries, it is estimated that more than 10% of adults present with various degrees of CKD. Irrespective of the initial cause, renal fibrosis is the hallmark of most progressive CKD, which is characterized by the excessive accumulation of fibroblasts and extracellular matrix (ECM). This is accompanied by glomerulosclerosis, tubular atrophy, tubulointerstitial inflammation, and irreversible loss of parenchymal cells. These pathological changes cause progressive deterioration of kidney function, and ultimately lead to end-stage renal disease (ESRD). Unfortunately, around 1–2% of the CKD patients will eventually succumb to the need of renal replacement therapy. Therefore, strategies for slowing or even preventing CKD to ESRD progression are of utmost importance.

In light of this, mechanistic studies of kidney fibrosis have been the focus of intensive research, and it is generally accepted that the critical steps for renal fibrogenesis are as follows: (1) activation of inflammatory response and inflammatory cell infiltration after renal injury; (2) release of profibrotic factors, including cytokines, growth factors, and chemokines; (3) excessive accumulation of fibroblasts and ECM in the interstitial compartment, due to the imbalance in ECM synthesis and degradation; (4) phenotypic change and irreversible loss of parenchymal cells; and (5) reduction in renal microvasculature. Among the above processes, ECM accumulation is the most critical step, as it causes renal scarring that leads to irreversible renal injury both structurally and functionally.

Recently, there are many elegant studies shedding light on the pathogenesis of renal fibrosis, arranging from the role of intrinsic kidney cells to the infiltrating inflammatory cells. New findings also uncover new molecular mechanisms mediating renal fibrogenesis. The development of reliable non-invasive biomarkers has also provided a new potential for diagnosis of CKD, and recent and ongoing advances in basic science research have also provided the necessary platform for new drugs development and novel therapies for renal fibrosis, which may alter the unfortunate fate of CKD patients down the road.

Currently, there are very limited books that systemically reviewed the research progress of renal fibrosis. In view of this, we have invited a group of experts in this field to compile a book that aims to systemically introduce the state-of-the-art research on renal fibrosis. The possible mechanisms, biomarkers, and strategies for prevention and treatment of renal fibrosis are to be elaborated in detail. It is hoped that this book will help readers to have a comprehensive understanding of the renal fibrosis. Finally, we would like to express our gratitude to the financial support of the Key National R&D Project of China Science and Technology Ministry (2018YFC1314000) and the Key International Cooperation Program of National Natural Science Foundation (81720108007).

Nanjing, China

Bi-Cheng Liu
Chair Professor & Head of Institute of Nephrology

Hong Kong, China

Hui-Yao Lan
Choh-Ming Li Professor of Biomedical Sciences

Contents

Contributors

Changlong An Division of Nephrology, Department of Medicine, University of Connecticut Health Center, Farmington, CT, USA;
Renal Section, Veterans Affairs Connecticut Healthcare System, West Haven, CT, USA;
Selzman Institute for Kidney Health and Section of Nephrology, Department of Medicine, Baylor College of Medicine, Houston, TX, USA

Ruo-Chen Che Department of Nephrology, Children's Hospital of Nanjing Medical University, Nanjing, China

Min Chen Key Laboratory of Renal Disease, Renal Division, Department of Medicine, Peking University First Hospital, Institute of Nephrology, Ministry of Health of China, Peking University, Beijing, China

Pei-Pei Chen Institute of Nephrology, Zhong Da Hospital, Southeast University School of Medicine, Nanjing, China

Ping-Sheng Chen Department of Pathology and Pathophysiology, School of Medicine, Southeast University, Nanjing, China;
Institute of Nephrology, Zhong Da Hospital, School of Medicine, Southeast University, Nanjing, China

Su-Fang Chen Key Laboratory of Renal Disease, Renal Division, Department of Medicine, Peking University First Hospital, Institute of Nephrology, Ministry of Health of China, Peking University, Beijing, China

Steven D. Crowley Division of Nephrology, Department of Medicine, Duke University and Durham VA Medical Centers, Durham, NC, USA

Zheng Dong Department of Cellular Biology and Anatomy, Medical College of Georgia, Augusta University and Charlie Norwood VA Medical Center, Augusta, GA, USA

Xiao-Gang Du Department of Nephrology, The First Affiliated Hospital of Chongqing Medical University of Medical Science, Chongqing, China

Syeda-Safoorah Fatima Department of Nephrology, Children's Hospital of Nanjing Medical University, Nanjing, China

Li-Jie He Department of Nephrology, The First Affiliated Hospital of Air Force Medical University, Xi'an, Shaanxi, China

Ze-Bo Hu Institute of Nephrology, Zhong Da Hospital, Southeast University School of Medicine, Nanjing, China

Li Jia Selzman Institute for Kidney Health and Section of Nephrology, Department of Medicine, Baylor College of Medicine, Houston, TX, USA

Hui-Yao Lan Department of Medicine and Therapeutics, Li Ka Shing Institute of Health Sciences, and Lui Chi Woo Institute of Innovative Medicine, The Chinese University of Hong Kong, Shatin, Hong Kong SAR, China

Bing Li Department of Nephrology, 2nd Affiliated Hospital, Harbin Medical University, Harbin, China

Jian-Si Li Department of Nephrology, 2nd Affiliated Hospital, Harbin Medical University, Harbin, China

Liang Li Department of Pharmacology, School of Basic Medical Sciences, Shandong University, Jinan, China

Yi-Ping Li Department of Pathology and Pathophysiology, School of Medicine, Southeast University, Nanjing, China

Zuo-Lin Li Institute of Nephrology, Zhong Da Hospital, Southeast University School of Medicine, Nanjing, China

Xin-Ling Liang Division of Nephrology, Guangdong Provincial People's Hospital, Guangdong Academy of Medical Sciences, Guangzhou, China

Bi-Cheng Liu Institute of Nephrology, Zhong Da Hospital, Southeast University School of Medicine, Nanjing, China

Dan Liu Institute of Nephrology, Zhong Da Hospital, Southeast University School of Medicine, Nanjing, China

Feng Liu Department of Nephrology, Shanghai East Hospital, Tongji University School of Medicine, Shanghai, China

Hong Liu Institute of Nephrology, Zhong Da Hospital, Southeast University School of Medicine, Nanjing, China

Youhua Liu State Key Laboratory of Organ Failure Research, National Clinical Research Center of Kidney Disease, Division of Nephrology, Nanfang Hospital, Southern Medical University, Guangzhou, China;
Department of Pathology, University of Pittsburgh School of Medicine, Pittsburgh, PA, USA

Man J. Livingston Department of Cellular Biology and Anatomy, Medical College of Georgia, Augusta University and Charlie Norwood VA Medical Center, Augusta, GA, USA

Chen-Chen Lu Institute of Nephrology, Zhong Da Hospital, Southeast University School of Medicine, Nanjing, China

Jian Lu Institute of Nephrology, Zhong Da Hospital, Southeast University School of Medicine, Nanjing, China

Ji-Cheng Lv Renal Division, Department of Medicine, Peking University First Hospital, Beijing, China;
KeyLaboratory of Renal Disease, Ministry of Health of China, Beijing, China;
Key Laboratory of Chronic Kidney Disease Prevention and Treatment (Peking University), Ministry of Education of China, Beijing, China

Lin-Li Lv Institute of Nephrology, Zhong Da Hospital, Southeast University School of Medicine, Nanjing, China

Kun-Ling Ma Institute of Nephrology, Zhong Da Hospital, Southeast University School of Medicine, Nanjing, China

Tao-Tao Ma Anhui Province Key Laboratory of Major Autoimmune Diseases, Anhui Institute of Innovative Drugs, School of Pharmacy, Anhui Medical University, Hefei, Anhui, China

Thomas Shiu-Kwong Mak Department of Medicine and Therapeutics, Li Ka Shing Institute of Health Sciences, and Lui Chi Woo Institute of Innovative Medicine, The Chinese University of Hong Kong, Shatin, Hong Kong SAR, China

Chang-Lin Mei Division of Nephrology, Kidney Institute of PLA, Changzheng Hospital, Second Military Medical University, Shanghai, China

Xiao-Ming Meng School of Pharmacy, Anhui Medical University, Hefei, Anhui, China;
Anhui Province Key Laboratory of Major Autoimmune Diseases, Anhui Institute of Innovative Drugs, School of Pharmacy, Anhui Medical University, Hefei, Anhui, China

Hai-Feng Ni Institute of Nephrology, Zhong Da Hospital, School of Medicine, Southeast University, Nanjing, China

Ruochen Qi Department of Urology, Zhongshan Hospital, Zhangjiang Technology Institute, Fudan University, Shanghai, China;
Shanghai Key Laboratory of Organ Transplantation, Shanghai, China

Yang Qiu Department of Nephrology, Union Hospital, Tongji Medical College, Huazhong University of Science and Technology, Wuhan, China

Mohammed Mazheruddin Quadri Department of Nephrology, Children's Hospital of Nanjing Medical University, Nanjing, China

Xiong-Zhong Ruan Department of Nephrology, The First Affiliated Hospital of Chongqing Medical University of Medical Science, Chongqing, China

Aanchal Shrestha Department of Nephrology, Children's Hospital of Nanjing Medical University, Nanjing, China

Anni Song Department of Nephrology, Union Hospital, Tongji Medical College, Huazhong University of Science and Technology, Wuhan, China

Hua Su Department of Nephrology, Union Hospital, Tongji Medical College, Huazhong University of Science and Technology, Wuhan, China

Hai-Jian Sun Department of Pharmacology, Yong Loo Lin School of Medicine, National University of Singapore, Singapore, Singapore

Lin Sun Department of Nephrology, Second Xiangya Hospital, Central South University, Changsha, Hunan, China

Shi-Ren Sun Department of Nephrology, The First Affiliated Hospital of Air Force Medical University, Xi'an, Shaanxi, China

Roderick J. Tan Department of Medicine, University of Pittsburgh School of Medicine, Pittsburgh, PA, USA

Tao-Tao Tang Institute of Nephrology, Zhong Da Hospital, Southeast University School of Medicine, Nanjing, China

Wei Tang Department of Pathogenic Biology, Basic Medical Sciences, Shandong University, Jinan, China

Cheng Wan Department of Nephrology, Union Hospital, Tongji Medical College, Huazhong University of Science and Technology, Wuhan, China

Gui-Hua Wang Institute of Nephrology, Zhong Da Hospital, Southeast University School of Medicine, Nanjing, China

Yanlin Wang Division of Nephrology, Department of Medicine, University of Connecticut Health Center, Farmington, CT, USA;
Renal Section, Veterans Affairs Connecticut Healthcare System, West Haven, CT, USA;
Selzman Institute for Kidney Health and Section of Nephrology, Department of Medicine, Baylor College of Medicine, Houston, TX, USA

Jia Wen Division of Nephrology, Department of Medicine, University of Connecticut Health Center, Farmington, CT, USA;
Selzman Institute for Kidney Health and Section of Nephrology, Department of Medicine, Baylor College of Medicine, Houston, TX, USA

Yi Wen Division of Nephrology, Zhongda Hospital, Southeast University, Nanjing, Jiangsu, China;
Division of Nephrology, Department of Medicine, Duke University and Durham VA Medical Centers, Durham, NC, USA

Ying Xiao Department of Nephrology, Second Xiangya Hospital, Central South University, Changsha, Hunan, China

Wei Xiong Department of Nephrology, Union Hospital, Tongji Medical College, Huazhong University of Science and Technology, Wuhan, China

Gang Xu Department of Nephrology, Tongji Hospital Affiliated with Tongji Medical College, Huazhong University of Science and Technology, Wuhan, Hubei, China

Cheng Xue Division of Nephrology, Kidney Institute of PLA, Changzheng Hospital, Second Military Medical University, Shanghai, China

Bin Yang Nantong-Leicester Joint Institute of Kidney Science, Department of Nephrology, Affiliated Hospital of Nantong University, Renal Group, Basic Medical Research Centre, Medical College of Nantong University, Nantong, Jiangsu, China;
Department of Cardiovascular Sciences, University of Leicester, University Hospitals of Leicester, Leicester, UK

Cheng Yang Department of Urology, Zhongshan Hospital, Zhangjiang Technology Institute, Fudan University, Shanghai, China;
Shanghai Key Laboratory of Organ Transplantation, Shanghai, China

Li Yang Renal Division, Department of Medicine, Peking University First Hospital, Beijing, China;
Key Laboratory of Renal Disease, Ministry of Health of China, Beijing, China;
Key Laboratory of Chronic Kidney Disease Prevention and Treatment (Peking University), Ministry of Education of China, Beijing, China

Zhen Yang Department of Nephrology, The First Affiliated Hospital of Air Force Medical University, Xi'an, Shaanxi, China

Fan Yi Department of Pharmacology, School of Basic Medical Sciences, Shandong University, Jinan, China

Qing Yin Institute of Nephrology, Zhong Da Hospital, Southeast University School of Medicine, Nanjing, China

Chen Yu Department of Nephrology, Tongji Hospital, Tongji University School of Medicine, Shanghai, China

Ying Yu Department of Nephrology, Tongji Hospital, Tongji University School of Medicine, Shanghai, China

Qian Yuan State Key Laboratory of Organ Failure Research, National Clinical Research Center of Kidney Disease, Division of Nephrology, Nanfang Hospital, Southern Medical University, Guangzhou, China

Ling-Feng Zeng Department of Nephrology, Second Xiangya Hospital, Central South University, Changsha, Hunan, China

Ai-Hua Zhang Department of Nephrology, Children's Hospital of Nanjing Medical University, Nanjing, China

Chun Zhang Department of Nephrology, Union Hospital, Tongji Medical College, Huazhong University of Science and Technology, Wuhan, China

Lu-Xia Zhang Renal Division, Department of Medicine, Peking University First Hospital, Beijing, China;
KeyLaboratory of Renal Disease, Ministry of Health of China, Beijing, China;
Key Laboratory of Chronic Kidney Disease Prevention and Treatment (Peking University), Ministry of Education of China, Beijing, China

Yang Zhang Institute of Nephrology, Zhong Da Hospital, Southeast University School of Medicine, Nanjing, China

Ying-Ying Zhang Department of Nephrology, Tongji Hospital, Tongji University School of Medicine, Shanghai, China

Jing-Hong Zhao Department of Nephrology, Xinqiao Hospital, Army Medical University, Chongqing, China

Xing-Chen Zhao Division of Nephrology, Guangdong Provincial People's Hospital, Guangdong Academy of Medical Sciences, Guangzhou, China

Le-Ting Zhou Institute of Nephrology, Zhong Da Hospital, Southeast University School of Medicine, Nanjing, China

Shougang Zhuang Department of Nephrology, Shanghai East Hospital, Tongji University School of Medicine, Shanghai, China;
Department of Medicine, Rhode Island Hospital, Alpert Medical School of Brown University, Providence, RI, USA

Part I
Renal Fibrosis in Chronic Kidney Disease

Chapter 1
Prevalence and Disease Burden of Chronic Kidney Disease

Ji-Cheng Lv and Lu-Xia Zhang

Abstract Chronic kidney disease (CKD) has been recognized as a leading public health problem worldwide. The global estimated prevalence of CKD is 13.4% (11.7–15.1%), and patients with end-stage kidney disease (ESKD) needing renal replacement therapy is estimated between 4.902 and 7.083 million. Through its effect on cardiovascular risk and ESKD, CKD directly affects the global burden of morbidity and mortality worldwide. The global increase in this disease is mainly driven by the increase in the prevalence of diabetes mellitus, hypertension, obesity, and aging. But in some regions, other causes such as infection, herbal and environmental toxins are still common. The large number of deaths for poor access to renal replacement therapy in developing countries, and also large increase of patients with ESKD in future, will produce substantial financial burden for even the most wealthy countries. Cost-effectiveness of preventive strategies to reduce the disease burden should be evaluated in relation to the local economic development and resource. Strategies reducing the cardiovascular risk in CKD still need further evaluation in large trials especially including patients with advanced kidney disease or end-stage kidney disease.

Keywords Chronic kidney disease · End-stage kidney disease · Prevalence · Cardiovascular disease

1.1 Introduction

Since the publication of the US Kidney Disease Outcomes Quality Initiative (KDQI) guidelines in 2002, chronic kidney disease (CKD) has been recognized as a leading

J.-C. Lv (✉) · L.-X. Zhang
Renal Division, Department of Medicine, Peking University First Hospital, Beijing, China
e-mail: jichenglv75@gmail.com

KeyLaboratory of Renal Disease, Ministry of Health of China, Beijing, China

Key Laboratory of Chronic Kidney Disease Prevention and Treatment (Peking University), Ministry of Education of China, Beijing, China

B.-C. Liu et al. (eds.), *Renal Fibrosis: Mechanisms and Therapies*,
Advances in Experimental Medicine and Biology 1165,
https://doi.org/10.1007/978-981-13-8871-2_1

public health problem that both increases the risk of end stage of kidney disease (ESKD) and cardiovascular disease as well as other complications (Bello et al. 2005; Go et al. 2004; Matsushita et al. 2010). The burden of chronic kidney disease is not restricted to its effect on clinically demands for renal replacement therapy; the disease has other major effects on the overall population. In the general population and peoples with diabetes or hypertension, clinical outcomes including mortality and cardiovascular events are strongly affected by the kidney involvement (Matsushita et al. 2010). Therefore, through its effect on cardiovascular risk and end-stage kidney disease, chronic kidney disease directly affects the global burden of mortality caused by cardiovascular disease, the most common cause of premature morbidity and mortality worldwide (GBD 2016 Causes of Death Collaborators 2017). Here, we summarized the disease burden, causes of chronic kidney disease outcomes and strategies for the disease preventions.

1.2 Epidemiology

1.2.1 Definition

CKD is defined as abnormalities of kidney structure or function, present for 3 months with implications for health, which was introduced by the US-based KDOQI in 2002 (National Kidney Foundation 2002). This definition was adopted internationally by the Kidney Disease: Improving Global Outcomes (KDIGO) in 2012 (Kidney Disease: Improving Global Outcomes (KDIGO) CKD Work Group 2013). According to the definition, markers of kidney damages include albuminuria, urine sediment abnormalities, electrolyte, or other abnormalities due to tubular disorders, histology abnormalities, and structure abnormalities by imaging. Decreased GFR is defined as GFR less than 60 ml/min per 1.73 m^2. In clinical practice, GFR is mainly estimated based on two equations, the Modification of Diet in Renal Disease (MDRD) study equation and the Chronic Kidney Disease Epidemiology Collaboration (CKD-EPI) equation (Levey et al. 2006, 2009; Stevens et al. 2011).

1.2.2 Prevalence of CKD

CKD is usually absent of symptoms until disease is advanced, and thus, precise calculation of the burden is difficult and accurate prevalence data are lacking (Jha et al. 2013). Observational studies estimating CKD prevalence in general populations from different parts of the world have been published (Coresh et al. 2007; Zhang et al. 2012). Based on the albuminuria and decreased estimated GFR, a national survey of the prevalence of CKD in USA rose from 10.0 to 13.1% between 1988–94 and 1999–2004 (Coresh et al. 2007). Using the similar method, a cross-section national survey in China showed the prevalence of CKD was 10.8% suggesting nearly 120 million adult having kidney disease in China (Zhang et al. 2012). When chronic kidney disease is defined solely by estimated GFR less than 60 ml/min per 1.73 m^2, approximate prevalence is 2.5–11.2% among the population across Europe, North America, Asia, and Australia (James et al. 2010). The prevalence of chronic kidney disease was 10.5–13.1% when defined by both the presence of albuminuria and decreased estimated GFR. A recent meta-analysis reviewed 100 studies including 6,908,440 participants (Hill et al. 2016). The global estimated prevalence of CKD stage 1–5 was 13.4% and stages 3–5 was 10.6%. CKD prevalence by stages was stage 1 (eGFR >90 ml/min per 1.73 m^2 and ACR >30 mg/g): 3.5% (95% CI: 2.8–4.2%); stage 2 (eGFR 60–89 + ACR >30): 3.9% (2.7–5.3%); stage 3 (eGFR 30–59): 7.6% (6.4–8.9%); stage 4 (eGFR 29–15): 0.4% (0.3–0.5%); and stage 5 (eGFR <15): 0.1% (0.1–0.1%). However, the reported prevalence of CKD varied widely amongst the studies. The prevalence of diabetes, hypertension and age of the participants were strongly associated with the CKD prevalence. The prevalence of CKD in developing country such as China (10.8%) (Zhang et al. 2012) was similar to that in developed countries to the USA (13.0%) (Coresh et al. 2007) and Norway (10.2%). However, the prevalence of stages 3 and 4 CKD in China was low compared with those in the developed countries. The prevalence of stage 3 CKD was 1.6% China, compared with 7.7% in the USA and 4.2% in Norway. One explanation might be that hypertension and diabetes have increased rapidly in the past 15–20 years in China, but for these diseases to affect chronic kidney disease at a population level might take another 10 years (Zhang et al. 2012). These findings will be observed in other developing countries and give us an opportunity of prevention of the disease progression to the end-stage kidney disease. Although CKD is common, the high prevalence reported in these studies was questioned that challenge the existence of a global epidemic of CKD. Concerns include using estimated GFR for identifying CKD, issues relating to the use of set GFR thresholds to define CKD in elderly populations and using of one-off testing for assessment of eGFR or albuminuria to define the prevalence of CKD in large-scale epidemiological studies. The pitfalls that exist in translating available epidemiological data hinder accurate assessment of the global burden of CKD (Glassock et al. 2017).

1.2.3 Prevalence of ESKD

Data on ESKD mainly reported by the country or region register systems. In a comprehensive systematic review, Liyanage et al. reported data on renal replacement therapy of ESKD from 123 countries or regions register system, representing 93% of the world population (Liyanage et al. 2015). The results showed there were 2.618 million people that received renal replacement therapy in 2010, 2.050 million received dialysis, and the remaining patients received renal transplant. The prevalence of renal replacement therapy varied widely across geographical regions, ranging from 80 per million people in Africa to 1840 per million people in North America, while the patients with end-stage kidney disease needing renal replacement therapy in 2010 were estimated between 4.902 and 7.083 million. This suggests that between 2.284 million (47%) to 7.083 million (73%) patients needing renal replacement therapy worldwide did not receive this therapy.

1.3 Changing Pattern in Spectrum of CKD

Diabetes and hypertension are the leading causes of chronic kidney disease in all developed countries. Glomerulonephritis and unknown causes are more common in countries of Asia and sub-Saharan Africa (Jha et al. 2013). These differences are related mainly to the burden of disease moving away from infections toward chronic lifestyle-related diseases, decreased birth rates, and increased life expectancy in developed countries. CKD is increasingly in developing countries due to the significant increase in non-infectious disease (particularly diabetes and hypertension). Furthermore, infectious diseases, such as HIV, schistosomiasis, and leishmaniasis, which also contribute to CKD, are highly prevalent in low-income to middle-income countries (George et al. 2017), while the spectrum of CKD and ESRD in middle-income countries is still undergoing transition. In the past three decades, China witnessed an economic boom and a considerable increase in metabolic diseases (Wang et al. 2018). Since the surge in the prevalence of hypertension and diabetes in the 1990s, metabolic diseases may eventually take over glomerulonephritis as the main cause of CKD and ESKD in China. Based on a national in-patient database of China, it was found that the proportion of hospital discharges from CKD due to diabetes mellitus had surpassed that from glomerulonephritis since 2010, and that the magnitude of this gap was more significant in urban cities than in rural areas (Zhang et al. 2016).

Another important reason for the CKD in Asia is the wide use of certain potentially toxic traditional herbal medicines (Wang et al. 2018; Debelle et al. 2008). The now well-known nephrotoxin, aristolochic acid, is contained in some commonly used traditional herbal products. According to the China National Survey of CKD, the prevalence of self-reported long-term intake of aristolochic acid-containing herbal medicine was 1.5% (Zhang et al. 2012). Balkan-endemic nephropathy affects people living along the tributaries of the Danube River and is characterized by chronic interstitial fibrosis with slow progression to end-stage kidney disease and urothelial malignant disease. In these regions, the exposure to aristolochic acid found in flour obtained from wheat contaminated with seeds of Aristolochia clematitis could be responsible for the so-called Balkan-endemic nephropathy. And, therefore, it is a deemed to be form of aristolochic acid nephropathy (Bamias and Boletis 2008).

Chronic kidney disease of unknown etiology (CKDu) is observed in several areas and among specific ethnic or occupational groups. Clusters of cases of chronic kidney disease of unknown origin have been reported in areas of Sri Lanka. The specific geographic distribution, preponderance among farming population, similar histology findings and absence of usual risk factors for kidney disease indicate undetected nephrotoxic agents playing a role in causation. Contamination of water, food, or both, by heavy metals, industrial chemicals, fertilizers, and pesticides has been suspected (Kumaresan and Seneviratne 2017). Mesoamerican nephropathy is a progressive, often fatal form of tubulointerstitial nephritis and chronic glomerular damage affecting young agricultural laborers in Central America (Wijkstrom et al. 2017). The cause of Mesoamerican nephropathy is not fully understood, but the leading hypothesis is occupational heat exposure with repeated episodes of volume and salt depletion.

1.4 Clinical Outcomes of CKD

Although much burden of chronic kidney disease is its effect on the demands of renal replacement therapy, it is estimated only −2% patients with CKD progress to ESKD. This suggests that most people with CKD will die before reaching the need for renal replacement therapy. Cardiovascular disease is reported as the leading cause of death in most studies of general population (Herzog et al. 2011). In the collaborative meta-analysis of Chronic Kidney Disease Prognosis Consortium, decreased estimated GFR and albuminuria were independently associated with all-cause death and cardiovascular death (Matsushita et al. 2010). In a large US cohort study including, Go et al. (2004) found an increasing risk for mortality and cardiovascular disease with lower levels of eGFR. In those with an eGFR <60 ml/min per 1.73 m^2, there were 25,621 deaths during follow-up, in contrast to 3171 individuals who began dialysis. In the Medicare population, the rate of death in individuals with CKD but without diabetes is >10 times greater than the risk for ESKD (Fried 2010). While in the African American Study of Kidney Disease and Hypertension (AASK) with a total of 11 years of follow-up, the rate of ESRD was 1.8 times greater than mortality (Alves et al. 2010). This suggests higher risk for ESRD in black individuals. Similar to the Chinese Cohort Study of Chronic Kidney Disease (C-STRIDE) (Gao et al. 2014), also more ESKD events observed than cardiovascular events (unpublished data). Data regarding the clinical outcomes in developing countries are scared especially for patients with glomerulonephritis. These studies also suggest the prognosis and timing of adverse clinical outcomes in CKD vary by patient characteristics including age, sex, race, eGFR, albumin-to-creatinine ratio, systolic blood pressure, smoking status, diabetes mellitus, and history of CVD. Risk models' estimates of risk and timing of these clinical outcomes were developed by the chronic renal insufficiency cohort (CRIC) study (Grams et al. 2017) and Chronic Kidney Disease Prognosis Consortium and Chronic Kidney Disease Epidemiology Collaboration (CKD-PC) (Grams et al. 2018). For example, the predicted 1-year probabilities for a hypothetical 60-year-old white woman with eGFR 30 ml/min/1.73 m^2, 1.8 g/d of proteinuria, and no diabetes or CVD, were 3.3, 4.1, and 0.3%, for first developing CVD, ESRD, and death, respectively. For a 40-year-old African–American man with similar characteristics but higher systolic blood pressure, the corresponding 1-year probabilities were 2.4, 13.2, and 0.1%. Thus, commonly measured clinical characteristics can predict the timing and occurrence of clinical outcomes in patients with CKD. Based on these models, effective measures should be considered to reduce the risk of clinical outcomes in the population with CKD.

1.5 Strategies in Facing the Challenge of CKD

1.5.1 Disease Burden in Future

Both the developing and developed countries face the challenge of an increasing burden of CKD. The aging population with the increase diabetes, obesity, and hypertension will steadily increase the background of susceptible population that will develop CKD and ESKD. Worldwide end-stage kidney disease burden will continue to increase in the developing as well as the developed countries (Liyanage et al. 2015). In developed countries such as USA, even the increase in ESRD incidence rates within age and race groups has leveled off and/or declined in recent years, the population changes in age and race distribution, obesity, and diabetes prevalence, and ESRD survival will result in a 11–18% increase in the crude incidence rate from 2015 to 2030. This incidence trend along with reductions in ESRD mortality will increase the number of patients with ESRD by 29–68% during the same period to between 971,000 and 1,259,000 in 2030 (McCullough et al. 2019). The projected number of people receiving renal replacement therapy will be more than double from 2.618 million people worldwide in 2010 to 5.439 million in 2030. On the other hand in developing countries, such as China, the transition from glomerulonephritis to metabolic disease-related CKD, as well as a large number of patients with early-stage CKD, implies an emerging source of subjects who could potentially progress to advanced CKD stages. These will impose an enormous pressure on the health system. The large absolute growth in the number of people receiving renal replacement therapy is projected for Asia, rising from 0.968 million people in 2010 to 2.162 million by 2030, in Africa, from 0.083 million to 0.236 million, and Latin America and the Caribbean, increasing from 0.373 million in 2010 to 0.903 million by 2030 (Liyanage et al. 2015).

1.5.2 Early Screening Program

Early detection of patients with chronic kidney disease and providing effective interventions would reduce risk of progression of kidney failure and major cardiovascular events (James et al. 2010). However, the role of general population-based screening remains controversial. Programs for early screening in patients with risk for CKD got some success. In Japan, glomerulonephritis like other Asian countries had ever been the leading cause of ESKD. The mandatory urinary analysis to early detect glomerulonephritis has been practiced in school children and all workers since the

1970s in Japan. The screening program has successfully detected milder forms of glomerulonephritis, early intervention and seems to have lowered the incidence of ESKD secondary to glomerulonephritis (Imai et al. 2007). Study in the US patients showed that microalbuminuria screening was cost-effective for patients with diabetes or hypertension, but not those without these conditions (Hoerger et al. 2010). Relative to no screening, targeted annual screening has cost-effectiveness ratios of $21,000/quality-adjusted life-year (QALY), $55,000/QALY, and $155,000/QALY for persons with diabetes, those with hypertension, and those with neither current diabetes nor current hypertension, respectively. Boulware et al. also reported that early detection of urine protein to slow progression of chronic kidney disease and decrease mortality was not cost-effective unless selectively directed toward high-risk groups (older persons and persons with hypertension) or conducted at an infrequent interval of 10 years (Boulware et al. 2003). Automated reporting of estimated GFR by laboratories worldwide boosts the number of referrals to nephrologists for early evaluation and therapy (Jain et al. 2009). However, whether this could improve the clinical outcome still remain investigation. Using the Canadian healthcare system, Manns and colleagues assessed the cost-effectiveness of using the estimated GFR for screening CKD (Manns et al. 2010). Findings suggest that it was not cost-effective overall or in subgroups of people with hypertension or older people and was cost-effective only in people with diabetes.

In the developing countries, such studies are still needed, since many patients are poorly access to therapy when progressing to ESKD needing renal replacement therapy. Screening for an early detection or treatment seems also feasible. In a program in Dharan, Nepal, 4471 high-risk individuals (with high blood pressure, high fasting glucose, proteinuria, and impaired renal function) were identified from 20,811 adult individuals by screening and refer to an early intervention (Sharma et al. 2014). After the 3-year follow-up, 63% of the individuals with proteinuria at baseline had normal values after an early and 48% of individuals with mild-to-moderate renal insufficiency at baseline had stable renal function. This study shows the feasibility of a screening and management program in a low-resource setting. But the cost-effectiveness of an intervention still needs further evaluation, and especially this depends on the local gross domestic product (GDP) of each region (George et al. 2017).

1.5.3 Strategies for Prevention CKD Progression

Clearly, diabetes, hypertension, and obesity will be the leading risk factors for CKD development and progression. The best long-term approach prevention for this process in the first place, perhaps through better obesity, diabetes, and hypertension prevention and management programs. Effective population-based approaches to prevention of CKD and ESKD including blood pressure control (Xie et al. 2016; Lv et al. 2013), and management of key risk factors, including diabetes and obesity—and acute kidney injury should be refined and tested. Large number studies have showed angiotensin-converting enzyme (ACE) inhibitors and angiotensin II receptor blockers (ARBs) in people with diabetic or non-diabetic kidney disease reduce the risk for kidney failure by 39 and 30%, respectively, and also risk of major cardiovascular events by 18 and 24%. ACE inhibitors also reduced the risk for all-cause mortality in this population (Xie et al. 2016). Both blood glucose control and renin–angiotensin system blockade could prevent new onset or worsening of albuminuria in patients with diabetes (Lv et al. 2012; Patel et al. 2008). A number of studies of sodium-glucose cotransporter 2 inhibitors among patients with diabetes have shown slowed progression to kidney disease end points, preventing albuminuria and/or substantial reduction in kidney function as measured by the eGFR (Neal et al. 2017; Wanner et al. 2016). Innovative models of preventive care should be piloted in low-income and middle-income countries, especially in areas where access to physicians is low. Evidence exists from places such as Chile, Taiwan, the UK, and Uruguay to suggest that multifaceted preventive strategies might stabilize or even reduce the incidence of people needing renal replacement therapy and lead to cost savings (Mazzuchi et al. 2006; Wei et al. 2010; Lin et al. 2013). These models should be implemented and evaluate their effect in clinical practice. The large number of deaths for poor access to renal replacement therapy in developing countries, and also large increase of patients with ESKD in future, will produce substantial financial burden for even the most wealthy countries. Clearly, cost-effective dialysis techniques should be developed.

1.5.4 Strategies for Reducing Cardiovascular Risk in CKD

Although cardiovascular events are the leading reason of death in patient with kidney disease, few trials have been undertaken specifically in populations with CKD. Until now, most evidence or strategies for cardiovascular risk intervention mainly comes from the subgroup analysis of the trial in general population, while most trials for cardiovascular risk in the general population have excluded patients with advanced kidney disease. This suggests we need studies to explore strategies reducing the risk of cardiovascular morbidity or mortality in patients with kidney disease. These strategies should include traditional (like hypertension, hyperlipidemia) and also non-traditional risk factors (like inflammatory factors, elevated plasma homocysteine, enhanced coagulability, anemia, increased arterial calcification) (Herzog et al. 2011; Muntner et al. 2004).

The Blood Pressure Lowering Treatment Trialists' Collaboration (BPLTTC) pooled 26 trials (152 290 participants), including 30,295 individuals with estimated GFR <60 mL/min/1.73 m^2 (Ninomiya et al. 2013). Blood pressure lowering effectively reduced the risk of cardiovascular events among people with moderately reduced estimated GFR. The relative benefits of blood pressure control were similar to those in patients without kidney disease. Blood pressure lowering agents also significantly reduce cardiovascular mortality in patient with end-stage kidney disease (Heerspink et al. 2009), while statin therapy did not reduce the cardiovascular events or death in patients receiving dialysis (Fellstrom et al. 2009; Wanner et al. 2005).

Meta-regression of the statin trials support statin benefits for cardiovascular events prevention were only observed in patients with mild or moderated decreased GFR and the relative or absolute benefits attenuated with the reduction of kidney function and not in the dialysis (Hou et al. 2013). Fibrate therapy also improved lipid profiles and prevent cardiovascular events in people with kidney disease. The risk of adverse events of statin or fibrate in patients with kidney disease does not increase as compared to those without CKD. In a post hoc subgroup analysis of Hypertension Optimal Treatment (HOT) study, aspirin therapy produces greater absolute reduction in major cardiovascular events and mortality in hypertensive patients with CKD, but with a non-significant greater risk of major bleeding events than with normal kidney function (Jardine et al. 2010). The increased risk of major bleeding appears to be outweighed by the substantial benefits. Among every 1000 persons with eGFR <45 ml/min/1.73 m^2 treated for 3.8 years, 76 major cardiovascular events and 54 all-cause deaths will be prevented while 27 excess major bleeds will occur. For the coronary heart disease with estimated GFR less than 60 ml/min/1.73 m^2, drug-eluting stents use in patients with eGFR <60 mL/min/1.73 m^2 is associated with a reduced rate of repeat revascularization and myocardial infarction without increased risk of stent thrombosis (Wang et al. 2013). Overall we still need large trials in CKD population especially including those with advanced kidney disease or ESKD to investigate these interventions.

Cardiovascular mortality and morbidity rates are highest in people with ESKD. Studied interventions include earlier initiation of dialysis, increased dialysis intensity driven by Kt/V targets or by hemofiltration volumes, the use of different dialysis membranes and hemodiafiltration. But randomized trials and systematic reviews have failed to consistently demonstrate any benefits for these interventions on cardiovascular outcomes (Jun et al. 2011). The FHN trial found the frequent daily dialysis (six sessions per week) improved the composite of increased LVM and death (HR for death or change in LVM 0.61; 95% CI 0.46–0.82) with a difference in adjusted mean change in LVM of −13.8 g (95% CI −21.8 to −5.8) (Chertow et al. 2010). Hyperphosphatemia is an independently associated risk factor of cardiovascular morbidity and mortality both in people with CKD. Analysis of the 11 randomized trials (4622 patients) showed that patients assigned to non-calcium-based binders had a 22% reduction in all-cause mortality compared with those assigned to calcium-based phosphate binders in chronic kidney disease (Jamal et al. 2013). In the China Stroke Primary Prevention Trial (CSPPT) study with a subgroup of 1671 patients with CKD, reducing homocysteine with folic acid and enalapril, compared with enalapril alone, significantly reduced the risk of cardiovascular death, myocardial infarction, and stroke by 20% (Huo et al. 2015). However, this was not observed in studies of other nations. Meta-analysis suggested that homocysteine lowering may be of benefit in populations without background supplementation or food fortification but not in populations with supplementation. Two small trials have reported a reduction in cardiovascular events from antioxidant therapy in ESKD patients (Tepel et al. 2003; Labarca 2000). These two trials (330 participants) suggest Vitamin E and acetylcysteine reduced cardiovascular events in patients with ESKD. However, the positive findings in these small studies are countered by the lack of benefit for the subgroup of 450 people with impaired kidney function in the HOPE-2 study (Steinberg and Witztum 2002). Counter to expectations, full correction of anemia with erythropoiesis-stimulating agents in CKD appears to lead to worse rather than better cardiovascular outcomes (Palmer et al. 2010).

In conclusion, most of the current evidence has been based on post hoc subgroup analyses of larger trials, predominantly testing pharmaceutical agents, which were generally not designed to specifically test benefits and harms in people with CKD and have few participants with ESKD. The trials conducted specifically in patients with CKD and ESKD are generally of relatively small size, restricting their capacity to determine effects in clinical outcomes. Based on this, patients with stage 3 CKD appear to derive similar or greater cardiovascular benefits from blood pressure lowering, lipid lowering, and aspirin-based antiplatelet therapy as the general population. Few other firm recommendations to alter current practice can be made on the basis of novel factors postulated to reduce cardiovascular events. Available strategies still need further evaluation in large trials especially including patients with advanced kidney disease or end-stage kidney disease.

Acknowledgements This study was supported by grants from the National Key Research and Development Program of China (2018YFC1314004).

References

Alves TP, Wang X, Wright JT Jr, Appel LJ, Greene T, Norris K et al (2010) Rate of ESRD exceeds mortality among African Americans with hypertensive nephrosclerosis. J Am Soc Nephrol 21:1361–1369

Bamias G, Boletis J (2008) Balkan nephropathy: evolution of our knowledge. Am J Kidney Dis 52:606–616

Bello AK, Nwankwo E, El Nahas AM (2005) Prevention of chronic kidney disease: a global challenge. Kidney Int Suppl 98:S11–S17

Boulware LE, Jaar BG, Tarver-Carr ME, Brancati FL, Powe NR (2003) Screening for proteinuria in us adults: a cost-effectiveness analysis. JAMA 290:3101–3114

Chertow GM, Levin NW, Beck GJ, Depner TA, Eggers PW, Gassman JJ et al (2010) In-center hemodialysis six times per week versus three times per week. N Engl J Med 363:2287–2300

Coresh J, Selvin E, Stevens LA, Manzi J, Kusek JW, Eggers P et al (2007) Prevalence of chronic kidney disease in the united states. JAMA 298:2038–2047

Debelle FD, Vanherweghem JL, Nortier JL (2008) Aristolochic acid nephropathy: a worldwide problem. Kidney Int 74:158–169

Fellstrom BC, Jardine AG, Schmieder RE, Holdaas H, Bannister K, Beutler J et al (2009) Rosuvastatin and cardiovascular events in patients undergoing hemodialysis. N Engl J Med 360:1395–1407

Fried LF (2010) Higher incidence of esrd than mortality in the AASK study. J Am Soc Nephrol 21:1244–1246

Gao B, Zhang L, Wang H, Zhao M (2014) Chinese cohort study of chronic kidney disease: design and methods. Chin Med J 127:2180–2185

GBD (2016) Causes of Death Collaborators (2017) Global, regional, and national age-sex specific mortality for 264 causes of death, 1980–2016: a systematic analysis for the global burden of disease study 2016. Lancet 390:1151–1210

George C, Mogueo A, Okpechi I, Echouffo-Tcheugui JB, Kengne AP (2017) Chronic kidney disease in low-income to middle-income countries: the case for increased screening. BMJ Glob Health 2:e000256

Glassock RJ, Warnock DG, Delanaye P (2017) The global burden of chronic kidney disease: estimates, variability and pitfalls. Nat Rev Nephrol 13:104–114

Go AS, Chertow GM, Fan D, McCulloch CE, Hsu CY (2004) Chronic kidney disease and the risks of death, cardiovascular events, and hospitalization. N Engl J Med 351:1296–1305

Grams ME, Sang Y, Ballew SH, Carrero JJ, Djurdjev O, Heerspink HJL et al (2018) Predicting timing of clinical outcomes in patients with chronic kidney disease and severely decreased glomerular filtration rate. Kidney Int 93:1442–1451

Grams ME, Yang W, Rebholz CM, Wang X, Porter AC, Inker LA et al (2017) Risks of adverse events in advanced CKD: the chronic renal insufficiency cohort (CRIC) study. Am J Kidney Dis 70:337–346

Heerspink HJ, Ninomiya T, Zoungas S, de Zeeuw D, Grobbee DE, Jardine MJ et al (2009) Effect of lowering blood pressure on cardiovascular events and mortality in patients on dialysis: a systematic review and meta-analysis of randomised controlled trials. Lancet 373:1009–1015

Herzog CA, Asinger RW, Berger AK, Charytan DM, Diez J, Hart RG et al (2011) Cardiovascular disease in chronic kidney disease. A clinical update from kidney disease: Improving global outcomes (KDIGO). Kidney Int 80:572–586

Hill NR, Fatoba ST, Oke JL, Hirst JA, O'Callaghan CA, Lasserson DS et al (2016) Global prevalence of chronic kidney disease—a systematic review and meta-analysis. PLoS ONE 11:e0158765

Hoerger TJ, Wittenborn JS, Segel JE, Burrows NR, Imai K, Eggers P et al (2010) A health policy model of CKD: 2. The cost-effectiveness of microalbuminuria screening. Am J Kidney Dis 55:463–473

Hou W, Lv J, Perkovic V, Yang L, Zhao N, Jardine MJ et al (2013) Effect of statin therapy on cardiovascular and renal outcomes in patients with chronic kidney disease: a systematic review and meta-analysis. Eur Heart J 34:1807–1817

Huo Y, Li J, Qin X, Huang Y, Wang X, Gottesman RF et al (2015) Efficacy of folic acid therapy in primary prevention of stroke among adults with hypertension in china: the CSPPT randomized clinical trial. JAMA 313:1325–1335

Imai E, Yamagata K, Iseki K, Iso H, Horio M, Mkino H et al (2007) Kidney disease screening program in japan: history, outcome, and perspectives. Clin J Am Soc Nephrol 2:1360–1366

Jain AK, McLeod I, Huo C, Cuerden MS, Akbari A, Tonelli M et al (2009) When laboratories report estimated glomerular filtration rates in addition to serum creatinines, nephrology consults increase. Kidney Int 76:318–323

Jamal SA, Vandermeer B, Raggi P, Mendelssohn DC, Chatterley T, Dorgan M et al (2013) Effect of calcium-based versus non-calcium-based phosphate binders on mortality in patients with chronic kidney disease: an updated systematic review and meta-analysis. Lancet 382:1268–1277

James MT, Hemmelgarn BR, Tonelli M (2010) Early recognition and prevention of chronic kidney disease. Lancet 375:1296–1309

Jardine MJ, Ninomiya T, Perkovic V, Cass A, Turnbull F, Gallagher MP et al (2010) Aspirin is beneficial in hypertensive patients with chronic kidney disease: a post-hoc subgroup analysis of a randomized controlled trial. J Am Coll Cardiol 56:956–965

Jha V, Garcia-Garcia G, Iseki K, Li Z, Naicker S, Plattner B et al (2013) Chronic kidney disease: global dimension and perspectives. Lancet 382:260–272

Jun M, Lv J, Perkovic V, Jardine MJ (2011) Managing cardiovascular risk in people with chronic kidney disease: a review of the evidence from randomized controlled trials. Ther Adv Chronic Dis 2:265–278

Kidney Disease: Improving Global Outcomes (KDIGO) CKD Work Group (2013) KDIGO 2012 clinical practice guideline for the evaluation and managementof chronic kidney disease. Kidney Int Suppl 3:1–150

Kumaresan J, Seneviratne R (2017) Beginning of a journey: unraveling the mystery of chronic kidney disease of unknown aetiology (CKDU) in sri lanka. Global Health 13:43

Labarca J (2000) Antibiotic cycling tested in nosocomial infections. Lancet 355:992

Levey AS, Coresh J, Greene T, Stevens LA, Zhang YL, Hendriksen S et al (2006) Using standardized serum creatinine values in the modification of diet in renal disease study equation for estimating glomerular filtration rate. Ann Intern Med 145:247–254

Levey AS, Stevens LA, Schmid CH, Zhang YL, Castro AF 3rd, Feldman HI et al (2009) A new equation to estimate glomerular filtration rate. Ann Intern Med 150:604–612

Lin CM, Yang MC, Hwang SJ, Sung JM (2013) Progression of stages 3b-5 chronic kidney disease–preliminary results of taiwan national pre-esrd disease management program in southern taiwan. J Formos Med Assoc 112:773–782

Liyanage T, Ninomiya T, Jha V, Neal B, Patrice HM, Okpechi I et al (2015) Worldwide access to treatment for end-stage kidney disease: a systematic review. Lancet 385:1975–1982

Lv J, Ehteshami P, Sarnak MJ, Tighiouart H, Jun M, Ninomiya T et al (2013) Effects of intensive blood pressure lowering on the progression of chronic kidney disease: a systematic review and meta-analysis. CMAJ 185:949–957

Lv J, Perkovic V, Foote CV, Craig ME, Craig JC, Strippoli GF (2012) Antihypertensive agents for preventing diabetic kidney disease. Cochrane Database Syst Rev 12: CD004136

Manns B, Hemmelgarn B, Tonelli M, Au F, Chiasson TC, Dong J et al (2010) Population based screening for chronic kidney disease: cost effectiveness study. BMJ 341:c5869

Matsushita K, van der Velde M, Astor BC, Woodward M, Levey AS, de Jong PE et al (2010) Association of estimated glomerular filtration rate and albuminuria with all-cause and cardiovascular mortality in general population cohorts: a collaborative meta-analysis. Lancet 375:2073–2081

Mazzuchi N, Schwedt E, Sola L, Gonzalez C, Ferreiro A (2006) Risk factors and prevention of end stage renal disease in uruguay. Ren Fail 28:617–625

McCullough KP, Morgenstern H, Saran R, Herman WH, Robinson BM (2019) Projecting esrd incidence and prevalence in the united states through 2030. J Am Soc Nephrol 30:127–135

Muntner P, Hamm LL, Kusek JW, Chen J, Whelton PK, He J (2004) The prevalence of nontraditional risk factors for coronary heart disease in patients with chronic kidney disease. Ann Intern Med 140:9–17

National Kidney Foundation (2002) K/DOQI clinical practice guidelines for chronic kidney disease: Evaluation, classification, and stratification. Am J Kidney Dis 39:S1–S266

Neal B, Perkovic V, Mahaffey KW, de Zeeuw D, Fulcher G, Erondu N et al (2017) Canagliflozin and cardiovascular and renal events in type 2 diabetes. N Engl J Med 377:644–657

Ninomiya T, Perkovic V, Turnbull F, Neal B, Barzi F, Cass A et al (2013) Blood pressure lowering and major cardiovascular events in people with and without chronic kidney disease: meta-analysis of randomised controlled trials. BMJ 347:f5680

Palmer SC, Navaneethan SD, Craig JC, Johnson DW, Tonelli M, Garg AX et al (2010) Meta-analysis: erythropoiesis-stimulating agents in patients with chronic kidney disease. Ann Intern Med 153:23–33

Patel A, MacMahon S, Chalmers J, Neal B, Billot L, Woodward M et al (2008) Intensive blood glucose control and vascular outcomes in patients with type 2 diabetes. N Engl J Med 358:2560–2572

Sharma SK, Ghimire A, Carminati S, Remuzzi G, Perico N (2014) Management of chronic kidney disease and its risk factors in eastern nepal. Lancet Glob Health 2:e506–e507

Steinberg D, Witztum JL (2002) Is the oxidative modification hypothesis relevant to human atherosclerosis? Do the antioxidant trials conducted to date refute the hypothesis? Circulation 105:2107–2111

Stevens LA, Claybon MA, Schmid CH, Chen J, Horio M, Imai E et al (2011) Evaluation of the chronic kidney disease epidemiology collaboration equation for estimating the glomerular filtration rate in multiple ethnicities. Kidney Int 79:555–562

Tepel M, van der Giet M, Statz M, Jankowski J, Zidek W (2003) The antioxidant acetylcysteine reduces cardiovascular events in patients with end-stage renal failure: a randomized, controlled trial. Circulation 107:992–995

Wang ZJ, Harjai KJ, Shenoy C, Gao F, Shi DM, Liu YY et al (2013) Drug-eluting stents versus bare-metal stents in patients with decreased GFR: a meta-analysis. Am J Kidney Dis 62:711–721

Wang J, Zhang L, Tang SC, Kashihara N, Kim YS, Togtokh A et al (2018) Disease burden and challenges of chronic kidney disease in north and east asia. Kidney Int 94:22–25

Wanner C, Inzucchi SE, Lachin JM, Fitchett D, von Eynatten M, Mattheus M et al (2016) Empagliflozin and progression of kidney disease in type 2 diabetes. N Engl J Med 375:323–334

Wanner C, Krane V, Marz W, Olschewski M, Mann JF, Ruf G et al (2005) Atorvastatin in patients with type 2 diabetes mellitus undergoing hemodialysis. N Engl J Med 353:238–248

Wei SY, Chang YY, Mau LW, Lin MY, Chiu HC, Tsai JC et al (2010) Chronic kidney disease care program improves quality of pre-end-stage renal disease care and reduces medical costs. Nephrology (Carlton) 15:108–115

Wijkstrom J, Gonzalez-Quiroz M, Hernandez M, Trujillo Z, Hultenby K, Ring A et al (2017) Renal morphology, clinical findings, and progression rate in mesoamerican nephropathy. Am J Kidney Dis 69:626–636

Xie X, Liu Y, Perkovic V, Li X, Ninomiya T, Hou W et al (2016) Renin-angiotensin system inhibitors and kidney and cardiovascular outcomes in patients with CKD: a bayesian network meta-analysis of randomized clinical trials. Am J Kidney Dis 67:728–741

Zhang L, Long J, Jiang W, Shi Y, He X, Zhou Z et al (2016) Trends in chronic kidney disease in china. N Engl J Med 375:905–906

Zhang L, Wang F, Wang L, Wang W, Liu B, Liu J et al (2012) Prevalence of chronic kidney disease in china: a cross-sectional survey. Lancet 379:815–822

Chapter 2
Morphology and Evaluation of Renal Fibrosis

Ping-Sheng Chen, Yi-Ping Li and Hai-Feng Ni

Abstract With continuing damage, both the indigenous cells of the cortex and medulla, and inflammatory cells are involved in the formation and development of renal fibrosis. Furthermore, interactions among the glomerular, tubular, and interstitial cells contribute to the process by excessive synthesis and decreased degradation of extracellular matrix. The morphology of kidney is different from pathological stages of diseases and changes with various causes. At the end stage of the disease, the kidneys are symmetrically contracted with diffuse granules. Most glomeruli show diffuse fibrosis and hyaline degeneration, and intervening tubules become atrophied. Renal interstitium shows obvious hyperplasia of fibrous tissues with marked infiltration of lymphocytes, mononuclear cells, and plasma cells. The renal arterioles are wall thickening frequently because of hyaline degeneration. Morphologic analysis based on Masson staining of the kidney tissues has been regarded as the golden standard to evaluate the visual fibrosis. However, the present studies have found that the evaluation system has poor repeatability. Several computer-aided image analysis techniques have been used to assess interstitial fibrosis. It is possible that the evaluation of renal fibrosis is carried out by the artificial intelligence renal biopsy pathological diagnosis system in the near future.

Keywords Morphology · Renal fibrosis · Evaluation

P.-S. Chen (✉) · Y.-P. Li
Department of Pathology and Pathophysiology, School of Medicine, Southeast University, Nanjing, China
e-mail: 101006254@seu.edu.cn

P.-S. Chen · H.-F. Ni
Institute of Nephrology, Zhong Da Hospital, School of Medicine, Southeast University, Nanjing, China

B.-C. Liu et al. (eds.), *Renal Fibrosis: Mechanisms and Therapies*,
Advances in Experimental Medicine and Biology 1165,
https://doi.org/10.1007/978-981-13-8871-2_2

2.1 Introduction

Renal fibrosis is the end stage of a variety of chronic progressive kidney diseases. It is a morphological change based on the combined responses of glomeruli, renal tubules, renal blood vessels, and/or renal interstitium exposed to continuous injury. Considering progressive structural damage, the glomerular sclerosis, tubulointerstitial inflammation, tubular atrophy, and even interstitial fibrosis will occur finally, whether the damage originates from glomeruli, renal tubules, renal interstitium, or renal blood vessels. As renal fibrosis processes, the kidney loses its function gradually (Satirapoj et al. 2012). Therefore, both appropriate morphological knowledge of renal fibrosis and objective analysis on the pathological degree will be valuable in directing treatment and estimating prognosis. This article will summarize renal fibrosis-related cells, histopathology, and evaluation methods.

2.2 Related Cells of Renal Fibrosis

Renal cortex and medullary intrinsic cells are involved in the formation and development of renal fibrosis under injury, furthermore, the interaction between glomerular cells and tubulointerstitial cells exists throughout the fibrosis process.

2.2.1 Related Cells of Glomerulosclerosis

Glomerular innate cells include endothelial cells, mesangial cells, visceral epithelial cells, and parietal epithelial cells; all the cells can produce varying amounts of extracellular matrix (ECM), which contains type IV and type VI collagen, laminin, and fibronectin. Under pathological condition, matrix synthesis increases and deposits in mesangial area, which squeezes the capillary lumen, then leads to the apoptosis of glomerular cells continuously. Apoptotic and exfoliated cells leave gaps in the glomeruli, which will be filled by ECM. Followed by endothelial injury, the basement membrane of capillaries is exposed and podocytes are damaged, further selective filtration of the glomerulus is deprived. The cytokines, chemokines, and growth factors, produced by the injured glomerular cells and inflammatory cells, promote interstitial inflammation and fibrosis (Eardley et al. 2006). Therefore, the renal tissues with chronic kidney diseases (CKDs) are lack of glomerular selective filtration, accompanied by glomerular cells apoptosis, inflammatory activation of glomerular residual cells, and leukocytes infiltration. In the end, the renal tissues present glomerular sclerosis. Thereby, the above changes reduce glomerular perfusion and filtration rate. Subsequently, the damage of renal tubule cells will inevitably come along with the injured glomeruli and albuminuria.

2.2.2 Related Cells of Interstitial Fibrosis

Interstitial fibrosis is associated with many other types of cells beside glomerular cells, which includes fibroblasts, myofibroblasts, fibrocyte, tubular epithelial cells, endothelial cells, capillary pericytes, mast cells, and inflammatory cells.

2.2.2.1 Fibroblasts and Myofibroblasts

Fibroblasts, the main cells of the renal interstitium, are the main cellular source of the ECM, and ECM constitutes renal connective tissue skeleton. It is challenging to study fibroblasts without specific markers. Fibroblasts are characterized by a large number of rough endoplasmic reticulum, abundant F-actin skeleton and membrane-expressed 5 'exonuclease, which differentiates from other interstitial cells. The fibroblasts interact with other cells, such as they interplay with the dendritic cells using cytoplasmic processes. Perceiving the basement membrane of the injured tubule cells, the fibroblasts will transform their phenotype to myofibroblasts by paracrine signaling, and myofibroblasts derived from fibroblasts will produce type III collagen (Dussaule et al. 2011; Meran and Steadman 2011).

Myofibroblasts express smooth muscle actin (SMA) and microfiber (stress fiber) with punctiform dense bodies (focal densities). Myofibril, as the hinge (fibronexi) connecting actin fibrils and extracellular fibronectin, densely distributes along the margin of myofibroblasts with round nuclei. Myofibroblasts are usually adjacent to basement membrane and produce plenty of ECM. Myofibroblasts also express vimentin, fibronectin containing A shear variants in the external functional region, and S100A4 (FSP-1). S100A4 is once thought to be a specific marker of myofibroblasts. Myofibroblasts are commonly accompanied by white blood cells. Interstitial myofibroblasts may originate from fibroblast, pericytes, perivascular cells, and endothelial cells (Lin et al. 2008; Kaissling and Le Hir 2008).

2.2.2.2 Fibrocytes

The fibrocyte is considered to be different from fibroblast although it can also synthesize ECM. It is spindle and derives from the white blood cells in peripheral blood. It possesses the hematogenous markers (CD45), interstitial cell marker (type I collagen) and chemokine receptor. Fibrocytes only exist in the diseased kidneys, and they may come from differentiation in situ or infiltration. T_H2-type cytokines can induce human fibrocyte differentiation, but T_H1-type cytokines inhibit the differentiation (Wada et al. 2011; Sakai et al. 2010).

2.2.2.3 Renal Tubular Epithelial Cells

Renal tubular epithelial cells show definite phenotypic changes in case of acute injury, which may provide a key clue for interstitial fibrogenesis. The intracellular signals will be activated and some specific genes will be expressed, which is induced by tubule extension, liquid shear stress, and biological mechanical force. These events will cause interstitial fibrosis. The damaged tubular epithelial cells present the increased intermediate filament (vimentin and nestin), up-regulated type I and III collagen, and down-regulated E-cadherin. Overexpression of vascular endothelial growth factor-A (VEGF-A) in renal tubular epithelial cells can raise the level of serum VEGF, then result in more capillaries with dilated lumens, type IV collagen deposition, and more fibroblasts and myofibroblasts in the interstitium. However, some studies reported that supplementing VEGF may relieve interstitial fibrosis (Hakroush et al. 2009; Lian et al. 2011).

Renal tubular epithelial cells enduring hypoxia can promote fibrosis through a variety of media including HIF-1α (Higgins et al. 2008; Kimura et al. 2008). Ras oncogene inhibitor RASAL1 can alleviate interstitial fibrosis by methylation of methyltransferase DNMT1. Another chromatin modifier, histone deacetylase, may modulate proinflammatory function and fibrotic process of renal tubulointerstitium injury (Sun et al. 2010; Yang et al. 2010).

2.2.2.4 Inflammation-Related Cells

Many kinds of inflammatory cells participate in tubulointerstitial fibrosis, which are elaborated as follows:

Lymphocytes

Lymphocytes may play an important role in the tubulointerstitial fibrosis, and $CD4^{+}T$ cells are considered to be particularly critical. $CD4^{+}T$ cells reconstructed mice (not $CD8^{+}T$ cells) displayed heavy interstitial fibrosis, and mice with reduced $CD4^{+}T$ cells showed relatively mild interstitial fibrosis (Nikolic-Paterson et al. 2010). Recently, microarray detection of transplanted kidney indicated that the increase of T cells as well natural killer cells accelerated the development of tubulointerstitial fibrosis. The infiltration of T cells and macrophages is often accompanied by low expression of IL-10, which is more likely to cause tubulointerstitial fibrosis than B lymphocyte infiltration.

Monocytes/Macrophages

Monocytes/macrophages are not homogeneous, and some subtypes, especially monocytes/macrophages expressing CD11b exhibit a pro-fibrotic effect. Galectin-3

(Gal-3), a galactosidase binding lectin, is secreted by macrophages, which is a profibrogenic factor. Renal tubular epithelial cells synthesize macrophage growth factor and CSF-1, then to regulate repair and relieve interstitial fibrosis. Macrophages play complicated roles in interstitial fibrosis, such as the monocyte subsets derived from bone marrow could actually delay the development of fibrosis (Anders and Ryu 2011; Vernon et al. 2010).

Dendritic Cells

Some dendritic cells distribute in the renal interstitium, and the present studies have shown that they play important roles during the process of interstitial fibrosis. The experiment demonstrated that the dendritic cells were decreased in number and the interstitial fibrosis could be alleviated in the transgenic mice with CD11c/diphtheria receptor were injected by diphtheria toxin. Other studies have reported that dendrite cells affected fibrosis indirectly by activating T cells (Snelgrove et al. 2012).

Mast Cells

Mast cell is a member of the innate immune system and is rare in the healthy kidneys. Most of them aggregate around the blood vessels and epithelial cells. Accompanying mast cells increased in number, the fibrosis is aggravated, and the renal function is reduced. Loss of the mast cells could relieve fibrosis. However, some studies reported that the cells could inhibit fibrosis (Kim et al. 2009).

Endothelial Cells

The relationship between endothelial cells and interstitial fibrosis is unclear. Some studies found that endothelial cells could transform into fibroblasts through endothelial–mesenchymal transdifferentiation (EndoMT) and promote fibrotic development. Both the capillary network and the lymphatic network are mainly composed of endothelial cells and are closely related to fibrosis. It was found that the decreased capillary network around renal tubule and renal hypofunction were aggravated with the time prolonging in allograft kidneys. The renal lymphatic network can assist the migration of inflammatory cells, and the interstitial fibrosis with lymphatic hyperplasia, which is partly due to the regulation of VEGF-C. Angiogenesis and inflammation inhibited by sirolimus can prevent interstitial fibrosis. The newborn lymphatic vessels directly close to the glomeruli with adherent vascular loops, which may be involved in the abnormal filtration of urine locally. Other studies have also shown that lymphangiogenesis is associated with tissue remodeling and differential expression of proteoglycan. As early as 72 h after transplantation, the lymphatic vessels appear, which may be related to inflammation (Piera-Velazquez et al. 2011; Stuht et al. 2007).

2.3 Histopathology of Renal Fibrosis

The healthy renal tissue contains ECM such as collagen, especially type I, III and IV collagen, which are the connective tissue structural framework of the kidney. The matrix deposit is the major cause of hypofunction in the kidneys with CKD.

Renal fibrosis/sclerosis includes two subtypes, glomerular sclerosis and interstitial fibrosis, induced by various causes. Therefore, it is hard to ascertain the subtype of renal fibrosis morphologically. Large scar lesions with tubular damage result from severe injury of focal tissues and parenchymal structure, which are commonly observed in pyelonephritis and renal infarction. The scar tissues contribute to maintain the intact structure and function of the damaged kidneys, the process is undoubtedly affiliated to wound healing. By contrast, diffuse or focal interstitial fibrosis around the normal and atrophic renal tubules is often be observed in the clinical renal biopsy, which is common in the glomerular diseases, renal tubular diseases, and renal vascular diseases.

2.3.1 General Morphological Characteristics of Renal Fibrosis

Renal fibrosis is continuously aggravated along with the development of the lesions in the diffusely progressive kidney diseases. The insulted kidneys present different morphological changes based on pathological stages and various causes at the end stage, the kidney will inevitably represent severe fibrosis. The basic pathological changes are described as follows.

2.3.1.1 Gross Morphology

The kidneys are symmetrically diminished in size and the mass is reduced (from 140 to 20 g). And the fibrotic kidneys show pale appearance, hard texture and diffusely granular surface (Fig. 2.1). On the cut surface, renal cortex is significantly thinner, and the corticomedullary demarcation is unclear, adipose tissues often proliferate around the renal pelvis, arterioles with narrowed lumina stand out because of thickened walls, renal capsule firmly attaches to the surface of the cortex. Such markedly injured kidneys often have been designated granular contracted kidney.

2.3.1.2 Microscopic Appearance

Most of the glomeruli (more than 75%) show diffuse sclerosis and hyaline degeneration, and the renal tubules implicated are atrophied and replaced by fibrous tissues. The hyaline and fibrotic glomeruli are drawn together due to the tightening of fibrous

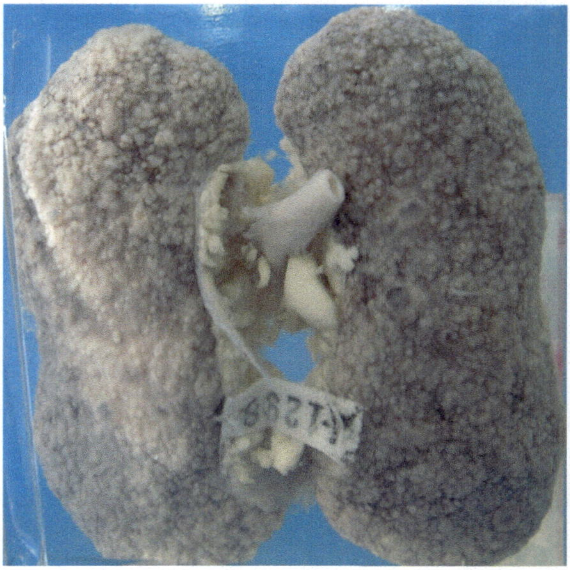

Fig. 2.1 Renal pathological morphology at end stage. The kidneys are symmetrically contracted with diffuse granules on the surface

tissues. The residual nephrons represent compensatory changes, characterized by the hypertrophic glomeruli and dilated renal tubules with various casts (mainly for protein casts). In the fibrotic interstitial tissues, chronic inflammatory cells (mainly for lymphocytes and monocytes, rarely for plasma cells) infiltration and hyalinized small and medium-sized arteries are present. The arterioles display intimal fibrosis with hyaline degeneration in the walls. Because most of the renal tissue is fibrotic and shrunk, the remaining nephrons protrude to the surface due to compensatory hypertrophy, the renal surface becomes finely granular. Once the damage progresses to the end-stage, all of the renal tissue structure, including glomeruli, renal tubules, and renal interstitium present nearly identical histopathological changes, which is difficult to differentiate the injuries from the sclerosis caused by blood vessels (such as hypertension kidney) or interstitium (such as chronic interstitial glomerulonephritis). All the kidneys show the same changes in morphology, a large number of nephrons damaged, interstitial fibrosis and chronic inflammatory cells infiltration, vascular intima fibrosis and hyaline degeneration (Fig. 2.2) (Chen 2017; Rubin and Strayer 2012).

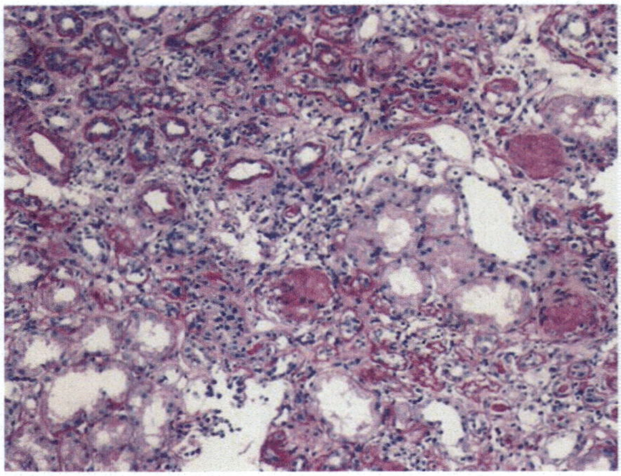

Fig. 2.2 Renal pathological morphology at end stage. Glomerulosclerosis, renal tubule atrophy, and renal interstitial fibrosis are shown with PAS staining. (PAS × 100)

2.3.2 Morphological Differences of Different Etiological Factors

As mentioned above, the morphology of the kidneys is different when the etiology is diverse, although it is difficult to differentiate at the end-stage. Herein, renal fibrosis considering several common causes will be discussed.

2.3.2.1 Glomerulosclerosis

Glomerulosclerosis mainly includes two types, ischemic glomerulosclerosis and proliferative glomerulosclerosis. Ischemic glomerulosclerosis is often seen in hypertension, characterized by protein-like substance deposition in the glomerular capsules (Bowman's capsules), without mesangial matrix accumulation in the glomeruli. But hypertension inevitably appears in the advanced stage of CKD, that is to say, hypertension-induced morphological changes exist in ischemic glomerulosclerosis, caused not only by primary hypertension but also by CKD. Therefore, combination with glomerular lesions and medical history is needed to differentiate them. Proliferative glomerulosclerosis is characterized by the increase in mesangial matrix and found in most primary or secondary CKD, such as diabetic nephropathy, lupus nephritis, membranous nephropathy, focal segmental glomerulosclerosis, membranoproliferative glomerulonephritis, renal vasculitis, Goodpasture's syndrome, diabetic nephropathy, and amyloid nephropathy (Fig. 2.3). In clinical practice, the coexistence of two scleroses is often observed (Zhou 2013). Sometimes, glomerulosclerosis may be glomerular segmental sclerosis or mesangial nodular sclerosis (Fig. 2.4).

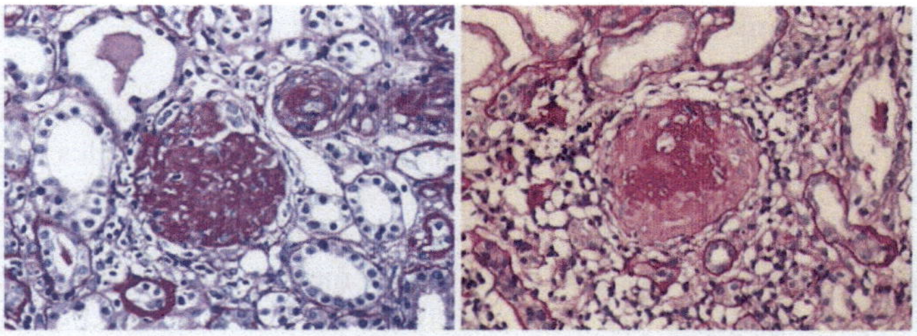

Fig. 2.3 Morphological difference between ischemic glomerulosclerosis and proliferative glomerulosclerosis. Left image for the representive images of ischemic glomerulosclerosis with arteriosclerosis, and right image for proliferative glomerulosclerosis (PAS × 200)

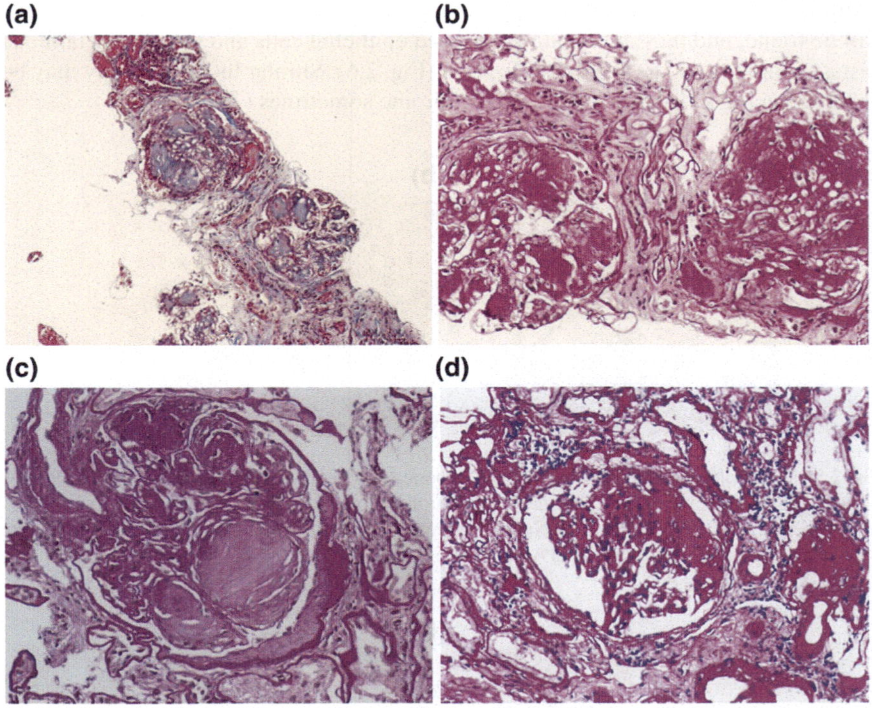

Fig. 2.4 Other types of glomerulosclerosis. **a** Glomerulosclerosis in diabetic nephropathy (Masson × 40); **b** Mesangial nodular glomerulosclerosis in diabetic nephropathy (PAS × 200); **c** The urinary pole showing two Kimmelstiel–Wilson (K–W) nodules in diabetic glomerulus (PAS × 400); **d** Focal segmental glomerulosclerosis (PAS × 200)

2.3.2.2 Renal Tubule Atrophy

The basic morphological features are that the epithelial cells become atrophic or lost in the renal tubules of fibrotic kidneys. Concretely, some tubules may become small in size, and the epithelial cells are lightly stained and surrounded by thick and shrunk basement membrane. Simultaneously, some tubules are obviously dilated with flattened epithelial cells, but the thickening of basement membrane seems unobvious. In diabetic nephropathy, the basement membrane thickened is found in the majority of renal tubules. Furthermore, the renal tubules with marked atrophy present more distinctly thickened basement membrane. It is necessary to distinguish nephropathy caused by diabetes mellitus and hypertension because both of them show arterioles wall thickening and hyaline degeneration (Fig. 2.5). In the chronic pyelonephritis, in addition to the loss of numerous renal tubules, atrophic and dilated renal tubules can still be found, and they are lined by flattened epithelial cells and filled by gelatinous casts, forming the so-called thyroidization (Fig. 2.6). Similar histopathology may be observed in the kidneys with Sjogren syndrome sometimes (Zhou 2013).

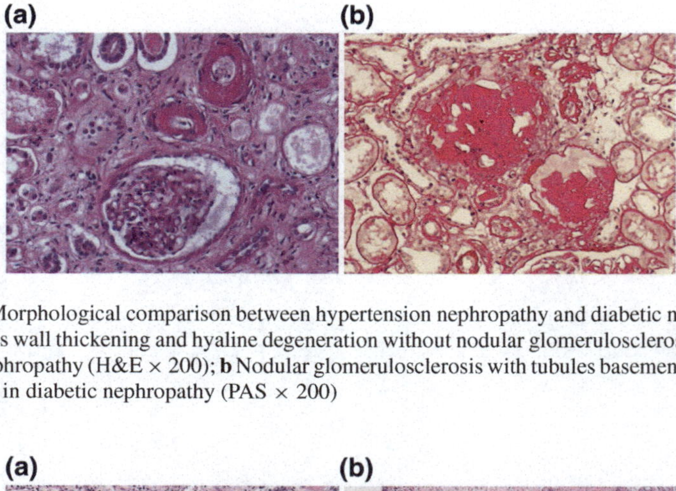

Fig. 2.5 Morphological comparison between hypertension nephropathy and diabetic nephropathy. **a** Arterioles wall thickening and hyaline degeneration without nodular glomerulosclerosis in hypertension nephropathy (H&E × 200); **b** Nodular glomerulosclerosis with tubules basement membrane thickening in diabetic nephropathy (PAS × 200)

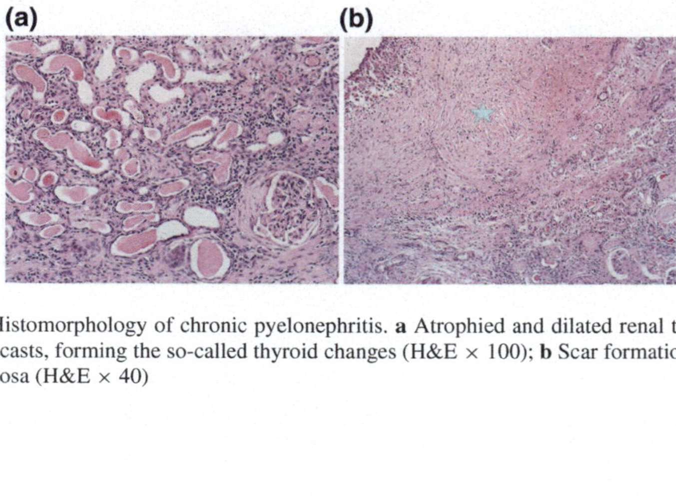

Fig. 2.6 Histomorphology of chronic pyelonephritis. **a** Atrophied and dilated renal tubules with gelatinous casts, forming the so-called thyroid changes (H&E × 100); **b** Scar formation under the pelvis mucosa (H&E × 40)

Renal tubular atrophy is usually accompanied by renal interstitial fibrosis and may be involved in the underlying mechanism related to renal blood flow, glomerular filtration rate or renal tubules loss. However, sometimes, renal tubular atrophy and interstitial fibrosis do not occur simultaneously, for example, fibrosis or inflammation response is hardly present in the renal artery stenosis with severe renal tubular atrophy.

2.3.2.3 Renal Interstitial Fibrosis

The fibrotic foci can be lumpy, flaky, focal, and latticed. At the advanced stage, the glomeruli were concentrated due to the contraction of the fibers. It is noteworthy that renal interstitial fibrosis is serious with storiform, but the destructive changes of the renal tubules including atrophy are relatively mild in IgG4-associated nephropathy (Yamaguchi et al. 2012). The renal tubules in the fibrous scars are thoroughly destroyed, and concave scars may be present on the surface of the kidneys in chronic pyelonephritis. In addition, the pelvis may be thickened even deformed in chronic pyelonephritis. Chronic pyelonephritis is often associated with urethral obstruction or urinary reflux. Tamm-Horsfall protein deposits as homogeneous fibrinoide lesions in renal interstitium, and it can be strongly stained by PAS and surrounded by inflammatory cells.

2.3.2.4 Inflammatory Cells Infiltration

More or less inflammatory cells, especially mononuclear lymphocytes infiltrate in the renal fibrotic areas. Plasma cells infiltration can be observed usually in IgG4-associated nephropathy or Sjogren syndrome renal injury (Yamaguchi et al. 2012); Lymphocytes, histocytes, and plasma cells constitute inflammatory cells infiltrated in chronic pyelonephritis (Fig. 2.7), and neutrophils infiltration can be found at

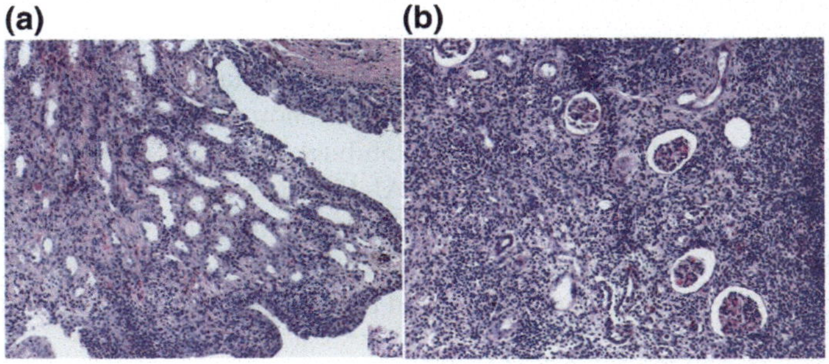

Fig. 2.7 Histomorphology of chronic pyelonephritis. **a** Abundant chronic inflammatory cells infiltration in the pelvis mucosa and submucosa renal tissue (H&E × 100); **b** The glomeruli showing mild lesions, but most renal tubules presenting complete destruction, and a large number of chronic inflammatory cells infiltration in the renal tissue (H&E × 100)

acute stage. Additionally, eosinophil infiltration may be observed in drug-induced renal injury and vasculitis. Epithelioid granulomas are commonly present in renal tuberculosis and sarcoidosis, but caseous necrosis may be found in tuberculosis, not in sarcoidosis.

2.3.2.5 Others

Atheromatous Thrombotic Nephropathy

This disease always caused by the rupture of atherosclerotic plaques, and the cholesterol crystals are released into the renal arterioles. In addition to the kidneys, the skin, the gastrointestinal tract, and the brain are also often involved. This disease can be noticed spontaneously, but it comes from vascular surgery, cardiac catheterization, or anticoagulation treatment in most cases. Atheromatous embolism is difficult to be diagnosed due to the involvement of multiple systems. The triad including treatment history, acute or chronic renal failure, and skin lesions will provide strong indication of the disease. Meanwhile, the diagnosis can be further confirmed by the increase of eosinophils in peripheral blood. Routinely, it can be finally diagnosed with the cholesterol crystals detected in the affected organs by biopsy or in retinal blood circulation system by ophthalmoscope. Renal arteriolar embolization of cholesterol crystallization contributes to renal infarct and fibrosis (Scolari and Ravani 2010). The patients with atheromatous thrombotic nephropathy are always with dismal prognosis.

High risk group of atherosclerotic thrombotic nephropathy is related to the following factors: male, over 60 years old, the white race, hypertension, smoking, diabetes mellitus, atherosclerotic vascular disease, ischemic heart disease, cerebrovascular disease, abdominal aortic aneurysm, peripheral vascular disease, and ischemic nephropathy (Scolari and Ravani 2010).

Autosomal Dominant Polycystic Renal Interstitial Fibrosis

Polycystic renal interstitial fibrosis is similar to the classic CKD in some ways, such as interstitial increased collagen deposition, abnormal matrix metalloproteinase (MMPs) activity, overexpression of tissue inhibitor of metalloproteinase-1 (TIMP-1) and plasminogen activator inhibitor-1 (PAI-1) and enhanced synthesis of TGF-β. However, the following changes are special, the sacculinization of tubules associated with tubular interstitial fibrosis, tubular epithelial change as the initiating factor of interstitial fibrosis, the periodical interaction between epithelial cells and interstitial fibroblasts (Norman 2011).

2.4 Fibrosis Evaluation

To treat the patients with kidney diseases, nephrologists often ask: Immunosuppressive drugs can be used for this patient? What will happen after medication? Since the risk of using immunosuppressive drugs is not excluded, it is necessary to consider its pros and cons before using. Here, the pathologist can provide helpful information by renal biopsy-guided drugs therapy. One is the relevant information to guide treatment (i.e., evidence of immunization activities), such as inflammation, necrosis, and active proliferation indications reflecting the serious degree of the disease. The other is which stage the disease is at (i.e., the signs of irreversible injury). The staging information is essential to predict whether the best therapeutic strategy is taken or not. The staging information is referred as chronic index, including the proportion of glomerulosclerosis, tubular atrophy and interstitial fibrosis in the biopsy tissues. Determining the degree of fibrosis will be beneficial to study of the efficacy of anti-fibrosis drugs and to compare the results of biopsies of transplant kidney of different batches. In the routine pathological work, it is not difficult to calculate the proportion of the sclerotic glomeruli in the renal tissues from puncture biopsy, but it is hard to determine the degree of interstitial fibrosis accurately. The morphological observation using Masson staining has now been used as the gold standard for visual determination of fibrosis, but the studies suggest that the method provide poor repeatability. Several computer-assisted image analyses have been applied to assess interstitial fibrosis, and the analyses based on special staining technique including Masson staining, Sirius red staining, and immunohistochemical staining (especially type III collagen), and observation of collagen I, collagen III using polarizing microscope (Fig. 2.8) (Farris and Colvin 2012; Robertsa et al. 2004).

Evaluation of fibrosis is to determine the amount of collagen deposit, which can be achieved by morphometry, biochemical methods, and fibrosis-related molecular markers detection.

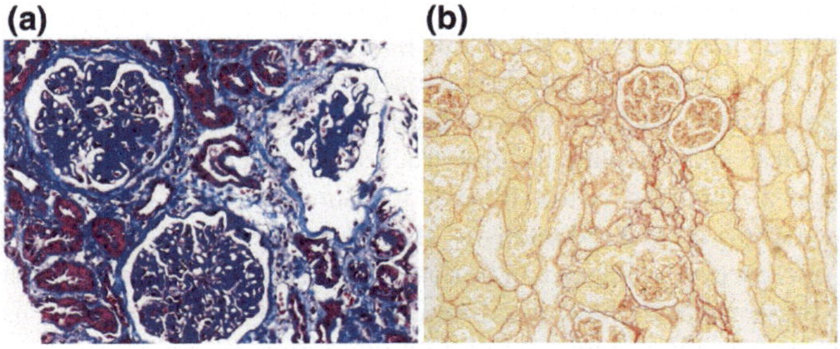

Fig. 2.8 Common staining methods to evaluate organ fibrosis. **a** Blue area showing the deposited collagen (Masson × 200); **b** Red area showing the deposited collagen (Sirius red × 100)

2.4.1 *Morphometry*

The morphometry can be divided into the microscope-assistant visual examination and computer-assistant image analysis. The tissues can be analyzed after special staining. The conventional staining methods are Masson's staining, Sirius red staining, V-G staining, and collagen immunohistochemical staining. Among them, Masson staining is the most popular method. The experienced pathologists can preliminarily evaluate the degree of fibrosis with conventional hematoxylin and eosin (H&E) staining under microscope.

2.4.1.1 Visual Examination

The fibrosis degree is determined by area percentage of fibrotic region in the whole tissue. Generally, grades are determined using the criteria: grade 0 for no fibrosis, grade 1 (mild) for fibrosis area ≤25%, grade 2 (moderate) for fibrosis area 25–50%, grade 3 (severe) for fibrosis area >50%. Although this method is simple and practical to be used extensively, it requires the observers with experienced renal pathological background. In addition, the standard of criterion is subjective, which leads to poor repeatability when the examination is done by different observers, even the same observer at the different time.

2.4.1.2 Computer Image Analysis

The method is to transmit the digital pictures of slices to the computer, then the operator determines the fibrosis area and non-fibrosis area, the computer will calculate the percentage of fibrosis areas using a small procedure. Judging whether an area is fibrotic is the first-phase preparation, which still needs professional knowledge. Another way to identify fibrosis is to mark interstitial type III or type IV collagen by immunohistochemistry, and the other methods have been described previously. If the figures have negligible and bright background, the digital photographs will have enough contrast. Thereby, the computer can recognize the positive region with a single threshold and calculate the area fraction. In recent years, Sirius red staining has also been gradually used for fibrosis measurement. The computer can assess collagen by identifying the red intensity of slices or observing Sirius red slices with polarizing microscope, collagen can be detected due to strong refraction. In the same laboratory, these methods have good repeatability, so they are widely used. Once the appropriate computer macro-instruction program runs, a slice can be completed within 2 min. It is worth noting that the reproducibility may be significantly different for various batches of staining slices. The Sirius red intensity is always quite inconsistent in different laboratories and the refraction intensity seems to be entirely dependent on the thickness of the slice. Therefore, the reproducibility of a laboratory does not represent the repeatability of different laboratories (Robertsa et al. 2004).

2.4.2 Biochemical Method

Hydroxyproline (Hyp) is one of the main components of collagen. Its content is about 13.4% in collagen, very little in elastin, and barely in other proteins under physiological condition. Therefore, the content of Hyp in different tissues can be used as an important indicator to examine the metabolism of collagen tissues. The detection principle is that the sample is hydrolyzed in hydrochloric acid solution to release hydroxyproline, subsequently oxidized by chloramine T to produce oxide containing pyrrole ring. The hydroxyproline oxide reacts with two methamidbenzaldehyde to produce a red compound, and its absorption value will be measured at the wavelength of 560 nm. However, it is generally recognized that the way to determine hydroxyproline is not suitable for renal biopsy examination because of too much tissue required. In addition, the following factors should be considered during the process of determination: ① Water content of living tissue: edema often occurs in the inflammation tissue, dehydration is needed before detection to avoid tricky judgments; ② Hydrolysis time and temperature: High pressure (120 °C) makes acidolysis to be easily completed in a short time to protect Hyp from excessive destruction; ③ The influence of pH in the measurement system: The buffer capacity of citric acid buffer is strong enough to neutralize pH difference of the sample to some extent, but the result will be influenced if pH of the sample exceeds the citric acid buffering capacity. Thus, the solution should be titrated to neutrality with NaOH after acidolysis; ④ Chromogenic temperature: the higher the chromogenic temperature, the shorter chromogenic time needed; ⑤ The effect of acid hydrolysate filtration on the reaction. The centrifugation is used to remove the residue produced by acidolysis. A small amount of the remaining residue will not affect the detection after further diluted in the following steps (Yang and Du 2004).

2.4.3 Molecular Measurement

With the development of modern biology, rapid and semi-automatic methods of gene or protein determination have been established to make the molecular measurement of renal fibrosis practicable. Fibrosis specific marker is the core of molecular measurement. In fact, collagen, as a fibrosis key molecule, is required to be detected to evaluate fibrosis. The above-mentioned image analysis with collagen immunohistochemical staining is one of the molecular measurements. However, it is not enough to measure collagen only because we need to understand not only the amount of collagen deposit at a certain time point but also the development trend of fibrosis.

Even if only collagen is selected as a detection index of fibrosis, it will not be simple and easy-on assessment strategy. Collagen has many types, and each type is a product of different alpha chains. Thus, any of these types are unable to reflect the whole fibrosis. Furthermore, the mechanism of collagen deposition is complicated, which is affected not only by collagen synthesis but also by the degradation. In addition, the activity of degrading enzyme is regulated by activating factors and inhibitory factors.

The main factors contributing to fibrosis are persistent tissue injury, increased number of activated myofibroblasts and continuous inflammatory response. The tissue injury is the initiating factor of fibrosis, myofibroblast activation is the major source of ECM, and inflammation can not only aggravate tissue injury but also promote other cells activation to myofibroblasts through a variety of inflammatory mediators. Hypoxia is the major factor to keep persistent tissue injury in the chronic progressive kidney disease, thus, hypoxia-inducible factor -1 α (HIF-1α), a specific marker of hypoxia, can reflect the extent of tissue hypoxia; α-smooth muscle actin (α-SMA), as one of the main molecular signs of myofibroblasts, is used to understand the number of myofibroblasts in the detected tissue; transforming growth factor -β (TGF-β) is the most important fibrosis promoting factor; most of inflammatory cells related to renal fibrosis tissue are lymphocytes and macrophages, among them, both $CD4^+$ T lymphocytes and $CD11b^+$ macrophages promote the development of fibrosis. Recent studies have found that toll-like receptor −4 (TLR4) not only mediates innate immunity but also accelerates renal fibroblast aggregation and tubulointerstitial fibrosis (Campbell et al. 2011). Therefore, the detection of HIF-1α, α-SMA, TGF-β, TLR4, CD4, and CD11b may help to understand the development trend of fibrosis. So far, these molecules can be examined by immunohistochemistry, in situ hybridization, and Western blot. Immunohistochemical method can represent not only the expression intensity but also the expressing site. Furthermore, it is more convenient and practicable among the described assays, although false positive and false negative occur sometimes.

To demonstrate the relationship between the fibrosis markers and chronic sclerosis/fibrosis, Bob group investigated the renal tissues from primary and secondary glomerulonephritis with immunohistochemistry using antibodies against SMA, vimentin, and TGF-β. Their data suggested a close correlation between the interstitial SMA marker index and the degree of sclerotic/fibrosis (chronic index). Specifically, the expression of vimentin and TGF-β is positively related to sclerosis/fibrosis (interstitial fibrosis, tubular atrophy, vascular hyalinosis/fibrosis, chronic index) but negatively associated with glomerular filtration rate. This study indicates that the other fibrosis-specific molecules except collagen can reflect not only the degree of fibrosis but also predicting prognosis (Bob et al. 2008).

2.4.4 *Artificial Intelligence-Assisted Pathological Diagnosis System*

Chronic kidney diseases are considered as a global public health challenge, the available latest data show that about 12% and 10.8% of adults are suffering from chronic kidney disease in the USA and China, respectively (Zhou et al. 2018). The therapy of such diseases mostly depends on accurate pathological diagnosis. The doctors engaged in the pathological diagnosis of kidney biopsy are very few in China. Such a large group of patients matched with a small team of professional doctors exhibit a significant gap, and it is inevitable to spawn a series of serious problems related to diagnostic work, such as overburdened work, high inaccurate report, and delayed diagnostic report. Therefore, it is urgent to break the deadlock, and it is lucky that the emerging of artificial intelligence pathological diagnosis system is expected to crack this dilemma (Yan et al. 2018; Miller and Brown 2018; Tizhoosh and Pantanowitz 2018).

The core of artificial intelligence is deep learning from virtual neural networks (Lecun et al. 2015). With the advent of the big data age and the development of a variety of more powerful computing devices including graphics handlers, deep learning can make full use of a variety of massive amounts of data, and automatically learn abstract knowledge expressing, that is, condensing raw data into a certain knowledge (Russell. 2017). CT, MRI, and pathological images are the best materials for artificial intelligence learning (Yu et al. 2017).

Artificial intelligence based on deep learning has been used in cytology diagnosis, as well as in the histopathological diagnosis including breast cancer, gastric cancer, lung cancer, and skin diseases, and the core roles are early tumor screening, disease grading, and diagnosis of benign and malignant diseases. It has been showing its unique advantages, although there is no extensively application at present (Savala et al. 2018; Qaiser et al. 2017; Haj-Hassan et al. 2017; Yoshida et al. 2017; Coudray et al. 2018; Olsen et al. 2018).

The pathological diagnosis of renal biopsy is quite different from that of routine surgical pathology. Although only a small amount of renal tissue from fine needle puncture is collected, it is required to be used for frozen slices, paraffin slices, and ultra-thin resin slices in each case, which are dyed by various staining techniques including immunofluorescence, H&E, Masson, PAS/PASM, ultra-thin slices double dyeing. Then, the tissues can be analyzed by fluorescence microscopy, optical microscopy, transmission electron microscopy, and polarizing microscopy, and finally combined with clinical manifestations to make a diagnosis. Compared with routine pathological diagnosis, renal biopsy diagnosis is more time-consuming and laborious, and the intervention of artificial intelligence is more urgent. Artificial intelligence has been applied to the management of clinical renal hemodialysis and prediction and risk stratification of kidney outcomes in IgA nephropathy (Hueso et al. 2018; Chen et al. 2019). American scholars have reported that the computer can identify and count the glomeruli, but so far there is no clinic use of artificial intelligence renal biopsy pathological diagnosis system based on deep learning (Rosenberg et al. 2016; Zee et al. 2018).

To establish an artificial intelligence pathological diagnosis system based on deep learning, it is the key to prepare sufficient and standardized training sets (Yu et al. 2017; Olsen et al. 2018). The training set is made up of a series of accurately labeled digital pathological images, thereby, accurate labeling is a prerequisite for making high-quality models.

The training set originates from the deviation or incomplete digital pathological images, which will lower the performance of the model to produce the distorted decision in the end. Renal biopsy pathological diagnosis mostly deals with a variety of acute and chronic inflammatory diseases, and the pathological changes include the increasing number of glomerular cells, the deposit of immune complexes and ECM, the thickening of the capillary wall, the narrowing of the lumen. And all of them are easy to be recognized by the machine, but it is complicated to analyze the lesions with inconsistent morphology at different stages, moreover, the tissues stained with numerous methods. Therefore, the accurately labeling lesions is a heavy workload and challenging task.

Currently, there is no public training set can be available globally. Compared with the judgment of glomerular, tubular, and vascular lesions, artificial intelligence is less difficult to assess renal fibrosis, due to the collagen deposition in the tissue can be revealed by Sirius Red staining or Masson staining or immunohistochemical staining. Image recognition can be performed by RGB Channel separation and convolution neural network model. After a lot of learning and training, the computer can automatically identify and make a conclusion. Today, the problem is that renal puncture tissue dyeing has not been standardized and made a stable staining protocol. It is hard to study when the dyeing quality is variable in different laboratories or different batches in the same laboratory. For this reason, there is long way to go before software could be used to assist the diagnosis of renal pathology.

2.5 Conclusion

The evaluation of renal fibrosis aims to determine the amount of collagen deposition, and the examining methods can be divided into morphometry, biochemical measurement, and molecular markers detection. Compared with clinical biological and histological data, immunohistochemistry plays an important role in evaluating glomerular diseases, but the classical semi-quantitative method of histology is still not out of date. Computer-assisted image analysis cannot completely replace artificial grading method currently, however, artificial intelligence-assisted renal fibrosis diagnosis system is urgently expected.

References

Anders HJ, Ryu M (2011) Renal microenvironments and macrophage phenotypes determine progression or resolution of renal inflammation and fibrosis. Kidney Int 80:915–925

Bob FR, Gluhovschi G, Herman D, Potencz E, Gluhovschi C, Trandafirescu V et al (2008) Histological, immunohistochemical and biological data in assessing interstitial fibrosis in patients with chronic glomerulonephritis. Acta Histochem 110:196–203

Campbell MT, Hile KL, Zhang H, Asanuma H, Vanderbrink BA, Rink RR et al (2011) Toll-like receptor 4: a novel signaling pathway during renal fibrogenesis. J Surg Res 168:e61–e69

Chen P (2017) Pathology, 2nd edn. Southeast University Publisher Co., Nanjing

Chen T, Li X, Li Y, Xia E, Qin Y, Liang S et al (2019) Prediction and risk stratification of kidney outcomes in IgA nephropathy. Am J Kidney Dis pii S0272–6386(19):30618–3
Coudray N, Ocampo PS, Sakellaropoulos T, Narula N, Snuderl M, Fenyö D et al (2018) Classification and mutation prediction from non-small cell lung cancer histopathology images using deep learning. Nat Med 24:1559–1567
Dussaule JC, Guerrot D, Huby AC, Chadjichristos C, Shweke N, Boffa JJ et al (2011) The role of cell plasticity in progression and reversal of renal fibrosis. Int J Exp Pathol 92:151–157
Eardley KS, Zehnder D, Quinkler M, Lepenies L, Bates RL, Savage CO et al (2006) The relationship between albuminuria, MCP-1/CCL2, and interstitial macrophages in chronic kidney disease. Kidney Int 69:1189–1197
Farris AB, Colvin RB (2012) Renal interstitial fibrosis: mechanisms and evaluation. Curr Opin Nephrol Hypertens 21:289–300
Haj-Hassan H, Chaddad A, Harkouss Y, Desrosiers C, Toews M, Tanougast C (2017) Classifications of multispectral colorectal cancer tissues using convolution neural network. J Pathol Inform 8:1–7
Hakroush S, Moeller MJ, Theilig F, Kaissling B, Sijmonsma TP, Jugold M et al (2009) Effects of increased renal tubular vascular endothelial growth factor (VEGF) on fibrosis, cyst formation, and glomerular disease. Am J Pathol 175:1883–1895
Higgins DF, Kimura K, Iwano M, Haase VH (2008) Hypoxia-inducible factor signaling in the development of tissue fibrosis. Cell Cycle 7:1128–1132
Hueso M, Vellido A, Montero N, Barbieri C, Ramos R, Angoso M et al (2018) Artificial tntelligence for the artificial kidney: pointers to the future of a personalized hemodialysis therapy. Kidney Dis 4:1–9
Kaissling B, Le Hir M (2008) The renal cortical interstitium: morphological and functional aspects. Histochem Cell Biol 130:247–262
Kim DH, Moon SO, Jung YJ, Lee AS, Kang KP, Lee TH et al (2009) Mast cells decrease renal fibrosis in unilateral ureteral obstruction. Kidney Int 75:1031–1038
Kimura K, Iwano M, Higgins DF, Yamaguchi Y, Nakatani K, Harada K et al (2008) Stable expression of HIF-1alpha in tubular epithelial cells promotes interstitial fibrosis. Am J Physiol Renal Physiol 295:F1023–F1029
Lecun Y, Bengio Y, Hinton G (2015) Deep learning. Nature 521:436
Lian YG, Zhou QG, Zhang YJ, Zheng FL (2011) VEGF ameliorates tubulointerstitial fibrosis in unilateral ureteral obstruction mice via inhibition of epithelialmesenchymal transition. Acta Pharmacol Sin 32:1513–1521
Lin SL, Kisseleva T, Brenner DA, Duffield JS (2008) Pericytes and perivascular fibroblasts are the primary source of collagenproducing cells in obstructive fibrosis of the kidney. Am J Pathol 173:1617–1627
Meran S, Steadman R (2011) Fibroblasts and myofibroblasts in renal fibrosis. Int J Exp Pathol 92:158–167
Miller DD, Brown EW (2018) Artificial intelligence in medical practice: the question to the answer? Am J Med 131:129–133
Nikolic-Paterson DJ (2010) $CD4^{+}$ T cells: A potential player in renal fibrosis. Kidney Int 78:333–335
Norman J (2011) Fibrosis and progression of Autosomal Dominant Polycystic Kidney Disease (ADPKD). Biochim Biophys Acta 1812:1327–1336
Olsen TG, Jackson BH, Feeser TA, Kent MN, Moad JC, Krishnamurthy S et al (2018) Diagnostic performance of deep learning algorithms applied to three common diagnoses in dermatopathology. J Pathol Inform 9:32
Piera-Velazquez S, Li Z, Jimenez SA (2011) Role of endothelial-mesenchymal transition (EndoMT) in the pathogenesis of fibrotic disorders. Am J Pathol 179:1074–1080
Qaiser T, Mukherjee A, Reddy PC, Munugoti SD, Tallam V, Pitkäaho T et al (2017) Her2 challenge contest: a detailed assessment of automated Her2 scoring algorithms in whole slide images of breast cancer tissues. Histopathol 72:227
Robertsa ISD, Furness PN, Cook HT (2004) Beyond diagnosis: stage and grade in inflammatory renal disease. Current Diagnostic Pathology 10:22–35

Rosenberg AZ, Palmer M, Merlino L, Troost JP, Gasim A, Bagnasco S et al (2016) The Application of digital pathology to improve accuracy in glomerular enumeration in renal biopsies. PLoS ONE 11:e0156441

Rubin R, Strayer DS (2012) Rubin's pathology: clinicopathologic foundations of medicine, 6th edn. Lippincott Williams & Wilkins, Philadelphia

Russell S (2017) Artificial intelligence: The future is superintelligent. Nature 548:520–521

Sakai N, Furuichi K, Shinozaki Y, Yamauchi H, Toyama T, Kitajima S et al (2010) Fibrocytes are involved in the pathogenesis of human chronic kidney disease. Human Pathol 41:672–678

Satirapoj B, Nast CC, Adler SG (2012) Novel insights into the relationship between glomerular pathology and progressive kidney disease. Adv Chronic Kidney Dis 19:93–100

Savala R, Dey P, Gupta N (2018) Artificial neural network model to distinguish follicular adenoma from follicular carcinoma on fine needle aspiration of thyroid. Diagn Cytopathol 46:244–249

Scolari F, Ravani P (2010) Atheroembolic renal disease. Lancet 375:1650–1660

Snelgrove SL, Kausman JY, Lo C, Lo C, Ooi JD, Coates PT et al (2012) Renal dendritic cells adopt a proinflammatory phenotype in obstructive uropathy to activate T cells, but do not directly contribute to fibrosis. Am J Pathol 180:91–103

Stuht S, Gwinner W, Franz I, Schwarz A, Jonigk D, Kreipe H et al (2007) Lymphatic neoangiogenesis in human renal allografts: results from sequential protocol biopsies. Am J Transplant 7:377–384

Sun G, Reddy MA, Yuan H, Lanting L, Kato M, Natarajan R (2010) Epigenetic histone methylation modulates fibrotic gene expression. J Am Soc Nephrol 21:2069–2080

Tizhoosh HR, Pantanowitz L (2018) Artificial intelligence and digital pathology: challenges and opportunities. J Pathol Inform 9:38

Vernon MA, Mylonas KJ, Hughes J (2010) Macrophages and renal fibrosis. Semin Nephrol 30:302–317

Wada T, Sakai N, Sakai Y, Matsushima K, Kaneko S, Furuichi K (2011) Involvement of bone-marrow-derived cells in kidney fibrosis. Clin Exp Nephrol 15:8–13

Yamaguchi Y, Kanetsuna Y, Honda K, Yamanaka N, Kawano M, Nagata M (2012) Japanese study group on IgG4-related nephropathy: Characteristic tubulointerstitial nephritis in IgG4-related disease. Hum Pathol 43:536–549

Yan W, Tang Y, Chang EI-C, Lai M, Xu Y (2018) Deep learning in digital pathology analysis. Chin J Biomed Eng 37:95–105

Yang XY, Du GH (2004) Microvolume assay method of hydroxyproline in animal tissues and application. Chin Pharmacol Bull 20:836–837

Yang L, Besschetnova TY, Brooks CR, Shah JV, Bonventre JV (2010) Epithelial cell cycle arrest in G2/M mediates kidney fibrosis after injury. Nat Med 6:535–543

Yoshida H, Shimazu T, Kiyuna T, Marugame A, Yamashita Y, Cosatto E et al (2017) Automated histological classification of whole-slide images of gastric biopsy specimens. Gastric Cancer 8:1–9

Yu G, Wei P, Chen Y, Zhu M (2017) Artificial Intelligence in pathological diagnosis and assessment of human solid tumor: application and thinking. Acad J Sec Mil Med Univ 38:1349–1354

Zee J, Hodgin JB, Mariani LH, Gaut JP, Palmer MB, Bagnasco SM et al (2018) Reproducibility and feasibility of strategies for morphologic assessment of renal biopsies using the nephrotic syndrome study network digital pathology scoring system. Arch Pathol Lab Med 142:613–625

Zhou W (2013) Real biopsy pathology, 3rd edn. Peiking University Publisher Co., Beijing

Zhou LT, Qiu S, Lv LL, Li ZL, Liu H, Tang RN et al (2018) Integrative bioinformatics analysis provides insight into the molecular mechanisms of chronic kidney disease. Kidney Blood Press Res 43:568–581

Chapter 3
Current Opinion for Hypertension in Renal Fibrosis

Hai-Jian Sun

Abstract Arterial hypertension remains to be a serious problem with considerable morbidity and mortality worldwide in the present age. Hypertension is a major risk factor for cardiovascular diseases such as stroke, myocardial infarction, renal failure, and heart failure. Hypertensive nephropathy is the second leading cause of death in chronic kidney disease (CKD) around the world. Long-time hypertension loading results in renal interstitial fibrosis, which is associated with aberrant activation of renal fibroblasts and excessive generation of extracellular matrix (ECM) proteins. Increasing evidence supported that proteinuria, tubular hypertrophy, oxidative stress, activation of renin–aldosterone–angiotensin system (RAAS), collagen turnover, chronic inflammation, and vasoactive substances synergistically contributed to the pathogenesis of hypertensive renal fibrosis. However, the mechanisms involving the pathogenesis of hypertensive renal fibrosis are complex and not fully understood. Also, the effective clinical therapy to halt or even reverse renal fibrosis in hypertension is still limited. In this chapter, we aimed to provide an overview of the main pathophysiologic and mechanistic features of renal fibrosis under hypertensive state. The completion of the studies in these directions would improve our understanding of renal fibrosis in hypertension and also help us better screen treatment strategies for preventing renal destruction associated with hypertension.

Keywords Hypertension · Kidney · Fibrosis · Chronic kidney disease

3.1 Introduction

A worldwide epidemic of chronic kidney disease (CKD) exists, and hypertensive nephropathy is a vital cause for end-stage renal disease (ESRD) in both developed and developing countries (Wei et al. 2013). With the increased morbidity and mortality in patients with ESRD, novel therapeutic strategies should focus to reduce

H.-J. Sun (✉)
Department of Pharmacology, Yong Loo Lin School of Medicine, National University of Singapore, Singapore 117600, Singapore
e-mail: phcsunh@nus.edu.sg

B.-C. Liu et al. (eds.), *Renal Fibrosis: Mechanisms and Therapies*, Advances in Experimental Medicine and Biology 1165,
https://doi.org/10.1007/978-981-13-8871-2_3

albuminuria, control blood pressure, restore renal function, and delay the progression of hypertensive nephropathy (Hart and Bakris 2010).

The kidney injury responses to hypertension are characterized by renal fibrosis, tubular hypertrophy, and glomerular alterations (Ghaderian and Beladi-Mousavi 2014). Dysregulation of renal homeostasis fundamentally triggers fibrotic scarring and ultimate loss of renal function (Grabias and Konstantopoulos 2014). Hypertensive renal fibrogenesis is a complex process involving a host of pathological scarring processes such as dysregulated extracellular matrix (ECM) assembly, anchoring or degradation, inflammatory cytokines formation, failed regeneration of tubular epithelium, microvascular rarefaction, and activated fibroblasts (Tampe and Zeisberg 2014). Moreover, endothelial-to-mesenchymal transition (EMT) in endothelial cells is also considered as an alternative mechanism for renal fibrosis (Harris and Neilson 2006). A comprehensive understanding of pro- and antifibrotic signaling pathways may identify crucial indications for therapeutic strategies to prevent and treat renal fibrosis in hypertension. In this chapter, we will focus on the current signaling pathways and targeted genes with high therapeutic potential in hypertensive renal fibrosis.

3.2 EMT in Hypertensive Renal Fibrosis

The transition from epithelial to mesenchymal phenotype is observed during embryonic development, and this phenomenon also takes place in hypertensive renal fibrosis (Dussaule et al. 2011). Initial studies suggest that about 30% of renal interstitial fibroblasts may be derived from tubular epithelial cells (Iwano et al. 2002). Currently, many researches support a notion that resident epithelial cells may dedifferentiate to a fibroblast-like phenotype through EMT process (Grabias and Konstantopoulos 2014). According to previous studies, the contribution of tubular epithelial cells to renal interstitial fibrosis is a subject of great interest.

In angiotensin II (Ang II)-infused hypertensive rats, the EMT in the peritubular renal capillaries is detected before the onset of significant proteinuria or renal fibrosis, further suggesting the close relationship between EMT and hypertensive renal fibrosis (Dussaule et al. 2011). Dietary salt intake induces hypertension, renal fibrosis, and tubular EMT, as evidenced by reduced expression of E-cadherin and enhanced expression of α-Smooth muscle actin (α-SMA) in rats (Wang et al. 2014). A toll-like receptor 4 antagonist counteracts aldosterone-triggered hypertension, renal fibrosis, and EMT in tubular epithelial cells (Zhang et al. 2015). These studies imply that EMT is a critical mechanism involving in the progression of renal fibrosis at least in experimental hypertensive models. Additional studies are needed to better recognize the molecular mechanisms of EMT in hypertensive renal fibrosis, and these investigations could lead to develop novel therapeutic strategies for hypertensive renal fibrosis.

3.3 Role of Ang II in Hypertensive Renal Fibrosis

3.3.1 RAAS, Ang II, and Hypertensive Renal Fibrosis

It is well known that renin–aldosterone–angiotensin system (RAAS) is crucially involved in hypertensive nephropathy (Udani et al. 2011). Within the RAAS, Ang II is regarded as an essential mediator for hypertension and hypertensive nephropathy (Wolf 2006). The level of renal Ang II is approximately one thousand-fold greater than the circulating Ang II (Klahr and Morrissey 2000). Most of the deleterious actions of Ang II are believed to be mediated by Ang II type 1 (AT1) receptors (Higuchi et al. 2007). Chronic infusion of Ang II causes hypertension, along with a fibrogenic phenotype involving the cortex and medulla in the kidney of murine models (Lombardi et al. 2001; Zhao et al. 2008). Pharmacological blockade of Ang II or antagonism of AT1 receptors retards or ameliorates renal fibrosis in several models of experimental or genetic hypertension (Chatziantoniou et al. 2004). In fact, Ang II receptor antagonists are widely used for the treatment of the hypertension in addition to its antifibrotic effects on the kidney (Bascands and Schanstra 2004).

3.3.2 EMT, Ang II, and Hypertensive Renal Fibrosis

It is documented that Ang II contributes to ECM production and renal fibrosis by activating many intracellular signaling pathways such as mitogen-activated protein kinases (MAPKs), transforming growth factor-β (TGF-β)/SMAD and the NFκB (nuclear factor κB) cascades (Ruiz-Ortega et al. 2006b; Mehta and Griendling 2007; Ruster and Wolf 2011). Ang II stimulates the migration of inflammatory cells and the production of chemokines and induces the synthesis of collagens by activation of platelet-derived growth factor (PDGF), nicotinamide adenine dinucleotide phosphate (NADPH) oxidase activity, connective tissue growth factor (CTGF), epidermal growth factor (EGF), and inflammation response (Fogo 2001a, b; el Nahas et al. 1997; McCarty 2006; Remuzzi et al. 2006; Klahr and Morrissey 2002; Modlinger et al. 2004; Ruiz-Ortega et al. 2006a). The effects of these factors on renal fibrosis have been well established in various models of hypertensive renal diseases (Klahr and Morrissey 2000) (Fig. 3.1).

3.3.3 Downstream Mediators of Ang II in Hypertensive Renal Fibrosis

The dysregulation of EMT signaling pathways is important in the early stages of interstitial fibrosis, and Ang II contributes to EMT through induction of profibrotic factors (Saad et al. 2010). Blockade of the MAPK cascade or RhoA-/Rho-associated kinase

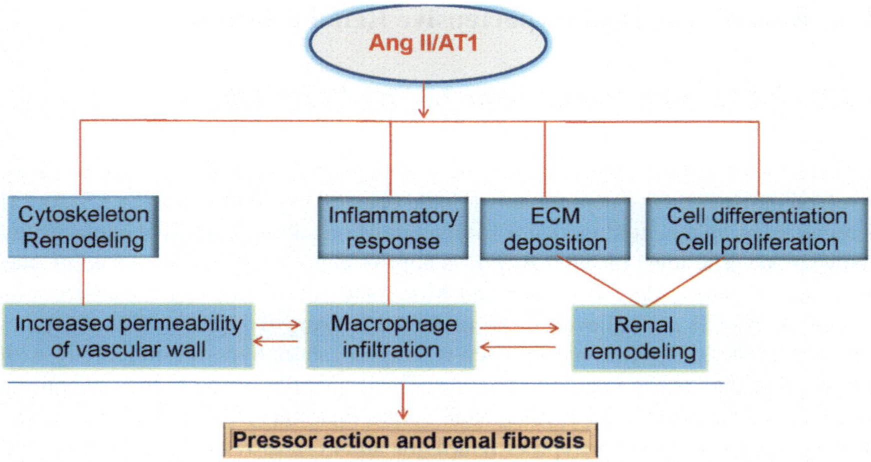

Fig. 3.1 Effects of Ang II/AT1 receptors axis on hypertensive renal damage. Angiotensin II (Ang II) acts on AT1 receptor to induce various biological effects including cytoskeleton remodeling, extracellular matrix generation, inflammatory response, macrophage infiltration, fibroblasts proliferation, tubules epithelial cell differentiation, and apoptosis in the kidney, all of which may interact with each other and eventually lead to hypertension and renal fibrosis

(ROCK) pathway diminishes Ang II-induced EMT in tubular cells (Rodrigues-Diez et al. 2008). Knockout of cytochrome P450 4A14 attenuates hypertension and renal fibrosis in Ang II-treated mice (Zhou et al. 2018). Ang II increases renal cytochrome P450 1B1 activity, which contributes to renal dysfunction and damages including renal fibrosis, tubular damage, and inflammation (Jennings et al. 2012). Infusion of Ang II increases blood pressure and aggravates renal fibrosis in mice, which were minimized in cytosolic phospholipase A2α−/− mice (Khan et al. 2016). It is revealed that interleukin 6 (IL-6) is a key downstream cytokine of Ang II, and it can directly induce fibrotic gene expressions in the kidney (Zhang et al. 2012). CC chemokine receptor 2 (CCR2) activation plays a central role in the development of hypertensive nephropathy including renal fibrosis in Ang II-treated mice (Liao et al. 2008). Moreover, tumor necrosis factor (TNF)-α is also involved in Ang II-evoked dendritic cell infiltration and maturation, and dexamethasone immunosuppression normalizes Ang II-induced hypertension and renal damage (Muller et al. 2002). Therefore, an effective anti-fibrotic therapy via blockade of Ang II/AT1 receptor signaling may develop attractive approaches for the treatment of hypertension and renal fibrosis. These current findings provide a better understanding of the mechanisms by which Ang II induces renal fibrosis in hypertension and open new therapeutic approaches to control the hypertensive renal fibrosis induced by Ang II.

3.4 Role of Oxidative Stress in Hypertensive Renal Fibrosis

3.4.1 NADPH, ROS, and Hypertensive Renal Fibrosis

A plethora of researches have verified that oxidative stress contributes to hypertensive kidney diseases associated with inflammation, endothelial dysfunction, tissue damage, and renal fibrosis (Lv et al. 2018). Excessive renal oxidative stress leads to renal damage and hypertension and reversal of oxidative stress by antioxidants decreases blood pressure and the related renal damage (Manning et al. 2005).

A number of studies have shown that chronic antioxidant treatment can effectively suppress renal oxidative stress and improve renal function in hypertension (Manning et al. 2005; Tian et al. 2006). NADPH oxidase is the main enzyme for ROS production, thereby participating in various organ damages including renal injury (Manning et al. 2005; Touyz et al. 2003). Among seven members of the NOX family of membrane-bound NADPH oxidases, NOX1, NOX2, and NOX4 are abundantly expressed in both human and rodent kidneys, and they exert a vital role in renal oxidative stress, inflammation, and fibrosis (Decleves and Sharma 2014). NOX1 is believed to mediate Ang II-induced hypertension (Jia et al. 2008), but the role of NOX1 in pathophysiology of hypertensive renal injury is incompletely known (Lee et al. 2015). An increase in NOX2 expression is found in the kidney of hypertensive rats, and simultaneous treatment with L-carnitine attenuates the renal fibrosis and oxidative stress in rats (Zambrano et al. 2014). There is still a dispute about the role of NOX4 in the renal injury. It is reported that NOX4 exerts a renoprotective action, and NOX4-deficient mice show the elevated tubulointerstitial fibrosis and oxidative stress after obstruction (Decleves and Sharma 2014). Quite a few documents demonstrate that NOX4-mediated ROS generation is responsible for unilateral ureteral obstruction or diabetes-induced renal fibrosis (Zhou et al. 2017; He et al. 2016). However, the functional role of NOX4 in hypertensive renal fibrosis is not well-defined.

3.4.2 PPAR-γ, Oxidative Stress, and Hypertensive Renal Fibrosis

Peroxisome proliferator-activated receptor-γ (PPAR-γ) belongs to a member of the ligand-activated nuclear transcriptional superfamily that is extensively expressed in the kidneys (Lv et al. 2018). The antifibrotic effects of PPAR-γ agonists are related to the reduction of mitochondrial ROS (Lv et al. 2018) (Fig. 3.2). Rosiglitazone treatment counteracts hypertension and exhibits anti-inflammatory and anti-fibrotic actions in the kidney of deoxycorticosterone acetate-salt hypertensive rats (Bae et al. 2010). Administration of irbesartan activates PPAR-γ to ameliorate renal fibrosis, glomerular injury, EMT, and oxidative stress in a mouse model of salt-sensitive hypertension (Kusunoki et al. 2013). L-carnitine attenuates the renal fibrosis in hyperten-

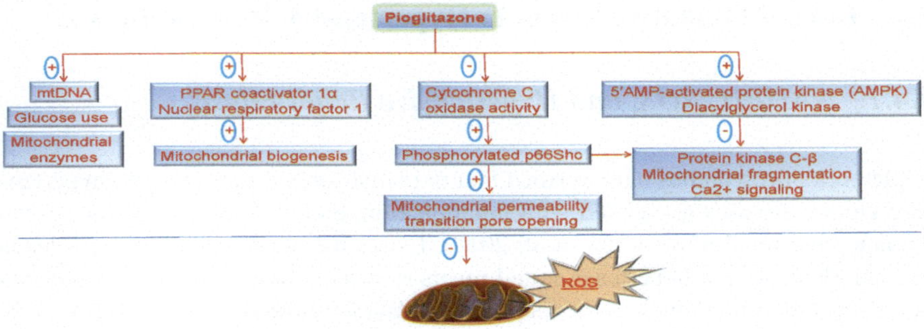

Fig. 3.2 Effects of pioglitazone on mitochondrial reactive oxygen species (ROS) during the development of renal fibrosis. Activation of PPAR-γ with pioglitazone may attenuate mitochondrial oxidative stress via several biological mechanisms: (1) upregulation of mitochondrial DNA, glucose utilization, mitochondrial enzymes; (2) increasing mitochondrial biogenesis by stimulating PPAR coactivator 1α and nuclear respiratory factor 1; (3) activating 5′AMP-activated protein kinase (AMPK) to inhibit protein kinase C-β or mitochondrial fragmentation Ca^{2+} signaling; (4) reducing cytochrome C oxidase activity. All these mechanisms are responsible for ameliorating effect of PPAR-γ on mitochondrial ROS

sive rats, and its effects may be dependent on PPAR-γ pathway activation (Zambrano et al. 2014). In short, a large body of evidence supports the protective role of PPAR-γ agonists in hypertensive renal fibrosis.

3.4.3 Immune Activation, Oxidative Stress, and Hypertensive Renal Fibrosis

Very recent studies have recognized a close link between immune activation and oxidative stress in hypertensive end-organ damages including renal fibrosis (Harwani et al. 2012). The renal effects of immune activation remain to be fully defined, but the roles of these cytokines in angiotensinogen production, sodium reabsorption, and renal fibrosis have been emerged (McMaster et al. 2015). The formation of ROS in dendritic cells leads to immune cell activation, cytokines release, and hypertensive renal fibrosis (McMaster et al. 2015) (Fig. 3.3). It has been revealed that T-cell suppressing agent mycophenolate mofetil retards blood pressure elevation and renal inflammation in experimental hypertensive models (Muller et al. 2002; Franco et al. 2007). High-dose infusion of Ang II into mice lacking interferon (IFN)-γ receptor 1 results in less renal fibrosis without changing hypertension (Marko et al. 2012). In conjunction with salt and water retention, T-cell-derived cytokines stimulate renal ROS formation, glomerular damage, renal fibrosis, and injury, all of which promote hypertensive renal dysfunction (McMaster et al. 2015). Substantial additional researches are needed to define the diagnostic and therapeutic tools based on T-cell-

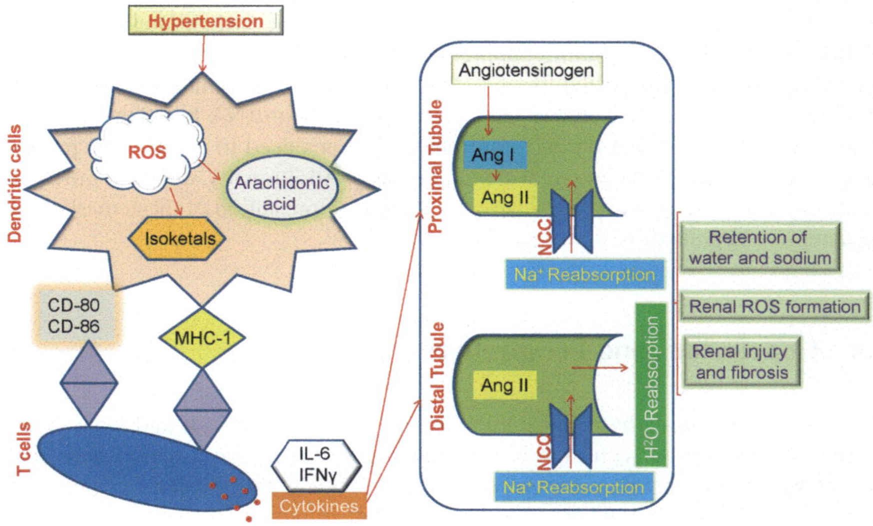

Fig. 3.3 Immune activation and hypertensive renal fibrosis. In hypertension, the excessive production of reactive oxygen species (ROS) in dendritic cells induces oxidation of arachidonic acid and isoketal formation. The forming proteins CD-80 and CD-86 by dendritic cells are presented to T cells, which polarize T cells to produce and release specific cytokines including interleukin-1β (IL-1β) and interferon γ (IFN-γ). These released cytokines promote the production of angiotensinogen, and angiotensin I (Ang I) is converted to angiotensin II (Ang II) by intrarenal angiotensin converting enzyme. Ang II upregulates transport channels such as sodium chloride cotransporter (NCC) in both proximal and distal convoluted tubules. In combination with salt and water retention, activation of T cells also triggers renal ROS generation, renal injury, and fibrosis. Certainly, the renal dysfunction response to T-cell activation may exacerbate hypertension

mediated local inflammation for the treatment of the end-organ damage associated with hypertension including renal fibrosis.

3.5 Role of MicroRNAs in Hypertensive Renal Fibrosis

MicroRNAs (miRNAs) are endogenously produced short RNAs with 21–25 nucleotides length, which play an important role in protein translation through binding to the 3′-untranslated region (UTR) of targeted genes (Cano and Nieto 2008). Profiling miRNA study of human hypertensive nephrosclerosis biopsy discloses that the levels of miR-141, miR-192, miR-200a, miR-200b, miR-205, and miR-429 are strikingly increased (Wei et al. 2013; Wang et al. 2010; Petrillo et al. 2017). Compared with control rats, the miR-29b expression in renal medulla of salt-sensitive hypertensive rats is obviously increased and increased miR-29b protects rats from hypertensive nephropathy (Liu et al. 2010). Low-dose paclitaxel treatment stimulates miR-192 to inhibit TGF-β-Smad2/3 signaling, thus ameliorating renal fibrosis in another animal model with 5/6 nephrectomy (Sun et al. 2011). Downregulation of miR-324-3p by ACE inhibitor suppresses hypertensive rat kidney damages, evidenced by less renal interstitial fibrosis (Macconi et al. 2012). Thus, a potential role of miRNAs in hypertensive nephropathy is emerging. For miRNA-based therapeutics against hypertensive renal fibrosis, more investigations need to identify the prominent responsible miRNAs and relevant target genes. Additionally, different miRNAs are discovered in renal diseases, whereas the clinical translation of these markers in hypertensive nephropathy is still lacking.

3.6 Conclusions and Perspectives

Hypertensive nephropathy is reflected by progressive renal fibrosis and renal insufficiency in association with high blood pressure, which is also a major complication of hypertension. From a therapeutic point of view, the imbalanced pro- and antifibrotic signaling pathways may be a guiding principle for developing novel antifibrotic strategies in hypertensive renal fibrosis and subsequent end-stage renal damage. Prospective therapeutics based on these findings will bring new hope for patients with hypertensive renal damage in the near future. Although our knowledge of mechanosensitive signaling events in hypertensive renal fibrosis is still in its infancy, a wide range of impressive studies focusing on hypertensive renal damage is growing rapidly in recent years. More works are needed to provide a full understanding of potential mediators in hypertensive renal fibrosis, which will undoubtedly help to identify novel therapeutic targets or propel the development of novel therapies that can halt or reverse hypertensive renal fibrosis.

Acknowledgements We would like to thank Professor J.S.B. (National University of Singapore) for helpful discussing and revising the overall the manuscript.

References

Bae EH, Kim IJ, Ma SK, Kim SW (2010) Rosiglitazone prevents the progression of renal injury in DOCA-salt hypertensive rats. Hypertens Res 33:255–262

Bascands JL, Schanstra JP (2004) Bradykinin and renal fibrosis: have we ACE'd it? J Am Soc Nephrol 15:2504–2506

Cano A, Nieto MA (2008) Non-coding RNAs take centre stage in epithelial- to- mesenchymal transition. Trends Cell Biol 18:357–359

Chatziantoniou C, Boffa JJ, Tharaux PL, Flamant M, Ronco P, Dussaule JC (2004) Progression and regression in renal vascular and glomerular fibrosis. Int J Exp Pathol 85:1–11

Decleves AE, Sharma K (2014) Novel targets of antifibrotic and anti-inflammatory treatment in CKD. Nat Rev Nephrol 10:257–267

Dussaule JC, Guerrot D, Huby AC, Chadjichristos C, Shweke N, Boffa JJ et al (2011) The role of cell plasticity in progression and reversal of renal fibrosis. Int J Exp Pathol 92:151–157

el Nahas AM, Muchaneta-Kubara EC, Essawy M, Soylemezoglu O (1997) Renal fibrosis: insights into pathogenesis and treatment. Int J Biochem Cell Biol 29:55–62
Fogo AB (2001a) Progression and potential regression of glomerulosclerosis. Kidney Int 59:804–819
Fogo AB (2001b) Renal fibrosis and the renin-angiotensin system. Adv Nephrol Necker Hosp 31:69–87
Franco M, Martinez F, Quiroz Y, Galicia O, Bautista R, Johnson RJ et al (2007) Renal angiotensin II concentration and interstitial infiltration of immune cells are correlated with blood pressure levels in salt- sensitive hypertension. Am J Physiol Regul Integr Comp Physiol 293:R251–R256
Ghaderian SB, Beladi-Mousavi SS (2014) The role of diabetes mellitus and hypertension in chronic kidney disease. J Renal Inj Prev 3:109–110
Grabias BM, Konstantopoulos K (2014) The physical basis of renal fibrosis: effects of altered hydrodynamic forces on kidney homeostasis. Am J Physiol Renal Physiol 306:F473–F485
Harris RC, Neilson EG (2006) Toward a unified theory of renal progression. Annu Rev Med 57:365–380
Hart PD, Bakris GL (2010) Hypertensive nephropathy: prevention and treatment recommendations. Expert Opin Pharmacother 11:2675–2686
Harwani SC, Chapleau MW, Legge KL, Ballas ZK, Abboud FM (2012) Neurohormonal modulation of the innate immune system is proinflammatory in the prehypertensive spontaneously hypertensive rat, a genetic model of essential hypertension. Circ Res 111:1190–1197
He T, Xiong J, Nie L, Yu Y, Guan X, Xu X et al (2016) Resveratrol inhibits renal interstitial fibrosis in diabetic nephropathy by regulating AMPK/NOX4/ROS pathway. J Mol Med (Berl) 94:1359–1371
Higuchi S, Ohtsu H, Suzuki H, Shirai H, Frank GD, Eguchi S (2007) Angiotensin II signal transduction through the AT1 receptor: novel insights into mechanisms and pathophysiology. Clin Sci (Lond) 112:417–428
Iwano M, Plieth D, Danoff TM, Xue C, Okada H, Neilson EG (2002) Evidence that fibroblasts derive from epithelium during tissue fibrosis. J Clin Invest 110:341–350
Jennings BL, Anderson LJ, Estes AM, Yaghini FA, Fang XR, Porter J et al (2012) Cytochrome P450 1B1 contributes to renal dysfunction and damage caused by angiotensin II in mice. Hypertension 59:348–354
Jia Z, Guo X, Zhang H, Wang MH, Dong Z, Yang T (2008) Microsomal prostaglandin synthase-1-derived prostaglandin E2 protects against angiotensin II-induced hypertension via inhibition of oxidative stress. Hypertension 52:952–959
Khan NS, Song CY, Thirunavukkarasu S, Fang XR, Bonventre JV, Malik KU (2016) Cytosolic phospholipase A2α is essential for renal dysfunction and end-organ damage associated with angiotensin II-induced hypertension. Am J Hypertens 29:258–265
Klahr S, Morrissey JJ (2000) The role of vasoactive compounds, growth factors and cytokines in the progression of renal disease. Kidney Int Suppl 75:S7–14
Klahr S, Morrissey J (2002) Obstructive nephropathy and renal fibrosis. Am J Physiol Renal Physiol 283:F861–F875
Kusunoki H, Taniyama Y, Rakugi H, Morishita R (2013) Cardiac and renal protective effects of irbesartan via peroxisome proliferator- activated receptorgamma- hepatocyte growth factor pathway independent of angiotensin II Type 1a receptor blockade in mouse model of salt-sensitive hypertension. J Am Heart Assoc 2:e000103
Lee SY, Kim SI, Choi ME (2015) Therapeutic targets for treating fibrotic kidney diseases. Transl Res 165:512–530
Liao TD, Yang XP, Liu YH, Shesely EG, Cavasin MA, Kuziel WA et al (2008) Role of inflammation in the development of renal damage and dysfunction in angiotensin II-induced hypertension. Hypertension 52:256–263
Liu Y, Taylor NE, Lu L, Usa K, Cowley AW Jr, Ferreri NR et al (2010) Renal medullary microRNAs in Dahl salt-sensitive rats: miR-29b regulates several collagens and related genes. Hypertension 55:974–982

Lombardi DM, Viswanathan M, Vio CP, Saavedra JM, Schwartz SM, Johnson RJ (2001) Renal and vascular injury induced by exogenous angiotensin II is AT1 receptor-dependent. Nephron 87:66–74

Lv W, Booz GW, Fan F, Wang Y, Roman RJ (2018) Oxidative stress and renal fibrosis: recent insights for the development of novel therapeutic strategies. Front Physiol 9:105

Macconi D, Tomasoni S, Romagnani P, Trionfini P, Sangalli F, Mazzinghi B et al (2012) MicroRNA-324-3p promotes renal fibrosis and is a target of ACE inhibition. J Am Soc Nephrol 23:1496–1505

Manning RD Jr, Tian N, Meng S (2005) Oxidative stress and antioxidant treatment in hypertension and the associated renal damage. Am J Nephrol 25:311–317

Marko L, Kvakan H, Park JK, Qadri F, Spallek B, Binger KJ et al (2012) Interferon- gamma signaling inhibition ameliorates angiotensin II- induced cardiac damage. Hypertension 60:1430–1436

McCarty MF (2006) Adjuvant strategies for prevention of glomerulosclerosis. Med Hypotheses 67:1277–1296

McMaster WG, Kirabo A, Madhur MS, Harrison DG (2015) Inflammation, immunity, and hypertensive end- organ damage. Circ Res 116:1022–1033

Mehta PK, Griendling KK (2007) Angiotensin II cell signaling: physiological and pathological effects in the cardiovascular system. Am J Physiol Cell Physiol 292:C82–C97

Modlinger PS, Wilcox CS, Aslam S (2004) Nitric oxide, oxidative stress, and progression of chronic renal failure. Semin Nephrol 24:354–365

Muller DN, Shagdarsuren E, Park JK, Dechend R, Mervaala E, Hampich F et al (2002) Immunosuppressive treatment protects against angiotensin II-induced renal damage. Am J Pathol 161:1679–1693

Petrillo F, Iervolino A, Zacchia M, Simeoni A, Masella C, Capolongo G et al (2017) MicroRNAs in renal diseases: a potential novel therapeutic target. Kidney Dis (Basel) 3:111–119

Remuzzi G, Benigni A, Remuzzi A (2006) Mechanisms of progression and regression of renal lesions of chronic nephropathies and diabetes. J Clin Invest 116:288–296

Rodrigues-Diez R, Carvajal-Gonzalez G, Sanchez-Lopez E, Rodriguez-Vita J, Rodrigues Diez R, Selgas R et al (2008) Pharmacological modulation of epithelial mesenchymal transition caused by angiotensin II. Role of ROCK and MAPK pathways. Pharm Res 25:2447–2461

Ruiz-Ortega M, Esteban V, Ruperez M, Sanchez-Lopez E, Rodriguez-Vita J, Carvajal G et al (2006a) Renal and vascular hypertension-induced inflammation: role of angiotensin II. Curr Opin Nephrol Hypertens 15:159–166

Ruiz-Ortega M, Ruperez M, Esteban V, Rodriguez-Vita J, Sanchez-Lopez E, Carvajal G et al (2006b) Angiotensin II: a key factor in the inflammatory and fibrotic response in kidney diseases. Nephrol Dial Transplant 21:16–20

Ruster C, Wolf G (2011) Angiotensin II as a morphogenic cytokine stimulating renal fibrogenesis. J Am Soc Nephrol 22:1189–1199

Saad S, Stanners SR, Yong R, Tang O, Pollock CA (2010) Notch mediated epithelial to mesenchymal transformation is associated with increased expression of the Snail transcription factor. Int J Biochem Cell Biol 42:1115–1122

Sun L, Zhang D, Liu F, Xiang X, Ling G, Xiao L et al (2011) Low-dose paclitaxel ameliorates fibrosis in the remnant kidney model by down-regulating miR-192. J Pathol 225:364–377

Tampe B, Zeisberg M (2014) Contribution of genetics and epigenetics to progression of kidney fibrosis. Nephrol Dial Transplant 29(Suppl 4): iv72–iv79

Tian N, Rose RA, Jordan S, Dwyer TM, Hughson MD, Manning RD Jr (2006) N-acetylcysteine improves renal dysfunction, ameliorates kidney damage and decreases blood pressure in salt-sensitive hypertension. J Hypertens 24:2263–2270

Touyz RM, Tabet F, Schiffrin EL (2003) Redox-dependent signalling by angiotensin II and vascular remodelling in hypertension. Clin Exp Pharmacol Physiol 30:860–866

Udani S, Lazich I, Bakris GL (2011) Epidemiology of hypertensive kidney disease. Nat Rev Nephrol 7:11–21

Wang G, Kwan BC, Lai FM, Choi PC, Chow KM, Li PK et al (2010) Intrarenal expression of miRNAs in patients with hypertensive nephrosclerosis. Am J Hypertens 23:78–84

Wang Y, Mu JJ, Liu FQ, Ren KY, Xiao HY, Yang Z et al (2014) Salt-induced epithelial-to-mesenchymal transition in Dahl salt-sensitive rats is dependent on elevated blood pressure. Braz J Med Biol Res 47:223–230

Wei Q, Mi QS, Dong Z (2013) The regulation and function of microRNAs in kidney diseases. IUBMB Life 65:602–614

Wolf G (2006) Renal injury due to renin-angiotensin-aldosterone system activation of the transforming growth factor-beta pathway. Kidney Int 70:1914–1919

Zambrano S, Blanca AJ, Ruiz-Armenta MV, Miguel-Carrasco JL, Arevalo M, Mate A et al (2014) L-carnitine attenuates the development of kidney fibrosis in hypertensive rats by upregulating PPAR-gamma. Am J Hypertens 27:460–470

Zhang W, Wang W, Yu H, Zhang Y, Dai Y, Ning C et al (2012) Interleukin 6 underlies angiotensin II-induced hypertension and chronic renal damage. Hypertension 59:136–144

Zhang Y, Peng W, Ao X, Dai H, Yuan L, Huang X et al (2015) TAK-242, a toll-like receptor 4 antagonist, protects against aldosterone-induced cardiac and renal injury. PLoS ONE 10:e0142456

Zhao W, Chen SS, Chen Y, Ahokas RA, Sun Y (2008) Kidney fibrosis in hypertensive rats: role of oxidative stress. Am J Nephrol 28:548–554

Zhou B, Mu J, Gong Y, Lu C, Zhao Y, He T et al (2017) Brd4 inhibition attenuates unilateral ureteral obstruction-induced fibrosis by blocking TGF-beta-mediated Nox4 expression. Redox Biol 11:390–402

Zhou Y, Yu J, Liu J, Cao R, Su W, Li S et al (2018) Induction of cytochrome P450 4A14 contributes to angiotensin II-induced renal fibrosis in mice. Biochim Biophys Acta 1864:860–870

Chapter 4 A Glimpse of the Mechanisms Related to Renal Fibrosis in Diabetic Nephropathy

Ling-Feng Zeng, Ying Xiao and Lin Sun

Abstract Diabetic nephropathy (DN) is a common kidney disease in people with diabetes, which is also a serious microvascular complication of diabetes and the main cause of end-stage renal disease (ESRD) in developed and developing countries. Renal fibrosis is a finally pathological change in DN. Nevertheless, the relevant mechanism of cause to renal fibrosis in DN is still complex. In this review, we summarized that the role of cell growth factors, epithelial–mesenchymal transition (EMT) in the renal fibrosis of DN, we also highlighted the miRNA and inflammatory cells, such as macrophage, T lymphocyte, and mastocyte modulate the progression of DN. In addition, there are certain other mechanisms that may yet be conclusively defined. Recent studies demonstrated that some of the new signaling pathways or molecules, such as Notch, Wnt, mTOR, Epac-Rap-1 pathway, may play a pivotal role in the modulation of ECM accumulation and renal fibrosis in DN. This review aims to elucidate the mechanism of renal fibrosis in DN and has provided new insights into possible therapeutic interventions to inhibit renal fibrosis and delay the development of DN.

Keywords Renal fibrosis · Diabetic nephropathy · TGF-β · Epithelial–mesenchymal transition · miRNA

4.1 Introduction

Diabetic nephropathy (DN) is the main cause of end-stage renal disease (ESRD), which is characterized by loss of normal nephrons, massive fibroblasts and myofibroblast hyperplasia, accumulation of extracellular matrix (ECM) proteins, formation of Kimmelstiel-Wilson nodules, thickening of the basement membrane and tubuloin-

---

Ling-Feng Zeng and Ying Xiao contributed equally to this work.

L.-F. Zeng · Y. Xiao · L. Sun (✉)
Department of Nephrology, Second Xiangya Hospital, Central South University, Changsha, Hunan, China
e-mail: sunlin@csu.edu.cn

B.-C. Liu et al. (eds.), *Renal Fibrosis: Mechanisms and Therapies*,
Advances in Experimental Medicine and Biology 1165,
https://doi.org/10.1007/978-981-13-8871-2_4

terstitial fibrosis (TIF) (Kanwar et al. 2011). Renal fibrosis result from DN is usually considered to be irreversible. And the pathogenesis of DN has not been fully elucidated. It is acknowledged that many cytokines, growth factors, and inflammatory mediators directly or indirectly this process. The chapter will be reviewed the mechanisms of pathogenetic and renal fibrosis in diabetic nephropathy.

4.2 The Cell Growth Factors and Diabetic Renal Fibrosis

4.2.1 Transforming Growth Factor

Transforming growth factor beta (TGF-β) was discovered in 1980 by De Larco and colleagues (de Larco and Todaro 1980), which has been linked with various organ fibrosis, including kidney. The TGF-β superfamily members include TGF-β family, activin, and bone morphogenetic protein (BMP). TGF-β is expressed widely in the body, especially in bone, lung, and kidney. Many cells, like parenchymal cells, lymphocytes, macrophages, and platelets, can express and release TGF-β. Abnormal activation of TGF-β and its receptors, as well as the downstream signaling pathways, can lead to increase extracellular matrix (ECM) accumulation and decreased degradation, thus causing renal fibrosis. Almost all types of renal cells can secrete TGF-β and express TGF-β receptors, which participate in the initiation and development of DN renal fibrosis through autocrine and paracrine pathways (Hu et al. 2018).

TGF-β1, a well-studied fibrogenic cytokine, exerts its fibrogenic effect by activating the downstream Smad signaling pathway in both experimental animal models and human kidney diseases (Hathaway et al. 2015). It is now evident that TGF-β1 activates Smad3 to mediate fibrosis, whereas overexpression of Smad7 prevents renal fibrosis (Meng et al. 2015, 2016; Feng et al. 2018).

Notably, TGF-β mRNA was significantly increased in the glomerulus of DN mice and was positively correlated with the increase of ECM accumulation and the degree of renal fibrosis (Fukasawa et al. 2004). In addition, the plasma TGF-β of diabetic mice is four times as high as that of normal mice (Wang et al. 2007). Furthermore, when TGF-β cDNA was introduced into the kidney of normal mice through glomerular arteriole by liposome, the expression of TGF-β in glomerulus increased and glomerulosclerosis occurred within one week (March et al. 2018). In vitro studies have also demonstrated that high glucose (HG) stimulated the expression of TGF-β in proximal tubular epithelial cells, glomerular endothelial cells, and mesangial cells (MCs), and the increase of ECM accumulation, whereas antisense oligonucleotides can antagonize these effects induced by HG (Meng et al. 2015; Tu et al. 2011; Mou et al. 2016). Therefore, TGF-β was an important stimulus factor for ECM accumulation and renal fibrosis.

Moreover, Tu et al. (2011) also found a significant increase of TGF-β expression in podocytes of DN patients, which attributing to DN renal fibrosis. Ziyadeh et al. (2000) found that anti-TGF-β neutralizing antibodies can rescue renal fibrosis in diabetic animal models, suggesting that the inhibition of TGF-β expression can delay the progression of renal fibrosis. However, the application of anti-TGF-β antibody may inhibit its anti-inflammatory effect, thereby aggravating the inflammatory damage of kidney and inducing tumors formation and various autoimmune diseases (Voelker et al. 2017). Therefore, the benefits and defects of anti-TGF-β therapy in DN still need more in-depth researches. Additionally, another member of TGF superfamily, BMP-7 is a pleiotropic protein, which plays a role in kidney development and many kidney diseases (Manson et al. 2015). Saika et al. found that BMP-7 can inhibit TGF-β1-induced epithelial–mesenchymal transition (EMT) formation and partly inhibits monocyte chemotactic protein-1 (MCP-1)-induced EMT (Meng et al. 2013).

Athinam et al. found that the level of TGF-β was gradually elevated in patients with impaired glucose tolerance, diabetes and DN, which suggested that TGF-β can be used as a potential indicator for DN (Vasanthakumar et al. 2015). Meanwhile, Verhave et al. (2013) found that TGF-β can independently predict the decline of renal function and act as an indicator of risk stratification in DN patients. Additionally, Sharma et al. demonstrated that inhibiting TGF-β by a small molecule might ameliorate the decline of estimated glomerular filtration rate (eGFR) (Dounousi et al. 2015).

4.2.2 Connective Tissue Growth Factor

The connective tissue growth factor (CTGF) is a member of the ECM protein family. It is a cysteine-rich heparin-binding protein, which contains 349 amino acids (36–38 KDa) (Cheng et al. 2014). Recent studies have revealed that CTGF was closely related to renal fibrosis of DN (Turner et al. 2018). Glomerular and renal interstitial CTGF expression was increased in animal models of type 1 and type 2 diabetes (Guha et al. 2007). Yokoi et al. (2008) demonstrated that overexpression of CTGF gene in diabetic mice significantly increased proteinuria, mesangial proliferation, and the production of collagens, and fibronectin (Fn). In addition, CTGF participates in renal fibrosis through binding to integrin receptors on the surface of the mesangial cells (MCs). Mason (2009) reported that CTGF bind to β-integrin and activated p42/p44 mitogen-activated protein kinase (MAPK) and phosphatidylinositol 3-kinase (PI3K), causing ECM accumulation in the mesangial region. Meanwhile, β-integrin neutralizing antibodies or specifically inhibiting the expression of MAPK or PI3K can reduce CTGF-induced ECM accumulation. Besides, Crean et al. (2006) found that CTGF can activate the PI3K-PKB pathway by the interaction with non-specific receptors and promote depolymerization of actin in glomerular MCs, leading to renal fibrosis. Furthermore, they confirmed that CTGF also can promote fibrosis by mediating the TGF-β-Smad signaling pathway. Consecutively, subcutaneous injection of TGF-β for 7 days only induced transient renal fibrosis, while simultaneous injection with CTGF-induced continuous renal fibrosis (Wang et al. 2011b). Further studies suggested that CTGF can bind BMP-7 with high affinity, thus interfering with BMP-7 signal through smad1/5 signaling pathway, and then enhancing TGF-β activity, causing structural and functional damage of the kidney in DN (Nguyen et al. 2008).

FG-3019, a human monoclonal antibody to CTGF, reduced microalbuminuria and renal fibrosis in DN patients safely and effectively in phase I clinical observation (Adler et al. 2010). Recent studies have shown that increased expression of CTGF and vascular endothelial growth factor-C (VEGF-C) in DN was associated with lymphangiogenesis in TIF, while CTGF knockout can decrease the expression of VEGF-C and reduced lymphangiogenesis (Kinashi et al. 2017). Analogously, by treating diabetic mice with a small molecule blocking CTGF, researchers can obviously reduce ECM accumulation, delay the EMT process, and improve glomerulosclerosis (Reddy et al. 2013; Zhang et al. 2016a, b). Overall, these data revealed a promising prospect of targeting CTGF in preventing DN, albuminuria, and renal fibrosis.

4.2.3 Angiotensin II

Angiotensin II (Ang II) is the most important bio-active substances in the renin–angiotensin system (RAS), the abnormal activation of Ang II is involved in the renal fibrosis of DN (Lv and Liu 2015). Ang II not only causes hemodynamic changes and leads to hypertension, but also acts as a growth-promoting factor that regulates the expression of many renal cytokines and plays a key role in the development of renal inflammation and fibrosis (Zha et al. 2017). The biological effects of Ang II are mediated by a specific membrane receptor, such as the Ang II Receptor. Recent studies have found that the level of Ang II was significantly increased in renal tissues of patients with DN and animal models (Aggarwal et al. 2017). Ang II not only participates in renal lesions through hemodynamic effects, but also promotes the proliferation and hypertrophy of MCs; it leads the ECM accumulation and renal fibrosis through non-hemodynamic effects (Rahimi 2016). Moreover, Ang II can stimulate the expression and activation of oxidative stress protein p66Shc in tubular epithelial cells (TECs) and participate in the oxidative damage of tubular cells through induces the overproduction of ROS in tubular cells mitochondria. In addition, high glucose can up-regulate Ang II expression in various renal cells, which can induce the accumulation of ECM and promoting the renal fibrosis of DN (Wang 2015). Generally, the mechanism of Ang II-induced renal fibrosis includes the promotion of hypertrophy and proliferation of MCs through the TGF-β signaling pathway (Meng et al. 2012). Ang II can induce EMT and the accumulation of ECM through TGF-β-independent or TGF-β-independent pathway. Furthermore, Ang II can also produce reactive oxygen species (ROS) (Verhave et al. 2013), increased the expression of nuclear factor kappa (NF-κB), TGF-β, IL-6, tumor necrosis factor-α (TNF-α) and chemokines, causing infiltration of macrophages and proliferation and differentiation of interstitial fibroblasts, resulting in damage of glomerular endothelial cells and tubulointerstitial fibrosis (TIF), resulting in a sustained decline in renal function and chronic renal failure (Yacoub and Campbell 2015).

At present, angiotensin-converting enzyme inhibitors (ACEI) and angiotensin receptor blocker (ARB) have been extensively used in treating DN and delaying renal fibrosis. It is known well that the decrease of blood pressure cannot fully explain the effect of ACEI and ARB in reducing proteinuria, indicating that Ang II has non-hemodynamic effects in renal fibrosis of DN (van der Sande et al. 2016). Because in non-DN with normal blood pressure, Ang II blockade also exerts renal protective effects (Umanath and Lewis 2018).

4.2.4 Hepatocyte Growth Factor

Hepatocyte growth factor (HGF) is a polypeptide growth factor that promotes hepatocyte regeneration and is widely expressed in organs derived from mesoderm (Mizuno et al. 2008). It has various biological effects, such as stimulating cell proliferation and differentiation, inhibiting cell apoptosis, and maintaining the structure and function of the organs (Gohda 2002). In renal tissue, HGF is mainly expressed in renal interstitial macrophages, glomerular epithelial cells, MCs and endothelial cells and exerts its biological effect on these cells by the form of autocrine or paracrine (Libetta et al. 2016). The elevated the level of HGF was involved in various renal injuries, like the nephropathy caused by nephrotoxic drugs and urinary tract obstruction, while HGF maintained its inactive form in normal tissues (Nlandu et al. 2016).

In recent years, the mechanism by which HGF induces renal fibrosis has been carried out. Mizuno and Nakamura (2004) found that exogenous HGF can reduce glomerular hypertrophy in streptozocin (STZ)-induced mice and inhibits the expression of collagen I, Fn and α-smooth muscle actin (α-SMA), whereas anti-HGF antibody can aggravate its pathological changes in the kidney. Similarly, in HG-treated MCs, the overexpression of HGF can reduce the expression of TGF-β, collagen IV, and α-SMA. Besides, Zhang et al. (2016a b) have revealed that HGF can ameliorate podocyte injury and proteinuria in hyperglycemia conditions by promoting autophagy via inhibition of the glycogen synthase kinase-3β (GSK-3β).

In addition, TGF-β inhibited the expression of HGF and led to renal fibrosis. In turn, exogenous HGF mitigated fibrosis by decreasing the level of TGF-β (Esposito et al. 2006). Current studies have been demonstrated that anti-fibrosis mechanisms of HGF mainly include the following two aspects: 1. Promoting the degradation of ECM: The degradation of ECM is mainly through two pathways, including the matrix metalloproteinases/tissue inhibitors of metalloproteinase (MMPs/TIMPs) and plasminogen activator/plasminogen activator inhibitor (PA/PAI) signaling pathway. In addition, HGF induced the expression of MMP-9 in tubule cells and antagonized the inhibitory effect of TGF-β on TIMP-2 and plasminogen activator inhibitor-1 (PAI-1) and then alleviating the tubulointerstitial fibrosis (TIF) (Esposito et al. 2006); 2. Inhibiting the formation of EMT in epithelial cells Yang et al. (2003) found that HGF can maintain the level of E-cadherin in renal TECs and inhibit the expression of α-SMA induced by TGF-β. At present, HGF has been used for the treatment of hepatic fibrosis, but its application in renal fibrosis caused by DN needs further identified.

4.3 Renal Epithelial–Mesenchymal Transition and Diabetic Renal Fibrosis

EMT is considered to be one of the initiating factors in the occurrence and development of renal TIF (Loeffler and Wolf 2015). In addition, the intrinsic fibroblasts of tubulointerstitium or exogenous fibroblasts derived from migration play an important role in the accumulation of ECM and the process of renal fibrosis (Hay and Zuk 1995). The EMT process of TECs includes: 1. the disappearance of cell polarity and the destruction of tight junctions between cells and cells, the loss of the original phenotype of TECs, such as the down-regulation of intercellular tight junction protein E-cadherin, and the transformation of cells into mesenchymal cells; 2. destruction of renal tubular basement membrane; 3. the renal TECs leave the renal tubules and enter the interstitium through the damaged basement membrane; 4. the epithelial cells of renal tubulointerstitium transform into myofibroblasts expressing α-SMA (Singh et al. 2018). Sanai et al. (2000) found that the expression of α-SMA significantly increased in the kidney of diabetic mice.

The main factors triggering EMT in DN include 1. increased production of advanced glycation end products (AGEs). Oldfield et al. (2001) proved that AGEs can bind to its receptor (RAGE) and activate TGF-β1 to trigger EMT in TECs. AGEs inhibitor (ALT711) can reduce the formation of EMT in tubular cells of STZ-induced diabetic mice accompanied by a significant decrease of TGF-β1. Besides, recent studies have revealed that AGEs can activate JAK/STAT and ERK1/2-MAPK signaling pathways by binding to RAGE, triggering EMT and promoting ECM synthesis (Huang et al. 2001). Moreover, AGEs also cause translocation and release of the high-mobility group protein 1 (HMGB1), contributing to promoting the expression of TGF-β and CTGF (Cheng et al. 2015). Meanwhile, after knocking out HMGB1 in mice, the expression of TGF-β is weakened through induction of AGE-BSA dependent RAGE pathway. And the high-mobility group protein 2 (HMGA 2) can be induced by AGEs in HMGA2-knockout mice, which plays a role in inhibiting the formation of reactive oxygen species (ROS) and the activation of p38MAPK (Hou et al. 2018). Furthermore, Huang et al. (2013) confirmed that neuroglioma associated protein-2 (CLIPR-2) participated in the TIF progress through ERK1/2 signaling pathway in the human proximal tubule cells (HK-2). On the other hand, epidermal growth factor (EGF) and fibroblast growth factor (FGF) can trigger EMT in renal TECs synergistic effect with TGF-β1 on EMT formation (Cheng et al. 2015). 2. Hyperlipidemia: Insufficient insulin secretion decreased intake of glucose by adipose tissue and the removal of triglycerides and cholesterol from plasma are the main causes for hyperlipidemia in diabetic patients (Yakush 2017). For example, Zeisberg et al. (2008) found that after 16 weeks of high-fat diet, the experimental mice displayed a significant increase in total blood cholesterol and the morphological change of the TECs was significantly changed compared with the control group fed with common diet. In addition, the expression of α-SMA and vimentin in the TECs was significantly increased, which verified that hyperlipidemia could induce EMT and renal fibrosis (Li and Bertram 2010). 3. Proteinuria: Proteinuria causes damage to the renal tubules when the urine protein exceeds the reabsorption capacity of TECs. It can enhance the activity of nuclear transcription factor NF-κB in TECs, which further activates the transcription of a variety of cytokine genes lead to after nuclear translocation, culminating in EMT and renal fibrosis (Wang et al. 1999).

4.4 Inflammatory Cells and Diabetic Renal Fibrosis

Recently, accumulated evidence has revealed that development of DN was associated with the many inflammatory cells, such as macrophages, T lymphocytes, neutrophils, and closely related to interstitial fibrosis as well (Moriya et al. 2004).

4.4.1 Macrophage

Macrophages are highly differentiated and mature cells in the mononuclear phagocyte system, which are differentiated after monocytes moving into tissues from blood and play a critical role in specific immunity and non-specific immunity in tissues and organs. Recent studies have demonstrated that macrophage infiltration of renal tissues was one of the characteristic features of inflammatory response in DN and played a central role in the development of DN (Tesch 2017). Macrophage infiltration was found in the glomerulus and tubulointerstitium in both diabetic rodent models and DN patients, and the degree of infiltration was positively correlated with interstitial fibrosis in DN (Tesch 2017). The mechanism of macrophage migrating to kidney in DN is still unclear, but inflammatory chemokines/chemokine receptors and integrins, such as intercellular adhesion molecule-1 (ICAM-1), mononuclear/macrophage chemoattractant protein-1 (MCP-1) and chemokine (CX3CL1), may involve in this process (Okada et al. 2003; Yang et al. 2018; Navarro-Gonzalez et al. 2011).

Some studies have proved that the expression of various inflammatory factors, intercellular adhesion molecules, and chemokines was obviously increased in diabetic kidney, which induced the adhesion between macrophage sand vascular endothelial cells, and then stimulated the production of MCP-1 and chronic inflammatory factors (IL-1, TNF-α, etc.),resulting in widespread aggregation of local inflammatory cells in kidney, especially macrophages (Lim and Tesch 2012; Donate-Correa et al. 2015; Aghadavod et al. 2016). Activated macrophages release nitric oxide, platelet-derived growth factor (PDGF), Interleukin-1 (IL-1), tumor necrosis factor-α (TNF-α), complement molecules, and metalloproteinases, etc. (Rousselle et al. 2017). These factors eventually result in the damage of renal vascular endothelial cells and proliferation of fibroblasts and MCs, glomerular hypertrophy as well as renal interstitial fibrosis. The clinical application of ACE inhibitors such as enalapril combined with immunosuppressants, like mycophenolate mofetil to delay the progression of DN may be related to the down-regulation of MCP-1 and ICAM-1 in kidney tissue of DN, the reduction of macrophage infiltration and renal interstitial inflammation, and the inhibition of renal interstitial fibrosis (Egido et al. 2017).

Additionally, macrophage migration inhibitory factor (MIF) which is elevated in diabetic kidney regulates the expression of many proinflammatory cytokines, including interleukin-6 (IL-6), interleukin-1β(IL-1β), tumor necrosis factor α (TNF-α), and plays a key role in macrophage accumulation and cell polarization in vitro (White et al. 2014). Wang et al. (2014b) found that there was a significant decrease in blood glucose, albuminuria, ECM accumulation, EMT, and macrophage activation in the kidney of db/db mice after treating with the MIF inhibitor ISO-1 for 8 weeks. Besides, they revealed that treatment with MIF to diabetic mice might activate macrophages and cause podocyte damage. In addition, MIF can induce EMT formation and ECM protein secretion in renal tubular cells (Wang et al. 2014b). Therefore, MIF inhibitors may be a potential therapeutic approach for DN by inhibiting the activation of macrophages (Chen et al. 2017; Hickey and Martin 2018).

4.4.2 T Lymphocyte

T lymphocytes, also known as thymus-dependent lymphocytes, are distributed in the thymus-dependent region of peripheral lymphoid tissues and recycled through lymphatic vessels, peripheral blood, and interstitial fluid (Bending et al. 1988). They can affect humoral immunity and simultaneously participate in cellular immune responses (Bruserud and Pawelec 1997). Both in DN animal model and patients, T lymphocyte infiltration can be found in the renal interstitium (Wu et al. 2011). Under the DN state with HG and activated RAS system, T lymphocytes migrate to kidney under the effects of cell adhesion molecules/inflammatory chemokines, such as LFA-1/ICAM-1, MCP-1. And then the activated T lymphocytes can secrete inflammatory molecules, like IFN-γ and TNF-α, thus activating endothelial cells and macrophages and inducing endothelial cell transdifferentiation and renal fibrosis81. It suggested that the abnormal activation and migration of T lymphocytes contributed to the development of renal fibrosis of DN. Moreover, studies proved that the tripterygium glycosides could reduce the infiltration of T lymphocytes in diabetic kidneys of animal models or DN patients by inhibiting the production of MCP-1 and the proliferation of T lymphocytes, but promoting their apoptosis, thereby exerting anti-inflammatory and antioxidant effects and finally reducing proteinuria and renal fibrosis (Gao et al. 2010).

Intriguingly, accumulated evidence demonstrated that the accumulation of IL-17A produced by T helper cells (Th17 cells) in diabetic kidneys may help delay the development of DN (Kim et al. 2015). $CD4^+$ $IL17A^+$ cells have been identified in monocytes isolated from kidneys of diabetic mice (Moon et al. 2012). In diabetic patients, urine level of IL-17A was increased in the presence of microalbuminuria but decreased in the presence of macroalbuminuria (Kim et al. 2015). Mice with IL-17A gene defect develop more serious renal injury of DN, while wild-type (WT) diabetic mice receive a low dose of IL-17A that protects against DN. Notably, the IL-17A treatment was related to the reduction of the macrophage infiltration, proinflammatory cytokines (MCP-1, IL-10, IL-6 and TNF-α) and STAT3 activation, thus revealing an anti-inflammatory effect in diabetic mice (Mohamed et al. 2016). Therefore, IL-17A can protect DN by alleviating the inflammatory response.

4.4.3 Mastocyte

Mastocytes originate from the bone marrow pluripotent progenitor cells. Under the stimulation of various immune or non-immune factors, mastocytes can migrate from bone marrow to renal tissue and develop into locally settled mast cells (He et al. 2017). The expression of MCP-1 in mast cells can be up-regulated by activation of stem cell factor (SCF)/c-kit signaling pathway in DN, which subsequently induce the aggregation of mast cells in renal tissue (Okayama and Kawakami 2006).

It has been proved that mastocytes were associated with renal interstitial fibrosis. For example, mastocytes were significantly increased in renal interstitial of rapidly progressive glomerulonephritis (RPGN) (Togawa et al. 2009), mesangial proliferative glomerulonephritis (Danilewicz and Wagrowska-Danilewicz 2005), and DN (Goto et al. 2002), and the number of mastocytes in renal interstitial were positively correlated with the ECM accumulation and the degree of TIF (Balakumar et al. 2009). Additionally, in DN, the mastocytes migrated to the kidney are activated to degranulate and release a series of inflammatory mediators, including TGF-β, renin, tryptase, chymotrypsin, heparin, histamine, and cathepsin G, which play direct or indirect role in glomerulosclerosis and TIF (Balakumar et al. 2009). Kidney biopsy in type 2 DN revealed mast cell aggregation and degranulation. They are mainly found around the glomerulus, paraperitoneal and perivascular areas, and their presence is related to tubulointerstitial injury and disease progression (Zheng et al. 2012). Consequently, it is reasonable to speculate that mastocytes may participate in the progression of DN with renal interstitial fibrosis through inflammatory mediators released by degranulation, though the specific molecular mechanism still needs to be further studied.

4.5 MicroRNA and Diabetic Renal Fibrosis

MicroRNAs (miRNAs) are small and highly conserved endogenous noncoding RNAs of 20–22 nucleotides in length, acting as post-transcriptional regulators of gene expression by binding to the 3′-untranslated regions of messenger RNA (mRNA) targets (Zhang et al. 2017; Chandrasekaran et al. 2012). In general, miRNAs lower the expression of the target genes through inhibiting the translation of related genes and/or promoting degradation of mRNA. Since the first miRNA, lin-4, was cloned in 1993 (Lee et al. 1993), over 2000 human mature miRNAs have been identified and at least 60% of all human protein-coding genes are currently estimated to be regulated by miRNAs (Kato and Natarajan 2015). Increasing evidence suggests that aberrant expression of miRNAs potentially underlies the development and progression of various diseases, including DN (Trionfini and Benigni 2017). MiRNA dysregulation may lead to disruption of podocyte homeostasis and EMT as well as accumulation of ECM proteins related to fibrosis and glomerular dysfunction (Trionfini et al. 2015; Zou et al. 2017; Assmann et al. 2018). Therefore, the function of miRNAs in the pathogenesis of renal fibrosis in DN is non-negligible (Fig. 4.1).

4.5.1 *MiR-192*

The pleiotropic effects of miR-192 in kidney are both fibrogenic and anti-fibrotic, which is highly expressed in kidney (Jenkins et al. 2012b). It is augmented by TGF-β or HG treatment in MCs, podocytes, and tubular cells (Deshpande et al. 2013). TGF-β1 increases the levels of miR-192 by directly up-regulating the expression of Smad3 or blocking a Smad7-dependent mechanism, and up-regulation of miR-192 in turn mediates activation of TGF-β/Smad signaling in the fibrous kidney (Zou et al. 2017). TGF-β also can trigger miRNA circuits involving miR-192 to increase TGF-β expression itself and accelerate DN (Chung et al. 2010; Kato et al. 2011).

In 2007, Kato et al. (2007) observed for the first time that miR-192 levels were enhanced significantly in glomeruli isolated from STZ-injected diabetic mice, diabetic db/db mice as well as MCs treated with TGF-β. MiR-192 increased the collagen I-α2 (Col1a2) levels by inducing the expression of Smad-interacting protein 1 (SIP1) (an E-box repressor), thereby leading to collagen deposition and renal fibrosis in diabetic glomeruli. Except SIP1, accumulating evidence suggested that by targeting the other two E-box repressors zinc finger E-box-binding homeobox 1 (Zeb1) and Zeb2, miR-192 can up-regulate the key fibrotic genes Col1a2 and collagen IV-α1 (Col4a1) in MCs related to early DN (Kato and Natarajan 2015). In 2013, Deshpande et al. (2013) demonstrated firstly that Zeb2 was even a target of miR-192 in human tissues. Meanwhile, diabetic conditions, including TGF-β and HG, can induce a reciprocal up-regulation of miR-192 and p53 in MCs and lead to enhanced expression of ECM genes, thereby creating an amplification loop via Zeb2 repression and accelerating the progression of DN. Besides, Hong et al. (2013) found that vascular endothelial growth factor (VEGF) suppressed Smad3 and miR-192 and subsequently inhibited TGF-β1-induced EMT. These data suggest that the elevation of miR-192 plays an important role in the deposition of ECM and renal fibrosis of DN.

Moreover, miR-192 can be induced by TGF-β1 through promoter Smad-binding elements and epigenetic regulation via acetylation of the transcription factor Ets-1 and histone H3, which were activated by the serine and threonine kinase Akt, indicating that miR-192 expression may be regulated by Akt inhibitors, such as MK-2206, or histone acetyltransferases (Kato et al. 2013). Except Smads and p53, the expression of miR-192 also regulated by hepatocyte nuclear factor-1 (HNF-1) in TECs (Chung et al. 2010; Jenkins et al. 2012a).

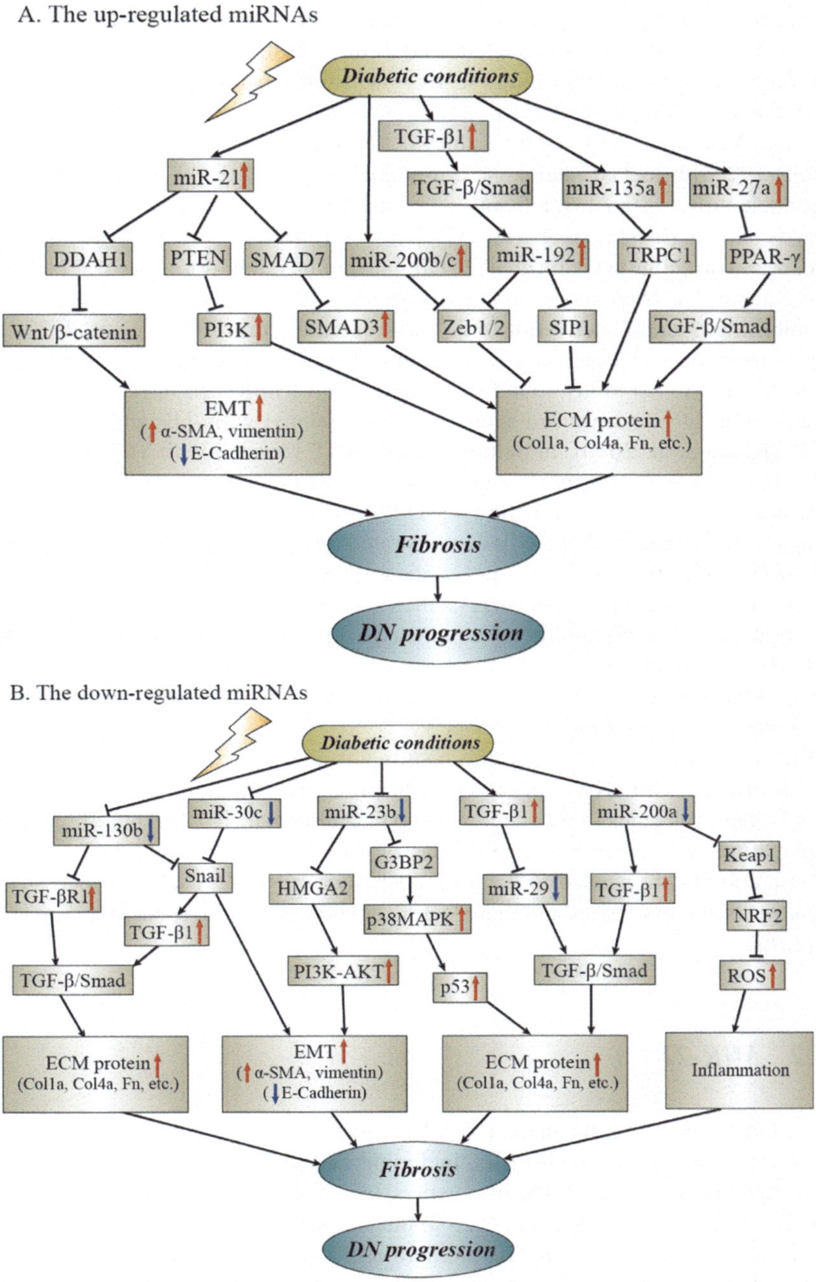

Fig. 4.1 Scheme of miRNAs and their potential roles in renal fibrosis of DN

Lower miR-192 gene expression showed less severe phenotypes of DN (glomerular hypertrophy, fibrosis, and proteinuria) compared to WT mice, suggesting that controlling miR-192 expression, and its downstream pathways may be beneficial for treating DN (Deshpande et al. 2013; Natarajan et al. 2012; Putta et al. 2012; Kato and Natarajan 2014). In the kidneys of STZ-induced diabetic mice, inhibition of miR-192 with locked nucleic acid-modified anti-miR-192 (LNA-anti-miR-192) oligonucleotides (oligos) significantly increased Zeb1/2 and decreased gene expression of collagen (Col1a2, Col4a1), TGF-β and fibronectin, thereby attenuating renal fibrosis and proteinuria (Putta et al. 2012). Intriguingly, low-dose paclitaxel, an anti-cancer agent, has been shown to ameliorate fibrosis in a rat remnant kidney model by inhibiting Smad2/3 activation and down-regulating miR-192 (Sun et al. 2011b). These results provided evidence for an anti-miRNA-based strategy to treat for DN.

However, another study found low expression of miR-192 in renal biopsies in patients with DN at advanced-stage, which correlated with TIF and reduction in eGFR (Krupa et al. 2010). In cultured human PTCs treated with TGF-β, the miR-192 expression was decreased, while expression of the E-Box repressors Zeb1 and Zeb2 was increased, thereby displaying TGF-β-mediated suppression of E-cadherin (Krupa et al. 2010). Further study demonstrated a similar association between miR-192 and E-cadherin in diabetic ApoE-deficient ($Apoe^{-/-}$) mice (Wang et al. 2010). To our knowledge, repression of E-cadherin is an early event that precedes other alterations during EMT and involves in early step of fibrogenesis in DN (Simpson et al. 2016). Therefore, loss of miR-192 expression promoted fibrogenesis in DN perhaps by enhancing TGF-β-mediated down-regulation of E-cadherin and miR-192 may exert an anti-fibrotic effect in PTCs.

The reasons for these conflicting results are unclear but could be due to cell-specific effects of miRNAs, differences studied in the animal models, as well as the role of other molecules in renal cells, like p53 (Putta et al. 2012). Therefore, further investigations are warranted to explore the role of miR-192 in renal fibrosis and larger diabetic cohorts are needed to determine whether miR-192 levels are increased in DN patients at early stages and then decline at later stages due to tubular damage or apoptosis.

4.5.2 MiR-21

MiR-21 is the most significant microRNA involved in fibrotic disorders, and its levels are up-regulated in kidney diseases, including DN (Noetel et al. 2012; Chau et al. 2012; Kolling et al. 2017). Despite expression in MCs and podocytes, the elevation of miR-21 was primarily detected in the tubular regions, which were positively correlated with the severity of TIF, glomerulosclerosis, and declining renal function (Zhong et al. 2013; Glowacki et al. 2013; Chung et al. 2013; McClelland et al. 2015). It has been demonstrated that the MiR-21 regulates downstream molecules of TGF-β1 and involves in TGF-β1-mediated fibrotic signaling pathways as a pro-fibrotic molecule in the kidney (Zhong et al. 2011).

Zhong et al. (2013) found that renal miR-21 was markedly increased in diabetic kidneys of db/db mice, which was associated with the development of microalbuminuria and renal fibrosis. In MCs and TECs from mouse, miR-21 was an important mediator of HG-induced tubular cells fibrosis by targeting Smad7, an inhibitor of TGF-β signaling, and overexpression of Smad7 was able to reverse the pro-fibrotic effects of miR-21 through blocking Smad3 activation. In consistence with these results, Lin et al. (2014) have recently identified Smad7 as a direct target of miR-21, thus preventing the proliferation of tubular cells in a rat DN model.

McClelland et al. (2015) focused on the role of miR-21 in PTCs and found that miR-21 promoted renal fibrosis in DN by targeting Smad7 and phosphatase and tensin homologue (PTEN), thereby causing the derepression of the Smad3-dependent and PI3K-dependent TGF-β1-signaling pathways, respectively, as reflected by increases in Smad3 and AKT phosphorylation, culminating in enhancing the expression of pro-fibrotic and ECM genes. Moreover, they found simultaneous repression of both Smad3 (miR-21–Smad7) and AKT (miR-21–PTEN) pathway via Smad3 knockdown and PI3K inhibitor LY294002 (LY), respectively, which revealed that both of these pathways contribute to the regulation of classic effects of TGF-β1 including modulation of Collagen I, PAI-1 and Fn1 expression in PTCs. Notably, they thought there still exists another miR-21 targeted pathway which is independent of Smad3 and AKT and mediates the pro-fibrotic effects of TGF-β1.

Furthermore, it has shown that miR-21 mimics enhanced TGF-β1-induced EMT in DN by directly down-regulating Smad7 and indirectly up-regulating Smad3, accompanied with an increase in collagen IV and Fn (Wang et al. 2014a). Conversely, in STZ-induced diabetic mice, miR-21 antagonism rescued various functional and structural parameters of DN, including TIF, mesangial matrix expansion, and albuminuria (Kolling et al. 2017). Intriguingly, by applying the LNA-200a (locked nucleic acid-modified anti-miR-200a) in nuclear factor-like 2 (Nrf2)-null mice, Wu et al. (2016) demonstrated that the curcumin analogue C66 ameliorated DN by both inhibiting miR-21 leading to down-regulation of Smad3 and activating Nrf2 through the stimulation of miR-200a targeting Keap1. Collectively, miR-21 plays a key role in the regulation of renal tubular ECM accumulation and renal fibrosis of DN. Silencing of miR-21 might be a novel efficient treatment strategy to halt the progression of DN and the novel curcumin analogue with dual effects urgently needs to be further tested.

In addition to TGF-β signaling, miR-21 triggered renal EMT by enhancing Wnt/β-catenin signaling via targeting a suppressor of this pathway, dimethylarginine dimethylaminohydrolase 1 (DDAH1), accompanied by an increase in α-SMA, collagen I and Fn expression and a reduction in E-cadherin expression (Zou et al. 2017; Liu et al. 2016b). Besides, miR-21 also promoted renal fibrosis by targeting MMP-9 and metalloproteinase inhibitor 1 in the KK-Ay mouse model of DN (Wang et al. 2013).

By contrast, miR-21 was also reported in one study to be down-regulated in db/db mice, and its overexpression blocked MCs proliferation by suppressing PTEN expression, thus suggesting a protective role during diabetic kidney injury (Zhang et al. 2009). In line with these data, genetic loss of miR-21 has also been shown to aggravate the progression of DN (Lai et al. 2015). Although these discrepancies are as yet unexplained, the differences in disease nature and conditions should be considered (Zhong et al. 2013).

4.5.3 MiR-200

The miR-200 family (miR-200a/b/c, miR-141 and miR-429) has been documented as regulators of the epithelial differentiation and mediates EMT formation in fibrotic diseases mainly by targeting TGF-β/Smad and Wnt/β-catenin signaling (Zou et al. 2017). Decreased of the miR-200 family plays a critical role in the repression of E-cadherin by Zeb1 and Zeb2 during TGF-β1-induced EMT (Bracken et al. 2015).

Recent studies indicated that miR-200a/b/c and miR-141 were decreased in DN as well as in renal TECs of rat treated with TGF-β (Zou et al. 2017). Wang et al. (2011a) demonstrated that miR-200a/141 prevented TGF-β-dependent EMT and renal fibrogenesis by repressing TGF-β1 and TGF-β2 expression, Smad-3 activation and decreasing matrix protein levels. In STZ-induced diabetic mice, Wei et al. (2014) revealed a novel mechanism whereby hyperglycemia-induced aldose reductase to regulate renal expression of miR-200a/141 to coordinately control hyperglycemia-induced renal oxidative stress, fibrogenesis, and EMT by posttranscriptionally targeting Kelch-like ECH-associated protein 1 (Keap 1), TGFβ2 and fibronectin. MiR-200a also ameliorates DN and promotes degradation of Keap1 mRNA. Additionally, Huang et al. (2015) reported that ectopic expression of miR-141 impeded the progression of TGF-β1-induced EMT through repression of homeodomain interacting protein kinase 2 (HIPK2) expression, which was considered as a tumor suppressor that activates Wnt, Notch and TGF-β-induced signaling while maintaining a high expression level of E-cadherin (Zou et al. 2017).

The increased expression of miR-200b and miR-200c in mouse MCs treated with TGF-β and renal glomeruli isolated from diabetic mice also has been reported (Kato et al. 2011; Putta et al. 2012). MiR-200b/c can up-regulate collagen expression and the autoregulation of TGF-β1 in mouse MCs by inhibiting Zeb1. And miR-200b precursor was capable of ameliorating renal tubulointerstitial fibrosis by inhibiting the synthesis of collagen I and III and Fn (Kato et al. 2011; Oba et al. 2010). Besides, miR-200b/c can activate Akt by targeting friend of GATA protein 2 (FOG2), an inhibitor of phosphoinositide 3-kinase (PI3K), leading to glomerular mesangial hypertrophy in DN (Park et al. 2013). Collectively, it is inferable that the miR-200 family is an important regulator of renal fibrosis in DN. And miR-200a/141 mainly acts as a protective molecule, while miR-200b/c primarily lies on promoting fibrosis and progression of DN, but the concrete mechanism is not clear.

4.5.4 MiR-29

MiR-29 family (miR-29a/b/c) has been increasingly noted to be a critical negative regulator of EMT in fibrotic renal diseases by regulating multiple signaling pathways, such as TGF-β/Smad, Wnt/β-catenin, and MAPK (Zou et al. 2017). In patients with type 2 diabetes or in cultured renal PTCs under HG conditions, levels of miR-29 family are down-regulated. Decrease of miR-29 family members leads to increasing collagen deposition/fibrosis by directly targeting multiple collagens (collagen I/III/IV) in TECs, MCs, and podocytes (Wang et al. 2010; Chen et al. 2014).

Wang et al. (2012) observed low levels of miR-29 in mice models with various pathologic similarities to the early and advanced stages of DN seen in humans. They identified an important downstream regulatory role of miR-29 family in TGF-β1-mediated fibrogenesis. Besides, they found that although administration of the Rho-associated kinase (ROCK) inhibitor fasudil prevented renal fibrosis and restored expression of renal cortical miR-29a/c, treatment with losartan resulted in a significant increase in miR-29b expression, which also attenuated the structural lesions in diabetic rats by reducing the ECM proteins. Chen et al. (2014) validated that under diabetic conditions, miR-29b was largely down-regulated in response to AGEs, which was related to up-regulation of collagen matrix (collagen I/IV) in MCs via the TGF-β/Smad3-dependent mechanism. Moreover, diabetic miR-29a transgenic mice also had lower fibrogenic factors expression, urinary protein secretion, and less glomerular fibrosis compared with diabetic WT mice (Lin et al. 2014). Additionally, Pan et al. (Pan et al. 2014) found that after treatment with Ang II to spontaneously hypertensive rats (SHRs) and human embryonic kidney epithelial cells (NRK-52E), the expression of miR-29b was decreased, while the expression of TGF-β, α-SMA, and collagen I was up-regulated, which leads to the EMT formation. Analogously, a recent report revealed that anti-diabetic drug linagliptin, a dipeptidyl peptidase-4 (DPP-4) inhibitor, can also confer renoprotection effects and ameliorate fibrosis in a mouse model of DN via inducing miR-29 by targeting DPP-4 (Kanasaki et al. 2014). Therefore, miR-29b may exert a protective effect on DN.

However, one study has shown that miR-29c was up-regulated in the kidney of db/db mice and in endothelial cells and podocytes under HG conditions (Chen et al. 2014). The elevated miR-29c can activate Rho kinase by directly decreasing the levels of Sprouty homolog 1 (Spry-1), which subsequently induced ECM protein accumulation and podocyte apoptosis. Furthermore, investigators also reported knockdown of miR-29c in vivo by a specific antisense oligonucleotide significantly attenuated albuminuria and kidney mesangial matrix accumulation in the db/db mice. Obviously, these results suggested a conflict role of miR-29 to the kidney with previous studies (Long et al. 2011). The disparities among the various studies may partially result from the different origin of cell lines and animal models, as well the functional differences among the miR-29 family members.

4.5.5 MiR-23b

Liu et al. (2016a) showed for the first time that miR-23b may act as a suppressor of EMT in mouse with DN through repressing PI3K-AKT signaling pathway activation by targeting HMGA2. They found that after treated with HG to HK-2 cells, a human proximal tubular cells line, a significant decrease in the expression of miR-23b was observed. Similar results were also seen in the kidney tissues of db/db mice. Conversely, they found that overexpression of miR-23b by administration of miR-23b agomir altered the expression levels of EMT-related genes, alleviated hyperglycemia-induced EMT formation, and improved renal functions in db/db mice. Additionally, HMGA2 knockdown or inhibition of the PI3K-AKT signaling pathway with LY294002 mimicked the effects of miR-23b overexpression on HG-mediated EMT, whereas HMGA2 overexpression or activation of the PI3K-AKT signaling pathway prevented the effects of miR-23b.

Similarly, Zhao et al. (2016) confirmed that miR-23b was a commonly decreased miRNA in the serum of patients with DN, in the kidneys of mouse with type 1 and 2 diabetes, and in HK-2 cells, human renal glomerular endothelial cells (HRGE) and mouse podocytes exposed to HG. They identified a binding site in the miR-23b promoter for p53. In vitro study inhibition of p53 through a 1-month injection with p53siRNA or the upstream p38MAPK by a 24-h treatment with p38MAPK inhibitor SB203580 enhances miR-23b expression. However, inhibition of Ras GTPase-activating protein SH3 domain-binding protein 2 (G3BP2) or overexpression of miR-23b decreases expression of p53 and p38MAPK. MiR-23b targets G3BP2 to alleviate fibrosis and albuminuria in DN.

4.5.6 MiR-30 Family

MiR-30 family (miR-30a/b/c/d/e), abundantly expressed in the kidney, has been shown to protect podocytes by targeting Notch1 and p53 and be required for renal development and homeostasis (Wu et al. 2014; Shi et al. 2008).

Peng et al. (2015) found that down-regulation of miR-30a in podocyte injury animal models and in patients with focal segmental glomerulosclerosis (FSGS). In addition, overexpression of miR-30a enhances epithelial markers but diminished these mesenchymal markers. They found that miR-30a targeted nuclear factor of activated T cells 3 (NFATc3) to protect cytoskeleton disorder or rearrangement in podocytes by inhibiting the nuclear translocation of NFATc3.

On the other hand, miR-30c has turned out to be essential for normal kidney homoeostasis and plays a protective role in DN independent of the benefits from reduced blood glucose (Zou et al. 2017; Jiang et al. 2013). In addition, Zhao et al. (2017b) found that in renal tubules of db/db mice and cultured HK2 cells exposed to HG ambience, loss of miR-30c accompanied with increased EMT. And they identified Snail1 as a direct target of miR-30c by which miR-30c regulated EMT in DN.

More importantly, they showed that the miR-30c inhibited Snail1-TGF-β1 axis in TECs undergoing EMT, thereby modulating the activation of fibroblasts and the fibrogenesis of myofibroblasts and finally protect against TIF in DN. In contrast, mouse with miR-30c gene deficiency enhanced the secretion of TGF-β1 from epitheliums and significantly promoted the progression of DN. This data indicates that miR-30c can protect against hyperglycemia-induced EMT and delay the development of DN via inhibiting Snail1-TGF-β1 pathway.

MiR-30e was down-regulated in tubular cells from obstruction-induced fibrotic kidneys and TGF-β1-treated NRK-52E cells (Zou et al. 2017). A miR-30e mimic increased the expression of miR-30e and significantly inhibited TGF-β1-induced the EMT formation of tubular cells partly by down-regulating the levels of mitochondrial uncoupling protein 2 (UCP2), whereas a miR-30e inhibitor imitated TGF-β1 effects, indicating there is a crosstalk between mitochondria dysfunction and renal fibrosis (Zou et al. 2017; Zhao et al. 2017a). Besides, Zhao et al. (2017a) found that in the renal tissues of db/db mice and the renal TECs treated with HG ambience, miR-30e was reduced, while Glioma pathogenesis-related-2 (GLIPR-2) was up-regulated, which was a direct target of miR-30e. More importantly, overexpression of miR-30e inhibits GLIPR-2, promotes the proliferation of renal TECs, and inhibits EMT formation by down-regulating vimentin, α-SMA and up-regulating E-cadherin, ultimately avoiding renal fibrosis in DN (Zhao et al. 2017a).

4.6 Signaling Pathways Involved in Diabetic Renal Fibrosis

4.6.1 Janus Kinase/Signal-Transducer and Activator of Transcription (JAK/STAT) Pathway

The JAK-STAT signaling pathway is a cytokine-stimulated signal transduction pathway, in which JAK proteins, a non-receptor tyrosine kinase, are activated by binding to cytokines such as interleukin and interferon. The activated JAK proteins can further activate STAT proteins, including STAT1, STAT2, STAT3, STAT4, STAT5a, STAT5b, and STAT6 and so on, which contain SH2 and SH3 domains and act as nuclear transcription factors (Jang and Baik 2013; Cai et al. 2015). STATs can combine with a specific peptide segment containing phosphorylated tyrosine and then initiate gene expression related to cell proliferation, differentiation, and migration, thus participating in a variety of physiological processes (Chuang and He 2010).

High glucose can activate JAK/STAT signaling pathway, induce EMT and TIF. Suppressor of cytokine signaling (SOCS) is the most important negative regulator of JAK/STAT signaling pathway (Liau et al. 2018). SOCS2 reduces the expression of TGF-β, Fn, and collagen by inhibiting the phosphorylation of STAT3 and JAK2 and alleviates the ECM deposition and TIF (Bao et al. 2015). Liu et al. (2014a) also found that STAT1 was activated in human renal TECs treated with HG and diabetic rats, which accompanied by the enhanced expression of α-SMA and inhibition of cytokeratin 18 (CK18) expression. However, the effect was blocked partially in that of treated with the AG-490, an inhibitor of JAK2. In addition, a clinical study showed that the patients with DN after oral administration of JAK1/JAK2 inhibitor baricitinib, the level of proteinuria was effectively reduced and renal function maintained stable (Liu et al. 2014b). Recent studies have found that the level of acetylation of STAT3 is significantly increased in the diabetic rats, while treatment with a STAT3 acetylation inhibitors, baricitinib, might effectively alleviate renal injury and proteinuria (Brosius et al. 2016).

4.6.2 Protein Kinase C (PKC) Pathway

PKC belongs to multi-functional serine and threonine kinase and is widely distributed in the body, which plays a key role in glucose metabolism, cell differentiation, gene expression regulation, and inflammatory process (Juan et al. 2012). It can be divided into 3 categories and 12 subtypes: traditional PKCs (PKC-α/β1/β2/γ), which can be activated by both Ca^{2+} and diacylglycerol (DAG); new PKCs (PKC-δ/ε/θ/μ/η), which can be activated by DAG; atypical PKCs (PKC-λ/ζ), which are activated by neither Ca^{2+} nor DAG (Zhang et al. 2014). The PKC pathway is considered to be the central link in the cross-reaction of multiple signaling pathways. In chronic hyperglycemia states, PKCs (including traditional PKCs and new PKCs) can be activated by the intermediate product DAG produced during glycolysis, the activated PKC will phosphorylate the serine or threonine residues in many proteins, and exerting a variety of biological functions (Teng et al. 2014).

It is well known that HG can activate PKC α, β1, β2 and then induces the production of various growth factors, such as TGF-β and VEGF (Teng et al. 2014). It further up-regulated the expression of ECM proteins, resulting in ECM accumulation. Therefore, the application of various inhibitors for PKC subtypes is an effective approach to the preventive the kidney injury of DN. Moreover, Menne et al. (2013) have found that the treatment with CGP41252 (PKC-α and PKC-β inhibitors) to STZ-induced mice, the activation of PKC-α and PKC-β were significantly inhibited, which accompanied with effectively alleviated of the glomerular hypertrophy and proteinuria.

4.6.3 Mitogen-Activated Protein Kinase (MAPK) Pathway

MAPKs include a group of serine/threonine protein kinases which can mediate the transmission of extracellular signals into cells and exert many biological effects, including cell division, proliferation, differentiation, migration, metabolic apoptosis, and gene expression (Burotto et al. 2014). The MAPK superfamily members are divided into four subfamilies: P38MAPK, extracellular regulated protein kinases (ERK), c-Jun N terminal kinase (JNK)/stress activated protein kinase (SAPK) and ERK5/big mitogen-activated protein kinase (BMK1) (Davis 1994). P38MARK includes four subtypes (P38α/β/γ/δ), which can be expressed in renal cells. JNK is a stress-activated protein kinase, including three subtypes (JNK1–3), of which JNK1 and JNK2 are normally expressed in the kidney (Ohta et al. 2018).

It is reported that HG stimulated the activation of p38 MAPK in the renal proximal TECs, which accompanied with up-regulated expression of MCP-1, TGF-β, VEGF, Fn, and collagen (Rane et al. 2010). At the same time, de Borst et al. (2009) found that treatment with TGF-β to fibroblasts can increase the expression of Fn and CTGF by activation of the JNK signaling pathway. Moreover, the activation of JNK has been correlated to the aggregation of interstitial macrophages, the expression of kidney injury molecule-1 (KIM-1), and the declining renal function in human DN. Liu et al. (2018) also reported that the use of baicalein (a specific inhibitor of P38MARK) can effectively reduce the expression of p38MAPK, p-p38MAPK, and NF-κB P65 in diabetic mouse, thereby alleviating the progression of DN and renal fibrosis.

4.6.4 Notch Signaling Pathway

The Notch signaling pathway is widely distributed in vertebrates and invertebrates. It is a highly conserved signaling pathway composed of Notch ligand, Notch receptor, and intracellular effector molecules. In mammals, the Notch signaling pathway includes three Delta-like ligands (DLL1, DLL3, DLL4), two ligands belonging to the Jagged family members (Jagged1, Jagged2), and four transmembrane receptors (Notchl ~ 4) (Ben-Shushan et al. 2015). The interaction between the Notch ligand and receptor leads to a change in the conformation of the Notch receptor, and the release of the Notch intracellular domain (NICD) under the action of the protease (Lin et al. 2018). The released NICD translocates to nucleus and combines with transcription factor CSL (CBF-l, Suppressor of hairless, and Lag-l) to form a NICD/CSL transcriptional complex. Subsequently, the complex activates the transcription of Notch target genes, including Hey and Hes genes, thus regulating cell proliferation, differentiation, apoptosis, and substance metabolism (Kopan and Ilagan 2009; Bedogni 2014; Lopez-Arribillaga et al. 2018). Sirin and Susztak (2012) confirmed the expression of Notch pathway-related proteins in DN was enhanced. The co-expression of Jagged1 and Hes1 in renal TECs of patients with DN was enhanced by in situ hybridization, and the expression of Notch1 protein in tubulointerstitium was closely connected with interstitial fibrosis. Lin et al. (2010a) found that STZ-induced diabetic mice treated with γ secretase, an inhibitor of Notch signaling pathway, the expression of VEGF and nephrin return to normal, accompanied by reducing renal TIF.

4.6.5 PI3K/Akt Signaling Pathway

PI3K is highly homologous with PKA and PKC, also called protein kinase B. PI3K/Akt signaling pathway is multi-functional, which can mediate cell proliferation, differentiation, metabolism, and other physiological processes (Xu et al. 2015). PI3K has a dual effect in both activities of serine/threonine protein kinase and phosphatidylinositol kinase, which is constituted by regulatory subunits (p85) and catalytic subunits (p110). The activation of the PI3K/Akt pathway is a multistep process. First, PI3K is activated by tyrosine kinase to generate PIP3. The binding of PIP3 to the pH domain of Akt causes Akt to polymerize on the cell membrane and undergo a conformational change. With the assistance of PDK1 and PDK2, the Thr308 and Ser473 loci of Akt are phosphorylated. Activated Akt affects a series of physiological processes in cells, including metabolism, protein synthesis, cell survival, and apoptosis (Xu et al. 2015; Jafari et al. 2019; Zhang et al. 2018).

Studies have shown that the PI3K/Akt signaling pathway is closely related to ECM accumulation and fibrosis in DN. Abnormal activation of Akt and overexpression of TGF-β1 participated in the EMT formation in renal tubular induced by HG ambience. Besides, activation of Akt was associated with as well as overproduction of interstitial ECM proteins and initiation of renal interstitial fibrosis in DN (Xie et al. 2015). Yano et al. (Lu et al. 2015) found that Ginkgo biloba extract, an inhibitor of Akt might alleviate renal interstitial fibrosis in DN. Additionally, Shemesh et al. (2014) also found that AS101, trichlorotellurate, can delay the progression of DN through pharmacological inhibition of Akt downstream pathway. As one of the inhibitors of Akt, PTEN alleviates the deposition of ECM and in diabetic mice by inhibiting the activation of Akt and the secretion of CTGF, thus alleviating the renal damage (Zhu et al. 2016). They also found that overexpression of CTMP (one of the endogenous inhibitors of Akt) in diabetic mouse model and human renal TECs cultured in HG reduced the excessive accumulation of ECM in renal tubulointerstitium by decreasing the expression of phospho-Akt (Ser473), TGF-β1, α-SMA, and the secretion of collagenI/II (Wu et al. 2011; Brosius et al. 2016).

4.6.6 Epac-Rap1 Signaling Pathway

Cyclic adenosine monophosphate (cAMP) is the key second messenger molecule that regulates various cellular functions including lipid metabolism, inflammation, and cell differentiation. PKA as the only effector of the cAMP signaling pathway exert an important role in the process (Wahlang et al. 2018). In 1998, de Rooij et al. (1998) identified a new cAMP effector called activated carbon acetate (Epac), which providing novel understanding of cAMP signaling pathway. Epac is a specific guanine nucleotide exchange factor for Ras-related GTPase Rap1 and Rap2 and acts as a "switch" in the activation of Rap molecules by converting the inactive Rap-GDP to the active Rap-GTP form (Bos 2003). Recent studies have verified that

Epac-Rap1 signaling pathway played an important role in renal fibrosis in DN. Lin et al. (2002) firstly discovered that Rap1b was related to the pathogenesis of DN and further studies demonstrated that Rap1b mediated Fn expression via the PKC pathway in HG-induced MCs, which consequently established the role of PKC-Rap1b-B-Raf signaling pathway in DN (Lin et al. 2002). Moreover, Yano et al. (2007) found that Epac-specific agonist 8-pHPT-2-O-Me-cAMP, rather than PKA inhibitors, significantly enhanced PI3K activity and inhibits the increase of mesangial matrix synthesis in MCs treated with Ang II. Besides, endothelin is an important vasoconstriction regulator and plays an important role in the pathogenesis of DN (Benz and Amann 2011). Kang et al. (1999) confirmed that in Rap1 participated in stimulating the proliferation and adhesion of MCs treated with endothelin through the ET-1-Pyk2-p130Cas/BCAR3-Rap1 signaling pathway. Additionally, it is reported that the expression of Epac protein was significantly increased in kidney tissue of STZ-induced DN mouse or HG-stimulated HK-2 cells, which was associated with cell hypertrophy and ECM accumulation.

4.6.7 *Wnt/β-Catenin Signaling Pathway*

The Wnt family of proteins belongs to a group of secreted lipid-modified glycoproteins which contain highly conserved cysteine residues with a molecular weight of 39–46 KDa (Tripurani et al. 2018). There are 19 different Wnt proteins that have been identified in humans and mice and most of which are composed of 350–380 amino acid residue (Hwang et al. 2009). The Wnt/β-catenin signaling pathway plays a key role in cell differentiation, proliferation, apoptosis, and migration through downstream key molecules such as GSK3β and β-catenin (Zhou and Liu 2016). Recent studies have linked Wnt/β-catenin signaling pathway associated with the progression of renal fibrosis in DN (Hwang et al. 2009; Mariappan et al. 2008; Sun et al. 2011a). Mariappan et al. (2008) demonstrated that the phosphorylation of GSK-3β, the key signaling molecule in the downstream of the Wnt in renal cortex of db/db mice was significantly increased, which was related to the expression level of laminin and Fn. In addition, GSK-3β promotes HG and high insulin-induced ECM synthesis process in renal TECs and plays an important role in renal hypertrophy and ECM accumulation in DN. Intriguingly, a study concluded that nicotine may enhance the MCs proliferation and Fn production under HG milieus partly through activating Wnt/β-catenin signaling pathway (Lan et al. 2018).

It has been demonstrated that the expression of Wnt and β-catenin in the kidney tissue of diabetic mouse and in vitro cultured proximal TECs exposed to HG were significantly increased, which consistent with the increased expression of CTGF and Fn (Mu et al. 2013). Moreover, MCs transformation plays an important role in the development of DN. Many studies have also shown that Wnt/β-catenin is associated with EMT formation of macrophages in DN. On the other hand, activation of β-catenin and the expression of its target genes can be detected in TECs exposed HG ambience (Zhou et al. 2012; Liu 2010). For example, Snail and Twist, two of β-catenin target genes, which can inhibit the expression of E-cadherin in tubule cells and increase the expression of Fn, α-SMA, and vimentin. Nevertheless, the inhibitor of Snail or β-catenin might reverse HG-induced EMT in tubule cells (Liu 2010). This data indicates that the Wnt/β-catenin signaling pathway was involved in the development of DN and renal fibrosis. Meanwhile, Wnt pathway interacts with TGF-β/Smad, Notch pathways and CTGF (Ho et al. 2012; Lin et al. 2010b; Xiao et al. 2013). Ho et al. (2012) reported that the expression of Wnt was down-regulated in macrophages after treated with HG, which accompanied by an increase in expression of TGF-β and Fn, and this effect was remission in macrophages transfected with Wnt4 and Wnt5a.

4.7 Conclusion

Renal fibrosis in DN is the pathological change. In this chapter, we discussed various cytokines, inflammatory cells, and a series of signaling pathways. Many signaling pathways, like the JAK/STAT, TGF-β-Smad, Wnt/β-catenin, and Epac-Rap1 pathway, are all involved in regulating the expression of ECM proteins or EMT in DN, which play an important role in renal fibrosis of DN. We also highlighted the potential role of miRNAs in aggravating or alleviating the process of renal fibrosis in DN. Despite these significant progresses, the detailed molecular mechanisms still need to be further studied. With the advancement of modern medical technology, more specific cytokines and key signaling molecules will be defined in the future, which is beneficial for developing the novel therapeutic strategies for the treatment of renal fibrosis in DN.

References

Adler SG, Schwartz S, Williams ME, Arauz-Pacheco C, Bolton WK, Lee T et al (2010) Phase 1 study of anti-CTGF monoclonal antibody in patients with diabetes and microalbuminuria. Clin J Am Soc Nephrol 5:1420–1428

Aggarwal N, Kare PK, Varshney P, Kalra OP, Madhu SV, Banerjee BD et al (2017) Role of angiotensin converting enzyme and angiotensinogen gene polymorphisms in angiotensin converting enzyme inhibitor-mediated antiproteinuric action in type 2 diabetic nephropathy patients. World J Diabetes 8:112–119

Aghadavod E, Khodadadi S, Baradaran A, Nasri P, Bahmani M, Rafieian-Kopaei M (2016) Role of oxidative stress and inflammatory factors in diabetic kidney disease. Iran J Kidney Dis 10:337–343

Assmann TS, Recamonde-Mendoza M, de Souza BM, Bauer AC, Crispim D (2018) MicroRNAs and diabetic kidney disease: systematic review and bioinformatic analysis. Mol Cell Endocrinol 477:90–102

Balakumar P, Reddy J, Singh M (2009) Do resident renal mast cells play a role in the pathogenesis of diabetic nephropathy? Mol Cell Biochem 330:187–192

Bao NN, Kong DY, Zhu D, Hao LR (2015) Influence of overexpression of SOCS2 on cells of DN rat. Asian Pac J Trop Med 8:583–589

Bedogni B (2014) Notch signaling in melanoma: interacting pathways and stromal influences that enhance Notch targeting. Pigment Cell Melanoma Res 27:162–168
Bending JJ, Lobo-Yeo A, Vergani D, Viberti GC (1988) Proteinuria and activated T-lymphocytes in diabetic nephropathy. Diabetes 37:507–511
Ben-Shushan E, Feldman E, Reubinoff BE (2015) Notch signaling regulates motor neuron differentiation of human embryonic stem cells. Stem Cells 33:403–415
Benz K, Amann K (2011) Endothelin in diabetic renal disease. Contrib Nephrol 172:139–148
Bos JL (2003) Epac: a new cAMP target and new avenues in cAMP research. Nat Rev Mol Cell Biol 4:733–738
Bracken CP, Khew-Goodall Y, Goodall GJ (2015) Network-based approaches to understand the roles of miR-200 and other microRNAs in cancer. Cancer Res 75:2594–2599
Brosius FC, Tuttle KR, Kretzler M (2016) JAK inhibition in the treatment of diabetic kidney disease. Diabetologia 59:1624–1627
Bruserud O, Pawelec G (1997) Interleukin-13 secretion by normal and posttransplant T lymphocytes; in vitro studies of cellular immune responses in the presence of acute leukaemia blast cells. Cancer Immunol Immunother 45:45–52
Burotto M, Chiou VL, Lee JM, Kohn EC (2014) The MAPK pathway across different malignancies: a new perspective. Cancer 120:3446–3456
Cai B, Cai JP, Luo YL, Chen C, Zhang S (2015) The specific roles of JAK/STAT signaling pathway in sepsis. Inflammation 38:1599–1608
Chandrasekaran K, Karolina DS, Sepramaniam S, Armugam A, Wintour EM, Bertram JF et al (2012) Role of microRNAs in kidney homeostasis and disease. Kidney Int 81:617–627
Chau BN, Xin C, Hartner J, Ren S, Castano AP, Linn G et al (2012) MicroRNA-21 promotes fibrosis of the kidney by silencing metabolic pathways. Sci Transl Med 4:118r–121r
Chen HY, Zhong X, Huang XR, Meng XM, You Y, Chung AC et al (2014) MicroRNA-29b inhibits diabetic nephropathy in db/db mice. Mol Ther 22:842–853
Chen H, Yang X, Lu K, Lu C, Zhao Y, Zheng S et al (2017) Inhibition of high glucose-induced inflammation and fibrosis by a novel curcumin derivative prevents renal and heart injury in diabetic mice. Toxicol Lett 278:48–58
Cheng M, Liu F, Peng Y, Chen J, Chen G, Xiao L et al (2014) Construction of a CTGF and RFP-coexpressed renal tubular epithelial cell and its application on evaluation of CTGF-specific siRNAs on epithelial-mesenchymal transition. Urology 83:1441–1443
Cheng M, Liu H, Zhang D, Liu Y, Wang C, Liu F et al (2015) HMGB1 enhances the AGE-induced expression of CTGF and TGF-beta via RAGE-dependent signaling in renal tubular epithelial cells. Am J Nephrol 41:257–266
Chuang PY, He JC (2010) JAK/STAT signaling in renal diseases. Kidney Int 78:231–234
Chung AC, Huang XR, Meng X, Lan HY (2010) miR-192 mediates TGF-beta/Smad3-driven renal fibrosis. J Am Soc Nephrol 21:1317–1325
Chung AC, Dong Y, Yang W, Zhong X, Li R, Lan HY (2013) Smad7 suppresses renal fibrosis via altering expression of TGF-beta/Smad3-regulated microRNAs. Mol Ther 21:388–398
Crean JK, Furlong F, Mitchell D, McArdle E, Godson C, Martin F (2006) Connective tissue growth factor/CCN2 stimulates actin disassembly through Akt/protein kinase B-mediated phosphorylation and cytoplasmic translocation of p27(Kip-1). FASEB J 20:1712–1714
Danilewicz M, Wagrowska-Danilewicz M (2005) Immunohistochemical analysis of the interstitial mast cells in rebiopsied patients with idiopathic mesangial proliferative glomerulonephritis. Pol J Pathol 56:63–68
Davis RJ (1994) MAPKs: new JNK expands the group. Trends Biochem Sci 19:470–473
de Borst MH, Prakash J, Sandovici M, Klok PA, Hamming I, Kok RJ et al (2009) c-Jun NH_2-terminal kinase is crucially involved in renal tubulo-interstitial inflammation. J Pharmacol Exp Ther 331:896–905
de Larco JE, Todaro GJ (1980) Sarcoma growth factor (SGF): specific binding to epidermal growth factor (EGF) membrane receptors. J Cell Physiol 102:267–277

de Rooij J, Zwartkruis FJ, Verheijen MH, Cool RH, Nijman SM, Wittinghofer A et al (1998) Epac is a Rap1 guanine-nucleotide-exchange factor directly activated by cyclic AMP. Nature 396:474–477
Deshpande SD, Putta S, Wang M, Lai JY, Bitzer M, Nelson RG et al (2013) Transforming growth factor-beta-induced cross talk between p53 and a microRNA in the pathogenesis of diabetic nephropathy. Diabetes 62:3151–3162
Donate-Correa J, Martin-Nunez E, Muros-de-Fuentes M, Mora-Fernandez C, Navarro-Gonzalez JF (2015) Inflammatory cytokines in diabetic nephropathy. J Diabetes Res 2015:948417
Dounousi E, Duni A, Leivaditis K, Vaios V, Eleftheriadis T, Liakopoulos V (2015) Improvements in the management of diabetic nephropathy. Rev Diabet Stud 12:119–133
Egido J, Rojas-Rivera J, Mas S, Ruiz-Ortega M, Sanz AB, Gonzalez PE et al (2017) Atrasentan for the treatment of diabetic nephropathy. Expert Opin Invest Drugs 26:741–750
Esposito C, Fasoli G, Rampino T, Dal Canton A (2006) Hepatocyte growth factor and kidney. G Ital Nefrol 23:381–388
Feng M, Tang PM, Huang XR, Sun SF, You YK, Xiao J et al (2018) TGF-beta Mediates renal fibrosis via the Smad3-Erbb4-IR long noncoding RNA axis. Mol Ther 26:148–161
Fukasawa H, Yamamoto T, Suzuki H, Togawa A, Ohashi N, Fujigaki Y et al (2004) Treatment with anti-TGF-beta antibody ameliorates chronic progressive nephritis by inhibiting Smad/TGF-beta signaling. Kidney Int 65:63–74
Gao Q, Shen W, Qin W, Zheng C, Zhang M, Zeng C et al (2010) Treatment of db/db diabetic mice with triptolide: a novel therapy for diabetic nephropathy. Nephrol Dial Transplant 25:3539–3547
Glowacki F, Savary G, Gnemmi V, Buob D, Van der Hauwaert C, Lo-Guidice JM et al (2013) Increased circulating miR-21 levels are associated with kidney fibrosis. PLoS ONE 8:e58014
Gohda E (2002) Function and regulation of production of hepatocyte growth factor (HGF). Nihon Yakurigaku Zasshi 119:287–294
Goto E, Honjo S, Yamashita H, Shomori K, Adachi H, Ito H (2002) Mast cells in human allografted kidney: correlation with interstitial fibrosis. Clin Transplant 16(Suppl 8):7–11
Guha M, Xu ZG, Tung D, Lanting L, Natarajan R (2007) Specific down-regulation of connective tissue growth factor attenuates progression of nephropathy in mouse models of type 1 and type 2 diabetes. FASEB J 21:3355–3368
Hathaway CK, Gasim AM, Grant R, Chang AS, Kim HS, Madden VJ et al (2015) Low TGFbeta1 expression prevents and high expression exacerbates diabetic nephropathy in mice. Proc Natl Acad Sci U S A 112:5815–5820
Hay ED, Zuk A (1995) Transformations between epithelium and mesenchyme: normal, pathological, and experimentally induced. Am J Kidney Dis 26:678–690
He J, Yuan G, Cheng F, Zhang J, Guo X (2017) Mast cell and M1 macrophage infiltration and local pro-inflammatory factors were attenuated with incretin-based therapies in obesity-related glomerulopathy. Metab Syndr Relat Disord 15:344–353
Hickey FB, Martin F (2018) Role of the immune system in diabetic kidney disease. Curr Diab Rep 18:20
Ho C, Lee PH, Hsu YC, Wang FS, Huang YT, Lin CL (2012) Sustained Wnt/beta-catenin signaling rescues high glucose induction of transforming growth factor-beta1-mediated renal fibrosis. Am J Med Sci 344:374–382
Hong JP, Li XM, Li MX, Zheng FL (2013) VEGF suppresses epithelial-mesenchymal transition by inhibiting the expression of Smad3 and miR192, a Smad3-dependent microRNA. Int J Mol Med 31:1436–1442
Hou M, Bao X, Luo F, Chen X, Liu L, Wu M (2018) HMGA2 modulates the TGFbeta/Smad, TGFbeta/ERK and Notch signaling pathways in human lens epithelial-mesenchymal transition. Curr Mol Med 18:71–82
Hu HH, Chen DQ, Wang YN, Feng YL, Cao G, Vaziri ND et al (2018) New insights into TGF-beta/smad signaling in tissue fibrosis. Chem Biol Interact 292:76–83
Huang JS, Guh JY, Chen HC, Hung WC, Lai YH, Chuang LY (2001) Role of receptor for advanced glycation end-product (RAGE) and the JAK/STAT-signaling pathway in AGE-induced collagen production in NRK-49F cells. J Cell Biochem 81:102–113

Huang S, Liu F, Niu Q, Li Y, Liu C, Zhang L et al (2013) GLIPR-2 overexpression in HK-2 cells promotes cell EMT and migration through ERK1/2 activation. PLoS ONE 8:e58574
Huang Y, Tong J, He F, Yu X, Fan L, Hu J et al (2015) miR-141 regulates TGF-beta1-induced epithelial-mesenchymal transition through repression of HIPK2 expression in renal tubular epithelial cells. Int J Mol Med 35:311–318
Hwang I, Seo EY, Ha H (2009) Wnt/beta-catenin signaling: a novel target for therapeutic intervention of fibrotic kidney disease. Arch Pharm Res 32:1653–1662
Jafari M, Ghadami E, Dadkhah T, Akhavan-Niaki H (2019) PI3k/AKT signaling pathway: erythropoiesis and beyond. J Cell Physiol 234:2373–2385
Jang YN, Baik EJ (2013) JAK-STAT pathway and myogenic differentiation. JAKSTAT 2:e23282
Jenkins RH, Martin J, Phillips AO, Bowen T, Fraser DJ (2012a) Pleiotropy of microRNA-192 in the kidney. Biochem Soc Trans 40:762–767
Jenkins RH, Martin J, Phillips AO, Bowen T, Fraser DJ (2012b) Transforming growth factor beta1 represses proximal tubular cell microRNA-192 expression through decreased hepatocyte nuclear factor DNA binding. Biochem J 443:407–416
Jiang L, Qiu W, Zhou Y, Wen P, Fang L, Cao H et al (2013) A microRNA-30e/mitochondrial uncoupling protein 2 axis mediates TGF-beta1-induced tubular epithelial cell extracellular matrix production and kidney fibrosis. Kidney Int 84:285–296
Juan YS, Chuang SM, Long CY, Lin RJ, Liu KM, Wu WJ et al (2012) Protein kinase C inhibitor prevents renal apoptotic and fibrotic changes in response to partial ureteric obstruction. BJU Int 110:283–292
Kanasaki K, Shi S, Kanasaki M, He J, Nagai T, Nakamura Y et al (2014) Linagliptin-mediated DPP-4 inhibition ameliorates kidney fibrosis in streptozotocin-induced diabetic mice by inhibiting endothelial-to-mesenchymal transition in a therapeutic regimen. Diabetes 63:2120–2131
Kang MJ, Wu X, Ly H, Thai K, Scholey JW (1999) Effect of glucose on stress-activated protein kinase activity in mesangial cells and diabetic glomeruli. Kidney Int 55:2203–2214
Kanwar YS, Sun L, Xie P, Liu FY, Chen S (2011) A glimpse of various pathogenetic mechanisms of diabetic nephropathy. Annu Rev Pathol 6:395–423
Kato M, Natarajan R (2014) Diabetic nephropathy—emerging epigenetic mechanisms. Nat Rev Nephrol 10:517–530
Kato M, Natarajan R (2015) MicroRNAs in diabetic nephropathy: functions, biomarkers, and therapeutic targets. Ann N Y Acad Sci 1353:72–88
Kato M, Zhang J, Wang M, Lanting L, Yuan H, Rossi JJ et al (2007) MicroRNA-192 in diabetic kidney glomeruli and its function in TGF-beta-induced collagen expression via inhibition of E-box repressors. Proc Natl Acad Sci U S A 104:3432–3437
Kato M, Arce L, Wang M, Putta S, Lanting L, Natarajan R (2011) A microRNA circuit mediates transforming growth factor-beta1 autoregulation in renal glomerular mesangial cells. Kidney Int 80:358–368
Kato M, Dang V, Wang M, Park JT, Deshpande S, Kadam S et al (2013) TGF-beta induces acetylation of chromatin and of Ets-1 to alleviate repression of miR-192 in diabetic nephropathy. Sci Signal 6:a43
Kim SM, Lee SH, Lee A, Kim DJ, Kim YG, Kim SY et al (2015) Targeting T helper 17 by mycophenolate mofetil attenuates diabetic nephropathy progression. Transl Res 166:375–383
Kinashi H, Falke LL, Nguyen TQ, Bovenschen N, Aten J, Leask A et al (2017) Connective tissue growth factor regulates fibrosis-associated renal lymphangiogenesis. Kidney Int 92:850–863
Kolling M, Kaucsar T, Schauerte C, Hubner A, Dettling A, Park JK et al (2017) Therapeutic miR-21 silencing ameliorates diabetic kidney disease in mice. Mol Ther 25:165–180
Kopan R, Ilagan MX (2009) The canonical Notch signaling pathway: unfolding the activation mechanism. Cell 137:216–233
Krupa A, Jenkins R, Luo DD, Lewis A, Phillips A, Fraser D (2010) Loss of MicroRNA-192 promotes fibrogenesis in diabetic nephropathy. J Am Soc Nephrol 21:438–447
Lai JY, Luo J, O'Connor C, Jing X, Nair V, Ju W et al (2015) MicroRNA-21 in glomerular injury. J Am Soc Nephrol 26:805–816

Lan X, Wen H, Aslam R, Shoshtari S, Mishra A, Kumar V et al (2018) Nicotine enhances mesangial cell proliferation and fibronectin production in high glucose milieu via activation of Wnt/beta-catenin pathway. Biosci Rep 38
Lee RC, Feinbaum RL, Ambros V (1993) The C. elegans heterochronic gene lin-4 encodes small RNAs with antisense complementarity to lin-14. Cell 75:843–854
Li J, Bertram JF (2010) Review: endothelial-myofibroblast transition, a new player in diabetic renal fibrosis. Nephrology (Carlton) 15:507–512
Liau N, Laktyushin A, Lucet IS, Murphy JM, Yao S, Whitlock E et al (2018) The molecular basis of JAK/STAT inhibition by SOCS1. Nat Commun 9:1558
Libetta C, Esposito P, Martinelli C, Grosjean F, Gregorini M, Rampino T et al (2016) Hepatocyte growth factor (HGF) and hemodialysis: physiopathology and clinical implications. Clin Exp Nephrol 20:371–378
Lim AK, Tesch GH (2012) Inflammation in diabetic nephropathy. Mediat Inflamm 2012:146154
Lin S, Sahai A, Chugh SS, Pan X, Wallner EI, Danesh FR et al (2002) High glucose stimulates synthesis of fibronectin via a novel protein kinase C, Rap1b, and B-Raf signaling pathway. J Biol Chem 277:41725–41735
Lin CL, Wang FS, Hsu YC, Chen CN, Tseng MJ, Saleem MA et al (2010a) Modulation of notch-1 signaling alleviates vascular endothelial growth factor-mediated diabetic nephropathy. Diabetes 59:1915–1925
Lin CL, Wang JY, Ko JY, Huang YT, Kuo YH, Wang FS (2010b) Dickkopf-1 promotes hyperglycemia-induced accumulation of mesangial matrix and renal dysfunction. J Am Soc Nephrol 21:124–135
Lin CL, Lee PH, Hsu YC, Lei CC, Ko JY, Chuang PC et al (2014) MicroRNA-29a promotion of nephrin acetylation ameliorates hyperglycemia-induced podocyte dysfunction. J Am Soc Nephrol 25:1698–1709
Lin Y, Zhao JL, Zheng QJ, Jiang X, Tian J, Liang SQ et al (2018) Notch signaling modulates macrophage polarization and phagocytosis through direct suppression of signal regulatory protein alpha expression. Front Immunol 9:1744
Liu Y (2010) New insights into epithelial-mesenchymal transition in kidney fibrosis. J Am Soc Nephrol 21:212–222
Liu Q, Xing L, Wang L, Yao F, Liu S, Hao J et al (2014a) Therapeutic effects of suppressors of cytokine signaling in diabetic nephropathy. J Histochem Cytochem 62:119–128
Liu R, Zhong Y, Li X, Chen H, Jim B, Zhou MM et al (2014b) Role of transcription factor acetylation in diabetic kidney disease. Diabetes 63:2440–2453
Liu H, Wang X, Liu S, Li H, Yuan X, Feng B et al (2016a) Effects and mechanism of miR-23b on glucose-mediated epithelial-to-mesenchymal transition in diabetic nephropathy. Int J Biochem Cell Biol 70:149–160
Liu XJ, Hong Q, Wang Z, Yu YY, Zou X, Xu LH (2016b) MicroRNA21 promotes interstitial fibrosis via targeting DDAH1: a potential role in renal fibrosis. Mol Cell Biochem 411:181–189
Liu D, Cao Y, Zhang X, Peng C, Tian X, Yan C et al (2018) Chemokine CC-motif ligand 2 participates in platelet function and arterial thrombosis by regulating PKCalpha-P38MAPK-HSP27 pathway. Biochim Biophys Acta 1864:2901–2912
Loeffler I, Wolf G (2015) Epithelial-to-mesenchymal transition in diabetic nephropathy: fact or fiction? Cells 4:631–652
Long J, Wang Y, Wang W, Chang BH, Danesh FR (2011) MicroRNA-29c is a signature microRNA under high glucose conditions that targets Sprouty homolog 1, and its in vivo knockdown prevents progression of diabetic nephropathy. J Biol Chem 286:11837–11848
Lopez-Arribillaga E, Rodilla V, Colomer C, Vert A, Shelton A, Cheng JH et al (2018) Manic Fringe deficiency imposes Jagged1 addiction to intestinal tumor cells. Nat Commun 9:2992
Lu Q, Zuo WZ, Ji XJ, Zhou YX, Liu YQ, Yao XQ et al (2015) Ethanolic Ginkgo biloba leaf extract prevents renal fibrosis through Akt/mTOR signaling in diabetic nephropathy. Phytomedicine 22:1071–1078

Lv LL, Liu BC (2015) Role of non-classical renin-angiotensin system axis in renal fibrosis. Front Physiol 6:117

Manson SR, Austin PF, Guo Q, Moore KH (2015) BMP-7 signaling and its critical roles in kidney development, the responses to renal injury, and chronic kidney disease. Vitam Horm 99:91–144

March JT, Golshirazi G, Cernisova V, Carr H, Leong Y, Lu-Nguyen N et al (2018) Targeting TGFbeta signaling to address fibrosis using antisense oligonucleotides. Biomedicines 6

Mariappan MM, Shetty M, Sataranatarajan K, Choudhury GG, Kasinath BS (2008) Glycogen synthase kinase 3beta is a novel regulator of high glucose- and high insulin-induced extracellular matrix protein synthesis in renal proximal tubular epithelial cells. J Biol Chem 283:30566–30575

Mason RM (2009) Connective tissue growth factor (CCN2), a pathogenic factor in diabetic nephropathy. What does it do? How does it do it? J Cell Commun Signal 3:95–104

McClelland AD, Herman-Edelstein M, Komers R, Jha JC, Winbanks CE, Hagiwara S et al (2015) miR-21 promotes renal fibrosis in diabetic nephropathy by targeting PTEN and SMAD7. Clin Sci (Lond) 129:1237–1249

Meng XM, Huang XR, Xiao J, Chen HY, Zhong X, Chung AC et al (2012) Diverse roles of TGF-beta receptor II in renal fibrosis and inflammation in vivo and in vitro. J Pathol 227:175–188

Meng X, Chung ACK, Lan HY (2013) Role of the TGF-β/BMP-7/Smad pathways in renal diseases. Clin Sci 124:243–254

Meng XM, Tang PM, Li J, Lan HY (2015) TGF-beta/Smad signaling in renal fibrosis. Front Physiol 6:82

Meng XM, Nikolic-Paterson DJ, Lan HY (2016) TGF-beta: the master regulator of fibrosis. Nat Rev Nephrol 12:325–338

Menne J, Shushakova N, Bartels J, Kiyan Y, Laudeley R, Haller H et al (2013) Dual inhibition of classical protein kinase C-alpha and protein kinase C-beta isoforms protects against experimental murine diabetic nephropathy. Diabetes 62:1167–1174

Mizuno S, Nakamura T (2004) Suppressions of chronic glomerular injuries and TGF-beta 1 production by HGF in attenuation of murine diabetic nephropathy. Am J Physiol Renal Physiol 286:F134–F143

Mizuno S, Matsumoto K, Nakamura T (2008) HGF as a renotrophic and anti-fibrotic regulator in chronic renal disease. Front Biosci 13:7072–7086

Mohamed R, Jayakumar C, Chen F, Fulton D, Stepp D, Gansevoort RT et al (2016) Low-dose IL-17 therapy prevents and reverses diabetic nephropathy, metabolic syndrome, and associated organ fibrosis. J Am Soc Nephrol 27:745–765

Moon JY, Jeong KH, Lee TW, Ihm CG, Lim SJ, Lee SH (2012) Aberrant recruitment and activation of T cells in diabetic nephropathy. Am J Nephrol 35:164–174

Moriya R, Manivel JC, Mauer M (2004) Juxtaglomerular apparatus T-cell infiltration affects glomerular structure in Type 1 diabetic patients. Diabetologia 47:82–88

Mou X, Zhou DY, Zhou DY, Ma JR, Liu YH, Chen HP et al (2016) Serum TGF-beta1 as a biomarker for type 2 diabetic nephropathy: a meta-analysis of randomized controlled trials. PLoS ONE 11:e149513

Mu J, Pang Q, Guo YH, Chen JG, Zeng W, Huang YJ et al (2013) Functional implications of microRNA-215 in TGF-beta1-induced phenotypic transition of mesangial cells by targeting CTNNBIP1. PLoS ONE 8:e58622

Natarajan R, Putta S, Kato M (2012) MicroRNAs and diabetic complications. J Cardiovasc Transl Res 5:413–422

Navarro-Gonzalez JF, Mora-Fernandez C, Muros DFM, Garcia-Perez J (2011) Inflammatory molecules and pathways in the pathogenesis of diabetic nephropathy. Nat Rev Nephrol 7:327–340

Nguyen TQ, Roestenberg P, van Nieuwenhoven FA, Bovenschen N, Li Z, Xu L et al (2008) CTGF inhibits BMP-7 signaling in diabetic nephropathy. J Am Soc Nephrol 19:2098–2107

Nlandu KS, Neelisetty S, Woodbury L, Green E, Harris RC, Zent R (2016) Deleting the TGF-beta receptor in proximal tubules impairs HGF signaling. Am J Physiol Renal Physiol 310:F499–F510

Noetel A, Kwiecinski M, Elfimova N, Huang J, Odenthal M (2012) microRNA are central players in anti- and profibrotic gene regulation during liver fibrosis. Front Physiol 3:49

Oba S, Kumano S, Suzuki E, Nishimatsu H, Takahashi M, Takamori H et al (2010) miR-200b precursor can ameliorate renal tubulointerstitial fibrosis. PLoS ONE 5:e13614
Ohta M, Chosa N, Kyakumoto S, Yokota S, Okubo N, Nemoto A et al (2018) IL-1β and TNF-α suppress TGF-β-promoted NGF expression in periodontal ligament-derived fibroblasts through inactivation of TGF-β- induced Smad2/3- and p38 MAPK-mediated signals. Int J Mol Med 42:1484–1494
Okada S, Shikata K, Matsuda M, Ogawa D, Usui H, Kido Y et al (2003) Intercellular adhesion molecule-1-deficient mice are resistant against renal injury after induction of diabetes. Diabetes 52:2586–2593
Okayama Y, Kawakami T (2006) Development, migration, and survival of mast cells. Immunol Res 34:97–115
Oldfield MD, Bach LA, Forbes JM, Nikolic-Paterson D, McRobert A, Thallas V et al (2001) Advanced glycation end products cause epithelial-myofibroblast transdifferentiation via the receptor for advanced glycation end products (RAGE). J Clin Invest 108:1853–1863
Pan J, Zhang J, Zhang X, Zhou X, Lu S, Huang X et al (2014) Role of microRNA-29b in angiotensin II-induced epithelial-mesenchymal transition in renal tubular epithelial cells. Int J Mol Med 34:1381–1387
Park JT, Kato M, Yuan H, Castro N, Lanting L, Wang M et al (2013) FOG2 protein down-regulation by transforming growth factor-beta1-induced microRNA-200b/c leads to Akt kinase activation and glomerular mesangial hypertrophy related to diabetic nephropathy. J Biol Chem 288:22469–22480
Peng R, Zhou L, Zhou Y, Zhao Y, Li Q, Ni D et al (2015) MiR-30a inhibits the epithelial-mesenchymal transition of podocytes through downregulation of NFATc3. Int J Mol Sci 16:24032–24047
Putta S, Lanting L, Sun G, Lawson G, Kato M, Natarajan R (2012) Inhibiting microRNA-192 ameliorates renal fibrosis in diabetic nephropathy. J Am Soc Nephrol 23:458–469
Rahimi Z (2016) The role of renin angiotensin aldosterone system genes in diabetic nephropathy. Can J Diabetes 40:178–183
Rane MJ, Song Y, Jin S, Barati MT, Wu R, Kausar H et al (2010) Interplay between Akt and p38 MAPK pathways in the regulation of renal tubular cell apoptosis associated with diabetic nephropathy. Am J Physiol Renal Physiol 298:F49–F61
Reddy MA, Tak PJ, Natarajan R (2013) Epigenetic modifications in the pathogenesis of diabetic nephropathy. Semin Nephrol 33:341–353
Rousselle A, Kettritz R, Schreiber A (2017) Monocytes promote crescent formation in anti-myeloperoxidase antibody-induced glomerulonephritis. Am J Pathol 187:1908–1915
Sanai T, Sobka T, Johnson T, El-Essawy M, Muchaneta-Kubara EC, Ben GO et al (2000) Expression of cytoskeletal proteins during the course of experimental diabetic nephropathy. Diabetologia 43:91–100
Shemesh II, Rozen-Zvi B, Kalechman Y, Gafter U, Sredni B (2014) AS101 prevents diabetic nephropathy progression and mesangial cell dysfunction: regulation of the AKT downstream pathway. PLoS ONE 9:e114287
Shi S, Yu L, Chiu C, Sun Y, Chen J, Khitrov G et al (2008) Podocyte-selective deletion of dicer induces proteinuria and glomerulosclerosis. J Am Soc Nephrol 19:2159–2169
Simpson K, Wonnacott A, Fraser DJ, Bowen T (2016) MicroRNAs in diabetic nephropathy: from biomarkers to therapy. Curr Diab Rep 16:35
Singh M, Yelle N, Venugopal C, Singh SK (2018) EMT: mechanisms and therapeutic implications. Pharmacol Ther 182:80–94
Sirin Y, Susztak K (2012) Notch in the kidney: development and disease. J Pathol 226:394–403
Sun L, Kondeti VK, Xie P, Raparia K, Kanwar YS (2011a) Epac1-mediated, high glucose-induced renal proximal tubular cells hypertrophy via the Akt/p21 pathway. Am J Pathol 179:1706–1718
Sun L, Zhang D, Liu F, Xiang X, Ling G, Xiao L et al (2011b) Low-dose paclitaxel ameliorates fibrosis in the remnant kidney model by down-regulating miR-192. J Pathol 225:364–377

Teng B, Duong M, Tossidou I, Yu X, Schiffer M (2014) Role of protein kinase C in podocytes and development of glomerular damage in diabetic nephropathy. Front Endocrinol (Lausanne) 5:179
Tesch GH (2017) Diabetic nephropathy—is this an immune disorder? Clin Sci (Lond) 131:2183–2199
Togawa H, Nakanishi K, Shima Y, Obana M, Sako M, Nozu K et al (2009) Increased chymase-positive mast cells in children with crescentic glomerulonephritis. Pediatr Nephrol 24:1071–1075
Trionfini P, Benigni A (2017) MicroRNAs as master regulators of glomerular function in health and disease. J Am Soc Nephrol 28:1686–1696
Trionfini P, Benigni A, Remuzzi G (2015) MicroRNAs in kidney physiology and disease. Nat Rev Nephrol 11:23–33
Tripurani SK, Wang Y, Fan YX, Rahimi M, Wong L, Lee MH et al (2018) Suppression of Wnt/β-catenin signaling by EGF receptor is required for hair follicle development. Mol Biol Cell 29:2784–2799
Tu Y, Wu T, Dai A, Pham Y, Chew P, de Haan JB et al (2011) Cell division autoantigen 1 enhances signaling and the profibrotic effects of transforming growth factor-beta in diabetic nephropathy. Kidney Int 79:199–209
Turner CA, Sharma V, Hagenauer MH, Chaudhury S, O'Connor AM, Hebda-Bauer EK et al (2018) Connective tissue growth factor is a novel prodepressant. Biol Psychiatry 84:555–562
Umanath K, Lewis JB (2018) Update on diabetic nephropathy: core curriculum 2018. Am J Kidney Dis 71:884–895
van der Sande NG, Dorresteijn JA, Visseren FL, Dwyer JP, Blankestijn PJ, van der Graaf Y et al (2016) Individualized prediction of the effect of angiotensin receptor blockade on renal and cardiovascular outcomes in patients with diabetic nephropathy. Diab Obes Metab 18:1120–1127
Vasanthakumar R, Mohan V, Anand G, Deepa M, Babu S, Aravindhan V (2015) Serum IL-9, IL-17, and TGF-beta levels in subjects with diabetic kidney disease (CURES-134). Cytokine 72:109–112
Verhave JC, Bouchard J, Goupil R, Pichette V, Brachemi S, Madore F et al (2013) Clinical value of inflammatory urinary biomarkers in overt diabetic nephropathy: a prospective study. Diabetes Res Clin Pract 101:333–340
Voelker J, Berg PH, Sheetz M, Duffin K, Shen T, Moser B et al (2017) Anti-TGF-beta1 antibody therapy in patients with diabetic nephropathy. J Am Soc Nephrol 28:953–962
Wahlang B, McClain C, Barve S, Gobejishvili L (2018) Role of cAMP and phosphodiesterase signaling in liver health and disease. Cell Signal 49:105–115
Wang S (2015) Role of upstream stimulatory factor 2 in diabetic nephropathy. Front Biol (Beijing) 10:221–229
Wang Y, Rangan GK, Tay YC, Wang Y, Harris DC (1999) Induction of monocyte chemoattractant protein-1 by albumin is mediated by nuclear factor kappaB in proximal tubule cells. J Am Soc Nephrol 10:1204–1213
Wang A, Ziyadeh FN, Lee EY, Pyagay PE, Sung SH, Sheardown SA et al (2007) Interference with TGF-beta signaling by Smad3-knockout in mice limits diabetic glomerulosclerosis without affecting albuminuria. Am J Physiol Renal Physiol 293:F1657–F1665
Wang B, Herman-Edelstein M, Koh P, Burns W, Jandeleit-Dahm K, Watson A et al (2010) E-cadherin expression is regulated by miR-192/215 by a mechanism that is independent of the profibrotic effects of transforming growth factor-beta. Diabetes 59:1794–1802
Wang B, Koh P, Winbanks C, Coughlan MT, McClelland A, Watson A et al (2011a) miR-200a prevents renal fibrogenesis through repression of TGF-beta2 expression. Diabetes 60:280–287
Wang Q, Usinger W, Nichols B, Gray J, Xu L, Seeley TW et al (2011b) Cooperative interaction of CTGF and TGF-beta in animal models of fibrotic disease. Fibrogenesis Tissue Repair 4:4
Wang B, Komers R, Carew R, Winbanks CE, Xu B, Herman-Edelstein M et al (2012) Suppression of microRNA-29 expression by TGF-beta1 promotes collagen expression and renal fibrosis. J Am Soc Nephrol 23:252–265
Wang J, Gao Y, Ma M, Li M, Zou D, Yang J et al (2013) Effect of miR-21 on renal fibrosis by regulating MMP-9 and TIMP1 in kk-ay diabetic nephropathy mice. Cell Biochem Biophys 67:537–546

Wang JY, Gao YB, Zhang N, Zou DW, Wang P, Zhu ZY et al (2014a) miR-21 overexpression enhances TGF-beta1-induced epithelial-to-mesenchymal transition by target smad7 and aggravates renal damage in diabetic nephropathy. Mol Cell Endocrinol 392:163–172

Wang Z, Wei M, Wang M, Chen L, Liu H, Ren Y et al (2014b) Inhibition of macrophage migration inhibitory factor reduces diabetic nephropathy in type II diabetes mice. Inflammation 37:2020–2029

Wei J, Zhang Y, Luo Y, Wang Z, Bi S, Song D et al (2014) Aldose reductase regulates miR-200a-3p/141-3p to coordinate Keap1-Nrf2, Tgfbeta1/2, and Zeb1/2 signaling in renal mesangial cells and the renal cortex of diabetic mice. Free Radic Biol Med 67:91–102

White DA, Su Y, Kanellakis P, Kiriazis H, Morand EF, Bucala R et al (2014) Differential roles of cardiac and leukocyte derived macrophage migration inhibitory factor in inflammatory responses and cardiac remodelling post myocardial infarction. J Mol Cell Cardiol 69:32–42

Wu CC, Sytwu HK, Lu KC, Lin YF (2011) Role of T cells in type 2 diabetic nephropathy. Exp Diabetes Res 2011:514738

Wu J, Zheng C, Fan Y, Zeng C, Chen Z, Qin W et al (2014) Downregulation of microRNA-30 facilitates podocyte injury and is prevented by glucocorticoids. J Am Soc Nephrol 25:92–104

Wu H, Kong L, Tan Y, Epstein PN, Zeng J, Gu J et al (2016) C66 ameliorates diabetic nephropathy in mice by both upregulating NRF2 function via increase in miR-200a and inhibiting miR-21. Diabetologia 59:1558–1568

Xiao L, Wang M, Yang S, Liu F, Sun L (2013) A glimpse of the pathogenetic mechanisms of Wnt/beta-catenin signaling in diabetic nephropathy. Biomed Res Int 2013:987064

Xie X, Xia W, Fei X, Xu Q, Yang X, Qiu D et al (2015) Relaxin inhibits high glucose-induced matrix accumulation in human mesangial cells by interfering with TGF-beta1 production and mesangial cells phenotypic transition. Biol Pharm Bull 38:1464–1469

Xu W, Yang Z, Lu N (2015) A new role for the PI3K/Akt signaling pathway in the epithelial-mesenchymal transition. Cell Adh Migr 9:317–324

Yacoub R, Campbell KN (2015) Inhibition of RAS in diabetic nephropathy. Int J Nephrol Renovasc Dis 8:29–40

Yakush WJ (2017) Management strategies for patients with diabetic kidney disease and chronic kidney disease in diabetes. Nurs Clin North Am 52:575–587

Yang J, Dai C, Liu Y (2003) Hepatocyte growth factor suppresses renal interstitial myofibroblast activation and intercepts Smad signal transduction. Am J Pathol 163:621–632

Yang Z, Guo Z, Dong J, Sheng S, Wang Y, Yu L et al (2018) miR-374a regulates inflammatory response in diabetic nephropathy by targeting MCP-1 expression. Front Pharmacol 9:900

Yano N, Suzuki D, Endoh M, Zhao TC, Padbury JF, Tseng YT (2007) A novel phosphoinositide 3-kinase-dependent pathway for angiotensin II/AT-1 receptor-mediated induction of collagen synthesis in MES-13 mesangial cells. J Biol Chem 282:18819–18830

Yokoi H, Mukoyama M, Mori K, Kasahara M, Suganami T, Sawai K et al (2008) Overexpression of connective tissue growth factor in podocytes worsens diabetic nephropathy in mice. Kidney Int 73:446–455

Zeisberg EM, Potenta SE, Sugimoto H, Zeisberg M, Kalluri R (2008) Fibroblasts in kidney fibrosis emerge via endothelial-to-mesenchymal transition. J Am Soc Nephrol 19:2282–2287

Zha D, Cheng H, Li W, Wu Y, Li X, Zhang L et al (2017) High glucose instigates tubulointerstitial injury by stimulating hetero-dimerization of adiponectin and angiotensin II receptors. Biochem Biophys Res Commun 493:840–846

Zhang Z, Peng H, Chen J, Chen X, Han F, Xu X et al (2009) MicroRNA-21 protects from mesangial cell proliferation induced by diabetic nephropathy in db/db mice. FEBS Lett 583:2009–2014

Zhang HT, Zhang D, Zha ZG, Hu CD (2014) Transcriptional activation of PRMT5 by NF-Y is required for cell growth and negatively regulated by the PKC/c-Fos signaling in prostate cancer cells. Biochim Biophys Acta 1839:1330–1340

Zhang C, Hou B, Yu S, Chen Q, Zhang N, Li H (2016a) HGF alleviates high glucose-induced injury in podocytes by GSK3beta inhibition and autophagy restoration. Biochim Biophys Acta 1863:2690–2699

Zhang H, Li A, Zhang W, Huang Z, Wang J, Yi B (2016b) High glucose-induced cytoplasmic translocation of Dnmt3a contributes to CTGF hypo-methylation in mesangial cells. Biosci Rep 36

Zhang Y, Sun X, Icli B, Feinberg MW (2017) Emerging roles for microRNAs in diabetic microvascular disease: novel targets for therapy. Endocr Rev 38:145–168

Zhang W, Hou X, Huang M, Zeng X, He X, Liao Y (2018) TDCPP protects cardiomyocytes from H_2O_2-induced injuries via activating PI3K/Akt/GSK3beta signaling pathway. Mol Cell Biochem

Zhao B, Li H, Liu J, Han P, Zhang C, Bai H et al (2016) MicroRNA-23b targets Ras GTPase-activating protein SH3 domain-binding protein 2 to alleviate fibrosis and albuminuria in diabetic nephropathy. J Am Soc Nephrol 27:2597–2608

Zhao D, Jia J, Shao H (2017a) miR-30e targets GLIPR-2 to modulate diabetic nephropathy: in vitro and in vivo experiments. J Mol Endocrinol 59:181–190

Zhao Y, Yin Z, Li H, Fan J, Yang S, Chen C et al (2017b) MiR-30c protects diabetic nephropathy by suppressing epithelial-to-mesenchymal transition in db/db mice. Aging Cell 16:387–400

Zheng JM, Yao GH, Cheng Z, Wang R, Liu ZH (2012) Pathogenic role of mast cells in the development of diabetic nephropathy: a study of patients at different stages of the disease. Diabetologia 55:801–811

Zhong X, Chung AC, Chen HY, Meng XM, Lan HY (2011) Smad3-mediated upregulation of miR-21 promotes renal fibrosis. J Am Soc Nephrol 22:1668–1681

Zhong X, Chung AC, Chen HY, Dong Y, Meng XM, Li R et al (2013) miR-21 is a key therapeutic target for renal injury in a mouse model of type 2 diabetes. Diabetologia 56:663–674

Zhou L, Liu Y (2016) Wnt/beta-catenin signaling and renin-angiotensin system in chronic kidney disease. Curr Opin Nephrol Hypertens 25:100–106

Zhou T, He X, Cheng R, Zhang B, Zhang RR, Chen Y et al (2012) Implication of dysregulation of the canonical wingless-type MMTV integration site (WNT) pathway in diabetic nephropathy. Diabetologia 55:255–266

Zhu L, Zhao S, Liu S, Liu Q, Li F, Hao J (2016) PTEN regulates renal extracellular matrix deposit via increased CTGF in diabetes mellitus. J Cell Biochem 117:1187–1198

Ziyadeh FN, Hoffman BB, Han DC, Iglesias-De LCM, Hong SW, Isono M et al (2000) Long-term prevention of renal insufficiency, excess matrix gene expression, and glomerular mesangial matrix expansion by treatment with monoclonal antitransforming growth factor-beta antibody in db/db diabetic mice. Proc Natl Acad Sci U S A 97:8015–8020

Zou XZ, Liu T, Gong ZC, Hu CP, Zhang Z (2017) MicroRNAs-mediated epithelial-mesenchymal transition in fibrotic diseases. Eur J Pharmacol 796:190–206

Chapter 5
Polycystic Kidney Disease and Renal Fibrosis

Cheng Xue and Chang-Lin Mei

Abstract Polycystic kidney disease (PKD) is a common genetic disorder characterized by formations of numerous cysts in kidneys and most caused by PKD1 or PKD2 mutations in autosomal dominant polycystic kidney disease (ADPKD). The interstitial inflammation and fibrosis is one of the major pathological changes in polycystic kidney tissues with an accumulation of inflammatory cells, chemokines, and cytokines. The immune response is observed across different stages and occurs prior to or coincident with cyst formation in ADPKD. Evidence for inflammation as an important contributor to cyst growth and fibrosis includes increased interstitial macrophages, upregulated expressions of pro-inflammatory cytokines, activated complement system, and activated pathways including NF-κB and JAK-STAT signaling in polycystic kidney tissues. Inflammatory cells are responsible for overproduction of several pro-fibrotic growth factors which promote renal fibrosis in ADPKD. These growth factors trigger epithelial mesenchymal transition and myofibroblast/fibrocyte activation, which stimulate the expansion of extracellular matrix (ECM) including collagen I, III, IV, V, and fibronectin, leading to renal fibrosis and reduced renal function. Besides, there are imbalanced ECM turnover regulators which lead to the increased ECM production and inadequate degradation in polycystic kidney tissues. Several fibrosis associated signaling pathways, such as TGFβ-SMAD, Wnt, and periostin–integrin-linked kinase are also activated in polycystic kidney tissues. Although the effective anti-fibrotic treatments are limited at the present time, slowing the cyst expansion and fibrosis development is very important for prolonging life span and improving the palliative care of ADPKD patients. The inhibition of pro-fibrotic cytokines involved in fibrosis might be a new therapeutic strategy for ADPKD in the future.

Keywords Polycystic kidney disease · Inflammation · Extracellular matrix · Fibrosis

---

C. Xue · C.-L. Mei (✉)
Division of Nephrology, Kidney Institute of PLA, Changzheng Hospital,
Second Military Medical University, Shanghai, China
e-mail: chlmei1954@126.com

B.-C. Liu et al. (eds.), *Renal Fibrosis: Mechanisms and Therapies*,
Advances in Experimental Medicine and Biology 1165,
https://doi.org/10.1007/978-981-13-8871-2_5

5.1 Introduction

Polycystic kidney disease (PKD) is a genetic disorder which is characterized by the formations of cysts in both kidneys and other organs. PKD can be divided into two forms according to genetic pattern, autosomal dominant polycystic kidney disease (ADPKD), and autosomal recessive polycystic kidney disease (ARPKD). ARPKD affects approximately 1/20,000 individuals (Bergmann 2017). Its causative gene is PKHD1 which encodes fibrocystin/polyductin. ADPKD affects about 1/1000 individuals (Wuthrich and Mei 2012; Xue et al. 2016). PKD1 and PKD2 are mainly two pathogenic genes which encode polycystin-1 (PC1) and polycystin-2 (PC2), respectively. Dysfunctional proteins of PC1/2 or fibrocystin influence the function of the primary cilia (Zimmerman and Yoder 2015), which is involved in mechanosensation to detect the fluid flow passing through the tubule lumen and regulate cell proliferation, oriented cell division, as well as cell polarity (Zimmerman and Yoder 2015). Because ADPKD is more common than ARPKD, the data about ADPKD and fibrosis is abundant, this chapter only focuses on ADPKD.

Cyst growth in ADPKD is associated with increases in epithelial cell proliferation, dedifferentiation, and fluid secretion. The enlargement of cysts leads to the compression and obstruction of surrounding nephrons which could significantly decrease the kidney function (Xue et al. 2018). At the late stage of ADPKD, cyst formation is always accompanied by extracellular matrix (ECM) deposition and fibrosis formation (Grantham et al. 2011). Fibrosis is characterized by excessive productions of collagen accompanied by decompositions of connective tissues. Fibrosis further reduces the renal function and eventually leads to end-stage renal disease (ESRD) (Grantham et al. 2011). Experimental studies found that disruptions of polycystins or primary cilia are associated with inflammation and fibrosis in a variety of polycystic kidney models (Song et al. 2017). In general, increased inflammation and fibrosis aggravates the disease progression. While inflammation and fibrosis are not primary causes of ADPKD, several cells like macrophages and molecules like transforming growth factor (TGF) β related to inflammation and fibrosis can influence renal function and ADPKD progression (Liu et al. 2014; Song et al. 2017).

5.2 Inflammation in ADPKD

Inflammation is an important process prior to or coincident with fibrosis in ADPKD (Fig. 5.1) (Harms et al. 2018). The immune response is observed across different stages of ADPKD progression. Perhaps the earliest immune changes are triggered by the loss of PC1/2 function, which may occur before the inflammation pathway activation by renal injury in ADPKD (Karihaloo et al. 2011).

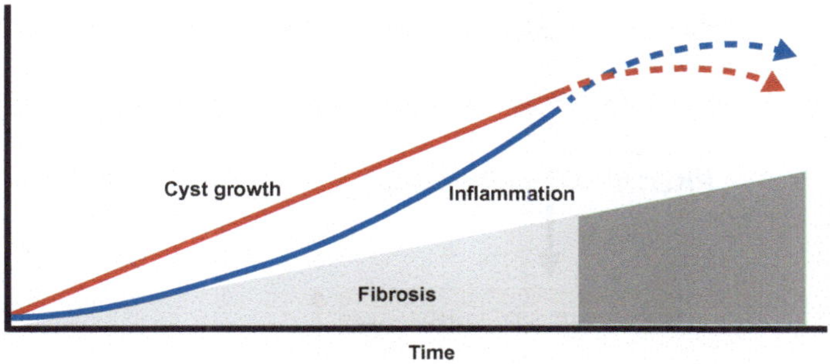

Fig. 5.1 Changes in renal cystic growth, fibrosis, and inflammation over time. Renal cystic growth increases over time in human studies (solid line). The kidney volume does not increase or may decrease in late stages, while the tissue becomes progressively more fibrotic in some animal studies (dashed line). Abnormal immune inflammation increases over time (solid line). However, in late stages, the progression magnitude of some immune responses may decrease due to the loss of functional renal parenchyma (dashed line)

5.2.1 Inflammatory Cells in ADPKD

Renal interstitial inflammation infiltrate is one of the most notable characteristics of ADPKD (Fig. 5.2). Among the inflammatory cells, macrophage is the most extensively studied cell. Macrophages consist of heterogeneous cell types which play specific roles in ADPKD progression (Karihaloo et al. 2011). Macrophage is involved in innate immunity, tissue development, repair, and homeostasis. Intriguingly, macrophages could become polarized and express pro-inflammatory or anti-inflammatory cytokines in response to signals in the tissue microenvironment (Gordon and Taylor 2005). On one hand, macrophages stimulated with interferon-γor LPS demonstrate a pro-inflammatory Th1-like phenotype, which is referred to as M1-like macrophage. M1 macrophage is characterized by expression of iNOS, interleukin (IL) 1β, and TNF-α (tumor necrosis factor α) (Sica and Mantovani 2012). On the other hand, macrophages treated with IL 4 or 13 produce an anti-inflammatory response and are referred to as M2-like macrophages. M2 macrophages express arginase 1 (Arg1) and IL10 and have wound healing and anti-inflammatory functions (Sica and Mantovani 2012). However, in vivo studies found that macrophages display a range of phenotypes that fall somewhere between the M1 and M2 spectrums and were able to rapidly switch the phenotypes based on external cues. The tissue microenvironment influenced the macrophage polarization (Swenson-Fields et al. 2013). For instance, M1 macrophages could transit into M2 like polarization following phagocytosis of apoptotic and necrotic cells or when cultured with renal epithelial cells.

In normal kidney, nearly half of the resident macrophages derive from the yolk sac while the rest derive from the hematopoietic lineage (Schulz et al. 2012). Resident macrophages play important roles in producing cytokines and monitoring surrounding cells in kidney. The second distinct source of macrophages is the infiltrating macrophage, which rapidly accumulate in response to kidneys injury (Schulz et al.

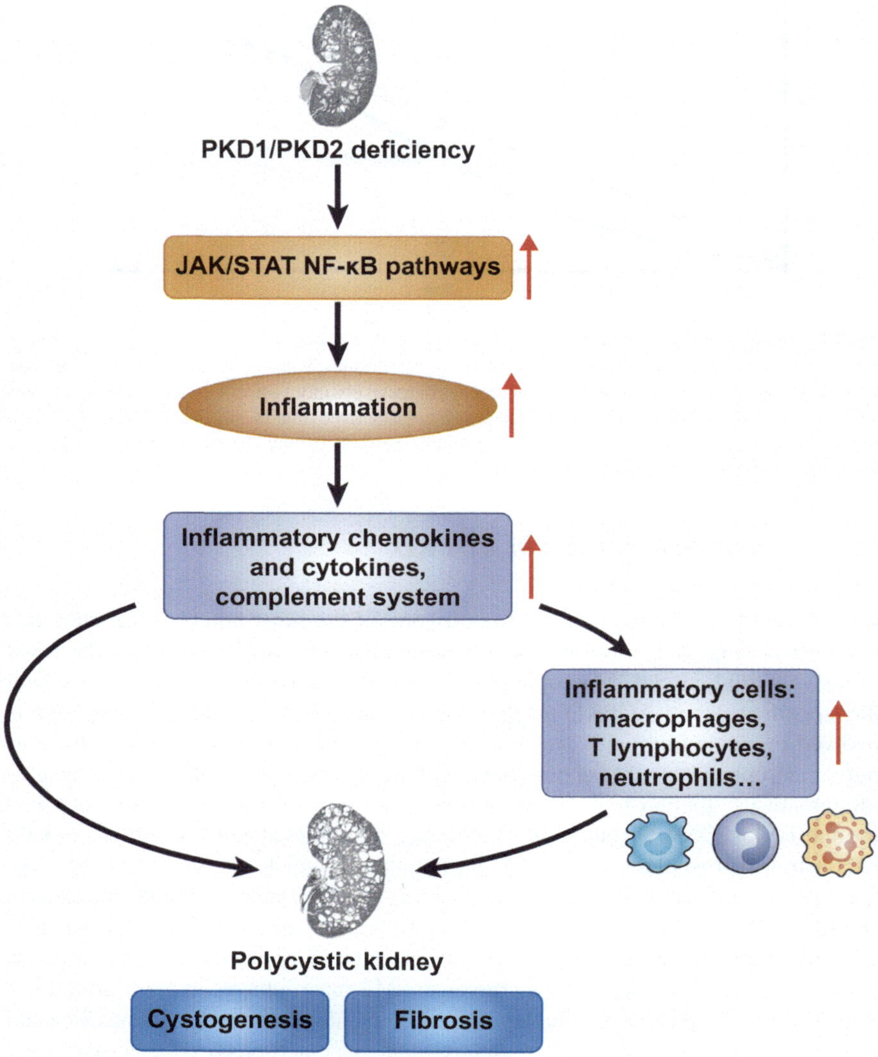

Fig. 5.2 A schematic summary of inflammation in ADPKD

2012). Infiltrating macrophages could produce pro-inflammatory cytokines and are associated with kidney injury and fibrosis.

Both resident and infiltrating macrophages increase in polycystic kidney tissue (Swenson-Fields et al. 2013). In rodent models, inflammatory cells including macrophages were present prior to or coincident with cyst initiation, and macrophages played important roles in promoting cyst formation (Bastos et al. 2009). Karihaloo et al. found that infiltrating macrophages contributed to the proliferation of the cystic lining cells (CLCs) and the progression of ADPKD in murine models (Karihaloo et al. 2011). They also verified that primary cyst epithelial cells from ADPKD promoted macrophages converting to the M2-like phenotype (Karihaloo et al. 2011). On the contrary, macrophage depletion using clodronate showed the reduction of cyst growth by influencing cell proliferation in ADPKD. Harris et al. established a model for the contribution of macrophages to PKD progression and regarded activated macrophages as a treatment target in ADPKD (Harris and Torres 2014). Recently, we revealed an interaction between macrophages and CLCs which promoted cyst growth in Pkd1−/− mice through the arginine-polyamine metabolic pathway (Yang et al. 2018). Arg1-encoded protein, arginase-1, was predominantly expressed in macrophages in a time-dependent manner in ADPKD (Yang et al. 2018). Multi-stage macrophage depletion verified that macrophages expressing high arginase-1 levels accounted for late-stage cyst enlargement, and inhibiting arginase-1 activity significantly retarded cyst growth and effectively lowered the cyst enlargement (Yang et al. 2018). In vitro experiments found that macrophages stimulated CLCs proliferation, and L-lactic acid, primarily generated by CLCs, significantly increased arginase-1 expression and polyamine synthesis in macrophages (Yang et al. 2018). Therefore, arginase-1 secreted by macrophages may be a key molecule involved in cyst formation process and may be a potential therapeutic target to delay ADPKD progression.

The involvement of other inflammatory cells like T lymphocytes, neutrophils, dendritic cells, and mast cells has been reported in ADPKD. The number of CD4-positive T cells was increased in the interstitial tissue of ADPKD (Zeier et al. 1992). Mast cells were found to be involved in PKD progression through the production of pro-inflammatory factors including chymase and Ang II (McPherson et al. 2004). Furthermore, elevated neutrophils were reported in human and canine PKD models (Bernhardt et al. 2007). While a wide variety of inflammatory cells increased in polycystic kidney tissue, future studies should focus on better understanding the respective roles of these cells in cyst formation and fibrosis.

5.2.2 Inflammatory Chemokines and Cytokines in ADPKD

Chemokines are responsible for infiltration, activation, and polarization of inflammatory cells and could regulate inflammatory cell behavior. One of the extensively studied chemokines in ADPKD is monocyte chemoattractant protein-1 (MCP-1, Ccl2), which binds to its cognate receptor CCR2. CCR2 belongs to the CC chemokine family of G-protein coupled receptors and typically expresses on the surface of T cells or monocytes (Li et al. 2008a). MCP1 was markedly increased in the cyst fluid of ADPKD patients and was associated with worsened renal function (Zheng et al. 2003). The increased expression of MCP-1 in rodent models of PKD paralleled the result observed in humans (Cowley et al. 2001). In vitro models, MCP-1 was produced by PKD1−/− tubule epithelial cells compared to the control cells (Song et al. 2017). Based on the increased MCP1 expression across different species, MCP-1 is proposed as a biomarker for ADPKD. Moreover, Cassini et al. (2018) found the upregulation of MCP-1 preceded macrophage infiltration in mouse polycystic kidney tissue. Macrophages accumulating around nascent cysts were pro-inflammatory and induced tubular cell injury and proliferation-independent cystic dilation (Cassini et al. 2018). One month later, macrophages switched to an alternative activation phenotype and further promoted cyst growth due to an additional threefold increase of tubular cell proliferative rates (Cassini et al. 2018). In PKD1-MCP1 double-knockout mice, there was a marked reduction in MCP-1 expression and macrophage numbers, resulting in slower cyst growth and improved renal function (Cassini et al. 2018). Treatment of PKD1−/− mice with Ccr2 inhibitor partially reproduced the improvement seen with MCP1 knockout (Cassini et al. 2018). Therefore, MCP-1 promoted macrophage accumulation and cyst growth via both proliferation-independent and -dependent mechanisms in ADPKD.

In addition to MCP-1, levels of TNF-α also increased in ADPKD cyst fluid with age and cyst severity (Li et al. 2008b). TNF-α is an immune cytokine which activates inflammatory signaling pathways and plays important roles in several biological processes. Furthermore, TNF-α plays a significant role in cyst formation where it interferes with processing and presentation of PC2 (Li et al. 2008b). Inhibition of TNF-α converting enzyme (TACE) led to a significant reduction in cyst volume (Dell et al. 2001). Taken together, TNF-α is a key contributor to cyst formation in ADPKD.

5.2.3 Complement System In ADPKD

The complement system is activated by three pathways: the classical complement pathway, alternative complement pathway, and lectin pathway. Three pathways converge on C3, then C3 convertase cleaves C5 which activating C6–9. The alternative pathway accounts for about 80–90% of the total complement activation, even when initially triggered by the classical pathway or lectin pathway (Ricklin et al. 2010). The end of the complement activation is the enhancement of antibodies and phagocytic cells to clear microbes and damaged cells from an organism, the promotion of inflammation, and the activation of membrane attack complex.

Growing evidence suggested that activation of the complement cascade contributed to ADPKD progression. Mrug et al. (2008) confirmed that innate immunity is involved in the PKD progression in mouse model, particularly abnormal C3 activation is a key element. Burtey et al. (Mrug et al. 2008) also confirmed the overexpression of nine complement–component genes (including C3) in Han: SPRD rat. The proteomic analysis of urine and cyst fluid from ADPKD patients with ESRD found 44 proteins including complement factors (Mason et al. 2009; Bakun et al. 2012. Mrug et al. 2014) further found that antigenic C3 was present in CLCs and that C3 activation fragments (iC3b) was present in renal cysts and urine from ADPKD patients, and iC3b might partly be responsible for cystogenic effects of M2 macrophages. Moreover, we found an excessive activation of alternative complement pathway in ADPKD progression (Su et al. 2014). Firstly, we screened the glycoproteome of urine samples from ADPKD patients, revealed that levels of complement factor B (CFB) and C9 increased along with disease progression. CFB is the key factor in complement alternate pathway, which is cleaved by factor D into two fragments: Ba and Bb. Bb, a serine protease, then combines with complement factor 3b to generate the C3 or C5 convertase which initiate and sustain the activation of alternative pathway. Then, we evaluated the effect of the complement inhibitor rosmarinic acid (RMA) in Pkd1−/− mice and Han: SPRD Cy/+ rats. RMA-treated models showed significantly lower cystic index, better renal function, lower inflammatory cells, and reduced fibrosis. We further explored the mechanism of CFB overexpression and alternative complement pathway activation in ADPKD (Ming Wu 2016.). We observed that the overexpression of CFB was associated with increased JAK2/STAT1 activity and an enhanced expression of PC1 C-terminal tail (PC1-CTT). Moreover, STAT1 inhibition by fludarabine in renal epithelial cells suppressed Arg1 expression induced by PC1-CTT-CM, which suggested that PC1-CTT-induced macrophage activation into M2 phenotype might be mediated by STAT1 and CFB. In conclusion, the above findings prove that the complement activation, especially the alternative complement pathway takes part in ADPKD progression.

5.2.4 Pathways of Inflammation in ADPKD

To date, NF-κB (Banzi et al. 2006; Qin et al. 2012) and JAK-STAT (Weimbs et al. 2013) pathways are the extensively studied pathways that correlate with inflammation in ADPKD. Pei et al. (Song et al. 2009) performed a global gene profiling in cysts from PKD1 mutant human kidneys. The analysis displayed a rich gene transcriptional profile for immune/inflammatory response including JAK-STAT and NF-κB signaling pathway.

Several studies have explored the mechanism of the NF-κB activation in ADPKD. NF-κB has been characterized as a key regulator of immune system for a long time and is responsible for the transcription of the genes encoding pro-inflammatory factors including TNF-α, IL1, CcL3, Ccl4, and MCP-1 (Hayden and Ghosh 2011, 2012; Pahl 1999). Qin et al. 2012 reported that c-Met and NF-κB-dependent overexpression of wnt signaling promoted cystogenesis in PKD. Park et al. (2010) reported that receptor of advanced glycation end product upregulated intracellular NF-κB signaling in PKD2 transgenic mice. Banzi et al. (2006) reported that PC1 activated a PKCα-mediated NK-κB signaling. Taken together, NF-κB is activated and induces inflammation responses in ADPKD.

The JAK-STAT pathway is the principal signaling mechanism for a wide array of cytokines and growth factors (Rawlings 2004), involving cell proliferation, differentiation, transcription, and immune response (John J. O'Shea 2013; Kaplan 2013). Anil Kumar Bhunia et al. (2002) first reported that PKD1 and PKD2 regulated activation of the JAK-STAT signaling pathway. The CTT of PC1 could undergo proteolytic cleavage (Qian et al. 2002; Wei et al. 2007) and nuclear translocation (Chauvet et al. 2004). In ADPKD kidneys, PC1 tail fragments are overexpressed, including both

30 kDa (a full-length) and 15 kDa fragments (a half-length). Weimbs et al. observed that cleaved PC1 tail interacted with STAT6 and P100, enhanced STAT6 activity (Low et al. 2006). They further reported that membrane-anchored PC1 activated STAT3, soluble CTT co-activated SATA1, 3, and 6, and STAT3 activated required JAK2 which interacted with PC1 tail (Talbot et al. 2011). In our laboratory, Ming Wu et al. (2016) showed that the PC1-CTT regulated CFB expression through JAK2/STAT1 pathway. We also showed that NF-κB acted as the downstream of PC1-CTT and might mediate PC1-CTT-induced CFB expression. The result suggested that targeting STAT1 and NF-κB might be a strategy to decrease inflammation in ADPKD.

5.3 ECM in ADPKD

The characteristic of tubule interstitial fibrosis is the accumulation of ECM proteins such as proteoglycan, collagen I, III, IV, V, elastin, fibronectin, and heparin sulfate proteoglycan (HSPG) (Wilson et al. 1996). ECM is a complex of proteins which fills the extracellular space in connective tissues. ECM mainly participates in cell supporting, cell adhesion, proliferation, differentiation, and fibrosis (Wilson et al. 1992). Abnormalities of ECM accumulation are found in human polycystic kidney tissue (pre-dialysis and ESRD) (Norman 2011; Wilson et al. 1992). ECM deposition and tissue remodeling are essential components of cyst progression (Wilson and Burrow 1999). The ADPKD-associated ECM abnormalities include disordered production, composition, and turnover (Wilson et al. 1996). Abnormalities most commonly appeared in cyst surrounding and interstitial basement membrane structures (Wilson et al. 1999).

Except the common factors which regulate the activity of pro-fibrotic pathways are same with the chronic kidney disease (CKD), recent animal studies indicated that polycystins might directly control ECM production (Liu et al. 2014). Upregulation of TGFβ, the major pro-fibrotic growth factor, was observed in PKD1 knock-out mice, and loss of PKD1 led to an increased responsiveness of cystic cells and fibroblasts to TGFβ (Hassane et al. 2010; Liu et al. 2014). Moreover, enhanced TGFβ pathway signaling and Smad2/3 nuclear localization were observed in ADPKD patients and rodent models (Hassane et al. 2010). Similarly, the loss of polycystin was associated with notochord collagen overexpression in zebrafish which was associated with fish body axis curvature defects (Mangos et al. 2010). Therefore, polycystin mutations play important roles in ADPKD-associated changes in ECM.

In addition, PKD-related changes in ECM were associated with other interstitial fibrosis factors like hepatocyte growth factor, epithelial growth factor (EGF), and fibroblast growth factor-1 (FGF1) (Du and Wilson 1995; Horie et al. 1994). Levels of these pro-fibrotic factors increased over time in ADPKD (Du and Wilson 1995). Pro-fibrotic factors might affect the pathobiology of early stages of ADPKD and the highest levels were found in near end-stage or end-stage ADPKD.

5.4 ECM Regulation and Turnover in ADPKD

When an imbalance of cytokines is generated in ADPKD, the abundance of pro-fibrotic factors such as TGFβ, PDGF, connective tissue growth factor, FGF2, and osteopontin are increased, whereas the abundance of anti-fibrotic factors such as bone morphogenic protein (BMP) 7 and hepatocyte growth factor are decreased (Song et al. 2017). More importantly, interstitial fibrosis is a consequence of both increased ECM production and inadequate matrix degradation which is another key regulator of fibrosis (Eddy 1996). Development of fibrosis in ADPKD and other fibrotic models depends on both the amount of produced ECM and the extent of matrix turnover during disease progression. Matrix metalloproteinases (MMPs) are responsible for degrading ECM proteins. Several classes of MMPs have been studied including collagenases such as MMP1, MMP8, gelatinases, metalloelastase, membrane-type MMPs, and others. Tissue inhibitors of metalloproteinases (TIMPs) inhibit MMPs (Catania et al. 2007). Consistent with the abnormal ECM change observed in polycystic kidney tissues, an imbalance renal expression of TIMPs and MMPs was found in kidneys of PKD (Norman 2011).

In kidneys, many types of MMPs and TIMPs are expressed, including MMP 2, 3, 9, 13, 14, 24, 25, 27, 28, and TIMP 1, 2, and 3 (Catania et al. 2007). Among them, MMP 1, 2, 9, 14, and TIMP1 are related to PKD (Catania et al. 2007). Increased MMP activity and collagen expression could stimulate the cyst formation in PKD. MMP2 is down-regulated in Han: SPRD rats, along with the upregulation of TIMP1 and TIMP2 (Schaefer et al. 1996). However, an overexpression of MMP2 was found in kidneys or tubules in ADPKD mice (e.g., Cys1cpk mice and PKD1−/− mice) (Hassane et al. 2010; Rankin et al. 1996). MMP-14 overexpression was found in kidneys of PKD1−/− mouse, Cys1cpk mice, and Han: SPRD rats (Grantham et al. 2011; Schieren et al. 2006). The suppression of MMP14 by batimastat was found to reduce cyst formation and kidney weight. Increased expressions of MMPs and TIMPs could induce remodeling and thickening of the cystic membrane, and the fibrosis was induced in CLCs (Song et al. 2017). Therefore, inhibitions of MMP 2, 14, and TIMP2 through sirolimus decreased the accumulation of ECM and alleviated PKD progression (Follonier Castella et al. 2010).

In humans, ADPKD was associated with increases of serum levels of MMP1, MMP9, and TIMP1 (Nakamura et al. 2000). There were increased levels of multiple MMPs, TIMPs (e.g., MMP2, MMP3, MMP9, TIMP1 and TIMP2), and plasminogen activator inhibitor 1 (PAI1) in both pre-dialysis and ESRD kidneys of ADPKD patients (Nakamura et al. 2000). The overexpression of PAI1 was reported in both human and mouse polycystic kidney tissues (Eddy 2009). It was believed that PAI1 was pro-fibrotic due to its ability to recruit macrophage and myofibroblasts to the tubulointerstitial area (Eddy 2009).

The functional consequences of these ECM turnover regulators are not fully understood in ADPKD. Consequences may include changes in several cellular functions (e.g., proliferation and differentiation) and immune regulation (breakdown collagen I into proline–glycine–proline (PGP) fragment) (Norman 2011; Snelgrove et al. 2010).

PGP is an important regulator of inflammatory neutrophil accumulation which is the key pathogenesis of chronic obstructive pulmonary disease (Snelgrove et al. 2010). MMP9 which participated in the generation of PGP was upregulated in ADPKD. PGP and other collagen fragments of ECM may play important immunoregulatory and modulating effects in polycystic kidney tissues.

Interactions between cells and ECM or its degradation products are mediated by matrix receptors such as integrins and syndecans (Geiger et al. 2001). Chemokine receptors often interact with ECM components at focal adhesion complexes which activate intracellular signaling to regulate major cellular processes (Ehrhardt et al. 2002). The complex interactions between ECM, receptors of ECM degradation fragments, signaling pathways, and ensuing transcriptional activation of specific genes form a unique microenvironment (Norman 2011). Studies demonstrated that several ECM receptors were increased in PKD patients, including integrins α2, β1, α8, β4, αv, syndecan-4, and integrin-associated focal adhesion kinase (Joly et al. 2003; Zeltner et al. 2008). ECM receptors (integrins α and β) may localize on primary cilia (McGlashan et al. 2006). Genetic disruption of integrin pathways could mitigate the renal cyst formation in mice (Desmouliere et al. 1993). Moreover, a hypomorphic mutation in laminin α5 gene could drive renal cyst formation (LeBleu et al. 2013). Therefore, the cell–ECM interactions may take part in PKD pathogenesis.

5.5 Cells in Regulation of ECM in ADPKD

5.5.1 Myofibroblasts

Myofibroblasts are specialized cells which can exert contractile forces, mediate wound healing, and substantially contribute to the ECM expansion and development of renal interstitial fibrosis (Fig. 5.3) (Qi et al. 2006). The hallmark of myofibroblasts is the expression of cytoskeletal α smooth muscle actin (αSMA) (Norman 2011). αSMA-positive cells were found in kidneys of PKD1−/− mice and end-stage kidneys of Han:SPRD rats (Hassane et al. 2010). There were also αSMA expressing interstitial cells in focal areas of early-stage kidneys and in widespread areas of end-stage kidneys in ADPKD patients. The origin of myofibroblasts in polycystic kidney may differentiate from different precursors such as resident interstitial fibroblasts (LeBleu et al. 2013). In a model of renal fibrosis, most myofibroblasts were derived through the proliferation of resident fibroblasts, although bone-marrow-derived fibrocytes also contributed about 35% of total myofibroblasts in the kidney (LeBleu et al. 2013). Myofibroblast transformation could be regulated by TGFβ or by the alteration of calcium flux in PKD (Desmouliere et al. 1993; Follonier Castella et al. 2010). Myofibroblasts also could differentiate from infiltrating inflammatory cells (Wada et al. 2011).

After appropriate stimulation, fibroblasts may differentiate into myofibroblasts. The stimulation may be enhanced in ADPKD in response to increased ECM-

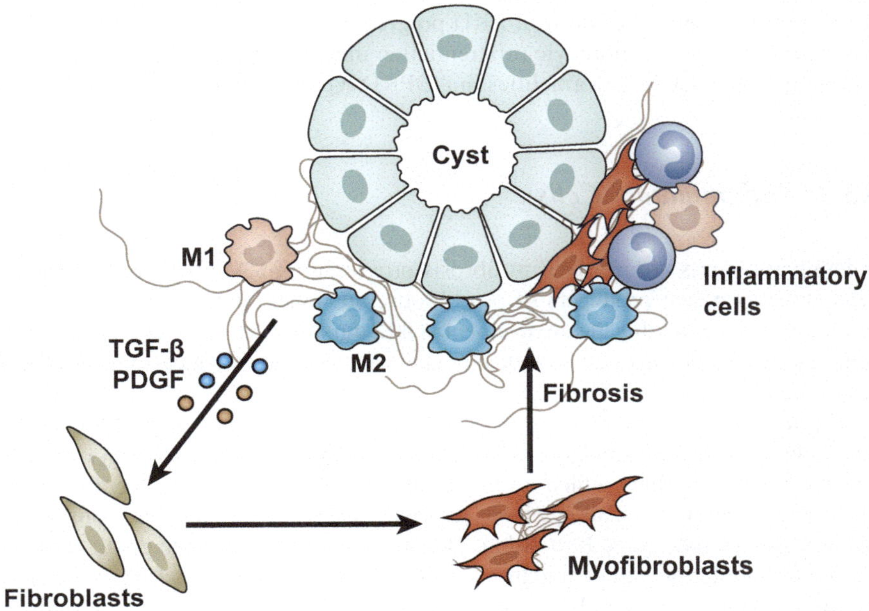

Fig. 5.3 Cells and pathways in the regulation of fibrosis in ADPKD. The persistent increase in inflammatory cells including macrophages leads to enhanced secretion of pro-inflammatory and pro-fibrotic cytokines which causes the transition of fibroblasts and fibrocytes to a myofibroblast phenotype. These myofibroblasts produce large amounts of ECM proteins leading to the deposition of collagen into ECM and fibrosis in ADPKD

promoting factors such as TGFβ, FGF1, platelet-derived growth factor (PDGF), and insulin-like growth factors (IGF) I and II (Kuo et al. 1997). These changes also included increased secretion of MMP2 and heat shock protein 47. Similar changes of TGFβ pathway activation and the proliferative response were observed in embryonic fibroblasts in PKD1−/− mice (Nishio et al. 2005). In addition, fibroblast differentiation to myofibroblasts could be modulated by macrophages that secrete growth factors including FGF2, PDGF, galectin 3, and IGF binding protein 5 (Huen and Cantley 2015).

5.5.2 *Fibrocytes*

Fibrocyte is another ECM producing cell. Fibrocytes differentiate from peripheral blood leukocytes and express both hematopoietic and stromal cell markers, as well as several chemokine receptors (Wada et al. 2011). The differentiation of fibrocytes is enhanced by cytokines associated with repair and pro-fibrogenic Th2-type immune response and is inhibited by pro-inflammatory Th1-type cytokines (Niedermeier et al. 2009). Because the speed of PKD progression was associated with the Th2-type immune response, fibrocyte differentiation may be enhanced in cystic kidneys and contribute to ECM abnormalities in ADPKD (Qi et al. 2006).

5.5.3 *Inflammatory Cells*

Inflammatory cells are involved in fibrosis progression of ADPKD. Inflammatory cells serve as potent producers of pro-fibrotic, pro-inflammatory, and pro-mitotic cytokines. Understanding the involvement of inflammatory cells and related signaling pathways in cyst fibrosis will provide innovative insights into the mechanism of PKD progression.

The most studied inflammatory cell associated with fibrosis in ADPKD is the macrophage. The macrophage accumulation accelerates the proliferation of CLCs and increases cytokines which cause fibrosis in ADPKD (Vernon et al. 2010). Macrophages include a heterogeneous group of cell types which differ from origin and activation (Anders and Ryu 2011). The involvement of macrophages in promoting the accumulation and degradation of ECM was well established. Macrophages secrete cytokines such as IL-10, MCP1, TNF-α, TGF-β, PDGF, or Arg1, which induce myofibroblast activation and ECM production (Anders and Ryu 2011). Then, myofibroblasts produce lots of ECM proteins which lead to fibrosis in polycystic kidneys. M1 macrophages produce TNF-α and encourage inflammation. M2 macrophage stimulates the tubular cell proliferation and fibrosis formation. In normal condition or ADPKD, renal epithelial cells could promote the differentiation of naïve macrophages into M2 macrophages (Vernon et al. 2010). Recently, M2 macrophages associated with Th2 cytokine-driven responses were found to promote renal tissue repair and fibrosis. M2 macrophages were also associated with ADPKD progression (Mrug et al. 2008). CD11b was used as the identification of the pro-fibrotic bone-marrow-derived macrophage (Lin et al. 2009). However, some macrophage types have fibrosis attenuating effects in ADPKD. Those macrophages produced several matrix-degrading proteases including MMP1, 2, 8, 9, and 13 (Semedo et al. 2010). In a UUO model, adoptive transfer of bone-marrow-derived macrophages into leukocyte-depleted mice could significantly attenuate fibrosis (Qi et al. 2006). Together, these data suggest that different macrophage subtypes could promote or regress fibrosis. In addition to macrophages, renal fibrosis and ECM abnormalities may be enhanced by lymphocytes, CD11c dendritic cells, and mast cells.

5.6 Epithelial Mesenchymal Transition

Renal vascular and tubular epithelial cells or macrophages can transit to a myofibroblast phenotype through epithelial mesenchymal transition (EMT). TGFβ is recognized as the main EMT inducer. In PCK rats, an orthologous model of human

PKD, e-cadherin and β-catenin in cystic tubules were attenuated and localized to lateral areas of cell–cell contact (Mun and Park 2016). Epithelial cells in cysts of PCK rats acquired mesenchymal features through EMT in response to cyst enlargement and participated in progressive renal fibrosis. However, EMT was found to make negligible contributions to the pathogenesis of renal cysts of ADPKD (Mun and Park 2016). Although EMT plays an important role in the formation of kidney myofibroblasts, the extent of EMT and its importance for the fibrosis in ADPKD patients is unknown.

5.7 Prognostic Value of ECM Abnormality in ADPKD

Recently, abnormal expression of collagen I and III derived fragments were identified in the urine of young pre-dialysis patients with ADPKD (Kistler et al. 2009). ECM remodeling may be an attractive predictive biomarker and a therapeutic target for development. A follow-up study based on the analyses of urinary collagen fragments provided further support for the hypothesis (Kistler et al. 2013). The study found that urinary proteomic score (mostly urinary collagen fragments) predicted ADPKD severity significantly (Kistler et al. 2013).

5.8 Fibrotic Signaling Pathways

Several intracellular signaling pathways in fibrosis are activated in ADPKD.

5.8.1 TGFβ-SMAD Signaling Pathway

TGFβ is highly expressed in CLCs of human and rodent models of PKD (Hassane et al. 2010). The TGFβ pathway is usually related to fibrosis, proliferation, cell—cell interaction, apoptosis, and cell differentiation (Sureshbabu et al. 2016). TGFβ is secreted by macrophages, lymphocytes, and dendritic cells. When TGFβ binds to its receptors, TGFβ receptor types I and II will assembly and activate receptor I. Then, the activated receptor I phosphorylates receptor-regulated SMAD which binds to the common-mediator SMAD (Leonhard et al. 2016). The SMAD complex then translocates to the nucleus and activates gene transcription. Increased TGFβ expression and SMAD signaling were found to correlate with the late-stage fibrosis of ADPKD and play an important role in cyst fibrosis and the disease progression rather than cyst initiation (Norman 2011). The expression of TGFR1 and 2 was found to be elevated in the PKD1 −/− mouse model (Chen et al. 2008). Moreover, the inhibition of TGFβ could decrease cyst formation and ADPKD progression. However,

TGFβ2 was found to alleviate ADPKD progression and cystogenesis by controlling the synthesis of ECM proteins and cell adhesion (Elberg et al. 2012).

Activin A is a cytokine belonging to the TGFβ family of growth factors. During kidney development, activin A is produced by ureteric bud (UB) and negatively regulates UB branching (Yamashita et al. 2004). Furthermore, activin A is involved in kidney regeneration following injury, suggesting an important role for activin A during kidney formation or regeneration. The inhibition of activin A signaling could slow the progression of PKD (Leonhard et al. 2016). In addition, the level of activin A increased in PKD1−/− mouse model and was associated with conditions wherein dysfunction of the cilia or polycystins caused rapid cyst formation (Yamashita et al. 2004). These studies indicated that activin A was a critical cytokine in cyst progression and ESRD since it was involved in epithelial regeneration and was capable of inducing ECM gene expression in kidney (Yamashita et al. 2004).

5.8.2 Wnt Signaling

Because β-catenin regulates the EMT process, Wnt signaling pathways are closely connected with fibrosis. There are three types of Wnt signaling pathways including the canonical Wnt pathway, the non-canonical Wnt/Ca2 + pathway, and the non-canonical planar cell polarity pathway. Wnt signaling regulates gene transcription, proliferation, cytoskeletal structure, and cell migration (Komiya and Habas 2008). When Wnt is absent, complexes that contain disheveled, glycogen synthase kinase-3β, axin, and adenomatous polyposis coli degrade β-catenin in the canonical Wnt pathway. When Wnt is present, the complex is inhibited, then β-catenin accumulates and functions as the transcription factor. A transgenic mouse model that constitutively expressed β-catenin developed PKD (Qi et al. 2006). The activity of Wnt signaling was increased in CLCs of patients with ADPKD (Mun and Park 2016). The expression of Wnt4 related to EMT increased in jck mice (Stark et al. 1994). In Gpr48−/− PKD mice, renal fibrosis was found to accompany the Wnt signaling pathway activation (Dang et al. 2014). Moreover, the Wnt signaling regulated primary cilia formation. Genetic knockout of ciliary proteins Bbs1 and Kif3a showed an activation of Wnt signaling in transgenic mice compared with normal mice (Corbit et al. 2008). Likewise, the upregulation of Wnt signaling increased the expression of fibronectin. Together, the overactivation of Wnt signaling results in an increased frequency of EMT and lead to fibrosis in ADPKD.

5.8.3 Periostin–Integrin-Linked Kinase Signaling

Periostin is a secreted protein which binds to the ECM components including collagen I and fibronectin and has been implicated in collagen fibrillogenesis (Wallace et al. 2014). Periostin was markedly overexpressed and could promote cyst growth and interstitial fibrosis in PKD mice (Wallace et al. 2014). Several integrins including α and β also were aberrantly expressed in CLCs. Integrin-linked kinase (ILK), which is stimulated by periostin, directly binds to β integrins and links the ECM to the actin cytoskeleton. Periostin-ILK signaling played a role in cytoskeleton reorganization by recruiting regulatory proteins such as parvin, paxillin, and kindlin2 (Raman et al. 2017). Raman et al. (Raman et al. 2017) found that knockdown of ILK and its downstream signaling strikingly reduced PKD fibrosis and extended the life of PKD mice.

5.9 Anti-fibrotic Therapies in ADPKD

Slowing the cyst expansion and development of fibrosis is very important to prolonging life span and improving palliative care of patients with ADPKD. However, the development of effective anti-fibrotic treatments in ADPKD patients is limited. Identifying the key molecular mechanism of fibrosis and how it contributes to ESRD will provide novel targets for anti-fibrotic therapies (Vilayur and Harris 2009). For example, targeting of αSMA-positive cells for anti-fibrotic therapy to help reduce scarring and retain renal function may be an attractive idea in the future (Song et al. 2017). B-type natriuretic peptide overexpression was found to ameliorate renal fibrocystic disease through the guanylyl cyclase A-cGMP axis in a rat model of ADPKD (Holditch et al. 2017). We previously found that rosiglitazone, a peroxisome-proliferator-activated receptor-γ agonist, was able to down-regulate the abnormally activated β-catenin signaling pathway and delay the progression of fibrosis in Han: SPRD rats (Dai et al. 2010). In addition, targeting other fibrotic factors may alter the pathobiology of cyst formation and ADPKD progression. For example, overexpression of BMP receptor activin receptor like kinase 3 or BMP7 knockout, could trigger renal cysts formation in mice through SMAD/WNT signaling (Hu et al. 2003). BMP7 treatment was found to attenuate the renal cystic disease progression in Nek8 jck mouse model, and soluble activin-type IIB receptor treatment could effectively block cyst formation in a mouse PKD model (Leonhard et al. 2016). Similarly, EGF is another important regulator of cystic epithelial cell proliferation through EGF pathway which is explored for therapeutic development of ADPKD (Du and Wilson 1995). Better understandings of mechanisms underlying the initiation and progression of fibrosis in ADPKD are urgently needed.

5.10 Conclusion

The fibrosis is increased in the late-stage and end-stage ADPKD. Inflammation can initiate fibrosis, causing thickening of tubular membrane and remodeling of the interstitium. However, inflammation and fibrosis alone cannot generate ADPKD, which also requires CLC proliferation. Following an increase in cystic cell proliferation, fibrotic cells and pro-inflammatory cytokines are elevated and then lead to the development of inflammation, fibrosis, and proliferation. Macrophage and its secreted cytokines accumulated in cystic fluids and urine of PKD. TGFβ, MMPs, and TIMPs could trigger fibrosis in ADPKD. MMPs and TIMPs could induce the accumulation of fibroblasts and collagen. The inhibition of cytokines involved in fibrosis could be a therapeutic strategy for ADPKD.

References

Anders HJ, Ryu M (2011) Renal microenvironments and macrophage phenotypes determine progression or resolution of renal inflammation and fibrosis. Kidney Int 80:915–925

Anil Kumar Bhunia KP, Boletta Alessandra, Liu Lijuan, Qian Feng, Pei-Ning Xu et al (2002) PKD1 induces p21waf1 and regulation of the cell cycle via direct activation of the JAK-STAT signaling pathway in a process requiring PKD2. Cell 109:157–168

Bakun M, Niemczyk M, Domanski D, Jazwiec R, Perzanowska A et al (2012) Urine proteome of autosomal dominant polycystic kidney disease patients. Clin Proteomics 9:13

Banzi M, Aguiari G, Trimi V, Mangolini A, Pinton P et al (2006) Polycystin-1 promotes PKCalpha-mediated NF-kappaB activation in kidney cells. Biochem Biophys Res Commun 350:257–262

Bastos AP, Piontek K, Silva AM, Martini D, Menezes LF et al (2009) Pkd1 haploinsufficiency increases renal damage and induces microcyst formation following ischemia/reperfusion. J Am Soc Nephrol 20:2389–2402

Bergmann C (2017) Genetics of autosomal recessive polycystic kidney disease and its differential diagnoses. Front Pediatr 5:221

Bernhardt WM, Wiesener MS, Weidemann A, Schmitt R, Weichert W et al (2007) Involvement of hypoxia-inducible transcription factors in polycystic kidney disease. Am J Pathol 170:830–842

Cassini MF, Kakade VR, Kurtz E, Sulkowski P, Glazer P et al (2018) Mcp1 promotes macrophage-dependent cyst expansion in autosomal dominant polycystic kidney disease. J Am Soc Nephrol

Catania JM, Chen G, Parrish AR (2007) Role of matrix metalloproteinases in renal pathophysiologies. Am J Physiol Renal Physiol 292:F905–F911

Chauvet V, Tian X, Husson H, Grimm DH, Wang T et al (2004) Mechanical stimuli induce cleavage and nuclear translocation of the polycystin-1 C terminus. J Clin Invest 114:1433–1443

Chen WC, Tzeng YS, Li H (2008) Gene expression in early and progression phases of autosomal dominant polycystic kidney disease. BMC Res Notes 1:131

Corbit KC, Shyer AE, Dowdle WE, Gaulden J, Singla V et al (2008) Kif3a constrains beta-catenin-dependent Wnt signalling through dual ciliary and non-ciliary mechanisms. Nat Cell Biol 10:70–76

Cowley BD Jr, Ricardo SD, Nagao S, Diamond JR (2001) Increased renal expression of monocyte chemoattractant protein-1 and osteopontin in ADPKD in rats. Kidney Int 60:2087–2096

Dai B, Liu Y, Mei C, Fu L, Xiong X et al (2010) Rosiglitazone attenuates development of polycystic kidney disease and prolongs survival in Han: SPRD rats. Clin Sci (Lond) 119:323–333

Dang Y, Liu B, Xu P, Zhu P, Zhai Y et al (2014) Gpr48 deficiency induces polycystic kidney lesions and renal fibrosis in mice by activating Wnt signal pathway. PLoS ONE 9:e89835

Dell KM, Nemo R, Sweeney WE Jr, Levin JI, Frost P et al (2001) A novel inhibitor of tumor necrosis factor-alpha converting enzyme ameliorates polycystic kidney disease. Kidney Int 60:1240–1248

Desmouliere A, Geinoz A, Gabbiani F, Gabbiani G (1993) Transforming growth factor-beta 1 induces alpha-smooth muscle actin expression in granulation tissue myofibroblasts and in quiescent and growing cultured fibroblasts. J Cell Biol 122:103–111

Du J, Wilson PD (1995) Abnormal polarization of EGF receptors and autocrine stimulation of cyst epithelial growth in human ADPKD. Am J Physiol 269:C487–C495

Eddy AA (1996) Molecular insights into renal interstitial fibrosis. J Am Soc Nephrol 7:2495–2508
Eddy AA (2009) Serine proteases, inhibitors and receptors in renal fibrosis. Thromb Haemost 101:656–664
Ehrhardt A, Ehrhardt GR, Guo X, Schrader JW (2002) Ras and relatives–job sharing and networking keep an old family together. Exp Hematol 30:1089–1106
Elberg D, Jayaraman S, Turman MA, Elberg G (2012) Transforming growth factor-beta inhibits cystogenesis in human autosomal dominant polycystic kidney epithelial cells. Exp Cell Res 318:1508–1516
Follonier Castella L, Gabbiani G, McCulloch CA, Hinz B (2010) Regulation of myofibroblast activities: calcium pulls some strings behind the scene. Exp Cell Res 316:2390–2401
Geiger B, Bershadsky A, Pankov R, Yamada KM (2001) Transmembrane crosstalk between the extracellular matrix–cytoskeleton crosstalk. Nat Rev Mol Cell Biol 2:793–805
Gordon S, Taylor PR (2005) Monocyte and macrophage heterogeneity. Nat Rev Immunol 5:953–964
Grantham JJ, Mulamalla S, Swenson-Fields KI (2011) Why kidneys fail in autosomal dominant polycystic kidney disease. Nat Rev Nephrol 7:556–566
Harms JC, Song CJ, Mrug M (2018) The role of inflammation and fibrosis in cystic kidney disease
Harris PC, Torres VE (2014) Genetic mechanisms and signaling pathways in autosomal dominant polycystic kidney disease. J Clin Invest 124:2315–2324
Hassane S, Leonhard WN, van der Wal A, Hawinkels LJ, Lantinga-van Leeuwen IS et al (2010) Elevated TGFbeta-Smad signalling in experimental Pkd1 models and human patients with polycystic kidney disease. J Pathol 222:21–31
Hayden MS, Ghosh S (2011) NF-kappaB in immunobiology. Cell Res 21:223–244
Hayden MS, Ghosh S (2012) NF-kappaB, the first quarter-century: remarkable progress and outstanding questions. Genes Dev 26:203–234
Holditch SJ, Schreiber CA, Harris PC, LaRusso NF, Ramirez-Alvarado M et al (2017) B-type natriuretic peptide overexpression ameliorates hepatorenal fibrocystic disease in a rat model of polycystic kidney disease. Kidney Int 92:657–668
Horie S, Higashihara E, Nutahara K, Mikami Y, Okubo A et al (1994) Mediation of renal cyst formation by hepatocyte growth factor. Lancet 344:789–791
Hu MC, Piscione TD, Rosenblum ND (2003) Elevated SMAD1/beta-catenin molecular complexes and renal medullary cystic dysplasia in ALK3 transgenic mice. Development 130:2753–2766
Huen SC, Cantley LG (2015) Macrophage-mediated injury and repair after ischemic kidney injury. Pediatr Nephrol 30:199–209
John J O'Shea, SMH, Louis M Staudt (2013) JAKs and STATs in immunity, immunodeficiency, and cancer. N Engl J Med 368:161–170
Joly D, Morel V, Hummel A, Ruello A, Nusbaum P et al (2003) Beta4 integrin and laminin 5 are aberrantly expressed in polycystic kidney disease: role in increased cell adhesion and migration. Am J Pathol 163:1791–1800
Kaplan MH (2013) STAT signaling in inflammation. Jak-Stat 2(1):e24198
Karihaloo A, Koraishy F, Huen SC, Lee Y, Merrick D et al (2011) Macrophages promote cyst growth in polycystic kidney disease. J Am Soc Nephrol 22:1809–1814
Kistler AD, Mischak H, Poster D, Dakna M, Wuthrich RP et al (2009) Identification of a unique urinary biomarker profile in patients with autosomal dominant polycystic kidney disease. Kidney Int 76:89–96
Kistler AD, Serra AL, Siwy J, Poster D, Krauer F et al (2013) Urinary proteomic biomarkers for diagnosis and risk stratification of autosomal dominant polycystic kidney disease: a multicentric study. PLoS ONE 8:e53016
Komiya Y, Habas R (2008) Wnt signal transduction pathways. Organogenesis 4:68–75
Kuo NT, Norman JT, Wilson PD (1997) Acidic FGF regulation of hyperproliferation of fibroblasts in human autosomal dominant polycystic kidney disease. Biochem Mol Med 61:178–191
LeBleu VS, Taduri G, O'Connell J, Teng Y, Cooke VG et al (2013) Origin and function of myofibroblasts in kidney fibrosis. Nat Med 19:1047–1053

Leonhard WN, Kunnen SJ, Plugge AJ, Pasternack A, Jianu SB et al (2016) Inhibition of activin signaling slows progression of polycystic kidney disease. J Am Soc Nephrol 27:3589–3599
Li L, Huang L, Sung SS, Vergis AL, Rosin DL et al (2008a) The chemokine receptors CCR47 and CX3CR47 mediate monocyte/macrophage trafficking in kidney ischemia-reperfusion injury. Kidney Int 74:1526–1537
Li X, Magenheimer BS, Xia S, Johnson T, Wallace DP et al (2008b) A tumor necrosis factor-alpha-mediated pathway promoting autosomal dominant polycystic kidney disease. Nat Med 14:863–868
Lin SL, Castano AP, Nowlin BT, Lupher ML Jr, Duffield JS (2009) Bone marrow Ly6Chigh monocytes are selectively recruited to injured kidney and differentiate into functionally distinct populations. J Immunol 183:6733–6743
Liu D, Wang CJ, Judge DP, Halushka MK, Ni J et al (2014) A Pkd1-Fbn1 genetic interaction implicates TGF-beta signaling in the pathogenesis of vascular complications in autosomal dominant polycystic kidney disease. J Am Soc Nephrol 25:81–91
Low SH, Vasanth S, Larson CH, Mukherjee S, Sharma N et al (2006) Polycystin-1, STAT6, and P100 function in a pathway that transduces ciliary mechanosensation and is activated in polycystic kidney disease. Dev Cell 10:57–69
Mangos S, Lam PY, Zhao A, Liu Y, Mudumana S et al (2010) The ADPKD genes pkd1a/b and pkd2 regulate extracellular matrix formation. Dis Model Mech 3:354–365
Mason SB, Lai X, Bacallao RL, Blazer-Yost BL, Gattone VH et al (2009) The biomarker enriched proteome of autosomal dominant polycystic kidney disease cyst fluid. Proteomics Clin Appl 3:1247–1250
McGlashan SR, Jensen CG, Poole CA (2006) Localization of extracellular matrix receptors on the chondrocyte primary cilium. J Histochem Cytochem 54:1005–1014
McPherson EA, Luo Z, Brown RA, LeBard LS, Corless CC et al (2004) Chymase-like angiotensin II-generating activity in end-stage human autosomal dominant polycystic kidney disease. J Am Soc Nephrol 15:493–500
Ming Wu MC, Jing Ying, Junhui Gu, Mei Shuqin (2016) The C-terminal tail of polycystin-1 regulates complement factor B expression by signal transducer and activator of transcription 1. Am J Physiol Renal Physiol 310:F1284–F1294
Mrug M, Zhou J, Woo Y, Cui X, Szalai AJ et al (2008) Overexpression of innate immune response genes in a model of recessive polycystic kidney disease. Kidney Int 73:63–76
Mrug M, Zhou J, Mrug S, Guay-Woodford LM, Yoder BK, Szalai AJ (2014) Complement C3 activation in cyst fluid and urine from autosomal dominant polycystic kidney disease patients. J Intern Med 276:539–540
Mun H, Park JH (2016) Inflammation and Fibrosis in ADPKD. Adv Exp Med Biol 933:35–44
Nakamura T, Ushiyama C, Suzuki S, Ebihara I, Shimada N et al (2000) Elevation of serum levels of metalloproteinase-1, tissue inhibitor of metalloproteinase-1 and type IV collagen, and plasma levels of metalloproteinase-9 in polycystic kidney disease. Am J Nephrol 20:32–36
Niedermeier M, Reich B, Rodriguez Gomez M, Denzel A, Schmidbauer K et al (2009) CD4 + T cells control the differentiation of Gr1 + monocytes into fibrocytes. Proc Natl Acad Sci U S A 106:17892–17897
Nishio S, Hatano M, Nagata M, Horie S, Koike T et al (2005) Pkd1 regulates immortalized proliferation of renal tubular epithelial cells through p53 induction and JNK activation. J Clin Invest 115:910–918
Norman J (2011) Fibrosis and progression of autosomal dominant polycystic kidney disease (ADPKD). Biochim Biophys Acta 1812:1327–1336
Pahl HL (1999) Activators and target genes of Rel/NF-kB transcription factors. Oncogene 18:6853–6866
Park EY, Seo MJ, Park JH (2010) Effects of specific genes activating RAGE on polycystic kidney disease. Am J Nephrol 32:169–178
Qi W, Chen X, Poronnik P, Pollock CA (2006) The renal cortical fibroblast in renal tubulointerstitial fibrosis. Int J Biochem Cell Biol 38:1–5

Qian F, Boletta A, Bhunia AK, Xu H, Liu L et al (2002) Cleavage of polycystin-1 requires the receptor for egg jelly domain and is disrupted by human autosomal-dominant polycystic kidney disease 1-associated mutations. Proc Nat Acad Sci U S A 99:16981–16986
Qin S, Taglienti M, Cai L, Zhou J, Kreidberg JA (2012) c-Met and NF-kappaB-dependent overexpression of Wnt7a and -7b and Pax2 promotes cystogenesis in polycystic kidney disease. J Am Soc Nephrol 23:1309–1318
Raman A, Reif GA, Dai Y, Khanna A, Li X et al (2017) Integrin-linked kinase signaling promotes cyst growth and fibrosis in polycystic kidney disease. J Am Soc Nephrol 28:2708–2719
Rankin CA, Suzuki K, Itoh Y, Ziemer DM, Grantham JJ et al (1996) Matrix metalloproteinases and TIMPS in cultured C57BL/6 J-cpk kidney tubules. Kidney Int 50:835–844
Rawlings Jason S, Rosler KM et al (2004) The JAK/STAT signaling pathway. J Cell Sci 117:1281–1283
Ricklin D, Hajishengallis G, Yang K, Lambris JD (2010) Complement: a key system for immune surveillance and homeostasis. Nat Immunol 11:785–797
Schaefer L, Han X, Gretz N, Hafner C, Meier K et al (1996) Tubular gelatinase A (MMP-2) and its tissue inhibitors in polycystic kidney disease in the Han: SPRD rat. Kidney Int 49:75–81
Schieren G, Rumberger B, Klein M, Kreutz C, Wilpert J et al (2006) Gene profiling of polycystic kidneys. Nephrol Dial Transplant 21:1816–1824
Schulz C, Gomez Perdiguero E, Chorro L, Szabo-Rogers H, Cagnard N et al (2012) A lineage of myeloid cells independent of Myb and hematopoietic stem cells. Science 336:86–90
Semedo P, Donizetti-Oliveira C, Burgos-Silva M, Cenedeze MA, Avancini Costa Malheiros DM et al (2010) Bone marrow mononuclear cells attenuate fibrosis development after severe acute kidney injury. Lab Invest 90:685–695
Sica A, Mantovani A (2012) Macrophage plasticity and polarization: in vivo veritas. J Clin Invest 122:787–795
Snelgrove RJ, Jackson PL, Hardison MT, Noerager BD, Kinloch A et al (2010) A critical role for LTA4H in limiting chronic pulmonary neutrophilic inflammation. Science 330:90–94
Song X, Di Giovanni V, He N, Wang K, Ingram A et al (2009) Systems biology of autosomal dominant polycystic kidney disease (ADPKD): computational identification of gene expression pathways and integrated regulatory networks. Hum Mol Genet 18:2328–2343
Song CJ, Zimmerman KA, Henke SJ, Yoder BK (2017) Inflammation and fibrosis in polycystic kidney disease. Results Probl Cell Differ 60:323–344
Stark K, Vainio S, Vassileva G, McMahon AP (1994) Epithelial transformation of metanephric mesenchyme in the developing kidney regulated by Wnt-4. Nature 372:679–683
Su Z, Wang X, Gao X, Liu Y, Pan C et al (2014) Excessive activation of the alternative complement pathway in autosomal dominant polycystic kidney disease. J Int Med 276:470–485
Sureshbabu A, Muhsin SA, Choi ME (2016) TGF-beta signaling in the kidney: profibrotic and protective effects. Am J Physiol Renal Physiol 310:F596–F606
Swenson-Fields KI, Vivian CJ, Salah SM, Peda JD, Davis BM et al (2013) Macrophages promote polycystic kidney disease progression. Kidney Int 83:855–864
Talbot JJ, Shillingford JM, Vasanth S, Doerr N, Mukherjee S et al (2011) Polycystin-1 regulates STAT activity by a dual mechanism. Proc Nati Acad Sci U S A 108:7985–7990
Vernon MA, Mylonas KJ, Hughes J (2010) Macrophages and renal fibrosis. Semin Nephrol 30:302–317
Vilayur E, Harris DC (2009) Emerging therapies for chronic kidney disease: what is their role? Nat Rev Nephrol 5:375–383
Wada T, Sakai N, Sakai Y, Matsushima K, Kaneko S, Furuichi K (2011) Involvement of bone-marrow-derived cells in kidney fibrosis. Clin Exp Nephrol 15:8–13
Wallace DP, White C, Savinkova L, Nivens E, Reif GA et al (2014) Periostin promotes renal cyst growth and interstitial fibrosis in polycystic kidney disease. Kidney Int 85:845–854
Wei W, Hackmann K, Xu H, Germino G, Qian F (2007) Characterization of cis-autoproteolysis of polycystin-1, the product of human polycystic kidney disease 1 gene. J Biol Chem 282:21729–21737

Weimbs T, Olsan EE, Talbot JJ (2013) Regulation of STATs by polycystin-1 and their role in polycystic kidney disease. Jak-Stat 2:e23650

Wilson PD, Burrow CR (1999) Cystic diseases of the kidney: role of adhesion molecules in normal and abnormal tubulogenesis. Exp Nephrol 7:114–124

Wilson PD, Hreniuk D, Gabow PA (1992) Abnormal extracellular matrix and excessive growth of human adult polycystic kidney disease epithelia. J Cell Physiol 150:360–369

Wilson PD, Norman JT, Kuo NT, Burrow CR (1996) Abnormalities in extracellular matrix regulation in autosomal dominant polycystic kidney disease. Contrib Nephrol 118:126–134

Wilson PD, Geng L, Li X, Burrow CR (1999) The PKD1 gene product, "polycystin-1," is a tyrosine-phosphorylated protein that colocalizes with alpha2beta1-integrin in focal clusters in adherent renal epithelia. Lab Invest 79:1311–1323

Wuthrich RP, Mei C (2012) Aquaretic treatment in polycystic kidney disease. N Engl J Med 367:2440–2442

Xue C, Zhou CC, Wu M, Mei CL (2016) The clinical manifestation and management of autosomal dominant polycystic kidney disease in China. Kidney Dis (Basel) 2:111–119

Xue C, Zhou C, Mei C (2018) Total kidney volume: the most valuable predictor of autosomal dominant polycystic kidney disease progression. Kidney Int 93:540–542

Yamashita S, Maeshima A, Kojima I, Nojima Y (2004) Activin A is a potent activator of renal interstitial fibroblasts. J Am Soc Nephrol 15:91–101

Yang Y, Chen M, Zhou J, Lv J, Song S et al (2018) Interactions between macrophages and cyst-lining epithelial cells promote kidney cyst growth in Pkd1-deficient mice. J Am Soc Nephrol 29:2310–2325

Zeier M, Fehrenbach P, Geberth S, Mohring K, Waldherr R, Ritz E (1992) Renal histology in polycystic kidney disease with incipient and advanced renal failure. Kidney Int 42:1259–1265

Zeltner R, Hilgers KF, Schmieder RE, Porst M, Schulze BD et al (2008) A promoter polymorphism of the alpha 8 integrin gene and the progression of autosomal-dominant polycystic kidney disease. Nephron Clin Pract 108:c169–c175

Zheng D, Wolfe M, Cowley BD Jr, Wallace DP et al (2003) Urinary excretion of monocyte chemoattractant protein-1 in autosomal dominant polycystic kidney disease. J Am Soc Nephrol 14:2588–2595

Zimmerman K, Yoder BK (2015) SnapShot: sensing and signaling by cilia. Cell 161(692–692):e691

Chapter 6
Pathogenesis of Chronic Allograft Dysfunction Progress to Renal Fibrosis

Cheng Yang, Ruochen Qi and Bin Yang

Abstract Kidney transplantation is a life-change measurement for the patients of end-stage renal disease (ESRD). However, the renal allograft cannot avoid initial acute kidney injury (AKI) and subsequent chronic allograft dysfunction (CAD), gradually develops fibrosis and eventually loses function. It is imperative to disclose the pathogenesis of AKI and CAD in order to facilitate interventions. We have studied the involvement of immunity, inflammation, and apoptosis in ischemia-reperfusion injury (IRI) and/or immunosuppressant induced AKI models, with associated chronic damage. Our research mainly focused on tubular epithelial cells (TECs) that are passive victims and also active participators in injury and mediate following repair or fibrosis. Targeting not only fibroblasts/myofibroblasts, but also TECs, might be a fundamental strategy to prevent and treat renal fibrosis. We have also evaluated the potential application of siRNA targeting caspase-3 and tissue protective erythropoietin derivatives, HBSP and CHBP, aiming to treat AKI and prevent CAD. Significant improvements have been obtained, but timely diagnosis and precise therapy of AKI and prevention of CAD progressing to ESRD are still very challenging. Modern technologies such as microarray and sequencing analysis have been used to identify biomarkers and potentially facilitate individual cell target treatment for transplant patients.

Keywords Fibrosis · Tubular epithelia cell · Kidney · Inflammation

---

C. Yang · R. Qi
Department of Urology, Zhongshan Hospital, Zhangjiang Technology Institute, Fudan University, Shanghai, China

Shanghai Key Laboratory of Organ Transplantation, Shanghai, China

B. Yang (✉)
Nantong-Leicester Joint Institute of Kidney Science, Department of Nephrology, Affiliated Hospital of Nantong University, Renal Group, Basic Medical Research Centre, Medical College of Nantong University, Nantong 226001, Jiangsu, China
e-mail: binyang_le@163.com; by5@le.ac.uk

Department of Cardiovascular Sciences, University of Leicester, University Hospitals of Leicester, Leicester LE1 7RH, UK

B.-C. Liu et al. (eds.), *Renal Fibrosis: Mechanisms and Therapies*,
Advances in Experimental Medicine and Biology 1165,
https://doi.org/10.1007/978-981-13-8871-2_6

6.1 Renal Tubulointerstitial Fibrosis

Fibrosis is the characteristics of end-stage renal disease (ESRD) regardless of etiology. How to prevent and reverse renal fibrosis in kidney diseases are seeking after by clinicians and scientists alike. The severity of fibrosis predicts the progression of disease, but does not correct well with glomerular filtration rate (GFR, Nicholson et al. 1996). Although some renal injuries may initially target the glomeruli rather than the tubulointerstitial compartment, the best predictor of prognosis in chronic kidney disease (CKD) or chronic allograft dysfunction (CAD) is the degree of tubulointerstitial fibrosis (TIF), often accompanied by tubular atrophy (TA), in kidney biopsies (Liu et al. 2018b). Immunosuppressive drugs have developed rapidly, but the long-term survival of transplant kidney is still challenging. Interstitial residential cells, infiltrated innate immune cells, endothelial cells (ECs), pericytes, and tubular epithelial cells (TECs) all actively contribute to CAD. Therefore, there are great interests in understanding the pathogenesis of progressive CAD to renal fibrosis, with more attentions on ECs, pericytes, and TECs (Fig. 6.1).

6.2 Tubulointerstitial Fibrosis in Transplanted Kidneys

Before 2005, chronic renal allograft dysfunction without typical manifestation of rejection was pathologically classified as chronic allograft nephropathy (CAN). In 2005, the 8th Banff Conference on Allograft Pathology held in Edmonton eliminated ‘CAN’ (Solez et al. 2007), since the use of the non-specific term ‘CAN’ tended to undermine recognition of morphological features that enable the diagnosis of specific causes of CAD. In the new Banff classification, the individual chronic lesions are defined as IF, TA, arterial fibrous intimal thickening, arteriolar hyalinosis, mesangial matrix increase, and transplant glomerulopathy (Solez et al. 2008). IF and TA, in particular, almost invariably occurs together and is often considered as the single parameter IF/TA (Sis et al. 2010). Because the tubulointerstitium comprises 90% volume of the kidney, IF/TA is generally the most prominent manifestation of structural allograft deterioration.

Fibrosis develops in many patients during the first year after transplantation, most of which is accumulated over the first 3 months. Afterward, the progressive rate of fibrosis slows down significantly. It is likely that this early accumulation of fibrosis mainly results from self-limiting inflammation related to transplantation stress. Nevertheless, fibrosis often progresses continuously (Vanhove et al. 2017). In a 10-year follow-up study, severe CAN (old definition for CAD) presented in 58.4% of patients, while glomerulosclerosis was shown in 37.3% of patients. The conclusion was made that IF/TA is irreversible and results in declining renal function and graft failure (Nankivell et al. 2003). Epithelial–mesenchymal transition (EMT) is a basic pathologic process during renal fibrosis, but we have a focus here to discuss the interaction between TECs and the immune system.

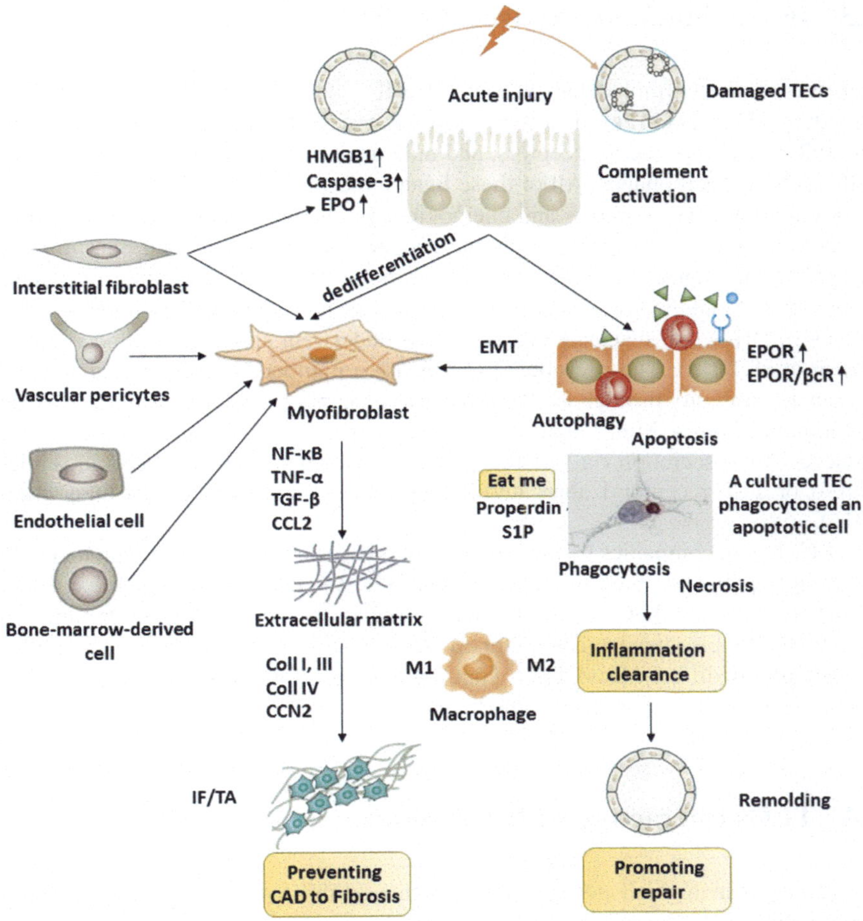

Fig. 6.1 Mechanisms of acute injury to repair or CAD to renal fibrosis. IR-related AKI induces immune responses, inflammation, and apoptotic and necrotic cell death. S1P and properdin are released by injured TECs, and also label damaged cells as 'fine me' signals, then facilitate the capture by phagocytes to clear inflammation and initiate repair. Erythropoietin (EPO) released by interstitial cells combines with EPO receptor that expressed on the phagocytes including macrophages and TECs. The Jak-2/STAT3 pathway is a typical and classical signaling pathway that involves in EPO receptor mediated repair and maybe also fibrosis

6.3 Innate Immunity Activated by TEC Injury

Damage-associated molecular patterns (DAMPs) are released from necrotic cells including TECs. DAMPs can activate Toll-like receptors (TLRs), NOD-like receptors (NLRs), C-type lectin receptors, and inflammasomes on resident inflammatory cells such as macrophages. All of these ligand–receptor actions drive the secretion of proinflammatory cytokines and chemokines (Ising and Heneka 2018). Complex chemokine gradients are formed via the cooperation of immune and non-immune cells to recruit innate immune cells. The number of studies including ours have demonstrated that the non-sterile inflammation cascade is induced by injured TECs (Hu et al. 2012). For instance, in AKI, TECs underwent apoptosis and necroptosis (Wang et al. 2016). The necrotic TECs release a number of DAMPs that activate the innate immunity through identical pattern recognition receptors including TLRs and inflammasomes. Mulay et al. showed that oxalate crystal formation inside TECs induced TNF-α secretion via TNF receptor 1-dependent mechanism. Blocking either TNF-α or TNF receptor 1 abrogated kidney injury and dysfunction (Mulay et al. 2016).

TECs loss accompanies by the initiation of the repair process. Although there are different hypotheses about the regeneration of TECs, the most acceptable concept is that the survived TECs dedifferentiate to regain the ability of proliferation. Up to now, the precise mechanism of this process is still unknown, but recent studies showed that dedifferentiated TECs contribute to myofibroblast activation and TIF via the secretion of inflammatory mediators.

6.4 TECs Interacting with Macrophages

Macrophages are highly heterogeneous cells that exhibit distinct phenotype and function depending on the microenvironment conditioned by the type and the stage of diseases. Classically, macrophages are classified into activated macrophages (M1) and alternatively activated macrophages (M2) in vitro. M2 are subdivided further into M2a, M2b and M2c according to their responses to different modulators. However, this classification does not truly reflect their phenotypes in tissue environment in vivo. Recently, Anders and Ryu proposed four types of macrophages in vivo, according to their predominant roles in the phases of wound healing, namely proinflammatory, anti-inflammatory, profibrotic, and fibrolytic macrophages (Anders and Ryu 2011). In a recent study, TECs were found to be associated with profibrotic macrophages. The exosome released from TECs packed with chemokine C-C motif chemokine ligand 2 (CCL2) mRNA and transferred CCL2 mRNA from TECs to macrophages. These signals provoked the activation and autocrine of macrophages and then recruited additional myeloid cells (Lv et al. 2018). Interestingly, the cilia in renal TECs can also activate chemokine signaling to recruit inflammatory cells. Viau et al. identified a complex of ciliary kinase liver kinase B1 (LKB1) and several ciliopathy-related proteins including NPHP1 and PKD1. For homeostasis, this ciliary module suppresses the expression of inflammatory chemokine CCL2 in TECs. Deletion of LKB1 or PKD1 in mouse renal tubules elevates CCL2 expression in a cell-autonomous manner and results in peritubular accumulation of $CCR2^{+}$ mononuclear phagocytes, promoting a ciliopathy phenotype. This study suggests that cilia-localized LKB1 regulates chemokine signaling, macrophage recruitment, and tissue homeostasis in the kidney (Viau et al. 2018).

6.5 Autophagy in Tubulointerstitial Fibrosis

Autophagy is a double-edged sword in TIF. It is an intracellular homeostatic mechanism and is important for the degradation of waste components from the cytoplasm in acidic lysosomal compartments (Deretic et al. 2013). In diabetic nephropathy, autophagy can degrade advanced glycation end products (AGEs) in TECs, while specifically inhibiting autophagy in TECs results in accumulation of AGEs along with worsened inflammation and fibrosis (Takahashi et al. 2017). Another in vitro study shows that overexpression of SK1 leads to enhanced autophagy activity and ameliorates fibrosis induced by TGF-β1 (Du et al. 2017). Despite the fact that autophagy serves as a protective mechanism upon injury, a few more studies focusing on the role of Atg5 reach controversial results. One study shows that genetic deletion of Atg5 results in Atg5-mediated autophagy deficiency specifically in TECs, which leads to marked cell cycle arrest at G2/M phase, increased COL1 deposition, and severe interstitial fibrosis in an unilateral urinary obstruction (UUO) model, which corresponds with the concept discussed above that autophagy plays as an antifibrotic role in TECs (Li et al. 2016). However, another similar study also focused on the role of Atg5 indicates that ablation of autophagy in TECs leads to significantly less tubular senescence and reduced interstitial fibrosis 30 days after IRI (Baisantry et al. 2016). This is also consistent with an additional study that also showed ameliorated tubular injury and tubulointerstitial fibrosis in the mouse kidneys subjected to proximal tubule-specific Atg7 knock out (KO). These results indicate the two-way effects of autophagy in renal fibrosis. Indeed, autophagy can degrade unnecessary or dysfunction components and prevent cell apoptosis, thus acting as a protective mechanism for cell survival. However, if severely damaged TECs managed to survive via autophagy, these compromised cells may undergo maladaptive repair and phenotype change, thereby secreting proinflammatory and profibrotic cytokines, leading to aggravated renal fibrosis (Livingston et al. 2016).

6.6 Phagocytosis of Apoptotic TECs by EPO Receptor-Related Pathways

Apoptosis is necessary for the clearance of inflammatory cells and injured TECs during repair/remodeling, while apoptotic cells have to be rapidly removed by phagocytosis to avoid further damage (Yang et al. 2015b). The previous study showed up-regulated caspase-3 in TECs is associated with apoptosis and inflammation in IRI kidneys (Yang et al. 2001, 2012, 2005). The anti-inflammation, anti-apoptosis, and pro-regeneration role of EPO have been broadly reported (Gilboa et al. 2017; Lee et al. 2015). We revealed that the renoprotective function of EPO was associated with increased caspase-3 activity and inflammatory cell clearance via apoptosis (Yang et al. 2011b). Several pieces of evidence revealed the role of EPO in the clearance of apoptotic cells by macrophage in renal IRI (Brooks et al. 2015; Humphreys et al. 2011; Kusaba et al. 2014; Poon et al. 2014). First, EPO receptor (EPOR) is expressed on macrophages and the treatment of macrophages with EPO suppresses inflammatory gene expression, which is associated with the engulfment of apoptotic cells (Lifshitz et al. 2010). Second, the 'find-me' signal S1P is known to induce hypoxia-inducible factor 1α (HIF-1α), a key regulator of EPO expression. Third, EPO has been shown to ameliorate autoimmune diseases by promoting dead cell clearance (Luo et al. 2013). Despite these published observations, how a hormone critical for erythropoiesis functions in the clearance of dying cells remained a mystery. In 2016, Luo et al. bridged this gap by demonstrating that EPO can prime phagocytes for effective efferocytosis (Elliott et al. 2017; Luo et al. 2016a).

6.7 Complement System in Renal Fibrosis

Renal IRI, as well as other injuries, activates the complement system by endogenous ligands that trigger proteolytic cleavage of complement components via three pathways: classical, lectin, and alternative pathways (CP, LP and AP). These eventually direct the formation of a membrane attack complex (MAC) (Gorsuch et al. 2012). The products of complement activation such as MAC (C5b-9), C3a and C5a play pivotal roles in cell damage and amplify inflammatory responses by complement receptor (CR)1 and CR3 recruiting and activating leukocytes (neutrophil/monocyte) (Ricklin and Lambris 2013; Thurman et al. 2006; Zipfel and Skerka 2009).

There are many reasons to believe that complement is involved in renal IRI and its inhibition might reduce the extent of injury. However, this is not always the case as complement in renal IRI is complicated by many factors such as the stage of injury and/or repair. Complement properdin might have a predominant role as a pattern recognition molecule to opsonizing damaged cells facilitating phagocytic clearance in repair, rather than amplifying complement activation in initial injury (Zwaini et al. 2017). Renal TECs actively participate in the generation of mediators including complement components upon renal IRI. Pharmacologic inhibitors, as well as complement component-deficient animals, have been used to dissect the role of complement in renal IRI and also to achieve therapeutic intervention. Fortunately, efforts in recent years have introduced the first complement inhibitor Eculizumab for clinical treatment and also added more than a dozen promising candidates including TT30 (AP inhibitor) and C1INH (recombinant human C1 Inhibitor) to the pipelines. fB (C3bBb promotor) deficiency reduces complement 3 (C3) deposition, decreases caspase-3 and tubular apoptosis, and protects the kidney from IRI (Thurman et al. 2003, 2006). These target complement components could also be potential biomarkers for the diagnosis of IRI in both native and transplanted kidneys. However, the time frame and long-term effect of complement-oriented therapy need to be further evaluated and optimized (Zwaini et al. 2017).

6.8 Metabolic Alteration of TEC Influence Fibrosis

In CAD or CKD, the number of functional nephrons decreases, creating more metabolic work for the remaining ones. Similar to declining renal function leading to glomerular hyperfiltration, the remaining TEC hypertrophy meets the increased demand of reabsorbing water and solutes. This is in accordance with the concept that protein intake correlates positively with proximal TEC hypertrophy (Johnston et al. 1987), and augmented protein consumption in the subtotal nephrectomy model in rats leads to more oxidative stress (Nath et al. 1990). TECs have a high level of energy demand of adenosine triphosphate (ATP) that is mostly produced by fatty acid oxidation. New findings indicate that dysregulation of fatty acid oxidation followed by intracellular lipid accumulation profoundly affects the fate of TECs by promoting EMT, inflammation, and eventually interstitial fibrosis (Kang et al. 2015). Other studies suggest that increased calories, rather than protein per se, increased the workload of TECs and, conversely, limiting calories intake protects against TIF.

Hyperammonemia, in addition, is the clinical characteristic of hepatorenal syndrome (HRS) in patients with end-stage liver disease. Tubular dysfunction is an important feature of renal injury in HRS. In a HRS rodent model, gene expression analyses signified proximal TEC injury, tissue hypoxia, inflammation, and activation of the fibrotic gene program. Marked changes in renal arginine metabolism (up-regulation of arginase-2 and downregulation of argininosuccinate synthase 1), resulted in decreased circulating arginine levels. The results from a study performed by Varga et al. suggest that hyperammonemia may contribute to impaired renal arginine metabolism, leading to decreased eNOS activity, impaired microcirculation, TEC death, tubulointerstitial nephritis, and fibrosis (Varga et al. 2018). Moreover, in diabetic kidney disease, extracellular vesicles (EVs), which contain microRNA, constitute a novel means of cell communication that may contribute to the inevitable expansion of renal fibrosis (Jia et al. 2018).

Changes in metabolism after injury may be better understood from the perspective of evolutionary trade-offs. Natural selection leads to changes that maintain kidney function through the reproductive years but may promote renal decline after this time period (Chevalier 2017). For example, nephron loss leads to hypertrophy and increased energy consumption by the remaining nephrons to maintain renal function in the short term, but this cannot be sustained and eventually leads to tubular loss and fibrosis. This evolutionary adaption may explain other responses in the CKD kidney that are initially adaptive but lead to dysfunction such as senescence and inflammation.

6.9 Impact of Immunosuppressive Agents on Fibrosis

The classical immunosuppressive agents include calcineurin inhibitors (CNI), mycophenolic acid, and prednisone. Among these three mainstay immunosuppressive agents, CNI has acute nephrotoxic effects, primarily resulting from hemodynamic alterations (vasoconstriction of the afferent arterioles) and reversible tubular dysfunction. The effects of different immunosuppressants on allograft fibrosis have been revealed. Our previous study compared typical CNI [cyclosporine A (CsA) and tacrolimus], mycophenolate mofetil (MMF), and rapamycin in long-term kidney IRI model. We found that CsA, neither tacrolimus, rapamycin nor MMF, increased interstitial inflammation and renal fibrosis (Yang et al. 2005). CNI, especially CsA, mediates chronic nephrotoxicity through a variety of mechanisms: chronic vasoconstriction and arteriolar narrowing resulting in persistent local hypoxia that stimulates reactive oxygen species production and renin–angiotensin–aldosterone system activation, both of which have profibrotic effects (Krauskopf et al. 2005; Wu et al. 2018).

Efforts have been made to achieve a better immunosuppressive protocol to prevent allograft fibrosis. The Symphony study established that the combination of low-dose tacrolimus, MMF, steroids and daclizumab resulted in a lower rate of acute rejection, better graft survival, and better graft function at 1 and 3 years after transplantation compared with low-dose CsA, standard-dose CsA, or low-dose sirolimus-based immunosuppression (Ekberg et al. 2007, 2009). The impact of CNI exposure on graft histology is not as clear. Studies have reported both high and low CsA exposures are independent predictors of IF/TA accumulation by 1–2 years (di Paolo et al. 2004; Krauskopf et al. 2005). In the BENEFIT study of belatacept versus CsA, both belatacept arms had better renal function and lower prevalence of IF/TA (mainly mild) on 1-year protocol biopsies (20–29%) compared with the CsA arm (44%) (Vincenti et al. 2005). Microarray analysis of a small subgroup of 1-year biopsies from the BENEFIT and BENEFIT-EXT showed that the CsA group was enriched for gene sets associated with fibrosis and chronic allograft injury (Vitalone et al. 2014). Comparative data with tacrolimus in that regard are currently lacking.

6.10 Intervention Using Caspase-3 siRNA

Caspase-3 is one of the intracellular executing proteases, which was found in an inactive phase in normal renal tissues. Caspase-3 could be stimulated through extrinsic death receptor-mediated pathways (such as TNF receptor) or intrinsic pathways via mitochondrial stimulation, as well endoplasmic reticulum, and result in the activation of caspases (Lifshitz et al. 2010; Yang et al. 2007). Active caspase-3 has the ability to regulate apoptosis during renal IRI, where it dismantles affected cells via its proteolytic activity, developing apoptotic bodies, and upregulating ligands to attract phagocytic cells (Luo et al. 2013, 2016b). Both human proximal TECs subjected to hypoxia/reoxygenation and mouse kidneys exposed to ischemia for 30 min followed by reperfusion for 24 h expressed higher cleaved caspase-3 protein, as well as more apoptotic bodies, compared to the sham groups, respectively (Elliott et al. 2017). We have previously shown up-regulated caspase-3 in TECs (Yang et al. 2001) associated with apoptosis and inflammation in different IR-induced acute and chronic kidney injury models (Yang et al. 2001, 2005, 2012). A recent study showed *caspase-3* deficiency had less microvascular congestion and activation in the early and later stage of AKI model. After 3 weeks of AKI, *caspase-3* KO reduced microvascular rarefaction and renal fibrosis, as well as decreased expression of α-smooth muscle actin and reduced collagen deposition within peritubular capillaries (Yang et al. 2018).

The modification of caspase-3 has also been evaluated in our series of studies using caspase-3 inhibitor (Yang et al. 2003, 2004), synthetic siRNA (Yang et al. 2015b), leukocyte deletion (Yang et al. 2010) and clinically used drugs atorvastatin (Haylor et al. 2011), EPO and its derivatives helix B surface peptide (HBSP) (Wu et al. 2013; Yang et al. 2013c; Zhang et al. 2017) and cyclic HBSP (CHBP, with longer half-life and stronger potency) (Liu et al. 2018a; Yang et al. 2015a).

The significant achievement in siRNA therapy has been obtained in different organ systems. The kidney is a comparatively easy target organ due to its unique structural and functional characteristics and has great potential for clinical applications in humans. Targeting C3 and caspase-3 using siRNA, respectively, improved renal function and histology in mouse IRI (Gobe et al. 2014; Sato and Yanagita 2013) and porcine kidney auto-transplant models (Kriz et al. 2011). We latterly utilized specific caspase-3 siRNA of different designs and delivery routes in a series of biological models of kidney transplant-related injuries (Yang et al. 2015b). A new version of chemically modified caspase-3 siRNA had a profound protection in TECs, isolated and auto-transplanted porcine kidneys, while its side effect and systemic complementary response were also thoroughly investigated including changes in innate immunity upon siRNA treatment. The most effective sequences of caspase-3 siRNA selected in porcine proximal tubular cells (LLC-PK1) cells (Yang et al. 2011a) were then validated in isolated porcine kidneys, in which caspase-3 siRNA silenced caspase-3, attenuated inflammation and apoptosis, improved renal histology and function (Yang et al. 2011c). Caspase-3 siRNA preserved porcine kidneys were further auto-transplanted for 2 days, but not protected (Yang et al. 2013a), as complementary systemic feedback activated the innate immunity (Yang et al. 2013b).

Moreover, new serum stabilized caspase-3 siRNA was administrated locally in kidney preservation and also intravenously in recipients. The 2-week auto-transplanted kidneys were finally protected without significant off-target effects (Yang et al. 2014). Although the siRNA target p53 has been used in a phase I clinical trial, and we paid attention to the serum stability and immune tolerance of siRNA, and also give siRNA once at the early stage of injury to protect kidneys for long-term, the future challenges of siRNA therapy include organ-/tissue-/cell-specific delivery and time-controlled silence, as well as selecting therapeutic targets in the different cause, stage, and location of injury.

6.11 Intervention Using EPO and Its Derivatives

EPO and its derivatives have been found to be renoprotective in different studies. EPO-derived HBSP only recognizes the tissue protective receptor, a heterodimer of EPOR and β common receptor (EPOR/βcR) that activates anti-inflammation and tissue repair pathways (Wu et al. 2013; Yang et al. 2013c; Zhang et al. 2017). EPOR/βcR is not expressed in normal tissues, but could be locally up-regulated earlier than EPO production upon injury (Collino et al. 2015; Dahan et al. 2016). We previously demonstrated that HBSP inhibited caspase-3 and improved renal IRI in mouse models (Wu et al. 2013; Yang et al. 2013c). Others also showed that HBSP reduces macrophage M1/M2 ratio, suppresses endothelial apoptosis (Ueba et al. 2013), and highly expressed EPOR/βcR mediates wound healing (Saqib et al. 2009). We recently demonstrated that the expression of EPOR/βcR was gradually increased by prolonged reperfusion time from 6 h to 1 w. In addition, 72-h renal IRI was more severe in properdin KO mice with further increased EPOR/βcR mainly located in TECs. Moreover, increased caspase-3 and apoptosis were greatly reduced by HBSP treatment, regardless of mouse genetic phenotype (unpublished data with abstracts accepted by meetings: Nephrology Dialysis Transplantation, Volume 33, Issue suppl_1, May 2018, Page i103, https://doi.org/10.1093/ndt/gfy104.FP214; UKKW2018, P212: https://www.ukkw.org.uk/abstracts-from-uk-kidney-week/).

The renoprotection and its underlying mechanism of EPO, HBSP and CHBP have also been studied. EPO decreased myeloperoxidase (MPO)+ neutrophil infiltration and proinflammatory mediator gene expression in rat IRI kidneys (Solez et al. 2008). EPO also induced a dose-dependent increase in caspase-3 precursor and stimulated caspase-3 cleavage in cisplatin-treated LLC-PK1. EPO increased activated caspase-3, apoptotic cells, and MPO+ cells in tubular lumens, but decreased apoptotic cells in tubular areas with decreased IL-1β activation and macrophage L1+ cells (Sis et al. 2010). HBSP and CHBP remained the renoprotective properties of EPO, but not its erythropoiesis function (Nankivell et al. 2003; Sato and Yanagita 2017; Vanhove et al. 2017). In addition, HBSP (ARA290) has been applied in a phase I clinical trial. Nevertheless, the mechanistic signaling pathways of EPOR/βcR and EPOR on TECs regulating phagocytosis and innate immunity in self-recovery post-renal IRI or progress to CKD/CAD need to be further explored. In addition, organ/cell target

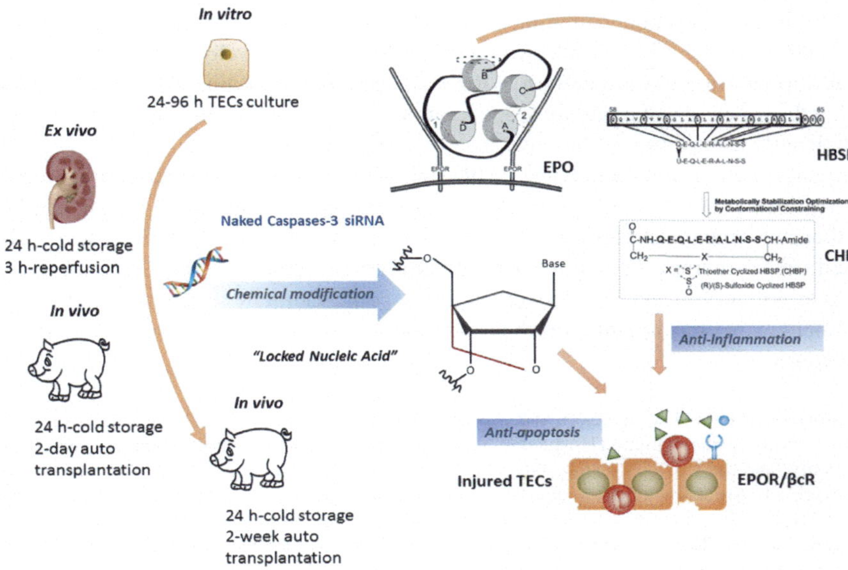

Fig. 6.2 Renoprotection of caspase-3 siRNA, EPO and its derivatives. The caspase-3 siRNA first protected porcine renal TECs against hydrogen peroxide-induced injury. The renoprotection of naked caspase-3 siRNA with the same sequences was further validated in an ex vivo isolated porcine kidney perfusion model, then shown that the siRNA was effective for cold preservation, but not in auto-transplanted kidneys without systematic siRNA treatment. Finally, the modified siRNA of caspase-3, by locked nucleic acid stabilized the siRNA in serum, significantly protected 2-week transplanted kidneys. In addition, the renoprotection of EPO and its derivatives were assessed in transplant-related renal injury models. Based on the essential structure of HBSP, we synthesized cyclic HBSP with more potent effects in vivo. CHBP is our patent drug for kidney protection with anti-inflammation and anti-apoptosis property

delivery of caspase-3 siRNA as a novel treatment might be achieved by caspase-3 siRNA conjugated with HBSP that is a ligand of EPOR/βcR. It should be possible to move biomarker-based individualized transplant care from a research hypothesis to clinical implementation (Fig. 6.2) (Menon et al. 2017).

6.12 Conclusion

Kidney transplantation is a mature therapy for ESRD patients. However, CAD remains a major concern for allograft failure. IF/TA is one of the most important manifestations of CAD. In recent years, TECs are drawing an increasing attention in the development of kidney injury and allograft failure. TECs actively mediate not only IR-induced AKI, but also the progression of AKI to CAD, even renal fibrosis. The inflammatory cytokines produced and released by injured TECs initiate complement activation and subsequent inflammation. This also further activates inflammatory cells and promotes TECs death, which might lead to the development of CAD and renal fibrosis. In addition, certain cytokines, such as properdin, secreted by TECs during the processes of repair and regeneration might result in TEC dedifferentiation significantly contribute to the clearance of inflammation, and the change of local microenvironment for remodeling. Although fibroblasts/myofibroblasts are executors of renal fibrosis, TECs seem to be major initiators. Targeting TECs might be another fundamental strategy to prevent and treat renal fibrosis. We believe that optimized siRNA therapy, in conjunction with advanced genetic screening technologies, together with erythropoietin derived peptides, could facilitate timely and individual treatment even pre-implantation to prevent and intervene CAD and renal fibrosis in near future.

Acknowledgements The studies cited in this chapter were supported by the Zhongshan Hospital of Fudan University; the Affiliated Hospital of Nantong University and the Medical School of Nantong University; and the University Hospitals of Leicester (UHL) and the University of Leicester. We also would like to acknowledge the support of various project grants such as the Kidney Care Appeal from UHL (to BY), the UK–China Fellowship for Excellence from the Department for Business Innovation and Skills (to BY), the National Natural Foundation of China (81400752 and 81770746 to CY and 81170689, 81570677 and 81873622 to BY), the National Key R&D Program of China (2018YFA0107502 to CY), the Medical and Health Talents Training Plan for the Excellent Youth of Shanghai Municipal (2018YQ50 to CY) and Shanghai Rising-Star Program (19QA1406300 to CY).

References

Anders HJ, Ryu M (2011) Renal microenvironments and macrophage phenotypes determine progression or resolution of renal inflammation and fibrosis. Kidney Int 80:915–925

Baisantry A, Bhayana S, Rong S, Ermeling E, Wrede C et al (2016) Autophagy induces prosenescent changes in proximal tubular S3 segments. J Am Soc Nephrol JASN 27:1609–1616

Brooks CR, Yeung MY, Brooks YS, Chen H, Ichimura T et al (2015) KIM-1-/TIM-1-mediated phagocytosis links ATG5-/ULK1-dependent clearance of apoptotic cells to antigen presentation. EMBO J 34:2441–2464

Chevalier RL (2017) Evolutionary nephrology. Kidney Int Rep 2:302–317

Collino M, Thiemermann C, Cerami A, Brines M (2015) Flipping the molecular switch for innate protection and repair of tissues: long-lasting effects of a non-erythropoietic small peptide engineered from erythropoietin. Pharmacol Ther 151:32–40

Dahan A, Brines M, Niesters M, Cerami A, van Velzen M (2016) Targeting the innate repair receptor to treat neuropathy. Pain Rep 1:e566

Deretic V, Saitoh T, Akira S (2013) Autophagy in infection, inflammation and immunity. Nat Rev Immunol 13:722–737

di Paolo S, Teutonico A, Stallone G, Infante B, Schena A et al (2004) Cyclosporin exposure correlates with 1 year graft function and histological damage in renal transplanted patients. Nephrol Dial Transplant 19:2107–2112

Du C, Ren Y, Yao F, Duan J, Zhao H et al (2017) Sphingosine kinase 1 protects renal tubular epithelial cells from renal fibrosis via induction of autophagy. Int J Biochem Cell Biol 90:17–28

Ekberg H, Tedesco-Silva H, Demirbas A, Vitko S, Nashan B et al (2007) Reduced exposure to calcineurin inhibitors in renal transplantation. N Engl J Med 357:2562–2575

Ekberg H, Bernasconi C, Tedesco-Silva H, Vitko S, Hugo C et al (2009) Calcineurin inhibitor minimization in the symphony study: observational results 3 years after transplantation. Am J Transplant (Official Journal of the American Society of Transplantation and the American Society of Transplant Surgeons) 9:1876–1885

Elliott MR, Koster KM, Murphy PS (2017) Efferocytosis signaling in the regulation of macrophage inflammatory responses. J Immunol 198:1387–1394

Gilboa D, Haim-Ohana Y, Deshet-Unger N, Ben-Califa N, Hiram-Bab S et al (2017) Erythropoietin enhances Kupffer cell number and activity in the challenged liver. Sci Rep 7:10379

Gobe GC, Bennett NC, West M, Colditz P, Brown L et al (2014) Increased progression to kidney fibrosis after erythropoietin is used as a treatment for acute kidney injury. Am J Physiol Renal Physiol 306:F681–F692

Gorsuch WB, Chrysanthou E, Schwaeble WJ, Stahl GL (2012) The complement system in ischemia-reperfusion injuries. Immunobiology 217:1026–1033

Haylor JL, Harris KP, Nicholson ML, Waller HL, Huang Q et al (2011) Atorvastatin improving renal ischemia reperfusion injury via direct inhibition of active caspase-3 in rats. Exp Biol Med (Maywood) 236:755–763

Hu L, Yang C, Zhao T, Xu M, Tang Q et al (2012) Erythropoietin ameliorates renal ischemia and reperfusion injury via inhibiting tubulointerstitial inflammation. J Surg Res 176:260–266

Humphreys BD, Czerniak S, DiRocco DP, Hasnain W, Cheema R et al (2011) Repair of injured proximal tubule does not involve specialized progenitors. Proc Natl Acad Sci USA 108:9226–9231

Ising C, Heneka MT (2018) Functional and structural damage of neurons by innate immune mechanisms during neurodegeneration. Cell Death Dis 9:120

Jia Y, Zheng Z, Guan M, Zhang Q, Li Y et al (2018) Exendin-4 ameliorates high glucose-induced fibrosis by inhibiting the secretion of miR-192 from injured renal tubular epithelial cells. Exp Mol Med 50:56

Johnston JR, Brenner BM, Hebert SC (1987) Uninephrectomy and dietary protein affect fluid absorption in rabbit proximal straight tubules. Am J Physiol 253:F222–F233

Kang HM, Ahn SH, Choi P, Ko YA, Han SH et al (2015) Defective fatty acid oxidation in renal tubular epithelial cells has a key role in kidney fibrosis development. Nat Med 21:37–46

Krauskopf A, Lhote P, Petermann O, Ruegg UT, Buetler TM (2005) Cyclosporin A generates superoxide in smooth muscle cells. Free Radic Res 39:913–919

Kriz W, Kaissling B, Le Hir M (2011) Epithelial-mesenchymal transition (EMT) in kidney fibrosis: fact or fantasy? J Clin Investig 121:468–474

Kusaba T, Lalli M, Kramann R, Kobayashi A, Humphreys BD (2014) Differentiated kidney epithelial cells repair injured proximal tubule. Proc Natl Acad Sci USA 111:1527–1532

Lee TS, Lu KY, Yu YB, Lee HT, Tsai FC (2015) β common receptor mediates erythropoietin-conferred protection on OxLDL-induced lipid accumulation and inflammation in macrophages. Mediators Inflamm 2015:439759

Li H, Peng X, Wang Y, Cao S, Xiong L et al (2016) Atg5-mediated autophagy deficiency in proximal tubules promotes cell cycle G2/M arrest and renal fibrosis. Autophagy 12:1472–1486

Lifshitz L, Tabak G, Gassmann M, Mittelman M, Neumann D (2010) Macrophages as novel target cells for erythropoietin. Haematologica 95:1823–1831

Liu A, Wu J, Yang C, Wu Y, Zhang Y et al (2018a) TRPM7 in CHBP-induced renoprotection upon ischemia reperfusion-related injury. Sci Rep 8:5510

Liu BC, Tang TT, Lv LL, Lan HY (2018b) Renal tubule injury: a driving force toward chronic kidney disease. Kidney Int 93:568–579

Livingston MJ, Ding HF, Huang S, Hill JA, Yin XM et al (2016) Persistent activation of autophagy in kidney tubular cells promotes renal interstitial fibrosis during unilateral ureteral obstruction. Autophagy 12:976–998

Luo B, Jiang M, Yang X, Zhang Z, Xiong J et al (2013) Erythropoietin is a hypoxia inducible factor-induced protective molecule in experimental autoimmune neuritis. Biochem Biophys Acta 1832:1260–1270

Luo B, Gan W, Liu Z, Shen Z, Wang J et al (2016a) Erythropoeitin signaling in macrophages promotes dying cell clearance and immune tolerance. Immunity 44:287–302

Luo B, Wang J, Liu Z, Shen Z, Shi R et al (2016b) Phagocyte respiratory burst activates macrophage erythropoietin signalling to promote acute inflammation resolution. Nat Commun 7:12177

Lv LL, Feng Y, Wen Y, Wu WJ, Ni HF et al (2018) Exosomal CCL2 from tubular epithelial cells is critical for albumin-induced tubulointerstitial inflammation. J Am Soc Nephrol JASN 29:919–935

Menon MC, Murphy B, Heeger PS (2017) Moving biomarkers toward clinical implementation in kidney transplantation. J Am Soc Nephrol 28:735–747

Mulay SR, Desai J, Kumar SV, Eberhard JN, Thomasova D et al (2016) Cytotoxicity of crystals involves RIPK3-MLKL-mediated necroptosis. Nat Commun 7:10274

Nankivell BJ, Borrows RJ, Fung CL, O'Connell PJ, Allen RD et al (2003) The natural history of chronic allograft nephropathy. N Engl J Med 349:2326–2333

Nath KA, Croatt AJ, Hostetter TH (1990) Oxygen consumption and oxidant stress in surviving nephrons. Am J Physiol 258:F1354–F1362

Nephrology Dialysis Transplantation, Volume 33, Issue suppl_1, May 2018, Page i103, https://doi.org/10.1093/ndt/gfy104.FP214

Nicholson ML, McCulloch TA, Harper SJ, Wheatley TJ, Edwards CM et al (1996) Early measurement of interstitial fibrosis predicts long-term renal function and graft survival in renal transplantation. Br J Surg 83:1082–1085

Poon IK, Lucas CD, Rossi AG, Ravichandran KS (2014) Apoptotic cell clearance: basic biology and therapeutic potential. Nat Rev Immunol 14:166–180

Ricklin D, Lambris JD (2013) Complement in immune and inflammatory disorders: therapeutic interventions. J Immunol 190:3839–3847

Saqib N, Duling L, Krier K, Howdieshell TR (2009) Temporal and spatial expression of erythropoietin, erythropoietin receptor, and common beta receptor in wound fluid and granulation tissue. Wounds 21:164–171

Sato Y, Yanagita M (2013) Renal anemia: from incurable to curable. Am J Physiol Renal Physiol 305:F1239–F1248

Sato Y, Yanagita M (2017) Resident fibroblasts in the kidney: a major driver of fibrosis and inflammation. Inflamm Regen 37:17

Sis B, Einecke G, Chang J, Hidalgo LG, Mengel M et al (2010) Cluster analysis of lesions in nonselected kidney transplant biopsies: microcirculation changes, tubulointerstitial inflammation and scarring. Am J Transplant (Official Journal of the American Society of Transplantation and the American Society of Transplant Surgeons) 10:421–430

Solez K, Colvin RB, Racusen LC, Sis B, Halloran PF et al (2007) Banff '05 meeting report: differential diagnosis of chronic allograft injury and elimination of chronic allograft nephropathy ('CAN'). Am J Transplant 7:518–526

Solez K, Colvin RB, Racusen LC, Haas M, Sis B et al (2008) Banff 07 classification of renal allograft pathology: updates and future directions. Am J Transplant (Official Journal of the American Society of Transplantation and the American Society of Transplant Surgeons) 8:753–760

Takahashi A, Takabatake Y, Kimura T, Maejima I, Namba T et al (2017) Autophagy inhibits the accumulation of advanced glycation end products by promoting lysosomal biogenesis and function in the kidney proximal tubules. Diabetes 66:1359–1372

Thurman JM, Ljubanovic D, Edelstein CL, Gilkeson GS, Holers VM (2003) Lack of a functional alternative complement pathway ameliorates ischemic acute renal failure in mice. J Immunol 170:1517–1523

Thurman JM, Royer PA, Ljubanovic D, Dursun B, Lenderink AM et al (2006) Treatment with an inhibitory monoclonal antibody to mouse factor B protects mice from induction of apoptosis and renal ischemia/reperfusion injury. J Am Soc Nephrol 17:707–715

Ueba H, Shiomi M, Brines M, Yamin M, Kobayashi T et al (2013) Suppression of coronary atherosclerosis by helix B surface Peptide, a nonerythropoietic, tissue-protective compound derived from erythropoietin. Mol Med 19:195–202

Vanhove T, Goldschmeding R, Kuypers D (2017) Kidney fibrosis: origins and interventions. Transplantation 101:713–726

Varga ZV, Erdelyi K, Paloczi J, Cinar R, Zsengeller ZK et al (2018) Disruption of renal arginine metabolism promotes kidney injury in hepatorenal syndrome. Hepatology 68(4):1519–1533

Viau A, Bienaime F, Lukas K, Todkar AP, Knoll M et al (2018) Cilia-localized LKB1 regulates chemokine signaling, macrophage recruitment, and tissue homeostasis in the kidney. EMBO J 37(15):e98615

Vincenti F, Larsen C, Durrbach A, Wekerle T, Nashan B et al (2005) Costimulation blockade with belatacept in renal transplantation. N Engl J Med 353:770–781

Vitalone MJ, Ganguly B, Hsieh S, Latek R, Kulbokas EJ et al (2014) Transcriptional profiling of belatacept and calcineurin inhibitor therapy in renal allograft recipients. Am J Transplant (Official Journal of the American Society of Transplantation and the American Society of Transplant Surgeons) 14:1912–1921

Wang S, Zhang C, Hu L, Yang C (2016) Necroptosis in acute kidney injury: a shedding light. Cell Death Dis 7:e2125

Wu Y, Zhang J, Liu F, Yang C, Zhang Y et al (2013) Protective effects of HBSP on ischemia reperfusion and cyclosporine a induced renal injury. Clin Dev Immunol 2013:758159

Wu Q, Wang X, Nepovimova E, Wang Y, Yang H et al (2018) Mechanism of cyclosporine A nephrotoxicity: oxidative stress, autophagy, and signalings. Food Chem Toxicol 118:889–907

Yang B, El Nahas AM, Thomas GL, Haylor JL, Watson PF et al (2001) Caspase-3 and apoptosis in experimental chronic renal scarring. Kidney Int 60:1765–1776

Yang B, Johnson TS, Haylor JL, Wagner B, Watson PF et al (2003) Effects of caspase inhibition on the progression of experimental glomerulonephritis. Kidney Int 63:2050–2064

Yang B, El Nahas AM, Fisher M, Wagner B, Huang L et al (2004) Inhibitors directed towards Caspase-1 and -3 are less effective than pan caspase inhibition in preventing renal proximal tubular cell apoptosis. Nephron Exp Nephrol 96:E39–E51

Yang B, Jain S, Pawluczyk IZ, Imtiaz S, Bowley L et al (2005) Inflammation and caspase activation in long-term renal ischemia/reperfusion injury and immunosuppression in rats. Kidney Int 68:2050–2067

Yang B, Harris KP, Jain S, Nicholson ML (2007) Caspase-7, Fas and FasL in long-term renal ischaemia/reperfusion and immunosuppressive injuries in rats. Am J Nephrol 27:397–408

Yang B, Hosgood SA, Harper SJ, Nicholson ML (2010) Leucocyte depletion improves renal function in porcine kidney hemoreperfusion through reduction of myeloperoxidase + cells, caspase-3, IL-1beta, and tubular apoptosis. J Surg Res 164:e315–e324

Yang B, Elias JE, Bloxham M, Nicholson ML (2011a) Synthetic small interfering RNA down-regulates caspase-3 and affects apoptosis, IL-1 beta, and viability of porcine proximal tubular cells. J Cell Biochem 112:1337–1347

Yang B, Hosgood SA, Bagul A, Waller HL, Nicholson ML (2011b) Erythropoietin regulates apoptosis, inflammation and tissue remodelling via caspase-3 and IL-1beta in isolated hemoperfused kidneys. Eur J Pharmacol 660:420–430

Yang B, Hosgood SA, Nicholson ML (2011c) Naked small interfering RNA of caspase-3 in preservation solution and autologous blood perfusate protects isolated ischemic porcine kidneys. Transplantation 91:501–507

Yang B, Hosgood SA, Da Z, Harper SJ, Waller HL et al (2012) Biomarkers assessing warm ischemic injury using an isolated porcine kidney hemoreperfusion model. Exp Biol Med 237:1462–1473

Yang C, Jia Y, Zhao T, Xue Y, Zhao Z et al (2013a) Naked caspase 3 small interfering RNA is effective in cold preservation but not in autotransplantation of porcine kidneys. J Surg Res 181:342–354

Yang C, Li L, Xue Y, Zhao Z, Zhao T et al (2013b) Innate immunity activation involved in unprotected porcine auto-transplant kidneys preserved by naked caspase-3 siRNA. J Transl Med 11:210

Yang C, Zhao T, Lin M, Zhao Z, Hu L et al (2013c) Helix B surface peptide administered after insult of ischemia reperfusion improved renal function, structure and apoptosis through beta common receptor/erythropoietin receptor and PI3K/Akt pathway in a murine model. Exp Biol Med 238:111–119

Yang C, Zhao T, Zhao Z, Jia Y, Li L et al (2014) Serum-stabilized naked caspase-3 siRNA protects autotransplant kidneys in a porcine model. Mol Ther J Am Soc Gene Ther 22:1817–1828

Yang C, Hosgood SA, Meeta P, Long Y, Zhu T et al (2015a) Cyclic helix B peptide in preservation solution and autologous blood perfusate ameliorates ischemia-reperfusion injury in isolated porcine kidneys. Transplant Direct 1:e6

Yang C, Zhang C, Zhao Z, Zhu T, Yang B (2015b) Fighting against kidney diseases with small interfering RNA: opportunities and challenges. J Transl Med 13:39

Yang B, Lan S, Dieude M, Sabo-Vatasescu JP, Karakeussian-Rimbaud A et al (2018) Caspase-3 is a pivotal regulator of microvascular rarefaction and renal fibrosis after ischemia-reperfusion injury. J Am Soc Nephrol 29:1900–1916

Zhang Y, Chen W, Wu Y, Yang B (2017) Renoprotection and mechanisms of erythropoietin and its derivatives helix B surface peptide in kidney injuries. Curr Protein Pept Sci 18:1183–1190

Zipfel PF, Skerka C (2009) Complement regulators and inhibitory proteins. Nat Rev Immunol 9:729–740

Zwaini Z, Dai H, Stover C, Yang B (2017) Role of complement properdin in renal ischemia-reperfusion injury. Curr Gene Ther 17:411–423

Chapter 7
How Acute Kidney Injury Contributes to Renal Fibrosis

Li Yang

Abstract Acute kidney injury (AKI) is a widespread clinical syndrome directly associated with patient short-term and long-term morbidity and mortality. During the last decade, the incidence rate of AKI has been increasing, the repeated and severe episodes of AKI have been recognized as a major risk factor chronic kidney diseases (CKD) and end-stage kidney disease (ESRD) leading to global disease burden. Proposed pathological processes and risk factors that add to the transition of AKI to CKD and ESRD include severity and frequency of kidney injury, older age, gender, genetics and chronic health conditions like diabetes, hypertension, and obesity. Therefore, there is a great interest in learning about the mechanism of AKI leading to renal fibrosis, the ultimate renal lesions of CKD. Over the last several years, a significant attention has been given to the field of renal fibrosis with impressive progression in knowing the mechanism of renal fibrosis to detailed cellular characterization and molecular pathways implicated in tubulointerstitial fibrosis. Research and clinical trial are underway for emerging biomarkers detecting early kidney injury, predicting kidney disease progression and developing strategies to efficiently treat AKI and to minimize AKI progression to CKD and ESRD. Specific interventions to prevent renal fibrosis are still experimental. Potential therapeutic advances based on those molecular mechanisms will hopefully offer promising insights into the development of new therapeutic interventions for patients in the near future.

Keywords Acute kidney injury · Fibrosis · Maladaptive repair · Proximal tubular cell · Inflammation

---

L. Yang (✉)
Renal Division, Department of Medicine, Peking University First Hospital, Beijing, China
e-mail: li.yang@bjmu.edu.cn

Key Laboratory of Renal Disease, Ministry of Health of China, Beijing, China

Key Laboratory of Chronic Kidney Disease Prevention and Treatment (Peking University), Ministry of Education of China, Beijing, China

B.-C. Liu et al. (eds.), *Renal Fibrosis: Mechanisms and Therapies*,
Advances in Experimental Medicine and Biology 1165,
https://doi.org/10.1007/978-981-13-8871-2_7

7.1 Introduction

Acute kidney injury (AKI) is a syndrome of reduced glomerular filtration rate (GFR) and reduced urine production, which can be caused by a large number of diseases, treatment side effects, and complications. Till this day, not a single therapy has been proven to recuperate the outcome of AKI. Traditionally, the renal function decline after AKI has been believed to be a reversible process, the resolution of AKI and the improvement of renal function and parameter have generally been considered to be efficient processes, and that in surviving patients, AKI would have no effect on long-term renal function. Owing to numerous recent studies, this ideology has changed now and its shown that there is a strong connection between AKI and the development to chronic kidney disease (CKD) (Basile et al. 2012).

Apart from reducing kidney function acutely, many diseases can damage the cells in the kidney, which has the ability to repair after mild injury, but if repair mechanisms are disordered, or if the injury-causing stimulus continues, severe damage results in scarring (fibrosis) and an associated progressive loss of kidney function.

Before the last decade, the exact mechanisms involved in CKD progression after AKI were still largely unknown and several different hypotheses were proposed, as this topic has acquired immense interest and studied extensively, now, there is growing evidence that AKI may lead to renal fibrosis when the adaptive responses and dedifferentiation go wrong after the injury (Ferenbach and Bonventre 2015). It is now been largely recognized that AKI is better prevented than being treated lately. As it is now largely known that even the milder form of AKI has an adverse outcome and may progress to renal fibrosis which is final common pathways to various terminal kidney diseases like CKD and ESRD. There is developing confirmation that patients who had survived a scene of AKI, even with clear recuperation in renal function, will have a significant future risk of progressing to chronic kidney disease (CKD) (Chawla et al. 2011; Chawla and Kimmel 2012; Leung et al. 2013). Albeit different clinical factors, for example, preexisting illness and old age assume a part, the severity of AKI is by all accounts the most critical risk factor for future CKD (Heung and Chawla 2014). Quantitative appraisal of the AKI severity may recognize patients who are at a high risk of AKI-CKD transition, so that proper follow-up and secondary prevention may be implemented or studied in this population (Farris and Alpers 2014).

7.2 Epidemiology of AKI to CKD Transition

There is a strong epidemiological indication of a causal clinical connection between the clinical syndrome of acute kidney disease (AKI) and the consequent development of chronic kidney disease (CKD) (Coca et al. 2009; Hsu 2012). AKI and CKD are an interconnected clinical syndrome as AKI leads to worsening of CKD and CKD predisposes to the clinical entity of AKI. The tubules of the kidney play an integral role in the fibrotic response, which ultimately leads to progressive kidney disease. CKD is gaining in prevalence worldwide and has a strong negative effect on an individual's quality of life both physically and mentally, apart from increasing the morbidity and mortality. In the course of the most recent two decades, the occurrence of CKD has increased more than three times; furthermore, as stated by the World Health Organization (WHO), it will be one of the three most common reasons for death and disability worldwide by 2020 (Lozano et al. 2012). A meta-analysis of observational data demonstrates that AKI increases the risk of CKD by almost eightfold (Coca et al. 2012).

A study in developed nations (USA, Canada, Western Europe, and Australia) estimated 1.5 million AKI survivors per year with approximately 15–20% who progress to advanced CKD within 24-month period resulting in approximately 300,000 cases of advanced CKD per year (Chawla and Kimmel 2012). According to a nationwide survey in China, approximately 3 million adult AKI patients are admitted to the hospitals across the country every year and of whom 50% develop CKD, which would result in 1.5 million cases of CKD per year (Chawla and Kimmel 2012).

The US Centers for Disease Control and Prevention project that 47% of 30-year-olds will develop CKD during their lifetime. Eleven percent of individuals with stage 3 CKD will eventually progress to end-stage renal disease (ESRD), requiring dialysis or kidney transplantation. CKD is also one of the strongest risk factors for cardiovascular disease. The most important AKI-CKD progressing factor identified by the public health system is lack of post-AKI period assessment guidelines, poor follow system, and cost factor.

7.3 Risk Factors and Etiologies Affecting the Trajectory of AKI to CKD

Developing evidences during the last several years from epidemiological and experimental findings have revealed that AKI itself is an important risk factor for the development of CKD and may also promote the CKD transition to end-stage renal disease (ESRD), including very few of the studies demonstrating the mechanism of AKI to CKD transition (Chawla et al. 2014; Lewington et al. 2013; Venkatachalam et al. 2010). Evidence has suggested that there exists a population heterogeneity in AKI to CKD continuum, various risk factors both extrinsic and intrinsic play an active role, in an excellent review on risk factors of AKI-CKD transition. Hewitson et al. (2017) have explained the concluded important factors affecting the AKI disease progression. Such factors are explained below. [A] Nature. The nature of the injury, in each case, the degree of fibrosis is driven in part by the extent of the tubule damage, and the duration of injury-causing stimuli, with severity and frequency of tubule injury being shown to determine prognosis (Takaori et al. 2016). Consistent with this, the severity of AKI predicts progression to CKD. [B] Age, a progressive reduction in renal function is common with aging, albeit with wide variability. Both hemodynamic and structural changes occur (Ishani et al. 2009; Zhou et al. 2008) and aging rats can be shown to have impaired redox homeostasis (Aydin et al. 2012) and angiogenesis (Kang et al. 2001). These changes predispose older kidneys to new acute organ injury as well as aggravating progression of CKD. [C] Genetics, it is estimated that about 25% of incident dialysis patients have close relatives with CKD (Freedman et al. 2005), and the distinct susceptibilities of different rodent strains to experimental CKD strongly suggest that genetic variations affect renal fibrogenesis (Kokeny et al. 2010). Similarly, familial clustering and disparities in the prevalence of CKD across race suggest a strong genetic component to progression. [D] Gender, many studies have also indicated a gender basis to the progression of senescence and CKD, with epidemiological studies showing that females have a lower prevalence and slower rate of progression than males (Yu et al. 2010). Nevertheless, other investigations suggest that the male dominance is due to damaging effects of testosterone (Baylis 1994; Hewitson et al. 2016) rather than the protective effects of estrogen. In addition, there is an acquired evidence that the pathophysiology, clinical characteristics, and prognosis of the cardiovascular and renal disease are totally unique among men and women, which is quite possible since the physiology of both the opposite gender is different. [E] Controllable risk factors like obesity, type 2 diabetes, and hypertension/ischemic nephropathy which are also considered the most common causes of CKD in developed nations are also an important factor which may complicate AKI and recovery from AKI (Goldstein et al. 2013; Moonen et al. 2018). There are numerous causes of AKI, also differing in frequency based on age, gender, environment, and geographical regions. But the most common one is recognized in three categories, such as prerenal associated with hypovolemia, renal acute tubular necrosis, and post-renal acute urinary obstruction. In many patients, AKI presents with an integrated etiology, often having the coexistence of sepsis, ischemia–reperfusion injury (IRI), and exposure to the nephrotoxic drugs. Although the initial injury mechanism is different for AKI with different clinical and biological characteristics, it is generally accepted that all primary causes of CKD share a common pathogenetic path to progressive injury leading to the destructive consequences of scarring (fibrosis) (Nogueira et al. 2017). In a very interesting review, Hultström et al. describe the in-common mechanisms behind tissue damage in AKI caused by different underlying diseases. Comparing six high-quality microarray studies of renal gene expression after AKI in disease models (gram-negative sepsis, gram-positive sepsis, ischemia–reperfusion injury, malignant hypertension, rhabdomyolysis, and cisplatin toxicity) identified 5254 differentially expressed genes in at least one of the AKI models; 66% of genes were found only in one model, showing that there are unique features to AKI depending on the underlying disease. There were in-common features in the form of four genes that were differentially expressed in all six models, 49 in at least five, and 215 were found in common between at least four models (Hultstrom et al. 2018).

7.4 The Proposed Mechanisms of Renal Fibrosis

Acute kidney injury in most circumstances is multifactorial in origin, whether it is associated with ischemia–reperfusion injury, sepsis caused by various infections, toxins or autoimmune disease. During AKI several distinct pathophysiological processes occurs at the same time and in sequence including microcirculatory dysfunction, endothelial cell dysfunction, formation of microvascular thrombi, inflammation and recruitment of different leukocytes and cytokines, tubular cell injury, tubular and renal venous congestion (Bellomo et al. 2017; Sharfuddin and Molitoris 2011). When it comes to AKI, tubular epithelial cell injury and peritubular interaction have great significance for renal fibrosis. Tubular epithelial cells injury occurs early during ischemia and involves alterations to the cytoskeleton and in surface membrane polarity. ATP depletion causes rapid disturbance of the actin cytoskeleton structure, which perturbs tight junctions and in turn leads to back leak of the tubular filtrate. Loss of cell–cell interaction and cell adhesion molecules results in flattened nonpolarized epithelial cells, denuded basement membranes, and expression of mesenchymal markers. Na^{+}/K^{+} -ATPase pumps, normally located at the basolateral membrane and tethered by the actin–spectrin cytoskeleton, redistribute to the apical membrane of the proximal tubular cell and are internalized into the cytosol during ischemic injury. Morphologically, proximal tubular cells lose their brush borders, undergo swelling, and blebbing of microvilli during injury, leading to cast formation (Sharfuddin and Molitoris 2011).

Under ideal conditions, injured proximal tubular cells undergo differentiation and subsequent re-epithelization. Recovery of proximal tubular cells begins with integrin reattachment, reassembly of the actin cytoskeleton, repolarization of the surface membranes, and redistribution of the sodium pumps back to their basolateral location and renal function recovers completely after an episode of AKI. The reversal of the pathophysiological process of AKI determines the timeline and the trajectory of renal recovery. However, it is observed under many circumstances that the severe forms of AKI or AKI on previously compromised renal function lead to epithelial dedifferentiation inducing profibrotic signals result in maladaptive repair and chronic kidney disease (CKD). Risk factors for maladaptive repair include increasing age, reduced baseline renal function, and prolonged duration and severity of AKI (Canaud and Bonventre 2015). If the repair process is maladaptive (which will be discussed in detail below) under the influence of profibrotic signals, matrix proteins accumulate in glomerulus which is termed as glomerulosclerosis, whereas tubulointerstitial fibrosis (TIF) is the presence of matrix protein replacing the tubules and/or the surrounding interstitium. Many of the renal injuries may involve both the glomeruli and tubular compartments, but the best predictor of renal survival in CKD of all the etiologies is the amount of TIF in kidney biopsies (Nath 1992).

Recently in laboratory science, several animal models were studied for the pathophysiology and therapy of AKI and for the prevention to renal fibrosis, although no animal model could exactly mimic all the alteration caused by progression of renal disease in humans such as renal fibrosis, because the factors associated with it are not completely understood; hence, the extrapolation of outcomes from animal models must be done very carefully. Most commonly investigated models in animals are ischemia/reperfusion injury induced by clamping of both the renal pedicles; other less commonly conducted models include toxic injury models, for example, cisplatin and folic acid injury and a sepsis model using cecal ligation and puncture, these models have given us some insights into the causal pathways and mechanism of renal fibrosis and therapeutics options (Nogueira et al. 2017; Schmiedt et al. 2016).

Over the last several years, with the help of the knowledge gained from this animal model's research experiments, many different theories and mechanism have been proposed on this abnormal recovery of AKI which may lead to renal fibrosis if the disease progresses.

The hypothesis of epithelial–mesenchymal transition (EMT) states that after injury, the kidney tubular cells undergo a phenotypic transition characterized by loss of epithelial markers and gain of mesenchymal features. Liu proposed four key events occur in tubular EMT in renal fibrogenesis, based on the studies of tubulointerstitial changes following experimental unilateral ureteric obstruction (UUO) in vivo and on the response of epithelial cells in vitro to TGFβ1, an inducer of EMT. These are the following events: loss of epithelial adhesion, cytoskeletal reorganization, de novo synthesis of α-SMA, disruption of the tubular basement membrane, and finally enhanced cell migration and invasion of the interstitium (Liu 2004, 2010). However, such concept has been debated; recent studies have demonstrated that genetic cell lineage tracing could not confirm a direct contribution of epithelial cells to the myofibroblast population in the fibrotic kidney (Kriz et al. 2011).

The idea of EMT was again supported by two studies and offered new insights into the potential role of tubular EMT in the development of renal fibrosis. These studies found out that tubular epithelial cells undergo a partial EMT; during the progression of the disease, the cell expresses the markers for both epithelial and mesenchymal cells and remains associated with the basement membrane. Lovisa et al. have further demonstrated that this partial EMT is sufficient to induce tubule function impairment through the arrest of cell cycle in the G2 phase leading to an abnormal recovery of tubular epithelial cells which will further progress to renal fibrosis (Lovisa et al. 2016).

Endothelial-to-mesenchymal transition (EndMT) is a subtype of EMT where specific epithelial cells known as endothelial which are a single layer of cells lining vessels serving as a permeable membrane transition into a mesenchymal cell type. EndMT has also been described in several publications as a contributor to the development and progression and renal fibrosis (Cruz-Solbes and Youker 2017). Zeisberg et al. carried out a landmark experiment to confirm the contribution of EndMT in renal fibrosis in three different types of animal models: unilateral ureteral obstruction nephropathy UUO, streptozotocin-induced diabetic nephropathy, and a surrogate for Alport syndrome (Zeisberg et al. 2008). The result showed that approximately 30–50% of fibroblasts co-expressed the endothelial marker CD31 and mesenchymal markers FSP-1 and αSMA; they confirmed this finding with endothelial lineage tracing using dual transgenic mice. Li et al. also confirmed that EndMT occurs and contributes to the early development of diabetic renal interstitial fibrosis. They did so by looking at a streptozotocin-induced diabetic nephropathy model using endothelial lineage-traceable mice to identify if EndMT occurred in diabetic renal interstitial fibrosis (Li et al. 2009).

However, recent development in the understanding of renal fibrosis has come up with new and reasonable evidence for the mechanism responsible for normal and abnormal repair and drew an important link between injury, abnormal repair, and development of fibrosis, stating that fibrosis is a characteristic of maladaptive repair (Fiorentino et al. 2018) that can result when the regenerating tubules become growth arrested, fail to differentiate, and undergo atrophy, these tubules also exhibit pathological signaling, paracrine, and autocrine activity that perturbs normal interaction between different renal and renal interstitial cells. There is a significant cellular cross talk between tubular epithelial–epithelial cell, tubular epithelial–fibroblast, epithelial–endothelial cells, and epithelial–inflammatory cells. One of the main changes is the interaction of pericytes and peritubular capillary endothelium. The maladaptive process results in dissociation of pericytes from capillaries is leading to capillary disintegration and microvascular rarefaction, pericyte to myofibroblast transformation, endothelial dysfunction, chronic inflammatory infiltrates, renin–angiotensin system (RAS) activation, mitochondrial dysfunction and epigenetic changes, all the pathological changes ultimately causing renal fibrosis (Chou et al. 2017). Maladaptive repair and progression of tubulointerstitial fibrosis is a complex process that involves several different types of cells and molecular pathways. Renal biopsies from the human with AKI and also CKD patients offer us a snapshot in time and usually shows all these pathological changes at once, making it difficult to decide the relative importance of each cellular compartment and sequence of events leading to renal fibrosis. In recent years a lot of attention was focused on myofibroblast and its origin (Chang et al. 2012), as this interstitial cell is thought to be the main producers of ECM, but new findings have highlighted the role of the proximal tubule and proximal tubular epithelial cell (PTEC) a specialized tubular segment just adjacent to the glomerulus, as not only the target of injury but also an important mediator of renal fibrosis progression (Bonventre 2014).

In the below section, we will discuss the maladaptive changes in cells and molecular mechanism implicated in tubulointerstitial fibrosis

7.5 The Maladaptive Repair

7.5.1 The Role of Proximal Tubular Epithelial Cells in Maladaptive Repair

The tubulointerstitium consists of multiple cell components including tubular epithelial, mesenchymal (fibroblast and pericytes), endothelial, and inflammatory cells all of which contribute to fibrosis progression.

Researches done over a period of several years have made clear that proximal tubular epithelial cells (PTECs) play an important role in the histopathology of renal injury, both in AKI and CKD. After an acute ischemic or toxic insult, PTECs (mostly those of proximal S3 segment) are most susceptible to injury (Bonventre and Yang 2011). Ischemia–reperfusion injury and drug-induced renal injury lead to a significant increase in production of reactive oxygen species (ROS) heavily contributing to PTEC injury (Granger and Kvietys 2015; Hosohata 2016). The PTECs are highly sensitive due to (i) the high metabolic activity/demand of these cells; (ii) medullary region is highly prone to hypoxia; and (iii) increased exposure to intra-tubular toxins due to upstream water absorption (Canaud and Bonventre 2015). The innate immune features of PTECs facilitate them to act as immune responders to a wide range of insults, with the subsequent production and release of bioactive mediators that drive the interstitial inflammation.

Senescent epithelial cells and G2/M cell cycle arrest of epithelial cells is an important mediator of CKD. Senescence is a cell fate decision when cell cycle arrest becomes permanent. Although this is generally the case, some senescent cells that do not express p16INK4a can resume growth after inactivation of p53 (Beausejour et al. 2003). Senescence has been studied extensively in response to DNA damage where some cell types undergo apoptosis, whereas others, particularly epithelial cells, undergo a senescence response. Besides serious levels of DNA damage, high levels of reactive oxygen species (ROS) are considered a major cause of senescence induction (Munoz-Espin and Serrano 2014; Sturmlechner et al. 2017). DNA damage leads to the activation of the ATM/ATR pathway, following CHK2 activation which phosphorylates cdc25 and p53, causing the G2/M arrest by CDK1 inhibition (Goodarzi et al. 2003; Yan et al. 2016). An important finding validating the involvement of G2/M arrest in renal injury was reported by Yang et al. in 2010 (Yang et al. 2010). By inspecting tubular epithelial cell cycle behavior after renal injury, they noted a causal association between epithelial cell cycle arrest at the G2/M phase and subsequent development of renal fibrosis due to maladaptive repair. They characterized the cell cycle profile of PTECs after an acute insult in five experimental mouse models of AKI: moderate bilateral ischemia–reperfusion injury (IRI), severe bilateral IRI, unilateral IRI, aristolochic acid nephropathy (AAN), and unilateral ureter obstruction (UUO). These models resemble the three most common causes of AKI seen in humans: ischemia, toxic exposure, and obstruction (Le Clef et al. 2016). All animal models, except moderate IRI, showed that the injury led to the development of severe fibrosis demonstrating the chronic fate of the kidney. They also found that many tubular cells arrested in the G2/M phase of the cell cycle result in activation of the DNA repair response with the resultant production and secretion of profibrotic factors, such as transforming growth factor beta (TGFβ), Collagen 1 (Col I), and CCN2, formerly known as connective tissue growth factor (CTGF) (Moonen et al. 2018). Mounting evidence has also shown that the injured PTEC that is transformed into a secretory phenotype and can recruit a variety of proinflammatory cytokines that drive the immune response in an autocrine manner or indirectly through infiltrating leukocyte in a paracrine manner, thus causing tubulointerstitial inflammation. A variety of cytokines and profibrotic factors have been shown to be produced by activated TECs, including IL-1b, IL-18, IL-15, IL-16, TNF-a, TWEAK, Fas ligand, CTGF, and vascular endothelial growth factor, platelet-derived growth factor (PDGF). Proinflammatory cytokines and profibrotic growth factor produced or recruited by injured TEC are listed in Table 7.1.

Transforming growth factor-β1 (TGF-β1) is one of the most crucial soluble mediators of renal fibrosis. Both angiotensin II exposure and Snail1 overexpression induce TGF-β1 production in tubular cells (Grande et al. 2016; Macconi et al. 2014), particularly when TEC is under arrest in the G2/M phase. Elevated TGF-β1 expression leads to a variety of paracrine and autocrine effects on target cells. TGF-β1 is a potent mitogen for fibroblasts and causes myofibroblast transformation (Meng et al. 2015). TGF-β1 also causes tubular cell apoptosis and induces the EMT program, leading to the production of an increased matrix. It also causes tubular cell hypertrophy, which may be closely linked to increased matrix production (Gewin et al. 2012; Lopez-Hernandez and Lopez-Novoa 2012). Stress-induced activation of the c-Jun amino-terminal kinase (JNK) pathway in cells of the glomerular and tubulointerstitium is a common feature of chronic kidney disease. JNK signaling acts to increase TGF-β1 expression, to promote activation of latent TGF-β1, and to promote transcription of profibrotic molecules via direct phosphorylation of the linker region of Smad3 (Grynberg et al. 2017). Ample progress has been made in identifying miRNA molecules which regulate TGF-β/Smad3 induced fibrosis. TGF-β specifically promotes fibrosis by increasing levels of miR21, miR-433, and miR-192 which amplify TGF-β signaling and stimulate dedifferentiation of tubular epithelial cells (Chung and Lan 2015). One of the main therapies for chronic kidney disease is inhibition of the production or action of angiotensin II (Ang II) as Ang II is an inducer of TGF-β production and activation (Macconi et al. 2014). Autocrine TGF-β signaling increases PTEC production of PDGF-β and CTGF/CCN2 that can signal neighboring fibroblast and induce TIF (Geng et al. 2012). CTGF is an essential cofactor for TGF-β signaling. CTGF interacts directly with the TGF-β receptor, downregulation of Smad7 activity, and inhibition of bone morphogenetic protein 7. CTGF can also bind LRP6, stimulating β-catenin activity (Kok et al. 2014). CTGF also promotes inflammatory cell infiltration in the renal interstitium by activating NF-kappaB pathway (Rodrigues-Diez et al. 2015). PDGF is another profibrotic growth factor produced by the epithelial cell that interacts with fibroblast. PDGF-β produced by injured TEC may increase PDGF-receptor phosphorylation and consequent signaling on neighboring fibroblast promotes TIF (Tang et al. 1996). The intrarenal renin–angiotensin system activation plays an important role in CKD progression; there is extensive research available that the major fraction of Ang II in renal tissues is produced from angiotensinogen by PTECs and it is converted to Ang I (Kobori et al. 2007). RAS production in the kidney has also shown to be Wnt/β-catenin dependent (Zhou et al. 2015). As discussed above, Ang II can stimulate TGF-β expression in cultured murine PTECs to promote inflammatory and fibrotic responses. Wnt/β-catenin signaling is a pathway involving the renal recovery from AKI. In the acute phase of injury, Wnt/β-catenin is likely to be protective. A lot of nephrotoxic events lead to marked accumulation of β-catenin in renal tubules (Zhou et al. 2016). In both renal ischemia/reperfusion injury and folic acid nephropathy, tubule-specific ablation of β-catenin has been shown

Table 7.1 Proinflammatory cytokines and profibrotic growth factors induced by injured TEC

Cytokines/factors	Effects	References
IL-6	Proinflammation	Yard et al. (1992)
IL-18	Triggers proinflammatory cytokines	Yang et al. (2015)
IL-34	Neutrophil and macrophage recruitment	Baek et al. (2015)
IL-6	Proinflammation	Yard et al. (1992)
IL-15	$CD103^+$ T-cell recruitment	Wong et al. (2003)
IL-1β	Triggers proinflammatory cytokines and initiates acute-phase responses	Leemans et al. (2014) Anders (2016)
IL-16	$CD4^+$ T-cell recruitment	Wang et al. (2008)
CSF-1	Macrophage recruitment and adhesion	Menke et al. (2009)
	Polarization into an M2 phenotype	Wang et al. (2015) Huen et al. (2015)
TNF-α	Triggers proinflammatory cytokines innate and adaptive immunity apoptosis	Al-Lamki and Mayadas (2015)
Fas ligand	Apoptosis	Lorz et al. (2000)
TWEAK	Cell death in the presence of TNF-α/IFN-γ	Sanz et al. (2014)
	Proinflammation	Sanz et al. (2011)
VEGF	Macrophage recruitment	Schrijvers et al. (2005)
	Endothelial cell proliferation and survival	Ninichuk et al. (2006)
CTGF	Triggers proinflammatory cytokines	Rodrigues-Diez et al. (2015)
	Profibrotic fibroblast proliferation	Geng et al. (2012)
PDGF	Profibrotic	Kimura et al. (2005)
	Fibroblast proliferation Fibroblast to myofibroblast differentiation	Chen et al. (2011)
TGF-β	Profibrotic	Wu et al. (2013)
	Tubular cell dedifferentiation	Meng et al. (2015)
	Fibroblast to myofibroblast differentiation	Lamouille et al. (2014)
Hh	Profibrotic	Fabian et al. (2012)
	Fibroblast to myofibroblast differentiation	Ding et al. (2012)
	Pericyte to myofibroblast differentiation	Zhou et al. (2014)
Notch	Profibrotic	Murea et al. (2010)

(continued)

Table 7.1 (continued)

Cytokines/factors	Effects	References
	Epithelial–mesenchymal transition	Bielesz et al. (2010)
Wnt	Fibroblast to myofibroblast differentiation	Maarouf et al. (2016)
	RAS↑	Tan et al. (2014)
RAS components	Proinflammatory	Liu et al. (2006)
	Triggers growth factors (TGF-b, CTGF)	Wolf et al. (1999) Chen et al. (2006)
ET-1	Proinflammatory	Gerstung et al. (2007)
	Profibrotic	Gomez-Garre et al. (2001) Zager et al. (2013)
Exosome	Fibroblast to myofibroblast differentiation	Borges et al. (2013)
	Epithelial–mesenchymal transition	Zhou et al. (2013b)
Complement	Epithelial–mesenchymal transition	Tang et al. (2009)
	RAS activation	Zhou et al. (2013a)

to exacerbate kidney injury by increasing TEC apoptosis (Zhou et al. 2012). However, persistent activation of Wnt signaling has a decisive role in driving AKI-CKD progression because sustained Wnt signaling causes uncontrolled fibroblast activation, RAS activation, inflammation, and excessive deposition of ECM (Maarouf et al. 2016). The relationship between mitochondrial dysfunction and CKD has been long suspected, but the exact mechanism connecting mitochondrial dysfunction to CKD has remained indefinable (Galvan et al. 2017). Also, mitochondrial dysfunction is thought to be pathogenic in AKI (Che et al. 2014; Hall and Schuh 2016). The proximal tubules are densely packed with mitochondria; PTEC generates ATP mainly via mitochondrial oxidative phosphorylation, whereas other renal cells like endothelial cells, podocytes, and mesangial cells exhibit more flexibility in their glycolytic capacity to generate energy (Wirthensohn and Guder 1986). Under stressful condition, mitochondrial fragmentation and permeability alteration contribute significantly to tubular dysfunction and cell death. Funk and Schnellmann demonstrated persistent disruption of mitochondrial homeostasis after AKI, which in turn may result in decreased cellular respiration accompanied by a reduction in cellular adenosine triphosphate, all contributing to the development of ensuing CKD (Funk and Schnellmann 2012). Plenty of evidence suggests that a mitoprotective drug like SS-31 can target the mitochondrial membrane to prevent the permeability transition and is very effective in diminishing ischemic AKI and the development of interstitial fibrosis (Liu et al. 2014).

7.5.2 *The Role of Endothelial Dysfunction in Maladaptive Repair*

AKI-induced endothelial injury has long-term chronic disease implications. Basile and colleagues documented that renal vascular network is significantly compromised following acute ischemic injury (Basile 2007). Hörbelt M and colleagues have also confirmed this finding; they found a nearly 45% drop in vascular density 4 weeks after an ischemic insult. This reduction in capillary vascular density is termed as a capillary rarefaction (Horbelt et al. 2007). Capillary rarefaction in the kidneys is thought to promote hypoxia, impair hemodynamic responses, and predispose to chronic kidney disease (CKD) progression and hypertension development. Consistent with decreased renal oxygenation in CKD is the increased expression of the oxygen-sensitive α-subunit of hypoxia-inducible factor (HIF). These heterodimeric basic helix-loop-helix transcription factors HIF-1 and HIF-2 are key mediators mediating a cellular adaptation to hypoxia by regulating glycolysis, angiogenesis, erythropoiesis, and cell survival decisions. HIF-1 and HIF-2 are furthermore involved in the regulation of biological processes that are relevant to wound healing, tissue repair, and fibrogenesis, such as extracellular matrix synthesis and turnover, cell adhesion and migration, and epithelial to mesenchymal transition (EMT) (Haase 2012). There is conflicting opinions regarding the effect of HIF on CKD pathophysiology, which seems to depend on the pathological discourse. At one point, stimulation of HIF and HIF-regulated genes by CoCl2 has been shown to exert renoprotective role in the hypoxic tubulointerstitium in rats with nephritis (Tanaka et al. 2005) and hypertensive type 2 diabetes (Ohtomo et al. 2008).

At another point, an inappropriate and prolonged activation of HIF is well known to play a pivotal role in initiating and promoting renal fibrogenesis via regulation of multiple signaling pathways in CKD. HIF activation can stimulate inflammatory cell proliferation and recruitment to the site of injury in animal models of CKD, which plays a role in setting up the fibrous scar formation. Additionally, activated HIF binds to its profibrogenic downstream genes and induces maladaptive expression of matrix modifying factors directly in hypoxic TECs, such as collagen I, plasminogen activator inhibitor 1 (PAI1), endothelin-1 (ET-1), connective tissue growth factor (CTGF), matrix metallopeptidase 2 (MMP-2), and tissue inhibitor of metalloproteinase 1 (TIMP1), which lead to increased production of interstitial collagen and decreased degradation of ECM (Liu et al. 2017).

There is a bidirectional relationship between tubular epithelial cells and capillary endothelial cells. Primary tubular epithelial cell injury promotes capillary rarefaction (Bonventre 2012) and capillary rarefaction further promotes hypoxic tubular cell injury, thus creating a vicious circle. There are several examples in which tubular injury precedes capillary rarefaction. Additionally, capillary rarefaction decreases tubular blood and oxygen supply, promoting the loss of tubular cell viability and tubular atrophy and interstitial fibrosis (Kida et al. 2014). Various mechanisms have been suggested to play a role in the development of capillary rarefaction; the very first hypothesis proposed nearly 20 years ago was that excessive collagen production by myofibroblasts reduces peritubular blood flow, causing tubular hypoxia and nephron dropout (Fine et al. 1998); following this hypothesis, a huge interest is shown in this field, and various other mechanisms were inflammation, an altered endothelial-tubular epithelial cell cross talk, a relative deficiency in angiogenic growth factors, increased activity of TGF-β1 and thrombospondin-1, loss of pericytes, vitamin D deficiency, a link to lymphatic neoangiogenesis and INK4a/ARF (cyclin-dependent kinase inhibitor 2a; CDKN2A) (Afsar et al. 2018).

Basile et al. have also shown that acute ischemia decreases the expression of angiogenic vascular endothelial growth factor (VEGF) while raising the expression of the VEGF antagonist, a disintegrin, and metalloproteinase with thrombospondin motif 1 (ADAMTS-1) (Basile et al. 2008). By contrast, administration of a VEGF fragment (VEGF-121) during repair preserved microcapillary density, indicating that VEGF agonism may be a therapeutic approach to prevent AKI-induced capillary rarefaction (Leonard et al. 2008). Pericyte–endothelial interactions also play an important role in capillary rarefaction. Pericytes produce angiopoietin-1, a growth factor that stabilizes the microvasculature by activating the endothelial Tie2 receptor. After renal injury, endothelium-derived angiopoietin-2, an antagonist of angiopoietin-1, increases, favoring capillary leakiness and pericyte loss (Tsai et al. 2014). Pericyte disintegration and loss lead to structural instability of blood vessels and to capillary rarefaction. Furthermore, detached pericytes are key precursors of myofibroblasts. Pericytes-turned myofibroblasts contribute to interstitial fibrosis that leads to further capillary rarefaction (Kramann and Humphreys 2014).

During hypoxia, TGF-β1 directly causes endothelial cell apoptosis and capillary pruning finally leading to vascular dropouts (Ballermann and Obeidat 2014). Thrombospondin-1 could enhance the fibrotic response by both activating TGF-β and employing antiangiogenic actions, thus leading to capillary rarefaction (Gewin et al. 2017). Inhibition of thrombospondin expression ameliorated tubulointerstitial fibrosis by promoting VEGF production and restoring peritubular capillary density.

7.5.3 The Role of Inflammatory Cells in the Maladaptive Repair

The interstitial inflammatory cells specifically macrophages and dendritic cells are important modulators of CKD. Generally, the macrophages and dendritic cells populate the uninjured renal interstitium, act as an antigen-presenting cell, and also carry out their phagocytic function; these inflammatory cells expand both through local proliferation and infiltration of circulating monocytes and subsequent differentiation after the renal injury. This persistent leukocyte accumulation and activation inside the kidney would promote extended periods of ischemia due to vascular congestion and may induce direct tubular and endothelial cell damage by the release of inflammatory mediators (Nelson et al. 2012; Weisheit et al. 2015).

Macrophages and dendritic cells in the injured kidney display appreciable plasticity and functional heterogeneity with different overlapping roles. These cells are identified with different markers by multicolor flow cytometry, for example, macrophages (CD11b, F4/80, and CD68) and dendritic cells (CD11c, MHCII, and CD80/86) (Weisheit et al. 2015). Macrophages can be classified into many subtypes, but they are broadly classified either as M1 (classically activated) or M2 (alternatively activated) macrophages. During AKI, M1 macrophages are known as proinflammatory and make cytokines such as IL-1, IL-6, and TNF-a, and they promote inflammation whereas M2 are mainly reparative and anti-inflammatory, they express arginase, mannose receptor, IL-10, and IL-4 receptor-a. However, in CKD M2 macrophages may stimulate tubulointerstitial fibrosis through the production of various profibrotic growth factors. A variety of experimental cell depletion strategies have shown that decreasing the number of interstitial macrophages reduces kidney fibrosis (Mosser and Edwards 2008).

Recent reviews and studies have shown that there is a significant cross talk between injured TEC and inflammatory cells. Injured renal epithelia induce the inflammatory response by the production of chemokines and chemoattractant cytokines that bind to receptors on inflammatory cells to promote migration of macrophages and dendritic cells to the site of injury. Injured proximal tubules can also promote local macrophage proliferation through the production of macrophage colony-stimulating factor (Gewin et al. 2017).

The macrophage phenotype, the timing relative to PTEC injury, and local microenvironment affect the epithelial/inflammatory cell cross talk leading to progression of CKD. Animal models of AKI, a risk factor for CKD progression, suggest that a proinflammatory, M1 macrophage phenotype predominates at 1–3 days post-injury. Depleting macrophages using clodronate or genetic methods may improve injury in this phase. However, at later time points, there is a switch to a reparative, M2 macrophage subtype that promotes recovery from injury. Depletion of macrophages at this later stage can impair renal recovery and lead to CKD (Vinuesa et al. 2008). In CKD, unlike AKI, the M2 phenotype is likely to promote TIF due to the production of factors like TGF-β, PDGF, and galectin-3 (Henderson et al. 2008). Therefore, a continuous epithelial injury may affect TIF progression through chemokine-dependent, increased M2 macrophage infiltration. Although injured epithelia also alter DC polarization through cytokine production, there are fewer data to support a role for DCs in TIF progression (Kitching 2014; Machida et al. 2010). Other inflammatory immune cells like (lymphocytes, natural killer cells) have also shown to play a role in CKD, but there are not many studies to mention here.

7.5.4 The Role of Myofibroblast in the Maladaptive Repair

Tubulointerstitial fibrosis is marked by an exuberant and pathological deposition of extracellular matrix (ECM) consisting of collagen I, III, and IV, laminin, fibronectin, perlecan, and heparin. Research on animal models of kidney disease has identified myofibroblasts as major matrix-producing cells contributing to fibrosis after injury. Myofibroblasts can be defined by being α-smooth muscle actin (α-SMA) positive and are mainly located in the renal interstitium and to a lesser extent in glomeruli in various animal models of renal fibrosis. The amount of myofibroblasts is proportionally related with severity of renal fibrosis. However, the diverse lineage and mixed phenotypic heterogeneity of myofibroblasts make them a difficult therapeutic target. Due to their central importance, the cellular origin of these cells is a critical question that has been extensively studied in animal models, but contradictory data leave this as a subject of an ongoing debate.

Interstitial myofibroblasts have been proposed to originate from one or more of five sources: fibroblasts, pericytes, bone marrow-derived myofibroblast, tubular epithelial–mesenchymal transition (EMT), or endothelial/mesenchymal transition EndMT. However, the role of EMT in renal fibrosis is rigorously challenged by lineage tracing studies and other recent scientific evidence (Sun et al. 2016).

Fibroblasts are inactive cells found within the interstitial space, important for maintaining the structural integrity of kidneys by producing a basal level of the extracellular matrix. Injured epithelia are potent producers of growth factors and cytokines such as TGF-β, PDGF, hedgehog, and Wnt ligands. Injured PTECs have paracrine effects on surrounding cells such as fibroblasts, causing them to transform into myofibroblasts. Activated fibroblasts have increased stress fibers, and they proliferate and produce ECM components such as collagens leading to progressive TIF (Baum and Duffy 2011).

The recent discovery of a functionally distinct subset of extracellular matrix-producing cells called pericytes might be another major source of myofibroblasts. Pericytes are stromal cells which stay in close contact with interstitial capillary endothelial cells and regulate capillary permeability. An interesting study by Kramann et al. identified a group of Gli1-positive perivascular mesenchymal-like pericytes such as myofibroblast progenitors, as shown by lineage tracing and cell ablation. Using transgenic Gli1-CreERt2/tdTomato mice in an experimental model of fibrosis, Gli1þ cells proliferated within the interstitial region and acquired expression of NG2 and α-SMA as myofibroblasts (Kramann et al. 2015).

Myofibroblasts may also originate from bone marrow-derived cells (BMDCs) which migrate into the kidney in response to injury. In the UUO model of kidney fibrosis, transplantation of transgenic bone marrow expressing the red fluorescent protein (RFP) under the control of α-SMA promoter (α-SMA-RFP) demonstrated 35% of the myofibroblasts was derived from bone marrow cells (LeBleu et al. 2013).

Fibrocytes are hematopoietic or bone marrow-derived collagen-producing cells expressing vimentin and CD34 marker; they infiltrate renal parenchymal tissue and may contribute to fibrogenesis. Patients with chronic allograft nephropathy with interstitial fibrosis had a large number of myofibroblasts derived from the recipient. Immunostaining of hematopoietic markers like CD11b, CD34, CD45, CD115, and Gr1, together with intracellular collagen I, was used to identify fibrocytes in kidneys. Specific cell ablation of αSMA-positive fibrocytes after bone marrow transplantation showed a significant reduction of renal fibrosis (Mack and Yanagita 2015).

TGF-β1 is synthesized by injured TEC in the kidneys as latent complexes covalently bound to the extracellular matrix (Huang et al. 2008). Active TGF-β1 is released via proteolytic cleavage from the extracellular matrix and binds to the TGF-βII receptor and starts Smad-dependent and Smad-independent signaling pathways. Phosphorylated Smad2/3 complex together with Smad4 translocates into the nucleus to modulate transcription of a number of target genes. TGFβ1 promotes progressive renal fibrosis by stimulating extracellular matrix synthesis, preventing its degradation, mediating tubular epithelial cells and endothelial cells to undergo EMT or EndoMT, respectively (Li et al. 2009).

7.6 Biomarkers Implicated in AKI—CKD Transition

New biomarkers of AKI are significant recent developments and typically indicates a specific component of AKI pathophysiology, such as epithelial tubular injury, cell cycle arrest, systemic inflammatory pathways, and glomerular filtration. AKI biomarkers will hopefully beneficial in different clinical processes, including diagnosing AKI earlier, prognostication of clinical outcomes, and prediction of response to therapy.

Biomarkers of AKI can be used to inference the short- and long-term adverse outcomes in various patient care settings. The commonly observed short-term outcomes are in-hospital mortality, need for renal replacement therapy (RRT), and length of stay. At present, there are evidences that biomarkers of AKI are related to long-term mortality, but the data are lacking to suggest that biomarkers of AKI are related to other important long-term patient outcomes, such as cardiovascular events and CKD, although prospective studies are ongoing like ASSESS-AKI study. The Assessment, Serial Evaluation, and Subsequent Sequelae of Acute Kidney Injury (ASSESS-AKI) study is prospectively studying the long-term outcomes in hospitalized patients, with or without chronic kidney disease, after an episode of acute kidney injury, to ascertain the natural history of acute kidney injury and define the risk factors for progression and for the development of complications, including cardiovascular disease, and the relationship between putative biomarkers and long-term outcomes (Go et al. 2010). Some of the AKI biomarkers may yield additional predictive information beyond that is provided by the AKI event itself. Several long-term follow-up studies have identified a subgroup of patients who suffer from "subclinical AKI." While these patients do not have AKI as defined by serum creatinine, they have elevated biomarkers of tubular injury and fare worse than patients without elevated biomarkers of AKI (Haase et al. 2011). This proposes that biomarkers of AKI may provide additional prognostic information beyond that offered by serum creatinine and serum creatinine-based current AKI definitions.

7.6.1 Serum and Urine Kidney Injury Molecule 1

KIM-1 expression is markedly upregulated after ischemia–reperfusion injury reflecting the proliferating dedifferentiated epithelial cells of the proximal tubules, and it appears to peak at approximately 48 h. It is known to promote epithelial cell regeneration and regulates tubule cell apoptosis. In murine AKI models, KIM-1 overexpression reduces kidney fibrosis and the development of ESRD (Humphreys et al. 2013). Sustained KIM-1 elevation in blood, however, indicates persistent tubular injury, which would be a risk for the development of CKD/ESRD. Urinary KIM-1 shows a similar correlation of kidney injury; hence, the persistent KIM-1 levels may be used to prognosticate development of ESRD.

7.6.2 Neutrophil Gelatinase-Associated Lipocalin (NGAL)

Intrarenal NGAL is highly upregulated following ischemic or nephrotoxic kidney injury. A urinary NGAL has been proposed as a potential marker of kidney disease progression to ESRD in animal models (Ko et al. 2010). Human studies, though small and few, have shown the possible utility of high baseline levels of plasma and urinary NGAL in predicting progression [reference]. The area under the receiver operating characteristic curve (AUC) is estimated at 0.78 for baseline urine NGAL and 0.70 for baseline serum NGAL (Bolignano et al. 2009), suggesting potential utility in prognostication post-AKI kidney disease progression.

7.6.3 Liver-Type Fatty Acid-Binding Protein

Liver-type fatty acid-binding protein is considered as a renal protective protein in general. It binds to and promotes the metabolism of fatty acids and has antioxidant properties. Urinary L-FABP levels are elevated almost immediately after AKI and peak within 6 h correlating strongly with renal ischemic time (Susantitaphong et al. 2013). L-FABP levels are also increased in patients with known renal disease risk factors of hypertension and diabetes in the absence of overt kidney damage, further improving its potential to be used to identify patients at increased risk.

7.6.4 Interleukin-18

Interleukin-18 (IL-18) is a 22-kD proinflammatory cytokine formed in the proximal tubular cells. Urinary IL-18 is elevated following renal injury (Parikh et al. 2004). The mature IL-18 induces inflammatory response through upregulating NF-kappaB pathway including TNF-α, iNOS, chemokines MCP-1, and MIP-2, which causes inflammation by macrophage and neutrophils induction (Wu et al. 2008). IL-18 worsens tubular necrosis in ischemia–reperfusion injury (Parikh et al. 2004) and animal models via Fas/Fas ligand pathways (Yano et al. 2015). Interrupting the IL-18 signaling has consistently shown to decrease kidney injury.

7.6.5 Renin–Angiotensin System Activation

Renin–angiotensin system (RAS) activation, especially intrarenal activation, causes progression of AKI and transition from acute to chronic kidney injury. Urinary angiotensinogen is considered a novel prognostic marker for AKI. AKI patients with elevated urinary angiotensinogen have been shown to progress to higher stages of AKI with higher mortality rates (Alge et al. 2013b). In patients with post-cardiac surgery AKI, elevated urinary angiotensinogen had an AUC of 0.75 for predicting progression of AKI to stage 3 and predicting mortality (Alge et al. 2013a). Measurement of urinary angiotensinogen can help with delineating AKI patients who are at risk of having accelerated CKD and could possibly benefit from RAS blockade agents.

7.7 Renal Antifibrotic Treatment

Renal fibrosis is the common final outcome of different progressive kidney diseases and, in the majority of chronic kidney diseases, becomes a leading target for antifibrotic interventions. The current therapeutic approaches for the reversal or inhibition of progression of many chronic kidney diseases are either not enough effective or safe to be used as treatment regimen in the clinical practice; because of this reason, drugs to cure kidney fibrosis are not in clinical yet. At present, the available medications for the treatment of renal fibrosis are angiotensin-converting enzyme (ACE) inhibitors (e.g., captopril), angiotensin II receptor blockers (ARBs), and renin inhibitors. They are used only to improve the symptoms and delay the progression of chronic kidney diseases, but this drug also comes with serious limitations. The safety and efficacy of these medications depend on different key factors such as the stage and type of chronic kidney disease, the causal fibrosis leading molecular mechanisms, and the patient medical history. The present research work done on the pathways and mechanisms of fibrogenesis has revealed impressive preclinical evidence showing that fibrosis can be slowed, arrested, or even be reversed. Research indicates that multiple ligands of TGF-β/Smads are the direct mediators for renal fibrosis; hence, inhibition of the TGF-β/Smads signaling pathway using various strategies can significantly reduce renal fibrotic lesions and ameliorate kidney injury, suggesting that targeting the TGF-β/Smads signaling pathway could be a new strategy for effective therapies.

Other drugs that are being developed for CKD-related fibrosis are targeting the key players that are known in the molecular mechanisms of fibrosis such as CTGF, endothelin-1, mothers against decapentaplegic homolog (SMAD) 3 and 4, phosphodiesterase type 5, bone morphogenetic protein (BMP)-7, and NADPH oxidase (NOX) 1 and 4 (Nastase et al. 2018).

7.8 Conclusion

Currently, there are not many therapeutic strategies for preventing progression after AKI. However, staggering advancements have been made in recent years to determine the risk factors, mechanisms, the role of cellular cross talk, molecular pathways involved, maladaptive repair under the influence of proinflammatory cytokines, profibrotic growth factor (Fig. 7.1), and the role of novel biomarkers in AKI-CKD transition. It is highly surprising that the planned clinical follow-up of the patients who survived AKI is very low, given the increasing incidence of AKI and its association with the progression to CKD. More information can be acquired with the proper clinical follow-up of AKI survivors, which can help with short- and long-term clinical trials and research by making a better animal model that can elaborate the understanding of AKI-CKD transition which is ultimately necessary for preventing TIF and developing more effective therapies to halt TIF progression to CKD.

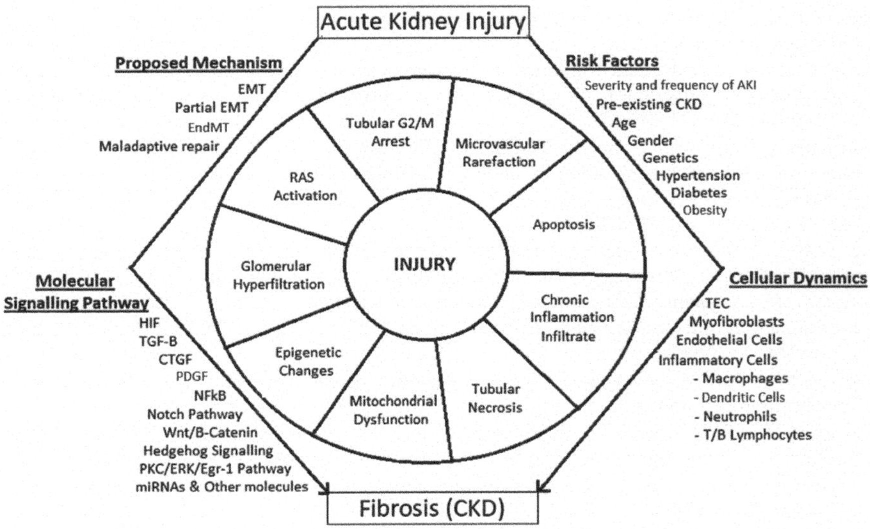

Fig. 7.1 Acute kidney injury to chronic kidney disease transition. Depicting the risk factors, proposed mechanisms, cellular dynamics, and molecular signaling pathway of fibrosis development after acute kidney injury

References

Afsar B, Afsar RE, Dagel T, Kaya E, Erus S et al (2018) Capillary rarefaction from the kidney point of view. Clin Kidney J 11:295–301

Alge JL, Karakala N, Neely BA, Janech MG, Tumlin JA et al (2013a) Association of elevated urinary concentration of renin-angiotensin system components and severe AKI. Clin J Am Soc Nephrol 8:2043–2052

Alge JL, Karakala N, Neely BA, Janech MG, Velez JC et al (2013b) Urinary angiotensinogen predicts adverse outcomes among acute kidney injury patients in the intensive care unit. Crit Care 17:R69

Al-Lamki RS, Mayadas TN (2015) TNF receptors: signaling pathways and contribution to renal dysfunction. Kidney Int 87:281–296

Anders HJ (2016) Of inflammasomes and alarmins: IL-1beta and IL-1alpha in kidney disease. J Am Soc Nephrol 27:2564–2575

Aydin S, Yanar K, Atukeren P, Dalo E, Sitar ME et al (2012) Comparison of oxidative stress biomarkers in renal tissues of D-galactose induced, naturally aged and young rats. Biogerontology 13:251–260

Baek JH, Zeng R, Weinmann-Menke J, Valerius MT, Wada Y et al (2015) IL-34 mediates acute kidney injury and worsens subsequent chronic kidney disease. J Clin Invest 125:3198–3214

Ballermann BJ, Obeidat M (2014) Tipping the balance from angiogenesis to fibrosis in CKD. Kidney Int Suppl 4:45–52 (2011)

Basile DP (2007) The endothelial cell in ischemic acute kidney injury: implications for acute and chronic function. Kidney Int 72:151–156

Basile DP, Fredrich K, Chelladurai B, Leonard EC, Parrish AR (2008) Renal ischemia reperfusion inhibits VEGF expression and induces ADAMTS-1, a novel VEGF inhibitor. Am J Physiol Renal Physiol 294:F928–F936

Basile DP, Anderson MD, Sutton TA (2012) Pathophysiology of acute kidney injury. Compr Physiol 2:1303–1353

Baum J, Duffy HS (2011) Fibroblasts and myofibroblasts: what are we talking about? J Cardiovasc Pharmacol 57:376–379

Baylis C (1994) Age-dependent glomerular damage in the rat. Dissociation between glomerular injury and both glomerular hypertension and hypertrophy. Male gender as a primary risk factor. J Clin Invest 94:1823–1829

Beausejour CM, Krtolica A, Galimi F, Narita M, Lowe SW et al (2003) Reversal of human cellular senescence: roles of the p53 and p16 pathways. EMBO J 22:4212–4222

Bellomo R, Kellum JA, Ronco C, Wald R, Martensson J et al (2017) Acute kidney injury in sepsis. Intensive Care Med 43:816–828

Bielesz B, Sirin Y, Si H, Niranjan T, Gruenwald A et al (2010) Epithelial Notch signaling regulates interstitial fibrosis development in the kidneys of mice and humans. J Clin Invest 120:4040–4054

Bolignano D, Lacquaniti A, Coppolino G, Donato V, Campo S et al (2009) Neutrophil gelatinase-associated lipocalin (NGAL) and progression of chronic kidney disease. Clin J Am Soc Nephrol 4:337–344

Bonventre JV (2012) Can we target tubular damage to prevent renal function decline in diabetes? Semin Nephrol 32:452–462

Bonventre JV (2014) Primary proximal tubule injury leads to epithelial cell cycle arrest, fibrosis, vascular rarefaction, and glomerulosclerosis. Kidney Int Suppl 4:39–44 (2011)

Bonventre JV, Yang L (2011) Cellular pathophysiology of ischemic acute kidney injury. J Clin Invest 121:4210–4221

Borges FT, Melo SA, Ozdemir BC, Kato N, Revuelta I et al (2013) TGF-beta1-containing exosomes from injured epithelial cells activate fibroblasts to initiate tissue regenerative responses and fibrosis. J Am Soc Nephrol 24:385–392

Canaud G, Bonventre JV (2015) Cell cycle arrest and the evolution of chronic kidney disease from acute kidney injury. Nephrol Dial Transplant 30:575–583

Chang FC, Chou YH, Chen YT, Lin SL (2012) Novel insights into pericyte-myofibroblast transition and therapeutic targets in renal fibrosis. J Formos Med Assoc 111:589–598
Chawla LS, Kimmel PL (2012) Acute kidney injury and chronic kidney disease: an integrated clinical syndrome. Kidney Int 82:516–524
Chawla LS, Amdur RL, Amodeo S, Kimmel PL, Palant CE (2011) The severity of acute kidney injury predicts progression to chronic kidney disease. Kidney Int 79:1361–1369
Chawla LS, Eggers PW, Star RA, Kimmel PL (2014) Acute kidney injury and chronic kidney disease as interconnected syndromes. N Engl J Med 371:58–66
Che R, Yuan Y, Huang S, Zhang A (2014) Mitochondrial dysfunction in the pathophysiology of renal diseases. Am J Physiol Renal Physiol 306:F367–F378
Chen L, Liu BC, Zhang XL, Zhang JD, Liu H, Li MX (2006) Influence of connective tissue growth factor antisense oligonucleotide on angiotensin II-induced epithelial mesenchymal transition in HK2 cells. Acta Pharmacol Sin 27:1029–1036
Chen YT, Chang FC, Wu CF, Chou YH, Hsu HL et al (2011) Platelet-derived growth factor receptor signaling activates pericyte-myofibroblast transition in obstructive and post-ischemic kidney fibrosis. Kidney Int 80:1170–1181
Chou YH, Huang TM, Chu TS (2017) Novel insights into acute kidney injury-chronic kidney disease continuum and the role of renin-angiotensin system. J Formos Med Assoc 116:652–659
Chung AC, Lan HY (2015) MicroRNAs in renal fibrosis. Front Physiol 6:50
Coca SG, Yusuf B, Shlipak MG, Garg AX, Parikh CR (2009) Long-term risk of mortality and other adverse outcomes after acute kidney injury: a systematic review and meta-analysis. Am J Kidney Dis 53:961–973
Coca SG, Singanamala S, Parikh CR (2012) Chronic kidney disease after acute kidney injury: a systematic review and meta-analysis. Kidney Int 81:442–448
Cruz-Solbes AS, Youker K (2017) Epithelial to mesenchymal transition (EMT) and endothelial to mesenchymal transition (EndMT): role and implications in kidney fibrosis. Results Probl Cell Differ 60:345–372
Ding H, Zhou D, Hao S, Zhou L, He W et al (2012) Sonic hedgehog signaling mediates epithelial-mesenchymal communication and promotes renal fibrosis. J Am Soc Nephrol 23:801–813
Fabian SL, Penchev RR, St-Jacques B, Rao AN, Sipila P et al (2012) Hedgehog-Gli pathway activation during kidney fibrosis. Am J Pathol 180:1441–1453
Farris AB, Alpers CE (2014) What is the best way to measure renal fibrosis?: a pathologist's perspective. Kidney Int Suppl 4:9–15 (2011)
Ferenbach DA, Bonventre JV (2015) Mechanisms of maladaptive repair after AKI leading to accelerated kidney ageing and CKD. Nat Rev Nephrol 11:264–276
Fine LG, Orphanides C, Norman JT (1998) Progressive renal disease: the chronic hypoxia hypothesis. Kidney Int Suppl 65:S74–S78
Fiorentino M, Grandaliano G, Gesualdo L, Castellano G (2018) Acute kidney injury to chronic kidney disease transition. Contrib Nephrol 193:45–54
Freedman BI, Volkova NV, Satko SG, Krisher J, Jurkovitz C et al (2005) Population-based screening for family history of end-stage renal disease among incident dialysis patients. Am J Nephrol 25:529–535
Funk JA, Schnellmann RG (2012) Persistent disruption of mitochondrial homeostasis after acute kidney injury. Am J Physiol Renal Physiol 302:F853–F864
Galvan DL, Green NH, Danesh FR (2017) The hallmarks of mitochondrial dysfunction in chronic kidney disease. Kidney Int 92:1051–1057
Geng H, Lan R, Singha PK, Gilchrist A, Weinreb PH et al (2012) Lysophosphatidic acid increases proximal tubule cell secretion of profibrotic cytokines PDGF-B and CTGF through LPA2- and Galphaq-mediated Rho and alphavbeta6 integrin-dependent activation of TGF-beta. Am J Pathol 181:1236–1249
Gerstung M, Roth T, Dienes HP, Licht C, Fries JW (2007) Endothelin-1 induces NF-kappaB via two independent pathways in human renal tubular epithelial cells. Am J Nephrol 27:294–300

Gewin L, Vadivelu S, Neelisetty S, Srichai MB, Paueksakon P et al (2012) Deleting the TGF-beta receptor attenuates acute proximal tubule injury. J Am Soc Nephrol 23:2001–2011
Gewin L, Zent R, Pozzi A (2017) Progression of chronic kidney disease: too much cellular talk causes damage. Kidney Int 91:552–560
Go AS, Parikh CR, Ikizler TA, Coca S, Siew ED et al (2010) The assessment, serial evaluation, and subsequent sequelae of acute kidney injury (ASSESS-AKI) study: design and methods. BMC Nephrol 11:22
Goldstein SL, Jaber BL, Faubel S, Chawla LS, Acute Kidney Injury Advisory Group of American Society of Nephrology (2013) AKI transition of care: a potential opportunity to detect and prevent CKD. Clin J Am Soc Nephrol 8:476–483
Gomez-Garre D, Largo R, Tejera N, Fortes J, Manzarbeitia F, Egido J (2001) Activation of NF-kappaB in tubular epithelial cells of rats with intense proteinuria: role of angiotensin II and endothelin-1. Hypertension 37:1171–1178
Goodarzi AA, Block WD, Lees-Miller SP (2003) The role of ATM and ATR in DNA damage-induced cell cycle control. Prog Cell Cycle Res 5:393–411
Grande MT, Sanchez-Laorden B, Lopez-Blau C, De Frutos CA, Boutet A et al (2016) Erratum: Snail1-induced partial epithelial-to-mesenchymal transition drives renal fibrosis in mice and can be targeted to reverse established disease. Nat Med 22:217
Granger DN, Kvietys PR (2015) Reperfusion injury and reactive oxygen species: the evolution of a concept. Redox Biol 6:524–551
Grynberg K, Ma FY, Nikolic-Paterson DJ (2017) The JNK signaling pathway in renal fibrosis. Front Physiol 8:829
Haase VH (2012) Hypoxia-inducible factor signaling in the development of kidney fibrosis. Fibrogenesis Tissue Repair 5:S16
Haase M, Devarajan P, Haase-Fielitz A, Bellomo R, Cruz DN et al (2011) The outcome of neutrophil gelatinase-associated lipocalin-positive subclinical acute kidney injury: a multicenter pooled analysis of prospective studies. J Am Coll Cardiol 57:1752–1761
Hall AM, Schuh CD (2016) Mitochondria as therapeutic targets in acute kidney injury. Curr Opin Nephrol Hypertens 25:355–362
Henderson NC, Mackinnon AC, Farnworth SL, Kipari T, Haslett C et al (2008) Galectin-3 expression and secretion links macrophages to the promotion of renal fibrosis. Am J Pathol 172:288–298
Heung M, Chawla LS (2014) Acute kidney injury: gateway to chronic kidney disease. Nephron Clin Pract 127:30–34
Hewitson TD, Boon WC, Simpson ER, Smith ER, Samuel CS (2016) Estrogens do not protect, but androgens exacerbate, collagen accumulation in the female mouse kidney after ureteric obstruction. Life Sci 158:130–136
Hewitson TD, Holt SG, Smith ER (2017) Progression of tubulointerstitial fibrosis and the chronic kidney disease phenotype—role of risk factors and epigenetics. Front Pharmacol 8:520
Horbelt M, Lee SY, Mang HE, Knipe NL, Sado Y, Kribben A et al (2007) Acute and chronic microvascular alterations in a mouse model of ischemic acute kidney injury. Am J Physiol Renal Physiol 293:F688–F695
Hosohata K (2016) Role of oxidative stress in drug-induced kidney injury. Int J Mol Sci 17:1826
Hsu CY (2012) Yes, AKI truly leads to CKD. J Am Soc Nephrol 23:967–969
Huang XR, Chung AC, Wang XJ, Lai KN, Lan HY (2008) Mice overexpressing latent TGF-beta1 are protected against renal fibrosis in obstructive kidney disease. Am J Physiol Renal Physiol 295:F118–F127
Huen SC, Huynh L, Marlier A, Lee Y, Moeckel GW, Cantley LG (2015) GM-CSF promotes macrophage alternative activation after renal ischemia/reperfusion injury. J Am Soc Nephrol 26:1334–1345
Hultstrom M, Becirovic-Agic M, Jonsson S (2018) Comparison of acute kidney injury of different etiology reveals in-common mechanisms of tissue damage. Physiol Genomics 50:127–141
Humphreys BD, Xu F, Sabbisetti V, Grgic I, Movahedi Naini S et al (2013) Chronic epithelial kidney injury molecule-1 expression causes murine kidney fibrosis. J Clin Invest 123:4023–4035

Ishani A, Xue JL, Himmelfarb J, Eggers PW, Kimmel PL, Molitoris BA et al (2009) Acute kidney injury increases risk of ESRD among elderly. J Am Soc Nephrol 20:223–228

Kang DH, Anderson S, Kim YG, Mazzalli M, Suga S et al (2001) Impaired angiogenesis in the aging kidney: vascular endothelial growth factor and thrombospondin-1 in renal disease. Am J Kidney Dis 37:601–611

Kida Y, Tchao BN, Yamaguchi I (2014) Peritubular capillary rarefaction: a new therapeutic target in chronic kidney disease. Pediatr Nephrol 29:333–342

Kimura M, Asano M, Abe K, Miyazaki M, Suzuki T, Hishida A (2005) Role of atrophic changes in proximal tubular cells in the peritubular deposition of type IV collagen in a rat renal ablation model. Nephrol Dial Transplant 20:1559–1565

Kitching AR (2014) Dendritic cells in progressive renal disease: some answers, many questions. Nephrol Dial Transplant 29:2185–2193

Ko GJ, Grigoryev DN, Linfert D, Jang HR, Watkins T et al (2010) Transcriptional analysis of kidneys during repair from AKI reveals possible roles for NGAL and KIM-1 as biomarkers of AKI-to-CKD transition. Am J Physiol Renal Physiol 298:F1472–F1483

Kobori H, Nangaku M, Navar LG, Nishiyama A (2007) The intrarenal renin-angiotensin system: from physiology to the pathobiology of hypertension and kidney disease. Pharmacol Rev 59:251–287

Kok HM, Falke LL, Goldschmeding R, Nguyen TQ (2014) Targeting CTGF, EGF and PDGF pathways to prevent progression of kidney disease. Nat Rev Nephrol 10:700–711

Kokeny G, Nemeth Z, Godo M, Hamar P (2010) The Rowett rat strain is resistant to renal fibrosis. Nephrol Dial Transplant 25:1458–1462

Kramann R, Humphreys BD (2014) Kidney pericytes: roles in regeneration and fibrosis. Semin Nephrol 34:374–383

Kramann R, Schneider RK, DiRocco DP, Machado F, Fleig S et al (2015) Perivascular Gli1$^+$ progenitors are key contributors to injury-induced organ fibrosis. Cell Stem Cell 16:51–66

Kriz W, Kaissling B, Le Hir M (2011) Epithelial-mesenchymal transition (EMT) in kidney fibrosis: fact or fantasy? J Clin Invest 121:468–474

Lamouille S, Xu J, Derynck R (2014) Molecular mechanisms of epithelial-mesenchymal transition. Nat Rev Mol Cell Biol 15:178–196

Le Clef N, Verhulst A, D'Haese PC, Vervaet BA (2016) Unilateral renal ischemia-reperfusion as a robust model for acute to chronic kidney injury in mice. PLoS ONE 11:e0152153

LeBleu VS, Taduri G, O'Connell J, Teng Y, Cooke VG et al (2013) Origin and function of myofibroblasts in kidney fibrosis. Nat Med 19:1047–1053

Leemans JC, Kors L, Anders HJ, Florquin S (2014) Pattern recognition receptors and the inflammasome in kidney disease. Nat Rev Nephrol 10:398–414

Leonard EC, Friedrich JL, Basile DP (2008) VEGF-121 preserves renal microvessel structure and ameliorates secondary renal disease following acute kidney injury. Am J Physiol Renal Physiol 295:F1648–F1657

Leung KC, Tonelli M, James MT (2013) Chronic kidney disease following acute kidney injury-risk and outcomes. Nat Rev Nephrol 9:77–85

Lewington AJ, Cerda J, Mehta RL (2013) Raising awareness of acute kidney injury: a global perspective of a silent killer. Kidney Int 84:457–467

Li J, Qu X, Bertram JF (2009) Endothelial-myofibroblast transition contributes to the early development of diabetic renal interstitial fibrosis in streptozotocin-induced diabetic mice. Am J Pathol 175:1380–1388

Liu Y (2004) Epithelial to mesenchymal transition in renal fibrogenesis: pathologic significance, molecular mechanism, and therapeutic intervention. J Am Soc Nephrol 15:1–12

Liu Y (2010) New insights into epithelial-mesenchymal transition in kidney fibrosis. J Am Soc Nephrol 21:212–222

Liu BC, Chen L, Sun J, Huang HQ, Ma KL, Liu H et al (2006) Connective tissue growth factor-mediated angiotensin II-induced hypertrophy of proximal tubular cells. Nephron Exp Nephrol 103:e16–e26

Liu S, Soong Y, Seshan SV, Szeto HH (2014) Novel cardiolipin therapeutic protects endothelial mitochondria during renal ischemia and mitigates microvascular rarefaction, inflammation, and fibrosis. Am J Physiol Renal Physiol 306:F970–F980

Liu M, Ning X, Li R, Yang Z, Yang X, Sun S et al (2017) Signalling pathways involved in hypoxia-induced renal fibrosis. J Cell Mol Med 21:1248–1259

Lopez-Hernandez FJ, Lopez-Novoa JM (2012) Role of TGF-beta in chronic kidney disease: an integration of tubular, glomerular and vascular effects. Cell Tissue Res 347:141–154

Lorz C, Ortiz A, Justo P, Gonzalez-Cuadrado S, Duque N et al (2000) Proapoptotic Fas ligand is expressed by normal kidney tubular epithelium and injured glomeruli. J Am Soc Nephrol 11:1266–1277

Lovisa S, Zeisberg M, Kalluri R (2016) Partial epithelial-to-mesenchymal transition and other new mechanisms of kidney fibrosis. Trends Endocrinol Metab 27:681–695

Lozano R, Naghavi M, Foreman K, Lim S, Shibuya K et al (2012) Global and regional mortality from 235 causes of death for 20 age groups in 1990 and 2010: a systematic analysis for the Global Burden of Disease Study 2010. Lancet 380:2095–2128

Maarouf OH, Aravamudhan A, Rangarajan D, Kusaba T, Zhang V et al (2016) Paracrine Wnt1 drives interstitial fibrosis without inflammation by tubulointerstitial cross-talk. J Am Soc Nephrol 27:781–790

Macconi D, Remuzzi G, Benigni A (2014) Key fibrogenic mediators: old players. Renin-angiotensin system. Kidney Int Suppl 4:58–64 (2011)

Machida Y, Kitamoto K, Izumi Y, Shiota M, Uchida J et al (2010) Renal fibrosis in murine obstructive nephropathy is attenuated by depletion of monocyte lineage, not dendritic cells. J Pharmacol Sci 114:464–473

Mack M, Yanagita M (2015) Origin of myofibroblasts and cellular events triggering fibrosis. Kidney Int 87:297–307

Meng XM, Tang PM, Li J, Lan HY (2015) TGF-beta/Smad signaling in renal fibrosis. Front Physiol 6:82

Menke J, Iwata Y, Rabacal WA, Basu R, Yeung YG et al (2009) CSF-1 signals directly to renal tubular epithelial cells to mediate repair in mice. J Clin Invest 119:2330–2342

Moonen L, D'Haese PC, Vervaet BA (2018) Epithelial cell cycle behaviour in the injured kidney. Int J Mol Sci 19:2038

Mosser DM, Edwards JP (2008) Exploring the full spectrum of macrophage activation. Nat Rev Immunol 8:958–969

Munoz-Espin D, Serrano M (2014) Cellular senescence: from physiology to pathology. Nat Rev Mol Cell Biol 15:482–496

Murea M, Park JK, Sharma S, Kato H, Gruenwald A et al (2010) Expression of Notch pathway proteins correlates with albuminuria, glomerulosclerosis, and renal function. Kidney Int 78:514–522

Nastase MV, Zeng-Brouwers J, Wygrecka M, Schaefer L (2018) Targeting renal fibrosis: Mechanisms and drug delivery systems. Adv Drug Deliv Rev 129:295–307

Nath KA (1992) Tubulointerstitial changes as a major determinant in the progression of renal damage. Am J Kidney Dis 20:1–17

Nelson PJ, Rees AJ, Griffin MD, Hughes J, Kurts C, Duffield J (2012) The renal mononuclear phagocytic system. J Am Soc Nephrol 23:194–203

Ninichuk V, Gross O, Segerer S, Hoffmann R, Radomska E et al (2006) Multipotent mesenchymal stem cells reduce interstitial fibrosis but do not delay progression of chronic kidney disease in collagen4A3-deficient mice. Kidney Int 70:121–129

Nogueira A, Pires MJ, Oliveira PA (2017) Pathophysiological mechanisms of renal fibrosis: a review of animal models and therapeutic strategies. Vivo 31:1–22

Ohtomo S, Nangaku M, Izuhara Y, Takizawa S, Strihou C, Miyata T (2008) Cobalt ameliorates renal injury in an obese, hypertensive type 2 diabetes rat model. Nephrol Dial Transplant 23:1166–1172

Parikh CR, Jani A, Melnikov VY, Faubel S, Edelstein CL (2004) Urinary interleukin-18 is a marker of human acute tubular necrosis. Am J Kidney Dis 43:405–414

Rodrigues-Diez RR, Garcia-Redondo AB, Orejudo M, Rodrigues-Diez R, Briones AM et al (2015) The C-terminal module IV of connective tissue growth factor, through EGFR/Nox1 signaling, activates the NF-kappaB pathway and proinflammatory factors in vascular smooth muscle cells. Antioxid Redox Signal 22:29–47

Sanz AB, Sanchez-Nino MD, Ortiz A (2011) TWEAK, a multifunctional cytokine in kidney injury. Kidney Int 80:708–718

Sanz AB, Izquierdo MC, Sanchez-Nino MD, Ucero AC, Egido J et al (2014) TWEAK and the progression of renal disease: clinical translation. Nephrol Dial Transplant 29(Suppl 1):i54–i62

Schmiedt CW, Brainard BM, Hinson W, Brown SA, Brown CA (2016) Unilateral renal ischemia as a model of acute kidney injury and renal fibrosis in cats. Vet Pathol 53:87–101

Schrijvers BF, Flyvbjerg A, Tilton RG, Rasch R, Lameire NH, De Vriese AS (2005) Pathophysiological role of vascular endothelial growth factor in the remnant kidney. Nephron Exp Nephrol 101:e9–e15

Sharfuddin AA, Molitoris BA (2011) Pathophysiology of ischemic acute kidney injury. Nat Rev Nephrol 7:189–200

Sturmlechner I, Durik M, Sieben CJ, Baker DJ, van Deursen JM (2017) Cellular senescence in renal ageing and disease. Nat Rev Nephrol 13:77–89

Sun YB, Qu X, Caruana G, Li J (2016) The origin of renal fibroblasts/myofibroblasts and the signals that trigger fibrosis. Differentiation 92:102–107

Susantitaphong P, Siribamrungwong M, Doi K, Noiri E, Terrin N, Jaber BL (2013) Performance of urinary liver-type fatty acid-binding protein in acute kidney injury: a meta-analysis. Am J Kidney Dis 61:430–439

Takaori K, Nakamura J, Yamamoto S, Nakata H, Sato Y et al (2016) Severity and frequency of proximal tubule injury determines renal prognosis. J Am Soc Nephrol 27:2393–2406

Tan RJ, Zhou D, Zhou L, Liu Y (2014) Wnt/beta-catenin signaling and kidney fibrosis. Kidney Int Suppl 4:84–90 (2011)

Tanaka T, Matsumoto M, Inagi R, Miyata T, Kojima I et al (2005) Induction of protective genes by cobalt ameliorates tubulointerstitial injury in the progressive Thy1 nephritis. Kidney Int 68:2714–2725

Tang WW, Ulich TR, Lacey DL, Hill DC, Qi M et al (1996) Platelet-derived growth factor-BB induces renal tubulointerstitial myofibroblast formation and tubulointerstitial fibrosis. Am J Pathol 148:1169–1180

Tang Z, Lu B, Hatch E, Sacks SH, Sheerin NS (2009) C3a mediates epithelial-to-mesenchymal transition in proteinuric nephropathy. J Am Soc Nephrol 20:593–603

Tsai YC, Chiu YW, Tsai JC, Kuo HT, Lee SC et al (2014) Association of angiopoietin-2 with renal outcome in chronic kidney disease. PLoS ONE 9:e108862

Venkatachalam MA, Griffin KA, Lan R, Geng H, Saikumar P, Bidani AK (2010) Acute kidney injury: a springboard for progression in chronic kidney disease. Am J Physiol Renal Physiol 298:F1078–F1094

Vinuesa E, Hotter G, Jung M, Herrero-Fresneda I, Torras J, Sola A (2008) Macrophage involvement in the kidney repair phase after ischaemia/reperfusion injury. J Pathol 214:104–113

Wang S, Diao H, Guan Q, Cruikshank WW, Delovitch TL et al (2008) Decreased renal ischemia-reperfusion injury by IL-16 inactivation. Kidney Int 73:318–326

Wang Y, Chang J, Yao B, Niu A, Kelly E et al (2015) Proximal tubule-derived colony stimulating factor-1 mediates polarization of renal macrophages and dendritic cells, and recovery in acute kidney injury. Kidney Int 88:1274–1282

Weisheit CK, Engel DR, Kurts C (2015) Dendritic cells and macrophages: sentinels in the kidney. Clin J Am Soc Nephrol 10:1841–1851

Wirthensohn G, Guder WG (1986) Renal substrate metabolism. Physiol Rev 66:469–497

Wolf G, Ziyadeh FN, Stahl RA (1999) Angiotensin II stimulates expression of transforming growth factor beta receptor type II in cultured mouse proximal tubular cells. J Mol Med (Berl) 77:556–564

Wong WK, Robertson H, Carroll HP, Ali S, Kirby JA (2003) Tubulitis in renal allograft rejection: role of transforming growth factor-beta and interleukin-15 in development and maintenance of CD103$^+$ intraepithelial T cells. Transplantation 75:505–514
Wu H, Craft ML, Wang P, Wyburn KR, Chen G et al (2008) IL-18 contributes to renal damage after ischemia-reperfusion. J Am Soc Nephrol 19:2331–2341
Wu CF, Chiang WC, Lai CF, Chang FC, Chen YT et al (2013) Transforming growth factor beta-1 stimulates profibrotic epithelial signaling to activate pericyte-myofibroblast transition in obstructive kidney fibrosis. Am J Pathol 182:118–131
Yan M, Tang C, Ma Z, Huang S, Dong Z (2016) DNA damage response in nephrotoxic and ischemic kidney injury. Toxicol Appl Pharmacol 313:104–108
Yang L, Besschetnova TY, Brooks CR, Shah JV, Bonventre JV (2010) Epithelial cell cycle arrest in G2/M mediates kidney fibrosis after injury. Nat Med 16:535–543
Yang Y, Zhang ZX, Lian D, Haig A, Bhattacharjee RN, Jevnikar AM (2015) IL-37 inhibits IL-18-induced tubular epithelial cell expression of pro-inflammatory cytokines and renal ischemia-reperfusion injury. Kidney Int 87:396–408
Yano T, Nozaki Y, Kinoshita K, Hino S, Hirooka Y et al (2015) The pathological role of IL-18Ralpha in renal ischemia/reperfusion injury. Lab Invest 95:78–91
Yard BA, Daha MR, Kooymans-Couthino M, Bruijn JA, Paape ME et al (1992) IL-1 alpha stimulated TNF alpha production by cultured human proximal tubular epithelial cells. Kidney Int 42:383–389
Yu M, Ryu DR, Kim SJ, Choi KB, Kang DH (2010) Clinical implication of metabolic syndrome on chronic kidney disease depends on gender and menopausal status: results from the Korean National Health and Nutrition Examination Survey. Nephrol Dial Transplant 25:469–477
Zager RA, Johnson AC, Andress D, Becker K (2013) Progressive endothelin-1 gene activation initiates chronic/end-stage renal disease following experimental ischemic/reperfusion injury. Kidney Int 84:703–712
Zeisberg EM, Potenta SE, Sugimoto H, Zeisberg M, Kalluri R (2008) Fibroblasts in kidney fibrosis emerge via endothelial-to-mesenchymal transition. J Am Soc Nephrol 19:2282–2287
Zhou XJ, Rakheja D, Yu X, Saxena R, Vaziri ND, Silva FG (2008) The aging kidney. Kidney Int 74:710–720
Zhou D, Li Y, Lin L, Zhou L, Igarashi P, Liu Y (2012) Tubule-specific ablation of endogenous beta-catenin aggravates acute kidney injury in mice. Kidney Int 82:537–547
Zhou X, Fukuda N, Matsuda H, Endo M, Wang X et al (2013a) Complement 3 activates the renal renin-angiotensin system by induction of epithelial-to-mesenchymal transition of the nephrotubulus in mice. Am J Physiol Renal Physiol 305:F957–F967
Zhou Y, Xiong M, Fang L, Jiang L, Wen P et al (2013b) miR-21-containing microvesicles from injured tubular epithelial cells promote tubular phenotype transition by targeting PTEN protein. Am J Pathol 183:1183–1196
Zhou D, Li Y, Zhou L, Tan RJ, Xiao L et al (2014) Sonic hedgehog is a novel tubule-derived growth factor for interstitial fibroblasts after kidney injury. J Am Soc Nephrol 25:2187–2200
Zhou L, Li Y, Hao S, Zhou D, Tan RJ et al (2015) Multiple genes of the renin-angiotensin system are novel targets of Wnt/beta-catenin signaling. J Am Soc Nephrol 26:107–120
Zhou D, Tan RJ, Fu H, Liu Y (2016) Wnt/beta-catenin signaling in kidney injury and repair: a double-edged sword. Lab Invest 96:156–167

Part II
Resident and infiltrating cell activation in renal fibrosis

Chapter 8
Role of Endothelial Cells in Renal Fibrosis

Zhen Yang, Li-Jie He and Shi-Ren Sun

Abstract Renal fibrosis has been regarded as the common pathway of end-stage renal failure. Understanding the fundamental mechanism that leads to renal fibrosis is essential for developing better therapeutic options for chronic kidney diseases. So far, the main abstractions are on the injury of tubular epithelial cells, activation of interstitial cells, expression of chemotactic factor and adhesion molecule, infiltration of inflammatory cells and homeostasis of ECM. However, emerging studies revealed that endothelial cells (ECs) might happen to endothelial-to-mesenchymal transition (EndMT) dependent and/or independent endothelial dysfunction, which were supposed to accelerate renal fibrosis and are identified as new mechanisms for the proliferation of myofibroblasts as well. In this chapter, we are about to interpret the role of ECs in renal fibrosis and analyze the related molecules and pathways of both EndMT and EndMT independent endothelial dysfunction.

Keywords Endothelial cell · Renal fibrosis · EndMT · Endothelial dysfunction

8.1 Introduction

The high prevalence and burden of chronic kidney diseases (CKD) have been well established, and it has emerged as a major threat to public health due to its 13% incidence rate (Coresh et al. 2007). Meanwhile, 45% of total deaths can be attributed to fibrotic disorders (Wynn 2008). Nowadays, it is well-known that all primary causes of CKD share a common pathogenetic pathway of progressive injury due to the destructive consequences of renal fibrosis, although numerous different kinds of diseases including glomerulonephritis, metabolic diseases, diabetes mellitus, atherosclerosis, obstructive nephropathy, interstitial nephritis, cystic nephropathies and polycystic kidney disease, can be major causes of CKD (Cho 2010). Since renal fibrosis has been the common final pathway to end-stage renal disease (ESRD), it is warranted

Z. Yang · L.-J. He · S.-R. Sun (✉)
Department of Nephrology, The First Affiliated Hospital of Air Force Medical University, Xi'an, Shaanxi, China
e-mail: sunshiren@medmail.com.cn

B.-C. Liu et al. (eds.), *Renal Fibrosis: Mechanisms and Therapies*,
Advances in Experimental Medicine and Biology 1165,
https://doi.org/10.1007/978-981-13-8871-2_8

to elucidate the mechanisms of renal fibrosis and develop better therapeutic agents for CKD.

So far, the scientists mainly focus on the injury of tubular epithelial cells, activation of interstitial cells, expression of chemotactic factor and adhesion molecule, infiltration of inflammatory cells and homeostasis of ECM (Iliescu et al. 2010; Kang et al. 2001). Emerging studies revealed that endothelial dysfunction is dependent or independent of endothelial-to-mesenchymal transition (EndMT), which has been supposed to accelerate renal fibrosis and identified as a new mechanism for the proliferation of myofibroblasts (Piera-Velazquez et al. 2011). Moreover, it has excited particular and extensive concern about the injury and/or loss of renal microvascular and subsequently hypoxia and ischemia, by which eventually leads to the renal fibrosis. In this chapter, we will focus on the role of ECs in renal fibrosis.

8.2 The Function of ECs

Generally, ECs compose the internal surface of the vessels and play the essential role in vascular homeostasis. Apart from its classical barrier role, the endothelium also takes part in many physiological processes including the control of vasomotor tone, the response of tissue inflammation and thrombosis (Deanfield et al. 2007; Feletou and Vanhoutte 2006). The endothelium structure is highly heterogeneous inside the renal microvessels environment, where ECs are multifunctional and highly specialized from the preglomerular arterioles to the peritubular capillary bed. Despite gradually rising interest in the identification of endothelial diseases associated with CKD, limited researches have investigated endothelial alterations during kidney injury. Fortunately, recent evidence has provided some novel insights into the pathophysiological role of intrarenal endothelium in the progression of CKD (Fig. 8.1) (Guerrot et al. 2012).

The main physiological functions of ECs include as follows: (1) Barrier function: The vascular endothelium can prevent the adherent and harmful substances such as thrombocyte and leukocyte from invading vessel while the intact structure of endothelium has the agonistic effects to the lipid deposition; (2) Endocrine function: ECs can synthesize and secrete a variety of vasoactive substances including endothelium-dependent contraction factor such as endothelin (ET), angiotensin II and thromboxane as well as the endothelium-dependent relaxing factor such as nitric oxide (NO), prostacyclin, bradykinin, endothelium-dependent hyperpolarizing factor and so on; (3) Regulation of vascular tone: ECs regulate the relaxation and contraction of blood vessels by releasing vasodilators and vasoconstrictors. Among these, ET and NO maintain a relative dynamic balance under physiological conditions to regulate renal vascular tone and kidney function; (4) Anticoagulation and fibrinolysis: Under vascular injury or stress, vascular ECs secrete vasoactive substances including thrombomodulin-1 (TM-1), voll Willebrand factor (vWF), tissue-type plasminogen activator (t-PA), etc., to coordinate the balance between procoagulant and anticoagulation to prevent thrombosis; (5) Involvement in inflammatory response:

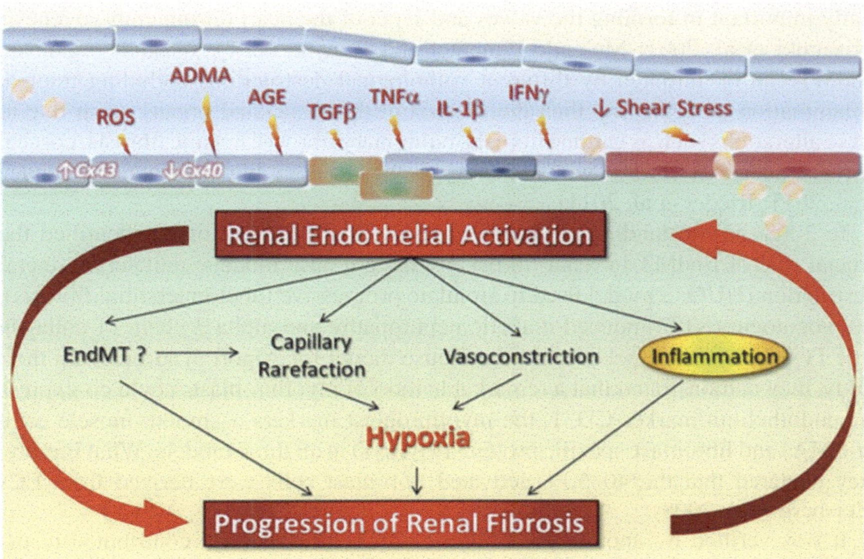

Fig. 8.1 Schematic view of the pathophysiological role of endothelial activation in CKD progression. ADMA, asymmetric dimethylarginine; ROS, reactive oxygen species; AGE, advanced glycation end products; TGF, transforming growth factor; TNF, tumor necrosis factor; IL, interleukin; IFN, interferon; EndMT, endothelial–mesenchymal transition; Cx40, Connexin 40: Cx43, Connexin 43 (Guerrot et al. 2012)

ECs express multiple adhesion molecules including selectin family, integrin family and immunoglobulin superfamily while tumor necrosis factor (TNF) can activate these adhesion molecules to exert the inflammatory effect; (6) Regulation of vascular remodeling: There are a large number of receptors on the ECs, which transmit the detected signals to the smooth muscles or secrete physiological factors through the smooth muscle-endothelial junction and cooperate with the smooth muscle to construct the vascular remodeling.

8.3 The EndMT and the Renal Fibrosis

In the past, epithelial–mesenchymal transition (EMT), characterizing by a phenotypical change occurred in epithelial cells which diminish their inherent cell–cell basement membrane linkage as well as their structural polarity to spindle-shaped and morphologically similar to mesenchymal/myofibroblast cells, was regarded as a crucial role to contribute to the tubular epithelial cells injury and renal fibrosis (Sun et al. 2009).

Recently, EndMT, first discovered in heart development (Markwald et al. 1975), was supposed to play a role in renal fibrosis as well. It was confirmed to be crucially important in forming the valves and septa of the heart during embryogenesis (Kisanuki et al. 2001; Mercado-Pimentel and Runyan 2007). In adult organism, EndMT can be initiated by different pathological destructions including trauma, inflammation or aging and then the fibrosis of the associated organs occur due to these alterations such as cardiac fibrosis, pulmonary fibrosis, hepatic fibrosis, corneal fibrosis, intestinal fibrosis, wound healing as well as renal fibrosis (Piera-Velazquez et al. 2011; Rieder et al. 2011).

In 2008, an outstanding work achieved by Zeisberg et al. firstly identified the crucial role of EndMT in renal fibrosis by three mouse models: unilateral ureteral obstruction (UUO; a model used to simulate progressive tubulointerstitial fibrosis), streptozotocin (STZ)-induced diabetic nephropathy and alpha 3 chain of collagen type IV (COL4A3) knockout mice (a mouse model for Alport syndrome). In their study, they demonstrated that a remarkable ratio of myofibroblasts could co-express the endothelium marker CD31, the myofibroblast markers α-smooth muscle actin (α-SMA) and fibroblast-specific protein-1 (FSP-1) in all three models. What is more, they declared that the 40–50% activated fibroblast cells were derived from ECs (Zeisberg et al. 2008).

It was verified by another group that EndMT occurred and contributed to the proliferation of myofibroblasts in the early stage of renal fibrosis in diabetic mouse models (STZ-induced diabetic nephropathy). And a considerable number of myofibroblasts of an endothelial origin in the fibrotic kidneys were recognized from the diabetic mouse models (Li et al. 2009). These data suggest that both EMT and EndMT contribute to the activated fibroblasts/myofibroblasts responsible for renal fibrosis in CKD.

8.4 The Mechanisms of EndMT in Renal Fibrosis

8.4.1 The Cellular Mechanisms of EndMT in Renal Fibrosis

The role of EndMT in kidney fibrosis is the differentiation and activation of fibroblasts or myofibroblasts from endothelia and supplying the matrix that generates mesenchymal cells. These cells implicate a disparate set of biomarkers including VE-cadherin, CD31, TIE1, TIE2, vWF and cytokeratins. Resembling EMT, ECs diminish their adhesion and apical–basal polarity to form highly invasive, migratory, spindle-shaped and prolonged mesenchymal cells during EndMT. Meanwhile, the changed biochemical process includes decreased expression of endothelial markers and elevated mesenchymal markers (α-SMA, N-cadherin, fibronectin, vimentin, types I and III collagen, MMP-2 and MMP-9, etc.) (Fig. 8.2) (Liu 2011; Medici and Kalluri 2012).

Nowadays, we have realized that cytokines secreted by stem cells would accumulate in the destructed kidney, which came along with the ECs proliferation or angiogenesis to accelerate renal regeneration (Ishikane et al. 2008; Togel et al. 2007). Then,

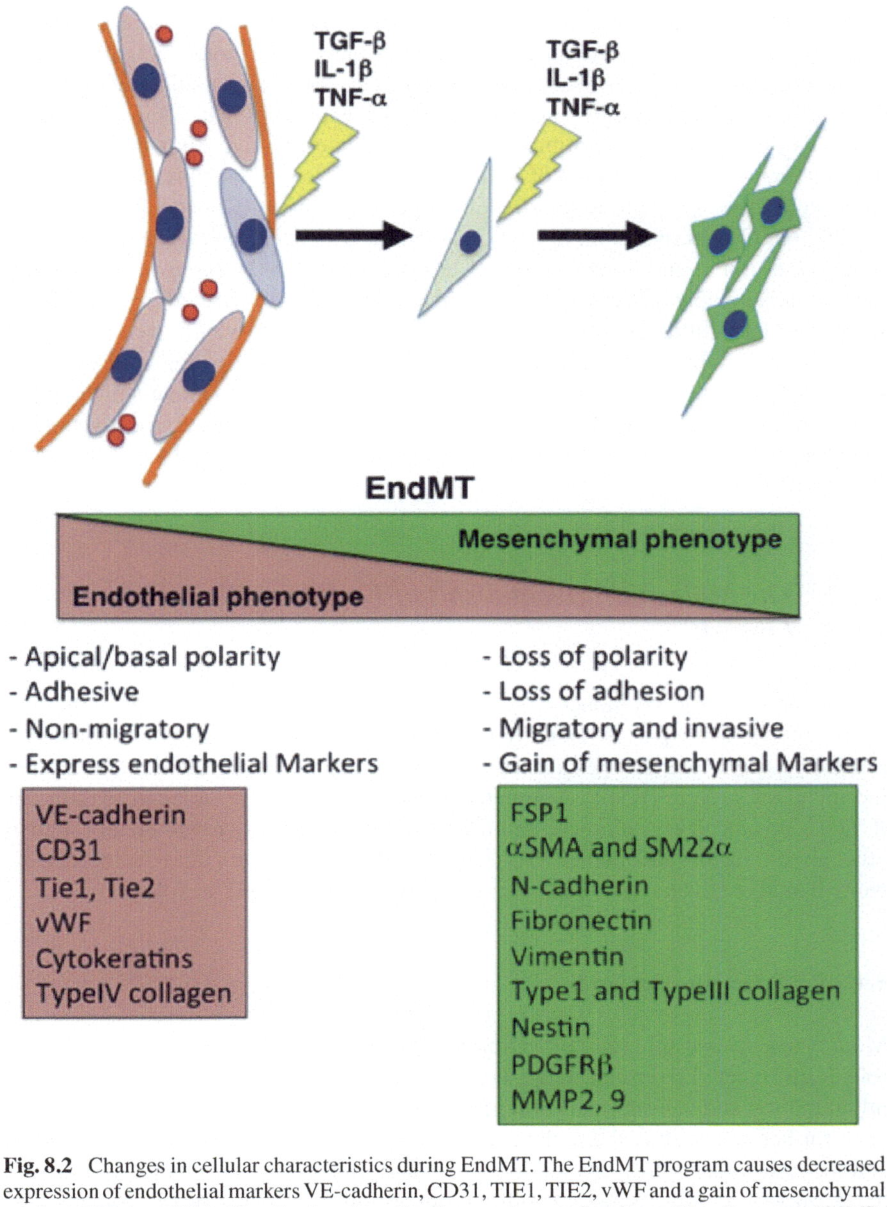

Fig. 8.2 Changes in cellular characteristics during EndMT. The EndMT program causes decreased expression of endothelial markers VE-cadherin, CD31, TIE1, TIE2, vWF and a gain of mesenchymal markers FSP-1, α-SMA, N-cadherin, vimentin, fibronectin, type I and type III collagen, and MMP-2 and MMP-9. Distinct changes in cell polarity and morphology accompany EndMT, as well as the loss of cell–cell junctions and increased motility. FSP-1, fibroblast-specific protein-1; α-SMA, α-smooth muscle actin; TIE, tyrosine kinase with immunoglobulin-like and EGF-like domains; PDGFRb, platelet-derived growth factor receptor b; MMP, matrix metalloproteinase (He et al. 2013)

a study showed that mouse kidney progenitor-like cells (MKPCs) were able to be isolated from Myh9-targeted mutant mice and displayed pluripotent activity in vitro and in vivo (Lee et al. 2010). Subsequently, scientists found that conditioned medium derived from MKPCs resulted in transforming growth factor beta (TGF-β)-induced EndMT in vitro due to its antifibrotic activity. The latest in vivo study demonstrated that intravenous administration of MKPCs might prevent 5/6 nephrectomized mice from renal fibrosis via preservation of angiogenic processes (Chen et al. 2015).

The role of hyperuricemia in CKD is still controversial. However, a recent study revealed an initial and essential role of uric acid (UA) in the progression of renal disease. It is notable that UA-induced renal disease could be resulted from oxidative stress and local inflammation during afferent arteriopathy, which eventually lead to glomerular hypertrophy and interstitial fibrosis due to the endothelial dysfunction. Moreover, the latest study showed that UA per se could mediate a phenotypic transition of epithelial and ECs via an induction of oxidative stress and glycocalyx shedding, which could be one of the mechanisms of UA-induced kidney disease (Kang 2018).

8.4.2 The Molecular Mechanisms of EndMT in Renal Fibrosis

Renal fibrosis could be induced by the activation of numerous cell types that act cooperatively and expose to an extremely complicated microenvironment under the regulation of various biological mediators including TGF-β/Smad signaling, bone morphogenic protein (BMP), angiotensin (Ang) II, connective tissue growth factor (CTGF), advanced glycation end products (AGEs), fibroblast growth factors (FGF), other growth factors and microRNAs (He et al. 2013). Apart from these factors above, we will analyze the newest molecules and pathways of EndMT as follows.

8.4.2.1 C3a/C5a-Aktsignal

More than a decade ago, scientists already identified the Akt pathway as a significant role in I/R injury (Loverre et al. 2004) and vascular homeostasis for its regulation on inflammation and cell survival (Shiojima and Walsh 2002). Then, Pasquale Ditonno et al. further demonstrated that the number of CD31+/α-SMA+ and CD31+/FPS-1+ cell increased during I/R injury, which indicated EndMT and eventually diminished the density of renal peritubular capillaries and promoted tissue fibrosis. Meanwhile, the progression of EndMT gradually vanished, with an intact density of peritubular capillaries and remarkable alleviation in renal fibrosis when the complements (C3a/C5a, etc.) were blocked. On the other hand, the complement (C3a/C5a) would enhance the EndMT with elevated expression of fibroblast markers and declined expression of specific endothelial markers when ECs were activated by anaphylatoxins in vitro, indicating the regulation of Akt pathway was crucial for the C3a and C5a-induced EndMT in vitro. Moreover, blocking complement in vivo led to inhibition of Akt signaling pathway, which alleviated EndMT and tissue fibrosis. Considerably, these results probably disclose a new role of complement in the EndMT via the Akt pathway and shed some light on inhibiting complement as a potential clinical therapy of tissue fibrosis (Curci et al. 2014).

8.4.2.2 SIRT3-Foxo3a-Catalase Pathway

Sirtuins (SIRTs) are nicotinamide adenine dinucleotide-dependent class III histone deacetylases, which consist of seven mammalian members (SIRT1-7). SIRT3 is localized in the mitochondrial matrix and is highly expressed in tissues with high metabolic turnover (Haigis and Guarente 2006; Newman et al. 2012; Onyango et al. 2002). A recent publication reported that SIRT3 depletion increased vascular oxidative stress and promoted endothelial dysfunction (Dikalova et al. 2017). But its role in EndMT and hypertensive renal injury still needs further elucidation. Transcription factor Foxo3a (forkhead box O3a) is a downstream target of SIRT3 and regulates the cellular antioxidant defense (Chen et al. 2014). Oxidative stress can promote EndMT (Sanchez-Duffhues et al. 2016). In this study, they used SIRT3$^{-/-}$ and SIRT3-TgEC (SIRT3 EC-specific transgenic) mice to investigate the possible role of SIRT3 in EndMT and hypertensive renal injury. They investigated the specific mechanism of SIRT3-mediated suppression of EndMT in Ang II-induced renal fibrosis and identified that Foxo3a-reactive oxygen species (ROS) may be a downstream target of SIRT3 (Lin et al. 2018).

8.4.2.3 DKK3-Wnt Pathway

Apart from C3a/C5a-Akt signal and SIRT3-Foxo3a-catalase pathway, the DKK3-Wnt pathway also plays a role in the EndMT and renal fibrosis. Notably, Dickkopf-3 (DKK3) is specifically expressed in the fibrogenic secretome, acting as the ligand of the Wnt/β-catenin pathway. In accordance with its localization, DKK3 could significantly promote myofibroblast activation and accelerate the transition from fibroblasts to myofibroblasts by antagonized effects of Wnt pathway inhibitor—DKK1. A new study showed that DKK3 could prohibit endothelial outgrowth, enhance myofibroblast formation and activate the EndMT, by which the angiogenic competence was hampered in RMVECs. Meanwhile, DKK3 could be downregulated by the administration of Wnt pathway suppressor—sulindac sulfide, along with the hampered renal fibrosis. Thus, this novel mechanism implicated that DKK3 could promote the formation of myofibroblasts and EndMT, resulting in the handicap of angiogenesis due to the DKK3's role of the agonist of the Wnt pathway. In addition, sulindac sulfide, a potent inhibitor of the Wnt pathway, should be noticed for its suppressing function of nephropathy-induced DKK3 expression and renal fibrosis (Lipphardt et al. 2019).

8.5 The Molecular Mechanisms of EndMT Independent Endothelial Dysfunction

Despite an increasing number of studies on EndMT, controversies still exist regarding the existence of EndMT in kidney disease, as well as its role in disease development (Cruz-Solbes and Youker 2017). On the other hand, in addition to the EndMT in renal fibrosis, there are other molecules and pathways participating in the development and progression of renal fibrosis and kidney diseases as well.

Therefore, we should also pay attention to the "EndMT independent endothelial dysfunction." Obviously, the "endothelial dysfunction" not only resulted from EndMT but also from other factors. However, it is still hard to exactly define the "endothelial dysfunction" so far because this versatile "homeostasis break" actually consists of a series of handicaps in vasomotor responses, antithrombogenic properties, vascular permeability, leukocyte recruitment and ECs proliferation. Indeed, it is supposed to include plenty of syndromes characterized by alterations in dissimilar endothelial functions (Joannides et al. 1995; Thuillez and Richard 2005). Thus, in this chapter, we attempt to give a comprehensive review on the "EndMT independent" pathways and to elaborate what has been clarified in favor and against these as follows.

8.5.1 Smad3 Linker Phosphorylation

Transforming growth factor (TGF)-β1/Smad3 signaling pathway, as the most significant one involved in endothelial dysfunction, contributes to tissue fibrosis via the phosphorylation of Smad2 and Smad3 (Matsuzaki 2012). It is reported that fibroblasts from $Smad3^{-/-}$ mice fail to auto-induced TGF-β1 expression (Piek et al. 2001). In addition, it is observed that Smad3 was indispensable of several models of tissue fibrosis (UUO, diabetic nephropathy and angiotensin II-induced renal and cardiac fibrosis), indicating a crucial role for TGF-β/Smad3 signaling in tissue fibrosis (Fujimoto et al. 2003; Huang et al. 2010; Liu et al. 2012; Sato et al. 2003; Wang et al. 2007). Phosphorylation is recognized as an important mechanism regulating Smad3 transcription factor activity and the fibrotic response (Flanders 2004; Kretzschmar and Massague 1998; Li et al. 2010; Zimmerman and Padgett 2000). One study found that endothelial dysfunction could exacerbate renal interstitial fibrosis via enhanced Smad3 linker phosphorylation, resulting in increased fibroblast proliferation and collagen production. While resolvin D1 (RvD1) could alleviate renal fibrosis in the mouse models, and further research revealed that RvD1 might prevent from endothelial dysfunction and inhibited Smad3/JNK pathway (Sun et al. 2013).

8.5.2 LPS-Binding Protein

In addition to the Smads, we should pay attention to the lipid A-containing lipopolysaccharide (LPS), a typical component of gram-negative bacteria and their walls, acting as a predominant role in the pathogenesis of sepsis (Ronco et al. 2000). As in gram-negative sepsis, during endotoxemia, LPS induces uncontrolled cytokines release, activation of coagulation on ECs (Zarjou and Agarwal 2011), which lead to shock, multiple organ damage and even death (Ramnath et al. 2008). Acute kidney injury (AKI) is a frequent complication of sepsis and endotoxemia (Zarjou and Agarwal 2011). LPS activated EC (Dunzendorfer et al. 2004) and tubular epithelial cells (Bussolati et al. 2002) through the toll-like receptor-4 (TLR-4), myeloid differentiation protein-2 (MD-2) and CD14 complex, subsequently triggering both proinflammatory and cytoprotective functions (Singla et al. 2011). In a swine model, the LPS administration would lead to the acute tubulointerstitial fibrosis which was related to dysfunctional α-SMA + ECs verified by Ki-67+ without apoptosis (caspase-3-). Additionally, the expression of vimentin and N-cadherin as well as the mRNA synthesis of collagen I were remarkably elevated in vitro due to the EC dysfunction led by LPS. In accordance, administration of citrate-based CPFA could apparently maintain the EC phenotype in both peritubular capillaries and renal arteries, which eventually extenuated the acute fibrosis in endotoxemic animals. Furthermore, it was observed that inhibiting LPS-induced collagen I production could diminish the effects of LPS on EC dysfunction by removal of LBP from plasma. These evidences implicate EC dysfunction plays a crucial role in the acute progression of tubulointerstitial fibrosis in LPS-mediated AKI. Moreover, it might be a future therapeutic option to protect against EC dysfunction and tissue fibrosis in endotoxemia-induced AKI by specifically removing the LPS adaptor protein LBP (Castellano et al. 2014).

8.5.3 IGF-1R-VE-Protein Tyrosine Phosphatase

Two decades ago, the insulin-like growth factor-1 receptor (IGF-1R) was already implicated in the regulation of cellular proliferation, differentiation and survival (Pollak 2000). Subsequently, there are an increasing number of studies showing that IGF-1R is a protective factor in ECs (Chisalita and Arnqvist 2004). Recently, it was reported that H_2O_2-mediated apoptosis was hampered by IGF-1 via improving mitochondrial dysfunction due to its protective role of maintaining the mitochondrial membrane potential and repressing caspase-3 activity in HUVECs. The mechanism was supposed to be the specific binding of IGF-1 to IGF-1R in kidney and retinal ECs (Pandini et al. 2002; Paneni et al. 2013; Spoerri et al. 1998), but whether IGF-1R plays a protective role in CKD-induced pathological responses is unknown. In this study, it is supposed that IGF-1R could stabilize the VE-protein tyrosine phosphatase/VE-cadherin complex, which is beneficial for maintaining endothelial barrier function.

In accordance, renal fibrosis would be aggravated by the block of IGF-1R due to the endothelial dysfunction (Ferrannini and Solini 2012).

8.5.4 Glucose Uptake

In the normal physiological conditions, approximately 90% of glucose reabsorption was facilitated by sodium-glucose cotransporter 2 (SGLT2) in the convoluted segments of renal proximal tubules where the sodium ion and glucose were transported at a 1:1 stoichiometry from the tubule lumen into the cells (Nakano et al. 2015). AKI is a common clinical syndrome defined as a sudden onset of reduced renal function (Munshi et al. 2011), leading the survivors an increased risk of developing CKD (Gammelager et al. 2014), myocardial infarction and heart failure which might influence AKI at the same time or after (Zhang et al. 2018). In this study, they found that treatment with luseogliflozin (A sodium-glucose cotransporter 2 inhibitor) after I/R attenuated the endothelial rarefaction, renal hypoxia and renal interstitial fibrosis, which implied a novel intervention strategy that the elevation of VEGF-A was partially beneficial for attenuating renal fibrosis by the administration of luseogliflozin which reduced the glucose absorption in the GLUT2-downregulated tubules (Bishop et al. 2009).

8.5.5 Renal Hedgehog Interacting Protein

The renal hedgehog interacting protein (Hhip), a component of the hedgehog pathway, also plays a significant role in the endothelial dysfunction and renal fibrosis. At the beginning, it is identified as a classical antagonist of all Hh ligands including Sonic, Indian and Desert (Bosanac et al. 2009; Chuang and McMahon 1999; Coulombe et al. 2004; Holtz et al. 2015; Kwong et al. 2014; Zhao et al. 2014). An early study in kidney diseases showed that Hhip could hamper nephrogenesis by its dramatical elevation in the kidneys of the diabetes mice models. Recently, a group declared that Hhip gene could be upregulated by high glucose (25 mM D-Glucose) in a time- and dose-dependent manner in cultured metanephric mesenchymal cells, while this elevation would be reversed by insulin, indicating the intimate relationship between glucose with Hhip. Along with the hyperglycemia-mediated renal Hhip expression, nephropathy development in diabetes was accelerated and eventually resulted in the glomerular endothelial dysfunction and renal fibrosis (Arfian et al. 2016).

8.5.6 *Vitamin D and Vascular Remodeling*

Vitamin D has been reported to attenuate kidney fibrosis in UUO model due to reduction of myofibroblast formation, inflammation and tubular cell apoptosis (Martínezmiguel et al. 2014). Vitamin D regulates endothelin-1 and NO production in ECs culture (Sureka et al. 2014). However, the role of their interactions in the pathology of diseases has not been fully understood. It was reported that ET-1/ETBR/eNOS upregulated under the renoprotective effects of vitamin D, which attenuated vascular remodeling and ischemic conditions in kidney fibrosis.

8.6 The Endothelial Repair from AKI to CKD

The kidney has a main function in the retention of total body homeostasis, which acts as a precise machine with many indispensable components including the glomeruli, tubular system, interstitium and vasculature (Sureka et al. 2014). In the kidney, the endothelium is the thin layer lining the interior surface of blood vessels and lymphatic vessels. Notably, the kidney has one of the organs containing the richest and most diversified ECs populations. ECs from renal arteries, arterioles, capillaries, venules, veins and glomerular capillaries have distinctive phenotypic features (Bernatova et al. 2014; Gimbrone and Garcia-Cardena 2013; Rajendran et al. 2013). Endothelium forms a passive barrier, dynamically regulating the permeability of the microvasculature. Phenotypic variations among ECs are mainly responsible for mediating adhesion and leukocytes recruitment (Persson 2015; Shen et al. 2016).

Epithelial–mesenchymal transition was characterized by a phenotypical change occurred in epithelial cells which diminish the inherent cell–cell basement membrane linkage as well as their structural polarity to turn to the spindle-shaped and morphologically similar to mesenchymal/myofibroblast cells and was supposed to contribute to the extracellular matrix (ECM) accumulation and renal fibrosis (Carew et al. 2012; Du et al. 2018; Shu et al. 2018). On the other hand, ECM deposition in the zone between peritubular capillaries and tubules hampered tubular normal function including removing toxins and inducing cellular transport. Therefore, the abundant accumulation of ECM in interstitial and glomeruli is the most remarkable hallmark of renal fibrosis and responsible for kidney destruction (Choi et al. 2015; Theocharis et al. 2016). Eventually, renal fibrosis can be labeled with abundant deposition of ECM composing of collagen I, III and IV, fibronectin, laminin, heparan and perlecan in tubulointerstitium, in particular the surrounding peritubular capillaries (O'Neill et al. 2013; Pushpakumar et al. 2015).

All types of progressive CKD inevitably induced renal fibrosis which finally leads to end-stage renal diseases (ESRD). Therefore, clarifying the pathogenesis of renal fibrosis might essentially promote the realization and intervention for renal failure. On the other hand, AKI represents a significant risk factor for the development of CKD (Demirjian et al. 2014; Drawz and Rahman 2015). Notably, the renal endothelial function may underlie impairment in renal perfusion which might contribute to the potential transition from AKI to CKD. How endothelial repair is mediated by different cell types following vascular injury in AKI is still unclear. Meanwhile, it was considered that endothelial dysfunction acts as a contributor from AKI to CKD (Basile et al. 2018). Some studies showed that tubular damage was exacerbated in renal ischemia-reperfusion, manifested as endothelial swelling (Yasuda et al. 2012), increased expression of adhesion molecules such as P-selectin and ICAM-1 (Boesen et al. 2012; Collett et al. 2017) and recruited various leukocyte populations (Fig. 8.3). Paracrine factors, such as cytokines and the danger signals from stressed epithelium, might be the potential link between the epithelial and vascular compartment, finally lead to the transition from AKI to CKD (Molitoris 2014). It has also been demonstrated that low nephron endowment at birth leads to not only CKD but also AKI development. Another study also revealed that the number of nephrons decreased with age (Suzuki and Arakawa 1991). In addition, although a canonical function of fibroblasts in pathological conditions is a driver of renal fibrosis, fibroblasts are also directly involved in the diverse pathogenesis from AKI to CKD progression (Zal et al. 2014; Zhu et al. 2014).

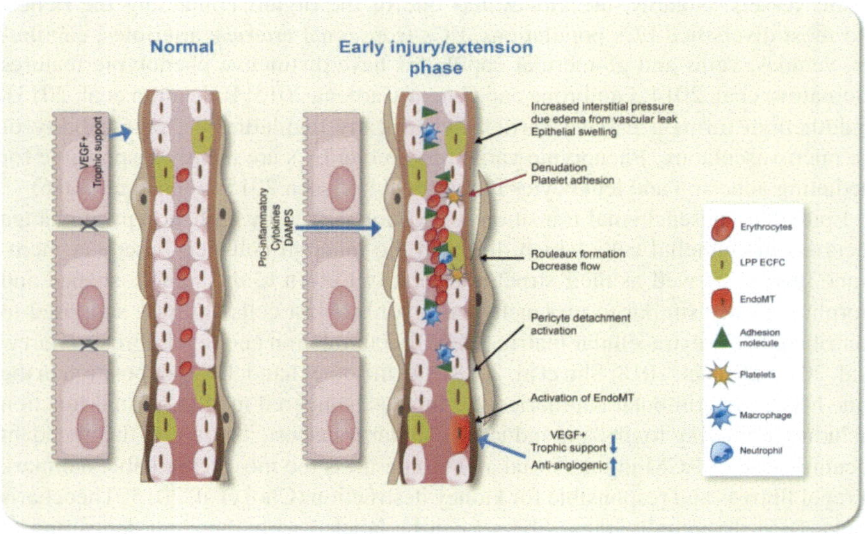

Fig. 8.3 Alterations in endothelial function contribute to the extension phase of AKI. On the left, a peritubular capillary is shown in close apposition to the tubular epithelium in a normal kidney. In response to injury, endothelial swelling narrows capillary space. Increased adhesion molecule expression facilitates leukocyte attachment, contributing to erythrocyte rouleaux formation and disrupting normal blood flow. Reduction in flow contributes to reduced shear stress and inhibition of NO formation, a potential trigger for endothelial–mesenchymal transition (EndMT). Addition potential contributors toward EndMT include a reduction in trophic support from damaged tubules or injury-activated pericytes (Basile et al. 2018)

One study showed that endothelial response might directly result in ischemia by altering the endothelial cytoskeletal structure, subsequently lead to reduced endothelial cell-adhesion complexes, endothelial mitochondrial damage in peritubular capillary and inflammation following ischemia (Fig. 8.4) (Liu et al. 2014; Mooren et al. 2014). Both ischemia and sepsis are known to induce glycocalyx shedding. Meanwhile, the casualty of endothelial cell barrier effect might trigger the coagulation cascades and consequently impair renal function through causing interstitial edema and elevated intratubular pressure (Basile 2007; Fu et al. 2015).

In addition, late outgrown endothelial cells were separated from culture of blood cells on collagen by exorcizing non-adherent monocytes and subsequent expansion, which were defined as endothelial colony forming cells (ECFCs) (Kawasaki et al. 2015). ECFCs express classic markers of ECs including CD31 and VEGFR2 as well as other markers. The peritubular capillary (PTCap) circulation comprises a network of capillary ECs consisting of cortical PTCaps. Up to now, the injury repair responses

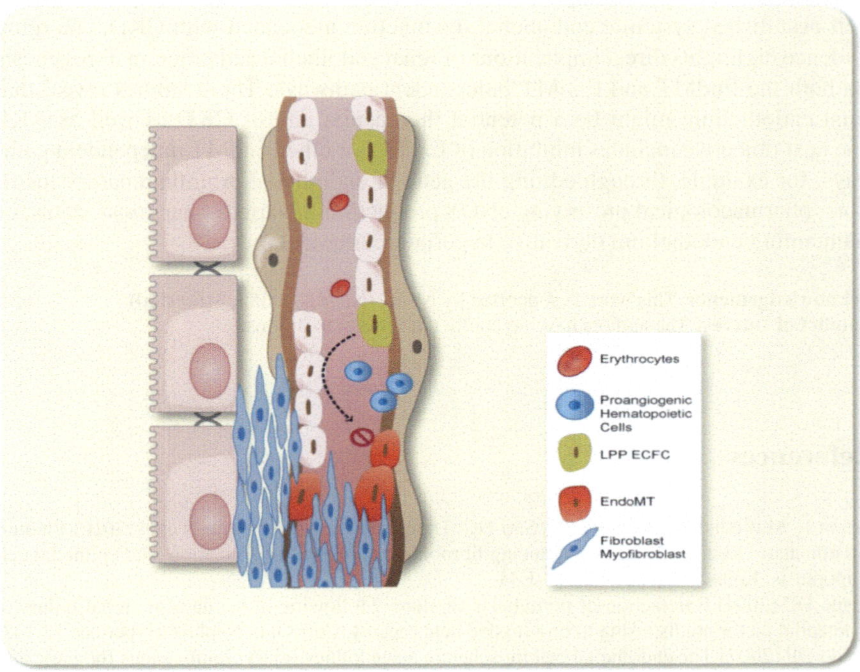

Fig. 8.4 Failed vascular recovery leads to peritubular capillary rarefaction following AKI. Concurrent with the resolution of GFR and tubular repair recovery of the capillary endothelium is inefficient due to a combination of EndMT and low endothelial proliferation. Infiltration of pro-angiogenic hematopoietic cells provides pro-angiogenic stimulation but renal endothelium is unresponsive due to the lack of HPP-ECFC activity intrinsic in the kidney. Expansion of fibroblasts or myofibroblasts, which derive from either pericyte activation or EndMT, may occlude blood vessels leading to a rarefied capillary bed (Basile et al. 2018)

of the ECs of PTCaps are still poorly understood, with a reduction in PTCap density in AKI (Basile 2004). The renal ECs seem to possess poor regenerative capacity, despite with patchy postinjury hypoxic domains that would promote angiogenesis. Generally, angiogenesis is indispensable for the maintenance of tissue function in serious ischemia and growth of new capillaries by sprouting and absorbing from pre-existing capillaries (Mentzer and Konerding 2014). After the injury, ECs exhibit activated caspase-3, suggesting that apoptosis may be responsible for vascular rarefaction (Horbelt et al. 2007).

8.7 Conclusion

Recently, endothelial dysfunction is supposed to be a multifaceted disorder that plays a crucial role in complications involved in kidney diseases in addition to the tubular epithelial cell damage and the interstitial cells activation. Apart from the well-established systemic endothelial dysfunction associated with CKD, emerging evidence highlights direct implications of renal endothelial activation in fibrogenesis via both the EndMT and EndMT independent pathways. These studies reveal that renal endothelium might be a potential therapeutic aim of CKD as well as AKI. The next objective includes inhibition of EndMT or other EndMT independent pathways, for example, through editing the gene of endothelial proinflammatory mediators, pharmacological preserving eNOS production in varied shear stress status, or maintaining endothelium derivative hyperpolarizing factors.

Acknowledgements This work is supported by NSFC (No: 81670655, 81600520).
Conflict of Interest The authors have no conflict of interest to disclose.

References

Arfian N, Muflikhah K, Soeyono SK, Sari DC, Tranggono U, Anggorowati N et al (2016) Vitamin d attenuates kidney fibrosis via reducing fibroblast expansion, inflammation, and epithelial cell apoptosis. Kobe J Med Sci 62:E38–E44

Basile DP (2004) Rarefaction of peritubular capillaries following ischemic acute renal failure: a potential factor predisposing to progressive nephropathy. Curr Opin Nephrol Hypertens 13:1–7

Basile DP (2007) The endothelial cell in ischemic acute kidney injury: implications for acute and chronic function. Kidney Int 72:151–156

Basile DP, Collett JA, Yoder MC (2018) Endothelial colony-forming cells and pro-angiogenic cells: clarifying definitions and their potential role in mitigating acute kidney injury. Acta Physiol (Oxf) 222(2):e12914

Bernatova I, Andriantsitohaina R, Arribas SM, Matchkov VV (2014) Endothelium in diseased states. Biomed Res Int 2014:810436

Bishop B, Aricescu AR, Harlos K, O'Callaghan CA, Jones EY, Siebold C (2009) Structural insights into hedgehog ligand sequestration by the human hedgehog-interacting protein hhip. Nat Struct Mol Biol 16:698–703

Boesen EI, Crislip GR, Sullivan JC (2012) Use of ultrasound to assess renal reperfusion and p-selectin expression following unilateral renal ischemia. Am J Physiol Renal Physiol 303:F1333–F1340

Bosanac I, Maun HR, Scales SJ, Wen X, Lingel A, Bazan JF et al (2009) The structure of SHH in complex with HHIP reveals a recognition role for the SHH pseudo active site in signaling. Nat Struct Mol Biol 16:691–697

Bussolati B, David S, Cambi V, Tobias PS, Camussi G (2002) Urinary soluble CD14 mediates human proximal tubular epithelial cell injury induced by LPS. Int J Mol Med 10:441–449

Carew RM, Wang B, Kantharidis P (2012) The role of EMT in renal fibrosis. Cell Tissue Res 347:103–116

Castellano G, Stasi A, Intini A, Gigante M, Di Palma AM, Divella C et al (2014) Endothelial dysfunction and renal fibrosis in endotoxemia-induced oliguric kidney injury: possible role of LPS-binding protein. Crit Care 18:520

Chen IC, Chiang WF, Chen PF, Chiang HC (2014) Stress-responsive deacetylase SIRT3 is up-regulated by areca nut extract-induced oxidative stress in human oral keratinocytes. J Cell Biochem 115:328–339

Chen CL, Chou KJ, Fang HC, Hsu CY, Huang WC, Huang CW et al (2015) Progenitor-like cells derived from mouse kidney protect against renal fibrosis in a remnant kidney model via decreased endothelial mesenchymal transition. Stem Cell Res Ther 6:239

Chisalita SI, Arnqvist HJ (2004) Insulin-like growth factor i receptors are more abundant than insulin receptors in human micro- and macrovascular endothelial cells. Am J Physiol Endocrinol Metab 286:E896–E901

Cho MH (2010) Renal fibrosis. Korean J Pediatr 53:735–740

Choi SH, Chun SY, Chae SY, Kim JR, Oh SH, Chung SK et al (2015) Development of a porcine renal extracellular matrix scaffold as a platform for kidney regeneration. J Biomed Mater Res A 103:1391–1403

Chuang PT, McMahon AP (1999) Vertebrate hedgehog signalling modulated by induction of a hedgehog-binding protein. Nature 397:617–621

Collett JA, Mehrotra P, Crone A, Shelley WC, Yoder MC, Basile DP (2017) Endothelial colony-forming cells ameliorate endothelial dysfunction via secreted factors following ischemia-reperfusion injury. Am J Physiol Renal Physiol 312:F897–F907

Coresh J, Selvin E, Stevens LA, Manzi J, Kusek JW, Eggers P et al (2007) Prevalence of chronic kidney disease in the united states. JAMA 298:2038–2047

Coulombe J, Traiffort E, Loulier K, Faure H, Ruat M (2004) Hedgehog interacting protein in the mature brain: membrane-associated and soluble forms. Mol Cell Neurosci 25:323–333

Cruz-Solbes AS, Youker K (2017) Epithelial to mesenchymal transition (EMT) and endothelial to mesenchymal transition (EndMT): role and implications in kidney fibrosis. Results Probl Cell Differ 60:345–372

Curci C, Castellano G, Stasi A, Divella C, Loverre A, Gigante M et al (2014) Endothelial-to-mesenchymal transition and renal fibrosis in ischaemia/reperfusion injury are mediated by complement anaphylatoxins and Akt pathway. Nephrol Dial Transplant 29:799–808

Deanfield JE, Halcox JP, Rabelink TJ (2007) Endothelial function and dysfunction: testing and clinical relevance. Circulation 115:1285–1295

Demirjian S, Lane BR, Derweesh IH, Takagi T, Fergany A, Campbell SC (2014) Chronic kidney disease due to surgical removal of nephrons: relative rates of progression and survival. J Urol 192:1057–1062

Dikalova AE, Itani HA, Nazarewicz RR, McMaster WG, Flynn CR, Uzhachenko R et al (2017) Sirt3 impairment and SOD2 hyperacetylation in vascular oxidative stress and hypertension. Circ Res 121:564–574

Drawz P, Rahman M (2015) Chronic kidney disease. Ann Intern Med 162:Itc1–Itc16

Du Y, Chen Q, Huang L, Wang S, Yin X, Zhou L et al (2018) VEGFR2 and VEGF-C suppresses the epithelial-mesenchymal transition via yap in retinal pigment epithelial cells. Curr Mol Med 18:273–286

Dunzendorfer S, Lee HK, Soldau K, Tobias PS (2004) Toll-like receptor 4 functions intracellularly in human coronary artery endothelial cells: roles of LBP and sCD14 in mediating lps responses. FASEB J 18:1117–1119

Feletou M, Vanhoutte PM (2006) Endothelial dysfunction: a multifaceted disorder (the Wiggers award lecture). Am J Physiol Heart Circ Physiol 291:H985–H1002

Ferrannini E, Solini A (2012) SGLT2 inhibition in diabetes mellitus: rationale and clinical prospects. Nat Rev Endocrinol 8:495–502

Flanders KC (2004) Smad3 as a mediator of the fibrotic response. Int J Exp Pathol 85:47–64

Fu J, Lee K, Chuang PY, Liu Z, He JC (2015) Glomerular endothelial cell injury and cross talk in diabetic kidney disease. Am J Physiol Renal Physiol 308:F287–F297

Fujimoto M, Maezawa Y, Yokote K, Joh K, Kobayashi K, Kawamura H et al (2003) Mice lacking Smad3 are protected against streptozotocin-induced diabetic glomerulopathy. Biochem Biophys Res Commun 305:1002–1007

Gammelager H, Christiansen CF, Johansen MB, Tonnesen E, Jespersen B, Sorensen HT (2014) Three-year risk of cardiovascular disease among intensive care patients with acute kidney injury: a population-based cohort study. Crit Care 18:492

Gimbrone MA Jr, Garcia-Cardena G (2013) Vascular endothelium, hemodynamics, and the pathobiology of atherosclerosis. Cardiovasc Pathol 22:9–15

Guerrot D, Dussaule JC, Kavvadas P, Boffa JJ, Chadjichristos CE, Chatziantoniou C (2012) Progression of renal fibrosis: the underestimated role of endothelial alterations. Fibrogenesis Tissue Repair 5:S15

Haigis MC, Guarente LP (2006) Mammalian sirtuins—emerging roles in physiology, aging, and calorie restriction. Genes Dev 20:2913–2921

He J, Xu Y, Koya D, Kanasaki K (2013) Role of the endothelial-to-mesenchymal transition in renal fibrosis of chronic kidney disease. Clin Exp Nephrol 17:488–497

Holtz AM, Griffiths SC, Davis SJ, Bishop B, Siebold C, Allen BL (2015) Secreted HHIP1 interacts with heparan sulfate and regulates hedgehog ligand localization and function. J Cell Biol 209:739–757

Horbelt M, Lee SY, Mang HE, Knipe NL, Sado Y, Kribben A et al (2007) Acute and chronic microvascular alterations in a mouse model of ischemic acute kidney injury. Am J Physiol Renal Physiol 293:F688–F695

Huang XR, Chung AC, Yang F, Yue W, Deng C, Lau CP et al (2010) Smad3 mediates cardiac inflammation and fibrosis in angiotensin II-induced hypertensive cardiac remodeling. Hypertension 55:1165–1171

Iliescu R, Fernandez SR, Kelsen S, Maric C, Chade AR (2010) Role of renal microcirculation in experimental renovascular disease. Nephrol Dial Transplant 25:1079–1087

Ishikane S, Ohnishi S, Yamahara K, Sada M, Harada K, Mishima K et al (2008) Allogeneic injection of fetal membrane-derived mesenchymal stem cells induces therapeutic angiogenesis in a rat model of hind limb ischemia. Stem Cells 26:2625–2633

Joannides R, Haefeli WE, Linder L, Richard V, Bakkali EH, Thuillez C et al (1995) Nitric oxide is responsible for flow-dependent dilatation of human peripheral conduit arteries in vivo. Circulation 91:1314–1319

Kang DH (2018) Hyperuricemia and progression of chronic kidney disease: role of phenotype transition of renal tubular and endothelial cells. Contrib Nephrol 192:48–55

Kang DH, Joly AH, Oh SW, Hugo C, Kerjaschki D, Gordon KL et al (2001) Impaired angiogenesis in the remnant kidney model: I Potential role of vascular endothelial growth factor and thrombospondin-1. J Am Soc Nephrol 12:1434–1447

Kawasaki T, Nishiwaki T, Sekine A, Nishimura R, Suda R, Urushibara T et al (2015) Vascular repair by tissue-resident endothelial progenitor cells in endotoxin-induced lung injury. Am J Respir Cell MolBiol 53:500–512

Kisanuki YY, Hammer RE, Miyazaki J, Williams SC, Richardson JA, Yanagisawa M (2001) Tie2-Cre transgenic mice: a new model for endothelial cell-lineage analysis in vivo. Dev Biol 230:230–242

Kretzschmar M, Massague J (1998) SMADs: mediators and regulators of TGF-beta signaling. Curr Opin Genet Dev 8:103–111

Kwong L, Bijlsma MF, Roelink H (2014) SHH-mediated degradation of HHIP allows cell autonomous and non-cell autonomous shhsignalling. Nat Commun 5:4849

Lee PT, Lin HH, Jiang ST, Lu PJ, Chou KJ, Fang HC et al (2010) Mouse kidney progenitor cells accelerate renal regeneration and prolong survival after ischemic injury. Stem Cells 28:573–584

Li J, Qu X, Bertram JF (2009) Endothelial-myofibroblast transition contributes to the early development of diabetic renal interstitial fibrosis in streptozotocin-induced diabetic mice. Am J Pathol 175:1380–1388

Li J, Qu X, Yao J, Caruana G, Ricardo SD, Yamamoto Y et al (2010) Blockade of endothelial-mesenchymal transition by a smad3 inhibitor delays the early development of streptozotocin-induced diabetic nephropathy. Diabetes 59:2612–2624

Lin JR, Zheng YJ, Zhang ZB, Shen WL, Li XD, Wei T et al (2018) Suppression of endothelial-to-mesenchymal transition by SIRT (sirtuin) 3 alleviated the development of hypertensive renal injury. Hypertension 72:350–360

Lipphardt M, Dihazi H, Jeon NL, Dadafarin S, Ratliff BB, Rowe DW et al (2019) Dickkopf-3 in aberrant endothelial secretome triggers renal fibroblast activation and endothelial-mesenchymal transition. Nephrol Dial Transplant 34:49–62

Liu Y (2011) Cellular and molecular mechanisms of renal fibrosis. Nat Rev Nephrol 7:684–696

Liu Z, Huang XR, Lan HY (2012) Smad3 mediates ANG II-induced hypertensive kidney disease in mice. Am J Physiol Renal Physiol 302:F986–F997

Liu S, Soong Y, Seshan SV, Szeto HH (2014) Novel cardiolipin therapeutic protects endothelial mitochondria during renal ischemia and mitigates microvascular rarefaction, inflammation, and fibrosis. Am J Physiol Renal Physiol 306:F970–F980

Loverre A, Ditonno P, Crovace A, Gesualdo L, Ranieri E, Pontrelli P et al (2004) Ischemia-reperfusion induces glomerular and tubular activation of proinflammatory and antiapoptotic pathways: differential modulation by rapamycin. J Am Soc Nephrol 15:2675–2686

Markwald RR, Fitzharris TP, Smith WN (1975) Sturctural analysis of endocardial cytodifferentiation. DevBiol 42:160–180

Martínezmiguel P, Valdivielso JM, Medranoandrés D, Romángarcía P, Canopeñalver JL, Rodríguezpuyol M et al (2014) The active form of vitamin d, calcitriol, induces a complex dual upregulation of endothelin and nitric oxide in cultured endothelial cells. Am J Physiol Endocrinol Metab 307:1085–1096

Matsuzaki K (2012) Smadphosphoisoform signals in acute and chronic liver injury: similarities and differences between epithelial and mesenchymal cells. Cell Tissue Res 347:225–243

Medici D, Kalluri R (2012) Endothelial-mesenchymal transition and its contribution to the emergence of stem cell phenotype. Semin Cancer Biol 22:379–384

Mentzer SJ, Konerding MA (2014) Intussusceptive angiogenesis: expansion and remodeling of microvascular networks. Angiogenesis 17:499–509

Mercado-Pimentel ME, Runyan RB (2007) Multiple transforming growth factor-beta isoforms and receptors function during epithelial-mesenchymal cell transformation in the embryonic heart. Cells Tissues Organs 185:146–156

Molitoris BA (2014) Therapeutic translation in acute kidney injury: the epithelial/endothelial axis. J Clin Invest 124:2355–2363

Mooren OL, Li J, Nawas J, Cooper JA (2014) Endothelial cells use dynamic actin to facilitate lymphocyte transendothelial migration and maintain the monolayer barrier. Mol Biol Cell 25:4115–4129

Munshi R, Hsu C, Himmelfarb J (2011) Advances in understanding ischemic acute kidney injury. BMC Med 9:11

Nakano D, Doi K, Kitamura H, Kuwabara T, Mori K, Mukoyama M et al (2015) Reduction of tubular flow rate as a mechanism of oliguria in the early phase of endotoxemia revealed by intravital imaging. J Am Soc Nephrol 26:3035–3044

Newman JC, He W, Verdin E (2012) Mitochondrial protein acylation and intermediary metabolism: regulation by sirtuins and implications for metabolic disease. J Biol Chem 287:42436–42443

O'Neill JD, Freytes DO, Anandappa AJ, Oliver JA, Vunjak-Novakovic GV (2013) The regulation of growth and metabolism of kidney stem cells with regional specificity using extracellular matrix derived from kidney. Biomaterials 34:9830–9841

Onyango P, Celic I, McCaffery JM, Boeke JD, Feinberg AP (2002) SIRT3, a human SIR2 homologue, is an nad-dependent deacetylase localized to mitochondria. Proc Natl Acad Sci USA 99:13653–13658

Pandini G, Frasca F, Mineo R, Sciacca L, Vigneri R, Belfiore A (2002) Insulin/insulin-like growth factor I hybrid receptors have different biological characteristics depending on the insulin receptor isoform involved. J Biol Chem 277:39684–39695

Paneni F, Beckman JA, Creager MA, Cosentino F (2013) Diabetes and vascular disease: pathophysiology, clinical consequences, and medical therapy: part I. Eur Heart J 34:2436–2443

Persson PB (2015) The multiple functions of the endothelium: More than just wallpaper. Acta Physiol (Oxf) 213:747–749

Piek E, Ju WJ, Heyer J, Escalante-Alcalde D, Stewart CL, Weinstein M et al (2001) Functional characterization of transforming growth factor beta signaling in SMAD2- and SMAD3-deficient fibroblasts. J Biol Chem 276:19945–19953

Piera-Velazquez S, Li Z, Jimenez SA (2011) Role of endothelial-mesenchymal transition (EndoMT) in the pathogenesis of fibrotic disorders. Am J Pathol 179:1074–1080

Pollak M (2000) Insulin-like growth factor physiology and cancer risk. Eur J Cancer 36:1224–1228

Pushpakumar S, Kundu S, Narayanan N, Sen U (2015) DNA hypermethylation in hyperhomocysteinemia contributes to abnormal extracellular matrix metabolism in the kidney. FASEB J 29:4713–4725

Rajendran P, Rengarajan T, Thangavel J, Nishigaki Y, Sakthisekaran D, Sethi G et al (2013) The vascular endothelium and human diseases. Int J Biol Sci 9:1057–1069

Ramnath RD, Ng SW, Guglielmotti A, Bhatia M (2008) Role of MCP-1 in endotoxemia and sepsis. Int Immunopharmacol 8:810–818

Rieder F, Kessler SP, West GA, Bhilocha S, de la Motte C, Sadler TM et al (2011) Inflammation-induced endothelial-to-mesenchymal transition: a novel mechanism of intestinal fibrosis. Am J Pathol 179:2660–2673

Ronco C, Brendolan A, Dan M, Piccinni P, Bellomo R, De Nitti C et al (2000) Adsorption in sepsis. Kidney Int Suppl 76:S148–S155

Sanchez-Duffhues G, Orlova V, Ten Dijke P (2016) In brief: endothelial-to-mesenchymal transition. J Pathol 238:378–380

Sato M, Muragaki Y, Saika S, Roberts AB, Ooshima A (2003) Targeted disruption of TGF-beta1/smad3 signaling protects against renal tubulointerstitial fibrosis induced by unilateral ureteral obstruction. J Clin Invest 112:1486–1494

Shen L, Evans IM, Souza D, Dreifaldt M, Dashwood MR, Vidya MA (2016) Adiponectin: an endothelium-derived vasoprotective factor? Curr Vasc Pharmacol 14:168–174

Shiojima I, Walsh K (2002) Role of Akt signaling in vascular homeostasis and angiogenesis. Circ Res 90:1243–1250

Shu DY, Wojciechowski M, Lovicu FJ (2018) Erk1/2-mediated EGFR-signaling is required for TGFbeta-induced lens epithelial-mesenchymal transition. Exp Eye Res 178:108–121

Singla S, Predescu D, Bardita C, Wang M, Zhang J, Balk RA et al (2011) Pro-inflammatory endothelial cell dysfunction is associated with intersectin-1s down-regulation. Respir Res 12:46

Spoerri PE, Ellis EA, Tarnuzzer RW, Grant MB (1998) Insulin-like growth factor: receptor and binding proteins in human retinal endothelial cell cultures of diabetic and non-diabetic origin. Growth Horm IGF Res 8:125–132

Sun S, Ning X, Zhang Y, Lu Y, Nie Y, Han S et al (2009) Hypoxia-inducible factor-1alpha induces twist expression in tubular epithelial cells subjected to hypoxia, leading to epithelial-to-mesenchymal transition. Kidney Int 75:1278–1287

Sun YB, Qu X, Li X, Nikolic-Paterson DJ, Li J (2013) Endothelial dysfunction exacerbates renal interstitial fibrosis through enhancing fibroblast smad3 linker phosphorylation in the mouse obstructed kidney. PLoS ONE 8:e84063
Sureka B, Mittal MK, Mittal A, Sinha M, Thukral BB (2014) Supernumerary kidneys—a rare anatomic variant. Surg Radiol Anat 36:199–202
Suzuki S, Arakawa M (1991) Age-related changes in the kidney of patients with mild mesangial proliferative glomerulonephritis. Nihon Ronen Igakkai Zasshi 28:311–317
Theocharis AD, Skandalis SS, Gialeli C, Karamanos NK (2016) Extracellular matrix structure. Adv Drug Deliv Rev 97:4–27
Thuillez C, Richard V (2005) Targeting endothelial dysfunction in hypertensive subjects. J Hum Hypertens 19(Suppl 1):S21–S25
Togel F, Weiss K, Yang Y, Hu Z, Zhang P, Westenfelder C (2007) Vasculotropic, paracrine actions of infused mesenchymal stem cells are important to the recovery from acute kidney injury. Am J Physiol Renal Physiol 292:F1626–F1635
Wang A, Ziyadeh FN, Lee EY, Pyagay PE, Sung SH, Sheardown SA et al (2007) Interference with TGF-beta signaling by Smad3-knockout in mice limits diabetic glomerulosclerosis without affecting albuminuria. Am J Physiol Renal Physiol 293:F1657–F1665
Wynn TA (2008) Cellular and molecular mechanisms of fibrosis. J Pathol 214:199–210
Yasuda K, Vasko R, Hayek P, Ratliff B, Bicer H, Mares J et al (2012) Functional consequences of inhibiting exocytosis of weibel-palade bodies in acute renal ischemia. Am J Physiol Renal Physiol 302:F713–F721
Zal F, Mostafavi-Pour Z, Moattari A, Sardarian A, Vessal M (2014) Altered expression of alpha 2 beta 1 integrin in kidney fibroblasts: a potential mechanism for CsA-induced nephrotoxicity. Arch Iran Med 17:556–562
Zarjou A, Agarwal A (2011) Sepsis and acute kidney injury. J Am SocNephrol 22:999–1006
Zeisberg EM, Potenta SE, Sugimoto H, Zeisberg M, Kalluri R (2008) Fibroblasts in kidney fibrosis emerge via endothelial-to-mesenchymal transition. J Am Soc Nephrol 19:2282–2287
Zhang Y, Nakano D, Guan Y, Hitomi H, Uemura A, Masaki T et al (2018) A sodium-glucose cotransporter 2 inhibitor attenuates renal capillary injury and fibrosis by a vascular endothelial growth factor-dependent pathway after renal injury in mice. Kidney Int 94:524–535
Zhao XP, Liao MC, Chang SY, Abdo S, Aliou Y, Chenier I et al (2014) Maternal diabetes modulates kidney formation in murine progeny: the role of hedgehog interacting protein (HHIP). Diabetologia 57:1986–1996
Zhu R, Zheng R, Deng Y, Chen Y, Zhang S (2014) Ergosterol peroxide from cordyceps cicadae ameliorates TGF-beta1-induced activation of kidney fibroblasts. Phytomedicine 21:372–378
Zimmerman CM, Padgett RW (2000) Transforming growth factor beta signaling mediators and modulators. Gene 249:17–30

Chapter 9
Mesangial Cells and Renal Fibrosis

Jing-Hong Zhao

Abstract The main cellular constituents in glomerular mesangium are mesangial cells, which account for approximately 30–40% of the total cells in the glomerulus. Together with the mesangial matrix, mesangial cells form the glomerular basement membrane (GBM) in the glomerulus, whose main function is to perform the filtration. Under the pathologic conditions, mesangial cells are activated, leading to hyperproliferation and excess extracellular matrix (ECM). Moreover, mesangial cells also secrete several kinds of inflammatory cytokines, adhesion molecules, chemokines, and enzymes, all of which participate in the process of renal glomerular fibrosis. During the past years, researchers have revealed the roles of mesangial cells and the associated signal pathways involved in renal fibrosis. In this section, we will discuss how mesangial cells are activated and its contributions to renal fibrosis, as well as the molecular mechanisms and novel anti-fibrotic agents. Full understanding of the contributions of mesangial cells to renal fibrosis will benefit the clinical drug developing.

Keyword Mesangial cell · Glomerulosclerosis · Renal fibrosis · TGF-β

9.1 Introduction

Renal fibrosis is a final common pathological consequence of most chronic progressive kidney diseases. Fibrosis is characterized by a loss of capillary networks and a deposit of fibrillary collagens in both renal interstitial and glomerular compartments. As known, the activities of the interstitial components, especially fibroblasts, are closely correlated with renal interstitial fibrosis. However, in recent years, more and more studies have found that renal mesangial cells also play important roles in the pathogenesis of glomerular fibrosis which ultimately leads to glomerular sclerosis.

Among renal cells, the glomerular mesangial cells share most phenotypical similarities with the fibroblasts. As one of the major matrix-producing cells, mesangial

J.-H. Zhao (✉)
Department of Nephrology, Xinqiao Hospital, Army Medical University, Chongqing, China
e-mail: zhaojinghong_cq@163.com

B.-C. Liu et al. (eds.), *Renal Fibrosis: Mechanisms and Therapies*,
Advances in Experimental Medicine and Biology 1165,
https://doi.org/10.1007/978-981-13-8871-2_9

cells can secrete mesangial matrix (MM) components, such as type IV and type V collagens and fibronectin, which contribute to the excess extracellular matrix (ECM). Moreover, mesangial cells also secrete several kinds of inflammatory cytokines, adhesion molecules, chemokines, and enzymes, all of which also participate in the process of renal fibrosis. An in-depth understanding of the mesangial dysfunction and the associated mechanisms will not only help us orchestrate multiple pathophysiological processes in renal fibrosis, but also provide novel insights into the development of novel therapeutic approaches. Thus, this chapter will spotlight on the basic knowledge and recent advances of researches about mesangial cells.

9.2 What Are Mesangium and Mesangial Cells?

The mesangium is a cluster of mononuclear stellate cells embedded in ECM. It occupies a pivotal anatomical position that located between the capillaries and the glomerular basement membrane (GBM) in the glomerulus. Mesangial cells are the main cellular constituents in mesangium. It is estimated that the mesangial cells account for nearly one-third of the total number of glomerulus cells (Olivetti et al. 1977, 1980). Morphologically, the mesangial cells are low cytoplasm-to-nucleus ratio, with many cytoplasmic processes containing fibrils (Makino 1988). Locationally, the endothelial cells are incompletely surrounded by GBM and epithelial cells (podocytes), leaving a gap for direct communication between mesangial cells and endothelial wall (Makino 1988). This area is considered as the juxtamesangium of the capillary wall. While the areas where there is no direct contact are defined as extraglomerular mesangium (Mené et al. 1989). The most significant characteristic of the extraglomerular mesangium is a separation between mesangial cells and the GBM by mesangial-derived ECM (Andrews and Coffey 1983). Based on the microscope, the ECM performs as a dense net with large amount of microfibrils, anchored to cell membranes via fibronectin, one of the most major components of the ECM (Silva et al. 1986; Makino 1988; Kriz et al. 1990). This extraordinarily structural arrangement may result in two major functional consequences: First, because of the luxuriant components of cytoskeleton, the mesangial cells can contract similar to smooth muscle cells and exert mechanical traction on the GBM and the endothelial capillary lining by mesangial processes and intermediary ECM. This contractile peculiarity not only can prevent an abnormal capillary hydrostatic pressure, but also subtly maintaining steady of capillary nerves. In addition, given that an abundant of mesangial processes stretching into the capillary lumen, it is conceivable that even a minor reduction or slight alteration of this area will sensibly affect the filtering proportion and ultrafiltration rate of the glomerulus. On the other hand, inflammatory swelling of the mesangium and endothelial cells may directly affect glomerular ultrafiltration (Johnston and Latta 1977; Latta and Fligiel 1985). Second, the contact between endothelial filtration surfaces and the mesangium provides a structural basis for the plasma ultrafiltration through the mesangium (Latta et al. 1960; Latta and Maunsbach 1962; Latta and Johnston 1978; Keane and Raij 1981). Resembling the hydrophilic and polyanionic composition of the GBM, this structure of the mesangium guarantees an ideal sieving for macromolecules of blood flow (Latta and Lee 1983).

In ultrastructural studies, the body of endothelial cells is often found associated with the juxtamesangial portion of the capillary (Sakai and Kriz 1987). The two cell types generally present wide contact surfaces without specialized membrane junctions that are often filled by a thin layer of ECM, indicating the absence of electrical linkage. The proximity of the two cells is interesting in view of the proposed role of endothelial cells in clearing deposits of filtered macromolecules by cytoplasmic envelopment and transfer to mesangial areas (Latta and Maunsbach 1962; Elema et al. 1976; Cattell et al. 1982). Numerous studies have suggested that two major routes exist for mesangial filtration, a rapid influx of material through fenestrations, which is rate limited by particle size and charge, and a slow, cell-mediated uptake and processing in which both endothelial and mesangial cells are implicated (Latta and Johnston 1978; Latta and Lee 1983; Latta and Fligiel 1985). The frequent accumulation of immunocyte deposited in this area in glomerulonephritis may represent overloading of these cell-dependent mechanisms. Efflux of macromolecules filtered across the mesangium seems to occur through three ways as follows (Michielsen and Creemers 1967; Latta 1970; Elema et al. 1976): (1) transfer across the GBM and epithelial foot processes into the urinary space, (2) flow through channels and the fibrillar matrix network into efferent capillaries, or (3) concentration at the glomerular hilus in proximity to the juxtaglomerular apparatus. This latter site, where tracers often accumulate at variable intervals after injection, is likely to provide access to the lymphatic drainage.

In conclusion, structural and functional characteristics justify the definition of mesangial cells as myofibroblasts—mesenchymal cells share features of both smooth muscle and fibroblasts (Michielsen and Creemers 1967). Marked analogies between these three cell types in culture support the concept of a multipotent interstitial cell of the kidney glomerulus, which is different at structural, mechanical, secretory, and metabolic functions. Altered environmental conditions, such as inflammation or metabolic disease states, may modify its (mesangial cells) phenotype to express characteristics better suited to perform specific tasks (Fig. 9.1).

9.3 Role of Mesangial Cell on Glomerular Fibrosis and its Response to Injury in Associated Diseases

Renal glomerular fibrosis is a process by which normal, functional glomerular tissue is replaced by accumulated deposits of ECM (Eddy 1996). It represents a final common pathway for glomerular lesions in all progressive forms of renal diseases, particularly associating with primary renal diseases such as immunoglobulin A (IgA) glomerulonephritis or metabolic diseases such as diabetic nephropathy (Hewitson and Becker 1995; Chen et al. 2003; Simonson 2007; Qian et al. 2008; Schmekel et al.

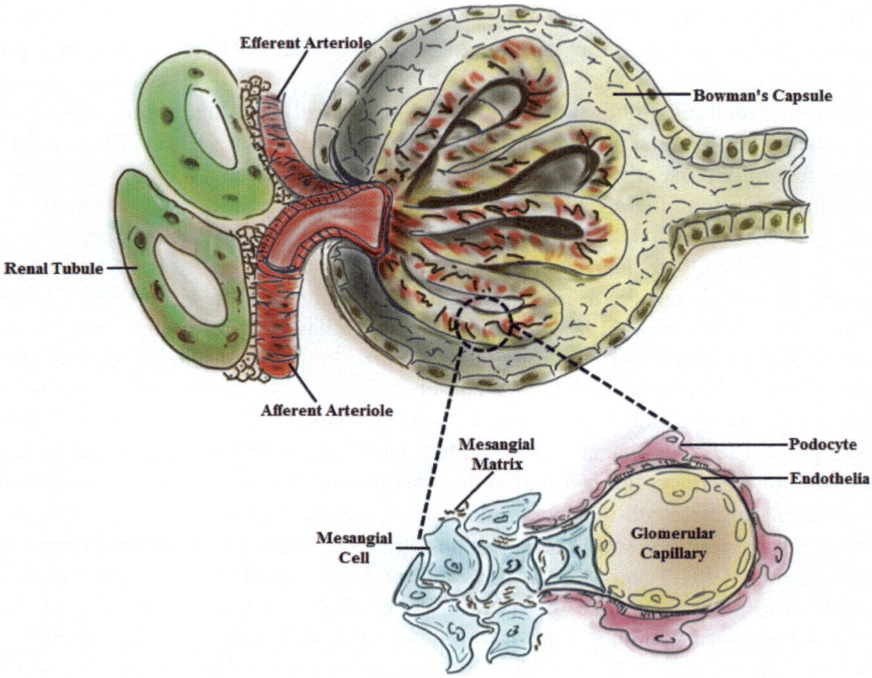

Fig. 9.1 Physiological sketch of mesangium

2010). To date, numerous studies have documented host of factors which play instrumental role in orchestrating fibrosis by multiple mechanisms. However, the initial pathogenesis of renal fibrosis is still uncertain. It is conceivable that three major cells of the glomerulus may participate in glomerular fibrotic process: podocyte, endothelial and mesangial cells (Eddy 2000; Liu 2006). The published genetic data of human nephrotic syndrome and the findings of transgenic murine podocyte abnormalities suggest that the damage of the visceral epithelial cell triggers glomerulosclerosis (Shirato et al. 1996; Wendt et al. 2003; Wharram et al. 2005; Wiggins et al. 2005; Shi et al. 2008). This assertion is supported by luxuriant data, implicating those potential epithelial cell stressors such as glomerular hypertension, hyperfiltration, or hypertrophy act as a detrimental role in sclerosis (Hostetter et al. 1981; Yoshida et al. 1989; Bidani et al. 1990; Miller et al. 1991; Ichikawa et al. 2005). Although podocytes play an important role in orchestrating glomerular sclerosis, it may not be the direct precursor of myofibroblasts which induce the major secretion of fibrillary collagens and accelerate progression of fibrosis. Still others suggest a role for glomerular endothelial cells (Liang et al. 2016a). Toyoda and his coworkers document that diminished endothelial cell fenestration can promote renal progression in diabetic nephropathy (Toyoda et al. 2007). Nevertheless, a reduction in the density and size of endothelial fenestrations seems prevailingly to lower glomerular hydraulic permeability, which directly correlates much closer with GFR than fibrosis. The only other morphometric variable to correlate remarkably with glomerular fibrosis is mesangial area. The last possibility is the most attractive because, in many models of glomerulosclerosis, a common major event of any origin is an increase of the mesangial compartment size, attributable to both mesangial matrix deposition and mesangial cell proliferation and hypertrophy (Ruef et al. 1992; Yuan et al. 2017). However, still very little is known about the mechanisms and functional understandings of mesangial cells in process from fibrosis to sclerosis.

Accumulating evidence demonstrate that mesangial cells are activated in numerous glomerular diseases, proliferate, and undergo a phenotypic alteration. This phenotypic modulation transforms the resting mesangial cell to express both smooth muscle-like (characterized by expression of α-smooth muscle actin) and fibroblast-like (characterized by generation of interstitial collagens such as type I and type III in addition to normal mesangial matrix constituents) features, supporting that the "activated" mesangial cell may be regarded as a type of myofibroblast (Johnson et al. 1992, Patel et al. 2003). It is presumable that proper response of glomerular injury allows its structure to restore. However, the organism often chooses an overcompensatory reaction after injury, which results in mesangial proliferation along with abnormal deposition of ECM, consequently inducing glomerular fibrosis or sclerosis.

For instance, large amounts of evidences have indicated that high glucose not only induced mesangial proliferation, but also stimulated the cell into expressing some pro-inflammatory factors (such as IL-1β, IL-6, TNF, CCL2/3, CXCL8) or pro-fibrotic constituents (such as TGF-β, fibronectin), mediated activation of reactive oxidation stress (ROS) to enhance oxidative stress level, and consequently triggered glomerular fibrosis (Nahman et al. 1992; Yuan et al. 2011; Zhang et al. 2012). Moreover, in diabetic rodent model, nodular expansion of glomerular mesangium is always along with increased quantities of ECM, and over 60% amount of type IV collagen, fibronectin and laminin are secreted by mesangial cells after glucose treatment (Reidy et al. 2014).

Another glomerular disease related closely to mesangial disorder is IgA nephropathy. A high level of circulating IgA which derived from a defective immune clearance or an increased production can increase risk of glomerulonephritis. The mesangial trapping of serum IgA immune complexes appears to be the most important event eliciting the IgA nephropathy (Pillebout and Nochy 2010). Both serum IgA and macromolecular IgA can react with mesangial matrix components (such as laminin, fibronectin, or type IV collagen), and the binding of IgA to them contributes to affinity between IgA and mesangium (Coppo et al. 1993). Owing to the high affinity, IgA prefers to deposit in mesangium area and targets to mesangial cells. In vitro study, the secretory IgA isolated from patients with IgA nephropathy can result in cultured human mesangial cells hypercellularity and stimulate the cell into secreting pro-inflammatory cytokines production (Liang et al. 2015). There is relentless accumulation of mesangial matrix with progressive IgA nephropathy, finally leading to glomerulosclerosis.

To date, a widespread recognized understanding of these unfortunate events is glomerular ECM accumulation. Actually, more than one paper support that the ECM accumulation often appears to begin in the mesangium (Pugliese et al. 1994; Fogo 1999). It has also been proposed that accumulation of mesangial-derived ECM in the mesangial interstitial space could alter the living conditions and further influence the cell status.

In another aspect, activated mesangial cells also secrete less matrix metalloproteinases (MMPs) which are capable to degrade matrix, assisting the maintenance of excess ECM (Chen et al. 2003; Mason and Wahab 2003; Alsaad and Herzenberg 2007). In addition, filtered macromolecules may be trapped in the mesangium, initiating an inflammatory response that could play a pivotal role in stimulating ECM synthesis (Santini et al. 2008; Min et al. 2009). A unifying hypothesis can be constructed that includes participation by all of the cellular elements of the glomerulus. Glomerular capillary hypertension or a genetic or acquired abnormality of podocyte adhesion or structure permits hyperfiltration of macromolecules. Paracrine signals from the injured podocyte stimulate endothelial cell expression of leukocyte adhesion molecules and impair endothelial cell fibrinolytic activity (Eremina et al. 2006). Signals from epithelial or endothelial cells to the mesangium, or direct delivery of pro-inflammatory substances through the glomerular filtrate, initiate a process that culminates in the accumulation of ECM (Eremina et al. 2006; Daehn et al. 2014).

The critical determinant of matrix accumulation is the balance between ECM synthesis and dissolution. This net matrix turnover reflects rates of matrix production (affected by transcription and translation) or degradation (determined by synthesis and activity of ECM proteases and their inhibitors) and factors that affect the susceptibility of the ECM proteins to degradation, such as MMP-2 and MMP-9, microRNAs (miR-21, miR-192, miR-377, miR-382, miR-491-5p, miR-29, and miR-200), glycosylation or the stability with which these proteins have been incorporated into the matrix (Chang et al. 2006; Kato et al. 2007, 2011; Howe et al. 2012). Recently, efforts have been directed toward modeling the cellular events regulating glomerular ECM dissolution. Meanwhile, ample approaches of physiological, pharmacological, and molecular medicine have been used to study how various mediators initiate or modify intracellular signaling pathways to cause mesangial cell matrix accumulation.

9.4 Role of Mesangial-Derived Biofactors on ECM

More than 30 years ago, Mauer and his colleagues documented a definite relationship between mesangial matrix expansion and renal diseases progression by demonstrating that the degree of mesangial expansion strongly predicts the clinical manifestations (Mauer et al. 1984). Since then, the major challenge to investigators has been to elucidate the underlying mechanisms that culminate in ECM accumulation.

As known, the mesangial cells generate their own matrix and embed themselves in it. There are two major functions of its matrix: to provide strong flexible support for glomerular capillaries and to create channels for filtering and addressing macromolecule (Latta and Fligiel 1985). Just resembling other extracellular matrixes, mesangial matrix is comprised of collagens; glycoproteins including fibronectin and

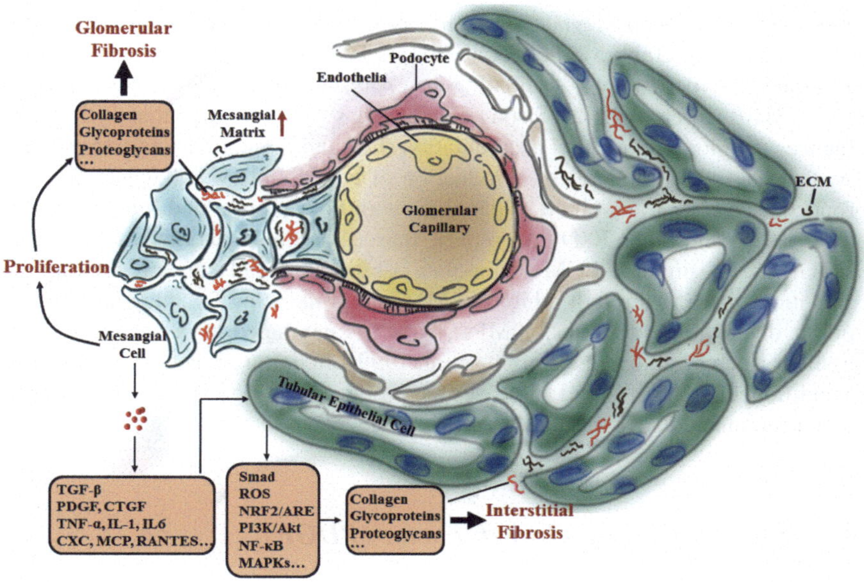

Fig. 9.2 Sketch of mechanisms associated with mesangial-induced fibrosis

laminin A, B1, and B2; proteoglycans including heparin sulfate and chondroitin sulfate proteoglycans; nidogen and so on (Fogo 1999). The amount of these constituents is carefully modulated by mesangial cells during health but can be significantly disordered in disease. Inversely, the matrix can influence the growth and proliferation of cells as well. In this part, we will review some vital biofactors that appear to be critical in the aberrant development of ECM (Fig. 9.2).

9.4.1 Transforming Growth Factor-β

As one of growth factors expressed in the glomerulus, transforming growth factor-β (TGF-β) has been confirmed to be associated with many cases of kidney diseases, including various glomerular diseases and diabetic nephropathy (Hoffman et al. 1998; Diamond-Stanic et al. 2012; López-Hernández and López-Novoa 2012). Although effective therapy for renal fibrosis is still lacking, a host of studies has well documented that TGF-β is the key mediator in CKD associated with progressive renal fibrosis. The transgenic rodents for over expression of TGF-β would ultimately develop into renal fibrosis (Liu 2006; Hewitson 2009). While considerable researches show that TGF-β exerts differently on different renal resident cells, in mesangial cell, upregulation of the TGF-β gene can promote the proliferation of mesangial cells so that produce excess matrix which may directly lead to a deterioration glomerulosclerosis and subsequently incur severe renal fibrosis (Schnaper et al. 2003). Conversely, intrarenal infusion of antisense oligonucleotides to decline the expression of TGF-β can alleviate glomerular sclerosis in rodent models of nephropathy (Lee and Song 2009; Mathew et al. 2011). TGF-β exerts a significant role for mesangial cell hypertrophy by mediating the expression of connective tissue growth factor (CTGF) (Li et al. 2012). Strong evidence has been provided that TGF-β also influences the glomerular ECM accumulation (Lee and Song 2009). In vitro studies show that TGF-β1 stimulates the production of type I and IV collagens, laminin, fibronectin, and heparan sulfate proteoglycans in both murine and human mesangial cells (Lee and Song 2009). Moreover, Baricos and his coworkers demonstrate that TGF-β can stimulate ECM accumulation by dramatically downregulating the degradation of ECM through inhibiting MMPs and inducing its antagonist (tissue inhibitor of metalloproteinases, TIMPs), which is paralleled by an increase of its ingredient, such as collagens (Baricos et al. 1999). In cultured mesangial cells and its conditioned media, TGF-β1 increases the production of type-I and -IV collagens at both protein and mRNA levels (Hubchak et al. 2009).

9.4.2 Platelet-Derived Growth Factor (PDGF)

As one of the most important mitogen factors for kidney, PDGF can be produced by platelets, endothelial cells, and mesangial cells (Shultz et al. 1988). PDGF can be induced by many factors including other two growth factors (EGF and FGF), TNF-α, TGF-α/β, and INFγ. Not only could be PDGF able to regulate mesangial cells proliferation but also it directly stimulated ultraproduction of ECM (Throckmorton et al. 1995). Recent observation indicated that specific inhibition of the PDGF-B chain by antisense mRNA significantly attenuated the overexpression of type I collagen and a-SMA (Throckmorton et al. 1995). Meanwhile, the antagonism of PDGF-D, an isoform of PDGF family, ameliorated epithelial–mesenchymal transition (EMT) in a rat model of mesangial proliferative glomerulonephritis (MsPGN), although it did not influence the loss of E-cadherin, but downregulated the tubular de novo vimentin expression (as we know, de novo expression of vimentin and a-SMA and the tubular epithelial loss of E-cadherin expression are major features of EMT) (Floege et al. 2007).

9.4.3 Pro-inflammatory Cytokines

The most related pro-inflammatory cytokines are tumor necrosis factor-α (TNF-α), interleukin-1 (IL-1), IL-6, and IL-8. These cytokines can not only stimulate the production of reactive oxygen species (ROS) and prostaglandin E2 (PGE2) in mesangial cells, but also induce the proliferation and differentiation of inflammatory cells, such as T- and B-cells, leading to mesangial dysfunction and immune response in kidneys (Dorsam et al. 2000). In the early 1990s, numerous studies have found that IL-1 could stimulate human mesangial cells to synthesize and release IL-6 and -8, assisting monocytes or neutrophils in adhering to vascular endothelium of glomeruli and regulating the remodeling of mesangial matrix (Radeke and Resch 1992). However, in the past few years, researchers put much more attention to the molecular mechanisms about how these cytokines influence cell functions. Liang et al. further discovered that the over-production of IL-1β, IL-8, and TGF-β1 induced by secretory IgA in mesangial cells was promoted by the downregulation of micro RNAs, such as miR-100-3p and -877-3p, which ultimately triggered the glomerular fibrosis (Liang et al. 2016b). The other groups additionally founded that a combination of IL-1β and TNF-α induces the expression of inducible NO synthase mainly through the NO/HIF axis, while hypoxia-driven signaling proteins, such as collagens, PDGF, and the secretory phospholipase A2, contribute to the proliferative and pro-fibrotic effects in renal cells (Kishida et al. 2005). Meanwhile, as the most vital inflammatory mediator, TNF-α has been proved to possess cytotoxicity and mesangial proliferation effects and induce mesangium contraction and cytoskeletal rearrangement, resulting in the alteration of GRF, which ultimately contribute to renal dysfunction and even fibrosis (Cooker et al. 2007). Moreover, mesangial-derived IL-6 can regulate the growth and differentiation of the mesangial cells themselves via inducing over-production of ROS and matrix components, and the levels of IL-6 are positive correlated with the degree of glomerular fibrosis (Eitner et al. 1997).

9.4.4 Chemokines

The present data imply that inflammation is responsible for the initiation and progress of primary renal injury, while the mesangial-derived chemokines are regarded as a vital accomplice of inflammation, mostly because they are secreted locally, acting on resident leukocyte as soon as the damage begins. These chemokines can further recruit inflammatory cells, as well as accelerate their adhesion and migration through the endothelium, thus promoting inflammatory cells to arrive at the injury site. Therefore, resembling the cytokines, chemokines and their receptors become the new center of attention.

On the basis of bioactivity, chemokine superfamilies are divided into four subclasses: C, CC, CXC, and CX_3C. The classical mesangium-derived members are monocyte chemotactic protein 1 (MCP-1), RANTES, and macrophage inflammatory protein 1α (MIP-1α), all of which belong to the CC-subclass. The pleiotropic cytokines, IL-8 and interferon-inducible protein 10 (IP-10), are belong to CXC-subclass. Murine glycoprotein TCA3, which is secreted from mesangial cells upon stimulation, is a member of CX_3C-subclass (Zlotnik and Yoshie 2000).

The functions of these chemokines are thoroughly studied in past decades. MCP-1 stimulated monocyte recruitment, retention, and proliferation within the mesangium after administration of oxidative modification of low-density lipoprotein, which played a critical role similar to that of systemic vascular cells in the pathobiological cellular events associated with glomerular fibrosis. In many cases, the excessive production of ROS represented a common mechanism of injury-induced chemokine and adhesion molecule activation (Yang et al. 2016). For example, Jia and colleagues suggested that glucose-induced MCP-1, TGF-β1 and fibronectin (FN) expression in mesangial cells related to the scathing (fibrotic) level of glomeruli and interstitial in diabetic nephropathy, whose early characteristics include glomerular hypertrophy and ECM accumulation (Xu et al. 2013). Presumably, the underlying mechanisms of high glucose-enhanced mesangium proliferation and MCP-1 expression are partly through activating the ROS/NF-κB signaling pathway. Meanwhile, Liu et al. further discovered that the mRNA expression of MCP-1 and the serum level of intracellular cell adhesion molecule-1 (ICAM-1) were both upregulated after treated with high glucose, and such increases were carried out most likely through hampering TGF-β signaling pathways (Liu et al. 2012). In addition, antisense MCP-1 could decrease mesangial cell proliferation and pathological injury in mesangial proliferative glomerulonephritis (MsPGN) model rats by decreasing expression of MCP-1 and CCR2; and suppressed mesangial cell proliferation and matrix accumulation in anti-Thy-1 MsPGN model rats, although which did not entirely depend on TGF-β1 (Liu et al. 2013). Additionally, mesangium also expresses chemokine receptors, such as CCR1 and 7, CXCR3, which means mesangial cells are both source and target of these bioactive factors secreted during injury. For example, human macrophage inflammatory protein-2 (MIP-2) could increase MCP-1 and RANTES expression to modulate mesangial cells migration and proliferation in vitro study (Uguccioni et al. 2010). Meanwhile, CCR7 is capable to regulate proliferation and apoptosis in mesangial cells (Banas et al. 2002).

9.5 Signal Pathways Associated with Glomerular Fibrosis

9.5.1 TGF-β Signaling Pathway

TGF-β is a master regulator of renal fibrosis (Meng et al. 2016). As mentioned above, numerous studies have manifested that increased expression and activation of pro-fibrotic TGF-β are associated with renal fibrosis. For the canonical TGF-β signaling pathway, TGF-β binds to TGF-β receptor 2 (TGFR2), which recruits and activates TGFR1 to activate receptor signaling. Subsequently, Smad2 and Smad3 are phosphorylated and translocate to the nucleus and activate the transcription of specific targets. It has been found that Serine-204 in the linker region of Smad3 mediated the collagen-I response to TGF-β in a cell phenotype-specific manner (Browne et al. 2013). Alternatively, binding of TGF-β with TGFR can also activate a wide variety of Smad-independent pathways (noncanonical signaling) to modify cell function. The involvement of TGF-β was evident by the findings that blockade of TGF-β1 in animal models of kidney disease reduces fibroblast activation and collagen deposition (Fig. 9.3).

Actually, TGF-β signaling is more complex than expected. Under the normal condition, TGF-β is synthesized in a precursor, which is kept as an inactive form with other two protein, the latency-associated peptide (LAP) and the latent TGF-β-

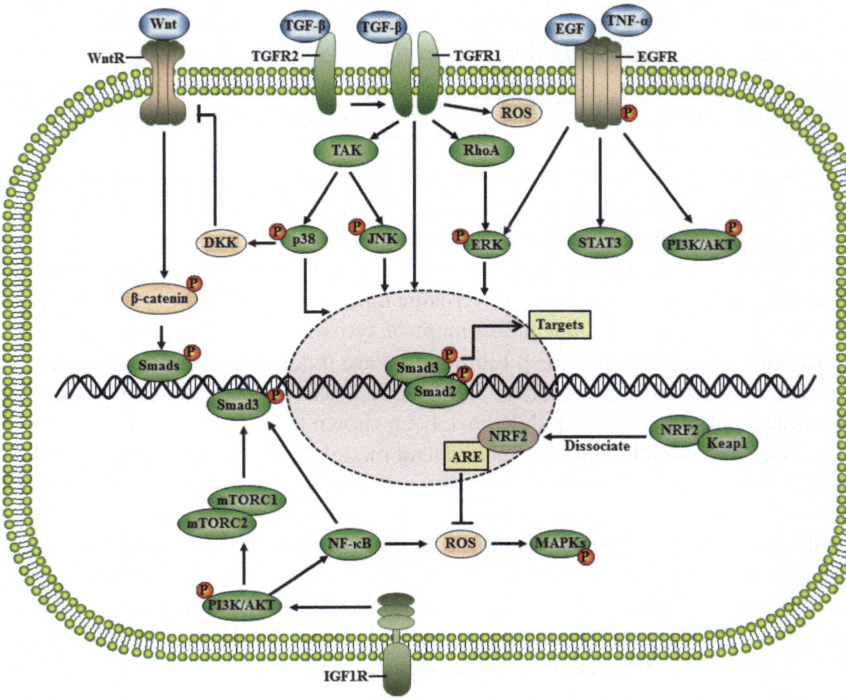

Fig. 9.3 Signaling pathways associated with mesangial activation

binding protein (LTBP). However, the latent TGF-β complex, which is secreted into the extracellular space, can be cleaved by a wide range of proteases to release active TGF-β. As for the role of Smad proteins, each member (Smad2, Smad3, or Smad4) exerts distinct and even opposing functions in the regulation of renal fibrosis. Smad3 is the sole member that can bind directly to Smad-binding elements within gene promoters, while Smad2 or Smad4 has no such domains (Piek et al. 2001).

The association with Smad4 is important for the translocation of the Smad2/3 complex into the nucleus (Meng et al. 2013). The Smad-mediated signaling induced by TGF-β can be controlled by a negative feedback mechanism: Smad7 is the inhibitor of TGF-β/Smad canonical signaling (Nakao et al. 1997). Smad7, whose transcription can be induced by the Smad3-containing complexes, competes with Smad2 and Smad3 for binding to activated TGFR1 and thus reduces TGF-β/Smad signaling (Shi and Massagué 2003). Under the condition of glomerular fibrosis, e.g., diabetic nephropathy, elevated TGF-β and phosphorylated Smad2/3 are found in many studies, suggesting the involvement of the canonical TGF-β signaling. On the other hand, many drugs with anti-fibrotic activity exert their biological functions by inhibiting this pathway.

For the Smad-independent TGF-β signaling, pathways associated with TGF-β activated kinase 1 (TAK1), phosphatidylinositol 3-kinase (PI3K)/AKT, Rho-like

GTPase, and MAPK are thought to mediate the biological functions of TGF-β. This noncanonical signaling mainly depends on the activation of other signal pathway induced by the TGF-β receptors, showing the cross talk with other pathways (Zhang 2009). Unlike those studies in cancer research, the noncanonical TGF-β signaling attracts relative less attention.

The most often studied cross-talking pathway is MAPK, including ERK, JNK, and p38-MAPK. TGF-β is shown to induce ERK activation in a Smad-independent manner. Upon TGF-beta stimulation, the activated TGFR recruits and directly phosphorylates ShcA proteins, which leads to the ShcA association with Grb2 and Sos, and subsequent activation of receptor tyrosine kinases with ERK MAP kinases (Lee et al. 1977). This suggests that recruitment of tyrosine kinase signaling pathways may account for aspects of TGF-β biology that are independent of Smad signaling. Moreover, all the MAPK kinases are activated in CKD, while genetic or pharmacologic strategies to block p38 or JNK have been shown to suppress renal fibrosis in a variety of animal models. For example, in a rat model of severe crescentic anti-GBM glomerulonephritis, JNK inhibitor CC-401 prevented severe glomerular and tubulointerstitial lesions demonstrated by the reduction in myofibroblast accumulation and collagen deposition (Ma et al. 2009). In addition, in vitro studies show that MAPK can upregulate TGF-β1 expression in various renal cells including fibroblasts and mesangial cells. In high glucose-treated mouse mesangial (MES13) cells, the gene transcription TGF-β was induced by high glucose and exogenous S100B through ERK1/2 and p38 MAPK pathway (Chuang et al. 2015). Interestingly, a 2004 study showed mesangial cells could produce thrombospondin-1 for activation of latent TGF-β1 via p38 MAPK and JNK, but not ERK1/2, which suggests an indirect link between MAPK and TGF-β signaling (Naito et al. 2004).

TAK1, also known as mitogen-activated protein kinase kinase kinase 7 (MAP3K7), is a member of the serine/threonine protein kinase family. This kinase mediates the signaling transduction induced by TGF-β and morphogenetic protein (BMP) and controls a variety of cell functions including transcription regulation and apoptosis. TGF-β can activate TAK1 directly, which plays a critical role in the progression of fibrosis. Besides TGF-β, TAK1 can also be activated by various stimuli including environmental stresses and inflammatory cytokines. Then the activated TAK1 transduces signals to several downstream targets, including the JNK, p38 MAPK, and IKK. In response to environmental stresses, it forms a complex with TAK1-binding protein 1 and 2 to activate pro-inflammatory signaling pathways. In a mouse model of ischemia-induced renal fibrosis, intraperitoneal injection of TAK1 inhibitor was found to ameliorate renal function and fibrosis, in which p38 MAPK was involved in (Zhou et al. 2018). In mouse mesangial cells, rapid activation of endogenous TAK1 activity and collagen induction were induced by TGF-β, which was companied with the activation of MKK3 and p38 MAPK (Kim et al. 2007). Similar effects of TGF-β-induced TAK1 activation were obtained in fibroblasts (Hocevar et al. 2005). These studies clearly indicated that TAK1 was a major regulator of TGF-β signaling and played an important role in renal pro-fibrotic response.

The Rho-family GTPase, including RhoA, Rac1, and Cdc42, is associated with cytoskeletal signaling. These molecules function as switch proteins, alternating between an active (GTP-bound) and an inactive (GDP-bound) state. This pathway has been implicated in fibrogenesis, which can be activated by TGF-β in a Smad-independent manner. Both RhoA and Rac1 are rapidly activated in response to TGF-β in human mesangial cells (Hubchak et al. 2009). Moreover, the induction of type I collagen expression stimulated by TGF-β was diminished by inhibiting Rac1 activity and was increased by a constitutively active Rac1 mutant. RhoA and its downstream target ROCK were activated in diabetic rat kidneys and high glucose-induced glomerular mesangial cells, in which TGF-β signaling was robust activated. However, the activation of RhoA/ROCK was inhibited by berberine treatment (Xie et al. 2013). More recently, TGR5 was found to suppress high glucose-induced upregulation of fibronectin in rat glomerular mesangial cells by inhibiting RhoA/ROCK signaling (Xiong et al. 2016).

9.5.2 Wnt/β-Catenin Signaling Pathway

The Wnt family consists of structurally related proteins, which have been implicated in oncogenesis and in several developmental processes, as well as in maintaining tissue homeostasis and initiating organ repair following injury. There are 19 secreted Wnt signaling proteins that have been identified in humans and mice. The secreted Wnt molecules can bind to cell-surface receptors, such as the seven transmembrane receptor frizzled protein and the low-density lipoprotein receptor-related protein (LRP). Similar to TGF-β signaling, there exist canonical and noncanonical Wnt pathways. The canonical Wnt pathway through β-catenin has been well established. Under inactive condition, β-catenin is bound to Axin and adenomatous polyposis coli (APC) and phosphorylated by glycogen synthase kinase-3β (GSK-3β) at N-terminal residues and then leads to the degradation of β-catenin. However, when Wnt signaling is activated by various stimulator, Wnt ligands activate FZD and LRP targeting APC and Axin, which leads to the dephosphorylation of GSK-3β and the recruitment of the cytosolic proteins Dishevelled (Dvl). Subsequently, the phosphorylation and the degradation of β-catenin are inhibited. Ultimately, β-catenin translocates to the nucleus and activates T-cell factor (TCF) and lymphoid enhancer factor (LEF) to regulate the gene expression of Wnt targets (Xiao et al. 2013). Extensive studies have shown Wnt/β-catenin signaling is involved in renal cell injury and renal fibrosis (Fig. 9.3).

The role of Wnt/β-catenin signaling in renal fibrosis remains controversial. Wnt/β-catenin signaling is silenced in the adult kidney. However, it can be activated following renal injury as part of the repair response. A protective role for Wnt/β-catenin signaling has been demonstrated in healing and repair following acute kidney injury (Peng and Dong 2012). Excessive activation of the Wnt/β-catenin signaling pathway results in chronic kidney disease (CKD) progression. Numerous studies have demonstrated that CKD rat kidney tissues exhibited moderate renal fibrosis and significantly increased expression levels of β-catenin and apoptosis-associated proteins. Moreover, response to Wnt/β-catenin signaling is found to be cell-type specific, as reverse effects on mesangial cell and tubular cells have been investigated during past decades. It has also been shown that high glucose or hyperglycemia can induce apoptosis of mesangial cells, which is one of the main causes of DN. A 2010 study found that high glucose induced the phosphorylation of Ser45-β-catenin and reduction of nuclear β-catenin levels, which was mediated by Dickkopf-1 (DKK-1), an endogenous inhibitor of Wnt/β-catenin signaling (Peng and Dong 2012). Further studies of this group confirmed that sustained Wnt/β-catenin signaling rescued high glucose induction of TGF-β-mediated renal fibrosis, and curcumin resumed HG depression of Wnt/β-catenin signaling (Ho et al. 2016). These studies suggest that modulation of Wnt/β-catenin signaling is a viable alternative strategy to rescue the TGF-β1-mediated fibrotic signaling pathway in diabetic renal injury. However, β-catenin was found increased in TNF-α-treated mesangial cells, another cell model of glomerular fibrosis. Silencing of β-catenin by siRNA could attenuate cell apoptosis induced by TNF-α and decrease the mRNA expression of various fibrosis markers (Lin et al. 2017). In injured renal tubular epithelia cells, Wnt/β-catenin signaling is upregulated in acute kidney injury and CKD, which is different from that of mesangial cells (Gewin 2018). Most importantly, inhibition of Wnt/β-catenin signaling can reduce fibrosis in experimental kidney disease including diabetic nephropathy, obstructive nephropathy, adriamycin nephropathy, chronic allograft nephropathy, and polycystic kidney disease. Thus, the roles of Wnt/β-catenin need further elucidation. Special attention should be paid to the animal models and the cell types involved.

Cross talk between the Wnt/β-catenin and TGF-β signaling in renal fibrosis attracts much attention. It has been demonstrated that knockdown of β-catenin by siRNA inhibited the mRNA expression TGF-β, which resulted in decreased expression of renal fibrosis markers (Lin et al. 2017). It has also been found that Wnt ligand can induce the production of TGF-β, suggesting a direct regulation of TGF-β (Peng and Dong 2012). In addition, β-catenin was found to interact with TGF-β to regulate EMT in tubular cells (Gewin 2018). On the other hand, TGF-β can activate Wnt/β-catenin signaling. TGF-β1 treatment induced Wnt1 expression, β-catenin activation, and stimulated the expression of Wnt/β-catenin downstream target genes (Zhou et al. 2012). Thus, TGF-β and β-catenin can counter-regulate each other, depending on the animal models or cell types.

9.5.3 EGFR Signaling

The epidermal growth factor receptor (EGFR) is a transmembrane glycoprotein that is a member of the tyrosine kinase superfamily. Binding of the protein to a ligand induces receptor dimerization and tyrosine autophosphorylation. Besides EGF, other ligands such as heparin-binding EGF-like growth factor (HB-EGF), TGF-α, and amphiregulin can activate EGFR. Upon activation by its ligands, the autophosphorylation of several tyrosine residues of EGFR occurs, which elicits downstream activation and signaling. The downstream signaling initiates several signal transduction cascades, including ERK and signal transducer and activator of transcription

3 (STAT3), and subsequently the induction of various cellular responses, such as proliferation, angiogenesis, and inhibition of apoptosis (Fig. 9.3).

Upregulation of the EGFR is involved in the regeneration of tubular epithelium after acute tubular necrosis, and exogenously administered EGF could attenuate the severity and duration of renal failure in early studies (Fine et al. 1992). However, activation of EGRF signaling is thought to be involved in renal fibrosis nowadays. Unlike other signal pathway, activation of EGFR signaling leads to renal fibrosis in all type of renal intrinsic cells (e.g., tubular epithelial cells, interstitial cells, and mesangial cells). In serum-induced mesangial cell, three growth factors in the EGF family were significantly induced as revealed by high-density oligonucleotide microarrays, which was thought to be associated with the proliferation of mesangial cell (Mishra et al. 2002). The author also confirmed the functional EGFR activation by using an EGFR-selective kinase inhibitor and a specific antibody, which suggest a role for EGFR signaling in control of mesangial cell growth. In cultured mesangial cells, inhibition of EGFR with erlotinib abrogated aldosterone-induced expression of ECM proteins, cell proliferation, and migration (Sheng et al. 2016). Similar results were observed in HB-EGF-treated mesangial cells. Moreover, activation of the EGFR was also investigated in several animal models, while either genetic or pharmacologic inhibition of EGFR significantly suppresses renal fibrosis. In Sprague Dawley rats with five-sixths nephrectomy, treatment of EGFR inhibitor erlotinib significantly blunted the progression of CKD, as evidenced by reduced levels of serum creatinine, proteinuria, and renal cortical profibrogenic genes and scores of glomerulosclerosis and tubulointerstitial damage (Yamamoto et al. 2018). Also, in a murine model of uninephrectomy, erlotinib-treated rats exhibited relieved structural lesion comparing with rats treated with aldosterone alone, as characterized by glomerular hypertrophy, mesangial cell proliferation, and expansion (Sheng et al. 2016). Similarly, the renoprotective effects of erlotinib were observed in a rat model of anti-Thy 1.1-induced experimental glomerulonephritis (Rintala et al. 2016). The role of EGFR signaling was also determined in diabetic nephropathy, which is characterized by glomerular hypertrophy. In primary mesangial cells, EGFR and downstream Akt activation was induced by high glucose, which was prevented by siRNA to either HB-EGF or ADAM17 (Uttarwar et al. 2011). Moreover, targeted deletion of EGFR attenuated diabetic nephropathy in streptozotocin-induced mouse diabetic nephropathy. These studies suggest that EGFR might be a potential therapeutic target for modulating renal fibrosis. Although the EGF level was increased in the serum of patients with diabetic nephropathy at all stages of CKD compared to healthy controls, a decreased expression of EGF mRNA was unexpectedly found in renal biopsy samples from DN patients (Lindenmeyer et al. 2007). The discrepancy might be explained by differences in disease stage or differences between species.

For mechanism, EGFR signaling always elicits downstream signal cascades. MAPK and PI3K-Akt are the major signaling cascades downstream of EGFR activation, which partially depends on the animal model or cell types. Moreover, activation of the EGFR can induce Smad3 phosphorylation and fibrotic response via ROS-dependent mechanisms, suggesting that EGFR signaling can crosstalk with

TGF-β signaling to regulate its activation and mediate its biological consequences (Samarakoon et al. 2013).

9.5.4 Akt Signaling

Proteins of the Akt family are serine/threonine kinases, which can be activated by growth factors, such as the insulin-like growth factor-1 receptor (IGFIR), PDGF, EGF, and TGF-β, or by other stimuli. The activation is rapid and specific, and it is abrogated by mutations in the pleckstrin homology domain of Akt. In mammalian cells, this family contains three closely related and highly conserved isoforms. The tissue distribution and function of each isoform are different. In kidney, Akt2 is the main isoform related to renal disease (Fig. 9.3).

Akt acts as a major signal node in signal transduction cascades. Under normal condition, Akt is maintained in an inactivated state. Growth factors and other stimuli are shown to activate Akt through phosphatidylinositol 3-kinase (PI3K). Thus, Akt is a critical mediator of growth factor-induced cell survival. Moreover, survival factors can suppress apoptosis in a transcription-independent manner by activating Akt, which in turn phosphorylates and inactivates components of the apoptotic machinery. For the signal transduction cascade of PI3K/Akt, PI3K is a heterodimeric protein complex, which can be activated by receptor tyrosine kinases, G-protein-coupled receptors, or Ras proteins. The biochemical function of PI3K is to convert phosphate dylinositol-4,5-diphosphate (PIP2) to phosphatidylinositol-3,4,5-triphosphate (PIP3), which form binding sites for PH domain-containing proteins, such as Akt and PDK1. Therefore, Akt and PDK1 are recruiting them to the membrane, where Akt is phosphorylated by PDK1 at Thr308 and by mammalian target of rapamycin complex-2 (mTORC2) at Ser473, leading to the activation of Akt. Subsequently, the activated Akt regulates its effectors, such as mTORC1, GSK3β, proapoptotic Bcl-2 family members, NF-κB, or Mdm2, which ultimately promote cell proliferation, survival, and tumor growth. This pathway can be negatively regulated by phosphatase and tensin homolog (PTEN) and tuberous sclerosis complex 1 (TSC1) and 2 (TSC2).

Numerous studies indicate that Akt plays a role in glomerular mesangial hypertrophy and ECM accumulation (Lan and Du 2014). As mentioned above, growth factor receptors with tyrosine kinases activity activate PI3K/Akt. This event can be found in TGF-β- and EGF-treated mesangial cells, which were accompanied by obvious cell hypertrophy (Ghosh-Choudhury and Abboud 2004). Moreover, the expression of fibrotic proteins induced by TGF-β is also regulated by PI3K/Akt signal transduction. In addition, PTEN, the negative regulator of PI3K/Akt, was found to be downregulated in high glucose-treated mesangial cells and in the streptozotocin-induced diabetic kidney cortex and glomeruli (Mahimainathan et al. 2006). Similarly, the effect of PTEN on AKT/mTORC1 was demonstrated in TGF-β-treated mesangial cells via a new mechanism, in which miR-21 was involved (Dey et al. 2012). It is reported that mesangial cell proliferation induced by PDGF and EGF was also coupled to the activation of Akt both in vitro and in vivo, which is inhibited by lipoxins, endogenously produced eicosanoids with a spectrum of bioactions (Mitchell et al. 2004). Autophagy, a process to maintain metabolic homeostasis, is found involved in diabetic nephropathy. More recently, triptolide was found to restore autophagy and alleviate fibrosis in DKD rats and high glucose-incubated human mesangial cells through the Akt/mTOR pathway (Li et al. 2017). All these studies clearly indicate that Akt activation is related to mesangial cell survival and glomerular fibrosis. In human, the finding that a dominant negative mutation of Akt (R274H) caused severe hyperinsulinemia and diabetes confirmed the role of Akt in such metabolic disorders. Impaired insulin-stimulated PI3K/Akt has also been implicated in human diabetic nephropathy.

One of the Akt effectors is NF-κB, which plays an important role in pro-inflammation. The role of Akt/NF-κB in mesangial cells and glomerular fibrosis will be discussed later.

9.5.5 Antioxidant NRF2/ARE Pathway

Like other organs, the kidney relies on aerobic metabolism and oxidative phosphorylation for energy production necessary for its normal physiological functions. Loss of ATP production due to mitochondrial dysfunction is associated with increased free radical generation and oxidative stress, which is one of the major mechanisms for the pathogenesis of renal fibrosis. Under normal physiological conditions, the production of ROS activates many beneficial effects for signaling pathways that maintain homeostasis, but excess production of ROS is detrimental. Generally, oxidative stress can result from an imbalance between the production of free radicals that are often increased by mitochondrial dysfunction and reduced antioxidant defenses. Especially, impairment of antioxidant systems disturbs downstream signaling events during the pathogenesis of CKD, contributing to renal senescence and cell apoptosis, as well as renal fibrosis. In cultured mesangial cells, high glucose can induce ROS production, which results in continuous oxidative stress. Many antioxidant systems that directly scavenge the excessive free radicals have been studied during past decades. Among them, nuclear factor erythroid 2-related factor 2 (NRF2) is a master regulator of cellular antioxidant activity. NRF2 is a transcription factor which is a member of a small family of basic leucine zipper (bZIP) proteins. It is ubiquitously expressed in a wide range of tissue and cell types, where it is held in the cytoplasm as an inactive complex bound to Kelch-like ECH-associated protein 1 (Keap1). Upon oxidative and electrophilic insults, Nrf2 dissociates from its cytoplasmic repressor Keap1 and then translocates to the nucleus, where it binds to antioxidant response element (ARE) and regulates gene expression (Lee and Johnson 2004). Many of these NRF2 targets oppose inflammatory and oxidative damage, and the targets include glutathione S-transferase, superoxide dismutase, glutamylcysteinyl synthetase, heme oxygenase-1 (HO-1), and so on. Currently, it is estimated that the total number of Nrf2 targets is about 250 (Fig. 9.3).

Accumulating evidence indicates that Nrf2/ARE signaling is protective in various models of renal disease, including renal fibrosis. Most studies investigating the role of Nrf2 were carried out in diabetic nephropathy. The glomeruli of human diabetic nephropathy patients were under oxidative stress and had elevated Nrf2 levels. The importance of Nrf2 was revealed by the fact that Nrf2 knockout (KO) diabetic mice exhibit more severe renal injury as compared with wild-type (WT) diabetic mice (Jiang et al. 1998). Moreover, impaired Nrf2 activity and decreased expression of its antioxidant targets have been observed in CKD animal models. On the contrary, NRF2 activation improves DN in mice. For example, Gong et al. found that polydatin promotes the Nrf2 activity to resist the upregulation of FN and ICAM-1 in diabetic mice kidneys (Gong et al. 2017). Similarly, hydrogen sulfide was found to alleviate diabetic nephropathy in a streptozotocin-induced diabetic rat model, as Nrf2 antioxidant pathway was activated by hydrogen sulfide (Zhou et al. 2014). More recently, MDM2 was found to control NRF2 antioxidant activity in prevention of DN via inhibition of P53 (Guo et al. 2018). Of interest, many natural products have been shown to reduce renal oxidative stress and slow the progression of CKD by activating Nrf2/ARE pathway. These Nrf2 activators include resveratrol, curcumin, oleanolic acid, sulforaphane, and cinnamic aldehyde. The role of antioxidant Nrf2/ARE pathway was also confirmed in cultured mesangial cells. Astaxanthin (AST) is a fat-soluble xanthophyll carotenoid with remarkable antioxidative capacity, and its renoprotective effects were demonstrated by the finding that it promoted Nrf2/ARE signaling to inhibit HG-induced renal fibrosis in glomerular mesangial cells (Huang et al. 2013). Another source of ROS in kidney is related to the overloading of advanced glycation end products (AGEs). AGEs can boost the generation of ROS in glomerular mesangial cells, which can be attenuated by Sirt1 through Nrf2/ARE antioxidative pathway (Hayden and Ghosh 2004). All these studies suggest activating Nrf2 signaling may be helpful in protecting against diabetic renal fibrosis.

9.5.6 NF-κB Pathway

As mentioned above, pro-inflammatory cytokines contribute to mesangial dysfunction and immune response in kidneys. On the one hand, inflammatory cytokines from other cells, e.g., infiltrated macrophage and T-cells, can lead to the proliferation of mesangial cells and matrix deposition. On the other hand, mesangial cells can also produce these cytokines under the condition of detriment insults. Here, we will discuss the NF-κB pathway associated with the latter situation (Fig. 9.3).

In general speaking, NF-κB participates in processes that lead to kidney damage and renal fibrosis. So far, five mammalian proteins of the Rel/NF-κB family have been cloned and characterized. These proteins include p50/NF-κB1, p52/NF-κB2, RelA/p65, RelB, and c-Rel. All Rel/NF-κB proteins can form either homo- or heterodimers, with the exception of RelB which is unable to homodimerize or heterodimerize to c-Rel and RelA (Hayden and Ghosh 2004). The most abundant form of NF-κB is NF-κB1 complexed with RelA, which is often designated "classical NF-κB." NF-κB is a ubiquitous transcription factor involved in many biological processes. It is held in the cytoplasm in an inactive state by specific inhibitors. Activation of the NF-κB system occurs in response to ligand binding to cell-surface receptors, which lead to the degradation of the inhibitor. Subsequently, NF-κB moves to the nucleus and activates transcription of specific genes. These cell-surface receptors include Toll-like receptor family, the TNF receptor family, the T- and B-cell receptor, and angiotensin II receptor (Massy et al. 1999). In addition, NF-κB can also be activated by cross talking with other signal transduction cascade. For example, activated MAPK is thought to be the physiological activator of NF-κB, which was demonstrated by the finding that rapeseed protein-derived antioxidant peptide RAP alleviated renal fibrosis through MAPK/NF-κB signaling pathways in diabetic nephropathy (Zhang et al. 2018). Likewise, Ji et al. found that andrographolide ameliorates diabetic nephropathy by attenuating hyperglycemia-mediated renal oxidative stress and inflammation via Akt/NF-κB pathway (Ji et al. 2016). As mentioned, activated NF-κB can induce gene expression of many targets, which harbor NF-κB-responsive sequence in their promoter regions. After NF-κB activation, the gene expression of inflammatory cytokines and pro-inflammatory mediators in mesangial cell will robust increase, which ultimately lead to cell proliferation.

All Rel proteins are expressed in mesangial cells. The detriment insults leading to NF-κB activation in mesangial cells include cytokines (IL-1β and TNF-α), immunoglobulins (IgA), angiotensin II, aldosterone, high glucose, AGEs, C-reactive protein, and growth factors such as TGF-β, anti-DNA antibodies, and adiponectins. It has been shown that high glucose can induce NF-κB activation as evident by IκB degradation and nuclear accumulation of NF-κB in mesangial cells (Ji et al. 2016). Recent studies focus on the regulation of NF-κB activity in diabetic nephropathy. Betulinic acid (BA), a pentacyclic triterpene derived from the bark of the white birch tree, has been demonstrated to have many pharmacological activities. Its inhibitory effect on NF-κB pathway was investigated in high glucose-induced mesangial cells and diabetic rat kidneys (Wang et al. 2016c). Liuwei Dihuang Pill (LDP) also has similar inhibitory effect on NF-κB pathway in mesangial cells (Xu et al. 2017). Of note, the epigenetic regulation of fibrogenic program in mesangial cells treated with high glucose was revealed by a recent study in which H3K9/14 hyperacetylation was found closely associated with NF-κB or CREB motifs, and was further confirmed by chromatin immunoprecipitation assays (Ji et al. 2016). All these studies suggest inhibition of NF-κB activation might be beneficial for mesangial cells and glomerular fibrosis, especially for diabetic nephropathy.

9.5.7 Mechanisms Based on Noncoding RNA

Epigenetic regulations are involved in the maintenance of renal cell functions in health and disease. It is a very interesting topic that how the "second genetic code" impacts the fate of mesangial cells and glomerular fibrosis. Among these epigenetic mechanisms, studies based on noncoding RNA gain much attention (Wanner and Bechtel-Walz 2017). Noncoding RNAs are arbitrarily separated into long noncoding RNAs (lncRNAs) (>200 nucleotides) and short ncRNAs (<200). And the latter can be further divided into three classes, namely microRNA (miRNA), small interfering RNA (siRNA), and piwi-interacting RNA (piRNA). Both lncRNAs and short ncRNAs influence gene expression but via different mechanisms. During the past decades, understanding of the functions and molecular mechanisms of miRNAs is better than that of other ncRNAs. MiRNAs are 21–25 nucleotides long and can cleave mRNA with the help of the Dicer-containing RNA-induced silencing complex. miRNAs have been intensely investigated and found to regulate various disease progresses including renal fibrosis, even some miRNAs are can be treated as therapeutic targets. However, the functional roles and molecular mechanisms of lncRNAs are less understood. It has been reported that lncRNAs can regulate gene expression and epigenetic mechanisms on the chromatin level (Mercer and Mattick 2013). As a large class of pervasive transcriptions in the genome, numerous lncRNAs are reported to be involved in various kinds of diseases, including renal fibrosis.

It has been found that miR-21 expression is correlated with tubulointerstitial injury in kidney biopsies of diabetic patients. In cultured mesangial cells, miR-21-mediated repression of Cdc25a and Cdk6 resulted in impaired cell cycle progression and subsequent mesangial cell hypertrophy (Kölling et al. 2017). A 2016 study showed that decreased miR-29a expression and attenuated Wnt/β-catenin signaling were concomitantly detected in glomeruli of STZ-induced diabetic mice, while gain of miR-29a function in diabetic mice substantially increased the expression of β-catenin and blocked the expressions of pro-fibrotic gene markers in glomerular mesangium (Hsu et al. 2016). This work suggested a protective effect of miR-29a on diabetic glomerular dysfunction by modulation of DKK1/Wnt/β-catenin signaling. Similarly, upregulation of miR-27a was found in high glucose-treated mesangial cells and in the kidney glomeruli of STZ-induced diabetic rats, while knockdown of miR-27a prevented high glucose-induced mesangial cell proliferation and also blocked the upregulation of ECM-associated pro-fibrotic genes (Wu et al. 2016). Then functions of other miRNAs such as let-7b, miRNA-21, miR-29b, miR-130b, miR-135a, miR-150, miR-192, miR-215, and miR-302 were also analyzed in mesangial cells and related animal models. These studies indicate miRNAs are highly involved in the maintenance of mesangial cell functions and can be treated as therapeutic targets.

The number of references about the involvement of lncRNAs in mesangial cells and glomerular fibrosis is constantly growing. By using lncRNA microarrays, the differential expressed lncRNAs were detected, and lncRNA CYP4B1-PS1-001 was significantly downregulated in response to early diabetic nephropathy in vitro and in vivo. Further analysis showed overexpression of CYP4B1-PS1-001 inhibited proliferation and fibrosis of mesangial cells (Wang et al. 2016a). Another work of this group showed that lncRNA ENSMUST00000147869 protected mesangial cells from proliferation and fibrosis induced by diabetic nephropathy (Wang et al. 2016b). Yi et al. showed that lncRNA Gm4419 knockdown could obviously inhibit the expressions of pro-inflammatory cytokines and renal fibrosis biomarkers, and reduce cell proliferation in MCs under high glucose condition, whereas overexpression of

Gm4419 could increase the inflammation, fibrosis, and cell proliferation in MCs under low-glucose condition (Yi et al. 2017). Mechanically, Gm4419 could activate the NF-κB pathway by directly interacting with p50, the subunit of NF-κB. Similarly, other lncRNAs such as ASncmtRNA-2, Erbb4-IR, and 1700020I14Rik were also functionally analyzed in mesangial cells, and their association with the expression of pro-fibrotic factors was confirmed. These suggest lncRNAs can function as novel targets for the research of mesangial cells and renal glomerular fibrosis.

9.6 Implication of Anti-fibrotic Agents

Treatment for renal fibrosis is limited and only partially successful. To date, the clinical trials which have been completed to evaluate the safety and efficacy of anti-fibrotic therapies in mesangial-associated glomerular fibrosis account for only a small portion. The current data from multiple studies provide us significance to pursue further researches to consummate a better understanding and application of these anti-fibrotic agents.

9.6.1 Drugs

As an oral pyridone derivative drug, pirfenidone (PFD) exerts anti-inflammatory and especially anti-fibrotic effects in various progressive fibrotic diseases. The clinical evaluation of PFD for its safety and efficacy has been pinpointed in renal fibrosis such as diabetic nephropathy and irreversible glomerulosclerosis and numerous extra-renal fibrotic disorders such as idiopathic pulmonary fibrosis and myelofibrosis. PFD not only ameliorates the morphological alterations of glomeruli and decreases proteinuria, but cuts off the accumulation of ECM in mesangium to prevent glomerular sclerosis. The most significant mechanism involved is that PFD can suppress the activity and production of TGF-β, which correlates closely with phenotypic alteration of mesangial cells to myofibroblasts causing an over-production of mesangial matrix. However, the side effects of PFD on mesangial cells remain to be elucidated by further rodent studies and clinical trials. Taken together, PFD is a promising agent for individuals with irreversible glomerulosclerosis.

In contrast to the Western medicine, there is accumulating evidence document that traditional Chinese medicine and the herb compounds isolated from medicinal plants are widely used in the treatment of renal fibrosis. Although some herbs such as aristolochic acids are considered to be renal pro-fibrotic, the others such as baicalein are widely used in China to treat patients with chronic renal failure and have been recognized as anti-fibrotic drug in kidneys. It has been reported that baicalein played a renoprotective role via inflammation-independent anti-fibrotic activities. Besides baicalein, quercetin, which share a similar core structure with baicalein, is one of the most abundant flavonoids in human diet, manifested a dose-dependent anti-fibrotic activity and relatively low cytotoxicity in diabetic nephropathy. In addition, quercetin can reduce oxidative stress and oxidative stress-induced signaling in cultured mesangial cells. More importantly, people can intake quercetin from fruits and vegetables instead of medicine.

9.6.2 HDAC Inhibitors

Acetylation of core histones is governed by opposing actions of a variety of histone acetyltransferases (HATs) and histone deacetylases (HDACs). Under pathological condition, the balance of normal histone acetylation can be disrupted, which leads to hyperacetylation or hypoacetylation of core histones and subsequently altered gene expression. Aberrant regulation of HDACs and HATs is found to disrupt the balance of normal histone acetylation. Likewise other disease model, deregulation of HDACs is found associated with renal fibrosis and many studies have demonstrated a role of histone acetylation in the progression of renal fibrosis (Chun 2017).

To recover from the setting of hypoacetylation, small molecular inhibitors of HDACs (termed as HDACi) have been tested for anti-fibrotic activity in various types of animal models. A number of structurally diverse HDAC inhibitors have been developed, many of which are currently being evaluated in clinical trials, especially for cancer therapy. In 2003, Mishra et al. found that TSA treatment could significantly decrease the urine-protein excretion and the proliferative hallmarks of glomerulonephritis associated with systemic lupus erythematosus (SLE)-induced lupus in mice for the first time (Mishra et al. 2003). Based on the chemical structure, HDACi can be classified into the following categories: hydroxamic acids [such as suberoylanilide hydroxamic acid (SAHA)]; cyclic peptide (such as FK228); short-chain and aromatic fatty acids (such as butyrate); benzamides (such as MS-275), and miscellaneous compounds. Many inhibitors are HDAC-type-specific, while some inhibitors are member-specific. For example, valproic acid (VPA) exerts inhibition of both class I and class II HDACs, with a high potency for class I HDACs, while Tubastatin A is a specific selective inhibitor of HDAC6. Increasing evidence has clearly indicated the inhibition of HDAC enzymes with specific inhibitors might become an important therapeutic approach for experimental renal fibrosis. A lot of biological effects of HDACi on mesangial cell have been confirmed, which include inhibition of cell proliferation and ECM accumulation, inhibition of inflammatory response and cytokines. For example, pretreatment with the HDAC inhibitor trichostatin A (TSA) or valproic acid (VPA) partially reversed human renal mesangial cells (Dai et al. 2016). SAHA, a Class I/II HDAC inhibitor approved by US FDA for cutaneous T-cell lymphoma (CTCL), was found to reduce EGFR protein and mRNA in cultured renal cells and in diabetic kidneys (Gilbert et al. 2011). Another study revealed that VPA, class I and II inhibitors, ameliorates STZ-induced diabetic nephropathy and experimental kidney fibrosis (Van Beneden et al. 2013). All these studies suggest a prospective clinical application of HDACi.

9.7 Conclusion

Increasing evidence has clearly indicated that mesangial cells play a crucial role in the pathogenesis of renal glomerular fibrosis. The increased cell numbers resulted from hyperproliferation and the accumulation of ECM induced by enhanced gene expression of pro-fibrotic markers ultimately lead to renal glomerular fibrosis. Therefore, maintenance of mesangial cell functions and reversal of aberrant gene expression benefit the normal renal function. This might become an important therapeutic approach for chronic fibrotic kidney diseases.

References

Alsaad KO, Herzenberg AM (2007) Distinguishing diabetic nephropathy from other causes of glomerulosclerosis: an update. J Clin Pathol 60:18–26

Andrews PM, Coffey AK (1983) Cytoplasmic contractile elements in glomerular cells. Fed Proc 42:3046–3052

Banas B, Wörnle M, Berger T, Nelson PJ, Cohen CD, Kretzler M et al (2002) Roles of SLC/CCL21 and CCR3 in human kidney for mesangial proliferation, migration, apoptosis, and tissue homeostasis. J Immunol 168:4301–4307

Baricos WH, Cortez SL, Deboisblanc M, Xin S (1999) Transforming growth factor-β is a potent inhibitor of extracellular matrix degradation by cultured human mesangial cells. J Am Soc Nephrol 10:790–795

Bidani AK, Mitchell KD, Schwartz MM, Navar LG, Lewis EJ (1990) Absence of glomerular injury or nephron loss in a normotensive rat remnant kidney model. Kidney Int 38:28–38

Browne JA, Liu X, Schnaper HW, Hayashida T (2013) Serine-204 in the linker region of Smad3 mediates the collagen-I response to TGF-β in a cell phenotype-specific manner. Exp Cell Res 319:2928–2937

Cattell V, Gaskin de Urdaneta A, Arlidge S, Collar JE, Roberts A, Smith J (1982) Uptake and clearance of ferritin by the glomerular mesangium. I. Phagocytosis by mesangial cells and blood monocytes. Lab Invest 47:296–303

Chang HR, Yang SF, Li ML, Lin CC, Hsieh YS, Lian JD (2006) Relationships between circulating matrix metalloproteinase-2 and -9 and renal function in patients with chronic kidney disease. Clin Chim Acta 366:243–248

Chen S, Jim B, Ziyadeh FN (2003) Diabetic nephropathy and transforming growth factor-beta: transforming our view of glomerulosclerosis and fibrosis build-up. Semin Nephrol 23:532–543

Chuang CT, Guh JY, Lu CY, Chen HC, Chuang LY (2015) S100B is required for high glucose-induced pro-fibrotic gene expression and hypertrophy in mesangial cells. Int J Mol Med 35:546–552

Chun P (2017) Therapeutic effects of histone deacetylase inhibitors on kidney disease. Arch Pharm Res 41:1–22

Cooker LA, Peterson D, Rambow J, Riser ML, Riser RE, Najmabadi F et al (2007) TNF-alpha, but not IFN-gamma, regulates CCN2 (CTGF), collagen type I, and proliferation in mesangial cells: possible roles in the progression of renal fibrosis. Am J Physiol Renal Physiol 293:F157–F165

Coppo R, Amore A, Gianoglio B, Reyna A, Peruzzi L, Roccatello D et al (1993) Serum IgA and macromolecular IgA reacting with mesangial matrix components. Contrib Nephrol 104:162–171

Daehn I, Casalena G, Zhang T, Shi S, Fenninger F, Barasch N et al (2014) Endothelial mitochondrial oxidative stress determines podocyte depletion in segmental glomerulosclerosis. J Clin Invest 124:1608–1621

Dai Q, Liu J, Du YL, Hao X, Ying J, Tan Y et al (2016) Histone deacetylase inhibitors attenuate P-aIgA1-induced cell proliferation and extracellular matrix synthesis in human renal mesangial cells in vitro. Acta Pharmacol Sin 37:228–234
Dey N, Ghosh-Choudhury N, Kasinath BS, Choudhury GG (2012) TGFβ-stimulated microRNA-21 utilizes PTEN to orchestrate AKT/mTORC1 signaling for mesangial cell hypertrophy and matrix expansion. PLoS ONE 7:e42316
Diamond-Stanic MK, You YH, Sharma K (2012) Sugar, sex, and TGF-β in diabetic nephropathy. Semin Nephrol 32:261–268
Dorsam G, Taher MM, Valerie KC, Kuemmerle NB, Chan JC, Franson RC (2000) Diphenyleneiodium chloride blocks inflammatory cytokine-induced up-regulation of group IIA phospholipase A(2) in rat mesangial cells. J Pharmacol Exp Ther 292:271–279
Eddy AA (1996) Molecular insights into renal interstitial fibrosis. J Am Soc Nephrol 7:2495–2508
Eddy AA (2000) Molecular basis of renal fibrosis. Pediatr Nephrol 15:290–301
Eitner F, Westerhuis R, Burg M, Weinhold B, Gröne HJ, Ostendorf T et al (1997) Role of interleukin-6 in mediating mesangial cell proliferation and matrix production in vivo. Kidney Int 51:69–78
Elema JD, Hoyer JR, Vernier RL (1976) The glomerular mesangium: uptake and transport of intravenously injected colloidal carbon in rats. Kidney Int 9:395–406
Eremina V, Cui S, Gerber H, Ferrara N, Haigh J, Nagy A et al (2006) Vascular endothelial growth factor a signaling in the podocyte-endothelial compartment is required for mesangial cell migration and survival. J Am Soc Nephrol 17:724–735
Fine LG, Hammerman MR, Abboud HE (1992) Evolving role of growth factors in the renal response to acute and chronic disease. J Am Soc Nephrol 2:1163–1170
Floege J, van Roeyen C, Boor P, Ostendorf T (2007) The role of PDGF-D in mesangioproliferative glomerulonephritis. Contrib Nephrol 157:153–158
Fogo AB (1999) Mesangial matrix modulation and glomerulosclerosis. Exp Nephrol 7:147–159
Gewin LS (2018) Renal tubule repair: is Wnt/β-catenin a friend or foe? Genes (Basel) 9:58
Ghosh-Choudhury G, Abboud HE (2004) Tyrosine phosphorylation-dependent PI 3 kinase/Akt signal transduction regulates TGFbeta-induced fibronectin expression in mesangial cells. Cell Sig 16:31–41
Gilbert RE, Huang Q, Thai K, Advani SL, Lee K, Yuen DA et al (2011) Histone deacetylase inhibition attenuates diabetes-associated kidney growth: potential role for epigenetic modification of the epidermal growth factor receptor. Kidney Int 79:1312–1321
Gong W, Li J, Chen Z, Huang J, Chen Q, Cai W et al (2017) Polydatin promotes Nrf2-ARE anti-oxidative pathway through activating CKIP-1 to resist HG-induced up-regulation of FN and ICAM-1 in GMCs and diabetic mice kidneys. Free Radic Biol Med 106:393–405
Guo W, Tian D, Jia Y, Huang W, Jiang M, Wang J et al (2018) MDM2 controls NRF2 antioxidant activity in prevention of diabetic kidney disease. Biochim Biophys Acta Mol Cell Res 1865:1034–1045
Hayden MS, Ghosh S (2004) Signaling to NF-kappaB. Genes Dev 18:2195–2224
Hewitson TD (2009) Renal tubulointerstitial fibrosis: common but never simple. Am J Physiol Renal Physiol 296:F1239–F1244
Hewitson TD, Becker GJ (1995) Interstitial myofibroblasts in IgA glomerulonephritis. Am J Nephrol 15:111–117
Ho C, Hsu YC, Lei CC, Mau SC, Shih YH, Lin CL (2016) Curcumin rescues diabetic renal fibrosis by targeting superoxide-mediated Wnt signaling pathways. Am J Med Sci 351:286–295
Hocevar BA, Prunier C, Howe PH (2005) Disabled-2 (Dab2) mediates transforming growth factor beta (TGFbeta)-stimulated fibronectin synthesis through TGFbeta-activated kinase 1 and activation of the JNK pathway. J Biol Chem 280:25920–25927
Hoffman BB, Sharma K, Ziyadeh FN (1998) Potential role of TGF-β in diabetic nephropathy. Miner Electrolyte Metab 24:190–196
Hostetter TH, Olson JL, Rennke HG, Venkatachalam MA, Brenner BM (1981) Hyperfiltration in remnant nephrons: a potentially adverse response to renal ablation. Am J Physiol 241:F85–F93

Howe EN, Cochrane DR, Richer JK (2012) The miR-200 and miR-221/222 microRNA families: opposing effects on epithelial identity. J Mammary Gland Biol Neoplasia 17:65–77

Hsu YC, Chang PJ, Ho C, Huang YT, Shih YH, Wang CJ et al (2016) Protective effects of miR-29a on diabetic glomerular dysfunction by modulation of DKK1/Wnt/β-catenin signaling. Sci Rep 6:30575

Huang K, Huang J, Xie X, Wang S, Chen C, Shen X et al (2013) Sirt1 resists advanced glycation end products-induced expressions of fibronectin and TGF-β1 by activating the Nrf2/ARE pathway in glomerular mesangial cells. Free Radic Biol Med 65:528–540

Hubchak SC, Sparks EE, Hayashida T, Schnaper HW (2009) Rac1 promotes TGF-beta-stimulated mesangial cell type I collagen expression through a PI3K/Akt-dependent mechanism. Am J Physiol Renal Physiol 297:F1316–F1323

Ichikawa I, Ma J, Motojima M, Matsusaka T (2005) Podocyte damage damages podocytes: autonomous vicious cycle that drives local spread of glomerular sclerosis. Curr Opin Nephrol Hypertens 14:205–210

Ji X, Li C, Ou Y, Li N, Yuan K, Yang G et al (2016) Andrographolide ameliorates diabetic nephropathy by attenuating hyperglycemia-mediated renal oxidative stress and inflammation via Akt/NF-κB pathway. Mol Cell Endocrinol 437:268–279

Jiang T, Huang Z, Lin Y, Zhang Z, Fang D, Zhang DD (1998) The protective role of Nrf2 in streptozotocin-induced diabetic nephropathy. Diabetes 59:850–860

Johnson RJ, Floege J, Yoshimura A, Iida H, Couser WG, Alpers CE (1992) The activated mesangial cell: a glomerular "myofibroblast"? J Am Soc Nephrol 2:S190–S197

Johnston WH, Latta H (1977) Glomerular mesangial and endothelial cell swelling following temporary renal ischemia and its role in the no-reflow phenomenon. Am J Pathol 89:153–166

Kato M, Zhang J, Wang M, Lanting L, Yuan H, Rossi JJ et al (2007) MicroRNA-192 in diabetic kidney glomeruli and its function in TGF-beta-induced collagen expression via inhibition of E-box repressors. Proc Natl Acad Sci U S A 104:3432–3437

Kato M, Arce L, Wang M, Putta S, Lanting L, Natarajan R (2011) A microRNA circuit mediates transforming growth factor-β1 autoregulation in renal glomerular mesangial cells. Kidney Int 80:358–368

Keane WF, Raij L (1981) Determinants of glomerular mesangial localization of immune complexes. Role of endothelial fenestrae. Lab Invest 45:366–371

Kim SI, Kwak JH, Zachariah M, He Y, Wang L, Choi ME (2007) TGF-beta-activated kinase 1 and TAK1-binding protein 1 cooperate to mediate TGF-beta1-induced MKK3-p38 MAPK activation and stimulation of type I collagen. Am J Physiol Renal Physiol 292:F1471–F1478

Kishida M, Urakaze M, Takata M, Nobata Y, Yamamoto N, Temaru R et al (2005) PGE1 inhibits the expression of PAI-1 mRNA induced by TNF-alpha in human mesangial cells. Exp Clin Endocrinol Diabetes 113:365–371

Kölling M, Kaucsar T, Schauerte C, Hübner A, Dettling A, Park JK et al (2017) Therapeutic miR-21 silencing ameliorates diabetic kidney disease in mice. Mol Ther 25:165–180

Kriz W, Elger M, Lemley K, Sakai T (1990) Structure of the glomerular mesangium: a biomechanical interpretation. Kidney Int Suppl 30:S2–S9

Lan A, Du J (2014) Potential role of Akt signaling in chronic kidney disease. Nephrol Dial Transplant 30:385–394

Latta H (1970) The glomerular capillary wall. J Ultrastruct Res 32:526–544

Latta H, Fligiel S (1985) Mesangial fenestrations, sieving, filtration, and flow. Lab Invest 52:591–598

Latta H, Johnston WH (1978) Granular and agranular cell counts in the juxtaglomerular apparatuses of rats with unilateral renovascular hypertension. Lab Invest 39:219–224

Latta H, Lee TC (1983) Effects of excessive sodium chloride on the juxtaglomerular apparatus and blood pressure of uninephrectomized rats. Lab Invest 49:99–106

Latta H, Maunsbach AB (1962) Relations of the centrolobular region of the glomerulus to the juxtaglomerular apparatus 1. J Ultrastruct Res 6:562–578

Latta H, Maunsbac AB, Madden SC (1960) The centrolobular region of the renal glomerulus studied by electron microscopy 1. J Ultrastruct Res 4:455–472

Lee JM, Johnson JA (2004) An important role of Nrf2-ARE pathway in the cellular defense mechanism. J Biochem Mol Biol 37:139–143

Lee HS, Song CY (2009) Differential role of mesangial cells and podocytes in TGF-beta-induced mesangial matrix synthesis in chronic glomerular disease. Histol Histopathol 24:901–908

Lee MK, Pardoux C, Hall MC, Lee PS, Warburton D, Qing J et al (1977) Uptake and transport of imposil by the glomerular mesangium in the mouse. Lab Invest 37:526–533

Li W, Cui M, Wei Y, Kong X, Tang L, Xu D (2012) Inhibition of the expression of TGF-β1 and CTGF in human mesangial cells by exendin-4, a glucagon-like peptide-1 receptor agonist. Cell Physiol Biochem 30:749–757

Li XY, Wang SS, Han Z, Han F, Chang YP, Yang Y et al (2017) Triptolide restores autophagy to alleviate diabetic renal fibrosis through the miR-141-3p/PTEN/Akt/mTOR pathway. Mol Ther Nucleic Acids 9:48–56

Liang Y, Zhang J, Zhou Y, Xing G, Zhao G, Liu Z (2015) Proliferation and cytokine production of human mesangial cells stimulated by secretory IgA isolated from patients with IgA nephropathy. Cell Physiol Biochem 36:1793–1808

Liang JS, Song WL, Yan Wu (2016a) The study on the molecular mechanism of endothelial cells in diabetic nephropathy. Chin J Diabetes 24:362

Liang Y, Zhao G, Tang L, Zhang J, Li T, Liu Z (2016b) MiR-100-3p and miR-877-3p regulate overproduction of IL-8 and IL-1β in mesangial cells activated by secretory IgA from IgA nephropathy patients. Exp Cell Res 347:312–321

Lin X, Zha Y, Zeng XZ, Dong R, Wang QH, Wang DT (2017) Role of the Wnt/β-catenin signaling pathway in inducing apoptosis and renal fibrosis in 5/6-nephrectomized rats. Mol Med Rep 15:3575–3582

Lindenmeyer MT, Kretzler M, Boucherot A, Berra S, Yasuda Y, Henger A et al (2007) Interstitial vascular rarefaction and reduced VEGF-A expression in human diabetic nephropathy. J Am Soc Nephrol 18:1765–1776

Liu Y (2006) Renal fibrosis: new insights into the pathogenesis and therapeutics. Kidney Int 69:213–217

Liu J, Li Y, Liu MD, Ren GW, Wang MM, Li YM, Cong B (2012) Effects of megsin gene transfection on the expressions of MCP-1 and ICAM-1 in mouse glomerular mesangial cells. Chin J Cell Mol Immunol 28:702–705

Liu H, Zhang XP, Yi ZW (2013) Efficacy of antisense monocyte chemoattractant protein-1 (MCP-1) in a rat model of mesangial proliferative glomerulonephritis. Ren Fail 35:1418–1428

López-Hernández FJ, López-Novoa JM (2012) Role of TGF-β in chronic kidney disease: an integration of tubular, glomerular and vascular effects. Cell Tissue Res 347:141–154

Ma FY, Flanc RS, Tesch GH, Bennett BL, Friedman GC, Nikolic-Paterson DJ (2009) Blockade of the c-Jun amino terminal kinase prevents crescent formation and halts established anti-GBM glomerulonephritis in the rat. Lab Invest 89:470–484

Mahimainathan L, Das F, Venkatesan B, Choudhury GG (2006) Mesangial cell hypertrophy by high glucose is mediated by downregulation of the tumor suppressor PTEN. Diabetes 55:2115–2125

Makino H (1988) Three-dimensional ultrastructure of rat acellular glomerulus by scanning electron microscopy. J Electron Microsc (Tokyo) 37:294–304

Mason RM, Wahab NA (2003) Extracellular matrix metabolism in diabetic nephropathy. J Am Soc Nephrol 14:1358–1373

Massy ZA, Guijarro C, O'Donnell MP, Kim Y, Kashtan CE, Egido J et al (1999) The central role of nuclear factor-kappaB in mesangial cell activation. Kidney Int Suppl 71:S76–S79

Mathew A, Cunard R, Sharma K (2011) Antifibrotic treatment and other new strategies for improving renal outcomes. Contrib Nephrol 170:217–227

Mauer SM, Steffes MW, Ellis EN, Sutherland DE, Brown DM, Goetz FC (1984) Structural-functional relationships in diabetic nephropathy. J Clin Invest 74:1143–1155

Mené P, Simonson MS, Dunn MJ (1989) Physiology of the mesangial cell. Physiol Rev 69:1347–1424

Meng XM, Chung AC, Lan HY (2013) Role of the TGF-β/BMP-7/Smad pathways in renal diseases. Clin Sci (Lond) 124:243–254

Meng XM, Nikolic-Paterson DJ, Lan HY (2016) TGF-β: the master regulator of fibrosis. Nat Rev Nephrol 12:325–338

Mercer TR, Mattick JS (2013) Structure and function of long noncoding RNAs in epigenetic regulation. Nat Struct Mol Biol 20:300–307

Michielsen P, Creemers J (1967) The structure and function of the glomerular mesangium. Ultrastruct Kidney 47:57–72

Miller PL, Rennke HG, Meyer TW (1991) Glomerular hypertrophy accelerates hypertensive glomerular injury in rats. Am J Physiol 261:459–465

Min D, Lyons JG, Bonner J, Twigg SM, Yue DK, McLennan SV (2009) Mesangial cell-derived factors alter monocyte activation and function through inflammatory pathways: possible pathogenic role in diabetic nephropathy. Am J Physiol Renal Physiol 297:1229–1237

Mishra R, Leahy P, Simonson MS (2002) Gene expression profiling reveals role for EGF-family ligands in mesangial cell proliferation. Am J Physiol Renal Physiol 283:F1151–F1159

Mishra N, Reilly CM, Brown DR, Ruiz P, Gilkeson GS (2003) Histone deacetylase inhibitors modulate renal disease in the MRL-lpr/lpr mouse. J Clin Invest 111:539–552

Mitchell D, Rodgers K, Hanly J, McMahon B, Brady HR, Martin F et al (2004) Lipoxins inhibit Akt/PKB activation and cell cycle progression in human mesangial cells. Am J Pathol 164:937–946

Nahman NS Jr, Leonhart KL, Cosio FG, Hebert CL (1992) Effects of high glucose on cellular proliferation and fibronectin production by cultured human mesangial cells. Kidney Int 41:396–402

Naito T, Masaki T, Nikolic-Paterson DJ, Tanji C, Yorioka N, Kohno N (2004) Angiotensin II induces thrombospondin-1 production in human mesangial cells via p38 MAPK and JNK: a mechanism for activation of latent TGF-beta1. Am J Physiol Renal Physiol 286:F278–F287

Nakao A, Afrakhte M, Morén A, Nakayama T, Christian JL, Heuchel R et al (1997) Identification of Smad7, a TGFbeta-inducible antagonist of TGF-beta signalling. Nature 389:631–635

Olivetti G, Anversa P, Rigamonti W, Vitali-Mazza L, Loud AV (1977) Morphometry of the renal corpuscle during normal postnatal growth and compensatory hypertrophy. A light microscope study. J Cell Biol 75:573–585

Olivetti G, Anversa P, Melissari M, Loud AV (1980) Morphometry of the renal corpuscle during postnatal growth and compensatory hypertrophy. Kidney Int 17:438–454

Patel K, Harding P, Haney LB, Glass WF 2nd (2003) Regulation of the mesangial cell myofibroblast phenotype by actin polymerization. J Cell Physiol 195:435–445

Peng J, Dong Z (2012) Role changes of β-catenin in kidney injury and repair. Kidney Int 82:509–511

Piek E, Ju WJ, Heyer J, Escalante-Alcalde D, Stewart CL, Weinstein M et al (2001) Functional characterization of transforming growth factor beta signaling in Smad2- and Smad3-deficient fibroblasts. J Biol Chem 276:19945–19953

Pillebout E, Nochy D (2010) Glomerulonephritis with IgA mesangial deposits. Nephrol Ther 6:545–557

Pugliese G, Pricci F, Pugliese F, Mene P, Lenti L, Andreani D et al (1994) Mechanisms of glucose-enhanced extracellular matrix accumulation in rat glomerular mesangial cells. Diabetes 43:478–490

Qian Y, Feldman E, Pennathur S, Kretzler M, Brosius FC 3rd (2008) From fibrosis to sclerosis: mechanisms of glomerulosclerosis in diabetic nephropathy. Diabetes 57:1439–1445

Radeke HH, Resch K (1992) The inflammatory function of renal glomerular mesangial cells and their interaction with the cellular immune system. Clin Investig 70:825–842

Reidy K, Kang HM, Hostetter T, Susztak K (2014) Molecular mechanisms of diabetic kidney disease. J Clin Invest 124:2333–2340

Rintala JM, Savikko J, Rintala SE, Palin N, Koskinen PK (2016) Epidermal growth factor receptor inhibition with erlotinib ameliorates anti-Thy 1.1-induced experimental glomerulonephritis. J Nephrol 29:359–365

Ruef C, Kashgarian M, Coleman DL (1992) Mesangial cell-matrix interactions. Effects on mesangial cell growth and cytokine secretion. Am J Pathol 141:429–439
Sakai T, Kriz W (1987) The structural relationship between mesangial cells and basement membrane of the renal glomerulus. Anat Embryol (Berl) 176:373–386
Samarakoon R, Dobberfuhl AD, Cooley C, Overstreet JM, Patel S, Goldschmeding R et al (2013) Induction of renal fibrotic genes by TGF-β1 requires EGFR activation, p53 and reactive oxygen species. Cell Sig 25:2198–2209
Santini E, Lupi R, Baldi S, Madec S, Chimenti D, Ferrannini E et al (2008) Effects of different LDL particles on inflammatory molecules in human mesangial cells. Diabetologia 51:2117–2125
Schmekel B, Svalander C, Bucht H, Westberg NG (2010) Mesangial IgA glomerulonephritis in adults. Clinical and histopathological observations. Acta Med Scand 210:363–372
Schnaper HW, Hayashida T, Hubchak SC, Poncelet AC (2003) TGF-beta signal transduction and mesangial cell fibrogenesis. Am J Physiol Renal Physiol 284:F243–F252
Sheng L, Yang M, Ding W, Zhang M, Niu J, Qiao Z et al (2016) Epidermal growth factor receptor signaling mediates aldosterone-induced pro-fibrotic responses in kidney. Exp Cell Res 346:99–110
Shi Y, Massagué J (2003) Mechanisms of TGF-beta signaling from cell membrane to the nucleus. Cell 113:685–700
Shi S, Yu L, Chiu C, Sun Y, Chen J, Khitrov G et al (2008) Podocyte-selective deletion of dicer induces proteinuria and glomerulosclerosis. J Am Soc Nephrol 19:2159–2169
Shirato I, Hosser H, Kimura K, Sakai T, Tomino Y, Kriz W (1996) The development of focal segmental glomerulosclerosis in masugi nephritis is based on progressive podocyte damage. Virchows Arch 429:255–273
Shultz PJ, DiCorleto PE, Silver BJ, Abboud HE (1988) Mesangial cells express PDGF mRNAs and proliferate in response to PDGF. Am J Physiol 255:674–684
Silva FG, Eigenbrodt EH, Glass M, Taft E (1986) An ultrastructural study of the renal juxtaglomerular apparatus and extraglomerular mesangium in patients with systemic lupus erythematosus. Am J Kidney Dis 7:47–57
Simonson MS (2007) Phenotypic transitions and fibrosis in diabetic nephropathy. Kidney Int 71:846–854
Throckmorton DC, Brogden AP, Min B, Rasmussen H, Kashgarian M (1995) PDGF and TGF-beta mediate collagen production by mesangial cells exposed to advanced glycosylation end products. Kidney Int 48:111–117
Toyoda M, Najafian B, Kim Y, Caramori ML, Mauer M (2007) Podocyte detachment and reduced glomerular capillary endothelial fenestration in human type 1 diabetic nephropathy. Diabetes 56:2155–2160
Uguccioni M, D'Apuzzo M, Loetscher M, Dewald B, Baggiolini M (2010) Actions of the chemotactic cytokines MCP-1, MCP-2, MCP-3, RANTES, MIP-1α and MIP-1β on human monocytes. Eur J Immunol 25:64–68
Uttarwar L, Peng F, Wu D, Kumar S, Gao B, Ingram AJ et al (2011) HB-EGF release mediates glucose-induced activation of the epidermal growth factor receptor in mesangial cells. Am J Physiol Renal Physiol 300:F921–F931
Van Beneden K, Geers C, Pauwels M, Mannaerts I, Wissing KM, Van den Branden C et al (2013) Comparison of trichostatin A and valproic acid treatment regimens in a mouse model of kidney fibrosis. Toxicol Appl Pharmacol 271:276–284
Wang M, Wang S, Yao D, Yan Q, Lu W (2016a) A novel long non-coding RNA CYP4B1-PS1-001 regulates proliferation and fibrosis in diabetic nephropathy. Mol Cell Endocrinol 426:136–145
Wang M, Yao D, Wang S, Yan Q, Lu W (2016b) Long non-coding RNA ENSMUST00000147869 protects mesangial cells from proliferation and fibrosis induced by diabetic nephropathy. Endocrine 54:1–12
Wang S, Yang Z, Xiong F, Chen C, Chao X, Huang J et al (2016c) Betulinic acid ameliorates experimental diabetic-induced renal inflammation and fibrosis via inhibiting the activation of NF-κB signaling pathway. Mol Cell Endocrinol 434:135–143
Wanner N, Bechtel-Walz W (2017) Epigenetics of kidney disease. Cell Tissue Res 369:75–92

Wendt TM, Tanji N, Guo J, Kislinger TR, Qu W, Lu Y et al (2003) RAGE drives the development of glomerulosclerosis and implicates podocyte activation in the pathogenesis of diabetic nephropathy. Am J Pathol 162:1123–1137

Wharram BL, Goyal M, Wiggins JE, Sanden SK, Hussain S, Filipiak WE et al (2005) Podocyte depletion causes glomerulosclerosis: diphtheria toxin-induced podocyte depletion in rats expressing human diphtheria toxin receptor transgene. J Am Soc Nephrol 16:2941–2952

Wiggins JE, Goyal M, Sanden SK, Wharram BL, Shedden KA, Misek DE et al (2005) Podocyte hypertrophy, "adaptation," and "decompensation" associated with glomerular enlargement and glomerulosclerosis in the aging rat: prevention by calorie restriction. J Am Soc Nephrol 16:2953–2966

Wu L, Wang Q, Guo F, Ma X, Ji H, Liu F et al (2016) MicroRNA-27a induces mesangial cell injury by targeting of PPARγ, and its in vivo knockdown prevents progression of diabetic nephropathy. Sci Rep 6:26072

Xiao L, Wang M, Yang S, Liu F, Sun L (2013) A glimpse of the pathogenetic mechanisms of Wnt/β-catenin signaling in diabetic nephropathy. Biomed Res Int 2013:987064

Xie X, Chang X, Chen L, Huang K, Huang J, Wang S et al (2013) Berberine ameliorates experimental diabetes-induced renal inflammation and fibronectin by inhibiting the activation of RhoA/ROCK signaling. Mol Cell Endocrinol 381:56–65

Xiong F, Li X, Yang Z, Wang Y, Gong W, Huang J et al (2016) TGR5 suppresses high glucose-induced upregulation of fibronectin and transforming growth factor-β1 in rat glomerular mesangial cells by inhibiting RhoA/ROCK signaling. Endocrine 54:657–670

Xu Z, Jia Q, Li H, Wang J, Zhao Y, Chen S et al (2013) Effects of parthenolide on high glucose-induced cell proliferation, NF-κB activation and MCP-1 expression in rat mesangial cells. Nan fang yi ke da xue xue bao 33:1471–1473

Xu ZJ, Shu S, Li ZJ, Liu YM, Zhang RY, Zhang Y (2017) Liuwei Dihuang pill treats diabetic nephropathy in rats by inhibiting of TGF-β/SMADS, MAPK, and NF-kB and upregulating expression of cytoglobin in renal tissues. Medicine (Baltimore) 96:e5879

Yamamoto Y, Iyoda M, Tachibana S, Matsumoto K, Wada Y, Suzuki T et al (2018) Erlotinib attenuates the progression of chronic kidney disease in rats with remnant kidney. Nephrol Dial Transplant 33:598–606

Yang X, Wang Y, Gao G (2016) High glucose induces rat mesangial cells proliferation and MCP-1 expression via ROS-mediated activation of NF-κB pathway, which is inhibited by eleutheroside E. J Recept Sig Transduct Res 36:152–157

Yi H, Peng R, Zhang LY, Sun Y, Peng HM, Liu HD et al (2017) LincRNA-Gm4419 knockdown ameliorates NF-κB/NLRP3 inflammasome-mediated inflammation in diabetic nephropathy. Cell Death Dis 8:e2583

Yoshida Y, Fogo A, Ichikawa I (1989) Glomerular hemodynamic changes vs. hypertrophy in experimental glomerular sclerosis. Kidney Int 35:654–660

Yuan P, Xue H, Zhou L, Qu L, Li C, Wang Z et al (2011) Rescue of mesangial cells from high glucose-induced over-proliferation and extracellular matrix secretion by hydrogen sulfide. Nephrol Dial Transplant 26:2119–2126

Yuan Y, Fan Q, Du SY, Yuan F, Xu L, Li L et al (2017) Effect of ursolic acid on hypertrophy and proliferation of renal glomerular mesangial cells and extracellular matrix production accumulation induced by high glucose. Chin J Clin Pharmacol 33:2158

Zhang YE (2009) Non-Smad pathways in TGF-β signaling. Cell Res 19:128–139

Zhang L, Pang S, Deng B, Qian L, Chen J, Zou J et al (2012) High glucose induces renal mesangial cell proliferation and fibronectin expression through JNK/NF-κB/NADPH oxidase/ROS pathway, which is inhibited by resveratrol. Int J Biochem Cell Biol 44:629–638

Zhang M, Yan Z, Bu L, An C, Wang D, Liu X et al (2018) Rapeseed protein-derived antioxidant peptide RAP alleviates renal fibrosis through MAPK/NF-κB signaling pathways in diabetic nephropathy. Drug Des Devel Ther 12:1255–1268

Zhou B, Liu Y, Kahn M, Ann DK, Han A, Wang H et al (2012) Interactions between β-catenin and transforming growth factor-β signaling pathways mediate epithelial-mesenchymal transition

and are dependent on the transcriptional co-activator cAMP-response element-binding protein (CREB)-binding protein (CBP). J Biol Chem 287:7026–7038
Zhou X, Feng Y, Zhan Z, Chen J (2014) Hydrogen sulfide alleviates diabetic nephropathy in a streptozotocin-induced diabetic rat model. J Biol Chem 289:28827–28834
Zhou J, Zhong J, Huang Z, Liao M, Lin S, Chen J et al (2018) TAK1 mediates apoptosis via p38 involve in ischemia-induced renal fibrosis. Artif Cells Nanomed Biotechnol 16:1–10
Zlotnik A, Yoshie O (2000) Chemokines. Immunity 12:121–127

Chapter 10
Role of Podocyte Injury in Glomerulosclerosis

Chen-Chen Lu, Gui-Hua Wang, Jian Lu, Pei-Pei Chen, Yang Zhang, Ze-Bo Hu and Kun-Ling Ma

Abstract Finding new therapeutic targets of glomerulosclerosis treatment is an ongoing quest. Due to a living environment of various stresses and pathological stimuli, podocytes are prone to injuries; moreover, as a cell without proliferative potential, loss of podocytes is vital in the pathogenesis of glomerulosclerosis. Thus, sufficient understanding of factors and underlying mechanisms of podocyte injury facilitates the advancement of treating and prevention of glomerulosclerosis. The clinical symptom of podocyte injury is proteinuria, sometimes with loss of kidney functions progressing to glomerulosclerosis. Injury-induced changes in podocyte physiology and function are actually not a simple passive process, but a complex interaction of proteins that comprise the anatomical structure of podocytes at molecular levels. This chapter lists several aspects of podocyte injuries along with potential mechanisms, including glucose and lipid metabolism disorder, hypertension, RAS activation, micro-inflammation, immune disorder, and other factors. These aspects are not technically separated items, but intertwined with each other in the pathogenesis of podocyte injuries.

Keywords Podocyte injury · Glomerular sclerosis

10.1 Introduction

10.1.1 The Structure and Physiology of Podocytes

The glomerular filtration membrane constituted by three components, porous endothelial cells, glomerular basement membrane (GBM), and epithelial cells in the GBM, which also called podocytes. Podocytes are highly differentiated epithelial cells consist of three distinct parts: cell body, major processes, and foot processes (FPs). Podocytes have a voluminous cell body, which is at the central position of the

C.-C. Lu · G.-H. Wang · J. Lu · P.-P. Chen · Y. Zhang · Z.-B. Hu · K.-L. Ma (✉)
Institute of Nephrology, Zhong Da Hospital, Southeast University School of Medicine, Nanjing, China
e-mail: klma05@163.com

B.-C. Liu et al. (eds.), *Renal Fibrosis: Mechanisms and Therapies*, Advances in Experimental Medicine and Biology 1165, https://doi.org/10.1007/978-981-13-8871-2_10

cell and lies in the urinary space. The cell body contains a nucleus, abundant endoplasmic reticulum, a well-developed Golgi system, lysosomes, and mitochondria. The densely distributed organelles in the cell body suggest a high level of anabolic and catabolic activity. Microtubules and intermediate filaments, such as vimentin and desmin, are the dominated cytoskeleton components in cell body to accounts for the unique shape of podocytes (Pavenstadt et al. 2003). From the cell body, podocytes give rise to primary processes that reach to glomerular capillary, forming an affixation by FPs.

Podocytes are polarized epithelial cells which contain a apical/luminal and a basal cell membrane. The apical surface domain forms a few fingerlike protrusions which bulge into Bowman's space. The apical domain is negatively charged, which limits the passage of albumin (also negatively charged) and maintain the separation of adjacent podocytes by anion charge. The basal cell membrane mediates the affixation to the GBM by $\alpha3\beta1$ integrin and α- and β-dystroglycans, which play the function by connecting to certain matrix proteins within the GBM (Kreidberg et al. 1996; Raats et al. 2000). Both of apical and basal membranes contain numerously distributed cholesterol-rich domains, and it was found that specific membrane proteins of podocytes are obviously arranged in rafts (Schwarz et al. 2001; Simons et al. 2001).

FPs functionally consist of a luminal or apical membrane domain and a basal cell membrane domain. FPs are characterized by a podosome-like cortical network of short, branched actin filaments and by the presence of highly ordered, parallel contractile actin filament bundles, which are thought to modulate the permeability of the filtration barrier through changes in FP morphology (Greka and Mundel 2012). The foot processes of neighboring podocytes are bridged by slit diaphragm (SD), which is the site of convective fluid flow through the visceral epithelium and the final barrier to urinary protein loss. Similar to the apical membrane domain of podocytes, the SD is also covered by a thick surface coat mainly constituted by sialoglycoproteins, including podoendin, podocalyxin, and others, which are responsible for the high negative surface charge of the podocytes. In addition, several molecules, including ZO-1 (zonula occludens protein), nephrin, CD2AP (CD2-associated protein), FAT, and P-cadherin, have all been shown to be expressed within the SD, and some of those molecules play a major role for its integrity (Pavenstadt et al. 2003).

The unique shape of podocyte and the maintenance of its processes are owing to a well-developed cytoskeleton, which serves as the podocyte's "backbone." And also, the actin-rich cytoskeleton makes podocytes to be able to alter shape continually and dynamically. The cytoskeleton is comprised by microfilaments (7–9 nm diameter), intermediate filaments (10 nm diameter), and microtubules (24 nm diameter), which are mainly defined by their diameter. Microtubules and intermediate filaments are predominant cytoskeletal constituents in the cell body and the primary processes. In the FPs, microfilaments are the main cytoskeletal component, which consist of a network with densely accumulated F-actin and myosin. FP actin cytoskeleton is extensively distributed in all three domains of FPs, resulting to an important role of actin for the function and dysfunction of podocytes. FP effacement requires the activation of actin filaments reorganization, a process which is regulated by multiple signaling events involving integrin activation, G protein-coupled receptor (GPCR) and growth factor receptor, and calcium (Ca^{2+}) influx pathways as upstream modulators of the actin cytoskeleton (Takeda et al. 2001).

The complex architecture of podocytes, in particular on the maintenance of highly ordered, parallel, contractile actin filament bundles in FPs, is required for the highly specialized functions of podocytes, which include (i) a size barrier to protein; (ii) charge barrier to protein; (iii) maintenance of the capillary loop shape; (iv) counteracting the intraglomerular pressure; (v) synthesis and maintenance of the GBM; (vi) production and secretion of vascular endothelial growth factor (VEGF) required for GEN integrity (Shankland 2006).

Podocyte is the most differentiated cell type in the glomerulus, which plays a crucial role in the glomerular filtration barrier. Podocyte foot processes with the interposed SD represent the last filtration barrier of GBM. The SD is a subtle signal transduction unit characterized by a modified adherens junction that bridges the 30–50-nm-wide filtration slits (Reiser et al. 2000). Transmembrane proteins such as nephrin and FAT constitute the rod-like units of SD which are connected by numerous linear bar, forming a network with pores the same size as or smaller than albumin (Mundel and Kriz 1995). Meanwhile, negatively charged apical domain of SD works as a charge barrier to prevent the albumin loss. Thus, the podocyte is the important size and charge barrier of GBM, and podocytes' damage leads to the disruption of GBM integrity and proteinuria.

Podocytes stabilize glomerular architecture owing to FPs counteract distensions of the glomerular basement membrane, which is regulated by vasoactive hormones. In this regard, they are responsible for 40% of the hydraulic resistance of the filtration barrier (Pavenstadt 2000). ANG II regulates the contractile state of their foot processes by activating a Cl^- conductance and increasing $[Ca^{2+}]_{i,}$ cAMP in podocytes, thereby modulating the ultrafiltration coefficient Kf. Other agonists such as AVP, oxytocin, norepinephrine, and parathormone have also been reported to modulate $[Ca^{2+}]_i$ in podocytes. Vasoactive hormones may also alter charge properties of the podocyte and thereby enhance urinary protein excretion (Pavenstadt 2000).

VEGF family consists of five secreted homodimeric glycoproteins: VEGF-A, VEGF-B, VEGF-C, VEGF-D, and placental growth factor. In human and murine kidneys, VEGF-A isoform is constitutively expressed in podocytes, while playing its role mainly by contact with VEGFR-1 and VEGFR-2 predominately localized on the glomerular endothelial cells. It was assumed that VEGF-A is critical for the regulation of endothelial cell survival, proliferation, differentiation, and migration as well as endothelium-dependent vasodilatation and vascular permeability (Advani 2014). The complex paracrine signaling pathway between podocytes and glomerular endothelial cells plays a central role in maintaining the structure and integrity of the kidney filtration barrier.

10.1.2 The Role of Podocyte Injury in the Progresses of Glomerulosclerosis

Podocyte injury is the common pathological process in many glomerular diseases such as minimal change disease, membranous glomerulopathy, focal segmental glomerulosclerosis (FSGS), diabetic nephropathy (DN), and lupus nephritis. Physiological stresses or pathological stimuli like mechanical stress, oxidative stress, and immunologic stress disrupt the homeostasis of glomerular filtration barrier. Transcapillary pressure increment is induced by glomerular hypertension/hyperfiltration, and podocyte processes' elongation is induced by capillary expansion contribute to cytoskeletal dysregulation and intrinsic stress (Neal et al. 2007). Pathological factors, such as ischemia–reperfusion, chemical/toxic substances from the primary urine, usually cause reactive oxygen species (ROS) production in podocytes (Chen et al. 2013). It was also reported that aldosterone and angiotensin II promoted receptor-mediated ROS generation in podocytes (Liu et al.2013). Immunologic stress is induced by cytokine/complement, such as CC chemokine receptor 2, tumor necrosis factor, and sublytic C5b-9-mediated intracellular stress in podocytes (Nagata 2016).

Typical electron microscopy manifestations of podocyte injury include microcystic, pseudocystic changes, vacuolization, the presence of cytoplasmic inclusion bodies, and detachment from the GBM. Besides those changes, foot process effacement is the most characteristic change in podocyte injury. The damage of SD proteins contributes to cytoskeleton disorganization, leading to podocyte effacement and proteinuria (Shankland 2006). SD between adjacent podocytes is constituted predominantly by SD proteins including nephrin, podocin, CD2AP, Neph1, and FAT1. Mutations/abnormalities of those proteins result proteinuria and kidney disease. Studies have shown that SD proteins regulate cytoskeleton organization and podocyte shape by interacting with proteins associated with actin cytoskeleton. FAT-1 is an organizer of actin polymerization. CD2AP connects the nephrin complex with the actin-modifying proteins WASP, CAPZ, cortactin, and the Arp2/3 complex.

Reduced podocyte number causes proteinuria and glomerulosclerosis. Podocyte detachment, podocyte apoptosis, and the lack of adequate podocyte proliferation are three main reasons leading to the decrease in podocyte number also called "podocytopenia." The lack of charge- and size-selective barriers induced by podocyte loss leads to proteinuria. Studies have demonstrated the correlation of podocytes reduction with the onset and progression of glomerulosclerosis. Because podocytes counteract the outward forces of glomerular pressures and maintain capillary loop shape, podocyte loss results to local bulging of the GBM when glomerular pressures increase in many renal diseases. The denuded GBM tends to form a synechia attachment by contacting with the parietal epithelial cells and Bowman's capsule, which is thought to be the first "committed step" of focal segmental glomerular sclerosis (FSGS) (Kriz et al. 1994, 1998a, b).

Podocytes maintain a healthy intraglomerular environment by cross talk with glomerular endothelial cells. Endothelial cell swelling and attenuation of fenestrae are observed in podocyte injury models by ultrastructural study (Kriz et al. 2013).

It was illustrated that podocyte injury disrupts intracapillary homeostasis, causing thrombotic micro-angiopathy and mesangial abnormalities by reducing VEGF signaling (Eremina et al. 2008; Kobayashi et al. 2015).

Glomerulosclerosis is a terminal consequence of podocyte injury. The classic type of glomerulosclerosis, as defined by segmental obliteration of glomerular capillaries by the extracellular matrix, has been believed to progress to complete sclerosis without regression (Nagata 2016). In early stage of FSGS, cellular lesions including transformed podocytes were accompanied by segmental sclerosis, supporting the fact that podocyte damage might be an early event of glomerulosclerosis. In a recent elegant review by Kim JS and his colleagues, the essential steps of glomerulosclerosis were suggested as follows: (1) increased glomerular capillary pressure and filtration flow through podocyte slits, (2) foot process effacement as an adaptive response, (3) podocyte hypertrophy and glomerulomegaly, (4) mismatch between glomerular tuft growth and podocyte hypertrophy, (5) stretching and attenuation of podocyte cell body, (6) pseudocysts formation by hindered flow of filtrates beneath the podocyte that is partially detached on bare areas of GBM, (7) complete podocyte detachment by enlarged pseudocysts and adhesion to Bowman's capsule, (8) glomerular tuft's adhesion to Bowman's capsule, (9) spreading of filtrates to interstitium out of nephron through adhesion structure, and (10) interstitial proliferation and nephron degeneration (Kim et al. 2016).

10.2 The Role of Glucose Metabolism Disorder in Podocyte Injury

Podocytes' injury and depletion was a crucial step in the development of albuminuria in DN. In DN, the number of podocyte-specific markers and podocytes number is decreased, which leads to the occurrence of albuminuria and further develops into glomerulosclerosis. Hyperglycemia is the main pathological change of diabetes and plays an important role in promoting the occurrence and development of DN. Increased intracellular glucose could induce multiple cell and molecular events in podocyte: (1) generation of reactive oxygen species (ROS) and advanced glycation end products (AGEs), (2) increased flux of polyols and hexosamines, (3) activation of protein kinase C (PKC), (4) increased cytokines and growth factors, (5) aberrant Notch signaling, and (6) activate the renal RAS. These abnormal molecular pathological changes mediate the functional and morphological changes of podocytes in a direct or indirect way, including podocyte hypertrophy, epithelial mesenchymal transition (EMT), podocyte detachment, and podocyte apoptosis.

Podocyte injury is a key factor in the development of DN. Recent studies in both type 1 and type 2 diabetes have proposed that a reduction in the number of podocytes may lead to the development of proteinuria. It is reported that the structure and function of podocytes are abnormal under high glucose conditions, such as podocyte fusion, septal injury, and podocyte loss. There has been evidence that podocytes possess a completely functional system for glucose uptake (Lewko et al. 2005). Coward et al. have revealed that the cultured human podocytes express glucose transporter (GLUTs) in two forms: GLUT1 and GLUT4, which participate in insulin-dependent glucose transport to the cell (Coward et al. 2005, 2007). In addition, the podocyte split protein, such as Nefin, is also involved in glucose transport. Schiffer et al. have demonstrated that podocytes also express another insulin-sensitive glucose transporter, GLUT8 (Schiffer et al. 2005). GLUT1 is the primary glucose transporter in most cells as well as in podocytes (Coward et al. 2005, 2007). In diabetes, hyperglycemia and alteration of glucose transporter cause increased intracellular glucose concentration in podocyte and lead to severe impairment of the glomerular filtration barrier. Conversely, Zhang et al. found that enhancement of GLUT1 expression in diabetic podocyte significantly reduced the mesangial expansion and fibronectin accumulation by inhibiting the expression of vascular endothelial growth factor (VEGF) (Zhang et al. 2010). Similarly to other cells, under high glucose condition, podocyte can undergo many pathological changes induced by aberrations in various cellular and molecular events. High glucose induces generation of advanced glycation end products (AGEs) and reactive oxygen species (ROS), increased flux of polyols and hexosamines, increased activity of protein kinase C (PKC), upregulated expression of cytokines and growth factors including vascular endothelial growth factor (VEGF), and transforming growth factor-beta (TGF-β), induces aberrant Notch signaling, and activates the renal RAS (Anil Kumar et al. 2014).

Pathomechanism of podocyte injury in DN mainly includes podocyte hypertrophy, EMT, podocyte detachment, and podocyte apoptosis. Podocyte hypertrophy is the initial stage of podocyte injury in early DN. Hyperglycemia upregulated the expression of cyclin-dependent kinase p27kip1, which leads to further cell cycle arrest and hypertrophy. It was found that p27kip1-/- mice had significantly improved renal damage in DN (Wolf et al. 2005). Several studies suggested high glucose-induced podocyte hypertrophy by activating mTORC1 pathway (Fantus et al. 2016). In addition, hyperglycemia increased expression of nuclear STAT3 via the activation of the upstream signal transduction element Gp130, which eventually leads to podocyte hypertrophy. Excessive hypertrophy could result in degenerative changes in podocyte structure and functions, leading to its detachment from glomerular basement membrane (GBM). Previous studies have shown that phenotype conversion of podocyte was involved in the early stage of podocyte deletion in DN by inducing podocyte detachment or podocyte apoptosis. Podocyte EMT is a manifestation of podocyte phenotype conversion and one of the initiating factors leading to a variety of glomerular diseases. When EMT occurs, the cells lose their original characteristics, resulting in disappearance of intercellular contact, impaired cell polarity, and expression of mesenchymal markers such as alpha smooth muscle actin (alpha-SMA) and fibroblast-specific protein 1 (FSP1). EMT is also an explanation for podocyte depletion in DN (Yamaguchi et al. 2009). Emerging evidence suggested that podocytes could undergo EMT in DN, characterized by loss of epithelial features such as nephrin and P-cadherin, while expressing mesenchymal markers such as FSP-1, type I collagen, and fibronectin (Reidy and Susztak 2009). Xing et al. (2015) demonstrated that stimulation with high glucose for 48 h could activate the PI3 K/AKT pathway in podocyte, and thereby induce the protein expressions of α-SMA and desmin. Dai et al. (2012) suggested that connective tissue growth factor (CTGF) and integrin-linked kinase (ILK) were involved in high glucose-induced phenotypic alterations of podocytes. Lv et al. (2013) findings elaborated that Rac1/PAK1 signaling contributed to high glucose-induced podocyte EMT via promoting β-catenin and Snail transcriptional activities, which could be a potential mechanism involved in podocytes injury in response to stimuli under diabetic conditions. Guo et al. indicated high glucose can also activate β-catenin and Snail expressions by upregulating GSK-3β. In addition, hyperglycemia-induced podocyte detachment by decreasing the expression of key proteins involved in the foot process actin cytoskeleton, split diaphragm (SD) integrity, and podocyte–GBM interactions. A3b1 integrins are the important transmembrane protein involved in anchoring foot processes in the GBM. High glucose regulates the expression of integrin subunits and inhibits the synthesis of agrin. Therefore, high glucose affects not only the structure of podocytes, but also their ability to adhere to GBM (Chen et al. 2000; Han et al. 2006; Yard et al. 2001). It is found that high glucose can alter podocyte adhesion by decreasing expression of integrin α3β1v which was an important receptor that could tightly connect podocyte with the GBM and participated in the adhesion function of podocyte. In addition, α-Actinin, an actin filament for protein crosslink, is also an important factor required for podocyte adhesions (Dandapani et al. 2007). High glucose and AGE treatment resulted in α-actinin-4 expression changes and induces cytoplasmatic translocation in podocyte (Ha 2006). There are some evidences that podocyte apoptosis played a role in reduction in density and number of glomerular in DN. High glucose led to podocyte apoptosis by increased production of ROS, activation of poly(ADP-ribose) polymerase, NF-kB, and p38 MAP kinase (Susztak et al. 2006; Szabo et al. 2006). In diabetes, the surface receptors of the AGEs are upregulated in the podocytes (Tanji et al. 2000). Binding AGEs to receptors activates activated transcription factor FOXO4, which also induced podocyte apoptosis via p38 protein kinase signaling pathways (Cohen et al. 2005). In addition, high glucose increased the protein expression of Nestin, which is a VI intermediate filament protein-related cell cytoskeleton, thereby increased podocyte apoptosis rate (Liu et al. 2012). High glucose increased the expression of TGF-β1 in podocyte. TGF-β1 could induce podocyte apoptosis by directly activate Smad7, p38 MAP kinase, and Notch pathway (Li et al. 2004).

In addition to its direct effects, elevated glucose may act indirectly, via the proinflammatory response, Ang II-dependent pathways, and lipid accumulation. Under high glucose conditions, secretion of the MCP-1 protein by cultured podocytes was increased rapidly (Han et al. 2004), and similar effect was observed in podocyte stimulated with AGEs (Gu et al. 2006). Podocyte can also express TNF-α, a cytokine produced by various immune cells, in response to high glucose stimulation and in diabetic conditions (Ikezumi et al. 2008; Ruster et al. 2009). High glucose could stimulate activity and expression of the local RAS components in podocyte, including Ang II and its AT1 receptors (Yoo et al. 2007). Following that, it was recently demonstrated that local RAS activation would lead to podocyte injury through a variety of pathways. Ang II could induce podocyte apoptosis through activation of NADPH oxidase and production of ROS, and upregulate the expression of GLUT transporters (Gill and Wilcox 2006; Nose et al. 2003). In addition, Ma et al. found that lipid accumulation in podocytes was increased under the high glucose stimulation, which is mediated through the disruption of low-density lipoprotein receptor (LDLr) pathway (Zhang et al. 2015a). Interestingly, reducing lipid accumulation in podocytes decreased the protein expression of SMA and increased the expression of nephrin in podocyte. These studies reveal that high glucose-induced lipid accumulation is involved in the podocyte injury in DN. Therefore, the above shows that high glucose could induce various other metabolic disorders and indirectly lead to podocyte injury.

10.3 Lipid Metabolism Disorder in Podocyte Injury

Lipid metabolism disorder is commonly observed in patients with chronic kidney disease (CKD), accompanied by increased fasting triglyceride levels and decreased high-density lipoprotein cholesterol (HDL-C) (Bianchi et al. 2016). It is increasingly recognized that dysregulation of lipid metabolism is involved in the development and progression of CKD, such as obesity-related renal disease and DN (de Vries et al. 2014). Podocytes, as specialized cells of glomerulus, play an important role in the pathologist of CKD when they are injured (Fiorina et al. 2014). And excessive lipid accumulation in podocytes can lead to cellular dysfunction and death, which is called lipotoxicity.

10.3.1 Cholesterol

Between neighboring podocytes, there is a unique interdigitating structure bridged by SD, maintaining the proper glomerular filtration (Ruotsalainen et al. 1999). Researches have revealed that SD is a lipid raft structure containing multiple podocyte-specific proteins, such as podocin and nephrin (Schermer and Benzing 2009). In particular, podocin can recruit and bind to cholesterol to form SD, and this binding can influence the composition of lipid membrane, allowing cholesterol to contact with the ion-channel transient receptor potential canonical 6 (TRPC6) (Huber et al. 2006). This suggests that cholesterol homeostasis is essential for glomerular functions. However, excessive cholesterol can also negatively disrupt the mutual binding of podocyte SD proteins, or interfere with the binding between podocyte SD proteins and caveolin-1, a lipid raft-associated protein, binding nephrin, and Cluster of Differentiation 2 (CD20)-associated protein (Sorensson et al. 2002)

The content and distribution of cellular cholesterol regulated by cholesterol synthesis and intracellular trafficking.

It is regulated by some functional proteins such as ATP-binding cassette transporter A1 (ABCA1) involving cholesterol efflux, 3-hydroxy-3-methyl-glutaryl CoA reductase (HMG-CoA reductase, HMGCR) regulating cholesterol synthesis and low-density lipoprotein receptor (LDLR) mediating cholesterol influx. The expression of HMGCR and LDLR is regulated by some transcription factors, such as the sterol regulatory element-binding protein (SREBP), under negative feedback loops. When cells are rich in cholesterol or its derivatives, the transcription of LDLR gene or other genes necessary to lipid synthesis are suppressed. As a result, the cells are not able to generate and uptake cholesterol, and then establish cholesterol homeostasis. In contrast, when intracellular sterols are exhausted, the transcriptions of SREBP target genes will be activated, increasing intracellular cholesterol (Zhang et al. 2016). This enables cellular cholesterol homeostasis despite physiological fluctuations in cholesterol requirements and exogenous supply.

However, it is demonstrated that the cellular cholesterol imbalance of podocytes can induce proteinuric glomerular diseases (Merscher et al. 2014). It is demonstrated that human glomerular podocytes express ABCA1, HMGCR, and LDLR (Merscher-Gomez et al. 2013). Ma et al. found that some pathogenic factors such as inflammation can disrupt LDLR pathway (Zhang et al. 2015b). Thus, excessive lipid accumulates in podocytes, resulting in effacement of the foot processes and epithelial mesenchymal transition of podocytes (Zhang et al. 2015b). It is recently demonstrated that human podocytes treated with the sera from diabetic kidney disease (DKD) patients had increased cholesterol accumulation compared with human podocytes exposed to the sera of patients with diabetes, but no DKD. This was associated with a reduction of ABCA1 and an impairment of cholesterol efflux (Merscher-Gomez et al. 2013). Besides, it is showed that c-x-c motif ligand 16 (CXCL16) is the main scavenger receptor for oxidized LDL (oxLDL) in human podocyte (Gutwein et al. 2009). The expression of glomerular CXCL16 was increased in patients with membranous nephropathy, accompanied with higher levels of oxLDL (Gutwein et al. 2009). And in diabetic db/db mice, CXCL16 pathway was activated, in parallel with increased cholesterol accumulation in kidney (Hu et al. 2018). In vitro, oxLDL can induce loss of nephrin expression from cultured podocytes (Bussolati et al. 2005).

In summary, cholesterol metabolism disorder can destroy the structure and function of podocytes, leading to the progression of CKD.

10.3.2 Fatty Acids and Triglycerides

In addition to hypercholesterolemia, free fatty acids (FFAs) can also affect podocyte function in kidney disease. The essential role of fatty acids is to form the phospholipid bilayers of the cell membranes and act as phospholipid messengers, transmitting vital intracellular signals (Lee 2011). Normal cellular fatty acid homeostasis reflects a balance between generation or delivery and utilization. SREBP-1c is involved in fatty acid and TG synthesis, targeting lipogenic enzymes including acetyl-CoA carboxylase (ACC) and fatty acid synthase (FAS) (Horton et al. 2002). FFAs can be transported into cells by the scavenger receptor platelet glycoprotein 4 (also called as CD36) or via the assistance of vascular endothelial growth factor B (VEGF-B) (Hagberg et al. 2010; Masuda et al. 2009). Cellular FFAs are esterified or transported into the mitochondria for oxidation and subsequent energy production (Lee 2011).

Palmitic and stearic acids, belonging to saturated FFAs (SFAs), and oleic acid, belonging to monounsaturated FFAs (MUFAs), account for 70–80% of plasma FFAs (Raclot et al. 1997). SFAs can induce insulin resistance and cell death, involving the pathogenesis of diabetes mellitus type 2 (T2DM) (Lennon et al. 2009; Sieber et al. 2010). In contrast, MUFAs can prevent SFA-induced lipotoxicity (Sieber et al. 2010). In human podocytes, insulin resistance can be induced by palmitic acid (Lennon et al. 2009). It is observed that insulin sensitivity in glomeruli of obese and diabetic rats is reduced (Mima et al. 2011). Podocyte-specific insulin receptor knockout mice develop albuminuria and glomerulosclerosis, indicating that normal insulin signaling is critical for podocyte function and survival (Welsh et al. 2010). These findings imply that FFAs play potential roles in insulin resistance, promoting the development and progression of obesity-related renal disease and DN.

In the tubulointerstitial and glomerular segment of renal biopsies obtained from patients with DN, endoplasmic reticulum (ER) stress is observed (Sieber et al. 2010). Importantly, in a T1D mouse model, the progression of DN can be attenuated by ameliorating ER stress (Qi et al. 2011). ER dyshomeostasis can decrease the ER folding capacity, thereby leading to accumulation of unfolded and misfolded proteins. This in turn initiates the unfolded protein response (UPR), adaptively maintaining proper ER function (Ma and Hendershot 2001). But if ER stress persists, apoptosis will be induced by the proapoptotic transcription factor C/EBP homologous protein (CHOP) (Rasheva and Domingos 2009). In podocytes, ER stress induced by palmitic acid results in the upregulation of several UPR markers/effectors, such as the ER chaperone heavy chain-binding protein (BiP), and CHOP, while monounsaturated palmitoleic and oleic acids only upregulated BiP but not CHOP (Sieber et al. 2010). As BiP can attenuate palmitic acid-induced apoptosis (Laybutt et al. 2007), the beneficial effect of MUFAs may own to the upregulation of BiP. In addition to the unfolded proteins, alterations in ER membrane lipid composition can also sensitively affect the expression of the ER stress sensor inositol requiring enzyme 1 (IRE-1) (Promlek et al. 2011). It is shown that small molecule compound 4m8C, specific IRE-1 inhibition, can attenuate palmitic acid-induced podocyte death (Sieber and Jehle 2014).

Enhanced FFA uptake by podocytes is induced by increased expression of CD36 and a decrease in fatty acid β-oxidation, leading to excessive intracellular lipid accumulation (Soetikno et al. 2013). In animal model of type 1 diabetes (T1D), increased expression of SREBP-1 in renal results in upregulation of enzymes responsible for FFA synthesis and as a consequence of a high level of triglyceride (TG) in renal (Hashizume and Mihara 2012). Accumulated lipids in podocytes limited mitochondrial fatty acids β-oxidation. It induced mitochondrial damage and inhibition of AMP kinase (AMPK) activity, leading to endoplasmic reticulum (ER) stress, autophagy, and apoptosis in podocytes. As a result, mitochondrial dysfunction caused decreased podocyte density and increased in foot process width, together with inflammation (Szeto et al. 2016). Renal accumulation of TG is associated with reduced expression of the ultrasensitive energy sensor AMPK strongly. This suggests that the imbalance between energy-generating and energy-consuming pathways might be related to podocyte dysfunction in DKD and other disorders in CKD, due to lipid accumulation (Wahl et al. 2016). And hypertriglyceridemia can also increase podocytic de novo expression of desmin, which represents podocyte injury (Joles et al. 2000).

10.3.3 Gangliosides and Sphingolipids

Since the first description that glycosphingolipid accumulation in the renal results in glomerular hypertrophy in streptozotocin (STZ)-induced diabetic mice, several studies have highlighted the role of sphingolipids and gangliosides in podocyte biology (Merscher-Gomez et al. 2013).

Analysis of kidney biopsy compartments from 14 patients with Fabry disease using unbiased quantitative stereology indicated age-dependent accumulation of globotriaosylceramide (Gb3) in podocytes (Najafian et al. 2011). In vitro, globotriaosylsphingosine (known as lysoglobotriaosylceramide) acts as a profibrotic metabolite in cultured human podocytes (Sanchez-Nino et al. 2011). Ganglioside GM3 (GM3) is a receptor for soluble Flt1, locating in lipid raft domains in the SD of podocytes. Binding of soluble Flt1 to GM3 plays essential roles in autocrine preservation of the podocyte actin cytoskeleton and in prevention of proteinuria (Jin et al. 2012). O-acetylated disialosyllactosylceramide (GD3), a sialic-acid-containing lipid, was identified as a podocyte-specific ganglioside in rat (Reivinen et al. 1992). Treating mice with an antibody against GD3 caused nephrin phosphorylation and dislocation from the podocyte SD (Simons et al. 2001).

It is an emerging concept that sphingolipids act as modulators of podocyte function in FSGS and other glomerular diseases. Patients with FSGS are more likely to have recurrence of proteinuria after kidney transplantation. And the number of acid sphingomyelinase-like phosphodiesterase 3b (SMPDL3b) positive podocytes is decreased in patients with recurrent proteinuria (Fornoni et al. 2011).

To sum up, lipid metabolism disorder is involved in the pathogenesis of podocyte injury. Cholesterol helps form SD between podocytes, maintaining the proper glomerular filtration. LDL-cholesterol uptake is mediated via the LDLR or CXCL16 and may cause ER stress. Cholesterol metabolism is regulated by several nuclear receptors and transcription factors, including SREBP. Excessive cholesterol accumulation in podocytes may contribute to kidney disease. Free fatty acids are primarily transported via CD36, causing oxidative and ER stress based on the degree of saturation. Sphingolipids and gangliosides also play a role in podocyte biology. Binding of soluble Flt1 to GM3 plays essential roles in autocrine preservation of the podocyte actin cytoskeleton and in prevention of proteinuria (Fig. 10.1).

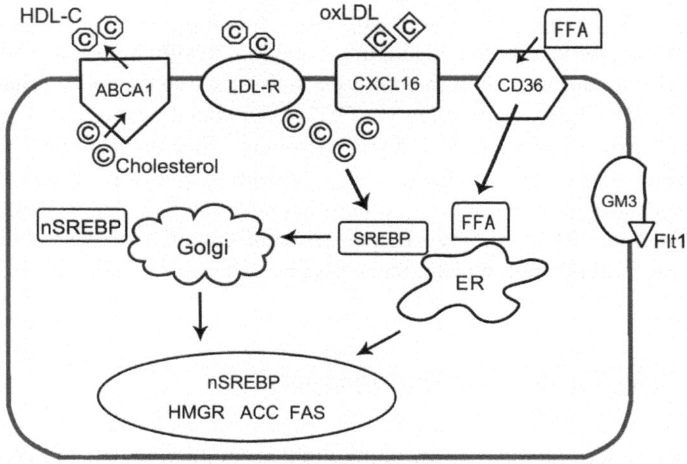

Fig. 10.1 Lipid metabolism disorder is involved in the pathogenesis of podocyte injury

10.4 Role of Hypertension in the Damage of Podocytes

Hypertension has become the second leading cause of end-stage renal disease (ESRD) after diabetes mellitus (Udani et al. 2011). High blood pressure can affect renal vessels, glomeruli, and tubulointerstitium. Recently, more and more studies have indicated that podocyte damage play an important role in hypertensive nephrosclerosis. Decreased intrarenal podocyte and increased urinary podocyte were observed in hypertensive nephrosclerosis (Wang et al. 2009). As terminally differentiated cells, podocyte loss leads to denudation of the glomerular basement membrane (GBM) and focal adhesion of the tufts to Bowman's capsule, which finally results in glomerulosclerosis and reduced filtration (Cellesi et al. 2015).

Podocyte loss in hypertension includes detachment of viable cells and apoptosis (Kriz et al. 2013). The major factor for podocyte loss in hypertension is the capillary hypertension, which cause glomerular hypertrophy and hyperfiltration (Kriz and Lemley 2015). Glomerular hypertrophy results in relatively decreased podocyte density. Puelles et al. (2016) examined the effect of hypertension on podocyte depletion using kidneys obtained from autopsy, and they did not observe a difference in total podocyte number solely driven by hypertension, while the relative podocyte depletion is associated with glomerular hypertrophy which resulted in the reductions in podocyte density. Hyperfiltration gives rise to increased shear stress by elevating driving force and augmenting GBM area. Podocytes cultured in vitro are sensitive to shear stress, which induces reorganization of cytoskeleton (Friedrich et al. 2006), and this helps them to cover an expanding GBM which further leads to foot process effacement. In desoxycorticosterone-trimethylacetate (DOCA) hypertensive mice, chloride intracellular channel 5A, which is highly enriched in podocytes foot process, protects against hypertension-induced podocyte injury through weakening the tensile strength of the actin cytoskeleton in Rac1-dependent manner (Tavasoli et al. 2016). This has been considered to be the protective response for podocyte to escape detachment. However, this strategy is not always successful and finally results in podocyte detachment from GBM, as seen in progressive stage of fawn-hooded hypertensive (FHH) (Kriz et al. 1998c) and DOCA hypertensive rat model (Kretzler et al. 1994). Apoptosis is another cause for podocyte loss under shear stress in hypertension, which is before or in conjunction with cell detachment (Kriz et al. 2013; Ying et al. 2000).

Besides mechanical stress, renin–angiotensin–aldosterone system (RAAS) plays central role in the pathogenesis of hypertensive nephrosclerosis, mainly through its actions on the subtype 1 receptor. Mechanical strain increased angiotensin II production and upregulation of angiotensin receptor 1 (AT1) in cultured podocytes, while the increased apoptosis induced by mechanical strain was also in an angiotensin II-dependent manner (Durvasula et al. 2004). Increased angiotensin II results in decreased expression of podocin and integrin β1, which are both vital in viable podocytes adhesion to the GBM and interaction of podocytes with other GBM components. This might elucidate that the elevated intraglomerular pressure is translated into a maladaptive response in podocyte probably due to the activation of local tissue angiotensin system. Furthermore, angiotensin II is also considered to be associated with the rearrangement of the actin cytoskeleton (Macconi et al. 2000). Aldosterone, an important mediator of the effect of angiotensin, has become a hot spot concern in hypertensive nephropathy. Using only the inhibition of aldosterone by eplerenone dramatically alleviated podocyte injury in Dahl salt-hypertensive rats, an animal model inclined to hypertensive glomerulosclerosis (Nagase et al. 2006). In a double-blind, randomized, placebo-controlled trial, additional use of low-dose eplerenone to renin–angiotensin system inhibitors has renoprotective effects in hypertensive patients with non-diabetic chronic kidney disease (Ando et al. 2014). These findings suggested that aldosterone plays an important role in hypertension-induced podocyte injury. The underlying mechanism is primarily due to aldosterone-induced mitochondrial dysfunction, which increased oxidative stress. In uninephrectomized rats infused with aldosterone and fed with high-salt diet, podocyte-associated proteins nephrin and podocin were dramatically decreased, along with reduced nicotinamide-adenine dinucleotide phosphate oxidase activation, increased oxidative stress, and enhanced aldosterone effector kinase Sgk1. Thus, podocyte is the prominent target for aldosterone by inducing oxidative stress and Sgk1 (Shibata et al. 2007). Selective mineralocorticoid receptor (MR) antagonist eplerenone also ameliorated the salt-induced proteinuria and podocyte injury in hypertensive rat model (Nagase et al. 2007).

After detachment from GBM, podocyte moves through meshes of Bowman's capsule to the urine and might keep alive. Unfortunately, the detection of viable podocytes in the urine is a complex procedure, which is still unavailable in all laboratories. However, elevated mRNA levels of podocin and nephrin can be examined in urine of hypertensive patients (Kelder et al. 2012). Recent studies suggested that increased podocyte-derived extracellular vesicles may predict podocyte stress and subsequent podocyte loss in hypertensive patients, which might provide a novel non-invasive detective method (Kwon et al. 2017). Podocytes, as the gatekeepers of protein in glomerular filtration barrier, are major targets of high blood pressure. In all, hypertension could cause mechanical stress and the activation of RAAS (mentioned below). Mechanical stress further induces capillary hypertension, promoting glomerular hypertrophy and hyperfiltration. These changes would lead to reduced podocyte density and the reorganization of cytoskeleton in podocytes, resulting in detachment of viable podocyte and podocyte apoptosis, progressing to final glomerulosclerosis. More studies are needed to prove that podocytes can be the detective marker for hypertensive nephrosclerosis and find the more specific method for early diagnosis and treatment.

10.5 Activation of RAS in Podocyte Injury

Hemodynamic changes and RAS of the glomeruli are key factors of CKD patients' persistent proteinuria and disease progression. Many investigations suggest that local intrarenal RAS activation contributes to kidney tissue injury (Gurley et al. 2011), and RAS activation accelerates renal injury by various mechanisms.

Angiotensinogen (AGT), the original of RAS, transforms into Ang II through the conversion of Ang I as a result of the enzymatic cleavage process by renin and ACE. As the most active peptide of RAS, Ang II was demonstrated to induce TGF-β expression and provoke oxidative stress and inflammation, which are main factors in the initiation, development, and progression of CKD (Ruggenenti et al. 2012).

Under a condition of continuous glomerular hypertension in CKD, podocytes may undergo actin cytoskeletal reorganization, compensatory hypertrophy, weakened local adhesion ability due to downregulation of adhesion molecules of basement membrane cells, and apoptosis of podocytes induced by local Ang II activation. The continuous increase of Ang II caused by mechanical stress further affects the capillary intraglomerular pressure, resulting in a vicious circle and contributing to the pathogenesis of glomerulosclerosis (Ruster and Wolf 2011). In addition to causing podocyte lesions by altering glomerular hemodynamics, Ang II also has a direct effect on the structure and functions of podocytes, which is mentioned later in this section.

Podocytes, in possession of a complete RAS (Marquez et al. 2015), can produce functional RAS elements themselves and participate in local RAS systems as well, playing an important role in not only its own physiological process but pathological status (Fig. 10.2). It has been reported that mechanical stress and high glucose could increase the production of local Ang II and AT1 receptor (AT_1R) in podocytes (Durvasula et al. 2004; Durvasula and Shankland 2008), with inducing the expression of other RAS elements (Sakoda et al. 2011).

There are also important elements of the RAS system expressing in human differentiated podocytes, including angiotensin, renin, ACE, AT_1R, and AT_2R subtype mRNA, but the related proteins were not detected (Liebau et al. 2006). Therefore, podocytes could not only be a target of the damage caused by Ang II, but a source of

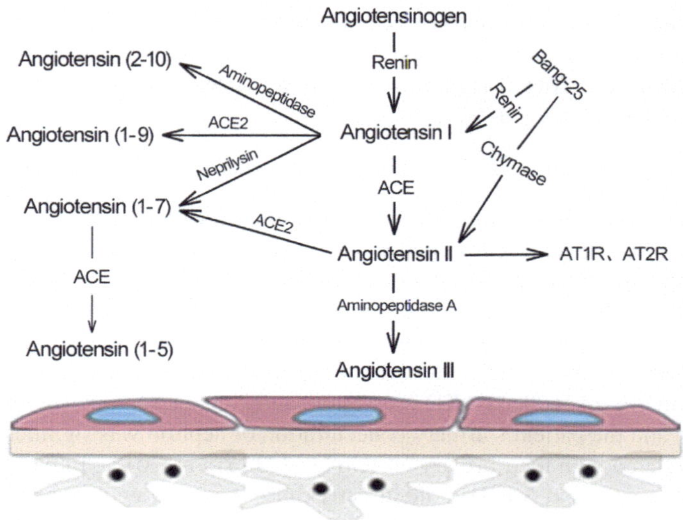

Fig. 10.2 Renin–angiotensin system (RAS) in podocytes

localized Ang II as well. However, it has been found that Ang II secreted by podocytes was not blocked by renin inhibitors, ACEI, and chymase inhibitors (Liebau et al. 2006), suggesting that there might be an unknown pathway for Ang II formation in podocytes.

Velez et al. (2007) used Matrix-Assisted Laser Desorption/Ionization Time of Flight Mass Spectrometry (MALDI-TOF-MS) to quantify the presence of RAS-related peptide chains in rat podocytes, in order to further explore the role of podocytes in the metabolism of RAS elements. As a result, after co-incubated with Ang I, mesangial cells mainly produced Ang II while the main product of podocytes was Ang (1–7) with almost no Ang II. Furthermore, it was confirmed that podocyte-producing Ang (1–7) is mainly through the neprilysin pathway, as ACE-mediated Ang II production did not result in an increase of Ang II concentration in podocytes, which might be related to podocytes' degradation of Ang II through ACE2 and aminopeptidase A pathways.

As a new member of RAS, ACE2 might have a negatively regulatory effect on ACE-produced Ang II of traditional RAS, mainly by accelerating the degradation of Ang II to attenuate its effect, and through the generation of Ang (1–7), which has the most expression in podocyte RAS. There has been no evidence that podocytes express the receptor of Ang (1–7), i.e., Mas, yet Ang (1–7) and its receptor seem to be involved in the renal protection for DN, such as regulating inflammation, oxidative stress, and retaining the progression of renal fibrosis.

Therefore, podocytes probably play an essential role in maintaining the balance of local RAS system in the kidney, similar to that between systemic Ang II and intrarenal RAS system, by degrading the systemic Ang II filtered from the glomeruli, and/or promoting the conversion of glomerular-filtrated Ang I and AGT to Ang (1–7), thereby regulating the damage caused by the whole systemic Ang II to the kidney.

The pathways of Ang II signaling mediating podocyte injury can be generally divided into the following aspect:

(1) To damage the function of the pore membrane and structure of the cytoskeleton

The SD is an essential structure of the glomerular filtration barrier, which is connected to the foot processes adjacent to podocytes. Nephrin and zonula occludens (ZO)-1 are main proteins of SD, preventing macromolecules from entering the urine. It has been found that SD is susceptible to damage, leading to decreased expression of nephrin and ZO-1, and cytoskeletal reorganization of podocytes.

It has been found that the expression of nephrin in renal biopsy specimens from patients with T2DM-induced DKD was significantly reduced compared with healthy volunteers, and the patient's urinary concentration of nephrin was significantly positively correlated with their urinary protein level (Jim et al. 2012). Ren et al. (2012) have found in vitro that Ang II could directly cause the downregulation and dephosphorylation of nephrin, which mediates podocyte injury. Besides, application of ACEI and ARB has been reported to inhibit the rearrangement of cytoplasmic ZO-1 and reduced the degree of proteinuria (Macconi et al. 2000).

(2) To induce podocyte apoptosis

One of the main causes of podocyte loss in CKD patients is podocyte apoptosis, and the occurrence of urinary podocyte plays an important role in glomerular sclerosis.

Ang II reportedly could induce the apoptosis of rat podocytes cultured in vitro in a dose- and time-dependent manner, and this process required cells to be exposed to TGF-β and TGF-β antibody could inhibit apoptosis of podocytes (Ding et al. 2002). After activation of TGF-β in diabetic glomeruli, the nuclear factor κB might be inhibited via the gene Smad7, resulting in podocyte apoptosis.

The advanced glycation end products (AGEs) were also found to activate the RAS system of podocytes, upregulate Ang II levels, and induce podocyte apoptosis via a AGEs receptor-PIK3/protein kinase B (Akt)-dependent signaling pathway; ARB could attenuate Ang II-induced podocyte apoptosis.

(3) To cause cell phenotypic transformation and hypertrophy

$p27^{Kip1}$ encodes a protein which belongs to cyclin-dependent kinase (Cdk) inhibitor proteins, which could control the cell cycle progression at G1 phrase, thereby inhibiting cell proliferation. It was found that Ang II could directly increase the levels of $p27^{Kip1}$ mRNA and protein in podocytes cultured in vitro and in vivo in DN, which was inhibited by ARB. Ang II-induced upregulation of $p27^{Kip1}$ expression might lead to podocyte hypertrophy (Xu et al. 2005). It was also observed that Ang II can upregulate the expression of $p27^{Kip1}$ protein, causing pathological podocyte hypertrophy similar to that in a DN text (Romero et al. 2010).

As an essential factor promoting the progression to renal fibrosis, EMT in podocytes will result in loss of epithelial markers with de novo expression of EMT markers; in more severe cases, it may lead to podocyte detachment from the glomerular basement membrane, thereby aggravating proteinuria and glomerulosclerosis (Li et al. 2015; Loeffler and Wolf 2015). A recent study reported that a high concentration of glucose and Ang II promoted EMT in podocytes, which could be reversed by silencing TCF8 (Bai et al. 2017).

(4) To induce podocyte membrane depolarization and damage the charge barrier

Studies by using patch clamp recording technique in isolated glomeruli in vitro have demonstrated that Ang II could cause sustained and irreversible depolarization of podocyte membranes. Stimulation of Ang II resulted in an immediate calcium influx of cultured podocytes (Greka and Mundel 2011). Studies have confirmed that TRPC6 colocalized with podocyte nephrin and podocin, and its functional mutation could disrupt the integrity of the pore membrane, leading to proteinuria and FSGS (Reiser et al. 2005; Winn et al. 2005). Numerous studies have found that abnormal calcium signaling may be the main cause of related podocyte diseases. For example, calcium increases evoked by Ang II are primarily mediated via TRPC6 channels and this pathway could be pharmacologically targeted to abate the development of DKD (Nijenhuis et al. 2011; Sonneveld et al. 2014).

(5) To induce podocyte autophagy

As terminally differentiated cells, podocytes mainly reduce intracellular accumulation of damaged DNA and macromolecular substances through autophagy rather than cell division (Pan et al. 2008). In vitro experiments, animal experiments, and human kidney biopsy indicate that podocytes have a high-level basis of autophagy, which plays an important role in maintaining the stability of podocytes. Recent research using a CKD animal model has demonstrated that autophagy is an essential intracellular process to encourage the survival of renal cells (Huber et al. 2012), while excessive and dysfunctional autophagy might result in podocyte injury (De Rechter et al. 2016). It has been found that Ang II could enhance the ROS production and increase oxidative stress in the renal system by enhancing the activity of systematic NADPH, leading to detrimental podocyte autophagy (De Rechter et al. 2016; Yadav et al. 2010), the underlying pathways of which is dependent or independent on mTOR (Mao et al. 2016). A recent study has found that autophagy could enhance the cell viability of Ang II-treated podocytes, suggesting improving autophagy may become a new targeted therapy to relieve Ang II-induced podocyte injury (Gao et al. 2017).

Traditionally concerned solely as an inactive precursor of renin, prorenin actually participates in the functional regulation of body through the hydrolysis of AGT to produce ANG I and can also bind to prorenin/renin receptor (PRR) (non-proteolytic pathway) to activate, like mitogen-activated protein kinases (MAPKs), initiating intracellular signal transductions. The plasma prorenin/renin ratio in diabetic patients was significantly higher, and the prorenin levels began to increase before the appearance of micro-albuminuria without changes in renin levels (Sakoda et al. 2011), suggesting that prorenin itself exerts somewhat important effects on DN.

Immunofluorescence double-labeling studies have showed that prorenin activated by non-proteolytic pathway coexisted with the podocyte marker nephrin, and electron microscopy also displayed that PRR was distributed on podocyte foot processes (Ichihara et al. 2006). Handle region peptide (HRP) is a polypeptide blocker of prorenin receptor. Ichihara et al. (2006) have found that gene deletion of AT_1R or using ACEI inhibitor could to some extent delay the occurrence of proteinuria and glomerular sclerosis in streptomycin-induced DN rats, while continuous instillation of HRP could almost completely block the progression of DN. It is noteworthy that the MAPK signaling pathway was activated in AT_1R-deficient mice, and HRP could significantly inhibit MAPK, indicative of an equally important role of prorenin coupling with PRR-induced angiotensin-independent pathway in diabetic kidney injuries. Besides, Sakoda et al. (2010) have confirmed that adding prorenin to human podocytes cultured in vitro could increase the intracellular level of Ang II and activate the MAPK intracellular signal transduction pathway, resulting in podocyte damage.

10.6 Roles of Micro-inflammation in Podocyte Injury

10.6.1 Definition of Micro-inflammation State

In mammalians, the acute-phase reaction is beneficial for eliminating acute insults for protection against microorganisms, limiting tissue damage, and maintaining homeostasis. This reaction would become disadvantageous under a chronic condition called micro-inflammation.

Micro-inflammation is a state with low-intensity, chronic persistent and dominant inflammation caused by the infection of non-pathogenic microorganisms, which is characterized by mild persistent elevation of inflammatory cytokines in the systemic circulation (Kaysen 2001; Schomig et al. 2000). Micro-inflammation is a continuous and relatively secretive action, the essence of which is immune inflammation.

10.6.2 Diagnosis and Detection of Micro-inflammation State

Micro-inflammation state has no obvious clinical symptoms, there is no specific diagnostic criteria, and the diagnosis of micro-inflammation relies mainly on the examination of circulating inflammatory biomarkers such as C-reactive protein (CRP) and serum amyloid A (SAA), tumor necrosis factor alpha (TNF-α), and interleukin-6 (IL-6). The acute-phase reactants including the above proteins are mainly synthesized by hepatocytes, such as complement components, coagulation proteins, and metal-binding proteins. It is important to note that when we are in the diagnosis of micro-inflammatory state, other causes and diseases of increased inflammatory markers must first be ruled out, such as connective tissue disease and recent microbial infection.

During acute-phase reaction, the concentration of CRP may increase over 1000-fold compared with normal levels (Kaysen 2001). In addition, CRP follows the course of a disease with little delay due to its short half-life. CRP is supposed to bind multiple other binding specificities such as opsonin of bacteria, immune complexes, and chromatin. CRP reflects not only the activity of inflammation, is also a sign of cytokine activation, its levels was positively associated with the degree of infection. The diagnosis of state of micro-inflammation based on CRP is the level of CRP > 8 mg/L but not more than 10–15 mg/L. SAA is a sensitive acute-phase reactant in micro-inflammatory state. The level of SAA obviously rises before other acute-phase reaction proteins.

10.6.3 The Mediators of Micro-inflammation State

A variety of inflammatory cytokines have emerged as being closely involved in the micro-inflammation state. Immune cells and intrinsic renal cells such as podocytes secrete proinflammatory cytokines including interleukin-1 (IL-1), IL-6, TNF-α, and monocyte chemoattractant protein-1 (MCP-1), which may contribute to the inflammatory process and aggravate diseases progression. For DN as an example, a strong induction of MCP-1 and keratinocyte chemoattractant (KC) by fetuin-A (FetA) or lipopolysaccharide (LPS) is associated with exacerbated palmitic acid-induced podocyte death. Moreover, the prevention of MCP-1 and KC secretion and inhibition of IL-1 attenuates the inflammatory and ultimate cell death response elicited by FetA alone or combined with palmitic acid. The study offers evidence that inflammation aggravates palmitic acid-induced podocyte death and the IL-1β signaling might be novel potential therapeutic targets for prevention and treatment of DN (Orellana et al. 2017).

Infiltrating macrophages/monocytes are associated with chronic, low-grade inflammation. The macrophages can interact with resident renal cells to generate a proinflammatory micro-environment that amplifies tissue injury and promotes scarring.

Macrophage-derived TNF-α had a direct role in the progression of DN. Conditional deletion of TNF-α from macrophages markedly reduced albuminuria, lessening the increase of plasma creatinine and histopathologic lesions (Awad et al. 2015). Likewise, tonicity-responsive enhancer-binding protein (TonEBP) in macrophages promotes hyperglycemia-mediated proinflammatory activation and chronic renal inflammation leading to DN and CKD (Choi et al. 2018).

10.6.4 Lipid-Related Inflammatory Signals

Lipids such as triglycerides and cholesterol may accumulate ectopically in the kidney, which contributes to a lipotoxicity process. Palmitic acid-treated podocytes had intracellular lipid accumulation and abnormal lipid metabolism, accompanied by the process of inflammation, insulin resistance, and rearrangements of the SD and actin cytoskeleton of podocyte. Thus, lipotoxicity accelerated podocyte damage through lipid accumulation related inflammation (Martinez-Garcia et al. 2015).

Lipoproteins including LDL, VLDL, and IDL might act as proinflammatory mediators, which promote the production of inflammatory cytokines, such as TGF-β, platelet-derived growth factor (PDGF), and IL-6 secreted from human mesangial cells. Lipoprotein-mediated cytokine production may cause recruitment of monocytes, lipid-mediated cell proliferation and apoptosis, and extracellular matrix production, thus contributing to podocyte injuries and glomerulosclerosis.

10.6.5 Micro-inflammation Promotes Podocyte Injuries

10.6.5.1 Micro-inflammation and Insulin Resistance of Podocytes

Chronic inflammation can reduce podocyte insulin sensitivity. Nucleotide-binding oligomerization domain-containing 2 (NOD2) is a subtype of intracellular pattern recognition receptor (PRR), playing functions in innate immunity. Of particular interest, increased levels of NOD2 were observed in DN patients and high fat diet (HFD)/STZ-induced mice models. Furthermore, HFD/STZ-induced diabetes mice models with NOD2 knock-out showed reduced podocyte injury and proteinuria compared with wild-type diabetic mice (Du et al. 2013). In vitro, NOD2 which was activated by bacterial component muramyl dipeptide in podocytes reduced insulin-induced glucose uptake and inhibited serine phosphorylation of IRS-1. Another study has explored the role of other PRR toll-like receptors (TLRs) in the *db/db* mice model of DN. Administration of a selective TLR2/4/6 inhibitor GIT27 improved insulin sensitivity, reduced albuminuria and urinary nephrin levels, indicative of reduced podocyte damage. TLR4 expression in podocytes was found to be highest expressed (Cha et al. 2013). Given the links between some specific PRRs activation and insulin stimulation in podocytes, how podocyte insulin responses are altered following PRRs activation and inhibition may need specifically investigated.

IKB/NF-κB is another important pathway of insulin resistance in podocyte, and NF-κB expression was increased in kidney tissues of patients with type 2 diabetes. NF-κB can increase the level of IRS serine phosphorylation and the expression of inflammatory MCP-1, IL-6, and TNF-α. Moreover, the increased expressed inflammatory factors can further activate the NF-κB. The inflammatory cytokines and the activation of NF-κB pathway form positive feedback to induce insulin resistance.

10.6.5.2 Micro-inflammation and Dyslipidemia Act Synergistically in Podocyte Injury

Xu et al. reported that chronic systemic inflammation exacerbates lipid accumulation in the kidney of ApoE knockout mice by diverting lipid from the plasma to the kidney via the SCAP-SREBP2-LDLr pathway and causing renal injury (Xu et al. 2011). Consisted with this, IL-1β stimulation in vitro increased the lipid accumulation in the podocytes by increasing the expression of lipid metabolism related proteins, for instance, LDLr, sterol regulatory element-binding protein-2 (SREBP-2) and SREBP cleavage-activating protein (SCAP), and through promoting translocation of the SCAP/SREBP-2 complex from the endoplasmic reticulum to the Golgi in the podocytes (Zhang et al. 2015b). Compared with db/db mice, podocyte injury was more severe in db/db mice with subcutaneous casein injections, which are supposed to induce inflammatory stress in vivo. Altogether, inflammation may be associated with high risk for chronic renal fibrosis.

10.6.5.3 Intrinsic Proinflammatory Signaling in Podocytes

Activation of intrinsic proinflammatory signaling in podocytes such as NF-κB signal pathway aggravates podocyte injury and proteinuria. In STZ-induced diabetic mice models with Ccr2 knock-out, transgenic CCR2 overexpression in the podocytes resulted in significantly increased albuminuria and podocyte loss, without concurrent increase in kidney macrophage infiltration or inflammatory cytokine production. These findings support that activation of CCR2 signaling cascade in podocytes mediates diabetic renal injury, which is independent of macrophage recruitment (You et al. 2017).

IL-20, a proinflammatory cytokine which is upregulated by high glucose and TGF-β1, can increase MCP-1 and TGF-β1 expression in podocytes and induce apoptosis in podocytes through activating caspase-8. In STZ-induced early DN mice models, anti-IL-20 monoclonal antibody (7E) treatment or IL-20R1-deficiency led to lower blood glucose and improved renal functions, and IL-20 is proved to be expressed in podocytes. Collectively, intrinsic proinflammatory signaling in podocytes contributes to podocyte damage (Fig. 10.3).

10.7 Immune Disorder in Podocyte Injury

Immune injuries are common causes of podocyte damage. Processes interfering with podocyte's structural or functional integrity lead to disruption of the glomerular filtration barrier.

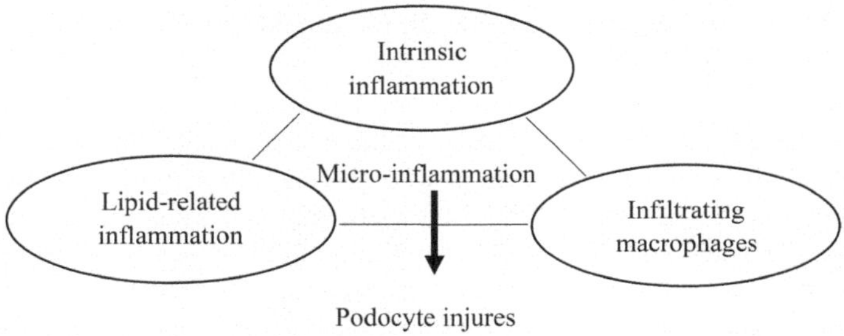

Fig. 10.3 Scheme pattern of micro-inflammation-mediated podocyte injury

10.7.1 Immunoactive Molecules Expressed at Podocytes

10.7.1.1 Complement and Complement Regulatory Protein

Primary-cultured human podocytes synthesize and secrete complement C3 physiologically, and the stimulation of inflammatory factor INF-γ could increase the production of C3. Under physiological conditions, C3 produced by glomerular podocytes can resist the invasion of foreign pathogens and protect local tissues. C3 activation can lead to decreased immune complex formation and increased disintegration. On the other hand, C3 activation leads to increased production of vasoactive molecules and chemokines, which in turn recruits more inflammatory mediators into the glomerulus. The activation of complement would produce proinflammatory components of complement, i.e., C5a. In immune complex diseases and ischemia-reperfusion injury, C5a is an important mediator that triggers an inflammatory cascade (Heller et al. 1999).

The kidney is one of the organs that are most susceptible to abnormally activated complement, which can be seen in various glomerulonephritis. The main pathogenesis of idiopathic membranous nephropathy (IMN) is caused by the binding of IgG to the intrinsic antigen on the basement membrane side of glomerular podocytes, which combine to form an antigen–antibody complex, thereby activating the complement-forming membrane attack complex (Takano et al. 2013). In IMN, the concentrations of complement cleavage products such as C3a, C5a and C5b-9 are significantly increased. C5b-9 is the final product of complement activation in three pathways of complement activation, causing podocyte injury not through conventional lysis, but probably via the mechanism related to the activation of corresponding intracellular signaling pathways in a subdissolved form. Ronco and Debiec have confirmed that the podocyte surface antigen megalin binded to the corresponding antibody underwent an immune complex reaction, activated the complement system, and promoted the formation of the membrane attack complex C5b-9 (Ronco and Debiec 2007). As a stimulant of podocytes, c5b-9 could destroy podocyte cytoskeletal proteins, inserting in the membrane to increase cell permeability, and activating a series of transduction pathways, resulting in the diffuse thickening of GBM and defects in glomerular filtration barrier, clinically leading to significant proteinuria.

In addition, podocytes begin to express complement receptor 1 (CR1, or C3bR, or CD35) during the capillary synthesis stage of renal development and are evenly distributed on the cell membrane and the membrane of foot processes. CR1 is expressed as a cofactor of the complement factor I and expressed in most circulating cells. CR1 is the only physiological blocker of complement synthesis in podocytes and inactivates the lysate of complement to promote the clearance of immune complexes, protecting podocytes from complement-mediated damage (Alexander et al. 2007). It has been reported that the production of CR1 was reduced in several glomerular diseases, making podocytes vulnerable to complement attacks.

Complement regulatory proteins include Crry, CD59, and decay acceleration factors (DAF or CD55), which are vital to limiting the activation of podocyte complement (Cheng et al. 2018). Podocyte expression of Crry and CD59 could inhibit C3 invertase and the synthesis of C5b-9, thus to protect podocytes from injuries induced by antibody-complement activation. In addition, podocytes both in vitro and in vivo could be detected of DAF. In a mouse model of nephritis, deficiency of DAF resulted in serious podocyte foot fusion, indicating that DAF might protect podocytes from complement-mediated injury (Bao et al. 2009).

10.7.1.2 Cytokines and Chemokines

In both physiological and pathological texts, podocytes of humans, rats, and mice all express the receptors of cytokines interleukin 4 (IL-4), IL-10, and IL-13. After stimulating podocytes cultured in vitro with IL-4 and IL-13, the skeletal structure and intercellular-link protein of podocytes were damaged and the permeability increased (Ha et al. 2017; Kim et al. 2017), suggesting that IL-4 and IL-13 could damage podocytes by binding to its receptors.

In early minimal change disease (MCD), FSGS, and MN, podocytes increasingly express inflammatory mediators IL-1 α/β along with IL-1 type 1 receptor (IL-1 RI), and IL-1 RI is decreasingly expressed at late stage of the disease when glomerular cell hyperplasia and sclerosis appear (Brahler et al. 2012), indicating that these molecules participate in podocyte damage and repair, glomerular local inflammation.

In addition, both podocytes cultured in vitro and renal tissue express receptors of functional CC chemokine receptor (CCR) and CXC chemokine receptors (CXCR), which could couple with corresponding chemokines to promote the production of cytoplasm Ca^{2+} and ROS and be involved in podocyte injuries (Huber et al. 2002). Moreover, it has been found that podocytes themselves could produce IL-8 (ligand of CXCR1/CXCR2), thus podocytes could be activated via autocrine.

CXCL16 might play an important role in the inflammatory response of kidney diseases. Podocytes overexpress CXCL16 under the stimulation of proinflammatory factors. Soluble CXCL16 plays a chemotactic role in inflammation and immune response, while transmembrane CXCL16 removes oxLDL (Gutwein et al. 2009), which is harmful to the kidney. Therefore, abnormal expression of CXCL16 in podocytes might cause renal damage due to excessive immune-inflammatory reaction or an accumulation of oxLDL. It has been found that the expression of CXCL16 and oxLDL in the glomeruli of MN patients increased not only significantly but consistently as well (Gutwein et al. 2009). The inflammatory factor IFN-γ is the strongest stimulator of CXCL16, which upregulates several forms and overall cell expression levels of CXCL16, consequently promoting podocyte damage (Wang et al. 2014).

10.7.1.3 Toll-like Receptors (TLRs)

Under physiological conditions, podocytes of humans and mice could express TLR4. Stimulating cultured murine podocytes in vitro with the ligand of TLR4-like LPS, lipid A, and fibrins (endogenous ligand), resulted in an increasing expression of CCL and CXCL. In the mouse model of cryoglobulinemia membrane proliferative glomerulonephritis, podocytes expressed more TLR4, promoting the synthesis and secretion of chemokines and further leukocyte recruitment and glomerular injury (Banas et al. 2008). It has been shown that under the stimulation of endogenous TLR4 ligand, podocytes upregulate TLR4, promote the production of proinflammatory chemokines, and actively participate in the recruitment of inflammatory cells, all leading to glomerular injuries (Banas et al. 2008).

Apart from TLR4, other members of the TLR family have also been proved to participate in podocyte injury. A recent study has pointed out that the overexpression of TLR-8 correlates with the progression of podocyte injury in glomerulonephritis, suggesting that altered levels of urinary Tlr8 mRNA might reflect the degree of podocyte injury in murine autoimmune GN (Kimura et al. 2014).

TLR-7 and TLR-9 expressed by B cells and dendritic cells have been considered as important molecules involved in the pathogenesis of systemic lupus. Recent study demonstrated that active LN onset in childhood expressed more TLR-9, accompanied by weakened expression of podocyte SD protein nephrin, podocin, and synaptopodin; in the meantime, patients showed proteinuria and high ds-DNA antibody and low complement (Machida et al. 2010). Therefore, under pathological conditions, TLRs link the innate immune system with podocyte and glomerular injuries.

10.7.1.4 Costimulatory Factors

B7-1(CD80) belongs to the immunoglobulin superfamily, mainly expressed in antigen-presenting cells, and provides a costimulatory signal by coupling with corresponding molecular receptors expressed on T cells, i.e., CD28 and CTLA 4, regulating the immune responses induced by activated T cells. It has been found that B7-1 was expressed on podocytes of lupus nephritis (LN) (Reiser et al. 2004), and the expression of podocyte B7-1 in LN patients and LN mouse models is positively correlated with the degree of proteinuria. However, new evidence has stricken up a discordant tune (Baye et al. 2016), leading to further mandatory studies of the application of B7-1 blockers in treating proteinuric patients (Novelli et al. 2016a).

Studies have shown that under the induction of hypoxia, high glucose, or bacteriocin lipopolysaccharide (LPS), the expression of B7-1 would be induced in podocytes which does not occur under physiological conditions, and participate in podocyte cytoskeletal reorganization and the pathogenesis of proteinuria (Chang et al. 2013; Fiorina et al. 2014; Shimada et al. 2012).

In the glomerulus of nephritis, podocyte-expressed B7-1 may also recruit T cells to where GBM is damaged and promote further inflammation. Podocytes from necrotic crescentic nephritis rat model and cultured rat podocytes in vitro could express both MHC I/II molecules and intercellular adhesion molecule 1 (ICAM-1) after stimulation of IFN-γ, suggesting that cytokines could present the antigen to infiltrating T cells (Goldwich et al. 2013). Recently, it has been pointed that compared to normal people, MCD patients but not FSGS patients excreted more urinary B7-1, while podocytes of relapsed MCD patients and FSGS patients did not express B7-1, thus B7-1 might be used to identify MCD and FSGS (Novelli et al. 2016b).

10.7.2 *Immune Disorder and Podocyte Injuries*

The glomerulus is a well-recognized target of miscellaneous immune-mediated injuries, and the pathogenesis of immune-mediated glomerular disease is multifactorial (Fig. 10.4).

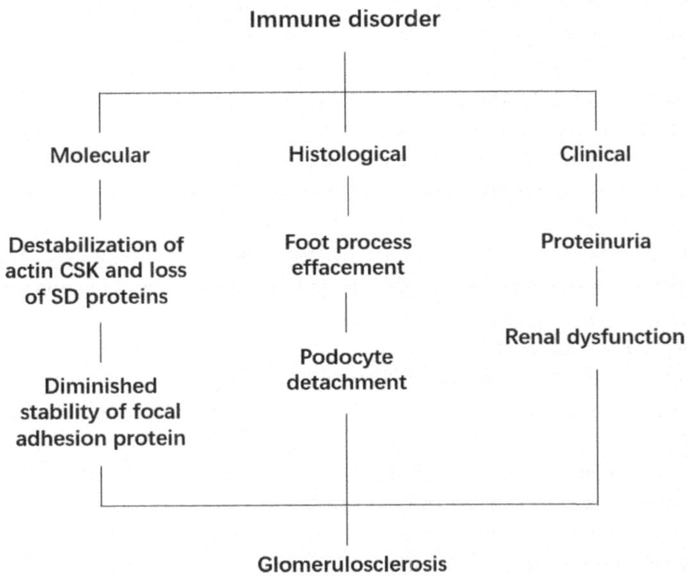

Fig. 10.4 Sequences of immune-mediated podocyte injury

10.7.2.1 Anti-podocyte Antibody

In MN, the surface molecules of glomerular podocyte act as antigens and trigger systematic immune responses, resulting in the formation of in situ immune complexes. The classic animal model of MN, Heymann nephritis, reproduces typical mesangial lesions by eliciting auto-antibodies against the podocyte membrane protein megalin in rats (Ronco and Debiec 2005).

It has been reported in vivo that the occurrence of human newborns MN was due to the production of auto-antibodies against glomerular podocyte membrane proteins. Neutral endopeptidase (NEP) is a membrane protein expressed on the surface of human podocytes. Studies have shown that neonatal MN occurs due to the presence of anti-NEP auto-antibodies in children (Herrmann et al. 2012). Its origin is due to the mother's carrying the relevant mutation gene and lacking NEP. If the mother bred a normal healthy fetus, the mother will produce an anti-NEP antibody against the fetus during pregnancy and the antibody enters the fetus through the placenta. Anti-NEP antibodies react with NEP antigens on fetal podocytes, forming an immune complex on the epithelial side, leading to neonatal MN. Although the incidence of this type of patients is very low, its pathogenesis confirms the role of anti-podocyte antigen antibodies in the development of human MN (Pozdzik et al. 2015).

10.7.2.2 T Cell Dysfunction

T cell dysfunction and the release of cytokines (circulatory factors) causing podocyte injury are associated with the formation of proteinuria in MCD patients. It is currently believed that the cytokines produced by Th1 and Th2 cells in T cells are involved in the occurrence of MCD, but the cytokines produced by Th2 cells (IL-4, IL-8, IL-13) might be more important (Mack 2009). Animal experiments have found that the injection of IL-8 to rats can reduce the content of heparin sulfate on the surface of podocytes, weaken the membrane filtration barrier of charge, and trigger proteinuria. There are also receptors for IL-4 and IL-13 on the podocyte, and the increase of circulating or local IL-4 and IL-13 can directly damage the podocyte through the receptors on the podocyte and increasing the permeability of the filtration membrane.

Shimada et al. (2011) have proposed that MCD is the result of a "two-hit" attack from podocyte immune dysfunction: The first hit is the effects of bacterial products, viruses, and various cytokines on podocytes, resulting in an abnormal expression of CD80 in podocytes, and further cytoskeleton reorganization and morphological changes of podocytes, increasing the permeability of GBM which might bring about proteinuria. However, due to the self-regulation of the body, podocytes can downregulate the expression of CD80. If the auto-regulatory function of podocytes and the body is defective, the sustained expression of CD80 would lead to proteinuria and even MCD. Moreover, Ishimoto et al. (2011) also observed increasing expression of CD80 in the urine of MCD patients; in view of the fact that the expression of CD80 in podocytes can be induced by IL-13 and bacterial products through the TLR pathway and regulated by CTLA4, suggesting that defective immune functions of podocytes is an essential cause of MCD.

10.7.2.3 Antigen–Antibody Immune Complex

Certain exogenous antigens (small molecular weight, positively charged) are implanted on the epithelial side and can also lead to the formation of in situ immune complexes. Hepatitis B virus (HBV)-associated nephropathy is often manifested as mesangial lesions (especially in children), and HBeAg plays an important role in its occurrence. The HBeAg molecule is of small mass and negatively charged and can be implanted across the glomerular basement membrane (GBM) on the epithelial side, triggering the formation of in situ immune complexes (Gupta and Quigg 2015).

Under inflammatory conditions, podocytes would inhibit the expression of MHC class II molecules, promoting the remove of immune complexes from the GBM. In some cases, podocytes might act as antigen-presenting cells themselves, taking up and processing antigens to initiate specific T cell responses. There has been evidence that transgenic mice with a loss of MHC class II exclusively in podocytes developed only a very moderate degree of nephrosclerosis and glomerular crescent formation compared to the control animals, indicative of their defective capacity to activate CD^{8+} T cells (Goldwich et al. 2013).

10.8 The Role of Other Factors in Podocyte Injury

Viral infection, such as human immunodeficiency virus (HIV)-1, parvovirus B19, cytomegalovirus (CMV), hepatitis B virus (HBV), and hepatitis C virus (HCV), is associated with podocyte injury. HIV-associated nephropathy (HIVAN) mostly manifests collapsing glomerulopathy or classic FSGS (Chandra and Kopp 2013). Podocyte infection is associated with podocyte injury and dedifferentiation and rapid loss of renal function. Studies have reported that HIV virus can be internalized by podocytes in vitro, which might be associated with receptors, such as viral coat protein gp120, and subsequent endocytosis, phagocytosis, or pinocytosis (Bruggeman 2017). Although the transmission of virus in vitro has been well documented, further studies are needed to demonstrate the definite mechanism by which the virus enters podocyte in vivo. Structural viral proteins, *gag* and *pol*, and non-structural proteins, *vpr*, *nef*, and *tat*, have been considered to be associated with HIVAN (Conaldi et al. 2002; Reid et al. 2001; Zuo et al. 2006). HBV is a major cause for membranous nephropathy and FSGS scarcely, which can be diagnosed by evidence of HBV antigen or antibodies on kidney biopsy. The possible mechanisms of HBV-induced podocyte injury might be as follows: detective infection of the cells by HBV, deposition of circulating immune complex in renal cells, effects of HBV-induced immunological mediators (Bhimma and Coovadia 2004; Sakai et al. 2011).

Podocytes are also targets of some toxicity drugs, which may further progress to glomerulosclerosis. For example, gold, bucillamine, and d-penicillamine, which are used for the treatment of rheumatoid arthritis, are confirmed to cause MN. The possible mechanism might be closely related to stress, energy metabolism, and inflammation (Fujiwara et al. 2011; Seguin et al. 2005). Other drugs, like non-steroid anti-inflammatory drugs and interferon, also can be inductor of podocyte injury. Organic solvents, like gasoline, dimethylbenzene, and formaldehyde, can induce podocyte injury including foot process fusion and decreased expression of nephrin and podocin (Qin et al. 2012).

Hypoxia can be induced by various pathogenic conditions including hypertension and diabetes. Chronic hypoxia can trigger endoplasmic reticulum (ER) stress, which result in increased ROS. Nephrin and alpha-actin-4, the structural components of SD, are subject to mutations, which cause defective protein folding in the ER of podocytes. The underling mechanism might include transient receptor protein 6 and complement complex and increased expression of MCP-1 (Chen et al. 2011; Cybulsky 2013; Maekawa and Inagi 2017). Targeting hypoxia and ER stress and the possible signal networks might be the novel target for intervention of podocyte injury in CKD.

10.9 Summary

Living in an environment of a variety of pathological stresses and stimuli, podocytes adapt to maintain the integrity and stability of the glomerular basement membrane, depending on their highly differentiated characteristics which also reflect the vulnerability of this barrier.

The different responses of podocytes to injury are associated with the pathology and prognosis of glomerular diseases. As a vital type of renal intrinsic cells, podocyte damage is an important cause of nephrotic proteinuria and glomerular sclerosis. However, as a highly differentiated terminal cell, podocyte has no proliferative potential, and loss of podocyte is associated with poor renal outcomes such as increased proteinuria, glomerulosclerosis, and renal disease progression.

Podocytes have different responses to injuries, including endoplasmic reticulum stress and autophagy reactions caused by abnormal energy metabolism. This chapter lists several aspects of podocyte injuries along with potential underlying mechanisms, including glucose and lipid metabolism disorder, hypertension, RAS activation, micro-inflammation, immune disorder, and other factors. These aspects are not technically separated items, but intertwined with each other in the pathogenesis of podocyte injuries.

Injured podocytes would undergo a series of morphological changes: FP disappearance, cellular shrink, pseudocysts form, cell hypertrophy, cytoplasmic lysosomal enrichment, etc. These changes eventually cause podocytes to detach from the GBM. Moreover, due to the lack of proliferative capacity, the number of glomerular podocytes would become less and less, until reduced by more than 20% when glomerulosclerosis occurs.

Glomerulosclerosis is not a specific disease but a state representing podocyte injury which is mediated by diverse causes. Podocytes interact with GBM and capillary loops tightly, dysfunction of which is an early event leading to glomerulosclerosis. Glomerulosclerosis seems like a station to stay in just before arriving to destination. Unanswered questions in the pathogenesis of podocyte injury and glomerulosclerosis are still ill-defined, and the causing list will continue to grow. Uncovering the selective targeting to pathogenesis and underlying mechanism of podocyte injury and glomerulosclerosis is bound to provide clues to answer for treatment and prevention of the disease in the future.

Acknowledgements This study was supported by grants from the National Key Research and Development Program of China (2018YFC1314004).

References

Advani A (2014) Vascular endothelial growth factor and the kidney: something of the marvellous. Curr Opin Nephrol Hypertens 23:87–92

Alexander JJ, Wang Y, Chang A, Jacob A, Minto AW, Karmegam M et al (2007) Mouse podocyte complement factor H: the functional analog to human complement receptor 1. J Am Soc Nephrol 18:1157–1166

Ando K, Ohtsu H, Uchida S, Kaname S, Arakawa Y, Fujita T et al (2014) Anti-albuminuric effect of the aldosterone blocker eplerenone in non-diabetic hypertensive patients with albuminuria: a double-blind, randomised, placebo-controlled trial. Lancet Diabetes Endocrinol 2:944–953

Anil Kumar P, Welsh GI, Saleem MA, Menon RK (2014) Molecular and cellular events mediating glomerular podocyte dysfunction and depletion in diabetes mellitus. Front Endocrinol (Lausanne) 5:151

Awad AS, You H, Gao T, Cooper TK, Nedospasov SA, Vacher J et al (2015) Macrophage-derived tumor necrosis factor-alpha mediates diabetic renal injury. Kidney Int 88:722–733

Bai C, Liang S, Wang Y, Jiao B (2017) Knocking down TCF8 inhibits high glucose- and angiotensin II-induced epithelial to mesenchymal transition in podocytes. Biosci Trends 11:77–84

Banas MC, Banas B, Hudkins KL, Wietecha TA, Iyoda M, Bock E et al (2008) TLR4 links podocytes with the innate immune system to mediate glomerular injury. J Am Soc Nephrol 19:704–713

Bao L, Haas M, Pippin J, Wang Y, Miwa T, Chang A et al (2009) Focal and segmental glomerulosclerosis induced in mice lacking decay-accelerating factor in T cells. J Clin Invest 119:1264–1274

Baye E, Gallazzini M, Delville M, Legendre C, Terzi F, Canaud G (2016) The costimulatory receptor B7-1 is not induced in injured podocytes. Kidney Int 90:1037–1044

Bhimma R, Coovadia HM (2004) Hepatitis B virus-associated nephropathy. Am J Nephrol 24:198–211

Bianchi S, Baronti A, Cominotto R, Bigazzi R (2016) Lipid metabolism abnormalities in Chronic Kidney Disease. G Ital Nefrol 33: S68

Brahler S, Ising C, Hagmann H, Rasmus M, Hoehne M, Kurschat C et al (2012) Intrinsic proinflammatory signaling in podocytes contributes to podocyte damage and prolonged proteinuria. Am J Physiol Renal Physiol 303:F1473–F1485

Bruggeman LA (2017) HIV-1 infection of renal cells in HIV-associated nephropathy. J Am Soc Nephrol 28:719–721

Bussolati B, Deregibus MC, Fonsato V, Doublier S, Spatola T, Procida S et al (2005) Statins prevent oxidized LDL-induced injury of glomerular podocytes by activating the phosphatidylinositol 3-kinase/AKT-signaling pathway. J Am Soc Nephrol 16:1936–1947

Cellesi F, Li M, Rastaldi MP (2015) Podocyte injury and repair mechanisms. Curr Opin Nephrol Hypertens 24:239–244
Cha JJ, Hyun YY, Lee MH, Kim JE, Nam DH, Song HK et al (2013) Renal protective effects of toll-like receptor 4 signaling blockade in type 2 diabetic mice. Endocrinology 154:2144–2155
Chandra P, Kopp JB (2013) Viruses and collapsing glomerulopathy: a brief critical review. Clin Kidney J 6:1–5
Chang JM, Hwang DY, Chen SC, Kuo MC, Hung CC, Hwang SJ et al (2013) B7-1 expression regulates the hypoxia-driven cytoskeleton rearrangement in glomerular podocytes. Am J Physiol Renal Physiol 304:F127–F136
Chen HC, Chen CA, Guh JY, Chang JM, Shin SJ, Lai YH (2000) Altering expression of alpha3beta1 integrin on podocytes of human and rats with diabetes. Life Sci 67:2345–2353
Chen S, He FF, Wang H, Fang Z, Shao N, Tian XJ et al (2011) Calcium entry via TRPC6 mediates albumin overload-induced endoplasmic reticulum stress and apoptosis in podocytes. Cell Calcium 50:523–529
Chen S, Meng XF, Zhang C (2013) Role of NADPH oxidase-mediated reactive oxygen species in podocyte injury. Biomed Res Int 2013:839761
Cheng L, Gou SJ, Qiu HY, Ma L, Fu P (2018) Complement regulatory proteins in kidneys of patients with anti-neutrophil cytoplasmic antibody (ANCA)-associated vasculitis. Clin Exp Immunol 191:116–124
Choi SY, Lim SW, Salimi S, Yoo EJ, Lee-Kwon W, Lee HH et al (2018) Tonicity-responsive enhancer-binding protein mediates hyperglycemia-induced inflammation and vascular and renal injury. J Am Soc Nephrol 29:492–504
Cohen MP, Chen S, Ziyadeh FN, Shea E, Hud EA, Lautenslager GT et al (2005) Evidence linking glycated albumin to altered glomerular nephrin and VEGF expression, proteinuria, and diabetic nephropathy. Kidney Int 68:1554–1561
Conaldi PG, Bottelli A, Baj A, Serra C, Fiore L, Federico G et al (2002) Human immunodeficiency virus-1 tat induces hyperproliferation and dysregulation of renal glomerular epithelial cells. Am J Pathol 161:53–61
Coward RJ, Welsh GI, Yang J, Tasman C, Lennon R, Koziell A et al (2005) The human glomerular podocyte is a novel target for insulin action. Diabetes 54:3095–3102
Coward RJ, Welsh GI, Koziell A, Hussain S, Lennon R, Ni L et al (2007) Nephrin is critical for the action of insulin on human glomerular podocytes. Diabetes 56:1127–1135
Cybulsky AV (2013) The intersecting roles of endoplasmic reticulum stress, ubiquitin-proteasome system, and autophagy in the pathogenesis of proteinuric kidney disease. Kidney Int 84:25–33
Dai HY, Zheng M, Lv LL, Tang RN, Ma KL, Liu D et al (2012) The roles of connective tissue growth factor and integrin-linked kinase in high glucose-induced phenotypic alterations of podocytes. J Cell Biochem 113:293–301
Dandapani SV, Sugimoto H, Matthews BD, Kolb RJ, Sinha S, Gerszten RE et al (2007) Alpha-actinin-4 is required for normal podocyte adhesion. J Biol Chem 282:467–477
De Rechter S, Decuypere JP, Ivanova E, van den Heuvel LP, De Smedt H, Levtchenko E et al (2016) Autophagy in renal diseases. Pediatr Nephrol 31:737–752
de Vries AP, Ruggenenti P, Ruan XZ, Praga M, Cruzado JM, Bajema IM et al (2014) Fatty kidney: emerging role of ectopic lipid in obesity-related renal disease. Lancet Diabetes Endocrinol 2:417–426
Ding G, Reddy K, Kapasi AA, Franki N, Gibbons N, Kasinath BS et al (2002) Angiotensin II induces apoptosis in rat glomerular epithelial cells. Am J Physiol Renal Physiol 283:F173–F180
Du P, Fan B, Han H, Zhen J, Shang J, Wang X et al (2013) NOD2 promotes renal injury by exacerbating inflammation and podocyte insulin resistance in diabetic nephropathy. Kidney Int 84:265–276
Durvasula RV, Shankland SJ (2008) Activation of a local renin angiotensin system in podocytes by glucose. Am J Physiol Renal Physiol 294:F830–F839
Durvasula RV, Petermann AT, Hiromura K, Blonski M, Pippin J, Mundel P et al (2004) Activation of a local tissue angiotensin system in podocytes by mechanical strain. Kidney Int 65:30–39

Eremina V, Jefferson JA, Kowalewska J, Hochster H, Haas M, Weisstuch J et al (2008) VEGF inhibition and renal thrombotic microangiopathy. N Engl J Med 358:1129–1136
Fantus D, Rogers NM, Grahammer F, Huber TB, Thomson AW (2016) Roles of mTOR complexes in the kidney: implications for renal disease and transplantation. Nat Rev Nephrol 12:587–609
Fiorina P, Vergani A, Bassi R, Niewczas MA, Altintas MM, Pezzolesi MG et al (2014) Role of podocyte B7-1 in diabetic nephropathy. J Am Soc Nephrol 25:1415–1429
Fornoni A, Sageshima J, Wei C, Merscher-Gomez S, Aguillon-Prada R, Jauregui AN et al (2011) Rituximab targets podocytes in recurrent focal segmental glomerulosclerosis. Sci Transl Med 3:85ra46
Friedrich C, Endlich N, Kriz W, Endlich K (2006) Podocytes are sensitive to fluid shear stress in vitro. Am J Physiol Renal Physiol 291:F856–F865
Fujiwara Y, Tsuchiya H, Sakai N, Shibata K, Fujimura A, Koshimizu TA (2011) Proximal tubules and podocytes are toxicity targets of bucillamine in a mouse model of drug-induced kidney injury. Eur J Pharmacol 670:208–215
Gao N, Wang H, Yin H, Yang Z (2017) Angiotensin II induces calcium-mediated autophagy in podocytes through enhancing reactive oxygen species levels. Chem Biol Interact 277:110–118
Gill PS, Wilcox CS (2006) NADPH oxidases in the kidney. Antioxid Redox Signal 8:1597–1607
Goldwich A, Burkard M, Olke M, Daniel C, Amann K, Hugo C et al (2013) Podocytes are non-hematopoietic professional antigen-presenting cells. J Am Soc Nephrol 24:906–916
Greka A, Mundel P (2011) Balancing calcium signals through TRPC5 and TRPC6 in podocytes. J Am Soc Nephrol 22:1969–1980
Greka A, Mundel P (2012) Cell biology and pathology of podocytes. Annu Rev Physiol 74:299–323
Gu L, Hagiwara S, Fan Q, Tanimoto M, Kobata M, Yamashita M et al (2006) Role of receptor for advanced glycation end-products and signalling events in advanced glycation end-product-induced monocyte chemoattractant protein-1 expression in differentiated mouse podocytes. Nephrol Dial Transplant 21:299–313
Gupta A, Quigg RJ (2015) Glomerular diseases associated with hepatitis B and C. Adv Chronic Kidney Dis 22:343–351
Gurley SB, Riquier-Brison ADM, Schnermann J, Sparks MA, Allen AM, Haase VH et al (2011) AT1A angiotensin receptors in the renal proximal tubule regulate blood pressure. Cell Metab 13:469–475
Gutwein P, Abdel-Bakky MS, Schramme A, Doberstein K, Kampfer-Kolb N, Amann K et al (2009) CXCL16 is expressed in podocytes and acts as a scavenger receptor for oxidized low-density lipoprotein. Am J Pathol 174:2061–2072
Ha TS (2006) High glucose and advanced glycosylated end-products affect the expression of alpha-actinin-4 in glomerular epithelial cells. Nephrology (Carlton) 11:435–441
Ha TS, Nam JA, Seong SB, Saleem MA, Park SJ, Shin JI (2017) Montelukast improves the changes of cytoskeletal and adaptor proteins of human podocytes by interleukin-13. Inflamm Res 66:793–802
Hagberg CE, Falkevall A, Wang X, Larsson E, Huusko J, Nilsson I et al (2010) Vascular endothelial growth factor B controls endothelial fatty acid uptake. Nature 464:917–921
Han SY, So GA, Jee YH, Han KH, Kang YS, Kim HK et al (2004) Effect of retinoic acid in experimental diabetic nephropathy. Immunol Cell Biol 82:568–576
Han SY, Kang YS, Jee YH, Han KH, Cha DR, Kang SW et al (2006) High glucose and angiotensin II increase beta1 integrin and integrin-linked kinase synthesis in cultured mouse podocytes. Cell Tissue Res 323:321–332
Hashizume M, Mihara M (2012) Atherogenic effects of TNF-alpha and IL-6 via up-regulation of scavenger receptors. Cytokine 58:424–430
Heller T, Hennecke M, Baumann U, Gessner JE, zu Vilsendorf AM, Baensch M et al (1999) Selection of a C5a receptor antagonist from phage libraries attenuating the inflammatory response in immune complex disease and ischemia/reperfusion injury. J Immunol 163:985–994
Herrmann SM, Sethi S, Fervenza FC (2012) Membranous nephropathy: the start of a paradigm shift. Curr Opin Nephrol Hypertens 21:203–210

Horton JD, Goldstein JL, Brown MS (2002) SREBPs: activators of the complete program of cholesterol and fatty acid synthesis in the liver. J Clin Invest 109:1125–1131
Hu ZB, Ma KL, Zhang Y, Wang GH, Liu L, Lu J et al (2018) Inflammation-activated CXCL16 pathway contributes to tubulointerstitial injury in mouse diabetic nephropathy. Acta Pharmacol Sin 39:1022–1033
Huber TB, Reinhardt HC, Exner M, Burger JA, Kerjaschki D, Saleem MA et al (2002) Expression of functional CCR and CXCR chemokine receptors in podocytes. J Immunol 168:6244–6252
Huber TB, Schermer B, Muller RU, Hohne M, Bartram M, Calixto A et al (2006) Podocin and MEC-2 bind cholesterol to regulate the activity of associated ion channels. Proc Natl Acad Sci U S A 103:17079–17086
Huber TB, Edelstein CL, Hartleben B, Inoki K, Jiang M, Koya D et al (2012) Emerging role of autophagy in kidney function, diseases and aging. Autophagy 8:1009–1031
Ichihara A, Suzuki F, Nakagawa T, Kaneshiro Y, Takemitsu T, Sakoda M et al (2006) Prorenin receptor blockade inhibits development of glomerulosclerosis in diabetic angiotensin II type 1a receptor-deficient mice. J Am Soc Nephrol 17:1950–1961
Ikezumi Y, Suzuki T, Karasawa T, Kawachi H, Nikolic-Paterson DJ, Uchiyama M (2008) Activated macrophages down-regulate podocyte nephrin and podocin expression via stress-activated protein kinases. Biochem Biophys Res Commun 376:706–711
Ishimoto T, Shimada M, Araya CE, Huskey J, Garin EH, Johnson RJ (2011) Minimal change disease: a CD80 podocytopathy? Semin Nephrol 31:320–325
Jim B, Ghanta M, Qipo A, Fan Y, Chuang PY, Cohen HW et al (2012) Dysregulated nephrin in diabetic nephropathy of type 2 diabetes: a cross sectional study. PLoS ONE 7:e36041
Jin J, Sison K, Li C, Tian R, Wnuk M, Sung HK et al (2012) Soluble FLT1 binds lipid microdomains in podocytes to control cell morphology and glomerular barrier function. Cell 151:384–399
Joles JA, Kunter U, Janssen U, Kriz W, Rabelink TJ, Koomans HA et al (2000) Early mechanisms of renal injury in hypercholesterolemic or hypertriglyceridemic rats. J Am Soc Nephrol 11:669–683
Kaysen GA (2001) The microinflammatory state in uremia: causes and potential consequences. J Am Soc Nephrol 12:1549–1557
Kelder TP, Penning ME, Uh HW, Cohen D, Bloemenkamp KW, Bruijn JA et al (2012) Quantitative polymerase chain reaction-based analysis of podocyturia is a feasible diagnostic tool in preeclampsia. Hypertension 60:1538–1544
Kim JS, Han BG, Choi SO, Cha SK (2016) Secondary focal segmental glomerulosclerosis: from podocyte injury to glomerulosclerosis. Biomed Res Int 2016:1630365
Kim AH, Chung JJ, Akilesh S, Koziell A, Jain S, Hodgin JB et al (2017) B cell-derived IL-4 acts on podocytes to induce proteinuria and foot process effacement. JCI Insight 2:pii: 81836
Kimura J, Ichii O, Miyazono K, Nakamura T, Horino T, Otsuka-Kanazawa S et al (2014) Overexpression of toll-like receptor 8 correlates with the progression of podocyte injury in murine autoimmune glomerulonephritis. Sci Rep 4:7290
Kobayashi N, Ueno T, Ohashi K, Yamashita H, Takahashi Y, Sakamoto K et al (2015) Podocyte injury-driven intracapillary plasminogen activator inhibitor type 1 accelerates podocyte loss via uPAR-mediated beta1-integrin endocytosis. Am J Physiol Renal Physiol 308:F614–F626
Kreidberg JA, Donovan MJ, Goldstein SL, Rennke H, Shepherd K, Jones RC et al (1996) Alpha 3 beta 1 integrin has a crucial role in kidney and lung organogenesis. Development 122:3537–3547
Kretzler M, Koeppen-Hagemann I, Kriz W (1994) Podocyte damage is a critical step in the development of glomerulosclerosis in the uninephrectomised-desoxycorticosterone hypertensive rat. Virchows Arch 425:181–193
Kriz W, Lemley KV (2015) A potential role for mechanical forces in the detachment of podocytes and the progression of CKD. J Am Soc Nephrol 26:258–269
Kriz W, Elger M, Nagata M, Kretzler M, Uiker S, Koeppen-Hageman I et al (1994) The role of podocytes in the development of glomerular sclerosis. Kidney Int Suppl 45:S64–S72
Kriz W, Gretz N, Lemley KV (1998a) Progression of glomerular diseases: is the podocyte the culprit? Kidney Int 54:687–697

Kriz W, Hosser H, Hahnel B, Gretz N, Provoost AP (1998b) From segmental glomerulosclerosis to total nephron degeneration and interstitial fibrosis: a histopathological study in rat models and human glomerulopathies. Nephrol Dial Transplant 13:2781–2798

Kriz W, Hosser H, Hahnel B, Simons JL, Provoost AP (1998c) Development of vascular pole-associated glomerulosclerosis in the Fawn-hooded rat. J Am Soc Nephrol 9:381–396

Kriz W, Shirato I, Nagata M, LeHir M, Lemley KV (2013) The podocyte's response to stress: the enigma of foot process effacement. Am J Physiol Renal Physiol 304:F333–F347

Kwon SH, Woollard JR, Saad A, Garovic VD, Zand L, Jordan KL et al (2017) Elevated urinary podocyte-derived extracellular microvesicles in renovascular hypertensive patients. Nephrol Dial Transplant 32:800–807

Laybutt DR, Preston AM, Akerfeldt MC, Kench JG, Busch AK, Biankin AV et al (2007) Endoplasmic reticulum stress contributes to beta cell apoptosis in type 2 diabetes. Diabetologia 50:752–763

Lee HS (2011) Mechanisms and consequences of hypertriglyceridemia and cellular lipid accumulation in chronic kidney disease and metabolic syndrome. Histol Histopathol 26:1599–1610

Lennon R, Pons D, Sabin MA, Wei C, Shield JP, Coward RJ et al (2009) Saturated fatty acids induce insulin resistance in human podocytes: implications for diabetic nephropathy. Nephrol Dial Transplant 24:3288–3296

Lewko B, Bryl E, Witkowski JM, Latawiec E, Golos M, Endlich N et al (2005) Characterization of glucose uptake by cultured rat podocytes. Kidney Blood Press Res 28:1–7

Li JH, Huang XR, Zhu HJ, Oldfield M, Cooper M, Truong LD et al (2004) Advanced glycation end products activate Smad signaling via TGF-beta-dependent and independent mechanisms: implications for diabetic renal and vascular disease. FASEB J 18:176–178

Li G, Li CX, Xia M, Ritter JK, Gehr TW, Boini K et al (2015) Enhanced epithelial-to-mesenchymal transition associated with lysosome dysfunction in podocytes: role of p62/Sequestosome 1 as a signaling hub. Cell Physiol Biochem 35:1773–1786

Liebau MC, Lang D, Bohm J, Endlich N, Bek MJ, Witherden I et al (2006) Functional expression of the renin-angiotensin system in human podocytes. Am J Physiol Renal Physiol 290:F710–F719

Liu W, Zhang Y, Hao J, Liu S, Liu Q, Zhao S et al (2012) Nestin protects mouse podocytes against high glucose-induced apoptosis by a Cdk5-dependent mechanism. J Cell Biochem 113:3186–3196

Liu Y, Hitomi H, Diah S, Deguchi K, Mori H, Masaki T et al (2013) Roles of Na(+)/H(+) exchanger type 1 and intracellular pH in angiotensin II-induced reactive oxygen species generation and podocyte apoptosis. J Pharmacol Sci 122:176–183

Loeffler I, Wolf G (2015) Epithelial-to-mesenchymal transition in diabetic nephropathy: fact or fiction? Cells 4:631–652

Lv Z, Hu M, Zhen J, Lin J, Wang Q, Wang R (2013) Rac1/PAK1 signaling promotes epithelial-mesenchymal transition of podocytes in vitro via triggering beta-catenin transcriptional activity under high glucose conditions. Int J Biochem Cell Biol 45:255–264

Ma Y, Hendershot LM (2001) The unfolding tale of the unfolded protein response. Cell 107:827–830

Macconi D, Ghilardi M, Bonassi ME, Mohamed EI, Abbate M, Colombi F et al (2000) Effect of angiotensin-converting enzyme inhibition on glomerular basement membrane permeability and distribution of zonula occludens-1 in MWF rats. J Am Soc Nephrol 11:477–489

Machida H, Ito S, Hirose T, Takeshita F, Oshiro H, Nakamura T et al (2010) Expression of toll-like receptor 9 in renal podocytes in childhood-onset active and inactive lupus nephritis. Nephrol Dial Transplant 25:2530–2537

Mack M (2009) Podocyte antigens, dendritic cells and T cells contribute to renal injury in newly developed mouse models of glomerulonephritis. Nephrol Dial Transplant 24:2984–2986

Maekawa H, Inagi R (2017) Stress signal network between hypoxia and ER stress in chronic kidney disease. Front Physiol 8:74

Mao N, Tan RZ, Wang SQ, Wei C, Shi XL, Fan JM et al (2016) Ginsenoside Rg1 inhibits angiotensin II-induced podocyte autophagy via AMPK/mTOR/PI3 K pathway. Cell Biol Int 40:917–925

Marquez E, Riera M, Pascual J, Soler MJ (2015) Renin-angiotensin system within the diabetic podocyte. Am J Physiol Renal Physiol 308:F1–F10

Martinez-Garcia C, Izquierdo-Lahuerta A, Vivas Y, Velasco I, Yeo TK, Chen S et al (2015) Renal lipotoxicity-associated inflammation and insulin resistance affects actin cytoskeleton organization in podocytes. PLoS ONE 10:e0142291
Masuda D, Hirano K, Oku H, Sandoval JC, Kawase R, Yuasa-Kawase M et al (2009) Chylomicron remnants are increased in the postprandial state in CD36 deficiency. J Lipid Res 50:999–1011
Merscher S, Pedigo CE, Mendez AJ (2014) Metabolism, energetics, and lipid biology in the podocyte-cellular cholesterol-mediated glomerular injury. Front Endocrinol (Lausanne) 5:169
Merscher-Gomez S, Guzman J, Pedigo CE, Lehto M, Aguillon-Prada R, Mendez A et al (2013) Cyclodextrin protects podocytes in diabetic kidney disease. Diabetes 62:3817–3827
Mima A, Ohshiro Y, Kitada M, Matsumoto M, Geraldes P, Li C et al (2011) Glomerular-specific protein kinase C-beta-induced insulin receptor substrate-1 dysfunction and insulin resistance in rat models of diabetes and obesity. Kidney Int 79:883–896
Mundel P, Kriz W (1995) Structure and function of podocytes: an update. Anat Embryol (Berl) 192:385–397
Nagase M, Shibata S, Yoshida S, Nagase T, Gotoda T, Fujita T (2006) Podocyte injury underlies the glomerulopathy of Dahl salt-hypertensive rats and is reversed by aldosterone blocker. Hypertension 47:1084–1093
Nagase M, Matsui H, Shibata S, Gotoda T, Fujita T (2007) Salt-induced nephropathy in obese spontaneously hypertensive rats via paradoxical activation of the mineralocorticoid receptor: role of oxidative stress. Hypertension 50:877–883
Nagata M (2016) Podocyte injury and its consequences. Kidney Int 89:1221–1230
Najafian B, Svarstad E, Bostad L, Gubler MC, Tondel C, Whitley C et al (2011) Progressive podocyte injury and globotriaosylceramide (GL-3) accumulation in young patients with Fabry disease. Kidney Int 79:663–670
Neal CR, Muston PR, Njegovan D, Verrill R, Harper SJ, Deen WM et al (2007) Glomerular filtration into the subpodocyte space is highly restricted under physiological perfusion conditions. Am J Physiol Renal Physiol 293:F1787–F1798
Nijenhuis T, Sloan AJ, Hoenderop JG, Flesche J, van Goor H, Kistler AD et al (2011) Angiotensin II contributes to podocyte injury by increasing TRPC6 expression via an NFAT-mediated positive feedback signaling pathway. Am J Pathol 179:1719–1732
Nose A, Mori Y, Uchiyama-Tanaka Y, Kishimoto N, Maruyama K, Matsubara H et al (2003) Regulation of glucose transporter (GLUT1) gene expression by angiotensin II in mesangial cells: involvement of HB-EGF and EGF receptor transactivation. Hypertens Res 26:67–73
Novelli R, Gagliardini E, Ruggiero B, Benigni A, Remuzzi G (2016a) Another piece of the puzzle of podocyte B7-1 expression: lupus nephritis. Nephron 133:129–138
Novelli R, Gagliardini E, Ruggiero B, Benigni A, Remuzzi G (2016b) Any value of podocyte B7-1 as a biomarker in human MCD and FSGS? Am J Physiol Renal Physiol 310:F335–F341
Orellana JM, Kampe K, Schulze F, Sieber J, Jehle AW (2017) Fetuin-A aggravates lipotoxicity in podocytes via interleukin-1 signaling. Physiol Rep 5:pii: e13287
Pan M, Maitin V, Parathath S, Andreo U, Lin SX, St Germain C et al (2008) Presecretory oxidation, aggregation, and autophagic destruction of apoprotein-B: a pathway for late-stage quality control. Proc Natl Acad Sci U S A 105:5862–5867
Pavenstadt H (2000) Roles of the podocyte in glomerular function. Am J Physiol Renal Physiol 278:F173–F179
Pavenstadt H, Kriz W, Kretzler M (2003) Cell biology of the glomerular podocyte. Physiol Rev 83:253–307
Pozdzik AA, Debiec H, Brocheriou I, Husson C, Rorive S, Broeders N et al (2015) Anti-NEP and anti-PLA2R antibodies in membranous nephropathy: an update. Rev Med Brux 36:166–171
Promlek T, Ishiwata-Kimata Y, Shido M, Sakuramoto M, Kohno K, Kimata Y (2011) Membrane aberrancy and unfolded proteins activate the endoplasmic reticulum stress sensor Ire1 in different ways. Mol Biol Cell 22:3520–3532

Puelles VG, Cullen-McEwen LA, Taylor GE, Li J, Hughson MD, Kerr PG et al (2016) Human podocyte depletion in association with older age and hypertension. Am J Physiol Renal Physiol 310:F656–F668
Qi W, Mu J, Luo ZF, Zeng W, Guo YH, Pang Q et al (2011) Attenuation of diabetic nephropathy in diabetes rats induced by streptozotocin by regulating the endoplasmic reticulum stress inflammatory response. Metabolism 60:594–603
Qin W, Xu Z, Lu Y, Zeng C, Zheng C, Wang S et al (2012) Mixed organic solvents induce renal injury in rats. PLoS ONE 7:e45873
Raats CJ, van den Born J, Bakker MA, Oppers-Walgreen B, Pisa BJ, Dijkman HB et al (2000) Expression of agrin, dystroglycan, and utrophin in normal renal tissue and in experimental glomerulopathies. Am J Pathol 156:1749–1765
Raclot T, Langin D, Lafontan M, Groscolas R (1997) Selective release of human adipocyte fatty acids according to molecular structure. Biochem J 324(Pt3):911–915
Rasheva VI, Domingos PM (2009) Cellular responses to endoplasmic reticulum stress and apoptosis. Apoptosis 14:996–1007
Reid W, Sadowska M, Denaro F, Rao S, Foulke J Jr, Hayes N et al (2001) An HIV-1 transgenic rat that develops HIV-related pathology and immunologic dysfunction. Proc Natl Acad Sci U S A 98:9271–9276
Reidy K, Susztak K (2009) Epithelial-mesenchymal transition and podocyte loss in diabetic kidney disease. Am J Kidney Dis 54:590–593
Reiser J, Kriz W, Kretzler M, Mundel P (2000) The glomerular slit diaphragm is a modified adherens junction. J Am Soc Nephrol 11:1–8
Reiser J, von Gersdorff G, Loos M, Oh J, Asanuma K, Giardino L et al (2004) Induction of B7-1 in podocytes is associated with nephrotic syndrome. J Clin Invest 113:1390–1397
Reiser J, Polu KR, Moller CC, Kenlan P, Altintas MM, Wei C et al (2005) TRPC6 is a glomerular slit diaphragm-associated channel required for normal renal function. Nat Genet 37:739–744
Reivinen J, Holthofer H, Miettinen A (1992) A cell-type specific ganglioside of glomerular podocytes in rat kidney: an O-acetylated GD3. Kidney Int 42:624–631
Ren Z, Liang W, Chen C, Yang H, Singhal PC, Ding G (2012) Angiotensin II induces nephrin dephosphorylation and podocyte injury: role of caveolin-1. Cell Signal 24:443–450
Romero M, Ortega A, Izquierdo A, Lopez-Luna P, Bosch RJ (2010) Parathyroid hormone-related protein induces hypertrophy in podocytes via TGF-beta(1) and p27(Kip1): implications for diabetic nephropathy. Nephrol Dial Transplant 25:2447–2457
Ronco P, Debiec H (2005) Molecular pathomechanisms of membranous nephropathy: from Heymann nephritis to alloimmunization. J Am Soc Nephrol 16:1205–1213
Ronco P, Debiec H (2007) Target antigens and nephritogenic antibodies in membranous nephropathy: of rats and men. Semin Immunopathol 29:445–458
Ruggenenti P, Cravedi P, Remuzzi G (2012) Mechanisms and treatment of CKD. J Am Soc Nephrol 23:1917–1928
Ruotsalainen V, Ljungberg P, Wartiovaara J, Lenkkeri U, Kestila M, Jalanko H et al (1999) Nephrin is specifically located at the slit diaphragm of glomerular podocytes. Proc Natl Acad Sci U S A 96:7962–7967
Ruster C, Wolf G (2011) Angiotensin II as a morphogenic cytokine stimulating renal fibrogenesis. J Am Soc Nephrol 22:1189–1199
Ruster C, Bondeva T, Franke S, Tanaka N, Yamamoto H, Wolf G (2009) Angiotensin II upregulates RAGE expression on podocytes: role of AT2 receptors. Am J Nephrol 29:538–550
Sakai K, Morito N, Usui J, Hagiwara M, Hiwatashi A, Fukuda K et al (2011) Focal segmental glomerulosclerosis as a complication of hepatitis B virus infection. Nephrol Dial Transplant 26:371–373
Sakoda M, Ichihara A, Kurauchi-Mito A, Narita T, Kinouchi K, Murohashi-Bokuda K et al (2010) Aliskiren inhibits intracellular angiotensin II levels without affecting (pro)renin receptor signals in human podocytes. Am J Hypertens 23:575–580

Sakoda M, Itoh H, Ichihara A (2011) Podocytes as a target of prorenin in diabetes. Curr Diabetes Rev 7:17–21
Sanchez-Nino MD, Sanz AB, Carrasco S, Saleem MA, Mathieson PW, Valdivielso JM et al (2011) Globotriaosylsphingosine actions on human glomerular podocytes: implications for Fabry nephropathy. Nephrol Dial Transplant 26:1797–1802
Schermer B, Benzing T (2009) Lipid-protein interactions along the slit diaphragm of podocytes. J Am Soc Nephrol 20:473–478
Schiffer M, Susztak K, Ranalletta M, Raff AC, Bottinger EP, Charron MJ (2005) Localization of the GLUT8 glucose transporter in murine kidney and regulation in vivo in nondiabetic and diabetic conditions. Am J Physiol Renal Physiol 289:F186–F193
Schomig M, Eisenhardt A, Ritz E (2000) The microinflammatory state of uremia. Blood Purif 18:327–332
Schwarz K, Simons M, Reiser J, Saleem MA, Faul C, Kriz W et al (2001) Podocin, a raft-associated component of the glomerular slit diaphragm, interacts with CD2AP and nephrin. J Clin Invest 108:1621–1629
Seguin B, Boutros PC, Li X, Okey AB, Uetrecht JP (2005) Gene expression profiling in a model of D-penicillamine-induced autoimmunity in the Brown Norway rat: predictive value of early signs of danger. Chem Res Toxicol 18:1193–1202
Shankland SJ (2006) The podocyte's response to injury: role in proteinuria and glomerulosclerosis. Kidney Int 69:2131–2147
Shibata S, Nagase M, Yoshida S, Kawachi H, Fujita T (2007) Podocyte as the target for aldosterone: roles of oxidative stress and Sgk1. Hypertension 49:355–364
Shimada M, Araya C, Rivard C, Ishimoto T, Johnson RJ, Garin EH (2011) Minimal change disease: a "two-hit" podocyte immune disorder? Pediatr Nephrol 26:645–649
Shimada M, Ishimoto T, Lee PY, Lanaspa MA, Rivard CJ, Roncal-Jimenez CA et al (2012) Toll-like receptor 3 ligands induce CD80 expression in human podocytes via an NF-kappaB-dependent pathway. Nephrol Dial Transplant 27:81–89
Sieber J, Jehle AW (2014) Free Fatty acids and their metabolism affect function and survival of podocytes. Front Endocrinol (Lausanne) 5:186
Sieber J, Lindenmeyer MT, Kampe K, Campbell KN, Cohen CD, Hopfer H et al (2010) Regulation of podocyte survival and endoplasmic reticulum stress by fatty acids. Am J Physiol Renal Physiol 299:F821–F829
Simons M, Schwarz K, Kriz W, Miettinen A, Reiser J, Mundel P et al (2001) Involvement of lipid rafts in nephrin phosphorylation and organization of the glomerular slit diaphragm. Am J Pathol 159:1069–1077
Soetikno V, Sari FR, Sukumaran V, Lakshmanan AP, Harima M, Suzuki K et al (2013) Curcumin decreases renal triglyceride accumulation through AMPK-SREBP signaling pathway in streptozotocin-induced type 1 diabetic rats. J Nutr Biochem 24:796–802
Sonneveld R, van der Vlag J, Baltissen MP, Verkaart SA, Wetzels JF, Berden JH et al (2014) Glucose specifically regulates TRPC6 expression in the podocyte in an AngII-dependent manner. Am J Pathol 184:1715–1726
Sorensson J, Fierlbeck W, Heider T, Schwarz K, Park DS, Mundel P et al (2002) Glomerular endothelial fenestrae in vivo are not formed from caveolae. J Am Soc Nephrol 13:2639–2647
Susztak K, Raff AC, Schiffer M, Bottinger EP (2006) Glucose-induced reactive oxygen species cause apoptosis of podocytes and podocyte depletion at the onset of diabetic nephropathy. Diabetes 55:225–233
Szabo C, Biser A, Benko R, Bottinger E, Susztak K (2006) Poly(ADP-ribose) polymerase inhibitors ameliorate nephropathy of type 2 diabetic Leprdb/db mice. Diabetes 55:3004–3012
Szeto HH, Liu S, Soong Y, Alam N, Prusky GT, Seshan SV (2016) Protection of mitochondria prevents high-fat diet-induced glomerulopathy and proximal tubular injury. Kidney Int 90:997–1011
Takano T, Elimam H, Cybulsky AV (2013) Complement-mediated cellular injury. Semin Nephrol 33:586–601

Takeda T, McQuistan T, Orlando RA, Farquhar MG (2001) Loss of glomerular foot processes is associated with uncoupling of podocalyxin from the actin cytoskeleton. J Clin Invest 108:289–301
Tanji N, Markowitz GS, Fu C, Kislinger T, Taguchi A, Pischetsrieder M et al (2000) Expression of advanced glycation end products and their cellular receptor RAGE in diabetic nephropathy and nondiabetic renal disease. J Am Soc Nephrol 11:1656–1666
Tavasoli M, Li L, Al-Momany A, Zhu LF, Adam BA, Wang Z et al (2016) The chloride intracellular channel 5A stimulates podocyte Rac1, protecting against hypertension-induced glomerular injury. Kidney Int 89:833–847
Udani S, Lazich I, Bakris GL (2011) Epidemiology of hypertensive kidney disease. Nat Rev Nephrol 7:11–21
Velez JC, Bland AM, Arthur JM, Raymond JR, Janech MG (2007) Characterization of renin-angiotensin system enzyme activities in cultured mouse podocytes. Am J Physiol Renal Physiol 293:F398–F407
Wahl P, Ducasa GM, Fornoni A (2016) Systemic and renal lipids in kidney disease development and progression. Am J Physiol Renal Physiol 310:F433–F445
Wang G, Lai FM, Kwan BC, Lai KB, Chow KM, Li PK et al (2009) Podocyte loss in human hypertensive nephrosclerosis. Am J Hypertens 22:300–306
Wang L, Sun S, Zhou A, Yao X, Wang Y (2014) oxLDL-induced lipid accumulation in glomerular podocytes: role of IFN-gamma, CXCL16, and ADAM10. Cell Biochem Biophys 70:529–538
Welsh GI, Hale LJ, Eremina V, Jeansson M, Maezawa Y, Lennon R et al (2010) Insulin signaling to the glomerular podocyte is critical for normal kidney function. Cell Metab 12:329–340
Winn MP, Conlon PJ, Lynn KL, Farrington MK, Creazzo T, Hawkins AF et al (2005) A mutation in the TRPC6 cation channel causes familial focal segmental glomerulosclerosis. Science 308:1801–1804
Wolf G, Schanze A, Stahl RA, Shankland SJ, Amann K (2005) p27(Kip1) Knockout mice are protected from diabetic nephropathy: evidence for p27(Kip1) haplotype insufficiency. Kidney Int 68:1583–1589
Xing L, Liu Q, Fu S, Li S, Yang L, Liu S et al (2015) PTEN inhibits high glucose-induced phenotypic transition in podocytes. J Cell Biochem 116:1776–1784
Xu ZG, Yoo TH, Ryu DR, Cheon Park H, Ha SK, Han DS et al (2005) Angiotensin II receptor blocker inhibits p27Kip1 expression in glucose-stimulated podocytes and in diabetic glomeruli. Kidney Int 67:944–952
Xu ZE, Chen Y, Huang A, Varghese Z, Moorhead JF, Yan F et al (2011) Inflammatory stress exacerbates lipid-mediated renal injury in ApoE/CD36/SRA triple knockout mice. Am J Physiol Renal Physiol 301:F713–F722
Yadav A, Vallabu S, Arora S, Tandon P, Slahan D, Teichberg S et al (2010) ANG II promotes autophagy in podocytes. Am J Physiol Cell Physiol 299:C488–C496
Yamaguchi Y, Iwano M, Suzuki D, Nakatani K, Kimura K, Harada K et al (2009) Epithelial-mesenchymal transition as a potential explanation for podocyte depletion in diabetic nephropathy. Am J Kidney Dis 54:653–664
Yard BA, Kahlert S, Engelleiter R, Resch S, Waldherr R, Groffen AJ et al (2001) Decreased glomerular expression of agrin in diabetic nephropathy and podocytes, cultured in high glucose medium. Exp Nephrol 9:214–222
Ying WZ, Wang PX, Sanders PW (2000) Induction of apoptosis during development of hypertensive nephrosclerosis. Kidney Int 58:2007–2017
Yoo TH, Li JJ, Kim JJ, Jung DS, Kwak SJ, Ryu DR et al (2007) Activation of the renin-angiotensin system within podocytes in diabetes. Kidney Int 71:1019–1027
You H, Gao T, Raup-Konsavage WM, Cooper TK, Bronson SK, Reeves WB et al (2017) Podocyte-specific chemokine (C-C motif) receptor 2 overexpression mediates diabetic renal injury in mice. Kidney Int 91:671–682
Zhang H, Schin M, Saha J, Burke K, Holzman LB, Filipiak W et al (2010) Podocyte-specific overexpression of GLUT1 surprisingly reduces mesangial matrix expansion in diabetic nephropathy in mice. Am J Physiol Renal Physiol 299:F91–F98

Zhang Y, Ma KL, Liu J, Wu Y, Hu ZB, Liu L et al (2015a) Dysregulation of low-density lipoprotein receptor contributes to podocyte injuries in diabetic nephropathy. Am J Physiol Endocrinol Metab 308:E1140–E1148

Zhang Y, Ma KL, Liu J, Wu Y, Hu ZB, Liu L et al (2015b) Inflammatory stress exacerbates lipid accumulation and podocyte injuries in diabetic nephropathy. Acta Diabetol 52:1045–1056

Zhang Y, Ma KL, Ruan XZ, Liu BC (2016) Dysregulation of the low-density lipoprotein receptor pathway is involved in lipid disorder-mediated organ injury. Int J Biol Sci 12:569–579

Zuo Y, Matsusaka T, Zhong J, Ma J, Ma LJ, Hanna Z et al (2006) HIV-1 genes vpr and nef synergistically damage podocytes, leading to glomerulosclerosis. J Am Soc Nephrol 17:2832–2843

Chapter 11
How Tubular Epithelial Cell Injury Contributes to Renal Fibrosis

Bi-Cheng Liu, Tao-Tao Tang and Lin-Li Lv

Abstract The renal tubules are the major component of the kidney and are vulnerable to a variety of injuries including ischemia, proteinuria, toxins, and metabolic disorders. It has long been believed that tubules are the victim of injury. In this review, we shift this concept to renal tubules as a driving force in the progression of kidney disease. In response to injury, tubular epithelial cells (TECs) can synthesize and secrete varieties of bioactive molecules that drive interstitial inflammation and fibrosis. Innate immune-sensing receptors on the TECs also aggravate immune responses. Necroinflammation, an auto-amplification loop between tubular cell death and interstitial inflammation, leads to the exacerbation of renal injury. Furthermore, TECs also play an active role in progressive renal injury via mechanisms associated with the conversion into collagen-producing fibroblast phenotype, cell cycle arrest at both G1/S and G2/M checkpoints, and metabolic disorder. Thus, a better understanding the mechanisms by which tubular injury drives AKI and CKD is necessary for the development of therapeutics to halt the progression of CKD.

Keywords Tubular epithelial cells · Renal fibrosis · Renal inflammation · Chronic kidney disease · Acute kidney injury

11.1 Introduction

The renal tubules and tubulointerstitium make up a significant portion of the kidney and are the major sites in response to injuries. Increasing evidence shows that tubular epithelial cells (TECs) play diverse roles in renal repair or progression to chronic kidney disease (CKD). The innate immune characteristics demonstrate TECs as immune responders to a wide range of insults, with the consequent production and release of bioactive molecules that drive interstitial inflammation and fibrosis. Accumulating evidence shows that renal function decline correlates better with tubu-

B.-C. Liu (✉) · T.-T. Tang · L.-L. Lv
Institute of Nephrology, Zhong Da Hospital, Southeast University School of Medicine, Nanjing, China
e-mail: liubc64@163.com

B.-C. Liu et al. (eds.), *Renal Fibrosis: Mechanisms and Therapies*, Advances in Experimental Medicine and Biology 1165,
https://doi.org/10.1007/978-981-13-8871-2_11

lointerstitial damage than that of glomerular injury (Risdon et al. 1968; Bohle et al. 1979; Mackensen-Haen et al. 1981). Maladaptive repair of injured tubules after acute kidney injury (AKI) also leads to the progression of CKD (Ferenbach and Bonventre 2015; Venkatachalam et al. 2015). Thus, TECs should be regarded not only as victims in the context of kidney disease but also as key inflammatory and fibrogenic cells that drive the progression from acute to chronic kidney disease, which will be the focus in this review. It should be noted that due to the length limitations, this review focuses on the emerging mechanisms by which TECs play a driving role in renal injury, whereas other potentially important factors/pathways not directly related to this topic are not discussed here.

11.2 Tubule-Derived Factors Associated with Tubulointerstitial Inflammation and Fibrosis

In response to stress and injury, TECs can be transformed into a secretory phenotype, with the consequent production and release of various bioactive molecules to favor the recruitment of inflammatory cells, the activation of fibroblasts, and the loss of endothelial cells, which eventually drive tubulointerstitial inflammation and fibrosis.

11.2.1 Pro-inflammatory Cytokines

In response to renal injury, TECs become activated and can actually facilitate the inflammatory response through induction of a variety of pro-inflammatory cytokines (e.g., interleukin, tumor necrosis factor, colony stimulating factor, and growth factor). After the first report of TNF-α and IL-6 produced by TEC following IL-1 stimulation (Jevnikar et al. 1991; Yard et al. 1992), a variety of cytokines produced by activated TECs are known including IL-1β, IL-18, IL-34, IL-16, CSF-1, TWEAK, VEGF, CTGF, and so on. In TECs, NLPR3 inflammasome activation causes the release of mature IL-1β and IL-18 during kidney injury (Leemans et al. 2014; Anders 2016). Observations by Menke and Wang showed that expression of CSF-1 is upregulated in TECs during kidney injury and may be responsible for the polarization of renal macrophages and recovery from AKI (Menke et al. 2009; Wang et al. 2015b; Huen et al. 2015). Baek et al. identified that TEC-derived IL-34 plays a key role in recruiting kidney macrophages and causing persistent kidney injury and the development of CKD (Baek et al. 2015).

11.2.2 Chemokines

Chemokines are a family of small molecular cytokines with chemotactic activity. TECs are rich sources of CCL subfamily (including MCP-1/CCL2, RANTES/CCL5, and MIP-1/CCL3) and CX3CL subfamily (fractalkine/CX3CL1), which have specific effects on monocytes and monocyte-derived lineages (Chung and Lan 2011). MCP-1/CCL2 is one of the most widely studied chemokines in AKI and CKD (Wang et al. 1997, 2000; Furuichi et al. 2003). CXCL8/IL-8 and CLCL12/SDF-1 are overexpressed after TECs injury and are chemotactic for a number of leukocyte populations (Li and Nord 2002, 2009; Zuk et al. 2014). A recent study reported that CXCL5 is increased in tubular cells following the induction of nephrotoxic nephritis and is responsible for the recruitment of neutrophils during acute renal tissue injury (Disteldorf et al. 2015).

11.2.3 ROS

It has become clear that oxidative stress contributes to CKD progression via myriad effects (Small et al. 2012; Massy et al. 2009; Nie and Hou 2012). Oxidative stress implies an increased production of reactive oxygen species (ROS), including superoxide anion (O_2^-), hydrogen peroxide(H_2O_2), and hydroxyl anion (OH^-). In response to multiple stimuli and agonists, mitochondrial dysfunction and NADPH oxidases have been recognized as the major contributors to ROS generation in TECs (Tang and Lai 2012; Sedeek et al. 2013). For instance, Ang II leads to tubular hypertrophy and TECs apoptosis via ROS-dependent mechanisms (Wolf et al. 2001; Leung et al. 2011). Albumin acts through epidermal growth factor receptor to stimulate NADPH oxidase and ROS production. ROS then activates NF-κB, which then ultimately leads to activation of ERK1/ERK2 pathway (Reich et al. 2005). In addition, albumin has also been shown to stimulate tubulointerstitial inflammation via the mROS-mediated activation of Nlrp3 inflammasome (Liu et al. 2014).

11.2.4 CRP

C-reactive protein (CRP) is an acute-phase protein, which is rapidly synthesized by the liver in response to infection, inflammation, and tissue damage. Besides its use as a biomarker of inflammation, CRP has been recognized as a pathogenic mediator in diabetic kidney disease (Liu et al. 2011), obstructive nephropathy (Li et al. 2011), and AKI (Pegues et al. 2013; Tang et al. 2014; Lai et al. 2016). CRP is also inducible by high glucose in human TECs and promotes renal inflammation and fibrosis through activation of TGF-β/SMAD and NF-κB signaling pathways under diabetic conditions and unilateral ureteral obstructive nephropathy (Liu et al. 2011; Li et al. 2011). Recent studies have demonstrated that CRP promotes AKI by causing TEC G1 cell-cycle arrest via CD32-Smad3–dependent p27-driven inhibition of the cyclin-dependent kinase 2/cyclin E mechanism (Tang et al. 2014; Lai et al. 2016).

11.2.5 Growth Factors

Transforming growth factor (TGF-β), connective tissue growth factor (CTGF), platelet-derived growth factor (PDGF), and vascular endothelial growth factor (VEGF) are the best-described growth factors involved in the tubulointerstitial fibrosis, of which TGF-β derived from injured TECs has long been considered as one of the most important pro-fibrotic growth factors (Yang et al. 2010; Lan et al. 2012; Geng et al. 2012; Meng et al. 2016; Wu et al. 2013; Grande et al. 2015). Both Ang II exposure and Snail 1 overexpression can induce TGF-β1 production by TECs (Grande et al. 2015; Macconi et al. 2014). After the hypoxic injury, TECs undergo cell cycle arrest. Particularly when cells are under arrest in the G2/M phase, these cells produce large amounts of TGF-β1 (Yang et al. 2010). Increased TGF-β production by TECs can promote TIF through paracrine signaling to activate adjacent fibroblasts and pericytes transforming into myofibroblast-type cells (Wu et al. 2013; Ignotz et al. 1987; Roberts et al. 1986). Interestingly, TEC is also a target of TGF-β1. TGF-β1 can induce cultured TECs to differentiate into cells with distinct myofibroblast morphology and marked upregulation of collagen production (Zeisberg et al. 2003; Fan et al. 1999). Meanwhile, autocrine TGF-β signaling increases TEC production of PDGF-β and CTGF/CCN2 that can signal on neighboring fibroblasts (Geng et al. 2012).

11.2.6 Intrarenal RAS

Renal local renin–angiotensin system (RAS) activation plays a pivotal role in the progression of CKD. Blockade of the RAS has become the mainstay therapy for the preservation of CKD (Hou et al. 2006). Ang II is the major bioactive product of the RAS driving renal fibrosis. There is substantial evidence that the major fraction of Ang II present in renal tissues is generated from angiotensinogen (AGT) and subsequently delivered to the kidney, as well as from AGT produced by the PTECs. Ang I delivered to the kidney can also be converted to Ang II (Kobori et al. 2007). Renin mRNA and renin-like activity have been observed in cultured PTECs (Henrich et al. 1996). The brush border membrane of proximal human kidney tubules also expresses abundant levels of angiotensin-converting enzyme (ACE) mRNA (Sibony et al. 1993) and protein (Vío and Jeanneret 2003). ACE has been detected in the proximal and distal tubular fluids (Casarini et al. 1997). Therefore, all of the major components required to generate Ang II are expressed within the renal tubules (Urushihara and Kagami 2017; Kobori and Urushihara 2013). And the upregulation of these RAS components may be in a Wnt/β-catenin-dependent manner (Zhou et al. 2015). Studies have demonstrated that Ang II stimulates TGF-β expression in cultured murine PTECs and upregulates specific receptors for TGF-β to further enhance its pro-inflammatory and fibrogenic action (Wolf et al. 1993, 1999; Liu et al. 2009). Ang II is also able to induce CTGF to mediate the fibrotic phenotype change (Liu et al. 2006, 2007; Chen et al. 2006). Moreover, we also proposed the interaction of Ang II and inflammation might be the critical node in the pathogenic tubuloglomerular feedback loop (Zhang and Liu 2011).

11.2.7 Wnt and Hh

The Wnt pathway has been implicated in the epithelial repair process, but an abundance of evidence also supports Wnt/β-catenin signaling in tubulointerstitial fibrosis (Kang et al. 2016; Kawakami et al. 2013; Tan et al. 2016; Edeling et al. 2016). There are 19 Wnt ligands, and all of them can bind to Frizzled and LRP5/6 receptors at the cell surface, leading to canonical signaling through β-catenin activation (Tan et al. 2014). Wnt proteins and receptors are upregulated after renal injury, and β-catenin activity appears to be increased in injured TECs (Zhou et al. 2012; He et al. 2009). Overexpression of Wnt1 in proximal tubules is sufficient to cause TIF and activate myofibroblasts to produce ECM, suggesting paracrine signaling (Maarouf et al. 2016). It is likely that injured TECs can produce Wnt ligands which then activate the neighboring fibroblasts to promote TIF (Gewin et al. 2017).

Hedgehog (Hh) signaling is a key mammalian developmental pathway and regulates tissue patterning, cell growth, and differentiation (Cain and Rosenblum 2011; Mao et al. 2010). Of three Hh ligands (Sonic Hh [Shh], Desert Hh [Dhh], and Indian Hh [Ihh]), Shh is well studied. Lineage tracing studies indicate that Shh and Ihh expression are upregulated in renal tubules after UUO (Fabian et al. 2012; Ding et al. 2012; Zhou et al. 2014). Interstitial fibroblasts and pericytes are the cells supposed to respond to these ligands. Shh induces fibroblast activation, manifested as an expression of α-SMA, fibronectin, collagen, and desmin (Ding et al. 2012).

11.2.8 Exosomes

Exosomes are small (30–100 nm in diameter), lipid bilayer membrane vesicles of endocytic origin. They can shuttle bioactive molecules including proteins, lipids, DNA, mRNA, and microRNAs (Zhang et al. 2016; Morrison et al. 2016). In kidneys, renal exosomes are produced and secreted by kidney cells which have been implicated in renal function and diseases via cell–cell communication (Krause et al. 2015). It is known that injured TECs can release exosomes containing TGF-β mRNA to activate fibroblasts, contributing to the development of renal fibrosis in post-AKI kidneys (Borges et al. 2013). We recently demonstrated that in the setting of proteinuric kidney disease, albumin triggered TECs to release exosomes packaged with CCL-2 mRNA, which was delivered to macrophages and led to interstitial inflammation (Lv et al. 2018). In addition, we also found that the HIF-1α-dependent release of miRNA-23a-enriched exosomes from hypoxic TECs activates macrophages to promote tubulointerstitial inflammation (Li et al. 2019).

11.3 Abnormal Repair of TECs: The Central Pathology Linking AKI to CKD

An increasing number of epidemiological studies have suggested that incomplete recovery from AKI can lead to progressive CKD (Waikar and Winkelmayer 2009; Okusa et al. 2009; Hsu 2012; Coca et al. 2012). This is supported by the finding that the incomplete tubular repair is tightly associated with persistent tubulointerstitial inflammation, proliferation of fibroblasts, and excessive deposition of extracellular matrix (Yang et al. 2010; Grgic et al. 2012). A number of recent studies have also demonstrated that tubule selective injury is sufficient to drive fibrosis, inflammation, and capillary rarefaction, which is making it to be a central link between AKI and CKD (Grgic et al. 2012; Takaori et al. 2016; Zhou et al. 2014; Humphreys et al. 2013).

In general, primary tubular injuries have a very good chance of recovery. The surviving cells dedifferentiate, migrate along the basement membrane, proliferate to restore cell number, and then restore the functional integrity of the nephron (Thadhani et al. 1996). However, some damaged TECs become atrophic or gain the fibrotic phenotype after AKI. This may be tightly associated with the abnormal repair process in response to the injuries. For example, in the initial repair phase after injury, TECs may become arrested in the G2/M phase, which may be associated with the activation of JNK signaling production of pro-fibrotic cytokine (Yang et al. 2010; Ferenbach and Bonventre 2015). This is confirmed by the ability of using pharmacological inhibition of G2/M-arrested cells with histone deacetylase inhibitors or p53 inhibition to block the process of fibrosis (Cianciolo Cosentino et al. 2013; Zhou et al. 2010). Recent studies also found that aging can sensitize TECs to be arrested at the cell cycle G2/M in response to cell stress and DNA damage, which provides a potential explanation for the increased risk of CKD progression after AKI in the elderly (Ferenbach and Bonventre 2015; Verzola et al. 2008; Liu et al. 2012; Yang and Fogo 2010). In addition, CRP-induced G1/S cell cycle arrest may also contribute to progressive TIF via the Smad3-p21/p27-dependent mechanism (Tang et al. 2014; Lai et al. 2016).

Wnt/β-catenin signaling is a pathway involving the recovery from AKI. In the acute phase of injury, Wnt/β-catenin is likely to be protective. In both IRI and folic acid nephropathy, tubule-specific ablation of β-catenin has been shown to aggravate kidney injury by increasing TEC apoptosis (Zhou et al. 2012). Activation of Wnt-4/β-catenin signaling allows entry into the cell cycle via the upregulation of cyclin D1 and cyclin A, two of the most crucial proteins in regulating cell proliferation and cell cycle progression (Terada et al. 2003; Angers and Moon 2009; Clevers and Nusse 2012). Therefore, an early and appropriate activation of Wnt/β-catenin signaling is required for minimizing the initial renal damages after AKI (Zhou et al. 2016). However, persistent activation of Wnt signaling has a decisive role in driving AKI to CKD progression because sustained Wnt signaling causes uncontrolled fibroblast activation, RAS activation, inflammation, and excessive deposition of ECM (Tan et al. 2014; Xiao et al. 2016). It is well known that tissue injury and inflammation are closely linked and interact with each other (Wallach et al. 2014). While initial renal inflammation may be protective in favoring the repair process in response to AKI, unresolved and prolonged renal inflammation may cause progressive renal fibrosis. Thus, better understanding the mechanisms by which tubular injury drives interstitial inflammation and renal fibrosis is of paramount importance.

11.4 Emerging Mechanisms of Tubule Injury Driving the Progression of CKD

11.4.1 Inflammation: Innate Immune-Sensing Receptors in TECs Activation

Uncontrolled or excessive inflammatory responses can lead to progressive kidney injury. In view of the immune characteristics of TECs, substantial information indicates that Toll-like receptors (TLRs), Nod-like receptors (NLRs) and the NACHT, LRR, and PYD domain-containing protein 3 (NLRP3) inflammasome have important roles in the pathogenesis of multiple renal disorders (Leemans et al. 2014). TLRs are a family of transmembrane receptors and the signal transduction initiated from TLRs activates effector cells via several kinases and NF-κB-dependent mechanisms (Gluba et al. 2010). TLRs are widely expressed in TECs. For instance, TECs are known to express both TLR2 and TLR4, and both TLR2 and TLR4 signaling are activated during IRI (Wu et al. 2010; Allam et al. 2012; Wolfs et al. 2002), sepsis-induced AKI (El-Achkar and Dagher 2006; El-Achkar et al. 2006; Dear et al. 2006), diabetic nephropathy (Lin et al. 2012, 2013; Mudaliar et al. 2013; Devaraj et al. 2011), unilateral ureter obstruction (Pulskens et al. 2010; Leemans et al. 2009; Campbell et al. 2011; Skuginna et al. 2011). Necrotic tubular cells release high-mobility group box1 protein (HMGB1), histones, heat-shock proteins, and other DAMPs that activate TLR2 and TLR4 on renal parenchymal cells and drive inflammation (Wu et al. 2010; Allam et al. 2012; Leemans et al. 2005). We also found that albumin might serve as an endogenous DAMP to trigger the activation of TLR2-MyD88-NF-κB pathway and pro-inflammatory cytokine TNF-α and IL-6 secretion (Ding et al. 2015) (Fig. 11.1). NLRs are cytoplasmic receptors. Shigeoka and co-workers showed that Nod1 and Nod2 are present in TECs in both mouse and human kidneys and that the absence of these receptors can protect the kidney from AKI by inhibiting TEC apoptosis and inflammation (Shigeoka et al. 2010).

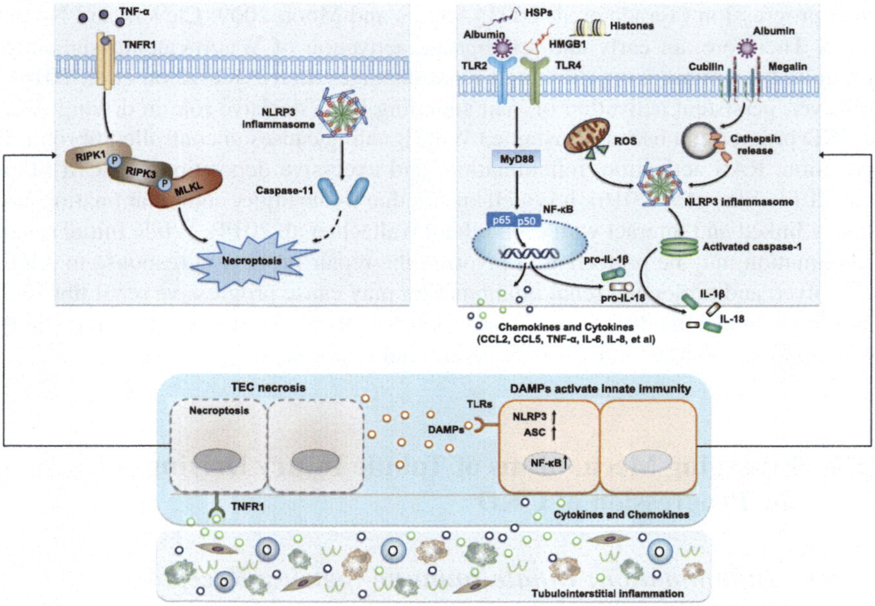

Fig. 11.1 Landscape of interstitial inflammation caused by damaged TECs. In response to injury, damaged TECs release various kinds of DAMPs that activate innate immunity through identical pattern recognition receptors including TLRs and inflammasomes, with the consequent production and release of cytokines and chemokines to recruit inflammatory cell infiltration in the interstitium, which eventually drive interstitial inflammation and fibrosis. Some injury factors can also be seen as DAMPs (such as albumin). In addition, TNF-α and possibly other cytokines drive necroptosis as a secondary cell death category contributing to tubular necrosis and renal dysfunction. This sets up the auto-amplification loop of necroinflammation

In addition, emerging evidence suggests an important role for NLRP3 inflammasome and IL-1β/IL-18 in the pathogenesis of acute and chronic inflammation and tissue remodeling in the kidney (Anders and Muruve 2011; Chang et al. 2014) (Fig. 11.1). Upregulation of the NLRP3 inflammasome is demonstrated in both classical immune cells as well as in TECs in a wide variety of tubulointerstitial disease (Anders and Muruve 2011; Chang et al. 2014). We recently found that proteinuria causes NLRP3 inflammasome activation and IL-1β/IL-18 maturation in a time course and dose-dependent manner in the proximal tubules (Liu et al. 2014). Further investigation indicated that megalin/cubilin-mediated albumin retention and lysosomal rupture are involved in the activation of NLRP3 inflammasome and interstitial inflammation (Liu et al. 2015). Moreover, Ang II has also been shown to induce NLRP3 inflammasome activation in TECs, which is associated with mitochondrial dysfunction or ER stress (Wang et al. 2015a; Wen et al. 2016). Thus, activation of the inflammasome pathway may represent a new mechanism of tubulointerstitial inflammation.

11.4.2 Necroinflammation: An Auto-Amplification Loop Between Tubular Injury and Tubulointerstitial Inflammation

Necroinflammation is a new pathological auto-amplification loop driven by necrosis (defined by cell death involving rupture of the plasma membrane) and inflammation (defined by cytokine release, increased vascular permeability, and recruitment of immune effector cells) (Linkermann et al. 2014; Mulay et al. 2016b). Following this pathological process, ischemia, toxins, and proteinuria can trigger tubulointerstitial inflammation, and in turn, tubulointerstitial inflammation causes TECs injury, which leads to an aggravation of interstitial inflammation (Fig. 11.1).

How TECs necrosis induces tubulointerstitial inflammation? In the last decade, it was unraveled that injured cells release DAMPs that activate innate immunity through identical pattern recognition receptors including TLRs and inflammasomes (Anders and Schaefer 2014). As mentioned above, this process is also involved in kidney inflammation and immunopathology (Anders and Muruve 2011; Anders et al. 2004; Anders 2010). AKI is most frequently associated with cell necrosis that implies DAMPs release. For example, ischemic, septic, or toxic forms of tubular necrosis can induce HMGB1, histones, heat-shock proteins, and other DAMPs release, which activate TLR2 and TLR4 on renal parenchymal cells and inflammatory cells to drive inflammation (Allam et al. 2012; Leelahavanichkul et al. 2011; Rabadi et al. 2012; Wu et al. 2010; Arumugam et al. 2009). Deficiency of receptor-interacting protein kinase 3 (RIPK3) or mixed lineage kinase domain-like (MLKL), two core proteins of the necroptosis pathway, blocks oxalate crystal-induced AKI and inflammation (Mulay et al. 2016a).

How tubulointerstitial inflammation induces TECs necrosis? DAMPs released by dying cells activate the pattern recognition receptors of infiltrating immune cells and intrinsic renal parenchymal cells and induce the release of numerous pro-inflammatory mediators. In particular, TNF-α and IFN-γ can induce necroptosis via two distinct pathways (Dannappel et al. 2014; Takahashi et al. 2014; Vanden Berghe et al. 2014). Mulay et al. showed that oxalate crystal formation inside tubules induced TNF-α secretion, which could activate the RIPK1, RIPK3, and MLKL pathway of necroptosis via TNFR1. And blocking either TNF-α or TNFR1 could abrogate kidney injury and dysfunction (Mulay et al. 2016a). Furthermore, the NLRP3 inflammasome activation not only triggers cytokine release but also pyroptosis, as a consequence of inflammasome-driven caspase-11 activation (Bergsbaken et al. 2009; Case et al. 2013). But if pyroptosis can occur in TECs is under debate (Krautwald and Linkermann 2014; Yang et al. 2014) (Fig. 11.1).

11.4.3 Partial Epithelial–Mesenchymal Transition (EMT)

TECs might directly contribute to renal fibrosis via EMT, a phenotypic conversion program that is characterized by the loss of epithelial markers (such as E-cadherin, zonula occludens-1 [ZO-1] and cytokeratin) and gain of mesenchymal features (including vimentin, α-smooth muscle actin [α-SMA], fibroblast-specific protein-1 [FSP1], interstitial matrix components type I collagen, and fibronectin) (Liu 2004; Strutz 2009). Historical data and recent new findings have suggested that renal fibrosis might occur as a result of the tubular epithelial cells injury. In response to this, TECs produce various chemokines and cytokines around peritubular compartments to attract and direct the influx of inflammatory cells to the tubulointerstitial space. Infiltrating cells in turn activate and produce a mixture of soluble factors, including pro-inflammatory, pro-fibrotic cytokines, and MMPs. Altered microenvironment contributes to the reshaping of the mesenchymal cell phenotype, and rendering TECs adaptable to changing cell phenotype for the sake of escaping apoptosis (Prunotto et al. 2012; Liu 2010).

However, the precise contribution of the EMT to kidney fibrosis remains a subject of debate, as studies using genetic cell lineage tracing could not find evidence of a direct contribution of epithelial cells to the myofibroblast population in the fibrotic kidney (Humphreys et al. 2010). Two studies recently addressed this dispute and offered new insights into the potential role of tubular EMT in the development and progression of renal fibrosis (Ovadya and Krizhanovsky 2015; Zhou and Liu 2016). The transcription factors Snail 1 and Twist are the main regulators of the EMT program. Grande et al. (2015) focus on Snail 1, whereas Lovisa et al. (2015) carried out experiments with both Snail 1 and Twist. By conditional deletion of Snail 1 or Twist in TECs, the EMT is specifically inhibited. As a result, fibrosis is reduced in several CKD models, including unilateral ureter obstruction, nephrotoxic serum-induced nephritis, and folic acid-induced nephropathy. And improvement of renal fibrosis also led to the preservation of tubular cell integrity and function. Interestingly, both studies found that TECs undergo incomplete EMT during renal fibrosis—the cells express markers of both epithelial and mesenchymal cells and remain associated with their basement membrane. In this respect, these observations are in harmony with earlier genetic cell lineage tracing studies and demonstrate that partial EMT is sufficient to induce tubular function impairment, triggering cell cycle arrest, and promoting the release of critical fibrogenic cytokines, although evidence for partial EMT in human CKD is rare.

11.4.4 Cell Cycle Arrest

A series of elegant studies have identified that G1/S and G2/M arrest in TECs is an important driver of maladaptive TECs repair and renal fibrosis, providing a link between AKI and CKD (Yang et al. 2010; Cianciolo Cosentino et al. 2013; Tang et al. 2013). Yang et al. demonstrated a causal association between epithelial cell cycle G2/M arrest and a fibrotic outcome in toxic and obstructive models of AKI. G2/M-arrested PTECs activate JNK signaling, which acts to upregulate pro-fibrotic cytokine (TGF-β1 and CTGF) production (Yang et al. 2010). Canaud et al. further identified PTECs in the G2/M phase form target of rapamycin–autophagy spatial coupling compartments, which facilitate pro-fibrotic secretion similar to the senescence-associated secretory phenotype (Canaud et al. 2019). Targeting the G2/M checkpoint to maintain the proper progression of TECs through the cell cycle during the injury phase has been proposed as an attractive therapeutic target to prevent the progression of CKD (Canaud and Bonventre 2015). Cianciolo Cosentino et al. reported that a histone acetylase inhibitor could reduce the number of cells in G2/M arrest and reduce post-injury tubular atrophy and interstitial fibrosis (Cianciolo Cosentino et al. 2013). Jenkins et al. suggested that miR-192 has an important role in aristolochic acid-induced G2/M arrest (Jenkins et al. 2014). Interestingly, the induction of a transient G0/G1 arrest in TECs with the CDK4/6 inhibitor PD0332991 before IRI ameliorated kidney injury by preventing apoptosis and pro-fibrotic cytokine production (DiRocco et al. 2014).

As previously discussed, the functional consequences of EMT during fibrotic injury are the induction of the G2 phase arrest of TECs (Lovisa et al. 2015). Genetic inhibition of EMT by knocking out Twist and Snail 1, resulted in a substantial decrease in the G2/M-arrested TECs. In vitro induction of EMT with TGF- β1 also induced G2/M arrest in TECs (Wu et al. 2013; Lovisa et al. 2015). Furthermore, it was found that the G2 arrest was mediated by the cell cycle inhibitor p21 (Lovisa et al. 2015). And it is in line with a finding that p21 in kidney proximal tubules mediates fibrosis (Megyesi et al. 2015).

11.4.5 Metabolic Disorder

The intracellular accumulation of excess non-esterified fatty acid (NEFA) and metabolites in TECs, namely lipotoxicity, can result in renal dysfunction, especially in the context of diabetic nephropathy (Schelling 2016; Kimmelstiel and Wilson 1936; Oliver et al. 1954; Herman-Edelstein et al. 2014). Several groups have shown that proximal tubule uptake of filtered NEFAs is the source of tubular toxicity in case of glomerular damage. Tubulointerstitial damage can be induced in rats by infusion of NEFA-loaded albumin and in vitro incubation with albumin-bound NEFAs stimulate PTEC apoptosis (Thomas et al. 2002; Kamijo et al. 2002; van Timmeren et al. 2005). Tubular cells have a high level of energy demand and the ATP that they use is mostly produced by fatty acid oxidation. New findings indicate that dysregulation of fatty acid oxidation followed intracellular lipid accumulation profoundly affects the fate of TECs, by promoting EMT, inflammation, and eventually interstitial fibrosis (Kang et al. 2015). They also investigated the mechanisms behind the depressed metabolic pathways in fibrotic kidney disease and further demonstrated that TGF-β1 inhibits the expression of carnitine palmitoyltransferase 1 (CPT1), the rate-limiting

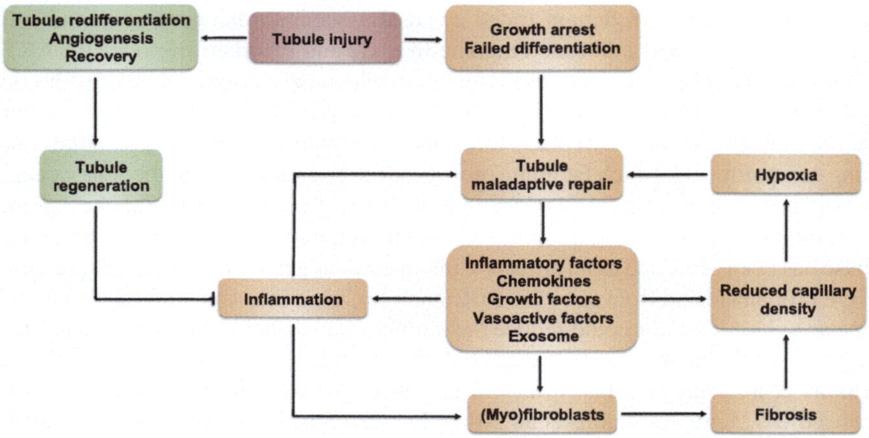

Fig. 11.2 Schematic diagram illustrating cycle feedback interactions between tubule pathology and interstitial pathology

enzyme in FAO, and thereby decreases fatty acid metabolism (Kang et al. 2015). Furthermore, miR-21 is shown to be implicated in the regulation of metabolic pathways recently (Trionfini et al. 2015; Chau et al. 2012). miR-21 promotes tubular injury and fibrosis by downregulating PPARα, with consequent alterations of TEC lipid metabolism. Inhibition of miR-21 reduces TGF-β-induced fibrogenesis and inflammation, preserves tubular integrity, as a result of enhanced PPARα/RXR activity and improved mitochondrial function (Gomez et al. 2015).

11.5 Conclusion

In this review, we shift TECs from the victim of injury to a driving force in the progression from AKI to CKD. Damaged TECs can contribute directly to interstitial inflammation and fibrosis through various kinds of mechanisms (Fig. 11.2). Thus, protecting tubules from repeated injury and restoring healthy tubular function may be the priority of treatment of kidney diseases. Although the mechanisms of tubular injury remain to be elucidated, the G1/S and G2/M cell cycle arrest may be a pivotal obstacle to the adaptive repair of injured TECs and targeting the G1/S and G2/M checkpoint to maintain the proper cell cycle transition may be an attractive therapeutic target to prevent the progression of CKD.

Acknowledgements This chapter was modified from a paper reported by our group in Kidney Int (Liu et al. 2018). The related contents are reused with permission.
This study was supported by grants from the National Key Research and Development Program of China (2018YFC1314004), the National Natural Scientific Foundation (No. 81720108007, 81130010, 81470997, and 81670696), and the Clinic Research Center of Jiangsu Province (No. BL2014080).

References

Allam R, Scherbaum CR, Darisipudi MN, Mulay SR, Hägele H, Lichtnekert J et al (2012) Histones from dying renal cells aggravate kidney injury via TLR2 and TLR4. J Am Soc Nephrol 23:1375–1388

Anders HJ (2010) Toll-like receptors and danger signaling in kidney injury. J Am Soc Nephrol 21:1270–1274

Anders HJ (2016) Of Inflammasomes and alarmins: IL-1β and IL-1α in kidney disease. J Am Soc Nephrol 27:2564–2575

Anders HJ, Muruve DA (2011) The inflammasomes in kidney disease. J Am Soc Nephrol 22:1007–1018

Anders HJ, Schaefer L (2014) Beyond tissue injury-damage-associated molecular patterns, toll-like receptors, and inflammasomes also drive regeneration and fibrosis. J Am Soc Nephrol 25:1387–1400

Anders HJ, Banas B, Schlöndorff D (2004) Signaling danger: toll-like receptors and their potential roles in kidney disease. J Am Soc Nephrol 15:854–867

Angers S, Moon RT (2009) Proximal events in Wnt signal transduction. Nat Rev Mol Cell Biol 10:468–477

Arumugam TV, Okun E, Tang SC, Thundyil J, Taylor SM, Woodruff TM (2009) Toll-like receptors in ischemia-reperfusion injury. Shock 32:4–16

Baek JH, Zeng R, Weinmann-Menke J, Valerius MT, Wada Y, Ajay AK et al (2015) IL-34 mediates acute kidney injury and worsens subsequent chronic kidney disease. J Clin Invest 125:3198–3214

Bergsbaken T, Fink SL, Cookson BT (2009) Pyroptosis: host cell death and inflammation. Nat Rev Microbiol 7:99–109

Bohle A, Christ H, Grund KE, Mackensen S (1979) The role of the interstitium of the renal cortex in renal disease. Contrib Nephrol 16:109–114

Borges FT, Melo SA, Özdemir BC, Kato N, Revuelta I, Miller CA et al (2013) TGF-β1-containing exosomes from injured epithelial cells activate fibroblasts to initiate tissue regenerative responses and fibrosis. J Am Soc Nephrol 24:385–392

Cain JE, Rosenblum ND (2011) Control of mammalian kidney development by the Hedgehog signaling pathway. Pediatr Nephrol 26:1365–1371

Campbell MT, Hile KL, Zhang H, Asanuma H, Vanderbrink BA, Rink RR et al (2011) Toll-like receptor 4: a novel signaling pathway during renal fibrogenesis. J Surg Res 168:e61–e69

Canaud G, Bonventre JV (2015) Cell cycle arrest and the evolution of chronic kidney disease from acute kidney injury. Nephrol Dial Transplant 30:575–583

Canaud G, Brooks CR, Kishi S, Taguchi K, Nishimura K, Magassa S et al (2019) Cyclin G1 and TASCC regulate kidney epithelial cell G2-M arrest and fibrotic maladaptive repair. Sci Transl Med 11

Casarini DE, Boim MA, Stella RC, Krieger-Azzolini MH, Krieger JE, Schor N (1997) Angiotensin I-converting enzyme activity in tubular fluid along the rat nephron. Am J Physiol 272:F405–F409

Case CL, Kohler LJ, Lima JB, Strowig T, de Zoete MR, Flavell RA et al (2013) Caspase-11 stimulates rapid flagellin-independent pyroptosis in response to *Legionella pneumophila*. Proc Natl Acad Sci U S A 110:1851–1856

Chang A, Ko K, Clark MR (2014) The emerging role of the inflammasome in kidney diseases. Curr Opin Nephrol Hypertens 23:204–210

Chau BN, Xin C, Hartner J, Ren S, Castano AP, Linn G et al (2012) MicroRNA-21 promotes fibrosis of the kidney by silencing metabolic pathways. Sci Transl Med 4:121ra18

Chen L, Liu BC, Zhang XL, Zhang JD, Liu H, Li MX (2006) Influence of connective tissue growth factor antisense oligonucleotide on angiotensin II-induced epithelial mesenchymal transition in HK2 cells. Acta Pharmacol Sin 27:1029–1036

Chung AC, Lan HY (2011) Chemokines in renal injury. J Am Soc Nephrol 22:802–809

Cianciolo Cosentino C, Skrypnyk NI, Brilli LL, Chiba T, Novitskaya T, Woods C et al (2013) Histone deacetylase inhibitor enhances recovery after AKI. J Am Soc Nephrol 24:943–953

Clevers H, Nusse R (2012) Wnt/β-catenin signaling and disease. Cell 149:1192–1205
Coca SG, Singanamala S, Parikh CR (2012) Chronic kidney disease after acute kidney injury: a systematic review and meta-analysis. Kidney Int 81:442–448
Dannappel M, Vlantis K, Kumari S, Polykratis A, Kim C, Wachsmuth L et al (2014) RIPK1 maintains epithelial homeostasis by inhibiting apoptosis and necroptosis. Nature 513:90–94
Dear JW, Yasuda H, Hu X, Hieny S, Yuen PS, Hewitt SM et al (2006) Sepsis-induced organ failure is mediated by different pathways in the kidney and liver: acute renal failure is dependent on MyD88 but not renal cell apoptosis. Kidney Int 69:832–836
Devaraj S, Tobias P, Kasinath BS, Ramsamooj R, Afify A, Jialal I (2011) Knockout of toll-like receptor-2 attenuates both the proinflammatory state of diabetes and incipient diabetic nephropathy. Arterioscler Thromb Vasc Biol 31:1796–1804
Ding H, Zhou D, Hao S, Zhou L, He W, Nie J et al (2012) Sonic hedgehog signaling mediates epithelial-mesenchymal communication and promotes renal fibrosis. J Am Soc Nephrol 23:801–813
Ding LH, Liu D, Xu M, Wu M, Liu H, Tang RN et al (2015) TLR2-MyD88-NF-κB pathway is involved in tubulointerstitial inflammation caused by proteinuria. Int J Biochem Cell Biol 69:114–120
DiRocco DP, Bisi J, Roberts P, Strum J, Wong KK, Sharpless N et al (2014) CDK4/6 inhibition induces epithelial cell cycle arrest and ameliorates acute kidney injury. Am J Physiol Renal Physiol 306:F379–F388
Disteldorf EM, Krebs CF, Paust HJ, Turner JE, Nouailles G, Tittel A et al (2015) CXCL5 drives neutrophil recruitment in TH17-mediated GN. J Am Soc Nephrol 26:55–66
Edeling M, Ragi G, Huang S, Pavenstädt H, Susztak K (2016) Developmental signalling pathways in renal fibrosis: the roles of Notch, Wnt and Hedgehog. Nat Rev Nephrol 12:426–439
El-Achkar TM, Dagher PC (2006) Renal Toll-like receptors: recent advances and implications for disease. Nat Clin Pract Nephrol 2:568–581
El-Achkar TM, Huang X, Plotkin Z, Sandoval RM, Rhodes GJ, Dagher PC (2006) Sepsis induces changes in the expression and distribution of Toll-like receptor 4 in the rat kidney. Am J Physiol Renal Physiol 290:F1034–F1043
Fabian SL, Penchev RR, St-Jacques B, Rao AN, Sipilä P, West KA et al (2012) Hedgehog-Gli pathway activation during kidney fibrosis. Am J Pathol 180:1441–1453
Fan JM, Ng YY, Hill PA, Nikolic-Paterson DJ, Mu W, Atkins RC et al (1999) Transforming growth factor-beta regulates tubular epithelial-myofibroblast transdifferentiation in vitro. Kidney Int 56:1455–1467
Ferenbach DA, Bonventre JV (2015) Mechanisms of maladaptive repair after AKI leading to accelerated kidney ageing and CKD. Nat Rev Nephrol 11:264–276
Furuichi K, Wada T, Iwata Y, Kitagawa K, Kobayashi K, Hashimoto H et al (2003) CCR39 signaling contributes to ischemia-reperfusion injury in kidney. J Am Soc Nephrol 14:2503–2515
Geng H, Lan R, Singha PK, Gilchrist A, Weinreb PH, Violette SM et al (2012) Lysophosphatidic acid increases proximal tubule cell secretion of profibrotic cytokines PDGF-B and CTGF through LPA2- and Gαq-mediated Rho and αvβ6 integrin-dependent activation of TGF-β. Am J Pathol 181:1236–1249
Gewin L, Zent R, Pozzi A (2017) Progression of chronic kidney disease: too much cellular talk causes damage. Kidney Int 91:552–560
Gluba A, Banach M, Hannam S, Mikhailidis DP, Sakowicz A, Rysz J (2010) The role of Toll-like receptors in renal diseases. Nat Rev Nephrol 6:224–235
Gomez IG, MacKenna DA, Johnson BG, Kaimal V, Roach AM, Ren S et al (2015) Anti-microRNA-21 oligonucleotides prevent Alport nephropathy progression by stimulating metabolic pathways. J Clin Invest 125:141–156
Grande MT, Sánchez-Laorden B, López-Blau C, De Frutos CA, Boutet A, Arévalo M et al (2015) Snail1-induced partial epithelial-to-mesenchymal transition drives renal fibrosis in mice and can be targeted to reverse established disease. Nat Med 21:989–997

Grgic I, Campanholle G, Bijol V, Wang C, Sabbisetti VS, Ichimura T et al (2012) Targeted proximal tubule injury triggers interstitial fibrosis and glomerulosclerosis. Kidney Int 82:172–183

He W, Dai C, Li Y, Zeng G, Monga SP, Liu Y (2009) Wnt/beta-catenin signaling promotes renal interstitial fibrosis. J Am Soc Nephrol 20:765–776

Henrich WL, McAllister EA, Eskue A, Miller T, Moe OW (1996) Renin regulation in cultured proximal tubular cells. Hypertension 27:1337–1340

Herman-Edelstein M, Scherzer P, Tobar A, Levi M, Gafter U (2014) Altered renal lipid metabolism and renal lipid accumulation in human diabetic nephropathy. J Lipid Res 55:561–572

Hou FF, Zhang X, Zhang GH, Xie D, Chen PY, Zhang WR et al (2006) Efficacy and safety of benazepril for advanced chronic renal insufficiency. N Engl J Med 354:131–140

Hsu CY (2012) Yes, AKI truly leads to CKD. J Am Soc Nephrol 23:967–969

Huen SC, Huynh L, Marlier A, Lee Y, Moeckel GW, Cantley LG (2015) GM-CSF promotes macrophage alternative activation after renal ischemia/reperfusion injury. J Am Soc Nephrol 26:1334–1345

Humphreys BD, Lin SL, Kobayashi A, Hudson TE, Nowlin BT, Bonventre JV et al (2010) Fate tracing reveals the pericyte and not epithelial origin of myofibroblasts in kidney fibrosis. Am J Pathol 176:85–97

Humphreys BD, Xu F, Sabbisetti V, Grgic I, Movahedi Naini S, Wang N et al (2013) Chronic epithelial kidney injury molecule-1 expression causes murine kidney fibrosis. J Clin Invest 123:4023–4035

Ignotz RA, Endo T, Massagué J (1987) Regulation of fibronectin and type I collagen mRNA levels by transforming growth factor-beta. J Biol Chem 262:6443–6446

Jenkins RH, Davies LC, Taylor PR, Akiyama H, Cumbes B, Beltrami C et al (2014) miR-192 induces G2/M growth arrest in aristolochic acid nephropathy. Am J Pathol 184:996–1009

Jevnikar AM, Brennan DC, Singer GG, Heng JE, Maslinski W, Wuthrich RP et al (1991) Stimulated kidney tubular epithelial cells express membrane associated and secreted TNF alpha. Kidney Int 40:203–211

Kamijo A, Kimura K, Sugaya T, Yamanouchi M, Hase H, Kaneko T et al (2002) Urinary free fatty acids bound to albumin aggravate tubulointerstitial damage. Kidney Int 62:1628–1637

Kang HM, Ahn SH, Choi P, Ko YA, Han SH, Chinga F et al (2015) Defective fatty acid oxidation in renal tubular epithelial cells has a key role in kidney fibrosis development. Nat Med 21:37–46

Kang HM, Huang S, Reidy K, Han SH, Chinga F, Susztak K (2016) Sox9-Positive progenitor cells play a key role in renal tubule epithelial regeneration in mice. Cell Rep 14:861–871

Kawakami T, Ren S, Duffield JS (2013) Wnt signalling in kidney diseases: dual roles in renal injury and repair. J Pathol 229:221–231

Kimmelstiel P, Wilson C (1936) Intercapillary lesions in the glomeruli of the kidney. Am J Pathol 12:83–98

Kobori H, Urushihara M (2013) Augmented intrarenal and urinary angiotensinogen in hypertension and chronic kidney disease. Pflugers Arch 465:3–12

Kobori H, Nangaku M, Navar LG, Nishiyama A (2007) The intrarenal renin-angiotensin system: from physiology to the pathobiology of hypertension and kidney disease. Pharmacol Rev 59:251–287

Krause M, Samoylenko A, Vainio SJ (2015) Exosomes as renal inductive signals in health and disease, and their application as diagnostic markers and therapeutic agents. Front Cell Dev Biol 3:65

Krautwald S, Linkermann A (2014) The fire within: pyroptosis in the kidney. Am J Physiol Renal Physiol 306:F168–F169

Lai W, Tang Y, Huang XR, Ming-Kuen Tang P, Xu A, Szalai AJ et al (2016) C-reactive protein promotes acute kidney injury via Smad3-dependent inhibition of CDK2/cyclin E. Kidney Int 90:610–626

Lan R, Geng H, Polichnowski AJ, Singha PK, Saikumar P, McEwen DG et al (2012) PTEN loss defines a TGF-β-induced tubule phenotype of failed differentiation and JNK signaling during renal fibrosis. Am J Physiol Renal Physiol 302:F1210–F1223

Leelahavanichkul A, Huang Y, Hu X, Zhou H, Tsuji T, Chen R et al (2011) Chronic kidney disease worsens sepsis and sepsis-induced acute kidney injury by releasing High Mobility Group Box Protein-1. Kidney Int 80:1198–1211

Leemans JC, Stokman G, Claessen N, Rouschop KM, Teske GJ, Kirschning CJ et al (2005) Renal-associated TLR2 mediates ischemia/reperfusion injury in the kidney. J Clin Invest 115:2894–2903

Leemans JC, Butter LM, Pulskens WP, Teske GJ, Claessen N, van der Poll T et al (2009) The role of Toll-like receptor 2 in inflammation and fibrosis during progressive renal injury. PLoS ONE 4:e5704

Leemans JC, Kors L, Anders HJ, Florquin S (2014) Pattern recognition receptors and the inflammasome in kidney disease. Nat Rev Nephrol 10:398–414

Leung JC, Chan LY, Tang SC, Lam MF, Chow CW, Lim AI et al (2011) Oxidative damages in tubular epithelial cells in IgA nephropathy: role of crosstalk between angiotensin II and aldosterone. J Transl Med 9:169

Li H, Nord EP (2002) CD40 ligation stimulates MCP-1 and IL-8 production, TRAF6 recruitment, and MAPK activation in proximal tubule cells. Am J Physiol Renal Physiol 282:F1020–F1033

Li H, Nord EP (2009) IL-8 amplifies CD40/CD154-mediated ICAM-1 production via the CXCR-1 receptor and p38-MAPK pathway in human renal proximal tubule cells. Am J Physiol Renal Physiol 296:F438–F445

Li ZI, Chung AC, Zhou L, Huang XR, Liu F, Fu P et al (2011) C-reactive protein promotes acute renal inflammation and fibrosis in unilateral ureteral obstructive nephropathy in mice. Lab Invest 91:837–851

Li ZL, Lv LL, Tang TT, Wang B, Feng Y, Zhou LT et al (2019) HIF-1α inducing exosomal microRNA-23a expression mediates the cross-talk between tubular epithelial cells and macrophages in tubulointerstitial inflammation. Kidney Int 95:388–404

Lin M, Yiu WH, Wu HJ, Chan LY, Leung JC, Au WS et al (2012) Toll-like receptor 4 promotes tubular inflammation in diabetic nephropathy. J Am Soc Nephrol 23:86–102

Lin M, Yiu WH, Li RX, Wu HJ, Wong DW, Chan LY et al (2013) The TLR4 antagonist CRX-526 protects against advanced diabetic nephropathy. Kidney Int 83:887–900

Linkermann A, Stockwell BR, Krautwald S, Anders HJ (2014) Regulated cell death and inflammation: an auto-amplification loop causes organ failure. Nat Rev Immunol 14:759–767

Liu Y (2004) Epithelial to mesenchymal transition in renal fibrogenesis: pathologic significance, molecular mechanism, and therapeutic intervention. J Am Soc Nephrol 15:1–12

Liu Y (2010) New insights into epithelial-mesenchymal transition in kidney fibrosis. J Am Soc Nephrol 21:212–222

Liu BC, Chen L, Sun J, Huang HQ, Ma KL, Liu H et al (2006) Connective tissue growth factor-mediated angiotensin II-induced hypertrophy of proximal tubular cells. Nephron Exp Nephrol 103:e16–e26

Liu XC, Liu BC, Zhang XL, Li MX, Zhang JD (2007) Role of ERK1/2 and PI3-K in the regulation of CTGF-induced ILK expression in HK-2 cells. Clin Chim Acta 382:89–94

Liu BC, Gao J, Li Q, Xu LM (2009) Albumin caused the increasing production of angiotensin II due to the dysregulation of ACE/ACE2 expression in HK2 cells. Clin Chim Acta 403:23–30

Liu F, Chen HY, Huang XR, Chung AC, Zhou L, Fu P et al (2011) C-reactive protein promotes diabetic kidney disease in a mouse model of type 1 diabetes. Diabetologia 54:2713–2723

Liu J, Yang JR, He YN, Cai GY, Zhang JG, Lin LR et al (2012) Accelerated senescence of renal tubular epithelial cells is associated with disease progression of patients with immunoglobulin A (IgA) nephropathy. Transl Res 159:454–463

Liu D, Xu M, Ding LH, Lv LL, Liu H, Ma KL et al (2014) Activation of the Nlrp3 inflammasome by mitochondrial reactive oxygen species: a novel mechanism of albumin-induced tubulointerstitial inflammation. Int J Biochem Cell Biol 57:7–19

Liu D, Wen Y, Tang TT, Lv LL, Tang RN, Liu H et al (2015) Megalin/cubulin-lysosome-mediated albumin reabsorption is involved in the tubular cell activation of NLRP3 inflammasome and tubulointerstitial inflammation. J Biol Chem 290:18018–18028

Liu BC, Tang TT, Lv LL, Lan HY (2018) Renal tubule injury: a driving force toward chronic kidney disease. Kidney Int 93:568–579

Lovisa S, LeBleu VS, Tampe B, Sugimoto H, Vadnagara K, Carstens JL et al (2015) Epithelial-to-mesenchymal transition induces cell cycle arrest and parenchymal damage in renal fibrosis. Nat Med 21:998–1009

Lv LL, Feng Y, Wen Y, Wu WJ, Ni HF, Li ZL et al (2018) Exosomal CCL2 from tubular epithelial cells is critical for albumin-induced tubulointerstitial inflammation. J Am Soc Nephrol 29:919–935

Maarouf OH, Aravamudhan A, Rangarajan D, Kusaba T, Zhang V, Welborn J et al (2016) Paracrine Wnt1 drives interstitial fibrosis without inflammation by tubulointerstitial cross-talk. J Am Soc Nephrol 27:781–790

Macconi D, Remuzzi G, Benigni A (2014) Key fibrogenic mediators: old players. Renin-angiotensin system. Kidney Int Suppl 4:58–64

Mackensen-Haen S, Bader R, Grund KE, Bohle A (1981) Correlations between renal cortical interstitial fibrosis, atrophy of the proximal tubules and impairment of the glomerular filtration rate. Clin Nephrol 15:167–171

Mao J, Kim BM, Rajurkar M, Shivdasani RA, McMahon AP (2010) Hedgehog signaling controls mesenchymal growth in the developing mammalian digestive tract. Development 137:1721–1729

Massy ZA, Stenvinkel P, Drueke TB (2009) The role of oxidative stress in chronic kidney disease. Semin Dial 22:405–408

Megyesi J, Tarcsafalvi A, Li S, Hodeify R, Seng NS, Portilla D et al (2015) Increased expression of p21WAF1/CIP1 in kidney proximal tubules mediates fibrosis. Am J Physiol Renal Physiol 308:F122–F130

Meng XM, Nikolic-Paterson DJ, Lan HY (2016) TGF-β: the master regulator of fibrosis. Nat Rev Nephrol 12:325–338

Menke J, Iwata Y, Rabacal WA, Basu R, Yeung YG, Humphreys BD et al (2009) CSF-1 signals directly to renal tubular epithelial cells to mediate repair in mice. J Clin Invest 119:2330–2342

Morrison EE, Bailey MA, Dear JW (2016) Renal extracellular vesicles: from physiology to clinical application. J Physiol 594:5735–5748

Mudaliar H, Pollock C, Komala MG, Chadban S, Wu H, Panchapakesan U (2013) The role of Toll-like receptor proteins (TLR) 2 and 4 in mediating inflammation in proximal tubules. Am J Physiol Renal Physiol 305:F143–F154

Mulay SR, Desai J, Kumar SV, Eberhard JN, Thomasova D, Romoli S et al (2016a) Cytotoxicity of crystals involves RIPK3-MLKL-mediated necroptosis. Nat Commun 7:10274

Mulay SR, Linkermann A, Anders HJ (2016b) Necroinflammation in kidney disease. J Am Soc Nephrol 27:27–39

Nie J, Hou FF (2012) Role of reactive oxygen species in the renal fibrosis. Chin Med J (Engl) 125:2598–2602

Okusa MD, Chertow GM, Portilla D, Acute Kidney Injury Advisory Group of the American Society of Nephrology (2009) The nexus of acute kidney injury, chronic kidney disease, and World Kidney Day 2009. Clin J Am Soc Nephrol 4:520–522

Oliver J, MacDowell M, Lee YC (1954) Cellular mechanisms of protein metabolism in the nephron. I. The structural aspects of proteinuria; tubular absorption, droplet formation, and the disposal of proteins. J Exp Med 99:589–604

Ovadya Y, Krizhanovsky V (2015) A new twist in kidney fibrosis. Nat Med 21:975–977

Pegues MA, McCrory MA, Zarjou A, Szalai AJ (2013) C-reactive protein exacerbates renal ischemia-reperfusion injury. Am J Physiol Renal Physiol 304:F1358–F1365

Prunotto M, Budd DC, Gabbiani G, Meier M, Formentini I, Hartmann G et al (2012) Epithelial-mesenchymal crosstalk alteration in kidney fibrosis. J Pathol 228:131–147

Pulskens WP, Rampanelli E, Teske GJ, Butter LM, Claessen N, Luirink IK et al (2010) TLR4 promotes fibrosis but attenuates tubular damage in progressive renal injury. J Am Soc Nephrol 21:1299–1308

Rabadi MM, Ghaly T, Goligorksy MS, Ratliff BB (2012) HMGB1 in renal ischemic injury. Am J Physiol Renal Physiol 303:F873–F885

Reich H, Tritchler D, Herzenberg AM, Kassiri Z, Zhou X, Gao W et al (2005) Albumin activates ERK via EGF receptor in human renal epithelial cells. J Am Soc Nephrol 16:1266–1278

Risdon RA, Sloper JC, De Wardener HE (1968) Relationship between renal function and histological changes found in renal-biopsy specimens from patients with persistent glomerular nephritis. Lancet 2:363–366

Roberts AB, Sporn MB, Assoian RK, Smith JM, Roche NS, Wakefield LM et al (1986) Transforming growth factor type beta: rapid induction of fibrosis and angiogenesis in vivo and stimulation of collagen formation in vitro. Proc Natl Acad Sci USA 83:4167–4171

Schelling JR (2016) Tubular atrophy in the pathogenesis of chronic kidney disease progression. Pediatr Nephrol 31:693–706

Sedeek M, Nasrallah R, Touyz RM, Hébert RL (2013) NADPH oxidases, reactive oxygen species, and the kidney: friend and foe. J Am Soc Nephrol 24:1512–1518

Shigeoka AA, Kambo A, Mathison JC, King AJ, Hall WF, da Silva Correia J et al (2010) Nod1 and nod2 are expressed in human and murine renal tubular epithelial cells and participate in renal ischemia reperfusion injury. J Immunol 184:2297–2304

Sibony M, Gasc JM, Soubrier F, Alhenc-Gelas F, Corvol P (1993) Gene expression and tissue localization of the two isoforms of angiotensin I converting enzyme. Hypertension 21:827–835

Skuginna V, Lech M, Allam R, Ryu M, Clauss S, Susanti HE et al (2011) Toll-like receptor signaling and SIGIRR in renal fibrosis upon unilateral ureteral obstruction. PLoS ONE 6:e19204

Small DM, Coombes JS, Bennett N, Johnson DW, Gobe GC (2012) Oxidative stress, anti-oxidant therapies and chronic kidney disease. Nephrology (Carlton) 17:311–321

Strutz FM (2009) EMT and proteinuria as progression factors. Kidney Int 75:475–481

Takahashi N, Vereecke L, Bertrand MJ, Duprez L, Berger SB, Divert T et al (2014) RIPK1 ensures intestinal homeostasis by protecting the epithelium against apoptosis. Nature 513:95–99

Takaori K, Nakamura J, Yamamoto S, Nakata H, Sato Y, Takase M et al (2016) Severity and frequency of proximal tubule injury determines renal prognosis. J Am Soc Nephrol 27:2393–2406

Tan RJ, Zhou D, Zhou L, Liu Y (2014) Wnt/β-catenin signaling and kidney fibrosis. Kidney Int Suppl 4:84–90

Tan RJ, Zhou D, Liu Y (2016) Signaling crosstalk between tubular epithelial cells and interstitial fibroblasts after kidney injury. Kidney Dis (Basel) 2:136–144

Tang SC, Lai KN (2012) The pathogenic role of the renal proximal tubular cell in diabetic nephropathy. Nephrol Dial Transplant 27:3049–3056

Tang J, Liu N, Tolbert E, Ponnusamy M, Ma L, Gong R et al (2013) Sustained activation of EGFR triggers renal fibrogenesis after acute kidney injury. Am J Pathol 183:160–172

Tang Y, Huang XR, Lv J, Chung AC, Zhang Y, Chen JZ et al (2014) C-reactive protein promotes acute kidney injury by impairing G1/S-dependent tubular epithelium cell regeneration. Clin Sci (Lond) 126:645–659

Terada Y, Tanaka H, Okado T, Shimamura H, Inoshita S, Kuwahara M et al (2003) Expression and function of the developmental gene Wnt-4 during experimental acute renal failure in rats. J Am Soc Nephrol 14:1223–1233

Thadhani R, Pascual M, Bonventre JV (1996) Acute renal failure. New Engl J Med 334:1448–1460

Thomas ME, Harris KP, Walls J, Furness PN, Brunskill NJ (2002) Fatty acids exacerbate tubulointerstitial injury in protein-overload proteinuria. Am J Physiol Renal Physiol 283:F640–F647

Trionfini P, Benigni A, Remuzzi G (2015) MicroRNAs in kidney physiology and disease. Nat Rev Nephrol 11:23–33

Urushihara M, Kagami S (2017) Role of the intrarenal renin-angiotensin system in the progression of renal disease. Pediatr Nephrol 32:1471–1479

van Timmeren MM, Bakker SJ, Stegeman CA, Gans RO, van Goor H (2005) Addition of oleic acid to delipidated bovine serum albumin aggravates renal damage in experimental protein-overload nephrosis. Nephrol Dial Transplant 20:2349–2357

Vanden Berghe T, Linkermann A, Jouan-Lanhouet S, Walczak H, Vandenabeele P (2014) Regulated necrosis: the expanding network of non-apoptotic cell death pathways. Nat Rev Mol Cell Biol 15:135–147
Venkatachalam MA, Weinberg JM, Kriz W, Bidani AK (2015) Failed tubule recovery, AKI-CKD transition, and kidney disease progression. J Am Soc Nephrol 26:1765–1776
Verzola D, Gandolfo MT, Gaetani G, Ferraris A, Mangerini R, Ferrario F et al (2008) Accelerated senescence in the kidneys of patients with type 2 diabetic nephropathy. Am J Physiol Renal Physiol 295:F1563–F1573
Vío CP, Jeanneret VA (2003) Local induction of angiotensin-converting enzyme in the kidney as a mechanism of progressive renal diseases. Kidney Int Suppl 86:S57–S63
Waikar SS, Winkelmayer WC (2009) Chronic on acute renal failure: long-term implications of severe acute kidney injury. JAMA 302:1227–1229
Wallach D, Kang TB, Kovalenko A (2014) Concepts of tissue injury and cell death in inflammation: a historical perspective. Nat Rev Immunol 14:51–59
Wang Y, Chen J, Chen L, Tay YC, Rangan GK, Harris DC (1997) Induction of monocyte chemoattractant protein-1 in proximal tubule cells by urinary protein. J Am Soc Nephrol 8:1537–1545
Wang Y, Rangan GK, Goodwin B, Tay YC, Harris DC (2000) Lipopolysaccharide-induced MCP-1 gene expression in rat tubular epithelial cells is nuclear factor-kappaB dependent. Kidney Int 57:2011–2022
Wang J, Wen Y, Lv LL, Liu H, Tang RN, Ma KL et al (2015a) Involvement of endoplasmic reticulum stress in angiotensin II-induced NLRP3 inflammasome activation in human renal proximal tubular cells in vitro. Acta Pharmacol Sin 36:821–830
Wang Y, Chang J, Yao B, Niu A, Kelly E, Breeggemann MC et al (2015b) Proximal tubule-derived colony stimulating factor-1 mediates polarization of renal macrophages and dendritic cells, and recovery in acute kidney injury. Kidney Int 88:1274–1282
Wen Y, Liu Y, Tang T, Lv L, Liu H, Ma K et al (2016) NLRP3 inflammasome activation is involved in Ang II-induced kidney damage via mitochondrial dysfunction. Oncotarget 7:54290–54302
Wolf G, Mueller E, Stahl RA, Ziyadeh FN (1993) Angiotensin II-induced hypertrophy of cultured murine proximal tubular cells is mediated by endogenous transforming growth factor-beta. J Clin Invest 92:1366–1372
Wolf G, Ziyadeh FN, Stahl RA (1999) Angiotensin II stimulates expression of transforming growth factor beta receptor type II in cultured mouse proximal tubular cells. J Mol Med (Berl) 77:556–564
Wolf G, Wenzel U, Hannken T, Stahl RA (2001) Angiotensin II induces p27(Kip1) expression in renal tubules in vivo: role of reactive oxygen species. J Mol Med (Berl) 79:382–389
Wolfs TG, Buurman WA, van Schadewijk A, de Vries B, Daemen MA, Hiemstra PS et al (2002) In vivo expression of Toll- like receptor 2 and 4 by renal epithelial cells: IFN-γ and TNF-α mediated up-regulation during inflammation. J Immunol 168:1286–1293
Wu H, Ma J, Wang P, Corpuz TM, Panchapakesan U, Wyburn KR et al (2010) HMGB1 contributes to kidney ischemia reperfusion injury. J Am Soc Nephrol 21:1878–1890
Wu CF, Chiang WC, Lai CF, Chang FC, Chen YT, Chou YH et al (2013) Transforming growth factor β-1 stimulates profibrotic epithelial signaling to activate pericyte-myofibroblast transition in obstructive kidney fibrosis. Am J Pathol 182:118–131
Xiao L, Zhou D, Tan RJ, Fu H, Zhou L, Hou FF et al (2016) Sustained activation of Wnt/β-catenin signaling drives AKI to CKD progression. J Am Soc Nephrol 27:1727–1740
Yang H, Fogo AB (2010) Cell senescence in the aging kidney. J Am Soc Nephrol 21:1436–1439
Yang L, Besschetnova TY, Brooks CR, Shah JV, Bonventre JV (2010) Epithelial cell cycle arrest in G2/M mediates kidney fibrosis after injury. Nat Med 16:535–543
Yang JR, Yao FH, Zhang JG, Ji ZY, Li KL, Zhan J et al (2014) Ischemia-reperfusion induces renal tubule pyroptosis via the CHOP-caspase-11 pathway. Am J Physiol Renal Physiol 306:F75–F84
Yard BA, Daha MR, Kooymans-Couthino M, Bruijn JA, Paape ME, Schrama E et al (1992) IL-1 alpha stimulated TNF alpha production by cultured human proximal tubular epithelial cells. Kidney Int 42:383–389

Zeisberg M, Hanai J, Sugimoto H, Mammoto T, Charytan D, Strutz F et al (2003) BMP-7 counteracts TGF-beta1-induced epithelial-to-mesenchymal transition and reverses chronic renal injury. Nat Med 9:964–968

Zhang JD, Liu BC (2011) Angiotensin II, a missing node in new pathogenic glomerulotubular feedback loop. Med Hypotheses 77:595–597

Zhang W, Zhou X, Zhang H, Yao Q, Liu Y, Dong Z (2016) Extracellular vesicles in diagnosis and therapy of kidney diseases. Am J Physiol Renal Physiol 311:F844–F851

Zhou D, Liu Y (2016) Renal fibrosis in 2015: understanding the mechanisms of kidney fibrosis. Nat Rev Nephrol 12:68–70

Zhou L, Fu P, Huang XR, Liu F, Lai KN, Lan HY (2010) Activation of p53 promotes renal injury in acute aristolochic acid nephropathy. J Am Soc Nephrol 21:31–41

Zhou D, Li Y, Lin L, Zhou L, Igarashi P, Liu Y (2012) Tubule-specific ablation of endogenous β-catenin aggravates acute kidney injury in mice. Kidney Int 82:537–547

Zhou D, Li Y, Zhou L, Tan RJ, Xiao L, Liang M et al (2014) Sonic hedgehog is a novel tubule-derived growth factor for interstitial fibroblasts after kidney injury. J Am Soc Nephrol 25:2187–2200

Zhou L, Li Y, Hao S, Zhou D, Tan RJ, Nie J et al (2015) Multiple genes of the renin-angiotensin system are novel targets of Wnt/β-catenin signaling. J Am Soc Nephrol 26:107–120

Zhou D, Tan RJ, Fu H, Liu Y (2016) Wnt/β-catenin signaling in kidney injury and repair: a double-edged sword. Lab Invest 96:156–167

Zuk A, Gershenovich M, Ivanova Y, MacFarland RT, Fricker SP, Ledbetter S (2014) $CXCR_4$ antagonism as a therapeutic approach to prevent acute kidney injury. Am J Physiol Renal Physiol 307:F783–F797

Chapter 12
Myofibroblast in Kidney Fibrosis: Origin, Activation, and Regulation

Qian Yuan, Roderick J. Tan and Youhua Liu

Abstract Renal fibrosis is characterized by excessive deposition of extracellular matrix (ECM), leading to destruction of normal kidney architecture and loss of renal function. The activation of α-smooth muscle actin-positive myofibroblasts plays a key role in this process. After kidney injury, profibrotic factors are secreted by injured tubular epithelia and infiltrated inflammatory cells to promote complex cascades of signaling events leading to myofibroblastic activation, proliferation, and ECM production. The origins of myofibroblasts remain controversial, and possibilities include resident fibroblasts, pericytes, bone marrow-derived cells, and endothelial cells. Recent evidence supports the existence of localized fibrogenic niches, which provides a specialized tissue microenvironment for myofibroblastic activation and expansion. Myofibroblasts often undergo epigenetic modifications, leading to their sustained activation and resistance to apoptosis. In this chapter, we discuss the origins, heterogeneity, and activation of myofibroblasts in diseased kidneys. We also highlight novel strategies for the treatment of patients with fibrotic kidney diseases.

Keywords Renal fibrosis · Myofibroblast · Fibrotic niche · EMT · Wnt signaling

Q. Yuan · Y. Liu (✉)
State Key Laboratory of Organ Failure Research, National Clinical Research Center of Kidney Disease, Division of Nephrology, Nanfang Hospital, Southern Medical University, Guangzhou, China
e-mail: yhliu@pitt.edu

R. J. Tan
Department of Medicine, University of Pittsburgh School of Medicine, Pittsburgh, PA, USA

Y. Liu
Department of Pathology, University of Pittsburgh School of Medicine, Pittsburgh, PA, USA

B.-C. Liu et al. (eds.), *Renal Fibrosis: Mechanisms and Therapies*,
Advances in Experimental Medicine and Biology 1165,
https://doi.org/10.1007/978-981-13-8871-2_12

12.1 Introduction

Chronic kidney disease (CKD) is a highly prevalent disorder affecting more than 10% of all adults worldwide (Snyder et al. 2009). Progressive CKD often leads to end-stage renal failure, for which the only options for survival are dialysis or renal transplantation. The unifying pathology of all advanced forms of CKD is tissue fibrosis characterized by excessive accumulation and deposition of extracellular matrix (ECM), leading to scar formation and progressive loss of kidney function.

Although many types of cells can produce ECM, it is a general belief that the major cell type contributing to fibrosis is the myofibroblast in diseased kidney. In this regard, central questions in the field are to understand where these myofibroblasts come from and how they are regulated in pathologic conditions. As such, much research in the past years has focused on the origin, activation, and regulation of myofibroblasts. Such work carries even greater significance considering the lack of effective therapies for the treatment of fibrotic CKD.

12.2 Characteristics of Myofibroblast

Myofibroblast is characterized by its signature protein α-smooth muscle actin (α-SMA). Although myofibroblasts are often considered to be equivalent to active fibroblasts in the literature, not all active fibroblasts express α-SMA. Myofibroblasts were originally discovered by utilizing electron microscopy in contracting granulation tissue (Gabbiani et al. 1971). Comparing to quiescent fibroblasts, which are spindle- or stellate-shaped, myofibroblasts often have large nucleus with abundant and extensive rough endoplasmic reticulum and are typically surrounded by collagen fibers. Myofibroblasts also have a greater capacity for generating collagen fibers than fibroblasts. Furthermore, myofibroblasts possess large bundles of contractile actin/myosin-containing stress fibers line up in parallel with the long axis of the cells (Hinz et al. 2012). The ECM protein fibronectin abundantly coats these cells (Sandbo and Dulin 2011). The α-SMA in myofibroblasts facilitates cell contractility. Other features of myofibroblast include electron dense focal adhesions in the cell membrane known as the fibronexus connecting intracellular myofilaments with the extracellular fibronectin filaments (Falke et al. 2015; Eyden 2008; Singer 1979). Similar characteristics are observed in smooth muscle cells. In some ways, myofibroblasts are often thought to be a mixed phenotype of smooth muscle cells and fibroblasts.

Myofibroblast contraction is unique and greatly contributes to tissue remodeling. When myofilament bundles in the cytoplasm contract, the mechanical force transmits to the extracellular fibrillar system via focal adhesions, leading to ECM contraction. Surprisingly, these contractures can last as long as years and consist of cyclic contractile events, in which strong contractions covering large areas (on the order of micrometers) alternate with weaker forces affecting smaller areas. The strong contractions are mediated by RhoA/Rho-associated kinase, while the weak contractions are a result of intracellular calcium signals (Hinz et al. 2012). While beneficial for wound healing, such chronic and unrelenting contraction is likely to be harmful in organ fibrosis. The mechanical stress further promotes ECM secretion, release of profibrotic cytokines, latent TGF-β1 activation, and other myofibroblastic activities (Van De Water et al. 2013; Frangogiannis 2017).

A variety of immunological methods to identify myofibroblasts have been developed. The α-SMA is deemed to be a classical marker of myofibroblasts because fibrotic activity is correlated closely with its expression (Liu 2011; Hinz et al. 2007). While it has been assumed that normal interstitial fibroblasts lack α-SMA, this marker is not as specific as previously thought. Data in collagen-GFP transgenic mice demonstrated that the majority of collagen-positive cells co-stained with α-SMA, but approximately 25% of the α-SMA-positive cells in tubular interstitial space did not express collagen (Lin et al. 2008). Further, leukocytes and pericytes may also express α-SMA (Falke et al. 2015). Some studies show that severely fibrotic areas of the kidney seemingly lack α-SMA-positive cells. In collagen-EGFP and α-SMA-RFP dual transgenic mice, only 26% of collagen-positive cells co-stained with α-SMA in the fibrotic kidney induced by unilateral ureteral obstruction (UUO). In bleomycin-induced lung fibrosis, this number was only 42%, while in carbon tetrachloride-induced hepatic fibrosis, the majority of collagen-positive cells express α-SMA (Sun et al. 2016a, b). The difference between the various studies could be due to experimental approach as well as organ studied, as immunofluorescence of α-SMA used in the former study is less convincing than Col-EGFP and SMA-RFP dual transgenic mice. Alternatives to α-SMA to label myofibroblasts include platelet-derived growth factor receptor-β (PDGFR-β), fibroblast-specific protein 1 (FSP-1), and CD73 (Falke et al. 2015; Liu 2011). Table 12.1 lists different markers of fibroblast, myofibroblast, vascular smooth muscle cell (VSMC), and pericytes. Since most individual markers are not specific, it may be ideal to use two or three markers together to define myofibroblasts with greater confidence.

There might be some distinct and novel myofibroblast markers that are organ-specific. For examples, studies by Bodmer et al. suggest that AOC3 (amine oxidase, copper containing 3), a target protein of the myofibroblast-reacting mAb PR2D3, is a novel myofibroblast marker in the colon (Hsia et al. 2016). Compared with α-SMA, AOC3 is more specific in distinguishing colorectal tumor-derived primary myofibroblast from skin-derived fibroblast. Whole-genome microarray mRNA-expression profiles reveal four additional genes that are significantly differentially expressed in these two cell types: NKX2-3 and LRRC17 in myofibroblasts and SHOX2 and TBX5 in skin fibroblasts (Hsia et al. 2016). Periostin appears only after myocardial infarction, but not in normal tissues, and restricts to myofibroblasts. Periostin positive cells are proven to include almost all cardiac myofibroblasts through single-cell RNA-Seq (Kanisicak et al. 2016). Similarly, single-cell RNA-Seq has discovered that Hhip, Aspn, and Mustn1 are superior to α-SMA in characterizing myofibroblast in mouse lung fibrotic tissue (Xie et al. 2018). The translation of these results to kidney myofibroblasts, however, awaits further investigation.

Table 12.1 Characteristic features and markers of various stromal cells in the kidney

Cell type	Marker					
	α-SMA	PDGFR-β	FSP-1	CD73	Desmin	Vimentin
Fibroblast	−	+	+	++	−	+
Myofibroblast	++	+	+	+	−	++
Pericyte	+	+	−	+	+	+
VSMC	+	+	+	−	+	+
Other cell types	Leukocyte, vascular smooth muscle cell, activated mesangial cells	Vascular smooth muscle cell, MSCs Mesangial cells, macrophages	Macrophage Leukocyte	Mesangial cell T cell Proximal tubular cell	Injured podocyte	Macrophage, MSCs Mesangial cells Podocytes Injured tubular cell

Abbreviations: *α-SMA* α-smooth muscle actin; *PDGFR-β* platelet-derived growth factor receptor β; *CD* cluster of differentiation; *FSP-1* fibroblast-specific protein 1; *VSMC* vascular smooth muscle cell; *MSCs* mesenchymal stem cells

12.3 Origins of Myofibroblast

Myofibroblasts are rare in normal kidneys but increase greatly in chronic fibrosis, and the origins of these cells are still an area of active research. Much of the current data relies upon lineage tracing techniques, in which cell types are permanently labeled and followed for differentiation into myofibroblasts (Mack and Yanagita 2015). Possible precursors include resident fibroblasts, pericytes, epithelial cells, endothelium, and circulating bone-marrow-derived cells (Fig. 12.1).

12.3.1 Resident Mesenchymal Cells

Resident mesenchymal cells, also known as stromal cells, refer to cells residing in the interstitial space of the kidney and exclude cells entering the kidney via the bloodstream. These cells include fibroblasts and mesenchymal stem cells, as well as specialized support cells for the peritubular capillaries known as pericytes. Current data provides strong support for these cells as myofibroblast precursors.

Resident fibroblasts form an extensive interconnected network with each other and with tubular epithelia and capillaries. Just one day after an obstructive kidney injury due to UUO, there was an increase in α-SMA co-staining with the fibroblast marker ecto-5′-nucleotidase (5′NT), suggesting myofibroblast transformation from resident fibroblasts (Picard et al. 2008). Using lineage tracing, it is found that myelin protein zero (P0)-positive cells differentiate specifically into fibroblasts. After kidney injury,

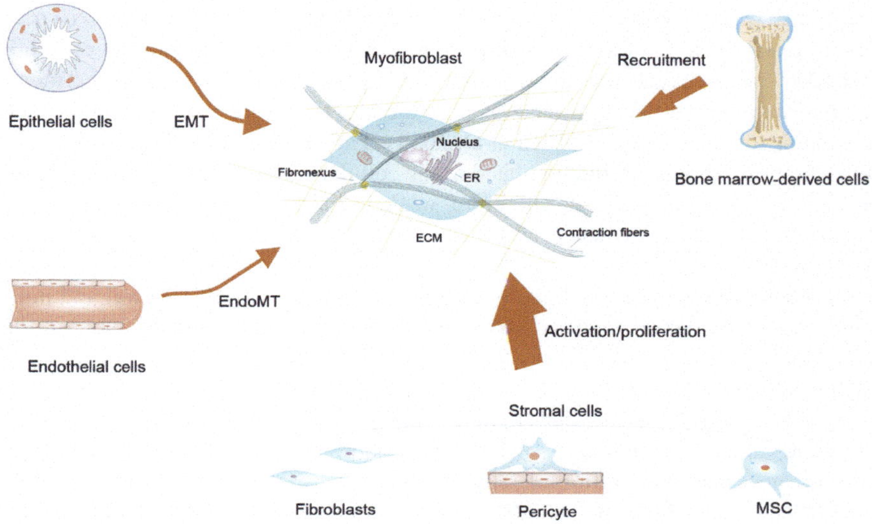

Fig. 12.1 Various origins of myofibroblasts. The origin of the myofibroblast includes several types of cells: interstitial stromal cells such as fibroblasts, pericytes and mesenchymal stem cells (MSC), bone marrow-derived cells, tubular epithelial cells, and endothelial cells. The exact contribution of each of these precursors to the total myofibroblasts pool is still controversial. Current evidence suggests stromal cells including fibroblasts, pericytes, and MSC are a major source of myofibroblasts. Bone marrow-derived cells infiltrate and contribute to the myofibroblast pool as well. "Partial EMT" is a theory suggesting that epithelial cells fall short of a full transition to fibroblasts but nonetheless promote renal fibrosis through the release of mediators to promote myofibroblast differentiation. (ECM—extracellular matrix; EMT—epithelial-mesenchymal transition; EndoMT—endothelial-mesenchymal transition; ER—endoplasmic reticulum; MSC—mesenchymal stem cell)

over 90% of PDGFR-β-expressing interstitial cells and α-SMA-expressing cells were derived from these P0-positive cells (Asada et al. 2011). Furthermore, about 50% of myofibroblasts in UUO is shown to be originated from local resident fibroblasts via proliferation (LeBleu et al. 2013).

While it seems intuitive to assume that the majority of myofibroblasts arise from resident fibroblasts, several studies have indicated that is not the case. Rather, pericytes are the primary source of myofibroblasts in diseased kidneys. Pericytes are vascular support cells that stabilize vessels and promote capillary basement membrane deposition, angiogenesis while preventing vessel rarefaction (Gomez and Duffield 2014). Using a lineage-tracing strategy to label all cells expressing FoxD1 during development, Humphreys and colleagues demonstrated that all pericytes derive from this cell population. After UUO and ischemic injury, they showed that all α-SMA-positive cells are originated from FoxD1-derived cells, providing strong evidence for a pericyte origin of myofibroblasts (Gomez and Duffield 2014; Humphreys et al. 2010). A caveat to this study is that vascular smooth muscle cells also derive from FoxD1-expressing cells and that available markers such as PDGFRβ do not reliably distinguish pericytes from resident fibroblasts (Table 12.1), which can also exist near blood vessels. Using a different marker (NG2), another study found that only 6% of α-SMA-positive cells are derived from $NG2^+$ pericytes after UUO. Furthermore, pericyte ablation does not significantly alter the fibrotic severity, suggesting the lack of a major functional role (LeBleu et al. 2013). The reasons behind the differences between these studies are likely related to the cellular markers chosen and the exact lineage tracing model utilized. Therefore, while initial data are promising and seem to provide compelling evidence, additional studies are required to clarify exact contributions of the pericyte.

Resident mesenchymal stem cells (MSCs) are capable of self-renewal and multilineage differentiation capacity and reside in the perivascular environment in the same "niche" as pericytes (Kramann et al. 2017; El Agha et al. 2017). Genetic fate-tracing experiments indicated that resident perivascular $Gli1^+$ cells, but not bone marrow-derived $Gli1^+$ cells, contribute to nearly 55% of myofibroblasts via proliferation. Conditional ablation of $Gli1^+$ cells by administration of DTX in transgenic mice or blocking proliferation of $Gli1^+$ cells by pharmacological GLI2 inhibition reduces nearly the same percentage of renal fibrosis induced by UUO (Kramann et al. 2015a, b). There also appears to be some overlap in $Gli1^+$ cells with FoxD1 precursors, both of which can ultimately express PDGFRβ (Humphreys 2018).

12.3.2 Epithelial Cells

The majority of the kidney parenchyma is comprised of tubular epithelium. Epithelial-to-mesenchymal transition (EMT) is a process in which epithelial cells undergo a transformation to a mesenchymal phenotype. This is characterized by loss of epithelial markers and function and acquisition of myofibroblastic features (Cruz-Solbes and Youker 2017). Lineage tracing experiments demonstrated that approximately 36% of $FSP1^+$ fibroblasts in UUO were derived from γGT^+ tubular epithelial cells (Iwano et al. 2002). However, more recent studies directly contradict these results and appear to show that epithelia contribute little to the myofibroblast population after injury (LeBleu et al. 2013; Humphreys et al. 2010; Koesters et al. 2010; Li et al. 2010). This may be due to the fact that EMT, if it occurs, appears to differ between experimental models and animal strains (Inoue et al. 2015). The in vivo demonstration of EMT appears to be much more challenging in comparison to the relative ease in which it can be demonstrated in vitro.

The coexpression of epithelial and mesenchymal markers, however, has been confirmed in a number of experimental and human kidney diseases (Sun et al. 2016a, b; Liu 2010), and therapeutics targeted at the process of EMT attenuates renal fibrosis (Li et al. 2009a, b; He et al. 2009; Chen et al. 2018; Sugimoto et al. 2012). Recent studies have popularized the idea of "partial EMT" in which epithelia acquire mesenchymal features but do not fully transform into a myofibroblast. In this process, injured epithelia undergo G2/M phase cell cycle arrest, downregulate epithelial markers and their characteristic solute transporters, and upregulate profibrotic cytokines and the expression of α-SMA. *Snail* and *Twist1* are shown to be key mediators of this process (Grande et al. 2015; Lovisa et al. 2015). Considering the difficulty demonstrating true and complete EMT in vivo, partial EMT appears to be a unifying hypothesis to explain much of the existing literature.

12.3.3 Endothelial Cells

Endothelial cells have also been reported to undergo transformation to a myofibroblast phenotype. Zeisberg and colleagues have reported that about 30–50% of the renal fibroblasts appearing in nephrotic models of kidney disease derive from the endothelial cell. This work was accomplished with a Tie2-Cre lineage tracing mouse and co-staining for α-SMA or FSP1 and the endothelial marker CD31 (Zeisberg et al. 2008). Another study conducted in Tie2-Cre mice showed that endothelial cells contributed to about 25% myofibroblasts in diabetic kidney (Li et al. 2009a, b). It should be acknowledged that Tie2 expression is not limited to endothelial cells but also other myeloid lineage cells. Thus, the percentages could be overestimated (Humphreys 2018). Another study also demonstrated that 10% of the endothelial cells co-localize with α-SMA in *Cdh5*-Cre; $YPF^{f/f}$ and α-SMA-RFP dual transgenic mice (LeBleu et al. 2013). Endothelial-to-mesenchymal transition (EndoMT) could play an important role in microvascular rarefaction, which causes worsened renal hypoxia after injury.

12.3.4 Bone Marrow-Derived Cells

Circulating monocytes and fibrocytes could contribute to the myofibroblasts pool. In one study, bone marrow-derived myofibroblasts are approximately 35% of the total (LeBleu et al. 2013), which is corroborated by other lineage tracing experiments (Li et al. 2007; Broekema et al. 2007). Recent studies also confirm the phenomenon of monocyte/macrophage-to-mesenchymal transition (Wang et al. 2017a, b). Directly or indirectly deleting bone marrow-derived cells decreases the percentage of renal fibrosis ranging from 15 to 50% (Mack and Yanagita 2015). However, other investigations suggest that only a small percentage, if any, of myofibroblasts originated from bone marrow-derived cells (Lin et al. 2008; Roufosse et al. 2006). It has also been proposed that bone marrow-derived myofibroblasts contribute differently to renal fibrosis, primarily through paracrine actions but not by directly producing large amounts of collagen (Sun et al. 2016a, b).

12.4 Heterogeneity of Myofibroblast

It is now clear that fibroblasts and myofibroblasts are not uniform populations and different types of these cells coexist in the kidney during injury (Fries et al. 1994). For instance, medullary kidney myofibroblasts are Wnt4-positive while cortical myofibroblasts are Wnt4-negative (DiRocco et al. 2013). The diverse origins of myofibroblast also contribute to their heterogeneity (Falke et al. 2015). It is logical to assume that myofibroblasts retain some features of their particular precursor cell (Boor and Floege 2012). As an example, it has been shown that erythropoietin (EPO)-producing fibroblasts that transform into myofibroblasts can still produce EPO under hypoxia or other stimulating conditions (Sato and Yanagita 2017).

Renal fibroblasts have various functions including producing ECM, secreting EPO and hepatocyte growth factor (HGF), and interactions with resident cells and with inflammatory cells (Sato and Yanagita 2017; Yang et al. 2003). Some fibroblasts contribute to the formation of tertiary lymphoid tissue in the kidney (Sato et al. 2016). Mannose receptor C-type 2 (Mrc2)-positive myofibroblasts play a paradoxical role in renal fibrosis through promoting collagen internalization and degradation by lysosomes (Lopez-Guisa et al. 2012). It has been shown that the ratio of expression of fibroblast activation protein (FAP) and α-SMA plays a role in myofibroblast phenotype. Gene expression and functional analysis exhibit that the main function of fibroblasts expressing high levels of FAP is to synthesize and degrade ECM in fibrosis. Meanwhile, high α-SMA-expressing myofibroblasts mediate contraction and possess greater proliferative capacity. Local composition and stiffness of the ECM, as well as TGF-β signaling, appear to govern the ratio of FAP^+ fibroblasts and $\alpha\text{-SMA}^+$ myofibroblasts (Avery et al. 2018).

The different features of kidney fibroblast and myofibroblast populations are still being described (Falke et al. 2015; Gladka et al. 2018; Tabib et al. 2018; Mizoguchi et al. 2018; Hu et al. 2018). Newer next-generation sequencing techniques such as single-cell RNA-Seq should be particularly useful in describing these subpopulations (Villani et al. 2017; Wu and Humphreys 2017). It is speculated that we may characterize different myofibroblast subpopulations in a way similar to documenting leukocyte subtypes in the future. This research carries added significance as we strive toward precision medicine.

12.5 Activation of Myofibroblast

One central issue in renal fibrosis field is to elucidate how tissue injury causes myofibroblast activation. After the initiation of injury, there is the generation of a profibrotic "niche" that then drives the recruitment, proliferation, and activation of myofibroblasts (Liu 2011). Injured tubules and infiltrated inflammatory cells produce profibrotic factors, which target to various myofibroblast precursors by paracrine or autocrine mechanisms, ultimately leads to myofibroblast activation. Here, we dis-

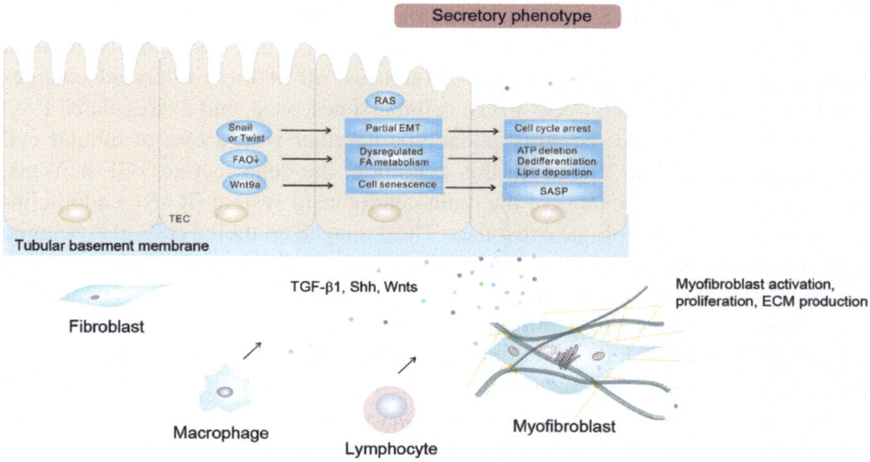

Fig. 12.2 The neighbor cells secret profibrotic factors triggering myofibroblasts activation. Partial EMT, cell senescence, and defective metabolism drive tubular epithelial cells (TEC) to secrete profibrotic factors. Chemokines produced by TEC, myofibroblast and resident immune cells induce more immune cells in circulation infiltrating to renal interstitial space. Inflammatory cells also are major sources of inflammatory and profibrotic cytokines. Myofibroblasts are also a donor of profibrotic factors

cuss the sources of profibrotic factors, as well as key signal pathways in mediating myofibroblast activation.

12.5.1 Sources of Profibrotic Factors

Profibrotic factors from tubular epithelia, inflammatory cells, and endothelium strongly affect the activation of myofibroblasts (Fig. 12.2). As the major component of renal parenchyma, tubular epithelia are particularly susceptible to injury induced by ischemic and toxic insults. Increasing evidence indicates that renal tubular epithelial cells are not only the victim of injury but also a driving force in the progression of kidney diseases (Liu et al. 2018).

How tubular injury drives fibroblast activation was an issue of controversy in the past. Several hypotheses have been postulated, such as EMT, cell cycle arrest, and cellular senescence. As described above, the activation of transcriptional factors *Snail* and *Twist1* in epithelia triggers a partial EMT leading to fibrosis (Grande et al. 2015; Lovisa et al. 2015). Studies also demonstrate that injured renal tubular cells are arrested in the G2/M phase of the cell cycle (Yang et al. 2010), and while in this state, they produce and secrete profibrotic cytokines. Abnormal upregulation of Wnt9a in tubular epithelia leads to cellular senescence, resulting in the release of the senescence-associated secretory phenotype (SASP), which cause myofibroblast activation (Luo et al. 2018). Other studies show that defective fatty acid oxidation (FAO) in tubular epithelia causes ATP depletion, cell death, and dedifferentiation (Kang et al. 2015). Regardless of the differences, all of these observed changes including partial EMT, cell cycle arrest, cellular senescence, and dysregulated FAO converge to a pathologic secretory phenotype together with a loss of tubular cell function (Zhou and Liu 2016a). In this condition, the cells secrete TGF-β, Wnts, hedgehog ligands, components of the renin–angiotensin system (RAS), endothelin-1, complement, and pathologic exosomes. The damaged epithelial cells also become proinflammatory through the release of chemokines, adhesion molecules, reactive oxygen species, and C-reactive protein (Liu et al. 2018; Gewin et al. 2017; Tan et al. 2016). Table 12.2 shows the profibrotic factors and their mechanism of action.

Injured endothelial cells also contribute to fibroblast activation via Notch and Wnt signaling (Lipphardt et al. 2017). Further, the destroyed integrity of peritubular capillary by EndoMT accelerates hypoxic injury (Nahrwold et al. 1974). In addition, perivascular pericytes lose their function when differentiating into myofibroblasts, which further aggravates the injury of endothelial cells, leading vascular rarefaction.

Macrophages, lymphocytes, dendritic cells, and mast cells are key contributors to injury (Rogers et al. 2014; Tapmeier et al. 2010; Heymann et al. 2009; Holdsworth and Summers 2008; Nikolic-Paterson et al. 2014). They produce cytokines that mediate myofibroblastic transformation and chemokines that attract for bone marrow-derived cells (Liu 2011; Kis et al. 2011). Co-culture with inflammatory cells induces tubular cells to undergo EMT via both direct cell-to-cell interactions and soluble factors (Nightingale et al. 2004; Li et al. 2011). Activated myofibroblasts express cytokines to affect normal fibroblasts, tubular epithelial cells, and endothelial cells in a paracrine fashion (Phan 2008), creating a vicious cycle.

12.5.2 Key Signal Pathways Mediating Myofibroblast Activation

There are several signal pathways playing key roles in mediating the activation and proliferation of various myofibroblast precursors. Among them, TGF-β is well established, classic signaling cascade that is believed to be essential in kidney fibrogenesis. Recent studies have pointed to a crucial role of developmental pathways such as Wnt/β-catenin, hedgehog, and Notch signaling in the activation of myofibroblast during kidney fibrosis.

12.5.2.1 TGF-β Signaling

TGF-β1 is regarded as the master regulator of myofibroblast activation. TGF-β1 is secreted in a latent form that is bound to latency-associated peptide (LAP) and ultimately latent TGF-β binding protein (LTBP) (Meng et al. 2015). Many factors

Table 12.2 Factors regulating myofibroblast activation

Factors	Mechanism	Cellular targets	Reference
Growth factors and cytokines			
TGF-β1	Ligand of TGF-β1 receptors	Fibroblast, EMT, EndoMT, MMT	Meng et al. (2016)
Wnt	Wnt ligand	Fibroblast, EMT	Tan et al. (2014), Zhou et al. (2017)
Hedgehog	Ligand of Hh pathway	Fibroblast	Zhou et al. (2014a, b)
Jagged	Ligand of Notch signaling	Fibroblast, EMT, EndoMT	Dees et al. (2011), Liu et al. (2009), Xiao et al. (2014)
Ang II	Stimulate TGF-β and ROS production	Fibroblast, EMT, EndoMT	Macconi et al. (2014)
CTGF	Activate Rho signaling; cofactor of TGF-β; interact with LRP6	Fibroblast, EMT EndoMT	Tsou et al. (2014)
PDGF	Stimulate TGF-β production	Fibroblast, EMT, EndoMT	Ostendorf et al. (2014)
FGF-2	Activate PI3K/AKT signaling	EMT	Masola et al. (2012)
IL-17	Induce oxidative stress	Fibroblast	Wang et al. (2017a, b)
IL-1β	Stimulate Th17 responses, TGF-β production	Fibroblast, EMT, EndoMT	Jones et al. (2009)
IL-4	Activate JAK3/STAT6 signaling	Fibrocyte	Yan et al. (2016), Liang et al. (2017)
IFN-γ	Activate JAK3/STAT6 signaling	Fibrocyte	Yan et al. (2016)
IL-12	Activate JAK3/STAT6 signaling	Fibrocyte	Yan et al. (2016)
IL-13	Activate STAT6 signaling; enhance TGF-β	Fibrocyte	Yan et al. (2015, 2016)
IL-6	Promote EMT via the Akt/GSK-3β/β-catenin pathway	EMT	Chen et al. (2017)
TNF-α	Increase TGF-β expression	Fibroblast, EMT, EndoMT	Liu (2011)
Adiponectin	Activate adiponectin/AMPK signaling	Fibrocyte	Yan et al. (2016)
ECM proteins			
TSP-1	Activate latent TGF-β	Fibroblast, EMT, EndoMT	Bige et al. (2012)
ED-A fibronectin	Activate TGF-β1 and store LTBP-1	Fibroblast, EMT, EndoMT	Serini et al. (1998)

(continued)

Table 12.2 (continued)

Factors	Mechanism	Cellular targets	Reference
TNC	Activate the integrin/FAK pathway	Fibroblast	Fu et al. (2017)
Oxidative stress			
AGEs	Activate NF-κB, MAPK, Smad2/3,small GTPase Ras, Cdc42, and Rac1	Fibroblast, EMT	He et al. (2013)
ROS	Activate tyrosine and serine/threonine kinases	Fibroblast, EMT	Barnes and Gorin (2011)
NOX	Mediate the generation of ROS	Fibroblast, EMT	Siani and Tirelli (2014)
Other intracellular factors			
tPA	Promotes fibroblast survival, proliferation by recruiting β1 integrin	Fibroblast	Hu et al. (2007)
Snail1	Induce partial EMT	Partial EMT	Grande et al. (2015)
ET1	Induces TGF-β and Snail1; activate Rho	Fibroblast, EMT, EndoMT	Tsou et al. (2014), Widyantoro et al. (2010)
Src	Regulate MMT; activate AKT, STAT3, EGFR	MMT, fibroblast	Zhou and Liu (2016b)
Integrin	Activate Rho signaling	Fibroblast	Tsou et al. (2014)
Antifibrotic factors			
BMP-7	Antagonizes TGF-β signaling	Fibroblast, EMT, EndoMT	Meng et al. (2013)
HGF	Antagonizes TGF-β1	Fibroblast, EMT, EndoMT	Liu (2004)
PPAR-α/γ	Reduce TGF-β expression	Fibroblast, EMT, EndoMT	Kawai et al. (2009), Li et al. (2013)
Smad7	Compete against Smad2/3	Fibroblast, EMT, EndoMT	Meng et al. (2016)

Abbreviation Advanced glycation end products (AGEs); Angiotensin II (Ang II); adenosine monophosphate-activated protein kinase (AMPK); bone morphogenetic protein-7 (BMP-7); connective tissue growth factor (CTGF); endothelin-1 (ET1); fibroblast growth factor-2 (FGF-2), glycogen synthase kinase 3β (GSK-3β); hypoxia inducible factor (HIF); interferon-γ (IFN-γ); interleukin (IL); Janus Kinase (JAK); low-density lipoprotein receptor-related protein 6 (LRP6); latent TGF-β binding proteins (LTBPs); NADPH oxidases (NOX); platelet-derived growth factor (PDGF); peroxisome proliferator-activated receptor (PPAR); protein kinase B (Akt); reactive oxygen species (ROS); signal transducers and activators of transcription (STAT); transforming growth factor-β1 (TGF-β1); TGF-β receptor (TGFR); tumor necrosis factor (TNF)-α; tenascin-C (TNC); tissue-type plasminogen activator (tPA); thrombospondin-1 (TSP-1) including αv integrin (Henderson et al. 2013), a strained ECM (Klingberg et al. 2014), and proteolytic cleavage by matrix metalloproteinases such as MMP-9 and MMP-2, plasmin, and thrombospondin (TSP) all can activate TGF-β1 from its latent form. In the canonical TGF-β signaling cascade, the direct binding of active TGF-β1 to TGF-β receptor II recruits TGF-β receptor I and results in Smad2 and Smad3 phosphorylation. Smad 4 associates with Smad2/3 and the entire complex translocates into the nucleus to induce profibrotic actions (Fig. 12.3) (Meng et al. 2015).

TGF-β signaling forms a complex network with other signaling pathways. In a noncanonical pathway, TGF-β1 activates the mitogen-activated protein kinase (MAPK) family and downstream extracellular signal-regulated kinase (ERK), c-Jun terminal kinase (JNK), and p38 MAPK. Activated MAPK can upregulate TGF-β1 and phosphorylate residues within the linker regions of Smad2 and Smad3 (Meng et al. 2016). TGF-β signaling can also mediate the oxidative stress injury caused by NADPH oxidases (NOX)-4 (Bondi et al. 2010) and activates MyoD which promotes myofibroblast differentiation mediated by TGF-β1 or PDGF (Hecker et al. 2011), respectively. Connexin43 positively regulates TGF-β signaling by competing with Smad2/3 for binding to microtubules and increasing the release of Smad2/3 (Dai et al. 2007).

12.5.2.2 Wnt Signaling

Wnt signal cascade is an evolutionarily conserved, developmental pathway. Wnt ligands transmit their signal via both canonical, β-catenin-dependent, and non-canonical, β-catenin-independent, mechanisms. Wnt/β-catenin signaling is silent in normal adult kidney, but reactivated after injury. When one of 19 of the known secreted Wnt proteins binds to cell surface Frizzled and LRP5/6 receptors, a cytoplasmic β-catenin degradation complex is inhibited. This results in the accumulation and nuclear translocation of β-catenin. Wnt signaling is a key promoter of renal myofibroblast activation as α-SMA, ECM components, Snail, EMT markers and RAS proteins including angiotensinogen, renin, angiotensin-converting enzyme and angiotensin receptor type 1 are all target genes of β-catenin (Tan et al. 2014; Zhou and Liu 2016c). The crosstalk between Wnt signaling and TGF-β signaling exists at various levels such as promoting ligand production reciprocally and synergetic regulation of target genes in the nucleus (He and Dai 2015; Piersma et al. 2015). Extensive studies show that Wnt signaling mediates myofibroblast activation (Luo et al. 2018; Zhou et al. 2017) and EMT (Zhou et al. 2013a, b). Blocked Wnt signaling greatly mitigates renal fibrosis (Zhou et al. 2013a, b).

12.5.2.3 Hedgehog Signaling

The ligands of hedgehog signaling are sonic hedgehog (Shh), Indian hedgehog (Ihh), and desert hedgehog (Dhh). Shh has been the primary focus of kidney disease studies. The binding of hedgehog ligands to the patched (Ptch) receptor relieves the

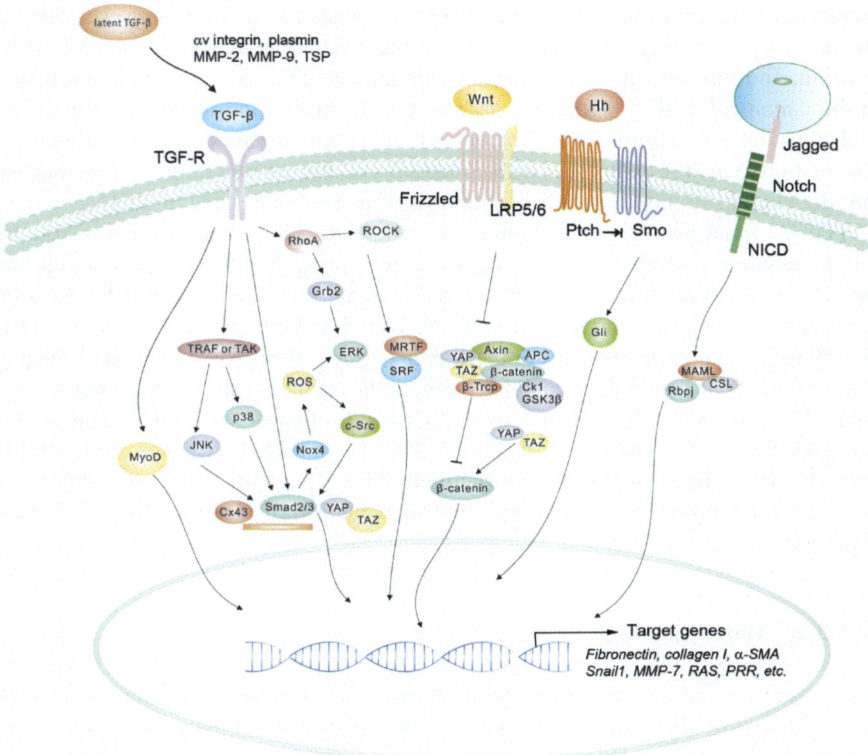

Fig. 12.3 Major signaling pathways involved in myofibroblast activation. TGF-β activates both Smad3 and MAPK. TGF-β1 activates three transcriptional factors: MyoD, Smad2/3 and MRTF/SRF. Nox4 is downstream of Smad3 and activate fibroblasts mediated by TGF-β signaling. YAP or TAZ facilitates Smad2/3 nuclear translocation. Smads are not only activated by TGF-β1, but also other molecules like TNF-α, CTGF, AGEs, MAPK, PDGF, ET1 and so on. The classical Wnt pathway is mediated by Frizzled and LRP5/6 receptor. The β-catenin degradation complex contains Axin, APC, GSK-3β, and CK1. Phosphorylated β-catenin degradation is mediated by the ubiquitin-proteasome system. Phosphorylated YAP and TAZ are necessary for docking of β-TrCP to the complex and accelerate β-catenin degradation. The binding of hedgehog ligands to Ptch receptor relieves the inhibition of Smo receptor. Active Smo promotes transcription factor Gli nuclear translocation. Notch signaling occurs as NICD binds to other transcriptional factors like MAML and Rbpj in nuclear after ligand–receptor binding. All these signaling pathways control their target genes in the nucleus. Common, fibrosis-related targets of these signals include fibronectin, collagen I, α-smooth muscle actin (α-SMA), Snail1, matrix metalloproteinase-7 (MMP-7), renin-angiotensin system (RAS), (pro)renin receptor (PRR), etc.

inhibition of the smoothened (Smo) protein. Active Smo promotes transcription factors Glis nuclear translocation. Shh is mainly expressed in tubular epithelia while it specifically targets to interstitial fibroblasts (Fabian et al. 2012). As a result, the Shh pathway is an example of intercellular crosstalk in which a tubule-derived profibrotic factor activates fibroblasts but not tubular epithelial cells (Zhou et al. 2014a, b). Shh upregulates the expression of Snail, α-SMA, fibronectin, and collagen I in fibroblasts (Ding et al. 2012). Recent data shows that $Gli1^{+}$ MSC are a major source of myofibroblasts and that conditional deletion of Gli2 attenuates renal fibrosis, highlighting the importance of Shh in chronic scarring (Kramann et al. 2015a, b).

12.5.2.4 Notch Signaling

The activation of Notch signaling requires cell–cell contact because the receptors, Notch 1–4, and the ligands, Jagged 1, 2, and Delta 1, 3, 4, are all transmembrane proteins. The binding of ligand and receptor causes cleavage of the Notch receptor, releasing the Notch intracellular domain (NICD) which then enters the nucleus. NICD binds to additional transcriptional factors such as MAML and Rbpj to upregulate target genes. The expression of Notch signaling components differs in different nephropathy (Edeling et al. 2016). Notch1 is mainly expressed in tubular epithelial cells and drives EMT due to *Snail* and *α-SMA* upregulation (Bielesz et al. 2010). TGF-β1 upregulates the expression of Notch3, and Notch3 global knockout mice demonstrate attenuated renal fibrosis and interstitial inflammation during injury (Djudjaj et al. 2012). Studies in other organs show that Notch signaling activates fibroblasts (Dees et al. 2011).

12.6 Fibrogenic Niche for Myofibroblast Activation

Kidney fibrotic lesions are not homogeneous across the kidney parenchyma. Rather, it typically initiates at certain focal sites, in which interstitial fibroblasts become activated, proliferate, and produce a large amount of ECM components. Although it remains enigmatic why fibroblasts at certain sites respond differentially after injury, it has been long hypothesized that there is a specialized niche/microenvironment, which dictates the activation of fibroblasts in their discrete locations.

12.6.1 Tenascin-C as an Organizer of Fibrogenic Niche

Tenascin-C (TNC) is a large oligomeric ECM glycoprotein with cell signaling properties. Structurally, TNC contains an N-terminal assembly domain that leads to the formation of hexamers, a variable number of tandem epidermal growth factor (EGF)-like repeats, numerous fibronectin type III domains, and a fibrinogen-like globe

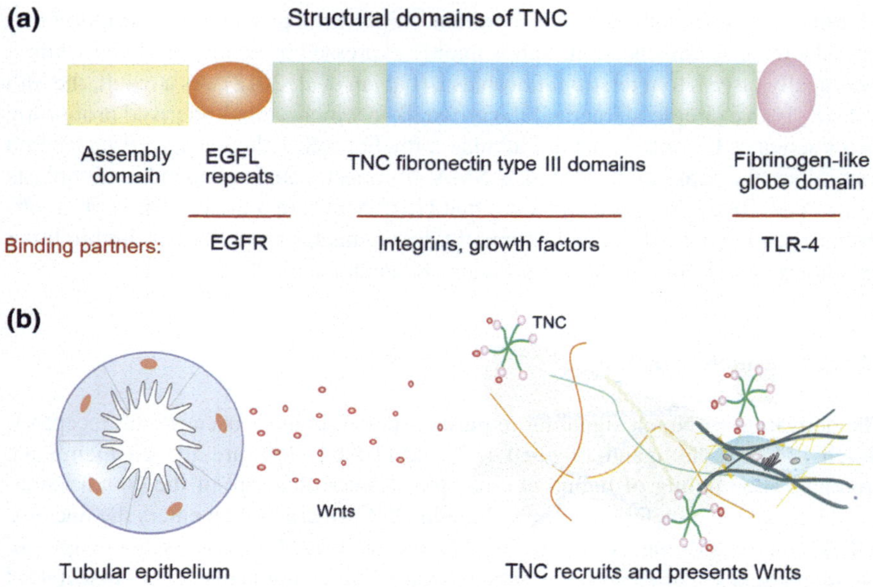

Fig. 12.4 Tenascin C (TNC) acts as an organizer of the fibrogenic niche. **a** Diagram shows the structural domains of TNC, which contains an N-terminal assembly domain that leads to the formation of hexamers, a variable number of tandem epidermal growth factor (EGF)-like repeats, numerous fibronectin type III domains, and a fibrinogen-like globe domain at the C-terminus. Different domains can have distinct binding partners such as EGF receptor (EGFR), integrins and other growth factors, as well as Toll-like receptor-4 (TLR-4). **b** TNC binds to Wnt ligands, and therefore is able to recruit them from surrounding environment and present them to responding cells such as fibroblasts

domain at the C-terminus (Fig. 12.4). In adults, little or no TNC is detected in the kidney and other organs. However, prominent de novo expression of TNC is reported in the fibrotic kidneys. As a so-called matricellular protein, TNC is not an obligatory structural element in the ECM, but it is able to bind to ECM structural proteins and cell surface receptors such as integrins and Toll-like receptor-4 (TLR-4). TNC binding to these receptors causes activation of their downstream pathways, thereby modulating cell adhesion, spreading, migration, and proliferation.

We recently show that TNC plays an important role in myofibroblast activation primarily by organizing a fibrogenic microenvironment (Fu et al. 2017). In vitro, profibrotic factors such as Shh, TGF-β1, Wnts stimulate TNC expression by fibroblasts. In vivo, TNC is rapidly upregulated in the fibrotic kidneys induced by UUO or ischemia/reperfusion injury (IRI) and predominantly localizes at the foci rich in fibroblasts in renal interstitium. TNC selectively promotes renal interstitial fibroblast proliferation and the expression of proliferation-related genes. TNC-rich decellularied ECM scaffold facilitates fibroblast proliferation, whereas TNC-deprived scaffold inhibits it. Matrix scaffold prepared from fibrotic kidney also promotes greater fibroblast proliferation ex vivo, and the effects can be reversed by deletion of TNC (Fu et al. 2017). These studies indicate a pivotal role for the TNC-rich microenvironment in kidney fibrogenesis. These observations are also consistent with a crucial role of TNC in building a specialized niche for stem or progenitor cells, as reported previously (von Holst 2008; Chiquet-Ehrismann et al. 2014).

12.6.2 Mechanism of Profibrotic Factors Enrichment

Fibrogenic niche should have a special mechanism for recruiting and concentrating profibrotic cues, building a microenvironment in which high levels of soluble fibrotic factors are present. In this aspect, TNC is quite unique in that it can bind a wide variety of profibrotic factors such as TGF-β1, PDGF, and FGF-2 (De Laporte et al. 2013). We recently show that TNC binds to and recruits Wnt ligands (Chen et al. 2019). As such, TNC is able to concentrate Wnt ligands from the surrounding milieu and present them to neighboring cells, which creates a specialized microenvironment in which Wnts are enriched. This conclusion is supported by an ex vivo approach using TNC-rich kidney tissue scaffold. Notably, TNC is a large hexameric protein that can potentially bind to multiple Wnts at the same time and therefore could possess a tremendous ability to recruit Wnt ligands. It is conceivable to propose that TNC functions as a "molecular sponge" that attracts and recruits Wnt ligands from the surrounding microenvironment.

12.6.3 Other Components of Fibrogenic Niche

The major components of fibrogenic niche include ECM and its associated proteins. ECM is a non-cellular, three-dimensional structure providing the physical support for parenchymal cells and controlling tissue homeostasis (Bonnans et al. 2014). ECM is dynamic, and its quantity and quality are changed in normal condition versus pathologic setting. The fibrogenic ECM is composed by aberrant fibrillar proteins, matricellular proteins, glycoproteins, and profibrotic cytokines (Fig. 12.5) (Walraven and Hinz 2018).

Matricellular proteins are not structural components but regulate the interaction of ECM and cells. Besides TNC, the most well-characterized matricellular proteins are the CCN family of proteins, which contains six members designated CCN1 to CCN6. The CCN represents the names of the first three members of the family to be discovered: Cyr61 (cysteine-rich protein 61), CTGF (connective tissue growth factor), and NOV (nephroblastoma overexpressed gene) (Jun and Lau 2011). Among them, CCN2 is widely known as CTGF, which is not only a growth factor but also an ECM protein that links ECM proteins such as collagen, fibronectin up with target cells (Yokoi and Mukoyama 2017). CTGF also serves as a necessary cofactor for TGF-β and can be pro-inflammatory on its own (Kok et al. 2014).

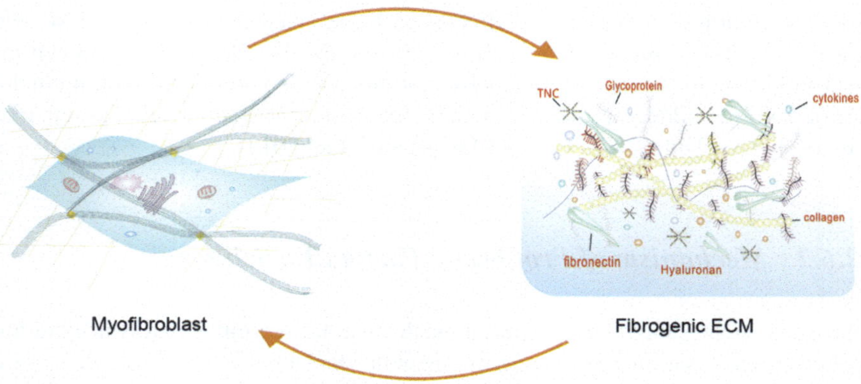

Fig. 12.5 Reciprocal stimulation of myofibroblast and fibrogenic ECM occurs in fibrotic kidney. Myofibroblast is the major contributor of ECM. Mechanical forces and constituent components residing in ECM promote the activation of myofibroblasts. Fibrillar proteins such as collagen and fibronectin are necessary for building the framework of ECM. Matricellular proteins such as TNC with multiple domains usually mediate the interaction of ECM proteins and target cells. Glycoproteins including HA contribute to renal fibrosis. The soluble profibrotic factors and inflammatory cells are also abundant in the fibrotic tissue microenvironment

Periostin functions as a scaffold protein by binding to many proteins due to multiple domains' structure. Proteins remain within close proximity by binding to adjacent domains of periostin and assemble easily into extracellular architectures (Kii and Ito 2017). Secreted protein acidic and rich in cysteine (SPARC) also exhibits diverse functions in renal fibrosis and glomerulosclerosis (Francki and Sage 2001).

Collagen is composed of three procollagen chains in a cylindrical structure with about 300 nm length and nearly 1.5 nm diameter. These specific features make it ideal for the basic building block of ECM. Collagen IV is a component of basement membrane (Bonnans et al. 2014). Collagen I and collagen III are the principal participators of interstitial ECM in fibrotic tissues, which are extensively reviewed elsewhere (Karsdal et al. 2017). Increased collagen crosslinking mediated by lysyl oxidase-like 2 (LOXL2) enzyme or transglutaminase 2 (TG2) accelerates collagen deposition and has been demonstrated in other tissues (Walraven and Hinz 2018) and in kidney (Cosgrove et al. 2018).

Fibronectin is a high-molecular-weight protein consisting of two nearly identical subunits linked by disulfide bonds. There are type I, II, and III fibronectins. Type III contains type III (FNIII) extra-domain A (ED-A) which is necessary for the activation of TGF-β1 and storage of LTBP-1 (Dallas et al. 2005). Fibroblasts initially produce fibronectin which directs the deposition of collagen I and then mature collagen preferentially interacts with relaxed fibronectin to align mechanical forces generated by the myofibroblast (Kubow et al. 2015).

Elastin is secreted by fibroblast and smooth muscle cells used to buffer strength induced by scar tissue. The decrease of elastin is harmful for vessels, however, and abundant elastin in ECM promotes tissue fibrosis (Klingberg et al. 2013).

Hyaluronan (HA) is an extracellular glycoprotein mediating cell adhesion, migration, and proliferation. It deposits in fibrotic kidneys such as diabetic nephropathy and allograft rejection, and attracts inflammatory cells by binding to CD44 receptors and TLR-4 (Colombaro et al. 2013). In addition to the aforementioned, multiple other glycoproteins have been found in ECM, like non-fibrillar collagen, LTBP, fibulins, small leucine-rich proteoglycans (SLRPs), verscian, and syndecan (Walraven and Hinz 2018).

Matricellular proteins, fibrillar proteins, glycoproteins, secreted profibrotic factors like TGF-β1, Wnts, Shh, TNF-α as well as exosomes constitute the backbone of the fibrogenic tissue microenvironment (Herrera et al. 2018). Such a niche is not only a breeding ground for fibroblast activation and proliferation, but also plays a critical role in mediating inflammatory cells infiltration, defective angiogenesis, and tubular epithelial cell atrophy in the fibrotic kidney.

12.7 Myofibroblast Activation and Epigenetic Modification

Fibroblast activation is a wound healing response after tissue injury. Chronic fibrosis differentiating from a successful wound healing is largely due to the lack of myofibroblast resolution. Studies show that sustained or irreversible myofibroblast activation is associated with altered epigenetic modification of myofibroblasts in fibrotic kidneys. Epigenetic modification mainly includes DNA methylation, histone modification, and small interfering RNAs (Hu and Phan 2013). These modifications can directly affect genes related to myofibroblast activation.

12.7.1 DNA Methylation

CpG island methylation is catalyzed by DNA methyl transferases (DNMTS) and generally tightens the DNA helix to represses genes transcription. Hyperactive Ras caused by hypermethylation of RASAL1, an inhibitor, contributes to fibroblast activation and renal fibrosis. Short-term exposure to TGF-β1 resulted in reversible transcriptional repression of RASAL1 while long-term exposure to TGF-β1 caused irreversible DNA methylation (Bechtel et al. 2010). DNA methylation can be inherited to progency cells, and this may be an explanation of sustained myofibroblast activation. Hypomethylation of profibrotic gene such as α-SMA (Hu and Phan 2013) or hypermethylation of antifibrotic genes like peroxisome proliferator-activated receptor γ (PPARγ), Klotho, or Krüppel-like factor 4 (KLF4) facilitates myofibroblasts activation (Zeisberg and Zeisberg 2013; Xiao et al. 2015). In theory, DNA methylation could be a therapeutic target, as hypermethylation of RASAL1 can be reversed by BMP7 (Tampe et al. 2014).

12.7.2 Histone Modification

Posttranslational histone modifications include methylation, acetylation, ubiquitylation, and phosphorylation. Among them, acetylation is the most well studied in myofibroblast activation. Histone acetylation and deacetylation are catalyzed by histone acetyltransferase (HAT) and histone deacetylase (HDAC) enzymes, respectively (Meng et al. 2016). A large amount of data shows that inhibition of different HDAC classes abolishes EMT and/or fibroblast activation in different disease models (Liu and Zhuang 2015). HDAC inhibitors are effective in reducing fibrosis (Novitskaya et al. 2014). TGF-β1 increases activating histone H3 lysine methylation (H3K4me1/2/3 levels) and decreases repressive H3K9me2/3 levels at ECM-associated gene promoters, the net effect being increase in profibrotic gene expression (Sun et al. 2010). Inhibition of histone methyltransferase inhibits EMT and renal fibrosis (Zhou et al. 2018a, b; Irifuku et al. 2016).

12.7.3 Small Interfering RNA Regulation

Small interfering RNAs are short noncoding RNAs encoded by genomic DNA. The function of these RNAs is silencing gene expression by degradation of mRNA and/or translational inhibition. Many microRNAs have been found to promote or inhibit the activation of myofibroblasts. MiR-30c suppresses EMT by targeting *Snail* (Zhao et al. 2017). MiR-200 family members including miR200a, b, and c and miR141 also inhibit EMT (Huang et al. 2015; Tang et al. 2013). MiR-221 blocks fibroblasts activation induced by Ang II (Di et al. 2014). However, MiR-21 promotes EMT induced by TGF-β1 (Liu et al. 2016), and microRNA-132 facilitates activation and proliferation of FoxD1-derivative myofibroblast (Bijkerk et al. 2016). Besides microRNA, recent studies indicate that long noncoding RNA (lncRNA) also plays a critical role in regulating myofibroblast activation and renal fibrosis (Zhou et al. 2014a, b; Wang et al. 2018). These findings provide a novel approach to inhibiting myofibroblast activation.

12.8 Therapies Directed at Myofibroblast

Given the importance of myofibroblasts in kidney fibrogenesis, inhibiting their activation and functions is a key avenue mitigating fibrotic lesions. Blocking myofibroblast activation from its diverse origins, abolishing the functions of differentiated myofibroblast, and inducing myofibroblast apoptosis are therapeutic approaches. Such therapies would theoretically be useful not only in the kidney but also in other fibrotic organs.

12.8.1 Therapies Based on Molecular Mechanism

We have summarized the profibrotic factors and antifibrotic factors in Table 12.2. The therapies inhibiting myofibroblast activation and ECM production include blocking profibrotic factors and their cascades; enhancing antifibrotic factors; and targeting epigenetic modification.

TGF-β signaling is a well-known major mediator of myofibroblast activation. Pirfenidone is a synthetic small molecule inhibiting TGF-β synthesis and shows promise to attenuate kidney fibrosis (Allinovi et al. 2018). TGF-β neutralizing antibodies such as Fresolimumab and LY2382770 have also been tested in clinical trials (Lee et al. 2015). Specific inhibitors of TGF-β downstream targets, including SIS3, GQ5 as well as a number of miRNAs regulating TGF-β signaling, have also been discovered (Ai et al. 2015). Restoration of BMP7 expression or treatment with recombinant protein is another way to antagonize TGF-β signaling (McVicker and Bennett 2017). THR-123 which is an agonist of BMP signaling suppresses EMT, inflammation, and reverses established fibrosis (Sugimoto et al. 2012). Similarly, HGF has been shown to ameliorate kidney fibrosis by antagonizing TGF-β signaling (Liu 2004). A comprehensive review summarizes the inhibitors and agonists in renal fibrosis based on TGF-β signaling and BMP signaling (Munoz-Felix et al. 2015). The glitazones, which are PPARγ agonists, are widely used in type 2 diabetes management. They may be effective in treating fibrosis by antagonizing of TGF-β (Bolignano and Zoccali 2012). Relaxin protein exhibits a similar role (Samuel and Hewitson 2009). However, TGF-β signaling is important in immune regulation and tumor suppression, and this raises the concerns that the strategy to block TGF-β1 could lead to adverse effects.

Outside of using TGF-β blockers, many other inhibitors also show promising effects. The Gli inhibitor GANT61 represses the proliferation of $Gli1^+$ myofibroblasts and attenuates renal fibrosis (Kramann et al. 2015a, b). RAS inhibitors including angiotensin-converting enzyme inhibitors (ACEI) and angiotensin receptor blockers (ARBs) have been widely used in the clinic and appear to prevent or slow the progression of CKD. A β-catenin signaling inhibitor, ICG-001, disrupts the interaction of β-catenin and cAMP response element binding (CREB) binding protein (CBP) to reduce EMT and renal fibrosis (Zhou and Liu 2016b). Klotho inhibits Wnt/β-catenin signaling via binding to Wnt ligands (Zhou et al. 2013a, b). The Notch γ-secretase inhibitor, Dibenzazepine, represses fibroblast proliferation and activation in UUO (Xiao et al. 2014). Inhibition of Shh signaling by smoothened inhibitor, cyclopamine, reduces the accumulation of myofibroblasts (Zhou et al. 2014a, b). Recent studies show that antagonizing type 2 cannabinoid receptor (CB2) by small molecule is also effective in preventing and reversing kidney fibrosis in preclinical setting (Zhou et al. 2018a, b).

Different from these inhibitors targeting specific proteins, gene therapy by delivering miRNA or small interfering RNA (siRNA) ameliorates disease by targeting genes. The therapies based on siRNA like Smad4 siRNA, Smoc2 siRNA, and microRNA like miR-let7c, miR-146a, all inhibit myofibroblast activation and attenuate renal fibrosis (Nastase et al. 2017). It will be important to ensure that RNA-based therapies are targeted, to avoid off-target effects of these biologicals.

12.8.2 *Therapies Targeting the Kidney*

Systemic drug delivery can be subject to numerous deleterious actions via off-target effects on normal organs or cells. Drug targeting strategies have the potential to solve the problem by accurately delivering drugs to the exact target cells or tissues. Passive targeting relies on blood flow, the pathophysiological characteristics of the organ and physicochemical properties of the drug. Active targeting usually modifies drugs with the ligands against a receptor or biomolecule expressed on targeted cells (Yazdani et al. 2017).

The antifibrotic drug, IFN-γ, has several side effects such as lipolysis and microglia activation. As discussed above, PDGFRβ is widely expressed by renal stromal cells. By conjugating PDGFRβ recognizing peptide to PEGylated IFN-γ (PPB-PEG-IFN-γ), this approach allows the peptide modified drugs to be specifically targeted to PDGFRβ positive myofibroblast in UUO kidney (Poosti et al. 2015, 2016). Nanocarrier systems have successfully delivered drugs to epithelial cells by targeting the low-density lipoprotein receptor-related protein 2 (commonly known as megalin/cubilin), and folate receptor 1α in the luminal brush border which cause proximal tubular epithelium to actively engulf them. The discovered nanocarrier materials include low-molecular-weight protein lysozyme (LZM), glucosamine, low-molecular-weight chitosan, and folate (Falke et al. 2015). Supramolecular hydrogels composed by biocompatible materials like HA are another carrier possessing the ability to release drugs at a controlled rate. Administration of IL-10 by local injecting of HA-hydrogels inhibits renal fibrosis in UUO model (Nastase et al. 2017).

12.8.3 *Induction of Myofibroblast Resolution*

In a perfect wound healing process, myofibroblasts ultimately revert to quiescent state or undergo apoptosis, leading to resolution after wound closure. However, in organ fibrosis, the myofibroblast becomes resistant to deactivation or exhibit prolonged lifespan. ECM stiffness upregulates the expression of the proapoptotic activator, Bim, and myofibroblasts are then primed for apoptosis. However, simultaneous overexpression of the antiapoptotic protein, Bcl-xL, maintains myofibroblast survival. ABT-263 terminates the antiapoptotic effects of Bcl-xL by releasing Bim to bind in activated myofibroblasts, but not quiescent fibroblasts (Lagares et al. 2017).

Converting myofibroblasts back to tubular epithelial cells after EMT is an ambitious goal and would not only decrease the matrix-producing cells but also aid in repair of denuded and injured tubules. Kaminski et al. found that the overexpression of four transcriptional factors, Emx2, Hnf1b, Hnf4a, and Pax8, in myofibroblasts led to this reverse transformation (Kaminski et al. 2016). The induced renal tubular epithelial cells (iRECs) had the global gene expression profile of primary renal epithelial cells. They also demonstrate epithelial properties and functions and integrate into decellularized tubules.

Although many antifibrotic drugs have proven effective in preclinical experiments, none have shown adequate efficacy in human clinical trials. Barriers to progress might include patient heterogeneity, differences in underlying pathogenesis, single-nucleotide polymorphisms, diverse epigenetic modifications (Allinovi et al. 2018) and side effects (Falke et al. 2015). The strategies outlined above provide new hope for the future of antifibrotic therapies in the kidney.

12.9 Conclusion

As the major effector cells, myofibroblasts play a fundamental role in the development and progression of kidney fibrosis. The identification of different origins, functional heterogeneity, and interplay of different signaling pathways and paracrine mediators highlight the complexity of the biology of these cells, and also present the challenge for therapeutic interventions. These findings, however, potentially offer new possibility for designing specific and precision therapy for millions of patients with fibrotic CKD. Future research must focus on resolving controversy, identifying local networks of cellular and molecular interactions, further characterizing fibrogenic niche, and ultimately translating these findings to the clinic.

Acknowledgements This work was supported by National Natural Science Foundation of China Grant 81521003 and 81770715, an American Society of Nephrology Gottschalk Award, an American Heart Association Fellow-to-Faculty award, and NIH grants DK079307, DK064005, and DK106049.

References

Ai J, Nie J, He J, Guo Q, Li M, Lei Y et al (2015) GQ5 hinders renal fibrosis in obstructive nephropathy by selectively inhibiting TGF-beta-induced Smad3 phosphorylation. J Am Soc Nephrol 26:1827–1838

Allinovi M, De Chiara L, Angelotti ML, Becherucci F, Romagnani P (2018) Anti-fibrotic treatments: a review of clinical evidence. Matrix Biol 69:333–354

Asada N, Takase M, Nakamura J, Oguchi A, Asada M, Suzuki N et al (2011) Dysfunction of fibroblasts of extrarenal origin underlies renal fibrosis and renal anemia in mice. J Clin Invest 121:3981–3990

Avery D, Govindaraju P, Jacob M, Todd L, Monslow J, Puré E (2018) Extracellular matrix directs phenotypic heterogeneity of activated fibroblasts. Matrix Biol 67:90–106

Barnes JL, Gorin Y (2011) Myofibroblast differentiation during fibrosis: role of NAD(P)H oxidases. Kidney Int 79:944–956

Bechtel W, McGoohan S, Zeisberg EM, Müller GA, Kalbacher H, Salant DJ et al (2010) Methylation determines fibroblast activation and fibrogenesis in the kidney. Nat Med 16:544–550

Bielesz B, Sirin Y, Si H, Niranjan T, Gruenwald A, Ahn S et al (2010) Epithelial Notch signaling regulates interstitial fibrosis development in the kidneys of mice and humans. J Clin Invest 120:4040–4054
Bige N, Shweke N, Benhassine S, Jouanneau C, Vandermeersch S, Dussaule J et al (2012) Thrombospondin-1 plays a profibrotic and pro-inflammatory role during ureteric obstruction. Kidney Int 81:1226–1238
Bijkerk R, de Bruin RG, van Solingen C, van Gils JM, Duijs JM, van der Veer EP et al (2016) Silencing of microRNA-132 reduces renal fibrosis by selectively inhibiting myofibroblast proliferation. Kidney Int 89:1268–1280
Bolignano D, Zoccali C (2012) Glitazones in chronic kidney disease: potential and concerns. Nutr Metab Cardiovasc Dis 22:167–175
Bondi CD, Manickam N, Lee DY, Block K, Gorin Y, Abboud HE et al (2010) NAD(P)H oxidase mediates TGF-beta1-induced activation of kidney myofibroblasts. J Am Soc Nephrol 21:93–102
Bonnans C, Chou J, Werb Z (2014) Remodelling the extracellular matrix in development and disease. Nat Rev Mol Cell Biol 15:786–801
Boor P, Floege J (2012) The renal (myo-)fibroblast: a heterogeneous group of cells. Nephrol Dial Transplant 27:3027–3036
Broekema M, Harmsen MC, van Luyn MJ, Koerts JA, Petersen AH, van Kooten TG et al (2007) Bone marrow-derived myofibroblasts contribute to the renal interstitial myofibroblast population and produce procollagen I after ischemia/reperfusion in rats. J Am Soc Nephrol 18:165–175
Chen Q, Yang D, Zong H, Zhu L, Wang L, Wang X et al (2017) Growth-induced stress enhances epithelial-mesenchymal transition induced by IL-6 in clear cell renal cell carcinoma via the Akt/GSK-3β/β-catenin signaling pathway. Oncogenesis 6:e375
Chen H, Yang T, Wang MC, Chen DQ, Yang Y, Zhao YY (2018) Novel RAS inhibitor 25-O-methylalisol F attenuates epithelial-to-mesenchymal transition and tubulo-interstitial fibrosis by selectively inhibiting TGF-beta-mediated Smad3 phosphorylation. Phytomedicine 42:207–218
Chen S, Fu H, Wu S, Zhu W, Liao J, Hong X et al (2019) Tenascin-C protects against acute kidney injury by recruiting Wnt ligands. Kidney Int 95:62–74
Chiquet-Ehrismann R, Orend G, Chiquet M, Tucker RP, Midwood KS (2014) Tenascins in stem cell niches. Matrix Biol 37:112–123
Colombaro V, Decleves AE, Jadot I, Voisin V, Giordano L, Habsch I et al (2013) Inhibition of hyaluronan is protective against renal ischaemia-reperfusion injury. Nephrol Dial Transplant 28:2484–2493
Cosgrove D, Dufek B, Meehan DT, Delimont D, Hartnett M, Samuelson G et al (2018) Lysyl oxidase like-2 contributes to renal fibrosis in Col4alpha3/Alport mice. Kidney Int 94:303–314
Cruz-Solbes AS, Youker K (2017) Epithelial to esenchymal transition (EMT) and endothelial to mesenchymal transition (EndMT): role and implications in kidney fibrosis. Results Probl Cell Differ 60:345–372
Dai P, Nakagami T, Tanaka H, Hitomi T, Takamatsu T (2007) Cx43 mediates TGF-beta signaling through competitive Smads binding to microtubules. Mol Biol Cell 18:2264–2273
Dallas SL, Sivakumar P, Jones CJ, Chen Q, Peters DM, Mosher DF et al (2005) Fibronectin regulates latent transforming growth factor-beta (TGF beta) by controlling matrix assembly of latent TGF beta-binding protein-1. J Biol Chem 280:18871–18880
De Laporte L, Rice JJ, Tortelli F, Hubbell JA (2013) Tenascin C promiscuously binds growth factors via its fifth fibronectin type III-like domain. PLoS ONE 8:e62076
Dees C, Tomcik M, Zerr P, Akhmetshina A, Horn A, Palumbo K et al (2011) Notch signalling regulates fibroblast activation and collagen release in systemic sclerosis. Ann Rheum Dis 70:1304–1310
Di J, Jiang L, Zhou Y, Cao H, Fang L, Wen P et al (2014) Ets-1 targeted by microRNA-221 regulates angiotensin II-induced renal fibroblast activation and fibrosis. Cell Physiol Biochem 34:1063–1074

Ding H, Zhou D, Hao S, Zhou L, He W, Nie J et al (2012) Sonic hedgehog signaling mediates epithelial-mesenchymal communication and promotes renal fibrosis. J Am Soc Nephrol 23:801–813

DiRocco DP, Kobayashi A, Taketo MM, McMahon AP, Humphreys BD (2013) Wnt4/beta-catenin signaling in medullary kidney myofibroblasts. J Am Soc Nephrol 24:1399–1412

Djudjaj S, Chatziantoniou C, Raffetseder U, Guerrot D, Dussaule J, Boor P et al (2012) Notch-3 receptor activation drives inflammation and fibrosis following tubulointerstitial kidney injury. J Pathol 228:286–299

Edeling M, Ragi G, Huang S, Pavenstädt H, Susztak K (2016) Developmental signalling pathways in renal fibrosis: the roles of Notch, Wnt and Hedgehog. Nat Rev Nephrol 12:426–439

El Agha E, Kramann R, Schneider RK, Li X, Seeger W, Humphreys BD et al (2017) Mesenchymal stem cells in fibrotic disease. Cell Stem Cell 21:166–177

Eyden B (2008) Translational medicine: the myofibroblast: phenotypic characterization as a prerequisite to understanding its functions in translational medicine. J Cell Mol Med 12:22–37

Fabian SL, Penchev RR, St-Jacques B, Rao AN, Sipila P, West KA et al (2012) Hedgehog-Gli pathway activation during kidney fibrosis. Am J Pathol 180:1441–1453

Falke LL, Gholizadeh S, Goldschmeding R, Kok RJ, Nguyen TQ (2015) Diverse origins of the myofibroblast-implications for kidney fibrosis. Nat Rev Nephrol 11:233–244

Francki A, Sage EH (2001) SPARC and the kidney glomerulus: matricellular proteins exhibit diverse functions under normal and pathological conditions. Trends Cardiovasc Med 11:32–37

Frangogiannis NG (2017) The extracellular matrix in myocardial injury, repair, and remodeling. J Clin Invest 127:1600–1612

Fries KM, Blieden T, Looney RJ, Sempowski GD, Silvera MR, Willis RA et al (1994) Evidence of fibroblast heterogeneity and the role of fibroblast subpopulations in fibrosis. Clin Immunol Immunopathol 72:283–292

Fu H, Tian Y, Zhou L, Zhou D, Tan RJ, Stolz DB et al (2017) Tenascin-C is a major component of the fibrogenic niche in kidney fibrosis. J Am Soc Nephrol 28:785–801

Gabbiani G, Ryan GB, Majno G (1971) Presence of modified fibroblasts in granulation tissue and their possible role in wound contraction. Experientia 27:549–550

Gewin L, Zent R, Pozzi A (2017) Progression of chronic kidney disease: too much cellular talk causes damage. Kidney Int 91:552–560

Gladka MM, Molenaar B, de Ruiter H, van der Elst S, Tsui H, Versteeg D et al (2018) Single-cell sequencing of the healthy and diseased heart reveals Ckap4 as a new modulator of fibroblasts activation. Circulation 138:166–180

Gomez IG, Duffield JS (2014) The FOXD1 lineage of kidney perivascular cells and myofibroblasts: functions and responses to injury. Kidney Int Suppl 4:26–33

Grande MT, Sánchez-Laorden B, López-Blau C, De Frutos CA, Boutet A, Arévalo M et al (2015) Snail1-induced partial epithelial-to-mesenchymal transition drives renal fibrosis in mice and can be targeted to reverse established disease. Nat Med 21:989–997

He W, Dai C (2015) Key fibrogenic signaling. Curr Pathobiol Rep 3:183–192

He W, Dai C, Li Y, Zeng G, Monga SP, Liu Y (2009) Wnt/beta-catenin signaling promotes renal interstitial fibrosis. J Am Soc Nephrol 20:765–776

He J, Xu Y, Koya D, Kanasaki K (2013) Role of the endothelial-to-mesenchymal transition in renal fibrosis of chronic kidney disease. Clin Exp Nephrol 17:488–497

Hecker L, Jagirdar R, Jin T, Thannickal VJ (2011) Reversible differentiation of myofibroblasts by MyoD. Exp Cell Res 317:1914–1921

Henderson NC, Arnold TD, Katamura Y, Giacomini MM, Rodriguez JD, McCarty JH et al (2013) Targeting of αv integrin identifies a core molecular pathway that regulates fibrosis in several organs. Nat Med 19:1617–1624

Herrera J, Henke CA, Bitterman PB (2018) Extracellular matrix as a driver of progressive fibrosis. J Clin Invest 128:45–53

Heymann F, Meyer-Schwesinger C, Hamilton-Williams EE, Hammerich L, Panzer U, Kaden S et al (2009) Kidney dendritic cell activation is required for progression of renal disease in a mouse model of glomerular injury. J Clin Invest 119:1286–1297

Hinz B, Phan SH, Thannickal VJ, Galli A, Bochaton-Piallat M, Gabbiani G (2007) Myofibroblast. Am J Pathol 170:1807–1816

Hinz B, Phan SH, Thannickal VJ, Prunotto M, Desmoulière A, Varga J et al (2012) Recent developments in myofibroblast biology. Am J Pathol 180:1340–1355

Holdsworth SR, Summers SA (2008) Role of mast cells in progressive renal diseases. J Am Soc Nephrol 19:2254–2261

Hsia L, Ashley N, Ouaret D, Wang LM, Wilding J, Bodmer WF (2016) Myofibroblasts are distinguished from activated skin fibroblasts by the expression of AOC3 and other associated markers. Proc Natl Acad Sci USA 113:E2162–E2171

Hu B, Phan SH (2013) Myofibroblasts. Curr Opin Rheumatol 25:71–77

Hu K, Wu C, Mars WM, Liu Y (2007) Tissue-type plasminogen activator promotes murine myofibroblast activation through LDL receptor-related protein 1-mediated integrin signaling. J Clin Invest 117:3821–3832

Hu MS, Moore AL, Longaker MT (2018) A fibroblast is not a fibroblast is not a fibroblast. J Invest Dermatol 138:729–730

Huang Y, Tong J, He F, Yu X, Fan L, Hu J et al (2015) MiR-141 regulates TGF-beta1-induced epithelial-mesenchymal transition through repression of HIPK2 expression in renal tubular epithelial cells. Int J Mol Med 35:311–318

Humphreys BD (2018) Mechanisms of renal fibrosis. Annu Rev Physiol 80:309–326

Humphreys BD, Lin S, Kobayashi A, Hudson TE, Nowlin BT, Bonventre JV et al (2010) Fate tracing reveals the pericyte and not epithelial origin of myofibroblasts in kidney fibrosis. Am J Pathol 176:85–97

Inoue T, Umezawa A, Takenaka T, Suzuki H, Okada H (2015) The contribution of epithelial-mesenchymal transition to renal fibrosis differs among kidney disease models. Kidney Int 87:233–238

Irifuku T, Doi S, Sasaki K, Doi T, Nakashima A, Ueno T, et al (2016) Inhibition of H3K9 histone methyltransferase G9a attenuates renal fibrosis and retains klotho expression. Kidney Int 89:147–157

Iwano M, Plieth D, Danoff TM, Xue C, Okada H, Neilson EG (2002) Evidence that fibroblasts derive from epithelium during tissue fibrosis. J Clin Invest 110:341–350

Jones LK, O'Sullivan KM, Semple T, Kuligowski MP, Fukami K, Ma FY et al (2009) IL-1RI deficiency ameliorates early experimental renal interstitial fibrosis. Nephrol Dial Transplant 24:3024–3032

Jun JI, Lau LF (2011) Taking aim at the extracellular matrix: CCN proteins as emerging therapeutic targets. Nat Rev Drug Discov 10:945–963

Kaminski MM, Tosic J, Kresbach C, Engel H, Klockenbusch J, Müller A et al (2016) Direct reprogramming of fibroblasts into renal tubular epithelial cells by defined transcription factors. Nat Cell Biol 18:1269–1280

Kang HM, Ahn SH, Choi P, Ko Y, Han SH, Chinga F et al (2015) Defective fatty acid oxidation in renal tubular epithelial cells has a key role in kidney fibrosis development. Nat Med 21:37–46

Kanisicak O, Khalil H, Ivey MJ, Karch J, Maliken BD, Correll RN et al (2016) Genetic lineage tracing defines myofibroblast origin and function in the injured heart. Nat Commun 7:12260

Karsdal MA, Nielsen SH, Leeming DJ, Langholm LL, Nielsen MJ, Manon-Jensen T et al (2017) The good and the bad collagens of fibrosis—their role in signaling and organ function. Adv Drug Deliv Rev 121:43–56

Kawai T, Masaki T, Doi S, Arakawa T, Yokoyama Y, Doi T, et al (2009) PPAR-gamma agonist attenuates renal interstitial fibrosis and inflammation through reduction of TGF-beta. Lab Invest 89:47–58

Kii I, Ito H (2017) Periostin and its interacting proteins in the construction of extracellular architectures. Cell Mol Life Sci 74:4269–4277

Kis K, Liu X, Hagood JS (2011) Myofibroblast differentiation and survival in fibrotic disease. Expert Rev Mol Med 13:e27

Klingberg F, Hinz B, White ES (2013) The myofibroblast matrix: implications for tissue repair and fibrosis. J Pathol 229:298–309

Klingberg F, Chow ML, Koehler A, Boo S, Buscemi L, Quinn TM et al (2014) Prestress in the extracellular matrix sensitizes latent TGF-β1 for activation. J Cell Biol 207:283–297

Koesters R, Kaissling B, LeHir M, Picard N, Theilig F, Gebhardt R et al (2010) Tubular overexpression of transforming growth tactor-β1 induces autophagy and fibrosis but not mesenchymal transition of renal epithelial cells. Am J Pathol 177:632–643

Kok HM, Falke LL, Goldschmeding R, Nguyen TQ (2014) Targeting CTGF, EGF and PDGF pathways to prevent progression of kidney disease. Nat Rev Nephrol 10:700–711

Kramann R, Fleig SV, Schneider RK, Fabian SL, DiRocco DP, Maarouf O et al (2015a) Pharmacological GLI2 inhibition prevents myofibroblast cell-cycle progression and reduces kidney fibrosis. J Clin Invest 125:2935–2951

Kramann R, Schneider RK, DiRocco DP, Machado F, Fleig S, Bondzie PA et al (2015b) Perivascular Gli1 + progenitors are key contributors to injury-induced organ fibrosis. Cell Stem Cell 16:51–66

Kramann R, Wongboonsin J, Chang-Panesso M, Machado FG, Humphreys BD (2017) Gli1(+) pericyte loss induces capillary rarefaction and proximal tubular injury. J Am Soc Nephrol 28:776–784

Kubow KE, Vukmirovic R, Zhe L, Klotzsch E, Smith ML, Gourdon D et al (2015) Mechanical forces regulate the interactions of fibronectin and collagen I in extracellular matrix. Nat Commun 6:8026

Lagares D, Santos A, Grasberger PE, Liu F, Probst CK, Rahimi RA et al (2017) Targeted apoptosis of myofibroblasts with the BH3 mimetic ABT-263 reverses established fibrosis. Sci Transl Med 9:3765

LeBleu VS, Taduri G, O'Connell J, Teng Y, Cooke VG, Woda C et al (2013) Origin and function of myofibroblasts in kidney fibrosis. Nat Med 19:1047–1053

Lee S, Kim SI, Choi ME (2015) Therapeutic targets for treating fibrotic kidney diseases. Transl Res 165:512–530

Li J, Deane JA, Campanale NV, Bertram JF, Ricardo SD (2007) The contribution of bone marrow-derived cells to the development of renal interstitial fibrosis. Stem Cells 25:697–706

Li J, Qu X, Bertram JF (2009a) Endothelial-myofibroblast transition contributes to the early development of diabetic renal interstitial fibrosis in streptozotocin-induced diabetic mice. Am J Pathol 175:1380–1388

Li Y, Tan X, Dai C, Stolz DB, Wang D, Liu Y (2009b) Inhibition of integrin-linked kinase attenuates renal interstitial fibrosis. J Am Soc Nephrol 20:1907–1918

Li L, Zepeda-Orozco D, Black R, Lin F (2010) Autophagy is a component of epithelial cell fate in obstructive uropathy. Am J Pathol 176:1767–1778

Li Q, Liu BC, Lv LL, Ma KL, Zhang XL, Phillips AO (2011) Monocytes induce proximal tubular epithelial-mesenchymal transition through NF-kappa B dependent upregulation of ICAM-1. J Cell Biochem 112:1585–1592

Li S, Mariappan N, Megyesi J, Shank B, Kannan K, Theus S et al (2013) Proximal tubule PPARalpha attenuates renal fibrosis and inflammation caused by unilateral ureteral obstruction. Am J Physiol Renal Physiol 305:F618–F627

Liang H, Zhang Z, Yan J, Wang Y, Hu Z, Mitch WE et al (2017) The IL-4 receptor alpha has a critical role in bone marrow-derived fibroblast activation and renal fibrosis. Kidney Int 92:1433–1443

Lin S, Kisseleva T, Brenner DA, Duffield JS (2008) Pericytes and perivascular fibroblasts are the primary source of collagen-producing cells in obstructive fibrosis of the kidney. Am J Pathol 173:1617–1627

Lipphardt M, Song JW, Matsumoto K, Dadafarin S, Dihazi H, Müller G et al (2017) The third path of tubulointerstitial fibrosis: aberrant endothelial secretome. Kidney Int 92:558–568

Liu Y (2004) Hepatocyte growth factor in kidney fibrosis: therapeutic potential and mechanisms of action. Am J Physiol Renal Physiol 287:F7–F16151

Liu Y (2010) New insights into epithelial-mesenchymal transition in kidney fibrosis. J Am Soc Nephrol 21:212–222
Liu Y (2011) Cellular and molecular mechanisms of renal fibrosis. Nat Rev Nephrol 7:684–696
Liu N, Zhuang S (2015) Treatment of chronic kidney diseases with histone deacetylase inhibitors. Front Physiol 6:121
Liu T, Hu B, Choi YY, Chung M, Ullenbruch M, Yu H et al (2009) Notch1 signaling in FIZZ1 induction of myofibroblast differentiation. Am J Pathol 174:1745–1755
Liu X, Hong Q, Wang Z, Yu Y, Zou X, Xu L (2016) Transforming growth factor-beta-sphingosine kinase 1/S1P signaling upregulates microRNA-21 to promote fibrosis in renal tubular epithelial cells. Exp Biol Med 241:265–272
Liu BC, Tang TT, Lv LL, Lan HY (2018) Renal tubule injury: a driving force toward chronic kidney disease. Kidney Int 93:568–579
Lopez-Guisa JM, Cai X, Collins SJ, Yamaguchi I, Okamura DM, Bugge TH et al (2012) Mannose receptor 2 attenuates renal fibrosis. J Am Soc Nephrol 23:236–251
Lovisa S, LeBleu VS, Tampe B, Sugimoto H, Vadnagara K, Carstens JL et al (2015) Epithelial-to-mesenchymal transition induces cell cycle arrest and parenchymal damage in renal fibrosis. Nat Med 21:998–1009
Luo C, Zhou S, Zhou Z, Liu Y, Yang L, Liu J et al (2018) Wnt9a promotes renal fibrosis by accelerating cellular senescence in tubular epithelial cells. J Am Soc Nephrol 29:1238–1256
Macconi D, Remuzzi G, Benigni A (2014) Key fibrogenic mediators: old players. Renin–angiotensin system. Kidney Int Suppl 4:58–64
Mack M, Yanagita M (2015) Origin of myofibroblasts and cellular events triggering fibrosis. Kidney Int 87:297–307
Masola V, Gambaro G, Tibaldi E, Brunati AM, Gastaldello A, D'Angelo A et al (2012) Heparanase and syndecan-1 interplay orchestrates fibroblast growth factor-2-induced epithelial-mesenchymal transition in renal tubular cells. J Biol Chem 287:1478–1488
McVicker BL, Bennett RG (2017) Novel anti-fibrotic therapies. Front Physiol 8:318
Meng X, Chung ACK, Lan HY (2013) Role of the TGF-β/BMP-7/Smad pathways in renal diseases. Clin Sci 124:243–254
Meng X, Tang PM, Li J, Lan HY (2015) TGF-β/Smad signaling in renal fibrosis. Front Physiol 6:82
Meng X, Nikolic-Paterson DJ, Lan HY (2016) TGF-β: the master regulator of fibrosis. Nat Rev Nephrol 12:325–338
Mizoguchi F, Slowikowski K, Wei K, Marshall JL, Rao DA, Chang SK et al (2018) Functionally distinct disease-associated fibroblast subsets in rheumatoid arthritis. Nat Commun 9:789
Munoz-Felix JM, Gonzalez-Nunez M, Martinez-Salgado C, Lopez-Novoa JM (2015) TGF-beta/BMP proteins as therapeutic targets in renal fibrosis. Where have we arrived after 25 years of trials and tribulations? Pharmacol Ther 156:44–58
Nahrwold ML, Lecky JH, Cohen PJ (1974) The effect of halothane on mitochondrial permeability to NADH. Life Sci 15:1261–1265
Nastase MV, Zeng-Brouwers J, Wygrecka M, Schaefer L (2017) Targeting renal fibrosis: mechanisms and drug delivery systems. Adv Drug Deliv Rev 129:295–307
Nightingale J, Patel S, Suzuki N, Buxton R, Takagi KI, Suzuki J et al (2004) Oncostatin M, a cytokine released by activated mononuclear cells, induces epithelial cell-myofibroblast transdifferentiation via Jak/Stat pathway activation. J Am Soc Nephrol 15:21–32
Nikolic-Paterson DJ, Wang S, Lan HY (2014) Macrophages promote renal fibrosis through direct and indirect mechanisms. Kidney Int Suppl 4:34–38
Novitskaya T, McDermott L, Zhang KX, Chiba T, Paueksakon P, Hukriede NA et al (2014) A PTBA small molecule enhances recovery and reduces postinjury fibrosis after aristolochic acid-induced kidney injury. Am J Physiol Renal Physiol 306:F496–F504
Ostendorf T, Boor P, van Roeyen CR, Floege J (2014) Platelet-derived growth factors (PDGFs) in glomerular and tubulointerstitial fibrosis. Kidney Int Suppl 4:65–69
Phan SH (2008) Biology of fibroblasts and myofibroblasts. Proc Am Thorac Soc 5:334–337

Picard N, Baum O, Vogetseder A, Kaissling B, Le Hir M (2008) Origin of renal myofibroblasts in the model of unilateral ureter obstruction in the rat. Histochem Cell Biol 130:141–155
Piersma B, Bank RA, Boersema M (2015) Signaling in Fibrosis: TGF-β, WNT, and YAP/TAZ Converge. Front Med 2:59
Poosti F, Bansal R, Yazdani S, Prakash J, Post E, Klok P et al (2015) Selective delivery of IFN-γ to renal interstitial myofibroblasts: a novel strategy for the treatment of renal fibrosis. FASEB J 29:1029–1042
Poosti F, Bansal R, Yazdani S, Prakash J, Beljaars L, van den Born J et al (2016) Interferon gamma peptidomimetic targeted to interstitial myofibroblasts attenuates renal fibrosis after unilateral ureteral obstruction in mice. Oncotarget 7:54240–54252
Rogers NM, Ferenbach DA, Isenberg JS, Thomson AW, Hughes J (2014) Dendritic cells and macrophages in the kidney: a spectrum of good and evil. Nat Rev Nephrol 10:625–643
Roufosse C, Bou-Gharios G, Prodromidi E, Alexakis C, Jeffery R, Khan S et al (2006) Bone marrow-derived cells do not contribute significantly to collagen I synthesis in a murine model of renal fibrosis. J Am Soc Nephrol 17:775–782
Samuel CS, Hewitson TD (2009) Relaxin and the progression of kidney disease. Curr Opin Nephrol Hypertens 18:9–14
Sandbo N, Dulin N (2011) Actin cytoskeleton in myofibroblast differentiation: ultrastructure defining form and driving function. Transl Res 158:181–196
Sato Y, Yanagita M (2017) Resident fibroblasts in the kidney: a major driver of fibrosis and inflammation. Inflamm Regen 37:17
Sato Y, Mii A, Hamazaki Y, Fujita H, Nakata H, Masuda K et al (2016) Heterogeneous fibroblasts underlie age-dependent tertiary lymphoid tissues in the kidney. JCI Insight 1:e87680
Serini G, Bochaton-Piallat ML, Ropraz P, Geinoz A, Borsi L, Zardi L et al (1998) The fibronectin domain ED-A is crucial for myofibroblastic phenotype induction by transforming growth factor-beta1. J Cell Biol 142:873–881
Siani A, Tirelli N (2014) Myofibroblast Differentiation: main features, biomedical relevance, and the role of reactive oxygen species. Antioxid Redox Signal 21:768–785
Singer II (1979) The fibronexus: a transmembrane association of fibronectin-containing fibers and bundles of 5 nm microfilaments in hamster and human fibroblasts. Cell 16:675–685
Snyder JJ, Foley RN, Collins AJ (2009) Prevalence of CKD in the United States: a sensitivity analysis using the National Health and Nutrition Examination Survey (NHANES) 1999–2004. Am J Kidney Dis 53:218–228
Sugimoto H, LeBleu VS, Bosukonda D, Keck P, Taduri G, Bechtel W et al (2012) Activin-like kinase 3 is important for kidney regeneration and reversal of fibrosis. Nat Med 18:396–404
Sun G, Reddy MA, Yuan H, Lanting L, Kato M, Natarajan R (2010) Epigenetic histone methylation modulates fibrotic gene expression. J Am Soc Nephrol 21:2069–2080
Sun K, Chang Y, Reed NI, Sheppard D (2016a) α-Smooth muscle actin is an inconsistent marker of fibroblasts responsible for force-dependent TGFβ activation or collagen production across multiple models of organ fibrosis. Am J Physiol Lung Cell Mol Physiol 310:L824–L836
Sun YBY, Qu X, Caruana G, Li J (2016b) The origin of renal fibroblasts/myofibroblasts and the signals that trigger fibrosis. Differentiation 92:102–107
Tabib T, Morse C, Wang T, Chen W, Lafyatis R (2018) SFRP2/DPP4 and FMO1/LSP1 define major fibroblast populations in human skin. J Invest Dermatol 138:802–810
Tampe B, Tampe D, Muller CA, Sugimoto H, LeBleu V, Xu X et al (2014) Tet3-mediated hydroxymethylation of epigenetically silenced genes contributes to bone morphogenic protein 7-induced reversal of kidney fibrosis. J Am Soc Nephrol 25:905–912
Tan RJ, Zhou D, Zhou L, Liu Y (2014) Wnt/β-catenin signaling and kidney fibrosis. Kidney Int Suppl 4:84–90
Tan RJ, Zhou D, Liu Y (2016) Signaling crosstalk between tubular epithelial cells and interstitial fibroblasts after kidney injury. Kidney Dis 2:136–144

Tang O, Chen X, Shen S, Hahn M, Pollock CA (2013) MiRNA-200b represses transforming growth factor-β1-induced EMT and fibronectin expression in kidney proximal tubular cells. Am J Physiol Renal Physiol 304:F1266–F1273
Tapmeier TT, Fearn A, Brown K, Chowdhury P, Sacks SH, Sheerin NS et al (2010) Pivotal role of CD4+ T cells in renal fibrosis following ureteric obstruction. Kidney Int 78:351–362
Tsou P, Haak AJ, Khanna D, Neubig RR (2014) Cellular mechanisms of tissue fibrosis. 8. Current and future drug targets in fibrosis: focus on Rho GTPase-regulated gene transcription. Am J Physiol Cell Physiol 307:C2–C13
Van De Water L, Varney S, Tomasek JJ (2013) Mechanoregulation of the myofibroblast in wound contraction, scarring, and fibrosis: opportunities for new therapeutic intervention. Adv Wound Care 2:122–141
Villani AC, Satija R, Reynolds G, Sarkizova S, Shekhar K, Fletcher J et al (2017) Single-cell RNA-seq reveals new types of human blood dendritic cells, monocytes, and progenitors. Science 356:4573
von Holst A (2008) Tenascin C in stem cell niches: redundant, permissive or instructive? Cells Tissues Organs 188:170–177
Walraven M, Hinz B (2018) Therapeutic approaches to control tissue repair and fibrosis: extracellular matrix as a game changer. Matrix Biol 72:205–224
Wang YY, Jiang H, Pan J, Huang XR, Wang YC, Huang HF et al (2017a) Macrophage-to-myofibroblast transition contributes to interstitial fibrosis in chronic renal allograft injury. J Am Soc Nephrol 28:2053–2067
Wang H, Qian J, Zhao X, Xing C, Sun B (2017b) β-Aminoisobutyric acid ameliorates the renal fibrosis in mouse obstructed kidneys via inhibition of renal fibroblast activation and fibrosis. J Pharmacol Sci 133:203–213
Wang P, Luo M, Song E, Zhou Z, Ma T, Wang J, et al (2018, in press) Long noncoding RNA lnc-TSI inhibits renal fibrogenesis by negatively regulating the TGF-β/Smad3 pathway. Sci Transl Med
Widyantoro B, Emoto N, Nakayama K, Anggrahini DW, Adiarto S, Iwasa N et al (2010) Endothelial cell-derived endothelin-1 promotes cardiac fibrosis in diabetic hearts through stimulation of endothelial-to-mesenchymal transition. Circulation 121:2407–2418105
Wu H, Humphreys BD (2017) The promise of single-cell RNA sequencing for kidney disease investigation. Kidney Int 92:1334–1342
Xiao Z, Zhang J, Peng X, Dong Y, Jia L, Li H et al (2014) The Notch γ-secretase inhibitor ameliorates kidney fibrosis via inhibition of TGF-β/Smad2/3 signaling pathway activation. Int J Biochem Cell Biol 55:65–71
Xiao X, Tang W, Yuan Q, Peng L, Yu P (2015) Epigenetic repression of Kruppel-like factor 4 through Dnmt1 contributes to EMT in renal fibrosis. Int J Mol Med 35:1596–1602
Xie T, Wang Y, Deng N, Huang G, Taghavifar F, Geng Y et al (2018) Single-cell deconvolution of fibroblast heterogeneity in mouse pulmonary fibrosis. Cell Rep 22:3625–3640
Yan J, Zhang Z, Yang J, Mitch WE, Wang Y (2015) JAK3/STAT6 Stimulates bone marrow-derived fibroblast activation in renal fibrosis. J Am Soc Nephrol 26:3060–3071
Yan J, Zhang Z, Jia L, Wang Y (2016) Role of bone marrow-derived fibroblasts in renal fibrosis. Front Physiol 7:61
Yang J, Dai C, Liu Y (2003) Hepatocyte growth factor suppresses renal interstitial myofibroblast activation and intercepts Smad signal transduction. Am J Pathol 163:621–632
Yang L, Besschetnova TY, Brooks CR, Shah JV, Bonventre JV (2010) Epithelial cell cycle arrest in G2/M mediates kidney fibrosis after injury. Nat Med 16:535–543
Yazdani S, Bansal R, Prakash J (2017) Drug targeting to myofibroblasts: implications for fibrosis and cancer. Adv Drug Deliv Rev 121:101–116
Yokoi H, Mukoyama M (2017) Analysis of pathological activities of CCN proteins in fibrotic diseases: kidney fibrosis. Methods Mol Biol 1489:431–443
Zeisberg EM, Zeisberg M (2013) The role of promoter hypermethylation in fibroblast activation and fibrogenesis. J Pathol 229:264–273

Zeisberg EM, Potenta SE, Sugimoto H, Zeisberg M, Kalluri R (2008) Fibroblasts in kidney fibrosis emerge via endothelial-to-mesenchymal transition. J Am Soc Nephrol 19:2282–2287

Zhao Y, Yin Z, Li H, Fan J, Yang S, Chen C et al (2017) MiR-30c protects diabetic nephropathy by suppressing epithelial-to-mesenchymal transition in db/db mice. Aging Cell 16:387–400

Zhou D, Liu Y (2016a) Renal fibrosis in 2015: understanding the mechanisms of kidney fibrosis. Nat Rev Nephrol 12:68–70

Zhou D, Liu Y (2016b) Therapy for kidney fibrosis: is the Src kinase a potential target? Kidney Int 89:12–14

Zhou L, Liu Y (2016c) Wnt/β-catenin signaling and renin–angiotensin system in chronic kidney disease. Curr Opin Nephrol Hypertens 25:100–106

Zhou D, Tan RJ, Zhou L, Li Y, Liu Y (2013a) Kidney tubular β-catenin signaling controls interstitial fibroblast fate via epithelial-mesenchymal communication. Sci Rep 3:1878

Zhou L, Li Y, Zhou D, Tan RJ, Liu Y (2013b) Loss of Klotho contributes to kidney injury by derepression of Wnt/beta-catenin signaling. J Am Soc Nephrol 24:771–785

Zhou D, Li Y, Zhou L, Tan RJ, Xiao L, Liang M et al (2014a) Sonic hedgehog is a novel tubule-derived growth factor for interstitial fibroblasts after kidney injury. J Am Soc Nephrol 25:2187–2200

Zhou Q, Chung AC, Huang XR, Dong Y, Yu X, Lan HY (2014b) Identification of novel long noncoding RNAs associated with TGF-beta/Smad3-mediated renal inflammation and fibrosis by RNA sequencing. Am J Pathol 184:409–417

Zhou D, Fu H, Zhang L, Zhang K, Min Y, Xiao L et al (2017) Tubule-derived Wnts are required for fibroblast activation and kidney fibrosis. J Am Soc Nephrol 28:2322–2336

Zhou L, Zhou S, Yang P, Tian Y, Feng Z, Xie XQ et al (2018a) Targeted inhibition of the type 2 cannabinoid receptor is a novel approach to reduce renal fibrosis. Kidney Int 4:756–772

Zhou X, Xiong C, Tolbert E, Zhao TC, Bayliss G, Zhuang S (2018) Targeting histone methyltransferase enhancer of zeste homolog-2 inhibits renal epithelial-mesenchymal transition and attenuates renal fibrosis. FASEB J:j201800237R

Chapter 13
Macrophages in Renal Fibrosis

Xiao-Ming Meng, Thomas Shiu-Kwong Mak and Hui-Yao Lan

Abstract Monocytes/macrophages are highly involved in the process of renal injury, repair and fibrosis in many aspects of experimental and human renal diseases. Monocyte-derived macrophages, characterized by high heterogeneity and plasticity, are recruited, activated, and polarized in the whole process of renal fibrotic diseases in response to local microenvironment. As classically activated M1 or $CD11b^+/Ly6C^{high}$ macrophages accelerate renal injury by producing pro-inflammatory factors like tumor necrosis factor-alpha (TNFα) and interleukins, alternatively activated M2 or $CD11b^+/Ly6C^{intermediate}$ macrophages may contribute to kidney repair by exerting anti-inflammation and wound healing functions. However, uncontrolled M2 macrophages or $CD11b^+/Ly6C^{low}$ macrophages promote renal fibrosis via paracrine effects or direct transition to myofibroblast-like cells via the process of macrophage-to-myofibroblast transition (MMT). In this regard, therapeutic strategies targeting monocyte/macrophage recruitment, activation, and polarization should be emphasized in the treatment of renal fibrosis.

Keywords Macrophage · Renal fibrosis · Macrophage-myofibroblast transition · Macrophage polarization

---

X.-M. Meng
School of Pharmacy, Anhui Medical University, Hefei 230032, Anhui, China
e-mail: mengxiaoming@ahmu.edu.cn

T. S.-K. Mak · H.-Y. Lan (✉)
Department of Medicine and Therapeutics, Li Ka Shing Institute of Health Sciences, and Lui Chi Woo Institute of Innovative Medicine, The Chinese University of Hong Kong, Shatin, Hong Kong SAR, China
e-mail: hylan@cuhk.edu.hk

T. S.-K. Mak
e-mail: thomak@hotmail.com.hk

B.-C. Liu et al. (eds.), *Renal Fibrosis: Mechanisms and Therapies*,
Advances in Experimental Medicine and Biology 1165,
https://doi.org/10.1007/978-981-13-8871-2_13

13.1 Introduction

Renal fibrosis is a common pathological feature of chronic kidney diseases (CKD) and characterized by excessive extracellular matrix (ECM) deposition and myofibroblast accumulation (Meng et al. 2016a). Macrophages, firstly identified by Metchnikoff over one hundred years ago, are highly diverse and exhibit a wide range of complex roles in host defense, tissue development, homeostasis, tissue injury and repair, and fibrosis (Wilson et al. 2004; Wynn and Vannella 2016). In the kidney, macrophages originate from yolk sac, fetal liver and bone marrow. It is noteworthy that bone marrow myeloid progenitors-derived monocytes are the major source of infiltrated macrophages (Huen and Cantley 2015). In the injured kidney, local production of chemokines induces the infiltration of neutrophils and naïve monocytes from which differentiate into phagocytic macrophages, and then they are polarized and activated in response to the local immune microenvironment (Yona et al. 2013). As the major mediator for inflammatory response, monocytes/macrophages are highly involved in the process of renal injury and repair in many aspects of experimental and human renal diseases (Duffield 2010). They are regarded as a critical link between renal inflammation and fibrosis (Meng et al. 2014). Macrophages, with high heterogeneity and plasticity, are activated and polarized into different phenotypes in the progression of renal disease, they then secrete various cytokines and growth factors accordingly, which may alter the microenvironment in diseased kidney in a feedback loop, the interplay between macrophages and neighboring cells such as immune cells and resident kidney cells may determine the fate of renal diseases (Anders and Ryu 2011; Duffield 2010; Ricardo et al. 2008). In this setting, this chapter highlighted recent progress in the understanding of the role of monocytes/macrophages in renal fibrosis, with a focus on the monocytes/macrophages recruitment, phenotypes, functions, and regulatory mechanisms in progression of renal fibrosis, then the therapeutic potential for macrophage-based or targeted therapy for renal fibrosis were also discussed.

13.2 Recruitment of Monocytes/Macrophages in the Kidney

Previous studies have shown that the recruitment of bone marrow-derived monocytes into kidney is a critical step for renal inflammation (Braga et al. 2018), with extensive discussion on the several key chemokines involved. CCR2 and its main ligand, CCL2 (also called MCP-1), are indicated in various types of kidney diseases; they are responsible for the recruitment of Ly6C^{High} monocytes and regulation of bone marrow-derived fibroblasts in injured kidney (Braga et al. 2018). Emerging evidence further shows that knockout of CCR2 and 4, instead of CCR3 and 5, attenuates renal fibrosis (Braga et al. 2018), these results are further confirmed by the finding that treatment of CCX140-B, a CCR2 inhibitor, protects against type 2 diabetic nephropathy (Weir 2015). Transforming growth factor-β (TGF-β) is reported to up-regulate the expression of CCL2 in macrophages and then promote monocyte recruitment and macrophage accumulation (Border and Noble 1994). The interaction between CX3CL1 and CX3CR1 is also responsible for the infiltration of Ly6C-CX3CR1^{high} macrophages, which contribute significantly to unilateral ureteral obstruction (UUO)-induced renal fibrosis (Peng et al. 2015). Additionally, chemokine CXCL16 and its receptor CXCR6 play important roles in recruiting monocytes from circulation to the injured kidney in UUO nephropathy, hypertensive nephropathy, and ischemia-reperfusion acute kidney injury (AKI) (Chen et al. 2011; Xia et al. 2013, 2014a, b). Tubular-derived IL-34, being one of the macrophage differentiation and growth factors, shares a common receptor with macrophage colony-stimulating factor (M-CSF). It fails to alter kidney macrophages' activation phenotypes but induces persistent tubular injury via macrophage recruitment and proliferation in the later stages of tubular repair and fibrosis (Baek et al. 2015). Newer evidence shows that the accumulation of B cells in the early stage of kidney injury enhances monocyte/macrophage mobilization and recruitment, thereby accelerates renal fibrosis in UUO nephropathy (Han et al. 2017).

13.3 Activation and Polarization of Monocytes/Macrophages in the Kidney

As aforementioned, bone marrow myeloid progenitors-derived monocytes are the major source for infiltrated macrophages (Duffield 2010; Wilson et al. 2004). Monocytes could be categorized into different subsets as defined by lymphocyte antigen 6C (Ly6C), an antigen representing the stages of a continuous maturation pathway, and chemokine receptor profiles like CCR2 and CX3CR1 (Ricardo et al. 2008; Sunderkötter et al. 2004). For example, CCR2^{+}Ly6C^{+} monocyte recruited to the site of inflammation has been identified as a specific monocyte subset that functions in immune response and tissue remodeling (Geissmann et al. 2003). Monocytes then differentiate into macrophages with distinct activation states in response to local microenvironment. To represent the Th1/Th2 paradigm, classification of M1/M2 macrophages has been widely used, although it may be a gross oversimplification of representing the expanded phenotype diversity accurately (Guilliams et al. 2014; Martinez and Gordon 2014; Murray et al. 2014; Wermuth and Jimenez 2015). Pro-inflammatory M1 macrophages, also termed as classically activated macrophages, are induced by interferon (IFN)-γ and lipopolysaccharide (LPS) in vitro, while wound healing/pro-fibrotic M2 macrophages, also called alternatively activated macrophages, are generated by interleukin (IL)-4 and IL-13 incubation. M2 macrophages could be further subcategorized based on different stimuli and functions: IL-4 and IL-13 trigger M2a macrophages; immune complexes induce M2b macrophages; IL-10 plus TGF-β or glucocorticoids induce anti-inflammatory M2c macrophages (Anders and Ryu 2011). In the injured kidney, macrophages are activated by multiple factors, which include other types of immune cells like T cells and NK cells, pathogen-associated molecular patterns (PAMPs), damage-associated molecular patterns (DAMPs), and immune complexes (Anders 2010; Duffield 2010). M1 macrophages are generally induced by pro-inflammatory cytokines like IFN-γ, LPS, and TNF-α while M2 macrophages are polarized by Th2 cytokines, and macrophages gained M2 phenotype after engulfing apoptotic cells (Anders and Ryu 2011; Swaminathan and Griffin 2008; Vinuesa et al. 2008). Evidence shows that high level of iNOS, instead of Arginase 1, is expressed in macrophages 24 h post-injury, indicating that pro-inflammatory M1 macrophages become predominant in the early stage of kidney diseases (Lee et al. 2011). Additionally, polarization between M1 and M2 is also detectable in vivo, which is supported by the finding that IFN-γ-stimulated M1 cells can switch to M2 in the repaired kidney after being injected in the early stage of AKI model (Lee et al. 2011).

Macrophages derived from circulating inflammatory Ly6C^{high} monocytes could also be divided into three subcategories depending on the level of Ly6C markers (Clements et al. 2016; Lin et al. 2009). CD11b^{+}/Ly6C^{high} macrophages are associated with the initiation of renal injury, they mimic the function of M1 macrophages by producing abundant pro-inflammatory cytokines (e.g., TNF-α) and chemokines (e.g., MIP-1) (Meng et al. 2015). Deletion of circulating monocytes and recruited Ly6C^{high} macrophages attenuates renal fibrosis (Lin et al. 2009). The number of the CD11b^{+}/Ly6C^{intermediate} macrophages is significantly increased during the repair stage. By contrast, CD11b^{+}/Ly6C^{low} macrophages are predominant in renal fibrosis through producing pro-fibrotic factors including platelet-derived growth factor (PDGF), insulin-like growth factor (IGF)-1, and CCL17, which are highly correlated to wound healing and fibrogensis (Duffield 2010). Additionally, gene signature in CD11b^{+}/Ly6C^{low} macrophages has been well defined and within the significantly altered genes, SPARC regulates the production of ECM while TIMP2 prevents MMPs-mediated ECM turnover and enhances matrix accumulation (Fan et al. 2014; Wang et al. 2010). Additionally, Macrophages-derived IGF-1 attenuates myofibroblast apoptosis and enhances collagen production (Wynes et al. 2004). In rhabdomyolysis-induced AKI mouse model, macrophage polarization was detected during the disease progression, an abundance of F4/80^{low}CD11b^{high}Ly6b^{high}CD206^{low} macrophages was found in kidney two days after rhabdomyolysis, whereas F4/80^{high}CD11b^{+}Ly6b^{low}CD206^{high} cells became predominant by day 8 (Belliere et al. 2015). All these evidences indicate the pro-fibrotic role of CD11b^{+}/Ly6C^{low} macrophages in renal fibrosis.

13.4 Role of Monocytes/Macrophages in Kidney Injury

Glomerular and interstitial macrophage infiltration is detectable in different types of AKI and progressive CKD of both experimental models and human biopsies (Wilson et al. 2004). Classically activated macrophages produce pro-inflammatory factors like IL-1, TNF-α, IL-6, IL-23, ROS, NO, and iNOS, overproduction of these factors induces severe kidney damage. Pro-inflammatory macrophages infiltration is highly correlated with the degree of renal damage in both AKI and CKD models. By using different macrophage depletion and transfer techniques, pathogenic roles of these pro-inflammatory macrophages have been determined in different kidney disease models (Cao et al. 2013). Liposomal clodronate-mediated macrophage depletion in early stage of ischemia-reperfusion injury (IRI) and rhabdomyolysis-induced AKI significantly reduces renal injury and long-term renal fibrosis, indicating the pathogenic role of M1 macrophages in the initiation of kidney injury (Belliere et al. 2015; Day et al. 2005; Jo et al. 2006; Ko et al. 2008). Additionally, depletion of macrophages with clodronate liposome or CCR2 deficiency attenuates renal injury and fibrosis in UUO nephropathy (Kitagawa et al. 2004; Kitamoto et al. 2009). Pro-inflammatory macrophages also mediate renal injury in CKD model, macrophages deletion or deactivation by clodronate, c-fms inhibitor, or JNK inhibitor prevents the progression of crescentic anti-GBM glomerulonephritis (D'Souza et al. 1999; Han et al. 2011; Ma et al. 2009). In contrast, adoptive transfer of bone marrow-derived macrophages in early stage of the same disease model enhances renal injury (Ikezumi et al. 2003). Taken together, pro-inflammatory M1 macrophages enhance renal injury possibly through mechanism as follows: First, accelerating renal inflammation by releasing an abundance of pro-inflammatory cytokines and chemokines (Cao et al. 2013); second, overproduction of ROS and TNF-α by macrophages induces apoptosis of renal resident cells, including tubular epithelial cells (TECs) and endothelial cells, and prevents their proliferation, thereby increases renal injury (Kluth et al. 2004); third, a plethora of pro-fibrotic cytokines and growth factors released from macrophages triggers abnormal wound healing and leads to renal fibrosis eventually (Anders and Ryu 2011).

13.5 Role of Monocytes/Macrophages in Kidney Repair

Anti-inflammatory and reparative roles of macrophages have been well studied (Day et al. 2005; Huen and Cantley 2015; Lee et al. 2011). M2 macrophages and $CD11b^{+}/Ly6C^{intermediate}$ macrophages become predominant in the repair stage of kidney disease models such as IRI and UUO nephropathy, and they serve as key regulators for renal inflammation resolution and wound healing (Cochrane et al. 2005; Lee et al. 2011). Fluorescence-labeled cell tracing study shows that 6 days after IRI, a majority of macrophages loss iNOS markers and gained high level of Arginase 1, showing the phenotypic switch of macrophages toward M2 in the repair phase of AKI (Lee et al. 2011). Depletion of macrophages in late stage of IRI model reduces TEC proliferation and delays renal repair, but transferring IL-4-polarized M2 macrophages induces the repair process (Vinuesa et al. 2008). IL-4/IL-13-polarized M2a macrophages are essential for the recovery from ischemic AKI (Zhang et al. 2017). Additionally, calcium-binding protein S100A8/A9 complex, as a typical DAMP, promotes M2 polarization, thereby increases renal repair following IRI (Dessing et al. 2015). M2 macrophages exhibit anti-inflammatory effect mainly through induction of anti-inflammatory factors and high endocytic capacities (Ricardo et al. 2008). M2 macrophages synthesize an abundance of IL-10 after engulfing unwanted cells and their debris. They produce other anti-inflammatory cytokines and trophic factors like TGF-β, IGF, and hepatocyte growth factor (HGF). M2 macrophages can deactivate T cells and macrophages to alleviate renal inflammation. It is noteworthy that M2c macrophages induce production of Tregs to exert more powerful anti-immunological effects compared with other subtypes (Lu et al. 2013; Mu et al. 2005). Moreover, M2 macrophages stimulate angiogenesis and promote endothelial repair (Mantovani et al. 2013). Failure of polarization from pro-inflammatory M1 to reparative M2 macrophages leads to progressive renal inflammation and fibrosis after IRI (Lech et al. 2014). Macrophage-derived Wnt7b signaling enhances epithelial response and accelerates renewal of stem cells or progenitor cells, thereby induces renal repair following IRI directly (Lin et al. 2010). BRP-39, a macrophage-produced chitinase-like protein, prevents tubular apoptosis in a PI3K/AKT-dependent manner (Schmidt et al. 2013). Macrophage-derived HO-1 also contributes to macrophage-mediated renoprotective effect (Ferenbach et al. 2010, 2011). Furthermore, cross talk between injured tubular cells and activated macrophages via retinoic acid signaling also coordinates tubular repair (Chiba et al. 2016).

13.6 Role of Monocytes/Macrophages in Kidney Fibrosis and Fibrolysis

Anti-inflammatory macrophages promote tubular re-epithelialization via the production of trophic factors. However, unresolved or severe inflammation initiates renal fibrosis (Anders and Ryu 2011). Evidence shows that depletion of macrophage attenuates renal fibrosis in most occasions, showing the pro-fibrotic effect of macrophages in various renal diseases (Meng et al. 2014; Vernon et al. 2010; Zeisberg and Duffield 2010). For example, depletion of monocytes/macrophages by liposome-encapsulated clodronate (LEC) lowers blood pressure and reduces hypertensive renal injury and fibrosis (Huang et al. 2018). Liposomal clodronate-mediated depletion of macrophages prevents renal fibrosis following IRI and UUO nephropathy (Ko et al. 2008; Sung et al. 2007), this is further evidenced by the finding that mutation of MCP-1 gene significantly suppresses renal fibrosis (Wada et al. 2004). Of note, large numbers of M2 macrophages, detected in the active fibrotic area in renal biopsy of IgA patients, are positively correlated with the severity of glomerulosclerosis and interstitial fibrosis (Ikezumi et al. 2011). Consistently, glucocorticoid treatment accelerates global glomerulosclerosis in rat thy-1 mesangial proliferative glomerulonephritis, and it is correlated with increased numbers of M2 macrophages (Ikezumi et al. 2010). Moreover, deficiency of macrophages in fibrotic phase prevents renal fibrosis via reducing TGF-β1 expression and capillary rarefaction (Han et al. 2013). Collectively, macrophages promote renal fibrosis possibly through mechanisms as followed: First, M2 macrophages produce numbers of pro-fibrotic factors, such as TGF-β1, fibroblast growth factor 2 (FGF-2), PDGF, and galectin-3, which promote myofibroblast proliferation, survival, and activation, and overproduction of ECM (Floege et al. 2008; Henderson et al. 2008; Wynes et al. 2004), although macrophage-derived TGF-β1 may not be essential for UUO-induced renal interstitial fibrosis (Huen et al. 2013); second, macrophage-derived cytokines and factors, such as IL-1, matrix metalloproteinases (MMP)-9, TGF-β1, angiotensin (Ang)-II, PDGF, IGF-1 and FGF-2, enhance myofibroblasts transdifferentiation or activation from tubular epithelial cells via epithelial-mesenchymal transition (EMT), endothelial cells via endothelial-mesenchymal transition (EndoMT), pericytes, local fibroblasts, and mesangial cells (Falke et al. 2015; LeBleu et al. 2013; Meng et al. 2013). Third, macrophages produce fibronectin and collagen in response to pro-fibrotic microenvironment (Gratchev et al. 2001; Schnoor et al. 2008). Emerging evidence indicates that monocytes/macrophages transdifferentiate into collagen-producing fibrocytes (Duffield 2010) or directly into myofibroblast-like cells (Bertrand et al. 1992; Chen et al. 2014; Mooney et al. 2010; Nikolic-Paterson et al. 2014; Pilling and Gomer 2012). Fourth, activated macrophages damage glomerular and peritubular capillaries, and thereby promote hypoxia-driven fibrosis (Fine and Norman 2008; Han et al. 2013). However, we should note that M2 macrophages might not definitely contribute to renal fibrosis (Anders and Ryu 2011). Inflammation and epithelial healing characterize the first-line danger response program for wound healing. Fibrosis, a major event in the second-line danger response program, only occurs when epithelial healing is incomplete or insufficient, such as in the cases of sustained injury and unresolved renal inflammation (Gurtner et al. 2008). During inflammatory response, bone marrow-derived macrophages are recruited into the inflamed kidney and further differentiate into collagen-producing myofibroblasts locally in the injured kidney via newly identified phenomenon termed macrophage-to-myofibroblast (MMT) (Wang et al. 2016, 2017; Meng et al. 2016b; Tang et al. 2018). The MMT cells can be recognized by their co-expression of macrophage (CD68) and myofibroblast (α-smooth muscle actin, α-SMA) markers in the diseased kidney and account for more than half of α-SMA-expressing macrophages in both human and experimental models of chronic kidney diseases including chronic renal allograph rejection (Wang et al. 2016, 2017; Meng et al. 2016b; Tang et al. 2018). However, some studies show that bone marrow-derived cells make only a small fraction of contribution to myofibroblasts directly; these conflicting results warrant further investigation (Lin et al. 2008; Roufosse et al. 2006; Kramann et al. 2018).

In the fibrolysis stage, macrophages could serve as a negative regulator for renal fibrosis (Anders and Ryu 2011). Evidence shows that fibrolytic macrophage promotes resolution of renal fibrosis through producing matrix metalloproteinases (MMPs), and thereby degrades ECM in fibrotic kidney (Anders and Ryu 2011; Ronco and Chatziantoniou 2008). However, the exact phenotype for fibrolytic macrophage is not fully understood. Regression of established fibrosis has been well studied in liver, depletion of macrophages in the late stage of CCL4-induced liver fibrosis prevents the clearance of liver scars, which may be caused by the loss of macrophage-triggered hepatic stellate cell (HSC) apoptosis (Duffield et al. 2005a), Moreover, macrophage-produced MMP-13 removes fibrotic scar in liver (Fallowfield et al. 2007). Transfer of bone marrow-derived macrophages reverses liver fibrosis and promotes liver recovery (Thomas et al. 2011). In kidney, deficiency of angiotensin II type 1 receptor (AngIIr1) reduces the phagocytic activity of macrophages, thereby promotes renal fibrosis as compared with mice transplanted with AngIIr1$^{+/+}$ bone marrow cells in the late phase of UUO nephropathy (Nishida et al. 2002). Additionally, urokinase-type plasminogen activator receptor (uPAR) enhances macrophage infiltration and scavenger receptor function, therefore increasing the resolution of renal fibrosis (Zhang et al. 2003). In addition, adoptive transfer of macrophages 14 days after UUO surgery attenuates renal fibrosis and enhances renal repair in a MMP-2-dependent manner (Nishida et al. 2005, 2007). Of note, functions of MMPs vary in different stages of renal diseases, for example, MMP-2 and MMP-9 are pathogenic by destroying glomerular and tubular basement membranes and inducing EMT in early stage of renal diseases (Cheng and Lovett 2003; Cheng et al. 2006; Rao et al. 2006; Ronco et al. 2007).

13.7 Regulatory Mechanisms of Macrophage Polarization in Renal Fibrosis

Molecular mechanisms underlying the activation and polarization of macrophages have been extensively investigated (Meng et al. 2015). Increasing evidence shows that macrophage polarization is regulated by various transcriptional factors like STATs, PPARs, KLFs, and C/EBP and multiple signaling pathways such as NF-κB, JNK, JAK/STAT, PI3K/AKT, Wnt/β-catenin, and Notch signals (Kapoor et al. 2015; Piccolo et al. 2017; Zhou et al. 2014). Some other mediators have also been identified, for example, high-mobility group box 1 (HMGB1) protein produced by TEC and infiltrated macrophages contribute to the M1 macrophage activation, as shown by the high level of iNOS and suppression of IL-10 in macrophages. Blocking HMGB1 production with a glycyrrhizic acid derivative reduced M1 polarization, kidney injury and fibrosis in UUO nephropathy (Tian et al. 2015). Knockout of suppressor of cytokine signaling-3 (SOCS-3), a critical intracellular negative regulator, enhances cell proliferation and M2 activation in a JAK/STAT-dependent mechanism while overexpression of SOCS-3 in TECs induces classical activation of the cocultured macrophages, indicating its role in macrophage polarization (Susnik et al. 2014). A recent study showed that myeloid-specific knockout of the transcription factor recombination signal binding protein-Jκ (RBP-J), a modulator essential for Notch activation, decreased monocyte infiltration and macrophage activation, thereby alleviated renal fibrosis (Jiang et al. 2018).

Mediators for M2 polarization have also been extensively reviewed. CSF-1 is an important inducer for macrophage polarization. Loss of CSF-1 reduces M2 macrophages, thereby inhibits TEC proliferation and tubular repair (Menke et al. 2009; Zhang et al. 2012). This is confirmed by the finding that CSF-1 promoted renal crystals clearance in hyperoxaluric mice via increasing the number of $CD11b^{+}F4/80^{+}CD163^{+}CD206^{high}$ M2 cells (Taguchi et al. 2014). Although granulocyte-macrophage (GM)-CSF usually induces the differentiation of peripheral $Ly6C^{high}$ monocytes to pro-inflammatory M1 macrophages (Lenzo et al. 2012; Murray and Wynn 2011), a recent in vivo study identified macrophages with a unique alternative activation state in response to GM-CSF, they were found in macrophages isolated from repair phase of injured kidneys in IRI model and promoted tubular proliferation and repair (Huen et al. 2015; Takeda et al. 1996). Additionally, treatment of IL-25, a novel cytokine for M2 polarization both in vivo and in vitro, prevents renal injury in adriamycin nephropathy via a IL-4/IL-13-dependent manner (Cao et al. 2011). Netrin-1 is an anti-inflammatory molecule induced in TECs from IRI model; it suppresses monocyte migration and function by targeting chemokines and NF-κB signaling. Netrin-1 transgenic mice show an increase in M2 macrophages infiltration with upregulation of IL-4, IL-13, and arginase-1 in a PPAR-dependent mechanism, showing that Netrin-1 is a critical inducer for M2 polarization (Ranganathan et al. 2013). Calcitriol, a bioactive 1,25-dihydroxyvitamin D3, promotes M2 polarization while inhibiting macrophage recruitment and activation, thereby attenuates proteinuria and renal injury in diabetic nephropathy (Zhang et al. 2014). In addition, loss of p53 from bone marrow accelerates renal injury and impairs renal repair caused by the deficiency of KLF4 expression and M2 polarization (Sutton et al. 2013). Moreover, paracrine effects of mesenchymal stem cells (MSCs) increases the infiltration of M2 macrophages which protects against renal acute injury, and the adoptive transfer of MSCs-cocultured macrophages in macrophage depletion mice induces much milder renal injury compared with control (Geng et al. 2014). The functions of MSCs on M2 polarization have also been reported in IRI injury (Wise et al. 2014). Additionally, recent in vivo studies showed that Wnt/β-catenin signaling promoted renal fibrosis by enhancing macrophage proliferation and M2 polarization in STAT3-dependent mechanisms (Feng et al. 2018a, b).

13.8 Monocyte/Macrophage-Based or Targeted Therapy in Treatment of Renal Fibrosis

Till now, therapeutic strategies by interfering with monocyte/macrophage recruitment, activation and polarization, or adoptive transfer of polarized macrophages have been extensively studied.

Previous studies showed that DNA vaccination or neutralized antibody-mediated inhibition of chemokines, like CCL2 and CCL5, prevents macrophage infiltration and renal damage in adriamycin nephropathy (Wu et al. 2005; Zheng et al. 2006), nephrotoxic serum nephritis (Lloyd et al. 1997; Tang et al. 1996; Wada et al. 1996), and anti-thy1.1 nephritis (Wenzel et al. 1997). Inhibition of CX3CR1 or intercellular adhesion molecule-1 (ICAM-1) protects against crescentic glomerulonephritis and nephrotoxic nephritis (Feng et al. 1999; Kawasaki et al. 1993). Additionally, anti-macrophage serum-induced depletion of macrophage prevents experimental glomerulonephritis (Holdsworth et al. 1981). Blocking c-fms, a receptor for CSF, protects against UUO and diabetic nephropathy by reducing the recruitment and proliferation of macrophages (Le Meur et al. 2002; Lim et al. 2009). Moreover, liposomal clodronate-mediated clearance of macrophage alleviates renal fibrosis (Kitamoto et al. 2009), this finding is further confirmed by the study showing that conditional depletion of $CD11b^{+}$ cells attenuates crescentic glomerulonephritis (Duffield et al. 2005b; Wang and Harris 2011). Notwithstanding, inconsistent evidence shows that blocking CCL2 or CCL5 fails to attenuate renal injury, indicating that the success of therapy by inhibiting macrophages recruitment might depend on the types and stages of kidney diseases (Anders et al. 2003; Clauss et al. 2009).

Accumulating evidence shows that modification of macrophage activation states could also prevent renal fibrosis. A recent study demonstrated that Beta-2 adrenergic receptor (β2AR) agonists increased the binding of β-arrestin2 and IκBα, leading to the down-regulation of NF-κB and deactivation of macrophages, thereby protected against diabetic renal complication (Noh et al. 2017). Blocking NF-κB signaling by antisense oligonucleotides or its natural inhibitor IκB suppresses the classical activation of macrophages but increases anti-inflammatory macrophages, thereby limits kidney injury (Tomita et al. 2000; Wilson et al. 2005). By increasing IL-4/IL-13-mediated M2 polarization, IL-25 protects against adriamycin nephropathy (Cao et al. 2011). Additionally, treatment of Quercetin reduced macrophage infiltration and M2 polarization by preventing ECM production and interstitial fibrosis in a TGF-β1/Smad-dependent mechanism in obstructive nephropathy (Lu et al. 2018).

Modified macrophages are directly used to treat renal diseases in some studies. IL-4/IL-13-polarized M2a spleen macrophages were transferred into SCID mice where functions of endogenous immune cells were excluded, results showed that renal histology and function were both restored in adriamycin nephropathy (Wang et al. 2007). The protective effect of ex vivo polarized macrophages was further confirmed in streptozotocin-induced type 1 diabetic nephropathy (Parsa et al. 2012). Of note, IL-10 and TGF-β-induced M2c macrophages show high efficiency in reducing renal damage and proteinuria compared with M2a, because they are capable of inducing immunosuppressing regulatory T cells differentiation via a B7-H4-dependant mechanism (Cao et al. 2010; Lu et al. 2013). IL-10/TGF-β or IL-4/IL-13-modified bone marrow-derived macrophages have limited protective effect due to the finite proliferation capacity of bone marrow cells, so it may confine the clinical application of macrophage-based therapy by modifying bone marrow cells from patients (Cao et al. 2014).

13.9 Conclusions and Perspective

Taken together, monocytes and macrophages are recruited into the injured kidney by chemokines released from kidney, and then they are activated and polarized into distinct phenotypes in response to the local microenvironment. Macrophages with different activation stages exert distinct or even diverse effects in the processes

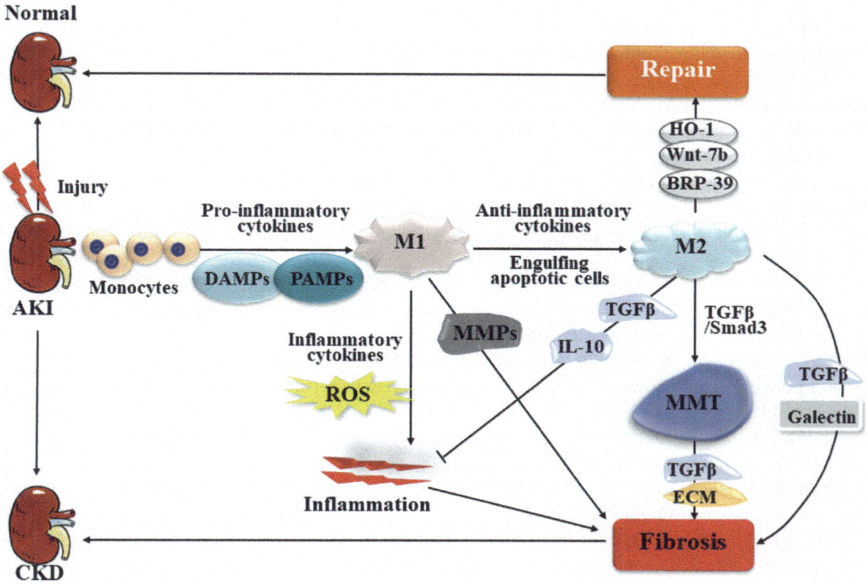

Fig. 13.1 Polarization and function of macrophages in renal injury, repair, and fibrosis

of renal injury, repair, and fibrosis (Fig. 13.1). Uncontrolled M2 macrophages or CD11b^{+}/Ly6C^{low} macrophages promote renal fibrosis via paracrine effects or direct transition to myofibroblast-like cells. In this regard, inhibiting monocyte/macrophage recruitment, modifying macrophage activation and polarization, or adoptive transfer of polarized macrophages may be promising therapies for renal fibrosis.

Acknowledgements This study is supported by grants from National Natural Science Foundation of China (National Science Foundation of China 81300580 and 81570623), Science and Technological Fund of Anhui Province for Outstanding Youth of China (Grant number: 1608085J07), Lui Chi Woo Institute of Innovative Medicine, and the Research Grants Council of Hong Kong (GRF 14121816, 14163317, C7018-16G, TRS T12-402/13N).

References

Anders HJ (2010) Toll-like receptors and danger signaling in kidney injury. J Am Soc Nephrol 21:1270–1274

Anders HJ, Ryu M (2011) Renal microenvironments and macrophage phenotypes determine progression or resolution of renal inflammation and fibrosis. Kidney Int 80:915–925

Anders HJ, Frink M, Linde Y, Banas B, Wörnle M, Cohen CD et al (2003) CC chemokine ligand 5/RANTES chemokine antagonists aggravate glomerulonephritis despite reduction of glomerular leukocyte infiltration. J Immunol 170:5658–5666

Baek JH, Zeng R, Weinmann-Menke J, Valerius MT, Wada Y, Ajay AK et al (2015) IL-34 mediates acute kidney injury and worsens subsequent chronic kidney disease. J Clin Investig 125:3198–3214

Belliere J, Casemayou A, Ducasse L, Zakaroff-Girard A, Martins F, Iacovoni JS et al (2015) Specific macrophage subtypes influence the progression of rhabdomyolysis-induced kidney injury. J Am Soc Nephrol 26:1363–1377

Bertrand S, Godoy M, Semal P, Van Gansen P (1992) Transdifferentiation of macrophages into fibroblasts as a result of Schistosoma mansoni infection. Int J Develop Biol 36:179–184

Border WA, Noble NA (1994) Transforming growth factor-beta in glomerular injury. Exp Nephrol 2:13–17

Braga TT, Correa-Costa M, Silva RC, Cruz MC, Hiyane MI, da Silva JS et al (2018) CCR8 contributes to the recruitment of monocytes and leads to kidney inflammation and fibrosis development. Inflammopharmacology 26:403–411

Cao Q, Wang Y, Zheng D, Sun Y, Wang Y, Lee VW et al (2010) IL-10/TGF-beta-modified macrophages induce regulatory T cells and protect against adriamycin nephrosis. J Am Soc Nephrol 21:933–942

Cao Q, Wang C, Zheng D, Wang Y, Lee VW, Wang YM et al (2011) IL-25 induces M2 macrophages and reduces renal injury in proteinuric kidney disease. J Am Soc Nephrol 22:1229–1239

Cao Q, Wang Y, Harris DC (2013) Pathogenic and protective role of macrophages in kidney disease. Am J Physiol Renal Physiol 305:F3–F11

Cao Q, Wang Y, Zheng D, Sun Y, Wang C, Wang XM et al (2014) Failed renoprotection by alternatively activated bone marrow macrophages is due to a proliferation-dependent phenotype switch in vivo. Kidney Int 85:794–806

Chen G, Lin SC, Chen J, He L, Dong F, Xu J et al (2011) CXCL16 recruits bone marrow-derived fibroblast precursors in renal fibrosis. J Am Soc Nephrol 22:1876–1886

Chen J, Xia Y, Lin X, Feng XH, Wang Y (2014) Smad3 signaling activates bone marrow-derived fibroblasts in renal fibrosis. Lab Invest J Tech Methods Pathol 94:545–556

Cheng S, Lovett DH (2003) Gelatinase A (MMP-2) is necessary and sufficient for renal tubular cell epithelial-mesenchymal transformation. Am J Pathol 162:1937–1949

Cheng S, Pollock AS, Mahimkar R, Olson JL, Lovett DH (2006) Matrix metalloproteinase 2 and basement membrane integrity: a unifying mechanism for progressive renal injury. FASEB J 20:1898–1900 (Official publication of the Federation of American Societies for Experimental Biology)

Chiba T, Skrypnyk NI, Skvarca LB, Penchev R, Zhang KX, Rochon ER et al (2016) Retinoic acid signaling coordinates macrophage-dependent injury and repair after AKI. J Am Soc Nephrol 27:495–508

Clauss S, Gross O, Kulkarni O, Avila-Ferrufino A, Radomska E, Segerer S et al (2009) Ccl2/Mcp-1 blockade reduces glomerular and interstitial macrophages but does not ameliorate renal pathology in collagen4A3-deficient mice with autosomal recessive Alport nephropathy. J Pathol 218:40–47

Clements M, Gershenovich M, Chaber C, Campos-Rivera J, Du P, Zhang M et al (2016) Differential Ly6C expression after renal ischemia-reperfusion identifies unique macrophage populations. J Am Soc Nephrol 27:159–170

Cochrane AL, Kett MM, Samuel CS, Campanale NV, Anderson WP, Hume DA et al (2005) Renal structural and functional repair in a mouse model of reversal of ureteral obstruction. J Am Soc Nephrol 16:3623–3630

Day YJ, Huang L, Ye H, Linden J, Okusa MD (2005) Renal ischemia-reperfusion injury and adenosine 2A receptor-mediated tissue protection: role of macrophages. Am J Physiol Renal Physiol 288:F722–F731

Dessing MC, Tammaro A, Pulskens WP, Teske GJ, Butter LM, Claessen N et al (2015) The calcium-binding protein complex S100A8/A9 has a crucial role in controlling macrophage-mediated renal repair following ischemia/reperfusion. Kidney Int 87:85–94

D'Souza MJ, Oettinger CW, Shah A, Tipping PG, Huang XR, Milton GV (1999) Macrophage depletion by albumin microencapsulated clodronate: attenuation of cytokine release in macrophage-dependent glomerulonephritis. Drug Dev Ind Pharm 25:591–596
Duffield JS (2010) Macrophages and immunologic inflammation of the kidney. Semin Nephrol 30:234–254
Duffield JS, Forbes SJ, Constandinou CM, Clay S, Partolina M, Vuthoori S et al (2005a) Selective depletion of macrophages reveals distinct, opposing roles during liver injury and repair. J Clin Investig 115:56–65
Duffield JS, Tipping PG, Kipari T, Cailhier JF, Clay S, Lang R et al (2005b) Conditional ablation of macrophages halts progression of crescentic glomerulonephritis. Am J Pathol 167:1207–1219
Falke LL, Gholizadeh S, Goldschmeding R, Kok RJ, Nguyen TQ (2015) Diverse origins of the myofibroblast-implications for kidney fibrosis. Nat Rev Nephrol 11:233–244
Fallowfield JA, Mizuno M, Kendall TJ, Constandinou CM, Benyon RC, Duffield JS et al (2007) Scar-associated macrophages are a major source of hepatic matrix metalloproteinase-13 and facilitate the resolution of murine hepatic fibrosis. J Immunol 178:5288–5295
Fan D, Takawale A, Basu R, Patel V, Lee J, Kandalam V et al (2014) Differential role of TIMP2 and TIMP3 in cardiac hypertrophy, fibrosis, and diastolic dysfunction. Cardiovasc Res 103:268–280
Feng L, Chen S, Garcia GE, Xia Y, Siani MA, Botti P et al (1999) Prevention of crescentic glomerulonephritis by immunoneutralization of the fractalkine receptor CX3CR30 rapid communication. Kidney Int 56:612–620
Feng Y, Liang Y, Ren J, Dai C (2018a) Canonical Wnt signaling promotes macrophage proliferation during kidney fibrosis. Kidney Dis (Basel) 4:95–103
Feng Y, Ren J, Gui Y, Wei W, Shu B, Lu Q et al (2018b) Wnt/beta-catenin-promoted macrophage alternative activation contributes to kidney fibrosis. J Am Soc Nephrol 29:182–193
Ferenbach DA, Ramdas V, Spencer N, Marson L, Anegon I, Hughes J et al (2010) Macrophages expressing heme oxygenase-1 improve renal function in ischemia/reperfusion injury. Mol Therapy J Am Soc Gene Therapy 18:1706–1713
Ferenbach DA, Nkejabega NC, McKay J, Choudhary AK, Vernon MA, Beesley MF et al (2011) The induction of macrophage hemeoxygenase-1 is protective during acute kidney injury in aging mice. Kidney Int 79:966–976
Fine LG, Norman JT (2008) Chronic hypoxia as a mechanism of progression of chronic kidney diseases: from hypothesis to novel therapeutics. Kidney Int 74:867–872
Floege J, Eitner F, Alpers CE (2008) A new look at platelet-derived growth factor in renal disease. J Am Soc Nephrol 19:12–23
Geissmann F, Jung S, Littman DR (2003) Blood monocytes consist of two principal subsets with distinct migratory properties. Immunity 19:71–82
Geng Y, Zhang L, Fu B, Zhang J, Hong Q, Hu J et al (2014) Mesenchymal stem cells ameliorate rhabdomyolysis-induced acute kidney injury via the activation of M2 macrophages. Stem Cell Res Therapy 5:80
Gratchev A, Guillot P, Hakiy N, Politz O, Orfanos CE, Schledzewski K et al (2001) Alternatively activated macrophages differentially express fibronectin and its splice variants and the extracellular matrix protein betaIG-H3. Scand J Immunol 53:386–392
Guilliams M, Ginhoux F, Jakubzick C, Naik SH, Onai N, Schraml BU et al (2014) Dendritic cells, monocytes and macrophages: a unified nomenclature based on ontogeny. Nat Rev Immunol 14:571–578
Gurtner GC, Werner S, Barrandon Y, Longaker MT (2008) Wound repair and regeneration. Nature 453:314–321
Han Y, Ma FY, Tesch GH, Manthey CL, Nikolic-Paterson DJ (2011) c-fms blockade reverses glomerular macrophage infiltration and halts development of crescentic anti-GBM glomerulonephritis in the rat. Lab Invest J Tech Methods Pathol 91:978–991
Han Y, Ma FY, Tesch GH, Manthey CL, Nikolic-Paterson DJ (2013) Role of macrophages in the fibrotic phase of rat crescentic glomerulonephritis. Am J Physiol Renal Physiol 304:F1043–F1053

Han H, Zhu J, Wang Y, Zhu Z, Chen Y, Lu L et al (2017) Renal recruitment of B lymphocytes exacerbates tubulointerstitial fibrosis by promoting monocyte mobilization and infiltration after unilateral ureteral obstruction. J Pathol 241:80–90

Henderson NC, Mackinnon AC, Farnworth SL, Kipari T, Haslett C, Iredale JP et al (2008) Galectin-3 expression and secretion links macrophages to the promotion of renal fibrosis. Am J Pathol 172:288–298

Holdsworth SR, Neale TJ, Wilson CB (1981) Abrogation of macrophage-dependent injury in experimental glomerulonephritis in the rabbit. Use of an antimacrophage serum. J Clin Investig 68:686–698

Huang L, Wang A, Hao Y, Li W, Liu C, Yang Z et al (2018) Macrophage depletion lowered blood pressure and attenuated hypertensive renal injury and fibrosis. Front Physiol 9:473

Huen SC, Cantley LG (2015) Macrophage-mediated injury and repair after ischemic kidney injury. Pediatr Nephrol 30:199–209

Huen SC, Moeckel GW, Cantley LG (2013) Macrophage-specific deletion of transforming growth factor-beta1 does not prevent renal fibrosis after severe ischemia-reperfusion or obstructive injury. Am J Physiol Renal Physiol 305:F477–F484

Huen SC, Huynh L, Marlier A, Lee Y, Moeckel GW, Cantley LG (2015) GM-CSF Promotes macrophage alternative activation after renal ischemia/reperfusion injury. J Am Soc Nephrol 26:1334–1345

Ikezumi Y, Hurst LA, Masaki T, Atkins RC, Nikolic-Paterson DJ (2003) Adoptive transfer studies demonstrate that macrophages can induce proteinuria and mesangial cell proliferation. Kidney Int 63:83–95

Ikezumi Y, Suzuki T, Karasawa T, Hasegawa H, Kawachi H, Nikolic-Paterson DJ et al (2010) Contrasting effects of steroids and mizoribine on macrophage activation and glomerular lesions in rat thy-1 mesangial proliferative glomerulonephritis. Am J Nephrol 31:273–282

Ikezumi Y, Suzuki T, Karasawa T, Hasegawa H, Yamada T, Imai N et al (2011) Identification of alternatively activated macrophages in new-onset paediatric and adult immunoglobulin A nephropathy: potential role in mesangial matrix expansion. Histopathology 58:198–210

Jiang Y, Wang Y, Ma P, An D, Zhao J, Liang S et al (2018) Myeloid-specific targeting of Notch ameliorates murine renal fibrosis via reduced infiltration and activation of bone marrow-derived macrophage. Protein Cell 10:196–210

Jo SK, Sung SA, Cho WY, Go KJ, Kim HK (2006) Macrophages contribute to the initiation of ischaemic acute renal failure in rats. Nephrol Dialysis Transplant 21:1231–1239 (Official publication of the European Dialysis and Transplant Association–European Renal Association)

Kapoor N, Niu J, Saad Y, Kumar S, Sirakova T, Becerra E et al (2015) Transcription factors STAT6 and KLF4 implement macrophage polarization via the dual catalytic powers of MCPIP. J Immunol 194:6011–6023

Kawasaki K, Yaoita E, Yamamoto T, Tamatani T, Miyasaka M, Kihara I (1993) Antibodies against intercellular adhesion molecule-1 and lymphocyte function-associated antigen-1 prevent glomerular injury in rat experimental crescentic glomerulonephritis. J Immunol 150:1074–1083

Kitagawa K, Wada T, Furuichi K, Hashimoto H, Ishiwata Y, Asano M et al (2004) Blockade of CCR58 ameliorates progressive fibrosis in kidney. Am J Pathol 165:237–246

Kitamoto K, Machida Y, Uchida J, Izumi Y, Shiota M, Nakao T et al (2009) Effects of liposome clodronate on renal leukocyte populations and renal fibrosis in murine obstructive nephropathy. J Pharmacol Sci 111:285–292

Kluth DC, Erwig LP, Rees AJ (2004) Multiple facets of macrophages in renal injury. Kidney Int 66:542–557

Ko GJ, Boo CS, Jo SK, Cho WY, Kim HK (2008) Macrophages contribute to the development of renal fibrosis following ischaemia/reperfusion-induced acute kidney injury. Nephrol Dialysis Transplant 23:842–852 (Official publication of the European Dialysis and Transplant Association–European Renal Association)

Kramann R, Machado F, Wu H, Kusaba T, Hoeft K, Schneider RK et al (2018) Parabiosis and single-cell RNA sequencing reveal a limited contribution of monocytes to myofibroblasts in kidney fibrosis. JCI Insight 3
Le Meur Y, Tesch GH, Hill PA, Mu W, Foti R, Nikolic-Paterson DJ et al (2002) Macrophage accumulation at a site of renal inflammation is dependent on the M-CSF/c-fms pathway. J Leukoc Biol 72:530–537
LeBleu VS, Taduri G, O'Connell J, Teng Y, Cooke VG, Woda C et al (2013) Origin and function of myofibroblasts in kidney fibrosis. Nat Med 19:1047–1053
Lech M, Gröbmayr R, Ryu M, Lorenz G, Hartter I, Mulay SR et al (2014) Macrophage phenotype controls long-term AKI outcomes-kidney regeneration versus atrophy. J Am Soc Nephrol 25:292–304
Lee S, Huen S, Nishio H, Nishio S, Lee HK, Choi BS et al (2011) Distinct macrophage phenotypes contribute to kidney injury and repair. J Am Soc Nephrol 22:317–326
Lenzo JC, Turner AL, Cook AD, Vlahos R, Anderson GP, Reynolds EC et al (2012) Control of macrophage lineage populations by CSF-1 receptor and GM-CSF in homeostasis and inflammation. Immunol Cell Biol 90:429–440
Lim AK, Ma FY, Nikolic-Paterson DJ, Thomas MC, Hurst LA, Tesch GH (2009) Antibody blockade of c-fms suppresses the progression of inflammation and injury in early diabetic nephropathy in obese db/db mice. Diabetologia 52:1669–1679
Lin SL, Kisseleva T, Brenner DA, Duffield JS (2008) Pericytes and perivascular fibroblasts are the primary source of collagen-producing cells in obstructive fibrosis of the kidney. Am J Pathol 173:1617–1627
Lin SL, Castaño AP, Nowlin BT, Lupher ML Jr, Duffield JS (2009) Bone marrow Ly6Chigh monocytes are selectively recruited to injured kidney and differentiate into functionally distinct populations. J Immunol 183:6733–6743
Lin SL, Li B, Rao S, Yeo EJ, Hudson TE, Nowlin BT et al (2010) Macrophage Wnt7b is critical for kidney repair and regeneration. Proc Natl Acad Sci USA 107:4194–4199
Lloyd CM, Dorf ME, Proudfoot A, Salant DJ, Gutierrez-Ramos JC (1997) Role of MCP-1 and RANTES in inflammation and progression to fibrosis during murine crescentic nephritis. J Leukoc Biol 62:676–680
Lu J, Cao Q, Zheng D, Sun Y, Wang C, Yu X et al (2013) Discrete functions of M2a and M2c macrophage subsets determine their relative efficacy in treating chronic kidney disease. Kidney Int 84:745–755
Lu H, Wu L, Liu L, Ruan Q, Zhang X, Hong W et al (2018) Quercetin ameliorates kidney injury and fibrosis by modulating M1/M2 macrophage polarization. Biochem Pharmacol 154:203–212
Ma FY, Flanc RS, Tesch GH, Bennett BL, Friedman GC, Nikolic-Paterson DJ (2009) Blockade of the c-Jun amino terminal kinase prevents crescent formation and halts established anti-GBM glomerulonephritis in the rat. Lab Invest 89:470–484
Mantovani A, Biswas SK, Galdiero MR, Sica A, Locati M (2013) Macrophage plasticity and polarization in tissue repair and remodelling. J Pathol 229:176–185
Martinez FO, Gordon S (2014) The M1 and M2 paradigm of macrophage activation: time for reassessment. F1000prime Rep 6:13
Meng XM, Chung AC, Lan HY (2013) Role of the TGF-beta/BMP-7/Smad pathways in renal diseases. Clin Sci (Lond) 124:243–254
Meng XM, Nikolic-Paterson DJ, Lan HY (2014) Inflammatory processes in renal fibrosis. Nat Rev Nephrol 10:493–503
Meng XM, Tang PM, Li J, Lan HY (2015) Macrophage phenotype in kidney injury and repair. Kidney Dis (Basel) 1:138–146
Meng XM, Nikolic-Paterson DJ, Lan HY (2016a) TGF-beta: the master regulator of fibrosis. Nat Rev Nephrol 12:325–338
Meng XM, Wang S, Huang XR, Yang C, Xiao J, Zhang Y et al (2016b) Inflammatory macrophages can transdifferentiate into myofibroblasts during renal fibrosis. Cell Death Dis 7:e2495

Menke J, Iwata Y, Rabacal WA, Basu R, Yeung YG, Humphreys BD et al (2009) CSF-1 signals directly to renal tubular epithelial cells to mediate repair in mice. J Clin Investig 119:2330–2342
Mooney JE, Rolfe BE, Osborne GW, Sester DP, van Rooijen N, Campbell GR et al (2010) Cellular plasticity of inflammatory myeloid cells in the peritoneal foreign body response. Am J Pathol 176:369–380
Mu W, Ouyang X, Agarwal A, Zhang L, Long DA, Cruz PE et al (2005) IL-10 suppresses chemokines, inflammation, and fibrosis in a model of chronic renal disease. J Am Soc Nephrol 16:3651–3660
Murray PJ, Wynn TA (2011) Protective and pathogenic functions of macrophage subsets. Nat Rev Immunol 11:723–737
Murray PJ, Allen JE, Biswas SK, Fisher EA, Gilroy DW, Goerdt S et al (2014) Macrophage activation and polarization: nomenclature and experimental guidelines. Immunity 41:14–20
Nikolic-Paterson DJ, Wang S, Lan HY (2014) Macrophages promote renal fibrosis through direct and indirect mechanisms. Kidney Int Suppl 4:34–38
Nishida M, Fujinaka H, Matsusaka T, Price J, Kon V, Fogo AB et al (2002) Absence of angiotensin II type 1 receptor in bone marrow-derived cells is detrimental in the evolution of renal fibrosis. J Clin Investig 110:1859–1868
Nishida M, Okumura Y, Fujimoto S, Shiraishi I, Itoi T, Hamaoka K (2005) Adoptive transfer of macrophages ameliorates renal fibrosis in mice. Biochem Biophys Res Commun 332:11–16
Nishida M, Okumura Y, Ozawa S, Shiraishi I, Itoi T, Hamaoka K (2007) MMP-2 inhibition reduces renal macrophage infiltration with increased fibrosis in UUO. Biochem Biophys Res Commun 354:133–139
Noh H, Yu MR, Kim HJ, Lee JH, Park BW, Wu IH et al (2017) Beta 2-adrenergic receptor agonists are novel regulators of macrophage activation in diabetic renal and cardiovascular complications. Kidney Int 92:101–113
Parsa R, Andresen P, Gillett A, Mia S, Zhang XM, Mayans S et al (2012) Adoptive transfer of immunomodulatory M2 macrophages prevents type 1 diabetes in NOD mice. Diabetes 61:2881–2892
Peng X, Zhang J, Xiao Z, Dong Y, Du J (2015) CX3CL1-CX3CR94 interaction increases the population of Ly6C(-)CX3CR94(hi) macrophages contributing to unilateral ureteral obstruction-induced fibrosis. J Immunol 195:2797–2805
Piccolo V, Curina A, Genua M, Ghisletti S, Simonatto M, Sabò A et al (2017) Opposing macrophage polarization programs show extensive epigenomic and transcriptional cross-talk. Nat Immunol 18:530–540
Pilling D, Gomer RH (2012) Differentiation of circulating monocytes into fibroblast-like cells. Methods Mol Biol 904:191–206
Ranganathan PV, Jayakumar C, Ramesh G (2013) Netrin-1-treated macrophages protect the kidney against ischemia-reperfusion injury and suppress inflammation by inducing M2 polarization. Am J Physiol Renal Physiol 304:F948–F957
Rao VH, Meehan DT, Delimont D, Nakajima M, Wada T, Gratton MA et al (2006) Role for macrophage metalloelastase in glomerular basement membrane damage associated with alport syndrome. Am J Pathol 169:32–46
Ricardo SD, van Goor H, Eddy AA (2008) Macrophage diversity in renal injury and repair. J Clin Investig 118:3522–3530
Ronco P, Chatziantoniou C (2008) Matrix metalloproteinases and matrix receptors in progression and reversal of kidney disease: therapeutic perspectives. Kidney Int 74:873–878
Ronco P, Lelongt B, Piedagnel R, Chatziantoniou C (2007) Matrix metalloproteinases in kidney disease progression and repair: a case of flipping the coin. Semin Nephrol 27:352–362
Roufosse C, Bou-Gharios G, Prodromidi E, Alexakis C, Jeffery R, Khan S et al (2006) Bone marrow-derived cells do not contribute significantly to collagen I synthesis in a murine model of renal fibrosis. J Am Soc Nephrol 17:775–782

Schmidt IM, Hall IE, Kale S, Lee S, He CH, Lee Y et al (2013) Chitinase-like protein Brp-39/YKL-40 modulates the renal response to ischemic injury and predicts delayed allograft function. J Am Soc Nephrol 24:309–319
Schnoor M, Cullen P, Lorkowski J, Stolle K, Robenek H, Troyer D et al (2008) Production of type VI collagen by human macrophages: a new dimension in macrophage functional heterogeneity. J Immunol 180:5707–5719
Sunderkötter C, Nikolic T, Dillon MJ, Van Rooijen N, Stehling M, Drevets DA et al (2004) Subpopulations of mouse blood monocytes differ in maturation stage and inflammatory response. J Immunol 172:4410–4417
Sung SA, Jo SK, Cho WY, Won NH, Kim HK (2007) Reduction of renal fibrosis as a result of liposome encapsulated clodronate induced macrophage depletion after unilateral ureteral obstruction in rats. Nephron Experiment Nephrol 105:e1–e9
Susnik N, Sörensen-Zender I, Rong S, von Vietinghoff S, Lu X, Rubera I et al (2014) Ablation of proximal tubular suppressor of cytokine signaling 3 enhances tubular cell cycling and modifies macrophage phenotype during acute kidney injury. Kidney Int 85:1357–1368
Sutton TA, Hato T, Mai E, Yoshimoto M, Kuehl S, Anderson M et al (2013) p53 is renoprotective after ischemic kidney injury by reducing inflammation. J Am Soc Nephrol 24:113–124
Swaminathan S, Griffin MD (2008) First responders: understanding monocyte-lineage traffic in the acutely injured kidney. Kidney Int 74:1509–1511
Taguchi K, Okada A, Kitamura H, Yasui T, Naiki T, Hamamoto S et al (2014) Colony-stimulating factor-1 signaling suppresses renal crystal formation. J Am Soc Nephrol 25:1680–1697
Takeda K, Tanaka T, Shi W, Matsumoto M, Minami M, Kashiwamura S et al (1996) Essential role of Stat6 in IL-4 signalling. Nature 380:627–630
Tang WW, Qi M, Warren JS (1996) Monocyte chemoattractant protein 1 mediates glomerular macrophage infiltration in anti-GBM Ab GN. Kidney Int 50:665–671
Tang PM, Zhou S, Li CJ, Liao J, Xiao J, Wang QM et al (2018) The proto-oncogene tyrosine protein kinase Src is essential for macrophage-myofibroblast transition during renal scarring. Kidney Int 93:173–187
Thomas JA, Pope C, Wojtacha D, Robson AJ, Gordon-Walker TT, Hartland S et al (2011) Macrophage therapy for murine liver fibrosis recruits host effector cells improving fibrosis, regeneration, and function. Hepatology 53:2003–2015
Tian S, Zhang L, Tang J, Guo X, Dong K, Chen SY (2015) HMGB1 exacerbates renal tubulointerstitial fibrosis through facilitating M1 macrophage phenotype at the early stage of obstructive injury. Am J Physiol Renal Physiol 308:F69–F75
Tomita N, Morishita R, Lan HY, Yamamoto K, Hashizume M, Notake M et al (2000) In vivo administration of a nuclear transcription factor-kappaB decoy suppresses experimental crescentic glomerulonephritis. J Am Soc Nephrol 11:1244–1252
Vernon MA, Mylonas KJ, Hughes J (2010) Macrophages and renal fibrosis. Semin Nephrol 30:302–317
Vinuesa E, Hotter G, Jung M, Herrero-Fresneda I, Torras J, Sola A (2008) Macrophage involvement in the kidney repair phase after ischaemia/reperfusion injury. J Pathol 214:104–113
Wada T, Yokoyama H, Furuichi K, Kobayashi KI, Harada K, Naruto M et al (1996) Intervention of crescentic glomerulonephritis by antibodies to monocyte chemotactic and activating factor (MCAF/MCP-1). FASEB journal 10:1418–1425
Wada T, Furuichi K, Sakai N, Iwata Y, Kitagawa K, Ishida Y et al (2004) Gene therapy via blockade of monocyte chemoattractant protein-1 for renal fibrosis. J Am Soc Nephrol 15:940–948
Wang Y, Harris DC (2011) Macrophages in renal disease. J Am Soc Nephrol 22:21–27
Wang Y, Wang YP, Zheng G, Lee VW, Ouyang L, Chang DH et al (2007) Ex vivo programmed macrophages ameliorate experimental chronic inflammatory renal disease. Kidney Int 72:290–299
Wang JC, Lai S, Guo X, Zhang X, de Crombrugghe B, Sonnylal S et al (2010) Attenuation of fibrosis in vitro and in vivo with SPARC siRNA. Arthritis Res Therapy 12:R60

Wang S, Meng XM, Ng YY, Ma FY, Zhou S, Zhang Y et al (2016) TGF-beta/Smad3 signalling regulates the transition of bone marrow-derived macrophages into myofibroblasts during tissue fibrosis. Oncotarget 7:8809–8822
Wang YY, Jiang H, Pan J, Huang XR, Wang YC, Huang HF et al (2017) Macrophage- to-myofibroblast transition contributes to interstitial fibrosis in chronic renal allograft injury. J Am Soc Nephrol 28:2053–2067
Weir MR (2015) CCR126 inhibition: a panacea for diabetic kidney disease? Lancet Diabetes Endocrinol 3:666–667
Wenzel U, Schneider A, Valente AJ, Abboud HE, Thaiss F, Helmchen UM et al (1997) Monocyte chemoattractant protein-1 mediates monocyte/macrophage influx in anti-thymocyte antibody-induced glomerulonephritis. Kidney Int 51:770–776
Wermuth PJ, Jimenez SA (2015) The significance of macrophage polarization subtypes for animal models of tissue fibrosis and human fibrotic diseases. Clin Transl Med 4:1–19
Wilson HM, Walbaum D, Rees AJ (2004) Macrophages and the kidney. Curr Opin Nephrol Hypertens 13:285–290
Wilson HM, Chettibi S, Jobin C, Walbaum D, Rees AJ, Kluth DC (2005) Inhibition of macrophage nuclear factor-kappaB leads to a dominant anti-inflammatory phenotype that attenuates glomerular inflammation in vivo. Am J Pathol 167:27–37
Wise AF, Williams TM, Kiewiet MB, Payne NL, Siatskas C, Samuel CS et al (2014) Human mesenchymal stem cells alter macrophage phenotype and promote regeneration via homing to the kidney following ischemia-reperfusion injury. Am J Physiol Renal Physiol 306:F1222–F1235
Wu H, Wang Y, Tay YC, Zheng G, Zhang C, Alexander SI et al (2005) DNA vaccination with naked DNA encoding MCP-1 and RANTES protects against renal injury in adriamycin nephropathy. Kidney Int 67:2178–2186
Wynes MW, Frankel SK, Riches DW (2004) IL-4-induced macrophage-derived IGF-I protects myofibroblasts from apoptosis following growth factor withdrawal. J Leukoc Biol 76:1019–1027
Wynn TA, Vannella KM (2016) Macrophages in tissue repair, regeneration, and fibrosis. Immunity 44:450–462
Xia Y, Entman ML, Wang Y (2013) Critical role of CXCL16 in hypertensive kidney injury and fibrosis. Hypertension 62:1129–1137
Xia Y, Jin X, Yan J, Entman ML, Wang Y (2014a) CXCR136 plays a critical role in angiotensin II-induced renal injury and fibrosis. Arterioscler Thromb Vasc Biol 34:1422–1428
Xia Y, Yan J, Jin X, Entman ML, Wang Y (2014b) The chemokine receptor CXCR137 contributes to recruitment of bone marrow-derived fibroblast precursors in renal fibrosis. Kidney Int 86:327–337
Yona S, Kim KW, Wolf Y, Mildner A, Varol D, Breker M et al (2013) Fate mapping reveals origins and dynamics of monocytes and tissue macrophages under homeostasis. Immunity 38:79–91
Zeisberg M, Duffield JS (2010) Resolved: EMT produces fibroblasts in the kidney. J Am Soc Nephrol 21:1247–1253
Zhang G, Kim H, Cai X, Lopez-Guisa JM, Carmeliet P, Eddy AA (2003) Urokinase receptor modulates cellular and angiogenic responses in obstructive nephropathy. J Am Soc Nephrol 14:1234–1253
Zhang MZ, Yao B, Yang S, Jiang L, Wang S, Fan X et al (2012) CSF-1 signaling mediates recovery from acute kidney injury. J Clin Investig 122:4519–4532
Zhang XL, Guo YF, Song ZX, Zhou M (2014) Vitamin D prevents podocyte injury via regulation of macrophage M1/M2 phenotype in diabetic nephropathy rats. Endocrinology 155:4939–4950
Zhang MZ, Wang X, Wang Y, Niu A, Wang S, Zou C et al (2017) IL-4/IL-13-mediated polarization of renal macrophages/dendritic cells to an M2a phenotype is essential for recovery from acute kidney injury. Kidney Int 91:375–386

Zheng G, Wang Y, Xiang SH, Tay YC, Wu H, Watson D et al (2006) DNA vaccination with CCL2 DNA modified by the addition of an adjuvant epitope protects against "nonimmune" toxic renal injury. J Am Soc Nephrol 17:465–474
Zhou D, Huang C, Lin Z, Zhan S, Kong L, Fang C et al (2014) Macrophage polarization and function with emphasis on the evolving roles of coordinated regulation of cellular signaling pathways. Cell Signal 26:192–197

Chapter 14
Targeting Bone Marrow-Derived Fibroblasts for Renal Fibrosis

Changlong An, Li Jia, Jia Wen and Yanlin Wang

Abstract Renal fibrosis is a major pathological feature of chronic kidney disease, which is characterized by massive fibroblast activation and excessive production and deposition of extracellular matrix (ECM). Renal fibrosis results in progressive loss of kidney function; however, there is currently no effective therapy available clinically to treat or even reverse renal fibrosis. Although activated fibroblasts/myofibroblasts are responsible for the production and deposition of ECM, their origin has been debatable. Recent studies have provided compelling evidence that bone marrow-derived fibroblast precursors contribute significantly to the population of myofibroblasts and the development of renal fibrosis. Therefore, targeting the molecular signaling mechanisms underlying the recruitment and activation of the bone marrow-derived fibroblast precursors may serve as novel therapeutic strategy for chronic kidney disease. In this review, we appraise recent advances in our understanding of the recruitment and activation of bone marrow-derived fibroblast precursors in the kidney and the development of renal fibrosis and highlight novel molecular signaling pathways that may lead to the development of new therapies for chronic kidney disease.

Keywords Chemokine · Cytokine · Bone marrow-derived fibroblast precursors · Fibroblasts · Renal fibrosis · Extracellular matrix · Chronic kidney disease · Monocyte-to-fibroblast transition

C. An · J. Wen · Y. Wang (✉)
Division of Nephrology, Department of Medicine, University of Connecticut Health Center, Farmington, CT, USA
e-mail: yanlwang@uchc.du

C. An · Y. Wang
Renal Section, Veterans Affairs Connecticut Healthcare System, West Haven, CT, USA

C. An · L. Jia · J. Wen · Y. Wang
Selzman Institute for Kidney Health and Section of Nephrology, Department of Medicine, Baylor College of Medicine, Houston, TX, USA

B.-C. Liu et al. (eds.), *Renal Fibrosis: Mechanisms and Therapies*, Advances in Experimental Medicine and Biology 1165,
https://doi.org/10.1007/978-981-13-8871-2_14

14.1 Introduction

Chronic kidney disease is a global public health problem that affects more than 20 million Americans who have chronic kidney disease and more than 450,000 Americans who suffer from end-stage renal disease (ESRD) requiring renal replacement therapy. Chronic kidney disease is the eighth leading cause of death in the USA and over 1 million people die from chronic kidney disease yearly worldwide. Renal fibrosis is a common pathological feature of chronic kidney disease regardless of various causes, which is associated with extensive accumulation of activated fibroblast/myofibroblasts and production and deposition of extracellular matrix including collagen type I, III, IV, fibronectin, vimentin, and proteoglycans at the site of tissue injury (Farris and Colvin 2012; Conway and Hughes 2012; Zeisberg and Neilson 2010; Yan et al. 2016).

Fibrosis is considered the consequence of dysregulated wound healing, in which multiple cell types such as resident fibroblasts and bone marrow-derived cells are recruited to the site of injury to participate in a wound-healing response. Renal interstitial fibrosis is characterized by massive fibroblast activation and excessive production and deposition of extracellular matrix. During the pathological development of renal fibrosis, extracellular matrix and fibroblasts replace normal kidney parenchyma, which leads to the destruction and collapse of kidney parenchyma resulting in progressive loss of kidney function. Fibrosis is a complex pathological process that is associated with the infiltration of inflammatory cells, accumulation of fibroblasts, and increase extracellular matrix deposition.

14.2 What Are Fibroblasts?

Fibroblasts are mesenchymal cells that exhibit spindle-shaped morphology. Due to the lack of specific markers, it is quite challenging to distinguish fibroblasts from other cells of mesenchymal origin such as pericytes, vascular smooth muscle cells, and resident mesenchymal stem cells (Strutz and Zeisberg 2006). These cells produce extracellular matrix including collagens, fibronectin, and other matrix proteins. Moreover, these cells synthesize matrix metalloproteinases and tissue inhibitors of metalloproteinases. Therefore, fibroblasts play an essential role in the maintenance of normal tissue structure and function (Strutz and Zeisberg 2006; Meran and Steadman 2011). Furthermore, interstitial fibroblasts in the kidney possess endocrine function through secretion of erythropoietin, which controls erythropoiesis by regulating the proliferation and differentiation of erythroid precursor cells (Zeisberg and Kalluri 2015).

14.3 What Are Myofibroblasts?

Activated fibroblasts termed myofibroblasts possess a unique contractile property that is characterized by expression of alpha-smooth muscle actin (α-SMA) (Tomasek et al. 2002). Myofibroblasts display an active phenotype with a large oval nucleus, abundant rough endoplasmic reticulum, and several sets of Golgi apparatuses (Zeisberg and Kalluri 2015). These features reflect their capacity to produce large amounts of extracellular matrix components. Fibroblasts activation into myofibroblasts is considered the principal cells that are responsible for the production of extracellular matrix during the pathogenesis of fibrotic kidney disease (Neilson 2006; Eddy 2005). However, the origin of the activated fibroblasts has been of intensive debate. It is traditionally thought that myofibroblasts are arisen from resident fibroblasts within the kidney. However, recent evidence indicates that myofibroblasts may originate from epithelial/endothelial transition (Iwano et al. 2002; Liu 2010; Sato et al. 2003; Zeisberg and Kalluri 2004; Zeisberg et al. 2008), pericytes (Lin et al. 2008; Humphreys et al. 2010), and bone marrow-derived progenitor cells (Iwano et al. 2002; Sakai et al. 2006; Grimm et al. 2001; Lebleu et al. 2013). Recent studies have shown that activated fibroblasts/myofibroblasts may originate from bone marrow-derived fibroblast precursors (Iwano et al. 2002; Sakai et al. 2006; Grimm et al. 2001; Broekema et al. 2007; Chen et al. 2011; Yan et al. 2016).

14.4 What Are Bone Marrow-Derived Fibroblasts?

Bone marrow-derived fibroblast precursors termed fibrocytes were first described as a subpopulation of leukocytes that express the hematopoietic stem cell/progenitor marker CD34, the leukocyte common antigen CD45 (a pan-hematopoietic cell marker), monocyte markers such as CD11b in conjunction with mesenchymal cell markers such as vimentin, collagen I, and collagen III (Bucala et al. 1994). The expression of the stem cell marker CD34 supports the notion that fibrocytes constitute a bone marrow-derived fibroblast population that circulates in the peripheral blood (Herzog and Bucala 2010). While CD45 shows more robust expression than CD34, both CD45 and CD34 downregulated in situ fibrocytes differentiate and mature (Mori et al. 2005). Fibrocytes can be easily isolated from peripheral blood mononuclear cells on either plastic or fibronectin-coated tissue culture plate in vitro. Under physiological condition, fibrocytes represent less than 1% of circulating leukocytes (Bucala et al. 1994). However, the number of fibrocytes increases significantly in response to certain chemokines or under pathological conditions such as fibrotic or inflammatory disorders (Herzog and Bucala 2010). The co-expression of CD45 or CD34 with collagen I or procollagen I is commonly used to identify bone marrow-derived fibroblasts/fibrocytes (Chen et al. 2011; Moeller et al. 2009). Fibrocytes also express other matrix proteins such as fibronectin and vimentin and additional hematopoietic markers such as CD11b, CD13, and leukocyte-specific protein 1 (LSP1) (Herzog and

Bucala 2010). Furthermore, fibrocytes possess antigen presentation property through expressing cell surface proteins including the class II major histocompatibility complex molecules and the costimulatory molecules such as CD80 and CD86. Therefore, fibrocytes are capable of activating naïve T cells (Chesney et al. 1997).

Recent studies have demonstrated that bone marrow-derived fibroblasts contribute to 30–60% of activated fibroblast/myofibroblast population during the development of renal fibrosis depending on experimental models, timing of detection, lineage tracing methods, and detection techniques (Lebleu et al. 2013; Broekema et al. 2007; Chen et al. 2011; Li et al. 2007; Wang et al. 2016; Xia et al. 2014b; Dong et al. 2016). In a clinical study of mismatched kidney transplantation in humans, the host-derived α-SMA-positive cells constitute 30% of α-SMA-positive myofibroblast population in allografts undergoing chronic rejection (Grimm et al. 2001). In rodent models of renal fibrosis, we and others have shown that bone marrow-derived fibroblast precursors migrate into the kidney in response to injury, contributing significantly to the population of myofibroblasts (Iwano et al. 2002; Broekema et al. 2007; Chen et al. 2011; Xia et al. 2014b; Roufosse et al. 2006). In an experimental study, Iwano et al. have demonstrated that 15% of fibroblasts derived bone marrow are present in the kidney 10 days after obstructive injury using bone marrow transplantation of transgenic mice that express fibroblast-specific protein 1 (FSP1) (Iwano et al. 2002). Another experimental study has shown that bone marrow-derived myofibroblasts contribute to 30% α-SMA-positive myofibroblasts 7 days after ischemia-reperfusion injury using bone marrow transplantation of transgenic rats that express human placental alkaline phosphatase (Broekema et al. 2007). More recently, LeBleu et al. have reported that 35% α-SMA-positive myofibroblasts are derived from bone marrow after obstructive injury 10 days using bone marrow transplantation of transgenic mice that express red fluorescence protein (RFP) that is driven by α-SMA promoter (Lebleu et al. 2013). Moreover, we have shown that bone marrow-derived hematopoietic fibroblasts proliferate and differentiate into myofibroblasts and migrate into the kidney using bone marrow transplantation of transgenic mice that express green fluorescence protein (GFP) that is driven by collagen α1(I) promoter (Xia et al. 2014b). One potential pitfall of bone marrow transplantation is the rate of engraftment of bone marrow cells. It is technically challenging to obtain 100% bone marrow reconstitution from donor mice. Therefore, this technique may underestimate the contribution of bone marrow-derived fibroblasts to the development of renal fibrosis. To objectively quantify the number of bone marrow-derived hematopoietic fibroblasts in the kidney, we have stained freshly isolated kidney cells with CD45, a hematopoietic cell marker, and collagen I, a mesenchymal cell marker, and analyzed with flow cytometry. Our results have demonstrated that $CD45^+$ and collagen I^+ cells constituted 45% of total collagen I^+ cells in the kidney 7 days after unilateral obstructive injury (Xia et al. 2014b). Consistent with our findings, a recent study has reported that hematopoietic, bone marrow-derived cells contribute to 38%–50% of the overall deposition of collagen I in the kidney in murine models of UUO and adenine-induced nephropathy (Buchtler et al. 2018). These results provide strong evidence that bone marrow-derived fibroblasts contribute significantly to the pathogenesis of renal fibrosis and suggest that targeting bone marrow-derived fibroblasts may represent a novel therapeutic strategy for chronic kidney disease.

14.5 How Are Bone Marrow-Derived Fibroblasts Recruited into the Kidney

The recruitment of circulating fibroblasts into sites of injury is mediated by locally produced chemokines. Chemokines are a family of small cytokines. Based on the relative position of cysteine residues near the NH2 terminus, chemokines can be classified into four major subfamilies: CC, CXC, C, and CX3C (Mackay 2001; Rollins 1997). In the CXC chemokines, one amino acid separates the first two cysteines, whereas in CC chemokines, these two cysteines are adjacent. A single member of the CX3C subfamily, CX3CL1 or fractalkine, has three amino acids between the two cysteines, whereas the first and third cysteines are missing in the (X)C subfamily. Chemokines exert their chemoattractant function through interactions with their respective receptors. Chemokine receptors belong to class A G protein-coupled receptors (GPCR) coupled with the Gαi class of the heterotrimeric G proteins. They are also grouped into four subfamilies according to the subfamily of their major chemokine ligands (Zlotnik and Yoshie 2012). Fibrocytes express chemokine receptors—CCR2, CCR7, CXCR4, and CXCR6 (Sakai et al. 2006; Chen et al. 2011; Xia et al. 2014b; Moore et al. 2005; Xia et al. 2013a; Xu et al. 2011; Xia et al. 2014a; Phillips et al. 2004).

The chemokine CXCL16 is a recently discovered chemokine that belongs to the CXC chemokine family (Matloubian et al. 2000; Wilbanks et al. 2001). It was originally discovered as a scavenger receptor for phosphatidylserine and oxidized low-density lipoprotein (SR-PSOX) (Shimaoka et al. 2000). Subsequently, CXCL16 was independently identified as a ligand for the CXC chemokine receptor, CXCR6 (BONZO, STRL33, TYMSTR) (Matloubian et al. 2000). CXCL16 contains six cysteines in the chemokine domain, a property only observed in a subfamily of CC chemokines. CXCL16 exists in two forms, a transmembrane form and a soluble form. The transmembrane form is a type I transmembrane glycoprotein consisting of an extracellular N-terminal chemokine domain, glycosylated mucin-like stalk, transmembrane-spanning region, and a short cytoplasmic domain with a YXPV motif, which is a potential tyrosine-phosphorylation and SH2-protein-binding site (Izquierdo et al. 2014). The transmembrane form can function as an adhesion molecule for CXCR6 expressing cells and scavenger receptor for oxidized low-density lipoprotein (Shimaoka et al. 2000, 2004). The soluble form is generated by cleavage at the cell surface and functions as a chemoattractant for cells expressing CXCR6 (Gough et al. 2004). CXCL16 is produced as an intracellular precursor that is processed and transported to the cell surface, where it undergoes metalloproteinase-dependent cleavage (Gough et al. 2004). The α-secretase-like activity of two disintegrins and metalloproteinases, ADAM10 and ADAM17/TNF-alpha converting enzyme (TACE) is responsible for the ectodomain release (Schulte et al. 2007). ADAM10 is capable of mediating both constitutive and inducible CXCL16 cleavage, whereas ADAM17 is only able to mediate inducible shedding of CXCL16 (Abel et al. 2004; Ludwig et al. 2005). ADAM10-mediated cleavage generates C-terminal CXCL16 fragments can be further cleaved by the γ-secretase complex (Schulte et al. 2007). It is speculated that the cleavage of CXCL16 by the γ-secretase complex may generate mediators of intracellular signaling as other proteins such as Notch and CD74 undergo intramembrane proteolysis (RIP) (Schulte et al. 2007; Ludwig and Weber 2007; Bielesz et al. 2010; Shachar 2017).

CXCL16 is induced in kidney tubular epithelial cells in vivo in a murine model of renal fibrosis induced by obstructive injury (Chen et al. 2011; Okamura et al. 2007). Tumor necrosis factor (TNF)-α and interferon (IFN)-γ upregulate CXCL16 expression in tubular epithelial cells in vitro (Xia et al. 2013a). In addition, the TNF superfamily cytokine TNF-like weak inducer of apoptosis (TWEAK) increases CXCL16 expression in kidney tubular epithelial cells in vitro and in vivo (Izquierdo et al. 2012). Angiotensin II, a key mediator of kidney injury and fibrosis, also augments CXCL16 expression in tubular epithelial cells via activation of NF-κB, a key signaling regulator of inflammation (Xia et al. 2013b). We have recently studied the functional role of CXCL16 in the pathogenesis of renal fibrosis in a murine model of obstructive nephropathy using CXCL16 knockout mice. Our results have shown that CXCL16 knockout mice display fewer bone marrow-derived fibroblast precursor accumulation and myofibroblast activation. Furthermore, genetic disruption of CXCL16 reduces extracellular matrix protein expression and inhibits collagen deposition in the obstructed kidneys compared with wild-type mice (Chen et al. 2011). These data indicate that CXCL16 promotes renal fibrosis by recruiting bone marrow-derived fibroblast precursors into the kidney in response to obstructive injury. Furthermore, we have examined the functional role of CXCL16 in angiotensin II-induced renal injury and fibrosis. The results have demonstrated that genetic disruption of CXCL16 protects the kidney from angiotensin II infusion-induced renal dysfunction, inhibits renal fibrosis, reduces proteinuria, suppresses bone marrow-derived fibroblast accumulation, myofibroblast formation, macrophage, and T cell infiltration, and pro-inflammatory cytokine expression without affecting blood pressure at baseline and in response to angiotensin II infusion (Xia et al. 2013b). Recently, we have shown that CXCL16 is induced in the kidney in deoxycorticosterone acetate (DOCA)-salt-induced hypertension. We further examine whether CXCL16 has a role in DOCA-salt-induced renal inflammation and fibrosis. Wild-type and CXCL16 knockout mice are subjected to DOCA-salt treatment for 3 weeks after uninephrectomy. Systolic blood pressure is similar at baseline between wild-type and CXCL16 knockout mice. DOCA-salt treatment increases systolic blood pressure significantly, which is comparable between wild-type and CXCL16 knockout mice. Compared with wild-type mice, CXCL16 deficient mice exhibit less renal dysfunction, proteinuria, and fibrosis after DOCA-salt treatment. CXCL16 deficiency reduces extracellular matrix protein production and inhibits bone marrow-derived fibroblast accumulation and myofibroblast formation in the kidneys in response to DOCA-salt treatment. Furthermore, CXCL16 deficiency reduces macrophage and T cell infiltration into the kidneys in response to DOCA-salt hypertension (Liang et al. 2016). More recently, we have shown that CXCL16 is induced in the kidney in a murine model of renal artery stenosis (RAS). We then determine whether CXCL16 is involved in renal injury and fibrosis using CXCL16 knockout mice. Both wild-type and CXCL16 knockout mice are subjected to renal artery stenosis induced by placing a cuff on the left renal artery. Wild-type and CXCL16 knockout mice have similar blood pressure at baseline. Renal artery stenosis results in an increase in systolic blood pressure that is comparable between wild-type and CXCL16 knockout mice. CXCL16 knockout mice are protected from renal injury and fibrosis induced by RAS. Genetic disruption of CXCL16 inhibits bone marrow-derived fibroblast accumulation and myofibroblast formation in the stenotic kidney, which is associated with less expression of extracellular matrix proteins. Furthermore, CXCL16 deficiency inhibits the infiltration of macrophages and T cells in the stenotic kidneys compared with wild-type mice (Ma et al. 2016). In support of the clinical relevance of these observations in experimental animal models, clinical studies have shown that circulating CXCL16 is elevated in patients with chronic kidney disease and diabetic nephropathy and high levels of CXCL16 are associated with chronic kidney disease progression and development of proteinuria (Lin et al. 2011; Zhao et al. 2014; Elewa et al. 2016).

CXCR6 is the receptor for CXCL16. CXCR6 was first cloned as an orphan receptor by three independent groups and was termed STRL33, BONZO, or TYMSTR (Alkhatib et al. 1997; Deng et al. 1997; Loetscher et al. 1997). We have recently shown that both circulating fibroblast precursors and bone marrow-derived fibroblasts in the kidney express CXCR6 (Chen et al. 2011; Xia et al. 2014b). We have studied the functional role of CXCR6 in mouse models of renal injury using CXCR6 knockout mice. Compared with wild-type mice, CXCR6 knockout mice accumulate fewer bone marrow-derived fibroblasts in response to injury induced by unilateral ureteral obstruction and ischemia-reperfusion. Moreover, CXCR6 knockout mice express less profibrotic molecules and exhibit fewer myofibroblasts in the obstructed kidneys compared with wild-type mice. CXCR6 deficiency inhibited total collagen deposition and suppressed the expression of collagen I and fibronectin in the kidneys in response to obstructive injury or ischemia-reperfusion injury. Furthermore, wild-type mice engrafted with $CXCR6^{-/-}$ bone marrow cells displayed fewer bone marrow-derived fibroblasts in the kidneys with obstructive injury and showed less severe renal fibrosis compared with wild-type mice engrafted with $CXCR6^{+/+}$ bone marrow cells. Transplant of wild-type bone marrow into $CXCR6^{-/-}$ recipients restored recruitment of myeloid fibroblasts and susceptibility to fibrosis (Xia et al. 2014b). We also examine the role of CXCR6 in angiotensin II-induced renal injury and fibrosis. Wild-type and CXCR6 knockout mice have similar systolic blood pressure at baseline. In response to angiotensin II, systolic blood pressure increases, which is comparable between wild-type and CXCR6 knockout mice. Compared with wild-type mice, CXCR6 knockout mice are protected from angiotensin II-induced renal dysfunction, proteinuria, and fibrosis. CXCR6 knockout mice accumulate fewer bone marrow-derived fibroblasts and myofibroblasts and generate less extracellular matrix protein in the kidneys after angiotensin II treatment. Furthermore, CXCR6 knockout mice display fewer macrophages and T cells and express less pro-inflammatory cytokines in the kidneys in response to angiotensin II treatment. Further, wild-type mice engrafted with CXCR6$^{-/-}$ bone marrow cells exhibit fewer bone marrow-derived fibroblasts, macrophages, and T cells in the kidney after angiotensin II treatment compared with wild-type mice engrafted with CXCR6$^{+/+}$ bone marrow cells (Xia et al. 2014a).

Chemokine (C-C motif) ligand 21 (CCL21) is a small cytokine belonging to the CC chemokine family. CCL21 contains six cysteines and is a potent chemoattractant for T cells, B cells, and dendritic cells (Ebert et al. 2005; Forster et al. 2008). Additionally, CCL21 can function as a chemoattractant for fibrocytes, which express CCR7 (Abe et al. 2001). Sakai et al. have examined the functional role of CCL21/CCR7 in the regulation of fibrocyte infiltration in a murine model of fibrosis induced by unilateral ureteral obstruction (Sakai et al. 2006). They show that wild-type mice exhibit significant fibrocyte infiltration into the kidney in response to obstructive injury. Blockade of CCL21 with a neutralizing antibody or genetic deletion of CCR7 markedly suppresses the infiltration of fibrocytes into the obstructive kidney with concomitant reduction of renal fibrosis. These results indicate that CCL21 recruits CCR7-positive fibrocytes into the obstructed kidney, thereby contributing to the development of renal fibrosis. Therefore, targeting CCL21/CCR7 axis may yield new therapeutic strategy for renal fibrosis.

Circulating fibroblast precursors/fibrocytes express the chemoattractant protein-1 (MCP-1) receptor—CCR2 (Xu et al. 2011). We have established a murine model of hypertension induced by continuous infusion of angiotensin II to study the mechanisms of reactive cardiac fibrosis (Xu et al. 2011; Haudek et al. 2010). We have found that the development of cardiac fibrosis is associated with induction of MCP-1 predominantly in endothelial and smooth muscle cells of the small vessels (Haudek et al. 2010). Angiotensin II infusion results in interstitial deposition of collagens that is mediated by infiltration of monocytic CD34$^+$ CD45$^+$ fibroblast precursors into the heart. These cells are of myeloid origin and contribute significantly to the development of cardiac fibrosis. Genetic disruption of MCP-1 prevents the development of angiotensin II-induced cardiac fibrosis and the accumulation of bone marrow-derived fibroblasts in the heart. In contrast, angiotensin II infusion leads to cardiac hypertrophy and increases systolic function and blood pressure in both wild-type and MCP-1-KO mice, which are not significantly different between the WT and MCP-1-KO mice over the 6-week course of infusion (Haudek et al. 2010). We have demonstrated that circulating fibroblast precursors express the MCP-1 receptor—CCR2 (Xu et al. 2011). We then determined whether CCR2 has a role in the uptake of bone marrow-derived fibroblast precursors into the heart and the development of cardiac fibrosis in response to angiotensin II-induced hypertension using CCR2 knockout mice. Angiotensin II infusion results in an elevation of blood pressure and development of cardiac hypertrophy that are not significantly different between wild-type and CCR2 knockout mice. However, the development of cardiac fibrosis and the accumulation of bone marrow-derived fibroblast precursors in the heart are significantly reduced in CCR2 knockout mice following Ang II infusion. These results indicate that MCP-1/CCR2 axis plays an important role in the recruitment of bone marrow-derived fibroblasts and the development of cardiac fibrosis in response to angiotensin II without affecting the hemodynamic response of angiotensin II. Further study has shown that the development of cardiac fibrosis in response to angiotensin II has two stages: A primary M1 inflammatory phase followed by a M2 fibrotic phase. The M1 cells produce tumor necrosis factor (TNF)-α. The M2 fibrotic phase and fibrogenesis are dependent on TNF-α receptor 1 (TNFR1) in myeloid cells. These results indicate that TNF-α produced by M1 cells is required for mediating a TNFR1-dependent maturation of M2 cells into collagen-producing fibroblasts (Duerrschmid et al. 2015).

The role of CCR2 in recruiting bone marrow-derived fibroblasts into the kidney during the development of renal fibrosis has been investigated. Reich et al. have shown that CCR2 does not regulate fibrocytes emigrate from bone marrow to the circulation because the number of circulating fibrocytes is comparable between wild-type mice and CCR2 knockout mice or wild-type mice treated with a CCR2 neutralizing antibody (Reich et al. 2013). However, genetic disruption of CCR2 reduces the number of bone marrow-derived fibrocytes and inhibits collagen expression and deposition in the kidney in response to obstructive injury. We have shown that bone marrow-derived collagen-expressing GFP^+ fibroblasts express platelet-derived growth factor receptor-β (PDGFR-β) and CCR2 in the obstructed kidneys of chimeric mice transplanted with donor bone marrow from collagen α1(I)-GFP reporter mice (Xia et al. 2013a). CCR2 knockout mice exhibit significantly fewer bone marrow-derived fibroblast precursors dual positive for CD45 and PDGFR-β or procollagen I in the obstructed kidneys compared with wild-type mice. Furthermore, CCR2 knockout mice accumulate fewer bone marrow-derived myofibroblasts and express less α-SMA or FSP-1 in the obstructed kidneys compared with wild-type mice. Furthermore, genetic deletion of CCR2 suppresses total collagen deposition and expression of collagen I and fibronectin in the obstructed kidney.

In these studies, we have noted that genetic disruption of a single chemokine/chemokine axis does not completely inhibit bone marrow-derived fibroblast precursor infiltration into the kidney and fibrosis development. These observations suggest that chemokine/receptor pairs may interact with each other in recruiting bone marrow-derived fibroblast precursors into the kidney. In support of this notion, the expression of CXCL16 in the kidney is reduced in CCR2 knockout mice in response to obstructive injury, suggesting the interactions of two distinct chemokine systems modulate renal tubular epithelial cell-initiated fibrosis (Xia et al. 2013a).

14.6 TGF-β1/Samd3 Signaling in Bone Marrow-Derived Fibroblast Activation

The activation of bone marrow-derived fibroblast precursors plays a crucial role in the pathogenesis of renal fibrosis (Chen et al. 2014; Yang et al. 2013). Myofibroblasts are a population of smooth muscle-like fibroblasts that express α-SMA (Powell et al. 1999). The activation of fibroblasts into myofibroblasts is generally accepted a crucial event in the pathogenesis of renal fibrosis (Nath 1992; Eddy 2013). Furthermore, experimental and clinical studies have shown that the number of interstitial myofibroblasts is associated closely with the severity of tubulointerstitial fibrosis and the rapidity of kidney disease progression (Zhang et al. 1995; Roberts et al. 1997; Essawy et al. 1997). The activation of bone marrow-derived fibroblast precursors is regulated by locally produced cytokines (Chen et al. 2014; Yang et al. 2013). TGF-β1 is a key profibrotic cytokine that plays an important role in the pathogenesis of renal fibrosis through activation of a cascade of intracellular signaling pathways (Border and Noble 1994; Border et al. 1990; Bottinger and Bitzer 2002; Lan 2011). Evidence suggests that activation of the canonical Smad signaling cascade plays an important role in stimulating ECM protein expression and tissue fibrosis (Sato et al. 2003; Latella et al. 2009; Huang et al. 2010; Verrecchia et al. 2001; Zhao et al. 2002; Meng et al. 2015). We have recently examined the functional role of Smad3 in the activation of bone marrow-derived fibroblast precursors in vitro and in vivo (Chen et al. 2014). In cultured monocytes, TGF-β1 activates Smad3. Smad3 deficient monocytes express less amounts of ECM proteins at baseline and Smad3 deficiency completely abolished TGF-β1-induced expression of α-SMA and extracellular matrix proteins in cultured monocytes in vitro. Smad3 knockout mice accumulate significantly fewer bone marrow-derived fibroblasts in the kidney after obstructive injury, exhibit less myofibroblast activation, and express less α-SMA in the obstructed kidney. Furthermore, genetic deletion of Smad3 reduces total collagen deposition and suppresses expression of extracellular matrix proteins. Additionally, wild-type mice engrafted with Smad3$^{-/-}$ bone marrow cells display fewer bone marrow-derived fibroblasts in the kidney with obstructive injury and show less severe renal fibrosis compared with wild-type mice engrafted with Smad3$^{+/+}$ bone marrow cells. These results indicate that Smad3 of bone marrow-derived cells plays an important role in bone marrow-derived fibroblast activation. However, we have observed that Samd3 deficiency does not completely abolish bone marrow-derived fibroblast activation, collagen deposition, and ECM protein expression in vivo. These results suggest that other factors may be involved in bone marrow-derived fibroblast activation.

14.7 IL-4Rα/JAK3/STAT6 Signaling in Bone Marrow-Derived Fibroblast Activation

The activation of bone marrow-derived fibroblasts is a dynamic process that is modulated by cells in the inflamed microenvironment. T cells have an important role in the development of renal fibrosis (Tapmeier et al. 2010), which have been shown to regulate bone marrow-derived fibrocyte activation (Niedermeier et al. 2009). Naïve CD4^{+} T cells can differentiate into two major distinct phenotypes, Th1 cells and Th2 cells, which are characterized by specific cytokine expression patterns (Wynn 2004). Th2 cells produce Th2 cytokines such as IL-4 and IL-13, which induce alternative activation of macrophage and promotes monocyte-to-fibroblast transition. Th1 cells make Th1 cytokines such as IFN-γ and IL-12, which promote classical activation of macrophages and inhibit fibrocyte differentiation (Wynn 2004; Shao et al. 2008). However, the molecular signaling mechanisms by which Th2 cytokines promote bone marrow-derived fibroblast activation are well understood. We have found that Janus kinase 3 (JAK3)/signal transducer and activator of transcription 6 (STAT6) signaling pathway is activated during the development of renal fibrosis and plays an important role in bone marrow-derived fibroblast activation, extracellular matrix protein production, and interstitial fibrosis development (Yan et al. 2015). Specifically, our results have shown that Th2 cytokine- IL-4 or IL-13 induces STAT6 activation and stimulates bone marrow monocytes to express ECM proteins and α-smooth muscle actin (α-SMA). CP690550, a specific JAK3 inhibitor, or STAT6 deficiency inhibits IL-4/IL-13-induced STAT6 activation and expression levels of ECM proteins and α-SMA in bone marrow monocytes in vitro. Furthermore, CP690550 treatment or STAT6 deficiency suppresses bone marrow-derived fibroblast activation and ECM protein production in the kidney in response to obstructive nephropathy. To further confirm the role of bone marrow STAT6 signaling in myeloid fibroblast activation, we performed bone marrow chimeric experiments. Wild-type mice transplanted with STAT6 null bone marrow cells exhibit fewer bone marrow-derived fibroblasts and develop a lesser degree of renal fibrosis. More recently, we have studied the role of IL-4 receptor α (IL-4Rα) in myeloid fibroblast activation in two experimental models of renal fibrosis induced by unilateral ureteral obstruction and folic acid (Liang et al. 2017). Compared with wild-type mice, IL-4Rα knockout mice accumulate fewer bone marrow-derived fibroblasts and myofibroblasts in the kidneys in response to obstructive injury or folic acid administration. Genetic disruption of IL-4Rα suppressed the expression of α-SMA, extracellular matrix proteins and collagen deposition in the injured kidney. Furthermore, genetic disruption of IL-4Rα inhibited the activation of STAT6 in the kidney in response to obstructive injury. To further examine the role of IL-4Rα in myeloid fibroblast activation and fibrosis in the kidney, we have performed bone marrow transplantation experiments. Wild-type mice engrafted with bone marrow cells from IL-4Rα knockout mice display fewer myeloid fibroblasts in the kidney and displayed less severe renal fibrosis following obstructive injury compared with wild-type mice engrafted with wild-type bone marrow cells. In cultured bone marrow monocytes, IL-4 treatment results in activation of STAT6 and stimulates protein expression of α-smooth muscle actin and fibronectin, whereas IL-4Rα deficiency abolishes IL-4 induced STAT6 activation and monocyte-to-fibroblast transition. These results indicate that IL-4Rα/JAK3/STAT6 signaling may serve as novel therapeutic targets for fibrotic kidney disease.

14.8 Adiponectin/AMPK Signaling in Bone Marrow-Derived Fibroblast Activation

Adiponectin is a multifunctional cytokine and an important factor that regulates lipid and carbohydrate metabolism. Accumulating evidence suggests that circulating adiponectin levels are elevated in patients with chronic kidney disease, and a high level of adiponectin is linked to increased cardiovascular mortality (Iwashima et al. 2006; Shimotomai et al. 2005; Zoccali et al. 2003; Zoccali and Mallamaci 2011; Mills et al. 2013). However, the mechanisms by which adiponectin regulates cardiovascular mortality and progression of chronic kidney disease are not understood. We have discovered that adiponectin is induced following ischemia-reperfusion and obstructive injury (Yang et al. 2013). Genetic disruption of adiponectin inhibits bone marrow-derived fibroblast accumulation and myofibroblast activation. Furthermore, adiponectin deficiency also reduces the expression of profibrotic chemokines and cytokines and the production of ECM proteins in the kidneys in response to obstructive injury or ischemia-reperfusion. These results indicate that adiponectin plays an important role in the activation and maturation of bone marrow-derived fibroblast precursors and the development of renal fibrosis. We further demonstrate that adiponectin stimulates α-SMA and extracellular matrix protein expression through activation of adenosine monophosphate-activated protein kinase (AMPK) in cultured bone marrow-derived monocytes in a time- and dose-dependent manner. Pharmacological inhibition of AMPK with compound C, or ectopic expression of dominant negative AMPK-α1 attenuates adiponectin-induced expression of α-SMA and extracellular matrix proteins in bone marrow-derived monocytes. AICAR is a cell-permeable adenosine analogue. Once inside the cell, AICAR is converted to 5-amino-4-imidazolecarboxamide ribotide (ZMP), which mimic AMP to activate AMPK (Corton et al. 1995). We have demonstrated that AMPK activation by AICAR results in increased expression of α-SMA and extracellular matrix proteins. Therefore, we have identified that adiponectin functions as a critical regulator of monocyte-to-fibroblast transition and renal fibrosis. More recently, we have discovered that AMPKα1 is induced in the kidney during the development of renal fibrosis (Wang et al. 2018). Mice with global or fibroblast-specific deletion of AMPKα1 accumulate fewer myofibroblasts, develop less fibrosis, and produce less extracellular matrix proteins in the kidney following unilateral ureteral obstruction or ischemia-reperfusion injury. We then explore the mechanisms by which AMPKα1 stimulates fibroblast activation. We have shown that AMPKα1 directly phosphorylates cofilin leading to cytoskeleton remodeling and myocardin-related transcription factor-A (MRTF-A) nuclear translocation resulting in fibroblast activation and extracellular matrix protein production. Therefore, inhibition of adiponectin/AMPK/MRTF-A signaling may serve as a novel therapeutic target for fibrotic kidney disease.

It is generally thought that macrophages do not produce extracellular matrix proteins. These cells promote fibrosis by producing molecules that activate fibroblasts (Wynn and Barron 2010). Recently, a model of two major macrophage classes has proposed. Classically activated macrophages exhibit a Th1-like phenotype and promote inflammation in response to Th1 cytokines; while alternatively activated macrophages or M2 macrophages display a Th2-like phenotype and stimulate ECM production in response to Th2 cytokines (Gordon and Martinez 2010). M2 macrophages are characterized by expressing MHC class II, mannose receptor (CD206), Ym1, Fizz1/Relm-α, and arginase. Alternatively activated macrophages have been implicated in the pathogenesis of fibrosis in other organs (Gordon and Martinez 2010) and repair after renal ischemia-reperfusion injury (Lee et al. 2011). We have recently demonstrated for the first time that alternatively activated macrophages produce procollagen I, suggesting a linking relationship between M2 macrophage polarization and bone marrow-derived fibroblast activation (Yang et al. 2013). Consistent with this novel concept, we have shown that adiponectin deficiency suppresses M2 macrophage polarization and inhibits the number of procollagen-expressing M2 macrophages in the injured kidneys (Yang et al. 2013).

14.9 Conclusion

Recent studies have provided compelling evidence that bone marrow-derived fibroblast precursors contribute significantly to the pathogenesis of renal fibrosis. Recruitment and activation of bone marrow-derived fibroblasts are mediated through the interactions between chemokines/cytokines and their receptors (Fig. 14.1). Therefore, targeting the signaling machinery of these chemokines/cytokines could rep-

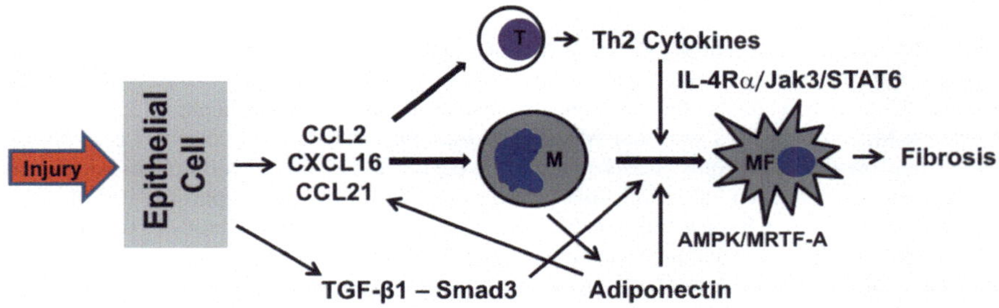

Fig. 14.1 A proposed model of signaling mechanisms underlying recruitment and activation of bone marrow-derived fibroblasts during the development of renal fibrosis. In response to injurious stimulation (obstruction, ischemia-reperfusion, folic acid, hypertension), tubular epithelial cells produce chemokines CXCL16/CCL2/CCL21 and cytokines TGF-β1. Chemokines—CXCL16/CCL2/CCL21—function concertedly to recruit bone marrow-derived cells (T cells, monocytes, and fibrocytes) via interaction with their respective receptors. TGF-β1 activates Smad3 to stimulate monocyte-to-fibroblast transition. Th2 cells produce Th2 cytokines (Il-4, IL-13), which activate IL-4Rα/JAK3/STAT6 signaling pathway to promote monocyte-to-fibroblast transition. Finally, adiponectin produced by inflammatory cells regulates chemokine and cytokine production and stimulates monocyte-to-fibroblast transition through activation of AMPK/MRTF-A signaling pathway (Yan et al. 2016). T—T cells; M—Monocytes

resent novel therapeutic strategy for the treatment of fibrotic kidney disease and possible fibrotic disorders of other organs.

Acknowledgements This work was supported by grants from the National Institutes of Health grant (R01DK95835) and the US Department of Veterans Administration grant (I01BX02650) to YW. This chapter was modified from an article authored by our group in Frontiers in Physiology (Yan et al. 2016). The related contents are reused with permission.

Conflict of Interest Statement
The authors declare that the research was conducted in the absence of any commercial or financial relationships that could be considered as a potential conflict of interest.

References

Abe R, Donnelly SC, Peng T, Bucala R, Metz CN (2001) Peripheral blood fibrocytes: differentiation pathway and migration to wound sites. J Immunol 166:7556–7562

Abel S, Hundhausen C, Mentlein R, Schulte A, Berkhout TA, Broadway N et al (2004) The transmembrane CXC-chemokine ligand 16 is induced by IFN-gamma and TNF-alpha and shed by the activity of the disintegrin-like metalloproteinase ADAM10. J Immunol 172:6362–6372

Alkhatib G, Liao F, Berger EA, Farber JM, Peden KW (1997) A new SIV co-receptor, STRL33. Nature 388:238

Bielesz B, Sirin Y, Si H, Niranjan T, Gruenwald A, Ahn S et al (2010) Epithelial Notch signaling regulates interstitial fibrosis development in the kidneys of mice and humans. J Clin Invest 120:4040–4054

Border WA, Noble NA (1994) Transforming growth factor beta in tissue fibrosis. N Engl J Med 331:1286–1292

Border WA, Okuda S, Languino LR, Sporn MB, Ruoslahti E (1990) Suppression of experimental glomerulonephritis by antiserum against transforming growth factor beta 1. Nature 346:371–374

Bottinger EP, Bitzer M (2002) TGF-beta signaling in renal disease. J Am Soc Nephrol 13:2600–2610

Broekema M, Harmsen MC, van Luyn MJ, Koerts JA, Petersen AH, van Kooten TG et al (2007) Bone marrow-derived myofibroblasts contribute to the renal interstitial myofibroblast population and produce procollagen I after ischemia/reperfusion in rats. J Am Soc Nephrol 18:165–175

Bucala R, Spiegel LA, Chesney J, Hogan M, Cerami A (1994) Circulating fibrocytes define a new leukocyte subpopulation that mediates tissue repair. Mol Med 1:71–81

Buchtler S, Grill A, Hofmarksrichter S, Stöckert P, Schiechl-Brachner G, Rodriguez Gomez M et al (2018) Cellular origin and functional relevance of collagen I production in the kidney. J Am Soc Nephrol 29:1859–1873

Chen G, Lin SC, Chen J, He L, Dong F, Xu J et al (2011) CXCL16 recruits bone marrow-derived fibroblast precursors in renal fibrosis. J Am Soc Nephrol 22:1876–1886

Chen J, Xia Y, Lin X, Feng XH, Wang Y (2014) Smad3 signaling activates bone marrow-derived fibroblasts in renal fibrosis. Lab Invest 94:545–556

Chesney J, Bacher M, Bender A, Bucala R (1997) The peripheral blood fibrocyte is a potent antigen-presenting cell capable of priming naive T cells in situ. Proc Natl Acad Sci U S A 94:6307–6312

Conway B, Hughes J (2012) Cellular orchestrators of renal fibrosis. QJM: Mon J Assoc Physicians 105:611–615

Corton JM, Gillespie JG, Hawley SA, Hardie DG (1995) 5-aminoimidazole-4-carboxamide ribonucleoside. A specific method for activating AMP-activated protein kinase in intact cells? Eur J Biochem 229:558–565

Deng HK, Unutmaz D, KewalRamani VN, Littman DR (1997) Expression cloning of new receptors used by simian and human immunodeficiency viruses. Nature 388:296–300

Dong Y, Yang M, Zhang J, Peng X, Cheng J, Cui T et al (2016) Depletion of $CD8^{+}$ T cells exacerbates $CD4^{+}$ T cell-induced monocyte-to-fibroblast transition in renal fibrosis. J Immunol 196:1874–1881

Duerrschmid C, Trial J, Wang Y, Entman ML, Haudek SB (2015) Tumor necrosis factor: a mechanistic link between angiotensin-II-induced cardiac inflammation and fibrosis. Circ Heart Fail 8:352–361

Ebert LM, Schaerli P, Moser B (2005) Chemokine-mediated control of T cell traffic in lymphoid and peripheral tissues. Mol Immunol 42:799–809

Eddy AA (2005) Progression in chronic kidney disease. Adv Chronic Kidney Dis 12:353–365

Eddy AA (2013) The origin of scar-forming kidney myofibroblasts. Nat Med 19:964–966

Elewa U, Sanchez-Nino MD, Mahillo-Fernandez I, Martin-Cleary C, Belen Sanz A, Perez-Gomez MV et al (2016) Circulating CXCL16 in diabetic kidney disease. Kidney Blood Press Res 41:663–671

Essawy M, Soylemezoglu O, Muchaneta-Kubara EC, Shortland J, Brown CB, el Nahas AM (1997) Myofibroblasts and the progression of diabetic nephropathy. Nephrol Dial Transpl Off Publ Eur Dial Transpl Assoc Eur Ren Assoc 12:43–50

Farris AB, Colvin RB (2012) Renal interstitial fibrosis: mechanisms and evaluation. Curr Opin Nephrol Hypertens 21:289–300

Forster R, Davalos-Misslitz AC, Rot A (2008) CCR25 and its ligands: balancing immunity and tolerance. Nat Rev Immunol 8:362–371

Gordon S, Martinez FO (2010) Alternative activation of macrophages: mechanism and functions. Immunity 32:593–604

Gough PJ, Garton KJ, Wille PT, Rychlewski M, Dempsey PJ, Raines EW (2004) A disintegrin and metalloproteinase 10-mediated cleavage and shedding regulates the cell surface expression of CXC chemokine ligand 16. J Immunol 172:3678–3685

Grimm PC, Nickerson P, Jeffery J, Savani RC, Gough J, McKenna RM et al (2001) Neointimal and tubulointerstitial infiltration by recipient mesenchymal cells in chronic renal-allograft rejection. N Engl J Med 345:93–97

Haudek SB, Cheng J, Du J, Wang Y, Hermosillo-Rodriguez J, Trial J et al (2010) Monocytic fibroblast precursors mediate fibrosis in angiotensin-II-induced cardiac hypertrophy. J Mol Cell Cardiol 49:499–507

Herzog EL, Bucala R (2010) Fibrocytes in health and disease. Exp Hematol 38:548–556

Huang XR, Chung AC, Yang F, Yue W, Deng C, Lau CP et al (2010) Smad3 mediates cardiac inflammation and fibrosis in angiotensin II-induced hypertensive cardiac remodeling. Hypertension 55:1165–1171

Humphreys BD, Lin SL, Kobayashi A, Hudson TE, Nowlin BT, Bonventre JV et al (2010) Fate tracing reveals the pericyte and not epithelial origin of myofibroblasts in kidney fibrosis. Am J Pathol 176:85–97

Iwano M, Plieth D, Danoff TM, Xue C, Okada H, Neilson EG (2002) Evidence that fibroblasts derive from epithelium during tissue fibrosis. J Clin Invest 110:341–350

Iwashima Y, Horio T, Kumada M, Suzuki Y, Kihara S, Rakugi H et al (2006) Adiponectin and renal function, and implication as a risk of cardiovascular disease. Am J Cardiol 98:1603–1608

Izquierdo MC, Sanz AB, Mezzano S, Blanco J, Carrasco S, Sanchez-Niño MD et al (2012) TWEAK (tumor necrosis factor-like weak inducer of apoptosis) activates CXCL16 expression during renal tubulointerstitial inflammation. Kidney Int 81:1098–1107

Izquierdo MC, Martin-Cleary C, Fernandez-Fernandez B, Elewa U, Sanchez-Niño MD, Carrero JJ et al (2014) CXCL16 in kidney and cardiovascular injury. Cytokine Growth Factor Rev 25:317–325

Lan HY (2011) Diverse roles of TGF-beta/Smads in renal fibrosis and inflammation. Int J Biol Sci 7:1056–1067

Latella G, Vetuschi A, Sferra R, Catitti V, D'Angelo A, Zanninelli G et al (2009) Targeted disruption of Smad3 confers resistance to the development of dimethylnitrosamine-induced hepatic fibrosis in mice. Liver Int 29:997–1009

Lebleu VS, Taduri G, O'Connell J, Teng Y, Cooke VG, Woda C et al (2013) Origin and function of myofibroblasts in kidney fibrosis. Nat Med 19:1047–1053

Lee S, Huen S, Nishio H, Nishio S, Lee HK, Choi BS et al (2011) Distinct macrophage phenotypes contribute to kidney injury and repair. J Am Soc Nephrol 22:317–326

Li J, Deane JA, Campanale NV, Bertram JF, Ricardo SD et al (2007) The contribution of bone marrow-derived cells to the development of renal interstitial fibrosis. Stem Cells 25:697–706

Liang H, Ma Z, Peng H, He L, Hu Z, Wang Y (2016) CXCL16 deficiency attenuates renal injury and fibrosis in salt-sensitive hypertension. Sci Rep 6:28715

Liang H, Zhang Z, Yan J, Wang Y, Hu Z, Mitch WE et al (2017) The IL-4 receptor alpha has a critical role in bone marrow-derived fibroblast activation and renal fibrosis. Kidney Int 92:1433–1443

Lin SL, Kisseleva T, Brenner DA, Duffield JS (2008) Pericytes and perivascular fibroblasts are the primary source of collagen-producing cells in obstructive fibrosis of the kidney. Am J Pathol 173:1617–1627

Lin Z, Gong Q, Zhou Z, Zhang W, Liao S, Liu Y et al (2011) Increased plasma CXCL16 levels in patients with chronic kidney diseases. Eur J Clin Invest 41:836–845

Liu Y (2010) New insights into epithelial-mesenchymal transition in kidney fibrosis. J Am Soc Nephrol 21:212–222

Loetscher M, Amara A, Oberlin E, Brass N, Legler D, Loetscher P et al (1997) TYMSTR, a putative chemokine receptor selectively expressed in activated T cells, exhibits HIV-1 coreceptor function. Curr Biol 7:652–660

Ludwig A, Weber C (2007) Transmembrane chemokines: versatile 'special agents' in vascular inflammation. Thromb Haemost 97:694–703

Ludwig A, Hundhausen C, Lambert MH, Broadway N, Andrews RC, Bickett DM et al (2005) Metalloproteinase inhibitors for the disintegrin-like metalloproteinases ADAM10 and ADAM17 that differentially block constitutive and phorbol ester-inducible shedding of cell surface molecules. Comb Chem High Throughput Screen 8:161–171

Ma Z, Jin X, He L, Wang Y (2016) CXCL16 regulates renal injury and fibrosis in experimental renal artery stenosis. Am J Physiol Heart Circ Physiol 311:H815–H821

Mackay CR (2001) Chemokines: immunology's high impact factors. Nat Immunol 2:95–101

Matloubian M, David A, Engel S, Ryan JE, Cyster JG (2000) A transmembrane CXC chemokine is a ligand for HIV-coreceptor Bonzo. Nat Immunol 1:298–304

Meng XM, Tang PM, Li J, Lan HY (2015) TGF-beta/Smad signaling in renal fibrosis. Front Physiol 6:82

Meran S, Steadman R (2011) Fibroblasts and myofibroblasts in renal fibrosis. Int J Exp Pathol 92:158–167

Mills KT, Hamm LL, Alper AB, Miller C, Hudaihed A, Balamuthusamy S et al (2013) Circulating adipocytokines and chronic kidney disease. PLoS ONE 8:e76902

Moeller A, Gilpin SE, Ask K, Cox G, Cook D, Gauldie J et al (2009) Circulating fibrocytes are an indicator of poor prognosis in idiopathic pulmonary fibrosis. Am J Respir Crit Care Med 179:588–594

Moore BB, Kolodsick JE, Thannickal VJ, Cooke K, Moore TA, Hogaboam C et al (2005) CCR57-mediated recruitment of fibrocytes to the alveolar space after fibrotic injury. Am J Pathol 166:675–684

Mori L, Bellini A, Stacey MA, Schmidt M, Mattoli S (2005) Fibrocytes contribute to the myofibroblast population in wounded skin and originate from the bone marrow. Exp Cell Res 304:81–90

Nath KA (1992) Tubulointerstitial changes as a major determinant in the progression of renal damage. Am J Kidney Dis 20:1–17

Neilson EG (2006) Mechanisms of disease: Fibroblasts–a new look at an old problem. Nat Clin Pract Nephrol 2:101–108

Niedermeier M, Reich B, Rodriguez Gomez M, Denzel A, Schmidbauer K, Göbel N et al (2009) $CD4^+$ T cells control the differentiation of $Gr1^+$ monocytes into fibrocytes. Proc Natl Acad Sci U S A 106:17892–17897

Okamura DM, Lopez-Guisa JM, Koelsch K, Collins S, Eddy AA (2007) Atherogenic scavenger receptor modulation in the tubulointerstitium in response to chronic renal injury. Am J Physiol Renal Physiol 293:F575–F585

Phillips RJ, Burdick MD, Hong K, Lutz MA, Murray LA, Xue YY et al (2004) Circulating fibrocytes traffic to the lungs in response to CXCL12 and mediate fibrosis. J Clin Invest 114:438–446

Powell DW, Mifflin RC, Valentich JD, Crowe SE, Saada JI, West AB (1999) Myofibroblasts. I. Paracrine cells important in health and disease. Am J Physiol 277:C1–C9

Reich B, Schmidbauer K, Rodriguez Gomez M, Johannes Hermann F, Göbel N, Brühl H et al (2013) Fibrocytes develop outside the kidney but contribute to renal fibrosis in a mouse model. Kidney Int 84:78–89

Roberts IS, Burrows C, Shanks JH, Venning M, McWilliam LJ (1997) Interstitial myofibroblasts: predictors of progression in membranous nephropathy. J Clin Pathol 50:123–127

Rollins BJ (1997) Chemokines. Blood 90:909–928

Roufosse C, Bou-Gharios G, Prodromidi E, Alexakis C, Jeffery R, Khan S et al (2006) Bone marrow-derived cells do not contribute significantly to collagen I synthesis in a murine model of renal fibrosis. J Am Soc Nephrol 17:775–782

Sakai N, Wada T, Yokoyama H, Lipp M, Ueha S, Matsushima K et al (2006) Secondary lymphoid tissue chemokine (SLC/CCL21)/CCR69 signaling regulates fibrocytes in renal fibrosis. Proc Natl Acad Sci U S A 103:14098–14103

Sato M, Muragaki Y, Saika S, Roberts AB, Ooshima A (2003) Targeted disruption of TGF-beta1/Smad3 signaling protects against renal tubulointerstitial fibrosis induced by unilateral ureteral obstruction. J Clin Invest 112:1486–1494

Schulte A, Schulz B, Andrzejewski MG, Hundhausen C, Mletzko S, Achilles J et al (2007) Sequential processing of the transmembrane chemokines CX3CL1 and CXCL16 by alpha- and gamma-secretases. Biochem Biophys Res Commun 358:233–240

Shachar I (2017) An essential MIF-CD74 signaling axis in kidney tubular regeneration, with prospects for precision medicine and pharmacological augmentation. Am J Physiol Renal Physiol 313:F1084–F1086

Shao DD, Suresh R, Vakil V, Gomer RH, Pilling D (2008) Pivotal Advance: Th-1 cytokines inhibit, and Th-2 cytokines promote fibrocyte differentiation. J Leukoc Biol 83:1323–1333

Shimaoka T, Kume N, Minami M, Hayashida K, Kataoka H, Kita T et al (2000) Molecular cloning of a novel scavenger receptor for oxidized low density lipoprotein, SR-PSOX, on macrophages. J Biol Chem 275:40663–40666

Shimaoka T, Nakayama T, Fukumoto N, Kume N, Takahashi S, Yamaguchi J et al (2004) Cell surface-anchored SR-PSOX/CXC chemokine ligand 16 mediates firm adhesion of CXC chemokine receptor 6-expressing cells. J Leukoc Biol 75:267–274

Shimotomai T, Kakei M, Narita T, Koshimura J, Hosoba M, Kato M et al (2005) Enhanced urinary adiponectin excretion in IgA-nephropathy patients with proteinuria. Ren Fail 27:323–328

Strutz F, Zeisberg M (2006) Renal fibroblasts and myofibroblasts in chronic kidney disease. J Am Soc Nephrol 17:2992–2998

Tapmeier TT, Fearn A, Brown K, Chowdhury P, Sacks SH, Sheerin NS et al (2010) Pivotal role of $CD4^+$ T cells in renal fibrosis following ureteric obstruction. Kidney Int 78:351–362

Tomasek JJ, Gabbiani G, Hinz B, Chaponnier C, Brown RA (2002) Myofibroblasts and mechano-regulation of connective tissue remodelling. Nat Rev Mol Cell Biol 3:349–363

Verrecchia F, Chu ML, Mauviel A (2001) Identification of novel TGF-beta/Smad gene targets in dermal fibroblasts using a combined cDNA microarray/promoter transactivation approach. J Biol Chem 276:17058–17062

Wang S, Meng XM, Ng YY, Ma FY, Zhou S, Zhang Y et al (2016) TGF-β/Smad3 signalling regulates the transition of bone marrow-derived macrophages into myofibroblasts during tissue fibrosis. Oncotarget 7:8809–8822

Wang Y, Jia L, Hu Z, Entman ML, Mitch WE, Wang Y (2018) AMP-activated protein kinase/myocardin-related transcription factor-A signaling regulates fibroblast activation and renal fibrosis. Kidney Int 93:81–94

Wilbanks A, Zondlo SC, Murphy K, Mak S, Soler D, Langdon P et al (2001) Expression cloning of the STRL33/BONZO/TYMSTRligand reveals elements of CC, CXC, and CX3C chemokines. J Immunol 166:5145–5154

Wynn TA (2004) Fibrotic disease and the T(H)1/T(H)2 paradigm. Nat Rev Immunol 4:583–594

Wynn TA, Barron L (2010) Macrophages: master regulators of inflammation and fibrosis. Semin Liver Dis 30:245–257

Xia Y, Entman ML, Wang Y (2013a) CCR2 regulates the uptake of bone marrow-derived fibroblasts in renal fibrosis. PLoS ONE 8:e77493

Xia Y, Entman ML, Wang Y (2013b) Critical role of CXCL16 in hypertensive kidney injury and fibrosis. Hypertension 62:1129–1137

Xia Y, Jin X, Yan J, Entman ML, Wang Y (2014a) CXCR88 plays a critical role in angiotensin II-induced renal injury and fibrosis. Arterioscler Thromb Vasc Biol 34:1422–1428

Xia Y, Yan J, Jin X, Entman ML, Wang Y (2014b) The chemokine receptor CXCR89 contributes to recruitment of bone marrow-derived fibroblast precursors in renal fibrosis. Kidney Int 86:327–337

Xu J, Lin SC, Chen J, Miao Y, Taffet GE, Entman ML et al (2011) CCR90 mediates the uptake of bone marrow-derived fibroblast precursors in angiotensin II-induced cardiac fibrosis. Am J Physiol Heart Circ Physiol 301:H538–H547

Yan J, Zhang Z, Yang J, Mitch WE, Wang Y (2015) JAK3/STAT6 stimulates bone marrow-derived fibroblast activation in renal fibrosis. J Am Soc Nephrol 26:3060–3071

Yan J, Zhang Z, Jia L, Wang Y (2016) Role of bone marrow-derived fibroblasts in renal fibrosis. Front Physiol 7:61

Yang J, Lin SC, Chen G, He L, Hu Z, Chan L et al (2013) Adiponectin promotes monocyte-to-fibroblast transition in renal fibrosis. J Am Soc Nephrol 24:1644–1659

Zeisberg M, Kalluri R (2004) The role of epithelial-to-mesenchymal transition in renal fibrosis. J Mol Med 82:175–181

Zeisberg M, Kalluri R (2015) Physiology of the renal interstitium. Clin J Am Soc Nephrol 10:1831–1840

Zeisberg M, Neilson EG (2010) Mechanisms of tubulointerstitial fibrosis. J Am Soc Nephrol 21:1819–1834

Zeisberg EM, Potenta SE, Sugimoto H, Zeisberg M, Kalluri R (2008) Fibroblasts in kidney fibrosis emerge via endothelial-to-mesenchymal transition. J Am Soc Nephrol 19:2282–2287

Zhang G, Moorhead PJ, el Nahas AM (1995) Myofibroblasts and the progression of experimental glomerulonephritis. Exp Nephrol 3:308–318

Zhao J, Shi W, Wang YL, Chen H, Bringas P Jr, Datto MB et al (2002) Smad3 deficiency attenuates bleomycin-induced pulmonary fibrosis in mice. Am J Physiol Lung Cell Mol Physiol 282:L585–L593

Zhao L, Wu F, Jin L, Lu T, Yang L, Pan X et al (2014) Serum CXCL16 as a novel marker of renal injury in type 2 diabetes mellitus. PLoS ONE 9:e87786

Zlotnik A, Yoshie O (2012) The chemokine superfamily revisited. Immunity 36:705–716

Zoccali C, Mallamaci F (2011) Adiponectin and leptin in chronic kidney disease: causal factors or mere risk markers? J Ren Nutr 21:87–91

Zoccali C, Mallamaci F, Panuccio V, Tripepi G, Cutrupi S, Parlongo S et al (2003) Adiponectin is markedly increased in patients with nephrotic syndrome and is related to metabolic risk factors. Kidney Int Suppl 84:S98–S102

Part III
Mediators and Cellular Processes in Renal Fibrosis

Chapter 15
Role of Aldosterone in Renal Fibrosis

Aanchal Shrestha, Ruo-Chen Che and Ai-Hua Zhang

Abstract Aldosterone is a mineralocorticoid hormone, as its main renal effect has been considered as electrolyte and water homeostasis in the distal tubule, thus maintaining blood pressure and extracellular fluid homeostasis through the activation of mineralocorticoid receptor (MR) in epithelial cells. However, over the past decade, numerous studies have documented the significant role of aldosterone in the progression of chronic kidney disease (CKD) which has become a subject of interest. It is being studied that aldosterone can affect cardiovascular and renal system, thereby contributing to tissue inflammation, injury, glomerulosclerosis, and interstitial fibrosis. Aldosterone acts on renal vessels, renal cells (glomerular mesangial cells, podocytes, vascular smooth muscle cells, tubular epithelial cells, and interstitial fibroblasts), and infiltrating inflammatory cells, inducing reactive oxygen species (ROS) production, upregulated epithelial growth factor receptor (EGFR), and type 1 angiotensin (AT1) receptor expressions, and activating nuclear factor kappa B (NF-κB), activator protein-1 (AP-1), and EGFR to further promote cell proliferation, apoptosis, and proliferation. Phenotypic transformation of epithelial cells stimulates the expression of transforming growth factor-β (TGF-β), connective tissue growth factor (CTGF), osteopontin (OPN), and plasminogen activator inhibitor-1 (PAI-1), eventually leading to renal fibrosis. MR antagonisms are related to inhibition of aldosterone-mediated pro-inflammatory and pro-fibrotic effect. In this review, we will summarize the important role of aldosterone in the pathogenesis of renal injury and fibrosis, emphasizing on its multiple underlying mechanisms and advances in aldosterone research along with the potential therapeutics for targeting MR in a renal fibrosis.

Keywords Aldosterone · Renal fibrosis · Chronic kidney disease

---

Aanchal Shrestha and Ruo-Chen Che contributed equally to this work.

A. Shrestha · R.-C. Che · A.-H. Zhang (✉)
Department of Nephrology, Children's Hospital of Nanjing Medical University, Nanjing, China
e-mail: zhaihua@njmu.edu.cn

B.-C. Liu et al. (eds.), *Renal Fibrosis: Mechanisms and Therapies*,
Advances in Experimental Medicine and Biology 1165,
https://doi.org/10.1007/978-981-13-8871-2_15

15.1 Introduction

In 1953, Simpson and Tait were the first to isolate a mineralocorticoid hormone, aldosterone (Simpson 1953). Aldosterone was initially found to have a physiological effect in regulating the electrolyte and water balance in the distal tubule, thus maintaining blood pressure and extracellular fluid homeostasis (Simpson 1953; Williams and Williams 2003) that is acted principally via cytosolic MR in epithelial cells (Rogerson and Fuller 2000). However, over the past two decades, aldosterone has also shown to act on the blood vessels, heart, and kidney. MRs are found in different tissues, such as vascular smooth muscle cells and endothelial cells, cardiomyocytes in the heart, mesangial cells and podocytes in the kidney, fibroblasts, adipocytes, hypothalamus in the brain, and macrophages expanding the distribution of aldosterone actions among non-epithelial tissues (Jaisser and Farman 2016; Funder 2010; Shibata et al. 2007; Brown 2013). Thus, there has been a shift of paradigm in our understanding of the more complex role of aldosterone, apart from the physiological role in renal sodium reabsorption. There are numerous studies which demonstrated aldosterone as a major deleterious hormone in the cardiovascular and renal system, thereby contributing to tissue inflammation, injury, glomerulosclerosis, and interstitial fibrosis (Epstein 2006; Hostetter and Ibrahim 2003). A study done by Greene et al. in 1996 first showed the role of aldosterone as an independent mediator in the progression of CKD, where aldosterone is infused intravenously in unilateral nephrectomized rat model which diminished the renoprotective effect of angiotensin-converting enzyme (ACE) inhibitor and angiotensin receptor blocker (ARB) (Greene et al. 1996). In the pathogenesis of CKD, the activation of renin–angiotensin–aldosterone system (RAAS) is an important event (Gonzalez et al. 2004). In RAAS, angiotensin II (Ang II) is one of the key elements which augments adrenal production of aldosterone and has appeared as a helper for renal injury by increasing the intraglomerular capillary pressure and ultrafiltration of plasma proteins and by promoting pro-inflammatory effects, pro-fibrogenic and growth stimulatory action (Rüster and Wolf 2006). CKD has become a massive public health concern which has increased the overall mortality by 31.7% over the last 10 years (Wang et al. 2016). Renal fibrosis, as the hallmark feature of CKD, is characterized by pathomorphologic changes comprising of glomerulosclerosis and tubulointerstitial fibrosis along with the process of excessive production and progressive extracellular matrix (ECM) protein accumulation such as collagen I and fibronectin leading to scar formation.

Herein, we will summarize the pivotal role of aldosterone in the pathogenesis of renal injury and fibrosis, emphasizing on its multiple underlying mechanisms and advances in aldosterone research along with the potential therapeutics for targeting MR in a renal fibrosis.

15.2 Aldosterone: Its Synthesis, Secretion, and Action

Aldosterone is synthesized in the body from cholesterol mainly in the cells of the zona glomerulosa layer of the adrenal cortex. However, extra-adrenal sites of aldosterone synthesis have also been identified in the heart, blood vessels, and brain (Bonvalet et al. 1995; Coirini et al. 1985; Kornel 1994). The enzyme aldosterone synthase, encoded by CYP11B2 gene, can convert deoxycorticosterone to aldosterone in a series of steps, including molecular arrangement and enzymatic reactions. Likewise, similar enzyme β-hydroxylase, encoded by the CYP11B1 gene, catalyzes the conversion of deoxycortisol to cortisol (Lenzini et al. 2007). There are many hormonal and paracrine factors that modify the regulation of aldosterone secretion. However, under physiological conditions, the important ones are Ang II, extracellular potassium, and adrenocorticotropic hormone (ACTH). Ang II, an octapeptide, is the main precursor of RAAS which is generated by the action of ACE on the inactive precursor Ang I. Ang II acts on aldosterone secretion mainly by binding to the G-protein-coupled receptor type 1 angiotensin (AT1), whereas type 2 angiotensin (AT2) receptor counteracts many of the AT1 receptor-mediated processes (Unger et al. 2011). An increased extracellular potassium concentration also augments aldosterone secretion leading to potassium excretion in the kidney. Both aldosterone and Ang II stimulate aldosterone secretion independently and equipotently (Williams 2005). The action of ACTH in the zona glomerulosa cells is mediated by cAMP which induces aldosterone secretion. In severe sodium or fluid loss, ACTH is also secreted and synergizes with Ang II or potassium in order to stimulate glomerulosa cells (Spat and Hunyady 2004). In the alteration of aldosterone synthesis, inhibitory factors such as atrial natriuretic peptides (ANP), dopamine, somatostatin, and extracellular calcium also participate. ANP is an effective inhibitor of aldosterone secretion. ANP secretion is increased in response to sodium and/or water load, thereby inhibiting the aldosterone secretion (Spat and Hunyady 2004).

At present, the varied actions of aldosterone seem to be related to two different pathways, such as genomic pathway and non-genomic pathway. In genomic pathway, aldosterone shows its physiological effects by binding to classical MR, a ligand-activated transcription factor which belongs to the nuclear receptor family with equal affinity for both mineralocorticoids and glucocorticoids (Arriza et al. 1987). It then results in a series of intracellular events beginning with the translocation of receptor ligand into the nucleus to promote protein synthesis and finally mediates the insertion of sodium channels into the epithelial cells of the collecting duct [epithelial sodium channel (ENaC)] to allow sodium reabsorption and potassium and hydrogen ions secretion (Porter and Edelman 1964; Porter et al. 1964). This process takes hours to produce the epithelial effect which can be blocked by protein synthesis inhibitor. The presence of isoenzymes 11β-hydroxysteroid dehydrogenase (11β-HSD) types 1 and 2, along with other factors independent of steroid receptor binding, seems to account for much of the mineralocorticoid selectivity seen biologically (Funder et al. 1988; Funder and Myles 1996). The specificity of the action of aldosterone in the cells containing MR (distal tubules and collecting ducts) is ensured by the expression and activity of 11β-HSD type 2 enzyme, which induces physiological responses by metabolizing circulating glucocorticoid hormones (cortisol in humans, corticosterone in rodents) into inactive 11-dehydro-derivatives (cortisone, 11-dehydrocorticosterone) with very low affinity for the MR (Farman and Rafestin-Oblin 2001). During the hypovolemic and hypoperfusion state of the kidney, RAAS activation leads to vasoconstriction and volume expansion.

Over the ensuing years, the non-genomic aldosterone effects have been demonstrated. In this, a non-classical type of MR is expressed mostly in nonelectrolyte transporting cells and the effects can be observed within minutes as nuclear binding or protein synthesis does not appear to take place (Christ and Wehling 1999; Boldyreff and Wehling 2003; Mihailidou and Funder 2005). The cell signaling pathway involves activation of protein kinase C along with release of intracellular calcium (Mihailidou and Funder 2005). Studies have shown that an alternative G-protein-coupled estrogen receptor, GPR30, may promote rapid, non-genomic effects in vascular endothelial cells (Gros et al. 2013) and vascular smooth cells (Gros et al. 2011).

15.3 Mechanism of Aldosterone-Mediated Renal Fibrosis

There has been a curious history regarding the relationship between aldosterone and renal disease which led to much researches conducted in the past. In 1964, Conn described the first 125 proven cases of primary hyperaldosteronism with hypertension in patients, among which 85% had gross proteinuria (Conn et al. 1964). The proteinuria was considered as a result of hypertension that accompanies the Conn syndrome, until 1996 when the experiment done by Greene et al. on a rat kidney model showed that mineralocorticoid hormones, particularly aldosterone, can induce proteinuria in the absence of hypertension (Greene et al. 1996). They showed that the treatment with ARB (losartan) and ACE inhibitor (enalapril) in remnant rat kidneys reduced proteinuria, hypertension, and glomerulosclerosis. They also demonstrated that rats injected with exogenous aldosterone in order to maintain very high aldosterone levels were subjected to losartan and enalapril, experienced proteinuria, hypertension, and glomerulosclerosis to the same level as the untreated rats with remnant kidneys did. Another study done on rats with complete unilateral ureteral obstruction (UUO) showed that the renal fibrosis was significantly reduced after the treatment with MR antagonist spironolactone with no changes in serum potassium, aldosterone, and urine sodium (Trachtman et al. 2004). In response to deoxycorticosterone acetate (DOCA) salt treatment, $p47^{phox-/-}$ and $gp91^{phox-/Y}$ mice were found to reduce blood pressure only within the first 2–3 days of treatment and later no significant difference was observed between these mice and wild-type (WT) control group (Zhang et al. 2011). On the contrary, four different kinds of mitochondrial ROS inhibitors (rotenone: mitochondrial respiratory chain complex I inhibitor, 5-hydroxydecanoate: specific mitochondrial ATP-sensitive potassium channel blocker, benzylguanidine: mitochondrial complexes I and III, and the cell-permeable manganese tetrakis (4-benzoic acid), and porphyrin: mitochondrial superoxide dismutase analogues) are found to lower the blood pressure and ROS production in renal tissue, indicating that mitochondria mediates aldosterone-induced ROS production (Zhang et al. 2011). This provides direct evidence that mitochondrial dysfunction contributes to aldosterone-dependent or MR-dependent ROS formation. Juknevicius and colleagues demonstrated that aldosterone infusion for 3 days in normal rats causes a more than twofold increase in the urinary excretion of TGF-β without changes in blood pressure or evidence of kidney damage through an MR-dependent posttranscriptional effect (Juknevicius et al. 2004).

Taken together, these data clearly indicate that aldosterone was involved in the pathogenesis of renal damage and has a direct histological mechanism besides mediating renal damage through hemodynamic pathways. In the following section, we will illustrate the cells, principal molecules, and other mediators involved in the contribution of aldosterone to the pathogenesis of renal fibrosis.

15.3.1 Induced Mesangial Injury

Mesangial cells are active intrinsic kidney cells which play a crucial role in maintaining the glomerular structural integrity of the microvascular bed which provides mesangial matrix homeostasis and modifies glomerular filtration (Abboud 2012; Brunskill and Potter 2012). Chronic aldosterone infusion increases ROS production through the activation of nicotinamide adenine dinucleotide phosphate (NADPH) oxidase (Nishiyama et al. 2004; Huang et al. 2009). Superoxide anion and hydrogen peroxide are ROS that implicates as important mediators in the mechanisms of aldosterone-mediated renal damage. There are a number of oxidant-generating systems that participate in ROS production such as NADPH oxidase, xanthine oxidase, mitochondrial respiratory chain, cyclooxygenase, lipoxygenase, cytochrome P-450, and nitric oxide synthase.

In rat mesangial cells, aldosterone could directly promote superoxide anion (O2-) generation through the activation of NADPH oxidase and translocation of the cytosolic components of NADPH enzyme p47phox and p67phox to the cell membrane (Miyata et al. 2005). Experiment done in cultured human mesangial cells shows that aldosterone stimulates mesangial cell apoptosis in a dose- and time-dependent manner. The administration of MR antagonist or antioxidant attenuates the proapoptotic effects of aldosterone. Aldosterone also induces dephosphorylation of Bcl-2-associated death promoter (BAD), a protein, which after dephosphorylation is linked to mitochondrial dysfunction leading to accumulation of cytochrome c into the cytosol and mesangial cell apoptosis (Mathew et al. 2008). Aldosterone-infused mesangial cells exhibited increased ROS production, which was blocked by MR antagonist, mitochondrial respiratory chain complex I inhibitor, or an NADPH oxidase inhibitor. These results indicate that mitochondria and NADPH oxidase mediate aldosterone-induced renal ROS production (Huang et al. 2009). Evidence points to aldosterone has a role in mesangial cell proliferation, which involved in MR expression and extracellular signal-regulated kinase 1 and 2 (ERK1/2) activation.

Similarly, we have shown that aldosterone-induced mesangial cell proliferation was mediated by Ki-RasA:GTP/c-Raf/MEK/ERK and PI3 K/Akt/mTOR/p70S6K1, an EGF-activated signaling pathway downstream from the EGFR (Huang et al. 2009). Recently, Zhang et al. revealed that Connexin 43 (Cx43), an important regulator of cell growth, showed decreased expression in mesangial cell proliferation and treatment with MR antagonist spironolactone, ERK1/2 inhibitor PD98059, and PKC inhibitor GF109203X attenuated the downregulation of Cx43 expression which suggests that this process might be mediated through (ERK1/2) and PKC pathways (Zhang et al. 2015). TGF-β is a pro-fibrogenic cytokine that has been extensively studied as a responsible factor for renal fibrosis. Aldosterone induces TGF-β1 expression via ERK1/2, c-Jun N-terminal kinase (JNK) and AP-1 pathways (Han et al. 2009).

Aldosterone also exhibited a pro-fibrotic effect by rapidly inducing mRNA and protein expression of CTGF in a time- and concentration-dependent manner (Gauer et al. 2007). The use of MR antagonist spironolactone, canrenoate, or eplerenone could not inhibit the rapid CTGF induction which suggests an MR-independent mechanism for this effect. However, selective inhibitor of glucocorticoid receptor (GR), RU-486, inhibited aldosterone-induced CTGF expression, which indicates that GR is important in aldosterone-induced regulation of CTGF (Gauer et al. 2007). CTGF can promote migration, hypertrophy, fibronectin production, and actin disassembly in mesangial cells and also increases type I, type III, and type IV collagen production by mesangial cells (Phanish et al. 2010). Aldosterone could upregulate CTGF expression in two ways: TGF-β dependent and TGF-β independent. In vivo study done by Terada et al. (2012) showed that in cultured rat mesangial cells, aldosterone promoted the expression of CTGF in TGF-β-dependent manner. In this study, aldosterone stimulates serum- and glucocorticoid-regulated kinase-1 (SGK1) expression and promotes expressions of CTGF and intercellular adhesion molecule-1 (ICAM-1) via NF-κB. However, in a mouse model of diabetic nephropathy, aldosterone significantly increases CTGF gene expression and protein synthesis even in the presence of TGF-β1 neutralizing antibody, which shows the TGF-β-independent pathway for aldosterone-induced CTGF production in cultured mesangial cells (Han et al. 2006). Moreover, treatment with spironolactone markedly decreases renal CTGF and collagen synthesis. Another possible mechanism through which aldosterone may involve in the development of fibrosis is by activating PAI-1 (Huang et al. 2008; Yuan et al. 2007). Huang et al. (2008) observed the increment in aldosterone-induced PAI-1 is partially facilitated by TGF-β1. TGF-β1 and aldosterone acts synergistically to stimulate the PAI-1 expression and decrease extracellular matrix degradation (Huang et al. 2008). Another study done by Yuan et al. (2007) also demonstrated upregulation of PAI-1 mRNA and protein expression by aldosterone, however, was independent of aldosterone-induced increases in TGF-β1 expression and intracellular ROS production. In a high glucose media, aldosterone produced an inflammatory response by upregulating NF-κB and monocyte chemoattractant protein-1 (MCP-1) via AT1a and AT2 pathways (Hao et al. 2015).

Evidences point to show that EGFR may take part in aldosterone-induced mesangial cell proliferation (Huang et al. 2009; Sheng et al. 2016). EGFR is a transmembrane protein with intrinsic tyrosine kinase activity. Upon ligand binding to the receptor induces dimerization and phosphorylation of tyrosine residues in its cytosolic domains. These phosphorylated residues act as docking sites for adaptor proteins and activate downstream signaling cascades (Forrester et al. 2016). The most widely studied signaling pathways are phosphatidylinositol-3 kinase (PI3 K)/AKT and Ras/mitogen-activated protein kinase (MAPK). Aldosterone, after binding to their receptor, can transactivate EGFR, via "a disintegrin and metalloproteases" (ADAMs), thereby regulating several processes including cellular functions, proliferation, hypertrophy, and migration. We have shown that aldosterone transactivates EGFR in cultured mesangial cells, because pretreatment with the EGFR antagonist AG1478 blocked mesangial cell proliferation as well as the activation of Ras/MAPK and PI3 K/Akt (Huang et al. 2009).

In addition, the mitogenic effect of aldosterone was shown to be mediated by Ki-RasA/c-Raf/MEK/ERK and PI3 K/Akt/mTOR/p70S6K1 signaling pathways (Huang et al. 2009). Similarly, Sheng et al. highlighted a role of EGFR in aldosterone-mediated renal fibrosis which relies on ROS-induced EGFR/ERK activation (Sheng et al. 2016).

15.3.2 Induced Podocyte Injury

The glomerular filtration barrier is composed of three layers: the capillary endothelium, the glomerular basement membrane (GBM), and the podocytes (specialized epithelial cells). Podocytes extend long primary processes from its cell body, the end of which contains foot processes (Wolf et al. 2005). The foot processes of adjacent podocytes are connected by a continuous membrane-like structure called the slit diaphragm and form an interdigitating pattern adhering to the glomerular basement membrane (Wolf et al. 2005; Pavenstadt et al. 2003). The proteins of the slit diaphragm such as nephrin, podocin, and CD2-associated protein (CD2AP) are not only structural, but also serve as a survival function for podocytes by interacting with each other within the body of the podocytes (Wolf et al. 2005). Podocytes thus act as a final filtration barrier against urinary protein loss (Wolf et al. 2005; Shankland 2006). Podocyte injury is the basic pathogenic mechanism that leads to the development of proteinuria and progression of glomerular sclerosis. Several studies have shown that podocytes express MR, which is the target site of aldosterone (Shibata et al. 2007; Kiyomoto et al. 2008; Nagase et al. 2006).

Our study provided direct evidence supporting the role of ROS in contributing to aldosterone-induced podocyte injury (Zhu et al. 2011). Aldosterone-infused mice showed podocyte injury with reduced glomerular expression of nephrin and increased ROS production in renal glomeruli. We examined the important role of mitochondria dysfunction in aldosterone-induced podocyte injury as pronounced by decreased mitochondrial membrane potential, ATP levels, and mitochondrial DNA copy number seen in podocytes and glomeruli. The study done by Nishiyama et al. (2004) showed that aldosterone/salt treatment induced hypertension and renal injury, as characterized by proteinuria, glomerular changes, and collagen accumulation. The renal injury was also associated with increase in renal cortical thiobarbituric acid reactive substance (TBARS) contents, a marker of ROS production, and mRNA expression of NADPH oxidase components, $p22^{phox}$, Nox-4, and $gp91^{phox}$. The use of MR antagonist eplerenone prevented the increases in TBARS levels and NADPH oxidase expression in the kidney, and tempol also reduced ROS levels and ameliorated renal injury (Nishiyama et al. 2004). Shibata et al. (2007) showed that continuous infusion of aldosterone in uninephrectomized rats induced massive proteinuria with hypertension and glomerular podocyte injury and decreased gene expressions of nephrin and podocin were closely related to podocyte injury. In the same study, NADPH oxidase activity and TBARS contents were enhanced and podocyte damage was ameliorated by tempol further supporting an important role for ROS in this rat model. It is found that aldosterone can induce the activation of SGK1 signaling pathway through oxidative stress and damage the podocyte (Shibata et al. 2007). Previously conducted studies have found that aldosterone promotes apoptosis of podocytes by inducing mitochondrial dysfunction (Su et al. 2013; Yuan, et al. 2012a). We showed that under normal conditions, endogenous peroxisome proliferator-activated receptor-g coactivator-1α (PGC-1α), a transcriptional coactivator of peroxisome proliferator-activated receptor-γ, and other nuclear hormone receptors are important for podocyte maintenance of mitochondrial function (Yuan, et al. 2012a). SIRT1/PGC-1α axis activation protects against aldosterone-induced podocyte injury likely by preventing mitochondrial dysfunction and SIRT1 agonist, resveratrol-blocked mitochondrial dysfunction, and podocyte injury by activating SIRT1/PGC-1 alpha (Yuan et al. 2012a). Aldosterone-induced podocyte apoptosis seems to be mediated through the activation of p38 MAPK signaling pathway because pretreatment with inhibitor of p38 MAPK suppressed apoptosis (Chen et al. 2009). We recently demonstrated the possible signaling pathways involved in mitochondrial dynamics (Yuan et al. 2017). In this study, aldosterone induced podocyte injury and increased the expression of p53, leading to mitochondrial dysfunction through activation of dynamin-related GTPase dynamin-related protein 1 (Drp1)-mediated mitochondrial fission. Pathogenic role of NOD-like receptor family, pyrin domain containing 3 (NLRP3) inflammasome in podocyte injury, has been identified in our another study (Bai et al. 2017). Herein, exposure of podocytes to aldosterone enhanced NLRP3, indicating an activation of NLRP3 inflammasome. NLRP3 deletion abolished aldosterone-induced NLRP3 inflammasome activation and also ameliorated aldosterone-induced podocyte injury evidenced by the improved proteinuria and silt diaphragm protein loss. In the ARHGDIA gene (encoding RhoGDI 1 alpha) knockout mice, the damage of foot processes is positively correlated with the intensity of Rac1 activation, and the Rac1 inhibitor NSC23766 can inhibit the activation of the SGK1 signaling pathway and reduce the proteinuria and the foot processes damage (Shibata et al. 2008).

15.3.3 Regulation of Vessels and Vascular Smooth Muscle Cells (VSMCs)

Aldosterone increases the VSMCs proliferation, increases the expression of vascular collagen components, decreases vascular compliance, and constricts the glomerular afferent and efferent arterioles. It increases the glomeruli pressure, perfusion, and filtration which stimulates vasculitis and hence results in renal fibrosis. The main mechanism of aldosterone in stimulation of VSMCs proliferation is as follows: (1) The combination of aldosterone/MR complex with the hormone response element (HRE) of AT1 receptor gene promoter upregulates the AT1 receptor expression which can enhance the effect of Ang II-induced vascular smooth muscle hypertrophy, hence stimulating the proliferation of VSMCs (Yamada et al. 2008) and (2) Aldosterone activates ERK1/2 signaling pathway and promotes the proliferation of VSMCs (Min et al. 2005). A study showed the synergistic signaling interaction between aldosterone and Ang II on the mitogenic action in VSMCs (Min et al. 2005). Here, aldosterone and Ang II increased the early activation of ERK1/2 involving the transactivation of the EGFR, whereas delayed activation involved increased MAPK-1 and Ki-ras2A expression. Other investigators used small interference RNA for AT1aR, AT1bR, and MR to explain the cross talk between aldosterone and angiotensin II in mouse VSMCs. Aldosterone and Ang II induced ERK1/2, JNK, and NF-κB phosphorylation (Bomback and Klemmer 2007). This study showed ERK1/2 activation through an AT1-dependent, MR-independent mechanism, and JNK and NF-κB activation required both AT1- and MR-dependent mechanisms after aldosterone stimulation.

Aldosterone increases expression of PAI-1 in VSMCs and endothelial cells (Brown et al. 2000a, b). PAI-1 is a member of the serine protease inhibitor (serpin) family, and it can promote fibrosis by preventing matrix metalloproteinases (MMPs) activation and ECM degradation by plasminogen activators and plasmin. PAI-1 is the major physiological inhibitor of tissue-type plasminogen activator (tPA) and urokinase-type plasminogen activator (uPA). It inhibits the activation of plasminogen to plasmin by tPA and also can also exert anti-fibrotic effects by retarding cellular infiltration and interfering with uPA or plasmin-mediated activation and the release of latent growth factors. By interacting with Ang II, aldosterone increases PAI-1 expression in a dose-dependent manner both in vitro cultured vascular smooth muscle cells and in vivo vascular endothelial cells (Brown et al. 2000b). This effect was blocked by MR antagonist spironolactone, indicating that aldosterone promotes PAI-1 expression through the classical salt mineralocorticoid receptor mechanism. In this study, plasma PAI-1 levels in patients with primary hyperaldosteronism increased significantly. In a radiation model of renal damage, inhibition of aldosterone also downregulates PAI-1 expression in vivo, and local PAI-1 expression is closely associated with sclerosis (Brown et al. 2000a). These collective evidences support the hypothesis that aldosterone induces renal fibrosis at least in part through its effects on PAI-1 expression. Grossmann et al. (2007) showed the aldosterone increases the expression of EGFR in vascular smooth muscle cells via an interaction with the EGFR promoter, which is MR specific.

Nitric oxide (NO) has several functions in the kidney including dilatation of the renal blood vessels, regulation of renal hemodynamics, blunting of tubuloglomerular feedback, modulation of renal sympathetic nerve activity, and inhibition of matrix protein aggregation, smooth muscle cell proliferation, and fibroblast proliferation. Nω-nitro-L-arginine methyl ester (L-NAME) is a potent inhibitor of NO synthase which can cause systemic hypertension, vascular injury, and vascular endothelial cell dysfunction. It thus plays an important role in renal fibrosis. Ikeda et al. (2009) demonstrated significant increase in blood pressure, proteinuria, renal inflammation, and fibrosis in L-NAME-treated rats. These effects were prevented significantly after the use of aldosterone antagonist, spironolactone, which may suggest the role of aldosterone in contributing to renal inflammation and fibrosis by inhibiting the formation of NO. Besides, Arima et al. (2003) examined the vascular action of aldosterone that can cause dose-dependent constriction in both afferent and efferent glomerular arterioles with more susceptibility in efferent arterioles. These effects are via non-genomic pathway as the mineralocorticoid receptor antagonist spironolactone was not able to block the vasoconstriction effect of aldosterone. Instead, the effect of aldosterone on both the arterioles was due to activation of phospholipase C and thus promoting calcium transport through L- and T-type calcium channels. In addition, aldosterone-induced vasculitis may also be involved in the process of renal fibrosis (Blasi et al. 2003). It was demonstrated that aldosterone/salt administration induces elevated expression of inflammatory components including MCP-1, IL-1β, and IL-6 in the blood vessels of rats. It was also accompanied with severe albuminuria, renal vascular injury, and histopathological changes in the kidney. The author further observed that eplerenone attenuated these inflammatory components, improved vasculitis, and reduced renal inflammation and fibrosis (Blasi et al. 2003).

15.3.4 Phenotypic Transition of Renal Tubular Epithelial Cells

Epithelial–mesenchymal transition (EMT) is defined as phenotypic conversion in epithelial cells which leads to loss of contacts between epithelial cell-to-cell basement membrane and hence gives rise to matrix-producing fibroblasts and myofibroblasts (Acloque et al. 2009; Thiery 2002; Kalluri and Weinberg 2009). EMT is known to play a pivotal role in the mechanism of tubulointerstitial fibrosis (Burns and Thomas 2011). We have previously demonstrated that the aldosterone induces EMT in renal epithelial cells through MR-mediated, mitochondria-derived ROS, and activation of ERK1/2 signaling pathway (Zhang et al. 2007). This effect was significantly blocked by MR antagonist eplerenone and mitochondrial respiratory chain complex I inhibitor rotenone. In the DOCA/salt mice model, EMT was evidenced by exhibiting the tubular epithelial cells as elongated and fibroblast-like morphology, expression of alpha-smooth muscle actin (fibroblast markers) was increased, and E-cadherin (epithelial markers) was decreased. We further extended our investigation demonstrating the role of mitochondrial function in aldosterone-induced EMT (Yuan et al. 2012b). PGC-1α and silent mating-type information regulation 2 homolog 1 (SIRT1) function together to control the mitochondrial biogenesis and function. In this study, aldosterone-infused human kidney cells showed decreased PGC1 expression and induced mitochondrial dysfunction leading to promoting EMT, whereas overexpression of PGC-1α prevented mitochondrial dysfunction and EMT. SIRT1/PGC-1α axis activation protected against aldosterone-induced EMT by inhibiting mitochondrial dysfunction (Yuan et al. 2012b). Aldosterone has been reported to contribute to normal renal tubular epithelial cell differentiation and proliferation from isolated renal stem cells (Minuth et al. 2005), and in vivo experiment done on juvenile animals can also cause dysregulation of renal cell growth and promote renal fibrosis (Heber et al. 2007). Human tubular epithelial cells treated with aldosterone activate EGFR via TGF-α/ADAM17 (Morgado-Pascual et al. 2015). Blockade of ADAM-17/TGF-α/EGFR pathway using the specific inhibitors, the ADAM17 inhibitor TAPI-2, an anti-TGF-α neutralizing antibody, and the EGFR kinase inhibitor AG1478, significantly diminished aldosterone-induced pro-inflammatory gene upregulation which supports the involvement of ADAM-17/TGF-α/EGFR axis in the regulation of pro-inflammatory factors by aldosterone (Morgado-Pascual et al. 2015).

15.3.5 Fibroblasts

In rat kidney fibroblast cells, aldosterone induces proliferation via MR-mediated kinase activity of the growth factor receptors and downstream cell signaling through P13 K/Akt, ERK, and JNK pathways (Huang et al. 2012). Aldosterone induced uninephrectomized rats in the presence of AT1 receptor blockade, increased the expression of TGF-β1 and collagen mRNAs, and diffused renal medullary and cortical fibrosis accompanied by the accumulation of abundant myofibroblasts at the sites of fibrosis. This suggests the independent role of aldosterone in renal fibrosis through TGF-β1 signaling pathways, thereby upregulating the collagen synthesis, downregulating the release of the extracellular matrix metalloproteinase collagenase, and hence promoting fibroblast proliferation (Sun et al. 2000). CTGF is recognized as a pro-fibrotic protein and plays an important role in the pathogenesis of chronic fibrotic disorders (Phanish et al. 2010). The CTGF gene promoter contains a TGF-β1-responsive element as well as other regulatory sequences and thus acts as an early responder during the onset of fibrotic diseases. In addition to its own pro-fibrotic effect, it is a downstream mediator of at least some of the pro-fibrotic effects of TGF-β1, in particular proliferation of fibroblasts and secretion of ECM proteins by fibroblasts (Phanish et al. 2010). OPN is a key mediator of aldosterone-induced renal fibrosis. It is a glycosylated phosphoprotein that is produced by osteoblasts, macrophages, endothelial cells, and epithelial cells and acts to facilitate cell adhesion and migration. In the normal kidney, it is mainly expressed in the loop of Henle and distal nephron and is thought to play a role in renal fibroblast proliferation and ECM synthesis. Aldosterone significantly increased OPN mRNA expression and that OPN-small inference RNA completely blocked aldosterone-induced collagen synthesis and renal fibroblast proliferation in renal fibroblasts (Irita et al. 2008). This study also showed the use of MR antagonist spironolactone abolished aldosterone-induced OPN expression through AP-1 and NF-κB binding activities, which suggests that OPN may be a downstream factor of aldosterone in renal fibrosis.

15.3.6 Infiltration of Inflammatory Cells

Infiltration of inflammatory cells is one of the major cellular events in renal fibrosis. In response to injury, inflammation is an essential part of host defense mechanism. However, inflammation which is non-resolving tends to develop fibrotic diseases (Liu 2011). Following renal fibrosis, the inflammatory cells become activated, which produce molecules such as ROS that damage tissues; hence, growth factors and fibrogenic cytokines are produced. Macrophage, a member of the mononuclear phagocyte family, has a key role in renal injury, inflammation, and fibrosis. They are pleiotropic inflammatory cells that take part in inflammatory reactions (Han et al. 2019). Blasi et al. (2003) showed the possible role of aldosterone in mediating renal fibrosis which is preceded by inflammatory components infiltration including macrophage, and increased the expression of pro-inflammatory cytokines OPN, interleukin-1 beta (IL-1β), and interleukin-6 (IL-6), MCP-1, suggesting that inflammation is an essential contributor of renal fibrosis. The heterogeneity of macrophage polarization has been recognized as an important feature of renal disease (Li et al. 2015). Aldosterone plus salt treated in rat kidney model contributes to classical macrophage activation to the inflammatory M1 phenotype, along with hypertension, renal damage, and fibrosis. These effects were effectively blocked by MR antagonism spironolactone, thereby showing the beneficial effect of MR antagonist in hypertensive renal disease (Martín-Fernández et al. 2016). NF-κB, a eukaryotic transcription factor, is one of the most important pro-inflammatory signal pathways which has been intensively investigated over the past 30 years because of its involvement of several biological programs. NF-κB can be activated by various molecules, such as lipopolysaccharide, cytokines, and ROS, and influence the transcription of several genes involved in the inflammatory response, cell growth, and adhesion (Baeuerle and Henkel 1994). Ding et al. conducted an experiment showing the role of NF-κB in the pathogenesis of aldosterone-induced renal injury (Ding et al. 2012). It was found that the activity of NF-κB and inflammation in the renal tissue of aldosterone-/salt-treated rats increased significantly along with the upregulated expression of ICAM-1, TGF-β, CTGF, and type IV collagen. Treatment with pyrrolidine dithiocarbamate (PDTC), a NF-κB inhibitor, significantly reduced the pathological damage of kidney tissue and reduced the expression of factors including CTGF, ICAM-1, TGF-β, and collagen IV (Ding et al. 2012).

15.4 Aldosterone Breakthrough

We have already mentioned studies that showed proteinuria induced by aldosterone is one of the important signs in the progression of renal disease and renal fibrosis. The renin–angiotensin system (RAS) plays a key role in the development of proteinuria in kidney diseases. Blockade of RAS with the use of ACE inhibitors decreases the levels of circulating Ang II, and ARBs inhibit the action of AT1R significantly reducing the urinary protein excretion which further delays the chronic progression of renal disease. However, with the use of ACE inhibitors or ARBs, Ang II inhibits release of renin and leads to increased plasma renal activity. Besides, after the initiation of ACE inhibitors or ARBs therapy for several weeks, plasma aldosterone concentration is increased to pretreatment level, following a decrease of aldosterone levels in the initial treatment phase. This phenomenon is known as aldosterone breakthrough, which also suggests the role of aldosterone in mediating proteinuria (Bomback and Klemmer 2007). Up to 10–53% CKD patients on ACE inhibitors or ARBs were found to have aldosterone breakthrough in previous studies using various definitions (Bomback and Klemmer 2007). It was shown that patients who experienced aldosterone breakthrough had enhanced decline in GFR compared with those without breakthrough (Schjoedt et al. 2004). Aldosterone breakthrough provides a theoretical basis for clinical application of aldosterone receptor antagonists. Many observational studies have confirmed that the combination of MR antagonist and ACE inhibitors or ARBs can effectively reduce proteinuria and delay renal fibrosis in patients with CKD (Schjoedt et al. 2004; Furumatsu et al. 2008; Schjoedt et al. 2006; Chrysostomou et al. 2006; Bianchi et al. 2006). Sato and Fukuda (2013) reported that aldosterone breakthrough occurred in 55% of hypertensive patients after the use of direct renin inhibitor (DRI) aliskiren, which acts by directly blocking the enzymatic action of renin without interfering with its production and interaction with renin receptors. Adding MR receptor antagonist eplerenone to DRI to those patients showed decrease in albuminuria and may benefit hypertensive patients by exerting a cardiovascular and/or renal protective effect.

The mechanism of aldosterone breakthrough is not fully elucidated. It may be related to the following factors: (1) Ang II is produced by alternative pathway and non-renin pathway. ACE inhibitor can block the classical Ang II generation pathway, but it does not play a role in the alternative pathway and non-renin pathway of Ang II generation (Urata et al. 1994a, b; Cicoira et al. 2001); (2) Long-term blockade of AT1 receptor leads to increase in Ang II which may mediate aldosterone synthesis through AT2 receptor (Naruse et al. 2002); (3) Some aldosterone stimulating factors that do not depend upon Ang II are endothelin, vasopressin, catecholamine, ACTH, glucocorticoid, hyperkalemia, hyperlipidemia, and reduced high-density lipoprotein.

15.5 Treatment

15.5.1 Clinical Trials

15.5.1.1 MR Antagonists

The two clinically approved and commonly used MR antagonists are spironolactone and eplerenone, both of which are steroidal compounds. They share similar molecular structure of aldosterone, natural MR ligands, and cortisol. Searle Laboratories (Skokie, Ill) introduced spironolactone as the first MR antagonist in 1959 (Jaisser and Farman 2016). It was recommended for the treatment of edema, hypertension, primary aldosteronism, and heart failure. It is commonly used as a diuretic. Spironolactone is a potent MR antagonist; however, due to its poorly selective nature toward MR, it inhibits the androgen and progesterone receptors (Kolkhof and Borden 2012). Therapeutic use of this drug has led to several side effects such as feminization, gynecomastia, impotence, and menstrual irregularities, which reflect its anti-androgenic and progestogenic activity. In the Randomized Aldactone Evaluation Study (RALES), 10% of men in the spironolactone-treated group had gynecomastia and breast pain and the incidence rate of hyperkalemia was only slightly higher than the placebo group (Pitt et al. 1999). Other side effects of spironolactone are hyperkalemia, hyponatremia, dizziness, abnormal renal function, and high creatinine concentration. These side effects limit the patients with spironolactone therapy. In recent years, spironolactone has shown to prevent renal fibrosis. In 2002, a second-generation MR antagonist eplerenone was launched for the treatment of hypertension and congestive heart failure. In comparison with spironolactone, eplerenone has more selectivity for MR but less potency (40-fold less affinity) and interacts less with other steroid receptors (Kolkhof and Borden 2012). Therefore, there are fewer hormonal-related side effects with eplerenone. In the Eplerenone Post-acute Myocardial Infarction Heart Failure Efficacy and Survival Study (EPHESUS), 15.6% of the patients with eplerenone therapy had serum potassium (K^+) >5.5 mEq/L in comparison with 11.2% with the placebo group (Pitt et al. 2008). Essentially, patients who are taking spironolactone and eplerenone need to monitor their serum potassium level and renal function carefully. The ratio of sodium/potassium in the urine is an accurate indicator of the appropriate dose of spironolactone.

Nonsteroidal compound dihydropyridine, which is also known as L-type calcium channel antagonists, can also act as MR antagonist (Kolkhof and Borden 2012). A potent and selective dihydropyridine-based MR antagonist, BR-4628, was generated due to optimization of antagonists without L-type calcium channel activation. Finerenone (BAY 94–8862) is a third-generation, novel dihydronaphthyridine class of MR antagonist formed by the optimization of BR-4628 which has shown better selectivity nature for MR over spironolactone and eplerenone (Jaisser and Farman 2016). The use of finerenone was investigated in a phase 2 clinical trial of Mineralocorticoid Receptor Antagonist Tolerability Study (ARTS), which showed equal effectiveness as spironolactone in reducing the levels of B-type natriuretic peptide

(BNP), amino-terminal proBNP, and albuminuria (Pitt et al. 2013). However, the rate of hyperkalemia and renal dysfunction was lowered in finerenone group demonstrating its safety profile than spironolactone in patients with chronic heart failure and kidney impairment. Another clinical trial, the ARTS-diabetic nephropathy (ARTS-DN), was launched in 823 patients suffering from type II diabetic mellitus or diabetic nephropathy in order to testify the efficacy of different oral doses of finerenone compared to placebo, who were initially receiving ACE inhibitors or ARBs (Bakris et al. 2015). At day 90 follow-up, finerenone (7.5–20 mg once daily) reduced the albuminuria (17.2–40.2%) in a dose-dependent manner as compared with placebo (13.6%). There was no significant correlation seen between changes in urinary albumin–creatinine ratio and changes in systolic blood pressure which suggests that lowering of blood pressure does not lower albumin excretion, and intraglomerular pressure is also independent of urinary albumin–creatinine ratio changes.

15.5.1.2 Aldosterone Synthase Inhibitor

Inhibition of aldosterone synthase is an alternative approach to antagonize the harmful aldosterone actions (both MR dependent and MR independent) by directly attenuating its production in cardiac, vascular, and renal target organs (Azizi et al. 2012). Therefore, enzymatic activity of the aldosterone synthase is inactivated resulting in decreased aldosterone concentrations in plasma and tissue. Aldosterone synthase inhibitor has been studied as a new treatment option for management of hypertension, cardiac failure, and renal diseases. LCI699 is the first orally active aldosterone synthase inhibitor drug that was discovered for humans. LCI699, when used in 14 patients with primary aldosteronism observed dose-dependent reduction in plasma and urinary aldosterone concentrations by 70–80%, decreased blood pressure, and plasma renin activity was increased mildly. Notably, LCI699 was shown to have limited selectivity to aldosterone synthase CYP11B2 (Azizi et al. 2012).

15.5.2 Experimental Studies

As mentioned earlier, various experiments were carried out in animal models which have shown a potential role of MR antagonists, along with other pathways inhibitors in delaying the progression of renal fibrosis through different mechanisms. In this section, we will briefly mention a few of them. We have examined the role of MR antagonist, mitochondrial respiratory chain complex I inhibitor, or an NADPH oxidase inhibitor in significantly mitigating the mesangial cell proliferation and ROS production which is induced by aldosterone (Huang et al. 2009). In this, eplerenone also inhibited EGFR transactivation which is a downstream effector of PI3 K/Akt and Ras/MAPK signaling pathway. Aldosterone-induced ERK1/2 phosphorylation in mesangial cells was significantly attenuated when using eplerenone as pretreatment (Nishiyama et al. 2005). Spironolactone can block the role of NO synthase inhibitor L-NAME in renal tissue which showed that reduced expression of TGF-beta 1, CTGF, OPN, and PAI-1 in renal tissue, proteinuria, and blood pressure was significantly decreased, hence alleviating renal fibrosis and renal function deterioration (Ikeda et al. 2009). In another study, we also showed that aldosterone-induced EMT was significantly blocked with the use of eplerenone and rotenone (Zhang et al. 2007). Treatment with eplerenone significantly improved aldosterone-induced renal injury and increased serum- and glucocorticoid-inducible protein kinase-1 (SGK1), ICAM-1, and CTGF expressions in the rat mesangial cells which may be considered as another therapeutic target in the treatment of renal fibrosis (Terada et al. 2012). Blockade of ADAM-17/TGF-α/EGFR pathway diminished aldosterone-induced pro-inflammatory gene upregulation (Morgado-Pascual et al. 2015) which may be another treatment option for preventing the progression of renal fibrosis. When treated with eplerenone, podocyte damage and proteinuria were markedly attenuated in the aldosterone-treated rats (Shibata et al. 2007). Besides, treatment with eplerenone or tempol also significantly reduced podocyte injury and proteinuria in other rat models of hypertensive glomerulosclerosis (Nagase et al. 2006, 2007) and type 2 diabetic rats (Nishiyama et al. 2010). Our study has shown activator of SIRT1, resveratrol-blocked mitochondrial dysfunction, and podocyte injury by activating SIRT1/PGC-1 alpha which may be therapeutically useful in renal diseases (Yuan et al. 2012a). Blockade of Drp1 inhibited mitochondrial fission and dysfunction, and podocyte apoptosis in an animal model of aldosterone-induced nephropathy which may provide another promising therapeutics for podocyte injury (Yuan et al. 2017). In order to discover highly selective aldosterone synthase inhibitor, an experimental study was conducted in cynomolgus monkey-based models which showed BI 689648 as a highly selective one (Weldon et al. 2016). It revealed an in vitro IC_{50} of 2 nM against aldosterone synthase and cortisol synthase and 150-fold selectivity. For in vivo selectivity profiling, an adrenocorticotropin-challenge model was used which exhibited BI 689648 to be more than 20-fold selective compared with FAD286 and LCI699. This study may highlight an important step forward in the identification of effective and high potential aldosterone synthase inhibitors for clinical setup in cardiometabolic diseases, diabetic nephropathy, and CKD.

15.6 Summary

Apart from its physiological action on salt and water homeostasis, herein, we have discussed aldosterone as a potent factor in the pathogenesis of renal fibrosis. The prevailing in vivo and in vitro experiments suggest the independent detrimental role of aldosterone along with its MR, eventually contributing to the development of renal injury and fibrosis. As shown in Fig. 15.1, aldosterone has multiple mechanisms underlying the pathogenic role in kidney ultimately promoting renal injury and fibrosis. Some small clinical studies have shown that the use of aldosterone antagonists in patients with CKD can reduce proteinuria and delay the progression

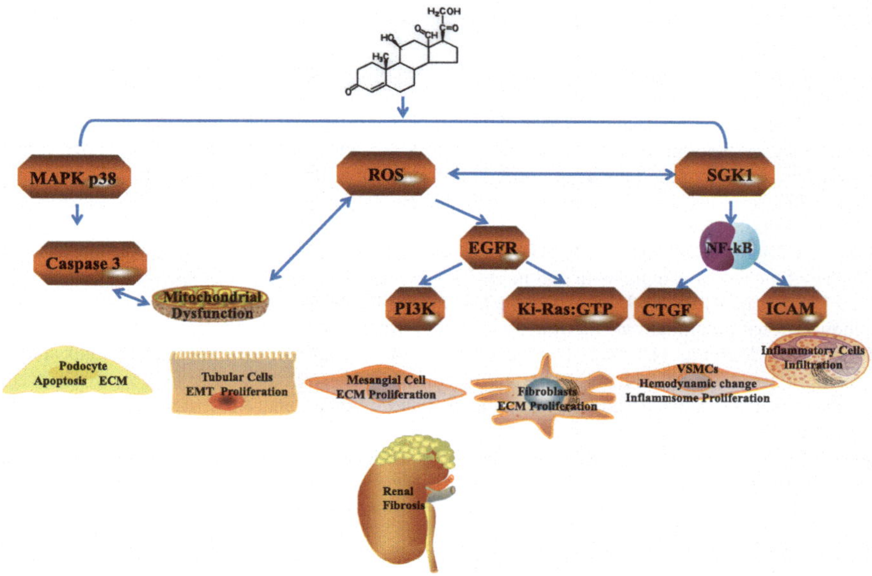

Fig. 15.1 Schematic diagram illustrating the possible mechanisms underlying the pathogenic role of aldosterone in renal fibrosis. Aldosterone stimulates the increase in ROS production which induces mitochondrial dysfunction. ROS also activates EGFR. EGFR transactivation activates PI3 K/Akt/mTOR/p70S6K1 and Ki-RasA/c-Raf/MEK/ERK and signaling pathways. Activation of p38 MAPK could trigger the Caspase 3 death pathway. Aldosterone induces SGK1 phosphorylation and promotes expressions of CTGF and ICAM-1 via NF-κB. In kidney, these changes lead to activation of pro-inflammatory and pro-fibrotic pathways ultimately promoting glomerulosclerosis and tubulointerstitial fibrosis

of renal fibrosis. However, large-scale clinical studies are needed to confirm the safety and effectiveness of aldosterone receptor antagonists in CKD patients.

Acknowledgement and Conflict of Interest The authors declare that they have no conflict of interest.

References

Abboud HE (2012) Mesangial cell biology. Exp Cell Res 318:979–985

Acloque H, Adams MS, Fishwick K, Bronner-Fraser M, Nieto MA (2009) Epithelial-mesenchymal transitions: the importance of changing cell state in development and disease. J Clin Invest 119:1438–1449

Arima S, Kohagura K, Xu H-L, Sugawara A, Abe T, Satoh F et al (2003) Nongenomic vascular action of aldosterone in the glomerular microcirculation. J Am Soc Nephrol 14:2255–2263

Arriza JL, Weinberger C, Cerelli G, Glaser TM, Handelin BL, Housman DE et al (1987) Cloning of human mineralocorticoid receptor complementary DNA: structural and functional kinship with the glucocorticoid receptor. Science 237:268–275

Azizi M, Amar L, Menard J (2012) Aldosterone synthase inhibition in humans. Nephrol Dial Transplant 28:36–43
Baeuerle PA, Henkel T (1994) Function and activation of NF-kappaB in the immune system. Annu Rev Immunol 12:141–179
Bai M, Chen Y, Zhao M, Zhang Y, He JC-J, Huang S et al (2017) NLRP3 inflammasome activation contributes to aldosterone-induced podocyte injury. Am J Physiol Renal Physiol 312:F556–F564
Bakris GL, Agarwal R, Chan JC, Cooper ME, Gansevoort RT, Haller H et al (2015) Effect of finerenone on albuminuria in patients with diabetic nephropathy: a randomized clinical trial. JAMA 314:884–894
Bianchi S, Bigazzi R, Campese VM (2006) Long-term effects of spironolactone on proteinuria and kidney function in patients with chronic kidney disease. Kidney Int 70:2116–2123
Blasi ER, Rocha R, Rudolph AE, Blomme EA, Polly ML, McMahon EG (2003) Aldosterone/salt induces renal inflammation and fibrosis in hypertensive rats. Kidney Int 63:1791–1800
Boldyreff B, Wehling M (2003) Non-genomic actions of aldosterone: mechanisms and consequences in kidney cells. Nephrol Dial Transplant 18:1693–1695
Bomback AS, Klemmer PJ (2007) The incidence and implications of aldosterone breakthrough. Nat Clin Pract Nephrol 3:486–492
Bonvalet J, Alfaidy N, Farman N, Lombes M (1995) Aldosterone: intracellular receptors in human heart. Eur Heart J 16:92–97
Brown NJ (2013) Contribution of aldosterone to cardiovascular and renal inflammation and fibrosis. Nat Rev Nephrol 9:459–469
Brown NJ, Kim KS, Chen YQ, Blevins LS, Nadeau JH, Meranze SG et al (2000a) Synergistic effect of adrenal steroids and angiotensin II on plasminogen activator inhibitor-1 production. J Clin Endocrinol Metab 85:336–344
Brown NJ, Nakamura S, Ma L, Nakamura I, Donnert E, Freeman M et al (2000b) Aldosterone modulates plasminogen activator inhibitor-1 and glomerulosclerosis in vivo. Kidney Int 58:1219–1227
Brunskill EW, Potter SS (2012) Changes in the gene expression programs of renal mesangial cells during diabetic nephropathy. BMC Nephrol 13:70
Burns W, Thomas M (2011) Angiotensin II and its role in tubular epithelial to mesenchymal transition associated with chronic kidney disease. Cells Tissues Organs 193:74–84
Chen C, Liang W, Jia J, Van Goor H, Singhal PC, Ding G (2009) Aldosterone induces apoptosis in rat podocytes: role of PI3-K/Akt and p38MAPK signaling pathways. Nephron Exp Nephrol 113:e26–e34
Christ M, Wehling M (1999) Rapid actions of aldosterone: lymphocytes, vascular smooth muscle and endothelial cells. Steroids 64:35–41
Chrysostomou A, Pedagogos E, MacGregor L, Becker GJ (2006) Double-blind, placebo-controlled study on the effect of the aldosterone receptor antagonist spironolactone in patients who have persistent proteinuria and are on long-term angiotensin-converting enzyme inhibitor therapy, with or without an angiotensin II receptor blocker. Clin J Am Soc Nephrol 1:256–262
Cicoira M, Zanolla L, Rossi A, Golia G, Franceschini L, Cabrini G et al (2001) Failure of aldosterone suppression despite angiotensin-converting enzyme (ACE) inhibitor administration in chronic heart failure is associated with ACE DD genotype. J Am Coll Cardiol 37:1808–1812
Coirini H, Mari A, De Nicola AF, Rainbow TC, McEwen BS (1985) Further studies of brain aldosterone binding sites employing new mineralocorticoid and glucocorticoid receptor markers in vitro. Brain Res 361:212–216
Conn JW, Knopf RF, Nesbit RM (1964) Clinical characteristics of primary aldosteronism from an analysis of 145 cases. Am J Surg 107:159–172
Ding W, Yang L, Zhang M, Gu Y (2012) Chronic inhibition of nuclear factor kappa B attenuates aldosterone/salt-induced renal injury. Life Sci 90:600–666
Epstein M (2006) Aldosterone blockade: an emerging strategy for abrogating progressive renal disease. Am J Med 119:912–919
Farman N, Rafestin-Oblin M-E (2001) Multiple aspects of mineralocorticoid selectivity. Am J Physiol Renal Physiol 280:F181–F192

Forrester SJ, Kawai T, O'Brien S, Thomas W, Harris RC, Eguchi S (2016) Epidermal growth factor receptor transactivation: mechanisms, pathophysiology, and potential therapies in the cardiovascular system. Annu Rev Pharmacol Toxicol 56:627–653
Funder J, Myles K (1996) Exclusion of corticosterone from epithelial mineralocorticoid receptors is insufficient for selectivity of aldosterone action: in vivo binding studies. Endocrinology 137:5264–5268
Funder JW (2010) Aldosterone and mineralocorticoid receptors in the cardiovascular system. Prog Cardiovasc Dis 52:393–400
Funder JW, Pearce PT, Smith R, Smith AI (1988) Mineralocorticoid action: target tissue specificity is enzyme, not receptor, mediated. Science 242:583–585
Furumatsu Y, Nagasawa Y, Tomida K, Mikami S, Kaneko T, Okada N et al (2008) Effect of renin-angiotensin-aldosterone system triple blockade on non-diabetic renal disease: addition of an aldosterone blocker, spironolactone, to combination treatment with an angiotensin-converting enzyme inhibitor and angiotensin II receptor blocker. Hypertens Res 31:59–67
Gauer S, Segitz V, Goppelt-Struebe M (2007) Aldosterone induces CTGF in mesangial cells by activation of the glucocorticoid receptor. Nephrol Dial Transplant 22:3154–3159
Gonzalez A, López B, Díez J (2004) Fibrosis in hypertensive heart disease: role of the renin-angiotensin-aldosterone system. Med Clin North Am 88:83–97
Greene E, Kren S, Hostetter T (1996) Role of aldosterone in the remnant kidney model in the rat. J Clin Invest 98:1063–1068
Gros R, Ding Q, Sklar LA, Prossnitz EE, Arterburn JB, Chorazyczewski J et al (2011) GPR30 expression is required for the mineralocorticoid receptor–independent rapid vascular effects of aldosterone. Hypertension 57:442–451
Gros R, Ding Q, Liu B, Chorazyczewski J, Feldman RD (2013) Aldosterone mediates its rapid effects in vascular endothelial cells through GPER activation. Am J Physiol Cell Physiol 304:C532–C540
Grossmann C, Krug AW, Freudinger R, Mildenberger S, Voelker K, Gekle M (2007) Aldosterone-induced EGFR expression: interaction between the human mineralocorticoid receptor and the human EGFR promoter. Am J Physiol Endocrinol Metab 292:E1790–E1800
Han K, Kang Y, Han SY, Jee Y, Lee M, Han J et al (2006) Spironolactone ameliorates renal injury and connective tissue growth factor expression in type II diabetic rats. Kidney Int 70:111–120
Han JS, Choi BS, Yang CW, Kim YS (2009) Aldosterone-induced TGF-β1 expression is regulated by mitogen-activated protein kinases and activator protein-1 in mesangial cells. J Korean Med Sci 24:S195–S203
Han HI, Skvarca LB, Espiritu EB, Davidson AJ, Hukriede NA (2019) The role of macrophages during acute kidney injury: destruction and repair. Pediatr Nephrol 34:1–9
Hao J, Ren L, Zhang L, Kong D, Hao L (2015) Aldosterone-induced inflammatory response of mesangial cells via angiotension II receptors. J Renin Angiotensin Aldosterone Syst 16:739–748
Heber S, Denk L, Hu K, Minuth WW (2007) Modulating the development of renal tubules growing in serum-free culture medium at an artificial interstitium. Tissue Eng 13:281–292
Hostetter TH, Ibrahim HN (2003) Aldosterone in chronic kidney and cardiac disease. J Am Soc Nephrol 14:2395–2401
Huang W, Xu C, Kahng KW, Noble NA, Border WA, Huang Y (2008) Aldosterone and TGF-β1 synergistically increase PAI-1 and decrease matrix degradation in rat renal mesangial and fibroblast cells. Am J Physiol Renal Physiol 294:F1287–F1295
Huang S, Zhang A, Ding G, Chen R (2009) Aldosterone-induced mesangial cell proliferation is mediated by EGF receptor transactivation. Am J Physiol Renal Physiol 296:F1323–F1333
Huang L, Nikolic-Paterson D, Ma F, Tesch G (2012) Aldosterone induces kidney fibroblast proliferation via activation of growth factor receptors and PI3K/MAPK signalling. Nephron Exp Nephrol 120:e115–e122
Ikeda H, Tsuruya K, Toyonaga J, Masutani K, Hayashida H, Hirakata H et al (2009) Spironolactone suppresses inflammation and prevents L-NAME–induced renal injury in rats. Kidney Int 75:147–155

Irita J, Okura T, Kurata M, Miyoshi K-i, Fukuoka T, Higaki J (2008) Osteopontin in rat renal fibroblasts: functional properties and transcriptional regulation by aldosterone. Hypertension 51:507–513

Jaisser F, Farman N (2016) Emerging roles of the mineralocorticoid receptor in pathology: toward new paradigms in clinical pharmacology. Pharmacol Rev 68:49–75

Juknevicius I, Segal Y, Kren S, Lee R, Hostetter TH (2004) Effect of aldosterone on renal transforming growth factor-β. Am J Physiol Renal Physiol 286:F1059–F1062

Kalluri R, Weinberg RA (2009) The basics of epithelial-mesenchymal transition. J Clin Invest 119:1420–1428

Kiyomoto H, Rafiq K, Mostofa M, Nishiyama A (2008) Possible underlying mechanisms responsible for aldosterone and mineralocorticoid receptor-dependent renal injury. J Pharmacol Sci 108:399–405

Kolkhof P, Borden SA (2012) Molecular pharmacology of the mineralocorticoid receptor: prospects for novel therapeutics. Mol Cell Endocrinol 350:310–317

Kornel L (1994) Colocalization of 11β-hydroxysteroid dehydrogenase and mineralocorticoid receptors in cultured vascular smooth muscle cells. Am J Hypertens 7:100–103

Lenzini L, Seccia TM, Aldighieri E, Belloni AS, Bernante P, Giuliani L et al (2007) Heterogeneity of aldosterone-producing adenomas revealed by a whole transcriptome analysis. Hypertension 50:1106–1113

Li C, Ding XY, Xiang DM, Xu J, Huang XL, Hou FF et al (2015) Enhanced M1 and impaired M2 macrophage polarization and reduced mitochondrial biogenesis via inhibition of AMP kinase in chronic kidney disease. Cell Physiol Biochem 36:358–372

Liu Y (2011) Cellular and molecular mechanisms of renal fibrosis. Nat Rev Nephrol 7:684–696

Martín-Fernández B, Rubio-Navarro A, Cortegano I, Ballesteros S, Alía M, Cannata-Ortiz P et al (2016) Aldosterone induces renal fibrosis and inflammatory M1-macrophage subtype via mineralocorticoid receptor in rats. PLoS ONE 11:e0145946

Mathew JT, Patni H, Chaudhary AN, Liang W, Gupta A, Chander PN et al (2008) Aldosterone induces mesangial cell apoptosis both in vivo and in vitro. Am J Physiol Renal Physiol 295:F73–F81

Mihailidou AS, Funder JW (2005) Nongenomic effects of mineralocorticoid receptor activation in the cardiovascular system. Steroids 70:347–351

Min LJ, Mogi M, Li JM, Iwanami J, Iwai M, Horiuchi M (2005) Aldosterone and angiotensin II synergistically induce mitogenic response in vascular smooth muscle cells. Circ Res 97:434–442

Minuth WW, Denk L, Heber S (2005) Growth of embryonic renal parenchyme at the interphase of a polyester artificial interstitium. Biomaterials 26:6588–6598

Miyata K, Rahman M, Shokoji T, Nagai Y, Zhang GX, Sun GP et al (2005) Aldosterone stimulates reactive oxygen species production through activation of NADPH oxidase in rat mesangial cells. J Am Soc Nephrol 16:2906–2912

Morgado-Pascual JL, Rayego-Mateos S, Valdivielso JM, Ortiz A, Egido J, Ruiz-Ortega M (2015) Paricalcitol inhibits aldosterone-induced proinflammatory factors by modulating epidermal growth factor receptor pathway in cultured tubular epithelial cells. Biomed Res Int 2015:783538

Nagase M, Shibata S, Yoshida S, Nagase T, Gotoda T, Fujita T (2006) Podocyte injury underlies the glomerulopathy of Dahl salt-hypertensive rats and is reversed by aldosterone blocker. Hypertension 47:1084–1093

Nagase M, Matsui H, Shibata S, Gotoda T, Fujita T (2007) Salt-induced nephropathy in obese spontaneously hypertensive rats via paradoxical activation of the mineralocorticoid receptor: role of oxidative stress. Hypertension 50:877–883

Naruse M, Tanabe A, Sato A, Takagi S, Tsuchiya K, Imaki T et al (2002) Aldosterone breakthrough during angiotensin II receptor antagonist therapy in stroke-prone spontaneously hypertensive rats. Hypertension 40(1):28–33

Nishiyama A, Yao L, Nagai Y, Miyata K, Yoshizumi M, Kagami S et al (2004) Possible contributions of reactive oxygen species and mitogen-activated protein kinase to renal injury in aldosterone/salt-induced hypertensive rats. Hypertension 43:841–848

Nishiyama A, Yao L, Fan Y, Kyaw M, Kataoka N, Hashimoto K et al (2005) Involvement of aldosterone and mineralocorticoid receptors in rat mesangial cell proliferation and deformability. Hypertension 45:710–716
Nishiyama A, Kobori H, Konishi Y, Morikawa T, Maeda I, Okumura M et al (2010) Mineralocorticoid receptor blockade enhances the antiproteinuric effect of an angiotensin II blocker through inhibiting podocyte injury in type 2 diabetic rats. J Pharmacol Exp Ther 332:1072–1080
Pavenstadt H, Kriz W, Kretzler M (2003) Cell biology of the glomerular podocyte. Physiol Rev 83:253–307
Phanish MK, Winn S, Dockrell M (2010) Connective tissue growth factor-(CTGF, CCN2)–a marker, mediator and therapeutic target for renal fibrosis. Nephron Exp Nephrol 114:e83–e92
Pitt B, Zannad F, Remme WJ, Cody R, Castaigne A, Perez A et al (1999) The effect of spironolactone on morbidity and mortality in patients with severe heart failure. N Engl J Med 341:709–717
Pitt B, Bakris G, Ruilope LM, DiCarlo L, Mukherjee R (2008) Serum potassium and clinical outcomes in the eplerenone post-acute myocardial infarction heart failure efficacy and survival study (EPHESUS). Circulation 118:1643–1650
Pitt B, Kober L, Ponikowski P, Gheorghiade M, Filippatos G, Krum H et al (2013) Safety and tolerability of the novel non-steroidal mineralocorticoid receptor antagonist BAY 94-8862 in patients with chronic heart failure and mild or moderate chronic kidney disease: a randomized, double-blind trial. Eur Heart J 34:2453–2463
Porter GA, Edelman IS (1964) The action of aldosterone and related corticosteroids on sodium transport across the toad bladder. J Clin Invest 43:611–620
Porter GA, Bogoroch R, Edelman IS (1964) On the mechanism of action of aldosterone on sodium transport: the role of RNA synthesis. Proc Natl Acad Sci U S A 52:1326–1333
Rogerson FM, Fuller PJ (2000) Mineralocorticoid action. Steroids 65:61–73
Rüster C, Wolf G (2006) Renin-angiotensin-aldosterone system and progression of renal disease. J Am Soc Nephrol 17:2985–2991
Sato A, Fukuda S (2013) Effect of aldosterone breakthrough on albuminuria during treatment with a direct renin inhibitor and combined effect with a mineralocorticoid receptor antagonist. Hypertens Res 36:879–884
Schjoedt K, Andersen S, Rossing P, Tarnow L, Parving H-H (2004) Aldosterone escape during blockade of the renin–angiotensin–aldosterone system in diabetic nephropathy is associated with enhanced decline in glomerular filtration rate. Diabetologia 47:1936–1939
Schjoedt K, Rossing K, Juhl T, Boomsma F, Tarnow L, Rossing P et al (2006) Beneficial impact of spironolactone on nephrotic range albuminuria in diabetic nephropathy. Kidney Int 70:536–542
Shankland S (2006) The podocyte's response to injury: role in proteinuria and glomerulosclerosis. Kidney Int 69:2131–2147
Sheng L, Yang M, Ding W, Zhang M, Niu J, Qiao Z et al (2016) Epidermal growth factor receptor signaling mediates aldosterone-induced profibrotic responses in kidney. Exp Cell Res 346:99–110
Shibata S, Nagase M, Yoshida S, Kawachi H, Fujita T (2007) Podocyte as the target for aldosterone: roles of oxidative stress and Sgk1. Hypertension 49:355–364
Shibata S, Nagase M, Yoshida S, Kawarazaki W, Kurihara H, Tanaka H et al (2008) Modification of mineralocorticoid receptor function by Rac1 GTPase: implication in proteinuric kidney disease. Nat Med 14:1370–1376
Simpson S (1953) Isolation from the adrenals of a new crystalline hormone with especially high effectiveness on mineral metabolism. Experientia 9(333–335):3
Spat A, Hunyady L (2004) Control of aldosterone secretion: a model for convergence in cellular signaling pathways. Physiol Rev 84:489–539
Su M, Dhoopun A-R, Yuan Y, Huang S, Zhu C, Ding G et al (2013) Mitochondrial dysfunction is an early event in aldosterone-induced podocyte injury. Am J Physiol Renal Physiol 305:F520–F531
Sun Y, Zhang J, Zhang JQ, Ramires FJ (2000) Local angiotensin II and transforming growth factor-β1 in renal fibrosis of rats. Hypertension 35:1078–1084
Terada Y, Ueda S, Hamada K, Shimamura Y, Ogata K, Inoue K et al (2012) Aldosterone stimulates nuclear factor-kappa B activity and transcription of intercellular adhesion molecule-1 and con-

nective tissue growth factor in rat mesangial cells via serum-and glucocorticoid-inducible protein kinase-1. Clin Exp Nephrol 16:81–88

Thiery JP (2002) Epithelial–mesenchymal transitions in tumour progression. Nat Rev Cancer 2:442–454

Trachtman H, Weiser AC, Valderrama E, Morgado M, Palmer LS (2004) Prevention of renal fibrosis by spironolactone in mice with complete unilateral ureteral obstruction. J Urol 172:1590–1594

Unger T, Paulis L, Sica DA (2011) Therapeutic perspectives in hypertension: novel means for renin–angiotensin–aldosterone system modulation and emerging device-based approaches. Eur Heart J 32:2739–2747

Urata H, Hoffmann S, Ganten D (1994a) Tissue angiotensin II system in the human heart. Eur Heart J 15:68–78

Urata H, Strobel F, Ganten D (1994b) Widespread tissue distribution of human chymase. J Hypertens Suppl 12:S17–S22

Wang H, Naghavi M, Allen C, Barber RM, Bhutta ZA, Carter A et al (2016) Global, regional, and national life expectancy, all-cause mortality, and cause-specific mortality for 249 causes of death, 1980–2015: a systematic analysis for the Global Burden of Disease Study 2015. Lancet 388:1459–1544

Weldon SM, Cerny MA, Gueneva-Boucheva K, Cogan D, Guo X, Moss N et al (2016) Selectivity of BI 689648, a novel, highly selective aldosterone synthase inhibitor: comparison with FAD286 and LCI699 in nonhuman primates. J Pharmacol Exp Ther 359:142–150

Williams GH (2005) Aldosterone biosynthesis, regulation, and classical mechanism of action. Heart Fail Rev 10:7–13

Williams JS, Williams GH (2003) 50th anniversary of aldosterone. J Clin Endocrinol Metab 88:2364–2372

Wolf G, Chen S, Ziyadeh FN (2005) From the periphery of the glomerular capillary wall toward the center of disease: podocyte injury comes of age in diabetic nephropathy. Diabetes 54:1626–1634

Yamada M, Kushibiki M, Osanai T, Tomita H, Okumura K (2008) Vasoconstrictor effect of aldosterone via angiotensin II type 1 (AT1) receptor: possible role of AT1 receptor dimerization. Cardiovasc Res 79:169–178

Yuan J, Jia R, Bao Y (2007) Aldosterone up-regulates production of plasminogen activator inhibitor-1 by renal mesangial cells. J Biochem Mol Biol 40:180–188

Yuan Y, Huang S, Wang W, Wang Y, Zhang P, Zhu C et al (2012a) Activation of peroxisome proliferator-activated receptor-γ coactivator 1α ameliorates mitochondrial dysfunction and protects podocytes from aldosterone-induced injury. Kidney Int 82:771–789

Yuan Y, Chen Y, Zhang P, Huang S, Zhu C, Ding G et al (2012b) Mitochondrial dysfunction accounts for aldosterone-induced epithelial-to-mesenchymal transition of renal proximal tubular epithelial cells. Free Radic Biol Med 53:30–43

Yuan Y, Zhang A, Qi J, Wang H, Liu X, Zhao M et al (2017) P53/Drp1-dependent mitochondrial fission mediates aldosterone-induced podocyte injury and mitochondrial dysfunction. Am J Physiol Renal Physiol 314:F798–F808

Zhang A, Jia Z, Guo X, Yang T (2007) Aldosterone induces epithelial-mesenchymal transition via ROS of mitochondrial origin. Am J Physiol Renal Physiol 293:F723–F731

Zhang A, Jia Z, Wang N, Tidwell TJ, Yang T (2011) Relative contributions of mitochondria and NADPH oxidase to deoxycorticosterone acetate-salt hypertension in mice. Kidney Int 80:51–60

Zhang A, Han Y, Wang B, Li S, Gan W (2015) Beyond gap junction channel function: the expression of Cx43 contributes to aldosterone-induced mesangial cell proliferation via the ERK1/2 and PKC pathways. Cell Physiol Biochem 36:1210–1222

Zhu C, Huang S, Yuan Y, Ding G, Chen R, Liu B et al (2011) Mitochondrial dysfunction mediates aldosterone-induced podocyte damage: a therapeutic target of PPARγ. Am J Pathol 178:2020–2031

Chapter 16
TGF-β/Smad and Renal Fibrosis

Tao-Tao Ma and Xiao-Ming Meng

Abstract Renal fibrosis is characterized by excessive deposition of extracellular matrix (ECM) that disrupts and replaces functional parenchyma, which leads to organ failure. It is known as the major pathological mechanism of chronic kidney disease (CKD). Although CKD has an impact on no less than 10% of the world population, therapeutic options are still limited. Regardless of etiology, elevated TGF-β levels are highly correlated with the activated pro-fibrotic pathways and disease progression. TGF-β, the key driver of renal fibrosis, is involved in a dynamic pathophysiological process that leads to CKD and end-stage renal disease (ESRD). It is becoming clear that epigenetics regulates renal programming, and therefore, the development and progression of renal disease. Indeed, recent evidence shows TGF-β1/Smad signaling regulates renal fibrosis via epigenetic-correlated mechanisms. This review focuses on the function of TGF-β/Smads in renal fibrogenesis, and the role of epigenetics as a regulator of pro-fibrotic gene expression.

Keywords TGF-β · Renal fibrosis · Smad · Non-coding RNA · Epigenetic modification

16.1 Introduction

Chronic kidney disease (CKD) is associated with excessive deposition of extracellular matrix (ECM), which causes loss of healthy renal structure and contributes to end-stage renal disease (ESRD). CKD affects 10% of the worldwide population with a high mortality rate due to limited effective treatments. Researches in animal and human studies confirm TGF-β1 is a predominant pathogenic factor that drives progressive forms of renal fibrosis (Meng et al. 2016). Collectively, these findings indicate TGF-β1 is a viable therapeutic target to prevent the onset or progression of fibrosis.

T.-T. Ma · X.-M. Meng (✉)
Anhui Province Key Laboratory of Major Autoimmune Diseases, Anhui Institute of Innovative Drugs, School of Pharmacy, Anhui Medical University, Hefei, Anhui, China
e-mail: mengxiaoming@ahmu.edu.cn

B.-C. Liu et al. (eds.), *Renal Fibrosis: Mechanisms and Therapies*, Advances in Experimental Medicine and Biology 1165,
https://doi.org/10.1007/978-981-13-8871-2_16

TGF-β is a member of transforming growth factor superfamily and consists of three isoforms TGF-β1, TGF-β2, and TGF-β3. All have been identified in mammals, sharing 70–82% amino acid homology. All three isoforms bind to TGF-β type 2 receptor (TGFR2) as homodimers that recruit and activate TGFR1 to initiate receptor signaling. Overexpression of active TGF-β1 in the liver of transgenic mice is sufficient to induce fibrotic disease in multiple organs, including kidneys (Meng et al. 2016). TGF-β is synthesized in a precursor form; latency-associated peptide (LAP) is released by cleavage near the N-terminus and binds with mature TGF-β homodimers that promote attachment with latent TGF-β binding protein (LTBP). This latent TGF-β/LAP/LTBP complex maintains TGF-β in an inactive form, preventing its interaction with receptors. Latent TGF-β complex is cleaved extracellularly by a broad range of proteases, including plasmin, matrix metalloproteinase (MMP) 2 and MMP9 that release active TGF-β. This mechanism is more complex than initially thought. LAP/TGF-β complex is capable to bind GARP, a receptor also known as LRRC32, which expresses in regulatory T cells and plays a significant role in regulating phenotypes and functions of these cells. This interaction may explain why latent TGF-β1 overexpression limits not only fibrosis, but also inflammation in models of kidney disease.

16.2 Role of TGF-β1 in Renal Fibrosis

Renal fibrosis, characterized by excessive accumulation of ECM, plays central roles in the development of ESRD that is caused by progressive renal dysfunction (Eddy and Neilson 2006). Although a large amount of evidence indicates TGF-β is the key mediator in CKD, an effective therapy for renal fibrosis sorely needed. TGF-β1 regulates multiple biological processes, including cell proliferation, apoptosis, and autophagy (Meng et al. 2013). Data show TGF-β is upregulated in injured kidneys in human patients and animal disease models (Yamamoto et al. 1996; Bottinger and Bitzer 2002). In addition, TGF-β is higher in patients with different kinds of renal diseases, and is positively related to the severity of fibrosis (Murakami et al. 1997). In addition, overexpression of TGF-β1 in rodent liver leads to the fibrotic response in kidneys while inhibiting TGF-β with neutralizing antibodies, inhibitors, or receptors depletion attenuates renal fibrosis in vivo and in vitro (Kopp et al. 1996; Border and Noble 1998; Moon et al. 2006; Petersen et al. 2008; Meng et al. 2012a, b). Compared with the active form of TGF-β1, the latent form prevents renal inflammation and fibrosis by upregulating Smad7, as observed in latent TGF-β transgenic mice with obstructive nephropathy or anti-GBM glomerulonephritis (Huang et al. 2008a, b).

Taken together, evidence shows TGF-β promotes renal fibrosis through several possible mechanisms: (1) TGF-β1 directly induces ECM synthesis via Smad3-dependent or Smad3-independent manners; (2) TGF-β1 suppresses the degradation of ECM by inhibiting MMPs and inducing natural inhibitors of MMPs like tissue inhibitor of metalloproteinases (TIMPs); (3) TGF-β1 plays a critical role in the trans-differentiation toward myofibroblasts from several cell types such as epithelial cells via epithelial–mesenchymal transition (EMT), endothelial cells via endothelial–mesenchymal transition (EndMT), and pericytes and bone marrow-derived macrophages via macrophage–myofibroblast transition (MMT) (Wu et al. 2013; Meng et al. 2016); (4) TGF-β1 acts directly on different types of renal resident cells to induce mesangial cells proliferation and elimination of tubular epithelial cells (TECs), podocytes, and endothelial cells, and this may lead to more severe renal damage and fibrosis (Bottinger and Bitzer 2002; Lopez-Hernandez and Lopez-Novoa 2012).

16.3 Role of Smads in Renal Fibrosis

Interestingly, although TGF-β1 plays a supporting role in renal fibrosis, its downstream Smads (like Smad2, Smad3, and Smad4) have distinct or even opposing effects while regulating fibrosis. One key difference among Smads is that Smad3 enhances target genes transcription by directly binding to promoters of Smad-binding elements (SBE). No DNA binding domain is detected in Smad2 and Smad4, so they only work as modulators for Smad3-based gene transcription (Dennler et al. 1998; Chen et al. 1999; Piek et al. 2001; Yuan and Varga 2001). Considering embryonic lethality of Smad2 or Smad4 mice, investigations into the role of Smad proteins have been limited due to a lack of available genetic models. (Goumans and Mummery 2000).

16.3.1 Smad2 and Smad3 in Renal Fibrosis

Both Smad2 and Smad3 are activated in the fibrotic kidney of CKD patients and animal models (Kim et al. 2003; Inazaki et al. 2004). Data show Smad3-deficient mice have less collagen deposition compared to wild-type mice with inductive kidney injury (Fujimoto et al. 2003; Sato et al. 2003; Zhou et al. 2010). In contrast, conditional knockout of Smad2 from TECs promotes renal fibrosis in obstructive nephropathy via inducing Smad3 activation, phosphor-Smad3 nuclear translocation and auto-induction of TGF-β1 (Meng et al. 2010). This indicates Smad2 exerts a negative effect on the pro-fibrotic function of Smad3.

16.3.2 Smad4 in Renal Fibrosis

Smad4 is a known player in both TGF-β and bone morphogenic protein (BMP) signaling pathways that facilitate nuclear translocation of Smad2/3 and Smad1/5/8 complexes, respectively (Meng et al. 2013). Smad4 deficiency in mesangial cells in vitro suppresses collagen I promoter activity (Tsuchida et al. 2003). Consistent with this finding, conditional deletion of Smad4 in tubular epithelial cells significantly reduced fibrosis in UUO-induced fibrosis mice without affecting Smad3 activation (Meng et al. 2012a, b).

16.3.3 Smad7 in Renal Fibrosis

Smad7 is reported as a negative feedback inhibitor of TGF-β/Smad signaling. Smad3 complexes induce Smad7 transcription in response to TGF-β1 (Yan and Chen 2011). Smad7 competes with Smad2 and Smad3 for binding sites on activated TGFR1. Thus, Smad7 negatively regulates TGF-β/Smad signaling (Shi and Massague 2003). It is important to note that during renal fibrosis, Smad7 mRNA transcription is countered by reduced protein by ubiquitin-dependent degradation via Smad ubiquitin regulatory factor 2 (Smurf2) (Fukasawa et al. 2004). Indeed, mice deficient in Smad7 show increased susceptibility to renal fibrosis (Chung et al. 2009; Chen et al. 2011). In comparison, overexpression of Smad7 attenuates renal fibrosis in kidney disease in animal models (Hou et al. 2005; Ka et al. 2007, 2012; Liu et al. 2014).

16.4 Role of TGF-β/Smad-Dependent Epigenetic Modification in Renal Fibrosis

Covalent changes to chromatin, such as DNA methylation or histone modifications, affect gene transcription. Modifications occur in a cell-type and microenvironment-specific manner. They universally regulate biological functions, including renal fibrosis. Evidence shows that TGF-β1/Smad regulates renal fibrosis via epigenetic modification (Meng et al. 2016).

16.4.1 DNA Methylation

In the mammalian genome, 60–90% of cytosine–phosphate–guanine (CpG) sites are methylated by DNA methyltransferases (DNMTs), which creates a 5-methylcytosine (5mC) site (Ehrlich et al. 1982). Counter to DNMTs, ten-eleven translocation (TET) proteins reduce 5mC sites to 5-hydroxy-methyl-cytosine (5hmC) (Tahiliani et al. 2009). Maintenance methylation is handled by DNA methyltransferase 1 (DNMT1), which establishes methylation patterns via preferential methylation of hemi-methylated DNA on newly synthesized DNA strands to solve the dilutive effect of cell proliferation (Bestor et al. 1992; Pradhan et al. 1999). De novo methylation is controlled by DNMT3A and DNMT3B (Okano et al. 1998; Hsieh 1999). DNA methylation occurs primarily at CpGd nucleotides, which are often enriched in the promoter region and the first five untranslated regions (Luczak and Jagodzinski 2006). Data show that some specific methylation patterns are correlated with kidney disease (Qiu et al. 2018). For example, TGF-β-mediated hypermethylation of RASAL1 via DNMT1 promotes fibroblast activation and fibrosis in kidney disease; this is resolved by anti-fibrotic factor BMP-7 (Tampe et al. 2014). Additionally, suppression of Krüppel-like factor 4 (KLF4) by hypermethylation of its promoter region contributes to TGF-β1-induced mesenchymal transition of TECs (Xiao et al. 2015). Inhibition of DNMT1 prevents TGF-β1-induced hypermethylation of the Smad7 gene promoter, which downregulates Smad7 expression and inhibits TGF-β1-induced α-SMA and collagen I production (Bian et al. 2014). In contrast, in cardiac fibroblasts, TGF-β1 downregulates DNMT1 expression and reduces methylation of the Col1a1 promoter and upregulates collagen I gene transcription. This suggests DNA methylation regulation by TGF-β1 works in a cell type-specific manner.

16.4.2 Histone Code

Histone modification is a universal genetic characteristic, and occurs in various ways, including acetylation, lysine and arginine methylation, phosphorylation, ubiquitylation, and sumoylation. All modifications are capable of conferring gene transcription regulation (Jenuwein and Allis 2001; Goldberg et al. 2007). Histone acetylation generally actives gene transcription, whereas histone methylation activates and inactivates promoters depending on the site of modified lysine. It is reported that active promoters are often decorated with H3K27ac and H3K4me3 marks, active enhancers with H3K27ac and H3K4me1 marks, in addition to silenced promoters and enhancers with H3K27me3 marks (Zhou et al. 2011; Kato and Natarajan 2014; Shlyueva et al. 2014). Interestingly, TGF-β1 represses H3K9me2/3-mediated transcription and enhances H3K4me1/2/3-mediated transcription on PAI-1, CTGF, and Col1a1 promoters, thereby induces fibrosis (Sun et al. 2010).

Histone acetylation and deacetylation are mediated by histone acetyltransferase (HAT) and histone deacetylase (HDAC) enzymes, respectively (Berger 2002). TGF-β1 induces expression of PAI-1 and p21 in mesangial cells, which also promotes glomerulosclerosis and hypertrophy. This requires histone acetyltransferase p300/CBP, which acetylates histone H3 at lysines 9 and 14, and transcription factors Smad3 and Sp1 (Yuan et al. 2013). In addition, p300/CBP acetylates Smad3 at Lys-378 in the MH2 domain to enhance its transcriptional activity (Inoue et al. 2007). Acetylation of Smad3 is induced in the early stage and sustained in the progression of severe fibrosis in UUO nephropathy (Li et al. 2010a, b).

It is important to note that the activities of HDAC inhibitors are increased in murine models of diabetic kidney disease, and treatment with non-selective HDAC inhibitor trichostatin A reduces ECM expression (Noh et al. 2009). The TGF-β1-induced EMT is countered by silencing HDAC-2 in TECs. MS-275, a class I HDAC inhibitor, inhibits TGF-β1/Smad3 signaling to suppress renal fibrosis in UUO models (Liu et al. 2013). Moreover, blocking HDAC-6 with inhibitors reduces binding of phosphor-Smad2/3 to SBE in promoter regions and suppresses angiotensin

II-induced hypertensive kidney fibrosis (Choi et al. 2015). This finding is further supported by studies in TECs showing that blockade of HDAC-6 suppresses TGF-β1-triggered EMT (Yoshikawa et al. 2007). Remarkably, a previous study showed that activation of TGF-β1 signaling decreased NR4A1 level via AKT and HDAC-dependent mechanisms, and thereby attenuated the inhibitory effect of NR4A1 in multiple models of tissue fibrosis, including kidney fibrosis (Palumbo-Zerr et al. 2015). Overall, considerable interest exists in exploiting the therapeutic potential of HDAC inhibitors in fibrotic diseases. However, given HDAC inhibitors have broad substrate specificity and deacetylate numerous proteins that are not associated with epigenetic regulation, this presents significant challenges to the field (Rius and Lyko 2012).

As a NAD^+-dependent protein deacetylase, Sirtuin 1 (SIRT1) may also play a role in renal fibrosis. SIRT1, which is activated by resveratrol, reduces acetylation but not the phosphorylation of Smad3, and attenuates UUO nephropathy. SIRT1 knockdown abolishes the protective effect of resveratrol. Co-IP results show a direct interaction between SIRT1 and acetylated Smad3. Consistently, resveratrol-induced SIRT1 activation attenuates renal dysfunction and fibrosis in remnant kidney model (Huang et al. 2014). However, contradictory findings have been reported. In one, administration of the SIRT1 activator, SRT1720, exacerbated renal fibrosis in the murine UUO model (Ponnusamy et al. 2015). In another, mice with fibroblast-specific knockdown of SIRT1 resisted bleomycin-induced lung fibrosis due to disrupted TGF-β/Smad signaling and collagen release in fibroblasts (Zerr et al. 2016). Thus, whether SIRT1 activation is a desirable strategy for renal fibrosis treatment remains unclear.

16.4.3 Noncoding RNA

Noncoding RNA is a modulator of cotranscriptional epigenetic silencing. It contains small (<200 nucleotides) and long (>200 nucleotides) noncoding RNA and regulates gene expression through various mechanisms. Accumulating evidence shows that noncoding RNAs, particularly microRNAs (miRNAs), play a significant role in the regulation of TGF-β/Smads on renal fibrotic genes (Meng et al. 2016). Long non-coding RNAs (lncRNA) affect gene expression and epigenetic mechanisms at the chromatin level (Mercer and Mattick 2013).

miRNAs are short RNA molecules of 20–22 nucleotides encoded by genomic DNA which could not be transcribed into peptides. These noncoding RNAs suppress gene expression by repressing and/or degradating targeted mRNAs (Leung 2015). miRNAs function in numerous biological processes including organ fibrosis. TGF-β1 induces transcription of several miRNA species, like miR-21, to exert pro-fibrotic effects. miR-21 is induced in the fibrotic kidney with levels correlated with the severity of fibrosis in the patients with diabetic nephropathy. TGF-β1 upregulates miR-21 level via Smad3-dependent mechanisms. miR-21 knockdown with oligonucleotides attenuates renal fibrosis in diabetic and UUO nephropathy (Zhong et al. 2011). Smad3-mediated upregulation of miR-21 promotes renal fibrosis (Zhong et al. 2013; Kolling et al. 2017). Compared with wild-type littermates, miR-21 knockout mice develop less fibrosis in response to injurious stimuli. This protective effect is mediated by the restoration of miR-21-targeted anti-fibrotic genes such as Smad7, Pparα, Spry1, and Pten (Chau et al. 2012; Zhong et al. 2013; Chung and Lan 2015). Upregulation of miR-17-5p is detected in fibrotic liver. In hepatic stellate cells (HSCs), TGF-β1-induced collagen I and α-SMA production is enhanced by miR-17-5p-mediated Smad7 suppression (Yu et al. 2015). In glomerular mesangial cells, TGF-β1 promotes cell survival and hypertrophy by upregulatingmiR-216a and miR-217, which targets the AKT inhibitor, PTEN, resulting in AKT activation (Kato et al. 2009). Additionally, TGF-β1 induces EMT via upregulating miR-382 which suppresses the adhesion molecule called E-cadherin (Kriegel et al. 2010). miR-433 is significantly induced by Smad3 in renal fibrosis, and miR-433 knockdown limits fibrosis via targeting the Azin1/TGF-β/Smad signaling in obstructive nephropathy (Li et al. 2013).

Several miRNAs regulating fibrosis molecules are suppressed by TGF-β1. For example, TGF-β1 downregulates miRNA let-7 in cultured mesangial cells, TECs, and diabetic and non-diabetic models of fibrotic kidney disease (Park et al. 2014; Wang et al. 2014). Let-7 exerts an anti-fibrotic response by downregulating TGFR1 expression, limiting TGF-β1-induced canonical and non-canonical signaling. miR-19b, miR-26a and miR-101, endogenous inhibitors of TGF-β1/Smad signaling, are suppressed by TGF-β1 in fibrotic organs (Lakner et al. 2012; Pan et al. 2012; Liang et al. 2014; Tu et al. 2014). A recent study showed loss of miR-30c in kidney of db/db mice; overexpression of miR-30c inhibited, but knockdown of miR-30c enhanced, the Snail1-TGF-β1 axis, and renal fibrosis in diabetic nephropathy (Zhao et al. 2017).

Together, these findings demonstrate that miRNAs play functional roles that affect Smad3 mechanisms related to regulating expression of pro-fibrotic and anti-fibrotic molecules. Challenges in targeting these miRNAs therapeutically included the inherent fact that each miRNA regulates many individual mRNAs, and the differences in how genes are regulated by individual miRNAs in mice compared with humans. Emerging evidence shows that miR-21 knockdown and miR-29 overexpression are the most effective anti-fibrotic therapies. Indeed, RG-012, an inhibitor of miR-21, is currently evaluated in a Phase I trial for treating Alport syndrome. And, MRG-201, a miR-29 mimic, is being tested in a Phase I trial for treating cutaneous scleroderma.

Thousands of lncRNAs have been explored by high-throughput sequencing and array analyses recently, some of them interplay with chromatin to regulate gene expression, although function of others remains unclear. In liver fibrosis, one group found that LINK-A lncRNA activated the TGF-β pathway in ovarian carcinoma cells (Ma and Xue 2018). Similarly, another found that LncRNA NKILA suppressed TGF-β-induced epithelial–mesenchymal transition (Wu et al. 2018). Of note, lncRNA MEG3 is reduced in liver fibrosis, and its overexpression limits TGF-β1-induced HSCs activation and experimental liver fibrosis (Wang et al. 2010; He et al. 2014). A recent study showed that lncRNA MEG3 was reduced in TGF-β1-treated HK2 cells; restoration of MEG3 blocked TGF-β1-induced promotion of EMT, cell viability, and proliferation (Xue et al. 2018). The function of lncRNA in kidney tissues has gained far more attention recently (Tang et al. 2017). A 2013 study found that a lncRNA, termed plasmacytoma variant translocation 1 (PVT1), increased TGF-β1 in mesangial cells and promoted renal fibrosis in diabetic nephropathy (Alvarez et al. 2013). No less than 21 lncRNAs are induced by TGF-β/Smad3 signaling in kidney by RNA sequencing techniques. It is noteworthy that lncRNAs np_5318 and 17,856 may function in TGF-β-induced renal fibrosis (Zhou et al. 2014). In TGF-β-stimulated HK-2 cells and a UUO-induced renal fibrosis model, lncRNA-H19 was significantly upregulated; knockdown of lncRNA-H19 prevented renal fibrosis both in vitro and in vivo (Xie et al. 2016). Most recently, a report showed Erbb4-IR, a novel lncRNA, is a key regulator in TGF-β/Smad3-mediated renal fibrosis. In it, kidney-specific knockdown of Erbb4-IR-induced renal Smad7, and inhibited TGF-β/Smad3-mediated renal fibrosis (Feng et al. 2018). These findings indicate Erbb4-IR may serve as a specific therapeutic target for chronic kidney disease. A pathological role for Erbb4-IR has also been suggested in type 2 diabetic nephropathy (Sun et al. 2018). Although data are currently limited, lncRNAs clearly play an important role in renal fibrosis and may present new therapeutic opportunities.

16.5 TGF-β/Smad and Mesenchymal Transition

Myofibroblasts are the major source of ECM production in organ fibrosis (Duffield et al. 2013; Falke et al. 2015). However, the cellular origins of myofibroblasts in renal fibrosis are still under debate. It is noteworthy that TGF-β/Smads plays a dominant role in the transdifferentiation or activation of myofibroblasts from different origins.

16.5.1 TGF-β and EMT

The phenomenon of tubular epithelial cells acquiring markers of myofibroblasts during interstitial fibrosis was first reported in 1995 (Strutz et al. 1995). EMT has been detected in human biopsies (Jinde et al. 2001; Oldfield et al. 2001) and animal models with fibrotic kidney diseases (Ng et al. 1998, 1999). Lines of evidence from in vitro studies show that TGF-β1 induces cultured TECs to differentiate into myofibroblast-like cells and produce a large amount of ECM, mediated primarily via Smad3-dependent manner, although non-canonical TGF-β1 pathways like RhoA and MAPK are also involved (Bhowmick et al. 2001; Bakin et al. 2002; Li et al. 2004; Mariasegaram et al. 2010; Wang et al. 2011). The concept of EMT has been challenged due to lineage tracing studies showing that tubular EMT contributes little to the myofibroblast population in renal fibrosis (Grgic et al. 2012; LeBleu et al. 2013; Mack and Yanagita 2015). Of note, novel evidence shows that tubular epithelial cells contribute to renal fibrosis without undergoing complete transition into α-SMA expressing myofibroblasts (Grande et al. 2015; Lovisa et al. 2015).

16.5.2 TGF-β and EndMT

EndMT is found in fibrotic kidneys and contributes to renal fibrosis (Piera-Velazquez et al. 2011). TGF-β promotes EndMT both in vitro and in vivo (van Meeteren and ten Dijke 2012). Conditional knockout of Tgfr2 from endothelial cells suppresses renal fibrosis (Xavier et al. 2015), and inhibition of Smad3 by SIS3 prevents EndMT and attenuates diabetic nephropathy (Li et al. 2010a, b). In addition, blocking dipeptidyl peptidase-4 limits EndMT and renal fibrosisin experimental diabetic nephropathy via TGF-β/Smad3/miR-29-dependent mechanisms (Shi et al. 2015).

16.5.3 TGF-β, Fibrocytes and Macrophages

Fibrocytes, a type of collagen I-producing cell derived from monocyte precursors in the bone marrow, are a source of myofibroblasts in models of renal fibrosis (Bucala et al. 1994; Wada et al. 2007; Strieter et al. 2009; Reich et al. 2013). TGF-β1 promotes fibrocyte-to-myofibroblast transition via Smad2/3 and JNK pathways (Marquardt and Muller-Hermelink 1990). Bone marrow-derived macrophages contribute significantly to myofibroblast populations and renal fibrosis via macrophage-to-myofibroblast transition via a TGF-β/Smad3-dependent mechanism (Nikolic-Paterson and Wang 2011; Wang et al. 2016).

16.5.4 TGF-β1 and Activation of Fibroblast-Type Cells

Activation of renal resident cells (glomerular mesangial cells, interstitial fibroblasts, and pericytes) is a key step for myofibroblast induction and renal fibrosis. TGF-β1 is a key factor during these processes. For example, TGF-β1 induces the production of α-SMA and collagen in mesangial cells (Schnaper et al. 2003; Mishra et al. 2008), and blocking TGF-β1 limits glomerulosclerosis (Border et al. 1990). Additionally, TGF-β1 activates intrinsic fibroblasts in kidney via a Nox4-dependent mechanism (Grande and Lopez-Novoa 2009; Barnes and Gorin 2011; Manickam et al. 2014). Previous studies showed that injured TECs produce a large amount of TGF-β1, which directly promotes the transition of adjacent pericytes into myofibroblast-type cells. This finding is supported by evidence showing TGF-β1 induces the production of platelet-derived growth factor (PDGF) from tubular epithelial cells, and promotes pericyte proliferation (Wu et al. 2013).

16.6 TGF-β/Smad-Targeted Therapy for Renal Fibrosis

16.6.1 Anti-fibrosis Therapy by General Blockade of TGF-β Signaling

Anti-TGF-β1 therapy has been thoroughly studied in fibrotic renal disease. Strategies to block it include TGF-β neutralizing antibodies, antisense oligodeoxynucleotides, soluble human TβRII (sTβRII.Fc), and specific inhibitors to TGF-β receptor (GW788388 and IN-1130) that protect against renal fibrosis in kidney disease models (Meng et al. 2015a, b). Importantly, TGF-β inhibitors have been applied in preclinical and clinical trials (Tampe and Zeisberg 2014). For example, Pirfenidone restored the estimated glomerular filtration rate (eGFR) in patients with diabetic nephropathy or focal segmental glomerulosclerosis (FSGS) by blocking TGF-β1 promoter (Cho et al. 2007; Sharma et al. 2011). Additional TGF-β neutralizers include Fresolimumab, identified in human FSGS, and LY2382770, tested in diabetic nephropathy (Tampe and Zeisberg 2014). Unfortunately, TGF-β-targeted therapy has failed due to generalized TGF-β signaling inhibition abolishes its anti-inflammatory and anti-tumorigenesis properties. For example, results from a randomized, double-blind, and phase II clinical study show that TGF-β1-specific humanized neutralizing monoclonal antibody failed to halt the progression of diabetic nephropathy (Voelker et al. 2017).

16.6.2 Anti-fibrosis Therapy Targeting Downstream Smads or Smad-Regulated MiRNAs

To avoid the side effects of global TGF-β1 signaling inhibition, anti-fibrosis therapy focuses on downstream effectors like Smad3, Smad7, and Smad-dependent miRNAs (Meng et al. 2016). Inhibition of Smad3 phosphorylation by SIS3, a Smad3 inhibitor, attenuates renal fibrosis in diabetic and obstructive nephropathy (Zhang et al. 2018). Treatment of BMP-7, a natural antagonist for TGF-β, prevents renal fibrosis in different types of renal diseases by inhibiting Smad3 activation (Zeisberg et al. 2003). GQ5, a small molecular phenolic compound isolated from dried resin of Toxicodendron vernicifluum, reduced the production of fibrotic indexes via inhibiting the interaction between TGF-β type I receptor and Smad3 (Ai et al. 2015). Additionally, as a natural inhibitor for Smad3, overexpression of Smad7 prevents both renal inflammation and fibrosis in different kidney disease models (Meng et al. 2016). A previous study showed that rebalancing Smad3/Smad7 ratio by combinational treatment of naringenin, a Smad3 inhibitor, and asiatic acid, a Smad7 agonist, provided added protection against UUO-induced renal fibrosis without enhancing cytotoxicity (Meng et al. 2015a, b). Accumulating evidence from in vivo studies also show that modification of certain Smad3-correlated miRNAs, like miR-21, miR-29, miR-192,

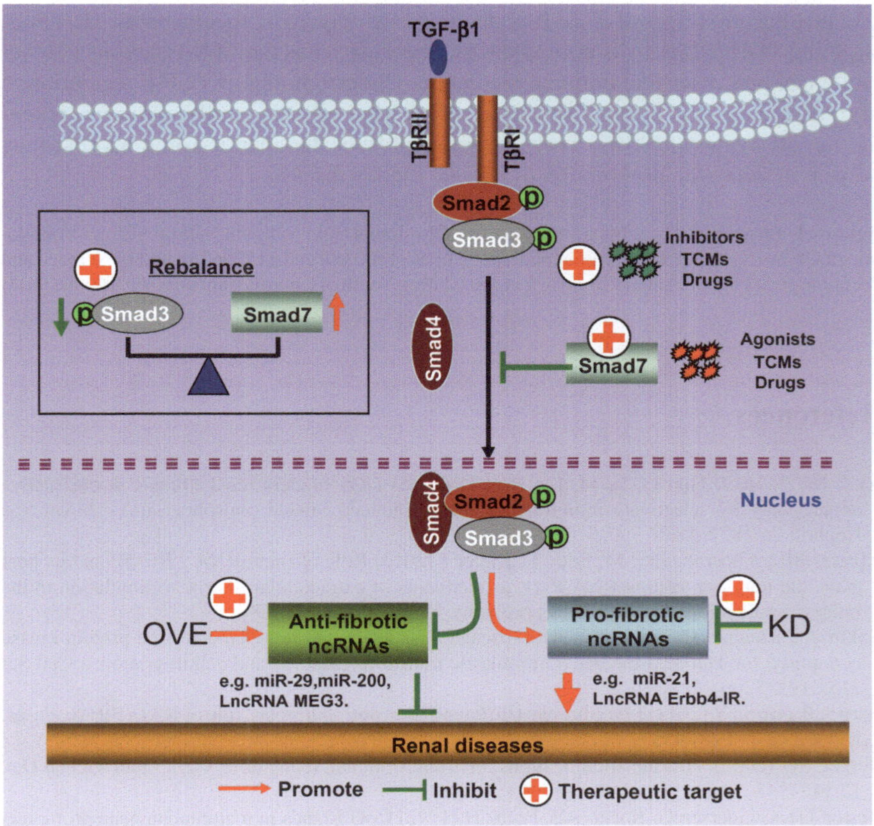

Fig. 16.1 Potential therapeutic targets in TGF-β/Smad signaling in renal fibrosis. As the downstream effectors of TGF-β1, Smad3 is pathogenic but Smad7 is protective in renal fibrosis. Targeting Smad3 with specific inhibitors or Smad3-dependent noncoding RNAs, and/or promoting Smad7 with gene therapy or specific agonists, are potential therapies for renal fibrotic disease. LncRNA: long noncoding RNA; TCM: Traditional Chinese Medicine

miR-200, and miR-433, represent promising therapeutic solutions for renal fibrosis (Zhong et al. 2011, 2013); (Qin et al. 2011; Chen et al. 2014; Meng et al. 2016) (shown in Fig. 16.1).

16.7 Conclusions and Perspectives

TGF-β signaling occurs through both canonical and non-canonical pathways to promote fibrosis. It has a significant influence on the progression of renal fibrosis, regulating crosstalk among a multitude of biological markers and signaling molecules.

The pro-fibrotic effect of TGF-β is mediated by epigenetic mechanisms. However, targeting TGF-β directly is not satisfactory because it has multiple functions in other biological processes like immune regulation. Understanding of TGF-β regulation in fibrotic diseases and mechanisms of its action have helped identify numerous potential targets against fibrosis. These targets represent therapeutic solutions that may prevent or delay the progression of fibrotic kidney disease.

Acknowledgements The laboratory is supported by grants from National Natural Science Foundation of China (National Science Foundation of China 81300580 and 81570623), and by Science and Technological Fund of Anhui Province for Outstanding Youth of China (Grant number: 1608085J07)

References

Ai J, Nie J, He J, Guo Q, Li M, Lei Y et al (2015) GQ5 hinders renal fibrosis in obstructive nephropathy by selectively inhibiting TGF-beta-Induced Smad3 phosphorylation. J Am Soc Nephrol 26:1827–1838

Alvarez ML, Khosroheidari M, Eddy E, Kiefer J (2013) Role of microRNA 1207-5P and its host gene, the long non-coding RNA Pvt1, as mediators of extracellular matrix accumulation in the kidney: implications for diabetic nephropathy. PLoS ONE 8:e77468

Bakin AV, Rinehart C, Tomlinson AK, Arteaga CL (2002) p38 mitogen-activated protein kinase is required for TGFbeta-mediated fibroblastic transdifferentiation and cell migration. J Cell Sci 115:3193–3206

Barnes JL, Gorin Y (2011) Myofibroblast differentiation during fibrosis: role of NAD(P)H oxidases. Kidney Int 79:944–956

Berger SL (2002) Histone modifications in transcriptional regulation. Curr Opin Genet Dev 12:142–148

Bestor TH, Gundersen G, Kolsto AB, Prydz H (1992) CpG islands in mammalian gene promoters are inherently resistant to de novo methylation. Genet Anal Tech Appl 9:48–53

Bhowmick NA, Ghiassi M, Bakin A, Aakre M, Lundquist CA, Engel ME et al (2001) Transforming growth factor-beta1 mediates epithelial to mesenchymal transdifferentiation through a RhoA-dependent mechanism. Mol Biol Cell 12:27–36

Bian EB, Huang C, Wang H, Chen XX, Zhang L, Lv XW et al (2014) Repression of Smad7 mediated by DNMT1 determines hepatic stellate cell activation and liver fibrosis in rats. Toxicol Lett 224:175–185

Border WA, Noble NA (1998) Evidence that TGF-beta should be a therapeutic target in diabetic nephropathy. Kidney Int 54:1390–1391

Border WA, Okuda S, Languino LR, Sporn MB, Ruoslahti E (1990) Suppression of experimental glomerulonephritis by antiserum against transforming growth factor beta 1. Nature 346:371–374

Bottinger EP, Bitzer M (2002) TGF-beta signaling in renal disease. J Am Soc Nephrol 13:2600–2610

Bucala R, Spiegel LA, Chesney J, Hogan M, Cerami A (1994) Circulating fibrocytes define a new leukocyte subpopulation that mediates tissue repair. Mol Med 1:71–81

Chau BN, Xin C, Hartner J, Ren S, Castano AP, Linn G et al (2012) MicroRNA-21 promotes fibrosis of the kidney by silencing metabolic pathways. Sci Transl Med 4:121ra118

Chen SJ, Yuan W, Mori Y, Levenson A, Trojanowska M, Varga J (1999) Stimulation of type I collagen transcription in human skin fibroblasts by TGF-beta: involvement of Smad 3. J Invest Dermatol 112:49–57

Chen HY, Huang XR, Wang W, Li JH, Heuchel RL, Chung AC et al (2011) The protective role of Smad7 in diabetic kidney disease: mechanism and therapeutic potential. Diabetes 60:590–601

Chen HY, Zhong X, Huang XR, Meng XM, You Y, Chung AC et al (2014) MicroRNA-29b inhibits diabetic nephropathy in db/db mice. Mol Ther 22:842–853

Cho ME, Smith DC, Branton MH, Penzak SR, Kopp JB (2007) Pirfenidone slows renal function decline in patients with focal segmental glomerulosclerosis. Clin J Am Soc Nephrol 2:906–913

Choi SY, Ryu Y, Kee HJ, Cho SN, Kim GR, Cho JY et al (2015) Tubastatin A suppresses renal fibrosis via regulation of epigenetic histone modification and Smad3-dependent fibrotic genes. Vascul Pharmacol 72:130–140

Chung AC, Lan HY (2015) MicroRNAs in renal fibrosis. Front Physiol 6:50

Chung AC, Huang XR, Zhou L, Heuchel R, Lai KN, Lan HY (2009) Disruption of the Smad7 gene promotes renal fibrosis and inflammation in unilateral ureteral obstruction (UUO) in mice. Nephrol Dial Transplant 24:1443–1454

Dennler S, Itoh S, Vivien D, ten Dijke P, Huet S, Gauthier JM (1998) Direct binding of Smad3 and Smad4 to critical TGF beta-inducible elements in the promoter of human plasminogen activator inhibitor-type 1 gene. EMBO J 17:3091–3100

Duffield JS, Lupher M, Thannickal VJ, Wynn TA (2013) Host responses in tissue repair and fibrosis. Annu Rev Pathol 8:241–276

Eddy AA, Neilson EG (2006) Chronic kidney disease progression. J Am Soc Nephrol 17:2964–2966

Ehrlich M, Gama-Sosa MA, Huang LH, Midgett RM, Kuo KC, McCune RA et al (1982) Amount and distribution of 5-methylcytosine in human DNA from different types of tissues of cells. Nucleic Acids Res 10:2709–2721

Falke LL, Gholizadeh S, Goldschmeding R, Kok RJ, Nguyen TQ (2015) Diverse origins of the myofibroblast-implications for kidney fibrosis. Nat Rev Nephrol 11:233–244

Feng M, Tang PM, Huang XR, Sun SF, You YK, Xiao J et al (2018) TGF-beta mediates renal fibrosis via the Smad3-Erbb4-IR long noncoding RNA axis. Mol Ther 26:148–161

Fujimoto M, Maezawa Y, Yokote K, Joh K, Kobayashi K, Kawamura H et al (2003) Mice lacking Smad3 are protected against streptozotocin-induced diabetic glomerulopathy. Biochem Biophys Res Commun 305:1002–1007

Fukasawa H, Yamamoto T, Togawa A, Ohashi N, Fujigaki Y, Oda T et al (2004) Down-regulation of Smad7 expression by ubiquitin-dependent degradation contributes to renal fibrosis in obstructive nephropathy in mice. Proc Natl Acad Sci U S A 101:8687–8692

Goldberg AD, Allis CD, Bernstein E (2007) Epigenetics: a landscape takes shape. Cell 128:635–638

Goumans MJ, Mummery C (2000) Functional analysis of the TGFbeta receptor/Smad pathway through gene ablation in mice. Int J Dev Biol 44:253–265

Grande MT, Lopez-Novoa JM (2009) Fibroblast activation and myofibroblast generation in obstructive nephropathy. Nat Rev Nephrol 5:319–328

Grande MT, Sanchez-Laorden B, Lopez-Blau C, De Frutos CA, Boutet A, Arevalo M et al (2015) Snail1-induced partial epithelial-to-mesenchymal transition drives renal fibrosis in mice and can be targeted to reverse established disease. Nat Med 21:989–997

Grgic I, Duffield JS, Humphreys BD (2012) The origin of interstitial myofibroblasts in chronic kidney disease. Pediatr Nephrol 27:183–193

He Y, Wu YT, Huang C, Meng XM, Ma TT, Wu BM et al (2014) Inhibitory effects of long noncoding RNA MEG3 on hepatic stellate cells activation and liver fibrogenesis. Biochim Biophys Acta 1842:2204–2215

Hou CC, Wang W, Huang XR, Fu P, Chen TH, Sheikh-Hamad D et al (2005) Ultrasound-microbubble-mediated gene transfer of inducible Smad7 blocks transforming growth factor-beta signaling and fibrosis in rat remnant kidney. Am J Pathol 166:761–771

Hsieh CL (1999) Evidence that protein binding specifies sites of DNA demethylation. Mol Cell Biol 19:46–56

Huang XR, Chung AC, Wang XJ, Lai KN, Lan HY (2008a) Mice overexpressing latent TGF-beta1 are protected against renal fibrosis in obstructive kidney disease. Am J Physiol Renal Physiol 295:F118–F127

Huang XR, Chung AC, Zhou L, Wang XJ, Lan HY (2008b) Latent TGF-beta1 protects against crescentic glomerulonephritis. J Am Soc Nephrol 19:233–242

Huang XZ, Wen D, Zhang M, Xie Q, Ma L, Guan Y et al (2014) Sirt1 activation ameliorates renal fibrosis by inhibiting the TGF-beta/Smad3 pathway. J Cell Biochem 115:996–1005

Inazaki K, Kanamaru Y, Kojima Y, Sueyoshi N, Okumura K, Kaneko K et al (2004) Smad3 deficiency attenuates renal fibrosis, inflammation, and apoptosis after unilateral ureteral obstruction. Kidney Int 66:597–604

Inoue Y, Itoh Y, Abe K, Okamoto T, Daitoku H, Fukamizu A et al (2007) Smad3 is acetylated by p300/CBP to regulate its transactivation activity. Oncogene 26:500–508

Jenuwein T, Allis CD (2001) Translating the histone code. Science 293:1074–1080

Jinde K, Nikolic-Paterson DJ, Huang XR, Sakai H, Kurokawa K, Atkins RC et al (2001) Tubular phenotypic change in progressive tubulointerstitial fibrosis in human glomerulonephritis. Am J Kidney Dis 38:761–769

Ka SM, Huang XR, Lan HY, Tsai PY, Yang SM, Shui HA et al (2007) Smad7 gene therapy ameliorates an autoimmune crescentic glomerulonephritis in mice. J Am Soc Nephrol 18:1777–1788

Ka SM, Yeh YC, Huang XR, Chao TK, Hung YJ, Yu CP et al (2012) Kidney-targeting Smad7 gene transfer inhibits renal TGF-beta/MAD homologue (SMAD) and nuclear factor kappaB (NF-kappaB) signalling pathways, and improves diabetic nephropathy in mice. Diabetologia 55:509–519

Kato M, Natarajan R (2014) Diabetic nephropathy–emerging epigenetic mechanisms. Nat Rev Nephrol 10:517–530

Kato M, Putta S, Wang M, Yuan H, Lanting L, Nair I et al (2009) TGF-beta activates Akt kinase through a microRNA-dependent amplifying circuit targeting PTEN. Nat Cell Biol 11:881–889

Kim JH, Kim BK, Moon KC, Hong HK, Lee HS (2003) Activation of the TGF-beta/Smad signaling pathway in focal segmental glomerulosclerosis. Kidney Int 64:1715–1721

Kolling M, Kaucsar T, Schauerte C, Hubner A, Dettling A, Park JK et al (2017) Therapeutic miR-21 silencing ameliorates diabetic kidney disease in mice. Mol Ther 25:165–180

Kopp JB, Factor VM, Mozes M, Nagy P, Sanderson N, Bottinger EP et al (1996) Transgenic mice with increased plasma levels of TGF-beta 1 develop progressive renal disease. Lab Invest 74:991–1003

Kriegel AJ, Fang Y, Liu Y, Tian Z, Mladinov D, Matus IR et al (2010) MicroRNA-target pairs in human renal epithelial cells treated with transforming growth factor beta 1: a novel role of miR-382. Nucleic Acids Res 38:8338–8347

Lakner AM, Steuerwald NM, Walling TL, Ghosh S, Li T, McKillop IH et al (2012) Inhibitory effects of microRNA 19b in hepatic stellate cell-mediated fibrogenesis. Hepatology 56:300–310

LeBleu VS, Taduri G, O'Connell J, Teng Y, Cooke VG, Woda C et al (2013) Origin and function of myofibroblasts in kidney fibrosis. Nat Med 19:1047–1053

Leung AKL (2015) The whereabouts of microRNA actions: cytoplasm and beyond. Trends Cell Biol 25:601–610

Li JH, Wang W, Huang XR, Oldfield M, Schmidt AM, Cooper ME et al (2004) Advanced glycation end products induce tubular epithelial-myofibroblast transition through the RAGE-ERK1/2 MAP kinase signaling pathway. Am J Pathol 164:1389–1397

Li J, Qu X, Ricardo SD, Bertram JF, Nikolic-Paterson DJ (2010a) Resveratrol inhibits renal fibrosis in the obstructed kidney: potential role in deacetylation of Smad3. Am J Pathol 177:1065–1071

Li J, Qu X, Yao J, Caruana G, Ricardo SD, Yamamoto Y et al (2010b) Blockade of endothelial-mesenchymal transition by a Smad3 inhibitor delays the early development of streptozotocin-induced diabetic nephropathy. Diabetes 59:2612–2624

Li R, Chung AC, Dong Y, Yang W, Zhong X, Lan HY (2013) The microRNA miR-433 promotes renal fibrosis by amplifying the TGF-beta/Smad3-Azin1 pathway. Kidney Int 84:1129–1144

Liang H, Xu C, Pan Z, Zhang Y, Xu Z, Chen Y et al (2014) The antifibrotic effects and mechanisms of microRNA-26a action in idiopathic pulmonary fibrosis. Mol Ther 22:1122–1133

Liu N, He S, Ma L, Ponnusamy M, Tang J, Tolbert E et al (2013) Blocking the class I histone deacetylase ameliorates renal fibrosis and inhibits renal fibroblast activation via modulating TGF-beta and EGFR signaling. PLoS ONE 8:e54001

Liu GX, Li YQ, Huang XR, Wei LH, Zhang Y, Feng M et al (2014) Smad7 inhibits AngII-mediated hypertensive nephropathy in a mouse model of hypertension. Clin Sci (Lond) 127:195–208

Lopez-Hernandez FJ, Lopez-Novoa JM (2012) Role of TGF-beta in chronic kidney disease: an integration of tubular, glomerular and vascular effects. Cell Tissue Res 347:141–154

Lovisa S, LeBleu VS, Tampe B, Sugimoto H, Vadnagara K, Carstens JL et al (2015) Epithelial-to-mesenchymal transition induces cell cycle arrest and parenchymal damage in renal fibrosis. Nat Med 21:998–1009

Luczak MW, Jagodzinski PP (2006) The role of DNA methylation in cancer development. Folia Histochem Cytobiol 44:143–154

Ma J, Xue M (2018) LINK-A lncRNA promotes migration and invasion of ovarian carcinoma cells by activating TGF-beta pathway. Biosci Rep 38:pii: BSR20180936

Mack M, Yanagita M (2015) Origin of myofibroblasts and cellular events triggering fibrosis. Kidney Int 87:297–307

Manickam N, Patel M, Griendling KK, Gorin Y, Barnes JL (2014) RhoA/Rho kinase mediates TGF-beta1-induced kidney myofibroblast activation through Poldip2/Nox4-derived reactive oxygen species. Am J Physiol Renal Physiol 307:F159–F171

Mariasegaram M, Tesch GH, Verhardt S, Hurst L, Lan HY, Nikolic-Paterson DJ (2010) Lefty antagonises TGF-beta1 induced epithelial-mesenchymal transition in tubular epithelial cells. Biochem Biophys Res Commun 393:855–859

Marquardt P, Muller-Hermelink HK (1990) Characteristics of T-lymphocytes infiltrating human B-cell lymphomas. Environ Health Perspect 88:233–235

Meng XM, Huang XR, Chung AC, Qin W, Shao X, Igarashi P et al (2010) Smad2 protects against TGF-beta/Smad3-mediated renal fibrosis. J Am Soc Nephrol 21:1477–1487

Meng XM, Huang XR, Xiao J, Chen HY, Zhong X, Chung AC et al (2012a) Diverse roles of TGF-beta receptor II in renal fibrosis and inflammation in vivo and in vitro. J Pathol 227:175–188

Meng XM, Huang XR, Xiao J, Chung AC, Qin W, Chen HY et al (2012b) Disruption of Smad4 impairs TGF-beta/Smad3 and Smad7 transcriptional regulation during renal inflammation and fibrosis in vivo and in vitro. Kidney Int 81:266–279

Meng XM, Chung AC, Lan HY (2013) Role of the TGF-beta/BMP-7/Smad pathways in renal diseases. Clin Sci (Lond) 124:243–254

Meng XM, Tang PM, Li J, Lan HY (2015a) TGF-beta/Smad signaling in renal fibrosis. Front Physiol 6:82

Meng XM, Zhang Y, Huang XR, Ren GL, Li J, Lan HY (2015b) Treatment of renal fibrosis by rebalancing TGF-beta/Smad signaling with the combination of asiatic acid and naringenin. Oncotarget 6:36984–36997

Meng XM, Nikolic-Paterson DJ, Lan HY (2016) TGF-beta: the master regulator of fibrosis. Nat Rev Nephrol 12:325–338

Mercer TR, Mattick JS (2013) Structure and function of long noncoding RNAs in epigenetic regulation. Nat Struct Mol Biol 20:300–307

Mishra R, Cool BL, Laderoute KR, Foretz M, Viollet B, Simonson MS (2008) AMP-activated protein kinase inhibits transforming growth factor-beta-induced Smad3-dependent transcription and myofibroblast transdifferentiation. J Biol Chem 283:10461–10469

Moon JA, Kim HT, Cho IS, Sheen YY, Kim DK (2006) IN-1130, a novel transforming growth factor-beta type I receptor kinase (ALK5) inhibitor, suppresses renal fibrosis in obstructive nephropathy. Kidney Int 70:1234–1243

Murakami K, Takemura T, Hino S, Yoshioka K (1997) Urinary transforming growth factor-beta in patients with glomerular diseases. Pediatr Nephrol 11:334–336

Ng YY, Huang TP, Yang WC, Chen ZP, Yang AH, Mu W et al (1998) Tubular epithelial-myofibroblast transdifferentiation in progressive tubulointerstitial fibrosis in 5/6 nephrectomized rats. Kidney Int 54:864–876

Ng YY, Fan JM, Mu W, Nikolic-Paterson DJ, Yang WC, Huang TP et al (1999) Glomerular epithelial-myofibroblast transdifferentiation in the evolution of glomerular crescent formation. Nephrol Dial Transplant 14:2860–2872

Nikolic-Paterson DJ, Wang S (2011) Lan HY (2014) Macrophages promote renal fibrosis through direct and indirect mechanisms. Kidney Int Suppl 4:34–38

Noh H, Oh EY, Seo JY, Yu MR, Kim YO, Ha H et al (2009) Histone deacetylase-2 is a key regulator of diabetes- and transforming growth factor-beta1-induced renal injury. Am J Physiol Renal Physiol 297:F729–F739

Okano M, Xie S, Li E (1998) Cloning and characterization of a family of novel mammalian DNA (cytosine-5) methyltransferases. Nat Genet 19:219–220

Oldfield MD, Bach LA, Forbes JM, Nikolic-Paterson D, McRobert A, Thallas V et al (2001) Advanced glycation end products cause epithelial-myofibroblast transdifferentiation via the receptor for advanced glycation end products (RAGE). J Clin Invest 108:1853–1863

Palumbo-Zerr K, Zerr P, Distler A, Fliehr J, Mancuso R, Huang J et al (2015) Orphan nuclear receptor NR4A1 regulates transforming growth factor-beta signaling and fibrosis. Nat Med 21:150–158

Pan Z, Sun X, Shan H, Wang N, Wang J, Ren J et al (2012) MicroRNA-101 inhibited postinfarct cardiac fibrosis and improved left ventricular compliance via the FBJ osteosarcoma oncogene/transforming growth factor-beta1 pathway. Circulation 126:840–850

Park JT, Kato M, Lanting L, Castro N, Nam BY, Wang M et al (2014) Repression of let-7 by transforming growth factor-beta1-induced Lin28 upregulates collagen expression in glomerular mesangial cells under diabetic conditions. Am J Physiol Renal Physiol 307:F1390–F1403

Petersen M, Thorikay M, Deckers M, van Dinther M, Grygielko ET, Gellibert F et al (2008) Oral administration of GW788388, an inhibitor of TGF-beta type I and II receptor kinases, decreases renal fibrosis. Kidney Int 73:705–715

Piek E, Ju WJ, Heyer J, Escalante-Alcalde D, Stewart CL, Weinstein M et al (2001) Functional characterization of transforming growth factor beta signaling in Smad2- and Smad3-deficient fibroblasts. J Biol Chem 276:19945–19953

Piera-Velazquez S, Li Z, Jimenez SA (2011) Role of endothelial-mesenchymal transition (EndoMT) in the pathogenesis of fibrotic disorders. Am J Pathol 179:1074–1080

Ponnusamy M, Zhuang MA, Zhou X, Tolbert E, Bayliss G, Zhao TC et al (2015) Activation of Sirtuin-1 promotes renal fibroblast activation and aggravates renal fibrogenesis. J Pharmacol Exp Ther 354:142–151

Pradhan S, Bacolla A, Wells RD, Roberts RJ (1999) Recombinant human DNA (cytosine-5) methyltransferase. I. Expression, purification, and comparison of de novo and maintenance methylation. J Biol Chem 274:33002–33010

Qin W, Chung AC, Huang XR, Meng XM, Hui DS, Yu CM et al (2011) TGF-beta/Smad3 signaling promotes renal fibrosis by inhibiting miR-29. J Am Soc Nephrol 22:1462–1474

Qiu C, Hanson RL, Fufaa G, Kobes S, Gluck C, Huang J et al (2018) Cytosine methylation predicts renal function decline in American Indians. Kidney Int 93:1417–1431

Reich B, Schmidbauer K, Rodriguez Gomez M, Johannes Hermann F, Gobel N, Bruhl H et al (2013) Fibrocytes develop outside the kidney but contribute to renal fibrosis in a mouse model. Kidney Int 84:78–89

Rius M, Lyko F (2012) Epigenetic cancer therapy: rationales, targets and drugs. Oncogene 31:4257–4265

Sato M, Muragaki Y, Saika S, Roberts AB, Ooshima A (2003) Targeted disruption of TGF-beta1/Smad3 signaling protects against renal tubulointerstitial fibrosis induced by unilateral ureteral obstruction. J Clin Invest 112:1486–1494

Schnaper HW, Hayashida T, Hubchak SC, Poncelet AC (2003) TGF-beta signal transduction and mesangial cell fibrogenesis. Am J Physiol Renal Physiol 284:F243–F252

Sharma K, Ix JH, Mathew AV, Cho M, Pflueger A, Dunn SR et al (2011) Pirfenidone for diabetic nephropathy. J Am Soc Nephrol 22:1144–1151

Shi Y, Massague J (2003) Mechanisms of TGF-beta signaling from cell membrane to the nucleus. Cell 113:685–700

Shi S, Srivastava SP, Kanasaki M, He J, Kitada M, Nagai T et al (2015) Interactions of DPP-4 and integrin beta1 influences endothelial-to-mesenchymal transition. Kidney Int 88:479–489

Shlyueva D, Stampfel G, Stark A (2014) Transcriptional enhancers: from properties to genome-wide predictions. Nat Rev Genet 15:272–286

Strieter RM, Keeley EC, Hughes MA, Burdick MD, Mehrad B (2009) The role of circulating mesenchymal progenitor cells (fibrocytes) in the pathogenesis of pulmonary fibrosis. J Leukoc Biol 86:1111–1118

Strutz F, Okada H, Lo CW, Danoff T, Carone RL, Tomaszewski JE et al (1995) Identification and characterization of a fibroblast marker: FSP1. J Cell Biol 130:393–405

Sun G, Reddy MA, Yuan H, Lanting L, Kato M, Natarajan R (2010) Epigenetic histone methylation modulates fibrotic gene expression. J Am Soc Nephrol 21:2069–2080

Sun SF, Tang PMK, Feng M, Xiao J, Huang XR, Li P et al (2018) Novel lncRNA Erbb4-IR promotes diabetic kidney injury in db/db mice by targeting miR-29b. Diabetes 67:731–744

Tahiliani M, Koh KP, Shen Y, Pastor WA, Bandukwala H, Brudno Y et al (2009) Conversion of 5-methylcytosine to 5-hydroxymethylcytosine in mammalian DNA by MLL partner TET1. Science 324:930–935

Tampe D, Zeisberg M (2014) Potential approaches to reverse or repair renal fibrosis. Nat Rev Nephrol 10:226–237

Tampe B, Tampe D, Muller CA, Sugimoto H, LeBleu V, Xu X et al (2014) Tet3-mediated hydroxymethylation of epigenetically silenced genes contributes to bone morphogenic protein 7-induced reversal of kidney fibrosis. J Am Soc Nephrol 25:905–912

Tang PM, Tang PC, Chung JY, Lan HY (2017) TGF-beta1 signaling in kidney disease: from Smads to long non-coding RNAs. Noncoding RNA Res 2:68–73

Tsuchida K, Zhu Y, Siva S, Dunn SR, Sharma K (2003) Role of Smad4 on TGF-beta-induced extracellular matrix stimulation in mesangial cells. Kidney Int 63:2000–2009

Tu X, Zhang H, Zhang J, Zhao S, Zheng X, Zhang Z et al (2014) MicroRNA-101 suppresses liver fibrosis by targeting the TGFbeta signalling pathway. J Pathol 234:46–59

van Meeteren LA, ten Dijke P (2012) Regulation of endothelial cell plasticity by TGF-beta. Cell Tissue Res 347:177–186

Voelker J, Berg PH, Sheetz M, Duffin K, Shen T, Moser B et al (2017) Anti-TGF-beta1 antibody therapy in patients with diabetic nephropathy. J Am Soc Nephrol 28:953–962

Wada T, Sakai N, Matsushima K, Kaneko S (2007) Fibrocytes: a new insight into kidney fibrosis. Kidney Int 72:269–273

Wang B, Herman-Edelstein M, Koh P, Burns W, Jandeleit-Dahm K, Watson A et al (2010) E-cadherin expression is regulated by miR-192/215 by a mechanism that is independent of the profibrotic effects of transforming growth factor-beta. Diabetes 59:1794–1802

Wang D, Dai C, Li Y, Liu Y (2011) Canonical Wnt/beta-catenin signaling mediates transforming growth factor-beta1-driven podocyte injury and proteinuria. Kidney Int 80:1159–1169

Wang B, Jha JC, Hagiwara S, McClelland AD, Jandeleit-Dahm K, Thomas MC et al (2014) Transforming growth factor-beta1-mediated renal fibrosis is dependent on the regulation of transforming growth factor receptor 1 expression by let-7b. Kidney Int 85:352–361

Wang S, Meng XM, Ng YY, Ma FY, Zhou S, Zhang Y et al (2016) TGF-beta/Smad3 signalling regulates the transition of bone marrow-derived macrophages into myofibroblasts during tissue fibrosis. Oncotarget 7:8809–8822

Wu CF, Chiang WC, Lai CF, Chang FC, Chen YT, Chou YH et al (2013) Transforming growth factor beta-1 stimulates profibrotic epithelial signaling to activate pericyte-myofibroblast transition in obstructive kidney fibrosis. Am J Pathol 182:118–131

Wu W, Chen F, Cui X, Yang L, Chen J, Zhao J et al (2018) LncRNA NKILA suppresses TGF-beta-induced epithelial-mesenchymal transition by blocking NF-kappaB signaling in breast cancer. Int J Cancer 43:2213–2224

Xavier S, Vasko R, Matsumoto K, Zullo JA, Chen R, Maizel J et al (2015) Curtailing endothelial TGF-beta signaling is sufficient to reduce endothelial-mesenchymal transition and fibrosis in CKD. J Am Soc Nephrol 26:817–829

Xiao X, Tang W, Yuan Q, Peng L, Yu P (2015) Epigenetic repression of Kruppel-like factor 4 through Dnmt1 contributes to EMT in renal fibrosis. Int J Mol Med 35:1596–1602

Xie H, Xue JD, Chao F, Jin YF, Fu Q (2016) Long non-coding RNA-H19 antagonism protects against renal fibrosis. Oncotarget 7:51473–51481

Xue R, Li Y, Li X, Ma J, An C, Ma Z (2018) miR-185 affected the EMT, cell viability and proliferation via DNMT1/MEG3 pathway in TGF-beta1-induced renal fibrosis. Cell Biol Int, 2018 Aug 10. https://doi.org/10.1002/cbin.11046 [Epub ahead of print]

Yamamoto T, Noble NA, Cohen AH, Nast CC, Hishida A, Gold LI et al (1996) Expression of transforming growth factor-beta isoforms in human glomerular diseases. Kidney Int 49:461–469

Yan X, Chen YG (2011) Smad7: not only a regulator, but also a cross-talk mediator of TGF-beta signalling. Biochem J 434:1–10

Yoshikawa M, Hishikawa K, Marumo T, Fujita T (2007) Inhibition of histone deacetylase activity suppresses epithelial-to-mesenchymal transition induced by TGF-beta1 in human renal epithelial cells. J Am Soc Nephrol 18:58–65

Yu F, Guo Y, Chen B, Dong P, Zheng J (2015) MicroRNA-17-5p activates hepatic stellate cells through targeting of Smad7. Lab Invest 95:781–789

Yuan W, Varga J (2001) Transforming growth factor-beta repression of matrix metalloproteinase-1 in dermal fibroblasts involves Smad3. J Biol Chem 276:38502–38510

Yuan H, Reddy MA, Sun G, Lanting L, Wang M, Kato M et al (2013) Involvement of p300/CBP and epigenetic histone acetylation in TGF-beta1-mediated gene transcription in mesangial cells. Am J Physiol Renal Physiol 304:F601–F613

Zeisberg M, Hanai J, Sugimoto H, Mammoto T, Charytan D, Strutz F et al (2003) BMP-7 counteracts TGF-beta1-induced epithelial-to-mesenchymal transition and reverses chronic renal injury. Nat Med 9:964–968

Zerr P, Palumbo-Zerr K, Huang J, Tomcik M, Sumova B, Distler O et al (2016) Sirt1 regulates canonical TGF-beta signalling to control fibroblast activation and tissue fibrosis. Ann Rheum Dis 75:226–233

Zhang Y, Meng XM, Huang XR, Lan HY (2018) The preventive and therapeutic implication for renal fibrosis by targeting TGF-beta/Smad3 signaling. Clin Sci (Lond) 132:1403–1415

Zhao Y, Yin Z, Li H, Fan J, Yang S, Chen C et al (2017) MiR-30c protects diabetic nephropathy by suppressing epithelial-to-mesenchymal transition in db/db mice. Aging Cell 16:387–400

Zhong X, Chung AC, Chen HY, Meng XM, Lan HY (2011) Smad3-mediated upregulation of miR-21 promotes renal fibrosis. J Am Soc Nephrol 22:1668–1681

Zhong X, Chung AC, Chen HY, Dong Y, Meng XM, Li R et al (2013) miR-21 is a key therapeutic target for renal injury in a mouse model of type 2 diabetes. Diabetologia 56:663–674

Zhou L, Fu P, Huang XR, Liu F, Chung AC, Lai KN et al (2010) Mechanism of chronic aristolochic acid nephropathy: role of Smad3. Am J Physiol Renal Physiol 298:F1006–F1017

Zhou VW, Goren A, Bernstein BE (2011) Charting histone modifications and the functional organization of mammalian genomes. Nat Rev Genet 12:7–18

Zhou Q, Chung AC, Huang XR, Dong Y, Yu X, Lan HY (2014) Identification of novel long noncoding RNAs associated with TGF-beta/Smad3-mediated renal inflammation and fibrosis by RNA sequencing. Am J Pathol 184:409–417

Chapter 17
Connective Tissue Growth Factor and Renal Fibrosis

Qing Yin and Hong Liu

Abstract CCN2, also known as connective tissue growth factor (CTGF), is one of important members of the CCN family. Generally, CTGF expresses at low levels in normal adult kidney, while increases significantly in various kidney diseases, playing an important role in the development of glomerular and tubulointerstitial fibrosis in progressive kidney diseases. CTGF is involved in cell proliferation, migration, and differentiation and can promote the progression of fibrosis directly or act as a downstream factor of transforming growth factor β (TGF-β). CTGF also regulates the expression and activity of TGF-β and bone morphogenetic protein (BMP), thereby playing an important role in the process of kidney repair. In patients with chronic kidney disease, elevated plasma CTGF is an independent risk factor for progression to end-stage renal disease and is closely related to glomerular filtration rate. Therefore, CTGF may be a potential biological marker of kidney fibrosis, but more clinical studies are needed to confirm this view. This section briefly describes the role and molecular mechanisms of CTGF in renal fibrosis and also discusses the potential value of targeting CCN2 for the treatment of renal fibrosis.

Keywords Connective tissue growth factor · Renal fibrosis · TGF-β

17.1 Introduction

Renal fibrosis is the final common pathway leading to end-stage renal failure. The pathogenesis of renal fibrosis is highly complex and is not known completely, but it involves the activation of tubular interstitial cells, inflammatory cell infiltration, renal tubular epithelial cell transdifferentiating into fibroblasts, apoptosis of renal epithelial cells, and excessive expression of cytokines, growth factors, and other factors. Histologically, renal fibrosis is characterized by excessive extracellular matrix accumulation in the glomerulus and interstitium, as well as vascular hyalinosis and

Q. Yin · H. Liu (✉)
Institute of Nephrology, Zhong Da Hospital, Southeast University School of Medicine, Nanjing, China
e-mail: jstzliu@sina.com

B.-C. Liu et al. (eds.), *Renal Fibrosis: Mechanisms and Therapies*, Advances in Experimental Medicine and Biology 1165,
https://doi.org/10.1007/978-981-13-8871-2_17

sclerosis and destruction of renal tissue structures, eventually leading to the loss of functioning nephrons. Transforming growth factor β (TGF-β) may be the most important cytokine that promotes renal fibrosis. CTGF is a downstream molecule of TGF-β and plays an important role in the renal fibrosis.

17.2 Connective Tissue Growth Factor (CTGF or CCN2)

CCN2, also known as CTGF, is one of the members of the CCN family (CCN 1-6). Human CTGF/CCN2 (hCTGF/CCN2) was initially described in 1991 by Bradham et al. in media conditioned with cultured human vascular endothelial cells (Bradham 1991). It was subsequently detected in fibroblasts, chondrocytes, tumor cells, endothelial cells, cartilaginous cells, and smooth muscle cells, and it is expressed in many human tissues, such as heart, lung, brain, kidney, liver, and placenta. It is most abundant in the kidney (Wu et al. 2017; Xing et al. 2018; Behnes et al. 2014). It is a member of the CCN family, which consists of six regulatory proteins, including cysteine-rich angiogenic inducer 61 (CYR 61 or CCN1), nephroblastoma overexpressed gene (NOV or CCN3), and the Wnt-inducible signaling pathway proteins 1–3 (WISP 1–3 or CCN 4–6) (Perbal and Perbal 2016). The CCN family is a group of polypeptide factors that regulate the growth of fibroblasts and epidermal cells and share a high degree of homology with each other.

CTGF is a 36- to 38-kD cysteine-rich peptide containing 349 amino acids. hCTGF gene maps to chromosome 6q23.1. It contains 5 exons and 4 introns of varying sizes, with an ATG start site, 3 ATTTA sites at its 3′ end, and a primary amino acid sequence showing 5 exons encoding its secretion signal peptide domains (Hall-Glenn and Lyons 2011). CTGF is a cysteine-rich matricellular protein that contains an N-terminal secretory peptide and has four multi-functional domains. These domains carry a diverse array of binding partners that potentially impact multiple signaling mechanisms. The first domain is homologous to insulin-like growth factor binding proteins (IGFBPs) but has low affinity for IGF. The second domain encodes a von Willebrand type C (VWC) repeat, which mediates the interaction of CCN2 with αvβ3, αvβ5, and growth factors such as bone morphogenic proteins (BMPs) and TGF-β. IGFBP and VWC domains bind to the major proteoglycan aggrecan produced by chondrocytes. The third domain is a type 1 thrombospondin (TSP) repeat protein. It mediates the ability of TSP to bind to ECM, matrix metalloproteinase (MMP), and integrin. Moreover, this domain regulates the interaction of CCN2 with VEGF and low-density lipoprotein receptor-related protein-1 (LRP1). The final, C-terminal (CT) motif contains cysteine knots similar to those found in many growth factors, including members of the TGF-β superfamily, platelet-derived growth factor (PDGF), and nerve growth factor (NGF). This domain also exists in other secreted proteins, including WISE, slit, and mucins. It mediates the interaction of CCN2 with LRP6, fibronectin, basement membrane proteoglycan, and fibulin-1. The secretion of CCN2 into the extracellular space depends on an N-terminal 37 amino acid signal sequence (Fig. 17.1).

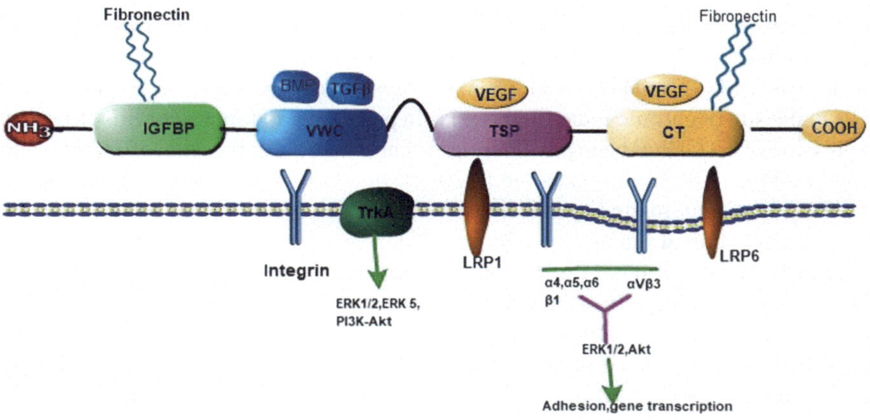

Fig. 17.1 Domain structure and amino acid sequence of the CTGF proteins. This figure summarizes the binding of various growth factors and receptors to CCN2, where the domains that interact with TrkA receptors are not known. IGFBP—insulin-like growth factor binding protein domain; VWC—von Willebrand factor C domain; TSP-1—thrombospondin type 1 repeat domain; CT—C-terminal domain with cysteine knot motif; and CTGF—connective tissue growth factor

17.3 Biological Characteristics of CTGF

The CCN2 domains bind to receptors to exert multiple biological functions, including promoting cell chemotaxis, migration, cellular adhesion, proliferation, differentiation, and extracellular matrix (ECM) synthesis (Xing et al. 2018; Hall-Glenn and Lyons 2011). The biological characteristics change with the target cell. For example, CTGF can promote fibroblast proliferation and ECM synthesis. It can participate in the process of endothelial cell migration, angiogenesis and lymphangiogenesis, as well as the proliferation and differentiation of chondrocytes (Montford and Furgeson 2017). CTGF can modify the expression of other molecules, such as TGF-β1, VEGF, BMP4, and BMP7 (Falke et al. 2017; Kiwanuka et al. 2017). Another important function of CTGF is to promote wound healing, mainly through its C-terminal domain binding to fibronectin (FN), enhancing the binding of FN to fibrin, thereby promoting wound healing (Mendes et al. 2015).

CTGF is mainly expressed in adult tissues and is elevated in pathological conditions such as fibrosis, atherosclerosis, osteoarthritis, and certain cancers (Chen and Lau 2009). CTGF is overexpressed in inflammatory bowel disease, hepatic fibrosis and idiopathic pulmonary fibrosis, skin lesions, and renal fibrosis, revealing that CTGF may participate in the pathogenesis of fibrosis (Dendooven et al. 2011). It is widely considered necessary for mediating TGF-β-induced fibrosis. Furthermore, urinary CTGF is elevated in many renal fibrotic disorders, including diabetic nephropathy (DN), chronic allograft nephropathy, IgA nephropathy, focal segmental glomerulosclerosis, crescentic nephritis, class IV lupus nephritis, and anti-Thy-1.1 nephritis (Wang et al. 2015; Tu et al. 2007; Nonaka Takahashi et al. 2008; Smeets et al. 2006; Hilhorst et al. 2015). CTGF is significantly upregulated in mesangial cells, epithelial cells, and endothelial cells in IgA nephropathy, focal stage glomerulosclerosis (FSGS), and DN. In addition, its expression increases in the mesenchymal cells at the site of chronic interstitial injury. In patients with lupus nephritis (LN), the expression of CTGF has been positively correlated with the expression of TGF-β1 and type I collagen (Tachaudomdach et al. 2012). In addition, the increased expression of CTGF is closely related to a decreased glomerular filtration rate and consequent deterioration of renal function. Therefore, CTGF expression in the kidney may serve as an early marker of renal disease progression in LN patients. It is speculated that CTGF may promote glomerular and renal interstitial fibrosis in various kidney diseases (Ito et al. 2010). In recent years, more and more studies on CTGF have shown that CTGF is both a mediator and a marker of tissue fibrosis. Therefore, CTGF may become a clinically useful, non-invasive tool for monitoring tissue fibrosis, and a therapeutic target for treating tissue fibrosis.

17.4 Connective Tissue Growth Factor and Renal Fibrosis

Multiple lines of evidence support the pro-fibrotic role of CCN2 in renal fibrosis: (i) CCN2 is upregulated in people with renal fibrosis; (ii) CCN2 inhibition can significantly improve the renal fibrosis in UUO model, the residual kidney model, and the experimental model of DN; and (iii) CCN2 mediates pro-fibrotic effects of advanced glycation end products (AGE) (Chung et al. 2010). Experimental studies in mice suggest that CTGF plays an important role in the development of renal fibrosis. CTGF has effects on many cell types in the kidney, including tubular epithelial mesenchymal transdifferentiation, stimulation of extracellular matrix deposition, induction of mesangial cell cycle arrest and hypertrophy, prolonging survival of activated cells, transition of quiescent renal cells to a myofibroblastic phenotype, and recruitment of inflammatory cells (Hall-Glenn and Lyons 2011). The glomerular CTGF mRNA and protein levels gradually increase with the progression of glomerulosclerosis. In addition, renal biopsies of CKD patients of different etiological origins showed that the expression of CTGF in these sites was increased in patients with glomerulosclerosis or interstitial fibrosis, indicating that CTGF plays an important role in renal fibrosis (Ito et al. 1998). Subsequently, Suzuki et al. indicated that CTGF may play an important role in the development and progression of glomerulosclerosis in DN, immunoglobulin A nephropathy, which are both accompanied by mesangial matrix expansion and end-stage renal failure (Suzuki et al. 2003). Animal models of kidney disease are also accompanied by changes in the expression of CTGF. For example, the expression of CTGF increases in diabetic rat kidney at the early stage, which might be an important mediator of renal hypertrophy through arresting cell cycling (Liu et al. 2006b). Rayego-Mateos et al. reported that CTGF regulates fibrosis and proliferation events of renal and in vitro fibroblasts and tubular epithelial cells via the EGFR pathway (Rayego-Mateos et al. 2018).

Renal interstitial fibrosis plays an important role in the progression of chronic kidney disease. Currently, it is believed that renal interstitial fibrosis is related to the occurrence and development of renal dysfunction and is an important indicator for predicting clinical prognosis. Ito et al. reported that at the site of chronic tubulointerstitial lesions, the expression of CTGF mRNA was increased and correlated with the degree of renal injury (Ito et al. 1998). Unilateral ureteric obstruction (UUO) in rodents is used as a model of CKD (Klahr and Morrissey 2002). After performing UUO for seven days, the number of myofibroblasts in the interstitium is obviously increased, accompanied by increased levels of type I collagen, EDA-fibronectin (a splice variant associated with fibrosis), fibronectin, and CTGF in these molecules. Administration of CTGF antisense oligonucleotide (ASO) via the renal vein could significantly attenuate these molecular events and improve the development of fibrosis in UUO (Chen et al. 2006). In another study, subcutaneous injection of CTGF-ASO in mice with streptozotocin-induced type 1 diabetes and mutant db/db mice with type 2 diabetes for 16 weeks. The mesangial matrix of type 1 mice is expanded in kidney cortex with increased mRNA levels of CTGF, fibronectin, collagen I, and TGF-β. The increase in the glomerular size of ASO-treated animals was attenuated, and the mRNA levels returned to those of controls. Similar effects have been found in the db/db mice, and improvements in kidney function tests were also noted (Guha et al. 2007). Yasuhiko Ito et al. found that tissue, urine, and plasma CTGF are important biomarkers for the progression of hypertensive glomerulosclerosis in humans. In addition, local overexpression of CTGF may be a key factor in hypertensive vascular injury and renal fibrosis (Ito et al. 2010). In general, these studies provide convincing evidence that CTGF plays a key role both in glomerulosclerosis and tubulointerstitial fibrosis.

CTGF mRNA in glomeruli is markedly increased in diabetic transgenic mice compared to controls, but mRNAs for matrix encoding genes are not significantly increased. However, the expression of matrix metalloproteinase 2 (MMP2) is increased in control diabetic mice, while the transgenic group shows a decrease. The difference in MMP2 expression is accompanied by a decrease in MMP2 activity in the transgenic mice. Therefore, the increased expression of CTGF associated with mesangial matrix expansion is likely to be due to decreased matrix degradation rather than increased matrix protein expression. Interestingly, not only is the CTGF protein expressed by the podocytes of diabetic transgenic mice increased, but the mesangial area is also increased. Thus, podocyte transgenic CTGF is likely to have paracrine effects on cells in the mesangial area (Yokoi et al. 2008).

17.5 Regulation of CTGF Expression

Many fibrogenic factors, such as TGF-β, angiotensin II (Ang II), endothelin, blood glucose, thrombin, traction and oxidative stress, can participate in the development and progression of renal fibrosis by inducing the expression of CTGF (Wong et al. 2018). TGF-β is thought to be the most important factor.

17.5.1 TGF-β and CTGF

TGF-β has long been recognized as a central mediator in the development of renal inflammation and fibrosis. In recent years, studies have shown that CTGF is an important downstream factor of TGF-β that induces fibrosis. There are three isoforms of TGF-β: TGF-β1, TGF-β2, and TGF-β3 (Ma et al. 2014; Kiwanuka et al. 2017; Wong et al. 2018). TGF-β exerts its various biological and immunological effects through complex signaling pathways. CTGF binds to TGF-β, promoting interaction with its receptor and increasing downstream signaling. CTGF and TGF-β act in a cooperative manner rather than in sequence to promote fibrosis (Yokoi et al. 2008). For instance, repeated daily subcutaneous injections of TGF-β1, TGF-β2, TGF-β3 or CTGF alone into the skin of neonatal mice failed to induce fibrosis. However, when TGF-β and CTGF were injected together, or when TGF-β was injected for 3 days, followed by CTGF for 4 days, persistent fibrosis developed (Mori et al. 1999). Another report indicated that when CTGF and TGF-β2 were injected intraperitoneally together, the peritoneal membranes of neonatal mice developed severe fibrosis, but injection of either TGF-β2 alone or CTGF alone did not induce fibrosis (Wang et al. 2011). Some scholars believe TGF-β1 can induce short-term fibrosis alone, while CTGF is necessary for the maintenance of chronic persistent fibrosis. Therefore, CTGF is not only a downstream factor of TGF-β but also acts synergistically with TGF-β, and its activity directly or indirectly regulates the expression of TGF-β in a positive-feedback manner.

The molecular mechanism by which these two cytokines interact with each other in driving fibrosis is not clear in vivo. In the TGF-β canonical signaling pathway, Smad2 and Smad3 are phosphorylated following TGF-β binding to its receptor (ALK5) and then combine with Smad4 and translocate to the nucleus in a complex, where they bind to the promoters of TGF-β response genes, stimulating CTGF transcription (Miyazawa et al. 2002). In addition, CTGF can accelerate the signal transduction by promoting the process of TGF-β binding to its receptors. CTGF mediates TGF-β's induction of glomerulosclerosis mainly by promoting the proliferation of mesangial cells and the synthesis of ECM proteins.

In a unilateral ureteral obstruction (UUO) rat model, TGF-β1 increased first and CTGF increased subsequently. During the course of disease progression, the expression of CTGF, PAI-1, and collagen increased continuously, and the CTGF protein was significantly increased in the fibrotic region and tubular epithelial cells. This was correlated with tubule-interstitial injury index, TGF-β1 and PAI-1 (Samarakoon and Higgins 2018). This suggested that CTGF may be induced by TGF-β1 in the tubule interstitium and mediate a variety of pro-fibrotic effects, such as increased ECM synthesis and reduced ECM degradation. In the case of persistent expression of TGF-β1, intravenous injection of CTGF antisense oligonucleotides can effectively block the expression of CTGF in proximal tubular epithelial cells and reduce the expression of ECM molecules and inhibit renal interstitial fibrosis eventually (Chen et al. 2006).

17.5.2 Hepatocyte Growth Factor and CTGF

Hepatocyte growth factor (HGF) is expressed widely in tissues, including in interstitial and endothelial cells of the kidney. Proximal tubular epithelial cells (PTECs) express the c-Met receptor for HGF. In 5/6-nephrectomized TGF-β1-transgenic mice, treatment with HGF reduces CTGF expression and alleviates glomerulosclerosis significantly. The addition of recombinant CTGF does not stimulate the expression of TGF-β1 mRNA in renal interstitial fibroblasts or mesangial cells, but inhibition of CTGF expression alleviates the degree of glomerulosclerosis. There are also reports that high-dose HGF can reduce the expression of CTGF induced by TGF-β1 and alleviate the progression of renal tubular interstitial fibrosis in the 5/6-nephrectomized model eventually. Treatment with HGF significantly reduces the CTGF expression and attenuates the expression of collagen and the development of fibrosis. This suggests that the protective effect of HGF on renal fibrosis relates to its regulation of TGF-β1 and CTGF. Furthermore, HGF is able to attenuate TGF-β-stimulated fibrogenic responses by downregulating CTGF.

17.5.3 Bone Morphogenetic Factor and CTGF

Bone morphogenetic factor-4 (BMP4) is a member of the TGF-β superfamily. CTGF binds directly to BMP4 at a higher affinity than it does for TGF-β. In the presence of CTGF, the interaction of BMP4 with its receptor and downstream signaling through Smad 1 phosphorylation are reduced, which are due to the direct binding between CTGF and BMP4 (Abreu et al. 2002). In fact, overexpression of BMP4 in transgenic mice causes significant glomerulosclerosis and proteinuria, similar to the changes in DN. Compared to $BMP4^{+/+}$ mice, glomerular sclerosis is reduced in diabetic $BMP4^{+/-}$ mice, confirming the role of BMP4 in this disease (Tominaga et al. 2011). The interaction between CTGF and BMP4 in DN remains to be explored, but based on the results of Abreu et al., CTGF seems unlikely to significantly inhibit the effects of BMP4 (Abreu et al. 2002).

Bone morphogenetic factor-7 is another member of the TGF-β superfamily. It also binds to CTGF (Nguyen et al. 2008). The administration of BMP7 reduces glomerulosclerosis in several different animal models of renal fibrosis. Wang and Hirschberg (2003) demonstrated that this factor antagonizes the TGF-β-induced mesangial cell fibrotic response, including reduced fibronectin and type IV collagen accumulation, CTGF expression, PAI-1 promoter activity, and MMP2 gelatinase activity (Wang and Hirschberg 2003). Similar to the expression of BMP type II receptors and ALK2 type I receptors, bone morphogenic factor-7 is normally expressed in renal tubular epithelial cells but is absent in STZ-induced diabetes. Therefore, in DN, as well as other CKDs, higher levels of CTGF are associated with decreased BMP7-signaling and disease progression.

17.5.4 Ang II and CTGF

Accumulating evidence has shown that the activation of the intrarenal RAS is involved in many kinds of renal diseases, which may play a key role in mediating the progression of chronic renal scarring. After systemic infusion of Ang II into normal rats for 3 days, the levels of renal CTGF mRNA and protein increase. At day 7, Ang II-infused rats overexpress CTGF in glomeruli, renal tubules, and renal arteries, and renal tubular injury and fibronectin deposition also increase (Ruperez et al. 2003). Liu et al. found that Ang II directly induces CTGF mRNA and protein expression in HK2 cells, and more interestingly, they demonstrated that the Ang II-induced increase of cell size, protein synthesis, and increase in total protein content can be significantly prevented by co-treatment with CTGF-AS. This strongly suggests that early expression of CTGF in kidney tissue is related to local activation of RAS (Liu et al. 2006a). Ang II can induce mesangial cell proliferation and increase the synthesis of extracellular matrix. Liu et al. demonstrated that Ang II induces the expression of CTGF mRNA and protein in a dose- and time-dependent manner, which suggests that Ang II could directly stimulate the expression and synthesis of CTGF in HK2 cells (Wong et al. 2018). AT1 receptor antagonists can reduce the overexpression of CTGF and FN and can alleviate renal damage.

17.5.5 High Glucose and Advanced Glycation End Products

High glucose–stimulated human mesangial cells can increase the expression of TGF-β1 and CTGF, while IFN-γ can activate STAT1 and phosphorylate it, inhibit the proliferation of mesangial cells, and downregulate the expression of TGF-β1 and CTGF mRNA and protein (Dai et al. 2016).

Podocytes undergo epithelial–mesenchymal transition (EMT) in DN. CTGF and integrin-linked kinase (ILK) are involved in the progression of DN. Under stimulation with high glucose, CTGF and ILK expression in podocytes increase in a dose- and time-dependent manner, whereas the increase does not occur in osmotic control cells. Morphological analysis of epithelial cells shows that inhibition of CTGF using anti-CTGF antibody can prevent the phenotypic transition (Dai et al. 2012).

A prospective study showed that in patients with type 1 diabetes, plasma CTGF was higher in patients with large amounts of proteinuria than in patients with normal proteinuria (Roestenberg et al. 2004). Elevated CTGF in circulating plasma is an independent risk factor for end-stage renal disease (ESRD) progression and death. Another study showed that the level of CTGF in urine was significantly higher in patients with type 1 diabetes with microalbuminuria or proteinuria than in patients with normal proteinuria, and there was a positive correlation between albuminuria and the level of CTGF secretion (Adler et al. 2010).

AGEs can stimulate the synthesis of ECM and play an important role in DN. In diabetic kidneys, CTGF is one of the important induction factors for the increased synthesis of the ECM (Twigg et al. 2001). After rats are intravenously injected with AGEs, the glomerular extracellular matrix accumulates and is associated with increased expression of CTGF (Wang et al. 2011). However, in vitro stimulation of rat mesangial cells with AGEs can increase the production of FN and type IV collagen, which can be completely blocked by pretreatment with anti-CTGF antibodies. These findings suggest that CTGF plays a key role in the progression of glomerulosclerosis in DN. Interestingly, inhibition of TGF-β1 mRNA expression with shRNA or neutralization of TGF-β1 protein with anti-TGF-β1 antibody does not completely block AGE-induced increases in CTGF expression, suggesting that AGE-induced increases in CTGF expression are not completely dependent on TGF-β1 (Chung et al. 2010).

17.5.6 Epithelial–Mesenchymal Transition and CTGF

Tubular epithelial–mesenchymal transdifferentiation (EMT) plays an important role in the development of tubulointerstitial fibrosis (Dai et al. 2016). During EMT, tubule epithelial cells lose their cell–cell junctions and epithelial cell phenotype, obtaining the phenotypes of α-SMA and myofibroblasts, coupled with the destruction of the basement membrane, make it easier for tubule epithelial cells to migrate into the stroma to cause matrix deposition and fibrosis.

Liu et al. used CTGF-AS to demonstrate the relationship between Ang II and CTGF in the process of Ang II-induced EMT. They demonstrated that Ang II induces the overexpression of CTGF mRNA and protein. Furthermore, co-treatment with CTGF-AS prevents the trans-differential responses induced by Ang II, including the expression of α-SMA in HK-2 cells and the expression of mesenchymal features in the ultrastructure of cells. Their data strongly suggest that CTGF might be an important mediator of the development of tubular EMT induced by Ang II (Liu et al. 2003). In vivo studies have found that in the progress of DN, CTGF is co-localized with EMT marker proteins in the kidney. Therefore, some scholars believe that the EMT induced by high glucose, AGEs, and TGF-β1 is partially mediated by CTGF (Twigg et al. 2001). In addition, stimulation of human tubular epithelial cells with recombinant human CTGF significantly increases the expression of α-SMA and tenascin C, while the expression of type IV collagen decreases, and CTGF antisense oligonucleotides significantly inhibit this effect. The above studies indicate that CTGF is an independent mediator of tubule EMT.

17.5.7 Stimulation of Extracellular Matrix Deposition

Tubulointerstitial fibrosis involves the loss of renal tubules and accumulation of ECM proteins, such as collagen (types I and III), fibronectin, and laminin. Recent studies have suggested that the major effector cells that are responsible for this excessive deposition by ECM are myofibroblasts. CTGF serves as an adaptor molecule connecting the cell surface and is considered to be the major protein of the extracellular matrix. CTGF also upregulates the expression of integrins on the cell surface and promotes the deposition of ECM (Liu et al. 2014). In mesangial cells, CTGF can induce the production of plasminogen activator inhibitor 1 (PAI-1), actin cytoskeletal rearrangement, and tissue metalloproteinase inhibitor 1 (TIMP1) increase, resulting in reduced matrix degradation (Zhang et al. 2015). CTGF can promote the transcription of α1 type I collagen, α5 integrin, and FN in renal interstitial fibroblasts and ultimately increase protein synthesis. In cultured human proximal tubular epithelial cells, CTGF antisense oligonucleotides can significantly inhibit the expression of PAI 1 and increase the degradation of ECM by plasminogen activators, thereby inhibiting ECM aggregation (Chen et al. 2006).

17.6 Cellular Receptors for CTGF and Downstream Signaling

Stimulation of cells with CTGF can cause the activation of multiple signaling cell molecules. Several CTGF receptors have been reported, but there is no specific cytokine. For example, CTGF can rapidly activate several intracellular signaling molecules in human mesangial cells (HMC), including extracellular signal-associated kinase 1/2, Jun NH2-terminal kinase, protein kinase B, CaMK II, protein kinase Cα, and protein kinase Cδ, indicating that it functions through signaling receptors (Wahab et al. 2005). These receptors include low-density lipoprotein receptor-related protein-1 (LRP-1; (Segarini et al. 2001)), low-density lipoprotein receptor-associated protein-6 [LRP-6; (Rooney et al. 2011)], tropomyosin-related kinases A (Wahab et al. 2005), and several adhesion receptors, such as β3 integrin (Crean et al. 2002), $\alpha_V\beta3$ integrins (Gao and Brigstock 2004), α5β1 integrins (Gao and Brigstock 2006), and αMβ2 integrins (Schober et al. 2002).

Convincing evidence has confirmed that LRP is a receptor for CTGF: (1) CTGF shows binding competition with many ligands that bind to LRP, including receptor-associated protein; (2) a CTGF-receptor complex is immunoprecipitated with LRP antibodies; (3) cells that are genetically deficient for LRP are unable to bind CTGF; and (4) CTGF is rapidly internalized and degraded, and this process is LRP-dependent (Segarini et al. 2001). LRP-1 is thought to be the main regulator of the plasma membrane proteome and binds to other plasma membrane proteins by bridging proteins and subsequently adapts to the main microenvironment through endocytosis and lysosomal degradation. CTGF but not TGF-β induces phosphorylation of LRP-1 in NRK-49F cells. RAP inhibits this activation and reduces the synergistic action of CTGF on TGF-β-induced α-SMA expression. Although whether this was downstream of LRP-1 phosphorylation has not been studied, the synergy of CTGF depends on its activation of the ERK1/2 signaling pathway (Yang et al. 2004). CTGF induces serine phosphorylation of LRP-6, and then can bind to the reporter gene TCF/LEF-luciferase to affect the transcription activity of β-catenin. Rooney et al.

identified several upregulated Wnt target genes in human nephropathy biopsy, UUO, and streptozotocin-induced diabetic animal models (Rooney et al. 2011). However, further studies are necessary to determine the exact role of the CTGF-activated LRP-6 pathway in mesangial cells.

Tropomyosin-related kinase A is a member of the neurotrophin receptor family (TrkA, TrkB, and TrkC). It is a receptor tyrosine kinase that is activated by ligand-induced formation of noncovalently associated receptor dimers (Schecterson and Bothwell 2010). After activation of some G protein-coupled receptors, such as A2a adenosine receptors and LRP-1 (Schecterson and Bothwell 2010), Trk receptors may be trans-activated by the Src-related kinase Fyn. In a subsequent study on TrkA involvement in DN, high glucose conditions were shown to induce CTGF in mesangial cells that activate downstream phosphorylation of TrkA and ERK1/2. These steps are blocked by knockdown of CTGF or TrkA siRNA, confirming the specific involvement and importance of CTGF-TrkA-primed signals in response to high glucose (Fragiadaki et al. 2012). CTGF induces TGF-β-inducible early response gene-1 (TIEG-1) in mesangial cells, and this action is inhibited by K252a and therefore may be mediated through TrkA activation (Wahab et al. 2005) (Fig. 17.2).

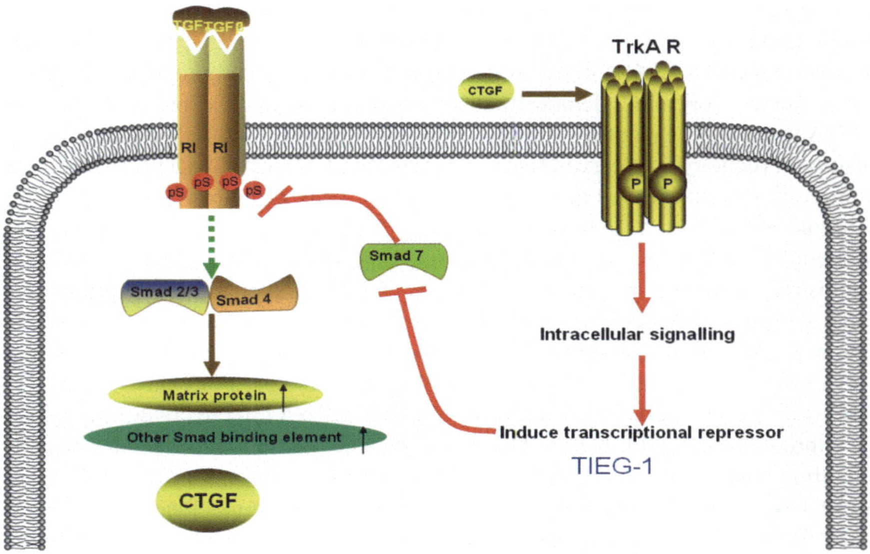

Fig. 17.2 Connective tissue growth factor (CTGF) stimulates the TGFβ-Smad signaling pathway. CTGF enhances TGF-β-induced mesangial cell phosphorylation and Smad2 and Smad3 nuclear translocation. Antisense oligonucleotides against TIEG-1 prevent CTGF-induced downregulation of Smad7. TIEG-1, TGF-β-inducible early response gene-1

17.7 Potential Therapeutic Value of CTGF

At present, the treatment fibrosis mainly focuses on anti-TGF-β approaches, but TGF-β has many types of target cells, and the efficacy is complex. Therefore, it is not feasible to completely block the expression or activity of TGF-β. For instance, mice depleted of TGF-β1 die of systemic inflammation soon after birth because they lose their inhibition of the inflammatory process. In contrast, CTGF expression is very low under normal conditions, and it is mainly expressed in mesenchymal cells. Its effect is limited to connective tissue. Therefore, blocking CTGF expression or inhibiting its biological activity may be more specific and more effective to treat fibrosis. The experiments in mice provide proof of principle that anti-CTGF therapies are successful in attenuating the development of renal fibrosis.

Hewitson et al. demonstrated that theobromine can reduce the expression of α-SMA and CTGF and inhibit the proliferation of fibroblasts in a dose-dependent manner, thereby reducing the progression of fibrosis and scar formation in rats with unilateral ureteral obstruction, suggesting that CTGF may become a new anti-fibrotic medium (Leask 2013). The HMG-CoA reductase inhibitor statin can interfere with the expression of CTGF mRNA and CTGF induced by LPA and TGF-β by blocking the isoprenylation of RhoA in human renal fibroblasts. This shows that statins are potent inhibitors of CTGF mRNA in mesangial cells. Therefore, in glomerular fibrosis, overexpression of CTGF by mesangial cells can be strongly regulated by statin drugs. Liu and colleagues demonstrated that Ang II induces the expression of CTGF mRNA and protein in a dose- and time-dependent manner, and AT1 receptor antagonists can reduce the overexpression of CTGF and FN and alleviate renal damage. Moreover, a phase 1 trial of the effects of an anti-CTGF antibody in human diabetics with microalbuminuria (indicative of early renal changes) showed not only that the antibody was safe to use but also that it reduced microalbuminuria (Adler et al. 2010). Decreased expression of systemic CTGF can ameliorate proteinuria and glomerular damage in anti-GBM nephritis by inhibiting inflammation and reducing extracellular matrix accumulation (Toda et al. 2017). Inhibition of CTGF can ameliorate peritoneal fibrosis by inhibiting the accumulation of fibroblasts and myofibroblasts and angiogenesis (Sakai et al. 2017). The delivery of CTGF siRNA effectively downregulates the expression of renal CTGF in UUO in a sequence-specific manner, significantly inhibiting renal tubular cell apoptosis and tubulointerstitial fibrosis (Ren et al. 2015). Anti-CCN2 therapy using fully humanized neutralizing monoclonal antibodies is currently in clinical trials in DN patients. The anti-fibrotic effects of CCN2 inhibition in patients with renal fibrosis will clearly require future clinical studies to further confirm, but it may be the first specific available anti-fibrotic treatment.

These findings suggest that CTGF and its receptors, as well as their activation of signaling pathways, provide a series of targets for new therapeutic interventions for fibrotic nephropathy.

17.8 Conclusions

The results of this study indicate that CTGF can participate in the process of renal fibrosis by promoting cell proliferation and ECM accumulation. A variety of cytokines, high glucose, Ang II, and other factors can promote the progression of renal fibrosis by inducing the expression of CTGF. CTGF is not only a mediator of tissue fibrosis but also a potential biological marker and a potential target for the treatment of renal fibrosis. More clinical studies are needed to assess the safety and efficacy of CTGF-neutralizing antibodies in patients with diabetes or non-diabetic kidney disease. Further studies of the role of CTGF in renal fibrosis are required, as they will have important implications for the prevention and treatment of renal fibrosis.

Acknowledgements This study was supported by grants from the National Key Research and Development Program of China (2018YFC1314004).

References

Abreu JG, Ketpura NI, Reversade B, De Robertis EM (2002) Connective-tissue growth factor (CTGF) modulates cell signalling by BMP and TGF-beta. Nat Cell Biol 4:599–604

Adler SG, Schwartz S, Williams ME, Arauz-Pacheco C, Bolton WK, Lee T et al (2010) Phase 1 study of anti-CTGF monoclonal antibody in patients with diabetes and microalbuminuria. Clin J Am Soc Nephrol 5:1420–1428

Behnes M, Brueckmann M, Lang S, Weiss C, Ahmad-Nejad P, Neumaier M et al (2014) Connective tissue growth factor (CTGF/CCN2): diagnostic and prognostic value in acute heart failure. Clin Res Cardiol 103:107–116

Bradham DM (1991) Connective tissue growth factor: a cysteine-rich mitogen secreted by human vascular endothelial cells is related to the SRC-induced immediate early gene product CEF-10. The Journal of Cell Biology 114:1285–1294

Chen CC, Lau LF (2009) Functions and mechanisms of action of CCN matricellular proteins. Int J Biochem Cell Biol 41:771–783

Chen L, Liu BC, Zhang XL, Zhang JD, Liu H, Li MX (2006) Influence of connective tissue growth factor antisense oligonucleotide on angiotensin II-induced epithelial mesenchymal transition in HK2 cells. Acta Pharmacol Sin 27:1029–1036

Chung AC, Zhang H, Kong YZ, Tan JJ, Huang XR, Kopp JB et al (2010) Advanced glycation end-products induce tubular CTGF via TGF-beta-independent Smad3 signaling. J Am Soc Nephrol 21:249–260

Crean JK, Finlay D, Murphy M, Moss C, Godson C, Martin F et al (2002) The role of p42/44 MAPK and protein kinase B in connective tissue growth factor induced extracellular matrix protein production, cell migration, and actin cytoskeletal rearrangement in human mesangial cells. J Biol Chem 277:44187–44194

Dai HY, Zheng M, Lv LL, Tang RN, Ma KL, Liu D et al (2012) The roles of connective tissue growth factor and integrin-linked kinase in high glucose-induced phenotypic alterations of podocytes. J Cell Biochem 113:293–301

Dai H, Zhang Y, Yuan L, Wu J, Ma L, Shi H (2016) CTGF mediates high-glucose induced epithelial-mesenchymal transition through activation of beta-catenin in podocytes. Ren Fail 38:1711–1716

Dendooven A, Gerritsen KG, Nguyen TQ, Kok RJ, Goldschmeding R (2011) Connective tissue growth factor (CTGF/CCN2) ELISA: a novel tool for monitoring fibrosis. Biomarkers 16:289–301

Falke LL, Leeuwis JW, Lyons KM, Mummery CL, Nguyen TQ, Goldschmeding R (2017) CCN2 reduction mediates protective effects of BMP7 treatment in obstructive nephropathy. J Cell Commun Signal 11:39–48

Fragiadaki M, Hill N, Hewitt R, Bou-Gharios G, Cook T, Tam FW et al (2012) Hyperglycemia causes renal cell damage via CCN2-induced activation of the TrkA receptor: implications for diabetic nephropathy. Diabetes 61:2280–2288

Gao R, Brigstock DR (2004) Connective tissue growth factor (CCN2) induces adhesion of rat activated hepatic stellate cells by binding of its C-terminal domain to integrin alpha(v)beta(3) and heparan sulfate proteoglycan. J Biol Chem 279:8848–8855

Gao R, Brigstock DR (2006) A novel integrin alpha5beta1 binding domain in module 4 of connective tissue growth factor (CCN2/CTGF) promotes adhesion and migration of activated pancreatic stellate cells. Gut 55:856–862

Guha M, Xu ZG, Tung D, Lanting L, Natarajan R (2007) Specific down-regulation of connective tissue growth factor attenuates progression of nephropathy in mouse models of type 1 and type 2 diabetes. FASEB J 21:3355–3368

Hall-Glenn F, Lyons KM (2011) Roles for CCN2 in normal physiological processes. Cell Mol Life Sci 68:3209–3217

Hilhorst M, Kok HM, Broekhuizen R, van Paassen P, van Breda Vriesman P, Goldschmeding R et al (2015) Connective tissue growth factor and the cicatrization of cellular crescents in ANCA-associated glomerulonephritis. Nephrol Dial Transplant 30:1291–1299

Ito Y, Aten J, Bende RJ, Oemar BS, Rabelink TJ, Weening JJ et al (1998) Expression of connective tissue growth factor in human renal fibrosis. Kidney Int 53:853–861

Ito Y, Goldschmeding R, Kasuga H, Claessen N, Nakayama M, Yuzawa Y et al (2010) Expression patterns of connective tissue growth factor and of TGF-beta isoforms during glomerular injury recapitulate glomerulogenesis. Am J Physiol Renal Physiol 299:F545–F558

Kiwanuka E, Junker JP, Eriksson E (2017) Transforming growth factor beta1 regulates the expression of CCN2 in human keratinocytes via Smad-ERK signalling. Int Wound J 14:1006–1018

Klahr S, Morrissey J (2002) Obstructive nephropathy and renal fibrosis. Am J Physiol Renal Physiol 283:F861–F875

Leask A (2013) CCN2: a novel, specific and valid target for anti-fibrotic drug intervention. Expert Opin Ther Targets 17:1067–1071

Liu BC, Sun J, Chen Q, Ma KL, Ruan XZ, Phillips AO (2003) Role of connective tissue growth factor in mediating hypertrophy of human proximal tubular cells induced by angiotensin II. Am J Nephrol 23:429–437

Liu BC, Chen L, Sun J, Huang HQ, Ma KL, Liu H et al (2006a) Connective tissue growth factor-mediated angiotensin II-induced hypertrophy of proximal tubular cells. Nephron Exp Nephrol 103:e16–e26

Liu BC, Huang HQ, Luo DD, Ma KL, Liu DG, Liu H (2006b) Connective tissue growth factor is associated with the early renal hypertrophy in uninephrectomized diabetic rats. Chin Med J 119:1010–1016

Liu Y, Li W, Liu H, Peng Y, Yang Q, Xiao L et al (2014) Inhibition effect of small interfering RNA of connective tissue growth factor on the expression of extracellular matrix molecules in cultured human renal proximal tubular cells. Ren Fail 36:278–284

Ma B, Kang Q, Qin L, Cui L, Pei C (2014) TGF-beta2 induces transdifferentiation and fibrosis in human lens epithelial cells via regulating gremlin and CTGF. Biochem Biophys Res Commun 447:689–695

Mendes FA, Coelho Aguiar JM, Kahn SA, Reis AH, Dubois LG, Romao LF et al (2015) Connective-tissue growth factor (CTGF/CCN2) induces astrogenesis and fibronectin expression of embryonic neural cells in vitro. PLoS ONE 10:e0133689

Miyazawa K, Shinozaki M, Hara T, Furuya T, Miyazono K (2002) Two major Smad pathways in TGF-beta superfamily signalling. Genes Cells 7:1191–1204
Montford JR, Furgeson SB (2017) A new CTGF target in renal fibrosis. Kidney Int 92:784–786
Mori T, Kawara S, Shinozaki M, Hayashi N, Kakinuma T, Igarashi A et al (1999) Role and interaction of connective tissue growth factor with transforming growth factor-beta in persistent fibrosis:a mouse fibrosis model. J Cell Physiol 181:153–159
Nguyen TQ, Roestenberg P, van Nieuwenhoven FA, Bovenschen N, Li Z, Xu L et al (2008) CTGF inhibits BMP-7 signaling in diabetic nephropathy. J Am Soc Nephrol 19:2098–2107
Nonaka Takahashi S, Fujita T, Takahashi T, Wada Y, Fuke Y, Satomura A et al (2008) TGF-beta1 and CTGF mRNAs are correlated with urinary protein level in IgA nephropathy. J Nephrol 21:53–63
Perbal A, Perbal B (2016) The CCN family of proteins: a 25th anniversary picture. J Cell Commun Signal 10:177–190
Rayego-Mateos S, Morgado-Pascual JL, Rodrigues-Diez RR, Rodrigues-Diez R, Falke LL, Mezzano S et al (2018) Connective tissue growth factor induces renal fibrosis via epidermal growth factor receptor activation. J Pathol 244:227–241
Ren Y, Du C, Yan L, Wei J, Wu H, Shi Y et al (2015) CTGF siRNA ameliorates tubular cell apoptosis and tubulointerstitial fibrosis in obstructed mouse kidneys in a Sirt1-independent manner. Drug Des Devel Ther 9:4155–4171
Roestenberg P, van Nieuwenhoven FA, Wieten L, Boer P, Diekman T, Tiller AM et al (2004) Connective tissue growth factor is increased in plasma of type 1 diabetic patients with nephropathy. Diabetes Care 27:1164–1170
Rooney B, O'Donovan H, Gaffney A, Browne M, Faherty N, Curran SP et al (2011) CTGF/CCN2 activates canonical Wnt signalling in mesangial cells through LRP6: implications for the pathogenesis of diabetic nephropathy. FEBS Lett 585:531–538
Ruperez M, Ruiz-Ortega M, Esteban V, Lorenzo O, Mezzano S, Plaza JJ et al (2003) Angiotensin II increases connective tissue growth factor in the kidney. Am J Pathol 163:1937–1947
Sakai N, Nakamura M, Lipson KE (2017) Inhibition of CTGF ameliorates peritoneal fibrosis through suppression of fibroblast and myofibroblast accumulation and angiogenesis. 7:5392
Samarakoon R, Higgins PJ (2018) The cytoskeletal network regulates expression of the profibrotic genes PAI-1 and CTGF in vascular smooth muscle cells. Adv Pharmacol 81:79–94
Schecterson LC, Bothwell M (2010) Neurotrophin receptors: old friends with new partners. Dev Neurobiol 70:332–338
Schober JM, Chen N, Grzeszkiewicz TM, Jovanovic I, Emeson EE, Ugarova TP et al (2002) Identification of integrin alpha(M)beta(2) as an adhesion receptor on peripheral blood monocytes for Cyr61 (CCN1) and connective tissue growth factor (CCN2): immediate-early gene products expressed in atherosclerotic lesions. Blood 99:4457–4465
Segarini PR, Nesbitt JE, Li D, Hays LG, Yates JR 3rd, Carmichael DF (2001) The low density lipoprotein receptor-related protein/alpha2-macroglobulin receptor is a receptor for connective tissue growth factor. J Biol Chem 276:40659–40667
Smeets B, Steenbergen ML, Dijkman HB, Verrijp KN, te Loeke NA, Aten J et al (2006) Angiotensin converting enzyme inhibition prevents development of collapsing focal segmental glomerulosclerosis in Thy-1.1 transgenic mice. Nephrol Dial Transplant 21:3087–3097
Suzuki D, Toyoda M, Umezono T, Uehara G, Zhang SY, Sakai T et al (2003) Glomerular expression of connective tissue growth factor mRNA in various renal diseases. Nephrology (Carlton) 8:92–97
Tachaudomdach C, Kantachuvesiri S, Changsirikulchai S, Wimolluck S, Pinpradap K, Kitiyakara C (2012) Connective tissue growth factor gene expression and decline in renal function in lupus nephritis. Exp Ther Med 3:713–718
Toda N, Mori K, Kasahara M, Ishii A, Koga K, Ohno S et al (2017) Crucial role of mesangial cell-derived connective tissue growth factor in a mouse model of anti-glomerular basement membrane glomerulonephritis. Sci Rep 7:42114
Tominaga T, Abe H, Ueda O, Goto C, Nakahara K, Murakami T et al (2011) Activation of bone morphogenetic protein 4 signaling leads to glomerulosclerosis that mimics diabetic nephropathy. J Biol Chem 286:20109–20116

Tu Z, Shi Y, Wang J, Bao J, Bu H (2007) Upregulation of connective tissue growth factor in a rat model of chronic allograft nephropathy. Nephrology (Carlton) 12:166–171

Twigg SM, Chen MM, Joly AH, Chakrapani SD, Tsubaki J, Kim HS et al (2001) Advanced glycosylation end products up-regulate connective tissue growth factor (insulin-like growth factor-binding protein-related protein 2) in human fibroblasts: a potential mechanism for expansion of extracellular matrix in diabetes mellitus. Endocrinology 142:1760–1769

Wahab NA, Weston BS, Mason RM (2005) Connective tissue growth factor CCN2 interacts with and activates the tyrosine kinase receptor TrkA. J Am Soc Nephrol 16:340–351

Wang S, Hirschberg R (2003) BMP7 antagonizes TGF-beta -dependent fibrogenesis in mesangial cells. Am J Physiol Renal Physiol 284:F1006–F1013

Wang X, McLennan SV, Twigg SM (2011) CCN-2 is up-regulated by and mediates effects of matrix bound advanced glycated end-products in human renal mesangial cells. J Cell Commun Signal 5:193–200

Wang S, Li B, Li C, Cui W, Miao L (2015) Potential renoprotective agents through inhibiting CTGF/CCN2 in diabetic nephropathy. J Diabetes Res 2015:962383

Wong CKS, Falkenham A, Myers T, Legare JF (2018) Connective tissue growth factor expression after angiotensin II exposure is dependent on transforming growth factor-beta signaling via the canonical Smad-dependent pathway in hypertensive induced myocardial fibrosis. J Renin Angiotensin Aldosterone Syst 19:1470320318759358

Wu YL, Li HY, Zhao XP, Jiao JY, Tang DX, Yan LJ et al (2017) Mesenchymal stem cell-derived CCN2 promotes the proliferation, migration and invasion of human tongue squamous cell carcinoma cells. Cancer Sci 108:897–909

Xing X, Li Z, Yu Z, Cheng G, Li D, Li Z (2018) Effects of connective tissue growth factor (CTGF/CCN2) on condylar chondrocyte proliferation, migration, maturation, differentiation and signalling pathway. Biochem Biophys Res Commun 495:1447–1453

Yang M, Huang H, Li J, Li D, Wang H (2004) Tyrosine phosphorylation of the LDL receptor-related protein (LRP) and activation of the ERK pathway are required for connective tissue growth factor to potentiate myofibroblast differentiation. FASEB J 18:1920–1921

Yokoi H, Mukoyama M, Mori K, Kasahara M, Suganami T, Sawai K et al (2008) Overexpression of connective tissue growth factor in podocytes worsens diabetic nephropathy in mice. Kidney Int 73:446–455

Zhang Q, Shu FL, Jiang YF, Huang XE (2015) Influence of expression plasmid of connective tissue growth factor and tissue inhibitor of metalloproteinase-1 shRNA on hepatic precancerous fibrosis in rats. Asian Pac J Cancer Prev 16:7205–7210

Chapter 18
Inflammatory Mediators and Renal Fibrosis

Xiao-Ming Meng

Abstract Renal inflammation is the initial, healthy response to renal injury. However, prolonged inflammation promotes the fibrosis process, which leads to chronic pathology and eventually end-stage kidney disease. There are two major sources of inflammatory cells: first, bone marrow-derived leukocytes that include neutrophils, macrophages, fibrocytes and mast cells, and second, locally activated kidney cells such as mesangial cells, podocytes, tubular epithelial cells, endothelial cells and fibroblasts. These activated cells produce many profibrotic cytokines and growth factors that cause accumulation and activation of myofibroblasts, and enhance the production of the extracellular matrix. In particular, activated macrophages are key mediators that drive acute inflammation into chronic kidney disease. They produce large amounts of profibrotic factors and modify the microenvironment via a paracrine effect, and they also transdifferentiate to myofibroblasts directly, although the origin of myofibroblasts in the fibrosing kidney remains controversial. Collectively, understanding inflammatory cell functions and mechanisms during renal fibrosis is paramount to improving diagnosis and treatment of chronic kidney disease.

Keywords Macrophage · Renal fibrosis · Myofibroblast · TGF-β

18.1 Introduction

Renal fibrosis is the consequence of a diverse range of kidney diseases and is characterized by excessive accumulation of fibroblasts and extracellular matrix (ECM) (Djudjaj and Boor 2018; Humphreys 2018; Liu 2006; Zeisberg and Neilson 2010). It is also the common pathological pathway leading to end-stage renal disease (ESRD) (Eddy and Neilson 2006; Meng et al. 2013). However, the mechanisms of renal fibrosis remain largely unclear. Indeed, current treatments are non-specific and ineffective. Thus, it is essential to better understand the inflammatory process from acute injury to chronic kidney disease (CKD).

X.-M. Meng (✉)
School of Pharmacy, Anhui Medical University, Hefei, Anhui, China
e-mail: mengxiaoming@ahmu.edu.cn

B.-C. Liu et al. (eds.), *Renal Fibrosis: Mechanisms and Therapies*,
Advances in Experimental Medicine and Biology 1165,
https://doi.org/10.1007/978-981-13-8871-2_18

Renal inflammation serves as the initial response to kidney stress or injury, and it protects the kidney from further damage (Meng et al. 2014). However, unresolved inflammation destroys kidney structure and function, leading to CKD that is characterized by progressive renal fibrosis (Grande et al. 2010; Lee and Kalluri 2010). Several key steps are involved in this process: first, the renal injury itself, followed by recruitment of inflammatory cells and release of profibrotic mediators that trigger accumulation and activation of myofibroblasts, then production and deposition of extracellular matrix (ECM), followed finally by glomerulosclerosis and tubular atrophy that is accompanied by microvascular rarefaction (Lee and Kalluri 2010; Liu 2011).

Regardless of the initial trigger, renal inflammation is characterized by glomerular and tubulointerstitial infiltration of inflammatory cells like neutrophils, T cells, macrophages, fibrocytes and mast cells (Meng et al. 2014). Activated intrinsic kidney cells, such as mesangial cells (MCs), podocytes, tubular epithelial cells (TECs) and endothelial cells, function as inflammatory mediators that produce proinflammatory cytokines, chemokines and growth factors and also participate in the repair process (Boor et al. 2010). After prolonged inflammation, histological and functional dysfunction occurs from excessive fibroblast and myofibroblast accumulation and ECM production. In fact, accumulation of myofibroblasts is the predominant source for collagen production and the dominant event in renal fibrosis progression. However, the origin of myofibroblasts remains largely controversial. Some research shows epithelial–mesenchymal transition (EMT) and endothelial–mesenchymal transition (EndoMT) are sources of myofibroblasts (Allison 2013; LeBleu et al. 2013; Liu 2011). But, resident fibroblasts also contribute to renal fibrosis (Duffield and Humphreys 2011; Lin et al. 2008; Strutz and Zeisberg 2006). In addition, accumulating evidence suggests pericytes are a major source of myofibroblasts (Humphreys et al. 2010; Lin et al. 2008). And, bone marrow-derived cells like fibrocytes and macrophages also contribute to the local accumulation of myofibroblasts (Broekema et al. 2007; Meng et al. 2016b; Wang et al. 2016b, 2017b). Thus, understanding the origin of myofibroblasts may help to clarify the pathway from renal inflammation to fibrosis. The current chapter focuses on key inflammatory cell types and pathways linking renal inflammation with fibrosis.

18.2 Bone Marrow-Derived Inflammatory Cells in Renal Fibrosis

When renal injury occurs, circulating leukocytes are recruited to the kidney (Chung and Lan 2011). Lymphocytes, monocytes, macrophages, mast cells and fibrocytes produce tissue damage factors like reactive oxygen species, cytokines and growth factors (Boor et al. 2010; Grande et al. 2010; Lee and Kalluri 2010). These profibrotic mediators cause myofibroblast accumulation and ECM production. Collectively, unresolved inflammation drives the profibrotic pathways that lead to fibrosis (Kanasaki et al. 2013; Lee and Kalluri 2010).

18.2.1 Macrophages in Renal Inflammation and Fibrosis

In the injured kidney, monocytes transdifferentiate into macrophages in response to oxidative stress, toxins and hypoxia. This represents a critical component of the mononuclear phagocyte system (Anders and Ryu 2011; Vernon et al. 2010). Macrophages are divided into two subtypes, classically activated (M1) and alternatively activated (M2). Macrophages adopt the M1 phenotype when stimulated by interferon-γ (IFN-γ) and lipopolysaccharide (LPS). M2 phenotype is induced by interleukin (IL)-4 or IL-13. M2 macrophages are further categorized into three subsets according to their response to stimuli. M2a, wound healing macrophages, are induced by IL-4 and IL-13; M2b are induced by immune complexes; M2c, regulatory macrophages, are induced by IL-10, TGF-β or glucocorticoid. During the progression from kidney injury to repair processes, inflammatory M1 macrophages switch toward M2 phenotype and exert distinct functions supporting the pathological process (Anders and Ryu 2011). Accordingly, the proinflammatory M1 macrophages generate TNF-α, IL-1β and reactive oxygen species in response to renal injury. By contrast, the M2 macrophages secrete anti-inflammatory cytokines like IL-10 and insulin-like growth factor-1 that promote tissue healing and angiogenesis (Ricardo et al. 2008).

Macrophages are key inflammatory cells in proliferative glomerulonephritis (GN), particularly in crescentic GN (Han et al. 2011; Ma et al. 2009a, 2010). Macrophages are recruited from either circulation or local environment, which correlates with the severity of glomerular and tubulointerstitial damage, and renal function impairment, and is prognostic of disease progression (Eardley et al. 2008; Lan et al. 1995; Yang et al. 1998a). The findings that show co-localization of α-SMA + myofibroblasts and proliferative macrophages in areas of severe renal damage indicate a correlation between macrophages and renal fibrosis (Yang et al. 1998b). This is underscored by results from biopsies of patients with CKD and fibrotic kidney of animal models showing a similar relationship between macrophage recruitment and disease severity (Eardley et al. 2008; Nishida and Hamaoka 2008).

Inflammatory macrophages are a major source of proinflammatory cytokines, such as TNF-α and IL-1β, and chemokines like MCP-1(Ma et al. 2010). They also secrete macrophage migration inhibitory factor (MIF), a critical inflammatory mediator that promotes the progression of kidney diseases (Lan et al. 1997). Moreover, macrophages produce a number of growth factors like TGF-β and PDGF that play key roles linking renal inflammation with fibrosis.

To date, the profibrotic role of macrophages has been studied exclusively by various depletion techniques. They showed macrophage deficiency attenuated renal fibrosis in various disease models, including ischemic acute kidney injury (Ko et al. 2008), crescentic glomerulonephritis (Han et al. 2013), membranoproliferative glomerulonephritis (Guo et al. 2011), obstructive nephropathy (Kitamoto et al. 2009) and diabetic nephropathy(Lin et al. 2009; You et al. 2013). In contrast, reconstitution of macrophages significantly aggravates existing fibrotic lesions. Much research has focused on the regulatory mechanisms of macrophages during the profibrotic effect. For example, a recent study revealed that genetic ablation of galectin-3, a β-galactosidase-binding lectin, protected against renal fibrosis in a unilateral ureteral obstruction (UUO) model without affecting the number of infiltrated macrophages. And, adoptive transfer of macrophages from wild-type mice, instead of galectin-3 KO mice, restored renal fibrosis (Henderson et al. 2008). Another study demonstrated that CpG-oligodeoxynucleotides-induced activation of TLR9 on macrophages accelerated interstitial fibrogenesis (Anders et al. 2004). This indicates the toll-like receptor-mediated signals on macrophages play critical roles in the progression of renal fibrosis. In addition, pharmacological inhibition of c-fms, the monocyte/macrophage receptor for CSF1, reduced the number of renal macrophages and tubular apoptosis, but had marginal impact on renal fibrosis (Han et al. 2013; Ma et al. 2009b). A recent study showed that macrophage-specific cyclooxygenase (Cox)-2 polarized and maintained a macrophage tissue-reparative M2 phenotype that protected against diabetic nephropathy (Wang et al. 2017a). This evidence shows that macrophages are a key bridge linking renal inflammation and renal fibrosis through several mechanisms. First, M1 macrophages secrete chemokines, cytokines and matrix metalloproteinases (MMPs) to accelerate the excessive infiltration of leukocytes into the injured region, destroying the original architecture of the kidney and aggravating renal injury (Ricardo et al. 2008). Second, cytokines and growth factors released by macrophages facilitate the proliferation and activation of resident fibroblasts. Macrophages promote transdifferentiation of other cells into myofibroblast-like cells through paracrine signaling. Macrophages are a major source of TGF-β1 in pathological conditions, and increased release enhances the number of infiltrated myofibroblasts by triggering the transdifferentiation of epithelial cells, endothelial cells and activation of pericytes and resident fibroblasts. M2 macrophages release IGF-1 and PDGF that promote myofibroblast survival (Floege et al. 2008; Wynes et al. 2004). Macrophage-derived PDGF also targets local fibroblasts and pericytes where the receptors for PDGF are highly expressed (Lin et al. 2008). And, macrophage-secreted MMP-9 induces EMT. Third, macrophages support extracellular matrix framework containing fibrinogen and collagens (Gratchev et al. 2001; Schnoor et al. 2008). Fourth, inflammatory macrophages induce vascular injury and capillary rarefaction, leading to tissue hypoxia and progression of renal fibrosis (Fine and Norman 2008). Finally, macrophages directly transdifferentiate into myofibroblast-like cells. Indeed, cells co-expressing the markers for both macrophage (CD68 +) and myofibroblast (α-SMA+) have been detected in patients with progressive renal fibrosis and animal models of obstructive nephropathy and chronic renal allograft injury (Wang et al. 2016b, 2017b). Furthermore, myeloid macrophages labeled with red fluorescence using lineage-tracing techniques show that bone marrow-derived macrophages, particularly M2, directly transdifferentiate into the collagen-producing myofibroblast in obstructive renal fibrosis model. This indicates macrophages contribute to renal fibrosis through direct mechanisms (Meng et al. 2016b). TGF-β/Smad3 signaling is a master regulator in this process (Wang et al. 2016b).

Several lines of evidence indicate that targeting the phenotypic alteration of macrophages has therapeutic potential in treating renal fibrosis. A recent study showed that transferring M2 macrophages significantly attenuated renal inflammation and fibrosis (Wang et al. 2007). Although M2a and M2c are both capable of attenuating renal injury by decreasing host macrophage and T cell infiltration, M2c is more effective at inducing Tregs and attenuating renal inflammation and fibrosis (Cao et al. 2013; Lu et al. 2013). Collectively, macrophages play a central role in linking renal inflammation and fibrosis (shown in Fig. 18.1). Inhibiting macrophage infiltration, modifying phenotype of macrophages by interfering with specific signaling molecules or transferring modified macrophages may represent a potential approach to treat renal fibrosis.

18.2.2 T Cells in Renal Fibrosis

T cells are found in kidneys of patients and experimental animals with chronic kidney disease (Harris and Neilson 2006; Robertson et al. 2004). In early stages of renal inflammation, T lymphocytes migrate to the injured region and are activated by antigen stimulation. Activated T cells produce a variety of cytokines and chemokines that recruit and activate macrophages, facilitating the inflammatory response (Lin et al. 2008). The functional role of T cells has been studied extensively in different renal disease models (Tipping and Holdsworth 2006; Zheng et al. 2005). For example, deletion of CD4+ T cells with a monoclonal antibody largely attenuated renal fibrosis. And, adoptive transfer of CD4+ T cells, but not CD8+ T cells, restored the severity of UUO-induced renal fibrosis without affecting the number of infiltrating macrophages (Tapmeier et al. 2010). This indicates CD4+ T cells directly contribute to renal fibrosis in a macrophage-independent manner. Additionally, T cells were shown to indirectly regulate fibrocyte differentiation (Niedermeier et al. 2009). In a study using chronic renal allograft, intratubular T cells triggered local epithelial cells to transition into proliferating fibroblast phenotype (Robertson et al. 2004). Among T cell subsets, Th2 cells play a more critical role in renal fibrosis. Reconstitution of Th2 cells in mice with CD4+ T cell deletion resulted in more severe renal fibrosis compared with adoptive transfer of Th1 cells (Liu et al. 2012). This indicates that targeting CD4+ T cells, particularly Th2 cells, may be a potential therapy for renal fibrosis. In another study, CD4+ Foxp3+ T cells mediated the repair of AKI and attenuated renal fibrosis via mTOR signaling (Chen et al. 2016). The function of Th17 cells and IL-17 in renal fibrosis is still under debate. IL-17 is mainly produced by Th17 and γδT cells (Kim et al. 2015). The Th17/IL-17 axis is profibrotic, and it induces the production of chemokine (C-X-C motif) ligand 5 (CXCL5) and recruits neutrophils that contribute to glomerular nephropathy (Disteldorf et al. 2015). Disruption of IL-17A reduces myofibroblast activation and extracellular matrix production in UUO nephropathy (Peng et al. 2015). Moreover, exposure to elevated dietary salt promotes

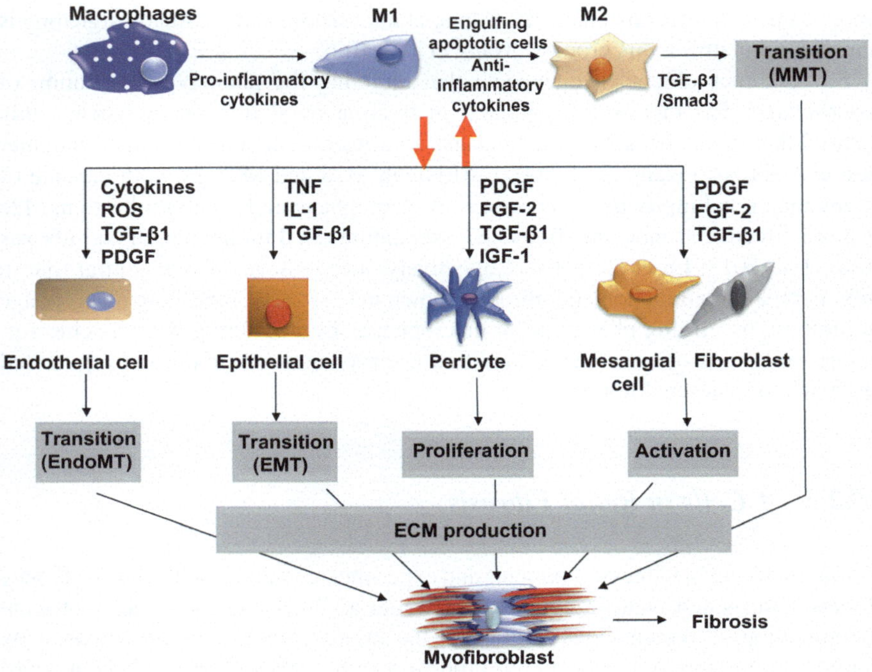

Fig. 18.1 Role of macrophages in linking renal inflammation and fibrosis. Monocyte/macrophages are recruited by chemokines released in response to microenvironment of injured kidney, and then they are activated by proinflammatory cytokines. These classically activated M1 macrophages activate intrinsic renal cells including tubular epithelial cells, endothelial cells and mesangial cells via producing factors such as cytokines (IL-1, TNF) and ROS. In the progression of renal disease, alternative activated M2 macrophages are induced by Th2 cytokines, and macrophages also gain M2 phenotype after engulfing apoptotic cells. These profibrotic macrophages promote renal fibrosis by producing an abundance of growth factors and ECM, and they may also undergo transition into myofibroblasts via TGF-β/Smad3-mediated MMT. In the fibrotic stage, the activated intrinsic renal cells proliferate, activate or directly transdifferentiate into myofibroblast-like cells directly via different mechanisms, and thereby participate in renal fibrosis. Abbreviations: ECM—extracellular matrix; EMT—epithelial–mesenchymal transition; EndoMT—endothelial–mesenchymal transition; MMT—macrophage–myofibroblast transition; FGF-2—fibroblast growth factor 2; IGF-1—insulin-like growth factor-1; PDGF—platelet-derived growth factor; ROS—reactive oxygen species; TGF-β—transforming growth factor β; and TNF—tumor necrosis factor

the progression from AKI to CKD through activation of Th17 cells (Mehrotra et al. 2015). This is consistent with the finding that disruption of IL-17 or IL-23 (a cytokine essential for the expansion and survival of Th17 cells) reduces albuminuria and glomerular crescent formation in experimental glomerulonephritis (Paust et al. 2009). In contrast, in a deoxycorticosterone acetate and angiotensin II-induced model of hypertensive nephropathy, loss of IL-17 accelerated hypertensive glomerular injury, which is correlated with reduced survival of TECs and more infiltration of γδT cells. Consistently, increased IL-17 suppressed renal inflammation and fibrosis in diabetic nephropathy by protecting against TECs and podocytes (Mohamed et al. 2016). All of these findings underscore Th17/IL-17 acts in a disease type- and dosage-dependent way. It is important to note that CD11c+ CD8+ T cells likely induce fibroblast apoptosis, thereby limiting renal fibrosis (Wang et al. 2016a). This is in agreement with a previous study that showed loss of CD8+ T cells increased CD4+ T cell-mediated monocyte-to-fibroblast transition and renal fibrosis (Dong et al. 2016).

18.2.3 Fibrocytes in Renal Fibrosis

Fibrocytes are bone marrow-derived circulating cells that originate from CD14+ monocytic lineage and share markers with leukocytes like CD45 and mesenchymal cells like type I collagen (Bucala et al. 1994; Chesney et al. 1998). In the injured kidney, fibrocytes are recruited via their highly expressed chemokine receptors CCR2, CCR7 and CXCR4 (Chen et al. 2011a; Reich et al. 2013; Sakai et al. 2006). The regulation of fibrocyte differentiation has been studied extensively. It is clear that profibrotic Th2 cytokines, like IL-4 and IL-13, induce the differentiation of fibrocytes into myofibroblasts in vitro. However, Th1 cytokines like IFN-γ exert inhibitory effects on this process (Shao et al. 2008). Further, inhibiting CCL21/CCR7 signaling with anti-CCL21 antibodies or genetic deletion inhibits fibrocyte recruitment and renal fibrosis (Wada et al. 2007). CCR2 is also a key mediator for migration of fibrocytes to the injured kidney (Reich et al. 2013). As monocyte-derived precursor cells, fibrocytes not only participate in renal inflammation but also contribute to renal fibrosis through several mechanisms. First, they produce a large amount of collagen and profibrotic growth factors such as TGF-β in response to stimuli (Buchtler et al., 2018). Second, they secrete a number of cytokines, such as TNF-α and MCP-1, that contribute to renal inflammation (Chesney et al. 1998; Reilkoff et al. 2011). Third, they directly convert into myofibroblast phenotype. Although this remains controversial, some studies report that a considerable ratio of collagen-producing fibroblasts in obstructive kidney diseases originates from fibrocytes or bone marrow (Broekema et al. 2007; Niedermeier et al. 2009).

18.2.4 Mast Cells in Renal Fibrosis

Mast cells are tissue-specific multifunctional cells found in low numbers in healthy kidneys, but are significantly increased in tubulointerstitial injury, regardless of the initiating disease (Holdsworth and Summers 2008). As granulated cells, mast cells produce an array of inflammatory cytokines, chemokines, growth factors and cell-specific neutral proteases that facilitate leukocyte recruitment and microbial destruction. Interestingly, infiltration of mast cells is positively correlated with disease severity (Kondo et al. 2001). However, the functional role of mast cells in fibrosis is still not clear (Mack and Rosenkranz 2009). It is possible that mast cells promote fibrosis by releasing profibrotic mediators, such as TGF-β and MMP, or by activating the AngII pathway through chymase-dependent mechanisms (Margulis et al. 2009; Wasse et al. 2012). Direct evidence of a role for mast cells in renal inflammation and fibrosis comes from a number of studies using mast cell-deficient mice. In one study, deletion of a key differentiation factor c-kit in $Kit^{W-sh}/^{W-sh}$ mice inhibited renal CD4+ T cells, macrophages and renal fibrosis. These were restored by adoptive transfer of bone marrow-derived mast cells from wild-type mice (Summers et al. 2012). In another, depletion of mast cells in the early phase of renal ischemia–reperfusion injury prevented CKD progression (Danelli et al. 2017). Consistent results were also found by deleting mouse mast cell protease 4 (mMCP4), the functional counterpart of human chymase, in mouse models of experimental anti-glomerular basement glomerulonephritis and UUO nephropathy. Results showed that mMCP4-deficient mice had significantly less renal inflammation and fibrosis; they were protected from progressive renal injury (Pons et al. 2017; Scandiuzzi et al. 2010). In contrast, results from other studies do not support the pathogenic role of mast cells in renal inflammation and fibrosis. For example, deletion of mMCP4 in UUO nephropathy increased kidney-infiltrating macrophages, T cells and local profibrotic TGF-β1 and CCL2 (Miyazawa et al. 2004). Similarly, in mast cell-deficient (Ws/Ws) rats, renal fibrosis in puromycin aminonucleoside nephrosis was measurably worse (Beghdadi et al. 2013; Miyazawa et al. 2004). This finding is further evidenced by a well-designed study that showed deficiency of mast cells in Kit(W)/Kit(W-v) mice enhanced renal fibrosis by increasing the infiltration of inflammatory cells (T cells and macrophages) and TGF-β production in obstructive kidney. The study also showed adoptive transfer of mast cells effectively suppressed renal fibrogenesis (Kim et al. 2009). Although the role of mast cells in renal inflammation and fibrosis remains unclear, there is no doubt that mast cells contribute to the pathological process from renal inflammation to fibrosis. The convincing and conflicting results from different animal models warrant further investigation.

18.3 Intrinsic Kidney Cells in Renal Inflammation and Fibrosis

Intrinsic kidney cells include tubular epithelial cells, podocytes, mesangial cells and endothelial cells. They actively participate in the disease process from inflammation to fibrosis after kidney injury. When activated, they produce a number of inflammatory cytokines, chemokines and growth factors, leading to progression of kidney diseases from acute inflammation to chronic fibrosing stage, resulting in ESRD (Gewin et al. 2017). In this regard, intrinsic kidney cells also function as inflammatory cells, like infiltrating leukocytes, and play critical roles in mediating fibrosis in CKD.

18.3.1 Tubular Epithelial Cells in Renal Inflammation and Fibrosis

Tubular epithelial cells (TECs) are the primary target of a variety of metabolic, immunologic, ischemic and toxic insults. TECs exhibit a wide range of responses including growth arrest, proliferation, apoptosis, autophagy and transdifferentiation that lead to tubular atrophy and renal fibrosis (Gewin 2018; Liu 2011). Several lines of evidence indicate TECs are not only common targets, but also major initiators of renal injury as producers of cytokines and chemokines (Hato et al. 2013; Liu et al. 2018; Nielsen et al. 2013). For instance, in a kidney transplantation model, activated TECs attracted all subsets of leukocytes and promoted their migration into the graft at early inflammatory stage of allograft rejection by releasing TEC-generated chemokines CCL2, CCL5, macrophage chemotactic osteopontin (OPN) and macrophage migration inhibitory factor (MIF) (Demmers et al. 2013; Nguan and Du 2009). Indeed, TEC itself is capable of producing abundant cytokines, including TNF-α, IL-1β, IL-15 and IL-6 to enhance local inflammation (Nguan and Du 2009). This proinflammatory property of TECs is highlighted by a study showing that deficiency of renal collecting duct epithelial cell-specific Klf5 largely attenuated renal inflammation by decreasing chemokine secretion (Fujiu et al. 2011). In addition, constitutive activation of TGF-β receptor type 1 (TβR1) kinase in tubular epithelial cells was sufficient to induce AKI, characterized by tubular cell apoptosis and necrosis, oxidative stress, interstitial accumulation of inflammatory cells and loss of renal function (Gentle et al. 2013). Consistently, blocking TGF-β signaling in tubular epithelial cells has significant impacts on renal inflammation, apoptosis and fibrogenesis in animal models of toxic nephropathy (Gewin et al. 2012) and obstructive nephropathy (Gewin et al. 2010; Meng et al. 2012a, b). Overactivation of TGF-β/Smad signaling in TEC promotes renal fibrosis in vivo and in vitro. In response to TGF-β1, TECs release a number of profibrotic factors like TGF-β1, which stimulate TECs to produce collagen extracellular matrix (Meng et al. 2010). Direct evidence comes from a number of studies that showed activated TEC transitions from epithelial into mesenchymal phenotype, the EMT process, which is mainly found in vitro in response to TGF-β, IL-1β and angiotensin II (AngII), in a number of animal models (Fan et al. 1999, 2001; Liu 2004, 2010; Yang et al. 2002). However, the EMT concept has been challenged by cell lineage-tracing studies showing myofibroblasts originate from TECs; it is rare to find EMT cells in renal biopsy samples from patients with chronic kidney disease (Allison 2013; LeBleu et al. 2013). It is worth noting that TECs undergo partial EMT, instead of completely transdifferentiating to myofibroblasts, and this is sufficient to drive renal interstitial fibrosis (Grande et al. 2015; Lovisa et al. 2015).

During inflammatory responses, several signaling pathways, including β-catenin (He et al. 2009), ILK (Li et al. 2003), Notch1 (Bielesz et al. 2010; Sharma et al. 2011; Ueno et al. 2013) and hypoxia-inducible factor 1 (HIF1) (Higgins et al. 2007), are activated in TECs and contribute to the progression of renal fibrosis. TEC-specific toll-like receptors, like TLR4, are key mediators in renal inflammation (Cheng et al. 2013; Correa-Costa et al. 2011; Lin et al. 2012; Zhang et al. 2008) and promote renal fibrosis (Campbell et al. 2011; Lin and Tang 2013). TLR4 suppressed BAMBI, the membrane-bound competitive inhibitor of the TGF-β type 1 receptor, increased the susceptibility of renal cells to TGF-β signaling and enhanced renal fibrogenesis (Pulskens et al. 2010). TLR4-mediated signaling promotes inflammatory cytokine IL-18-induced α-SMA and collagen production, and also decreases E-cadherin levels through AP-1 activation (Meldrum et al. 2012). Interestingly, the inflammatory phase of TECs also affects the repair process after AKI. It is now clear that the c-reactive protein (CRP) pathway is induced in TECs and plays an active role in promoting renal inflammation and fibrosis in diabetic kidney injury (Liu et al. 2011). Activation of the CRP pathway in TECs also delays the recovery from acute ischemic renal injury by impairing the G1/S cell cycle (Liu et al. 2011). Moreover, a previous study highlighted a link between TECs and renal fibrosis by demonstrating that G2/M-arrested proximal tubular cells activated c-jun NH(2)-terminal kinase (JNK) signaling, thereby inducing profibrotic cytokine production (Yang et al. 2010).

18.3.2 Mesangial Cells in Renal Inflammation and Fibrosis

Glomerulosclerosis pathogenesis is correlated with activation and proliferation of MCs, abnormalities of podocytes and endothelial cell dysfunction (Abboud 2012; Schnaper et al. 2003). As the key cell type for glomerulosclerosis, MCs are the primary target of immune-mediated glomerular diseases like IgA nephropathy or metabolic diseases like diabetic nephropathy (Gomez-Guerrero et al. 2005). When challenged by stimuli, MCs produce inflammatory mediators, including chemokines (e.g., CCL2), cytokines (e.g., TNF-α and IL-6) and various oxygen species (Schlondorff and Banas 2009). Results of transcriptomic and proteomic profiling reveal that multiple inflammatory pathways, such as LXR/RXR, FXR/RXR, and acute-phase response signaling were activated in MCs from IgA nephropathy (Liu et al. 2017). Evidence from a recent study showed that parathyroid hormone-related protein (PTHrP) served as a critical modulator of inflammatory cytokine production in MCs (Hochane et al. 2018). Proinflammatory mediators released by MCs damage the endothelial barrier, exposing the mesangium to macromolecules that trigger a positive feedback loop of inflammatory cascades (Schlondorff and Banas 2009). Leukocyte migration and infiltration into the glomerulus cause initiation and amplification of glomerular injury and are mediated by adhesion molecules and chemokines, which can be locally synthesized by MCs (Lu et al. 2018). Thus, MCs function to amplify inflammatory processes in the inflamed glomerulus. Growing evidence shows that MC-derived cytokines induce podocyte injury, resulting in proteinuria (Lai et al. 2008, 2009). It is also worth noting that inflammatory mediators released by MCs promote activation and proliferation of MC themselves via an autocrine mechanism. Furthermore, activated MCs produce growth factors, including TGF-β, PDGF, FGF, HGF, EGF and CTGF (Crean et al. 2004; Floege et al. 1998, 2008; Laping et al. 2000; Schnaper et al. 2003). These induce MC proliferation, α-SMA + phenotype transformation and ECM production, leading to glomerulosclerosis (Taniguchi et al.

2013). The findings that ECM accumulation often appears to begin in the mesangium of glomerulosclerosis suggest a critical role for mesangial cells in glomerulosclerosis (Schnaper et al. 2003).

18.3.3 Podocytes in Renal Inflammation and Fibrosis

Podocytes are terminally differentiated epithelial cells with limited proliferative capacity. They play an important role in renal inflammation and fibrosis, especially in glomerular diseases (Fogo 2011; Grahammer et al. 2013). Functionally, one study showed podocyte-specific ablation of NEMO, an NF-κB essential modulator, decreased secretion of proinflammatory chemokines and increased remission of proteinuria, restoring podocyte morphology in a mouse model of nephrotoxic nephritis (Brahler et al. 2012). Consistently, conditional knockout of signal transducer and activator of transcription 3 (STAT3) from podocytes attenuated inflammatory response and development of crescentic glomerulonephritis (Dai et al. 2013). Further, a recent study showed that podocyte-specific chemokine receptor (CCR) 2 overexpression enhanced inflammatory response in diabetic nephropathy (You et al. 2017).

Podocytes also play pivotal roles in the development of glomerulosclerosis and tubulointerstitial fibrosis, which lead to progressive proteinuric kidney disease (Deelman and Sharma 2009; Shih et al. 1999). A previous study demonstrated that deficiency of Akt2 in podocytes exacerbated glomerulosclerosis and albuminuria (Canaud et al. 2013). In another, increased caspase-8-mediated apoptosis in podocytes significantly increased kidney damage, foot process effacement, mesangial expansion and glomerulosclerosis (Rutkowski et al. 2013). In addition, conditional knockout of yes-associated protein (YAP) from podocytes accelerated FSGS and progressive renal failure (Schwartzman et al. 2016).

Podocyte-related mechanisms in glomerular diseases, especially diabetic nephropathy (DN), are a hot topic. In diabetic models, Notch signaling plays a critical role in podocyte injury; overexpression of the intracellular domain of Notch (ICN) in podocytes induces proteinuria and glomerulosclerosis. This demonstrates a prominent role in podocyte dysfunction as a key step in glomerulosclerosis and renal fibrosis (Niranjan et al. 2008). Evidence also shows that vitamin D/vitamin D receptor (VDR) signaling in podocytes protects against podocyte loss and glomerular fibrosis in diabetic kidneys (Wang et al. 2012). The role of podocyte-specific mTOR is also important, as it is highly expressed in human DN samples. Indeed, increased activation of mTORC1 enhances proteinuria and progressive glomerulosclerosis in DN, but genetic reduction of mTORC1 decreases symptoms (Inoki et al. 2011). Interestingly, another study demonstrated that knockout of mTOR accelerated glomerular damage. However, inhibiting mTORC1 signaling by genetically reducing mTORC1 copy number prevented glomerulosclerosis and progression of glomerular disease in DN. This highlights the importance of a critical balance in mTOR activity as a regulator of the pathology of diabetic kidney diseases (Fogo 2011; Godel et al. 2011).

Podocyte-secreted growth factors also induce renal fibrosis by building new microenvironments in glomeruli. Vascular endothelial growth factor (VEGF) regulates endothelial functions, including endothelial cell migration, differentiation and survival. Podocytes are a major source of VEGF in the kidney (Mathieson 2009). Selective knockdown of VEGF in podocytes caused endothelial and mesangial cell dysfunction and progressive glomerulosclerosis (Eremina et al. 2006; Siddiqi and Advani 2013). In contrast, overexpression of VEGF in a mouse model of diabetes led to advanced diabetic glomerulopathy, characterized by proteinuria, glomerulomegaly, glomerular basement membrane thickening and excessive mesangial expansion (Veron et al. 2010, 2011). This is consistent with previous findings that showed doxycycline-induced overexpression of soluble VEGF receptor-1 (sFlt-1) in podocytes ameliorated glomerulopathy in diabetic mice (Ku et al. 2008). In addition, podocyte-specific PDGF overexpression induced mesangial cell proliferation and led to mesangioproliferative disease, glomerulosclerosis and crescentic glomerulonephritis (van Roeyen et al. 2011). CTGF, a matricellular protein, is also a critical pathogenic factor in podocytes. One study showed CTGF overexpression induced expansion of the mesangial matrix and exacerbated albuminuria in a mouse model of DN (Yokoi et al. 2008).

Renin–angiotensin system (RAS) is an important target in renal fibrosis (Mezzano et al. 2001). Several studies have shown that RAS components are highly expressed in podocytes (Durvasula and Shankland 2006, 2008), indicating a potential role in podocyte-mediated renal fibrosis. Moreover, podocyte-specific deletion of the mammalian homologue of yeast vacuolar protein sorting defective 34 (mVps34), a critical regulator in autophagy, led to glomerulosclerosis and interstitial fibrosis accompanied by podocyte vacuolization and proteinaceous casts (Chen et al. 2013).

A final note, EMT is found in podocytes in animal models (Li et al. 2008; Sam et al. 2006) and human biopsy samples of diabetic nephropathy, IgA nephropathy and lupus nephritis (Yamaguchi et al. 2009). Clearly, podocytes participate in disease states and regulate pathology when environmental conditions are primed.

18.3.4 Endothelial Cells in Renal Inflammation and Fibrosis

Endothelial cells play critical roles in the pathophysiological processes of regional blood flow regulation and leukocyte recruitment (Pober and Sessa 2007). When the kidney is injured, the interaction between endothelial cells and leukocytes is initiated. Recruited leukocytes roll along the activated endothelium that is mediated by selectins, E-selectin and P-selectin, and leukocyte surface antigens. Leukocyte adhesion is facilitated by endothelial expression of ICAM-1, VCAM-1 and PECAM; leukocytes transmigrate to injured tissue and exert proinflammatory functions (Guerrot et al. 2012). Evidence of this comes from a study that showed blocking P-selectin increased renal blood flow and promoted recovery of renal function in an ischemic model of renal failure (Bojakowski et al. 2001). Consistently, another study showed deficiency of CD147, a ligand for E-selectin, impaired neutrophil recruitment and attenuated kidney damage (Kato et al. 2009). Similarly, inhibition of ICAM-1 with monoclonal antibody or genetic knockdown techniques also protected against renal injury (Kelly et al. 1994, 1996). Taken together, these findings underscore the importance of endothelial activation in local inflammation at an early stage.

Endothelial activation and subsequent leukocyte adhesion create hemodynamic resistance that reduces regional blood flow (Guerrot et al. 2012). In such conditions, ischemia and oxidative stress trigger endothelial apoptosis that leads to rarefaction of peritubular capillaries (Kelly et al. 2009; Venkatachalam et al. 2010). A pathogenic role for endothelial dysfunction was described in a study that found Crim1 deficiency, a protein involved in endothelial maintenance and integrity, resulted in excessive deposition of collagen, dysfunction and permeability of peritubular capillaries, and eventually renal fibrosis (Wilkinson et al. 2009). Importantly, rarefaction of peritubular capillaries induces hypoxia, which has significant impacts on progression of renal fibrosis by triggering EMT and/or promoting matrix production from myofibroblasts (Higgins et al. 2007). Compelling evidence for the pathogenic role of hypoxia in renal fibrosis was provided by studies showing that degradation or inhibition of hypoxia-induced factor1α (HIF1α), a central regulator of cellular responses to hypoxia (Higgins et al. 2008), largely reduced renal fibrosis (Higgins et al. 2007; Kimura et al. 2008). The correlation between HIF1α and tubulointerstitial fibrosis was also confirmed in biopsy samples from patients (Higgins et al. 2007). Chronic hypoxia is accompanied by increased oxidative stress and generation of ROS, which trigger advanced glycation end products, advanced oxidative protein products and advanced lipoperoxidation end products (D'Agati and Schmidt 2010; Negre-Salvayre et al. 2008). These may target kidney cells as pathogenic mediators and enhance renal inflammation and fibrosis (Daroux et al. 2010; Shanmugam et al. 2008; Shi et al. 2008; Zhou et al. 2009).

Finally, inflammatory endothelial cells also contribute to renal fibrosis through transdifferentiation into collagen-producing cells via EndoMT in response to stimuli (Zeisberg et al. 2008). Of note, TGF-β1 is the key mediator inducing EndoMT via Smad3-dependent mechanisms (Li et al. 2010; Xavier et al. 2015). A recent study showed EndoMT is regulated by SIRT (sirtuin)3/Foxo3a axis and integrin β1/dipeptidyl peptidase (DPP)-4 in vivo (Lin et al. 2018). Moreover, loss of heparin-binding EGF-like factor (HB-EGF) in endothelial cells attenuates angiotensin II (Ang II)-induced renal inflammation and fibrosis (Zeng et al. 2016). This emphasizes the critical role endothelial cells play in linking inflammation and fibrosis.

18.4 Key Fibrosis-Correlated Growth Factors Released by Inflammatory Cells

Resident and recruited inflammatory cells secrete growth factors such as TGF-β, PDGF, FGF, HGF, EGF and CTGF that exert pro- or anti-fibrotic roles in renal fibrosis (Boor and Floege 2011; Lv et al. 2018). TGF-β1 is the most abundant isoform of TGF-β family members and is secreted by all types of resident renal cells and infiltrated inflammatory cells. It is secreted as a latent precursor, called latent TGF-β1 that binds to latent TGF-β-binding protein (LTBP). When exposed to stimuli, including ROS, plasmin and acid (Meng et al. 2015, 2016a), TGF-β1 is released from latency-associated peptide (LAP) and LTBP (Lyons et al. 1990; Meng et al. 2013; Munger et al. 1999). Mature TGF-β1 then binds to its type II receptor to recruit type I receptors and activates downstream signals. The major source of TGF-β1 includes macrophages, tubular epithelial cells and myofibroblasts (Fukuda et al. 2001; Lee and Kalluri 2010; Ma et al. 2003). After release and activation, TGF-β1 activates Smad and non-Smad signaling in renal fibrosis, inflammation, cell growth, apoptosis and differentiation (Bottinger and Bitzer 2002). In general, TGF-β1 has both pro- and anti-inflammation properties (Huang et al. 2008a, b; Kitamura and Suto 1997; Zhang et al. 2009). In renal fibrosis, TGF-β and downstream Smad3 are pathogenic, while Smad2 and Smad7 seem to be renoprotective (Meng et al. 2013).

TGF-β1 behaves in renal disease in a cell-specific and dosage-dependent manner (Hathaway et al. 2015; Meng et al. 2016a). TGF-β1 recruits macrophages by secreting chemokines, including MCP-1 and OPN (Lan 2011; Zhang et al. 2009), and promotes macrophage polarization toward M2 phenotype (Sica and Mantovani 2012). Our recent results also show TGF-β1 initiates transdifferentiation of macrophages into myofibroblast-like cells (Wang et al. 2016b). In addition, TGF-β1 induces the production of Foxp3+ T regulatory cells (Tregs) (Fu et al. 2004) and inhibits the progression of autoimmune kidney disease (Wang et al. 2006). A recent study showed that combination treatment of rhTGF-β and ICG-001, an inhibitor of β-catenin/TCF, suppressed both renal inflammation and fibrosis in UUO nephropathy and kidney ischemia/reperfusion (Qiao et al. 2018). Several other studies confirm TGF-β1 plays regulatory roles in other types of T cells (Gorelik and Flavell 2002; Kitching and Holdsworth 2011; Turner et al. 2010). In addition, TGF-β mediates mast cell chemotaxis (Gruber et al. 1994) and production of mast cell protease (Funaba et al. 2006).

TGF-β1 also functions in resident inflammatory cells by enhancing ECM production from TECs and promoting EMT. TGF-β1 also induces apoptosis in tubular epithelial cells (Lopez-Hernandez and Lopez-Novoa 2012). In vitro, data indicate that podocytes produce ECM in response to TGF-β1 (Gruden et al. 2005). In TGF-β1 transgenic mice, apoptosis of podocytes is induced by overactivation of Smad7 (Schiffer et al. 2001). Further, a previous study indicated that TGF-β-induced apoptosis in cultured mouse podocytes acted through upregulation of mitochondrial Nox4 via Smad2/3-dependent mechanisms (Das et al. 2014). As terminally differentiated epithelial cells, podocytes also undergo EMT in response to TGF-β1 (Li et al. 2008). Furthermore, emerging evidence indicates that TGF-β1 stimulates mesangial cells to secrete type I, III and IV collagen, laminin and fibronectin, supporting glomerular ECM accumulation (Gruden et al. 2005; Lopez-Hernandez and Lopez-Novoa 2012). TGF-β1 also induces hypertrophy and proliferation of mesangial cells to accelerate glomerulosclerosis (Gruden et al. 2005). Additionally, TGF-β1 has an impact on apoptosis, proliferation and migration of endothelial cells (Lebrin et al. 2005). Although TGF-β1 has a proapoptotic effect on endothelial cells in most scenarios (Loeffler and Wolf 2013), it also promotes the release of VEGF from podocytes and

TECs which protect endothelial cells from apoptosis (Kang et al. 2002). Notably, as a downstream molecule of TGF-β1, Smad2 triggers expression of antiangiogenic factors like TSP-1 and VEGF-A antagonist. In comparison, TGF-β1-induced VEGF mRNA and protein expression are Smad3-dependent (Nakagawa et al. 2004a, b). TGF-β1 is known classically to stimulate EndoMT (Zeisberg et al. 2008). Thus, data clearly show TGF-β1 could be secreted by inflammatory cells and may be one of the key growth factors that drives renal fibrogenesis.

The PDGF family contains four isoforms, PDGF-A, -B, -C and -D, and two receptor chains, PDGFR-α and -β, that are constitutively or inducibly expressed in most renal cells (Floege et al. 2008). PDGF exerts numerous biological functions in renal diseases, including production of pro- and anti-inflammatory mediators, ECM accumulation, cell proliferation and migration (Ostendorf et al. 2012). PDGF signaling is a crucial mediator in renal fibrosis (LeBleu and Kalluri 2011), as evidenced by the findings that show blocking PDGF-D attenuates tubulointerstitial fibrosis in both early and late stages of glomerulonephritis (Boor et al. 2007; Ostendorf et al. 2006). PDGF-D has a direct profibrotic effect on the tubular interstitium by promoting EMT (Kong et al. 2009) or by triggering expression of PDGFR-β from resident fibroblasts or pericytes, which are the major source of interstitial myofibroblasts (Humphreys et al. 2010; Lin et al. 2008). Inhibition of PDGFR signaling reduces the number of myofibroblasts and ameliorates renal damage in obstructive nephropathy (Chen et al. 2011b). This indicates a key role for PDGFR signaling in myofibroblast generation from pericytes. Compelling evidence for a role of PDGFRα and its ligand PDGF-CC in renal fibrosis is provided by results showing that inhibition of PDGF-CC suppresses renal fibrosis in obstructive nephropathy (Eitner et al. 2008). PDGF-CC also exerts profibrotic effects by directly enhancing fibroblast proliferation, as well as accelerating leukocyte infiltration (Eitner et al. 2008). Moreover, PDGF-CC serves as a pro-angiogenic mediator in glomeruli (Boor et al. 2015). In contrast to PDGF-A, which has limited effects in renal fibrosis (Tang et al. 1996), PDGF-B and PDGFR-β mediate mesangial proliferation both in vitro and in vivo; high doses of PDGF-BB induce proliferation of tubulointerstitial cells and promote the myofibroblast generation and fibrosis (Boor et al. 2014; Buhl et al. 2016; Das et al. 2017; Tang et al. 1996). Collectively, renal inflammatory cells produce plenty of PDGF, driving the progression of renal fibrosis.

18.5 Conclusion

Inflammation is the initial response to cellular injuries and is the key process in wound healing and renal repair. However, unresolved renal inflammation leads to fibrosis with loss of renal function. In this process, both circulating leukocytes and intrinsic kidney cells are key inflammatory cell types and play critical roles in driving acute renal inflammation to chronic fibrosis, leading to the end-stage kidney disease. There are two primary mechanisms regulating the pathway from renal inflammation to fibrosis. First, fibrogenetic growth factors produced by inflammatory cells promote local fibroblasts, or inflammatory cells themselves, to proliferate and produce ECM. Second, inflammatory cells convert into collagen-producing fibroblasts or myofibroblasts directly from EMT, EndoMT, pericytes, fibrocytes or MMT. Many inflammatory cytokines and growth factors, like TGF-β, are involved in the inflammation–fibrosis process. Collectively, blocking excessive inflammatory responses may represent an effective therapy for renal fibrotic diseases.

Acknowledgements This laboratory is supported by grants from National Natural Science Foundation of China (National Science Foundation of China 81300580 and 81570623) and by Science and Technological Fund of Anhui Province for Outstanding Youth of China (Grant number: 1608085J07).

References

Abboud HE (2012) Mesangial cell biology. Exp Cell Res 318:979–985

Allison SJ (2013) Fibrosis: the source of myofibroblasts in kidney fibrosis. Nat Rev Nephrol 9:494

Anders HJ, Ryu M (2011) Renal microenvironments and macrophage phenotypes determine progression or resolution of renal inflammation and fibrosis. Kidney Int 80:915–925

Anders HJ, Vielhauer V, Eis V, Linde Y, Kretzler M et al (2004) Activation of toll-like receptor-9 induces progression of renal disease in MRL-Fas(lpr) mice. FASEB J 18:534–536

Beghdadi W, Madjene LC, Claver J, Pejler G, Beaudoin L, Lehuen A, Daugas E, Blank U (2013) Mast cell chymase protects against renal fibrosis in murine unilateral ureteral obstruction. Kidney Int 84:317–326

Bielesz B, Sirin Y, Si H, Niranjan T, Gruenwald A et al (2010) Epithelial Notch signaling regulates interstitial fibrosis development in the kidneys of mice and humans. J Clin Invest 120:4040–4054

Bojakowski K, Abramczyk P, Bojakowska M, Zwolinska A, Przybylski J, Gaciong Z (2001) Fucoidan improves the renal blood flow in the early stage of renal ischemia/reperfusion injury in the rat. J Physiol Pharmacol 52:137–143

Boor P, Floege J (2011) Chronic kidney disease growth factors in renal fibrosis. Clin Exp Pharmacol Physiol 38:441–450

Boor P, Konieczny A, Villa L, Kunter U, van Roeyen CR et al (2007) PDGF-D inhibition by CR10 ameliorates tubulointerstitial fibrosis following experimental glomerulonephritis. Nephrol Dial Transplant 22:1323–1331

Boor P, Ostendorf T, Floege J (2010) Renal fibrosis: novel insights into mechanisms and therapeutic targets. Nat Rev Nephrol 6:643–656

Boor P, Ostendorf T, Floege J (2014) PDGF and the progression of renal disease. Nephrol Dial Transplant 29(Suppl 1):i45–i54

Boor P, Babickova J, Steegh F, Hautvast P, Martin IV et al (2015) Role of platelet-derived growth factor-CC in capillary rarefaction in renal fibrosis. Am J Pathol 185:2132–2142

Bottinger EP, Bitzer M (2002) TGF-beta signaling in renal disease. J Am Soc Nephrol 13:2600–2610

Brahler S, Ising C, Hagmann H, Rasmus M, Hoehne M et al (2012) Intrinsic proinflammatory signaling in podocytes contributes to podocyte damage and prolonged proteinuria. Am J Physiol Renal Physiol 303:F1473–F1485

Broekema M, Harmsen MC, van Luyn MJ, Koerts JA, Petersen AH et al (2007) Bone marrow-derived myofibroblasts contribute to the renal interstitial myofibroblast population and produce procollagen I after ischemia/reperfusion in rats. J Am Soc Nephrol 18:165–175

Bucala R, Spiegel LA, Chesney J, Hogan M, Cerami A (1994) Circulating fibrocytes define a new leukocyte subpopulation that mediates tissue repair. Mol Med 1:71–81

Buchtler S, Grill A, Hofmarksrichter S, Stockert P, Schiechl-Brachner G et al (2018) Cellular origin and functional relevance of collagen i production in the kidney. J Am Soc Nephrol 29:1859–1873

Buhl EM, Djudjaj S, Babickova J, Klinkhammer BM, Folestad E et al (2016) The role of PDGF-D in healthy and fibrotic kidneys. Kidney Int 8:848–861

Campbell MT, Hile KL, Zhang H, Asanuma H, Vanderbrink BA et al (2011) Toll-like receptor 4: a novel signaling pathway during renal fibrogenesis. J Surg Res 168:e61–e69

Canaud G, Bienaime F, Viau A, Treins C, Baron W et al (2013) AKT2 is essential to maintain podocyte viability and function during chronic kidney disease. Nat Med 19:1288–1296

Cao Q, Wang Y, Harris DC (2013) Pathogenic and protective role of macrophages in kidney disease. Am J Physiol Renal Physiol 305:F3–F11

Chen G, Lin SC, Chen J, He L, Dong F et al (2011a) CXCL16 recruits bone marrow-derived fibroblast precursors in renal fibrosis. J Am Soc Nephrol 22:1876–1886

Chen YT, Chang FC, Wu CF, Chou YH, Hsu HL et al (2011b) Platelet-derived growth factor receptor signaling activates pericyte-myofibroblast transition in obstructive and post-ischemic kidney fibrosis. Kidney Int 80:1170–1181

Chen J, Chen MX, Fogo AB, Harris RC, Chen JK (2013) mVps34 deletion in podocytes causes glomerulosclerosis by disrupting intracellular vesicle trafficking. J Am Soc Nephrol 24:198–207

Chen G, Dong Z, Liu H, Liu Y, Duan S, Liu F, Chen H (2016) mTOR signaling regulates protective activity of transferred CD4+ Foxp3+ T Cells in repair of acute kidney injury. J Immunol 197:3917–3926

Cheng A, Dong Y, Zhu F, Liu Y, Hou FF, Nie J (2013) AGE-LDL activates Toll like receptor 4 pathway and promotes inflammatory cytokines production in renal tubular epithelial cells. Int J Biol Sci 9:94–107

Chesney J, Metz C, Stavitsky AB, Bacher M, Bucala R (1998) Regulated production of type I collagen and inflammatory cytokines by peripheral blood fibrocytes. J Immunol 160:419–425

Chung AC, Lan HY (2011) Chemokines in renal injury. J Am Soc Nephrol 22:802–809

Correa-Costa M, Braga TT, Semedo P, Hayashida CY, Bechara LR et al (2011) Pivotal role of Toll-like receptors 2 and 4, its adaptor molecule MyD88, and inflammasome complex in experimental tubule-interstitial nephritis. PLoS ONE 6:e29004

Crean JK, Furlong F, Finlay D, Mitchell D, Murphy M et al (2004) Connective tissue growth factor [CTGF]/CCN2 stimulates mesangial cell migration through integrated dissolution of focal adhesion complexes and activation of cell polarization. FASEB J 18:1541–1543

D'Agati V, Schmidt AM (2010) RAGE and the pathogenesis of chronic kidney disease. Nat Rev Nephrol 6:352–360

Dai Y, Gu L, Yuan W, Yu Q, Ni Z et al (2013) Podocyte-specific deletion of signal transducer and activator of transcription 3 attenuates nephrotoxic serum-induced glomerulonephritis. Kidney Int 84:950–961

Danelli L, Madjene LC, Madera-Salcedo I, Gautier G, Pacreau E et al (2017) Early phase mast cell activation determines the chronic outcome of renal ischemia-reperfusion injury. J Immunol 198:2374–2382

Daroux M, Prevost G, Maillard-Lefebvre H, Gaxatte C, D'Agati VD et al (2010) Advanced glycation end-products: implications for diabetic and non-diabetic nephropathies. Diabetes Metab 36:1–10

Das R, Xu S, Quan X, Nguyen TT, Kong ID et al (2014) Upregulation of mitochondrial Nox4 mediates TGF-beta-induced apoptosis in cultured mouse podocytes. Am J Physiol Renal Physiol 306:F155–F167

Das F, Ghosh-Choudhury N, Venkatesan B, Kasinath BS, Ghosh Choudhury G (2017) PDGF receptor-beta uses Akt/mTORC1 signaling node to promote high glucose-induced renal proximal tubular cell collagen I (alpha2) expression. Am J Physiol Renal Physiol 313:F291–F307

Deelman L, Sharma K (2009) Mechanisms of kidney fibrosis and the role of antifibrotic therapies. Curr Opin Nephrol Hypertens 18:85–90

Demmers MW, Baan CC, van Beelen E, Ijzermans JN, Weimar W, Rowshani AT (2013) Differential effects of activated human renal epithelial cells on T-cell migration. PLoS ONE 8:e64916

Disteldorf EM, Krebs CF, Paust HJ, Turner JE, Nouailles G et al (2015) CXCL5 drives neutrophil recruitment in TH17-mediated GN. J Am Soc Nephrol 26:55–66
Djudjaj S, Boor P (2018) Cellular and molecular mechanisms of kidney fibrosis. Mol Aspects Med
Dong Y, Yang M, Zhang J, Peng X, Cheng J, Cui T, Du J (2016) Depletion of CD8+ T cells exacerbates CD4+ T cell-induced monocyte-to-fibroblast transition in renal fibrosis. J Immunol 196:1874–1881
Duffield JS, Humphreys BD (2011) Origin of new cells in the adult kidney: results from genetic labeling techniques. Kidney Int 79:494–501
Durvasula RV, Shankland SJ (2006) The renin-angiotensin system in glomerular podocytes: mediator of glomerulosclerosis and link to hypertensive nephropathy. Curr Hypertens Rep 8:132–138
Durvasula RV, Shankland SJ (2008) Activation of a local renin angiotensin system in podocytes by glucose. Am J Physiol Renal Physiol 294:F830–F839
Eardley KS, Kubal C, Zehnder D, Quinkler M, Lepenies J et al (2008) The role of capillary density, macrophage infiltration and interstitial scarring in the pathogenesis of human chronic kidney disease. Kidney Int 74:495–504
Eddy AA, Neilson EG (2006) Chronic kidney disease progression. J Am Soc Nephrol 17:2964–2966
Eitner F, Bucher E, van Roeyen C, Kunter U, Rong S et al (2008) PDGF-C is a proinflammatory cytokine that mediates renal interstitial fibrosis. J Am Soc Nephrol 19:281–289
Eremina V, Cui S, Gerber H, Ferrara N, Haigh J et al (2006) Vascular endothelial growth factor a signaling in the podocyte-endothelial compartment is required for mesangial cell migration and survival. J Am Soc Nephrol 17:724–735
Fan JM, Ng YY, Hill PA, Nikolic-Paterson DJ, Mu W et al (1999) Transforming growth factor-beta regulates tubular epithelial-myofibroblast transdifferentiation in vitro. Kidney Int 56:1455–1467
Fan JM, Huang XR, Ng YY, Nikolic-Paterson DJ, Mu W et al (2001) Interleukin-1 induces tubular epithelial-myofibroblast transdifferentiation through a transforming growth factor-beta1-dependent mechanism in vitro. Am J Kidney Dis 37:820–831
Fine LG, Norman JT (2008) Chronic hypoxia as a mechanism of progression of chronic kidney diseases: from hypothesis to novel therapeutics. Kidney Int 74:867–872
Floege J, Burg M, Hugo C, Gordon KL, Van Goor H et al (1998) Endogenous fibroblast growth factor-2 mediates cytotoxicity in experimental mesangioproliferative glomerulonephritis. J Am Soc Nephrol 9:792–801
Floege J, Eitner F, Alpers CE (2008) A new look at platelet-derived growth factor in renal disease. J Am Soc Nephrol 19:12–23
Fogo AB (2011) The targeted podocyte. J Clin Invest 121:2142–2145
Fu S, Zhang N, Yopp AC, Chen D, Mao M et al (2004) TGF-beta induces Foxp3+ T-regulatory cells from CD4+ CD25-precursors. Am J Transplant 4:1614–1627
Fujiu K, Manabe I, Nagai R (2011) Renal collecting duct epithelial cells regulate inflammation in tubulointerstitial damage in mice. J Clin Invest 121:3425–3441
Fukuda K, Yoshitomi K, Yanagida T, Tokumoto M, Hirakata H (2001) Quantification of TGF-beta1 mRNA along rat nephron in obstructive nephropathy. Am J Physiol Renal Physiol 281:F513–F521
Funaba M, Ikeda T, Murakami M, Ogawa K, Nishino Y et al (2006) Transcriptional regulation of mouse mast cell protease-7 by TGF-beta. Biochem Biophys Acta 1759:166–170
Gentle ME, Shi S, Daehn I, Zhang T, Qi H et al (2013) Epithelial cell TGFbeta signaling induces acute tubular injury and interstitial inflammation. J Am Soc Nephrol 24:787–799
Gewin LS (2018) Renal fibrosis: primacy of the proximal tubule. Matrix Biol 68–69:248–262
Gewin L, Bulus N, Mernaugh G, Moeckel G, Harris RC et al (2010) TGF-beta receptor deletion in the renal collecting system exacerbates fibrosis. J Am Soc Nephrol 21:1334–1343
Gewin L, Vadivelu S, Neelisetty S, Srichai MB, Paueksakon P et al (2012) Deleting the TGF-beta receptor attenuates acute proximal tubule injury. J Am Soc Nephrol 23:2001–2011
Gewin L, Zent R, Pozzi A (2017) Progression of chronic kidney disease: too much cellular talk causes damage. Kidney Int 91:552–560
Godel M, Hartleben B, Herbach N, Liu S, Zschiedrich S et al (2011) Role of mTOR in podocyte function and diabetic nephropathy in humans and mice. J Clin Invest 121:2197–2209

Gomez-Guerrero C, Hernandez-Vargas P, Lopez-Franco O, Ortiz-Munoz G, Egido J (2005) Mesangial cells and glomerular inflammation: from the pathogenesis to novel therapeutic approaches. Curr Drug Targets Inflamm Allergy 4:341–351

Gorelik L, Flavell RA (2002) Transforming growth factor-beta in T-cell biology. Nat Rev Immunol 2:46–53

Grahammer F, Schell C, Huber TB (2013) The podocyte slit diaphragm–from a thin grey line to a complex signalling hub. Nat Rev Nephrol 9:587–598

Grande MT, Perez-Barriocanal F, Lopez-Novoa JM (2010) Role of inflammation in tubulo-interstitial damage associated to obstructive nephropathy. J Inflamm (Lond) 7:19

Grande MT, Sanchez-Laorden B, Lopez-Blau C, De Frutos CA, Boutet A et al (2015) Snail1-induced partial epithelial-to-mesenchymal transition drives renal fibrosis in mice and can be targeted to reverse established disease. Nat Med 21:989–997

Gratchev A, Guillot P, Hakiy N, Politz O, Orfanos CE, Schledzewski K, Goerdt S (2001) Alternatively activated macrophages differentially express fibronectin and its splice variants and the extracellular matrix protein betaIG-H3. Scand J Immunol 53:386–392

Gruber BL, Marchese MJ, Kew RR (1994) Transforming growth factor-beta 1 mediates mast cell chemotaxis. J Immunol 152:5860–5867

Gruden G, Perin PC, Camussi G (2005) Insight on the pathogenesis of diabetic nephropathy from the study of podocyte and mesangial cell biology. Curr Diab Rev 1:27–40

Guerrot D, Dussaule JC, Kavvadas P, Boffa JJ, Chadjichristos CE, Chatziantoniou C (2012) Progression of renal fibrosis: the underestimated role of endothelial alterations. Fibrogenesis Tissue repair 5(Suppl 1):S15

Guo S, Wietecha TA, Hudkins KL, Kida Y, Spencer MW et al (2011) Macrophages are essential contributors to kidney injury in murine cryoglobulinemic membranoproliferative glomerulonephritis. Kidney Int 80:946–958

Han Y, Ma FY, Tesch GH, Manthey CL, Nikolic-Paterson DJ (2011) c-fms blockade reverses glomerular macrophage infiltration and halts development of crescentic anti-GBM glomerulonephritis in the rat. Lab Invest 91:978–991

Han Y, Ma FY, Tesch GH, Manthey CL, Nikolic-Paterson DJ (2013) Role of macrophages in the fibrotic phase of rat crescentic glomerulonephritis. Am J Physiol Renal Physiol 304:F1043–F1053

Harris RC, Neilson EG (2006) Toward a unified theory of renal progression. Annu Rev Med 57:365–380

Hathaway CK, Gasim AM, Grant R, Chang AS, Kim HS et al (2015) Low TGFbeta1 expression prevents and high expression exacerbates diabetic nephropathy in mice. Proc Natl Acad Sci USA 112:5815–5820

Hato T, El-Achkar TM, Dagher PC (2013) Sisters in arms: myeloid and tubular epithelial cells shape renal innate immunity. Am J Physiol Renal Physiol 304:F1243–F1251

He W, Dai C, Li Y, Zeng G, Monga SP, Liu Y (2009) Wnt/beta-catenin signaling promotes renal interstitial fibrosis. J Am Soc Nephrol 20:765–776

Henderson NC, Mackinnon AC, Farnworth SL, Kipari T, Haslett C et al (2008) Galectin-3 expression and secretion links macrophages to the promotion of renal fibrosis. Am J Pathol 172:288–298

Higgins DF, Kimura K, Bernhardt WM, Shrimanker N, Akai Y et al (2007) Hypoxia promotes fibrogenesis in vivo via HIF-1 stimulation of epithelial-to-mesenchymal transition. J Clin Invest 117:3810–3820

Higgins DF, Kimura K, Iwano M, Haase VH (2008) Hypoxia-inducible factor signaling in the development of tissue fibrosis. Cell Cycle 7:1128–1132

Hochane M, Raison D, Coquard C, Beraud C, Bethry A et al (2018) Parathyroid hormone-related protein modulates inflammation in mouse mesangial cells and blunts apoptosis by enhancing COX-2 expression. Am J Physiol Cell Physiol 314:C242–C253

Holdsworth SR, Summers SA (2008) Role of mast cells in progressive renal diseases. J Am Soc Nephrol 19:2254–2261

Huang XR, Chung AC, Wang XJ, Lai KN, Lan HY (2008a) Mice overexpressing latent TGF-beta1 are protected against renal fibrosis in obstructive kidney disease. Am J Physiol Renal Physiol 295:F118–F127
Huang XR, Chung AC, Zhou L, Wang XJ, Lan HY (2008b) Latent TGF-beta1 protects against crescentic glomerulonephritis. J Am Soc Nephrol 19:233–242
Humphreys BD (2018) Mechanisms of renal fibrosis. Annu Rev Physiol 80:309–326
Humphreys BD, Lin SL, Kobayashi A, Hudson TE, Nowlin BT et al (2010) Fate tracing reveals the pericyte and not epithelial origin of myofibroblasts in kidney fibrosis. Am J Pathol 176:85–97
Inoki K, Mori H, Wang J, Suzuki T, Hong S et al (2011) mTORC1 activation in podocytes is a critical step in the development of diabetic nephropathy in mice. J Clin Invest 121:2181–2196
Kanasaki K, Taduri G, Koya D (2013) Diabetic nephropathy: the role of inflammation in fibroblast activation and kidney fibrosis. Front Endocrinol 4:7
Kang DH, Kanellis J, Hugo C, Truong L, Anderson S et al (2002) Role of the microvascular endothelium in progressive renal disease. J Am Soc Nephrol 13:806–816
Kato N, Yuzawa Y, Kosugi T, Hobo A, Sato W et al (2009) The E-selectin ligand basigin/CD147 is responsible for neutrophil recruitment in renal ischemia/reperfusion. J Am Soc Nephrol 20:1565–1576
Kelly KJ, Williams WW Jr, Colvin RB, Bonventre JV (1994) Antibody to intercellular adhesion molecule 1 protects the kidney against ischemic injury. Proc Nati Acad Sci USA 91:812–816
Kelly KJ, Williams WW Jr, Colvin RB, Meehan SM, Springer TA et al (1996) Intercellular adhesion molecule-1-deficient mice are protected against ischemic renal injury. J Clin Invest 97:1056–1063
Kelly KJ, Burford JL, Dominguez JH (2009) Postischemic inflammatory syndrome: a critical mechanism of progression in diabetic nephropathy. Am J Physiol Renal Physiol 297:F923–F931
Kim DH, Moon SO, Jung YJ, Lee AS, Kang KP et al (2009) Mast cells decrease renal fibrosis in unilateral ureteral obstruction. Kidney Int 75:1031–1038
Kim SM, Lee SH, Lee A, Kim DJ, Kim YG et al (2015) Targeting T helper 17 by mycophenolate mofetil attenuates diabetic nephropathy progression. Transl Res 166:375–383
Kimura K, Iwano M, Higgins DF, Yamaguchi Y, Nakatani K et al (2008) Stable expression of HIF-1alpha in tubular epithelial cells promotes interstitial fibrosis. Am J Physiol Renal Physiol 295:F1023–F1029
Kitamoto K, Machida Y, Uchida J, Izumi Y, Shiota M et al (2009) Effects of liposome clodronate on renal leukocyte populations and renal fibrosis in murine obstructive nephropathy. J Pharmacol Sci 111:285–292
Kitamura M, Suto TS (1997) TGF-beta and glomerulonephritis: anti-inflammatory versus prosclerotic actions. Nephrol Dial Transplant 12:669–679
Kitching AR, Holdsworth SR (2011) The emergence of TH17 cells as effectors of renal injury. J Am Soc Nephrol 22:235–238
Ko GJ, Boo CS, Jo SK, Cho WY, Kim HK (2008) Macrophages contribute to the development of renal fibrosis following ischaemia/reperfusion-induced acute kidney injury. Nephrol Dial Transplant 23:842–852
Kondo S, Kagami S, Kido H, Strutz F, Muller GA, Kuroda Y (2001) Role of mast cell tryptase in renal interstitial fibrosis. J Am Soc Nephrol 12:1668–1676
Kong D, Li Y, Wang Z, Banerjee S, Ahmad A et al (2009) miR-200 regulates PDGF-D-mediated epithelial-mesenchymal transition, adhesion, and invasion of prostate cancer cells. Stem Cells 27:1712–1721
Ku CH, White KE, Dei Cas A, Hayward A, Webster Z et al (2008) Inducible overexpression of sFlt-1 in podocytes ameliorates glomerulopathy in diabetic mice. Diabetes 57:2824–2833
Lai KN, Leung JC, Chan LY, Saleem MA, Mathieson PW et al (2008) Activation of podocytes by mesangial-derived TNF-alpha: glomerulo-podocytic communication in IgA nephropathy. Am J Physiol Renal Physiol 294:F945–F955
Lai KN, Leung JC, Chan LY, Saleem MA, Mathieson PW et al (2009) Podocyte injury induced by mesangial-derived cytokines in IgA nephropathy. Nephrol Dial Transplant 24:62–72

Lan HY (2011) Diverse roles of TGF-beta/Smads in renal fibrosis and inflammation. Int J Biol Sci 7:1056–1067

Lan HY, Nikolic-Paterson DJ, Mu W, Atkins RC (1995) Local macrophage proliferation in the progression of glomerular and tubulointerstitial injury in rat anti-GBM glomerulonephritis. Kidney Int 48:753–760

Lan HY, Bacher M, Yang N, Mu W, Nikolic-Paterson DJ et al (1997) The pathogenic role of macrophage migration inhibitory factor in immunologically induced kidney disease in the rat. J Exp Med 185:1455–1465

Laping NJ, Olson BA, Ho T, Ziyadeh FN, Albrightson CR (2000) Hepatocyte growth factor: a regulator of extracellular matrix genes in mouse mesangial cells. Biochem Pharmacol 59:847–853

LeBleu VS, Kalluri R (2011) Blockade of PDGF receptor signaling reduces myofibroblast number and attenuates renal fibrosis. Kidney Int 80:1119–1121

LeBleu VS, Taduri G, O'Connell J, Teng Y, Cooke VG et al (2013) Origin and function of myofibroblasts in kidney fibrosis. Nat Med 19:1047–1053

Lebrin F, Deckers M, Bertolino P, Ten Dijke P (2005) TGF-beta receptor function in the endothelium. Cardiovascular Res 65:599–608

Lee SB, Kalluri R (2010) Mechanistic connection between inflammation and fibrosis. Kidney Int Suppl S22–S26

Li Y, Yang J, Dai C, Wu C, Liu Y (2003) Role for integrin-linked kinase in mediating tubular epithelial to mesenchymal transition and renal interstitial fibrogenesis. J Clin Invest 112:503–516

Li Y, Kang YS, Dai C, Kiss LP, Wen X, Liu Y (2008) Epithelial-to-mesenchymal transition is a potential pathway leading to podocyte dysfunction and proteinuria. Am J Pathol 172:299–308

Li J, Qu X, Yao J, Caruana G, Ricardo SD et al (2010) Blockade of endothelial-mesenchymal transition by a Smad3 inhibitor delays the early development of streptozotocin-induced diabetic nephropathy. Diabetes 59:2612–2624

Lin M, Tang SC (2013) Toll-like receptors: sensing and reacting to diabetic injury in the kidney. Nephrol Dial Transplant

Lin SL, Kisseleva T, Brenner DA, Duffield JS (2008) Pericytes and perivascular fibroblasts are the primary source of collagen-producing cells in obstructive fibrosis of the kidney. Am J Pathol 173:1617–1627

Lin SL, Castano AP, Nowlin BT, Lupher ML Jr, Duffield JS (2009) Bone marrow Ly6Chigh monocytes are selectively recruited to injured kidney and differentiate into functionally distinct populations. J Immunol 183:6733–6743

Lin M, Yiu WH, Wu HJ, Chan LY, Leung JC et al (2012) Toll-like receptor 4 promotes tubular inflammation in diabetic nephropathy. J Am Soc Nephrol 23:86–102

Lin JR, Zheng YJ, Zhang ZB, Shen WL, Li XD et al (2018) Suppression of endothelial-to-mesenchymal transition by SIRT (Sirtuin) 3 alleviated the development of hypertensive renal injury. Hypertension

Liu Y (2004) Epithelial to mesenchymal transition in renal fibrogenesis: pathologic significance, molecular mechanism, and therapeutic intervention. J Am Soc Nephrol 15:1–12

Liu Y (2006) Renal fibrosis: new insights into the pathogenesis and therapeutics. Kidney Int 69:213–217

Liu Y (2010) New insights into epithelial-mesenchymal transition in kidney fibrosis. J Am Soc Nephrol 21:212–222

Liu Y (2011) Cellular and molecular mechanisms of renal fibrosis. Nat Rev Nephrol 7:684–696

Liu F, Chen HY, Huang XR, Chung AC, Zhou L et al (2011) C-reactive protein promotes diabetic kidney disease in a mouse model of type 1 diabetes. Diabetologia 54:2713–2723

Liu L, Kou P, Zeng Q, Pei G, Li Y et al (2012) CD4+ T Lymphocytes, especially Th2 cells, contribute to the progress of renal fibrosis. Am J Nephrol 36:386–396

Liu P, Lassen E, Nair V, Berthier CC, Suguro M et al (2017) Transcriptomic and proteomic profiling provides insight into mesangial cell function in IgA nephropathy. J Am Soc Nephrol 28:2961–2972

Liu BC, Tang TT, Lv LL, Lan HY (2018) Renal tubule injury: a driving force toward chronic kidney disease. Kidney Int 93:568–579

Loeffler I, Wolf G (2013) Transforming growth factor-beta and the progression of renal disease. Nephrol Dial Transplant

Lopez-Hernandez FJ, Lopez-Novoa JM (2012) Role of TGF-beta in chronic kidney disease: an integration of tubular, glomerular and vascular effects. Cell Tissue Res 347:141–154

Lovisa S, LeBleu VS, Tampe B, Sugimoto H, Vadnagara K et al (2015) Epithelial-to-mesenchymal transition induces cell cycle arrest and parenchymal damage in renal fibrosis. Nat Med 21:998–1009

Lu J, Cao Q, Zheng D, Sun Y, Wang C et al (2013) Discrete functions of M2a and M2c macrophage subsets determine their relative efficacy in treating chronic kidney disease. Kidney Int 84:745–755

Lu Y, Mei Y, Chen L, Wu L, Wang X et al (2018) The role of transcriptional factor D-site-binding protein in circadian CCL2 gene expression in anti-Thy1 nephritis. Cell Mol Immunol

Lv W, Booz GW, Wang Y, Fan F, Roman RJ (2018) Inflammation and renal fibrosis: recent developments on key signaling molecules as potential therapeutic targets. Eur J Pharmacol 820:65–76

Lyons RM, Gentry LE, Purchio AF, Moses HL (1990) Mechanism of activation of latent recombinant transforming growth factor beta 1 by plasmin. J Cell Biol 110:1361–1367

Ma LJ, Yang H, Gaspert A, Carlesso G, Barty MM et al (2003) Transforming growth factor-beta-dependent and -independent pathways of induction of tubulointerstitial fibrosis in beta6(−/−) mice. Am J Pathol 163:1261–1273

Ma FY, Flanc RS, Tesch GH, Bennett BL, Friedman GC et al (2009a) Blockade of the c-Jun amino terminal kinase prevents crescent formation and halts established anti-GBM glomerulonephritis in the rat. Lab Inves 89:470–484

Ma FY, Liu J, Kitching AR, Manthey CL, Nikolic-Paterson DJ (2009b) Targeting renal macrophage accumulation via c-fms kinase reduces tubular apoptosis but fails to modify progressive fibrosis in the obstructed rat kidney. Am J Physiol Renal Physiol 296:F177–F185

Ma FY, Ikezumi Y, Nikolic-Paterson DJ (2010) Macrophage signaling pathways: a novel target in renal disease. Semin Nephrol 30:334–344

Mack M, Rosenkranz AR (2009) Basophils and mast cells in renal injury. Kidney Int 76:1142–1147

Margulis A, Nocka KH, Brennan AM, Deng B, Fleming M et al (2009) Mast cell-dependent contraction of human airway smooth muscle cell-containing collagen gels: influence of cytokines, matrix metalloproteases, and serine proteases. J Immunol 183:1739–1750

Mathieson PW (2009) Update on the podocyte. Curr Opin Nephrol Hypertens 18:206–211

Mehrotra P, Patel JB, Ivancic CM, Collett JA, Basile DP (2015) Th-17 cell activation in response to high salt following acute kidney injury is associated with progressive fibrosis and attenuated by AT-1R antagonism. Kidney Int 88:776–784

Meldrum KK, Zhang H, Hile KL, Moldower LL, Dong Z, Meldrum DR (2012) Profibrotic effect of interleukin-18 in HK-2 cells is dependent on stimulation of the Toll-like receptor 4 (TLR4) promoter and increased TLR4 expression. J Biol Chem 287:40391–40399

Meng XM, Huang XR, Chung AC, Qin W, Shao X et al (2010) Smad2 protects against TGF-beta/Smad3-mediated renal fibrosis. J Am Soc Nephrol 21:1477–1487

Meng XM, Huang XR, Xiao J, Chen HY, Zhong X, Chung AC, Lan HY (2012a) Diverse roles of TGF-beta receptor II in renal fibrosis and inflammation in vivo and in vitro. J Pathol 227:175–188

Meng XM, Huang XR, Xiao J, Chung AC, Qin W et al (2012b) Disruption of Smad4 impairs TGF-beta/Smad3 and Smad7 transcriptional regulation during renal inflammation and fibrosis in vivo and in vitro. Kidney Int 81:266–279

Meng XM, Chung AC, Lan HY (2013) Role of the TGF-beta/BMP-7/Smad pathways in renal diseases. Clin Sci (Lond) 124:243–254

Meng XM, Nikolic-Paterson DJ, Lan HY (2014) Inflammatory processes in renal fibrosis. Nat Rev Nephrol 10:493–503

Meng XM, Tang PM, Li J, Lan HY (2015) TGF-beta/Smad signaling in renal fibrosis. Front Physiol 6:82

Meng XM, Nikolic-Paterson DJ, Lan HY (2016a) TGF-beta: the master regulator of fibrosis. Nat Rev Nephrol 12:325–338

Meng XM, Wang S, Huang XR, Yang C, Xiao J et al (2016b) Inflammatory macrophages can transdifferentiate into myofibroblasts during renal fibrosis. Cell Death Dis 7:e2495

Mezzano SA, Ruiz-Ortega M, Egido J (2001) Angiotensin II and renal fibrosis. Hypertension 38:635–638

Miyazawa S, Hotta O, Doi N, Natori Y, Nishikawa K (2004) Role of mast cells in the development of renal fibrosis: use of mast cell-deficient rats. Kidney Int 65: 2228–2237

Mohamed R, Jayakumar C, Chen F, Fulton D, Stepp D et al (2016) Low-dose IL-17 therapy prevents and reverses diabetic nephropathy, metabolic syndrome, and associated organ fibrosis. J Am Soc Nephrol 27:745–765

Munger JS, Huang X, Kawakatsu H, Griffiths MJ, Dalton SL et al (1999) The integrin alpha v beta 6 binds and activates latent TGF beta 1: a mechanism for regulating pulmonary inflammation and fibrosis. Cell 96:319–328

Nakagawa T, Lan HY, Zhu HJ, Kang DH, Schreiner GF, Johnson RJ (2004a) Differential regulation of VEGF by TGF-beta and hypoxia in rat proximal tubular cells. Am J Physiol Renal Physiol 287:F658–F664

Nakagawa T, Li JH, Garcia G, Mu W, Piek E et al (2004b) TGF-beta induces proangiogenic and antiangiogenic factors via parallel but distinct Smad pathways. Kidney Int 66:605–613

Negre-Salvayre A, Coatrieux C, Ingueneau C, Salvayre R (2008) Advanced lipid peroxidation end products in oxidative damage to proteins. Potential role in diseases and therapeutic prospects for the inhibitors. Br J Pharmacol 153:6–20

Nguan CY, Du C (2009) Renal tubular epithelial cells as immunoregulatory cells in renal allograft rejection. Transplant Rev (Orlando) 23:129–138

Niedermeier M, Reich B, Rodriguez Gomez M, Denzel A, Schmidbauer K et al (2009) CD4 + T cells control the differentiation of Gr1+ monocytes into fibrocytes. Proc Nati Acad Sci USA 106:17892–17897

Nielsen R, Mollet G, Esquivel EL, Weyer K, Nielsen PK et al (2013) Increased lysosomal proteolysis counteracts protein accumulation in the proximal tubule during focal segmental glomerulosclerosis. Kidney Int 84:902–910

Niranjan T, Bielesz B, Gruenwald A, Ponda MP, Kopp JB et al (2008) The Notch pathway in podocytes plays a role in the development of glomerular disease. Nat Med 14:290–298

Nishida M, Hamaoka K (2008) Macrophage phenotype and renal fibrosis in obstructive nephropathy. Nephron Exp Nephrol 110:e31–e36

Ostendorf T, Rong S, Boor P, Wiedemann S, Kunter U et al (2006) Antagonism of PDGF-D by human antibody CR170 prevents renal scarring in experimental glomerulonephritis. J Am Soc Nephrol 17:1054–1062

Ostendorf T, Eitner F, Floege J (2012) The PDGF family in renal fibrosis. Pediatr Nephrol 27:1041–1050

Paust HJ, Turner JE, Steinmetz OM, Peters A, Heymann F et al (2009) The IL-23/Th17 axis contributes to renal injury in experimental glomerulonephritis. J Am Soc Nephrol 2:969–979

Peng X, Xiao Z, Zhang J, Li Y, Dong Y, Du J (2015) IL-17A produced by both gammadelta T and Th17 cells promotes renal fibrosis via RANTES-mediated leukocyte infiltration after renal obstruction. J Pathol 235:79–89

Pober JS, Sessa WC (2007) Evolving functions of endothelial cells in inflammation. Nat Rev Immunol 7:803–815

Pons M, Ali L, Beghdadi W, Danelli L, Alison M et al (2017) Mast cells and MCPT4 chymase promote renal impairment after partial ureteral obstruction. Front Immunol 8:450

Pulskens WP, Rampanelli E, Teske GJ, Butter LM, Claessen N et al (2010) TLR4 promotes fibrosis but attenuates tubular damage in progressive renal injury. J Am Soc Nephrol 21:1299–1308

Qiao X, Rao P, Zhang Y, Liu L, Pang M et al (2018) Redirecting TGF-beta signaling through the beta-catenin/Foxo complex prevents kidney fibrosis. J Am Soc Nephrol 29:557–570

Reich B, Schmidbauer K, Rodriguez Gomez M, Johannes Hermann F, Gobel N et al (2013) Fibrocytes develop outside the kidney but contribute to renal fibrosis in a mouse model. Kidney Int 84:78–89
Reilkoff RA, Bucala R, Herzog EL (2011) Fibrocytes: emerging effector cells in chronic inflammation. Nat Rev Immunol 11:427–435
Ricardo SD, van Goor H, Eddy AA (2008) Macrophage diversity in renal injury and repair. J Clin Invest 118:3522–3530
Robertson H, Ali S, McDonnell BJ, Burt AD, Kirby JA (2004) Chronic renal allograft dysfunction: the role of T cell-mediated tubular epithelial to mesenchymal cell transition. J Am Soc Nephrol 15:390–397
Rutkowski JM, Wang ZV, Park AS, Zhang J, Zhang D et al (2013) Adiponectin promotes functional recovery after podocyte ablation. J Am Soc Nephrol 24:268–282
Sakai N, Wada T, Yokoyama H, Lipp M, Ueha S, Matsushima K, Kaneko S (2006) Secondary lymphoid tissue chemokine (SLC/CCL21)/CCR182 signaling regulates fibrocytes in renal fibrosis. Proc Nati Acad Sci USA 103:14098–14103
Sam R, Wanna L, Gudehithlu KP, Garber SL, Dunea G et al (2006) Glomerular epithelial cells transform to myofibroblasts: early but not late removal of TGF-beta1 reverses transformation. Transl Res 148:142–148
Scandiuzzi L, Beghdadi W, Daugas E, Abrink M, Tiwari N et al (2010) Mouse mast cell protease-4 deteriorates renal function by contributing to inflammation and fibrosis in immune complex-mediated glomerulonephritis. J Immunol 185:624–633
Schiffer M, Bitzer M, Roberts IS, Kopp JB, ten Dijke P et al (2001) Apoptosis in podocytes induced by TGF-beta and Smad7. J Clin Invest 108:807–816
Schlondorff D, Banas B (2009) The mesangial cell revisited: no cell is an island. J Am Soc Nephrol 20:1179–1187
Schnaper HW, Hayashida T, Hubchak SC, Poncelet AC (2003) TGF-beta signal transduction and mesangial cell fibrogenesis. Am J Physiol Renal Physiol 284:F243–F252
Schnoor M, Cullen P, Lorkowski J, Stolle K, Robenek H et al (2008) Production of type VI collagen by human macrophages: a new dimension in macrophage functional heterogeneity. J Immunol 180:5707–5719
Schwartzman M, Reginensi A, Wong JS, Basgen JM, Meliambro K et al (2016) Podocyte-specific deletion of yes-associated protein causes FSGS and progressive renal failure. J Am Soc Nephrol 27:216–226
Shanmugam N, Figarola JL, Li Y, Swiderski PM, Rahbar S, Natarajan R (2008) Proinflammatory effects of advanced lipoxidation end products in monocytes. Diabetes 57:879–888
Shao DD, Suresh R, Vakil V, Gomer RH, Pilling D (2008) Pivotal Advance: Th-1 cytokines inhibit, and Th-2 cytokines promote fibrocyte differentiation. J Leukoc Biol 83:1323–1333
Sharma S, Sirin Y, Susztak K (2011) The story of Notch and chronic kidney disease. Curr Opin Nephrol Hypertens 20:56–61
Shi XY, Hou FF, Niu HX, Wang GB, Xie D (2008) Advanced oxidation protein products promote inflammation in diabetic kidney through activation of renal nicotinamide adenine dinucleotide phosphate oxidase. Endocrinology 149:1829–1839
Shih NY, Li J, Karpitskii V, Nguyen A, Dustin ML, Kanagawa O et al (1999) Congenital nephrotic syndrome in mice lacking CD2-associated protein. Science 286:312–315
Sica A, Mantovani A (2012) Macrophage plasticity and polarization: in vivo veritas. J Clin Invest 122:787–795
Siddiqi FS, Advani A (2013) Endothelial-podocyte crosstalk: the missing link between endothelial dysfunction and albuminuria in diabetes. Diabetes 62:3647–3655
Strutz F, Zeisberg M (2006) Renal fibroblasts and myofibroblasts in chronic kidney disease. J Am Soc Nephrol 17:2992–2998
Summers SA, Gan PY, Dewage L, Ma FT, Ooi JD et al (2012) Mast cell activation and degranulation promotes renal fibrosis in experimental unilateral ureteric obstruction. Kidney Int 82:676–685

Tang WW, Ulich TR, Lacey DL, Hill DC, Qi M et al (1996) Platelet-derived growth factor-BB induces renal tubulointerstitial myofibroblast formation and tubulointerstitial fibrosis. Am J Pathol 148:1169–1180
Taniguchi K, Xia L, Goldberg HJ, Lee KW, Shah A et al (2013) Inhibition of Src kinase blocks high glucose-induced EGFR transactivation and collagen synthesis in mesangial cells and prevents diabetic nephropathy in mice. Diabetes 62:3874–3886
Tapmeier TT, Fearn A, Brown K, Chowdhury P, Sacks SH et al (2010) Pivotal role of CD4+ T cells in renal fibrosis following ureteric obstruction. Kidney Int 78:351–362
Tipping PG, Holdsworth SR (2006) T cells in crescentic glomerulonephritis. J Am Soc Nephrol 17:1253–1263
Turner JE, Paust HJ, Steinmetz OM, Panzer U (2010) The Th17 immune response in renal inflammation. Kidney Int 77:1070–1075
Ueno T, Kobayashi N, Nakayama M, Takashima Y, Ohse T et al (2013) Aberrant Notch1-dependent effects on glomerular parietal epithelial cells promotes collapsing focal segmental glomerulosclerosis with progressive podocyte loss. Kidney Int 83:1065–1075
van Roeyen CR, Eitner F, Boor P, Moeller MJ, Raffetseder U et al (2011) Induction of progressive glomerulonephritis by podocyte-specific overexpression of platelet-derived growth factor-D. Kidney Int 80:1292–1305
Venkatachalam MA, Griffin KA, Lan R, Geng H, Saikumar P, Bidani AK (2010) Acute kidney injury: a springboard for progression in chronic kidney disease. Am J Physiol Renal Physiol 298:F1078–F1094
Vernon MA, Mylonas KJ, Hughes J (2010) Macrophages and renal fibrosis. Semin Nephrol 30:302–317
Veron D, Reidy KJ, Bertuccio C, Teichman J, Villegas G et al (2010) Overexpression of VEGF-A in podocytes of adult mice causes glomerular disease. Kidney Int 77:989–999
Veron D, Bertuccio CA, Marlier A, Reidy K, Garcia AM et al (2011) Podocyte vascular endothelial growth factor (Vegf(1)(6)(4)) overexpression causes severe nodular glomerulosclerosis in a mouse model of type 1 diabetes. Diabetologia 54:1227–1241
Wada T, Sakai N, Matsushima K, Kaneko S (2007) Fibrocytes: a new insight into kidney fibrosis. Kidney Int 72:269–273
Wang YM, Zhang GY, Wang Y, Hu M, Wu H et al (2006) Foxp3-transduced polyclonal regulatory T cells protect against chronic renal injury from adriamycin. J Am Soc Nephrol 17:697–706
Wang Y, Wang YP, Zheng G, Lee VW, Ouyang L et al (2007) Ex vivo programmed macrophages ameliorate experimental chronic inflammatory renal disease. Kidney Int 72:290–299
Wang Y, Deb DK, Zhang Z, Sun T, Liu W et al (2012) Vitamin D receptor signaling in podocytes protects against diabetic nephropathy. J Am Soc Nephrol 23:1977–1986
Wang H, Wang J, Bai Y, Li J, Li L, Dong Y (2016a) CD11c(+) CD8(+) T Cells Reduce Renal Fibrosis Following Ureteric Obstruction by Inducing Fibroblast Apoptosis. Int J Mol Sci 18
Wang S, Meng XM, Ng YY, Ma FY, Zhou S et al (2016b) TGF-beta/Smad3 signalling regulates the transition of bone marrow-derived macrophages into myofibroblasts during tissue fibrosis. Oncotarget 7:8809–8822
Wang X, Yao B, Wang Y, Fan X, Wang S et al (2017a) Macrophage cyclooxygenase-2 protects against development of diabetic nephropathy. Diabetes 66:494–504
Wang YY, Jiang H, Pan J, Huang XR, Wang YC et al (2017b) Macrophage-to- myofibroblast transition contributes to interstitial fibrosis in chronic renal allograft injury. J Am Soc Nephrol 28:2053–2067
Wasse H, Naqvi N, Husain A (2012) Impact of mast cell chymase on renal disease progression. Curr Hypertens Rev 8:15–23
Wilkinson L, Gilbert T, Sipos A, Toma I, Pennisi DJ et al (2009) Loss of renal microvascular integrity in postnatal Crim1 hypomorphic transgenic mice. Kidney Int 76:1161–1171
Wynes MW, Frankel SK, Riches DW (2004) IL-4-induced macrophage-derived IGF-I protects myofibroblasts from apoptosis following growth factor withdrawal. J Leukoc Biol 76:1019–1027

Xavier S, Vasko R, Matsumoto K, Zullo JA, Chen R et al (2015) Curtailing endothelial TGF-beta signaling is sufficient to reduce endothelial-mesenchymal transition and fibrosis in CKD. J Am Soc Nephrol 26:817–829

Yamaguchi Y, Iwano M, Suzuki D, Nakatani K, Kimura K et al (2009) Epithelial-mesenchymal transition as a potential explanation for podocyte depletion in diabetic nephropathy. Am J Kidney Dis 54:653–664

Yang N, Isbel NM, Nikolic-Paterson DJ, Li Y, Ye R et al (1998a) Local macrophage proliferation in human glomerulonephritis. Kidney Int 54:143–151

Yang N, Wu LL, Nikolic-Paterson DJ, Ng YY, Yang WC et al (1998b) Local macrophage and myofibroblast proliferation in progressive renal injury in the rat remnant kidney. Nephrol Dial Transplant 13:1967–1974

Yang J, Dai C, Liu Y (2002) Hepatocyte growth factor gene therapy and angiotensin II blockade synergistically attenuate renal interstitial fibrosis in mice. J Am Soc Nephrol 13:2464–2477

Yang L, Besschetnova TY, Brooks CR, Shah JV, Bonventre JV (2010) Epithelial cell cycle arrest in G2/M mediates kidney fibrosis after injury. Nat Med 16:535–543, 531p following 143

Yokoi H, Mukoyama M, Mori K, Kasahara M, Suganami T et al (2008) Overexpression of connective tissue growth factor in podocytes worsens diabetic nephropathy in mice. Kidney Int 73:446–455

You H, Gao T, Cooper TK, Brian Reeves W, Awad AS (2013) Macrophages directly mediate diabetic renal injury. Am J Physiol Renal Physiol 305:F1719–F1727

You H, Gao T, Raup-Konsavage WM, Cooper TK, Bronson SK et al (2017) Podocyte-specific chemokine (C-C motif) receptor 2 overexpression mediates diabetic renal injury in mice. Kidney Int 91:671–682

Zeisberg M, Neilson EG (2010) Mechanisms of tubulointerstitial fibrosis. J Am Soc Nephrol 21:1819–1834

Zeisberg EM, Potenta SE, Sugimoto H, Zeisberg M, Kalluri R (2008) Fibroblasts in kidney fibrosis emerge via endothelial-to-mesenchymal transition. J Am Soc Nephrol 19:2282–2287

Zeng F, Kloepfer LA, Finney C, Diedrich A, Harris RC (2016) Specific endothelial heparin-binding EGF-like growth factor deletion ameliorates renal injury induced by chronic angiotensin II infusion. Am J Physiol Renal Physiol 311:F695–F707

Zhang B, Ramesh G, Uematsu S, Akira S, Reeves WB (2008) TLR4 signaling mediates inflammation and tissue injury in nephrotoxicity. J Am Soc Nephrol 19:923–932

Zhang F, Tsai S, Kato K, Yamanouchi D, Wang C et al (2009) Transforming growth factor-beta promotes recruitment of bone marrow cells and bone marrow-derived mesenchymal stem cells through stimulation of MCP-1 production in vascular smooth muscle cells. J Biol Chem 284:17564–17574

Zheng G, Wang Y, Mahajan D, Qin X, Alexander SI, Harris DC (2005) The role of tubulointerstitial inflammation. Kidney Int Suppl S96–S100

Zhou LL, Hou FF, Wang GB, Yang F, Xie D, Wang YP, Tian JW (2009) Accumulation of advanced oxidation protein products induces podocyte apoptosis and deletion through NADPH-dependent mechanisms. Kidney Int 76:1148–1160

Chapter 19
Role of Inflammasome in Chronic Kidney Disease

Liang Li, Wei Tang and Fan Yi

Abstract The inflammasome is a multiprotein complex assembled by intracytoplasmic pattern recognition receptors and is a key component of the innate immune system for host defense. Inflammasome recruits and activates the proinflammatory protease caspase-1 by recognizing pathogen-associated molecular patterns (PAMPs) or host-derived damage-associated molecular patterns (DAMPs). Activated caspase-1 cleaves the precursors of IL-1β and IL-18 to produce the corresponding mature cytokines. Several types of inflammasomes have been identified, such as NLRP3, NLRP1, IPAF (NLRC4) and AIM2. NLRP3 has recently been reported as a central pathogenic mechanism of chronic kidney disease (CKD). In this chapter, we briefly summarize the current knowledge about the roles of inflammasomes in the pathogenesis of CKD. A better understanding of the function of inflammasomes will provide unexpected opportunities to develop new therapies for kidney diseases by modulation of the innate immune system.

Keywords Inflammasome · Chronic kidney disease · Pattern recognition receptors · Inflammation · Fibrosis

19.1 Introduction

Chronic kidney disease (CKD) is a leading cause of death in the world. Although numerous studies have shown that persistent inflammation is a hallmark of chronic kidney injury and the activation of the innate immune system is of importance in the development of CKD (Granata et al. 2016), the underlying mechanisms of inflammation-induced CKD still keep largely unknown. Pattern recognition receptors

L. Li · F. Yi (✉)
Department of Pharmacology, School of Basic Medical Sciences, Shandong University, Jinan 250012, China
e-mail: fanyi@sdu.edu.cn

W. Tang
Department of Pathogenic Biology, Basic Medical Sciences, Shandong University, Jinan 250012, China

B.-C. Liu et al. (eds.), *Renal Fibrosis: Mechanisms and Therapies*,
Advances in Experimental Medicine and Biology 1165,
https://doi.org/10.1007/978-981-13-8871-2_19

(PRRs) act as sensors of the innate immune system and detect pathogen-associated molecular patterns (PAMPs) or damage-associated molecular patterns (DAMPs), which initiate immune responses to resolve infections and repair damaged tissues. Inappropriate or chronic PRR activation results in excessive inflammation and exacerbates tissue damage. During this process, the involvement of inflammatory cells in the damaged renal interstitium is a universal finding in failing kidneys and correlates inversely with renal function. In addition to immune cells, kidney intrinsic cells such as endothelial cells, podocytes and renal tubular epithelial cells are also involved in the process of inflammation, resulting in the loss of glomerular function and tubulointerstitial fibrosis (El-Nahas 2003; Nakagawa et al. 2012).

PRRs are classified into several subfamilies, which consist of toll-like receptors, nucleotide oligomerization domain (NOD)-like receptors (NLRs), C-type lectin-like receptors and retinoic acid-inducible gene I-like receptors (Wang and Yi 2015). Among them, NLRs are recently identified as intracellular PRRs that are essential to innate immune responses and homeostasis. A better understanding of the function of NLRs will provide unexpected opportunities to develop new therapies for CKD by modulation of the innate immune system. In this chapter, we summarize the current knowledge about the role and the molecular mechanisms underlying the activation of inflammasomes, one major subtype of NLRs, in the development of CKD and discuss the opportunities of pharmacological targeting of inflammasome-mediated signaling pathways at multiple levels for the treatment of CKD.

19.2 The Structure and Function of NLRs

NLRs are recently identified intracellular PRRs and currently 22 members have been identified in humans and 34 in mice which contain three different domains (Fritz et al. 2006): a highly conserved nucleotide binding and oligomerization (NACHT) domain that is required for self-oligomerization of the proteins upon ligand recognition, a C-terminal leucine-rich repeat that acts as a sensing domain and a N-terminal effector domain that mediates downstream signaling such as caspase activation and recruitment domain (CARD) or a pyrin domain (PYD) (Caruso et al. 2014). On the basic of their N-terminal effector domains, NLRs are divided into five subfamilies consisting of NLRA, NLRB, NLRC, NLRP and NLRX. NLRC subfamily is composed of five members including NOD1 (NLRC1), NOD2 (NLRC2 or CARD15), NLRC3, NLRC4 and NLRC5 (Fig. 19.1). NOD1 and NOD2 are two well-studied NLRs that mediate the activation of NF-κB and mitogen-activated protein kinases (MAPKs) pathways in response to peptidoglycan-related molecules (Keestra-Gounder and Tsolis 2017). By recruiting receptor-interacting protein 2 (Rip2) (Magalhaes et al. 2011), NOD1 and NOD2 undergo homophilic CARD–CARD interaction with Rip2. Subsequently, Rip2 interacts with IKK complexes (IKKs), and thereby inducing NOD1 or NOD2 to interact with IKKs, leading to the activation of NF-κB signaling pathway and promoting the transcription of proinflammatory cytokines such as TNF-α, IL-1β, IL-6 and IL-18 (Magalhaes et al. 2011). NOD1 and NOD2 are firstly found to be

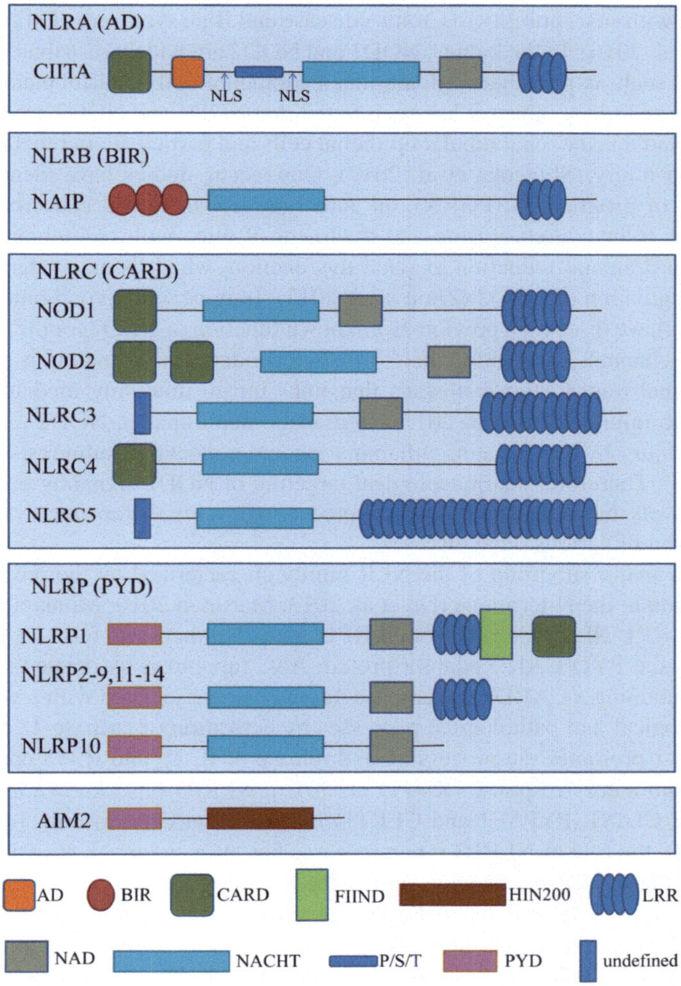

Fig. 19.1 Classification and molecular structures of NLRs and AIM2. *AD,* Activator domain; *AIM2,* Absent In Melanoma 2; *BIR,* baculovirus inhibitor of an apoptosis protein repeat; *CARD,* Caspase activation and recruitment domains; *FIIND,* Function to find domain; *HIN200* haematopoietic interferon-inducible nuclear 200; *LRR,* leucine-rich repeat; *NACHT,* neuronal apoptosis inhibitor protein (NAIP); *NAD,* NACHT-associated domain; *NLS,* nuclear localization sequence; *PYD,* pyrin domain

associated with susceptibility to Crohn's disease and Blau syndrome (Gu et al. 2018; Iwasaki et al. 2016). In the kidney, NOD1 and NOD2 are widely distributed in many cell types, such as podocytes, glomerular endothelial cells and tubular epithelial cells. In acute kidney injury, it has been found that NOD1 and NOD2 are expressed in human and murine renal tubular epithelial cells and participate in renal ischemia-reperfusion injury (Shigeoka et al. 2010). Our recent studies have identified that deficiency of progranulin (PGRN), an autocrine growth factor, exacerbates renal injury after ischemia/reperfusion and treatment of mice with recombinant PGRN results in a dramatic reduction in renal dysfunction, which is associated with the negative regulation of NOD2 (Zhou et al. 2015). In mice with hyperhomocysteinemia (hHcys), we discover a previously unknown function of NOD2 for the regulation of TRPC6 channels, suggesting that TRPC6-dependent Ca^{2+}signaling is one of the critical signal transduction pathways that links innate immunity mediator NOD2 to podocyte injury (Han et al. 2013). In diabetic nephropathy, NOD2 exacerbates podocyte injury by exacerbating inflammation and podocyte insulin resistance (Du et al. 2013). Therefore, pharmacological targeting of NOD2 signaling pathways at multiple levels may help design a new approach to develop therapeutic strategies for the treatment of various renal diseases.

Another major subgroup of the NLR family characterized by the presence of a PYD domain at the N-terminus (Du et al. 2013; Martinon 2010; Monk et al. 2017), such as NLRP1, NLRP3, NLRP6 or NLRP12, can oligomerize to form an inflammasome with the PYD-CARD adaptor protein ASC (apoptosis-associated speck-like protein containing a CARD) and caspase-1, which are associated with a wide range of physiological and pathological processes by activation of caspase-1. Activation of caspase-1 promotes the processing and release of IL-1β and IL-18 contributing to the inflammatory response (Kim et al. 2016). NLRP3 (also known as NALP3, Cryopyrin, CIAS1, PYPAF1 and CLR11) is located on chromosome 1q44 (Shen et al. 2018), the role of NLRP3 inflammasome has attracted much attention. External and endogenous danger signals induce the assembly and activation of inflammasome, which are closely related to the pathogenesis of multiple autoimmune diseases, aseptic inflammation and chronic inflammatory diseases (Broderick et al. 2015). In addition, NLRP3 inflammasome is one of the best characterized members in the kidney. Studies have also shown that NLRP3 inflammasome plays a vital role in the pathogenesis of CKD.

19.3 The Role of NLRP3 Inflammasome in CKD

Numerous studies have demonstrated that inflammation is a hallmark of CKD and the innate immune system participates in many inflammatory processes during the development of CKD. The expression of NLRP3 in the kidney and the activity of NLRP3 inflammasome are significantly increased in different types of CKDs. In this section, we review the current knowledge regarding the functional roles of NLRP3 inflammasome in CKD including renal fibrosis, diabetic nephropathy, obesity-related kidney

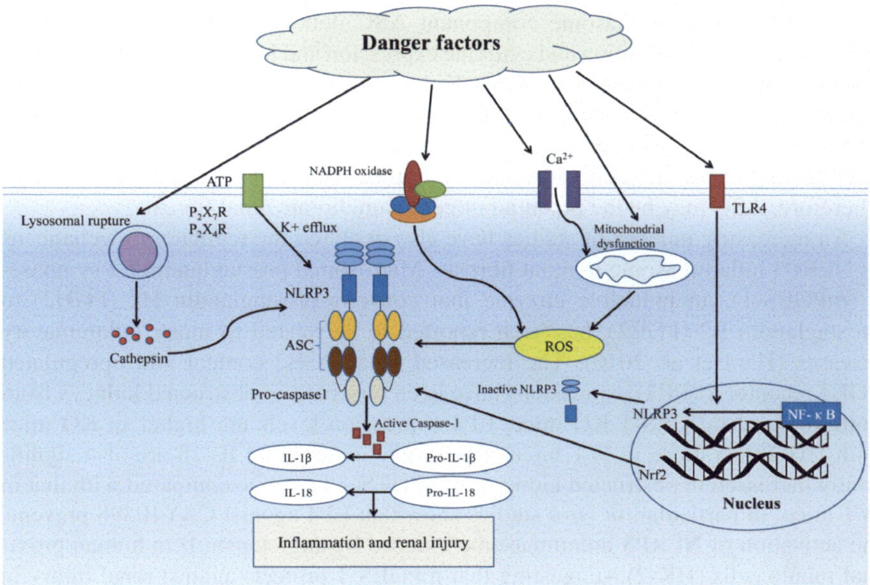

Fig. 19.2 Summarized recent findings on the mechanisms of NLRP3 inflammasome activation in CKDs

disease, chronic glomerulonephritis, IgA nephropathy, crystal-related nephropathy and hyperhomocysteinemia-induced renal injury (Fig. 19.2).

19.4 NLRP3 Inflammasome and Renal Fibrosis

Renal fibrosis is the final common pathway of numerous progressive kidney diseases, which is characterized by the activation and proliferation of renal interstitial fibroblasts as well as accumulation of extracellular matrix components. The unilateral ureteric obstruction (UUO) mouse model is a well-established animal model for the study of renal interstitial fibrosis. It has been found that NLRP3$^{-/-}$ mice have less tubular injury, inflammation and fibrosis after UUO compared with wild-type (WT) mice, accompanied by reduced caspase-1 activation and secretion of IL-1β and IL-18 (Vilaysane et al. 2010). Further bone marrow chimeras studies reveal that NLRP3 mediates the inflammatory processes in both hematopoietic and non-hematopoietic cellular compartments (Vilaysane et al. 2010). However, Pulskens et al. reported the non-canonical effects of NLRP3 following progressive renal injury induced by UUO. They have found that the deficiency of NLRP3 leads to early tubular damage and interstitial edema, suggesting a protective role of NLRP3 which is against progressive renal injury by preserving the vascular barrier and epithelial integrity in the kidney. Further studies need to be clarified in this issue.

The NLRP3 inflammasome component ASC deficiency significantly reduces inflammatory cell infiltration and cytokine expression and improves subsequent renal injury and fibrosis (Komada et al. 2014). Kidney-specific silencing of ASC attenuated proteinuria, albuminuria and glomerulosclerosis in mouse models of renal diseases (Abais et al. 2013; Zhang et al. 2012). Furthermore, ASC is specifically upregulated in collecting duct (CD) epithelial cells from mice with UUO (Komada et al. 2014). Therefore, ASC may be the potential target to ameliorate renal fibrosis.

An increasing number of studies have also explored the regulatory mechanisms of NLRP3 inflammasome in renal fibrosis. Microsomal prostaglandin E2 synthase-1 (mPGES-1), an inducible enzyme that converts prostaglandin H2 (PGH2) to prostaglandin E2 (PGE2), has been reported to be related to many inflammatory diseases (Hara et al. 2010). The increased renal PGE2 content and upregulated PGE2 receptor 4 (EP4) expressions have been observed in obstructed kidneys from both WT and mPGES-1 KO mice. EP4 expression levels are higher in KO mice with UUO than those in WT mice. The levels of ASC and IL-1β are also significantly increased in obstructed kidneys of mPGES-1 KO mice compared with that in WT mice. In particular, *in vitro* studies show that EP4 agonist CAY10598 prevents the activation of NLRP3 inflammasome induced by angiotensin II in human proximal tubule cells (HK-2), suggesting that mPGES-1 protects against renal injury in obstructive nephropathy at least in part, through the negative regulation of NLRP3 inflammasome (Luo et al. 2017). Nuclear factor erythroid 2-related factor 2 (Nrf2), a master transcription factor for antioxidant and detoxification responses (Ma 2013), has been reported to be associated with the NLRP3 inflammasome activation (Zhao et al. 2014). In the mouse UUO model, Nrf2 is necessary for the induction of IL-1β and caspase-1 within M1 macrophages (Sogawa et al. 2017). Loss of Nrf2 suppresses fibrosis and inflammation in the mouse UUO model by the reduction of the expression of NLRP3 and IL-1β. Some studies have also reported that activating renal tubular Wnt/β-catenin signaling triggers renal inflammation and fibrosis *via* the TLR-4/NLRP3 inflammasome axis (Wong et al. 2018).

19.5 NLRP3 Inflammasome and Diabetic Nephropathy

Diabetic nephropathy (DN) is one of the major microvascular complications of diabetes mellitus and the most common cause of the end-stage renal disease (Kanwar et al. 2011). Dysregulation of the innate immune response via NLRP3 has been implicated in the development and progression of diabetic nephropathy. In patients with DN, not only is the expression of NLRP3 in renal tubules significantly increased (Chen et al. 2013), but also the level of urinary protein is negatively correlated with the expression of caspase-1, IL-1β and IL-18 in renal tubules (Fang et al. 2013). NLRP3 inflammasome is also activated in animal models of DN (Shahzad et al. 2015). Bakker et al. have further demonstrated that NLRP3 plays a key role in diet-induced nephropathy and renal cholesterol accumulation (Bakker et al. 2014).

In addition to immune cells, non-bone marrow-derived renal intrinsic cells such as podocytes, endothelial cells and renal tubular epithelial cells can also initiate inflammatory responses (Shahzad et al. 2015). Recent studies have unraveled the role of the activated NLRP3 inflammasome in renal parenchymal cells and addressed the importance of NLRP3 inflammasome activation in non-myeloid-derived cells in diabetic nephropathy. Shahzad et al. demonstrated that NLRP3 inflammasome activation in non-myeloid-derived cells aggravates diabetic nephropathy. They found that abolishing NLRP3 or expression in bone marrow-derived cells fails to protect the mice against diabetic nephropathy. Conversely, NLRP3-deficient mice are protected against diabetic nephropathy despite transplantation of wild-type bone marrow. These results provide direct evidence that NLRP3 in renal parenchymal cells significantly contributes to the pathogenesis of diabetic nephropathy. Consistently, in vitro studies further confirm the upregulation of NLRP3 under high glucose condition in podocytes (Gao et al. 2014) and in renal tubular epithelial cells (Fang et al. 2013).

19.6 NLRP3 Inflammasome and Obesity-Related Kidney Disease

Obesity is becoming a serious health problem worldwide. Mounting evidence indicates that high body mass index (BMI) and obesity are important risk factors for CKD (Kovesdy et al. 2017). Previous studies have identified that visceral fat generates bioactive substances that contribute to the development of obesity-related kidney disease (ORKD). Numerous studies have indicated that the NLRP3 inflammasome participates in ORKD development and may serve as a key modulator of ORKD. Inhibition of the ASC protects mice from high fat diet (HFD)-induced obesity, glomerular injury and podocyte damage (Boini et al. 2014). The associated molecular of inflammasome activation is related to the production of fatty acid metabolites ceramide and palmitate, since their abundance in adipose tissue positively correlates with the development of obesity and type 2 diabetes (Ke et al. 2018). Studies have also reported inflammatory response to obesity, such as TLR4 and NLRP3 inflammasome activation as well as IL-1β secretion, attenuates β3-adrenoreceptor-induced beige adipocyte formation via oxidative stress and mitochondrial dysfunction, providing insights into targeting innate inflammatory system for enhancement of the adaptive thermogenesis against obesity (Okla et al. 2018).

Mitochondrial dysfunction can directly activate the NLRP3 inflammasome by itself or indirectly activate the NLRP3 inflammasome via ROS production. The activation of NLRP3 inflammasome can also induce mitochondrial damage in ORKD. Mitochondrial dysfunction is involved in obesity and obesity-related disorders which can initiate autophagy, a cellular degradation pathway essential for survival, then remove dysfunctional mitochondria and maintain cellular homeostasis (Conley et al. 2017). Yamamoto et al. found that autophagy ablation exaggerates HFD-induced mitochondrial dysfunction and inflammasome activation and proposed that HFD-impaired autophagic flux contributes to kidney lipotoxicity.

19.7 NLRP3 Inflammasome and Chronic Glomerulonephritis

Chronic glomerulonephritis (GN) is a series of immunological diseases with similar symptoms, different patho-manifestations and prognoses. Nearly all forms of acute glomerulonephritis have a tendency to progress to chronic glomerulonephritis. The condition is characterized by irreversible and progressive glomerular and tubulointerstitial fibrosis, ultimately leading to a reduction in the glomerular filtration rate (GFR) and retention of uremic toxins. Although some studies have shown that anti-GBM glomerulonephritis involves IL-1 but is independent of NLRP3/ASC inflammasome-mediated activation of caspase-1 (Lichtnekert et al. 2011), recent studies have demonstrated that a selective inhibitor of NLRP3 inflammasome activity, Bay11-7082, prevents the assembling and activation of the inflammasome, and decreases proteinuria, blood urea nitrogen and glomerular damage during GN. Bay11-7082 treatment can also decrease renal immune complex deposition and the levels of IL-1β, TNF-α and CCL2 as well as the amount of infiltration of macrophages (Zhao et al. 2013).

19.8 NLRP3 Inflammasome and IgA Nephropathy

Immunoglobulin IgA nephropathy (IgAN) is the leading form of primary glomerulonephritis associated with end-stage renal disease. Deposition of predominantly glomerular IgA immune complexes (ICs) or IgA immune aggregates the activation of innate immunity (Mestecky et al. 2013; Yu and Chiang 2014). The diagnosis of IgA nephropathy currently only depends on renal biopsy. The characteristic pathological features of IgAN are glomerular mesangial cells proliferation, mesangial matrix increasing and IgA-based immunoglobulins. Recent studies have found that NLRP3 inflammasome serves as a key regulator in the pathogenesis of IgAN (Liu et al. 2014; Tsai et al. 2017). IgA ICs can activate the NLRP3 inflammasome in macrophages and enhance the levels of NLRP3 and pro-IL-1β in macrophages (Tsai et al. 2017). IgA ICs can also induce DC activity, resulting in $CD4^+$ T cells activation and differentiation/polarization (Tsai et al. 2017). These findings indicate that NLRP3 inflammasome plays an important role in the development of IgAN (He et al. 2015; Liu et al. 2014). In IgA nephropathy, IgA ICs induce NLRP3 inflammasome activation through ROS-mediated pathways. IgA ICs increase mitochondrial ROS generation and mitochondrial DNA release into the cytosol of macrophages. Using mito-TEMPO, an inhibitor of mitochondrial ROS, the levels of NLRP3 and pro-IL-1β and IL-1β secretion in IgA ICs-activated macrophages are significantly reduced by mito-TEMPO (Tsai et al. 2017), indicating that NLRP3 inflammasome contributes to the development of IgAN, at least in part by inducing mitochondrial damage and ROS production. Recent studies have also found that some compounds for the potential treatment of IgAN, which are associated with the NLRP3 activation. Icariin, a major constituent of flavonoids isolated from plants of genus *Epimedium*, has been reported to suppress inflammatory responses and alleviate glomerulosclerosis in IgAN rats by inhibiting IgA deposition and NLRP3 inflammasome activation (Zhang et al. 2017). Antroquinonol, a pure active compound from *Antrodia camphorata* mycelium, inhibits renal inflammation and reduces oxidative stress in a mouse model of renal fibrosis (Yang et al. 2013). Antroquinonol has been shown to inhibit the activation of NLRP3 inflammasome and reduce ROS production in IgA-IC-primed macrophages (Yang et al. 2013).

19.9 NLRP3 Inflammasome and Crystalline Nephropathies

Inflammasome activation is closely associated with crystal formation including calcium oxalate, cholesterol emboli, uric acid, free light chains, myoglobin and cysteine (Hutton et al. 2016; Mulay et al. 2016; Mulay et al. 2014). Crystals can trigger renal injury when deposited or formed inside the kidney. The high osmolarity in certain renal segments, especially in the medulla, predisposes this organ to crystal formation (Kurts 2013; Mulay et al. 2014). Monosodium urate and calcium oxalate are well-characterized activators of the NLRP3 inflammasome in the kidney, which can induce lysosomal rupture when phagocytosed and subsequently cause damage to mitochondria, generating ROS (Emmerson et al. 1990). NLRP3 inflammasome-mediated IL-1β secretion is recognized to be the essential pathophysiological element of crystal- and particle-induced inflammation and has been demonstrated to contribute to crystalline nephropathies. Studies from Mulay SR et al. have found that calcium oxalates have a similar agonistic potential on NLRP3 and trigger IL-1β secretion, and this process contributes to calcium oxalates-induced renal inflammation in nephrocalcinosis (Mulay and Anders 2017; Mulay et al. 2013). Further studies confirm that NLRP3-mediated inflammation is a principal cause of progressive renal failure in oxalate nephropathy. In a mouse model of progressive oxalate nephropathy, NLRP3 expression is significantly increased in the kidney. NLRP3-null mice are completely protected from the progressive renal failure in oxalate nephropathy. Meanwhile, NLRP3 deficiency does not affect oxalate homeostasis. Thus, progressive renal failure in oxalate nephropathy results primarily from NLRP3-mediated inflammation (Knauf et al. 2013).

19.10 NLRP3 Inflammasome and Hyperhomocysteinemia-Induced Renal Injury

Hyperhomocysteinemia (hHcys) is an important independent risk factor for the development of cardiovascular disease and end-stage renal disease (Abais et al. 2013). Recent studies have found that NLRP3 inflammasome activation is importantly involved in podocyte dysfunction and glomerulosclerosis induced by hyperhomocysteinemia (Cavalca et al. 2001). Inhibition of NLRP3 inflammasome by either ASC silencing or inhibition of caspase-1 has recently prevented the development of hyperhomocysteinemia-induced glomerular injury and podocyte injury (Zhang et al. 2012). In addition, TXNIP is a negative regulator of the antioxidant thioredoxin that links the changes in oxidative stress to NLRP3 activation. When ROS accumulates, TXNIP can sense ROS and time-dependently dissociate from thioredoxin to bind with NLRP3, leading to inflammasome formation and activation. Abais et al. have also reported that inhibition of TXNIP prevents homocysteine-induced TXNIP protein recruitment to form NLRP3 inflammasome and reduces caspase-1 activity in glomeruli of mice with hyperhomocysteinemia, indicating that TXNIP binding to NLRP3 is a key signaling mechanism for homocysteine-induced NLRP3 inflammasome activation.

19.11 Other Inflammasomes and CKD

Besides NLRP3, other inflammasomes are also involved in CKD, including NLRP1, NLRC4 and the human interferon-inducible 200 protein AIM2 (Fig. 19.1). NLRP1 contains a PYD domain and a C-terminal CARD domain, and therefore interacts directly with caspase-1 and 5 (Boyden and Dietrich 2006). The NLRP1 inflammasome can also be activated by MDP (Faustin et al. 2007). Recent studies have elucidated the involvement of NLRP1 in the development of diabetic kidney disease (DKD). Two NLRP1 gene variants: rs2670660 and rs11651270 are significantly associated with a decreased risk to develop DKD, suggesting that NLRP1 may play an important role in DKD (Soares et al. 2018). However, the molecular mechanisms of NLRP1 inflammasome in CKD keep unknown.

NLRC4 responds to bacterial flagellin and bacteria containing type III/IV secretion systems such as *Salmonella typhimurium* and *Pseudomonas aeruginosa*. NLRC4 can activate caspase-1 through a direct CARD–CARD interaction in the absence of the adaptor ASC (Duncan and Canna 2018). In the aging kidney tissue, the expression levels of NLRP3 and NLRC4 are significantly increased compared with the young group, indicating that age-associated renal diseases are related to NLRC4 inflammasome.

The AIM2 inflammasome is one of the major regulators in innate immune responses. Unlike other three inflammasomes, AIM2 is a PYD containing human interferon-inducible 200 proteins that does not belong to the NLR family (Hu et al.

2016; Man et al. 2015) (Fig. 19.1). AIM2 forms a DNA-sensing inflammasome with ASC and caspase-1 that mediates cytokine processing and pyroptosis. AIM2 is involved in antiviral host defense (Fernandes-Alnemri et al. 2010). In the kidney, AIM2 is mainly activated by renal macrophages. AIM2 is also expressed in glomeruli, tubules and infiltrating leukocytes (Hu et al. 2016). In a mouse UUO model, AIM2 deficiency attenuates renal injury and inflammation. Moreover, in HBV-associated glomerulonephritis (HBV-GN) model, the expression of AIM2 is also significantly induced, which is related to the inflammatory cytokine expression (Komada et al. 2018; Zheng et al. 2018). Furthermore, AIM2 can be activated by cytosolic ds-DNA, which may play an important role in the pathogenesis of lupus in mice (Zhang et al. 2013).

19.12 Conclusion/Perspectives

There is an increasing number of evidence for an association between NLRP3 inflammasome activation and CKD, which leads to caspase-1-mediated IL-1β/IL-18 production. This chapter highlights the current findings regarding the role of NLRP3-mediated signaling pathways in different types of CKDs. Despite remarkable progress in inflammasome, there are still numerous aspects that need to be understood in the kidney. First of all, in the kidney, the molecular mechanisms for individual PRRs to induce pleiotropic outcomes remain largely unknown. In addition, controversies exist on the certainty of detrimental or beneficial effects of some inflammasomes in different disease states or different experimental animal models. For instance, a protective role of NLRP3 against progressive renal injury by preventing early renal interstitial edema and vascular permeability in UUO is recently observed (Pulskens et al. 2014). Second, recognition and clarification of the non-canonical effects of NLRP3 inflammasome activation in the kidney and alternative pathways to activate this inflammasome may also be important, because it may result in a combination of injurious actions independent of typical inflammation, which may cause direct damage to kidney cells, suggesting that direct targeting of the NLRP3 inflammasome may serve as a new therapeutic strategy for treatment of CKD (Yu et al. 2017). Third, the strategies for therapeutic potential of targeting inflammasomes in CKD are still very limited, despite several studies have shown that some components can significantly inhibit the activity of inflammasome and protect against renal injury in CKD. For instance, Danggui Buxue Tang (DBT) is a kind of traditional Chinese medicine, with two main ingredients, Danggui (*Radix Angelicae Sinensis*, RAS) and Huangqi (*Radix Astragali*, RA), with a weight ratio of 1:5 (Song et al, 2004). Studies have discovered that DBT alleviates the progression of diabetic nephropathy induced with STZ through reducing the level of IL-1 by inhibiting NLRP3 inflammasome expression and activation. Growing evidence reveals that recombinant thrombomodulin domain 1 (THBDD1), one of the components of anticoagulant pathway, binds to thrombin on the cell surface to prevent the formation from thrombus (Conway 2012), which can inhibit the activity of NLRP3 inflammasome and ameliorate renal injury. Collectively, a better understanding of the function of individual inflammasome and the development-related inhibitors will provide unexpected opportunities for the treatment of CKD.

References

Abais JM, Zhang C, Xia M, Liu Q, Gehr TW, Boini KM et al (2013) NADPH oxidase-mediated triggering of inflammasome activation in mouse podocytes and glomeruli during hyperhomocysteinemia. Antioxid Redox Signal 18:1537–1548

Bakker PJ, Butter LM, Kors L, Teske GJ, Aten J, Sutterwala FS et al (2014) Nlrp3 is a key modulator of diet-induced nephropathy and renal cholesterol accumulation. Kidney Int 85:1112–1122

Boini KM, Xia M, Abais JM, Li G, Pitzer AL, Gehr TW et al (2014) Activation of inflammasomes in podocyte injury of mice on the high fat diet: effects of ASC gene deletion and silencing. Biochim Biophys Acta 1843:836–845

Boyden ED, Dietrich WF (2006) Nalp1b controls mouse macrophage susceptibility to anthrax lethal toxin. Nat Genet 38:240–244

Broderick L, De Nardo D, Franklin BS, Hoffman HM, Latz E (2015) The inflammasomes and autoinflammatory syndromes. Annu Rev Pathol 10:395–424

Caruso R, Warner N, Inohara N, Nunez G (2014) NOD1 and NOD2: signaling, host defense, and inflammatory disease. Immunity 41:898–908

Cavalca V, Cighetti G, Bamonti F, Loaldi A, Bortone L, Novembrino C et al (2001) Oxidative stress and homocysteine in coronary artery disease. Clin Chem 47:887–892

Chen K, Zhang J, Zhang W, Zhang J, Yang J, Li K et al (2013) ATP-P2X4 signaling mediates NLRP3 inflammasome activation: a novel pathway of diabetic nephropathy. Int J Biochem Cell Biol 45:932–943

Conley SM, Abais JM, Boini KM, Li PL (2017) Inflammasome activation in chronic glomerular diseases. Curr Drug Targets 18:1019–1029

Conway EM (2012) Thrombomodulin and its role in inflammation. Semin Immunopathol 34:107–125

Du P, Fan B, Han H, Zhen J, Shan J, Wang X et al (2013) NOD2 promotes renal injury by exacerbating inflammation and podocyte insulin resistance in diabetic nephropathy. Kidney Int 84:265–276

Duncan JA, Canna SW (2018) The NLRC4 inflammasome. Immunol Rev 281:115–123

El-Nahas AM (2003) Plasticity of kidney cells: role in kidney remodeling and scarring. Kidney Int 64:1553–1563

Emmerson BT, Cross M, Osborne JM, Axelsen RA (1990) Reaction of MDCK cells to crystals of monosodium urate monohydrate and uric acid. Kidney Int 37:36–43

Fang L, Xie D, Wu X, Cao HD, Su WF, Yang JW (2013) Involvement of endoplasmic reticulum stress in albuminuria induced inflammasome activation in renal proximal tubular cells. PLoS ONE 8:e72344

Faustin B, Lartigue L, Bruey JM, Luciano F, Sergienko E, Bailly-Maitre B et al (2007) Reconstituted NALP1 inflammasome reveals two-step mechanism of caspase-1 activation. Mol Cell 25:713–724

Fernandes-Alnemri T, Yu JW, Juliana C, Solorzano L, Kang S, Wu J et al (2010) The AIM2 inflammasome is critical for innate immunity to Francisella tularensis. Nat Immunol 11:385–393

Fritz JH, Ferrero RL, Philpott DJ, Girardin SE (2006) Nod-like proteins in immunity, inflammation and disease. Nat Immunol 7:1250–1257

Gao P, Meng XF, Su H, He FF, Chen S, Tang H et al (2014) Thioredoxin-interacting protein mediates NALP3 inflammasome activation in podocytes during diabetic nephropathy. Biochim Biophys Acta 1843:2448–2460

Granata S, Dalla Gassa A, Bellin G, Lupo A, Zaza G (2016) Transcriptomics: a step behind the comprehension of the polygenic influence on oxidative stress, immune deregulation, and mitochondrial dysfunction in chronic kidney disease. Biomed Res Int 2016:9290857

Gu J, Liu G, Xing J, Song H, Wang Z (2018) Fecal bacteria from Crohn's disease patients more potently activated NOD-like receptors and Toll-like receptors in macrophages, in an IL-4-repressible fashion. Microb Pathog 121:40–44

Han H, Wang Y, Li X, Wang PA, Wei X, Liang W et al (2013) Novel role of NOD2 in mediating Ca2+ signaling: evidence from NOD2-regulated podocyte TRPC6 channels in hyperhomocysteinemia. Hypertension 62:506–511

Hara S, Kamei D, Sasaki Y, Tanemoto A, Nakatani Y, Murakami M (2010) Prostaglandin E synthases: understanding their pathophysiological roles through mouse genetic models. Biochimie 92:651–659

He L, Peng X, Liu G, Tang C, Liu H, Liu F et al (2015) Anti-inflammatory effects of triptolide on IgA nephropathy in rats. Immunopharmacol Immunotoxicol 37:421–427

Hu B, Jin C, Li HB, Tong J, Ouyang X, Cetinbas NM et al (2016) The DNA-sensing AIM2 inflammasome controls radiation-induced cell death and tissue injury. Science 354:765–768

Hutton HL, Ooi JD, Holdsworth SR, Kitching AR (2016) The NLRP3 inflammasome in kidney disease and autoimmunity. Nephrology 21:736–744

Iwasaki T, Kaneko N, Ito Y, Takeda H, Sawasaki T, Heike T et al (2016) Nod2-Nodosome in a cell-free system: implications in pathogenesis and drug discovery for blau syndrome and early-onset sarcoidosis. Sci World J 2016:2597376

Kanwar YS, Sun L, Xie P, Liu FY, Chen S (2011) A glimpse of various pathogenetic mechanisms of diabetic nephropathy. Annu Rev Pathol 6:395–423

Ke B, Shen W, Fang X, Wu Q (2018) The NLPR3 inflammasome and obesity-related kidney disease. J Cell Mol Med 22:16–24

Keestra-Gounder AM, Tsolis RM (2017) NOD1 and NOD2: beyond peptidoglycan sensing. Trends Immunol 38:758–767

Kim YK, Shin JS, Nahm MH (2016) NOD-like receptors in infection, immunity, and diseases. Yonsei Med J 57:5–14

Knauf F, Asplin JR, Granja I, Schmidt IM, Moeckel GW, David RJ et al (2013) NALP3-mediated inflammation is a principal cause of progressive renal failure in oxalate nephropathy. Kidney Int 84:895–901

Komada T, Chung H, Lau A, Platnich JM, Beck PL, Benediktsson H et al (2018) Macrophage uptake of necrotic cell DNA activates the AIM2 inflammasome to regulate a proinflammatory phenotype in CKD. J Am Soc Nephrol 29:1165–1181

Komada T, Usui F, Shirasuna K, Kawashima A, Kimura H, Karasawa T et al (2014) ASC in renal collecting duct epithelial cells contributes to inflammation and injury after unilateral ureteral obstruction. Am J Pathol 184:1287–1298

Kovesdy CP, Furth SL, Zoccali C (2017) Obesity and kidney disease: hidden consequences of the epidemic. Braz J Med Res 50:e6075

Kurts C (2013) A crystal-clear mechanism of chronic kidney disease. Kidney Int 84:859–861

Lichtnekert J, Kulkarni OP, Mulay SR, Rupanagudi KV, Ryu M, Allam R et al (2011) Anti-GBM glomerulonephritis involves IL-1 but is independent of NLRP3/ASC inflammasome-mediated activation of caspase-1. PLoS ONE 6:e26778

Liu D, Xu M, Ding LH, Lv LL, Liu H, Ma KL et al (2014) Activation of the Nlrp3 inflammasome by mitochondrial reactive oxygen species: a novel mechanism of albumin-induced tubulointerstitial inflammation. Int J Biochem 57:7–19

Luo R, Kakizoe Y, Wang F, Fan X, Hu S, Yang T et al (2017) Deficiency of mPGES-1 exacerbates renal fibrosis and inflammation in mice with unilateral ureteral obstruction. Am J Physiol Renal Physiol 312:F121–F133

Ma Q (2013) Role of nrf2 in oxidative stress and toxicity. Annu Rev Pharmacol Toxicol 53:401–426

Magalhaes JG, Sorbara MT, Girardin SE, Philpott DJ (2011) What is new with nods? Curr Opin Immunol 23:29–34

Man SM, Karki R, Malireddi RK, Neale G, Vogel P, Yamamoto M et al (2015) The transcription factor IRF1 and guanylate-binding proteins target activation of the AIM2 inflammasome by Francisella infection. Nat Immunol 16:467–475

Martinon F (2010) Signaling by ROS drives inflammasome activation. Eur J Immunol 40:616–619

Mestecky J, Raska M, Julian BA, Gharavi AG, Renfrow MB, Moldoveanu Z et al (2013) IgA nephropathy: molecular mechanisms of the disease. Annu Rev Pathol 8:217–240

Monk D, Sanchez-Delgado M, Fisher R (2017) NLRPs, the subcortical maternal complex and genomic imprinting. Reproduction 154:R161–R170

Mulay SR, Anders HJ (2017) Crystal nephropathies: mechanisms of crystal-induced kidney injury. Nat Rev Nephrol 13:226–240

Mulay SR, Desai J, Kumar SV, Eberhard JN, Thomasova D, Romoli S et al (2016) Cytotoxicity of crystals involves RIPK3-MLKL-mediated necroptosis. Nat Commun 7:10274

Mulay SR, Evan A, Anders HJ (2014) Molecular mechanisms of crystal-related kidney inflammation and injury Implications for cholesterol embolism, crystalline nephropathies and kidney stone disease. Nephrol Dial Transplant 29:507–514

Mulay SR, Kulkarni OP, Rupanagudi KV, Migliorini A, Darisipudi MN, Vilaysane A et al (2013) Calcium oxalate crystals induce renal inflammation by NLRP3-mediated IL-1beta secretion. J Clin Invest 123:236–246

Nakagawa N, Yuhki K, Kawabe J, Fujino T, Takahata O, Kabara M et al (2012) The intrinsic prostaglandin E2-EP4 system of the renal tubular epithelium limits the development of tubulointerstitial fibrosis in mice. Kidney Int 82:158–171

Okla M, Zaher W, Alfayez M, Chung S (2018) Inhibitory effects of toll-like Receptor 4, NLRP3 inflammasome, and interleukin-1beta on white adipocyte browning. Inflammation 41:626–642

Pulskens WP, Butter LM, Teske GJ, Claessen N, Dessing MC, Flavell RA et al (2014) Nlrp3 prevents early renal interstitial edema and vascular permeability in unilateral ureteral obstruction. PLoS ONE 9:e85775

Shahzad K, Bock F, Dong W, Wang H, Kopf S, Kohli S et al (2015) Nlrp3-inflammasome activation in non-myeloid-derived cells aggravates diabetic nephropathy. Kidney Int 87:74–84

Shen HH, Yang YX, Meng X, Luo XY, Li XM, Shuai ZW et al (2018) NLRP3: a promising therapeutic target for autoimmune diseases. Autoimmun Rev 17:694–702

Shigeoka AA, Kambo A, Mathison JC, King AJ, Hall WF, da Silva Correia J et al (2010) Nod1 and nod2 are expressed in human and murine renal tubular epithelial cells and participate in renal ischemia reperfusion injury. J Immunol 184:2297–2304

Soares JLS, Fernandes FP, Patente TA, Monteiro MB, Parisi MC, Giannella-Neto D et al (2018) Gain-of-function variants in NLRP1 protect against the development of diabetic kidney disease: NLRP1 inflammasome role in metabolic stress sensing? Clin Immunol 187:46–49

Sogawa Y, Nagasu H, Iwase S, Ihoriya C, Itano S, Uchida A et al (2017) Infiltration of M1, but not M2, macrophages is impaired after unilateral ureter obstruction in Nrf2-deficient mice. Sci Rep 7:8801

Song ZH, Ji ZN, Lo CK, Dong TT, Zhao KJ, Li OT et al (2004) Chemical and biological assessment of a traditional chinese herbal decoction prepared from Radix Astragali and Radix Angelicae Sinensis: orthogonal array design to optimize the extraction of chemical constituents. Planta Med 70:1222–1227

Tsai YL, Hua KF, Chen A, Wei CW, Chen WS, Wu CY et al (2017) NLRP3 inflammasome: pathogenic role and potential therapeutic target for IgA nephropathy. Sci Rep 7:41123

Vilaysane A, Chun J, Seamone ME, Wang W, Chin R, Hirota S et al (2010) The NLRP3 inflammasome promotes renal inflammation and contributes to CKD. J Am Soc Nephrol 21:1732–1744

Wang X, Yi F (2015) Implication of pattern-recognition receptors in cardiovascular diseases. Antioxid Redox Signal 22:1130–1145

Wong DWL, Yiu WH, Chan KW, Li Y, Li B, Lok SWY et al (2018) Activated renal tubular Wnt/beta-catenin signaling triggers renal inflammation during overload proteinuria. Kidney Int 93:1367–1383

Yang SM, Ka SM, Hua KF, Wu TH, Chuang YP, Lin YW et al (2013) Antroquinonol mitigates an accelerated and progressive IgA nephropathy model in mice by activating the Nrf2 pathway and inhibiting T cells and NLRP3 inflammasome. Free Radic Bio Med 61:285–297
Yu G, Bai Z, Chen Z, Chen H, Wang G, Wang G et al (2017) The NLRP3 inflammasome is a potential target of ozone therapy aiming to ease chronic renal inflammation in chronic kidney disease. Int Immunopharmacol 43:203–209
Yu HH, Chiang BL (2014) Diagnosis and classification of IgA nephropathy. Autoimmun Rev 13:556–559
Zhang C, Boini KM, Xia M, Abais JM, Li X, Liu Q et al (2012) Activation of Nod-like receptor protein 3 inflammasomes turns on podocyte injury and glomerular sclerosis in hyperhomocysteinemia. Hypertension 60:154–162
Zhang L, Wang XZ, Li YS, Zhang L, Hao LR (2017) Icariin ameliorates IgA nephropathy by inhibition of nuclear factor kappa b/Nlrp3 pathway. FEBS Open Bio 7:54–63
Zhang W, Cai Y, Xu W, Yin Z, Gao X, Xiong S (2013) AIM2 facilitates the apoptotic DNA-induced systemic lupus erythematosus via arbitrating macrophage functional maturation. J Clin Immunol 33:925–937
Zhao C, Gillette DD, Li X, Zhang Z, Wen H (2014) Nuclear factor E2-related factor-2 (Nrf2) is required for NLRP3 and AIM2 inflammasome activation. J Biol Chem 289:17020–17029
Zhao J, Zhang H, Huang Y, Wang H, Wang S, Zhao C et al (2013) Bay11-7082 attenuates murine lupus nephritis via inhibiting NLRP3 inflammasome and NF-κB activation. Int Immunopharmacol 17:116–122
Zheng L, Zhang J, Yuan X, Tang J, Qiu S, Peng Z et al (2018) Fluorofenidone attenuates interleukin-1β production by interacting with NLRP3 inflammasome in unilateral ureteral obstruction. Nephrology 23:573–584
Zhou M, Tang W, Fu Y, Xu X, Wang Z, Lu Y et al (2015) Progranulin protects against renal ischemia/reperfusion injury in mice. Kidney Int 87:918–929

Chapter 20
Complement Activation in Progression of Chronic Kidney Disease

Su-Fang Chen and Min Chen

Abstract Chronic kidney disease (CKD) is a public health problem worldwide, with increasing incidence and prevalence. The mechanisms underlying the progression to end-stage renal disease (ESRD) is not fully understood. The complement system was traditionally regarded as an important part of innate immunity required for host protection against infection and for maintaining host hemostasis. However, compelling evidence from both clinical and experimental studies has strongly incriminated complement activation as a pivotal pathogenic mediator of the development of multiple renal diseases and progressive replacement of functioning nephrons by fibrosis. Both anaphylatoxins, i.e., C3a and C5a, and membrane attack complex (MAC) contribute to the damage that occurs during chronic renal progression through various mechanisms including direct proinflammatory and fibrogenic activity, chemotactic effect, activation of the renal renin–angiotensin system, and enhancement of T-cell immunity. Evolving understanding of the mechanisms of complement-mediated renal injury has led to the emergence of complement-targeting therapeutics. A variety of specific antibodies and inhibitors targeting complement components have shown efficacy in reducing disease in animal models. Moreover, building on these advances, targeting complement has gained encouraging success in treating patients with renal diseases such as atypical hemolytic uremic syndrome (aHUS). Nevertheless, it still requires a great deal of effort to develop inhibitors that can be applied to treat more patients effectively in routine clinical practice.

Keywords Complement · Chronic kidney disease · Glomerular diseases · Renal fibrosis

---

S.-F. Chen · M. Chen (✉)
Key Laboratory of Renal Disease, Renal Division, Department of Medicine, Peking University First Hospital, Institute of Nephrology, Ministry of Health of China, Peking University, Beijing, China
e-mail: chenmin74@sina.com

B.-C. Liu et al. (eds.), *Renal Fibrosis: Mechanisms and Therapies*,
Advances in Experimental Medicine and Biology 1165,
https://doi.org/10.1007/978-981-13-8871-2_20

20.1 Introduction

Chronic kidney disease (CKD) is a public health problem worldwide, with increasing incidence and prevalence (Couser et al. 2011; Wang et al. 2018). Disease progression to end-stage renal disease (ESRD) leads to dramatic increases in morbidity and mortality, and it is associated with progressive glomerulosclerosis, tubulointerstitial and vascular fibrosis. Possible mechanisms responsible for the development of renal diseases and the progression of renal fibrosis leading to nephron loss have been explored in the past decades. The complement system, traditionally regarded as an important component of innate immunity required for host defense, has now been identified as a critical pathogenic mediator of the development of multiple renal diseases. Besides an important role in driving renal inflammation, cumulative evidence demonstrates that activation of complement system contributes to progressive renal fibrosis and loss of renal function. Herein, we provide a review of current understanding regarding the role of complement activation in the development and progression of kidney diseases, and highlight potential therapeutic targets.

20.2 Overview of the Complement System

The complement system is an enzyme cascade comprising of more than 30 proteins. It serves as one of the first lines of defense in innate immunity by protecting against invading microorganisms and facilitating the elimination of immune complexes and apoptotic cells, as well as bridging the innate and acquired immunity. Complement activation can be initiated through three different pathways—the classical pathway, the lectin pathway, and the alternative pathway (Fig. 20.1).

The classical pathway is typically triggered by binding of IgM or IgG containing immune complexes to C1, leading to the activation of serine proteases C1r and C1s. The latter cleaves complement components C4 and C2 to form the C4bC2a enzyme complex, the C3 convertase in the classical pathway.

The lectin pathway is activated by the recognition of carbohydrates on microbial surfaces by mannose-binding lectin (MBL), ficolins, and several collections. Following the process, MBL-associated serine proteases (MASP-1 and MASP-2) are activated, resulting in the formation of classical pathway C3 convertase C4bC2a by cleaving C4 and C2.

Different from the classical and lectin pathways, activation of the alternative pathway is initiated by spontaneously hydrolysis of C3 (referred to as tick-over), yielding C3(H_2O). By recruiting factor B (FB), which is enzymatically cleaved by factor D (FD), it leads to the formation of C3(H_2O) Bb, which continuously cleaves C3 into C3a and C3b at a low rate. This pathway is progressed and amplified on activating surfaces (e.g., bacterial surface) where C3b is protected from inactivation by complement regulatory proteins, such as factor I (FI) and factor H (FH). C3b associates with Bb to form the alternative pathway C3 convertase (C3bBb). Properdin

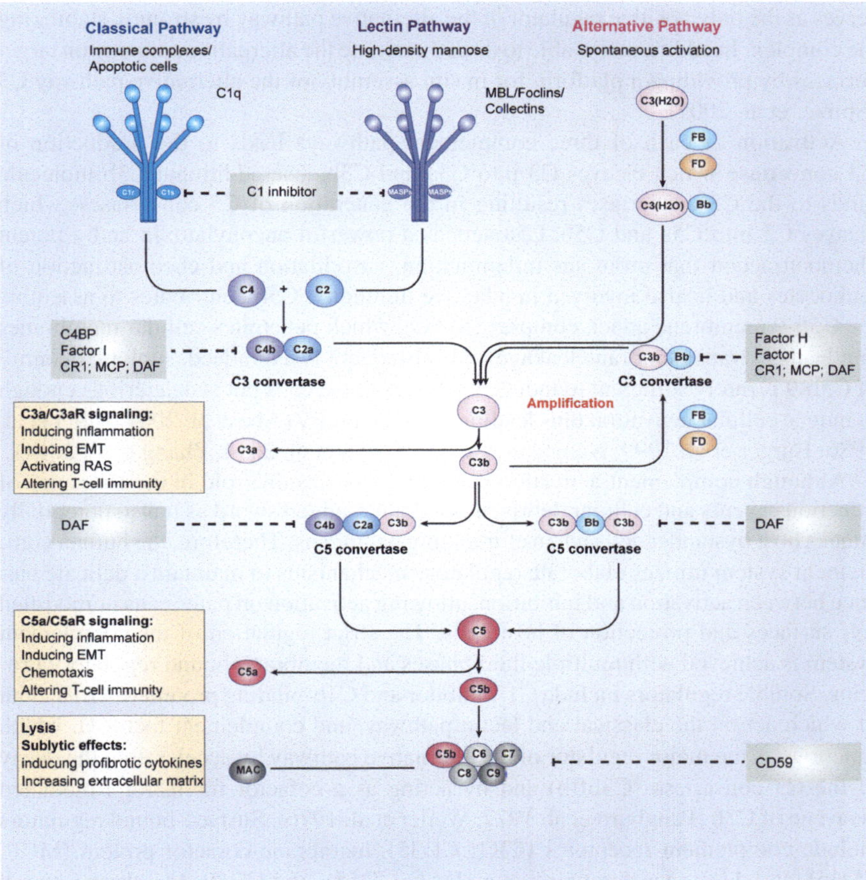

Fig. 20.1 Complement cascade. Complement activation can be activated through three different pathways—the classical pathway, the lectin pathway, and the alternative pathway. The classical pathway is induced by the binding of C1q to immune complexes or apoptotic cells. The lectin pathway is triggered by the recognition of mannose residues on microbial surface by MBL/ficolins/collections. The alternative pathway is spontaneously and continuously activated. Activation of any of the three pathways leads to activation of C3, initiate the formation of C3 convertases, which cleaves C3 to generate C3a and C3b. The additional C3b molecule binds to the C3 convertases resulting in the generation of C5 convertases, which cleaves C5 into C5a and C5b. By recruiting the components C6, C7, C8, and C9, C5b contributes to assemble membrane attack complex. Physiologically, the progression of the cascade is strictly controlled at each step by multiple fluid phase and membrane-bound regulatory proteins (shown in gray boxes) to protect host cells and tissues from complement-mediated injury. Uncontrolled complement activation leads to the generation of multiple effector compounds including C3a, C5a, and MAC, which are detrimental to the host. The effect mechanism involved in the progression of chronic kidney diseases has been highlighted in the yellow boxes. *Abbreviations* C4BP, C4b-binding protein; CR1, complement receptor 1; DAF, decay-accelerating factor; EMT, epithelial-to-mesenchymal transition; FB, factor B; FD, factor D; MAC, membrane attack complex; MASP, mannose-binding lectin-associated serine protease; MBL, mannose-binding lectin; MCP, membrane cofactor protein; RAS, renin–angiotensin system

serves as the only positive regulator of the alternative pathway by strongly stabilizing the complex. In addition, it is able to directly initiate the alternative pathway on target surfaces by providing a platform for in situ assembly of the alternative pathway C3 (Spitzer et al. 2007).

Activation of each of three complement pathways leads to the production of C3 convertase which cleaves C3 into C3a and C3b. The additional C3b molecule binds to the C3 convertases resulting in the generation of C5 convertases, which cleaves C5 into C5a and C5b. C5a acts as a powerful anaphylatoxin and a potent chemoattractant that promotes inflammation, vasodilation and chemoattraction of leukocytes and is also involved in adaptive immunity. C5b contributes to assemble the C5b-9 membrane attack complex (MAC), which penetrates cellular membranes and leads to transmembrane leakage and subsequent lysis. Indeed, sublytic quantity of C5b-9 is rarely sufficient to induce the lysis of host cells but is deleterious enough to induce cellular activation thus leading to tissue injury (Abe et al. 2004; Adler et al. 1986; Burger et al. 1999; Nangaku et al. 2005; Qiu et al. 2014; Zhang et al. 2014).

Although complement activation plays an indispensable role in the clearance of infectious agents and cellular debris, it is a double-edged sword as it also potentially attacks host bystander cells when eliminating pathogens. Therefore, the human complement system utilizes elaborate regulatory mechanisms to maintain a delicate balance between activation and inhibition, allowing activation on pathogens or modified self surfaces and protection of host cells. The strict regulation of the complement system is achieved with multiple fluid phases and membrane-bound regulatory proteins. Soluble regulators include C1-inhibitor and C4b-binding protein (C4BP), both of which act on the classical and lectin pathway, and complement factor H, which functions as the major regulator of the alternative pathway by accelerating the decay of the C3 convertase (C3bBb) and by acting as a cofactor for factor I-mediated cleavage of C3b (Pangburn et al. 1977; Weiler et al. 1976). Surface-bound regulators include complement receptor 1 (CR1; CD35), membrane cofactor protein (MCP; CD46), and decay-accelerating factor (DAF; CD55), and CD59. Membrane-bound regulators control all the three major complement activation pathways through various way, including preventing assemble of C3 convertase by inactivation of both C3b and C4b (CR1 and MCP) (Masaki et al. 1992), accelerating the decay of C3 and C5 convertases by rapidly promoting disassociation of Bb and C2a from their binding sites (DAF) (Fujita et al. 1987), and inhibiting the formation of MAC by CD59. Once activation of the complement cascade is not appropriately controlled or inhibited at host surfaces, it can progress resulting in excessive complement activation and tissue injury.

20.3 Complement in Glomerular Diseases

Activation of the complement cascade has been implicated in the development of many types of glomerulonephritis, including atypical hemolytic uremic syndrome (aHUS), C3 glomerulopathy (C3G), lupus nephritis (LN), anti-neutrophil cytoplasmic autoantibody (ANCA)-associated glomerulonephritis, IgA nephropathy (IgAN), and diabetic nephropathy (DN).

20.3.1 Atypical Hemolytic Uremic Syndrome

Atypical hemolytic uremic syndrome, accounting for approximately 10% of hemolytic uremic syndrome, is used to describe the cases not related to infection with Shiga toxin-producing bacteria. It is a life-threatening thrombotic microangiopathy characterized by microangiopathic hemolytic anemia, thrombocytopenia, and acute renal injury (Noris and Remuzzi 2009), with about half of patients progressing to ESRD (Noris and Remuzzi 2009). The strong association between uncontrolled complement activation and aHUS has been well recognized, with genetic or acquired abnormalities resulting in dysregulation of the alternative complement pathway have been found in 50–60% of cases (George and Nester 2014; Noris and Remuzzi 2009). Inherited loss of function mutations in complement regulators (factor H, I, and CD46) and gain of function mutations in activation proteins (C3 and factor B) have all been reported (Caprioli et al. 2001; Fremeaux-Bacchi et al. 2004, 2008; Goicoechea de Jorge et al. 2007; Noris et al. 2003). Acquired anti-FH autoantibodies disturbing functional activity of FH have also been reported to be associated with aHUS (Dragon-Durey et al. 2005; Jozsi et al. 2007; Moore et al. 2010). The understanding was further advanced by recent discoveries that genomic rearrangements of complement factor H-related proteins (CFHRs) leading to the production of fusion proteins that competitively inhibit FH activity contributed to the development of aHUS (Challis et al. 2016; Eyler et al. 2013; Francis et al. 2012; Valoti et al. 2015; Venables et al. 2006). Additional evidence derives from an animal model that the mice with mutated FH lacking surface recognition domains spontaneously developed aHUS (Pickering et al. 2007). Using this model, the pivotal role of C5 activation in the development of spontaneous aHUS has been demonstrated (de Jorge et al. 2011). Based on these findings, complement inhibition targeting C5 by the anti-C5 monoclonal antibody eculizumab has revolutionized the treatment of aHUS (Legendre et al. 2013). Nevertheless, patients those are refractory to eculizumab may suffer from diseases driven by C3 cleavage products, indicating that C3 and/or C3a/C3aR may be a possible effective target for therapeutic intervention.

20.3.2 C3 Glomerulopathy

C3 glomerulopathy describes a spectrum of disorders sharing the pathological finding of predominant C3 deposition within the glomerulus in the absence of immunoglobulins, including dense deposit disease (DDD), C3 glomerulonephritis (C3GN), and CFHR5 nephropathy (Fakhouri et al. 2010b). Progression to ESRD and post-transplant recurrence is common in all forms of C3 glomerulopathy. A large number of studies have highlighted a strong link between dysregulation of the complement alternative pathway and the development of these diseases. In patients with C3G, various genetic and acquired complement abnormalities have been identified, including inherited mutations in complement regulator genes (factor H, I, and CD46) and C3 convertase genes (C3 and factor B) (Iatropoulos et al. 2016; Levy et al. 1986; Schmidt et al. 1999; Servais et al. 2012), and the development of C3 nephritic factors (C3Nefs), which are autoantibodies that stabilize the C3 convertase leading to uncontrolled C3 activation. Besides, important developments in the understanding of complement dysregulation in C3G have been made by recent findings that a number of genomic rearrangements across different CFHR genes resulting in deregulation of FH have been associated with C3G (Athanasiou et al. 2011; Chen et al. 2014; Malik et al. 2012; Medjeral-Thomas et al. 2014; Togarsimalemath et al. 2017; Tortajada et al. 2013; Xiao et al. 2016). Animal models with complement gene abnormalities further provide supports for complement involvement in this setting. FH deficiency in pigs was found to be associated with DDD (Hogasen et al. 1995). Moreover, FH-deficient mice created in Pickering's group spontaneously develop diseases sharing clinical and laboratory characteristics of human C3GN (Pickering et al. 2002). Whereas, restoration of alternative pathway regulation through administration of human FH or engineered FH constructs ameliorate the experimental C3GN (Fakhouri et al. 2010a; Michelfelder et al. 2018; Nichols et al. 2015; Yang et al. 2018). Inactivating the terminal complement C5, but not C6, in the FH-deficient mice significantly reverses glomerular injury, indicating a critical role for C5 activation in the induction of renal lesions (Pickering et al. 2006). The C5 inhibitor eculizumab has been reported in treating several cases with C3G, but with variable results (Bomback et al. 2012; Daina et al. 2012; Gurkan et al. 2013; Le Quintrec et al. 2015; Vivarelli et al. 2012). Identifying patients who will benefit from terminal complement blockade versus upstream blockade may be of great potential to improve patients' outcomes. Other complement inhibition options targeting upstream complement activation such as C3 convertase should be further explored.

20.3.3 Lupus Nephritis

Lupus nephritis (LN) is a serious complication of systemic lupus erythematosus (SLE), characterized by autoimmune disorder targeting multiple systems. It is generally recognized that the role of complement in the mechanism of lupus is paradoxical, both beneficial and deleterious. The evidence for complement-mediated protection against SLE derives from the fact that inherited homozygous deficiencies of classical pathway components (C1q, C1r, C1 s, C4, and C2) predispose individuals to develop lupus (Manderson et al. 2004). On the other hand, compelling evidence from both clinical observation and animal study has incriminated complement activation as pathogenic in lupus nephritis. Firstly, decreased serum C3 and C4 levels and accumulation of immune complex in the kidney are frequently observed in patients with LN. It is further evidenced by the findings that complement split products such as

C4d, Bb and C3d are sensitive indicators of disease activity (Manzi et al. 1996). In addition to those clinical observations, further evidence was supported by animal models. Using the MRL/lpr mice model of lupus, Bao et al. demonstrated that factor H deficiency resulted in dysregulation of the alternative complement pathway accelerates the development of lupus nephritis (Bao et al. 2011). In contrast, MRL/lpr mice genetically deficient in factor B developed less severe renal disease and had improved survival (Watanabe et al. 2000). Similarly, factor D deficiency significantly reduced glomerular C3 deposition, serum creatinine levels, and pathological injury in MRL/lpr mice (Elliott et al. 2004). The important role of complement activation in the development of lupus nephritis is further confirmed by the efficacy of complement inhibition. Complement inhibition targeting C3b deposition using a CR2-Crry fusion protein significantly improved survival and renal function, and also dramatically reduced glomerulonephritis and renal vasculitis in MRL/lpr mice (Atkinson et al. 2008). Blockade of C5a receptor prevented the progressive impairment in renal function of MRL/lpr mice (Bao et al. 2005). Similarly, blocking C5 also markedly ameliorated the disease and increased survival in experimental lupus nephritis using the murine NZB/W(F1) lupus model (Wang et al. 1996). In the past few years, reports have emerged describing the successful use of the anti-C5 monoclonal antibody, eculizumab in severe resistant lupus nephritis, especially cases with thrombotic microangiopathy (Coppo et al. 2015; El-Husseini et al. 2015; Pickering et al. 2015).

20.3.4 Anti-neutrophil Cytoplasmic Autoantibody (ANCA)-Associated Glomerulonephritis

ANCA-associated vasculitis (AAV) is a group of potentially life-threatening autoimmune diseases. Kidneys are frequently involved, histologically characterized by pauci-immune necrotizing crescentic glomerulonephritis (NCGN) with little immunoglobulin deposition. Nevertheless, increasing evidence has demonstrated an indispensable role of complement activation, particularly, via the alternative pathway, in the development of AAV. Although with a relative paucity of glomerular staining for C3 and C4, various complement components including Bb, properdin, C3c, C3d, and C5b-9 could be detected in renal specimens of patients with AAV (Chen et al. 2009; Gou et al. 2013b; Hilhorst et al. 2017). Deposition of Bb and properdin, components of the alternative pathway, was associated with renal injury. Moreover, complement involvement is further supported by the observation that elevated levels of Bb, C3a, C5a, and soluble C5b-9 were detected in urine and plasma samples of patients with active AAV, and the correlation of Bb levels with disease activity (Gou et al. 2013a, b). Using the anti-MPO antibodies induced necrotizing crescentic glomerulonephritis mouse model, Xiao et al. demonstrated that complement depletion with cobra venom factor protected mice from developing NCGN. FB deficiency or C5 deficiency was protective, whereas C4 deficiency was not, highlighting the critical role of the alternative pathway in disease development (Xiao et al. 2007). Blocking C5 using anti-C5 mAb could also prevent the development of ANCA-associated glomerulonephritis in mice (Huugen et al. 2007). The critical role of C5a in disease induction was further demonstrated by that deficiency or blockade C5aR effectively ameliorates anti-MPO-induced NCGN (Schreiber et al. 2009; Xiao et al. 2014). By contrast, C6 deficiency has no effect on the disease development (Xiao et al. 2014), indicating that C5 cleavage but not MAC formation is required to induce injury. Building on the documentation that complement, especially C5a/C5aR, acts as a critical mediator in developing AAV, targeting C5aR emerged as a potentially effective therapy for AAV. As demonstrated by a multicenter phase II randomized double-blinded placebo-controlled trial led by EUVAS, CCX168 (avacopan), a C5aR antagonism, was effective and safe in replacing high-dose glucocorticoids in treating AAV patients (Jayne et al. 2017).

20.3.5 IgA Nephropathy

IgA nephropathy (IgAN), characterized by dominant or co-dominant IgA deposition in the glomeruli, is one of the most common glomerulopathy worldwide, leading to ESRD in up to 40% of patients within 20 years after diagnostic biopsy (Berthoux et al. 2008). It is generally accepted that the activation of complement participates in the development of IgAN (Maillard et al. 2015). This process is likely to be initiated by IgA-containing circulating immune complexes, inducing activation of the alternative and lectin pathways (Maillard et al. 2015). The involvement of the alternative complement activation in IgAN is suggested by frequently deposition of C3, properdin and factor H in diseased glomeruli (McCoy et al. 1974), accompanied by the presence of C3 cleavage products in the circulation. Moreover, mesangial C3 deposition and circulating C3 activation products were correlated with severity of the histological lesions and a higher risk of progression to ESRD (Kim et al. 2012; Wyatt and Julian 1988; Zwirner et al. 1997). A strong linkage between the alternative pathway and IgAN was given by the genome-wide association study identifying that a single nucleotide polymorphism rs6677604 in CFH which linked with CFHR3-1 deletion confers a significant protective effect in IgAN (Gharavi et al. 2011). Significant associations of rs6677604 in CFH and CFHR3-1 deletion with circulating factor H levels and mesangial C3 deposition in patients with IgAN have been observed (Zhu et al. 2015), qualifying activators and regulators of the alternative pathway as important players in the development of the disease. The involvement of alternative complement pathway in IgAN is further supported by recent discoveries that FHR-1 and FHR-5, which antagonize the regulatory ability of factor H, were found to be elevated in IgAN and associated with the risk of renal dysfunction and developing ESRD (Medjeral-Thomas et al. 2017; Tortajada et al. 2017; Zhu et al. 2018). Regarding the role of the lectin pathway, glomerular deposition of MBL, L-ficolin, and MBL-associated serine proteases was observed in approximately 25% of patients with IgAN, correlating with more pronounced histologic damage (Roos et al. 2006). Moreover, suggestive deleterious role of MBL in IgAN progression was revealed by a recent study showing that MBL deficiency and MBL excess were both associated with worse renal outcome (Guo et al. 2017). Therefore, complement may be a possible targeted approach for the treatment of human IgAN, which requires further study.

20.3.6 Diabetic Nephropathy

Diabetic nephropathy (DN) is one of the most common and serious consequences of diabetes mellitus (DM) and is the leading cause of CKD worldwide (Jha et al. 2013; Zhang et al. 2016b). There is growing evidence indicating that complement activation may contribute to the susceptibility and progression of DN (Flyvbjerg 2017). In clinical studies, circulating levels of MBL have emerged as a robust predictor of the development of diabetic nephropathy and progression to ESRD (Hansen et al. 2003, 2004, 2010; Hovind et al. 2005; Ostergaard et al. 2015). In mice with streptozotocin-induced diabetes, MBL deficiency attenuates renal damage, supporting a pathological effect of MBL on the development of DN (Ostergaard et al. 2007). In addition, as indicated by a prospective study, high levels of H-ficolin are associated with an increased risk of progression to microalbuminuria in patients with type 1 diabetes mellitus (Ostergaard et al. 2014). Downstream complement components such as C3, C4, C4d, Bb, C3a, C5a, and MAC in the pathogenesis of DN have also been highlighted by a number of studies (Barnett et al. 1984; Falk et al. 1983a, b; Fujita et al. 2013; Li et al. 2019; Woroniecka et al. 2011). Therefore, these findings provide the foundation for considering targeting complement as a potential approach to halt or slow the progression of diabetic nephropathy.

20.4 Complements in Renal Fibrosis

For various renal disease associated with CKD, the loss of functioning nephrons resulted in renal fibrosis is the final common pathway of developing ESRD. Increasing evidence links complement activation to the pathogenesis of progressive kidney fibrosis. Previous findings in transplanted kidneys have significantly contributed to our understanding of the role of complement in chronic renal injury, as interstitial fibrosis and tubular atrophy (IFTA) serves as the most common cause of kidney transplant failure. Associative evidence derives from the findings that specific C5 polymorphisms in both the donor and recipient were associated with late worse graft function and poorer transplant survival (Jeong et al. 2011). In addition, proteomic analysis of kidney transplant biopsies revealed strong associations between chronic progressive injury and overexpression of complement components (Nakorchevsky et al. 2010). Studies in focal segmental glomerulosclerosis (FSGS) which characterized by focal and segmental glomerular sclerosis showed that IgM and C3 deposition are frequently detected in the affected glomeruli and associated with worse renal outcomes (Zhang et al. 2016a). Collectively, these clinical observations indicate that complement is a mediator of chronic kidney injury and fibrosis.

The role of complement in renal fibrosis is further supported by multiple animal models. The mouse model induced by adriamycin shows typical features of chronic progressive renal disease in humans, resembling human FSGS. Either C3 or FD deficiency protected mice from developing glomerulosclerosis, tubulointerstitial injury, and renal dysfunction induced by adriamycin (Turnberg et al. 2006). In addition, inhibition of FB using a monoclonal antibody delayed the development of renal failure (Lenderink et al. 2007). It indicates an important role of alternative pathway activation in mediating chronic renal injury. Similarly, C3aR deficiency in mice significantly reduced adriamycin-induced glomerular sclerosis and tubulointerstitial injury (Tang et al. 2009). Consistently, death due to renal failure was greatly reduced in C3aR-deficient mice. Furthermore, the kidneys of C3aR-deficient mice had significantly less accumulation of α-smooth muscle actin (α-SMA) expressing myofibroblast and interstitial collagen I (Tang et al. 2009). Therefore, it highlights a pathogenic role of C3aR in inducing tubular epithelial-to-mesenchymal transition (EMT), an important process in renal fibrosis. In contrast, CD59-deficient mice lacking the major regulator of MAC formation, developed significantly more glomerulosclerosis and tubulointerstitial injury, as well as poor renal function and greater mortality than control mice (Turnberg et al. 2006). In parallel, increased myofibroblasts accumulation and increased interstitial collagen deposition was observed in CD59 deficient mice (Turnberg et al. 2006), indicating an important role of MAC in driving renal fibrosis. Using a rat model of adriamycin-induced FSGS, Rangan et al. showed that although glomerulosclerosis and accumulation of myofibroblast within glomeruli were not altered in C6-deficient rats, interstitial fibrosis and peritubular myofibroblast accumulation were significantly reduced (Rangan et al. 2004). And, in the absence of tubulointerstitial MAC formation, peritubular a-SMA accumulation and interstitial fibronectin deposition were dramatically reduced in C6 deficient rats (Rangan et al. 2004). Consistent with these findings, C6-deficient rats were protected from tubulointerstitial damage and progressive interstitial fibrosis in another proteinuric nephropathy model induced by aminonucleoside of puromycin (Nangaku et al. 1999). Additionally, in a rat model of chronic progressive renal disease secondary to nephron loss, performed by 5/6 nephrectomies, C6 deficiency ameliorated tubulointerstitial injury, attenuated progressive interstitial fibrosis with decreased interstitial extracellular matrix accumulation and macrophage infiltration, and improved renal function (Nangaku et al. 2002). Other animal models investigating the involvement of complement activation in the development of tubulointerstitial injury include the model of unilateral ureteric obstruction (UUO), induced by the ligation of one ureter. The injury is characterized by the progressive development of interstitial inflammation and fibrosis, infiltration of macrophage, T cells and fibroblast, and ultimately loss of functional nephrons, closely resembling the pathology of human CKD. C3 deficiency significantly reduced interstitial fibrosis and tubular atrophy in UUO mice. Moreover, epithelial-to-mesenchymal transition within the kidney was suppressed by C3 deficiency (Zhou et al. 2013). Another study using C5-deficient mice showed that markers of renal fibrosis, including α-SMA, vimentin, type I collagen and fibronectin, and macrophage infiltration were significantly reduced after five days of UUO (Boor et al. 2007). Significant reductions in mRNA of transforming growth factor (TGF-β) and platelet-derived growth factor (PDGF), all of which have been shown to be crucially involved in renal scarring (Chevalier 2006; Eitner and Floege 2005; Ostendorf et al. 2001, 2006), were also observed in C5-deficient mice (Boor et al. 2007). In consistence, treatment of UUO mice with a C5aR antagonist also ameliorates renal scarring (Boor et al. 2007). Altogether, a large body of evidence from both clinical and experimental studies using animal models has strongly identified complement activation as an important mediator of chronic glomerular and tubulointerstitial injury.

In vitro studies have advanced our understanding of how complement components contribute to the development of progressive renal fibrosis. Exposure of tubular epithelial cells to serum proteins or the C3a induced the cells to undergo an EMT process, as confirmed by a reduction in E-cadherin expression and increase in α-SMA and collagen I expression (Tang et al. 2009). This effect could be prevented by treating with a C3aR antagonist (Tang et al. 2009). This finding was confirmed and extended by another study showing that C3a induced EMT of tubular epithelial cells by increasing expression of TGF-β1, Krüppel-like factor 5 (KLF5) (Zhou et al. 2013). Moreover, a subsequent expression of renin in epithelial cells was induced by C3a through an increased expression of the transcription factor LXRα (Zhou et al. 2013). Therefore, it demonstrated that C3a is involved in the mechanism of progressive renal injury by inducing EMT and activating the renal renin–angiotensin systems.

The complement activation product C5a has been suggested to mediate renal fibrosis by several mechanisms. An important profibrotic mechanism of C5a was demonstrated by the finding that stimulated C5a stimulated production TGF-β1 in cultured murine tubular cells, and this was inhibited by a C5aR antagonist (Boor et al. 2007). Another effective mechanism may be attributed to its well-characterized chemotactic properties on leukocytes including neutrophils, monocytes, and macrophages. Monocyte/macrophage influx is widely recognized as a contributor to the development of renal tubulointerstitial fibrosis (Sean Eardley and Cockwell 2005). Moreover, a third mechanism by which C5a contributes to renal tubulointerstitial fibrosis may be mediated by its regulatory activity in T-cell immunity. It has been reported that C5a is able to inhibit the polarization of T-helper cells to Th1 cells (Kohl et al. 2006; Wenderfer et al. 2005). The subsequent shift of T-helper cells to Th2 cells, together with their cytokine response pattern, such as TGF-β release, acts in a profibrotic manner (Wynn 2004). In addition, both C5a and C3a serve as regulators of regulatory T cells (Tregs), which play an indispensable role in suppressing excessive immune responses deleterious to the host. Tregs have been demonstrated to have a protective role against renal ischemia-reperfusion injury by increasing tubular cell proliferation, improving renal function, and reducing renal fibrosis (Kim et al. 2013). And, it is instrumental for long-term transplant renal survival in humans (Sakaguchi et al. 2008). Genetic deficiency or pharmacological blockade of C3aR/C5aR signaling in natural regulatory T cells (nTregs) enhances their in vitro and in vivo suppression activity, indicating that signaling through C5aR and C3aR diminishes function of nTregs (Kwan et al. 2013). Additional evidence supporting a crucial regulatory role of C3aR/C5aR signaling on Tregs was derived from other sets of data showing that absence or blockade of C3aR/C5aR signaling augments induced regulatory T cells (iTregs) generation, and inhibits their conversion to IFN-γ/TNF-a-producing effector T cells (Strainic et al. 2013; van der Touw et al. 2013). Antagonizing C3aR and C5aR signaling also facilitate generation and stability of human iTregs from naive precursors in vitro (Strainic et al. 2013; van der Touw et al. 2013).

C5b-9 is another important mediator in inducing progressive renal injury. It has been reported that sublytic amount of C5b-9 could induce the production of proinflammatory and profibrotic cytokines by a variety of renal cells. In vitro complement activation on the proximal tubular cell surface triggers the generation of proinflammatory mediators including IL-6 and TNF-α in a C6-dependent manner (David et al. 1997). Sublytic C5b-9 also stimulated IL-6 and TGF-β1 production in rat glomerular mesangial cells (Zhang et al. 2014). Besides, treating endothelial cells with C5b-9 induced the release of profibrotic factors including fibroblast growth factor and PGDF (Benzaquen et al. 1994). Furthermore, effects of C5b-9 on the production of extracellular matrix components have been observed. Activation of tubular epithelial cells by sublytic concentration of C5b-9 rapidly increased the expression of fibronectin mRNA as well as the synthesis of fibronectin protein (Burger et al. 1999). Besides, stimulating proximal tubular epithelial cells with C5b-9 led to increased expression of collagen type IV (Abe et al. 2004). Additionally, incubation with human glomerular epithelial cells with sublytic doses of C5b-9 significantly increased the collagen synthesis (Torbohm et al. 1990). Collectively, these in vitro evidences supported that C5b-9 can increase the profibrotic process associated with progressive renal injury.

Therefore, evidence from studies both in vivo and in vitro suggests that both anaphylatoxins and MAC may be responsible for the damage that occurs during chronic renal progression. Either complement depletion or inhibition in experimental progressive renal disease can reduce the progression of renal fibrosis and loss of functioning nephrons.

20.5 Conclusion

The complement system is a pathogenic mediator of renal diseases in humans. Besides an important role in mediating of a broad range of glomerular disease, complement activation contributes to progressive glomerulosclerosis and interstitial fibrosis. The evolving understanding of complement activation in renal diseases has significantly contributed to the elucidation of the mechanisms of kidney injury and the development of effective complement-targeted therapeutics. A variety of specific antibodies and inhibitors targeting complement components have shown efficacy in reduce disease in animal models. Moreover, building on these advances, targeting complement in patients has gained great success in treating renal diseases such as aHUS. It indicates that inhibition of complement has the potential to abrogate disease progression and improve patient health. Nevertheless, it still requires a great deal of effort to develop inhibitors that can be applied to treat more patients effectively in routine clinical practice.

References

Abe K, Li K, Sacks SH, Sheerin NS (2004) The membrane attack complex, C5b-9, up regulates collagen gene expression in renal tubular epithelial cells. Clin Exp Immunol 136:60–66

Adler S, Baker PJ, Johnson RJ, Ochi RF, Pritzl P, Couser WG (1986) Complement membrane attack complex stimulates production of reactive oxygen metabolites by cultured rat mesangial cells. J Clin Invest 77:762–767

Athanasiou Y, Voskarides K, Gale DP, Damianou L, Patsias C et al (2011) Familial C3 glomerulopathy associated with CFHR5 mutations: clinical characteristics of 91 patients in 16 pedigrees. Clin J Am Soc Nephrol 6:1436–1446

Atkinson C, Qiao F, Song H, Gilkeson GS, Tomlinson S (2008) Low-dose targeted complement inhibition protects against renal disease and other manifestations of autoimmune disease in MRL/lpr mice. J Immunol 180:1231–1238

Bao L, Osawe I, Puri T, Lambris JD, Haas M, Quigg RJ (2005) C5a promotes development of experimental lupus nephritis which can be blocked with a specific receptor antagonist. Eur J Immunol 35:2496–2506

Bao L, Haas M, Quigg RJ (2011) Complement factor H deficiency accelerates development of lupus nephritis. J Am Soc Nephrol 22:285–295

Barnett AH, Mijovic C, Fletcher J, Chesner I, Kulkuska-Langlands BM (1984) Low plasma C4 concentrations: association with microangiopathy in insulin dependent diabetes. Br Med J (Clin Res Ed) 289:943–945

Benzaquen LR, Nicholson-Weller A, Halperin JA (1994) Terminal complement proteins C5b-9 release basic fibroblast growth factor and platelet-derived growth factor from endothelial cells. J Exp Med 179:985–992

Berthoux FC, Mohey H, Afiani A (2008) Natural history of primary IgA nephropathy. Semin Nephrol 28:4–9

Bomback AS, Smith RJ, Barile GR, Zhang Y, Heher EC et al (2012) Eculizumab for dense deposit disease and C3 glomerulonephritis. Clin J Am Soc Nephrol 7:748–756

Boor P, Konieczny A, Villa L, Schult AL, Bucher E et al (2007) Complement C5 mediates experimental tubulointerstitial fibrosis. J Am Soc Nephrol 18:1508–1515

Burger A, Wagner C, Hug F, Hansch GM (1999) Up-regulation of intracellular calcium, cyclic adenosine monophosphate and fibronectin synthesis in tubuar epithelial cells by complement. Eur J Immunol 29:1188–1193

Caprioli J, Bettinaglio P, Zipfel PF, Amadei B, Daina E et al (2001) The molecular basis of familial hemolytic uremic syndrome: mutation analysis of factor H gene reveals a hot spot in short consensus repeat 20. J Am Soc Nephrol 12:297–307

Challis RC, Araujo GS, Wong EK, Anderson HE, Awan A et al (2016) A De Novo deletion in the regulators of complement activation cluster producing a hybrid complement factor H/complement factor H-related 3 gene in atypical hemolytic uremic syndrome. J Am Soc Nephrol 27:1617–1624

Chen M, Xing GQ, Yu F, Liu G, Zhao MH (2009) Complement deposition in renal histopathology of patients with ANCA-associated pauci-immune glomerulonephritis. Nephrol Dial Transplant 24:1247–1252

Chen Q, Wiesener M, Eberhardt HU, Hartmann A, Uzonyi B et al (2014) Complement factor H-related hybrid protein deregulates complement in dense deposit disease. J Clin Invest 124:145–155

Chevalier RL (2006) Obstructive nephropathy: towards biomarker discovery and gene therapy. Nat Clin Pract Nephrol 2:157–168

Coppo R, Peruzzi L, Amore A, Martino S, Vergano L et al (2015) Dramatic effects of eculizumab in a child with diffuse proliferative lupus nephritis resistant to conventional therapy. Pediatr Nephrol 30:167–172
Couser WG, Remuzzi G, Mendis S, Tonelli M (2011) The contribution of chronic kidney disease to the global burden of major noncommunicable diseases. Kidney Int 80:1258–1270
Daina E, Noris M, Remuzzi G (2012) Eculizumab in a patient with dense-deposit disease. N Engl J Med 366:1161–1163
David S, Biancone L, Caserta C, Bussolati B, Cambi V, Camussi G (1997) Alternative pathway complement activation induces proinflammatory activity in human proximal tubular epithelial cells. Nephrol Dial Transplant 12:51–56
de Jorge EG, Macor P, Paixao-Cavalcante D, Rose KL, Tedesco F et al (2011) The development of atypical hemolytic uremic syndrome depends on complement C5. J Am Soc Nephrol 22:137–145
Dragon-Durey MA, Loirat C, Cloarec S, Macher MA, Blouin J et al (2005) Anti-factor H autoantibodies associated with atypical hemolytic uremic syndrome. J Am Soc Nephrol 16:555–563
Eitner F, Floege J (2005) Therapeutic targets for prevention and regression of progressive fibrosing renal diseases. Curr Opin Investig Drugs 6:255–261
El-Husseini A, Hannan S, Awad A, Jennings S, Cornea V, Sawaya BP (2015) Thrombotic microangiopathy in systemic lupus erythematosus: efficacy of eculizumab. Am J Kidney Dis 65:127–130
Elliott MK, Jarmi T, Ruiz P, Xu Y, Holers VM, Gilkeson GS (2004) Effects of complement factor D deficiency on the renal disease of MRL/lpr mice. Kidney Int 65:129–138
Eyler SJ, Meyer NC, Zhang Y, Xiao X, Nester CM, Smith RJ (2013) A novel hybrid CFHR1/CFH gene causes atypical hemolytic uremic syndrome. Pediatr Nephrol 28:2221–2225
Fakhouri F, de Jorge EG, Brune F, Azam P, Cook HT, Pickering MC (2010a) Treatment with human complement factor H rapidly reverses renal complement deposition in factor H-deficient mice. Kidney Int 78:279–286
Fakhouri F, Fremeaux-Bacchi V, Noel LH, Cook HT, Pickering MC (2010b) C3 glomerulopathy: a new classification. Nat Rev Nephrol 6:494–499
Falk RJ, Dalmasso AP, Kim Y, Tsai CH, Scheinman JI, Gewurz H, Michael AF (1983a) Neoantigen of the polymerized ninth component of complement. Characterization of a monoclonal antibody and immunohistochemical localization in renal disease. J Clin Invest 72:560–573
Falk RJ, Scheinman JI, Mauer SM, Michael AF (1983b) Polyantigenic expansion of basement membrane constituents in diabetic nephropathy. Diabetes 32(Suppl 2):34–39
Flyvbjerg A (2017) The role of the complement system in diabetic nephropathy. Nat Rev Nephrol 13:311–318
Francis NJ, McNicholas B, Awan A, Waldron M, Reddan D et al (2012) A novel hybrid CFH/CFHR3 gene generated by a microhomology-mediated deletion in familial atypical hemolytic uremic syndrome. Blood 119:591–601
Fremeaux-Bacchi V, Dragon-Durey MA, Blouin J, Vigneau C, Kuypers D et al (2004) Complement factor I: a susceptibility gene for atypical haemolytic uraemic syndrome. J Med Genet 41:e84
Fremeaux-Bacchi V, Miller EC, Liszewski MK, Strain L, Blouin J et al (2008) Mutations in complement C3 predispose to development of atypical hemolytic uremic syndrome. Blood 112:4948–4952
Fujita T, Inoue T, Ogawa K, Iida K, Tamura N (1987) The mechanism of action of decay-accelerating factor (DAF). DAF inhibits the assembly of C3 convertases by dissociating C2a and Bb. J Exp Med 166:1221–1228
Fujita T, Hemmi S, Kajiwara M, Yabuki M, Fuke Y et al (2013) Complement-mediated chronic inflammation is associated with diabetic microvascular complication. Diabetes Metab Res Rev 29:220–226
George JN, Nester CM (2014) Syndromes of thrombotic microangiopathy. N Engl J Med 371:654–666
Gharavi AG, Kiryluk K, Choi M, Li Y, Hou P et al (2011) Genome-wide association study identifies susceptibility loci for IgA nephropathy. Nat Genet 43:321–327

Goicoechea de Jorge E, Harris CL, Esparza-Gordillo J, Carreras L, Arranz EA et al (2007) Gain-of-function mutations in complement factor B are associated with atypical hemolytic uremic syndrome. Proc Natl Acad Sci U S A 104:240–245

Gou SJ, Yuan J, Chen M, Yu F, Zhao MH (2013a) Circulating complement activation in patients with anti-neutrophil cytoplasmic antibody-associated vasculitis. Kidney Int 83:129–137

Gou SJ, Yuan J, Wang C, Zhao MH, Chen M (2013b) Alternative complement pathway activation products in urine and kidneys of patients with ANCA-associated GN. Clin J Am Soc Nephrol 8:1884–1891

Guo WY, Zhu L, Meng SJ, Shi SF, Liu LJ et al (2017) Mannose-binding lectin levels could predict prognosis in IgA nephropathy. J Am Soc Nephrol 28:3175–3181

Gurkan S, Fyfe B, Weiss L, Xiao X, Zhang Y, Smith RJ (2013) Eculizumab and recurrent C3 glomerulonephritis. Pediatr Nephrol 28:1975–1981

Hansen TK, Thiel S, Knudsen ST, Gravholt CH, Christiansen JS et al (2003) Elevated levels of mannan-binding lectin in patients with type 1 diabetes. J Clin Endocrinol Metab 88:4857–4861

Hansen TK, Tarnow L, Thiel S, Steffensen R, Stehouwer CD et al (2004) Association between mannose-binding lectin and vascular complications in type 1 diabetes. Diabetes 53:1570–1576

Hansen TK, Forsblom C, Saraheimo M, Thorn L, Waden J et al (2010) Association between mannose-binding lectin, high-sensitivity C-reactive protein and the progression of diabetic nephropathy in type 1 diabetes. Diabetologia 53:1517–1524

Hilhorst M, van Paassen P, van Rie H, Bijnens N, Heerings-Rewinkel P et al (2017) Complement in ANCA-associated glomerulonephritis. Nephrol Dial Transplant 32:1302–1313

Hogasen K, Jansen JH, Mollnes TE, Hovdenes J, Harboe M (1995) Hereditary porcine membranoproliferative glomerulonephritis type II is caused by factor H deficiency. J Clin Invest 95:1054–1061

Hovind P, Hansen TK, Tarnow L, Thiel S, Steffensen R et al (2005) Mannose-binding lectin as a predictor of microalbuminuria in type 1 diabetes: an inception cohort study. Diabetes 54:1523–1527

Huugen D, van Esch A, Xiao H, Peutz-Kootstra CJ, Buurman WA et al (2007) Inhibition of complement factor C5 protects against anti-myeloperoxidase antibody-mediated glomerulonephritis in mice. Kidney Int 71:646–654

Iatropoulos P, Noris M, Mele C, Piras R, Valoti E et al (2016) Complement gene variants determine the risk of immunoglobulin-associated MPGN and C3 glomerulopathy and predict long-term renal outcome. Mol Immunol 71:131–142

Jayne DRW, Bruchfeld AN, Harper L, Schaier M, Venning MC et al (2017) Randomized trial of C5a receptor inhibitor avacopan in ANCA-associated vasculitis. J Am Soc Nephrol 28:2756–2767

Jeong JC, Hwang YH, Kim H, Ro H, Park HC et al (2011) Association of complement 5 genetic polymorphism with renal allograft outcomes in Korea. Nephrol Dial Transplant 26:3378–3385

Jha V, Garcia-Garcia G, Iseki K, Li Z, Naicker S et al (2013) Chronic kidney disease: global dimension and perspectives. Lancet 382:260–272

Jozsi M, Strobel S, Dahse HM, Liu WS, Hoyer PF et al (2007) Anti factor H autoantibodies block C-terminal recognition function of factor H in hemolytic uremic syndrome. Blood 110:1516–1518

Kim SJ, Koo HM, Lim BJ, Oh HJ, Yoo DE et al (2012) Decreased circulating C3 levels and mesangial C3 deposition predict renal outcome in patients with IgA nephropathy. PLoS ONE 7:e40495

Kim MG, Koo TY, Yan JJ, Lee E, Han KH et al (2013) IL-2/anti-IL-2 complex attenuates renal ischemia-reperfusion injury through expansion of regulatory T cells. J Am Soc Nephrol 24:1529–1536

Kohl J, Baelder R, Lewkowich IP, Pandey MK, Hawlisch H et al (2006) A regulatory role for the C5a anaphylatoxin in type 2 immunity in asthma. J Clin Invest 116:783–796

Kwan WH, van der Touw W, Paz-Artal E, Li MO, Heeger PS (2013) Signaling through C5a receptor and C3a receptor diminishes function of murine natural regulatory T cells. J Exp Med 210:257–268

Le Quintrec M, Lionet A, Kandel C, Bourdon F, Gnemmi V et al (2015) Eculizumab for treatment of rapidly progressive C3 glomerulopathy. Am J Kidney Dis 65:484–489

Legendre CM, Licht C, Muus P, Greenbaum LA, Babu S et al (2013) Terminal complement inhibitor eculizumab in atypical hemolytic-uremic syndrome. N Engl J Med 368:2169–2181
Lenderink AM, Liegel K, Ljubanovic D, Coleman KE, Gilkeson GS et al (2007) The alternative pathway of complement is activated in the glomeruli and tubulointerstitium of mice with adriamycin nephropathy. Am J Physiol Renal Physiol 293:F555–F564
Levy M, Halbwachs-Mecarelli L, Gubler MC, Kohout G, Bensenouci A et al (1986) H deficiency in two brothers with atypical dense intramembranous deposit disease. Kidney Int 30:949–956
Li XQ, Chang DY, Chen M, Zhao MH (2019) Complement activation in patients with diabetic nephropathy. Diabetes Metab 45:248–253
Maillard N, Wyatt RJ, Julian BA, Kiryluk K, Gharavi A et al (2015) Current understanding of the role of complement in IgA nephropathy. J Am Soc Nephrol 26:1503–1512
Malik TH, Lavin PJ, Goicoechea de Jorge E, Vernon KA, Rose KL et al (2012) A hybrid CFHR3-1 gene causes familial C3 glomerulopathy. J Am Soc Nephrol 23:1155–1160
Manderson AP, Botto M, Walport MJ (2004) The role of complement in the development of systemic lupus erythematosus. Annu Rev Immunol 22:431–456
Manzi S, Rairie JE, Carpenter AB, Kelly RH, Jagarlapudi SP et al (1996) Sensitivity and specificity of plasma and urine complement split products as indicators of lupus disease activity. Arthritis Rheum 39:1178–1188
Masaki T, Matsumoto M, Nakanishi I, Yasuda R, Seya T (1992) Factor I-dependent inactivation of human complement C4b of the classical pathway by C3b/C4b receptor (CR70, CD35) and membrane cofactor protein (MCP, CD46). J Biochem 111:573–578
McCoy RC, Abramowsky CR, Tisher CC (1974) IgA nephropathy. Am J Pathol 76:123–144
Medjeral-Thomas N, Malik TH, Patel MP, Toth T, Cook HT et al (2014) A novel CFHR5 fusion protein causes C3 glomerulopathy in a family without Cypriot ancestry. Kidney Int 85:933–937
Medjeral-Thomas NR, Lomax-Browne HJ, Beckwith H, Willicombe M, McLean AG et al (2017) Circulating complement factor H-related proteins 1 and 5 correlate with disease activity in IgA nephropathy. Kidney Int 92:942–952
Michelfelder S, Fischer F, Waldin A, Horle KV, Pohl M et al (2018) The MFHR1 fusion protein is a novel synthetic multitarget complement inhibitor with therapeutic potential. J Am Soc Nephrol 29:1141–1153
Moore I, Strain L, Pappworth I, Kavanagh D, Barlow PN et al (2010) Association of factor H autoantibodies with deletions of CFHR1, CFHR3, CFHR4, and with mutations in CFH, CFI, CD46, and C3 in patients with atypical hemolytic uremic syndrome. Blood 115:379–387
Nakorchevsky A, Hewel JA, Kurian SM, Mondala TS, Campbell D et al (2010) Molecular mechanisms of chronic kidney transplant rejection via large-scale proteogenomic analysis of tissue biopsies. J Am Soc Nephrol 21:362–373
Nangaku M, Pippin J, Couser WG (1999) Complement membrane attack complex (C5b-9) mediates interstitial disease in experimental nephrotic syndrome. J Am Soc Nephrol 10:2323–2331
Nangaku M, Pippin J, Couser WG (2002) C6 mediates chronic progression of tubulointerstitial damage in rats with remnant kidneys. J Am Soc Nephrol 13:928–936
Nangaku M, Shankland SJ, Couser WG (2005) Cellular response to injury in membranous nephropathy. J Am Soc Nephrol 16:1195–1204
Nichols EM, Barbour TD, Pappworth IY, Wong EK, Palmer JM et al (2015) An extended mini-complement factor H molecule ameliorates experimental C3 glomerulopathy. Kidney Int 88:1314–1322
Noris M, Remuzzi G (2009) Atypical hemolytic-uremic syndrome. N Engl J Med 361:1676–1687
Noris M, Brioschi S, Caprioli J, Todeschini M, Bresin E et al (2003) Familial haemolytic uraemic syndrome and an MCP mutation. Lancet 362:1542–1547
Ostendorf T, Kunter U, Grone HJ, Bahlmann F, Kawachi H et al (2001) Specific antagonism of PDGF prevents renal scarring in experimental glomerulonephritis. J Am Soc Nephrol 12:909–918
Ostendorf T, Rong S, Boor P, Wiedemann S, Kunter U et al (2006) Antagonism of PDGF-D by human antibody CR84 prevents renal scarring in experimental glomerulonephritis. J Am Soc Nephrol 17:1054–1062

Ostergaard J, Thiel S, Gadjeva M, Hansen TK, Rasch R, Flyvbjerg A (2007) Mannose-binding lectin deficiency attenuates renal changes in a streptozotocin-induced model of type 1 diabetes in mice. Diabetologia 50:1541–1549

Ostergaard JA, Thiel S, Hovind P, Holt CB, Parving HH et al (2014) Association of the pattern recognition molecule H-ficolin with incident microalbuminuria in an inception cohort of newly diagnosed type 1 diabetic patients: an 18 year follow-up study. Diabetologia 57:2201–2207

Ostergaard JA, Thiel S, Lajer M, Steffensen R, Parving HH et al (2015) Increased all-cause mortality in patients with type 1 diabetes and high-expression mannan-binding lectin genotypes: a 12-year follow-up study. Diabetes Care 38:1898–1903

Pangburn MK, Schreiber RD, Muller-Eberhard HJ (1977) Human complement C3b inactivator: isolation, characterization, and demonstration of an absolute requirement for the serum protein beta1H for cleavage of C3b and C4b in solution. J Exp Med 146:257–270

Pickering MC, Cook HT, Warren J, Bygrave AE, Moss J et al (2002) Uncontrolled C3 activation causes membranoproliferative glomerulonephritis in mice deficient in complement factor H. Nat Genet 31:424–428

Pickering MC, Warren J, Rose KL, Carlucci F, Wang Y et al (2006) Prevention of C5 activation ameliorates spontaneous and experimental glomerulonephritis in factor H-deficient mice. Proc Natl Acad Sci U S A 103:9649–9654

Pickering MC, de Jorge EG, Martinez-Barricarte R, Recalde S, Garcia-Layana A et al (2007) Spontaneous hemolytic uremic syndrome triggered by complement factor H lacking surface recognition domains. J Exp Med 204:1249–1256

Pickering MC, Ismajli M, Condon MB, McKenna N, Hall AE et al (2015) Eculizumab as rescue therapy in severe resistant lupus nephritis. Rheumatology (Oxford) 54:2286–2288

Qiu W, Zhou J, Zhu G, Zhao D, He F et al (2014) Sublytic C5b-9 triggers glomerular mesangial cell apoptosis via XAF1 gene activation mediated by p300-dependent IRF-1 acetylation. Cell Death Dis 5:e1176

Rangan GK, Pippin JW, Couser WG (2004) C5b-9 regulates peritubular myofibroblast accumulation in experimental focal segmental glomerulosclerosis. Kidney Int 66:1838–1848

Roos A, Rastaldi MP, Calvaresi N, Oortwijn BD, Schlagwein N et al (2006) Glomerular activation of the lectin pathway of complement in IgA nephropathy is associated with more severe renal disease. J Am Soc Nephrol 17:1724–1734

Sakaguchi S, Yamaguchi T, Nomura T, Ono M (2008) Regulatory T cells and immune tolerance. Cell 133:775–787

Schmidt BZ, Fowler NL, Hidvegi T, Perlmutter DH, Colten HR (1999) Disruption of disulfide bonds is responsible for impaired secretion in human complement factor H deficiency. J Biol Chem 274:11782–11788

Schreiber A, Xiao H, Jennette JC, Schneider W, Luft FC, Kettritz R (2009) C5a receptor mediates neutrophil activation and ANCA-induced glomerulonephritis. J Am Soc Nephrol 20:289–298

Sean Eardley K, Cockwell P (2005) Macrophages and progressive tubulointerstitial disease. Kidney Int 68:437–455

Servais A, Noel LH, Roumenina LT, Le Quintrec M, Ngo S et al (2012) Acquired and genetic complement abnormalities play a critical role in dense deposit disease and other C3 glomerulopathies. Kidney Int 82:454–464

Spitzer D, Mitchell LM, Atkinson JP, Hourcade DE (2007) Properdin can initiate complement activation by binding specific target surfaces and providing a platform for de novo convertase assembly. J Immunol 179:2600–2608

Strainic MG, Shevach EM, An F, Lin F, Medof ME (2013) Absence of signaling into CD4(+) cells via C3aR and C5aR enables autoinductive TGF-beta1 signaling and induction of Foxp3(+) regulatory T cells. Nat Immunol 14:162–171

Tang Z, Lu B, Hatch E, Sacks SH, Sheerin NS (2009) C3a mediates epithelial-to- mesenchymal transition in proteinuric nephropathy. J Am Soc Nephrol 20:593–603

Togarsimalemath SK, Sethi SK, Duggal R, Le Quintrec M, Jha P et al (2017) A novel CFHR1-CFHR5 hybrid leads to a familial dominant C3 glomerulopathy. Kidney Int 92:876–887

Torbohm I, Schonermark M, Wingen AM, Berger B, Rother K, Hansch GM (1990) C5b-8 and C5b-9 modulate the collagen release of human glomerular epithelial cells. Kidney Int 37:1098–1104

Tortajada A, Yebenes H, Abarrategui-Garrido C, Anter J, Garcia-Fernandez JM et al (2013) C3 glomerulopathy-associated CFHR1 mutation alters FHR oligomerization and complement regulation. J Clin Invest 123:2434–2446

Tortajada A, Gutierrez E, Goicoechea de Jorge E, Anter J, Segarra A, Espinosa M, Blasco M, Roman E, Marco H, Quintana LF et al (2017) Elevated factor H-related protein 1 and factor H pathogenic variants decrease complement regulation in IgA nephropathy. Kidney Int 92:953–963

Turnberg D, Lewis M, Moss J, Xu Y, Botto M, Cook HT (2006) Complement activation contributes to both glomerular and tubulointerstitial damage in adriamycin nephropathy in mice. J Immunol 177:4094–4102

Valoti E, Alberti M, Tortajada A, Garcia-Fernandez J, Gastoldi S et al (2015) A novel atypical hemolytic uremic syndrome-associated hybrid CFHR1/CFH gene encoding a fusion protein that antagonizes factor H-dependent complement regulation. J Am Soc Nephrol 26:209–219

van der Touw W, Cravedi P, Kwan WH, Paz-Artal E, Merad M, Heeger PS (2013) Cutting edge: receptors for C3a and C5a modulate stability of alloantigen-reactive induced regulatory T cells. J Immunol 190:5921–5925

Venables JP, Strain L, Routledge D, Bourn D, Powell HM et al (2006) Atypical haemolytic uraemic syndrome associated with a hybrid complement gene. PLoS Med 3:e431

Vivarelli M, Pasini A, Emma F (2012) Eculizumab for the treatment of dense-deposit disease. N Engl J Med 366:1163–1165

Wang Y, Hu Q, Madri JA, Rollins SA, Chodera A, Matis LA (1996) Amelioration of lupus-like autoimmune disease in NZB/WF1 mice after treatment with a blocking monoclonal antibody specific for complement component C5. Proc Natl Acad Sci U S A 93:8563–8568

Wang J, Zhang L, Tang SC, Kashihara N, Kim YS et al (2018) Disease burden and challenges of chronic kidney disease in North and East Asia. Kidney Int 94:22–25

Watanabe H, Garnier G, Circolo A, Wetsel RA, Ruiz P et al (2000) Modulation of renal disease in MRL/lpr mice genetically deficient in the alternative complement pathway factor B. J Immunol 164:786–794

Weiler JM, Daha MR, Austen KF, Fearon DT (1976) Control of the amplification convertase of complement by the plasma protein beta1H. Proc Natl Acad Sci U S A 73:3268–3272

Wenderfer SE, Ke B, Hollmann TJ, Wetsel RA, Lan HY, Braun MC (2005) C5a receptor deficiency attenuates T cell function and renal disease in MRLlpr mice. J Am Soc Nephrol 16:3572–3582

Woroniecka KI, Park AS, Mohtat D, Thomas DB, Pullman JM, Susztak K (2011) Transcriptome analysis of human diabetic kidney disease. Diabetes 60:2354–2369

Wyatt RJ, Julian BA (1988) Activation of complement in IgA nephropathy. Am J Kidney Dis 12:437–442

Wynn TA (2004) Fibrotic disease and the T(H)1/T(H)2 paradigm. Nat Rev Immunol 4:583–594

Xiao H, Schreiber A, Heeringa P, Falk RJ, Jennette JC (2007) Alternative complement pathway in the pathogenesis of disease mediated by anti-neutrophil cytoplasmic autoantibodies. Am J Pathol 170:52–64

Xiao H, Dairaghi DJ, Powers JP, Ertl LS, Baumgart T et al (2014) C5a receptor (CD88) blockade protects against MPO-ANCA GN. J Am Soc Nephrol 25:225–231

Xiao X, Ghossein C, Tortajada A, Zhang Y, Meyer N et al (2016) Familial C3 glomerulonephritis caused by a novel CFHR5-CFHR2 fusion gene. Mol Immunol 77:89–96

Yang Y, Denton H, Davies OR, Smith-Jackson K, Kerr H et al (2018) An engineered complement factor H construct for treatment of C3 glomerulopathy. J Am Soc Nephrol 29:1649–1661

Zhang J, Li Y, Shan K, Wang L, Qiu W et al (2014) Sublytic C5b-9 induces IL-6 and TGF-beta1 production by glomerular mesangial cells in rat Thy-1 nephritis through p300-mediated C/EBPbeta acetylation. FASEB J 28:1511–1525

Zhang YM, Gu QH, Huang J, Qu Z,Wang X et al (2016a) Clinical significance of IgMand C3 glomerular deposition in primary focal segmental glomerulosclerosis. Clin J AmSoc Nephrol 11:1582–1589

Zhang L, Long J, Jiang W, Shi Y, He X et al (2016b) Trends in chronic kidney disease in China. N Engl J Med 375:905–906

Zhou X, Fukuda N, Matsuda H, Endo M, Wang X et al (2013) Complement 3 activates the renal renin-angiotensin system by induction of epithelial-to-mesenchymal transition of the nephrotubulus in mice. Am J Physiol Renal Physiol 305:F957–F967

Zhu L, Zhai YL, Wang FM, Hou P, Lv JC et al (2015) Variants in complement factor H and complement factor H-related protein genes, CFHR3 and CFHR1, affect complement activation in IgA nephropathy. J Am Soc Nephrol 26:1195–1204

Zhu L, Guo WY, Shi SF, Liu LJ et al (2018) Circulating complement factor H-related protein 5 levels contribute to development and progression of IgA nephropathy. Kidney Int 94:150–158

Zwirner J, Burg M, Schulze M, Brunkhorst R, Gotze O, Koch KM, Floege J (1997) Activated complement C3: a potentially novel predictor of progressive IgA nephropathy. Kidney Int 51:1257–1264

Chapter 21
Renal Effects of Cytokines in Hypertension

Yi Wen and Steven D. Crowley

Abstract Preclinical studies point to a key role for immune cells in hypertension via augmenting renal injury and/or hypertensive responses. Blood pressure elevation in rheumatologic patients is attenuated by anti-inflammatory therapies. Both the innate and adaptive immune systems contribute to the pathogenesis of hypertension by modulating renal sodium balance, blood flow, and functions of the vasculature and epithelial cells in the kidney. Monocytes/macrophages and T lymphocytes are pivotal mediators of hypertensive responses, while dendritic cells and B lymphocytes can regulate blood pressure indirectly by promoting T lymphocytes activation. Pro-inflammatory cytokines, such as tumor necrosis factor-α (TNF), interleukin-1 (IL-1), interleukin-17 (IL-17), and interferon-γ (IFN), amplify blood pressure elevation and/or renal injury. By contrast, interleukin-10 (IL-10) protects against renal and vascular function when produced by T helper 2 cells (Th2) and regulatory T cells (Treg). Thus, understanding the renal effects of cytokines in hypertension will provide targets for precise immunotherapies to inhibit targeted organ damage while preserving necessary immunity.

Keywords Immune system · Cytokine · Chemokines · Kidney · Hypertension

21.1 Introduction

Hypertension is a prominent risk factor for severe kidney and cardiovascular disease, afflicting more than 1 billion people worldwide (NCD Risk Factor Collaboration (NCD-RisC) 2017). Kidney and cardiovascular damage can amplify hypertensive responses, resulting in a high risk of mortality (Lionakis et al. 2012). Despite the

Y. Wen
Division of Nephrology, Zhongda Hospital, Southeast University, Nanjing, Jiangsu, China

Y. Wen · S. D. Crowley (✉)
Division of Nephrology, Department of Medicine, Duke University and Durham VA Medical Centers, Durham, NC, USA
e-mail: steven.d.crowley@duke.edu

B.-C. Liu et al. (eds.), *Renal Fibrosis: Mechanisms and Therapies*,
Advances in Experimental Medicine and Biology 1165,
https://doi.org/10.1007/978-981-13-8871-2_21

availability of pharmacologic therapies targeting the vascular system, cardiac function, renal sodium balance, and sympathetic outflow, blood pressure in up to half of the hypertensive patients remains poorly controlled (Egan et al. 2010; Lionakis et al. 2012). Preclinical studies have revealed that both innate and adaptive immune systems contribute to the pathogenesis of hypertension, and immunotherapies may represent a novel approach to regulate blood pressure and target organ damage (Coffman 2011; Crowley and Jeffs 2016; Crowley et al. 2011). Thus, understanding the precise mechanisms through which the immune system regulates blood pressure is important to develop novel therapies without the risks of immunosuppression. Inflammatory cells affect functions of the cardiovascular system and kidney by altering cellular damage or repair and elaborating cytokines within the vasculature and kidney. This chapter will discuss the role that cytokines play in hypertension and target organ damage by direct actions in the kidney.

21.2 Inflammatory Cells in Hypertension

Autopsy studies demonstrated that inflammatory cells infiltrate the kidneys of patients with severe hypertension (Sommers et al. 1958). Preclinical studies demonstrated the distinct role of myeloid cells and lymphocytes in the pathogenesis of hypertension. Activated monocytes/macrophages express lysozyme M (LysM), and deletion of LysM positive macrophages decreases blood pressure by inhibiting endothelial damage with the consequent preservation of sodium excretion (Wenzel et al. 2011). CC chemokine receptor 2 (CCR2) on monocytes/macrophages binds to monocyte chemoattractant protein (MCP-1 or CCL2), facilitating their recruitment into the kidney during hypertension (Chan et al. 2012; Elmarakby et al. 2007), while CCL5 attenuates target organ damage and fibrosis in hypertension by interrupting CCR2-mediated infiltration of inflammatory macrophages into the kidney (Rudemiller et al. 2016). In contrast, VEGF-C-expressing macrophages in the skin inhibit blood pressure elevation by promoting lymphatic drainage of sodium from interstitial reservoirs to the circulating system. Thus, the exact effect of macrophages on blood pressure and renal function depends on their phenotype and distribution (Machnik et al. 2009; Wiig et al. 2013).

Dendritic cells (DCs) are another subset of myeloid cells that connect innate and adaptive immune systems by processing and presenting antigens to T lymphocytes that in turn stimulate reactive oxygen species and cytokine production that contributes to hypertensive responses. DCs inside the kidney facilitate infiltration of T lymphocytes (Yatim et al. 2016), and the requirement for antigen presentation by DCs to activate T lymphocytes in hypertensive responses has been demonstrated in multiple models (Vinh et al. 2010). Therefore, activated DCs can promote the susceptibility to hypertension only in the presence of functional T lymphocytes (Kirabo et al. 2014).

The activated adaptive immune system plays a major role in blood pressure elevation and target organ damage (Mattson et al. 2006). Guzik and colleagues demonstrated that adoptive transfer of T lymphocytes but not B lymphocytes restores hypertensive responses and vascular dysfunction in Ang II-infused Rag1−/− mice (Guzik et al. 2007). Similarly, Crowley and colleagues found that lymphocyte deficiency promotes sodium excretion during hypertension, possibly via stimulation of eNOS and COX-2 in the kidney (Crowley et al. 2010). In addition, CD8+ rather than CD4+ T lymphocytes are the T cell subset that contributes to hypertensive responses by impairing natriuresis and endothelial function. Although the adoptive transfer of B lymphocytes alone does not affect blood pressure in Rag1−/− mice (Guzik et al. 2007), B lymphocytes may drive blood pressure elevation by promoting T lymphocyte activation and cytokines production in WT mice (Chan et al. 2015; Mathis et al. 2014). In conclusion, cells in both the innate and adaptive immune systems contribute to hypertension via actions in the vasculature and the kidney. Next, we focus on the specific actions of individual cytokines produced by these cells to drive renovascular dysfunction and/or renal sodium retention.

21.3 Renal Effects of Cytokines in Hypertension

21.3.1 Tumor Necrosis Factor-α (TNF)

TNF is known as a macrophage cytokine, which is also generated by T lymphocytes and resident kidney cells (Majid 2011). The effects of TNF on kidney function and blood pressure are comprehensive (Ramseyer and Garvin 2013). While exogenous TNF promotes natriuresis by ligation of TNF receptor 1 (TNFR1), the effects of endogenous TNF during brisk activation of the renin–angiotensin system (RAS) are to stimulate blood pressure and kidney injury (Castillo et al. 2012; Chen et al. 2010). Accordingly, the Ang II-induced hypertensive responses are inhibited by TNF-deletion or blockade in rodent models (Guzik et al. 2007; Sriramula et al. 2008; Zhang et al. 2014). NOS3-dependent natriuresis is suppressed by TNF in the thick ascending limb (Ramseyer et al. 2012), and kidney cross-transplant studies demonstrated the net effect of endogenous TNF in the kidney is to augment hypertensive responses in vivo (Zhang et al. 2014). TNF induces glomerular epithelial cell injury in vivo (Bertani et al. 1989; Gomez-Chiarri et al. 1994). In rats, TNF blockade attenuates target organ damage in several hypertensive models (Elmarakby et al. 2006, 2008; Venegas-Pont et al. 2010), and salt-dependent hypertension can be prevented by direct infusion of a TNF blocker into kidney interstitium (Huang et al. 2016). Thus, in vivo studies highlight a role for TNF in augmenting sodium retention via TNF receptor 2 during hypertension (Singh et al. 2013). The effects of TNF on blood pressure in humans are complex. In patients with chronic heart failure, TNF blockade did not affect blood pressure (Chung et al. 2003; Mann et al. 2004). However, in hypertensive patients with frank immune activation, urinary TNF expression correlated with levels of blood pressure (Herrera et al. 2006), and TNF inhibition with infliximab reduces 24-h ambulatory blood pressure in rheumatoid arthritis patients (Yoshida et al. 2014).

Future studies will need to identify approaches for more specific TNF blockade which can prevent sodium retention and avoid off-target immunosuppression.

21.3.2 Interleukin-1 (IL-1)

A prototypical pro-inflammatory cytokine, IL-1, is generated by hematopoietic cells and resident kidney cells (Sims and Smith 2010). IL-1 exerts the recruitment of Myd88 and activation of IL-1 receptor associated kinases (IRAKs) via ligation of IL-1 receptors (IL-R1), facilitating the transcription of pro-inflammatory cytokines, such as TNF. The NLRP3 inflammasome drives the maturation of pro-IL-1 to active IL-1. NLRP3 deficiency in mice blunts the hypertensive responses and/or target organ damage during Ang II or mineralocorticoid infusion (Krishnan et al. 2016; Shirasuna et al. 2015; Wen et al. 2016). Although infused IL-1 exerts a natriuretic effect (Kohan et al. 1989; Schreiner and Kohan 1990; Takahashi et al. 1992), the net effect of endogenous IL-1 is to potentiate hypertensive responses. Intracisternal infusion of IL-1 augments sympathetic activation leading to systemic vasoconstriction, which impairs sodium natriuresis (Takahashi et al. 1992). Similarly, infusion of exogenous IL-1 in the systemic or pulmonary vasculature augments hypertensive responses (Shi et al. 2010; Voelkel et al. 1994). Zhang and colleagues found that IL-1R1 deficiency or blockade blunts NKCC-dependent sodium retention in the thick ascending limb and thereby attenuates Ang II-induced hypertension. IL-1R1 activation stimulates the maturation of myeloid cells which inhibits sodium retention by producing nitric oxide (NO) (Zhang et al. 2016). However, the direct activation of IL-1R1 on macrophage phenotype remains unclear, as inflammasome-derived IL-1R1 activation on myeloid cells facilitates NO production during infection (Lima-Junior et al. 2013). Future studies will need to investigate how to precisely disrupt the actions of IL-1 to attenuate cardiovascular damage without augmenting the susceptibility to lethal infection (Ridker et al. 2017).

21.3.3 Interferon-γ (IFN)

IFN is produced by T cells and macrophages, regulating and marking Th1 differentiation and activating myeloid cells and B lymphocytes. IFN limits natriuresis by stimulating the NHE3 transporter in the proximal tubule and by enhancing NKCC2 and NCC activity in the distal nephron (Kamat et al. 2015). IFN deficiency blunts Ang II-dependent hypertensive responses in rodent models (Saleh et al. 2015). However, blockade targeting IFN receptor 1 (IFNR1) did not reduce Ang II-dependent hypertensive responses (Marko et al. 2012), suggesting that IFN receptor 2 (IFNR2) plays a major role in regulating sodium retention. Nevertheless, inhibition of IFNR1 does prevent the progression of tubulointerstitial inflammation during Ang II-dependent hypertension (Marko et al. 2012).

21.3.4 *Transforming Growth Factor-β (TGF-β)*

TGF-β is a key driver of kidney fibrosis under RAS-associated hypertension (Kagami et al. 1994; Schreiner and Kohan 1990). TGF-β augments kidney fibrosis by inhibiting the activation of matrix metalloproteinases (MMPs) that increase the deposition of extracellular matrix (Border 1994; Douthwaite et al. 1999; Mozes et al. 1999). Infusion of exogenous TGF-β1 or TGF-β2 promotes kidney fibrosis, albuminuria, and increases levels of blood pressure, possibly by inducing vascular dysfunction and/or augmented sodium retention (Ledbetter et al. 2000). Chronic infusion of Ang II increases the levels of TGF-β in serum (Noble and Border 1997). Dietary salt intake augments the levels of TGF-β in the kidney (Sanders 2009; Ying et al. 2008). In Dahl SS rats, TGF-β blockade significantly attenuates blood pressure elevation, proteinuria, and kidney fibrosis (Dahly et al. 2002; Murphy et al. 2012). Inversely, TGF-β generated by T regulatory lymphocytes (Tregs) cooperates with IL-10 to attenuate hypertensive responses by suppressing the activation of T effector lymphocytes (Barhoumi et al. 2011). In hypertension, the exact effects of TGF-β on renal function may depend on origin and concentration. Future studies will need to investigate how to accurately disrupt the TGF-β signaling, given the complex and redundant pathways downstream of TGF-β (Wei et al. 2013).

21.3.5 *Interleukin-17 (IL-17)*

Interleukin-17A is produced by RORrt + CD4+ T cells and plays an important role in infection and autoimmune diseases (Chen and Kolls 2017). IL-17A augments the generation of pro-inflammatory cytokines and chemokines that stimulate cell-mediated immune response (Kim et al. 2016; Korn et al. 2009). IL-17A is produced by T cells that infiltrate the arterial wall and drive damage to vascular smooth muscle cells by increasing production of ROS and pro-inflammatory cytokines (Eid et al. 2009; Pietrowski et al. 2011). In hypertensive patients, the levels of serum IL-17 are significantly increased compared to healthy people (Madhur et al. 2010). Similarly, chronic infusion of Ang II stimulates the production of IL-17 and increases the expression of IL-17 inside the vessel wall (Madhur et al. 2010). Infusion of exogenous IL-17 augments hypertensive responses and endothelial dysfunction by stimulating the Rho-kinase pathway (Nguyen et al. 2013). Deficiency or blockade targeting IL-17A but not IL-17F reduces blood pressure elevation and kidney inflammation in Ang II-dependent hypertension (Madhur et al. 2010; Saleh et al. 2016). Deletion of γδ T cells, another source of IL-17, significantly attenuates Ang II-dependent blood pressure elevation and vascular injury (Caillon et al. 2017). IL-17A stimulates sodium retention by activating the NHE3 sodium transporter in the proximal tubule and the NCC exchanger in the distal nephron (Norlander et al. 2016). However, the non-specific inhibition of IL-17 causes neutral or detrimental effects on kidney function during hypertension. Thus, future studies will need to clarify the tissue- and isoform-specific actions of IL-17 (Krebs et al. 2014; Marko et al. 2012).

21.3.6 Interleukin-10 (IL-10)

IL-10 is an anti-inflammatory cytokine and produced by Th2 lymphocytes, Tregs, mast cells, and monocytes. IL-10 attenuates the generation of pro-inflammatory cytokines and chemokines by inhibiting NF-κB activation (Kim and Kim 2014; Rodriguez-Iturbe et al. 2017). In the rat, infusion of exogenous IL-10 attenuates proteinuria, endothelial damage, and blood pressure elevation during pregnancy-induced hypertension (Chatterjee et al. 2015; Harmon et al. 2015; Tinsley et al. 2010). In TLR3-induced preeclampsia, IL-10 deficiency augments endothelial damage and hypertensive responses, which can be blunted by supplementation of exogenous IL10 (Chatterjee et al. 2011). IL-10 deficiency exacerbates microvascular endothelial damage and blood pressure elevation by stimulating NADPH oxidase or RhoA/Rho kinase signaling (Didion et al. 2009; Kassan et al. 2011; Lima et al. 2016). Thus, the actions of IL-10 in protecting vascular function are consistent, but future studies will need to investigate the effects of IL-10 on systemic vascular resistance and/or renal sodium handling.

21.4 Conclusion

Guyton and colleagues demonstrated that increased sodium and water retention are the foundation for blood pressure elevation (Guyton 1991). Even when salt is non-osmotically stored in the dermis, blood pressure is still regulated by the kidney via titration of circulating volume (Machnik et al. 2009; Wiig et al. 2013). Thus, understanding the actions of inflammatory cytokines in modulating renal function is critical to understand the role of the immune system in the pathogenesis of hypertension. Sympathetic tone and renal nerve activation, renal blood flow-dependent endothelial dysfunction, and sodium transport in the thick ascending limb are the major targets for inflammatory cytokines (Zhang et al. 2016). The effects of TGF-β on renal function and hypertension are complex due to the pro-fibrotic and immunosuppressive functions. Generally, polarized inflammatory cells alter sodium retention and blood pressure elevation by secreting individual cytokines. Thus, pro-inflammatory M1 macrophages, Th1 and Th17 cells augment kidney injury and hypertensive responses by producing TNF, IL-17A, IL-1, and IFN. In contrast, Treg-derived IL-10 inhibits blood pressure elevation and target organ damage. Macrophage-derived nitric oxide (NO) and VEGF-C also perform antihypertensive actions by facilitating natriuresis (Machnik et al. 2009; Zhang et al. 2016). Considering the concentration, tissue distribution, and subtype of certain cytokines, the animal studies cited above certainly include exceptions to the template in which inflammation augments hypertensive responses. Nevertheless, the effects of inflammatory cytokines in stimulating blood pressure elevation may reflect the protective actions of the immune system in preventing circulatory collapse during an overwhelming infection.

Future studies will be needed to define the actions of cytokines in human hypertension. The serum levels of inflammatory cytokines correlate with blood pressure levels in some clinical studies (Bautista et al. 2005), and blockade targeting certain cytokines can reduce blood pressure hypertensive patients with the rheumatologic disease (Herrera et al. 2006; Yoshida et al. 2014). However, the risk of infection from immunomodulatory therapy cannot be dismissed in patients with cardiovascular disease (Ridker et al. 2017). Thus, understanding the precise renal effects of cytokines in hypertension is paramount to prevent cytokine-dependent sodium retention while

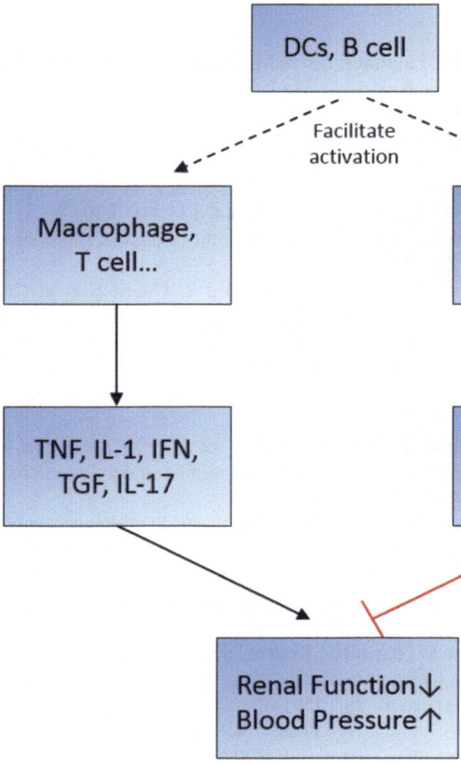

Fig. 21.1 Renal effects of cytokines in hypertension. Both B cells and dendritic cells (DC) participate in the pathogenesis of hypertension through facilitating the activation of inflammatory cells. Pro-inflammatory cytokines including tumor necrosis factor-α (TNF), interleukin-1 (IL-1), interferon-gamma (IFN), transforming growth factor-β (TGF-β), and interleukin-17 (IL-17) are produced by macrophages and T cells and augment renal function decline and the hypertensive response. By contrast, interleukin-10 (IL-10) is generated by T helper 2 cells (Th2) and regulatory T cells (Treg) and attenuates renal injury and blood pressure elevation

avoiding immunosuppression and tumor escape. Meanwhile, given the recent findings that sodium retention stimulates the pro-inflammatory polarization of T lymphocytes and macrophages (Hernandez et al. 2015; Jantsch et al. 2015; Kleinewietfeld et al. 2013; Wu et al. 2013), therapy that combines diuretics together with an anti-inflammatory agent offers potential for attenuating renal and cardiovascular damage in certain hypertensive patients (Fig. 21.1).

Acknowledgements NIH grants DK087893, HL128355; Veterans Health Administration, Office of Research and Development, Biomedical Laboratory Research and Development Grant BX000893.

References

Barhoumi T, Kasal DA, Li MW, Shbat L, Laurant P, Neves MF et al (2011) T regulatory lymphocytes prevent angiotensin II-induced hypertension and vascular injury. Hypertension 57:469–476

Bautista LE, Vera LM, Arenas IA, Gamarra G (2005) Independent association between inflammatory markers (C-reactive protein, interleukin-6, and TNF-alpha) and essential hypertension. J. Hum. Hypertens. 19:149–154

Bertani T, Abbate M, Zoja C, Corna D, Perico N, Ghezzi P, Remuzzi G (1989) Tumor necrosis factor induces glomerular damage in the rabbit. Am. J. Pathol. 134:419–430

Border WA (1994) Transforming growth factor-beta and the pathogenesis of glomerular diseases. Curr. Opin. Nephrol. Hypertens. 3:54–58

Caillon A, Mian M, Fraulob-Aquino JC, Huo KG, Barhoumi T, Ouerd S et al (2017) Gammadelta T cells mediate angiotensin II-induced hypertension and vascular injury. Circulation 135:2155–2162

Castillo A, Islam MT, Prieto MC, Majid DS (2012) Tumor necrosis factor-alpha receptor type 1, not type 2, mediates its acute responses in the kidney. Am. J. Physiol. Renal Physiol. 302:F1650–F1657

Chan CT, Moore JP, Budzyn K, Guida E, Diep H, Vinh A et al (2012) Reversal of vascular macrophage accumulation and hypertension by a CCR7 antagonist in deoxycorticosterone/salt-treated mice. Hypertension 60:1207–1212

Chan CT, Sobey CG, Lieu M, Ferens D, Kett MM, Diep H et al (2015) Obligatory role for B cells in the development of angiotensin II-dependent hypertension. Hypertension 66:1023–1033

Chatterjee P, Chiasson VL, Kopriva SE, Young KJ, Chatterjee V, Jones KA et al (2011) Interleukin 10 deficiency exacerbates toll-like receptor 3-induced preeclampsia-like symptoms in mice. Hypertension 58:489–496

Chatterjee P, Chiasson VL, Seerangan G, Tobin RP, Kopriva SE, Newell-Rogers MK et al (2015) Cotreatment with interleukin 4 and interleukin 10 modulates immune cells and prevents hypertension in pregnant mice. Am. J. Hypertens. 28:135–142

Chen K, Kolls JK (2017) Interluekin-17A (IL17A). Gene 614:8–14

Chen CC, Pedraza PL, Hao S, Stier CT, Ferreri NR (2010) TNFR1-deficient mice display altered blood pressure and renal responses to ANG II infusion. Am. J. Physiol. Renal Physiol. 299:F1141–F1150

Chung ES, Packer M, Lo KH, Fasanmade AA, Willerson JT (2003) Randomized, double-blind, placebo-controlled, pilot trial of infliximab, a chimeric monoclonal antibody to tumor necrosis factor-alpha, in patients with moderate-to-severe heart failure: results of the anti-TNF therapy against congestive heart failure (ATTACH) trial. Circulation 107:3133–3140

Coffman TM (2011) Under pressure: the search for the essential mechanisms of hypertension. Nat. Med. 17:1402–1409

Crowley SD, Jeffs AD (2016) Targeting cytokine signaling in salt-sensitive hypertension. Am. J. Physiol. Renal Physiol. 311:F1153–F1158

Crowley SD, Song YS, Lin EE, Griffiths R, Kim HS, Ruiz P (2010) Lymphocyte responses exacerbate angiotensin II-dependent hypertension. Am. J. Physiol. Regul. Integr. Comp. Physiol. 298:R1089–R1097

Crowley SD, Zhang J, Herrera M, Griffiths R, Ruiz P, Coffman TM (2011) Role of AT(1) receptor-mediated salt retention in angiotensin II-dependent hypertension. Am. J. Physiol. Renal Physiol. 301:F1124–F1130

Dahly AJ, Hoagland KM, Flasch AK, Jha S, Ledbetter SR, Roman RJ (2002) Antihypertensive effects of chronic anti-TGF-beta antibody therapy in Dahl S rats. Am. J. Physiol. Regul. Integr. Comp. Physiol. 283:R757–R767

Didion SP, Kinzenbaw DA, Schrader LI, Chu Y, Faraci FM (2009) Endogenous interleukin-10 inhibits angiotensin II-induced vascular dysfunction. Hypertension 54:619–624

Douthwaite JA, Johnson TS, Haylor JL, Watson P, Nahas El (1999) Effects of transforming growth factor-beta1 on renal extracellular matrix components and their regulating proteins. J. Am. Soc. Nephrol. 10:2109–2119

Egan BM, Zhao Y, Axon RN (2010) US trends in prevalence, awareness, treatment, and control of hypertension, 1988-2008. JAMA 303:2043–2050

Eid RE, Rao DA, Zhou J, Lo SF, Ranjbaran H, Gallo A et al (2009) Interleukin-17 and interferon-gamma are produced concomitantly by human coronary artery-infiltrating T cells and act synergistically on vascular smooth muscle cells. Circulation 119:1424–1432

Elmarakby AA, Quigley JE, Pollock DM, Imig JD (2006) Tumor necrosis factor alpha blockade increases renal Cyp2c23 expression and slows the progression of renal damage in salt-sensitive hypertension. Hypertension 47:557–562

Elmarakby AA, Quigley JE, Olearczyk JJ, Sridhar A, Cook AK, Inscho EW et al (2007) Chemokine receptor 2b inhibition provides renal protection in angiotensin II - salt hypertension. Hypertension 50:1069–1076

Elmarakby AA, Quigley JE, Imig JD, Pollock JS, Pollock DM (2008) TNF-alpha inhibition reduces renal injury in DOCA-salt hypertensive rats. Am. J. Physiol. Regul. Integr. Comp. Physiol. 294:R76–R83

Gomez-Chiarri M, Ortiz A, Lerma JL, Lopez-Armada MJ, Mampaso F, Gonzalez E et al (1994) Involvement of tumor necrosis factor and platelet-activating factor in the pathogenesis of experimental nephrosis in rats. Lab. Invest. 70:449–459

Guyton AC (1991) Blood pressure control—special role of the kidneys and body fluids. Science 252:1813–1816

Guzik TJ, Hoch NE, Brown KA, McCann LA, Rahman A, Dikalov S et al (2007) Role of the T cell in the genesis of angiotensin II induced hypertension and vascular dysfunction. J. Exp. Med. 204:2449–2460

Harmon A, Cornelius D, Amaral L, Paige A, Herse F, Ibrahim T et al (2015) IL-10 supplementation increases Tregs and decreases hypertension in the RUPP rat model of preeclampsia. Hypertens Pregnancy 34:291–306

Hernandez AL, Kitz A, Wu C, Lowther DE, Rodriguez DM, Vudattu N et al (2015) Sodium chloride inhibits the suppressive function of $FOXP3^{+}$ regulatory T cells. J Clin Invest 125:4212–4222

Herrera J, Ferrebuz A, MacGregor EG, Rodriguez-Iturbe B (2006) Mycophenolate mofetil treatment improves hypertension in patients with psoriasis and rheumatoid arthritis. J. Am. Soc. Nephrol. 17:S218–S225

Huang B, Cheng Y, Usa K, Liu Y, Baker MA, Mattson DL et al (2016) Renal tumor necrosis factor alpha contributes to hypertension in Dahl salt-sensitive rats. Sci Rep 6:21960

Jantsch J, Schatz V, Friedrich D, Schroder A, Kopp C, Siegert I et al (2015) Cutaneous Na+ storage strengthens the antimicrobial barrier function of the skin and boosts macrophage-driven host defense. Cell Metab. 21:493–501

Kagami S, Border WA, Miller DE, Noble NA (1994) Angiotensin II stimulates extracellular matrix protein synthesis through induction of transforming growth factor-beta expression in rat glomerular mesangial cells. J Clin Invest 93:2431–2437

Kamat NV, Thabet SR, Xiao L, Saleh MA, Kirabo A, Madhur MS et al (2015) Renal transporter activation during angiotensin-II hypertension is blunted in interferon-gamma$^{-/-}$ and interleukin-17A$^{-/-}$ mice. Hypertension 65:569–576

Kassan M, Galan M, Partyka M, Trebak M, Matrougui K (2011) Interleukin-10 released by CD4(+)CD25(+) natural regulatory T cells improves microvascular endothelial function through inhibition of NADPH oxidase activity in hypertensive mice. Arterioscler. Thromb. Vasc. Biol. 31:2534–2542

Kim HY, Kim HS (2014) IL-10 up-regulates CCL5 expression in vascular smooth muscle cells from spontaneously hypertensive rats. Cytokine 68:40–49

Kim BS, Park YJ, Chung Y (2016) Targeting IL-17 in autoimmunity and inflammation. Arch Pharm Res 39:1537–1547

Kirabo A, Fontana V, de Faria AP, Loperena R, Galindo CL, Wu J et al (2014) DC isoketal-modified proteins activate T cells and promote hypertension. J Clin Invest 124:4642–4656

Kleinewietfeld M, Manzel A, Titze J, Kvakan H, Yosef N, Linker RA et al (2013) Sodium chloride drives autoimmune disease by the induction of pathogenic TH17 cells. Nature 496:518–522

Kohan DE, Merli CA, Simon EE (1989) Micropuncture localization of the natriuretic effect of interleukin 1. Am. J. Physiol. 256:F810–F813

Korn T, Bettelli E, Oukka M, Kuchroo VK (2009) IL-17 and Th17 Cells. Annu. Rev. Immunol. 27:485–517

Krebs CF, Lange S, Niemann G, Rosendahl A, Lehners A, Meyer-Schwesinger C et al (2014) Deficiency of the interleukin 17/23 axis accelerates renal injury in mice with deoxycorticosterone acetate + angiotensin ii-induced hypertension. Hypertension 63:565–571

Krishnan SM, Dowling JK, Ling YH, Diep H, Chan CT, Ferens D et al (2016) Inflammasome activity is essential for one kidney/deoxycorticosterone acetate/salt-induced hypertension in mice. Br. J. Pharmacol. 173:752–765

Ledbetter S, Kurtzberg L, Doyle S, Pratt BM (2000) Renal fibrosis in mice treated with human recombinant transforming growth factor-beta2. Kidney Int. 58:2367–2376

Lima VV, Zemse SM, Chiao CW, Bomfim GF, Tostes RC, Clinton R, Giachini FR (2016) Interleukin-10 limits increased blood pressure and vascular RhoA/Rho-kinase signaling in angiotensin II-infused mice. Life Sci. 145:137–143

Lima-Junior DS, Costa DL, Carregaro V, Cunha LD, Silva AL, Mineo TW et al (2013) Inflammasome-derived IL-1beta production induces nitric oxide-mediated resistance to Leishmania. Nat. Med. 19:909–915

Lionakis N, Mendrinos D, Sanidas E, Favatas G, Georgopoulou M (2012) Hypertension in the elderly. World J Cardiol 4:135–147

Machnik A, Neuhofer W, Jantsch J, Dahlmann A, Tammela T, Machura K et al (2009) Macrophages regulate salt-dependent volume and blood pressure by a vascular endothelial growth factor-C-dependent buffering mechanism. Nat. Med. 15:545–552

Madhur MS, Lob HE, McCann LA, Iwakura Y, Blinder Y, Guzik TJ et al (2010) Interleukin 17 promotes angiotensin II-induced hypertension and vascular dysfunction. Hypertension 55:500–507

Majid DS (2011) Tumor necrosis factor-alpha and kidney function: experimental findings in mice. Adv. Exp. Med. Biol. 691:471–480

Mann DL, McMurray JJ, Packer M, Swedberg K, Borer JS, Colucci WS et al (2004) Targeted anticytokine therapy in patients with chronic heart failure: results of the randomized etanercept worldwide evaluation (RENEWAL). Circulation 109:1594–1602

Marko L, Kvakan H, Park JK, Qadri F, Spallek B, Binger KJ et al (2012) Interferon-gamma signaling inhibition ameliorates angiotensin II-induced cardiac damage. Hypertension 60:1430–1436

Mathis KW, Wallace K, Flynn ER, Maric-Bilkan C, LaMarca B, Ryan MJ (2014) Preventing autoimmunity protects against the development of hypertension and renal injury. Hypertension 64:792–800

Mattson DL, James L, Berdan EA, Meister CJ (2006) Immune suppression attenuates hypertension and renal disease in the Dahl salt-sensitive rat. Hypertension 48:149–156

Mozes MM, Bottinger EP, Jacot TA, Kopp JB (1999) Renal expression of fibrotic matrix proteins and of transforming growth factor-beta (TGF-beta) isoforms in TGF-beta transgenic mice. J. Am. Soc. Nephrol. 10:271–280

Murphy SR, Dahly-Vernon AJ, Dunn KM, Chen CC, Ledbetter SR, Williams JM et al (2012) Renoprotective effects of anti-TGF-beta antibody and antihypertensive therapies in Dahl S rats. Am. J. Physiol. Regul. Integr. Comp. Physiol. 303:R57–R69

NCD Risk Factor Collaboration (NCD-RisC) (2017) Worldwide trends in blood pressure from 1975 to 2015: a pooled analysis of 1479 population-based measurement studies with 191 million participants. Lancet 389:37–55

Nguyen H, Chiasson VL, Chatterjee P, Kopriva SE, Young KJ, Mitchell BM (2013) Interleukin-17 causes Rho-kinase-mediated endothelial dysfunction and hypertension. Cardiovasc. Res. 97:696–704

Noble NA, Border WA (1997) Angiotensin II in renal fibrosis: should TGF-beta rather than blood pressure be the therapeutic target? Semin. Nephrol. 17:455–466

Norlander AE, Saleh MA, Kamat NV, Ko B, Gnecco J, Zhu L et al (2016) Interleukin-17A regulates renal sodium transporters and renal injury in angiotensin II-induced hypertension. Hypertension 68:167–174

Pietrowski E, Bender B, Huppert J, White R, Luhmann HJ, Kuhlmann CR (2011) Pro-inflammatory effects of interleukin-17A on vascular smooth muscle cells involve NAD(P)H-oxidase derived reactive oxygen species. J. Vasc. Res. 48:52–58

Ramseyer VD, Garvin JL (2013) Tumor necrosis factor-alpha: regulation of renal function and blood pressure. Am. J. Physiol. Renal Physiol. 304:F1231–F1242

Ramseyer VD, Hong NJ, Garvin JL (2012) Tumor necrosis factor alpha decreases nitric oxide synthase type 3 expression primarily via Rho/Rho kinase in the thick ascending limb. Hypertension 59:1145–1150

Ridker PM, Everett BM, Thuren T, MacFadyen JG, Chang WH, Ballantyne C et al (2017) Antiinflammatory therapy with canakinumab for atherosclerotic disease. N. Engl. J. Med. 377:1119–1131

Rodriguez-Iturbe B, Pons H, Johnson RJ (2017) Role of the immune system in hypertension. Physiol. Rev. 97:1127–1164

Rudemiller NP, Patel MB, Zhang JD, Jeffs AD, Karlovich NS, Griffiths R et al (2016) C-C motif chemokine 5 attenuates angiotensin II-dependent kidney injury by limiting renal macrophage infiltration. Am. J. Pathol. 186:2846–2856

Saleh MA, McMaster WG, Wu J, Norlander AE, Funt SA, Thabet SR et al (2015) Lymphocyte adaptor protein LNK deficiency exacerbates hypertension and end-organ inflammation. J Clin Invest 125:1189–1202

Saleh MA, Norlander AE, Madhur MS (2016) Inhibition of interleukin 17-A but not Interleukin-17F signaling lowers blood pressure and reduces end-organ Inflammation in angiotensin II-induced hypertension. JACC Basic Transl Sci 1:606–616

Sanders PW (2009) Vascular consequences of dietary salt intake. Am. J. Physiol. Renal Physiol. 297:F237–F243

Schreiner GF, Kohan DE (1990) Regulation of renal transport processes and hemodynamics by macrophages and lymphocytes. Am. J. Physiol. 258:F761–F767

Shi P, Diez-Freire C, Jun JY, Qi Y, Katovich MJ, Li Q et al (2010) Brain microglial cytokines in neurogenic hypertension. Hypertension 56:297–303

Shirasuna K, Karasawa T, Usui F, Kobayashi M, Komada T, Kimura H et al (2015) NLRP3 deficiency improves angiotensin II-induced hypertension but not fetal growth restriction during pregnancy. Endocrinology 156:4281–4292

Sims JE, Smith DE (2010) The IL-1 family: regulators of immunity. Nat. Rev. Immunol. 10:89–102

Singh P, Bahrami L, Castillo A, Majid DS (2013) TNF-alpha type 2 receptor mediates renal inflammatory response to chronic angiotensin II administration with high salt intake in mice. Am. J. Physiol. Renal Physiol. 304:F991–F999

Sommers SC, Relman AS, Smithwick RH (1958) Histologic studies of kidney biopsy specimens from patients with hypertension. Am. J. Pathol. 34:685–715

Sriramula S, Haque M, Majid DS, Francis J (2008) Involvement of tumor necrosis factor-alpha in angiotensin II-mediated effects on salt appetite, hypertension, and cardiac hypertrophy. Hypertension 51:1345–1351

Takahashi H, Nishimura M, Sakamoto M, Ikegaki I, Nakanishi T, Yoshimura M (1992) Effects of interleukin-1 beta on blood pressure, sympathetic nerve activity, and pituitary endocrine functions in anesthetized rats. Am. J. Hypertens. 5:224–229

Tinsley JH, South S, Chiasson VL, Mitchell BM (2010) Interleukin-10 reduces inflammation, endothelial dysfunction, and blood pressure in hypertensive pregnant rats. Am. J. Physiol. Regul. Integr. Comp. Physiol. 298:R713–R719

Venegas-Pont M, Manigrasso MB, Grifoni SC, LaMarca BB, Maric C, Racusen LC et al (2010) Tumor necrosis factor-alpha antagonist etanercept decreases blood pressure and protects the kidney in a mouse model of systemic lupus erythematosus. Hypertension 56:643–649

Vinh A, Chen W, Blinder Y, Weiss D, Taylor WR, Goronzy JJ et al (2010) Inhibition and genetic ablation of the B7/CD28 T-cell costimulation axis prevents experimental hypertension. Circulation 122:2529–2537

Voelkel NF, Tuder RM, Bridges J, Arend WP (1994) Interleukin-1 receptor antagonist treatment reduces pulmonary hypertension generated in rats by monocrotaline. Am. J. Respir. Cell Mol. Biol. 11:664–675

Wei LH, Huang XR, Zhang Y, Li YQ, Chen HY, Heuchel R et al (2013) Deficiency of Smad7 enhances cardiac remodeling induced by angiotensin II infusion in a mouse model of hypertension. PLoS ONE 8:e70195

Wen Y, Liu Y, Tang T, Lv L, Liu H, Ma K, Liu B (2016) NLRP3 inflammasome activation is involved in ang II-induced kidney damage via mitochondrial dysfunction. Oncotarget 7:54290–54302

Wenzel P, Knorr M, Kossmann S, Stratmann J, Hausding M, Schuhmacher S et al (2011) Lysozyme M-positive monocytes mediate angiotensin II-induced arterial hypertension and vascular dysfunction. Circulation 124:1370–1381

Wiig H, Schroder A, Neuhofer W, Jantsch J, Kopp C, Karlsen TV et al (2013) Immune cells control skin lymphatic electrolyte homeostasis and blood pressure. J Clin Invest 123:2803–2815

Wu C, Yosef N, Thalhamer T, Zhu C, Xiao S, Kishi Y et al (2013) Induction of pathogenic TH17 cells by inducible salt-sensing kinase SGK1. Nature 496:513–517

Yatim KM, Gosto M, Humar R, Williams AL, Oberbarnscheidt MH (2016) Renal dendritic cells sample blood-borne antigen and guide T-cell migration to the kidney by means of intravascular processes. Kidney Int. 90:818–827

Ying WZ, Aaron K, Sanders PW (2008) Mechanism of dietary salt-mediated increase in intravascular production of TGF-beta1. Am. J. Physiol. Renal Physiol. 295:F406–F414

Yoshida S, Takeuchi T, Kotani T, Yamamoto N, Hata K, Nagai K et al (2014) Infliximab, a TNF-alpha inhibitor, reduces 24-h ambulatory blood pressure in rheumatoid arthritis patients. J. Hum. Hypertens. 28:165–169

Zhang J, Patel MB, Griffiths R, Mao A, Song YS, Karlovich NS et al (2014) Tumor necrosis factor-alpha produced in the kidney contributes to angiotensin II-dependent hypertension. Hypertension 64:1275–1281

Zhang J, Rudemiller NP, Patel MB, Karlovich NS, Wu M, McDonough AA et al (2016) Interleukin-1 receptor activation potentiates salt reabsorption in angiotensin II-induced hypertension via the NKCC2 co-transporter in the nephron. Cell Metab. 23:360–368

Chapter 22
Role of Extracellular Vesicles in Renal Inflammation and Fibrosis

Lin-Li Lv

Abstract Extracellular vesicles (EVs) are the membrane-surrounded structures released by almost all types of cells. Accumulating evidences have suggested that EVs secretion is enhanced under stress conditions and have been associated with a large wide of cellular physiological and pathological processes. In this part, recent understanding about the generation and biological function of EVs was reviewed. Moreover, the role of EVs in renal inflammation and fibrosis and future challenges of EVs study in kidney disease were discussed.

Keywords Extracellular vesicles · Renal fibrosis · Renal inflammation · Biomarker

22.1 Generation of Extracellular Vesicles (EVs)

Extracellular vesicles (EVs) are the membrane-surrounded structures released by almost all types of cells under study. Currently, EVs are classified into three categories based on their biogenesis, exosomes, microvesicles (MVs) and apoptotic bodies. Apoptotic bodies, with a diameter range from 200 nm to 5 mm, are shed from the plasma membrane of dying cells undergoing programmed cell death. Microvesicles are shed from the plasma membrane of viable cells with a size of 100–800 nm. Exosomes are 30–150 nm in size and are released into the extracellular space when multi-vesicular bodies (MVBs) fuse with the plasma membrane (Lasser et al. 2018; Mathieu et al. 2019).

Exosome release may increase under cellular stress conditions, such as a dysfunctional lysosomal pathway, endoplasmic reticulum stress, hypoxia or irradiation. It is demonstrated that inflammatory conditions promote the release of MVs with increased procoagulant activity from tumor cell lines (Gieseler et al. 2018). Accumulating evidence showed that cancer cells produce more exosomes under hypoxic conditions than do parental cells under normoxic conditions. The increasing secre-

L.-L. Lv (✉)
Institute of Nephrology, Zhong Da Hospital, Southeast University School of Medicine, Nanjing, China
e-mail: lvlinli@seu.edu.cn

B.-C. Liu et al. (eds.), *Renal Fibrosis: Mechanisms and Therapies*,
Advances in Experimental Medicine and Biology 1165,
https://doi.org/10.1007/978-981-13-8871-2_22

tion of exosomes may be a way of eliminating waste products accumulated in the cells under stress conditions. However, it is also likely that cells subjected to stress communicate with adjacent cells via the release of EVs (Dusso et al. 2018).

During secretion, EVs contain some endogenous substances from the parent cells including membrane traffic proteins (i.e., annexins, flotillin), multivesicular bodies (i.e., TSG101, Alix), integrins and tetraspanins (CD9, CD63, CD81, CD29). Most importantly, RNAs, DNAs, proteins and lipids from parent cells are also found in the released EVs (Escrevente et al. 2011; Svensson et al. 2013). Comparative study also revealed that the component varied in different populations of EVs. While large EVs (L-EV) and small EVs (S-EV) (exosomes) purified from the same cells contained similar amounts of protein, the DNA was more abundant in L-EV, despite S-EVs being more numerous (Vagner and Spinelli 2018).

22.2 Biological Function of EVs

Initially, EVs were proposed to release for getting rid of cellular waste. However, recent data support the idea of exosomes as an alternative way of maintaining cellular homeostasis. Importantly, these vesicles are believed to play a role in intercellular communication and have been associated with a large wide of cellular physiological and pathological processes. For example, secretion of exosomes from late endosomes is required for directional cell movement. Fibronectin is found on the outside of exosomes to support cellular adhesion and migration (Sung et al. 2015). Moreover, exosomes contain numerous cargoes that could impact cell motility, proliferation, phenotypic change and maturation. However, the total array of activities of the EVs is still under study. Generally, increasing evidence supported that EVs played important roles in immune modulation, tissue injury and repair and cellular homeostasis.

22.2.1 EVs in Immune Modulation

Exosomes and microvesicles have been shown to participate in antigen presentation, immune modulation, antitumor immunity and autoimmunity. EVs may exhibit immune suppressing or activation depending on the specific circumstances and the content (Caruso and Poon 2018). EVs have been shown to regulate molecule pathways in recipient cells through receptor–ligand interaction or function cargo transfer via endocytosis as well as membrane fusion. EVs can modulate immune responses by transporting damage-associated molecular patterns (DAMPs), cytokines such as IL-1β and functional microRNAs. Alternatively, EVs could regulate immunological memory through the surface expression of antigen-presenting MHC I and MHC II molecules.

22.2.1.1 EVs and DAMPs

Cells upon stress or injury release EVs containing damage-associated molecular patterns (DAMPs), which can contribute to tissue inflammation. Newly identified potential DAMPs include extracellular heat shock proteins (eHsp72), uric acid crystals, mitochondrial DNA (mtDNA), endogenous RNAs, high mobility group box (HMGB)1 and ATP (Fleshner and Crane 2017). Histones are the protein component of nucleosomes, which are the important DAMPs in tissue injury. Circulating histones contribute to inflammation by interacting with specific receptors, notably toll-like receptor 4 (TLR4). Recent study showed histones are actively released within EVs by LPS-activated macrophages. And histones are present on the outer surface of vesicles and can interact with TLR4 (Nair et al. 2018). Exosomes could also transfer mitochondria from airway myeloid-derived regulatory cells to T cells and participate in intercellular communication within the airways of human subjects with asthma (Hough et al. 2018).

Besides, under pathological conditions, endogenous RNAs act as DAMPs for pattern recognition receptors (PRRs). RN7SL1 is an endogenous RNA that is normally shielded by RNA-binding proteins. Interestingly, triggering of stromal NOTCH-MYC by breast cancer cells results in the increase of RN7SL1 and unshielded RN7SL1 in stromal exosomes. After exosome transfer to immune cells, unshielded RN7SL1 drives an inflammatory response (Nabet et al. 2017).

22.2.1.2 EVs and Cytokines

In addition to be secreted in soluble free format, cytokines are also imported into EVs and release to extracellular space. For instance, IL-1β is a secreted protein that lacks a signal peptide and cannot be secreted in a traditional pathway. Thus, IL-1β has been the most studied cytokines that could be secreted in a protected form by being packaged and secreted via both exosomes and MVs (Lopez-Castejon and Brough 2011; MacKenzie et al. 2001). In activated macrophages, the packaging of IL-1β into exosomes is mediated by ATP-gated P2X7 receptors (P2X7R) (Lopez-Castejon and Brough 2011). A recent report found that a wide variety of cytokines were encapsulated into EVs as observed in different in vitro, ex vivo and in vivo systems. Exosomes produced by hypoxic tumor cells are highly enriched in immunomodulatory proteins and chemokines (Park et al. 2019). Importantly, EVs carrying cytokines are more stable than free cytokines and are biologically active upon interacting with sensitive cells (Fitzgerald et al. 2018). Free cytokines are usually unstable and have very short half-life in plasma (Kudo et al. 1990); EVs-associated cytokines are protected by the vesicles which may destine for signaling processes at sites distant to the local inflammatory lesion.

22.2.1.3 EVs and microRNA

Among EVs, exosomes are the fraction that is enriched in genetic material, mostly non-coding RNAs. Circulating exosomes are the major mechanism for miRNA transport. In addition to bounding to protective proteins, such as high-density lipoprotein and Argonaute protein, miRNAs were packaged into protective microvesicles, such as exosomes (Fleshner and Crane 2017). Since the first study reported in 2007 (Valadi et al. 2007), increasing evidence showed that small non-coding RNAs within EVs are the major contributors to the molecular events in the recipient cell (Turchinovich et al. 2019). Exosome miRNAs could reprogramme recipient cells upon internalization. Adipose tissue macrophages secreted exosomes containing miRNA cargo, which can be transferred to insulin target cell types with robust effects on cellular insulin action (Ying et al. 2017). Another function of EV-associated miRNAs is that miRNA in exosomes may activate TLRs as paracrine agonists and contribute to inflammation. TLR7 and TLR8 are located in intracellular endosomes; Fabbri et al. demonstrated miRNAs in cancer-released exosomes could reach and bind TLR7 and TLR8 in a "receiving" cell (Fabbri et al. 2012).

22.2.2 EVs in Tissue Regeneration and Repair

Studies have revealed that EVs may contribute to tissue integrity and repair. Administration of EVs released from healthy cells, especially stem cells, has been shown to promote tissue regeneration in a variety of tissue injury models. Stem/progenitor cell-derived EVs have been demonstrated as a regenerative therapy for the acceleration of wound healing in a range of clinically relevant animal models of cutaneous wounds (Fabbri et al. 2012). A number of possible mechanisms involving EV-mediated transfer of functional molecules that trigger pro-repair pathways in target cells have been proved (Cabral et al. 2018). Gupta et al. identified a distinct type of early apoptotic extracellular vesicle with specific mitogenic activity, which are found in damaged mouse glomeruli and thus might have regenerative effects in the kidney (Gupta et al. 2017).

Interestingly, exosomes from human umbilical cord MSC inhibited STZ-induced β-cell apoptosis and restored the insulin secreting function of T2DM. In the rat models, exosomes from hucMSC can alleviate T2DM through reversing peripheral insulin resistance and relieving β-cell destruction (Sun et al. 2018). Endogenous annexin A1 (ANXA1) is released as a component of EVs derived from intestinal epithelial cells, and these ANXA1-containing EVs-activated wound repair circuits (Leoni et al. 2015).

22.2.3 EVs in Cell Homeostasis

EVs were initially proposed as cellular wastes garbage (Johnstone et al. 1987); however, the role of EVs secretion in maintaining cellular homeostasis is largely unknown. Some recent data indicate that EVs were an alternative approach to eliminate waste products to maintain cellular homeostasis.

EVs were a protein quality control pathway that, unlike degradation-based approach, promotes protein homeostasis by exporting misfolded proteins through excretion route (Desdin-Mico and Mittelbrunn 2017). Takahashi A et al. reported that exosome secretion maintains cellular homeostasis by removing harmful cytoplasmic DNA from cells. The inhibition of exosome secretion results in the accumulation of nuclear DNA and consequently senescence-like cell cycle arrest or apoptosis in normal human cells (Takahashi et al. 2017). However, the effect of secreted EVs with DNA needs further clarification. Indeed, a recent study reports that T cell EVs that contain genomic and mitochondrial DNA can be transferred to dendritic cells (DC), inducing antiviral responses (Torralba et al. 2018). Interestingly, MSC eliminated depolarized mitochondria by the release of EVs to enhance MSCs' cell survival (Phinney et al. 2015).

Multivesicular bodies (MVB) can either be directed to lysosomes for degradation or transported to the plasma membrane for exosome release. However, it is still unclear that how this balance is regulated in different cellular status. Autophagy is another pathway in the maintenance of protein homeostasis and the preservation of proper organelle function by selective removal of damaged organelles. Thus, exosome biogenesis and autophagy are linked by the endolysosomal pathway to preserve intracellular protein and RNA homeostasis. Under conditions that stimulate autophagy, MVBs are directed to the autophagic pathway that consequently inhibits exosome release (Hessvik and Llorente 2018; Xu et al. 2018). In neuronal cells, autophagy stimulation with the mTOR inhibitor rapamycin strongly inhibited exosomal prion release. Autophagy modulation can control the lateral transfer of prions by interfering with their exosomal release (Abdulrahman et al. 2018). Interestingly, autophagy, lysosome and exosomes are coordinated to maintain cell homeostasis. When uropathogenic E. coli (UPEC) infects bladder epithelial cells (BECs), they are targeted by autophagy but avoid degradation because they can neutralize lysosomal pH. This change is detected by mucolipin TRP channel 3 (TRPML3) in lysosomes, initiating lysosome exocytosis and exosome-encased bacteria (Miao et al. 2015).

However, since EVs were released into extracellular space, which also mediate the spreading of signals to surrounding cells in addition to preserve the parent cell homeostasis.

22.3 Functional Role of EVs in Renal Inflammation and Fibrosis

Preliminary studies showed that EVs secreted into the circulation and extracellular fluid have roles in renal physiology and pathophysiology through intra-nephron communication. The cross talk meditated by EVs especially among cell types with their plasma membranes facing glomerular filtration tract such as epithelial cells or in direct contact with the vascular compartment such as endothelial cells may reasonably be common in the kidney. The signaling carried by EVs may participate in renal inflammation and fibrosis. Besides, EVs from particular cells may also contribute to the repair of kidney injury.

22.3.1 Role of EVs in Tubulointerstitial Inflammation

Proximal tubule epithelial cells are the most populous cell type in the kidney and carry out diverse regulatory and endocrine functions in normal kidney physiology as well as the pathogenesis of kidney disease (Liu et al. 2018). Interestingly, recent studies indicated that external insults such as hypoxia, exposure to proteinuria or physical wounding triggered the release of EVs from injured TECs carrying specific cargo. In the condition of hypoxia, hypoxia-inducible factor-1 (HIF-1) promoted exosome production in tubular epithelial cells (Zhang et al. 2017).

Wang et al. reported that the expression of soluble epoxide hydrolase in renal tubular epithelial cells regulates macrophage infiltration and polarization in IgA nephropathy (Wang et al. 2018). Indeed, EVs pass from injured TECs to interstitial space via a damaged basement membrane also contributed to macrophage activation. Upon exposure to proteinuria, TECs produced increasing exosomes loading with CCL2 mRNA which could be transferred to macrophages and promoted macrophage activation. It may constitute a critical mechanism of albumin-induced tubulointerstitial inflammation (Lv et al. 2018). Similarly, hypoxia TECs produced more exosomes loaded with miR-23a which was transferred to interstitial macrophages and promoted the pro-inflammatory response (Li et al. 2019). Interestingly, in the tumor microenvironment, exosome-mimetic nanovesicles derived from M1 macrophages could induce polarization of M2 macrophages to M1 macrophages in vitro and in vivo. Thus, exosome may represent a novel mediator for inducing macrophage polarization (Choo et al. 2018).

22.3.2 *Role of EVs in Renal Fibrosis*

It has been shown that miR-21 was packaged into microvesicles released by tubular epithelial cells, which then entered recipient tubular cells, and promoted tubular phenotype transition (Zhou et al. 2013).

Borges FT et al. reported that exosomes released by injured epithelial cells promote fibroblast activation, including proliferation, a-smooth muscle actin expression, F-actin expression and type I collagen production, that is dependent on exosomes delivering of TGF-β1 mRNA. The study indicated the potential utility of exosome-targeted therapies to control tissue fibrosis (Borges et al. 2013). Since EV-associated MMPs can contribute to the degradation of extracellular matrix surrounding cells, and sometimes stimulate critical signaling pathways (Sanderson et al. 2017; Shimoda and Khokha 2017), whether EV-associated MMPs participated in the development remains an interesting question that deserves future investigation.

Podocyte exosomes were secreted in the urine and might pass through the renal tubule and transmit information to tubular epithelial cells (Prunotto et al. 2013). Given its location adjacent to the glomerulus, the proximal tubule represents a possible site of interaction for podocyte EVs. It has been demonstrated in in vitro study that podocyte microparticles (MPs) did communicate with proximal tubule epithelial cells (PTECs) and induced the cell fibrotic responses. MPs were isolated from the media of differentiated, untreated human podocytes (hPODs) and administered to cultured PTECs. Treatment with podocyte MPs promoted proximal tubule fibrotic signaling via p38 MAPK and CD36 (Munkonda et al. 2018).

22.4 EVs Biomarkers in Kidney Disease

EVs are found in different biofluids and carry multiple kinds of molecules, including proteins, lipids and nucleic acids, providing useful information of the parental cells. Accumulating evidence demonstrated that EVs' cargo changed in disease states, positioning EVs as potential sources for the discovery of novel biomarkers. Recent study found in patients with metastatic melanoma, the level of circulating exosomal PD-L1 positively correlated with that of IFN-γ, and was a predictive marker for stratifying clinical responders from non-responders for anti-PD-1 therapy (Chen et al. 2018).

RNAs (mRNA and miRNA) and proteins from kidney resident or infiltrating cells can be loaded into urinary extracellular vesicles (uEVs) and thus protected from degradation. Urinary EVs might be useful sources of novel biomarker discovery for renal disease. Urinary EV uromodulin (UMOD) mRNA levels are progressively elevated from T2DM to DKD groups and correlate with eGFR and ACR levels (Yamamoto et al. 2018). We have recently isolated exosome released from podocyte in the urine, and the structure was positive for the markers of both exosome and podocyte that is CD9, AQP2 and Nephrin. Further study showed that CD2AP mRNA from exosome was correlated with both kidney function and severity of fibrosis (Lv et al. 2014).

Moreover, the stability and enrichment of miRNA in exosome make it a promising candidate biomarker for kidney disease. Exosome miRNA was stable despite repeated frozen-thaw cycles and long-term storage (Lv et al. 2013a). Interestingly, miRNAs were extremely enriched in the urinary exosome subpopulation, but not MVs in hypertensive patients. Low exosomal miR-146a was associated with the presence of albuminuria (Perez-Hernandez et al. 2018). MiR-29c from urinary exosome was significantly reduced in CKD patients and inversely correlated with renal fibrosis scores (Lv et al. 2013b). Besides, the diagnostic potential of exosomal miRNA has also been demonstrated in autosomal dominant polycystic kidney disease (ADPKD), streptozotocin (STZ)-induced diabetic nephropathy animal models, patients with minimal change disease (MCD) and focal segmental glomerulosclerosis (FSGS) (Ben-Dov et al. 2014; Mohan et al. 2016; Ramezani et al. 2015). However, the findings of exosomal biomarker in different types of kidney disease warrant further validation studies.

22.5 Future Challenges and Directions of EVs Study in Kidney Disease

Despite all the promising data regarding the EVs-mediated injury signal amplifying, disease diagnosis and therapy, more research is still needed to extend these findings. Since it is technically challenging to obtain a totally pure EV fraction free from non-vesicular components, it is needed to establish comprehensive studies when EV function was studied. Recently, International Society for Extracellular Vesicles (ISEV) provides researchers with minimal experimental requirements that should be used to attribute any specific biological cargo or functions to EVs (Lotvall et al. 2014). And an update of the MISEV guidelines was published in 2018 (Thery et al. 2018). It is suggested to follow the minimal requirement of experiments when performing EV studies.

As for the functional study, nearly all results describing EVs-mediated cellular cross talk are based on the experimental data in vitro or partly in rodent models. Mechanisms concerning to the specific loading of bioactive molecules into EVs, their release and uptake by target renal cells are poorly understood. The major impediment to mechanistic studies is to clarify how EVs are formed and how to track their origin and destination in vivo (Shah et al. 2018). Identifying the tissues or specific cell of origin for circulating EVs may be especially important in understanding their relevance to disease. However, methods for identifying tissue or cell-specific EVs require development (Shah et al. 2018). We have reported the purification of exosomes from tubule. However, it is still difficult to identify other cellular origin of EVs in the kidney. Evidence from in vivo study is needed to support the intra-nephron communication.

Besides, van Balkom et al. speculated that Tamm–Horsfall protein (uromodulin) may limit exosomal fusion in downstream nephron segments, because urinary exosomes are usually surrounded by polymeric fibers formed from Tamm–Horsfall protein, which would prevent them from getting contact with surfaces of target cells (Fernandez-Llama et al. 2010; van Balkom et al. 2011). The effect of Tamm–Horsfall protein on urinary exosome communication with downstream tubule segment needs further investigation.

For urinary EV biomarker study, more efforts are necessary to standardize the EVs isolation methods and the nomenclature of EVs, which might help to get more comparable and repeatable data. Importantly, the low purity of EVs obtained from different isolation methods is common. A recent study report that ultracentrifugation followed by size exclusion chromatography (UC-SEC) yielded the most homogenous population of exosomes and less non-vesicle co-precipitated proteins from urine, compared to polyethylene glycol precipitation (PEG)-precipitation and ultracentrifugation (UC) method (Gheinani et al. 2018). The optimal method for EV purification is important especially for urine samples from proteinuric nephropathy patients. Besides, few reports have so far investigated systemic RNA cargo in urinary EVs, and the next-generation sequencing might be promising approaching for revealing the RNA transcriptome of urinary EVs in different kidney diseases (Turchinovich et al. 2019).

In conclusions, EVs play important roles in renal inflammation and fibrosis through conveying injury and repair signals between neighbor and distant cells. Cargoes of EVs might represent promising biomarkers for renal pathological injury and deterioration of kidney function. However, further studies are still needed to explore the role and potential diagnostic applications of EVs in renal disease.

Acknowledgements This chapter was modified from a paper reported by our group in J Cell Mol Med (Lv et al. 2019).

References

Abdulrahman BA, Abdelaziz DH, Schatzl HM (2018) Autophagy regulates exosomal release of prions in neuronal cells. J Biol Chem 293:8956–8968

Ben-Dov IZ, Tan YC, Morozov P, Wilson PD, Rennert H, Blumenfeld JD et al (2014) Urine microRNA as potential biomarkers of autosomal dominant polycystic kidney disease progression: description of miRNA profiles at baseline. PLoS ONE 9:e86856

Borges FT, Melo SA, Ozdemir BC, Kato N, Revuelta I, Miller CA et al (2013) TGF-beta1-containing exosomes from injured epithelial cells activate fibroblasts to initiate tissue regenerative responses and fibrosis. J Am Soc Nephrol 24:385–392

Cabral J, Ryan AE, Griffin MD, Ritter T (2018) Extracellular vesicles as modulators of wound healing. Adv Drug Deliv Rev 129:394–406

Caruso S, Poon IKH (2018) Apoptotic cell-derived extracellular vesicles: more than just debris. Front Immunol 9:1486

Chen G, Huang AC, Zhang W, Zhang G, Wu M, Xu W et al (2018) Exosomal PD-L1 contributes to immunosuppression and is associated with anti-PD-1 response. Nature 560:382–386

Choo YW, Kang M, Kim HY, Han J, Kang S, Lee JR et al (2018) M1 macrophage-derived nanovesicles potentiate the anticancer efficacy of immune checkpoint inhibitors. ACS Nano 12:8977–8993

Desdin-Mico G, Mittelbrunn M (2017) Role of exosomes in the protection of cellular homeostasis. Cell Adh Migr 11:127–134

Dusso A, Colombo MI, Shanahan CM (2018) Not all vascular smooth muscle cell exosomes calcify equally in chronic kidney disease. Kidney Int 93:298–301

Escrevente C, Keller S, Altevogt P, Costa J (2011) Interaction and uptake of exosomes by ovarian cancer cells. BMC Cancer 11:108

Fabbri M, Paone A, Calore F, Galli R, Gaudio E, Santhanam R (2012) MicroRNAs bind to Toll-like receptors to induce prometastatic inflammatory response. Proc Natl Acad Sci U S A 109:E2110–E2116

Fernandez-Llama P, Khositseth S, Gonzales PA, Star RA, Pisitkun T, Knepper MA (2010) Tamm-Horsfall protein and urinary exosome isolation. Kidney Int 77:736–742

Fitzgerald W, Freeman ML, Lederman MM, Vasilieva E, Romero R, Margolis L (2018) A system of cytokines encapsulated in extracellular vesicles. Sci Rep 8:8973

Fleshner M, Crane CR (2017) Exosomes, DAMPs and miRNA: features of stress physiology and immune homeostasis. Trends Immunol 38:768–776

Gheinani AH, Vogeli M, Baumgartner U, Vassella E, Draeger A, Burkhard FC et al (2018) Improved isolation strategies to increase the yield and purity of human urinary exosomes for biomarker discovery. Sci Rep 8:3945

Gieseler F, Plattfaut C, Quecke T, Freund A, Ungefroren H, Ender F (2018) Heterogeneity of microvesicles from cancer cell lines under inflammatory stimulation with TNF-alpha. Cell Biol Int 42:1533–1544

Gupta KH, Goldufsky JW, Wood SJ, Tardi NJ, Moorthy GS, Gilbert DZ (2017) Apoptosis and compensatory proliferation signaling are coupled by CrkI-containing microvesicles. Dev Cell 41(674–684):e5

Hessvik NP, Llorente A (2018) Current knowledge on exosome biogenesis and release. Cell Mol Life Sci 75:193–208

Hough KP, Trevor JL, Strenkowski JG, Wang Y, Chacko BK, Tousif S (2018) Exosomal transfer of mitochondria from airway myeloid-derived regulatory cells to T cells. Redox Biol 18:54–64

Johnstone RM, Adam M, Hammond JR, Orr L, Turbide C (1987) Vesicle formation during reticulocyte maturation. Association of plasma membrane activities with released vesicles (exosomes). J Biol Chem 262:9412–9420

Kudo S, Mizuno K, Hirai Y, Shimizu T (1990) Clearance and tissue distribution of recombinant human interleukin 1 beta in rats. Cancer Res 50:5751–5755

Lasser C, Jang SC, Lotvall J (2018) Subpopulations of extracellular vesicles and their therapeutic potential. Mol Aspects Med 60:1–14

Leoni G, Neumann PA, Kamaly N, Quiros M, Nishio H, Jones HR et al (2015) Annexin A1-containing extracellular vesicles and polymeric nanoparticles promote epithelial wound repair. J Clin Invest 125:1215–1227

Li ZL, Lv LL, Tang TT, Wang B, Feng Y, Zhou LT et al (2019) HIF-1alpha inducing exosomal microRNA-23a expression mediates the cross-talk between tubular epithelial cells and macrophages in tubulointerstitial inflammation. Kidney Int 95:388–404

Liu BC, Tang TT, Lv LL, Lan HY (2018) Renal tubule injury: a driving force toward chronic kidney disease. Kidney Int 93:568–579

Lopez-Castejon G, Brough D (2011) Understanding the mechanism of IL-1beta secretion. Cytokine Growth Factor Rev 22:189–195

Lotvall J, Hill AF, Hochberg F, Buzas EI, Di Vizio D, Gardiner C et al (2014) Minimal experimental requirements for definition of extracellular vesicles and their functions: a position statement from the International Society for Extracellular Vesicles. J Extracell Vesicles 3:26913

Lv LL, Cao Y, Liu D, Xu M, Liu H, Tang RN et al (2013a) Isolation and quantification of microRNAs from urinary exosomes/microvesicles for biomarker discovery. Int J Biol Sci 9:1021–1031

Lv LL, Cao YH, Ni HF, Xu M, Liu D, Liu H et al (2013b) MicroRNA-29c in urinary exosome/microvesicle as a biomarker of renal fibrosis. Am J Physiol Renal Physiol 305:F1220–F1227
Lv LL, Cao YH, Pan MM, Liu H, Tang RN, Ma KL et al (2014) CD2AP mRNA in urinary exosome as biomarker of kidney disease. Clin Chim Acta 428:26–31
Lv LL, Feng Y, Tang TT, Liu BC (2019) New insight into the role of extracellular vesicles in kidney disease. J Cell Mol Med 23:731–739
Lv LL, Feng Y, Wen Y, Wu WJ, Ni HF, Li ZL (2018) Exosomal CCL2 from tubular epithelial cells is critical for albumin-induced tubulointerstitial inflammation. J Am Soc Nephrol 29:919–935
MacKenzie A, Wilson HL, Kiss-Toth E, Dower SK, North RA, Surprenant A (2001) Rapid secretion of interleukin-1beta by microvesicle shedding. Immunity 15:825–835
Mathieu M, Martin-Jaular L, Lavieu G, Thery C (2019) Specificities of secretion and uptake of exosomes and other extracellular vesicles for cell-to-cell communication. Nat Cell Biol 21:9–17
Miao Y, Li G, Zhang X, Xu H, Abraham SN (2015) A TRP channel senses lysosome neutralization by pathogens to trigger their expulsion. Cell 161:1306–1319
Mohan A, Singh RS, Kumari M, Garg D, Upadhyay A, Ecelbarger CM et al (2016) Urinary exosomal microRNA-451-5p is a potential early biomarker of diabetic nephropathy in rats. PLoS ONE 11:e0154055
Munkonda MN, Akbari S, Landry C, Sun S, Xiao F, Turner M et al (2018) Podocyte-derived microparticles promote proximal tubule fibrotic signaling via p38 MAPK and CD36. J Extracell Vesicles 7:1432206
Nabet BY, Qiu Y, Shabason JE, Wu TJ, Yoon T, Kim BC (2017) Exosome RNA unshielding couples stromal activation to pattern recognition receptor signaling in cancer. Cell 170:352–366 (e13)
Nair RR, Mazza D, Brambilla F, Gorzanelli A, Agresti A, Bianchi ME (2018) LPS-challenged macrophages release microvesicles coated with histones. Front Immunol 9:1463
Park JE, Dutta B, Tse SW, Gupta N, Tan CF, Low JK (2019) Hypoxia-induced tumor exosomes promote M2-like macrophage polarization of infiltrating myeloid cells and microRNA-mediated metabolic shift. Oncogene
Perez-Hernandez J, Olivares D, Forner MJ, Ortega A, Solaz E, Martinez F et al (2018) Urinary exosome miR-146a is a potential marker of albuminuria in essential hypertension. J Transl Med 16:228
Phinney DG, Di Giuseppe M, Njah J, Sala E, Shiva S, St Croix CM (2015) Mesenchymal stem cells use extracellular vesicles to outsource mitophagy and shuttle microRNAs. Nat Commun 6:8472
Prunotto M, Farina A, Lane L, Pernin A, Schifferli J, Hochstrasser DF et al (2013) Proteomic analysis of podocyte exosome-enriched fraction from normal human urine. J Proteomics 82:193–229
Ramezani A, Devaney JM, Cohen S, Wing MR, Scott R, Knoblach S et al (2015) Circulating and urinary microRNA profile in focal segmental glomerulosclerosis: a pilot study. Eur J Clin Invest 45:394–404
Sanderson RD, Bandari SK, Vlodavsky I (2017) Proteases and glycosidases on the surface of exosomes: newly discovered mechanisms for extracellular remodeling. Matrix Biol 75–76:160–169
Shah R, Patel T, Freedman JE (2018) Circulating extracellular vesicles in human disease. N Engl J Med 379:958–966
Shimoda M, Khokha R (2017) Metalloproteinases in extracellular vesicles. Biochim Biophys Acta 1864:1989–2000
Sun Y, Shi H, Yin S, Ji C, Zhang X, Zhang B (2018) Human mesenchymal stem cell derived exosomes alleviate type 2 diabetes mellitus through reversing peripheral insulin resistance and relieving beta-cell destruction. ACS Nano 12:7613–7628
Sung BH, Ketova T, Hoshino D, Zijlstra A, Weaver AM (2015) Directional cell movement through tissues is controlled by exosome secretion. Nat Commun 6:7164
Svensson KJ, Christianson HC, Wittrup A, Bourseau-Guilmain E, Lindqvist E, Svensson LM (2013) Exosome uptake depends on ERK1/2-heat shock protein 27 signaling and lipid Raft-mediated endocytosis negatively regulated by caveolin-1. J Biol Chem 288:17713–17724
Takahashi A, Okada R, Nagao K, Kawamata Y, Hanyu A, Yoshimoto S (2017) Exosomes maintain cellular homeostasis by excreting harmful DNA from cells. Nat Commun 8:15287

Thery C, Witwer KW, Aikawa E, Alcaraz MJ, Anderson JD, Andriantsitohaina R et al (2018) Minimal information for studies of extracellular vesicles 2018 (MISEV2018): a position statement of the International Society for Extracellular Vesicles and update of the MISEV2014 guidelines. J Extracell Vesicles 7:1535750
Torralba D, Baixauli F, Villarroya-Beltri C, Fernandez-Delgado I, Latorre-Pellicer A, Acin-Perez R (2018) Priming of dendritic cells by DNA-containing extracellular vesicles from activated T cells through antigen-driven contacts. Nat Commun 9:2658
Turchinovich A, Drapkina O, Tonevitsky A (2019) Transcriptome of extracellular vesicles: state-of-the-art. Front Immunol 10:202
Vagner T, Spinelli C (2018) Large extracellular vesicles carry most of the tumour DNA circulating in prostate cancer patient plasma 7:1505403
Valadi H, Ekstrom K, Bossios A, Sjostrand M, Lee JJ, Lotvall JO (2007) Exosome-mediated transfer of mRNAs and microRNAs is a novel mechanism of genetic exchange between cells. Nat Cell Biol 9:654–659
van Balkom BW, Pisitkun T, Verhaar MC, Knepper MA (2011) Exosomes and the kidney: prospects for diagnosis and therapy of renal diseases. Kidney Int 80:1138–1145
Wang Q, Liang Y, Qiao Y, Zhao X, Yang Y, Yang S (2018) Expression of soluble epoxide hydrolase in renal tubular epithelial cells regulates macrophage infiltration and polarization in IgA nephropathy. Am J Physiol Renal Physiol 315:F915–F926
Xu J, Camfield R, Gorski SM (2018) The interplay between exosomes and autophagy—partners in crime. J Cell Sci 131
Yamamoto CM, Murakami T, Oakes ML, Mitsuhashi M, Kelly C, Henry RR et al (2018) Uromodulin mRNA from urinary extracellular vesicles correlate to kidney function decline in type 2 diabetes mellitus. Am J Nephrol 47:283–291
Ying W, Riopel M, Bandyopadhyay G, Dong Y, Birmingham A, Seo JB (2017) Adipose tissue macrophage-derived exosomal miRNAs can modulate in vivo and in vitro insulin sensitivity. Cell 171:372–384
Zhang W, Zhou X, Yao Q, Liu Y, Zhang H, Dong Z (2017) HIF-1-mediated production of exosomes during hypoxia is protective in renal tubular cells. Am J Physiol Renal Physiol 313:F906–F913
Zhou Y, Xiong M, Fang L, Jiang L, Wen P, Dai C et al (2013) miR-21-containing microvesicles from injured tubular epithelial cells promote tubular phenotype transition by targeting PTEN protein. Am J Pathol 183:1183–1196

Chapter 23
Hypoxia and Renal Tubulointerstitial Fibrosis

Zuo-Lin Li and Bi-Cheng Liu

Abstract Hypoxia, one of the most common causes of kidney injury, is a key pathological condition in various kidney diseases. Renal fibrosis is the terminal pathway involved in the continuous progression of chronic kidney disease (CKD), characterized by glomerulosclerosis and tubulointerstitial fibrosis (TIF). Recent studies have shown that hypoxia is a key factor promoting the progression of TIF. Loss of microvasculature, reduced oxygen dispersion, and metabolic abnormality of cells in the kidney are the main causes of the hypoxic state. Hypoxia can, in turn, profoundly affect the tubular epithelial cells, endothelial cells, pericytes, fibroblasts, inflammatory cells, and progenitor cells. In this chapter, we reviewed the critical roles of hypoxia in the pathophysiology of TIF and discussed the potential of anti-hypoxia as its promising therapeutic target.

Keywords Hypoxia · Tubulointerstitial fibrosis · Hypoxia-inducible factor

23.1 Characteristics of Kidney Oxygenation

It is well known that cells require adenosine triphosphate (ATP) for normal physiological functions of the intracellular organelles. Both oxidative phosphorylation and aerobic glycolysis produce ATP, but the oxidative phosphorylation pathway is more efficient. Hence, most cells need adequate supply of oxygen (O_2) to generate ATP via mitochondrial oxidative phosphorylation. Lack of O_2 supply is a definite cellular stress. Therefore, the balance between O_2 delivery and its consumption through various mechanisms to maintain O_2 homeostasis is vital for normal physiological functions.

The cardiovascular system, which is responsible for regulating blood supply, is the critical governor that maintains O_2 homeostasis in all organs. Majority of the organs have multiple autoregulatory mechanisms that help maintain their normal physio-

Z.-L. Li · B.-C. Liu (✉)
Institute of Nephrology, Zhong Da Hospital, Southeast University School of Medicine, Nanjing, China
e-mail: liubc64@163.com

B.-C. Liu et al. (eds.), *Renal Fibrosis: Mechanisms and Therapies*, Advances in Experimental Medicine and Biology 1165,
https://doi.org/10.1007/978-981-13-8871-2_23

logical functions by regulating blood flow (Davis and Kuo 2008). In the kidney, the major mechanisms mediating autoregulation of renal blood flow (RBF) depend on the myogenic and tubuloglomerular feedback system (Carlström et al. 2015). Additionally, various pathways for the metabolic regulation of RBF are found in the kidney (Pittman 2011). Unlike other organs where increments in the blood flow improve oxygenation uniformly, increments in the RBF enhance the glomerular filtration rate (GFR) and filtered reabsorptive load, which further augments O_2 consumption (Blantz et al. 2007). Therefore, in the kidney, an increase in the RBF simultaneously increases O_2 consumption and may not improve oxygenation.

Generally, oxygen in the arterial blood, RBF, GFR, oxygen consumed by the cells and arterial-to-venous oxygen shunting, which is not available to the cells for consumption, determine the tissue oxygen tension (pO_2) in the kidneys (Evans et al. 2008). In order to sustain GFR, the kidney possesses a very high RBF, which constitutes approximately 25% of the cardiac output. However, the low level of oxygen extracted from the blood and the operating mechanisms of the arterial-to-venous oxygen shunting provide an explanation about why intrarenal pO_2 is remarkably lower than the arterial pO_2.

23.1.1 A Heterogeneous Environment

Even though the kidney cortex is remarkably perfused, ischemic renal injury is a complication far more common than corresponding injuries to the other organs. A heterogeneous blood supply within the kidney is the major explanation for this apparent paradox. The cortex receives more than 25% of the total cardiac output with a tissue pO_2 of 50–60 mmHg while the medulla receives much less, the tissue pO_2 being 10–20 mmHg. Although the low blood flow leaves the medullary tissue at the brink of hypoxia, the cortico-medullary oxygen gradient is necessary to preserve the medullary osmotic gradients for urinary concentration, and compelling evidence has demonstrated that salt depletion leads to a change in the medullary oxygenation and size of the thick ascending limb, causing marked cortical hypoxia (Stillman et al. 1994).

23.1.2 Regulation of Renal Oxygenation

There are several mechanisms that maintain pO_2 homeostasis within the kidney, which include altering the O_2 delivery, either by altering the RBF or extraction of O_2 from the blood, and inducing changes in the O_2 consumption due to altered oxygen demand (Evans et al. 2010). However, O_2 consumption is a primary determinant of the intrarenal pO_2 when the RBF is unaltered and within the normal range. Approximately 20% of the consumption of total renal O_2 is related to the basal cellular O_2 demanding processes and the remaining 80% to the tubular sodium transport (T_{Na}).

Thus, T_{Na} is a major factor influencing the total renal O_2 consumption (Layton et al. 2016).

Besides, nitric oxide (NO) is a major regulator of microvascular O_2 supply and consumption. NO could increase the RBF and hence O_2 delivery via vasodilation (Laycock et al. 1998), suggesting a modulatory role of NO on O_2 consumption. Additionally, decreased bioavailability of NO is considered to contribute to the high renal O_2 consumption in various pathophysiological conditions, like diabetes, hypertension, and progressive renal disease (Singh et al. 2013; Tessari 2015; Welch et al. 2003). Angiotensin II (Ang II) also impacts renal oxygenation significantly and appears to be antagonistic to the roles of NO. In various conditions, it has been observed that high Ang II could lower O_2 delivery and increase its consumption by inducing renal vasoconstriction (Emans et al. 2016; van der Bel et al. 2016), and inhibition of Ang II with angiotensin-converting enzyme inhibition (ACEI) and angiotensin receptor blockade (ARB) lowering the elevated renal O_2 consumption also illustrates the key roles played by Ang II in regulating microvascular O_2 supply and consumption.

Recently, the hypoxia-inducible factor (HIF), a key factor of nuclear transcription regulating O_2 homeostasis, has received significant attention. HIF has been found to regulate the renal pathophysiological function by regulating multiple target genes including those responsible for vasomotor regulation (inducible NO synthase and heme oxygenase-1), angiogenic growth (vascular endothelial growth factor, VEGF), energy metabolism (glucose transporters and key glycolytic enzymes), and cell apoptosis (Haase 2017a). Compelling evidence shows that the function of HIF is pivotal in the regulation of renal oxygenation under various pathophysiological conditions. Hence, it is considered as a new target of investigation and is being explored increasingly.

23.2 Hypoxia in Kidney Diseases

23.2.1 Acute Kidney Injury

Acute kidney injury (AKI) is linked with increased morbidity and the risk of death (Rewa and Bagshaw 2014). Furthermore, patients who recover from AKI are at increased risk of developing CKD and consequent ESRD (Leung et al. 2013). Renal tissue hypoxia has been proposed as a critical mediator in the manifestation of multiple forms of AKI.

Ischemia/reperfusion (I/R) injury is the leading cause of AKI in the hospital setting (Lameire et al. 2013). The renal vasculature being temporarily occluded during surgical procedures such as renal transplantation or reparation of an abdominal aneurysm, is one of the most common causes of AKI. Consequentially, delivery of O_2 to the kidney ceases in making it anoxic (Bonventre and Yang 2011; Abdelkader et al. 2014). Thus, renal hypoxia caused by hypoperfusion is likely to be a major pathological cause of kidney injury. Cardiac surgery, requiring cardiopulmonary bypass (CPB) in particular, is a major cause of AKI in hospitals. Renal hypoxia induced by CPB might be associated with abnormalities in both delivery and consumption of renal O_2. A study has demonstrated that the delivery of renal O_2 had decreased by ~20% during CPB in humans (Lannemyr et al. 2017) and the persistent of hypoxia in kidney after CPB may be a major pathological mechanism (Sgouralis et al. 2017). Administration of contrast agents can lead to AKI. Renal tissue hypoxia is a key contributor to the pathogenesis of contrast-induced nephropathy. Hypoxia of the kidney has been consistently observed in animals administered with contrast agents (Zhang et al. 2012); however, the exact mechanisms that cause AKI induced by contrast agents still need to be unequivocally defined. A body of evidence has shown that administration of contrast agents could induce constriction of the vasa recta and consequent tubular dilation. Increased tubular reabsorption of sodium, and thus O_2 consumption, could also play a role in its manifestation (Fahling et al. 2017). Besides, sepsis is becoming one of the leading causes of AKI in hospitals (Uchino et al. 2005). Since rodent models of septic AKI are characterized by global vasoconstriction consequently resulting in a marked reduction of the renal blood flow, it is not surprising that renal hypoxia has been observed in this setting (Holthoff et al. 2012; Wang et al. 2012). Clinically, almost sepsis-induced AKI is caused by bacterial infection, which is likely to be associated with a hyperdynamic state. The evidence of hypoxia in sepsis-induced AKI was also elaborated by infusion of live *Escherichia coli*, a clinically relevant ovine model for sepsis, characterized by a hyperdynamic state and well-maintained cortical oxygenation (Calzavacca et al. 2015). Thus, even in the absence of ischemia, hypoxia is likely to be a common pathologic feature of septic AKI. Collectively, hypoxia appears to be a common characteristic of AKI. Renal hypoxia can lead to capillary rarefaction and thus propagates a vicious cycle of tissue hypoxia and kidney injury.

23.2.2 Chronic Kidney Disease

CKD is a complex, multifactorial disease. In 1998, Fine et al (1998) proposed the "chronic hypoxia hypothesis" for the pathogenesis of CKD for the first time. Since then, the fascinating concept has been investigated intensively by Eckardt, Johnson, and many other investigators (Eckardt et al. 2003; Kang et al. 2002). Despite various underlying pathological mechanisms involved across the different etiologies of CKD, renal hypoxia appears to be the key feature leading to the progression of kidney disease.

Hypoxia is a common finding in diabetic nephropathy (DN). Oxygen microelectrodes, which have been widely used in studies involving oxygenation in the kidneys, have demonstrated the presence of hypoxia in DN (Palm et al. 2003). Several studies using blood-oxygen-level-dependent (BOLD)-MRI have provided evidence of hypoxia in patients with DN (Yin et al. 2012). In remnant kidney models, hypoxia of the renal cortex and medulla has been observed (Pruijm et al. 2018). Meanwhile, there is evidence suggesting that, 6–8 weeks following subtotal nephrectomy, pO_2 of the kidney was found to be greater (Priyadarshi et al. 2002). The contradiction between these evidences may be related to the time course of progressive kidney diseases. Nevertheless, the balance of evidence suggests that hypoxia is a common characteristic of CKD. In polycystic kidney disease (PKD), which is the most common genetic form of CKD, vascular remodeling and expansion of the tubules have the potential to not only decrease tissue perfusion, but also increase the distance over which O_2 must diffuse to reach the renal tissue. There is also evidence suggesting that O_2 demand, at least proportional to the level of sodium reabsorption, is augmented in PKD. Thus, a decreased supply of O_2 to the tissues and its inappropriately high consumption are likely to contribute to the severe tissue hypoxia (Ow et al. 2014). The evidence of hypoxia in lupus nephritis (LN) has also been explored. Using BOLD-MIR, it was determined that the renal hypoxia was a critical pathologic feature in patients with LN. Moreover, upregulation of glomerular HIF-1α, which could evidence the phenomenon of hypoxia, was observed in renal biopsies obtained from LN patients (Deng et al. 2014). These data demonstrated that renal hypoxia might be a common mechanism for LN. However, the exact mechanisms of renal oxygenation in LN patients remain unclear (Shi et al. 2017). Compelling evidence indicates that the presence of inflammatory cytokines, a critical feature of LN, mediates fibrogenesis (Davidson 2016), in which the accumulation of ECM could promote hypoxia by increasing the distance of O_2 diffusion.

23.3 Mechanisms of Hypoxia in Renal Tissue

Due to its typical structure, the kidney, which plays a critical role in maintaining physiological functions by performing complex roles during transport, is susceptible to hypoxic injury. Hence, a variety of molecular mechanisms that are adaptive to renal hypoxia have evolved. Compelling evidence illustrates that these mechanisms play a vital role in maintaining renal hemostasis, and hence, their wide-ranging implications for the pathogenesis and treatment of renal diseases are being investigated.

23.3.1 Hypoxia-Inducible Factor

Recently, the HIF, a critical factor of nuclear transcription factor involved in maintaining O_2 homeostasis, has received extensive attention from scientists as well as clinicians. Due to its significant role in cellular adaptation to hypoxia, the HIF pathway has become a promising therapeutic target in areas of hypoxic conditions. Here, the key mechanisms that integrate the HIF pathway with other responses of hypoxia in the kidney are summarized.

HIF is a major contributor in cellular adaptation to hypoxia. It is a heterodimeric transcription factor consisting of alpha (HIF-α) and beta (HIF-β) subunits. Under normoxic conditions, HIF-α could be hydroxylated by prolyl hydroxylase domain-containing proteins (PHDs), which is the crucial enzyme in regulating HIF. Hydroxylated HIF-α could be degraded immediately by the polyubiquitination–proteasome system following von Hippel–Lindau tumor suppressor protein (vHL) activation (Ivan et al. 2001). Contrarily, under hypoxic conditions, HIF-α cannot be hydroxylated by the PHDs and is able to translocate to the nucleus where it forms a heterodimer with HIF-β. Being a critical transcription factor, the HIF heterodimers bind to hypoxia response elements resulting in the transactivation of multiple target genes (Nangaku et al. 2013) (Fig. 23.1). Indeed, HIF could upregulate the transcription of more than 100 target genes controlling hematopoiesis (e.g., erythropoietin, EPO), angiogenesis (e.g., VEGF), and anaerobic metabolism (e.g., glucose transporters and glycolytic enzymes) (Mole et al. 2009). HIF-α consists of two major active isoforms, HIF-1α and HIF-2α. In the hypoxic kidney, HIF-1α is expressed predominantly in tubular epithelial cells and works as a master regulator of hypoxic stress, whereas HIF-2α is expressed mainly in endothelial cells and interstitial cells. However, HIF is not expressed in the normal renal medulla (Rosenberger et al. 2002). These interesting observations are likely to be associated with their distinct metabolic characteristics and expression patterns of the PHD isoform (Schödel et al. 2009).

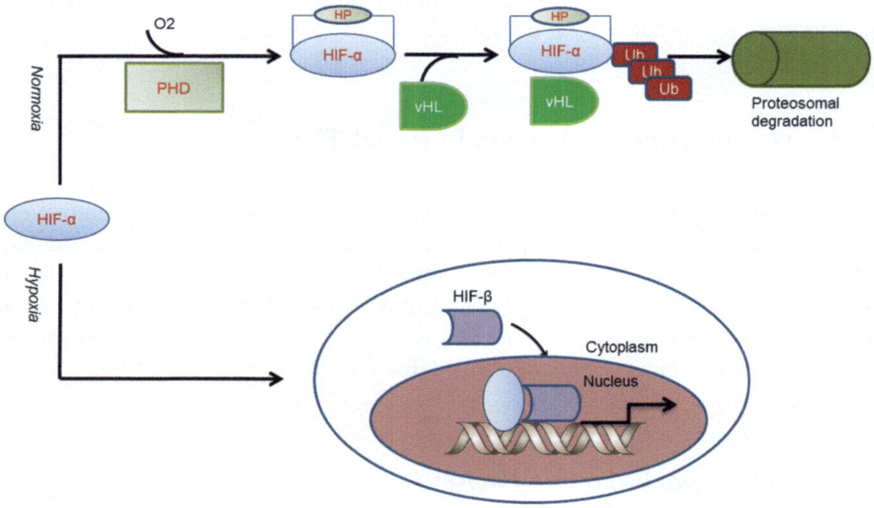

Fig. 23.1 Regulation of hypoxia-inducible factors. In normoxia, certain proline residues of the HIF subunits are hydroxylated by prolyl hydroxylases (PHDs) using oxygen. The hydroxyprolines (HP) allow von Hippel–Lindau (vHL) protein to bind and ubiquitinate HIF-α, which results in their proteasomal degradation. In contrast, HIF-α escapes posttranslational modification by PHDs under conditions of hypoxia. HIF-α forms a heterodimer with a β subunit (HIF-β) and then binds to a hypoxia response element (HRE) in the regulatory region of a target gene and promotes its transcription. Ub, ubiquitin

23.3.2 The Kidney in Control of O_2 Carrying Capacity

It is well recognized that the stimulation to produce red blood cells through increased synthesis of EPO is one of the most important mechanisms mediating systemic adaptation to hypoxia. Because kidneys are the main sources of EPO in adults, an extensive understanding of its physiologic and molecular mechanisms extensively in the kidneys is of extreme clinical significance in order to pave the way for discovery of the oxygen sensing machinery, namely the PHD/HIF pathway. It is well known that peritubular interstitial fibroblasts are the EPO-producing cells in the kidney (Asada et al. 2011). Accumulating evidence shows that HIF-2α, not HIF-1, is a major regulator for renal EPO synthesis (Haase 2013; Haase 2017b). In contrast to the nonrenal EPO-producing cell types, genetic studies in mice have determined that the activity of HIF-2 in renal EPO-producing cells (REPCs) is mainly controlled by PHD2 (Haase 2017b). It has been observed that the EPO transcription is highly responsive to O_2 in order to increase its carrying capacity and represents one of the most sensitive hypoxia responses in the kidney. It is important to understand the exact mechanisms of this response to explore the capacity of the kidneys in the regulation of O_2.

23.3.3 Metabolic Reprogramming in the Kidney

Increasing evidence has uncovered that changes occur in the metabolic pathways involved in various biological functions and is referred to as metabolic reprogramming. Recent evidence demonstrates that metabolic reprogramming has a critical role in the progression of kidney injury. In fact, reprogramming of metabolism can determine the extent of organ dysfunction, including that of the kidney, suggesting a novel conceptual model of the cellular response to hypoxic injury.

The mitochondrion uses O_2 to generate ATP via the mitochondrial respiratory chain, which fuels multiple cellular processes. Efficient metabolic adaptation to low pO_2 is therefore imperative for the maintenance of normal renal functioning. A body of evidence demonstrates that the PHD/HIF pathway has a central role in metabolic reprogramming. Indeed, HIF-1α is required for metabolic reprogramming in human diseases. HIF is able to shift the metabolism from oxidative phosphorylation to anaerobic glycolysis by regulating cellular energy metabolism at multiple levels. Further investigation revealed that HIF-1 could reprogram metabolism by blocking the conversion of pyruvate to acetyl-CoA through transcriptional upregulation of pyruvate dehydrogenase kinase and regulating the expression of proteins that compose the mitochondrial respiratory chain (Weinberg 2011). Moreover, activation of HIF-1 could also keep intracellular pH in the slightly alkaline range by regulating the expression of the membrane-bound ectoenzyme carbonic anhydrase IX (Parks et al. 2011). These effects of HIF activation may be of relevance for the renal adaptation to hypoxia.

23.4 Hypoxia and Tubulointerstitial Fibrosis

Regardless of the initiating etiology, tubulointerstitial fibrosis (TIF) is a characteristic feature of chronic and progressive kidney diseases and can be used as the best predictive indicator of renal functional decline in glomerular diseases, DN, and tubulointerstitial nephritis. A number of characteristic features could be observed during TIF, which are as follows: infiltration of inflammatory cells; increase in the number of fibroblasts due to proliferation of resident fibroblasts and recruitment of cells to the tubulointerstitium; the appearance of myofibroblasts, which arise by differentiation of various resident or infiltrated cells (Webster et al. 2017); the accumulation of ECM as the result of increased production of matrix proteins, including type I collagen, which is the most abundant matrix protein found during TIF; tubular atrophy as a consequence of epithelial–mesenchymal transdifferentiation (EMT) and apoptosis; and rarefaction of the peritubular capillaries. Thus, tubulointerstitial injury has become a leading candidate to target for CKD therapies.

Many factors are involved in the progression of TIF. One of the most important occurrences leading to TIF is hypoxia (Schnaper 2017). A large body of experimental and clinical evidence supports the occurrence of multiple mechanisms resulting in hypoxia of the tubulointerstitium, which are as follows: (a) A significant reduction of blood flow in peritubular capillaries due to glomerulosclerosis was observed. (b) Distortion and loss of peritubular capillaries decrease blood perfusion. (c) Constriction of the efferent arterioles and peritubular capillaries due to activation of the renin-angiotensin system induces a decrease in the blood flow. (d) Pericytes, a constituent of peritubular capillaries, detach from vessels and compromises the capillaries (Starling 2017). Furthermore, ECM deposition produced by various fibrogenetic cells widens the distance between the capillaries and tubules, diminishing the efficiency of O_2 diffusion. At least six different cells, exhibiting diverse mechanisms, have been proposed as contributors to the progression of tubulointerstitial injury induced by hypoxia (Fig. 23.2).

23.4.1 Tubular Epithelial Cells

The tubules and tubulointerstitium make up a significant portion of the kidney and are the major sites of response for injuries, such as hypoxia, proteinuria, toxins, and metabolic disorders. Increasing evidence shows that tubular epithelial cells play diverse roles in either renal repair or progression to CKD.

Proximal tubular epithelial (PTE) cells are the predominant epithelial cell type in the cortex that is packed with mitochondria and dependent on oxidative phosphorylation. Hence, they are particularly vulnerable to hypoxic injury and have become the focus of research in the related field. In PTE cells, hypoxia induces a complex transcriptional response resulting in changes in the expression of many genes (Liu et al. 2018). The potential role of PTE cells in tubulointerstitial injury is demonstrated by

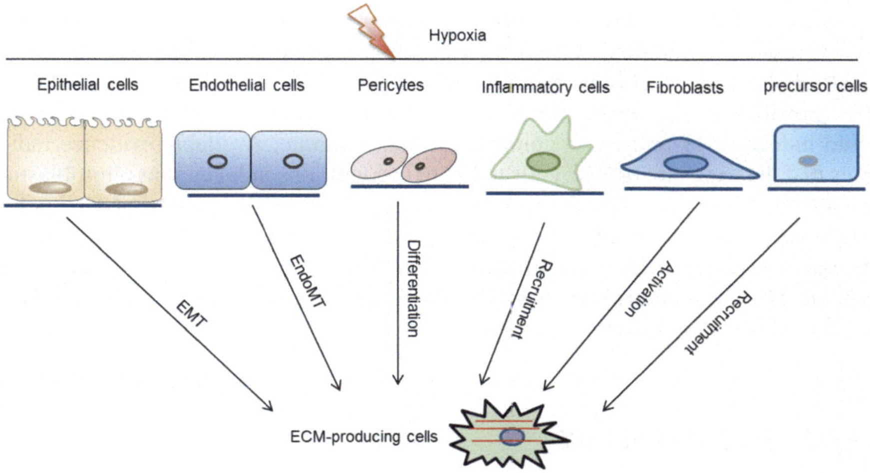

Fig. 23.2 Multiple cells have been proposed in renal fibrosis. ECM-producing cells can be derived from at least six different sources as follows: tubular EMT; capillary EndoMT; differentiation by vascular pericytes; recruitment from circulating inflammatory cells; phenotypic activation of fibroblasts and recruitment from circulating precursor cells. The relative contribution of each source to the ECM-producing cells in renal fibrosis is controversial. Conceivably, local activation of resident fibroblasts remains the major route for the generation of ECM-producing cells in diseased kidneys, at least in the early stage. By contrast, EMT could be a late event and may contribute to the irreversible progression of fibrosis. Abbreviations: ECM, extracellular matrix; EMT, epithelial–mesenchymal transition; EndoMT, endothelial–mesenchymal transition

the preferential activation of inflammatory and fibrogenic signaling. In response to hypoxia, PTE cells can transform into a secretory phenotype and elicit the response of pro-inflammatory mediators. In injured PTE cells, NF-κB signaling is activated immediately, which triggers the production and release of a variety of inflammatory cytokines, chemokines, and adhesion molecules, and initiates peritubular inflammation. Meanwhile, after hypoxic injury, tubular epithelial cells undergo changes in their structure and phenotype by expressing and producing multiple profibrotic factors. Compelling evidence shows that tubular epithelial cells are susceptible to the action of TGF-β1 which is produced by various cells and instigates a fibrogenic program. In PTE cells, hypoxia-induced injury could activate ILK (Li et al. 2003), Wnt (Zhou et al. 2017), Notch-1 (Edeling et al. 2016) and HIF-1 pathway (Higgins et al. 2007), all of which play critical role in renal fibrogenesis. Furthermore, there is a considerable literature implicating the activation of major fibrogenic signaling in PTE cells, such as the NF-κB, Smad and β-catenin signaling pathways which are characteristic features of various fibrotic kidney diseases.

Additionally, PTE cells' injury induced by hypoxia could also alter matrix metabolism, thereby promoting ECM accumulation and suppression of matrix degradation. Hypoxic PTE cells might contribute directly to TIF by transforming into myofibroblasts, a process known as the EMT (Zhou and Liu 2016). The last decade has witnessed that the HIF pathway is a master regulator, capable of controlling the expression of hundreds of genes during hypoxia-induced injury. By exposing renal tubular cells to chronic hypoxia, Manotham et al. (2004) demonstrated that HIF activation could induce fibrosis. In 2007, Higgins et al. reported that HIF-1α could induce EMT and cause epithelial cells to migrate. Available evidence indicates that the exposure of PTE cells to hypoxic conditions induces a myofibroblastic phenotype. Consistent with the results in vivo, a prolonged exposure of PTE cells to hypoxic conditions could lead to apoptosis. The tubular atrophy accompanying fibrosis is possibly a result of EMT and apoptosis. It is therefore concluded that hypoxic TECs can contribute directly to the occurrence of interstitial inflammation and fibrosis through various mechanisms.

23.4.2 Endothelial Cell

Mounting evidence supports the finding that vascular dysfunction plays a primary role in creating a hypoxic environment and triggering a fibrotic response in the tubulointerstitium. The fibrotic response, in turn, could exacerbate injury and enlarge the regions in which hypoxia occurs, by impacting the adjacent previously unaffected capillaries, and setting up an inexorable cycle of destruction to organ failure. A strong body of evidence gathered from experiments and patients has established a close correlation between peritubular capillary rarefaction and development of glomerular and tubulointerstitial scarring (Kang et al. 2002).

Renal TIF is typically associated with peritubular microvascular rarefaction. Multiple mechanisms could account for the loss of peritubular capillaries in the fibrotic kidneys, in which pathologic changes of endothelial cells are one of the most common causes of microvascular rarefaction. Although endothelial cells are a primary target for hypoxic injury and numerous evidences have indicated the complex response of endothelial cells to hypoxia, the exact mechanisms of hypoxic response in endothelial cells are largely unclear. It is well known that endothelial dysfunction contributes to impairment in the integrity of peritubular capillaries. Available evidence indicates that exposure of endothelial cells to hypoxia could activate the receptor for advanced glycation end products, which is a common finding in TIF, and activate the p38 mitogen-activated protein kinase and NF-κB signaling pathway, leading to the progression of the renal disease (Matsui et al. 2015). Additionally, there is increasing evidence that the endothelial cells possess the capacity to transdifferentiate to myofibroblasts (EndoMT) under hypoxic environments (Xavier et al. 2015). Thus, it is reasonable to speculate that the EndoMT, in turn, induced the dysfunction of endothelial cells and increased the number of ECM-producing fibroblasts, exacerbating the hypoxia of kidneys. Accordingly, blocking EndoMT by a specific inhibitor prevents injury to the endothelial cells, retarding the progression of DN (Li et al. 2010). Furthermore, the EndoMT not only translates endothelium cells into the matrix-producing (myo)fibroblasts, but also leads to the apoptosis of endothelial cells directly. Loss of endothelial cells in fibrotic kidneys implies that apoptosis may be the predominant response of renal endothelial cells to hypoxia. Indeed, ischemia and oxidative stress are the major apoptotic stimuli for endothelial cells, which could be a potential mechanism resulting in the rarefaction of peritubular capillaries (Tanaka et al. 2014).

23.4.3 Pericytes

Recently, emerging evidence illustrates that the pericyte, another hypoxia-sensitive component of the peritubular microvasculature, plays a vital role in the progression of TIF (Kramann and Humphreys 2014). Pericytes carry out the important function constricting the capillaries and stabilizing the vessels. Indeed, studies have determined that the pericyte could be a regulator of renal blood flow. And more interestingly, the number of pericytes in the kidneys is very large.

Pericytes not only contribute to pathological vasoconstriction, but also have the potential to promote ECM accumulation by differentiating into synthetic myofibroblast-like cells (Falke et al. 2015). Additionally, under hypoxic conditions, myofibroblast-like cells derived from pericytes could induce deficiency of pericytes and dysfunction of the endothelium cells, eventually leading to a rarefaction of the peritubular capillaries. Increasing evidence indicates that the PDGFRβ signaling in pericytes plays critical roles in kidney injury. When PDGFRβ signaling in pericytes is blocked, capillary rarefaction and TIF are attenuated in an animal model of progressive CKD (Lin et al. 2011). Furthermore, pericyte–myofibroblast differentiation triggers a switch in the secretion of VEGF isomers (from VEGF-164 to VEGF-120 and VEGF-188), resulting in endothelial loss and microvascular rarefaction in TIF (Lin et al. 2011).

23.4.4 Fibroblasts

In the fibrotic kidney, fibroblasts are the major ECM-producing cells, which is predominantly located in the tubulointerstitium. Exposure of the fibroblasts to hypoxic conditions promotes a fibrogenic phenotype with increased proliferation and enhances differentiation of the myofibroblast. Consistent with the results in PTE cells, exposure to hypoxia also increases ECM production, upregulating a variety of matrix proteins (Gilkes et al. 2013). Additionally, hypoxia could also upregulate various enzymes involved in the posttranslational modification of collagen, resulting in the development of resistance to degradation (Norman et al. 2000). Although hypoxia could induce changes of the fibrogenic phenotype in fibroblasts, the exact mechanisms underlying this process remain unexplored. There is evidence that the fibrogenic phenotype induced by hypoxia is likely to occur due to the functions of hypoxia-induced growth factors. Additionally, in parallel with increased ECM production, hypoxia also suppresses matrix degradation by decreasing the expression and activity of multiple matrix metalloproteinases, which are likely to demonstrate an HIF-dependent expression. However, these proteases have not been widely studied in renal pathophysiology. Meanwhile, the exact roles of other molecular pathways, which are regulated by hypoxia in renal fibroblasts, are currently unexplored in hypoxia-induced fibrosis.

23.4.5 *Inflammatory Cells*

Inflammation is a well-known component of the fibrotic response (Lv et al. 2018a, b), and infiltration of inflammatory cells in the tubulointerstitium can be correlated with deteriorating renal function (Kitching 2014). Among the complex pathological mechanisms, hypoxia is a part of the process of normal immune cell development and function, and the HIF pathway is a critical regulator of these. Hypoxia could also induce accumulation of pro-inflammatory cells at the site of injury, resulting in tubulointerstitium injury (Rama et al. 2008). Higgins et al (2007) found that, in the absence of HIF-1α, TIF was inhibited along with reduced infiltration of inflammatory cells and collagen deposition in kidneys subjected to unilateral ureteral obstruction (UUO), suggesting the key roles of the HIF-1 pathway. Hypoxia also causes the production of pro-inflammatory factors for the immune cells. In 2008, Rama and colleagues reported that dendritic cell differentiation and stimulated lymphocyte proliferation could be induced by hypoxia and HIF-1α-dependent, providing a link between hypoxia and immune activation (Rama et al. 2008). Also, I/R injured kidneys showed clear signs of dendritic cell maturation, and the evidences also highlighted the pivotal role of the HIF-1 pathway in regulating the immune response.

Among the inflammatory cell populations, macrophages exhibit heterogeneous phenotypes and have incredible functional plasticity, and play a key role in the development and recovery of kidney diseases (Huen and Cantley 2017; Lv et al. 2017, 2018a, b). Macrophages can switch phenotypes in response to their microenvironment and alter their effect from pro-inflammatory to anti-inflammatory. For example, in contrast to M1 macrophages, subsets of M2 macrophages can resolve inflammation and repair injury. Compelling evidence shows that TIF is accompanied by the accumulation of macrophages, which are derived from both resident and recruited cell populations. Resident macrophages maintain tissue homeostasis under physiological conditions and react upon tissue insults. Although macrophages have a broad spectrum of functions, as a whole, they are considered to have a tendency to induce TIF. Macrophages produce profibrotic and pro-inflammatory factors, which cause activation of myofibroblasts and their fibrogenesis, creating a positive feedback loop and developing TIF.

23.4.6 *Progenitor Cell*

In CKD, a proportion of ECM-producing cells are derived from circulating precursor cells (Xia et al. 2014). There is evidence that perivascular Gli1+ progenitors are key contributors to injury-induced organ fibrosis, including the kidney. Although the exact mechanisms of progenitor cell recruitment are unclear, the role for hypoxia seems plausible as progenitor cells preferentially home to ischemic sites, which are mediated by the HIF (Ceradini and Gurtner 2005). Additionally, in ischemic disease, overexpression of the HIF-1α-mediated stromal cell-derived factor-1 is also involved in the ischemic injury through bone marrow-derived progenitor cell recruitment. However, the exact mechanism of the recruitment of precursor cells during hypoxia-induced kidney injury is not clear. The IL-4 receptor potentially has a critical role in bone marrow-derived fibroblast activation and the TIF promoting. Moreover, hypoxia may alter the function of progenitor cells, which potentially have the ability to promote fibrogenesis.

23.5 Hypoxia Is a Therapeutic Target in CKD

With hypoxia having a primary role in the development of TIF, it is interesting to speculate that anti-hypoxia may be a novel avenue that can be employed in preventing TIF. Normalization of microvascular perfusion; preservation and repair of the microvasculature; modulation of the HIF; manipulation of progenitor cell homing; correction of anemia and repair of the PTE (not discussed here) are potential approaches to ameliorate hypoxia-mediated profibrotic changes.

Normalization of microvascular perfusion by alleviating vasoconstriction and enhancing vasodilation is a logical goal. Therapeutic manipulation of the renin–angiotensin system (ACEI and ARB) has been well established. Administration of ACEI could restore cortical perfusion and reduce tissue hypoxia thereby extending a range of renoprotective mechanisms (Efrati et al. 2012). Endothelin (ET), another potent vasoconstrictor, is activated during CKD and is a critical factor in the transition from AKI to CKD. Preclinical and clinical studies have demonstrated that selective ET receptors' antagonists have shown renoprotective effects. Particularly, antagonism of ET-A receptors has maximal benefit in decreasing vasoconstriction and thus blocks direct fibrogenic effects of the ET (Kasztan et al. 2017). On the other hand, it may also be a potential therapeutic target to enhance NO signaling which is compromised in the setting of CKD.

Microvascular rarefaction is a hallmark of renal fibrosis. It is well known that hypoxia is a critical angiogenic regulatory stimulus and one of the anomalies in hypoxia as a fibrogenic stimulus is the failure of angiogenic repair in hypoxia-induced kidney injury. In models of CKD, levels of VEGF expression are significantly reduced in the advanced stages of the disease, (Kang et al. 2002) the potential mechanism for which is unclear. However, reduction in the levels of VEGF can be correlated with tubular atrophy and infiltration of inflammatory cells. Interestingly, a recent study demonstrated that under hypoxic conditions, albumin could suppress VEGF production in tubular epithelial cells, suggesting a possible link between proteinuria and hypoxia in CKD (Katavetin et al. 2008). Expression of angiopoietin-1, another key angiogenic factor, is also suppressed in the model of TIF, suggesting an overall reduction in the pro-angiogenic factors (Kim et al. 2006.) Parallelly, an increase in the anti-angiogenic factors may be a good explanation for failure of the angiogenic repair. One of the important endogenous inhibitors of angiogenesis is endostatin, the levels of which are increased by hypoxia (Paddenberg et al. 2006). Thus, preservation and repair of the tubulointerstitial microvasculature by administration of exogenous pro-angiogenic factors or inhibition of endogenous anti-angiogenic factors may present promising therapeutic strategies.

HIF is a critical transcription factor mediating adaptive response to hypoxia. A large body of preclinical evidence in a various kidney injury models supports the observation that primary role for the HIF activation is protection from injuries, such as I/R injury, cisplatin-induced nephropathy, acute and progressive glomerulonephritis. Consistently, genetic deficiency in the HIF was found to exacerbate ischemic injury (Hill et al. 2008). Collectively, these data illustrate that HIF activation protects the kidney against injury. In terms of clinical applications, stabilizing the HIF by blocking PHD is a more promising and feasible approach and number of inhibitors are under development with initial promising results (Bonomini et al. 2016). Although the therapeutic strategies in CKD by activating HIF pathway have elicited considerable interest, a note of caution needs to be sounded that cell-specific activation of the HIF in tubular drives EMT and that HIF activates multiple profibrotic genes (Baumann et al. 2016) and in podocytes induces rapidly progressive glomerulonephritis (Ding et al. 2006). Furthermore, deletion of HIF-1α in tubular epithelial cells has been shown to reduce fibrosis in a CKD model (Higgins et al. 2007). Given that activation of the HIF pathway appears to have both renoprotective and profibrotic effects, potential therapeutic strategies that target this pathway will require careful evaluation.

A more speculative therapeutic aspect of the hypoxic response is the manipulation of progenitor cell homing. However, the exact functions of these cells are still unclear and need to be further explored.

23.6 Conclusion

The "chronic hypoxia hypothesis" was first put forward in 1998. Since then, the role of hypoxia in the pathogenesis of CKD has been investigated and validated intensively using in vitro studies, in vivo models, and studies in patients. Increasing evidence indicates that hypoxia plays a critical role in the progression of kidney disease. A thorough understanding of the role of hypoxia in TIF and its interaction with other factors opens the door to a multitude of novel therapeutic strategies aimed at preventing or retarding various kidney diseases.

Acknowledgements This study was supported by grants from the National Key Research and Development Program of China (2018YFC1314004) and the National Natural Scientific Foundation (No. 81720108007, 81130010, 81470997 and 81670696), the Clinic Research Center of Jiangsu Province (No. BL2014080) to Professor Bi-Cheng Liu as PI.

References

Abdelkader A, Ho J, Ow CP, Eppel GA, Rajapakse NW, Schlaich MP, Evans RG (2014) Renal oxygenation in acute renal ischemia-reperfusion injury. Am J Physiol Renal Physiol 306:F1026–F1038

Asada N, Takase M, Nakamura J, Oguchi A, Asada M, Suzuki N, Yamamura K, Nagoshi N, Shibata S, Rao TN, Fehling HJ, Fukatsu A, Minegishi N, Kita T, Kimura T, Okano H, Yamamoto M, Yanagita M (2011) Dysfunction of fibroblasts of extrarenal origin underlies renal fibrosis and renal anemia in mice. J Clin Invest 121:3981–3990

Baumann B, Hayashida T, Liang X, Schnaper HW (2016) Hypoxia-inducible factor-1α promotes glomerulosclerosis and regulates COL1A2 expression through interactions with Smad3. Kidney Int 90:797–808

Blantz RC, Deng A, Miracle CM, Thomson SC (2007) Regulation of kidney function and metabolism: a question of supply and demand. Trans Am Clin Climatol Assoc 118:23–43

Bonomini M, Del Vecchio L, Sirolli V, Locatelli F (2016) New treatment approaches for the anemia of CKD. Am J Kidney Dis 67:133–142

Bonventre JV, Yang L (2011) Cellular pathophysiology of ischemic acute kidney injury. J Clin Invest 121:4210–4221

Carlström M, Wilcox CS, Arendshorst WJ (2015) Renal autoregulation in health and disease. Physiol Rev 95:405–511

Calzavacca P, Evans RG, Bailey M, Bellomo R, May CN (2015) Cortical and medullary tissue perfusion and oxygenation in experimental septic acute kidney injury. Crit Care Med 43:e431–e439

Ceradini DJ, Gurtner GC (2005) Homing to hypoxia: HIF-1 as a mediator of progenitor cell recruitment to injured tissue. Trends Cardiovasc Med 15:57–63

Davidson A (2016) What is damaging the kidney in lupus nephritis? Nat Rev Rheumatol 12:143–153

Davis MJHM, Kuo L (2008) Local regulation of microvascular perfusion. In: Tuma RFDW, Ley K (eds). Handbook of physiology: microcirculation, 2nd edn. Elsevier, San Diego, CA

Deng W, Ren Y, Feng X, Yao G, Chen W, Sun Y, Wang H, Gao X, Sun L (2014) Hypoxia inducible factor-1 alpha promotes mesangial cell proliferation in lupus nephritis. Am J Nephrol 40:507–515

Ding M, Cui S, Li C, Jothy S, Haase V, Steer BM, Marsden PA, Pippin J, Shankland S, Rastaldi MP, Cohen CD, Kretzler M, Quaggin SE (2006) Loss of tumour suppressor Vhlh leads to upregulation of Cxcr4 and rapidly progressive glomerulonephritis in mice. Nat Med 12:1081–1087

Eckardt KU, Rosenberger C, Jurgensen JS, Wiesener MS (2003) Role of hypoxia in the pathogenesis of renal disease. Blood Purif 21:253–257

Edeling M, Ragi G, Huang S, Pavenstädt H, Susztak K (2016) Developmental signalling pathways in renal fibrosis: the roles of Notch, Wnt and Hedgehog. Nat Rev Nephrol 12:426–439

Efrati S, Berman S, Hamad RA, Siman-Tov Y, Ilgiyaev E, Maslyakov I, Weissgarten J (2012) Effect of captopril treatment on recuperation from ischemia/reperfusion-induced acute renal injury. Nephrol Dial Transplant 27:136–145

Emans TW, Janssen BJ, Pinkham MI, Ow CP, Evans RG, Joles JA, Malpas SC, Krediet CT, Koeners MP (2016) Exogenous and endogenous angiotensin-II decrease renal cortical oxygen tension in conscious rats by limiting renal blood flow. J Physiol 594:6287–6300

Evans RG, Eppel GA, Michaels S, Burke SL, Nematbakhsh M, Head GA, Carroll JF, O'Connor PM (2010) Multiple mechanisms act to maintain kidney oxygenation during renal ischemia in anesthetized rabbits. Am J Physiol Renal Physiol 298:F1235–F1243

Evans RG, Gardiner BS, Smith DW, O'Connor PM (2008) Intrarenal oxygenation: unique challenges and the biophysical basis of homeostasis. Am J Physiol Renal Physiol 295:F1259–F1270
Fahling M, Seeliger E, Patzak A, Persson PB (2017) Understanding and preventing contrast-induced acute kidney injury. Nat Rev Nephrol 13:169–180
Falke LL, Gholizadeh S, Goldschmeding R, Kok RJ, Nguyen TQ (2015) Diverse origins of the myofibroblast-implications for kidney fibrosis. Nat Rev Nephrol 11:233–244
Fine LG, Orphanides C, Norman JT (1998) Progressive renal disease: the chronic hypoxia hypothesis. Kidney Int 65(Suppl):S74–S78
Gilkes DM, Bajpai S, Chaturvedi P, Wirtz D, Semenza GL (2013) Hypoxia-inducible factor 1 (HIF-1) promotes extracellular matrix remodeling under hypoxic conditions by inducing P4HA1, P4HA2, and PLOD2 expression in fibroblasts. J Biol Chem 288:10819–10829
Haase VH (2013) Regulation of erythropoiesis by hypoxia-inducible factors. Blood Rev 27:41–53
Haase VH (2017a) Oxygen sensors as therapeutic targets in kidney disease. Nephrol Ther 13(Suppl 1):S29–S34
Haase VH (2017b) HIF-prolyl hydroxylases as therapeutic targets in erythropoiesis and iron metabolism. Hemodial Int 21(Suppl 1):S110–S124
Higgins DF, Kimura K, Bernhardt WM, Shrimanker N, Akai Y, Hohenstein B, Saito Y, Johnson RS, Kretzler M, Cohen CD, Eckardt KU, Iwano M, Haase VH (2007) Hypoxia promotes fibrogenesis in vivo via HIF-1 stimulation of epithelial-tomesenchymal transition. J Clin Invest 117:3810–3820
Hill P, Shukla D, Tran MG, Aragones J, Cook HT, Carmeliet P, Maxwell PH (2008) Inhibition of hypoxia inducible hydroxylases protects against renal ischemia-reperfusion injury. J Am Soc Nephrol 19:39–46
Holthoff JH, Wang Z, Seely KA, Gokden N, Mayeux PR (2012) Resveratrol improves renal microcirculation, protects the tubular epithelium, and prolongs survival in a mouse model of sepsis-induced acute kidney injury. Kidney Int 81:370–378
Huen SC, Cantley LG (2017) Macrophages in renal injury and repair. Annu Rev Physiol 79:449–469
Ivan M, Kondo K, Yang H, Kim W, Valiando J, Ohh M, Salic A, Asara JM, Lane WS, Kaelin WG (2001) HIFalpha targeted for VHL-mediated destruction by proline hydroxylation: implications for O2 sensing. Science 292:464–468
Kang DH, Kanellis J, Hugo C, Truong L, Anderson S, Kerjaschki D, Schreiner GF, Johnson RJ (2002) Role of the microvascular endothelium in progressive renal disease. J Am Soc Nephrol 13:806–816
Kasztan M, Fox BM, Speed JS, De Miguel C, Gohar EY, Townes TM, Kutlar A, Pollock JS, Pollock DM (2017) Long-term Endothelin-A receptor antagonism provides robust renal protection in humanized sickle cell disease Mice. J Am Soc Nephrol 28:2443–2458
Katavetin P, Inagi R, Miyata T, Tanaka T, Sassa R, Ingelfinger JR, Fujita T, Nangaku M (2008) Albumin suppresses vascular endothelial growth factor via alteration of hypoxia-inducible factor/hypoxia-responsive element pathway. Biochem Biophys Res Commun 367:305–310
Kim W, Moon SO, Lee SY, Jang KY, Cho CH, Koh GY, Choi KS, Yoon KH, Sung MJ, Kim DH, Lee S, Kang KP, Park SK (2006) COMP-angiopoietin-1 ameliorates renal fibrosis in a unilateral ureteral obstruction model. J Am Soc Nephrol 17:2474–2483
Kitching AR (2014) Dendritic cells in progressive renal disease: some answers, many questions. Nephrol Dial Transplant 29:2185–2193
Kramann R, Humphreys BD (2014) Kidney pericytes: roles in regeneration and fibrosis. Semin Nephrol 34:374–383
Lameire NH, Bagga A, Cruz D, De Maeseneer J, Endre Z, Kellum JA, Liu KD, Mehta RL, Pannu N, Van Biesen W, Vanholder R (2013) Acute kidney injury: an increasing global concern. Lancet 382:170–179
Lannemyr L, Bragadottir G, Krumbholz V, Redfors B, Sellgren J, Ricksten SE (2017) Effects of cardiopulmonary bypass on renal perfusion, filtration, and oxygenation in patients undergoing cardiac surgery. Anesthesiology 126:205–213

Layton AT, Laghmani K, Vallon V, Edwards A (2016) Solute transport and oxygen consumption along the nephrons: effects of Na+ transport inhibitors. Am J Physiol Renal Physiol 311:F1217–F1229

Laycock SK, Vogel T, Forfia PR, Tuzman J, Xu X, Ochoa M, Thompson CI, Nasjletti A, Hintze TH (1998) Role of nitric oxide in the control of renal oxygen consumption and the regulation of chemical work in the kidney. Circ Res 82:1263–1271

Leung KC, Tonelli M, James MT (2013) Chronic kidney disease following acute kidney injury-risk and outcomes. Nat Rev Nephrol 9:77–85

Li J, Qu X, Yao J, Caruana G, Ricardo SD, Yamamoto Y, Yamamoto H, Bertram JF (2010) Blockade of endothelial-mesenchymal transition by a Smad3 inhibitor delays the early development of streptozotocin-induced diabetic nephropathy. Diabetes 59:2612–2624

Li Y, Yang J, Dai C, Wu C, Liu Y (2003) Role for integrin-linked kinase in mediating tubular epithelial to mesenchymal transition and renal interstitial fibrogenesis. J Clin Invest 112:503–516

Lin SL, Chang FC, Schrimpf C, Chen YT, Wu CF, Wu VC, Chiang WC, Kuhnert F, Kuo CJ, Chen YM, Wu KD, Tsai TJ, Duffield JS (2011) Targeting endothelium-pericyte cross talk by inhibiting VEGF receptor signaling attenuates kidney microvascular rarefaction and fibrosis. Am J Pathol 178:911–923

Liu BC, Tang TT, Lv LL, Lan HY (2018) Renal tubule injury: a driving force toward chronic kidney disease. Kidney Int 93:568–579

Lv LL, Tang PM, Li CJ, You YK, Li J, Huang XR, Ni J, Feng M, Liu BC, Lan HY (2017) The pattern recognition receptor, Mincle, is essential for maintaining the M1 macrophage phenotype in acute renal inflammation. Kidney Int 91:587–602

Lv LL, Feng Y, Wen Y, Wu WJ, Ni HF, Li ZL, Zhou LT, Wang B, Zhang JD, Crowley SD, Liu BC (2018a) Exosomal CCL2 from tubular epithelial cells is critical for albumin-induced tubulointerstitial inflammation. J Am Soc Nephrol 29:919–935

Lv W, Booz GW, Wang Y, Fan F, Roman RJ (2018b) Inflammation and renal fibrosis: recent developments on key signaling molecules as potential therapeutic targets. Eur J Pharmacol 820:65–76

Manotham K, Tanaka T, Matsumoto M, Ohse T, Inagi R, Miyata T, Kurokawa K, Fujita T, Ingelfinger JR, Nangaku M (2004) Transdifferentiation of cultured tubular cells induced by hypoxia. Kidney Int 65:871–880

Matsui T, Oda E, Higashimoto Y, Yamagishi S (2015) Glyceraldehyde-derived pyridinium (GLAP) evokes oxidative stress and inflammatory and thrombogenic reactions in endothelial cells via the interaction with RAGE. Cardiovasc Diabetol 14:1

Mole DR, Blancher C, Copley RR, Pollard PJ, Gleadle JM, Ragoussis J, Ratcliffe PJ (2009) Genome-wide association of hypoxia-inducible factor (HIF)-1alpha and HIF-2alpha DNA binding with expression profiling of hypoxia-inducible transcripts. J Biol Chem 284:16767–16775

Nangaku M, Rosenberger C, Heyman SN, Eckardt KU (2013) Regulation of hypoxia-inducible factor in kidney disease. Clin Exp Pharmacol Physiol 40:148–157

Norman JT, Clark IM, Garcia PL (2000) Hypoxia promotes fibrogenesis in human renal fibroblasts. Kidney Int 58:2351–2366

Ow CPC, Abdelkader A, Hilliard LM, Phillips JK, Evans RG (2014) Determinants of renal tissue hypoxia in a rat model of polycystic kidney disease. Am J Physiol Regul Integr Comp Physiol 307:R1207–R1215

Paddenberg R, Faulhammer P, Goldenberg A, Kummer W (2006) Hypoxia-induced increase of endostatin in murine aorta and lung. Histochem Cell Biol 125:1–12

Palm F, Cederberg J, Hansell P, Liss P, Carlsson PO (2003) Reactive oxygen species cause diabetes-induced decrease in renal oxygen tension. Diabetologia 46:1153–1160

Parks SK, Chiche J, Pouyssegur J (2011) pH control mechanisms of tumor survival and growth. J Cell Physiol 226:299–308

Pittman RN (2011) Regulation of tissue oxygenation, 1st edn. Morgan & Claypool Life Science, San Rafael

Pruijm M, Milani B, Pivin E, Podhajska A, Vogt B, Stuber M, Burnier M (2018) Reduced cortical oxygenation predicts a progressive decline of renal function in patients with chronic kidney disease. Kidney Int 93:932–940
Priyadarshi A, Periyasamy S, Burke TJ, Britton SL, Malhortra D, Shapiro JI (2002) Effects of reduction of renal mass on renal oxygen tension and erythropoietin production in the rat. Kidney Int 61:542–546
Rama I, Bruene B, Torras J, Koehl R, Cruzado JM, Bestard O, Franquesa M, Lloberas N, Weigert A, Herrero-Fresneda I, Gulias O, Grinyó JM (2008) Hypoxia stimulus: an adaptive immune response during dendritic cell maturation. Kidney Int 73:816–825
Rewa O, Bagshaw SM (2014) Acute kidney injury-epidemiology, outcomes and economics. Nat Rev Nephrol 10:193–207
Rosenberger C, Mandriota S, Jürgensen JS, Wiesener MS, Hörstrup JH, Frei U, Ratcliffe PJ, Maxwell PH, Bachmann S, Eckardt KU (2002) Expression of hypoxia-inducible factor-1alpha and -2alpha in hypoxic and ischemic rat kidneys. J Am Soc Nephrol 13:1721–1732
Schnaper HW (2017) The tubulointerstitial pathophysiology of progressive kidney disease. Adv Chronic Kidney Dis 24:107–116
Schödel J, Klanke B, Weidemann A, Buchholz B, Bernhardt W, Bertog M, Amann K, Korbmacher C, Wiesener M, Warnecke C, Kurtz A, Eckardt KU, Willam C (2009) HIF-prolyl hydroxylases in the rat kidney: physiologic expression patterns and regulation in acute kidney injury. Am J Pathol 174:1663–1674
Sgouralis I, Evans RG, Layton AT (2017) Renal medullary and urinary oxygen tension during cardiopulmonary bypass in the rat. Math Med Biol 34:313–333
Shi H, Yan T, Li D, Jia J, Shang W, Wei L, Zheng Z (2017) Detection of renal hypoxia in lupus nephritis using blood oxygen level-dependent MR imaging: a multiple correspondence analysis. Kidney Blood Press Res 42:123–135
Singh P, Ricksten SE, Bragadottir G, Redfors B, Nordquist L (2013) Renal oxygenation and haemodynamics in acute kidney injury and chronic kidney disease. Clin Exp Pharmacol Physiol 40:138–147
Starling S (2017) Renal fibrosis: pericytes activate complement in fibrosis. Nat Rev Nephrol 13:262
Stillman IE, Brezis M, Heyman SN, Epstein FH, Spokes K, Rosen S (1994) Effects of salt depletion on the kidney: changes in medullary oxygenation and thick ascending limb size. J Am Soc Nephrol 4:1538–1545
Tanaka S, Tanaka T, Nangaku M (2014) Hypoxia as a key player in the AKI-to-CKD transition. Am J Physiol Renal Physiol 307:F1187–F1195
Tessari P (2015) Nitric oxide in the normal kidney and in patients with diabetic nephropathy. J Nephrol 28:257–268
Uchino S, Kellum JA, Bellomo R, Doig GS, Morimatsu H, Morgera S, Schetz M, Tan I, Bouman C, Macedo E, Gibney N, Tolwani A, Ronco C, Beginning and Ending Supportive Therapy for the Kidney (BEST Kidney) Investigators (2005) Acute renal failure in critically ill patients: a multinational, multicenter study. JAMA 294:813–818
van der Bel R, Coolen BF, Nederveen AJ, Potters WV, Verberne HJ, Vogt L, Stroes ES, Krediet CT (2016) Magnetic resonance imaging-derived renal oxygenation and perfusion during continuous, steady-state Angiotensin-II infusion in healthy humans. J Am Heart Assoc 5:e003185
Wang Z, Holthoff JH, Seely KA, Pathak E, Spencer HJ 3rd, Gokden N, Mayeux PR (2012) Development of oxidative stress in the peritubular capillary microenvironment mediates sepsis-induced renal microcirculatory failure and acute kidney injury. Am J Pathol 180:505–516
Webster AC, Nagler EV, Morton RL, Masson P (2017) Chronic kidney disease. Lancet 389(10075):1238–1252
Weinberg JM (2011) Mitochondrial biogenesis in kidney disease. J Am Soc Nephrol 22:431–436
Welch WJ, Baumgartl H, Lubbers D, Wilcox CS (2003) Renal oxygenation defects in the spontaneously hypertensive rat: role of AT1 receptors. Kidney Int 63:202–208

Xavier S, Vasko R, Matsumoto K, Zullo JA, Chen R, Maizel J, Chander PN, Goligorsky MS (2015) Curtailing endothelial TGF-β signaling is sufficient to reduce endothelial-mesenchymal transition and fibrosis in CKD. J Am Soc Nephrol 26:817–829

Xia Y, Yan J, Jin X, Entman ML, Wang Y (2014) The chemokine receptor CXCR82 contributes to recruitment of bone marrow-derived fibroblast precursors in renal fibrosis. Kidney Int 86:327–337

Yin W, Liu F, Li X, Yang L, Zhao S, Huang ZX, Huang YQ, Liu RB (2012) Noninvasive evaluation of renal oxygenation in diabetic nephropathy by BOLD-MRI. Eur J Radiol 81:1426–1431

Zhang Y, Wang J, Yang X, Wang X, Zhang J, Fang J, Jiang X (2012) The serial effect of iodinated contrast media on renal hemodynamics and oxygenation as evaluated by ASL and BOLD MRI. Contrast Media Mol Imaging 7:418–425

Zhou D, Fu H, Zhang L, Zhang K, Min Y, Xiao L, Lin L, Bastacky SI, Liu Y (2017) Tubule-derived Wnts are required for fibroblast activation and kidney fibrosis. J Am Soc Nephrol 28:2322–2336

Zhou D, Liu Y (2016) Renal fibrosis in 2015: understanding the mechanisms of kidney fibrosis. Nat Rev Nephrol 12:68–70

Chapter 24
New Understanding on the Role of Proteinuria in Progression of Chronic Kidney Disease

Dan Liu and Lin-Li Lv

Abstract Proteinuria is identified as an important marker and risk factor of progression in chronic kidney disease. However, the precise mechanism of action in the progress of chronic kidney disease is still unclear. Mesangial toxicity from specific filtered compounds such as albumin-bound fatty acids and transferrin/iron, tubular overload and hyperplasia, and induction of proinflammatory molecules such as MCP-1 and inflammatory cytokines are some of the proposed mechanisms. Reversing intraglomerular hypertension with protein restriction or antihypertensive therapy may be beneficial both by diminishing hemodynamic injury to the glomeruli and by reducing protein filtration. Therefore, understanding proteinuria and its role in renal tubular interstitial inflammation and fibrosis is of great significance for the study of renal protective therapy, such as antiproteinuric treatments, and delaying the progression of chronic renal disease.

Keywords Proteinuria · Chronic kidney disease · End-stage renal disease

24.1 Introduction

Chronic kidney disease (CKD) arises from many heterogeneous disease and pathways that alter the function and structure of the kidney irreversibly, over months to years. The diagnosis of CKD is based on a chronic reduction in kidney function and structural kidney damage. There are studies showing the increasing prevalence of global CKD. Among the first world countries such as USA and Australia, the prevalence of CKD is reported to be around 11% (Webster et al. 2017), while the adult prevalence rate in China is as high as 10.8% (Zhang et al. 2012). There is evidence of varying incidence, prevalence, and progression of CKD in different countries among ethnicity and social class. Multiple CKD such as chronic glomerulonephritis, dia-

D. Liu · L.-L. Lv (✉)
Institute of Nephrology, Zhong Da Hospital, Southeast University School of Medicine, Nanjing, China
e-mail: lvlinli@seu.edu.cn

B.-C. Liu et al. (eds.), *Renal Fibrosis: Mechanisms and Therapies*, Advances in Experimental Medicine and Biology 1165,
https://doi.org/10.1007/978-981-13-8871-2_24

betic nephropathy, and polycystic kidney disease could progress to ESRD. Therefore, CKD has become an important public health issue of global concern.

Proteinuria is an important marker and the stronger predicator of CKD outcomes. There are studies with the evidence supporting a strong association between proteinuria and the risk of CKD progression among nondiabetic and diabetic patients. A mass screening of 107,192 members studied among over 10 years of the general population, in Okinawa, Japan identified proteinuria as the most prevailing risk predictor of ESRD (Iseki et al. 1996). In the Ramipril Efficacy in Nephropathy (REIN) trial (Ruggenenti et al. 1997), urinary protein excretion was the only baseline variable that correlated with the rate of glomerular filtration rate (GFR) decline progressing to ESRD while studied among 274 patients with nondiabetic chronic nephropathies. Among the patients stratified according to baseline proteinuria levels, the lowest tertile consistently had the lowest rate of renal disease progression and ESRD, as compared with the middle and in the highest tertiles (Ruggenenti et al. 1998). The same was observed for diabetic patients. Angiotensin II Antagonist Losartan (RENAAL) study and in the Irbesartan in Diabetic Nephropathy Trial (IDNT) showed the baseline urinary albumin-to-creatinine ratio was a strong independent predictor of ESRD progression in the Reduction of Endpoints in NIDDM (Keane et al. 2006; Atkins et al. 2005).

Thus, among CKD patients and the general population, proteinuria proves to be a potential marker for the renal outcome. Furthermore, the RENAAL study found the evidence of baseline albuminuria as the most important independent predictor for the risk of ESRD, especially among ethnic white, Black, Asian, and Hispanic individuals (de Zeeuw et al. 2006; Cravedi et al. 2012).

Cellular mechanisms of proteinuria for the determination and progression of CKD are the proteins that access to the glomerular filtrate, which is a consequence of distorted glomerular permeability further reaching the tubular lumen. The further tubular synthesis of bioactive mediators causes it to act as trigger of interstitial inflammatory and fibrotic reactions. This chapter focuses on the pathophysiology of proteinuria and its pathogenic influence on tubulointerstitial inflammation and fibrosis.

24.2 Pathophysiology of Proteinuria

24.2.1 Alteration of Glomerular Filtration Barrier

Serum albumin is reabsorbed by renal tubules after glomerular filtration, so the appearance of urine albumin is caused by abnormal function of glomerular and renal tubular, which increases albumin excretion. The injury of glomerular and renal tubules is involved in the initial process of proteinuria formation. A chronic proteinuric glomerulopathies are defined as common manifestation of the sustained or permanent loss of selectivity of the glomerular barrier to protein filtration. The glomerular filtration barrier consists of triple-layer membrane structure, the innermost fenestrated glomerular endothelial, middle glomerular basement membrane, and outermost podocytes. Increased hydrostatic pressure or a damaged glomerular filtration barrier may cause overload in the glomerular proteinuria. The underlying pathogenesis of glomerular disease related to proteinuria is hyperglycemia causing glycated hemoglobin concentration, dyslipidemia, inflammatory cytokines, reactive oxygen species, the activation of angiotensin system (renin–angiotensin system, RAS), causing increased endothelial permeability (Jefferson et al. 2008). Furthermore, some studies reported the association of glycocalyx for glomerular endothelial defects with proteinuria highlighting the importance of this complex structure (Toblli et al. 2012). The negatively charged glomerular charge barrier destruction, consisting of collagen and laminin, which leads to negatively charged molecules into urine (Huxley and Williams 2000). Besides that, inflammatory cell infiltration, glomerular mesangial cell growth, and extracellular matrix production were also reported as the underlying mechanisms for proteinuria. Podocytes are the largest and the most differentiated cell type of the glomerulus and an essential part of the filtration unit. In the experimental model, it is thought that podocyte-related molecular dysfunction, such as nephrin and podocin, leads to the alterations of the slit diaphragm and proteinuria (Greka and Mundel 2012). The transient receptor potential cation (TRPC) is a member of proteins involved in the regulation of Ca^{2+} influx. Recently, the study demonstrated by increasing TRPC6 expression via an NFAT-mediated positive feedback signaling pathway, the Ang II participates in the podocyte injury. Therefore, the finding has highlighted the crucial involvement of TRPC6 in the pathogenesis of podocyte injury and proteinuria (Nijenhuis et al. 2011). Klotho is a type-1 membrane protein produced mainly in the kidney. In a study by Kim et al., secreted Klotho suppressed TRPC6-mediated Ca^{2+} influx in cultured mouse podocytes through inhibiting phosphoinositide 3-kinase-dependent exocytosis of the channel. This finding may offer a new therapeutic strategy for the treatment and management of proteinuria (Kim et al. 2017).

24.2.2 Protein Reabsorption of Renal Tubules

The filtered albumin through the glomerulus is reabsorbed by the receptor-mediated endocytosis in the proximal renal tubular cells which is further transferred into lysosomes for degradation. The receptor megalin and cubilin are responsible for the reabsorption of albumin. Furthermore, new studies on mice have found the reabsorption of albumin in the proximal renal tubules depends on the simultaneous expression of megalin and cubilin (Weyer et al. 2011; Amsellem et al. 2010). Clinical study on extremely rare Donnai-Barrow/Facio-oculoacoustico-renal (DB/FOAR) syndrome, which is characterized by low-molecular-weight proteinuria, found that the dysfunction of endocytic uptake of albumin was associated with the absence of megalin expression (Storm et al. 2013). With renal tubular proteinuria, albumin via binding to its receptor induces inflammation further releasing cytokines to promote inflammation and fibrosis eventually leading to a decreased renal function. TGF-β is currently recognized as a proinflammatory and fibrotic medium induced by albumin exposure. This may be through a negative feedback mechanism which increases the albumin filtration, meanwhile inhibits the megalin- and cubilin-mediated albumin endocytic uptake, resulting in increased urine albumin concentration.

24.3 Proteinuria and Chronic Kidney Disease Progression

It was thought that proteinuria as the result of glomerular filtration barrier or renal tubular interstitial injury, whereas in recent years, number of experimental studies show that the proteinuria is not only a renal lesion index, but also with a close relation for the progression of chronic kidney disease. Therefore, proteinuria could be an independent pathogenic factor in the pathological process of kidney damage.

24.3.1 Proteinuria and Progression of Glomerulosclerosis

In chronic renal disease, the persistent proteinuria and the leakage of plasma protein which accumulates in glomerular could be an independent risk factor exacerbating glomerulosclerosis. Early studies have concluded that clustered plasma proteins in the glomerular mesangial region can cause mesangial cell damage, proliferation, with an increase in mesangial matrix production eventually aggravating glomerulosclerosis. Lipoprotein, especially low-density lipoprotein (LDL), is one of the glomerular filtered protein-damaging glomerular mesangial cells (GMs) through the following mechanisms: ① LDL binding to mesangial cell surface receptor, which stimulates proto-oncogene such as c-fos, c-jun expression, resulting in mesangial cell proliferation. ② LDL promotes the production of the extracellular matrix protein fibronectin in mesangial cells as well as induces production of macrophage-specific chemotactic protein-1 (MCP-1) and platelet-derived growth factor (PDGF). ③ The presence of LDL may form more oxidized cytotoxic LDL in macrophages and mesangial cells further stimulating macrophages to produce growth factors, cytokines, and other transmitters that stimulate collagen synthesis and cell proliferation, eventually supporting the propagation of glomerulosclerosis (Burton and Harris 1996). In recent years, further study has found the similarity between podocytes to renal tubular epithelial cells along with the expression of megalin, on the contrary, do not contain cubilin. In the PAN nephritis model, the detection of FITC-labeled albumin in podocyte cytoplasm could suggest podocyte having a non-megalin/cubulin-mediated albumin absorption mechanism, which is different from renal tubular epithelial cells. This could be another cause of for podocyte morphology dysfunction which increases the progression of glomerulosclerosis (Tojo et al. 2008). Peired et al. observed that on adriamycin nephropathy induced mice and transgenic mice, the model of human FSGS, albuminuria reduced retinoic acid bioavailability, and impaired retinoic acid response element (RARE)-mediated transcription of podocyte-specific genes, thus inhibiting podocyte differentiation suggesting proteinuria contributes directly to progressive glomerulosclerosis (Peired et al. 2013).

24.3.2 Proteinuria and Tubulointerstitial Inflammation

Recent studies have recognized that urinary protein per se has a toxic effect which leads to tubulointerstitial injury. Filtered protein stimulates the abnormal regulation in the signaling pathway of the proximal tubular epithelial cells (PTECs) causing a change in the tubular cell growth, apoptosis, gene transcription and further induces the production of inflammatory factors inducing inflammation and fibrosis. This is one of the major reasons for the progression of chronic renal disease to ESRD (Gorriz and Martinez-Castelao 2012). The pathology related to proteinuria causing renal tubulointerstitial damage has always been a hot topic and is summed up as follows.

24.3.2.1 Direct Toxicity of Proteinuria on Renal Tubules

A widely accepted underlying mechanism for the progressive kidney injury is tubulointerstitial damage induced through direct toxicity of the filtered proteins (Liu et al. 2018). There are studies showing the association of plasma protein has a direct effect on the regulation of gene expression levels in renal tubular cells and also plays a very important role in chronic renal tubular interstitial injury (Abbate et al. 2006).

The cell apoptosis can contribute to a loss of functional PTEC in proteinuric nephropathy. The role of albumin in regulating an apoptotic signaling pathway involving protein kinase B (PKB), phosphoinositide 3 (PI-3) kinase, and megalin could be a crucial factor (Caruso-Neves et al. 2006). The low concentrations of albumin lead to anchoring of megalin and PKB in cell membrane area, and this stimulates PI-3 K-dependent cell protective pathway further promoting the growth of PTECs. Conversely, high concentrations of albumin downregulate megalin expression at plasma membrane side further decreasing PKB activation. This initiates the apoptotic pathway leading to the death of tubular epithelial cells (Baines and Brunskill 2011). It has recently been found that albumin through inducing PKC-δ pathway causes renal tubular injury and apoptosis, but there is still need for the clarification of the specific mechanism (Li et al. 2010).

Through cell endocytosis, PTEC handling of filtered proteins is mainly dependent on the megalin/cubilin complex. In addition to the role of the "scavenger" receptor, further new researches found megalin may also be involved in signal transduction before albumin endocytosis in PTECs. The mouse model study of albumin overload confirmed glomerular filtered albumin with competitive inhibition of renal tubular resorption. It was mainly derived from the lysosomal enzymatic degradation of albumin inflow into the renal tubular epithelial cells, rather than the megalin receptor-mediated albumin endocytosis. (Lee et al. 2013). Zou et al. were the first to find the involvement of megalin in the regulated intramembrane proteolysis (RIP) which is an evolutionarily conserved process that links cell surface receptor function to gene transcription (Zou et al. 2004). The ectodomain cleavage of the megalin receptor is mediated initially by PKC-dependent matrix metalloproteinase. Subsequently, γ-secretase family of proteases cleaves the remaining transmembrane–cytosolic receptor fragment, and the released cytosolic fragment is translocated to the nucleus for gene transcription (Biemesderfer 2006). The RIP process of megalin could have important physiological functions explaining the phenotypic changes of proximal tubules in proteinuric kidney disease. Evidence contributing to megalin being involved in early activation of proximal tubule cells during non-selective proteinuria was explained through experiments in megalin-knock-out/NEP25 mice treated with the immunotoxin LMB2. This was a model for nephrotic syndrome, focal segmental glomerulosclerosis, and tubulointerstitial injury. Here, megalin-deficient cells in proximal tubule resorbed less proteins and expressed less tubular cell injury markers, such as MCP-1 and heme-oxygenase 1(Motoyoshi et al. 2008). Although it is hypothesized that albumin could stimulate RIP of megalin, it is yet to be confirmed whether albumin regulates the gene transcription levels of renal tubular epithelial cells with RIP process. In the opossum kidney cells, the overexpression of the transmembrane—cytosolic fragment of megalin has shown to inhibit transcription of the megalin and Na^+, H^+-exchanger-3 genes. Further, inhibition of γ-secretase attenuated this specific effect (Baines and Brunskill 2011). Thrailkill et al. as compared with nonalbuminuric patients with diabetes reported patients with diabetic nephropathy and albuminuria display enhanced urinary shedding of the extracellular fragment of megalin. The study further speculated this finding could be mediated by increased matrix metalloproteinase activity in the albuminuric diabetic kidney (Thrailkill et al. 2009). A study by Fernandez et al. confirmed albumin associated with direct decreases in Klotho expression in cultured tubular cells. This suggests it could contribute to the higher risk of CKD progression in human CKD Category G1-2 and premature death (Fernandez-Fernandez et al. 2018). These studies may provide a new evidential direction for the mechanism of albumin toxicity among renal tubular epithelial cells.

24.3.2.2 Proteinuria and Tubular Interstitial Inflammation

Many studies which linked albumin exposure to PTEC dysfunction found the production of inflammatory mediators and tubulointerstitial damage. Vitro experiments have shown that exposure of proximal tubular cells to plasma proteins, such as IgG, transferrin, and albumin, leads to the release of proinflammatory and profibrotic molecules, including C–C motif chemokine ligand 5 (also known as RANTES), osteopontin, monocyte chemoattractant protein-1 (MCP-1), and vasoconstrictor peptide endothelin-1 (Abbate et al. 2006). Urine protein upregulates tubular gene expression and overexpresses various chemokines, recruiting inflammatory cells such as monocytes and T cells to accumulate in the tubulointerstitium causing tubulointerstitial inflammation and releasing interleukins which attract neutrophils aggregation, also, promoting the synthesis of a variety of profibrotic molecules, such as angiotensin II, endothelin, and TGF-β. These substances damage the structure of the renal tubular basement membrane and facilitate the passage of tubular-derived products into the interstitium and peritubular capillary spaces causing tubulointerstitial fibrosis which makes profibrotic molecules an important etiological factor (Gorriz and Martinez-Castelao 2012). In the rat model of proteinuria overload, the expression of RANTES, IL-8, MCP-1, and osteopontin in PTECs was significantly increased and the expression of NF-κB was also up-regulated which was found to be associated with chemotaxis and inflammatory cells infiltration. Thus, the series of alteration were responsible for tubulointerstitial inflammation. After anti-MCP-1 treatment, tubulointerstitial inflammation, fibrosis, and tubular damage were significantly reduced (Eardley et al. 2006; Shimizu et al. 2003). Additionally, chemokine-induced macrophage and lymphocyte infiltration can regulate the synthesis of tubule matrix by synthesizing TGF-β, PDGF, ET-1, and plasminogen activator inhibitor 1 (PAI-1) which lead to tubular epithelial–mesenchymal transition (EMT) which is one of the important mechanisms for renal interstitial fibrosis. The mechanism of EMT depends on three main pathways, including Integrin/ILK, TGF-β/Smad, and Wnt/β catenin (Boor et al. 2010). TGF-β induces Smad phosphorylation to activate Smad pathway that is involved in EMT of tubular cells. However, bone morphogenetic protein-7 (BMP-7) can inhibit TGF-β-induced transdifferentiation process, reversing the progression of fibrosis (Strutz 2009).

24.3.2.3 Proteinuria-Induced Tubular Cell Oxidative Stress

Excessive production of superoxide plays a crucial role in the pathogenesis of various diseases. The reactive oxygen species (ROS) served as the second messenger is a highly cytotoxic cytokine that directly induces tubulointerstitial damage. Some studies also confirmed the evidence of a new mechanism of tubulointerstitial inflammation by demonstrating that albumin and IgG cause a rapid and sustained generation of hydrogen peroxide (H_2O_2) in human proximal tubular cells which further regulates the activation of NF-κB. Then this activation releases various inflammatory factors such as MCP-1 and IL-8 which act in tubulointerstitial inflammation. However, antioxidants inhibit H_2O_2 release which results in inhibition of NF-κB pathway (Morigi et al. 2002; Souma et al. 2011). Nakajima et al. found that albumin activates STAT signaling pathway in renal tubular epithelial cells, resulting in increased ROS-induced renal tissue damage, and ROS clearance abnormalities can also lead to tubulointerstitial inflammation (Nakajima et al. 2004). In vitro study by Souma et al. in 2011 demonstrated that via activation of a Pyk2 kinase signaling pathway in cultured proximal tubule cells and acidic media significantly promotes oleic acid-bound albumin-induced radical oxygen production. As in vivo, alkalized luminal feeding of mice with protein-overload nephropathy selectively attenuated superoxide-induced DNA damage and proximal tubule injury and alkalinized the urine (Souma et al. 2011). As a class B scavenger receptor, CD36 is distributed in widespread tissue including renal tubular epithelial cells and is mediated in the uptake of oxidized lipoproteins and long chain fatty acids by macrophages. Advanced oxidation protein products (AOPPs) modified human serum albumin is subject to CD36dependent endocytosis by immortalized human PTECs and results in increased production of intracellular ROS that may take place via a mechanism that involves NADPH oxidase and PKC and upregulate TGF-β1 expression causing tubulointerstitial inflammation and fibrosis (Iwao et al. 2008).

24.3.2.4 Proteinuria Causes Inflammasome Activation in Tubule Cells

The inflammasome-forming NOD-like receptor (NLR) genes are key regulators of the innate immune response, which integrate various danger signals into caspase-1-activating platforms leading to the processing and secretion of the proinflammatory cytokines IL-1β and IL-18. Recent evidence suggests a crucial role for NLRP3, the most understood inflammasome and IL-1β/IL-18 in the pathogenesis of acute and chronic inflammation and tissue remodeling in the kidney (Anders and Muruve 2011; Chang et al. 2014). Various in vitro and in vivo studies are present which show that proteinuria could cause inflammasome activation in the proximal tubules (Liu et al. 2014; Nishi et al. 2013). This association was further supported by in vitro data showing that bovine serum albumin was induced in a human tubule cell line caspase-1 activation and maturation of IL-1β and IL-18 in a time- and dose-dependent manner. Recent data showed albumin overload resulted in NLRP3 inflammasome activation mediated by cathepsin B-related lysosomal rupture which ultimately induced tubulointerstitial inflammation (Liu et al. 2015).

24.3.2.5 Proteinuria Triggers Immune Response

Damaged tubules after protein overload secrete some inflammatory mediators which may represent danger signals that may elicit an immune response directed against self-protein. Actually, dendritic Cells (DCs) are able to promote local immunity via $CD8^+$ T cells that are activated in regional lymph nodes and recruited in the renal interstitium. A study by Macconi et al. demonstrated that proximal tubular cells transport albumin to DCs, where the protein was degraded into N-terminal 24-residue fragment of albumin (ALB1-24) via proteolytic cleavage. This peptide was taken up by DC for further processing by proteasomes into antigen peptides that had the binding motif for MHC Class I and were capable of activating $CD8^+$ T cells. In vivo, DCs from renal lymph nodes loaded with the albumin peptide ALB1-24 activated syngeneic $CD8^+$ T cells in primary culture (Macconi et al. 2009; Zoja et al. 2015). Thus, the above-mentioned studies showed a link among proteinuria, autoimmunity, and renal disease progression.

24.3.2.6 Proteinuria and Renin–Angiotensin System (RAS)

Alvarez et al. gave Lewis rats protein overload and found that not only it leads to proteinuria, but also cause local RAS activation and inflammatory cell infiltration in the kidney, and promote the formation of salt-sensitive hypertension (Alvarez et al. 2002). The renin–angiotensin system was activated by exposure of proximal tubule cells to high albumin doses via a PKC–NADPH oxidase-dependent pathway (Cao et al. 2011). Our in vitro studies also found that albumin induces an increase of ACE expression in HK-2 cells, while ACE2 expression was downregulated and angiotensin II production was increased. Captopril, however, attenuated the expression of ACE but increased expression of ACE2 induced by BSA, confirming that proteinuria-induced activation of intrarenal RAS by upregulation of ACE and downregulation of ACE2 (Liu et al. 2009).

24.3.2.7 Proteinuria Induces Cellular Crosstalk via Exosome

Beyond classic signaling through soluble cytokines and inflammatory mediators, extracellular vesicles were reported to have the potential role to deliver genetic information (e.g., mRNAs and miRNAs) (Pitt et al. 2016). Exosomes from injured tubular cells transfer TGF-β1 mRNA into fibroblasts and induce their proliferation, type I collagen production, a-smooth muscle actin expression, and F-actin-dependent cytoskeletal rearrangements (Borges et al. 2013). New evidence indicated that exosomes, as a messenger which are loaded with lipids, proteins, and RNAs and increasingly produced in the supernatant when TECs were treated with albumin. Transfer of exosomal CCL2 mRNA from tubular epithelial cells to macrophages after exposure of TECs to albumin is a critical mechanism through which albuminuria induces tubulointerstitial inflammation (Lv et al. 2018).

24.3.2.8 Proteinuria and Serum Metabolomic Alterations

Latest study conducted clinical trials with per-protocol measures of 24-hour proteinuria, eGFR and 58 different types of identified metabolites. Untargeted metabolomic profiling demonstrated a strong association of proteinuria in CKD with several serum metabolites (including 4-hydroxychlorthalonil and 1,5-AG) and a metabolic pathway (PEs). These findings focus on the need of future investigations into the pathophysiology of proteinuria, prediction of risks, and the development of novel preventive or treatment measures in CKD (Luo et al. 2019).

24.4 Reducing Proteinuria Retards Renal Disease Progression

To reverse intraglomerular hypertension, protein restriction or antihypertensive therapy has been beneficial by diminishing hemodynamic injury to the glomeruli and by reducing protein filtration, thus, lowering proteinuria. Some clinical trials have consistently shown the renoprotective effect of proteinuria reduction (Remuzzi et al. 2006). With no doubt, the inhibition of the renin–angiotensin–aldosterone system (RAAS) has been shown to be associated with a maximum reduction in proteinuria and long-term renal risk reduction as well as renoprotection. Numerous clinical and experimental studies have discussed the effect of either an angiotensin-converting enzyme inhibitor (ACEI) or angiotensin II receptor blocker (ARB) as a single therapy or in a dual blockade of RAAS to reduce or control proteinuria (Lewis et al. 2001; Ruggenenti et al. 1999). In the REIN study, patients were randomly assigned to receive ramipril or other conventional antihypertensive therapy to maintain diastolic blood pressure at 90 mmHg or below. It was assessed the hypothesis that through a specific antiproteinuric effect, ACEI could be superior to other antihypertensive drugs in limiting the GFR decline and preventing ESRD in patients with chronic nephropathies. Patients who had more proteinuria to start with ACEI therapy benefited more than those who had less proteinuria. Of note, independent of the initial level of proteinuria, the extent of short-term proteinuria reduction significantly correlated with the reduction in the rate of GFR decline and progression to ESRD than in the long-term proteinuria (Ruggenenti et al. 2003). Two large trials with overt nephropathy showed that proteinuria reduction by ARB treatment was associated with a lower incidence of serum creatinine increase and risk of progression to ESRD (Keane et al. 2006; Atkins et al. 2005). The latest study found that the median duration of CKD stage 3a to deteriorate was approximately 8.2 years shorter, and CKD stage 3b was 5.6 years shorter in those with ≥ 1 g/g (urine protein-to-creatinine ratio) of proteinuria compared with those with <1 g/g of proteinuria (Ku et al. 2018).

24.5 Summary

CKD is an important high-risk factor for many chronic noncommunicable diseases and cancer. Proteinuria is a marker of renal disease and an independent risk factor for CKD progression. There are variety of mechanisms involved in renal inflammation and fibrosis formation and progression, but glomerular damage leading to proteinuria may form a critical loopback effect between proteinuria and tubulointerstitial damage (Fig. 24.1). If proteinuria persistent, it causes tubular cell damage, leading to tubulointerstitial inflammation and fibrosis, and ultimately renal function declines and progresses from CKD to ESRD. Therefore, further elucidation of the mechanism of proteinuria has great significance for the early prevention and treatment of CKD as it leads to tubulointerstitial inflammation and fibrosis formation.

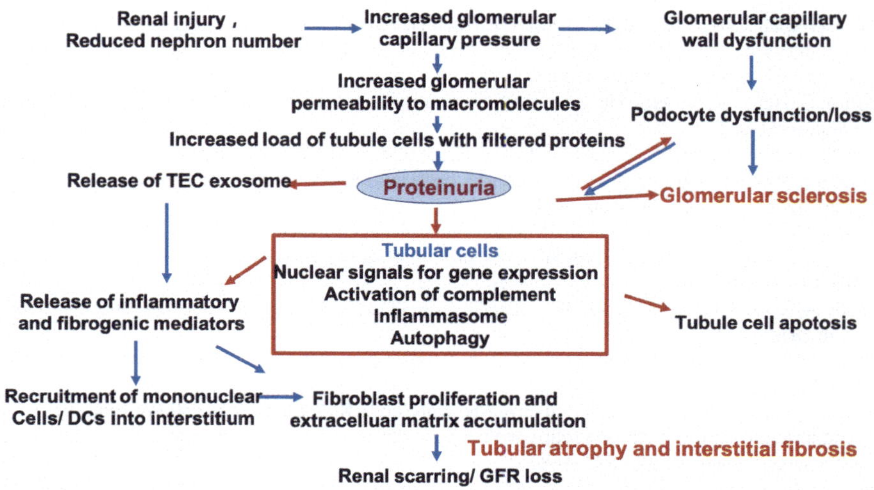

Fig. 24.1 Schematic representation of events underlying progressive glomerular and tubulointerstitial injury of proteinuric nephropathies

Acknowledgements This study was supported by grants from the National Key Research and Development Program of China (2018YFC1314004), the National Natural Science Foundation of China (No. 81720108007, 81670696, 81470922, 81600513 and 31671194), the Clinic Research Center of Jiangsu Province (No. BL2014080), and the Postgraduate Research and Practice Innovation Program of Jiangsu Province (No. KYCX18_0171).

References

Abbate M, Zoja C, Remuzzi G (2006) How does proteinuria cause progressive renal damage? J Am Soc Nephrol 17:2974–2984

Alvarez V, Quiroz Y, Nava M, Pons H, Rodriguez-Iturbe B (2002) Overload proteinuria is followed by salt-sensitive hypertension caused by renal infiltration of immune cells. Am J Physiol Renal Physiol 283:F1132–F1141

Amsellem S, Gburek J, Hamard G, Nielsen R, Willnow TE, Devuyst O et al (2010) Cubilin is essential for albumin reabsorption in the renal proximal tubule. J Am Soc Nephrol 21:1859–1867

Anders HJ, Muruve DA (2011) The inflammasomes in kidney disease. J Am Soc Nephrol 22:1007–1018

Atkins RC, Briganti EM, Lewis JB, Hunsicker LG, Braden G, Champion de Crespigny PJ et al (2005) Proteinuria reduction and progression to renal failure in patients with type 2 diabetes mellitus and overt nephropathy. Am J Kidney Dis 45:281–287

Baines RJ, Brunskill NJ (2011) Tubular toxicity of proteinuria. Nat Rev Nephrol 7:177–180

Biemesderfer D (2006) Regulated intramembrane proteolysis of megalin: linking urinary protein and gene regulation in proximal tubule? Kidney Int 69:1717–1721

Boor P, Ostendorf T, Floege J (2010) Renal fibrosis: novel insights into mechanisms and therapeutic targets. Nat Rev Nephrol 6:643–656

Borges FT, Melo SA, Ozdemir BC, Kato N, Revuelta I, Miller CA et al (2013) TGF-beta1-containing exosomes from injured epithelial cells activate fibroblasts to initiate tissue regenerative responses and fibrosis. J Am Soc Nephrol 24:385–392

Burton C, Harris KP (1996) The role of proteinuria in the progression of chronic renal failure. Am J Kidney Dis 27:765–775

Cao W, Zhou QG, Nie J, Wang GB, Liu Y, Zhou ZM et al (2011) Albumin overload activates intrarenal renin-angiotensin system through protein kinase C and NADPH oxidase-dependent pathway. J Hypertens 29:1411–1421

Caruso-Neves C, Pinheiro AA, Cai H, Souza-Menezes J, Guggino WB (2006) PKB and megalin determine the survival or death of renal proximal tubule cells. Proc Natl Acad Sci U S A 103:18810–18815

Chang A, Ko K, Clark MR (2014) The emerging role of the inflammasome in kidney diseases. Curr Opin Nephrol Hypertens 23:204–210

Cravedi P, Ruggenenti P, Remuzzi G (2012) Proteinuria should be used as a surrogate in CKD. Nat Rev Nephrol 8:301–306

de Zeeuw D, Ramjit D, Zhang Z, Ribeiro AB, Kurokawa K, Lash JP et al (2006) Renal risk and renoprotection among ethnic groups with type 2 diabetic nephropathy: a post hoc analysis of RENAAL. Kidney Int 69:1675–1682

Eardley KS, Zehnder D, Quinkler M, Lepenies J, Bates RL, Savage CO et al (2006) The relationship between albuminuria, MCP-1/CCL2, and interstitial macrophages in chronic kidney disease. Kidney Int 69:1189–1197

Fernandez-Fernandez B, Izquierdo MC, Valino-Rivas L, Nastou D, Sanz AB, Ortiz A et al (2018) Albumin Downregulates Klotho in tubular cells. Nephrol Dial Transplant 33:1712–1722

Gorriz JL, Martinez-Castelao A (2012) Proteinuria: detection and role in native renal disease progression. Transplant Rev (Orlando) 26:3–13

Greka A, Mundel P (2012) Cell biology and pathology of podocytes. Annu Rev Physiol 74:299–323

Huxley VH, Williams DA (2000) Role of a glycocalyx on coronary arteriole permeability to proteins: evidence from enzyme treatments. Am J Physiol Heart Circ Physiol 278:H1177–H1185

Iseki K, Iseki C, Ikemiya Y, Fukiyama K (1996) Risk of developing end-stage renal disease in a cohort of mass screening. Kidney Int 49:800–805

Iwao Y, Nakajou K, Nagai R, Kitamura K, Anraku M, Maruyama T et al (2008) CD36 is one of important receptors promoting renal tubular injury by advanced oxidation protein products. Am J Physiol Renal Physiol 295:F1871–F1880

Jefferson JA, Shankland SJ, Pichler RH (2008) Proteinuria in diabetic kidney disease: a mechanistic viewpoint. Kidney Int 74:22–36

Keane WF, Zhang Z, Lyle PA, Cooper ME, de Zeeuw D, Grunfeld JP et al (2006) Risk scores for predicting outcomes in patients with type 2 diabetes and nephropathy: the RENAAL study. Clin J Am Soc Nephrol 1:761–767

Kim JH, Xie J, Hwang KH, Wu YL, Oliver N, Eom M et al (2017) Klotho may ameliorate proteinuria by targeting TRPC6 channels in podocytes. J Am Soc Nephrol 28:140–151

Ku E, Johansen KL, McCulloch CE (2018) Time-centered approach to understanding risk factors for the progression of CKD. Clin J Am Soc Nephrol 13:693–701

Lee D, Gleich K, Fraser SA, Katerelos M, Mount PF, Power DA (2013) Limited capacity of proximal tubular proteolysis in mice with proteinuria. Am J Physiol Renal Physiol 304:F1009–F1019

Lewis EJ, Hunsicker LG, Clarke WR, Berl T, Pohl MA, Lewis JB et al (2001) Renoprotective effect of the angiotensin-receptor antagonist irbesartan in patients with nephropathy due to type 2 diabetes. N Engl J Med 345:851–860

Li X, Pabla N, Wei Q, Dong G, Messing RO, Wang CY et al (2010) PKC-delta promotes renal tubular cell apoptosis associated with proteinuria. J Am Soc Nephrol 21:1115–1124

Liu BC, Gao J, Li Q, Xu LM (2009) Albumin caused the increasing production of angiotensin II due to the dysregulation of ACE/ACE2 expression in HK2 cells. Clin Chim Acta 403:23–30

Liu BC, Tang TT, Lv LL, Lan HY (2018) Renal tubule injury: a driving force toward chronic kidney disease. Kidney Int 93:568–579

Liu D, Wen Y, Tang TT, Lv LL, Tang RN, Liu H et al (2015) Megalin/cubulin-lysosome-mediated albumin reabsorption is involved in the tubular cell activation of NLRP3 inflammasome and tubulointerstitial inflammation. J Biol Chem 290:18018–18028

Liu D, Xu M, Ding LH, Lv LL, Liu H, Ma KL et al (2014) Activation of the Nlrp3 inflammasome by mitochondrial reactive oxygen species: a novel mechanism of albumin-induced tubulointerstitial inflammation. Int J Biochem Cell Biol 57:7–19

Luo S, Coresh J, Tin A, Rebholz CM, Appel LJ, Chen J et al (2019) Serum metabolomic alterations associated with proteinuria in CKD. Clin J Am Soc Nephrol 14:342–353

Lv LL, Feng Y, Wen Y, Wu WJ, Ni HF, Li ZL et al (2018) Exosomal CCL2 from tubular epithelial cells is critical for albumin-induced tubulointerstitial inflammation. J Am Soc Nephrol 29:919–935

Macconi D, Chiabrando C, Schiarea S, Aiello S, Cassis L, Gagliardini E et al (2009) Proteasomal processing of albumin by renal dendritic cells generates antigenic peptides. J Am Soc Nephrol 20:123–130

Morigi M, Macconi D, Zoja C, Donadelli R, Buelli S, Zanchi C et al (2002) Protein overload-induced NF-kappaB activation in proximal tubular cells requires H(2)O(2) through a PKC-dependent pathway. J Am Soc Nephrol 13:1179–1189

Motoyoshi Y, Matsusaka T, Saito A, Pastan I, Willnow TE, Mizutani S et al (2008) Megalin contributes to the early injury of proximal tubule cells during nonselective proteinuria. Kidney Int 74:1262–1269

Nakajima H, Takenaka M, Kaimori JY, Hamano T, Iwatani H, Sugaya T et al (2004) Activation of the signal transducer and activator of transcription signaling pathway in renal proximal tubular cells by albumin. J Am Soc Nephrol 15:276–285

Nijenhuis T, Sloan AJ, Hoenderop JG, Flesche J, van Goor H, Kistler AD et al (2011) Angiotensin II contributes to podocyte injury by increasing TRPC6 expression via an NFAT-mediated positive feedback signaling pathway. Am J Pathol 179:1719–1732

Nishi Y, Satoh M, Nagasu H, Kadoya H, Ihoriya C, Kidokoro K et al (2013) Selective estrogen receptor modulation attenuates proteinuria-induced renal tubular damage by modulating mitochondrial oxidative status. Kidney Int 83:662–673

Peired A, Angelotti ML, Ronconi E, la Marca G, Mazzinghi B, Sisti A et al (2013) Proteinuria impairs podocyte regeneration by sequestering retinoic acid. J Am Soc Nephrol 24:1756–1768

Pitt JM, Kroemer G, Zitvogel L (2016) Extracellular vesicles: masters of intercellular communication and potential clinical interventions. J Clin Invest 126:1139–1143

Remuzzi G, Benigni A, Remuzzi A (2006) Mechanisms of progression and regression of renal lesions of chronic nephropathies and diabetes. J Clin Invest 116:288–296

Ruggenenti P, Perna A, Gherardi G, Garini G, Zoccali C, Salvadori M et al (1999) Renoprotective properties of ACE-inhibition in non-diabetic nephropathies with non-nephrotic proteinuria. Lancet 354:359–364

Ruggenenti P, Perna A, Mosconi L, Matalone M, Pisoni R, Gaspari F et al (1997) Proteinuria predicts end-stage renal failure in non-diabetic chronic nephropathies. The "Gruppo Italiano di Studi Epidemiologici in Nefrologia" (GISEN). Kidney Int 63(Suppl):S54–S57

Ruggenenti P, Perna A, Mosconi L, Pisoni R, Remuzzi G (1998) Urinary protein excretion rate is the best independent predictor of ESRF in non-diabetic proteinuric chronic nephropathies. "Gruppo Italiano di Studi Epidemiologici in Nefrologia" (GISEN). Kidney Int 53:1209–1216

Ruggenenti P, Perna A, Remuzzi G, Investigators GG (2003) Retarding progression of chronic renal disease: the neglected issue of residual proteinuria. Kidney Int 63:2254–2261

Shimizu H, Maruyama S, Yuzawa Y, Kato T, Miki Y, Suzuki S et al (2003) Anti-monocyte chemoattractant protein-1 gene therapy attenuates renal injury induced by protein-overload proteinuria. J Am Soc Nephrol 14:1496–1505

Souma T, Abe M, Moriguchi T, Takai J, Yanagisawa-Miyazawa N, Shibata E et al (2011) Luminal alkalinization attenuates proteinuria-induced oxidative damage in proximal tubular cells. J Am Soc Nephrol 22:635–648

Storm T, Tranebjaerg L, Frykholm C, Birn H, Verroust PJ, Neveus T et al (2013) Renal phenotypic investigations of megalin-deficient patients: novel insights into tubular proteinuria and albumin filtration. Nephrol Dial Transplant 28:585–591

Strutz FM (2009) EMT and proteinuria as progression factors. Kidney Int 75:475–481

Thrailkill KM, Nimmo T, Bunn RC, Cockrell GE, Moreau CS, Mackintosh S et al (2009) Microalbuminuria in type 1 diabetes is associated with enhanced excretion of the endocytic multiligand receptors megalin and cubilin. Diab Care 32:1266–1268

Toblli JE, Bevione P, Di Gennaro F, Madalena L, Cao G, Angerosa M (2012) Understanding the mechanisms of proteinuria: therapeutic implications. Int J Nephrol 2012:546039

Tojo A, Onozato ML, Kitiyakara C, Kinugasa S, Fukuda S, Sakai T et al (2008) Glomerular albumin filtration through podocyte cell body in puromycin aminonucleoside nephrotic rat. Med Mol Morphol 41:92–98

Webster AC, Nagler EV, Morton RL, Masson P (2017) Chronic kidney disease. Lancet 389:1238–1252

Weyer K, Storm T, Shan J, Vainio S, Kozyraki R, Verroust PJ et al (2011) Mouse model of proximal tubule endocytic dysfunction. Nephrol Dial Transplant 26:3446–3451

Zhang L, Wang F, Wang L, Wang W, Liu B, Liu J et al (2012) Prevalence of chronic kidney disease in China: a cross-sectional survey. Lancet 379:815–822

Zoja C, Abbate M, Remuzzi G (2015) Progression of renal injury toward interstitial inflammation and glomerular sclerosis is dependent on abnormal protein filtration. Nephrol Dial Transplant 30:706–712

Zou Z, Chung B, Nguyen T, Mentone S, Thomson B, Biemesderfer D (2004) Linking receptor-mediated endocytosis and cell signaling: evidence for regulated intramembrane proteolysis of megalin in proximal tubule. J Biol Chem 279:34302–34310

Chapter 25
Mitochondria and Renal Fibrosis

Mohammed Mazheruddin Quadri, Syeda-Safoorah Fatima, Ruo-Chen Che and Ai-Hua Zhang

Abstract Mitochondria are important organelles in eukaryotic cells and perform a variety of biosynthetic and metabolic functions. Many human diseases are closely related to mitochondrial dysfunction. Kidney is an organ with high-energy requirements, which is distributed with a large number of mitochondria. Mitochondrial dysfunction plays a crucial role in the pathogenesis of kidney disease, and studies have shown that mitochondrial dysfunction is involved in the physiological process of renal fibrosis. This review introduced the biogenesis and pathophysiology of mitochondria, illustrated the involvement of mitochondrial dysfunction in renal fibrosis based on various kinds of cells, and finally summarized the latest mitochondria-targeted therapies.

Keywords Mitochondria · Renal fibrosis · Therapies

25.1 Basic Information of Mitochondria

25.1.1 Mitochondrial Structure

Mitochondria are ubiquitous and dynamic cellular organelles with many fundamental roles in maintaining normal cellular function and mainly responsible for production of adenosine triphosphate (ATP), which is oxygen dependent, utilized as energy currency in many cellular processes. Over two billion years ago, mitochondria originated via endosymbiotic effect in which a prokaryotic cell related to alpha-proteobacteria were engulfed by a eukaryotic predecessor cell. In this process of phylogenesis, the endosymbiont was merged into the cellular network (Becker and Wagner 2018). Due to this reason, several mitochondrial proteins also have bacterial origin. Similar to bacteria, each human mitochondrion is a double membrane which exists in

Mohammed Mazheruddin Quadri and Syeda-Safoorah Fatima contributed equally to this work.

M. M. Quadri · S.-S. Fatima · R.-C. Che · A.-H. Zhang (✉)
Department of Nephrology, Children's Hospital of Nanjing Medical University, Nanjing, China
e-mail: zhaihua@njmu.edu.cn

B.-C. Liu et al. (eds.), *Renal Fibrosis: Mechanisms and Therapies*,
Advances in Experimental Medicine and Biology 1165,
https://doi.org/10.1007/978-981-13-8871-2_25

most eukaryotic cells except for mature erythrocytes which is bound by an outer mitochondrial membrane (OMM) and an inner mitochondrial membrane (IMM). Between the OMM and IMM, there is an intermembrane space. IMM is folded to form cristae that envelopes the matrix. The OMM has pores that allow the exchange of metabolites between the IMM and cytosol, molecules smaller than 5000 Da pass through passive diffusion, and larger molecules pass through the mitochondrion via translocases on the OMM. OMM also seals the mitochondria from releasing harmful agents into the cytosol such as ROS and mitochondrial DNA (mtDNA). When the cell incurs an irreversible injury, the OMM permeability expands and proteins' molecules present inside in the intermembrane space like cytochrome c move out and begins the process of programmed cell death. There are current studies that indicate that the OMM channels and their transportation process are more varied than considered earlier. Transportation of well-defined sets of precursor proteins' crossways and into the OMM is assisted by four protein-conducting channels. For small hydrophilic particles, the main transportation is voltage-dependent anion channel (VDAC). Additionally, there are three more channels whose substrate specificity which is yet to be known is present in the OMM. The IMM is comprised of well-defined structural regions which include the membrane lining, the junctions of cristae, and the cristae. Cristae are IMM foldings which substantially increases the surface area; also, the electron transport chain (ETC) proteins are situated here. The mitofilins which are accumulated between the IMM and OMM constitute in the cristae morphology. The ample number of proteins in the IMM carries out a lot of functions such as redox reactions, ATP production, blocks transportation of ions, and governs the mitochondrial dynamics. The IMM also wraps the matrix in which the oxidative phosphorylation (OXPHOS) enzyme and genetic material of mitochondria exist (Che et al. 2014).

25.1.2 Mitochondria Genome

Almost in all aspects, the mtDNA differs substantially from the nuclear genome, although a majority of mitochondrial proteins are encoded by the nuclear genome and imported into mitochondria. As opposed to nuclear genome (nDNA), mtDNA of humans is a circular structure made up of double strands which are 16.6 kb long heavy and light strands consisting of purines and pyrimidines. In total, there are 37 genes in mtDNA and there are short non-coding sequences and also one large non-coding region which has triple strands known as D-loop, consisting of transcriptional promoters and minimum of one origins' replication. mtDNA is formed as nucleoid by TFAM (mitochondrial transcription factor A), which condenses the DNA by bending and then wrapping it into compact nucleoid structures. Several different kinds of replications occur in mammalian cells as stated by Holt et al. In one of the model called as "bootlace," when the leading DNA strand is being replicated preformed RNA fragments hybridize the lagging DNA strand. The process of replication initiates at the site of genes cytb, nad5, and nad6 and continues bidirectionally till the D-loop region. Until now, several mitochondrial polymerases have been mentioned other than DNA PolG, which are PrimPol, PolQ, PolB, PolH, and PolZ, but their functions are still unknown (Volobueva et al. 2018). There are several genes involved in the initiation (POLRMT, TFAM, TFB2B) and elongation (TEFM, MTERF1) in the transcription of mtDNA; however, only mutations in TFAM have been shown to cause human diseases to date (Stiles et al. 2016).

A lot of factors in mtDNA maintenance and OXPHOS components (complexes I–IV), and the ATP synthase (complex V) are encrypted within the nucleus. The mtDNA encodes 13 polypeptides, two ribosomal RNAs (mt-rRNAs), and 22 transfer RNAs (mt-tRNAs). Furthermore, these polypeptides consists of seven subunits (ND1–ND6 and ND4L) of complex I (NADH dehydrogenase), three subunits [cyclooxygenase (COX) I–III] of complex IV (cytochrome c oxidase), cytochrome b (Cyt b) of complex III (ubiquinol cytochrome c oxidoreductase), and two subunits (ATPase 6 and ATPase 8) of complex V (ATP synthase). There are multiple copies of mtDNA within each human cell and depending on the cell type the total amount varies. Inheritance of mtDNA is strictly maternal.

The electron transport chain is a significant source of reactive oxygen species (ROS), in particular by complex I and complex III. Excessive ROS might damage nucleic acids and have a role in the pathogenesis of mitochondrial diseases. Increasing number of mutations in the nuclear genes that encode mitochondrial proteins has been shown to cause mitochondrial diseases. Other causes of defective mitochondrial function caused by nDNA mutations include apoptosis, mitochondrial chaperones, and mitochondrial metabolism (Gorman et al. 2016).

25.1.3 Mitochondrial Energy Production

More than 90% of energy production in the human body takes place in mitochondria. Oxygen in the cells is utilized by mitochondria to produce energy in the form of ATP through a process called as oxidative phosphorylation (OXPHOS). The enzymes for the metabolic paths such as fatty acid beta-oxidation, tricarboxylic acid (TCA) cycle, part of the urea cycle are present in the mitochondrial matrix. Electron transport chain complexes I–IV of the respiratory chain use the reducing potential of NADH and FADH2 which is generated during the process of glycolysis, TCA cycle, and fatty acid oxidation to pump protons in the mitochondrial intermembrane space, hence generating an electrochemical gradient which is then utilized by complex V to assist the phosphorylation of ADP to ATP. The oxidative phosphorylation electron transport chain (OXPHOS-ETC) complexes are arranged into supercomplexes (SCs) of defined ratio: Respirasome (SC I + III2 + IV), a supercomplex formed by CI in combination with CIII2 and CIV, also CI combines with CIII2 alone (SC I + III2). A super complex (SC III2 + IV) is formed by the combination of CIII2 and CIV. CV forms a dimer CV2. Recent cryo-electron microscopy studies have shown the structures of SC I + III2 and SC I + III2 + IV. Evidence for the existence and importance of respiratory SCs shows that (i) isolated respirasomes are capable of transferring electrons from NADH to O2, (ii) CIII2 and CIV affect the stability of CI in the inner mitochondrial membrane, (iii) SCs decrease the number of reactive oxygen species (ROS) formed during electron transport, and (iv) also there are differing opinions that SC improvises the transfer efficiency of the electron carriers cyt c27 and Q and may also trap a small amount of Q, hence dividing the membrane pool into two distinct functional groups. It has been suggested that SC formation is important for the stability of the individual OXPHOS-ETC complexes. There is evidence to suggest that the formation of SCs may be adaptive to an increased energy demand in the cell (Letts and Sazanov 2017). In recent studies of biomarkers for mitochondrial energy metabolism diseases, serum fibroblast growth factor 21 (sFGF21) and serum growth differentiation factor 15 (sGDF15) appear to be promising molecules (Boenzi and Diodato 2018).

25.1.4 Mitochondrial Dynamics

Mitochondria were viewed as organelles that are isolated and static for a long time, until it was revealed by advances in live cell imaging and genetic screening that mitochondria are highly dynamic. Mitochondria can exhibit a change of their position inside cells and also its architecture continuously by processes of fission and fusion reactions; this is referred as "mitochondrial dynamics" which has become a center of attention recently as they are necessary not only for mitochondrial morphology maintenance, but also for maintaining mtDNA integrity, regulate mtROS levels, calcium homeostasis, and oxidative phosphorylation and participating in metabolic processes.

Mitochondrial fusion involves fusion of both outer mitochondrial membrane (OMM) and inner mitochondrial membrane (IMM). Mitochondrial fusion of the OMM is induced by MitoPLD, a member of the phospholipase D family, which converts cardiolipin into phosphatidic acid, and by homo- or hetero-dimerization of the guanosine triphosphate hydrolases (GTPases) mitofusin (Mfn) 1 and 2. Mitochondrial fusion of IMM requires optic atrophy 1 (OPA1) anchored in the inner membrane and exposed to the intermembrane space. On the other hand, GTPase dynamin-related protein-1(Drp1) is responsible for mitochondrial fission which occurs when Drp1 translocates to the OMM to bind its receptors, such as FIS1, MFF, MID49, and MID51, and forms a multimeric structure around the fission site of the mitochondrion. While the precise reason is not yet known, mitochondrial protein 18 kDa (MTP18) gene ablation induces fusion of mitochondria and expression of cleaved OPA1 (S-OPA1) induces fission. Recently, researchers have discovered that an additional GTPase protein, dynamin-2 (Dyn2/Dnm2), directly coordinates mitochondrial fission with Drp1. Many new indirect mitochondrial regulatory proteins have also revealed which have been implicated in mitochondrial-related diseases. Vacuolar protein sorting-35 (VPS35) is a central protein that indirectly modulates mitochondrial homeostasis through the trafficking of several mitochondrial proteins. It is a subunit of the retromer cargo-selective complex (CSC). The chief function of this complex was to recover mannose 6-phosphate receptor from peripheral endosomes to the Golgi complex in the cells. Studies implicate VPS35 and the retromer complex in mitochondrial partial dysfunction and in the regulation of mitochondrial fusion and fission (Farmer et al. 2018). An increase in mitochondrial fission or decreased fusion process may be harmful to mitochondrial functions and cell survival. For instance, Drp1 and Fis1 are augmented in cyclosporine A and rhabdomyolysis-induced tubular apoptosis, and impediment of these proteins can recover tubular functions (Che et al. 2014).

25.1.5 Mitochondrial Nanotunnels

Mitochondria are intracellular organelles that move, fuse, and divide responding to biochemical stimulus. They also influence a broad spectrum of cellular and physiological roles through their signals. It is now known that mitochondria form a dynamic network of signaling organelles not just being a powerhouse functioning in isolation from one another as thought earlier. They can communicate with each other through nanotunnels which are thin double-membrane projections that involve both the IMM and OMM and connect the matrices of non-adjacent mitochondria. Emerging evidence suggests that mitochondrial nanotunnels are generated by transport proteins and they are in immotile mitochondria; this immotility encourages the formation of membrane projections in different systems. The mitochondrial nanotunnels were first reported about 10 years ago using electron microscopy and confocal imaging of GFP-labeled mitochondria in cultured African green monkey renal cells. The evidence of mitochondrial nanotunnels comes from imaging studies in mammalian cells, other nanotunnel-like structures which exchange molecular information between bacteria, the mitochondrial ancestor; and recent literature on specific cell projections that enable distant communication in mammalian cells. Because nanotunnels have only recently been observed, several important questions still remain unanswered (Vincent et al. 2017).

25.2 The Pathophysiology of Mitochondria

25.2.1 Mitochondrial Oxidative Stress

Molecules that are derived from oxygen and which can readily oxidize other molecules are known as ROS. At low levels, ROS have a physiological function as critical signaling molecules in which they take part in evoking proliferation and survival in response to stressful situations but at increased levels, mitochondrial ROS could be harmful causing cellular damage (Galvan et al. 2017). Mitochondria, apart from being a powerhouse, are also cellular sites for ROS production while hosting the processes of oxidative phosphorylation. Free radicals like the superoxide anion (O_2^-) may be generated in the respiratory chain when electrons derived from NADH or succinate oxidation may leak and react with O_2 or other electron acceptors. Superoxide anion (O_2^-) can be further reduced to the hydroxyl radical (OH^-) and hydrogen peroxide (H_2O_2). Major sites for ROS production are predicted to be respiratory chain complexes I and III. ROS can harm lipids, proteins which include nucleic acids (DNA, RNA). There exist various defense mechanisms against ROS within the cell. At first, superoxide is converted to H_2O_2 by two intracellular superoxide dismutases (SODs): Cu–Zn SOD (SOD1) and MnSOD (SOD2), and H_2O_2 can be changed into H_2O by an enzyme catalase or glutathione peroxidase. A lot of non-enzymatic cellular ROS scavengers are reported, such as carotenoids, ascorbate, glutathione, and flavonoids. Hence, oxidative damage is caused by oxidative stress resulting from an imbalance between ROS generation and disposal (Lagouge and Larsson 2013). As mentioned above, lesser levels of ROS encourage proliferation and higher levels of ROS lead to apoptosis. It was stated that a "ROS window" is needed for normal or cancerous cellular functions: Cell death is activated above this level and proliferation is blocked below it (Panieri and Santoro 2016). Ample evidence based on experimental models of kidney injury and patients has suggested that in failing kidneys generation of ROS is significantly increased. Consolidative proposals suggest that excessive production of mitochondrial ROS is linked to mitochondrial dysfunction causing cellular damage and progression of renal disease.

25.2.2 Mitochondrial Autophagy

Autophagy is defined as a natural regulated lysosome-dependent intracellular degradative process that recycles cytoplasmic components into bioenergetic and biosynthetic materials for maintenance of cellular homeostasis. As the function of autophagy is critically important in several stressful conditions, disruption of autophagy can result in abnormal cell function and diseases. Mitophagy is a type of autophagy that targets mitochondrial degradation by removing damaged mitochondria and recycling and reallocating useful components. Recent advances in mitophagy researches are characterization of Parkin–phosphoubiquitin complex (Kumar et al. 2017) discovery and study of molecular structure of PINK–ubiquitin complex (Schubert et al. 2017) as well as advancement in learning that the inner mitochondrial membrane protein prohibitin 2 serves as mitophagy receptor (Wei et al. 2017). Classification of mitophagy regulatory pathways is based on ubiquitin-dependent or independent. In the PINK1–Parkin-mediated pathway of mitophagy, after stress, Parkin recruitment is promoted when PINK1 is stabilized on the OMM after which Parkin ubiquitinates several outer membrane components and leads to phosphorylation of Poly-Ub (ubiquitin) by PINK1 which serves as an "eat me" signal for the autophagic machinery. Phosphorylated poly-Ub chains on mitochondrial proteins are recognized by adaptor proteins like p62, optineurin (OPTN), nuclear dot protein 52 (NDP52) and initiate autophagosome formation by binding with autophagosomal light chain 3 (LC3). OPTN is phosphorylated by the serine/threonine-protein kinase TBK1 (TANK-binding kinase 1), enhancing its binding affinity to Ub chains. A feed-forward mechanism promoting mitochondrial clearance is established by the OPTN–TBK1 complex. Gp78 (Glycoprotein 78), SMURF1 (SMAD-specific ubiquitin protein ligase 1), MUL1 (mitochondrial E3 ubiquitin protein ligase 1), SIAH1 (synphilin-1 recruited seven in absentia homolog), and ARIH1 (Ariadne RBR E3 ubiquitin protein ligase 1) represent as several other E3 ubiquitin ligases targeting OMM proteins prior to mitophagy. Mitochondrial dynamics and motility are regulated by the PINK1–Parkin pathway for proteasomal degradation targeting mitofusins (MFN) and Miro (outer mitochondrial membrane protein). In receptor-mediated mitophagy, several receptors like NIX (NIP3-like protein X), BNIP3 (BCL2 interacting protein 3), and FUNDC1 (FUN14 domain-containing protein 1) which are located on OMM mitophagy receptors are localized to the OMM and interacts with LC3 to facilitate mitochondrial elimination. Following mitochondrial dysfunction PHB2 (prohibitin 2) and cardiolipin which are localized externally to OMM interacts with LC3 impairment. The specificity of the process in different tissues is ensured by different receptors and followed by diverse stimuli. Their association with LC3 is increased by NIX and BNIP3 phosphorylation. The regulation of mitochondrial dynamic during hypoxia which is mainly modulated by FUNDC1 phosphorylation happens under the influence of both CK2 (casein kinase 2) and Sc kinases and PGAM5 (phosphoglycerate mutase family member 5) phosphatase. Through the breakup and release of OPA1 and the recruitment of DRP1 (dynamin-related protein 1) on the mitochondrial surface, these mitophagy receptors encourage fission of damaged organelles (Palikaras et al. 2018).

Numerous studies have shown that in the animal model of diabetic nephropathy, a large amount of mitochondrial debris accumulation were observed in the renal tissue, suggesting that the clearance process of abnormal mitochondria might be defective in diabetic nephropathy. Calpain10 is a mitochondrial and cytosolic Ca^{2+}-regulated cysteine protease. In STZ-induced early diabetic (four-week) rat models, calpain10 protein and mRNA levels are reduced after STZ injection, and mitochondrial autophagy regulator PINK1 was upregulated. After transfection with calpain10 siRNA, mitochondrial fusion was reduced, mitochondrial division and mitochondrial autophagy were increased, indicating that mitochondrial autophagy activity may be elevated in early diabetic stage, and calpain10 can negatively regulate mitochondria autophagy. However, the insulin-injected diabetic rats did not show upregulation of PINK, indicating that mitochondrial autophagy is induced by a high glucose environment. However, the expression level of PINK1 protein in renal tubular epithelial cells of diabetic nephropathy mice was significantly down-regulated, and the mitochondrial fragment content increased. The researchers speculated that this may be the disorder of abnormal mitochondria clearance. Some scholars speculate that in the early stage of diabetes, to remove abnormal mitochondria, the mitochondrial autophagy activity is compensatory increased, but with the progress of diabetic nephropathy, abnormal mitochondrial content exceeds compensatory capacity. Mitochondrial autophagy activity decreases, and damage mitochondria cannot be removed in time, eventually leading to cell death (Deng et al. 2017).

25.2.3 Mitochondria and Inflammasome

The inflammatory response plays an important role in chronic kidney disease. In recent years, it has been found that NLRP3 inflammatory bodies regulate the secretion of IL-1β and IL-18, which are the core of the inflammatory response. The NLRP3 inflammatory corpuscle is a polyprotein polymer complex that has been found to be composed of the NLRP3 backbone, apoptosis-associated spot-like protein (ASC), and caspase-1 (caspase-1). NLRP3, also known as cryopyrin, PYPAF1 or Nalp3, is encoded by the cold-induced autoinflammatory syndrome (CIAS)-1 gene. It is a member of the NLR family and belongs to pathogen-associated receptors (PRRs). It consists of an N-terminal thermoprotein-like domain (PYD), a C-terminal leucine repeat and an intermediate NACHT domain. ASC, also known as TMS1, encoded by the PYCARD gene, is a connexin consisting of the N-terminal PYD domain and the C-terminal caspase recruitment domain CARD (caspase recruitment domain). After NLRP3 is activated, oligos are formed by the interaction between homologous molecules, and the PYD domain can be linked to the PYD domain of ASC. Then, the CARD of ASC is linked to the CARD of caspase-1, which forms an inflammatory body that activates caspase-1. The activated caspase-1 promotes the maturation of IL-1β and IL-18 by cleaving pro-IL-1β and pro-IL-18. These mature interleukins participate in the inflammatory response and the body's innate immune response. Studies have shown that mitochondria dysfunction activates NLRP3 inflammatory bodies by releasing ROS or mtDNA, while mitochondrial autophagy inhibits NLRP3 inflammatory bodies, and mitochondria regulate NLRP3 inflammatory body activation through self-change (Zhuang et al. 2014, 2015).

25.2.4 Mitochondria and Endoplasmic Reticulum Stress

As the factory for the folding and trafficking of proteins, endoplasmic reticulum (ER) is highly sensitive to injury. Stimuli beyond control could lead to aggregation of misfolded proteins in the endoplasmic reticulum that immensely alters several cellular signaling processes, which includes energy production, reduction–oxidation (redox) homeostasis, inflammation, differentiation, and programmed cell death. Varied cellular stressors like changes in calcium homeostasis, redox imbalance, impaired protein glycosylation, or faulty protein folding results in misfolded or unfolded proteins to accumulate in the ER lumen, a condition called as ER stress (ERS). ER–mitochondria contact sites are involved in autophagosome biogenesis which is the main organelle of the autophagy degradation pathway. The ER–mitochondria contact sites have been shown to be a place for the recruitment of the ATG (autophagy-related, such as ATG1, ATG5, ATG8, ATG12, ATG16L1) machinery and autophagosome assembly (Molino et al. 2017). ER membrane-derived structures that form physical contact sites between the ER and mitochondria are called mitochondria-associated membranes (MAMs); they control the chief functions of these two organelles in forming

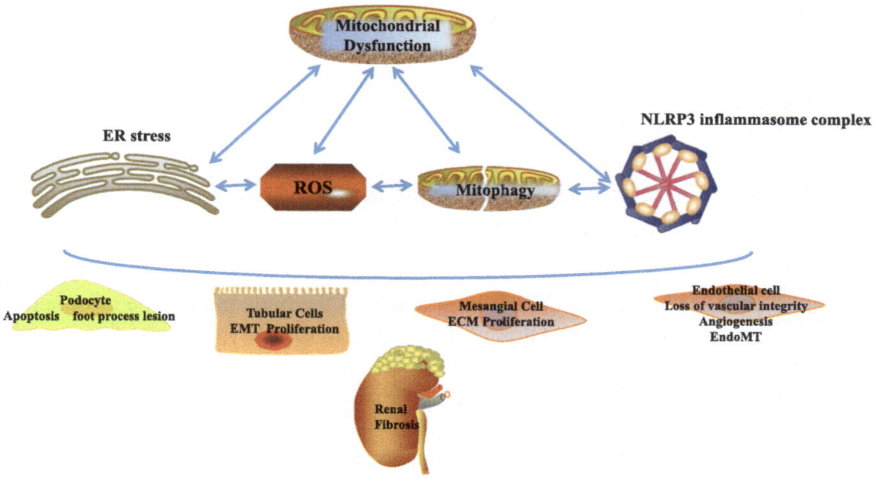

Fig. 25.1 Interaction between mitochondria dysfunction, endoplasmic reticulum stress, reactive oxygen species, mitophagy, and inflammasome leads to various cell lesions in kidney, which ultimately induces renal fibrosis.ER: endoplasmic reticulum, ROS: reactive oxygen species, ECM: extracellular matrix, EMT: epithelial to mesenchymal transition, EndoMT: endothelial-to-mesenchymal transition

a unique subcompartment at the junction between ER and mitochondria. MAMs are crucial for lipid biosynthesis in normal cellular conditions, ER–mitochondria lipid or Ca^{2+} transfer, mitochondrial dynamics and bioenergetics, apoptosis, or autophagy. MAMs are now known to be the popular sites for the transmission of stress signals from the endoplasmic reticulum to mitochondria, specifically during the of loss of ER proteostasis, by involving the unfolded protein response (Van Vliet and Agostinis 2018). In MCD patients, ER stress in parallel with mitochondrial dysfunction was found to trigger tubule-interstitial fibrosis and inflammation linked to heavy proteinuria. ER stress triggered a sterile inflammatory cascade, activating NLRP3 and caspase 1 in proteinuric patients. Therefore, it can be concluded that the crosstalk among ROS, mitochondrial dysfunction, ER stress, and inflammasome is one of the major culprits for renal fibrosis (Fig. 25.1).

25.3 Mitochondria and Renal Fibrosis

Renal fibrosis, including glomerular and tubular interstitial fibrosis, is the common pathologic pathway that various chronic kidney diseases (CKD) progressing to end-stage renal diseases. The underlying mechanism remains unclear. Early in 2000, French experts Doleris LM first time reported four cases of mitochondrial cytopathies suffering from focal segmental glomerulosclerosis. Renal fibrosis induced by mitochondrial dysfunction has started drew much attention. Recent studies show that in peripheral blood monocytes of ESRD patients, ROS production, COX I and IV increase significantly while respiratory complex IV decreases, which indicated that mitochondrial dysfunction closely correlates with progression of CKD.

25.3.1 Mitochondrial Dysfunction-Induced Mesangial Cell Proliferation and Extracellular Matrix (ECM) Deposit

Mesangial cells are located in the mesangial area of the bulb and are the most active reactive cells in the glomerulus. Glomerular disease induced by inflammation is often accompanied by significant mesangial cell proliferation in the early stage. Abnormal proliferation of mesangial cells, secondary release of inflammatory mediator and ECM accumulation leads to glomerular sclerosis, which is one of the critical risk factors progressing to ESRD. Oxidative damage plays an important role in mesangial cell proliferation and ECM aggregation. The main sources of ROS in mesangial cells are mitochondria (electron leaking from the respiratory chain), NADPH oxidase, xanthine oxidase, cyclooxygenase, lipoxygenase, cytochrome P450 oxidase, and nitric oxide synthase. In a high glucose environment, the mitochondrial respiratory chain of the human mesangial cell line produces excessive ROS, accompanied by a decrease in MnSOD activity, mtDNA copy number, mitochondrial membrane potential, and ATP production (Xu et al. 2012). ROS accumulation activates nuclear factor-κB (NF-κB) and activated protein-1 (AP-1), upregulate TGF-β1 expression, which promotes inflammation, ECM secretion, and subsequently glomerular sclerosis. Recently, it has also been found that aldosterone can also induce an increase in mitochondrial ROS production in mesangial cells. Mitochondria-derived ROS can promote mesangial cell proliferation by activating Ras/extracellular signal-regulated kinase (ERK1/2) signaling pathway. Mitochondrial Cyt bc1 complex inhibitor stigmatellin (Akool et al. 2012) and the respiratory chain complex I inhibitor rotenone (Yuan et al. 2012) inhibit mesangial cells proliferation by reducing mitochondrial-derived ROS production and blocking ERK1/2 signaling pathway activation. SIRT1 (silent information regulator 2 associated enzyme-1) is a NAD^+-dependent histone deacetylase that activates PGC-1α and regulates mitochondrial function by promoting deacetylation of PGC-1α. The SIRT1 agonist resveratrol blocks high glucose-induced mitochondrial lesions and inhibits mesangial cell proliferation by activating SIRT1.

25.3.2 Mitochondrial Dysfunction-Mediated Endothelial Cell Injury

The kidney is rich in blood vessels. The integrity of the glomerular capillary plexus is critical for maintaining glomerular filtration. Glomerular endothelial cells, which are one of the intrinsic cells of the glomerulus, are covered on the side of the glomerular capillary wall and have many small holes. These small holes are of different sizes with negative charge. As the first barrier of glomerular filtration membrane, it maintains the selective filtration functions. Glomerular endothelial cells can also adhere to bacteria and white blood cells, repair the basement membrane, and have anticoagulant and antithrombotic effects. Due to direct contact with blood flow, glomerular endothelial cells stand in the frontier against hemodynamic abnormalities, immune inflammation, excessive ROS, lipid metabolism disorders, and insulin resistance, etc. If the glomerular endothelial cells undergo progressive damage, it will affect the integrity of the filtration membrane. Plasma protein and blood cells leak out. Proliferation or apoptosis after endothelial cell injury leads to microvascular occlusion, glomerular capillary function loss, and eventually glomerular sclerosis. Recently, it is recognized that endothelial cell dropout after acute injury like hypoxia takes part in the transition from acute ischemic kidney injury to renal fibrosis. It has shown that mitochondrial swollen and loss of cristae membranes in endothelial cells after 45-min ischemia in the rat led to loss of vascular integrity. Measures preventing mitochondria from damage could effectively reduce the loss of peritubular capillaries and cortical arterioles (Liu et al. 2014). In addition, endothelial cell injury is an important cause of microalbuminuria in early diabetic nephropathy. It was found that high glucose can induce mitochondrial lesions in glomerular capillary endothelial cells, which are characterized by a significant increase in mitochondrial superoxide anion production and mitochondrial membrane permeability transition pores open with decreased membrane potential, respiratory chain enzyme complex I deactivation, and RCR reduction. 5/6 nephrectomy rat model is a classic model of chronic progressive renal injury with glomerular sclerosis and interstitial fibrosis as basic pathological changes. In the early stage of the rat model of remnant kidney, glomerular capillary endothelial cells proliferated, with the length and density of glomerular capillary plexus increased, and there was a phenomenon called "angiogenesis." However, with the deterioration of renal function, the proliferation of endothelial cells is weakened with inflammation and apoptosis. Inflammatory cells adhere to endothelial cells, and NF-κB activity in renal tissues is enhanced. Tumor necrosis factor-α (TNF-α) and TGF-β expression were upregulated and finally caused renal fibrosis (Soetikno et al. 2013; Fedulov et al. 2005). Recent studies have found that TNF-α can induce glomerular endothelial cell apoptosis through mitochondrial pathway. After TNF-α stimulation of endothelial cells, mitochondrial membrane permeability transition pores open with membrane permeability increased, inducing cytoplasm enter of cytochrome c, increase of proapoptotic protein Bak expression and decrease of anti-apoptotic protein Bcl-xL (Messmer et al. 1999). In addition, endothelial-to-mesenchymal transition (EndoMT) has recently emerged as a potentially important contributor in promoting fibrosis in chronic kidney disease. Evidence showed that ROS plays a critical role in EndoMT and ROS scavenger to maintain the homeostasis of endothelial function (Lin et al. 2018).

25.3.3 Mitochondrial Dysfunction-Mediated Podocyte Injury

Podocyte, a kind of glomerular visceral epithelial cell, attaches to the outside of the glomerular basement membrane (GBM). Podocyte, together with GBM and capillary endothelial cells, constitute the glomerular hemofiltration barrier. The podocyte cell body protrudes a lot of pseudopods, also known as the foot process (FP), which are finger-crossed over the outer surface of the GBM and are attached to the GBM by adhesion molecules and proteoglycan molecules. The fissure between two adjacent foot processes is called a slit, with a diameter of about 40 nm, and its surface is covered with a zipper-like structure–slit diaphragm (SD), which is the last barrier of plasma protein filtration. The integrity of the ruptured septum is the key to determining the permeability of the glomerular filtration barrier, and the membrane protein on the rupture membrane is closely related to the occurrence of proteinuria. A large body of evidence indicates that the complex formed by nephrin, CD2AP, and podocin will anchor the ruptured membrane to the podocyte actin backbone, which plays an important role in maintaining the normal function of the glomerular filtration barrier. In 2000, Doleris et al. (2000) reported for the first time that four patients with mitochondrial cell disease were concurrent with focal segmental glomerulosclerosis (FSGS). Subsequent studies showed that renal pathology in patients with mitochondrial disease was mainly FSGS. For example, the podocyte injury is particularly obvious in the patients with mitochondrial gene A3243G mutation, including the abnormalities of podocyte mitochondria size and morphological structure, mitochondria swelling, irregular shape, increased sputum plateau structure, smaller podocyte cell body, and foot process fusion. In the experimental animal model of FSGS (puromycin aminonucleoside nephropathy), mitochondrial oxidative phosphorylation dysfunction, decreased mtDNA copy number and mutation, and decreased expression of mitochondrial respiratory chain enzyme complex subunit were observed in renal tissues, and then apoptosis of podocytes occurred (Shoubridge 2001). In aldosterone-infused mice, we find that mitochondrial lesions are the initiating factor of early podocyte injury. In the early stage, ahead of the occurrence of proteinuria and foot process fusion, the mice have mitochondrial dysfunction, mitochondrial swelling, sputum disappearance, increased ROS production and decline in mitochondrial membrane potential, mtDNA copy number, ATP production and enzyme complex I activity. Overexpression of the SIRT1 agonist (resveratrol), SIRT1, or PGC-1 can alleviate podocyte injury by blocking mitochondrial dysfunction (Zhu et al. 2011). In addition, podocyte injury is a process with disturbance of mitochondrial dynamics and quality control. Mitochondrial fission induced by high glucose conditions leads to podocyte foot process effacement, probably through phosphorylation of Drp-1 by Rho-associated coiled-coil-containing protein kinase 1 (ROCK1). The fission talked above could be reversed by podocyte-specific silencing of ROCK1 (Wang et al. 2012). Furthermore, mitophagy is responsible for scavenging abnormal mitochondria with debility. As it is said above, suppressed autophagy by inhibition of mTOR maintains damaged mitochondria, leading to proteinuria. Unfortunately, podocytes exhibit high levels of autophagy while be unable to regenerate. As the compensation of podocyte loss, parietal cells in Bowman's capsule gradually develop fibrosis, showing the double-edged sword effect of autophagy.

25.3.4 Mitochondria-Induced Tubular Interstitial Fibrosis

Renal interstitial fibrosis (RIF) is a common outcome in various chronic kidney diseases. The role of renal tubular epithelial cells (RTEC) in the pathogenesis of tubulointerstitial fibrosis is of increasing concern. Normal RTEC has potent metabolic activity and potential proliferative capacity and can secrete a variety of cytokines. In the disease state, RTEC are very sensitive and highly susceptible to damage, such as proteinuria, inflammatory mediators, ischemia, hypoxia, poisoning, or glucose. RTEC are not only target cells for injury, but also active "participants," which are the main source of inflammatory factors and ECM production in kidney tissues. Injured tubular epithelial cells can undergo phenotypic transformation, transforming epithelial cell phenotype into mesenchymal cell phenotype (EMT), secreting a large number of inflammatory factors, chemokines, profibrotic factors and matrix proteins, destroying normal tubulointerstitial structure, and leading to tubulointerstitial inflammation and fibrosis. We demonstrated that with mtDNA depletion and mitochondrial dysfunction, epithelial cells are susceptible to mesenchymal cells transition with risk factors. While with the recovery of mtDNA, EMT was reversed in association with reexpression of endogenous E-cadherin, downregulation of α-SMA expression, and restoration of an epithelial phenotype. In folic acid-induced renal fibrosis model, mitochondria structure and function disturbance led to mitophagy and apoptotic necrosis, which induced fatty acid metabolism and accelerate interstitial fibrosis (Liu et al. 2014). Transforming growth factor (TGF)-β is involved in the pathogenesis of renal cell injury in progressive CKD (Casalena et al. 2012), inducing both renal cell apoptosis and renal fibrosis. On the one hand, TGF-β overexpression in various renal cells directly leads to mitochondrial dysfunction. Mitochondrial fragmentation, cell apoptosis, and expansion were observed in podocytes, mesangial cells, and proximal tubular epithelial cells, which further result in progressive glomerulosclerosis, extracellular matrix accumulation, glomerular basement membrane (GBM) thickening, and renal fibrosis (Schiffer et al. 2001) (Casalena et al. 2012). On the other hand, TGF-β takes part in the molecular pathway of pro-fibrogenic process with many other factors. The TGF-β/SMAD pathway is known to be one of the key pathways of EMT and endothelial–mesenchymal transition (EndoMT). TGF-β deteriorates oxidative stress and stimulates SMAD3 expression in a positive feedback loop through impairing mitochondrial antioxidant systems with concomitant enhancing prooxidant NADPH oxidase (Nox4). To verify this, TGF-β-induced profibrotic gene expression could be downregulated by the disruption of mitochondria-derived ROS. (Jain et al. 2013). Of note, in the program of fibrosis mitochondrial biogenesis is not inhibited. The study from Hickey et al. (2011) showed higher expression of genes encoding key mitochondrial proteins from renal biopsies of diabetic nephropathy patients, thus demonstrating the existence of mitochondrial biogenesis in renal proliferation and fibrosis. Induced in high glucose 1 (IHG1), mitochondrial protein amplifies TGF-β1 signaling, decreased the TGF-β1 inhibitor SMAD7, and maintains PGC-1α expression, thereby promoting mitochondrial biogenesis and further contributing to renal fibrosis. As we know, cAMP signaling is an indispensable pathway of mitochondrial homeostasis regulation and dynamics. The level of cytosolic AMP is positively associated with mitochondrial number and the level of ATP in the tubule. Ding et al. found that restoring cAMP by the phosphodiesterases (PDE4) inhibitor rolipram may ameliorate renal fibrosis by targeting C/EBP-β/PGC1-α and mitochondrial biogenesis (Ding et al. 2018). Recent study examined the genetic change of five families with autosomal dominant renal Fanconi syndrome and kidney failure. It turned out that monoallelic mutations happened in the gene encoding glycine amidinotransferase (GATM), a renal proximal tubular enzyme in the creatine biosynthetic pathway. The mutation induced the abnormal aggregation of GATM. Intramitochondrial fibrillary deposition of GATM leads to elongated and abnormal mitochondria, which further activate inflammasome and cell death. Therefore, renal proximal tubular mitochondrial pathology initiates the development of fibrosis (Reichold et al. 2018).

25.3.5 Mitochondrial Cytopathy and Renal Fibrosis

Hereditary mitochondrial cytopathy is a type of disease caused by the presence of one or more genetic defects in the oxidative respiratory chain complex. The cause can be divided into two major categories: mtDNA mutations and nuclear DNA (nDNA) mutations. Since both mtDNA and nDNA are involved in the coding of the respiratory chain complex, mutations in both mtDNA and nDNA may lead to mitochondrial cytopathies. Genetic patterns are also diverse, such as sporadic, mtDNA maternal genetic mutation, autosomal recessive inheritance, autosomal dominant inheritance, and X-linked genetic nDNA mutations. In addition, unlike nDNA, mtDNA mutations are heterogeneous, that is, because there are hundreds or even thousands of mtDNA copies in tissue cells, mutant mtDNA and normal mtDNA can coexist in one cell in different proportions. The phenotypic effect is determined by the relative ratio of mutant mtDNA to normal mtDNA and the extent to which this tissue is dependent on ATP produced by mitochondria (Dimauro and Schon 2003). Mutant mtDNA needs to reach a certain proportion, which is enough to cause changes in tissue and organ function, so the disease phenotype caused by mtDNA mutation has diverse characteristics. Kidney involvement is only part of the multisystemic lesion of mitochondrial cell disease, mainly characterized by focal segmental glomerulosclerosis (FSGS), tubular dysfunction, and cystic nephropathy. These pathological changes eventually lead to renal fibrosis and ESRD. FSGS is one of renal lesions frequently correlated with genetic defects. Podocytes have limited potency for regeneration. In addition, in these cases arteriolar hyalinosis is commonly seen (Seidowsky et al. 2013). Therefore, mitochondrial cytopathies could result in uncontrollable podocyte damage and microvascular lesions, which arouse a desperate strike to the kidney. Besides

FSGS, mitochondrial gene mutations can lead to tubulointerstitial lesions, renal tubulointerstitial inflammation, and progressive renal insufficiency. Renal biopsy of the patient with mtDNA depletion showed chronic tubulointerstitial lesions, tubule atrophy, interstitial fibrosis, mitochondrial swelling, and abnormal morphology of RTEC under electron microscope. As one of the major victims of MCs, proximal tubular cells are relatively vulnerable to oxidative stress. Fanconi syndrome and Bartter-like syndrome are frequently observed since the renal tubule defects mainly manifest as loss of electrolytes and low-molecular-weight proteins. Additionally, myoclonic epilepsy and ragged red muscle fibers (MERRFs), and Kearns-Sayre, Pearson's, and Leigh syndromes are also described in patients with mitochondrial tubulopathy. The majority of genetic mutations detected in these diseases are fragment deletions of mtDNA.

Based on different type of mutation, the following are illustrated.

25.3.5.1 mtDNA Point Mutation

The most common one is the mitochondrial gene A3243G mutation. This mutation can lead to mitochondrial myopathy with hyperlactosis and stroke-like episodes (MELAS), as well as diabetes, renal failure, and deafness. MELAS syndrome is caused by a point mutation in the MTTL1 gene encoding mitochondrial tRNALEU, of which 80% are A3243G mutations, 7.5–10% are A3253G mutations, and 7.5% are T3271C mutations (Ireland et al. 2004). Guery et al. found that in nine patients with kidney disease containing this mutation site, the pathological changes were mostly focal segmental glomerulosclerosis (FSGS), but the clinical types varied greatly, from tubulointerstitial disease to polycystic kidney disease (Guery et al. 2003). Among 9 patients, 5 developed ESRD within a median follow-up 5 years. Although the pathogenesis of MELAS-associated FSGS is unknown, more morphological mitochondria are seen in the glomerular podocytes of these patients, the podocyte foot process fusion disappears (Gucer et al. 2005), and hyaline degeneration is observed in small arteries. Therefore, irreversible podocyte injury and microvascular disease may be the cause of renal damage in MELAS syndrome.

25.3.5.2 mtDNA Deletion Rearrangement

There are reports suggesting that there is a 2–10 kb mtDNA deletion rearrangement in some patients with tubular disease. Such patients often have neurological symptoms such as Pearson and Kearns-Sayre syndrome. The size and location of mtDNA deletion rearrangements are not related to the clinical manifestations or severity of the disease.

25.3.5.3 Reduction in mtDNA Copy Number

This mutation can lead to lethal infant breathing, muscle, liver, and kidney failure. In this type of patient, there is one or more mtDNA copy number deletions in the tissue, and there is dysfunction of the affected tissue oxidative respiratory chain (Carrozzo et al. 2003).

Mitochondrial diseases caused by mutations in nuclear genes are rare, including mutations in genes encoding coenzyme Q10. Coenzyme Q10-related gene mutations, including PDSS1, PDSS2, COQ2, COQ4, COQ6, COQ7, COQ9, ADCK3, ADCK4, can lead to hormone-resistant nephrotic syndrome. COX10, SURF1, BCS1L, UQCC2, TMEM70, and other respiratory chain protein gene mutations have been reported in patients with proximal tubular disease. For example, the BCS1 gene is responsible for encoding a mitochondrial inner membrane protein involved in the assembly of complex III in the respiratory chain. It has been reported that four families with BCS1 mutations exhibit a series of clinical manifestations of liver failure, lactic acidosis, hepatocyte dysfunction, tubular disease, encephalopathy, hearing, and visual loss (De Lonlay et al. 2001). COX10 mutations lead to multi-system involvement in fatal neonatal encephalopathy, which is mainly characterized by renal tubular disease (Antonicka et al. 2003).

25.4 Treatment

25.4.1 Gene Therapy

miRNAs, conserved, clustered, and specifically expressed, are a group of non-coding regulatory RNAs of approximately 22 nucleotides in length, which are widely present in viruses, plants, and humans. At present, there are nearly 800 identified miRNAs in the human genome, which are mainly involved in the regulation of gene expression by pairing with the 3′ untranslated region (3′-UTR) of the target gene mRNA sequence, and participate in various biological processes. Previous studies have shown that miRNAs are mainly involved in biological processes such as growth and development, cell differentiation, proliferation, apoptosis, and tumorigenesis. A number of in vitro and in vivo studies on the involvement of miRNAs in the pathogenesis of kidney disease have been confirmed, some of which play a role in mitochondria (Gomez et al. 2016). MiR-30e is down-regulated in renal tissue fibrosis, and its analog can exert anti-fibrotic effect by acting on mitochondrial protein UCP2. MiR-21 participates in the fibrosis process of various organs through a critical pathway that acts on energy metabolism. In the Alport syndrome mouse model, miR-21 silencing enhances mitochondrial function, reduces mitochondrial ROS production, significantly promotes survival of diseased mice, and reduces glomerular sclerosis, interstitial fibrosis, tubular damage, and inflammatory response. MiR-17 is capable of mitochondrial metabolism and promotes the growth of polycystic kidney cysts.

In addition, studies have reported that transfection of tRNALys into cells mutated into the A8344G MERRF gene partially restores mitochondrial function, and similar results were observed in MELAS cells bearing the A3243G mutation (Hajarnis et al. 2017). However, no gene therapy was found to be the most effective up to now. The treatment modification of the mitochondrial genome is more difficult than the treatment of nuclear DNA disorders. The heterogeneity of mitochondria and the presence of multiple genomes in a single cell are extremely complex.

25.4.2 Medication

25.4.2.1 Coenzyme Q10

Coenzyme Q10 is a biological element of the mitochondrial respiratory chain that allows electrons to be transferred from complex I/II to complex III. Endogenous coenzyme Q10 is derived from tyrosine in body tissues or is ingested from the diet. Coenzyme Q10 has the characteristics of large molecular weight, strong lipophilicity, and poor water solubility, so its bioavailability is relatively low. Coenzyme Q10 prevents membrane lipid peroxidation and exerts anti-apoptotic effects by inhibiting the opening of mitochondrial permeability transition pores and mitochondrial membrane potential depolarization. It promotes oxygen consumption, ATP synthesis, and mitochondrial protein synthesis.

The total and oxidized forms of coenzyme Q10 in the renal cortex of db/db mice were reduced. Supplementation of coenzyme Q10 improved mitochondrial dysfunction and reduced the amount of collagen in diabetic kidney tissue. Increased coenzyme Q10 in type 2 diabetic mice reduces oxygen consumption, inhibits mitochondrial division, and reduces glomerular hyperfiltration and proteinuria. In a nephrectomized rat model, dietary supplementation of coenzyme Q10 reduced reactive oxygen species levels and improved renal function. In kidney damage caused by chronic nicotine exposure, coenzyme Q10 activates the antioxidant system by activating the non-mitochondrial fork protein p66shc, thereby counteracting nicotine-induced oxidative stress damage in renal tubular epithelial cells. The level of coenzyme Q10 in hemodialysis patients was lower than that in healthy controls, and the level of coenzyme Q10 was negatively correlated with the thickness of epicardial adipose tissue (19891905). Another study reported that supplementation of coenzyme Q10 in hemodialysis patients for 6 months reduced the level of oxidative stress in the body. The latest clinical randomized, double-blind, controlled trial showed that regular hemodialysis patients were safe to take 1200 mg of coenzyme Q10 per day and were able to reduce the level of oxidative stress index F2-isoprostanes (Rivara et al. 2017).

25.4.2.2 Thiazolidinediones

Thiazolidinedione hypoglycemic agents, such as rosiglitazone, pioglitazone, are synthetic PPAR gamma agonists and are now widely used in the treatment of clinical type 2 diabetes. Some experts have found that these drugs also have a protective effect on the kidney, which can reduce proteinuria and relieve renal fibrosis. It is speculated that the mechanism may be related to inhibition of oxidative stress and inflammatory response. Studies have shown that PPARγ can be involved in the regulation of mitochondrial biogenesis and oxidative phosphorylation in a variety of cells. Thiazolidinediones have potential therapeutic effects by blocking mitochondrial dysfunction. PPARγ mutation can increase the production of reactive oxygen species in the cell, leading to the decrease of mitochondrial membrane potential, and overexpression of PPARγ can maintain the normal function of mitochondria.

In the rat model of type 2 diabetes, pioglitazone significantly inhibited glomerular sclerosis, tubulointerstitial fibrosis, tubule atrophy, and podocyte apoptosis. PPARγ agonists can effectively relieve glomerular sclerosis and tubulointerstitial fibrosis in animal models of multiple kidney diseases such as DOCA-salt hypertensive rats, unilateral ureteral obstruction rats, 5/6 nephrectomized rats, Zucker obese rats, and cyclosporine A nephropathy rats. In the renal injury of puromycin nucleoside, the expression of PPARγ mRNA in podocytes was significantly down-regulated, and the PPARγ agonist pioglitazone inhibited podocyte injury and alleviated glomerular sclerosis by regulating the expression of cyclin. In aldosterone-induced podocyte injury, it was confirmed that the PPAR gamma agonist rosiglitazone protected podocytes from damage by blocking mitochondrial dysfunction (Zhu et al. 2011).

25.4.2.3 Resveratrol

Resveratrol, also known as quinol, is a plant polyphenolic compound found in plants and fruits such as *Polygonum cuspidatum*, grapes, peanuts, mulberry, and the highest content of fresh grape skin. Studies have confirmed that resveratrol has many biological activities such as cardiovascular protection, anti-cancer, anti-inflammatory, and anti-oxidation. Mitochondria are one of the main sites for the production of reactive oxygen species in cells. Studies have found that resveratrol inhibits high glucose-induced mitochondrial superoxide anion production in a variety of cell types. The inhibition of mitochondrial oxidative stress by resveratrol may be partly related to the activation of the mitochondrial antioxidant system. In addition, resveratrol promotes mitochondrial biogenesis in multiple tissues; resveratrol is the most potent activator of SIRT1 currently found. Resveratrol has been shown to act by increasing the activity of SIRT1, thereby regulating the transcription of PGC-1α and its target genes, and promoting mitochondrial function protection by promoting mitochondrial biosynthesis (Yuan et al. 2012).

25.4.2.4 Omega 3 Polyunsaturated Fatty Acids

Omega 3 is a member of the polyunsaturated fatty acid family and plays an important role in regulating cell and organ membrane structure and function. Omega 3 is mainly derived from fish oil. Omega 3 blocks the activation of NF-κB by inhibiting the phosphorylation of IKB, thereby reducing the expression of adhesion molecules, cytokines, and pro-inflammatory factors. Omega 3 has anti-platelet effects as a precursor to prostaglandins. In addition, they are involved in membrane fluidity, ion channel transport, and regulation of mitochondrial biosynthesis. Moreover, Omega 3 also has antioxidant properties that reduce the production of reactive oxygen species by enhancing the endogenous antioxidant system (An et al. 2009). Clinical supplementation of Omega 3 can be beneficial for lipid metabolism, blood pressure, redox status, and the cardiovascular system.

After 12 weeks of supplementation with Omega 3 in a rat model of chronic kidney disease, the proapoptotic, pro-inflammatory, and proapoptotic signals were down-regulated. In UUO, Omega 3 reduces fibroblast activation and kidney fibrosis. Studies have reported that the level of Omega 3 unsaturated fatty acids in hemodialysis patients is lower than in the normal control group, which may be related to increased cardiovascular risk. In addition, supplementation of fish oil in hemodialysis patients can significantly reduce systolic blood pressure and diastolic blood pressure. Omega 3 treatment reduces the activity of peripheral blood leukocyte apoptosis-related enzymes in patients with end-stage renal disease. A recent meta-analysis of Omega 3 treatment of chronic kidney disease has shown that supplementation with Omega 3 significantly reduces the risk of end-stage renal disease and delays disease progression.

Although these studies on Omega 3 bring hope to kidney disease treatment, there are some inconsistent results that require attention. This may require more clinical randomized controlled trials to confirm the role of Omega 3 in the chronic kidney patient population and further confirm the dose and interval of supplementation of Omega 3 in a special population such as hemodialysis to achieve a reasonable blood concentration. In turn, avoid side effects

25.4.2.5 Vitamin E

The inhibition of oxidative factors by vitamin E alleviates the development of a variety of cardiovascular diseases, aging, and other chronic disease degenerative diseases, including CKD. Dietary vitamin E supplementation delays the progression of chronic diseases. Studies have confirmed that vitamin E supplementation plays a beneficial role in dialysis patients with chronic kidney disease. The SPACE study investigated the effects of daily supplementation of 800 IU of vitamin E on cardiovascular events in hemodialysis patients. After a long follow-up, vitamin E reduced the incidence of cardiovascular events and myocardial occlusion. After the dialysis membrane is coated with vitamin E, it can reduce oxidative stress and inflammation indicators in hemodialysis patients, improve hemoglobin levels, and reduce the use of erythropoietin (Yang et al. 2006).

25.4.2.6 MitoQ

MitoQ is the most widely studied and widely used mitochondria-targeted antioxidant. It consists of a triphenylphosphine cation and a benzoquinone moiety of coenzyme Q10 covalently bonded through a ten carbon fatty chain. MitoQ can be concentrated in the mitochondria by mitochondrial membrane potential and then degraded into a form that can exert antioxidant activity. As the ubiquinone moiety is inserted into the hydrophobic interior of the mitochondrial membrane, the matrix within the membrane is its primary absorbing moiety, allowing it to exert a protective effect on lipid peroxidation (Kelso et al. 2001). MitoQ has been used in a variety of animal disease models to test its protective effects against mitochondrial oxidative stress damage. In mice with ischemia–reperfusion-induced renal injury, intravenous injection of MitoQ 15 min before renal vascular occlusion can reduce oxidative stress damage and protect renal function (Dare et al. 2015). In Akita mice with type 1 diabetic nephropathy, it was confirmed that MitoQ has anti-fibrosis and anti-glomerative chronic damage. In type 2 diabetic nephropathy db/db mice, MitoQ inhibits cellular oxidative stress by upregulating autophagy, preventing mitochondrial membrane potential and decreasing mtDNA copy number in renal tubular epithelial cells, down-regulating cleavage protein Drp1 expression, restoring fusion protein Mfn2 expression, as well as inhibiting cellular oxidative stress and apoptosis. In stroke-susceptible spontaneously hypertensive rats, MitoQ could prevent the development of hypertension, promote the bioavailability of endothelial NO, and reduce cardiac hypertrophy. The addition of MitoQ to renal tissue prevents mitochondrial dysfunction and improves cell viability and kidney morphology. Oral MitoQ has entered Phase I and Phase II clinical trials.

25.4.2.7 SS-31

SS-31 is a small molecule peptide that is capable of freely penetrating cells and targeting the mitochondria. In 5/6 nephrectomized rats, SS-31 can clear excess ROS in mitochondria, reduce ROS production, stabilize mitochondrial membrane potential, prevent mitochondrial permeability changes, reduce cytochrome c release, and prevent kidney tissue inflammation and fibrosis. In the UUO rat model, administration of SS-31 at a dose of 1 or 3 mg/kg 1 day before surgery significantly reduced tubular apoptosis, macrophage infiltration, and renal fibrosis, while increasing tubular proliferation. In the rat model of ischemia–reperfusion, subcutaneous administration of SS-31 30 min before bilateral renal ischemia can reduce the apoptosis of tubular cells caused by reperfusion, maintain the structure of renal tubules, and maintain the integrity and function of mitochondria. ATP synthesis can be restored upon reperfusion. Oxidative stress and inflammatory response are alleviated, and the rate of tubular cell regeneration is accelerated. In addition, SS-31 also has a protective effect on capillary endothelial cells during ischemia and perfusion. Similar to renal tubular epithelial cells, SS-31 protects mitochondrial structure. In a recent study of aging-related glomeruli disease, SS-31 improved mitochondrial damage in podocytes and parietal epithelial cells, reducing glomerular sclerosis (Sweetwyne et al. 2017).

25.4.2.8 Sirolimus

Drugs that regulate autophagy and mitochondrial autophagy include sirolimus (rapamycin), an inhibitor of mTORC1. A variety of diabetic animal models have demonstrated that it can inhibit the activation of the mTOR signaling pathway, thereby preventing kidney hypertrophy, glomerular sclerosis, reducing proteinuria, and inhibiting mesangial proliferation. However, rapamycin is considered to be an immunosuppressive agent widely used in kidney transplant patients, and its clinical treatment is limited due to its side effects. Rapamycin not only inhibits mTORC1 but also inhibits mTORC2. Studies in type 2 diabetic mice have found that rapamycin reduces high glucose tolerance and insulin activation by inhibiting mTORC2. mTORC1 is regulated by mTOR's rapamycin-insensitive companion of mTOR, which is regulated by mTOR's raptor rapamycin-associated protein of mTOR. mTORC2 regulates the expression of autophagy genes by phosphorylating FOXO3a by activating the Akt signaling pathway. FOXO3a regulates LC3, Bnip3, Nix, Atg4b, and Atg12l, and these genes play an important role in mitochondrial autophagy. This also explains the side effects of knockout mTOR in podocyte conditions, resulting in autophagic flow disorders and renal failure (Cina et al. 2012). Rapamycin and its analogs are still in the experimental research stage as effective drugs for clinical chronic kidney diseases such as diabetic nephropathy.

25.5 Conclusion

Mitochondria are the key energy supply organelle in human bodies, which make it the vulnerable target of detrimental stimulus. No molecule is an island. The crosstalk among mitochondrial dysfunction, ROS, ER stress, mitophagy, and inflammasome induces cytokines' thunderstorm, which induce various cell lesions in kidney. The ECM deposit, EMT, EndoMT, and several other decompensated changes ultimately lead to renal fibrosis. To date, there are few adoptive therapies targeted mitochondrial dysfunction. More research and clinical trials are needed.

References

Akool ES, Gauer S, Osman B, Doller A, Schulz S, Geiger H et al (2012) Cyclosporin A and tacrolimus induce renal Erk1/2 pathway via ROS-induced and metalloproteinase-dependent EGF-receptor signaling. BiochemPharmacol 83:286–295

An WS, Kim HJ, Cho KH, Vaziri ND (2009) Omega-3 fatty acid supplementation attenuates oxidative stress, inflammation, and tubulointerstitial fibrosis in the remnant kidney. Am J Physiol Renal Physiol 297:F895–F903

Antonicka H, Leary SC, Guercin GH, Agar JN, Horvath R, Kennaway NG et al (2003) Mutations in COX10 result in a defect in mitochondrial heme A biosynthesis and account for multiple, early-onset clinical phenotypes associated with isolated COX deficiency. Hum Mol Genet 12:2693–2702

Becker T, Wagner R (2018) Mitochondrial outer membrane channels: emerging diversity in transport processes. BioEssays 40:e1800013

Boenzi S, Diodato D (2018) Biomarkers for mitochondrial energy metabolism diseases. Essays Biochem 62:443–454

Carrozzo R, Bornstein B, Lucioli S, Campos Y, De La Pena P, Petit N et al (2003) Mutation analysis in 16 patients with mtDNA depletion. Hum Mutat 21:453–454

Casalena G, Daehn I, Bottinger E (2012) Transforming growth factor-beta, bioenergetics, and mitochondria in renal disease. SeminNephrol 32:295–303

Che R, Yuan Y, Huang S, Zhang A (2014) Mitochondrial dysfunction in the pathophysiology of renal diseases. Am J Physiol Renal Physiol 306:F367–F378

Cina DP, Onay T, Paltoo A, Li C, Maezawa Y, De Arteaga J et al (2012) Inhibition of MTOR disrupts autophagic flux in podocytes. J Am SocNephrol 23:412–420

Dare AJ, Bolton EA, Pettigrew GJ, Bradley JA, Saeb-Parsy K, Murphy MP (2015) Protection against renal ischemia-reperfusion injury in vivo by the mitochondria targeted antioxidant MitoQ. Redox Biol 5:163–168

De Lonlay P, Valnot I, Barrientos A, Gorbatyuk M, Tzagoloff A, Taanman JW et al (2001) A mutant mitochondrial respiratory chain assembly protein causes complex III deficiency in patients with tubulopathy, encephalopathy and liver failure. Nat Genet 29:57–60

Deng X, Xie Y, Zhang A (2017) Advance of autophagy in chronic kidney diseases. Ren Fail 39:306–313

Dimauro S, Schon EA (2003) Mitochondrial respiratory-chain diseases. N Engl J Med 348:2656–2668

Ding H, Bai F, Cao H, Xu J, Fang L, Wu J et al (2018) PDE/cAMP/Epac/C/EBP-beta signaling cascade regulates mitochondria biogenesis of tubular epithelial cells in renal fibrosis. Antioxid Redox Signal 29:637–652

Doleris LM, Hill GS, Chedin P, Nochy D, Bellanne-Chantelot C, Hanslik T et al (2000) Focal segmental glomerulosclerosis associated with mitochondrial cytopathy. Kidney Int 58:1851–1858

Farmer T, Naslavsky N, Caplan S (2018) Tying trafficking to fusion and fission at the mighty mitochondria. Traffic 19:569–577

Fedulov AV, Ses TP, Gavrisheva NA, Rybakova MG, Vassilyeva JG, Tkachenko SB et al (2005) Serum TGF-beta 1 and TNF-alpha levels and cardiac fibrosis in experimental chronic renal failure. Immunol Invest 34:143–152

Galvan DL, Green NH, Danesh FR (2017) The hallmarks of mitochondrial dysfunction in chronic kidney disease. Kidney Int 92:1051–1057

Gomez IG, Nakagawa N, Duffield JS (2016) MicroRNAs as novel therapeutic targets to treat kidney injury and fibrosis. Am J Physiol Renal Physiol 310:F931–F944

Gorman GS, Chinnery PF, Dimauro S, Hirano M, Koga Y, Mcfarland R et al (2016) Mitochondrial diseases. Nat Rev Dis Primers 2:16080

Gucer S, Talim B, Asan E, Korkusuz P, Ozen S, Unal S et al (2005) Focal segmental glomerulosclerosis associated with mitochondrial cytopathy: report of two cases with special emphasis on podocytes. Pediatr Dev Pathol 8(6):710–717

Guery B, Choukroun G, Noel LH, Clavel P, Rotig A, Lebon S et al (2003) The spectrum of systemic involvement in adults presenting with renal lesion and mitochondrial tRNA(Leu) gene mutation. J Am SocNephrol 14:2099–2108
Hajarnis S, Lakhia R, Yheskel M, Williams D, Sorourian M, Liu X et al (2017) microRNA-17 family promotes polycystic kidney disease progression through modulation of mitochondrial metabolism. Nat Commun 8:14395
Hickey FB, Corcoran JB, Docherty NG, Griffin B, Bhreathnach U, Furlong F et al (2011) IHG-1 promotes mitochondrial biogenesis by stabilizing PGC-1alpha. J Am SocNephrol 22:1475–1485
Ireland J, Rossetti S, Haugen E, Ireland J, MichelsV Harris P (2004) Mitochondrial causes of renal insufficiency and hearing loss. Kidney Int 65:2444–2445
Jain M, Rivera S, Monclus EA, Synenki L, Zirk A, Eisenbart J et al (2013) Mitochondrial reactive oxygen species regulate transforming growth factor-beta signaling. J BiolChem 288:770–777
Kelso GF, Porteous CM, Coulter CV, Hughes G, Porteous WK, Ledgerwood EC et al (2001) Selective targeting of a redox-active ubiquinone to mitochondria within cells: antioxidant and antiapoptotic properties. J BiolChem 276:4588–4596
Kumar A, Chaugule VK, Condos TEC, Barber KR, Johnson C, Toth R et al (2017) Parkin-phosphoubiquitin complex reveals cryptic ubiquitin-binding site required for RBR ligase activity. Nat Struct MolBiol 24:475–483
Lagouge M, Larsson NG (2013) The role of mitochondrial DNA mutations and free radicals in disease and ageing. J Intern Med 273:529–543
Letts JA, Sazanov LA (2017) Clarifying the supercomplex: the higher-order organization of the mitochondrial electron transport chain. Nat Struct MolBiol 24:800–808
Lin JR, Zheng YJ, Zhang ZB, Shen WL, Li XD, Wei T et al (2018) Suppression of endothelial-to-mesenchymal transition by SIRT (Sirtuin) 3 alleviated the development of hypertensive renal injury. Hypertension 72:350–360
Liu S, Soong Y, Seshan SV, Szeto HH (2014) Novel cardiolipin therapeutic protects endothelial mitochondria during renal ischemia and mitigates microvascular rarefaction, inflammation, and fibrosis. Am J Physiol Renal Physiol 306:F970–F980
Messmer UK, Briner VA, Pfeilschifter J (1999) Tumor necrosis factor-alpha and lipopolysaccharide induce apoptotic cell death in bovine glomerular endothelial cells. Kidney Int 55:2322–2337
Molino D, Nascimbeni AC, Giordano F, Codogno P, Morel E (2017) ER-driven membrane contact sites: Evolutionary conserved machineries for stress response and autophagy regulation? CommunIntegrBiol 10:e1401699
Palikaras K, Lionaki E, Tavernarakis N (2018) Mechanisms of mitophagy in cellular homeostasis, physiology and pathology. Nat Cell Biol 20:1013–1022
Panieri E, Santoro MM (2016) ROS homeostasis and metabolism: a dangerous liason in cancer cells. Cell Death Dis 7:e2253
Reichold M, Klootwijk ED, Reinders J, Otto EA, Milani M, Broeker C et al (2018) Glycine amidinotransferase (GATM), renal fanconi syndrome, and kidney failure. J Am SocNephrol 29:1849–1858
Rivara MB,Yeung CK, Robinson-Cohen C, Phillips BR, Ruzinski J, Rock D et al (2017) Effect of coenzyme Q10 on biomarkers of oxidative stress and cardiac function in hemodialysis patients: the CoQ10 biomarker trial. Am J Kidney Dis 69:389–399
Schiffer M, Bitzer M, Roberts IS, Kopp JB, Ten Dijke P, Mundel P et al (2001) Apoptosis in podocytes induced by TGF-beta and Smad7. J Clin Invest 108:807–816
Schubert AF, Gladkova C, Pardon E, Wagstaff JL, Freund SMV, Steyaert J et al (2017) Structure of PINK1 in complex with its substrate ubiquitin. Nature 552:51–56
Seidowsky A, Hoffmann M, Glowacki F, Dhaenens CM, Devaux JP, De Sainte Foy CL et al (2013) Renal involvement in MELAS syndrome—a series of 5 cases and review of the literature. ClinNephrol 80:456–463
Shoubridge EA (2001) Nuclear genetic defects of oxidative phosphorylation. Hum Mol Genet 10:2277–2284

Soetikno V, Sari FR, Lakshmanan AP, Arumugam S, Harima M, Suzuki K et al (2013) Curcumin alleviates oxidative stress, inflammation, and renal fibrosis in remnant kidney through the Nrf2-keap1 pathway. MolNutr Food Res 57:1649–1659

Stiles AR, Simon MT, Stover A, Eftekharian S, Khanlou N, Wang HL et al (2016) Mutations in TFAM, encoding mitochondrial transcription factor A, cause neonatal liver failure associated with mtDNA depletion. Mol Genet Metab 119:91–99

Sweetwyne MT, Pippin JW, Eng DG, Hudkins KL, Chiao YA, Campbell MD et al (2017) The mitochondrial-targeted peptide, SS-31, improves glomerular architecture in mice of advanced age. Kidney Int 91:1126–1145

Van Vliet AR, Agostinis P (2018) Mitochondria-associated membranes and ER stress. Curr Top Microbiol Immunol 414:73–102

Vincent AE, Turnbull DM, Eisner V, Hajnoczky G, Picard M (2017) Mitochondrial nanotunnels. Trends Cell Biol 27:787–799

Volobueva AS, Melnichenko AA, GrechkoAV Orekhov AN (2018) Mitochondrial genome variability: the effect on cellular functional activity. Ther Clin Risk Manag 14:237–245

Wang W, Wang Y, Long J, Wang J, Haudek SB, Overbeek P et al (2012) Mitochondrial fission triggered by hyperglycemia is mediated by ROCK1 activation in podocytes and endothelial cells. Cell Metab 15:186–200

Wei Y, Chiang WC, Sumpter R Jr, Mishra P, Levine B (2017) Prohibitin 2 is an inner mitochondrial membrane mitophagy receptor. Cell 168:224–238

Xu Y, Nie L, Yin YG, Tang JL, Zhou JY, Li DD et al (2012) Resveratrol protects against hyperglycemia-induced oxidative damage to mitochondria by activating SIRT1 in rat mesangial cells. Toxicol Appl Pharmacol 259:395–401

Yang CC, Hsu SP, Wu MS, Hsu SM, Chien CT (2006) Effects of vitamin C infusion and vitamin E-coated membrane on hemodialysis-induced oxidative stress. Kidney Int 69:706–714

Yuan Y, Huang S, Wang W, Wang Y, Zhang P, Zhu C et al (2012) Activation of peroxisome proliferator-activated receptor-gamma coactivator 1alpha ameliorates mitochondrial dysfunction and protects podocytes from aldosterone-induced injury. Kidney Int 82:771–789

Zhu C, Huang S, Yuan Y, Ding G, Chen R, Liu B et al (2011) Mitochondrial dysfunction mediates aldosterone-induced podocyte damage: a therapeutic target of PPARgamma. Am J Pathol 178:2020–2031

Zhuang Y, Ding G, Zhao M, Bai M, Yang L, Ni J et al (2014) NLRP3 inflammasome mediates albumin-induced renal tubular injury through impaired mitochondrial function. J BiolChem 289:25101–25111

Zhuang Y, Yasinta M, Hu C, Zhao M, Ding G, Bai M et al (2015) Mitochondrial dysfunction confers albumin-induced NLRP3 inflammasome activation and renal tubular injury. Am J Physiol Renal Physiol 308:F857–F866

Chapter 26
Lipid Metabolism Disorder and Renal Fibrosis

Xiao-Gang Du and Xiong-Zhong Ruan

Abstract Since the lipid nephrotoxicity hypothesis was proposed in 1982, increasing evidence has supported the hypothesis that lipid abnormalities contributed to the progression of glomerulosclerosis. In this chapter, we will discuss the general promises of the original hypothesis, focusing especially on the role of lipids and metabolic inflammation accompanying CKD in renal fibrosis and potential new strategies of prevention.

Keywords Lipids · Inflammation · Glomerular atherosclerosis · Tubulo-interstitial diseases · Statin

26.1 Introduction

Dyslipidemia is frequently observed in nephrotic syndrome (NS) and each stage of chronic kidney disease (CKD). The lipid profile of CKD patients is typified by high-circulating triglycerides, very low-density lipoprotein (VLDL), intermediate density lipoprotein (IDL), chylomicron remnants (CM), small dense LDL (sdLDL), and low-plasma high-density lipoprotein (HDL) cholesterol. Patients with proteinuria and patients treated with peritoneal dialysis (PD) have higher levels of LDL-C than non-proteinuric CKD or hemodialysis (HD) patients. The LDL cholesterol level in HD patients is not usually elevated and may even be reduced. A higher risk of death from cardiovascular disease is associated with low plasma cholesterol ("reverse epidemiology") (Liu et al. 2004; Lowrie and Lew 1990). Increased glomerular filtration membrane permeability is related to proteinuria and loss of lipoprotein lipase (LPL) activators ApoC-II and ApoA1. In addition, CKD patients present an upsurge of apolipoprotein C-III, a competitive inhibitor of LPL (Ginsberg et al. 1986), consequently leading to a delay of CM and VLDL removal which increases VLDL triglycerides (Moorhead et al. 1982). The loss of ApoA1 in urine results in decrease

X.-G. Du · X.-Z. Ruan (✉)
Department of Nephrology, The First Affiliated Hospital of Chongqing Medical University of Medical Science, Chongqing, China
e-mail: xiongzruan@gmail.com

B.-C. Liu et al. (eds.), *Renal Fibrosis: Mechanisms and Therapies*,
Advances in Experimental Medicine and Biology 1165,
https://doi.org/10.1007/978-981-13-8871-2_26

of plasma HDL levels. Furthermore, abnormal lecithin–cholesterol acyltransferase (LCAT) and cholesteryl ester transfer protein activities aggravate lipid abnormalities.

In addition to the changes in plasma levels, the composition and size distribution of all lipoproteins have also been modified in patients with CKD. The best example is the shift in the composition and size distribution of LDL to the increased content of sdLDL, which can undergo oxidation and enter into arterial wall easily, causing vascular damage (Berneis and Krauss 2002). In addition, the ratios of cholesterol-to-triglyceride, free cholesterol-to-cholesterol ester, and phospholipid-to-protein are changed. It has been strongly suggested that the elevated serum triglyceride-to-HDL ratio is not only a feature of dyslipidemia in patients with renal diseases but also an independent and highly predictive parameter for disease progression. Several clinical investigations have confirmed that the increased serum triglyceride-to-HDL ratio significantly influences the development of CKD and the decline of the estimated glomerular filtration rate (Tsuruya et al. 2015). In addition, CKD results in deficiency of HDL-associated enzymes (paraoxonase, glutathione peroxidase, and LCAT) and conversion of HDL from an antioxidant/anti-inflammatory agent to a pro-oxidant and pro-inflammatory agent (Moradi et al. 2009; Vaziri et al. 2009). These abnormalities can compound the effects of HDL deficiency in promoting an atherogenic diathesis in CKD. Lp(a) and apolipoprotein (apo)A-IV are also increased in this population. This lipid profile is similar to the atherogenic dyslipidemia of diabetics, and may sometimes be observed in early stages of primary kidney disease when measured glomerular filtration rate (GFR) is normal (Wanner and Ritz 2008).

Dyslipidemia is not only a consequence of CKD, but also a cause of kidney injury. Kidney injury on the molecular level involves complex interactions between all of the resident cell types and infiltrating cells from circulation. While tubular cells are often the initial site of injury, fibroblasts are subsequently recruited and activated. Initial responses are likely an attempt to repair kidneys and recapitulate embryologic developmental pathways. However, should repair processes persist, a profibrotic environment is promoted, which leads to the replacement of normal kidney with scar tissue composed of a variety of extracellular matrix (ECM) components, including collagens and fibronectin (Liu 2011). Renal fibrosis is a crucial determinant of progressive CKD. The pathogenesis of renal fibrosis is poorly understood but is commonly associated with dyslipidemia, suggesting that dyslipidemia plays an important role in renal fibrosis.

In this chapter, we will discuss the scientific evidence of lipid-mediated renal injury and potential mechanisms by which lipid causes renal fibrosis.

26.2 Lipid Nephrotoxicity

In 1982, Moorhead and colleagues published the "Lipid Nephrotoxicity Hypothesis" in the Lancet that stimulated lipid studies in the context of kidney diseases (Moorhead et al. 1982). In this "two-hit" model, the original disease could coexist or to be replaced by lipid-mediated damage. Persistent albuminuria stimulates excess lipoprotein synthesis by the liver, hence maintaining the lipid injury cycle. It also proposed that numerous of the features of progressive glomerular and tubulo-interstitial diseases share biological mechanisms with those of atherosclerosis including dyslipidemia, oxidative stress, inflammatory stress, and genetic factors. The term "glomerular atherosclerosis" was proposed. Since then, many laboratory and clinical studies (Grone et al. 1989; Moorhead et al. 1982) have supported the hypothesis that hyperlipidemia resulting from compensatory hepatic synthesis of lipoproteins in response to urinary loss of albumin contributed to the progression of glomerulosclerosis and tubulo-interstitial fibrosis.

Lipids are attributed to 3% of the wet weight of the normal human kidney. The major lipid components in the kidney are phospholipids (>50%), triglycerides (20%), and FFAs (10%). It seems that dyslipidemia per se might not be sufficient for renal injury, but it is one of the essential components of the multiple-hit mechanisms. It can affect the kidney by inducing lipotoxicity, inflammation, and oxidative stress, as well as signaling events and hormones with renal activity. Intensive laboratory studies have demonstrated that dyslipidemia in CKD can be both a consequence and a cause of the progression of CKD.

26.3 Impaired Cholesterol Homeostasis and Kidney Injury

Normally, there is a fine balance between cholesterol uptake, de novo synthesis and consumption in cells. LDL cholesterol uptake is mediated by multiple factors, such as LDL receptors, scavenger receptors, and lipid synthesis-related enzymes which express in kidney. Physiologically, the key biological function of LDL is to carry cholesterol from the liver to peripheral tissues, such as the kidney for sterol or phospholipid synthesis and energy supply. HDL is believed to play a beneficial role in the kidney and cardiovascular system by extracting cholesterol from peripheral organs and reverse transporting it to the liver. Renal dysfunction may disrupt the balance of lipid homeostasis, causing lipid accumulation which has been found in all kinds of renal resident cells such as podocytes, mesangial cells, endothelial cells, tubular epithelial cells, and interstitial cells. In addition, HDL has antioxidative, anti-inflammatory, and endothelium-protective properties. All of the above characteristics of HDL are critical for maintenance of normal lipid metabolism and for protecting the kidney from excessive lipid accumulation under conditions of lipid overloading.

It has long been established that cholesterol supplementation in the diets of several animal species leads to focal and segmental glomerulosclerosis (FSGS). French et al. showed that feeding guinea pigs with a 1% cholesterol diet caused severe glomerular disease (Brenner 1985; Feng et al. 2012). Peric-Golia et al. have demonstrated that feeding normal male Sprague-Dawley rats a diet containing 3–4% cholesterol resulted in hypercholesterolemia accompanied by aortic damage and renal glomerular abnormalities including lipid droplets, hyalinosis, glomerulosclerosis and interstitial fibrosis (Kasiske et al. 1990; Peric-Golia and Peric-Golia 1983). The severity of the glomerular injury is greatly increased if dietary induced hyperlipidemia is combined with either a loss of functioning nephrons, partial nephrectomy, or hypertension. Rats fed a diet consisting of 4% cholesterol which had a unilateral nephrectomy at 1 month developed significantly higher glomerular scarring than cholesterol-fed rats with two kidneys. Chronic renal failure induced by 5/6 nephrectomy results in abnormal lipid accumulation in the remnant kidney which is associated with upregulation of receptors involved in the influx of oxidized lipids and lipoproteins, activation of fatty acid biosynthesis and inhibition of pathways involved in fatty acid oxidation (Kim et al. 2009). Studies using the puromycin amino nucleoside (PAN) nephrotic rat model have also shown that cholesterol feeding increases the severity of proteinuria and FSGS (Kasiske et al. 1990; Lee et al. 1997). Apo B and apo E were found in increased amounts in the mesangium and co-localized with oil red O-positive lipid deposits (van Goor et al. 1993). Animals with endogenous hyperlipidemia (Kasiske et al. 1985) also develop progressive glomerular damage. Such models include the hyperlipidemic Sprague-Dawley rat developed by Imai (Imai et al. 1977), the spontaneously hypertensive rat described by Koletsky (Koletsky 1975), and the obese Zucker rat (Kasiske et al. 1985). Glomerular injury is also greater when systemic hypertension is combined with hyperlipidemia (Grone et al. 1993).

Several clinical studies have documented an association between dyslipidemia and the progression of CKD. Atherosclerosis risk in communities (Muntner et al. 2000) with low HDL cholesterol and increased non-HDL cholesterol was associated with increased risk of developing a reduced GFR ($\leq$55 ml/min/1.73 m^2). In the Atherosclerosis Risk in Communities (ARIC) study, higher HDL cholesterol levels were associated with a lower risk of CKD progression although one study showed an association between high LDL cholesterol levels and progression of kidney disease (Samuelsson et al. 1998). The weight of evidence therefore suggests that hypertriglyceridemia, accumulation of LDL cholesterol, and low HDL cholesterol are associated with higher risk of CKD progression. Survival statistics in renal transplant patients have also demonstrated that survival despite declining renal function is far superior in patients with normalized lipid profiles (Kobashigawa and Kasiske 1997; Pascual et al. 2002).

Foam cells and lipid deposits are found in FSGS in human renal biopsies (Lee et al. 1991). In these individuals, lipid deposition in the glomerulus is associated with progressive renal insufficiency. In patients with hepatorenal syndrome, the presence of lipoproteins with abnormal compositions have been reported to have progressive glomerular damage. A unique form of the nephrotic syndrome was reported in Japanese patients, where mesangial proliferation, mesangial expansion, glomerular deposition of lipoproteins, and FSGS were associated with high levels of circulating apoE (Koitabashi et al. 1990). Lee et al. found that 8.4% of 631 CKD patients had ultra-structurally detectable extracellular lipid in non-sclerotic glomeruli, which suggests that there may be an early pre-sclerotic stage of lipoprotein-mediated damage (Lee et al. 1991). Takemura also demonstrated that predominant deposition of apo B and apo E in the mesangial area in mesangial proliferative glomerulonephritis and the distribution and staining intensity of these apolipoproteins correlated with the grade of mesangial proliferation and proteinuria, but were independent of plasma lipid levels (Takemura et al. 1993).

It seems that hypercholesterolemia and cholesterol mainly affect glomerular mesangial cells which have been shown to bind LDL and Ox-LDL, leading to more cell proliferation via multiple downstream effects. LDL also stimulates the expression of extracellular matrix proteins including fibronectin. Furthermore, glomerular macrophages obtained from hypercholesterolemic animals displayed higher TGF-β mRNA expression which contributes to glomerular matrix expansion and glomerulosclerosis (Ding et al. 1994).

26.4 Deficiency of Fatty Acid Metabolism and Tubular Fibrosis

The kidney is an organ with a high-energy demand, but it has a relatively low glycolytic capacity. Therefore, beta-oxidation of free fatty acids (FFAs) in the mitochondria is the major source of energy, especially in proximal tubule cells. Due to the low level of fatty acid synthase (FAS), the kidney has a limited ability to synthesize fatty acids de novo; thus, it mainly relies on the uptake of triglycerides or FFAs from the circulation for energy.

Albumin carries >99% of plasma long-chain fatty acids (LCFAs). In CKD (especially diabetic kidney disease), the circulating levels of LCFAs are markedly increased, leading to an increase in the LCFA load per albumin molecule, with a molar FFA:albumin ratio of approximately 6 compared to ≤ 1 in health (Arici et al. 2003). Serum LCFAs bound to albumin in glomerular filtrate are reabsorbed by the proximal tubular cells via fatty acid transport CD36 and mediate renal tubular fibrosis (Ruggiero et al. 2014), highlighting the roles of hypertriglyceridemia and high level of FFAs on tubular fibrosis (Yang et al. 2017).

In CKD, the accumulation of triacylglycerols and FFAs with lower fatty acid oxidation (FAO) has critical roles in foam cell formation and the pathogenesis of kidney fibrosis (Yang et al. 2017). Lipid accumulation, reduced FAO, and ROS are reciprocal causations. Many factors can induce disbalance between fatty acid uptake and consumption in CKD, such as oxidative damage of mitochondria, which results in reduction in FAO (Jiang et al. 2017). Gene ontology analysis indicated that genes related to fatty acid metabolism and their key transcriptional regulator complex (PPARα/PGC-1α) were markedly lower in human chronic kidney disease (CKD) samples with higher lipid accumulation in diseased renal TECs. Mouse models of kidney fibrosis showed similar metabolic changes including low fatty acid and high intracellular lipid accumulation. Genetic or pharmacological improvement of FAO protected animals from kidney fibrosis development in vivo, suggesting restoring FAO could offer new therapeutic approaches for the treatment and prevention of kidney fibrosis.

As a specific membrane protein mediating kidney uptake of circulating fatty acids, cluster of differentiation 36 (CD36) modulates multiple pathways (Stremmel et al. 2001), which are closely linked to CKD and renal fibrosis. CD36 can be bound with context-specific binding partners (such as Toll-like receptor 2 (TLR2), TLR4, TLR6, and Na +/K + ATPase) to activate multiple signal pathways (Brown et al. 2015; Li et al. 2013; Stewart et al. 2010). The interactions of CD36 with TLR2 (Brown et al. 2015; Li et al. 2013; Stewart et al. 2010), tetraspanin (Huang et al. 2011), and integrin (Antonov et al. 2004; Yakubenko et al. 2011) induces uptake of ox-LDL and formation of foam cells in atherosclerosis. In vitro, CD36-dependent uptake of palmitic acid led to lipid accumulation in podocytes together with a dose-dependent increase in the levels of mitochondrial ROS, depolarization of mitochondria, ATP depletion, and apoptosis (Baranova et al. 2005; Hua et al. 2015). Knockout of CD36 in monocytes in unilateral ureteral obstruction (UUO)-induced kidney fibrosis mouse models and ischaemia–reperfusion injury models reduced the severity of fibrosis and ameliorated kidney function (Pennathur et al. 2015).

It is also reported that oxidized HDL enhanced the production of reactive oxygen species (ROS) and upregulated the expression of pro-inflammatory factors, impairing the function of human renal proximal tubule epithelial cells (HK-2) mainly through the scavenger receptor, CD36 (Gao et al. 2014). Zhang M also found that that oxidized HDL enhanced pro-inflammatory properties in mesangial cells partly via CD36 (Zhang et al. 2010). A CKD mouse model on high-fat diet shows CD36 is a key modulator of pro-inflammatory and oxidative pathways that promote fibrogenesis (Okamura et al. 2009). In proximal tubular epithelial cells, CD36-dependent signaling was involved in the activation of protein kinase Cα (PKCα) and NADPH oxidase (Cao et al. 2013). NADPH oxidase-dependent ROS production can lead to multiple downstream events such as NLRP3 inflammasome priming, activation of NF-κB, secretion of TGF-β, and downregulation of PPARα, PPARGC1α, and CPT1 (Yang et al. 2017). Anti-CD36 monoclonal antibody significantly inhibited secretion of TGF-β1 induced by AOPPs-HSA in HK-2 cells (Iwao et al. 2008). In fact, CD36 promoted renal fibrosis not only by increasing oxidative stress and activating pro-inflammatory pathways (Kim et al. 2009), but also by regulation the expression of TGF-β1 (Iwao et al. 2008) and activation of intrarenal renin–angiotensin system (RAS), playing a critical role in the progression of CKDs and promoting renal fibrogenesis (Kobori et al. 2007).

Hypertriglyceridemia is also regarded as an independent risk factor for developing proteinuria and CKD (Lee et al. 2009; Tozawa et al. 2002). Adipose tissue is typically the appropriate organ for lipid storage. However, once the entire storage capacity is exceeded, the lipids will be deposited ectopically, including in the kidney. In obesity, gross observations have revealed a large fat accumulation in the renal sinus and an altered appearance of renal parenchyma. Aided by microscopy and various staining techniques, ectopic lipid deposition in kidney has been confirmed. In fact, in addition to obesity and metabolic diseases, ectopic lipid aggregation is not uncommon in diverse renal diseases such as diabetic kidney disease (DKD), minimal change disease (especially tip variant), lupus nephritis, membranous nephropathy, Alport syndrome, Fabry disease (FD), hypertensive glomerular injury, and the aging kidney. An association between local lipid accumulation and kidney damage has been confirmed in multiple animal models and clinical studies (Wang et al. 2012; Weinberg 2006).

26.5 Pathophyiological Changes Driven by Lipid Overloading in Kidney

Lipid loaded foam cells in the kidney and atherosclerotic plaques support pathophysiological roles for lipids in the progression of both CKD.

26.5.1 *Oxidative Stress*

Though initial events involved in lipid-mediated renal damage are unclear, oxidative stress is thought to be especially important. Hyperlipidemia causes significantly higher rates of monocyte reactive oxygen species (ROS) production, which is strongly associated with impairment of endothelium-dependent relaxation and elevated plasma Ox-LDL level. Arteries from hypercholesterolemic animals produced significantly higher rates of oxygen radical than control arteries.

The mechanisms by which hyperlipidemia contributes to systemic oxidative stress in CKD remain unclear. Plasma HDL cholesterol with its important antioxidant function is reduced in CKD (Vaziri 2008). Inflammatory mediators, including TNFα and IL-1β, are ROS activating factors in the kidney and may induce oxygen radical production by MCs. Immune-mediated mesangial injury causes increased oxygen radical and eicosanoid production (Oberle et al. 1992). An important source of ROS is NAD(P)H-oxidase (NOX). The NOX family includes seven members: Nox1–Nox7. Nox1 and Nox2 (gp91phox-containing NADPH oxidase), together with Nox4 and Nox5 have been identified in the cardiovascular–renal systems and also have been implicated in oxidative stress in kidney (Sedeek et al. 2009) disease. Additionally, the leukocyte-derived enzymes myeloperoxidase (MPO) and xanthine oxidoreductase (XOR) may contribute to oxidative stress pathways in end-stage renal disease (ESRD) with a role in cardiovascular dysfunction (Kaysen and Eiserich 2004).

Inappropriate ROS production may contribute to tissue dysfunction in three ways: (i) dysregulation of redox-sensitive signaling pathways, (ii) oxidative damage to biological structures including DNA, proteins, and lipids, and (iii) activation of macrophages. Lipid peroxidation is the first step in the generation of Ox-LDL, which can accumulate in renal mesangial cells (Ruan et al. 1999). The process of lipid peroxidation itself generates free radicals and ROS.

Ox-LDL, produced in vitro by incubating LDL with CuSO4, could induce podocyte (Bussolati et al. 2005) and endothelial cells apoptosis, which may influence cellular turnover in vascular and renal injuries. Furthermore, all major cell types in the artery wall and kidney including endothelial cells, SMCs, monocyte–macrophages and MCs have been shown to cause oxidative modification of LDL in vitro (Fernando et al. 1993; Heinecke et al. 1986). Oxidative stress decreases renal NO production and availability (Rahman et al. 1999) and stimulates angiotensin II synthesis, suggesting that renin–angiotensin system (RAS) activation may contribute to lipid-induced renal injury. It has been demonstrated that Angiotensin II increases the expression of

TGF-β and plasminogen activator inhibitor-1 (PAI-1), thereby propagating glomerular fibrosis (Chalmers et al. 2006). Oxidized LDL has also been identified in the lesions of FSGS in vivo (Lee 1999).

26.5.2 Endoplasmic Reticulum (ER) Stress

Metabolic stress within the ER induces a coordinated unfolded protein response (UPR), which helps the ER to cope with the accumulation of misfolded proteins. UPR is initiated by three ER transmembrane proteins [namely PKR-like ER-regulated kinase (PERK), inositol-requiring enzyme-1 (IRE-1), and activating transcription factor-6 (ATF-6) (Ron and Walter 2007). Recent studies report that intracellular accumulation of saturated fatty acids and cholesterol results in ER stress, resulting in apoptosis of macrophages. Macrophage scavenger receptor type A is essential in regulating ER stress-induced apoptosis (Tabas 2002). Palmitate also induces ER stress by increasing IRE1 protein levels and activating the c-Jun NH_2-terminal kinase (JNK) pathway (Bachar et al. 2009). In both cultured cells and whole animals, ER stress leads to activation of the JNK and IKK/NF-κB pathways, promoting an inflammatory response. ER stress, in turn, leads to dysregulation of the endogenous sterol response mechanism and concordantly activates oxidative stress pathways (Kovacs et al. 2009).

26.5.3 Inflammatory Stress

The presence of oxidative and ER stress activates the NF-κB pathway which has been reported associated with inflammatory events in glomerulonephritis, as well the progression of CKD (Guijarro and Egido 2001). In addition, lipids may act as pro-inflammatory mediators. At certain concentrations, LDL, VLDL, and IDL enhanced the secretion of inflammatory cytokines by MCs, including IL-6, PDGF, and TGFβ. Since HDL could downregulate VCAM-1 and E-selectin on endothelial surfaces and reduce NF-κB, low HDL cholesterol levels may augment inflammatory responses (Ashby et al. 1998). In Apo E KO mice (Tomiyama-Hanayama et al. 2009) blocking IL-6 receptor prevented the progression of proteinuria and renal lipid deposition, as well as the mesangial cell proliferation associated with severe hyperlipoproteinemia. These results strongly support the role of pro-inflammatory cytokines in the pathogenesis of hyperlipidemia-induced glomerular injury. Inflammation also enhances both medial and intimal calcifications which contribute to vascular, and perhaps also renal injury (Al-Aly 2008; Moe and Chen 2005).

Ox-LDL preferentially binds to the glomerulus when injected intra-arterially in the rat and to mesangial cells in vitro (Coritsidis et al. 1991). Ox-LDL is a potent pro-inflammatory chemoattractant for macrophages and T lymphocytes with a role in recruiting circulating monocytes either directly or by inducing SMC, MCs, and/or endothelial cells to produce chemotactic and adhesive factors such as MCP-1, monocyte colony stimulating factor (m-CSF), and IL-1β (Berliner et al. 1990; Cases and Coll 2005). Modified LDL may also inhibit the motility of resident monocytes once they have differentiated into macrophages within the site (Quinn et al. 1987). Both oxidized LDL and minimally oxidized LDL stimulated TNFα secretion by MCs by activating the NF-κB pathway (Guijarro and Egido 2001).

26.5.4 Renal Fibrosis

Renal fibrosis is the key pathological change in the process of end-stage renal disease, and it is the final pathological outcome of various chronic progressive renal diseases.

26.5.4.1 Wnt/β-Catenin Signaling and Renal Fibrosis

Wnt/β-catenin signaling is an evolutionarily conserved, outside-in signal pathway that has a fundamental role in development and homeostasis of tissues, regulating cell morphology, proliferation, motility, and cell fate. Canonical and non-canonical Wnt ligands regulate key metabolic signaling pathways such as mTOR and insulin signaling. Given the importance of Wnt signaling in homeostasis and metabolism, aberrant Wnt signaling may contribute to chronic metabolic diseases. Although relatively silent in normal adult kidneys, Wnt/β-catenin signaling is re-activated after renal injury in a variety of animal models and in human kidney disorders such as obstructive nephropathy, diabetic nephropathy, adriamycin nephropathy, remnant kidneys after 5/6 nephrectomy, polycystic kidney disease, and chronic allograft nephropathy (Dai et al. 2009; He et al. 2009; Kato et al. 2011; Surendran et al. 2005; von Toerne et al. 2009). This re-activated canonical Wnt signaling can contribute to adult kidney fibrosis. Inhibition of canonical signaling also reduced total renal fibrosis and matrix deposition (He et al. 2009; Iglesias et al. 2007). In kidney cells, Wnt/β-catenin promotes the expression of numerous fibrosis-related genes such as Snail1, plasminogen activator inhibitor-1, and matrix metalloproteinase-7 (Dai et al. 2009; He et al. 2010).

Wnt/β-catenin has also been known as a molecule switch promoting adipogenesis and lipid accumulation (Dai et al. 2009; He et al. 2012). Wnt signaling maintains preadipocytes in an undifferentiated state through inhibition of the adipogenic transcription factors PPARγ. When Wnt signaling in preadipocytes is prevented, these cells differentiate into adipocytes, while disruption of Wnt signaling also causes adipogenesis. In fact, non-canonical Wnts, Wnt5a, and Wnt5b, have been shown to promote adipogenesis by increasing PPAR-γ expression which is critical for adipose tissue (van Tienen et al. 2009). Meanwhile, blocking Wnt/β-catenin also can effectively reduce serum triglyceride levels (Feng et al. 2016; Malliou et al. 2018), and the Wnt/β-catenin pathway was activated in aortas of mice fed with a high-cholesterol diet (Borrell-Pages et al. 2015).

In CKD, mounting evidence indicates that Wnt/β-catenin signaling plays a key role in promoting podocyte injury and/or dysfunction through dedifferentiation and mesenchymal transition, which led to proteinuria and glomerulosclerosis (Heikkila et al. 2010; Liu 2010; Zeisberg and Neilson 2009). It is also found that canonical Wnt/β-catenin signaling mediates TGF-β1-driven podocyte injury and proteinuria (Kato et al. 2011). Fenofibrate, a PPARα agonist, has been shown to increase the expression of lipolytic enzymes and reduce lipid accumulation in glomeruli (Tanaka et al. 2011). It can inhibit aberrant activation of the canonical Wnt pathway in the kidneys of diabetic rats (Cheng et al. 2016). In view of the importance of Wnt/b-catenin signaling in renal fibrosis and lipid metabolism, blockade of this signaling might be beneficial in fibrotic CKD (Zhou and Liu 2015), especially with lipid metabolism disorders.

26.5.4.2 TGF-β

In addition to Wnt/β-catenin signaling, researchers also found that TGF-β plays a role in lipid metabolism disorders and renal fibrosis. TGF-β is one of most widely potent regulators of many fibrosis diseases such as various chronic liver diseases and CKD, which may induce epithelial–mesenchymal transition (EMT) of renal tubular epithelial cells (Ding et al. 1993; Liu et al. 2018). In animal model, blockade of TGF-β signaling ameliorated ECM accumulation in anti-Thy-1 nephritis (Kasuga et al. 2001), reduced proteinuria, renal fibrosis, and glomerular damage in hypertension-induced renal injury (Murphy et al. 2012), suggesting that anti-TGF-β therapy might afford renoprotection opposing the progression of renal disease.

Disorders of lipid metabolism may increase the expression of TGF-β. Studies indicated that LCFA upregulated the mRNA expression of TGF-β and key downstream transcription factors Snail, Twist and Zeb1 of hepatocyte (Liu et al. 2018). Hypercholesterolemia induced by cholesterol feeding upregulated glomerular TGF-β1 and fibronectin gene expression (Ding et al. 1993). It is shown that Ox-LDL increased fibronectin expression in glomerular epithelial cells by a mechanism involving expression of TGF-β, suggesting that abnormal lipid accumulation in human glomerular epithelial cells may contribute to the pathogenesis of glomerulosclerosis through TGF-β-mediated mechanism (Ding et al. 1994). A study also found a significant increase in renal cortical TGF-β1 mRNA levels with the appearance of TGF-β1-positive interstitial cells in rats with diet-induced hypercholesterolemia (Eddy 1996). All of these showed that TGF-β is an important mediator of the kidney pathological effects induced by lipids.

Furthermore, Wnt/β-catenin signaling works in combination with TGF-β signaling in the process of fibrosis, and TGF-β signaling can induce the expression of Wnt/β-catenin superfamily members and vice versa (Guo et al. 2012). Inhibition of Wnt is effective in blocking TGF-β-mediated β-catenin activation in vivo, resulting in diminished podocyte injury and albuminuria induced by TGF-β1 (Wang et al. 2011). Taken together, TGF-β is a potential regulator underlying lipid metabolism disorders and renal fibrosis.

26.6 Lipid Lowering: A Potential Approach for Prevention of Renal Fibrosis

Mounting evidence shows that hyperlipidemia can lead to glomerulosclerosis and renal fibrosis, and therefore, lipid lowering may act as a potential therapeutic approach for prevention of renal fibrosis. A research (Chade et al. 2005) showed a simultaneous increase in extracellular matrix deposition with an increased collagen IV expression and fibrosis in renal tissues of pigs fed a high-cholesterol diet and a lipid-lowering dietary intervention partly regressed interstitial and vascular fibrosis, renal functional decline, and structural injury. Another study showed that lipid-lowering therapy with rosuvastatin significantly decreased mesangial expansion and exerted a protective effect on renal function and morphology in young Zucker rats fed a HFD (Reisin et al. 2009). In a study including 43 patients with idiopathic nephrotic syndrome, lipid-lowering therapy with fluvastatin decreased the interstitial fibrosis and renal fat deposits and played a protective effect on kidney function (Gheith et al. 2002).

However, several lines of evidence demonstrated that statins may have pleiotropic beneficial effects including antioxidant properties, anti-inflammatory properties, myofibroblast function downregulation and decreasing extracellular matrix (ECM) production, fibrinolytic effects through downregulation of plasminogen activator inhibitor-1 (PAI-1), inhibition of epithelial-to-mesenchymal transition (EMT), and suppressed mesangial and monocyte/macrophage proliferation, which are all independent of their lipid-lowering effect. It is shown that epithelial-to-mesenchymal transition plays a crucial role in contributing to the developmental and pathological processes of tissue fibrosis, including renal fibrosis. Clark et al. (2016) demonstrated that simvastatin could inhibit EMT through activation heme oxygenase-1 (HO-1) gene in renal proximal tubule cells, which may play an important role in protection of the kidney from EMT and consequent fibrosis. In STZ-induced diabetic nephropathy rat models, lovastatin significantly ameliorated micro-albuminuria and renal hypertrophy, inhibited EMT, and prevented renal fibrosis without affecting blood glucose and blood lipids level, the mechanism of which might be partly through suppression of oxidative stress and downregulated TGF-β1-Smad signaling pathway (Ma et al. 2017). Johnson et al. (1999) found that simvastatin may abrogate cyclosporin (CsA)-augmented interstitial collagen synthesis and further ameliorate interstitial fibrosis via direct actions on renal cortical fibroblasts in a human model of cyclosporin nephrotoxicity. Another study showed that simvastatin significantly abated development of renal fibrosis, probably by upregulating inhibitors of TGF-β signaling and thereby attenuating EMT, fibrogenic activity, and remodeling. In high-cholesterol diet-induced hypercholesterolemia pig models, simvastatin attenuated inflammatory and oxidative injury as well as fibrosis in kidneys, with an associated increase in renal perfusion (Chade et al. 2008). Hamasaki et al. (2012) found that simvastatin contributed to preventing the progression of renal fibrosis by upregulating BMP-7-mediated anti-fibrotic signaling in a mouse tubulo-interstitial injury model. The mechanism is involved in regulating HOXA13-USAG-1 pathway. Nam et al. (2013) showed that rosuvastatin supplementation attenuated renal inflammation and cell apoptosis and inhibited the fibrotic processes via the Smad-dependent and Smad-independent pathways in a rat model of CsA-induced nephropathy.

26.7 Conclusion

Clinical and experimental evidence suggest that dyslipidemia is not only a consequence of CKD, but also a cause for the progression of CKD by activating inflammatory, oxidative and ER stress, and renal fibrosis. Renal fibrosis mediated by Wnt/β-catenin signaling and TGF-β signaling is the final pathological outcome of various chronic progressive renal diseases. Lipid lowering may act as a potential therapeutic approach for prevention of renal fibrosis. The beneficial effect of statins is not solely derived from their lipid-lowering effect, but also its pleiotropic beneficial effects including antioxidant, anti-inflammatory, and anti-fibrotic properties, which may play a crucial role in prevention of renal fibrotic disease.

Acknowledgements "Lipid Nephrotoxicity: New Concept for an Old Disease" was reported by our group in Dyslipidemias in Kidney Disease (Ruan et al. 2014). The related contents are reused with permission.

References

Al-Aly Z (2008) Arterial calcification: a tumor necrosis factor-alpha mediated vascular Wnt-opathy. Transl Res 151:233–239

Antonov AS, Kolodgie FD, Munn DH, Gerrity RG (2004) Regulation of macrophage foam cell formation by alphaVbeta3 integrin: potential role in human atherosclerosis. Am J Pathol 165:247–258

Arici M, Chana R, Lewington A, Brown J, Brunskill NJ (2003) Stimulation of proximal tubular cell apoptosis by albumin-bound fatty acids mediated by peroxisome proliferator activated receptor-gamma. J Am Soc Nephrol 14:17–27

Ashby DT, Rye KA, Clay MA, Vadas MA, Gamble JR, Barter PJ (1998) Factors influencing the ability of HDL to inhibit expression of vascular cell adhesion molecule-1 in endothelial cells. Arterioscler Thromb Vasc Biol 18:1450–1455

Bachar E, Ariav Y, Ketzinel-Gilad M, Cerasi E, Kaiser N et al (2009) Glucose amplifies fatty acid-induced endoplasmic reticulum stress in pancreatic beta-cells via activation of mTORC1. PLoS ONE 4:e4954

Baranova IN, Vishnyakova TG. Bocharov AV, Kurlander R, Chen Z et al. (2005) Serum amyloid A binding to CLA-1 (CD36 and LIMPII analogous-1) mediates serum amyloid A protein-induced activation of ERK1/2 and p 38 mitogen-activated protein kinases. J Biol Chem 280:8031–8040

Berliner JA, Territo MC, Sevanian A, Ramin S, Kim JA et al (1990) Minimally modified low density lipoprotein stimulates monocyte endothelial interactions. J Clin Invest 85:1260–1266

Berneis KK, Krauss RM (2002) Metabolic origins and clinical significance of LDL heterogeneity. J Lipid Res 43:1363–1379

Borrell-Pages M, Romero JC, Badimon L (2015) LRP5 deficiency down-regulates Wnt signalling and promotes aortic lipid infiltration in hypercholesterolaemic mice. J Cell Mol Med 19:770–777
Brenner BM (1985) Nephron adaptation to renal injury or ablation. Am J Physiol 249:F324–F337
Brown PM, Kennedy DJ, Morton RE, Febbraio M (2015) CD36/SR-B2-TLR2 dependent pathways enhance *Porphyromonas gingivalis* mediated atherosclerosis in the Ldlr KO mouse model. PLoS ONE 10:e0125126
Bussolati B, Deregibus MC, Fonsato V, Doublier S, Spatola T et al (2005) Statins prevent oxidized LDL-induced injury of glomerular podocytes by activating the phosphatidylinositol 3-kinase/AKT-signaling pathway. J Am Soc Nephrol 16:1936–1947
Cao W, Xu J, Zhou ZM, Wang GB, Hou FF et al (2013) Advanced oxidation protein products activate intrarenal renin-angiotensin system via a CD36-mediated, redox-dependent pathway. Antioxid Redox Signal 18:19–35
Cases A, Coll E (2005) Dyslipidemia and the progression of renal disease in chronic renal failure patients. Kidney Int Suppl S87–S93
Chade AR, Mushin OP, Zhu X, Rodriguez-Porcel M, Grande JP et al (2005) Pathways of renal fibrosis and modulation of matrix turnover in experimental hypercholesterolemia. Hypertension 46:772–779
Chade AR, Zhu XY, Grande JP, Krier JD, Lerman A, Lerman LO (2008) Simvastatin abates development of renal fibrosis in experimental renovascular disease. J Hypertens 26:1651–1660
Chalmers L, Kaskel FJ, Bamgbola O (2006) The role of obesity and its bioclinical correlates in the progression of chronic kidney disease. Adv Chronic Kidney Dis 13:352–364
Cheng R, Ding L, He X, Takahashi Y, Ma JX (2016) Interaction of PPARalpha with the canonic Wnt pathway in the regulation of renal fibrosis. Diabetes 65:3730–3743
Clark JS, Carter AJ, Dixit M, Arany I (2016) Simvastatin inhibits epithelial-to-mesenchymal transition through induction of HO-1 in cultured renal proximal tubule cells. Vivo 30:407–411
Coritsidis G, Rifici V, Gupta S, Rie J, Shan ZH et al (1991) Preferential binding of oxidized LDL to rat glomeruli in vivo and cultured mesangial cells in vitro. Kidney Int 39:858–866
Dai C, Stolz DB, Kiss LP, Monga SP, Holzman LB, Liu Y (2009) Wnt/beta-catenin signaling promotes podocyte dysfunction and albuminuria. J Am Soc Nephrol 20:1997–2008
Ding G, Pesek-Diamond I, Diamond JR (1993) Cholesterol, macrophages, and gene expression of TGF-beta 1 and fibronectin during nephrosis. Am J Physiol 264:F577–F584
Ding G, van Goor H, Frye J, Diamond JR (1994) Transforming growth factor-beta expression in macrophages during hypercholesterolemic states. Am J Physiol 267:F937–F943
Eddy AA (1996) Interstitial inflammation and fibrosis in rats with diet-induced hypercholesterolemia. Kidney Int 50:1139–1149
Feng XL, Theodoratou E, Liu L, Chan KY, Hipgrave D et al (2012) Social, economic, political and health system and program determinants of child mortality reduction in China between 1990 and 2006: a systematic analysis. J Glob Health 2:010405
Feng X, Gao X, Jia Y, Zhang H, Xu Y, Wang G (2016) PPAR-alpha agonist fenofibrate decreased RANTES levels in type 2 diabetes patients with hypertriglyceridemia. Med Sci Monit 22:743–751
Fernando RL, Varghese Z, Moorhead JF (1993) Oxidation of low-density lipoproteins by rat mesangial cells and the interaction of oxidized low-density lipoproteins with rat mesangial cells in vitro. Nephrol Dial Transplant 8:512–518
Gao X, Wu J, Qian Y, Fu L, Wu G et al (2014) Oxidized high-density lipoprotein impairs the function of human renal proximal tubule epithelial cells through CD36. Int J Mol Med 34:564–572
Gheith OA, Sobh MA, Mohamed Kel S, El-Baz MA, El-Husseini F et al (2002) Impact of treatment of dyslipidemia on renal function, fat deposits and scarring in patients with persistent nephrotic syndrome. Nephron 91:612–619
Ginsberg HN, Le NA, Goldberg IJ, Gibson JC, Rubinstein A et al (1986) Apolipoprotein B metabolism in subjects with deficiency of apolipoproteins CIII and AI. Evidence that apolipoprotein CIII inhibits catabolism of triglyceride-rich lipoproteins by lipoprotein lipase in vivo. J Clin Invest 78:1287–1295

Grone HJ, Walli A, Grone E, Niedmann P, Thiery J et al (1989) Induction of glomerulosclerosis by dietary lipids. A functional and morphologic study in the rat. Lab Invest 60:433–446
Grone HJ, Walli AK, Grone EF (1993) Arterial hypertension and hyperlipidemia as determinants of glomerulosclerosis. Clin Investig 71:834–839
Guijarro C, Egido J (2001) Transcription factor-kappa B (NF-kappa B) and renal disease. Kidney Int 59:415–424
Guo Y, Xiao L, Sun L, Liu F (2012) Wnt/beta-catenin signaling: a promising new target for fibrosis diseases. Physiol Res 61:337–346
Hamasaki Y, Doi K, Okamoto K, Ijichi H, Seki G et al (2012) 3-Hydroxy-3-methylglutaryl-coenzyme A reductase inhibitor simvastatin ameliorates renal fibrosis through HOXA13-USAG-1 pathway. Lab Invest 92:1161–1170
He W, Dai C, Li Y, Zeng G, Monga SP et al (2009) Wnt/beta-catenin signaling promotes renal interstitial fibrosis. J Am Soc Nephrol 20:765–776
He W, Tan R, Dai C, Li Y, Wang D et al (2010) Plasminogen activator inhibitor-1 is a transcriptional target of the canonical pathway of Wnt/beta-catenin signaling. J Biol Chem 285:24665–24675
He W, Tan RJ, Li Y, Wang D, Nie J et al (2012) Matrix metalloproteinase-7 as a surrogate marker predicts renal Wnt/beta-catenin activity in CKD. J Am Soc Nephrol 23:294–304
Heikkila E, Juhila J, Lassila M, Messing M, Perala N et al (2010) beta-Catenin mediates adriamycin-induced albuminuria and podocyte injury in adult mouse kidneys. Nephrol Dial Transplant 25:2437–2446
Heinecke JW, Baker L, Rosen H, Chait A (1986) Superoxide-mediated modification of low density lipoprotein by arterial smooth muscle cells. J Clin Invest 77:757–761
Hua W, Huang HZ, Tan LT, Wan JM, Gui HB et al (2015) CD36 mediated fatty acid-induced podocyte apoptosis via oxidative stress. PLoS ONE 10:e0127507
Huang W, Febbraio M, Silverstein RL (2011) CD9 tetraspanin interacts with CD36 on the surface of macrophages: a possible regulatory influence on uptake of oxidized low density lipoprotein. PLoS ONE 6:e29092
Iglesias DM, Hueber PA, Chu L, Campbell R, Patenaude AM et al (2007) Canonical WNT signaling during kidney development. Am J Physiol Renal Physiol 293:F494–F500
Imai Y, Matsumura H, Shino A, Oka K, Suzuoki Z (1977) Induction of aortic lipid deposition in a high-response (ExHC) rat fed a diet containing cholesterol and cholic acid. Atherosclerosis 28:453–461
Iwao Y, Nakajou K, Nagai R, Kitamura K, Anraku M et al (2008) CD36 is one of important receptors promoting renal tubular injury by advanced oxidation protein products. Am J Physiol Renal Physiol 295:F1871–F1880
Jiang XS, Chen XM, Wan JM, Gui HB, Ruan XZ et al (2017) Autophagy protects against palmitic acid-induced apoptosis in podocytes in vitro. Sci Rep 7:42764
Johnson DW, Saunders HJ, Field MJ, Pollock CA (1999) In vitro effects of simvastatin on tubulointerstitial cells in a human model of cyclosporin nephrotoxicity. Am J Physiol 276:F467–F475
Kasiske BL, Cleary MP, O'Donnell MP, Keane WF (1985) Effects of genetic obesity on renal structure and function in the Zucker rat. J Lab Clin Med 106:598–604
Kasiske BL, O'Donnell MP, Schmitz PG, Kim Y, Keane WF (1990) Renal injury of diet-induced hypercholesterolemia in rats. Kidney Int 37:880–891
Kasuga H, Ito Y, Sakamoto S, Kawachi H, Shimizu F et al (2001) Effects of anti-TGF-beta type II receptor antibody on experimental glomerulonephritis. Kidney Int 60:1745–1755
Kato H, Gruenwald A, Suh JH, Miner JH, Barisoni-Thomas L et al (2011) Wnt/beta-catenin pathway in podocytes integrates cell adhesion, differentiation, and survival. J Biol Chem 286:26003–26015
Kaysen GA, Eiserich JP (2004) The role of oxidative stress-altered lipoprotein structure and function and microinflammation on cardiovascular risk in patients with minor renal dysfunction. J Am Soc Nephrol 15:538–548
Kim HJ, Moradi H, Yuan J, Norris K, Vaziri ND (2009) Renal mass reduction results in accumulation of lipids and dysregulation of lipid regulatory proteins in the remnant kidney. Am J Physiol Renal Physiol 296:F1297–F1306

Kobashigawa JA, Kasiske BL (1997) Hyperlipidemia in solid organ transplantation. Transplantation 63:331–338
Kobori H, Nangaku M, Navar LG, Nishiyama A (2007) The intrarenal renin-angiotensin system: from physiology to the pathobiology of hypertension and kidney disease. Pharmacol Rev 59:251–287
Koitabashi Y, Ikoma M, Miyahira T, Fujita R, Mio H et al (1990) Long-term follow-up of a paediatric case of lipoprotein glomerulopathy. Pediatr Nephrol 4:122–128
Koletsky S (1975) Pathologic findings and laboratory data in a new strain of obese hypertensive rats. Am J Pathol 80:129–142
Kovacs WJ, Tape KN, Shackelford JE, Wikander TM, Richards MJ et al (2009) Peroxisome deficiency causes a complex phenotype because of hepatic SREBP/Insig dysregulation associated with endoplasmic reticulum stress. J Biol Chem 284:7232–7245
Lee HS (1999) Oxidized LDL, glomerular mesangial cells and collagen. Diabetes Res Clin Pract 45:117–122
Lee HS, Lee JS, Koh HI, Ko KW (1991) Intraglomerular lipid deposition in routine biopsies. Clin Nephrol 36:67–75
Lee HS, Jeong JY, Kim BC, Kim YS, Zhang YZ et al (1997) Dietary antioxidant inhibits lipoprotein oxidation and renal injury in experimental focal segmental glomerulosclerosis. Kidney Int 51:1151–1159
Lee PH, Chang HY, Tung CW, Hsu YC, Lei CC et al (2009) Hypertriglyceridemia: an independent risk factor of chronic kidney disease in Taiwanese adults. Am J Med Sci 338:185–189
Li Y, Qi X, Tong X, Wang S (2013) Thrombospondin 1 activates the macrophage Toll-like receptor 4 pathway. Cell Mol Immunol 10:506–512
Liu Y (2010) New insights into epithelial-mesenchymal transition in kidney fibrosis. J Am Soc Nephrol 21:212–222
Liu Y (2011) Cellular and molecular mechanisms of renal fibrosis. Nat Rev Nephrol 7:684–696
Liu Y, Coresh J, Eustace JA, Longenecker JC, Jaar B et al (2004) Association between cholesterol level and mortality in dialysis patients: role of inflammation and malnutrition. JAMA 291:451–459
Liu J, Yang P, Zuo G, He S, Tan W et al (2018) Long-chain fatty acid activates hepatocytes through CD36 mediated oxidative stress. Lipids Health Dis 17:153
Lowrie EG, Lew NL (1990) Death risk in hemodialysis patients: the predictive value of commonly measured variables and an evaluation of death rate differences between facilities. Am J Kidney Dis 15:458–482
Ma Z, Zhu L, Liu Y, Wang Z, Yang Y et al (2017) Lovastatin alleviates endothelial-to-mesenchymal transition in glomeruli via suppression of oxidative stress and TGF-beta1 signaling. Front Pharmacol 8:473
Malliou F, Andreadou I, Gonzalez FJ, Lazou A, Xepapadaki E et al (2018) The olive constituent oleuropein, as a PPARalpha agonist, markedly reduces serum triglycerides. J Nutr Biochem 59:17–28
Moe SM, Chen NX (2005) Inflammation and vascular calcification. Blood Purif 23:64–71
Moorhead JF, Chan MK, El-Nahas M, Varghese Z (1982) Lipid nephrotoxicity in chronic progressive glomerular and tubulo-interstitial disease. Lancet 2:1309–1311
Moradi H, Pahl MV, Elahimehr R, Vaziri ND (2009) Impaired antioxidant activity of high-density lipoprotein in chronic kidney disease. Transl Res 153:77–85
Muntner P, Coresh J, Smith JC, Eckfeldt J, Klag MJ (2000) Plasma lipids and risk of developing renal dysfunction: the atherosclerosis risk in communities study. Kidney Int 58:293–301
Murphy SR, Dahly-Vernon AJ, Dunn KM, Chen CC, Ledbetter SR et al (2012) Renoprotective effects of anti-TGF-beta antibody and antihypertensive therapies in Dahl S rats. Am J Physiol Regul Integr Comp Physiol 303:R57–R69
Nam HK, Lee SJ, Kim MH, Rho JH, Son YK et al (2013) Rosuvastatin attenuates inflammation, apoptosis and fibrosis in a rat model of cyclosporine-induced nephropathy. Am J Nephrol 37:7–15
Oberle GP, Niemeyer J, Thaiss F, Schoeppe W, Stahl RA (1992) Increased oxygen radical and eicosanoid formation in immune-mediated mesangial cell injury. Kidney Int 42:69–74

Okamura DM, Pennathur S, Pasichnyk K, Lopez-Guisa JM et al (2009) CD36 regulates oxidative stress and inflammation in hypercholesterolemic CKD. J Am Soc Nephrol 20:495–505
Pascual M, Theruvath T, Kawai T, Tolkoff-Rubin N, Cosimi AB (2002) Strategies to improve long-term outcomes after renal transplantation. N Engl J Med 346:580–590
Pennathur S, Pasichnyk K, Bahrami NM, Zeng L, Febbraio M et al (2015) The macrophage phagocytic receptor CD36 promotes fibrogenic pathways on removal of apoptotic cells during chronic kidney injury. Am J Pathol 185:2232–2245
Peric-Golia L, Peric-Golia M (1983) Aortic and renal lesions in hypercholesterolemic adult, male, virgin sprague-dawley rats. Atherosclerosis 46:57–65
Quinn MT, Parthasarathy S, Fong LG, Steinberg D (1987) Oxidatively modified low density lipoproteins: a potential role in recruitment and retention of monocyte/macrophages during atherogenesis. Proc Natl Acad Sci USA 84:2995–2998
Rahman MM, Varghese Z, Fuller BJ, Moorhead JF (1999) Renal vasoconstriction induced by oxidized LDL is inhibited by scavengers of reactive oxygen species and L-arginine. Clin Nephrol 51:98–107
Reisin E, Ebenezer PJ, Liao J, Lee BS, Larroque M et al (2009) Effect of the HMG-CoA reductase inhibitor rosuvastatin on early chronic kidney injury in obese zucker rats fed with an atherogenic diet. Am J Med Sci 338:301–309
Ron D, Walter P (2007) Signal integration in the endoplasmic reticulum unfolded protein response. Nat Rev Mol Cell Biol 8:519–529
Ruan XZ, Varghese Z, Powis SH, Moorhead JF (1999) Human mesangial cells express inducible macrophage scavenger receptor. Kidney Int 56:440–451
Ruggiero C, Elks CM, Kruger C, Cleland E, Addison K et al (2014) Albumin-bound fatty acids but not albumin itself alter redox balance in tubular epithelial cells and induce a peroxide-mediated redox-sensitive apoptosis. Am J Physiol Renal Physiol 306:F896–F906
Samuelsson O, Attman PO, Knight-Gibson C, Larsson R, Mulec H et al (1998) Complex apolipoprotein B-containing lipoprotein particles are associated with a higher rate of progression of human chronic renal insufficiency. J Am Soc Nephrol 9:1482–1488
Sedeek M, Hebert RL, Kennedy CR, Burns KD, Touyz RM (2009) Molecular mechanisms of hypertension: role of Nox family NADPH oxidases. Curr Opin Nephrol Hypertens 18:122–127
Stewart CR, Stuart LM, Wilkinson K, van Gils JM, Deng J et al (2010) CD36 ligands promote sterile inflammation through assembly of a Toll-like receptor 4 and 6 heterodimer. Nat Immunol 11:155–161
Stremmel W, Pohl L, Ring A, Herrmann T (2001) A new concept of cellular uptake and intracellular trafficking of long-chain fatty acids. Lipids 36:981–989
Surendran K, Schiavi S, Hruska KA (2005) Wnt-dependent beta-catenin signaling is activated after unilateral ureteral obstruction, and recombinant secreted frizzled-related protein 4 alters the progression of renal fibrosis. J Am Soc Nephrol 16:2373–2384
Tabas I (2002) Consequences of cellular cholesterol accumulation: basic concepts and physiological implications. J Clin Invest 110:905–911
Takemura T, Yoshioka K, Aya N, Murakami K, Matumoto A et al (1993) Apolipoproteins and lipoprotein receptors in glomeruli in human kidney diseases. Kidney Int 43:918–927
Tanaka Y, Kume S, Araki S, Isshiki K, Chin-Kanasaki M et al (2011) Fenofibrate, a PPARalpha agonist, has renoprotective effects in mice by enhancing renal lipolysis. Kidney Int 79:871–882
Tomiyama-Hanayama M, Rakugi H, Kohara M, Mima T, Adachi Y et al (2009) Effect of interleukin-6 receptor blockage on renal injury in apolipoprotein E-deficient mice. Am J Physiol Renal Physiol 297:F679–F684
Tozawa M, Iseki K, Iseki C, Oshiro S, Ikemiya Y et al (2002) Triglyceride, but not total cholesterol or low-density lipoprotein cholesterol levels, predict development of proteinuria. Kidney Int 62:1743–1749
Tsuruya K, Yoshida H, Nagata M, Kitazono T, Iseki K et al (2015) Impact of the triglycerides to high-density lipoprotein cholesterol ratio on the incidence and progression of CKD: A longitudinal study in a large Japanese Population. Am J Kidney Dis 66:972–983

van Goor H, van der Horst ML, Atmosoerodjo J, Joles JA, van Tol A et al (1993) Renal apolipoproteins in nephrotic rats. Am J Pathol 142:1804–1812
van Tienen FH, Laeremans H, van der Kallen CJ, Smeets HJ et al (2009) Wnt5b stimulates adipogenesis by activating PPARgamma, and inhibiting the beta-catenin dependent Wnt signaling pathway together with Wnt5a. Biochem Biophys Res Commun 387:207–211
Vaziri ND (2008) Causal link between oxidative stress, inflammation, and hypertension. Iran J Kidney Dis 2:1–10
Vaziri ND, Moradi H, Pahl MV, Fogelman AM, Navab M (2009) In vitro stimulation of HDL anti-inflammatory activity and inhibition of LDL pro-inflammatory activity in the plasma of patients with end-stage renal disease by an apoA-1 mimetic peptide. Kidney Int 76:437–444
von Toerne C, Schmidt C, Adams J, Kiss E, Bedke J et al (2009) Wnt pathway regulation in chronic renal allograft damage. Am J Transplant 9:2223–2239
Wang D, Dai C, Li Y, Liu Y (2011) Canonical Wnt/beta-catenin signaling mediates transforming growth factor-beta1-driven podocyte injury and proteinuria. Kidney Int 80:1159–1169
Wang W, He B, Shi W, Liang X, Ma J et al (2012) Deletion of scavenger receptor A protects mice from progressive nephropathy independent of lipid control during diet-induced hyperlipidemia. Kidney Int 81:1002–1014
Wanner C, Ritz E (2008) Reducing lipids for CV protection in CKD patients-current evidence. Kidney Int Suppl S24–S28
Weinberg JM (2006) Lipotoxicity. Kidney Int 70:1560–1566
Yakubenko VP, Bhattacharjee A, Pluskota E, Cathcart MK (2011) alphaMbeta(2) integrin activation prevents alternative activation of human and murine macrophages and impedes foam cell formation. Circ Res 108:544–554
Yang X, Okamura DM, Lu X, Chen Y, Moorhead J et al (2017) CD36 in chronic kidney disease: novel insights and therapeutic opportunities. Nat Rev Nephrol 13:769–781
Zeisberg M, Neilson EG (2009) Biomarkers for epithelial-mesenchymal transitions. J Clin Invest 119:1429–1437
Zhang M, Gao X, Wu J, Liu D, Cai H et al (2010) Oxidized high-density lipoprotein enhances inflammatory activity in rat mesangial cells. Diabetes Metab Res Rev 26:455–463
Zhou L, Liu Y (2015) Wnt/beta-catenin signalling and podocyte dysfunction in proteinuric kidney disease. Nat Rev Nephrol 11:535–545

Chapter 27
Renal Interstitial Lymphangiogenesis in Renal Fibrosis

Gang Xu

Abstract The basic physiological functions of the lymphatic system include absorption of water and macromolecular substances in the interstitial fluid to maintain the fluid homeostasis, promoting the intestinal absorption of nutrients such as lipids and vitamins from food. Recent studies have found that lymphangiogenesis is associated with some pathological conditions, such as tumor metastasis, injury repair, and chronic inflammation. For a long time, the study of lymphatic vessels (LVs) has been stagnant because of the lack of lymphatic-specific cytology and molecular markers. Renal interstitial lymphangiogenesis is found in patients with chronic kidney disease (CKD) and a series of animal models of renal fibrosis. Intervention of the formation or maturation of LVs in renal tissue of CKD may reduce the drainage of inflammatory cells, attenuate chronic inflammation, delay the progression of renal fibrosis, and improve renal function. This review will summarize the latest findings on renal interstitial lymphangiogenesis in CKD.

Keywords Lymphangiogenesis · Kidney disease · Renal fibrosis · Immune responses

27.1 Introduction

There are two major circulatory systems in the body: the blood circulation system and the lymphatic circulation system, first described by Hippocrates. Although the two have many similarities in anatomy, structure, and function, only the blood circulation system has been widely and deeply studied, and lymphatic systems have been neglected for their elusive morphological structure and unique pathophysiological effects. It was not until the last decade that a series of landmark discoveries changed this situation (Choi et al. 2012). The first and most important finding is the lymphatic endothelial cells (LECs)-specific lymphatic endothelial growth factor

G. Xu (✉)
Department of Nephrology, Tongji Hospital Affiliated with Tongji Medical College, Huazhong University of Science and Technology, Wuhan, Hubei, China
e-mail: gangxu99@163.com

B.-C. Liu et al. (eds.), *Renal Fibrosis: Mechanisms and Therapies*,
Advances in Experimental Medicine and Biology 1165,
https://doi.org/10.1007/978-981-13-8871-2_27

receptor 3 (VEGFR-3) (Kaipainen et al. 1995). VEGFR-3 is one of the members of the VEGF receptor family. It is structurally related to VEGFR-1 and VEGFR-2. It is expressed in most vascular endothelial cells at the early stage of development, but is fixed in the lymphatic system at the late stage of development and after completion of development. Wigle et al. found that knocking out the homologous transcription factor PROX-1 leads to early blockade of lymphatic development and could not form the lymphatic system. After the completion of the development, the persistence of the LECs phenotype depends on the sustained expression of PROX-1, and overexpression of PROX-1 in vascular endothelial cells can also reconstitute it into LVs. Due to these important functions, PROX-1 is called the "master gene for lymphatic development" (Hong and Detmar 2003). Therefore, PROX-1 can be considered a marker of LECs. Following the discovery of VEGFR-3 and PROX-1, two other important markers of LVs, LYVE-1 and podoplanin (PDPN), were simultaneously reported in 1999. LYVE-1 is a receptor for hyaluronic acid (HA), which is specifically expressed in LECs and is homologous to CD44 (Banerji et al. 1999). Except as a lymphatic vessel marker, LYVE-1 is also expressed in activated tissue macrophages and hepatic and spleen sinusoidal endothelial cells. PDPN, a mucin-type transmembrane sialoglycoprotein, is expressed in endothelial cells of all major veins at 11.5 days of embryonic development in mice and gradually decreased in the expression of venous endothelial in late embryo, but it is continuously expressed in LECs during the whole development period. It is worth mentioning that D2-40, as a novel PDPN monoclonal antibody, can react with antigenic determinants of lymphatic rather than with vascular endothelium. This antibody has been widely used in the prediction of tumor metastasis and prognosis by lymphatic density. Besides, lymphatic biomarkers such as COUP-TF and FOXC2 also play an important role in the development and formation of LVs (Oliver 2004). With the progress of the research, the lymphatic system is no longer an adjunct to the blood system, but also has independent research value.

The lymphatic system is composed of LVs and secondary lymphoid organs. The tissue fluid, extracellular macromolecules, and cells present in the interstitial tissues, enter the primary capillary lymphatic vessels, merge into a larger collection of lymphatic vessels, pass through the lymph nodes, and finally enter the subclavian vein through the thoracic duct to return to the blood circulation system (Choi et al. 2012). The main function of primary capillary lymphatic vessels is drainage of interstitial fluid; it is composed of monolayer lymphatic endothelial cells, with no continuous basement membrane and pericytes surrounding, and a gap that can lead to the surrounding tissue (Johnson and Jackson 2008); endothelial cells are connected by a special discontinuous button-like joint, forming a primary valve to make lymph flow in one direction. At the same time, the capillary lymphatic endothelial cells are adhered to the outer layer of the filament and are directly anchored to the extracellular matrix. When the interstitial pressure rises, the interstitial filaments will prevent the collapse of the capillary lymphatic vessels by pulling the endothelial cells, opening the cell connections to allow interstitial fluid to enter (Choi et al. 2012; Schulte-Merker et al. 2011). The primary function of the collecting lymphatics is to transport lymph; the endothelial cells are tightly connected by a zipper-like joint, and there are basement membranes and pericardial cells, smooth muscle cells to promote lymphatic transport (Mäkinen et al. 2007). Similar to the vein structure, the collecting lymphatic vessels have pairs of valves that prevent lymphatic reflux (Johnson and Jackson 2008).

In addition to transporting interstitial fluid and cells to maintain tissue homeostasis, the regulation of the immune response by the lymphatic system has been a hot topic in the past decade. Recent studies have found that lymphangiogenesis is often associated with some pathological conditions, such as tumor metastasis (Martínez-Corral et al. 2012; Xiong et al. 2018), injury repair (Güç et al. 2017), and chronic inflammation (D'Alessio et al. 2014; Huggenberger et al. 2011; Kim et al. 2014). Immune cells enter the lymphatics through the gap between the endothelial cells and then transported to the regional lymph nodes to react with the antigen causing an immune response to remove pathogens from the body (Khawar et al. 2017). In addition, LECs can directly regulate immune response: Studies have shown that TLR-4 expression in LECs is an important part of LPs-induced lymphangiogenesis (Kang et al. 2009); inflammatory LECs can regulate immune response by directly inhibiting the maturation and function of DC cells through Mac-1/ICAM-1 pathway (Podgrabinska et al. 2009); these findings make the role of lymphatic vessels in inflammation and repair of damage increasingly prominent. Chronic kidney disease (CKD) is an immune inflammatory disease, and renal fibrosis is the end stage of all chronic renal diseases. The relationship between the progression of CKD and lymphangiogenesis is being paid more and more attention. This chapter will summarize the present status of research progress on renal interstitial lymphangiogenesis in chronic kidney disease.

27.2 Lymphangiogenesis and Kidney Disease

In healthy renal tissue, relatively few lymphatic capillaries are distributed around the interlobular artery and interlobular vein, and the glomerulus and renal interstitium are not distributed (Yazdani et al. 2014). In minimal change nephropathy and focal glomerulosclerosis without interstitial fibrosis, D2-40-positive LVs in the renal interstitial region are rare. However, some chronic kidney diseases such as lupus nephritis, antineutrophil cytoplasmic antibody-related glomerulonephritis, tubulointerstitial nephritis, focal segmental glomerulosclerosis, crescentic glomerulonephritis, type II diabetic nephropathy, and IgA nephropathy show markedly increased populations of LVs compared to controls (Heller et al. 2007; Khawar et al. 2017; Sakamoto et al. 2009; Zimmer et al. 2010). A number of studies have shown that the formation of proteinuria can aggravate kidney damage, and large amounts of proteinuria will cause renal tubular epithelial cells to express and release chemokines and mediators, causing inflammatory cells recruitment and kidney impairment (Bakris 2008; Eddy 2004; Moreno et al. 2014). Proteinuria can trigger renal lymphangiogenesis before the development of interstitial fibrosis via these chemokines promoting the recruitment of circulating leukocytes, such as regulatory T-cells (Yazdani et al. 2012).

It is noteworthy that lymphatic hyperplasia is significantly less in acute tubulointerstitial nephritis than in chronic tubulointerstitial nephritis. This suggests that chronic inflammation of the kidney may promote lymphangiogenesis. There is a significant correlation between the extent of renal interstitial damage and lymphatic vessel density in different types of chronic kidney disease. The density of renal interstitial LVs in type II diabetic nephropathy is higher than non-diabetic nephropathy. The density of renal interstitial lymphatics is positively correlated with the density of fibroblasts and macrophages in the renal tissue of non-diabetic nephropathy, while the density of renal interstitial LVs in diabetic nephropathy is not correlated with the density of fibroblasts and macrophages. The density of interstitial LVs in interstitial nephritis is significantly positively correlated with the density of fibroblasts, but not with the density of macrophages (Sakamoto et al. 2009). This indicates that the mechanism of renal interstitial lymphangiogenesis in different pathological types of chronic kidney disease varies.

Even so, pathological lymphangiogenesis requires some common factors. The best known factors to promote lymphangiogenesis are vascular endothelial growth factor (VEGF)-C and VEGF-D. VEGF-C binds to VEGFR-2 and VEGFR-3, which plays a crucial role in survival, proliferation, and migration of lymphatic endothelial cells (LECs) (Veikkola et al. 2001). VEGF-C or VEGF-D can stimulate protein kinase C-dependent activation of the ERK1 or ERK2 signaling cascade and phosphorylation of AKT (Mäkinen et al. 2001; Shibuya 2013). A previous study showed that the induction of lymphangiogenesis ameliorates inflammation and fibrosis in the renal interstitium. Enhancement of the VEGF-C signaling pathway in LECs may be a therapeutic strategy for renal fibrosis (Hasegawa et al. 2017). On the contrary, VEGF-C deficiency results in embryonic death in mice (Karkkainen et al. 2004). In addition, VEGF-D is another major inducer of lymphangiogenesis, also signaling via VEGFR3, but its ability to promote lymphangiogenesis is less important than that of VEGF-C. Both VEGF-C and VEGF-D are also able to induce angiogenesis (Benest et al. 2008; Rissanen et al. 2003). The VEGF-C/VEGFR-3 signaling pathway is currently considered to be the most important signaling pathway in lymphangiogenesis. The budding response of LECs to VEGF-C is mediated by VEGFR-3 and its co-receptor NRP-2. Although LECs budding requires VEGFR-3, knocking out VEGFR-3 does not alter lymphoid sac formation, suggesting possible VEGF-C also works with VEGFR-2. The activation of VEGFR-2 alone will only cause lymphatic vessel expansion without causing budding, indicating that VEGFR-2 and VEGFR-3 also have effects on lymphatic sprouting (Zhang et al. 2010). Beyond that, Tie1 and Tie2, which are vital to vascular remodeling, maturation, and stability, are also worth mentioning. Tie1 deficiency leads to lymphatic hypertrophy and abnormal remodeling, while the absence of Tie2 ligand can form functional lymphatic and valve-free lymphatics (D'Amico et al. 2010), which indicates that Tie1 and Tie2 also play an important role in lymphatic development. Recent findings suggest that basic fibroblast growth factor (FGF-2) also promotes lymphangiogenesis by interacting with the

LYVE-1 transmembrane domain on LVs, and this process does not depend on the activation of VEGFR-3 signaling (Platonova et al. 2013).

In the rat unilateral ureteral obstruction (UUO) model, the expression of TGF-β1 and VEGF-C was detected in renal tubular epithelial cells and monocytes, and the expression level increased gradually with the time prolongation of ureteral obstruction and reached the highest value at 14 days. Serial sections showed that the expression site and expression intensity of VEGF-C were consistent with TGF-β1. TGF-β1 up-regulated the expression of VEGF-C in tubular epithelial cells, fibroblasts, and macrophages, and TGF-β1 receptor inhibitor LY364947 inhibited the up-regulation of VEGF-C in vitro. LY364947 also inhibited the up-regulation of VEGF-C and lymphangiogenesis in the kidney in vivo (Suzuki et al. 2012). This indicates that in the UUO model, tubular cell-derived TGF-β1 ultimately induces lymphangiogenesis by up-regulating the expression of VEGF-C in renal parenchymal cells and macrophages. In addition, the connective tissue growth factor (CTGF) is another important factor in the production of VEGF-C (Kinashi et al. 2017). The study reported in this issue of Kidney International demonstrates that lymphangiogenesis to be associated with increased expression of CTGF and VEGF-C in human obstructed nephropathy as well as in diabetic kidney disease. Consistent with the assumptions, the increase in LVs and VEGF-C was significantly reduced in CTGF knockout compared to wild-type mice after UUO and ischemia–reperfusion injury. This conclusion was also verified in vitro using HK-2 cells. These two reports confirmed that renal tubules are damaged not only as victims but also as a promoter of the progression of kidney disease, which promoted lymphangiogenesis and the development of chronic kidney disease by producing TGF-β1, CTGF, and other cytokines.

As discussed, renal lymphangiogenesis has been reported to occur in numerous chronic kidney diseases. However, the origin of the newborn LVs is not completely understood. Our research team found that lymphangiogenesis occurs in the kidney after renal injury mainly results from the local proliferation of preexisting lymphatic endothelium (unpublished data). In order to confirm this hypothesis, we detected in situ immunostaining of Ki67 with LYVE-1. We found more intrarenal proliferating LVs (LYVE-1+ Ki67+ vessels) in sections from UUO or IRI kidneys compared with controls. Furthermore, we found lymphangiogenesis-related cytokines VEGF-C, VEGF-D, and FGF-2 were up-regulated in renal tubular epithelial cells (TECs). We used two different methods, parabiosis and bone marrow (BM) chimera, to determine whether bone marrow-derived cells also directly contribute to lymphangiogenesis in the UUO model. Parabiosis surgery was performed between wild-type mice and green fluorescence protein (GFP) mice, and chimera was done by BM transplantation from GFP mice to wild-type mice. We found few co-localizations of LYVE-1 and GFP in UUO kidneys after parabiosis or BM chimera. This finding indicates preexisting lymphatic endothelium proliferation, but not direct trans-differentiation of bone marrow-derived cells, which are the main source of lymphatic endothelium cells during renal lymphangiogenesis.

27.3 Lymphangiogenesis and Renal Fibrosis

The mechanisms of renal injury in different pathological types are not the same, but fibrosis is a common pathway for all types of kidney disease progression to end-stage nephropathy. Lesions of the renal tubule basement membrane and the glomerular capsule of the glomerular capsule during fibrosis will cause the primary urine to be filtered back to the renal interstitial (Kriz et al. 2001) so that the pressure of renal interstitial fluid increased, coupled with a large number of renal interstitial inflammatory cells infiltration. Therefore, objectively, more output channels are needed to drain these interstitial fluids and inflammatory cells, but at the same time, the perivascular capillary network is gradually destroyed as the fibrosis progresses and the inflammatory cells infiltrate, so from the basic functions of LVs (drainage interstitial fluid and inflammatory cells), the rebirth of renal interstitial lymphatics can be seen as a compensatory response to the kidney itself to drain excess interstitial fluid and inflammatory cells.

It is generally known that CKD is an immune inflammatory disease, and renal fibrosis is the end stage of all chronic kidney diseases. A recent review highlights the close relationship between inflammation and fibrosis (Wick et al. 2013). Lymphangiogenesis has also been previously correlated with renal fibrosis in substantial studies (Lee et al. 2013; Matsui et al. 2003; Sakamoto et al. 2009; Suzuki et al. 2012; Zampell et al. 2012). In animal research, 5/6 nephrectomized rat model as a more common model of renal fibrosis, its pathological morphology, and human end-stage renal disease have many similarities, such as loss of nephrons, compensatory hypertrophy of residual renal tubules, interstitial inflammatory cells infiltration, and progressive renal fibrosis. In the 5/6 nephrectomized group, the capillaries around the renal tubules were damaged compared with the normal group, and the renal interstitial LVs were significantly increased. LVs are distributed around the dilated tubules and fibrotic areas, and there are inflammatory cells in the individual lumen. In situ hybridization shows that lymphangiogenesis factor VEGF-C mRNA is expressed in renal interstitial infiltrating mononuclear cells (Matsui et al. 2003). Yazdani S et al. established a model of chronic adriamycin nephropathy from unilateral renal artery infusion of doxorubicin. The experiment lasted for 30 weeks. Within the first 12 weeks, accompanied by the increase in proteinuria, lymphatic density increased by 3 times, while renal tubular osteopontin expression was up-regulated, the number of mesenchymal fibroblasts increased significantly, but collagen deposition and macrophage infiltration did not increase significantly, while VEGF-C was mainly derived from renal tubular epithelium cell. In the ACEI intervention group, the renal damage decreased, while the lymphatic vessel formation was reduced correspondingly (Yazdani et al. 2012). This indicates that lymphangiogenesis is not synchronized with renal fibrosis in the adriamycin-induced nephropathy model, and VEGF-C derived from damaged renal parenchymal cells, rather than macrophage-derived VEGF-C, is involved in lymphangiogenesis. The above animal models have initially explored the mechanism of interstitial lymphangiogenesis during renal fibrosis and their foothold is VEGF-C. But in different animal models, due to various types and degrees of mass damage, there are also differences in the mechanism of lymphangiogenesis.

It is well known that LVs participate in the inflammatory response by regulating lymphocyte infiltration. Afferent LVs attract activated fibrocytes, dendritic, T, and B cells expressing the chemokine receptor 7 (CCR7) to secondary lymphoid organs by producing CC chemokine ligand 21 (CCL21) (Förster et al. 2008; Tammela and Alitalo 2010). Therefore, these neogenetic LVs are able to promote chronic kidney inflammation and renal fibrosis. Different degrees of lymphangiogenesis exist in almost all specimens of renal transplant rejection (Adair et al. 2007). A study on renal transplant rejection found that active lymphangiogenesis, with more than 50-fold increase in LV numbers being associated with nodular mononuclear infiltrates, which including a huge number of Ki-67^{+} CD4^{+} and CD8^{+} T lymphocytes, S100^{+} dendritic cells, and Ki-67^{+} CD20^{+} B lymphocytes and λ- and κ-chain-expressing plasmacytoid cells. In the same way, these CCR7+ cells within the nodular infiltrates seemed to be attracted by CCL21 that is produced and released by LECs (Kerjaschki et al. 2004). This elegant study provides direct evidence that CCR7/CCL21 signaling pathway can promote chronic inflammation and accelerate the progression of renal fibrosis. Another clinical study on kidney transplantation is also a good confirmation of this view. The authors collected 165 biopsy specimens from patients receiving kidney transplantation, histological examination using hematoxylin–eosin, periodic acid–Schiff, and Masson's trichrome staining, and CCR7, CCL21 were determined by immunohistochemistry. The results show that CCR7-positive expression was a strong protective factor for kidney transplant rejection, while CCL21-positive expression led to high susceptibility to recurrent nephropathy (Zhou et al. 2013). This experiment also demonstrated the important role of a CCR7/CCL21 signaling pathway in renal fibrosis induced by kidney transplantation. Animal experiments have also demonstrated that lymphangiogenesis regulates the progression of renal fibrosis through this signaling pathway. A previous study published in PNAS showed that a large number of CD45-positive inflammatory cells and type 1 collagen deposition reached the highest peak in 7 days in the UUO-induced renal fibrosis model. Most of these inflammatory cells express CCR7, and blocking CCR7/CCL21 signaling by anti-CCL21 antibody can reduce renal inflammatory cells infiltration and ameliorate renal fibrosis. This study shows that a CCL21-positive lumen (i.e., LVs) can cause a large amount of CCR7 inflammatory cells to be recruited in the kidney, therefore promoting renal fibrosis (Sakai et al. 2006). Similarly, we also found that UUO-induced renal fibrosis can also be alleviated by injecting mice with CCR7-neutralizing antibodies (unpublished data).

So far, in addition to CCR7/CCL21 signaling pathway, the role of hyaluronan in the regulation of renal fibrosis by lymphangiogenesis has also been emphasized. Hyaluronan (HA), a ubiquitous component of the extracellular matrix, is a variable length, long-chain polysaccharide containing repeating disaccharide units of glucuronic acid and n-acetylglucosamine. A recent study reported the relationship between hyaluronan and lymphangiogenesis of UUO. Researchers found that HA accumulation is correlated with the number of LYVE-1-positive LVs in the UUO kidney because HA could stimulate lymphangiogenesis, cooperating with VEGF-

C. Depletion of macrophages with clodronate decreased UUO-induced HA accumulation and lymphangiogenesis. Additionally, hyaluronan synthase (HAS) mRNA expression and HA production were increased in bone marrow-derived macrophages upon stimulation with TGF-b1. Additionally, VEGF-C expression and LYVE-1-positive lymphatic area were significantly lower from TLR4 null mice than that from TLR4 wild-type mice after UUO (Jung et al. 2015). This study demonstrated that hyaluronan-induced VEGF-C promotes fibrosis-induced lymphangiogenesis via Toll-like receptor 4-dependent signal pathway and has opened up a new way for us to study the role of lymphangiogenesis in renal fibrosis.

As discussed above, there is a well-established interaction between inflammatory cells and lymphangiogenesis via CCR7/CCL21 signaling pathway or some other undiscovered mechanisms. It means that an increase in LVs number might be detrimental by exacerbating the inflammatory response and thereby promoting renal fibrosis.

Even so, the causal relationship between lymphangiogenesis and renal fibrosis has not been fully elucidated and lymphangiogenesis may be a double-edged sword in some kidney diseases. In addition, whether lymphangiogenesis delays or accelerates the progression of renal fibrosis remains controversial. At present, the opinion accepted by most scholars is that lymphangiogenesis is closely related to fibrosis and the beneficial or detrimental effects of lymphangiogenesis depending on the difference in the inflammatory microenvironment of each experimental model. A recent study used VEGF-C to promote the proliferation of LV and found that it can alleviate renal fibrosis. Researchers found that lymphangiogenesis increased significantly compared with the control group after sustained injection of VEGF-C into UUO model mice. Interestingly, the inflammatory cells and inflammatory cytokines in the interstitium of the kidneys are greatly reduced (Hasegawa et al. 2017). This discovery makes us have a further understanding of the function of the new LVs. Also of concern is that disturbance of lymphatic circulation by ligation was shown to induce renal fibrosis by enhanced activation of the TGF-beta1/Smad signaling (Zhang et al. 2008).

In other organ studies, such as idiopathic pulmonary fibrosis, soluble factors such as short-fragment hyaluronic acid and macrophages contribute to lymphangiogenesis (El-Chemaly et al. 2009). A recent study found that a large number of LVs in fibrotic lungs show aberrant association with mural cells and excessive basement membrane deposition, which is driven by ectopic PDGF-B expression in LECs and signaling through PDGFR-β. Because of impaired lymphatic drainage, aberrant mural cell coverage fostered the accumulation of fibrogenic molecules and the attraction of fibroblasts to the perilymphatic space, promoting the progression of pulmonary fibrosis. On the contrary, blocking the PDGF-B/PDGFR-ß signaling prevents the attraction of fibroblasts to the perilymphatic space, to restore lymphatic drainage, and to ameliorate pulmonary fibrosis (Meinecke et al. 2012). This study describes the close relationship between lymphangiogenesis and pulmonary fibrosis. However, it is unclear whether this mechanism is also present in renal fibrosis.

27.4 Management and Therapeutic Targets of Lymphangiogenesis-related Renal Fibrosis

Studies have shown that lymphangiogenesis induced by exogenous pathways can help to alleviate chronic inflammatory reactions in arthritis and skin (Huggenberger et al. 2010; Zhou et al. 2011). After tumor lymph node dissection, the LVs that are transfected with the adenovirus coated with VEGF-C can mature and become functional after a period of time, thereby reducing lymphedema and increasing the removal of inflammatory cells (Tammela et al. 2007). This suggests that promoting lymphangiogenesis or lymphatic functional maturation is conducive to the reduction of lesions. Conversely, inhibition of immature lymphangiogenesis after heart and islet transplantation reduces lymphocytes infiltration, inhibits rejection, and prolongs graft survival (Nykänen et al. 2010; Yin et al. 2011). Therefore, both blocking and promoting lymphangiogenesis have been suggested to be advantageous and imply to be highly context-dependent and organ-specific. On the one hand, the newborn LVs can serve as an excessive inflammatory cell output channel, thereby attenuating chronic inflammatory response. A recent study showed that promoting lymphangiogenesis by VEGF-C156S increased clearance of interstitial hyaluronan and had a promise for improving lung graft outcomes. Numerous reports confirm that inducing lymphangiogenesis has been shown to decrease interstitial fluid accumulation and attenuate inflammatory response (Cheung et al. 2006; Kim et al. 2012; Szuba et al. 2002). On the other hand, lymphangiogenesis as a "fast channel" for inflammatory cell migration accelerate the recruitment of inflammatory cells to the injured site and participate in the local inflammatory response amplification. It is possible to improve the drainage of inflammatory cells, weaken the chronic inflammation, and delay the progression of renal fibrosis through exogenous measures to intervene in the formation or maturation of LVs in the renal tissue of patients with CKD. A report confirmed that VitD is anti-lymphangiogenic through VDR-dependent anti-proliferative and pro-apoptotic mechanisms. The VEGF-C/VEGF-D–VEGFR-3 signaling axis is the most prominent that induces lymphangiogenesis as we discussed above. Niina K. Palin et al. showed that lymphangiogenesis is associated with chronic kidney allograft injury in rat UUO model, and sirolimus is a potent inhibitor of lymphangiogenesis in renal allografts via inhibition VEGF-C/VEGFR-3 pathway (Palin et al. 2013). Using sirolimus to block the VEGF-C/VEGFR-3 pathway and inhibit lymphangiogenesis, thereby delaying renal fibrosis is expected to become a new therapeutic target. In cardiac allograft rejection and arteriosclerosis, VEGFR-3 inhibition may as a novel lymphatic vessel-targeted immunomodulatory therapy by reducing allograft lymphatic vessel CCL21 production (Nykänen et al. 2010).

The CCR7/CCL21 signaling pathway is another important mechanism for lymphatics involved in renal fibrosis. CCR7/CCL21 signaling of fibrocytes may provide therapeutic targets for combating renal fibrosis (Sakai et al. 2006). We also found that blocking CCR7+ cells recruitment by inhibition of LVs via genetic and pharmacologic approaches or using CCR7 neutralizing antibody attenuated intrarenal inflammation and fibrosis (unpublished data).

Despite the long history of research on the mechanisms and functions of lymphangiogenesis involved in renal fibrosis, there is still a long way to go for the clinical application of Intervention lymphangiogenesis. As far as the current research is concerned, both promoting lymphangiogenesis—for example, by using VEGF-C156s—and inhibiting it—for example, by using sVEGFR3-FC—could be served as a therapeutic strategy. Only through in-depth experiments to understand the functions of the newborn LVs and their relationship with kidney disease, can we find the exact appropriate intervention targets to delay or even reverse the progression of renal fibrosis.

References

Adair A, Mitchell DR, Kipari T, Qi F, Bellamy CO, Robertson F et al (2007) Peritubular capillary rarefaction and lymphangiogenesis in chronic allograft failure. Transplantation 83:1542–1550
Bakris GL (2008) Slowing nephropathy progression: focus on proteinuria reduction. Clin J Am Soc Nephrol 3(Suppl 1):S3–10
Banerji S, Ni J, Wang SX, Clasper S, Su J, Tammi R et al (1999) LYVE-1, a new homologue of the CD44 glycoprotein, is a lymph-specific receptor for hyaluronan. J Cell Biol 144:789–801
Benest AV, Harper SJ, Herttuala SY, Alitalo K, Bates DO (2008) VEGF-C induced angiogenesis preferentially occurs at a distance from lymphangiogenesis. Cardiovasc Res 78:315–323
Cheung L, Han J, Beilhack A, Joshi S, Wilburn P, Dua A et al (2006) An experimental model for the study of lymphedema and its response to therapeutic lymphangiogenesis. BioDrugs 20:363–370
Choi I, Lee S, Hong YK (2012) The new era of the lymphatic system: no longer secondary to the blood vascular system. Cold Spring Harb Perspect Med 2:a006445
D'Alessio S, Correale C, Tacconi C, Gandelli A, Pietrogrande G, Vetrano S et al (2014) VEGF-C-dependent stimulation of lymphatic function ameliorates experimental inflammatory bowel disease. J Clin Invest 124:3863–3878
D'Amico G, Korhonen EA, Waltari M, Saharinen P, Laakkonen P, Alitalo K (2010) Loss of endothelial Tie1 receptor impairs lymphatic vessel development-brief report. Arterioscler Thromb Vasc Biol 30:207–209
Eddy AA (2004) Proteinuria and interstitial injury. Nephrol Dial Transplant 19:277–281
El-Chemaly S, Malide D, Zudaire E, Ikeda Y, Weinberg BA, Pacheco-Rodriguez G et al (2009) Abnormal lymphangiogenesis in idiopathic pulmonary fibrosis with insights into cellular and molecular mechanisms. Proc Natl Acad Sci U S A 106:3958–3963
Förster R, Davalos-Misslitz AC, Rot A (2008) CCR11 and its ligands: balancing immunity and tolerance. Nat Rev Immunol 8:362–371
Güç E, Briquez PS, Foretay D, Fankhauser MA, Hubbell JA, Kilarski WW, Swartz MA (2017) Local induction of lymphangiogenesis with engineered fibrin-binding VEGF-C promotes wound healing by increasing immune cell trafficking and matrix remodeling. Biomaterials 131:160–175
Hasegawa S, Nakano T, Torisu K, Tsuchimoto A, Eriguchi M, Haruyama N et al (2017) Vascular endothelial growth factor-C ameliorates renal interstitial fibrosis through lymphangiogenesis in mouse unilateral ureteral obstruction. Lab Invest 97:1439–1452
Heller F, Lindenmeyer MT, Cohen CD, Brandt U, Draganovici D, Fischereder M et al (2007) The contribution of B cells to renal interstitial inflammation. Am J Pathol 170:457–468
Hong YK, Detmar M (2003) Prox1, master regulator of the lymphatic vasculature phenotype. Cell Tissue Res 314:85–92
Huggenberger R, Siddiqui SS, Brander D, Ullmann S, Zimmermann K, Antsiferova M et al (2011) An important role of lymphatic vessel activation in limiting acute inflammation. Blood 117:4667–4678

Huggenberger R, Ullmann S, Proulx ST, Pytowski B, Alitalo K, Detmar M (2010) Stimulation of lymphangiogenesis via VEGFR-3 inhibits chronic skin inflammation. J Exp Med 207:2255–2269
Johnson LA, Jackson DG (2008) Cell traffic and the lymphatic endothelium. Ann N Y Acad Sci 1131:119–133
Jung YJ, Lee AS, Nguyen-Thanh T, Kang KP, Lee S, Jang KY et al (2015) Hyaluronan-induced VEGF-C promotes fibrosis-induced lymphangiogenesis via Toll-like receptor 4-dependent signal pathway. Biochem Biophys Res Commun 466:339–345
Kaipainen A, Korhonen J, Mustonen T, van Hinsbergh VW, Fang GH, Dumont D et al (1995) Expression of the fms-like tyrosine kinase 4 gene becomes restricted to lymphatic endothelium during development. Proc Natl Acad Sci U S A 92:3566–3570
Kang S, Lee SP, Kim KE, Kim HZ, Mémet S, Koh GY (2009) Toll-like receptor 4 in lymphatic endothelial cells contributes to LPS-induced lymphangiogenesis by chemotactic recruitment of macrophages. Blood 113:2605–2613
Karkkainen MJ, Haiko P, Sainio K, Partanen J, Taipale J, Petrova TV et al (2004) Vascular endothelial growth factor C is required for sprouting of the first lymphatic vessels from embryonic veins. Nat Immunol 5:74–80
Kerjaschki D, Regele HM, Moosberger I, Nagy-Bojarski K, Watschinger B, Soleiman A et al (2004) Lymphatic neoangiogenesis in human kidney transplants is associated with immunologically active lymphocytic infiltrates. J Am Soc Nephrol 15:603–612
Khawar MB, Mukhtar M, Abbasi MH, Sheikh N (2017) IL-32θ: a recently identified anti-inflammatory variant of IL-32 and its preventive role in various disorders and tumor suppressor activity. Am J Transl Res 9:4726–4737
Kim H, Kataru RP, Koh GY (2012) Regulation and implications of inflammatory lymphangiogenesis. Trends Immunol 33:350–356
Kim H, Kataru RP, Koh GY (2014) Inflammation-associated lymphangiogenesis: a double-edged sword? J Clin Invest 124:936–942
Kinashi H, Falke LL, Nguyen TQ, Bovenschen N, Aten J, Leask A et al (2017) Connective tissue growth factor regulates fibrosis-associated renal lymphangiogenesis. Kidney Int 92:850–863
Kriz W, Hartmann I, Hosser H, Hähnel B, Kränzlin B, Provoost A et al (2001) Tracer studies in the rat demonstrate misdirected filtration and peritubular filtrate spreading in nephrons with segmental glomerulosclerosis. J Am Soc Nephrol 12:496–506
Lee AS, Lee JE, Jung YJ, Kim DH, Kang KP, Lee S et al (2013) Vascular endothelial growth factor-C and -D are involved in lymphangiogenesis in mouse unilateral ureteral obstruction. Kidney Int 83:50–62
Mäkinen T, Norrmén C, Petrova TV (2007) Molecular mechanisms of lymphatic vascular development. Cell Mol Life Sci 64:1915–1929
Mäkinen T, Veikkola T, Mustjoki S, Karpanen T, Catimel B, Nice EC et al (2001) Isolated lymphatic endothelial cells transduce growth, survival and migratory signals via the VEGF-C/D receptor VEGFR-3. EMBO J 20:4762–4773
Martínez-Corral I, Olmeda D, Diéguez-Hurtado R, Tammela T, Alitalo K, Ortega S (2012) In vivo imaging of lymphatic vessels in development, wound healing, inflammation, and tumor metastasis. Proc Natl Acad Sci U S A 109:6223–6228
Matsui K, Nagy-Bojarsky K, Laakkonen P, Krieger S, Mechtler K, Uchida S et al (2003) Lymphatic microvessels in the rat remnant kidney model of renal fibrosis: aminopeptidase p and podoplanin are discriminatory markers for endothelial cells of blood and lymphatic vessels. J Am Soc Nephrol 14:1981–1989
Meinecke AK, Nagy N, Lago GD, Kirmse S, Klose R, Schrödter K et al (2012) Aberrant mural cell recruitment to lymphatic vessels and impaired lymphatic drainage in a murine model of pulmonary fibrosis. Blood 119:5931–5942
Moreno JA, Moreno S, Rubio-Navarro A, Gómez-Guerrero C, Ortiz A, Egido J (2014) Role of chemokines in proteinuric kidney disorders. Expert Rev Mol Med 16:e3
Nykänen AI, Sandelin H, Krebs R, Keränen MA, Tuuminen R, Kärpänen T et al (2010) Targeting lymphatic vessel activation and CCL21 production by vascular endothelial growth factor receptor-

3 inhibition has novel immunomodulatory and antiarteriosclerotic effects in cardiac allografts. Circulation 121:1413–1422
Oliver G (2004) Lymphatic vasculature development. Nat Rev Immunol 4:35–45
Palin NK, Savikko J, Koskinen PK (2013) Sirolimus inhibits lymphangiogenesis in rat renal allografts, a novel mechanism to prevent chronic kidney allograft injury. Transpl Int 26:195–205
Platonova N, Miquel G, Regenfuss B, Taouji S, Cursiefen C, Chevet E et al (2013) Evidence for the interaction of fibroblast growth factor-2 with the lymphatic endothelial cell marker LYVE-1. Blood 121:1229–1237
Podgrabinska S, Kamalu O, Mayer L, Shimaoka M, Snoeck H, Randolph GJ et al (2009) Inflamed lymphatic endothelium suppresses dendritic cell maturation and function via Mac-1/ICAM-1-dependent mechanism. J Immunol 183:1767–1779
Rissanen TT, Markkanen JE, Gruchala M, Heikura T, Puranen A, Kettunen MI et al (2003) VEGF-D is the strongest angiogenic and lymphangiogenic effector among VEGFs delivered into skeletal muscle via adenoviruses. Circ Res 92:1098–1106
Sakai N, Wada T, Yokoyama H, Lipp M, Ueha S, Matsushima K et al (2006) Secondary lymphoid tissue chemokine (SLC/CCL21)/CCR42 signaling regulates fibrocytes in renal fibrosis. Proc Natl Acad Sci U S A 103:14098–14103
Sakamoto I, Ito Y, Mizuno M, Suzuki Y, Sawai A, Tanaka A et al (2009) Lymphatic vessels develop during tubulointerstitial fibrosis. Kidney Int 75:828–838
Schulte-Merker S, Sabine A, Petrova TV (2011) Lymphatic vascular morphogenesis in development, physiology, and disease. J Cell Biol 193:607–618
Shibuya M (2013) Vascular endothelial growth factor and its receptor system: physiological functions in angiogenesis and pathological roles in various diseases. J Biochem 153:13–19
Suzuki Y, Ito Y, Mizuno M, Kinashi H, Sawai A, Noda Y et al (2012) Transforming growth factor-beta induces vascular endothelial growth factor-C expression leading to lymphangiogenesis in rat unilateral ureteral obstruction. Kidney Int 81:865–879
Szuba A, Skobe M, Karkkainen MJ, Shin WS, Beynet DP, Rockson NB et al (2002) Therapeutic lymphangiogenesis with human recombinant VEGF-C. FASEB J 16:1985–1987
Tammela T, Alitalo K (2010) Lymphangiogenesis: molecular mechanisms and future promise. Cell 140:460–476
Tammela T, Saaristo A, Holopainen T, Lyytikkä J, Kotronen A, Pitkonen M et al (2007) Therapeutic differentiation and maturation of lymphatic vessels after lymph node dissection and transplantation. Nat Med 13:1458–1466
Veikkola T, Jussila L, Makinen T, Karpanen T, Jeltsch M, Petrova TV et al (2001) Signalling via vascular endothelial growth factor receptor-3 is sufficient for lymphangiogenesis in transgenic mice. EMBO J 20:1223–1231
Wick G, Grundtman C, Mayerl C, Wimpissinger TF, Feichtinger J, Zelger B et al (2013) The immunology of fibrosis. Annu Rev Immunol 31:107–135
Xiong Y, Liu Z, Zhao X, Ruan S, Zhang X, Wang S et al (2018) CPT1A regulates breast cancer-associated lymphangiogenesis via VEGF signaling. Biomed Pharmacother 106:1–7
Yazdani S, Navis G, Hillebrands JL, van Goor H, van den Born J (2014) Lymphangiogenesis in renal diseases: passive bystander or active participant? Expert Rev Mol Med 16:e15
Yazdani S, Poosti F, Kramer AB, Mirković K, Kwakernaak AJ, Hovingh M et al (2012) Proteinuria triggers renal lymphangiogenesis prior to the development of interstitial fibrosis. PLoS ONE 7:e50209
Yin N, Zhang N, Xu J, Shi Q, Ding Y, Bromberg JS (2011) Targeting lymphangiogenesis after islet transplantation prolongs islet allograft survival. Transplantation 92:25–30
Zampell JC, Yan A, Elhadad S, Avraham T, Weitman E, Mehrara BJ (2012) CD4(+) cells regulate fibrosis and lymphangiogenesis in response to lymphatic fluid stasis. PLoS ONE 7:e49940
Zhang L, Zhou F, Han W, Shen B, Luo J, Shibuya M et al (2010) VEGFR-3 ligand-binding and kinase activity are required for lymphangiogenesis but not for angiogenesis. Cell Res 20:1319–1331

Zhang T, Guan G, Liu G, Sun J, Chen B, Li X et al (2008) Disturbance of lymph circulation develops renal fibrosis in rats with or without contralateral nephrectomy. Nephrology (Carlton) 13:128–138

Zhou HL, Wang YT, Gao T, Wang WG, Wang YS (2013) Distribution and expression of fibroblast-specific protein chemokine CCL21 and chemokine receptor CCR59 in renal allografts. Transplant Proc 45:538–545

Zhou Q, Guo R, Wood R, Boyce BF, Liang Q, Wang YJ et al (2011) Vascular endothelial growth factor C attenuates joint damage in chronic inflammatory arthritis by accelerating local lymphatic drainage in mice. Arthritis Rheum 63:2318–2328

Zimmer JK, Dahdal S, Mühlfeld C, Bergmann IP, Gugger M, Huynh-Do U (2010) Lymphangiogenesis is upregulated in kidneys of patients with multiple myeloma. Anat Rec 293:1497–1505 (Hoboken)

Chapter 28
Cell Apoptosis and Autophagy in Renal Fibrosis

Xing-Chen Zhao, Man J. Livingston, Xin-Ling Liang and Zheng Dong

Abstract Renal fibrosis is the final common pathway of all chronic kidney diseases progressing to end-stage renal diseases. Autophagy, a highly conserved lysosomal degradation pathway, plays important roles in maintaining cellular homeostasis in all major types of kidney cells including renal tubular cells as well as podocytes, mesangial cells and endothelial cells in glomeruli. Autophagy dysfunction is implicated in the pathogenesis of various renal pathologies. Here, we analyze the pathological role and regulation of autophagy in renal fibrosis and related kidney diseases in both glomeruli and tubulointerstitial compartments. Further research is expected to gain significant mechanistic insights and discover pathway-specific and kidney-selective therapies targeting autophagy to prevent renal fibrosis and related kidney diseases.

Keywords Autophagy · Renal fibrosis · Focal segmental glomerulosclerosis · Diabetic kidney disease · Acute kidney injury · Podocytes · Proximal tubular epithelial cells

28.1 Introduction: Basics of Autophagy

First proposed by Christian de Duve in 1963, autophagy is a term derived from Greek and refers to "self-eating" (Klionsky 2008). It is an evolutionarily conserved catabolic process from yeasts to mammals, by which portions of cytosolic components and organelles are delivered to the lysosomes for degradation and recycling (Mizushima et al. 2008; Mizushima and Komatsu 2011). There are three types of

Co-first authors with equal contributions.

X.-C. Zhao · X.-L. Liang (✉)
Division of Nephrology, Guangdong Provincial People's Hospital, Guangdong Academy of Medical Sciences, Guangzhou, China
e-mail: xinlingliang_ggh@163.com

M. J. Livingston · Z. Dong (✉)
Department of Cellular Biology and Anatomy, Medical College of Georgia, Augusta University and Charlie Norwood VA Medical Center, Augusta, GA 30912, USA
e-mail: zdong@augusta.edu

B.-C. Liu et al. (eds.), *Renal Fibrosis: Mechanisms and Therapies*, Advances in Experimental Medicine and Biology 1165,
https://doi.org/10.1007/978-981-13-8871-2_28

autophagy in mammalian cells: macroautophagy, microautophagy and chaperone-mediated autophagy, which differ in the type of cargo to be degraded and the way of delivering them to lysosomes. Macroautophagy (herein referred to as autophagy), the best characterized form and the focus of this chapter, begins with the enveloping of large cytosolic structures by double-membraned autophagosomes and then fuses with the lysosome. Microautophagy involves the direct engulfment of small cytoplasmic cargos within the invagination of the lysosomal membrane. Chaperone-mediated autophagy is a process of direct transport of selective unfolded protein via chaperonin across the lysosome membrane for degradation (Levine and Kroemer 2008; Mizushima et al. 2008; Ravikumar et al. 2010).

The process of autophagy consists of a cascade of cellular events (Fig. 28.1). It is initiated by the formation of a double-membrane, cup-shaped structure termed phagophore around sequestered cytoplasmic targets, followed by expansion and closure of the phagophore to form an autophagosome. The autophagosome then docks and fuses with a lysosome to form an autolysosome, in which the autophagosome inner membrane and the cytoplasmic substrates are degraded by acidic lysosomal hydrolases. Eventually, the resultant degradation products are released for recycling (Mizushima et al. 2008; Mizushima and Komatsu 2011). The complete dynamic flow of autophagy is thus called autophagic flux. Although the origin of the autophagosome membrane has been a matter of debate for decades, it is now generally agreed that the initiation of the phagophore is associated with the phosphatidylinositol 3-phosphate (PtdIns3P)-enriched membrane compartment that appears to be linked to the endoplasmic reticulum (ER). An omega-shaped protrusion (also called omegasome) forms on ER at the initiation site of phagophore and functions as a scaffold for autophagosome biogenesis (Yang and Klionsky 2010; Rubinsztein et al. 2012). Furthermore, the phagophore is in contact with many surrounding organelles either by vesicular transport to and from the phagophore or through transient membrane contacts and exchange of proteins and lipids to grow and expand (Yang and Klionsky 2010; Rubinsztein et al. 2012; Yu et al. 2018).

In mammals, autophagosome biogenesis is regulated coordinately at different steps by the core machinery that consists of more than 36 autophagy-related (*Atg*) genes (Yang and Klionsky 2010; Klionsky et al. 2011; Mizushima et al. 2011; Rubinsztein et al. 2012; Hurley and Young 2017). The initiation of phagophore formation is orchestrated by the ULK1/2 (Unc-51-like kinase 1/2) complex, which is composed of ULK1/2 serine–threonine kinase, ATG13, FIP200 (FAK family kinase-interacting protein of 200 kDa, also known as RB1CC1, RB1-inducible coiled-coil protein 1) and ATG101. The ULK1/2 complex senses upstream signals and relays them to the downstream central autophagy pathway via the coordination and interaction with other ATG proteins or complexes (Yang and Klionsky 2010; Rubinsztein et al. 2012). Phagophore nucleation is dependent on the class III phosphatidylinositol 3-kinase (PtdIns3K) complex that consists of PIK3C3/VPS34 (vacuolar protein sorting 34) lipid kinase, PIK3R4/VPS15, beclin 1/BECN1 and ATG14L. This complex produces PtdIns3P at the site of phagophore nucleation to stabilize membrane curvature and to recruit more downstream factors for phagophore growth (Yang and Klionsky 2010; Rubinsztein et al. 2012). The delivery of membrane from other sources to the forming

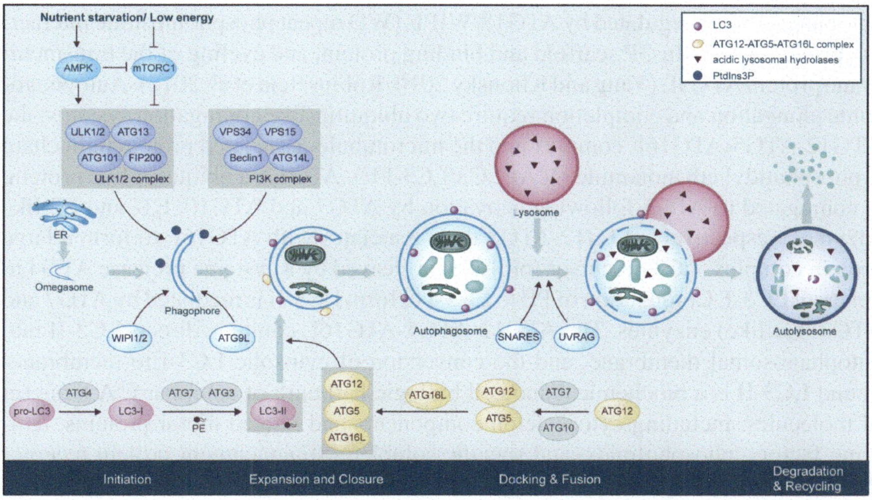

Fig. 28.1 Basics of autophagy. The autophagy pathway and its molecular regulation are shown. Autophagy is initiated by the formation of omegasome, an omega-shaped protrusion from the ER, which is enriched in phosphatidylinositol 3-phosphate (PtdIns3P) and further forms the phagophore. Then, the autophagosome is formed by the expansion and closure of phagophore, which then docks and fuses with a lysosome to form an autolysosome. The autolysosome degrades the inner membrane of autophagosome and the cytoplasmic substrates by acidic lysosomal hydrolases to recycle the nutrients. The initiation of autophagy is regulated by the ULK1/2 complex and the class III phosphatidylinositol 3-kinase (PI3K) complex. The ULK1/2 complex senses upstream signals and relays them to the downstream central autophagy pathway via the coordination and interaction with other ATG proteins or complexes. The PI3K complex produces PtdIns3P at the site of phagophore nucleation to stabilize membrane curvature and to recruit more downstream factors for phagophore growth. ATG18/WIPIs (WD repeat phosphoinositide interacting proteins) and ATG9L contribute to delivery of membrane to the phagophore. The expansion and closure of phagophore to form autophagosome require two ubiquitin-like conjugation systems: the ATG12-ATG5-ATG16L complex and the microtubule-associated protein light chain 3-phosphatidyl ethanolamine (MAPLC3/LC3-PE). LC3 precursor (proLC3) is cleaved by a cysteine protease ATG4 to generate LC3-I. The PE then conjugated to LC3-I to form LC3-II by ATG7 and ATG3 (E2-like) enzymes. The ATG12-ATG5-ATG16L complex is formed by the ATG7 (E1-like) and ATG10 (E2-like) enzymes, which directs LC3-II into autophagosomal membrane. Specific soluble NSF attachment protein receptor (SNARE) complexes contribute to the fusion of autophagosomes with lysosomes. The mTORC1 and AMP-activated protein kinase (AMPK) are major autophagy regulators, sensing and integrating multiple signaling pathways. mTORC1 inhibits autophagy by phosphorylating ULK1 to inhibit its activity, while AMPK induces autophagy by either directly phosphorylating ULK1 or suppressing mTORC1. Nutrient starvation or low energy induces autophagy by suppressing mTORC1 and activating AMPK pathway

autophagosome is regulated by ATG18/WIPIs (WD repeat phosphoinositide interacting proteins), a PtdIns3P scaffold and binding protein, and cycling of the transmembrane protein ATG9L (Yang and Klionsky 2010; Rubinsztein et al. 2012). Autophagosome elongation and completion require two ubiquitin-like conjugation systems: the ATG12-ATG5-ATG16L complex and the microtubule-associated protein light chain 3-phosphatidyl ethanolamine (MAPLC3/LC3-PE). ATG12, a ubiquitin-like protein, is conjugated to ATG5 following activation by ATG7 and ATG10, E1- and E2-like enzymes, respectively. ATG12-ATG5 then associates with ATG16L to form a large protein complex. LC3 precursor (proLC3) is cleaved by a cysteine protease ATG4 to generate LC3-I. Conjugation of PE to LC3-I to form LC3-II is mediated by ATG7 and ATG3 (E2-like) enzymes. The ATG12-ATG5-ATG16L complex directs LC3-II into autophagosomal membrane, and the conversion of cytosolic LC3-I to membrane-bound LC3-II is a biochemical and cell biologic hallmark of autophagy. A large set of molecules, including cytoskeleton components and related motor proteins, tethering factors, phospholipids, and specific soluble NSF attachment protein receptor (SNARE) complexes have been identified as important players in the maturation and fusion of autophagosome with lysosome (Yang and Klionsky 2010; Rubinsztein et al. 2012; Yu et al. 2018). Dynein, a microtubule motor, mediates the centripetal movement of autophagosomes to ensure a close spatial positioning between autophagosomes and lysosomes. UV radiation resistance-associated gene (UVRAG), via interaction with PtdIns3 K complex, activates the GTPase Rab7 (Ras-related protein 7) for tethering. As the core components of the fusion machinery, SNAREs can mediate membrane fusion on their own as well as via the regulation and interaction with other fusion factors. Upstream of the core machinery, autophagy is tightly regulated by a complex signaling network (He and Klionsky 2009; Mehrpour et al. 2010; Klionsky et al. 2016). The mechanistic target of rapamycin (mTOR) pathway, specifically mTOR complex 1 (mTORC1), serves as a sensor and a master negative regulator of autophagy. Multiple signaling pathways stimulated by nutrients, growth factors and energy may integrate and merge at mTORC1 to regulate autophagy. mTOR-independent mechanisms have also been implicated in autophagy regulation. A variety of cellular stress, including hypoxic stress, oxidative stress, ER stress and DNA damage, may also induce autophagy through various signaling pathways (Klionsky et al. 2016).

Multiple techniques have been developed to monitor autophagy in cells or tissues (Klionsky et al. 2016). Electron microscopy (EM) is a classic method for detecting and analyzing various autophagic structures including the phagophore, autophagosome and autolysosome. However, EM is less quantitative and could be problematic due to sampling artifacts. LC3-based assays have been widely used for the analysis of autophagy in cells and tissues. Especially, the conversion of LC3-I to LC3-II or expression of LC3-II measured by immunoblotting is a well-accepted biomarker for autophagy. In addition, LC3-I resides in cytosol whereas LC3-II is membrane-bound on autophagosomes. The change of LC3 localization can be revealed by microscopic examination of endogenous LC3 following immunostaining or exogenous green fluorescence protein (GFP)-LC3 following transfection. Considering the dynamic change of autophagic flux, several methods are utilized to detect the dynamic flow: (1) LC3 turnover assay that compares LC3-II abundance in the presence and absence of lysosome inhibitors; (2) examination of the turnover of autophagic substrates such as p62/sequestosome 1 (SQSTM1); and (3) the use of a tandem red fluorescent protein (mRFP)/mCherry-GFP-LC3 reporter for the monitoring of autophagosomes (red- and green-costained LC3 puncta) and autolysosomes (red-only LC3 puncta) (Klionsky et al. 2016).

Autophagy can non-selectively break down bulk cytosol and also selectively recognize and digest specific organelles such as mitochondria, ER and lysosomes, protein aggregates, lipid droplets and intracellular pathogens (Mizushima et al. 2008; Mizushima and Komatsu 2011; Zaffagnini and Martens 2016). Under physiological conditions, a basal level of autophagy in most cells is an important mechanism of removing potentially harmful or simply unneeded cytoplasmic materials, which is essential to the maintenance of cellular homeostasis. During starvation or nutrient deprivation, autophagy is activated to break down and recycle cytosolic contents to replenish pools of biosynthetic precursors (amino acids, lipids, nucleotides) and energy sources (Sionov et al. 2015; Dikic and Elazar 2018). Emerging studies have further shown the role of autophagy in a much broader range of cell stress and pathological conditions. Under these conditions, induction of autophagy primarily serves as an adaptive and defensive strategy for cell to deal with stress for survival (Sionov et al. 2015; Dikic and Elazar 2018). Conversely, dysregulation of autophagy may contribute to the pathogenesis of various diseases, such as cancer and cardiovascular disease (Huber et al. 2012; Choi et al. 2013; Sionov et al. 2015; Dikic and Elazar 2018).

28.2 Autophagy in Renal Resident Cells

Autophagy is an important mechanism for maintaining cellular homeostasis in all major types of renal resident cells including podocytes, mesangial cells, glomerular endothelial cells and renal tubular epithelial cells.

28.2.1 Autophagy in Podocytes

Podocytes, also called glomerular visceral epithelial cells, are highly specialized epithelial cells with a large cell body and primary processes that further branch into fine secondary foot processes. The foot processes from adjacent podocytes interdigitate and wrap around the outside of glomerular basement membrane (GBM) that surrounds the capillaries of glomeruli. Podocytes play a critical role in maintaining the selective permeability and structural integrity of the glomerular filtration barrier. Terminally differentiated, podocytes are incapable of proliferation, and the mechanism of podocyte replacement is limited (Pavenstädt et al. 2003). As a result, podocytes are the most vulnerable component of the glomerulus and can be irreversibly injured by various insults, leading to proteinuria and glomerulosclerosis that contribute to the pathogenesis of many glomerular diseases. As long-lived cells, podocytes rely on cellular quality control mechanisms to maintain their structural and functional homeostasis, and autophagy is one such mechanism under both normal and disease conditions (Zhang et al. 2014a).

Compared to other types of cells in the kidney, podocytes, especially differentiated podocytes, exhibit a high level of constitutive autophagy. In normal adult rats, abundant LC3 staining was detected in glomeruli but barely seen in proximal tubules. Punctate LC3 staining co-localized with podocalyxin, a podocyte marker. LC3-II accumulated in tissue lysates of isolated glomeruli. In conditionally immortalized mouse podocytes, a high level of basal autophagy was also primarily seen in differentiated podocytes (Asanuma et al. 2003). In *GFP-LC3* transgenic mice, compared with tubular cells, podocytes displayed clearly detectable levels of GFP-LC3 puncta under basal conditions (Mizushima et al. 2004; Hartleben et al. 2010; Fang et al. 2013). However, in newborn *GFP-LC3* mice, GFP-LC3 puncta were not seen until more mature or differentiated podocytes appeared in the late capillary loop stage (Hartleben et al. 2010; Fang et al. 2013). The formation of GFP-LC3 puncta was enhanced upon starvation (Mizushima et al. 2004). LC3 turnover assay further revealed remarkably increases in both the number of GFP-LC3 puncta and the accumulation of LC3-II by lysosomal inhibitor chloroquine, suggesting a complete process of autophagic flux in podocytes (Hartleben et al. 2010; Fang et al. 2013).

In doxycycline-inducible podocyte-specific *Atg5* knockout mice, acute induction of *Atg5* deletion in 12-week-old mice triggered a rapid onset of albuminuria, suggesting that autophagy is a fundamental mechanism for podocyte homeostasis (Hartleben et al. 2010; Fang et al. 2013). In contrast to this inducible knockout model, mice with constitutive podocyte-specific *Atg5* knockout were indistinguishable from their wild-type littermates for up to 2–4 months after birth, suggesting that other adaptive pathways may compensate for the constitutive loss of *Atg5* (Hartleben et al. 2010; Fang et al. 2013). Indeed, proteasome activity was significantly increased in glomerular lysates of 8-month-old podocyte-specific *Atg5* knockout mice, preventing the accumulation of polyubiquitinated proteins. Inhibition of proteasome activity led to increased albuminuria in 6-month-old podocyte-specific *Atg5* knockout mice. In differentiated podocytes in vitro, inhibition of proteasome activated autophagy. These results suggest that basal autophagy in podocytes acts in concert with the proteasome pathway to maintain podocyte homeostasis (Hartleben et al. 2010; Fang et al. 2013). Using paracellular permeability influx assay, Fang et al. found that inhibition of autophagy by 3-methyladenine in cultured mouse podocytes led to an increased leakage of albumin across the podocyte monolayer. The expression of podocyte slit diaphragm proteins such as nephrin and podocin was suppressed by 3-methyladenine and *BECN1* siRNA, further suggesting an essential role of high basal autophagy in the maintenance of podocyte health (Hartleben et al. 2010; Fang et al. 2013). Since lysosome is crucial for autophagic degradation, lysosomal dysfunction in podocytes by gene ablation of *mTOR*, *prorenin* or *VPS34* also resulted in severe glomerulosclerosis and proteinuria (Oshima et al. 2011; Cinà et al. 2012a; Chen et al. 2013). These results highlight the importance of an intact autophagic flux pathway in maintaining podocyte homeostasis.

In humans, Zeng et al. demonstrated a negative correlation between the podocyte autophagic activity and the progression of glomerular diseases (Zeng et al. 2014). In experimental models, autophagy was induced by adriamycin in cultured podocytes in vitro and in podocytes in mice. Importantly, inducible ablation of autophagy-related gene 7(ATG7) in podocytes exacerbated podocyte injury, glomerulopathy and proteinuria during adriamycin treatment, supporting a protective role of autophagy in podocytes (Yi et al. 2017). Recent studies have further suggested defective autophagy in podocytes of aged kidneys. Compared with young rats, the expression of LC3 and ATG7 was significantly decreased in old rats. p62/SQSTM1 and polyubiquitin aggregates accumulated in aged kidneys, accompanied with an accumulation of damaged mitochondria and induction of oxidative stress (Cui et al. 2012). Similar findings were also shown in aging mice (Wanner et al. 2014). In 20- to 24-month-old podocyte-specific *Atg5* knockout mice, *Atg5* deletion accelerated podocyte aging, leading to massive ER stress and oxidative stress in the knockout mice. The compensatory proteasome pathway that clears protein aggregates in young *Atg5* knockout mice was also significantly reduced in the old mice. Ultimately, the Atg5-deficient old mice developed podocyte loss and progressive glomerulosclerosis (Hartleben et al. 2010; Fang et al. 2013). These results highlight the fundamental role of autophagy for the longtime maintenance of glomerular podocytes.

Notably, podocytes also display higher mTORC1 activity compared with other glomerular cells, which appears to be required for postnatal growth (Narita et al. 2011; Fukuda et al. 2012). Podocytes stop cell division in the capillary loop stage; thus, an increasing glomerular volume must be accompanied by mTORC1-dependent growth of every single podocyte to cover the glomerular capillaries (Hartleben et al. 2014; Inoki 2014). The high levels of both basal autophagy and mTORC1 in podocytes seem contradictory to the concept that mTORC1 negatively regulates autophagy; however, it may suggest the existence of a unique mechanism involving a mutual function and coordination of mTORC1 and autophagy in these cells. TOR-autophagy spatial coupling compartment (TASCC), a distinct cytoplasmic compartment, has been identified in podocytes (Narita et al. 2011). Located at the trans side of Golgi apparatus, TASCC is mostly occupied by mTORC1, lysosomes and autolysosomes but largely excludes autophagosomes. The formation of a spatial mTORC1 gradient within a podocyte by TASCC sequestration thus allows simultaneous activation of mTORC1 and autophagy in different areas of the same cell (Narita et al. 2011). Functionally, this system plays a beneficial role in generating sufficient secretory proteins with a constant energy and source supply derived from autophagy. More importantly, it also creates a self-regulating mechanism, in which the autolysosomal degradation products reinforce mTOR enrichment and activity to in turn suppress autophagy and recycle lysosomes. This feedback regulation, called autophagic lysosome reformation (ALR), is very important for a balance and fine-tuning between mTOR pathway and autophagy-lysosomal pathway (Yu et al. 2010, 2018). In line with this, mice with podocyte-specific *mTOR* knockout developed severe proteinuria and renal failure 3–5 weeks after birth. *mTOR* deficiency led to impaired autophagic flux in podocytes, and the failure of ALR appeared to account for this podocyte dysfunction (Cinà et al. 2012b).

28.2.2 Autophagy in Mesangial Cells

Glomerular mesangial cells are specialized contractile cells that are located within the mesangium. They provide structural support for the glomerular tufts and also form a functional unit together with adjacent podocytes and glomerular endothelial cells to regulate glomerular filtration. Mesangial cells produce ECM components in the mesangium and play an important role in maintaining mesangial matrix homeostasis.

The role of autophagy in mesangial cells is poorly understood. Transforming growth factor (TGF)-β1 induced autophagy in mouse mesangial cells and protected against serum deprivation-induced apoptosis. The induction of autophagy by TGF-β1 in mesangial cells was shown to be mediated by TGF-β-activated kinase 1 (TAK1) and PI3 K-protein kinase B(PKB)/Akt pathway. TGF-β1 failed to rescue autophagy-deficient mesangial cells from serum deprivation-induced apoptosis, further supporting a pro-survival role of autophagy in mesangial cells (Kim et al. 2012a). The protective effect of mesangial autophagy was also associated with its role in the maintenance of matrix protein homeostasis (Kim et al. 2012a). Primary mouse mesangial cells isolated from autophagy-deficient mice expressed a higher basal level of collagen I protein. In response to TGF-β1, both protein and mRNA levels of collagen I were induced, and notably, the increased collagen I protein was co-localized with LC3 and lysosomal marker lysosome-associated membrane protein 1 (LAMP1). Inhibition of autophagy by *BECN1* knockdown or lysosomal inhibitors further increased collagen I protein accumulation without affecting mRNA expression. Upregulation of autophagy reduced collagen I protein in wild-type but not in autophagy-deficient mesangial cells. These results suggest a critical role of autophagy in negatively regulating to limit excessive ECM deposition in mesangial cells by promoting collagen I degradation (Kim et al. 2012a).

28.2.3 Autophagy in Glomerular Endothelial Cells

Glomerular endothelial cells localize in the inner side of GBM and are important components of the glomerular filtration barrier. The renal microvasculature also plays a key role in renal physiology by regulating vasomotor tone, vascular permeability, leukocyte recruitment and antithrombogenic responses. Glomerular endothelial dysfunction is associated with the progression of CKD and renal fibrosis; however, the underlying mechanisms remain largely unknown. Thus far, very few studies have examined the role of autophagy in glomerular endothelial cells. Xavier et al. showed that bone morphogenetic protein and activin membrane-bound inhibitor (BAMBI), a competitive receptor antagonist for the TGF-β receptor family, was increased in cultured mouse glomerular endothelial cells treated with TGF-β. By contrast, the decrease of BAMBI by serum starvation or rapamycin was completely impeded by lysosomal inhibitor bafilomycin A1 and partially inhibited by 3-methyladenine, but not by proteasome inhibitors. These results suggest a role of autophagy in regulating BAMBI turnover in endothelial cells, which may influence endothelial function via BAMBI-mediated regulation of TGF-β pathway (Xavier et al. 2010).

28.2.4 Autophagy in Proximal Tubular Epithelial Cells (PTECs)

PTECs are the key targets in both acute kidney injury (AKI) and chronic kidney diseases (CKDs). Under physiological conditions, PTECs exhibit a relatively low level of autophagy. Mice with proximal tubule-specific knockout of *Atg5* or *Atg7* showed progressive renal damage and also developed premature renal aging, as indicated by an accumulation of deformed mitochondria, p62/SQSTM1 and polyubiquitin-positive inclusion bodies as well as increased tubular cell apoptosis and renal interstitial fibrosis. These results suggest that a low but sufficient level of basal autophagy is needed to maintain cellular homeostasis in PTECs under normal conditions and a higher level of autophagy is required for the cells to deal with age-related stress (Kimura et al. 2011; Liu et al. 2012). Autophagy is remarkably activated in PTECs under various stress conditions and plays a renoprotective role against tubular injury and cell death (Huber et al. 2012; Jiang et al. 2012; Havasi and Dong 2016; Zhang et al. 2016; Tang et al. 2018).

28.3 Autophagy in Renal Fibrosis and Related Kidney Diseases

Renal fibrosis, characterized by the excessive deposition of extracellular matrix (ECM) in glomeruli and tubulointerstitium, is the common pathological hallmark of progressive CKD regardless of the initial causes. The pathogenesis of renal fibrosis involves an extremely complex interaction of multiple cellular events including excessive proliferation and activation of fibroblasts, increased deposition of ECM, infiltration of inflammatory cells, tubular atrophy, glomerulosclerosis and microvascular rarefaction (Liu 2011; Duffield 2014; Humphreys 2018). Recently, increasing evidence has demonstrated that dysregulated autophagy may also contribute to the pathogenesis of renal fibrosis and related kidney diseases.

28.3.1 Autophagy in Focal Segmental Glomerulosclerosis (FSGS)

FSGS is a heterogeneous fibrotic kidney disease with poor clinical outcome due to progressive loss of kidney function. It is either idiopathic or secondary to a number of other disorders. The cause of idiopathic or primary FSGS remains unclear. Although FSGS is characterized by lesions in the glomerulus, particularly podocytes, early functional defects in proximal tubular cells are also found in some patients with FSGS (Löwik et al. 2009; Kawakami et al. 2015). A pathological feature in the development of FSGS is podocyte injury and loss, leading to the adhesion of glomerular capillary tuft to Bowman's capsule followed by occlusion of the capillaries and eventual total nephron degeneration (D'Agati 2012; Hartleben et al. 2014).

Using in vivo and in vitro models of puromycin aminonucleoside (PAN) nephrosis (an experimental model of human FSGS), Asanuma et al. showed that autophagy induction in podocytes was correlated with the recovery from PAN nephrosis, and provided the first evidence that podocyte autophagy may prevent the development of FSGS (Asanuma et al. 2003). Upregulation of autophagy in podocytes was further shown in human renal biopsy specimens from patients with FSGS, as indicated by increased *Atg3* mRNA and LC3 puncta. Accordingly, podocyte-specific *Atg5* knockout significantly sensitized mice to glomerulosclerosis induced by PAN, further suggesting that autophagy defends the integrity of podocytes against the development of FSGS (Hartleben et al. 2010). In an adriamycin-induced experimental model of FSGS, our recent work demonstrated an induction of autophagy in cultured podocytes and the podocytes in mice (Yi et al. 2017). In cultured podocytes, activation of autophagy by rapamycin suppressed adriamycin-induced apoptosis, whereas inhibition of autophagy by chloroquine sensitized the cells to apoptosis. In inducible podocyte-specific Atg7 knockout mice, adriamycin induced more severe podocyte injury, glomerulopathy and proteinuria as compared with wild-type mice, further supporting a protective role of podocyte autophagy in FSGS (Yi et al. 2017).

Using renal biopsies from patients with minimal change disease (MCD) or FSGS, a recent study examined the role of podocyte autophagy in controlling the progression of podocytopathies (Zeng et al. 2014). It showed that podocytes from MCD patients had higher levels of BECN1-mediated autophagy than podocytes from FSGS patients.

Tracking of podocyte autophagy in MCD patients by repeat renal biopsies showed that the patients with decreased autophagy activity in podocyte progressed to FSGS, whereas those with relatively high levels of podocyte autophagy maintained MCD status. In PAN-treated cultured podocytes, apoptosis was enhanced by autophagy inhibition through *BECN1* knockdown or autophagy inhibitors (3-methyladenine or chloroquine), while alleviated by autophagy induction through rapamycin. In PAN-treated rats, inhibition of autophagy led to severe renal dysfunction and podocyte injuries with earlier onset and greater proteinuria, and more extensive foot process effacement and reduction in podocyte markers. Conversely, restoration of autophagy by rapamycin resulted in the attenuation of proteinuria and foot process effacement, and better preservation of podocyte markers. These results suggest that a gradually impaired autophagy in podocytes contributes to the progression of podocytopathies from MCD to FSGS (Zeng et al. 2014). In addition, compared with the podocytes from patients with MCD, the ubiquitin-proteasome activity was also compromised in podocytes from patients with FSGS, leading to abnormal protein accumulation and compensatory upregulation of autophagy (Beeken et al. 2014). These results further demonstrate that the two proteolytic systems may act in concert to protect podocytes and prevent the progression of FSGS.

To further determine the role of autophagy in the pathogenesis of FSGS, Kawakami et al. generated a mouse model in which *Atg5* or *Atg7* gene was mutated in the embryonic progenitor cells (Humphreys et al. 2008; Kawakami et al. 2015). The mutant mice showed mild albuminuria with normal kidney function and widespread podocyte foot process effacement without significant mesangial matrix deposition at 2 months old. At 4 months of age, these mice developed severe albuminuria, tubulointerstitial pathologies and glomerular changes. The mice also showed histologic features of FSGS in the kidneys, such as segmental lesions of scarring with capillary loop obliteration, tuft to capsule adhesion, lesions at the tubular pole and glomeruli with tuft collapse. The mutant mice died from kidney failure by 6 months. Ultrastructurally, podocytes and tubular cells from the mutant mice displayed vacuolization, abnormal mitochondria, evidence of ER stress and increased production of ROS, which recapitulated the features of human idiopathic FSGS kidney biopsy specimens. This over-production of ROS in podocyte and tubules, appearing early before any histological pathology, may be a result of autophagy deficiency-mediated impaired clearance of dysfunctional mitochondria and its proteins. These findings indicate that mitochondrial dysfunction and ER stress due to impairment of autophagic turnover may play a central role in FSGS development (Carney 2015; Kawakami et al. 2015).

28.3.2 Autophagy in Diabetic Kidney Disease (DKD)

DKD is a serious complication of diabetes mellitus and a leading cause of CKD and end-stage renal disease (ESRD) worldwide (Levin et al. 2017). The pathogenesis of DKD is extremely complex, involving multifactorial interactions between hyperglycemia-mediated metabolic alterations, hemodynamic abnormalities and intracellular stress (Brownlee 2005; Forbes and Cooper 2013). The clinical hallmark pathology of DKD is persistent albuminuria or proteinuria followed by decreased glomerular filtration rate (GFR), tubular cell damage and tubulointerstitial lesions, eventually leading to renal failure. Additional pathological features of DKD include an accumulation of ECM components, thickening of both GBM and tubular basement membrane, mesangial expansion, glomerulosclerosis, podocyte effacement, tubular atrophy, and afferent and efferent arteriolar hyalinosis (Abbate et al. 2006).

Emerging evidence has suggested that autophagy is impaired in diabetic kidneys (Fig. 28.2). The defective autophagy in DKD is associated with the abnormalities of multiple nutrient-sensing pathways including mTOR, AMP-activated protein kinase

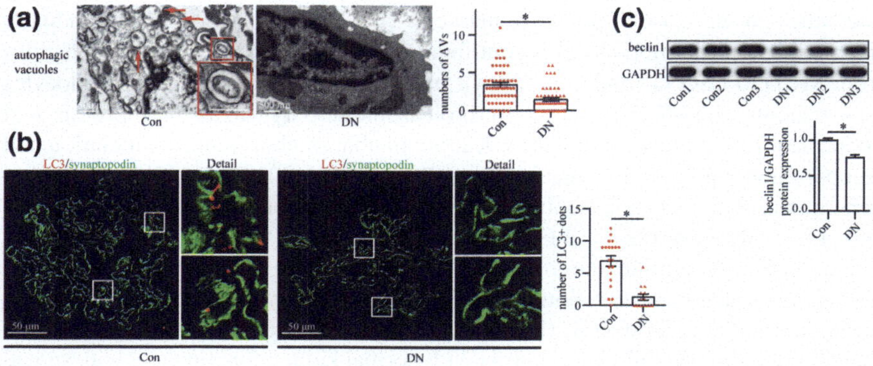

Fig. 28.2 Autophagic insufficiency in diabetic kidneys. **a** TEM of kidney specimens from control subjects (normal renal tissues from patients with renal cell carcinoma) and patients with diabetic nephropathy (DN). The number of autophagic vacuoles (arrows) was reduced in podocytes from patients with DN (means ± SEM, $n = 4$ for control subjects, $n = 5$ for patients with DN; 51–57 images were selected from each group). Student's *t*-test, $*P < 0.05$ versus control group. Scale bar = 500 nm. **b** Immunofluorescence staining for LC3 (red) and synaptopodin (green) in the renal cortex from control subjects and patients with DN. Quantification of the number of LC3—positive dots (yellow) indicated a reduction in the number of autophagosomes in podocytes from patients with DN (means ± SEM, $n = 3$, and 14–20 images from each group). Student's *t*-test, $*P < 0.05$ versus control. Scale bar = 50 μm. **c** Western blot assay of beclin 1 expression in renal cortex from control subjects and patients with DN. Beclin 1 expression normalized against GAPDH was reduced in the renal cortex from patients with DN (means ± SEM, $n = 3$). Student's *t*-test, $*P < 0.05$ versus control

(AMPK) and sirtuins (SIRTs) (Kume et al. 2012; Levin et al. 2017; Yang et al. 2018). mTOR, especially mTORC1, is activated under excessive nutrient conditions by increased levels of glucose, amino acid and growth factors (Wellen and Thompson 2010; Zoncu et al. 2011). mTORC1 negatively regulates autophagy by phosphorylating ULK1 to inhibit its activity (Hosokawa et al. 2009; Jung et al. 2009). Upon nutrient/energy depletion, AMPK and SIRTs are activated, respectively, in response to increasing levels of intracellular AMP and nicotinamide adenine dinucleotide (NAD+) levels (Steinberg and Kemp 2009; Imai and Guarente 2010). In contrast to mTORC1, both AMPK and SIRTs are positive regulators of autophagy. AMPK either directly phosphorylates ULK1 to promote autophagy or inhibits mTORC1 for autophagy induction (Lee et al. 2010; Kim et al. 2011; Alers et al. 2012). SIRT1, the most studied member of the SIRTs family, promotes autophagy by deacetylating ATG5, ATG7 and LC3 (Lee et al. 2008). SIRT1 also deacetylates the transcriptional factor Forkhead box O3a (FoxO3a), leading to activation of BNIP3 (BCL2/adenovirus E1B 19-kDa interacting protein 3) (Kume et al. 2010). In addition, SIRT1 cross talks with AMPK and mTOR to regulate autophagy (Cantó et al. 2009; Ghosh et al. 2010). Dysregulation of these nutrient-sensing pathways under diabetic conditions contributes to the defective autophagy and the pathogenesis of DKD (Kume et al. 2012; Levin et al. 2017; Yang et al. 2018).

Hyperactivation of mTORC1 was frequently seen in animal models and patients of both type 1 and 2 DKDs (Nagai et al. 2005; Mori et al. 2009; Zhang et al. 2014b). In non-diabetic mice, activation of mTORC1 specifically in podocytes induced renal damage that recapitulates features of DKD including GBM thickening, ECM expansion, podocyte loss and proteinuria (Inoki et al. 2011). The causative link between the hyperactivation of mTORC1 and the development of DKD was further verified in both mouse models and human DKD samples (Gödel et al. 2011). In diabetic PTECs, hyperactivation of mTORC1 also induced apoptosis and tubular hypertrophy (Sakaguchi et al. 2006; Velagapudi et al. 2011). On the contrary, inhibition of mTORC1 exerted a renoprotective role against DKD. Inhibiting mTORC1 pharmacologically by rapamycin alleviated kidney injury and attenuated expression of pro-inflammatory and profibrotic cytokines in STZ-induced diabetic rats (Yang et al. 2007; Wittmann et al. 2009). Rapamycin also reduced proteinuria, glomerulosclerosis, mesangial expansion and renal hypertrophy in both STZ-induced diabetic rats and db/db mice (Lloberas et al. 2006; Sakaguchi et al. 2006; Sataranatarajan et al. 2007; Mori et al. 2009; Stridh et al. 2015). Rapamycin rescued autophagy inhibition in cultured podocytes during prolonged high glucose treatment (Fang et al. 2013). The protective effects of rapamycin via inhibiting mTORC1 for autophagy activation were also shown in STZ-induced diabetic mice (Xiao et al. 2014). Pharmacological inhibition of mTORC1 by Torin1 also restored autophagy in db/db mice with high levels of advanced glycation end products (AGEs) and in AGEs-stimulated cultured podocytes (Zhao et al. 2018). In diabetic Wistar fatty rats, inhibition of mTORC1 by a very low protein diet restored autophagy in PTECs and protected against tubular cell injury, inflammation and interstitial fibrosis (Kitada et al. 2016). These findings suggest that hyperactivation of mTOR signaling pathway, via negatively regulating autophagy, plays a critical role in the pathogenesis of DKD.

The activity of AMPK was suppressed in both type 1 and 2 diabetic kidneys, which, importantly, could be reversed by several AMPK activators, leading to the restoration of autophagy and the attenuation of diabetic kidney injury. For example, berberine suppressed high glucose-induced podocyte apoptosis and protected diabetic mice by reactivating AMPK and autophagy (Zhao et al. 2014; Jin et al. 2017). Cinacalcet attenuated diabetic injury in db/db mice by activating autophagy via AMPK pathway (Lim et al. 2018). Consistently, other AMPK activators, such as resveratrol, metformin and AICAR (5-aminoimidazole-4-carboxamide-1β-riboside), also afforded renoprotective effects against kidney injury in both type 1 and 2 models of DKD (Kim et al. 2012b, 2013; Dugan et al. 2013). Similar to AMPK, SIRT1 was downregulated in renal cells in human and animal models of DKD, and activation of SIRT1 protected the kidney from diabetic injury. Mice with overexpressed SIRT1 specifically in proximal tubules were resistant to diabetes-related progression of podocyte damage and subsequent proteinuria (Hasegawa et al. 2013). Resveratrol, via restoring SIRT1 activity, provided beneficial effects in both podocytes and mesangial cells (Chuang et al. 2011; Wu et al. 2012). Recently, Ma et al. demonstrated a role of SIRT1 and consequent autophagy reactivation in the renoprotective effects of resveratrol in type 2 diabetic rats and in hypoxia-treated PTECs (Ma et al. 2016). Similarly, calorie restriction, via increasing SIRT1 activity, activated autophagy in a mouse model of type 2 diabetes and ameliorated glomerular and tubular injury of DKD (Kitada et al. 2011). These findings suggest that inactivation of AMPK and SIRT1 pathways under diabetic conditions suppresses autophagy and facilitates the development of DKD.

Although dysregulated nutrient-sensing pathways suppress autophagy in diabetic kidneys, as a compensatory response, autophagy is also induced by stress signaling to maintain cell integrity. The failure of this adaptive response may lead to abnormal accumulation of damaged organelles, such as mitochondria and ER, and progression of DKD. In this regard, both oxidative stress via reactive oxygen species (ROS) and ER stress have been suggested to regulate autophagy under diabetic conditions (Levin et al. 2017; Yang et al. 2018). Autophagy was induced in podocytes by high glucose within 24 h along with ROS generation, which was inhibited by antioxidant *N*-acetylcysteine (Ma et al. 2013). ROS-mediated autophagy was also seen in podocytes treated with palmitic acid, a saturated free fatty acid (FFA), for 24 h (Jiang et al. 2017). Oxidative stress stimulated autophagy to remove damaged mitochondria (mitophagy). This autophagy-mediated mechanism of mitochondrial quality control and subsequent reduction of ROS is indispensable for protecting against kidney injury under diabetic conditions (Higgins and Coughlan 2014). Autophagy also plays an important role in maintaining ER health during DKD. Defective autophagy under diabetic conditions may lead to prolonged ER stress, and activation of autophagy for ER degradation (ERphagy) is required to protect kidneys from cytotoxic ER stress. Tauroursodeoxycholic acid (TUDCA), a chemical chaperone, inhibited AGEs-induced podocyte apoptosis by reducing ER stress (Chen et al. 2008). Autophagy was restored in TUDCA-treated diabetic mice, accompanied by reduced podocyte injury and proteinuria (Fang et al. 2013; Cao et al. 2016a). In STZ-induced diabetic rats and db/db mice, 4-phenylbutyric acid (4-PBA), a chemical chaperone and ER stress inhibitor, suppressed ER stress-associated inflammation and kidney injury (Qi et al. 2011; Cao et al. 2016b). Both TUDCA and 4-PBA reactivated autophagy in db/db mice and in high glucose-treated podocytes, thus preventing ER stress-induced podocyte apoptosis under diabetic conditions (Cao et al. 2016b).

The changes of autophagy in podocyte seem to be time-dependent and correlated with the severity of podocyte injury and the progression of DKD. Autophagy was induced in immortalized murine podocytes within 24 h of high glucose treatment (Ma et al. 2013), but inhibited at 48 h (Fang et al. 2013). Similarly, in primary podocytes, autophagic flux was prompted by 24-h treatment of high glucose. Genetical inhibition of autophagy by *Atg5* ablation sensitized the podocytes to high glucose-induced apoptosis. In a stable podocyte cell line (SVI), autophagy was activated at 48 h of high glucose treatment, while inhibited after 15 days of high glucose treatment (Lenoir et al. 2015). In diabetic mice, autophagy was induced in podocytes at 4 weeks after STZ injection, when mice were hyperglycemic but had not yet developed glomerular lesions. At 8-week post-STZ injection, along with the occurrence of glomerular lesions autophagy in podocytes was inhibited. Podocyte-specific *Atg5* deletion accelerated glomerular injury in these mice. TEM further confirmed the presence of mesangial expansion and glomerulosclerosis in podocyte-specific autophagy-deficient mice, underscoring the communication between podocytes and mesangial cells under diabetic conditions (Lenoir et al. 2015). In renal biopsy samples from diabetic patients, autophagy was also decreased in podocytes (Fang et al. 2013). These results suggest that autophagy is induced for renoprotection by short term of high glucose (early stage diabetes), but it is suppressed by long term of high glucose exposure (late stage diabetes) contributing to aggravated glomerular injury and progression of DKD. Interestingly in cultured podocytes, along with autophagy impairment, ER stress was induced by prolonged high glucose treatment for up to 60 h (Fang et al. 2013). There was a switch from an adaptive unfolded protein response (UPR) to a cytotoxic ER stress response. ER stress inhibitors such as salubrinal or TUDCA restored autophagy and reduced podocyte injury, indicating that this adaptive-to-cytotoxic switch may be correlated to deficient autophagic turnover of damaged ER (Fang et al. 2013). Tagawa et al. further confirmed that autophagy deficiency in podocytes is pivotal for the progression of advanced DKD, particularly the development of massive proteinuria (Tagawa et al. 2016). Autophagy was impaired in the podocytes of type 2 diabetic patients and Otsuka Long-Evans Tokushima Fatty (OLETF) rats with massive proteinuria, but not in those with absent or minimal proteinuria. Compared with high-fat diet (HFD)-fed control mice that had minimal proteinuria, HFD-fed mice with podocyte-specific *Atg5* deletion displayed more severe podocyte injury and tubulointerstitial fibrosis, damaged lysosomes accumulated in podocytes. These autophagy-deficient diabetic mice showed massive proteinuria, suggesting that autophagy is crucial for the clearance of damaged lysosomes in DKD and its impairment is involved in the progression of podocyte injury and proteinuria in DKD. The sera from diabetic patients and OLETF rats with massive proteinuria impeded autophagy–lysosomal pathway in cultured podocytes and prompted apoptosis, suggesting that serum-derived factors may negatively regulate podocyte autophagy in the progression of DKD (Tagawa et al. 2016).

The mechanism underlying diabetes-associated autophagy impairment in podocytes is under investigation. Nutrient-sensing pathways and intracellular stress signaling pathways are most explored. Several recent studies have provided novel insights. β-arrestins were upregulated in the kidney of diabetic mice and kidney biopsies from diabetic patients as well as in cultured podocytes exposed to high glucose. β-arrestins interacted with ATG7 to downregulate ATG12-ATG5 conjugation, thereby suppressing autophagy in podocytes (Liu et al. 2016). Sun et al. further demonstrated a negative regulation of podocyte autophagy by miR-217. In high glucose-induced podocytes, miR-217 was upregulated to induce podocyte injury and insulin resistance. Inhibition of miR-217 expression reactivated autophagy and reduced podocyte injury. Phosphatase and tensin homolog (PTEN) was a target of miR-217 in podocytes (Sun et al. 2017). A role of FoxO1 in regulating mitophagy in podocytes has also been suggested. Overexpression of FoxO1 promoted mitophagy via PTEN-induced putative kinase 1 (PINK1)/Parkin pathway, leading to the clearance of aberrant mitochondria and the amelioration of podocyte injury (Li et al. 2017). Furthermore, HDAC4 was induced in cultured podocytes by high glucose, AGEs and transforming growth factor (TGF)-β. By deacetylating signal transducers and activators of transcription factor 1 (STAT1), HDAC4 inhibited podocyte autophagy and induced podocyte injury. Inhibition of HDAC4 reactivated autophagy and alleviated podocyte injury. These results suggest that HDAC4 may suppress podocyte autophagy via STAT1 to accelerate the development of DKD (Wei and Dong 2014).

Hyperglycemia inhibits autophagy in proximal and distal tubules of diabetic animals. Sodium-glucose cotransporter 2 (SGLT2), mainly expressed in proximal tubules, regulates the high capacity reabsorption of glucose in PTECs via sodium gradient produced by sodium/potassium ATPase pumps. Inhibition of SGLT2 reduced the reabsorption of glucose in PTECs and decreased blood glucose concentrations (Nair and Wilding 2010). Deletion of *Sglt2* alleviated STZ-induced abnormal accumulation of p62/SQSTM1 in kidney, indicating that SGLT2-mediated glucose uptake contributes to autophagy impairment in PTECs under diabetic conditions (Vallon et al. 2013). High glucose also induced p53 to upregulate the transcription and expression of miR-155 in PTECs. Overexpression of miR-155 targeted SIRT1 by binding to the SIRT1 3′UTR region to reduce its expression and activity on essential ATG proteins such as ATG5 and LC3, leading to autophagy impairment in PTECs (Wang et al. 2018). In addition, hyperactivated mTORC1 was involved in obesity-related autophagy inhibition in PTECs in type 2 diabetic mice and patients, as diet restriction or rapamycin restored autophagy under these conditions (Moruno-Manchon et al. 2018). Similarly in diabetic Wistar fatty rats, dietary restriction rescued autophagy deficiency in PTECs via suppression of mTORC1 and activation of SIRT1 pathways, which subsequently resulted in the protection against diabetes-induced tubular injury and renal dysfunction (Kitada et al. 2011, 2016). Recent studies further reveal an interconnection between autophagy impairment and AGEs accumulation in PTECs and its contribution to tubular injury in DKD. AGEs overload led to lysosomal dysfunction and disruption of autophagic flux in PTECs (Liu et al. 2015; Takahashi et al. 2017). AGEs overload upregulated LAMP1 in *Atg5*-competent primary PTECs, while it was suppressed in *Atg5*-deficient cells with AGEs overload, indicating that autophagy contributes to inducing lysosomal biogenesis and function under diabetic conditions. Similarly, the upregulation of lysosome was suppressed in PTEC-specific *Atg5* deletion of diabetic mice compared to the diabetic control mice. Along with autophagy deficiency, the PTEC-specific *Atg5* deletion of diabetic mice displayed enhanced accumulation of AGEs in PTECs, glomeruli and renal vasculature, and increased inflammation and renal fibrosis (Takahashi et al. 2017). These results suggest that autophagy may degrade accumulated AGEs by promoting lysosomal biogenesis and function in PTECs. As such, impaired autophagy under diabetic conditions may lead to failure in the turnover of lysosomes and consequent AGE accumulation, which further disrupts lysosomal function and blocks lysosomal degradation of AGEs. This vicious cycle subsequently results in irreversible injury in both glomeruli and tubulointerstitium for the progression of DKD (Takahashi et al. 2017).

The role of autophagy in mesangial cells and glomerular endothelial cells remains largely unclear. The expression of tissue inhibitors of metalloproteinase 3 (TIMP3) was reduced in STZ-induced diabetic mice and diabetic patients. *Timp3*-deficient mice had more severe diabetic kidney injury compared with control mice. *Timp* deletion in cultured mesangial cells suppressed autophagy via FoxO1/STAT pathway, and re-expression of TIMP in these mesangial cells reversed autophagy impairment (Fiorentino et al. 2013). In AGEs-treated mesangial cells, autophagy/mitophagy was activated via ROS and protected against AGEs-induced mitochondrial dysfunction and cell apoptosis (Xu et al. 2016a). In contrast, autophagy was suppressed in cultured rat mesangial cells in response to high glucose, accompanied with collagen I accumulation and cell hypertrophy. The inhibition of autophagy was mediated by miR-21/PTEN/Akt/mTOR pathway, and ursolic acid restored autophagy activity and attenuated mesangial injury and collagen I production (Lu et al. 2015). Using endothelial-specific *Atg5* knockout mice, a recent study demonstrated direct evidence on the role of glomerular endothelial cell autophagy in DKD (Lenoir et al. 2015). Endothelial-specific *Atg5* knockout non-diabetic mice exhibited mild lesions to the glomerular filtration barrier. These mice also had more severe glomerular endothelial injuries following STZ injection, showing GBM thickening and podocyte effacement. These results suggest that via a crosstalk between glomerular endothelial cells and surrounding podocytes autophagy in glomerular endothelial cells preserves both endothelial integrity and podocyte homeostasis (Lenoir et al. 2015).

28.3.3 Autophagy in Acute Kidney Injury (AKI) and Kidney Repair

AKI, mostly caused by nephrotoxic drugs, renal ischemia-reperfusion and sepsis, is a major renal disease associated with poor clinical outcomes in both short term (high morbidity and mortality) and long term (the development of CKD and ESRD) (Bellomo et al. 2012; Zuk and Bonventre 2016). The pathogenesis of AKI is multifactorial and involves a complex interplay among microvascular, tubular and inflammatory factors. Tubular cell injury and death are the key pathological features of this disorder (Bellomo et al. 2012; Linkermann et al. 2014; Zuk and Bonventre 2016). The activation of autophagy in AKI was initially demonstrated in experimental models of cisplatin-induced nephrotoxicity (Periyasamy-Thandavan et al. 2008; Yang et al. 2008). While Yang et al. showed autophagy activation in cultured renal tubular cells (Yang et al. 2008), we demonstrated it using both cell culture and mouse models (Periyasamy-Thandavan et al. 2008). Interestingly, both studies suggested a protective role of autophagy in renal tubular cells. Our follow-up study further verified autophagy activation and its protective role in renal ischemic/hypoxic AKI (Jiang et al. 2010). In 2012, we and other two groups independently demonstrated the protective role of tubular cell autophagy in AKI using renal tubule-specific autophagy gene knockout mouse models (Liu et al. 2012; Takahashi et al. 2012; Cheng et al. 2015). These findings have been summarized in recent reviews (Livingston and Dong 2014; Havasi and Dong 2016).

After injury, renal tubular cells have the capacity to regenerate for kidney repair. This process involves the activation of multiple signaling pathways in injured and regenerating tubular cells leading to the production and secretion of growth factors, cytokines and inflammatory mediators. Normal tubular repair begins with dedifferentiation, migration and proliferation of surviving cells to replace injured cells, followed by re-differentiation to restore normal epithelial structure and function. However, tubular repair following severe or multiple episodes of AKI is often incomplete and maladaptive, leading to renal interstitial fibrosis and CKD (Ferenbach and Bonventre 2015; Venkatachalam et al. 2015; Basile et al. 2016; He et al. 2017). Although the mechanisms underlying AKI to CKD transition remain to be elucidated, emerging evidence suggests a central role of proximal tubules in the disease progression. After severe or episodic AKI normal tubular epithelial cells undergo a phenotype change with persistent production and secretion of profibrotic proteins. These tubule-derived molecules may drive renal interstitial fibrosis and AKI to CKD transition via autocrine and paracrine functions (Ferenbach and Bonventre 2015; Venkatachalam et al. 2015; Basile et al. 2016).

Autophagy is induced in tubular cells during AKI and protects against kidney injury (Livingston and Dong 2014; Kaushal and Shah 2016). During recovery following AKI, resolution of autophagy in tubular cells may promote cell proliferation for tubular regeneration and repair (Li et al. 2014). Using autophagy reporter mice with a tandem RFP-EGFP-LC3 fusion protein expressing ubiquitously under the CAG promoter, Li et al. revealed dynamic changes of autophagy in renal tubules during ischemic AKI and the following recovery phase. At 1 day of reperfusion, there was increased formation of both autophagosomes and autolysosomes in proximal tubule cells. At 3 days of reperfusion, autolysosomes appeared, suggesting the resolution of autophagy during the recovery phase. Mechanistically, mTOR was activated after ischemic AKI. Some renal tubules showed defective mTOR activity and persistent RFP-LC3 puncta, and interestingly tubular proliferation was also inhibited, indicating that autophagic cells are less likely to divide for tubular repair (Li et al. 2014). These findings are consistent with an earlier study suggesting that rapamycin delays recovery from ischemic AKI (Lieberthal et al. 2006).

A recent study by Brooks et al. further revealed a novel mechanism of epithelial biology linking phagocytosis, autophagy and antigen presentation to regulation of the inflammatory response after injury (Brooks et al. 2015). Kidney injury molecule-1 (KIM-1) expressed on proximal tubular cells transformed the cells into phagocytes for the uptake of luminal apoptotic cell debris. This KIM-1-mediated phagocytosis was subsequently processed through autophagy for efficient clearance of apoptotic cells and autophagic degradation of phagosomes, leading to major histocompatibility complex (MHC) restricted antigen presentation that suppressed CD4 T cell proliferation but increased the percentage of regulatory T cells in an autophagy-dependent manner. These results highlight the role of autophagy in downregulating the inflammatory response after injury and maintaining self-tolerance in proximal tubular cells, both of which would contribute to an improved tubular repair (Brooks et al. 2015).

The role of autophagy in maladaptive repair and AKI to CKD transition needs to be further elucidated. In this setting, autophagy may be less effective and thereby fail to mediate intracellular degradation of newly synthesized fibrotic proteins (Zuk and Bonventre 2016). Instead, autophagy, coordinated with several other proximal tubular responses such as dedifferentiation, cell cycle changes and metabolic changes, may be adaptive initially, but ultimately lead to maladaptive responses that promote interstitial fibrosis and AKI to CKD transition (He et al. 2014; Gewin 2018). Along this line, a profibrotic role of autophagy has been shown in a mouse model of post-ischemic kidney fibrosis (Baisantry et al. 2016). Sustained activation of autophagy (30 days) following ischemic AKI induced prosenescent changes in proximal tubules of the S3 segments during recovery phase. Selective deletion of Atg5 in these proximal tubules suppressed autophagy, which in turn inhibited the development of a senescent phenotype and AKI progression to CKD (Baisantry et al. 2016). Of interest, compared with wild-type mice, selective Atg5 knockout mice showed more tubular cell death at the S3 segment at 2 h after reperfusion but less tubular damage and inflammation at day 3, suggesting that autophagy inhibition may enhance cell death in severely damaged tubular cells during injury phase but is beneficial for adaptive tubular repair during recovery phase. When such compromised cells with intact autophagy escape from cell death pathway and persist, they may develop into a senescent phenotype and promote renal interstitial fibrosis (Baisantry et al. 2016).

28.3.4 Autophagy in Renal Interstitial Fibrosis Induced by Unilateral Ureteral Obstruction (UUO) or TGF-β1

So far, most of the studies on the role of autophagy in renal interstitial fibrosis were performed in models of UUO or TGF-β1 and the findings are controversial. In mice subjected to UUO, autophagy was activated in renal tubules along with tubular apoptosis (Li et al. 2010; Forbes et al. 2011; Xu et al. 2013). Under this condition, autophagy and apoptosis acted in concert to induce tubular atrophy and nephron loss (Li et al. 2010). Oxidative stress-mediated mitochondrial damage was likely to promote autophagy and apoptosis in renal tubules, which may play a role in facilitating tubular decomposition in UUO (Xu et al. 2013). Using a tetracycline-controlled mouse model with TGF-β1 overexpression specifically in renal tubules, Koesters et al. showed that persistent expression of TGF-β1 promoted autophagy in renal tubules, leading to tubular dedifferentiation with widespread peritubular fibrosis. Notably, such degenerating cells were not positive for TUNEL staining for apoptosis, indicating that autophagy could be a key driver of tubular atrophy in TGF-β1-induced renal fibrosis (Koesters et al. 2010). Using pharmacological and genetic inhibitory approaches, we further demonstrated a profibrotic role of autophagy in a mouse model of UUO and in TGF-β1-treated PTECs (Livingston et al. 2016). Autophagy was persistently activated in proximal tubules following UUO. Pharmacological and genetic blockade of autophagy attenuated interstitial fibrosis, accompanied with the alleviation of tubular cell apoptosis, interstitial macrophage infiltration and production of fibroblast growth factor 2. In primary culture of PTECs, TGF-β1 induced fibronectin accumulation and cell death in an autophagy-dependent method (Livingston et al. 2016). A recent study by Yan et al. further demonstrated a connection between sustained activation of autophagy and lipid accumulation in tubular epithelial cells during kidney fibrosis (Yan et al. 2018). UUO-induced lipid accumulation in tubular cells was significantly reduced by autophagy inhibitors, along with the attenuation of renal interstitial fibrosis, tubular cell apoptosis and tubular cell dedifferentiation. This fibrosis-related lipid accumulation was not associated with lipophagy–lysosome pathway but was dependent on BECN1. These results highlight a role of autophagy in regulation of the lipid metabolism in renal tubular cells (Yan et al. 2018). Of interest, during thioacetamide or CCl4-induced liver injury, autophagy was activated in hepatic stellate cells, which further broke down lipids to fuel the activation of these cells to promote liver fibrosis (Thoen et al. 2011; Hernández-Gea et al. 2012). Therefore, autophagy may involve in lipid metabolism through a bidirectional mechanism of inducing lipolysis to activate liver fibroblasts as well as promoting lipid accumulation to induce lipotoxicity in renal tubular cells, both of which contribute to the development of fibrosis in these organs. A role for autophagy in the activation of fibroblasts and its profibrotic effects were also shown in UUO kidneys and in cultured renal fibroblast cells exposed to TGF-β1. Protein kinase C (PKC)-α, via stimulating autophagic flux, drove renal fibroblast activation and kidney fibrosis (Xue et al. 2018).

On the contrary, several studies have demonstrated an antifibrotic role of autophagy in UUO-associated renal interstitial fibrosis. In a rat model of UUO, inhibition of autophagy by 3-methyladenine aggravated tubular cell apoptosis and interstitial fibrosis, indicating that autophagy may inhibit fibrosis by suppressing tubular apoptosis (Kim et al. 2012c). In primary culture of mouse kidney mesangial cells, both protein and mRNA levels of collagen I were induced by TGF-β1. Inhibition of autophagy by BECN1 knockdown or lysosomal inhibitors further increased collagen I protein accumulation without affecting its mRNA level, indicating a potential effect for autophagy in regulating ECM deposition in mesangial cells by facilitating the degradation of collagen I (Kim et al. 2012a). MAP1S, via interacting with LC3, activated autophagy to suppress fibrosis by mediating fibronectin turnover. Defective MAP1S impaired autophagic clearance of fibronectin and induced renal fibrosis in aged mice. Reduced expression of MAP1S in renal biopsies from patients with kidney fibrosis was also accompanied with elevated accumulation of fibronectin (Xu et al. 2016b). Ding Y et al. further showed the role for autophagy in promoting the degradation of mature TGF-β1 in UUO kidneys and in TGF-β1-treated PTECs, further suggesting that autophagy can inhibit renal interstitial fibrosis by negative regulation of TGF-β1 (Ding et al. 2014). Using a proximal tubule-specific *Atg5* knockout mouse model, Li et al. suggested that autophagy deficiency promoted cell cycle G2/M arrest and accelerated renal interstitial fibrosis following UUO. Stimulation of *Atg5*-deficient primary proximal tubular cells with angiotensin II also led to G2/M arrest and increased production of collagen I, which was rescued by the restoration of autophagy-competent *Atg5* in these cells (Li et al. 2016). Elucidation of the mechanisms of autophagy regulation in renal interstitial fibrosis is pivotal for identifying potential therapeutic targets to treat progressive CKD.

28.4 Conclusions and Perspectives

Basal level of autophagy is essential to the maintenance of cellular homeostasis in renal resident cells including podocytes, renal tubular cells, mesangial cells and glomerular endothelial cells. Autophagy defects in these cells have been implicated in the development of CKD such as FSGS and DKD. Autophagy is induced in response to AKI for renoprotection. After kidney injury, tightly regulated autophagy may participate in adaptive renal repair, whereas dysregulated autophagy may lead to maladaptive repair contributing to AKI to CKD transition. The role of autophagy in renal interstitial fibrosis is multifaceted and complex. Further research is needed to gain significant insights into the role of autophagy in the pathogenesis of renal fibrosis and related kidney diseases as well as the regulatory mechanisms of autophagy in these disease settings. A comprehensive understanding of the regulation and pathological roles of autophagy in renal fibrosis will facilitate the discovery of novel therapeutic strategy that can target autophagy for the prevention and treatment of fibrosis-related CKD.

References

Abbate M, Zoja C, Remuzzi G (2006) How does proteinuria cause progressive renal damage. J Am Soc Nephrol 17:2974–2984

Alers S, Löffler AS, Wesselborg S, Stork B (2012) Role of AMPK-mTOR-Ulk1/2 in the regulation of autophagy: cross talk, shortcuts, and feedbacks. Mol Cell Biol 32:2–11

Asanuma K, Tanida I, Shirato I, Ueno T, Takahara H, Nishitani T et al (2003) MAP-LC3, a promising autophagosomal marker, is processed during the differentiation and recovery of podocytes from PAN nephrosis. FASEB J 17:1165–1167

Baisantry A, Bhayana S, Rong S, Ermeling E, Wrede C, Hegermann J et al (2016) Autophagy induces prosenescent changes in proximal tubular S3 segments. J Am Soc Nephrol 27:1609–1616

Basile DP, Bonventre JV, Mehta R, Nangaku M, Unwin R, Rosner MH et al (2016) Progression after AKI: understanding maladaptive repair processes to predict and identify therapeutic treatments. J Am Soc Nephrol 27:687–697

Beeken M, Lindenmeyer MT, Blattner SM, Radón V, Oh J, Meyer TN et al (2014) Alterations in the ubiquitin proteasome system in persistent but not reversible proteinuric diseases. J Am Soc Nephrol 25:2511–2525

Bellomo R, Kellum JA, Ronco C (2012) Acute kidney injury. Lancet 380:756–766

Brooks CR, Yeung MY, Brooks YS, Chen H, Ichimura T, Henderson JM et al (2015) KIM-1-/TIM-1-mediated phagocytosis links ATG5-/ULK1-dependent clearance of apoptotic cells to antigen presentation. EMBO J 34:2441–2464

Brownlee M (2005) The pathobiology of diabetic complications: a unifying mechanism. Diabetes 54:1615–1625

Cantó C, Gerhart-Hines Z, Feige JN, Lagouge M, Noriega L, Milne JC et al (2009) AMPK regulates energy expenditure by modulating NAD+ metabolism and SIRT1 activity. Nature 458:1056–1060

Cao A, Wang L, Chen X, Guo H, Chu S, Zhang X et al (2016a) Ursodeoxycholic acid ameliorated diabetic nephropathy by attenuating hyperglycemia-mediated oxidative stress. Biol Pharm Bull 39:1300–1308

Cao AL, Wang L, Chen X, Wang YM, Guo HJ, Chu S et al (2016b) Ursodeoxycholic acid and 4-phenylbutyrate prevent endoplasmic reticulum stress-induced podocyte apoptosis in diabetic nephropathy. Lab Invest 96:610–622

Carney EF (2015) Glomerular disease: autophagy failure and mitochondrial dysfunction in FSGS. Nat Rev Nephrol 11:66

Chen Y, Liu CP, Xu KF, Mao XD, Lu YB, Fang L et al (2008) Effect of taurine-conjugated ursodeoxycholic acid on endoplasmic reticulum stress and apoptosis induced by advanced glycation end products in cultured mouse podocytes. Am J Nephrol 28:1014–1022

Chen J, Chen MX, Fogo AB, Harris RC, Chen JK (2013) mVps34 deletion in podocytes causes glomerulosclerosis by disrupting intracellular vesicle trafficking. J Am Soc Nephrol 24:198–207

Cheng H, Fan X, Lawson WE, Paueksakon P, Harris RC (2015) Telomerase deficiency delays renal recovery in mice after ischemia-reperfusion injury by impairing autophagy. Kidney Int 88:85–94

Choi AM, Ryter SW, Levine B (2013) Autophagy in human health and disease. N Engl J Med 368:651–662

Chuang PY, Dai Y, Liu R, He H, Kretzler M, Jim B et al (2011) Alteration of forkhead box O (foxo4) acetylation mediates apoptosis of podocytes in diabetes mellitus. PLoS ONE 6:e23566

Cinà DP, Onay T, Paltoo A, Li C, Maezawa Y, De Arteaga J et al (2012a) MTOR regulates autophagic flux in the glomerulus. Autophagy 8:696–698

Cinà DP, Onay T, Paltoo A, Li C, Maezawa Y, De Arteaga J et al (2012b) Inhibition of MTOR disrupts autophagic flux in podocytes. J Am Soc Nephrol 23:412–420

Cui J, Bai XY, Shi S, Cui S, Hong Q, Cai G et al (2012) Age-related changes in the function of autophagy in rat kidneys. Age (Dordr) 34:329–339

D'Agati VD (2012) Pathobiology of focal segmental glomerulosclerosis: new developments. Curr Opin Nephrol Hypertens 21:243–250

Dikic I, Elazar Z (2018) Mechanism and medical implications of mammalian autophagy. Nat Rev Mol Cell Biol 19:349–364

Ding Y, Sl K, Lee SY, Koo JK, Wang Z, Choi ME (2014) Autophagy regulates TGF-β expression and suppresses kidney fibrosis induced by unilateral ureteral obstruction. J Am Soc Nephrol 25:2835–2846

Duffield JS (2014) Cellular and molecular mechanisms in kidney fibrosis. J Clin Invest 124:2299–2306

Dugan LL, You YH, Ali SS, Diamond-Stanic M, Miyamoto S, DeCleves AE et al (2013) AMPK dysregulation promotes diabetes-related reduction of superoxide and mitochondrial function. J Clin Invest 123:4888–4899

Fang L, Zhou Y, Cao H, Wen P, Jiang L, He W et al (2013) Autophagy attenuates diabetic glomerular damage through protection of hyperglycemia-induced podocyte injury. PLoS ONE 8:e60546

Ferenbach DA, Bonventre JV (2015) Mechanisms of maladaptive repair after AKI leading to accelerated kidney ageing and CKD. Nat Rev Nephrol 11:264–276

Fiorentino L, Cavalera M, Menini S, Marchetti V, Mavilio M, Fabrizi M et al (2013) Loss of TIMP3 underlies diabetic nephropathy via FoxO1/STAT1 interplay. EMBO Mol Med 5:441–455

Forbes JM, Cooper ME (2013) Mechanisms of diabetic complications. Physiol Rev 93:137–188

Forbes MS, Thornhill BA, Chevalier RL (2011) Proximal tubular injury and rapid formation of atubular glomeruli in mice with unilateral ureteral obstruction: a new look at an old model. Am J Physiol Renal Physiol 301:F110–F117

Fukuda A, Chowdhury MA, Venkatareddy MP, Wang SQ, Nishizono R, Suzuki T et al (2012) Growth-dependent podocyte failure causes glomerulosclerosis. J Am Soc Nephrol 23:1351–1363

Gewin LS (2018) Renal fibrosis: primacy of the proximal tubule. Matrix Biol 68–69:248–262

Ghosh HS, McBurney M, Robbins PD (2010) SIRT1 negatively regulates the mammalian target of rapamycin. PLoS ONE 5:e9199

Gödel M, Hartleben B, Herbach N, Liu S, Zschiedrich S, Lu S et al (2011) Role of mTOR in podocyte function and diabetic nephropathy in humans and mice. J Clin Invest 121:2197–2209

Hartleben B, Godel M, Meyer-Schwesinger C, Liu S, Ulrich T, Kobler S et al (2010) Autophagy influences glomerular disease susceptibility and maintains podocyte homeostasis in aging mice. J Clin Invest 120:1084–1096
Hartleben B, Wanner N, Huber TB (2014) Autophagy in glomerular health and disease. Semin Nephrol 34:42–52
Hasegawa K, Wakino S, Simic P, Sakamaki Y, Minakuchi H, Fujimura K et al (2013) Renal tubular Sirt1 attenuates diabetic albuminuria by epigenetically suppressing Claudin-1 overexpression in podocytes. Nat Med 19:1496–1504
Havasi A, Dong Z (2016) Autophagy and tubular cell death in the kidney. Semin Nephrol 36:174–188
He C, Klionsky DJ (2009) Regulation mechanisms and signaling pathways of autophagy. Annu Rev Genet 43:67–93
He L, Livingston MJ, Dong Z (2014) Autophagy in acute kidney injury and repair. Nephron Clin Pract 127:56–60
He L, Wei Q, Liu J, Yi M, Liu Y, Liu H et al (2017) AKI on CKD: heightened injury, suppressed repair, and the underlying mechanisms. Kidney Int 92:1071–1083
Hernández-Gea V, Ghiassi-Nejad Z, Rozenfeld R, Gordon R, Fiel MI, Yue Z et al (2012) Autophagy releases lipid that promotes fibrogenesis by activated hepatic stellate cells in mice and in human tissues. Gastroenterology 142:938–946
Higgins GC, Coughlan MT (2014) Mitochondrial dysfunction and mitophagy: the beginning and end to diabetic nephropathy. Br J Pharmacol 171:1917–1942
Hosokawa N, Hara T, Kaizuka T, Kishi C, Takamura A, Miura Y et al (2009) Nutrient-dependent mTORC1 association with the ULK1-Atg13-FIP200 complex required for autophagy. Mol Biol Cell 20:1981–1991
Huber TB, Edelstein CL, Hartleben B, Inoki K, Jiang M, Koya D et al (2012) Emerging role of autophagy in kidney function, diseases and aging. Autophagy 8:1009–1031
Humphreys BD (2018) Mechanisms of Renal Fibrosis. Annu Rev Physiol 80:309–326
Humphreys BD, Valerius MT, Kobayashi A, Mugford JW, Soeung S, Duffield JS et al (2008) Intrinsic epithelial cells repair the kidney after injury. Cell Stem Cell 2:284–291
Hurley JH, Young LN (2017) Mechanisms of autophagy initiation. Annu Rev Biochem 86:225–244
Imai S, Guarente L (2010) Ten years of NAD-dependent SIR2 family deacetylases: implications for metabolic diseases. Trends Pharmacol Sci 31:212–220
Inoki K (2014) mTOR signaling in autophagy regulation in the kidney. Semin Nephrol 34:2–8
Inoki K, Mori H, Wang J, Suzuki T, Hong S, Yoshida S et al (2011) mTORC1 activation in podocytes is a critical step in the development of diabetic nephropathy in mice. J Clin Invest 121:2181–2196
Jiang M, Liu K, Luo J, Dong Z (2010) Autophagy is a renoprotective mechanism during in vitro hypoxia and in vivo ischemia-reperfusion injury. Am J Pathol 176:1181–1192
Jiang M, Wei Q, Dong G, Komatsu M, Su Y, Dong Z (2012) Autophagy in proximal tubules protects against acute kidney injury. Kidney Int 82:1271–1283
Jiang XS, Chen XM, Wan JM, Gui HB, Ruan XZ, Du XG (2017) Autophagy protects against palmitic acid-induced apoptosis in podocytes in vitro. Sci Rep 7:42764
Jin Y, Liu S, Ma Q, Xiao D, Chen L (2017) Berberine enhances the AMPK activation and autophagy and mitigates high glucose-induced apoptosis of mouse podocytes. Eur J Pharmacol 794:106–114
Jung CH, Jun CB, Ro SH, Kim YM, Otto NM, Cao J et al (2009) ULK-Atg13-FIP200 complexes mediate mTOR signaling to the autophagy machinery. Mol Biol Cell 20:1992–2003
Kaushal GP, Shah SV (2016) Autophagy in acute kidney injury. Kidney Int 89:779–791
Kawakami T, Gomez IG, Ren S, Hudkins K, Roach A, Alpers CE et al (2015) Deficient autophagy results in mitochondrial dysfunction and FSGS. J Am Soc Nephrol 26:1040–1052
Kim J, Kundu M, Viollet B, Guan KL (2011) AMPK and mTOR regulate autophagy through direct phosphorylation of Ulk1. Nat Cell Biol 13:132–141
Kim J, Shon E, Kim CS, Kim JS (2012a) Renal podocyte injury in a rat model of type 2 diabetes is prevented by metformin. Exp Diabetes Res 2012:210821

Kim SI, Na HJ, Ding Y, Wang Z, Lee SJ, Choi ME (2012b) Autophagy promotes intracellular degradation of type I collagen induced by transforming growth factor (TGF)-β1. J Biol Chem 287:11677–11688

Kim WY, Nam SA, Song HC, Ko JS, Park SH, Kim HL et al (2012c) The role of autophagy in unilateral ureteral obstruction rat model. Nephrology (Carlton) 17:148–159

Kim MY, Lim JH, Youn HH, Hong YA, Yang KS, Park HS et al (2013) Resveratrol prevents renal lipotoxicity and inhibits mesangial cell glucotoxicity in a manner dependent on the AMPK-SIRT1-PGC1α axis in db/db mice. Diabetologia 56:204–217

Kimura T, Takabatake Y, Takahashi A, Kaimori JY, Matsui I, Namba T et al (2011) Autophagy protects the proximal tubule from degeneration and acute ischemic injury. J Am Soc Nephrol 22:902–913

Kitada M, Takeda A, Nagai T, Ito H, Kanasaki K, Koya D (2011) Dietary restriction ameliorates diabetic nephropathy through anti-inflammatory effects and regulation of the autophagy via restoration of Sirt1 in diabetic Wistar fatty (fa/fa) rats: a model of type 2 diabetes. Exp Diabetes Res 2011:908185

Kitada M, Ogura Y, Suzuki T, Sen S, Lee SM, Kanasaki K et al (2016) A very-low-protein diet ameliorates advanced diabetic nephropathy through autophagy induction by suppression of the mTORC1 pathway in Wistar fatty rats, an animal model of type 2 diabetes and obesity. Diabetologia 59:1307–1317

Klionsky DJ (2008) Autophagy revisited: a conversation with Christian de Duve. Autophagy 4:740–743

Klionsky DJ, Baehrecke EH, Brumell JH, Chu CT, Codogno P, Cuervo AM et al (2011) A comprehensive glossary of autophagy-related molecules and processes (2nd edition). Autophagy 7:1273–1294

Klionsky DJ, Abdelmohsen K, Abe A, Abedin MJ, Abeliovich H, Acevedo AA et al (2016) Guidelines for the use and interpretation of assays for monitoring autophagy (3rd edition). Autophagy 12:1–222

Koesters R, Kaissling B, Lehir M, Picard N, Theilig F, Gebhardt R et al (2010) Tubular overexpression of transforming growth factor-beta1 induces autophagy and fibrosis but not mesenchymal transition of renal epithelial cells. Am J Pathol 177:632–643

Kume S, Uzu T, Horiike K, Chin-Kanasaki M, Isshiki K, Araki S et al (2010) Calorie restriction enhances cell adaptation to hypoxia through Sirt1-dependent mitochondrial autophagy in mouse aged kidney. J Clin Invest 120:1043–1055

Kume S, Thomas MC, Koya D (2012) Nutrient sensing, autophagy, and diabetic nephropathy. Diabetes 61:23–29

Lee IH, Cao L, Mostoslavsky R, Lombard DB, Liu J, Bruns NE et al (2008) A role for the NAD-dependent deacetylase Sirt1 in the regulation of autophagy. Proc Natl Acad Sci U S A 105:3374–3379

Lee JW, Park S, Takahashi Y, Wang HG (2010) The association of AMPK with ULK1 regulates autophagy. PLoS ONE 5:e15394

Lenoir O, Jasiek M, Hénique C, Guyonnet L, Hartleben B, Bork T et al (2015) Endothelial cell and podocyte autophagy synergistically protect from diabetes-induced glomerulosclerosis. Autophagy 11:1130–1145

Levin A, Tonelli M, Bonventre J, Coresh J, Donner JA, Fogo AB et al (2017) Global kidney health 2017 and beyond: a roadmap for closing gaps in care, research, and policy. Lancet

Levine B, Kroemer G (2008) Autophagy in the pathogenesis of disease. Cell 132:27–42

Li L, Zepeda-Orozco D, Black R, Lin F (2010) Autophagy is a component of epithelial cell fate in obstructive uropathy. Am J Pathol 176:1767–1778

Li L, Wang ZV, Hill JA, Lin F (2014) New autophagy reporter mice reveal dynamics of proximal tubular autophagy. J Am Soc Nephrol 25:305–315

Li H, Peng X, Wang Y, Cao S, Xiong L, Fan J et al (2016) Atg5-mediated autophagy deficiency in proximal tubules promotes cell cycle G2/M arrest and renal fibrosis. Autophagy 12:1472–1486

Li W, Du M, Wang Q, Ma X, Wu L, Guo F et al (2017) FoxO1 promotes mitophagy in the podocytes of diabetic male mice via the PINK1/parkin pathway. Endocrinology 158:2155–2167
Lieberthal W, Fuhro R, Andry C, Patel V, Levine JS (2006) Rapamycin delays but does not prevent recovery from acute renal failure: role of acquired tubular resistance. Transplantation 82:17–22
Lim JH, Kim HW, Kim MY, Kim TW, Kim EN, Kim Y et al (2018) Cinacalcet-mediated activation of the CaMKKβ-LKB1-AMPK pathway attenuates diabetic nephropathy in db/db mice by modulation of apoptosis and autophagy. Cell Death Dis 9:270
Linkermann A, Chen G, Dong G, Kunzendorf U, Krautwald S, Dong Z (2014) Regulated cell death in AKI. J Am Soc Nephrol 25:2689–2701
Liu Y (2011) Cellular and molecular mechanisms of renal fibrosis. Nat Rev Nephrol 7:684–696
Liu S, Hartleben B, Kretz O, Wiech T, Igarashi P, Mizushima N et al (2012) Autophagy plays a critical role in kidney tubule maintenance, aging and ischemia-reperfusion injury. Autophagy 8:826–837
Liu WJ, Shen TT, Chen RH, Wu HL, Wang YJ, Deng JK et al (2015) Autophagy-lysosome pathway in renal tubular epithelial cells is disrupted by advanced glycation end products in diabetic nephropathy. J Biol Chem 290:20499–20510
Liu J, Li QX, Wang XJ, Zhang C, Duan YQ, Wang ZY et al (2016) β-Arrestins promote podocyte injury by inhibition of autophagy in diabetic nephropathy. Cell Death Dis 7:e2183
Livingston MJ, Dong Z (2014) Autophagy in acute kidney injury. Semin Nephrol 34:17–26
Livingston MJ, Ding HF, Huang S, Hill JA, Yin XM, Dong Z (2016) Persistent activation of autophagy in kidney tubular cells promotes renal interstitial fibrosis during unilateral ureteral obstruction. Autophagy 12:976–998
Lloberas N, Cruzado JM, Franquesa M, Herrero-Fresneda I, Torras J, Alperovich G et al (2006) Mammalian target of rapamycin pathway blockade slows progression of diabetic kidney disease in rats. J Am Soc Nephrol 17:1395–1404
Löwik MM, Groenen PJ, Levtchenko EN, Monnens LA, van den Heuvel LP (2009) Molecular genetic analysis of podocyte genes in focal segmental glomerulosclerosis—a review. Eur J Pediatr 168:1291–1304
Lu X, Fan Q, Xu L, Li L, Yue Y, Xu Y et al (2015) Ursolic acid attenuates diabetic mesangial cell injury through the up-regulation of autophagy via miRNA-21/PTEN/Akt/mTOR suppression. PLoS ONE 10:e0117400
Ma T, Zhu J, Chen X, Zha D, Singhal PC, Ding G (2013) High glucose induces autophagy in podocytes. Exp Cell Res 319:779–789
Ma L, Fu R, Duan Z, Lu J, Gao J, Tian L et al (2016) Sirt1 is essential for resveratrol enhancement of hypoxia-induced autophagy in the type 2 diabetic nephropathy rat. Pathol Res Pract 212:310–318
Mehrpour M, Esclatine A, Beau I, Codogno P (2010) Overview of macroautophagy regulation in mammalian cells. Cell Res 20:748–762
Mizushima N, Komatsu M (2011) Autophagy: renovation of cells and tissues. Cell 147:728–741
Mizushima N, Yamamoto A, Matsui M, Yoshimori T, Ohsumi Y (2004) In vivo analysis of autophagy in response to nutrient starvation using transgenic mice expressing a fluorescent autophagosome marker. Mol Biol Cell 15:1101–1111
Mizushima N, Levine B, Cuervo AM, Klionsky DJ (2008) Autophagy fights disease through cellular self-digestion. Nature 451:1069–1075
Mizushima N, Yoshimori T, Ohsumi Y (2011) The role of Atg proteins in autophagosome formation. Annu Rev Cell Dev Biol 27:107–132
Mori H, Inoki K, Masutani K, Wakabayashi Y, Komai K, Nakagawa R et al (2009) The mTOR pathway is highly activated in diabetic nephropathy and rapamycin has a strong therapeutic potential. Biochem Biophys Res Commun 384:471–475
Moruno-Manchon JF, Uzor NE, Kesler SR, Wefel JS, Townley DM, Nagaraja AS et al (2018) Peroxisomes contribute to oxidative stress in neurons during doxorubicin-based chemotherapy. Mol Cell Neurosci 86:65–71
Nagai K, Matsubara T, Mima A, Sumi E, Kanamori H, Iehara N et al (2005) Gas6 induces Akt/mTOR-mediated mesangial hypertrophy in diabetic nephropathy. Kidney Int 68:552–561

Nair S, Wilding JP (2010) Sodium glucose cotransporter 2 inhibitors as a new treatment for diabetes mellitus. J Clin Endocrinol Metab 95:34–42
Narita M, Young AR, Arakawa S, Samarajiwa SA, Nakashima T, Yoshida S et al (2011) Spatial coupling of mTOR and autophagy augments secretory phenotypes. Science 332:966–970
Oshima Y, Kinouchi K, Ichihara A, Sakoda M, Kurauchi-Mito A, Bokuda K et al (2011) Prorenin receptor is essential for normal podocyte structure and function. J Am Soc Nephrol 22:2203–2212
Pavenstädt H, Kriz W, Kretzler M (2003) Cell biology of the glomerular podocyte. Physiol Rev 83:253–307
Periyasamy-Thandavan S, Jiang M, Wei Q, Smith R, Yin XM, Dong Z (2008) Autophagy is cytoprotective during cisplatin injury of renal proximal tubular cells. Kidney Int 74:631–640
Qi W, Mu J, Luo ZF, Zeng W, Guo YH, Pang Q et al (2011) Attenuation of diabetic nephropathy in diabetes rats induced by streptozotocin by regulating the endoplasmic reticulum stress inflammatory response. Metabolism 60:594–603
Ravikumar B, Sarkar S, Davies JE, Futter M, Garcia-Arencibia M, Green-Thompson ZW et al (2010) Regulation of mammalian autophagy in physiology and pathophysiology. Physiol Rev 90:1383–1435
Rubinsztein DC, Shpilka T, Elazar Z (2012) Mechanisms of autophagosome biogenesis. Curr Biol 22:R29–R34
Sakaguchi M, Isono M, Isshiki K, Sugimoto T, Koya D, Kashiwagi A (2006) Inhibition of mTOR signaling with rapamycin attenuates renal hypertrophy in the early diabetic mice. Biochem Biophys Res Commun 340:296–301
Sataranatarajan K, Mariappan MM, Lee MJ, Feliers D, Choudhury GG, Barnes JL et al (2007) Regulation of elongation phase of mRNA translation in diabetic nephropathy: amelioration by rapamycin. Am J Pathol 171:1733–1742
Sionov RV, Vlahopoulos SA, Granot Z (2015) Regulation of bim in health and disease. Oncotarget 6:23058–23134
Steinberg GR, Kemp BE (2009) AMPK in health and disease. Physiol Rev 89:1025–1078
Stridh S, Palm F, Takahashi T, Ikegami-Kawai M, Hansell P (2015) Inhibition of mTOR activity in diabetes mellitus reduces proteinuria but not renal accumulation of hyaluronan. Ups J Med Sci 120:233–240
Sun J, Li ZP, Zhang RQ, Zhang HM (2017) Repression of miR-217 protects against high glucose-induced podocyte injury and insulin resistance by restoring PTEN-mediated autophagy pathway. Biochem Biophys Res Commun 483:318–324
Tagawa A, Yasuda M, Kume S, Yamahara K, Nakazawa J, Chin-Kanasaki M et al (2016) Impaired podocyte autophagy exacerbates proteinuria in diabetic nephropathy. Diabetes 65:755–767
Takahashi A, Kimura T, Takabatake Y, Namba T, Kaimori J, Kitamura H et al (2012) Autophagy guards against cisplatin-induced acute kidney injury. Am J Pathol 180:517–525
Takahashi A, Takabatake Y, Kimura T, Maejima I, Namba T, Yamamoto T et al (2017) Autophagy inhibits the accumulation of advanced glycation end products by promoting lysosomal biogenesis and function in the kidney proximal tubules. Diabetes 66:1359–1372
Tang C, Han H, Yan M, Zhu S, Liu J, Liu Z et al (2018) PINK1-PRKN/PARK2 pathway of mitophagy is activated to protect against renal ischemia-reperfusion injury. Autophagy 1–18
Thoen LF, Guimarães EL, Dollé L, Mannaerts I, Najimi M, Sokal E et al (2011) A role for autophagy during hepatic stellate cell activation. J Hepatol 55:1353–1360
Vallon V, Rose M, Gerasimova M, Satriano J, Platt KA, Koepsell H et al (2013) Knockout of Na-glucose transporter SGLT2 attenuates hyperglycemia and glomerular hyperfiltration but not kidney growth or injury in diabetes mellitus. Am J Physiol Renal Physiol 304:F156–F167
Velagapudi C, Bhandari BS, Abboud-Werner S, Simone S, Abboud HE, Habib SL (2011) The tuberin/mTOR pathway promotes apoptosis of tubular epithelial cells in diabetes. J Am Soc Nephrol 22:262–273
Venkatachalam MA, Weinberg JM, Kriz W, Bidani AK (2015) Failed tubule recovery, AKI-CKD transition, and kidney disease progression. J Am Soc Nephrol 26:1765–1776

Wang Y, Zheng ZJ, Jia YJ, Yang YL, Xue YM (2018) Role of p53/miR-155-5p/sirt1 loop in renal tubular injury of diabetic kidney disease. J Transl Med 16:146

Wanner N, Hartleben B, Herbach N, Goedel M, Stickel N, Zeiser R et al (2014) Unraveling the role of podocyte turnover in glomerular aging and injury. J Am Soc Nephrol 25:707–716

Wei Q, Dong Z (2014) HDAC4 blocks autophagy to trigger podocyte injury: non-epigenetic action in diabetic nephropathy. Kidney Int 86:666–668

Wellen KE, Thompson CB (2010) Cellular metabolic stress: considering how cells respond to nutrient excess. Mol Cell 40:323–332

Wittmann S, Daniel C, Stief A, Vogelbacher R, Amann K, Hugo C (2009) Long-term treatment of sirolimus but not cyclosporine ameliorates diabetic nephropathy in the rat. Transplantation 87:1290–1299

Wu L, Zhang Y, Ma X, Zhang N, Qin G (2012) The effect of resveratrol on FoxO1 expression in kidneys of diabetic nephropathy rats. Mol Biol Rep 39:9085–9093

Xavier S, Gilbert V, Rastaldi MP, Krick S, Kollins D, Reddy A et al (2010) BAMBI is expressed in endothelial cells and is regulated by lysosomal/autolysosomal degradation. PLoS ONE 5:e12995

Xiao T, Guan X, Nie L, Wang S, Sun L, He T et al (2014) Rapamycin promotes podocyte autophagy and ameliorates renal injury in diabetic mice. Mol Cell Biochem 394:145–154

Xu Y, Ruan S, Wu X, Chen H, Zheng K, Fu B (2013) Autophagy and apoptosis in tubular cells following unilateral ureteral obstruction are associated with mitochondrial oxidative stress. Int J Mol Med 31:628–636

Xu G, Yue F, Huang H, He Y, Li X, Zhao H et al (2016a) Defects in MAP1S-mediated autophagy turnover of fibronectin cause renal fibrosis. Aging (Albany NY) 8:977–985

Xu L, Fan Q, Wang X, Zhao X, Wang L (2016b) Inhibition of autophagy increased AGE/ROS-mediated apoptosis in mesangial cells. Cell Death Dis 7:e2445

Xue X, Ren J, Sun X, Gui Y, Feng Y, Shu B et al (2018) Protein kinase Cα drives fibroblast activation and kidney fibrosis by stimulating autophagic flux. J Biol Chem 293:11119–11130

Yan Q, Song Y, Zhang L, Chen Z, Yang C, Liu S et al (2018) Autophagy activation contributes to lipid accumulation in tubular epithelial cells during kidney fibrosis. Cell Death Discov 5:2

Yang Z, Klionsky DJ (2010) Mammalian autophagy: core molecular machinery and signaling regulation. Curr Opin Cell Biol 22:124–131

Yang Y, Wang J, Qin L, Shou Z, Zhao J, Wang H et al (2007) Rapamycin prevents early steps of the development of diabetic nephropathy in rats. Am J Nephrol 27:495–502

Yang C, Kaushal V, Shah SV, Kaushal GP (2008) Autophagy is associated with apoptosis in cisplatin injury to renal tubular epithelial cells. Am J Physiol Renal Physiol 294:F777–F787

Yang D, Livingston MJ, Liu Z, Dong G, Zhang M, Chen JK et al (2018) Autophagy in diabetic kidney disease: regulation, pathological role and therapeutic potential. Cell Mol Life Sci 75:669–688

Yi M, Zhang L, Liu Y, Livingston MJ, Chen JK, Nahman NS et al (2017) Autophagy is activated to protect against podocyte injury in adriamycin-induced nephropathy. Am J Physiol Renal Physiol 313:F74–F84

Yu L, McPhee CK, Zheng L, Mardones GA, Rong Y, Peng J et al (2010) Termination of autophagy and reformation of lysosomes regulated by mTOR. Nature 465:942–946

Yu L, Chen Y, Tooze SA (2018) Autophagy pathway: cellular and molecular mechanisms. Autophagy 14:207–215

Zaffagnini G, Martens S (2016) Mechanisms of selective autophagy. J Mol Biol 428:1714–1724

Zeng C, Fan Y, Wu J, Shi S, Chen Z, Zhong Y et al (2014) Podocyte autophagic activity plays a protective role in renal injury and delays the progression of podocytopathies. J Pathol 234:203–213

Zhang L, Livingston MJ, Chen JK, Dong Z (2014a) Autophagy in podocytes. Contrib Nephrol 183:83–100

Zhang MZ, Wang Y, Paueksakon P, Harris RC (2014b) Epidermal growth factor receptor inhibition slows progression of diabetic nephropathy in association with a decrease in endoplasmic reticulum stress and an increase in autophagy. Diabetes 63:2063–2072

Zhang D, Pan J, Xiang X, Liu Y, Dong G, Livingston MJ et al (2016) Protein kinase Cδ suppresses autophagy to induce kidney cell apoptosis in cisplatin nephrotoxicity. J Am Soc Nephrol

Zhao L, Sun LN, Nie HB, Wang XL, Guan GJ (2014) Berberine improves kidney function in diabetic mice via AMPK activation. PLoS ONE 9:e113398

Zhao X, Chen Y, Tan X, Zhang L, Zhang H, Li Z et al (2018) Advanced glycation end-products suppress autophagic flux in podocytes by activating mammalian target of rapamycin and inhibiting nuclear translocation of transcription factor EB. J Pathol 245:235–248

Zoncu R, Efeyan A, Sabatini DM (2011) mTOR: from growth signal integration to cancer, diabetes and ageing. Nat Rev Mol Cell Biol 12:21–35

Zuk A, Bonventre JV (2016) Acute kidney injury. Annu Rev Med 67:293–307

Chapter 29
Oxidative Stress and Renal Fibrosis: Mechanisms and Therapies

Hua Su, Cheng Wan, Anni Song, Yang Qiu, Wei Xiong and Chun Zhang

Abstract Oxidative stress results from the disruption of the redox system marked by a notable overproduction of reactive oxygen species. There are four major sources of reactive oxygen species, including NADPH oxidases, mitochondria, nitric oxide synthases, and xanthine oxidases. It is well known that renal abnormalities trigger the production of reactive oxygen species by diverse mechanisms under various pathologic stimuli, such as acute kidney injury, chronic kidney disease, nephrotic syndrome, and metabolic disturbances. Mutually, accumulating evidences have identified that oxidative stress plays an essential role in tubulointerstitial fibrosis by myofibroblast activation as well as in glomerulosclerosis by mesangial sclerosis, podocyte abnormality, and parietal epithelial cell injury. Given the involvement of oxidative stress in renal fibrosis, therapies targeting oxidative stress seem promising in renal fibrosis management. In this review, we sketch the updated knowledge of the mechanisms of oxidative stress generation during renal diseases, the pathogenic processes of oxidative stress elicited renal fibrosis and treatments targeting oxidative stress during tubulointerstitial fibrosis and glomerulosclerosis.

Keywords Oxidative stress · Tubulointerstitial fibrosis · Glomerulosclerosis

29.1 Introduction

The concept of oxidative stress (OS) was proposed in 1985 as "a disturbance in the prooxidant–antioxidant balance in favor of the former" (Sies 1985) and was then updated in 2006 to a more precise definition "a disruption of redox signaling and control" (Jones 2006). In terms of its definition, OS results from increased levels of endogenous and exogenous free radicals and other oxidants and decreased levels of antioxidants along with the depressed activity of antioxidant enzymes.

H. Su · C. Wan · A. Song · Y. Qiu · W. Xiong · C. Zhang (✉)
Department of Nephrology, Union Hospital, Tongji Medical College, Huazhong University of Science and Technology, Wuhan 430022, China
e-mail: drzhangchun@hust.edu.cn

B.-C. Liu et al. (eds.), *Renal Fibrosis: Mechanisms and Therapies*,
Advances in Experimental Medicine and Biology 1165,
https://doi.org/10.1007/978-981-13-8871-2_29

Attempts have been made to classify OS based on its intensity as basal, low intensity, intermediate, and high level theoretically. As we knew, reactive oxygen species (ROS) are generated in living organism abidingly, and however, at basal OS, it is too negligible to observe or measure. At low-intensity OS, expression of antioxidant enzymes is up-regulated with commonly normal prooxidant–antioxidant balance. As for intermediate-intensity OS, heat shock proteins (HSPs) and inflammatory proteins are up-regulated along with enhanced antioxidant enzymes, at this time, bodies are struggling to sustain the equilibrium. High-intensity OS leads to perturbation of mitochondrial permeability transition pore and destruction of electron transporters, resulting in activation of apoptosis and necrosis (Lushchak 2014). Thus far, due to lacking in well-defined assays to evaluate, this classification is too complicated to apply (Lichtenberg and Pinchuk 2015).

OS results from overproduction of ROS and reactive nitrogen species (RNS). It is well known that ROS/RNS is a double-edged sword, and under physiologic status, it is indispensable for cell signaling conduction, hormonal effects, ion channel activity, cell growth, senescence, and so on (Sedeek et al. 2013). However, when encountering improper stimuli, redox homeostasis may be destroyed, thus inducing a series of injurious reactions.

OS is one of the major components contributing to renal fibrosis which is the common pathway leading to end-stage renal disease (ESRD), while diverse renal abnormalities trigger OS and accelerate the disorders of kidney, such as acute kidney injury (AKI), chronic kidney disease (CKD), nephrotic syndrome (NS), and metabolic disturbances. In this review, we address the updated knowledge of the generation of OS during above-mentioned pathologic status and further elucidate the mechanisms and treatments about OS-associated tubulointerstitial fibrosis and glomerulosclerosis.

29.2 The Generation of OS in Renal Diseases

There are four major pathways for ROS generation in the organisms, all of which are implicated in renal diseases. Generally, the first mentioned and the most important resource of ROS is nicotinamide adenine dinucleotide phosphate (NADPH) oxidase (Nox) (Gill and Wilcox 2006; Lassègue and Clempus 2003). Nox consists of a membrane subunit flavocytochrome b558 and cytosolic subunits which include $p40^{phox}$, $p47^{phox}$, $p67^{phox}$, and Rac. The assembly of these subunits induces NADPH to transfer electrons to O_2 to generate $O_2^{\cdot-}$ (Krause 2007). Secondly, in mitochondria, NADH-ubiquinone oxidoreductase (complex I) and ubiquinol-cytochrome c reductase (complex III) transfer electrons to O_2 to produce $O_2^{\cdot-}$ and generate H_2O_2 by reacting with manganese superoxide dismutase (MnSOD) (Handy and Loscalzo 2012). Thirdly, nitric oxide synthase (NOS) is the enzyme which catalyzes *L*-arginine to produce nitric oxide (NO), and the uncoupling of endothelial NOS (eNOS) with its cofactors such as tetrahydrobiopterin and calmodulin generates $O_2^{\cdot-}$ by transferring electrons to O_2 (Forstermann and Munzel 2006). Subsequently, $O_2^{\cdot-}$ reacts with NO forming peroxynitrite ($ONOO^-$) which is a potent RNS that converts tyrosine residues to 3-nitrotyrosine and may induce oxidative damage to other macromolecules. Fourthly, xanthine oxidase (XO) is also a ROS-producing enzyme which oxidizes hypoxanthine to uric acid and xanthine (Pacher et al. 2006).

29.2.1 The Generation of OS in AKI

AKI is defined as a sudden reduction in glomerular filtration rate (GFR), which leads to azotemia and/or decreased urine production. It is mainly caused by renal ischemia, nephrotoxic agents, sepsis, and the reduction in effective intravascular volume (Bonventre and Yang 2011), and various sources of ROS are implicated in AKI as we illustrated below (Ratliff et al. 2016; Araujo and Welch 2006).

For ischemia- or nephrotoxicity-induced AKI, mitochondrial dysfunction is the major source of ROS (Lushchak 2014), which occurs before the rise of serum creatinine level and is an early pathologic event in AKI (Ralto and Parikh 2016). Simultaneously, the activity of antioxidant enzymes, like superoxide dismutase (SOD), glutathione reductase, and catalase, is depressed with the declined antioxidants such as the enzymic cofactor glutathione (GSH) (Chirino and Pedraza-Chaverri 2009).

In sepsis-mediated AKI, besides the dysfunction of mitochondria (Brealey et al. 2002), the inducible NOS (iNOS) is up-regulated significantly with escalating NO production (Brealey et al. 2002), which not only induces the uncoupling of eNOS and generates $O_2^{\cdot-}$, but also competes with SOD and generates peroxynitrite (Radi 2013). Moreover, the infiltration of inflammatory cells with myeloperoxidase and Nox activation are the predominant sources of ROS generation in sepsis. The extensive inflammatory response also induces vascular endothelium to release chemokines (Andrades et al. 2011) and damage-associated molecular patterns (DAMPs) such as high-mobility group box-1 and HSPs (Gill et al. 2010). Chemokines recruit immune cells and DAMPs activate the toll-like receptor with ensuing ROS generation (Andrades et al. 2011; Basile et al. 2012). In severe burns and trauma, rhabdomyolysis is one of the major causes for AKI, during which massive heme-containing myoglobin (Mb) is released (Baliga et al. 1999). Consequently, Mb degrades low-molecular-weight peroxides to generate the redox-active irons.

29.2.2 The Generation of OS in CKD

CKD is a progressive decline of renal function over time, which requires renal replacement therapies such as hemodialysis and peritoneal dialysis when developing to ESRD. OS in CKD is in association with uremic toxins and dialysis, caused by increased oxidase activities and/or diminished antioxidants.

The accumulation of uremic toxins induces OS in CKD patients (Oberg et al. 2004). For instance, indoxyl sulfate activates Nox (Tumur et al. 2010) and XO (Dou et al. 2007), which generates ROS by catalyzing the oxidation of hypoxanthine to xanthine and further to uric acid. Homocysteine (Hcy) generates ROS through auto-oxidation and inhibition of SOD and GSH peroxidase (GPx) (Massy et al. 2001).

During hemodialysis, the contact with dialytic membranes, such as polysulfone and regenerated cellulose (Ward and McLeish 2003), and the challenge of dialysate contaminants, such as endotoxins, trigger ROS generation by activating neutrophils and peripheral blood mononuclear cells. Under peritoneal dialysis, the high concentration of glucose in dialysate induces human peritoneal mesothelial cells to produce ROS (Noh et al. 2006). Additionally, hydrophilic nonenzymatic antioxidants are lost due to the highly permeable dialytic membranes (Ward et al. 2003).

In fact, even from the early stage of CKD, there is an obvious decline of both enzymatic and nonenzymatic antioxidants including thiol, vitamins E and C, GPx, arylesterase, and SOD (Tbahriti et al. 2013). For instance, thiol, a nonenzymatic antioxidant reacting with most oxidants, is depleted in CKD. Vitamin C, which reacts with ROS or RNS to generate semidehydroascorbic acid, is insufficient in CKD patients who are under dietary restriction of vegetables and fruits. GPx, an enzymatic antioxidant which degrades H_2O_2 and other peroxides, is also profoundly deficient in CKD patients (Fassett et al. 2015).

29.2.3 The Generation of OS in NS

NS is a constellation of clinical syndromes which manifests as massive proteinuria, hypoalbuminemia, edema, and hyperlipidemia. Although the underlying etiologies and pathogenesis are dissimilar, however, there is a strong link between excessive protein trafficking and intensity of OS in the kidney. Urinary albumin elicits significant ROS generation largely through mechanisms listed as below.

Firstly, urinary albumin is reabsorbed in proximal tubular cells, by the scavenger receptors including CD36, megalin, and cubilin (Baines et al. 2012), and this process activates protein kinase C (PKC). PKC is a family of serine/threonine protein kinases which is implicated in Nox activation and ROS generation. For instance, PKCδ phosphorylates p47phox, which then translocates to the membrane-bound cytochrome to assemble Nox, thus facilitating the activation of Nox and ROS production (Brown et al. 2003).

Secondly, albumin induces proximal tubular epithelial cells to secrete complement C3 and chemokines, such as interleukin-8 (IL-8), RANTES, endothelin-1 and monocyte chemotactic protein 1 (MCP-1). These chemokines recruit inflammatory cells which are major sources of ROS. For instance, RANTES and MCP-1 recruit monocytes and T cells, while IL-8 recruits neutrophils, monocytes, and T cells.

Thirdly, hypoalbuminemia aggravates OS as well, since albumin performs a major antioxidant role in plasma, such as providing thiol groups, scavenging free radicals, for example, hypochlorous acid and $ONOO^-$, and binding copper ions which enhance ROS-induced oxidative damage (Sitar et al. 2013).

29.2.4 The Generation of OS in Metabolic Disturbances

Hyperglycemia, hyperlipidemia, and hyperuricemia are three major metabolic disturbances which cause OS in kidney diseases.

As we knew, diabetes is the leading cause of ESRD. Approximately 50% of diabetes patients develop diabetic nephropathy (DN), many of whom progress to ESRD (Tuttle et al. 2014). Under the diabetic state, ROS production is augmented via diverse mechanisms. Firstly, hyperglycemia and other stimuli such as angiotensin II (AngII), transforming growth factor-β (TGF-β), and advanced glycation end products (AGEs) up-regulate Nox expression in intrinsic renal cells (Wan et al. 2016). For instance, Nox4 expression is enhanced in mesangial cells in response to AngII, leading to excessive ROS generation directly and Nox4-mediated uncoupling of eNOS which further exacerbates OS indirectly (Lee et al. 2013). Secondly, mitochondria are speculated to be the main source of ROS in DN. Hyperglycemia provides excessive electrons to the Krebs's cycle in mitochondria, resulting in mitochondrial membrane hyperpolarization and redundant ROS generation. Besides, SOD is diminished in DN, while XO is up-regulated.

Hyperlipidemia is suggested to mediate renal injury, which is predominantly attributed to the dysregulation of redox signaling (Su et al. 2017). The overmuch fatty acid induces ROS production probably due to elevated β-oxidation and the ensuing electron overflow in the mitochondrial electron transport chain (Seifert et al. 2010). Meanwhile, the elevation of fatty acids in plasma activates monocytes and neutrophils to generate ROS and release various cytokines which recruit more inflammatory cells and amplify OS (Tripathy et al. 2003). Furthermore, unsaturated fatty acids, triglycerides and phospholipids, are sensitive to oxidation, while peroxidation products of the lipids are more liable to activate Nox via AngII and TGF-β1, forming a vicious cycle (Lee and Song 2009).

Uric acid (UA) is the end product of the purine nucleotide metabolism in human. The decline of GFR causes decreased UA clearance and hence results in hyperuricemia, which is prevalent in patients with advanced stages of CKD (Madero et al. 2009). Excessive UA leads to its ionized formation urates, which participate in aggravating OS by activating inflammasome (Martinon et al. 2006), or by being converted to urate radical (Meotti et al. 2011). In addition, hyperuricemia up-regulates the expression of Nox4 and AngII and decreases NO bioavailability, along with mitochondrial alterations and reduced adenosine triphosphate (ATP) production (Sánchez-Lozada et al. 2012), all of which contribute to OS.

29.3 The Mechanisms of Renal Fibrosis Triggered by OS

Renal fibrosis is a common pathophysiologic process shared among diverse renal diseases which gradually develop to ESRD. Renal fibrosis is marked with certain pathological features including myofibroblast activation, extracellular matrix (ECM) deposit as well as glomerulosclerosis. Notably, OS is one of the important components contributing to renal fibrosis, and OS triggers tubulointerstitial fibrosis and glomerulosclerosis via multiple mechanisms, which will be illustrated in the following parts.

29.3.1 *OS-Induced Tubulointerstitial Fibrosis*

Tubulointerstitial fibrosis is the utmost recognized pathological feature in renal fibrosis, while OS is implicated in the pathogenesis of tubulointerstitial fibrosis by myofibroblast activation through various pathways as listed below (Fig. 29.1).

29.3.1.1 OS and Fibroblast/Pericyte Transdifferentiation

Under normal conditions, renal tubulointerstitial fibroblasts are rather quiescent. It is a kind of low metabolic and inactive cells. However, various stimuli, such as inflammation, hypoxia, and hyperglycemia, can strongly elicit fibroblast activation, proliferation, phenotypic changes, and transforming into myofibroblasts. It is well

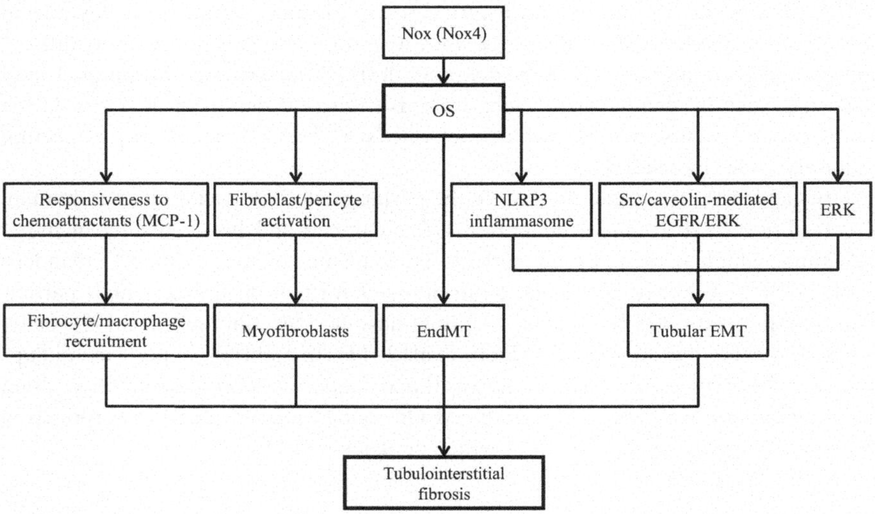

Fig. 29.1 Correlation between OS and tubulointerstitial fibrosis. OS mainly produced by Nox, especially by Nox4, leads to fibroblast/pericyte transformation, tubular EMT, fibrocyte/macrophage recruitment, and EndMT, which are four crucial events of tubulointerstitial fibrosis. OS usually induces fibroblast/pericyte activation and EndMT directly, whereas it triggers EMT by activation of Src/caveolin-mediated EGFR/ERK pathway, NLRP3 inflammasome, and so on. And OS frequently enhances macrophage recruitment by increasing the responsiveness of macrophages to chemoattractants, especially to MCP-1

known that the myofibroblasts expressing α-smooth muscle actin (α-SMA) are the main source of ECM accumulation in the tubulointerstitium during fibrosis. Nox is a well-recognized mediator to promoting the transition of fibroblasts to myofibroblasts, thus increasing the synthesis of ECM and aggravating renal fibrosis. It is reported that inhibition of Nox4 prevents ROS production and myofibroblast differentiation and consequently reduces tubulointerstitial fibrosis (Bondi et al. 2010).

Pericytes are another main source of myofibroblasts which are embedded in the basement membrane of the vessel. As pericytes stabilize endothelial cells structurally and functionally, the loss of pericytes destabilizes endothelial cells and further leads to the rarefication of microvasculature which results in tissue hypoxia and ensuing fibrosis (Kida et al. 2014). It has been discovered that ROS has a potential ability to trigger the transition from pericytes to myofibroblast phenotype in bleomycin-induced pulmonary fibrosis (Andersson-Sjoland et al. 2016), but the mechanisms of ROS-elicited pericyte-myofibroblasts transformation remain elusive.

29.3.1.2 OS and Tubular EMT

Renal tubular epithelial–mesenchymal transition (EMT) means mature tubular epithelial cells (TECs) lose the epithelial phenotype and acquire mesenchymal phenotype. When exposed to OS, inflammation, and hypoxia, tubular EMT occurs commonly, namely TECs transforming into myofibroblasts. Evidences have shown that Noxs participate in this process. It has been found that high glucose (HG)-induced EMT can be attenuated by suppression of Nox1 and Nox4 (He et al. 2015). Also, the possible involving signaling pathways in EMT, including extracellular signal-regulated kinase (ERK), Src/caveolin-mediated epidermal growth factor receptor (EGFR)/ERK (Kim et al. 2012; Chen et al. 2012a), and NOD-like receptor family, pyrin domain containing 3 (NLRP3) inflammasome, and so on, are frequently activated by mitochondrial ROS (Lorenz et al. 2014; Shahzad et al. 2015).

29.3.1.3 OS and Fibrocyte/Macrophage Recruitment

Fibrocytes, a class of monocytes with the surface markers of hematopoietic cells such as CD34\CD45\MHC-II molecules, function similarly to fibroblasts, such as secreting collagen and other matrix proteins. Sakai's study showed that in the model of progressive renal fibrosis in unilateral ureteral obstruction (UUO) mice, the interstitial infiltration of CD45 and type I collagen dual-positive ($CD45^+/ColI^+$) fibrocytes were found, particularly in the cortex–medulla junction (Sakai et al. 2006). $CD45^+/ColI^+$ fibrocytes were present in DN renal interstitium as well, and the number of fibrocytes was correlated to the degree of renal fibrosis.

Many studies have suggested the correlation between ROS and the activation of fibrocytes. In the kidney, the decreased expression of bone morphogenetic protein-6 (BMP-6) is associated with the increase of fibrocytes, in which BMP-6 deficiency aggravates tubulointerstitial damage and fibrosis due to indirect activation of ROS

(Dendooven et al. 2011). Consistently, Wagner et al. demonstrated that fibrocytes were increased in contrast-treated animals in association with increased expression of fibrotic markers, Nox4, and ROS (Wagner et al. 2012). However, the exact role of ROS in fibrocyte activation during renal fibrosis remains obscure, and more evidences are needed.

Macrophages, also derived from monocytes, are rapidly recruited in glomeruli or tubulointerstitium after acute renal injury, to regulate immune response, and phagocytize debris and apoptotic cells as a protective role in an early stage. However, persistent renal damage induces macrophage to infiltrate into renal tissue and to produce various growth factors, which eventually lead to the destruction of the normal kidney and irreversible fibrosis.

In renal diseases, macrophage recruitment is always accompanied by the activation of OS. Lee et al. found that aryl hydrocarbon receptor (AhR) was increased in DN, while AhR deficiency by genetic knockout or pharmacological inhibition decreased macrophage infiltration, which was associated with the attenuated Nox-derived ROS (Lee et al. 2016). In cyclosporine A-induced chronic nephrotoxicity, macrophage depletion could attenuate the reduction in GFR, renal blood flow, and the development of tubulointerstitial fibrosis, likely due to the inhibition of NO pathway and OS (Carlos et al. 2010). The underlying association between macrophage recruitment and OS has been explored. A number of studies have shown that OS enhances the responsiveness of macrophages to chemoattractants, such as MCP-1, osteopontin, intercellular adhesion molecule 1 (ICAM-1), and RANTES, which subsequently recruit macrophages to the tubulointerstitium (Qiao et al. 2009).

29.3.1.4 OS and EndMT

Recently, endothelial–mesenchymal transition (EndMT) is found to induce expression of fibroblast markers as well, and its pathological role in the process of fibrosis has been gradually recognized and emphasized. EndMT is defined as a cellular process in which endothelial cells undergo morphological changes with loss of their endothelial phenotype and gain mesenchymal properties, characterized by the loss of CD31 and von Willebrand factor along with up-regulation of α-SMA and fibroblast-specific protein 1. EndMT induced by HG, inflammation, and growth factors as well as OS eventually leads to renal fibrosis.

ROS has been proved to participate in the induction of EndMT directly. Treatment with H_2O_2 triggered EndMT in primary endothelial cells (Montorfano et al. 2014). On the other hand, antioxidant treatment is protective for EndMT. *N*-acetyl-*L*-cysteine, a ROS scavenger compound, has been verified effective in preventing endotoxin-induced EndMT. Moreover, apocynin, a Nox inhibitor, was found sufficient to inhibit the development of fibrotic markers, indicating Nox activation as a crucial step in endotoxin-induced EndMT (Echeverria et al. 2013). Taken together, OS is a precipitating cause of EndMT, and inhibition of ROS may provide new insight into the treatment of renal fibrosis.

29.3.2 OS-Induced Glomerulosclerosis

Mesangial sclerosis, podocyte abnormality, and parietal epithelial cell (PEC) injury are three crucial events and key diagnostic points for glomerulosclerosis, which is not a certain specific glomerular disease but a kind of pathological feature in renal fibrosis, especially in focal segmental glomerular sclerosis (FSGS). There is a consensus that OS plays an essential role in the pathogenesis of glomerulosclerosis, and some features labeling glomerulosclerosis may affect OS in turn (Fig. 29.2).

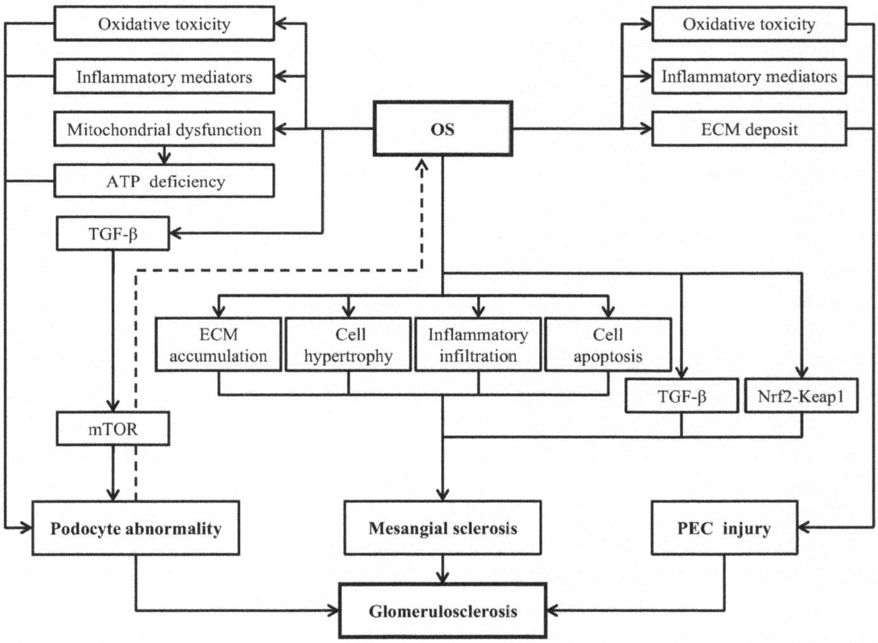

Fig. 29.2 Correlation between OS and glomerulosclerosis. OS leads to mesangial sclerosis, podocyte abnormality, and PEC injury, which are three essential processes of glomerulosclerosis. Mesangial sclerosis as the central event is attributed to ECM accumulation, mesangial cell hypertrophy, inflammatory infiltration, and mesangial cell apoptosis. Diverse signaling pathways are also involved, such as Nrf2-Keap1 and TGF-β pathways. Podocyte abnormality is triggered via oxidative toxicity, increased inflammatory mediators, mitochondrial dysfunction and ensuing ATP deficiency with TGF-β/mTOR signaling activation. In turn, podocyte abnormalities enhance ROS production via mTOR signaling. Finally, OS initiates PEC injury due to oxidative toxicity, increased inflammatory mediators, and ECM deposit

29.3.2.1 OS and Mesangial Sclerosis

Mesangial cells mainly offer structural support for capillary loops by generating ECM and alter glomerular filtration by regulating smooth muscle activity and diverse ion channels. Mesangial sclerosis manifests as ECM accumulation, mesangial cell hypertrophy, as well as recently recognized inflammatory infiltration and mesangial cell apoptosis, all of which are linked with OS.

Mark et al. discovered that ECM production, such as fibronectin and collagen IV, increased along with ROS production in HG treated mesangial cells, while it could be reversed by addition of antioxidants, suggesting OS might potentiate ECM accumulation in mesangial sclerosis.

There are a few evidences illustrating the underlying mechanisms about OS-induced ECM deposit in the mesangium. Yang et al. found that mesangial cells treated with *L*-homocysteins (*L*-Hcys) presented a raise in Nox activity, $O_2^{\cdot-}$ generation, cell proliferation, collagen I formation, and tissue inhibitors of metalloproteinase-1 (TIMP-1), which could be blocked by Nox inhibitors (Yang and Zou 2003). As TIMP-1 inhibits matrix metalloproteinase, a main enzyme responsible for degrading collagens and reducing ECM deposit, it indicates that Hcys may not only trigger mesangial cell proliferation but also induce collagen deposit by enhancing TIMP-1 expression associated with Nox activity. Similarly, hyperglycemia may induce mesangial fibrosis by activating PKC, producing AGEs, regulating sorbitol accumulation and nuclear factor kappa-light-chain-enhancer of activated B cell (NF-κB) activity, which are mainly mediated by ROS (Nishikawa et al. 2000; Catherwood et al. 2002). In addition, in terms of abundance and activity, TGF-β and connective tissue growth factor (CTGF) are also considered to be implicated in OS-induced ECM deposit (Mason 2003).

Furthermore, Nishikimi et al. identified that monocyte/macrophage infiltration along with OS in hypertensive glomerulosclerosis rats and treatment of Rho-kinase inhibitor could reduce OS and inflammation (Nishikimi and Matsuoka 2006). Consistently, Jha et al. noticed depressed renal expression of NF-κB and MCP-1 and attenuated glomerular macrophage infiltration in diabetic mice with deletion or inhibition of Nox4 (Jha et al. 2014). Thus, inflammatory infiltration takes part in OS-induced mesangial sclerosis probably via NF-κB and MCP-1.

Moreover, Wang et al. observed mesangial cell hypertrophy in DN and revealed the possible mechanism that HG-induced OS could affect mesangial cell cycle by preventing cells from progressing to G1/S transition, ultimately keeping cells in G1 phase. During G1 phase, the RNA and protein synthesis are enhanced, and mesangial cells undergo hypertrophy (Wang et al. 2006; Tang et al. 2011). Furthermore, the hypertrophic mesangial cell produces more ECM via autocrine action and further exacerbates ECM accumulation. Thereby, OS-induced mesangial cell hypertrophy and ECM deposit collectively contribute to glomerulosclerosis.

Recently, Lu et al. revealed that under HG, the apoptotic markers along with ROS levels were increased, which were tightly associated with mesangial sclerosis, and mammalian target of rapamycin (mTOR) inhibitor rapamycin could not only alleviate OS but also reduce mesangial cell apoptosis (Lu et al. 2017).

Diverse cell signaling pathways are involved in OS-induced mesangial sclerosis, for instance, the classic TGF-β-Smad pathways. TGF-β-Smad2/3 accelerates ECM accumulation, while TGF-β-Smad7 slows it down (Wang et al. 2006; Tang et al. 2011). Besides, other pathways, such as mitogen-activated protein kinase (MAPK), Janus kinase (JAK)–signal transducer and activator of transcription (STAT), and a number of transcription factors, are also involved in glomerular ECM deposition (Tang et al. 2011; Mason 2003).

In adriamycin-induced FSGS, antioxidant tetrahydroxystilbene glucoside reduces OS and collagen level, with a raise of nuclear factor erythroid-2-related factor 2 (Nrf2) in glomeruli. Meanwhile, silencing Nrf2 and its coupling molecule Kelch-like erythroid-cell-derived protein 1 (Keap1) in mesangial cells accelerates sclerosis. Thus, the antioxidant axis Nrf2-Keap1 plays a crucial role in the protection against OS-mediated mesangiosclerosis (Lin et al. 2018).

29.3.2.2 The Interplay Between OS and Podocyte Abnormalities

Podocyte is a kind of mono-layer cell with foot processes covering the outside of the glomerular basement membrane, which functions as one of the major components of the glomerular filtration barrier. The involvement of OS in podocyte impairment has already been verified, although the underlying mechanisms remain obscured.

Due to its direct oxidative toxicity, OS causes podocyte foot process effacement and fusion (Okamura and Pennathur 2015). Besides, OS induces podocyte impairment and depletion partially by inflammatory mediators, such as cyclooxygenase-2 (COX-2), iNOS, IL-6, and tumor necrosis factor-α (TNF-α) (Lin et al. 2018). Moreover, excessive ROS promotes depolarization of the mitochondrial membrane which consequently impairs the electron transport chain and leads to the ATP depletion. Morphologically, podocytes present cell body attenuation, foot process effacement and fusion, pseudocyst formation, and microvillous transformation and then gradually advance to necrosis and apoptosis to the loss of podocyte (Jefferson and Shankland 2014). Notably, it has been recognized that OS-stimulated podocyte apoptosis is closely related with the progression of glomerulosclerosis in DN (Chen et al. 2012b).

It is reported that quite a lot of signaling cascades are associated with OS-induced podocyte injuries, for example, TGF-β-Smad2/3 pathway, protein 38 (p38)/MAPK, c-Jun *N*-terminal kinase (JNK), ERK1/2, and so on (Kim et al. 2003).

In turn, podocyte abnormalities evidently enhance OS generation. When podocytes undergo pathological stimuli, mTOR is activated and exerts damage in a dose-dependent manner. Explanatorily, low-degree mTOR activity rescues podocyte from deletion; however, intensive and persistent activation of mTOR increases the level and activity of Nox leading to the deletion of podocyte. Theoretically, persistent mTOR-Nox activation disturbs mitochondrial respiration and ATP synthesis by ROS overproduction which results in podocyte injuries (Zschiedrich et al. 2017).

Together, there is a vicious cycle between ROS and podocyte damage, which jointly worsened podocyte associated disorders including glomerulosclerosis.

29.3.2.3 OS and PEC Injury

Initially, PEC is mainly considered just as a component of Bowman's capsule without well-defined function. However, emerging evidences show that PEC also serves as podocyte ancestral cells with differentiation capability toward podocytes or TECs (Eng et al. 2015; Shankland et al. 2013). Several mechanisms are attributed to the OS-induced PEC injury that including oxidative toxicity, increased inflammatory mediators, and ECM deposit which are similar to podocytes (Su et al. 2015; Smeets et al. 2011). In addition, ERK1/2 pathway and renin–angiotensin–aldosterone system (RAAS), which are always mentioned under OS status, are found to be associated with PEC proliferation and necrosis during crescent formation, suggesting their critical role in OS-induced PEC injury.

29.4 Targeting OS in Renal Fibrosis Management

OS is extensively implicated in renal fibrosis as we elucidated above, and thus, targeting OS seems to be a promising strategy in renal fibrosis management. Numerous antioxidants have been proposed to slow down CKD progression by reducing renal fibrosis in preclinical or clinical trials.

29.4.1 Targeting Nox

Since Nox family is vital in renal fibrosis, GKT137831 and GKT136901, two dual Nox1 and Nox4 inhibitors, have been demonstrated to reduce cytosolic and mitochondrial ROS production, macrophage infiltration, ECM accumulation, and albuminuria in different renal disease models (Decleves and Sharma 2014; Jha et al. 2014; Gorin et al. 2015).

A novel pan-Nox-inhibitor, APX-115 which decreases Nox1, Nox2, and Nox4 protein expression and reduces ROS generation, has been found to attenuate albuminuria and preserve GFR in addition to ameliorating mesangial expansion and macrophage infiltration in the diabetic kidney. Moreover, APX-115 has been validated to reduce HG-mediated ROS and fibronectin production in vitro (Cha et al. 2017).

Nox mediates TGF-β1-induced OS and fibrosis in part. P144, a synthetic peptide from TGF-β1 type III receptor competitively binding TGF-β1, could prevent TGF-β1-induced Nox activation and OS and down-regulate expression of collagen and CTGF, exerting antioxidative and antifibrotic effects (Baltanas et al. 2013).

There are many other signaling pathways regulating Nox, such as phosphatidylinositol 3-kinase (PI3K)/AKT and adenosine monophosphate-activated protein kinase (AMPK). Fluorofenidone has been reported to inhibit Nox and ECM deposition via PI3K/AKT pathway, and resveratrol has been verified to inhibit renal fibroblast proliferation and activation by regulating AMPK/Nox/ROS signaling; therefore, they are both considered as potential novel therapeutic agents against renal fibrosis (Qin et al. 2013; He et al. 2016).

29.4.2 *Targeting Mitochondria*

Since pirfenidone is a newly identified antifibrotic agent, it may preserve the function and structure of mitochondria by maintaining ATP production, increasing mitochondrial DNA (mtDNA) and stabilizing mitochondrial membrane potential, as well as reducing OS by enhancing MnSOD activity (Chen et al. 2013).

Pioglitazone is also found to attenuate renal fibrosis partially by improving mitochondrial function. Pioglitazone corrects the dysfunction of mitochondria via promoting mitochondrial biogenesis and reducing mitochondrial ROS by reducing the activity of cytochrome c oxidase and phosphorylation of $p66^{Shc}$, in addition to increasing mtDNA, maintaining ATP production, and stabilizing mitochondrial membrane potential (Lv et al. 2018).

Similar mitochondrial improvement has been seen in renal fibrosis promoted by hypochlorite-modified albumin treated by SS-31, a mitochondrial-targeted antioxidant peptide (Zhao et al. 2017).

$p66^{Shc}$, which is a recently recognized mediator of mitochondrial ROS production, may be suppressed by probucol. Since it reverses OS and ameliorates renal injury in DN, probucol is considered to be a potent antioxidant agent (Yang et al. 2017).

29.4.3 *Targeting NOS and XO*

Relaxin, a member of the insulin-like growth factor family of hormones, has been revealed to provide antifibrotic effects since it may up-regulate the expression of NOS1 and MnSOD, hence increasing the production of NO and attenuating OS (Leo et al. 2017; Sasser 2013). In addition, it has been shown to increase NO production by activating eNOS, thus improving endothelial function and alleviating inflammation.

Besides relaxin, curcumin and pyrolidium dithiocarbamate are also reported to targeting NOS, as they may protect against shock-wave lithotripsy-induced renal injury by decreasing expressions of iNOS and serum NO level (Bas et al. 2009; Tugcu et al. 2008).

XO is also a critical source of ROS. Febuxostat, an XO inhibitor approved for treating hyperuricemia, has been verified to prominently reduce XO activity and thereby alleviate OS in addition to endoplasmic reticulum stress, contributing to preserving renal function and reducing histopathological changes such as tubular injury, interstitial fibrosis, and macrophage infiltration (Tsuda et al. 2012).

29.4.4 Targeting Nrf2 Signaling Pathway

Nrf2 regulates expression of a wide array of genes encoding thiol molecules, antioxidant proteins, and enzymes. Many natural Nrf2 activators, such as curcumin, sulforaphane, astaxanthin, and cinnamic aldehyde, have been shown to alleviate OS and renal fibrosis (Soetikno et al. 2013; Choi et al. 2014; Liu et al. 2015). However, there were other unsatisfactory clinical findings that curcumin had no effect on Nrf2 activation, antioxidant enzyme activities, proteinuria, or GFR (Jimenez-Osorio et al. 2016).

Mycophenolate mofetil (MMF) has also been proposed to exert renoprotective effect by regulating Nrf2-Keap1 pathway and thereby attenuate OS (Tapia et al. 2016), as well as dioscin, which adjusts TGF-β1/Smad to inhibit renal fibrosis in addition (Qiao et al. 2018).

Fimasartan, which is a novel angiotensin II receptor blocker, has been found to up-regulate the Nrf2 pathway, accompanied by decreased expression of Nox family and increased expression of antioxidant enzymes, protecting against renal fibrosis induced by UUO (Kim et al. 2015).

In addition to the up-regulation of Nrf2, the enhanced nuclear translocation of Nrf2 by treatment with oleanolic acid has also been reported (Hong et al. 2014).

Moreover, sitagliptin, a dipeptidyl peptidase-4 (DPP-4) inhibitor, has been proved beneficial in DN therapy, by reason that it may attenuate OS by depressing miR-200a, which is an inhibitor of Keap1 (Civantos et al. 2017).

29.4.5 Other Antioxidants

Cysteamine bitartrate, another antioxidant, has been found to reduce renal fibrosis in a UUO model accompanied by attenuation of OS, in parallel with the decreased generation of ROS in cultured macrophages. Moreover, cysteamine bitartrate reduces myofibroblast differentiation and proliferation in vivo and in vitro via TGF-β-independent pathway (Okamura et al. 2014).

Semicarbazide-sensitive amine oxidases (SSAOs) produce H_2O_2 via oxidatively deaminating primary amines. Inhibiting SSAO by PXS-4728A reduces ROS generation, profibrotic cytokine secretion, and ECM deposition (Katagiri et al. 2016).

Besides, the SOD mimetic tempol and the angiotensin converting enzyme inhibitor ramipril have been tested to have an antioxidative effect by blunting diabetes-induced up-regulation of Noxs and ameliorating renal endoplasmic reticulum stress (De Blasio et al. 2017).

29.5 Conclusion and Perspectives

Renal diseases trigger OS marked by excessive ROS generation via different predominant mechanisms according to different etiologies. There are four major sources of OS including Nox, mitochondria, NOS, and XO in general. OS contributes critically to the pathophysiology of renal fibrosis which is the common pathway of renal diseases leading to ESRD. OS leads to fibroblast/pericyte activation, tubular EMT, fibrocyte/macrophage recruitment, and EndMT for tubulointerstitial fibrosis, along with mesangial sclerosis, podocyte abnormality, and PEC injury for glomerulosclerosis. The targeting OS therapy seems promising in renal fibrosis management as it may inhibit ROS generation, improve mitochondrial dysfunction, and regulate the antioxidative system. Many antioxidants have been tested to attenuate renal fibrosis, some of which are satisfying but some are not. A better understanding of the underlying mechanisms by which OS induces renal fibrosis and renal fibrosis triggers OS will provide new insights into the development of novel therapies to hold back the progression of renal diseases.

References

Andersson-Sjoland A, Karlsson JC, Rydell-Tormanen K (2016) ROS-induced endothelial stress contributes to pulmonary fibrosis through pericytes and Wnt signaling. Lab Invest 96:206–217

Andrades ME, Morina A, Spasic S, Spasojevic I (2011) Bench-to-bedside review: sepsis-from the redox point of view. Critical care (London, England) 15:230

Araujo M, Welch WJ (2006) Oxidative stress and nitric oxide in kidney function. Curr Opin Nephrol Hypertens 15:72–77

Baines RJ, Chana RS, Hall M, Febbraio M, Kennedy D, Brunskill NJ (2012) CD36 mediates proximal tubular binding and uptake of albumin and is upregulated in proteinuric nephropathies. Am J Physiol Renal Physiol 303:F1006–F1014

Baliga R, Ueda N, Walker PD, Shah SV (1999) Oxidant mechanisms in toxic acute renal failure. Drug Metab Rev 31:971–997

Baltanas A, Miguel-Carrasco JL, San Jose G, Cebrian C, Moreno MU, Dotor J et al (2013) A synthetic peptide from transforming growth factor-beta(1) type III receptor inhibits NADPH oxidase and prevents oxidative stress in the kidney of spontaneously hypertensive rats. Antioxid Redox Signal 19:1607–1618

Bas M, Tugcu V, Kemahli E, Ozbek E, Uhri M, Altug T et al (2009) Curcumin prevents shock-wave lithotripsy-induced renal injury through inhibition of nuclear factor kappa-B and inducible nitric oxide synthase activity in rats. Urol Res 37:159–164

Basile DP, Anderson MD, Sutton TA (2012) Pathophysiology of acute kidney injury. Compr. Physiol. 2:1303–1353

Bondi CD, Manickam N, Lee DY, Block K, Gorin Y, Abboud HE et al (2010) NAD(P)H oxidase mediates TGF-beta1-induced activation of kidney myofibroblasts. J Am Soc Nephrol 21:93–102

Bonventre JV, Yang L (2011) Cellular pathophysiology of ischemic acute kidney injury. J Clin Investig 121:4210–4221

Brealey D, Brand M, Hargreaves I, Heales S, Land J, Smolenski R et al (2002) Association between mitochondrial dysfunction and severity and outcome of septic shock. Lancet 360:219–223

Brown GE, Stewart MQ, Liu H, Ha VL, Yaffe MB (2003) A novel assay system implicates PtdIns(3,4)P(2), PtdIns(3)P, and PKC delta in intracellular production of reactive oxygen species by the NADPH oxidase. Mol Cell 11:35–47

Carlos CP, Mendes GEF, Miquelin AR, Luz MAM, da Silva CGA, van Rooijen N et al (2010) Macrophage depletion attenuates chronic cyclosporine a nephrotoxicity. Transplantation 89:1362–1370

Catherwood MA, Powell LA, Anderson P, McMaster D, Sharpe PC, Trimble ER (2002) Glucose-induced oxidative stress in mesangial cells. Kidney Int 61:599–608

Cha JJ, Min HS, Kim KT, Kim JE, Ghee JY, Kim HW et al (2017) APX-115, a first-in-class pan-NADPH oxidase (Nox) inhibitor, protects db/db mice from renal injury. Lab Invest 97:419

Chen J, Chen JK, Harris RC (2012a) Angiotensin II induces epithelial-to-mesenchymal transition in renal epithelial cells through reactive oxygen species/Src/caveolin-mediated activation of an epidermal growth factor receptor-extracellular signal-regulated kinase signaling pathway. Mol Cell Biol 32:981–991

Chen YQ, Wang XX, Yao XM, Zhang DL, Yang XF, Tian SF et al (2012b) Abated microRNA-195 expression protected mesangial cells from apoptosis in early diabetic renal injury in mice. J Nephrol 25:566–576

Chen JF, Liu H, Ni HF, Lv LL, Zhang MH, Zhang AH et al (2013) Improved mitochondrial function underlies the protective effect of pirfenidone against tubulointerstitial fibrosis in 5/6 nephrectomized rats. PLoS ONE 8:e83593

Chirino YI, Pedraza-Chaverri J (2009) Role of oxidative and nitrosative stress in cisplatin-induced nephrotoxicity. Exp Toxicol Pathol: Off J Ges Fur Toxikol Pathol 61:223–242

Choi BH, Kang KS, Kwak MK (2014) Effect of redox modulating NRF2 activators on chronic kidney disease. Molecules (Basel, Switzerland) 19:12727–12759

Civantos E, Bosch E, Ramirez E, Zhenyukh O, Egido J, Lorenzo O et al (2017) Sitagliptin ameliorates oxidative stress in experimental diabetic nephropathy by diminishing the miR-200a/Keap-1/Nrf2 antioxidant pathway. Diabetes, Metablc Syndr Obes: Targets Ther 10:207–222

De Blasio MJ, Ramalingam A, Cao AH, Prakoso D, Ye JM, Pickering R et al (2017) The superoxide dismutase mimetic tempol blunts diabetes-induced upregulation of NADPH oxidase and endoplasmic reticulum stress in a rat model of diabetic nephropathy. Eur J Pharmacol 807:12–20

Decleves AE, Sharma K (2014) Novel targets of antifibrotic and anti-inflammatory treatment in CKD. Nature reviews Nephrology 10:257–267

Dendooven A, van Oostrom O, van der Giezen DM, Leeuwis JW, Snijckers C, Joles JA et al (2011) Loss of endogenous bone morphogenetic protein-6 aggravates renal fibrosis. Am J Pathol 178:1069–1079

Dou L, Jourde-Chiche N, Faure V, Cerini C, Berland Y, Dignat-George F et al (2007) The uremic solute indoxyl sulfate induces oxidative stress in endothelial cells. J Thromb Haemost: JTH 5:1302–1308

Echeverria C, Montorfano I, Sarmiento D, Becerra A, Nunez-Villena F, Figueroa XF et al (2013) Lipopolysaccharide induces a fibrotic-like phenotype in endothelial cells. J Cell Mol Med 17:800–814

Eng DG, Sunseri MW, Kaverina NV, Roeder SS, Pippin JW, Shankland SJ (2015) Glomerular parietal epithelial cells contribute to adult podocyte regeneration in experimental focal segmental glomerulosclerosis. Kidney Int 88:999–1012

Fassett RG, Robertson IK, Ball MJ, Geraghty DP, Coombes JS (2015) Effects of atorvastatin on oxidative stress in chronic kidney disease. Nephrology (Carlton, Vic) 20:697–705

Forstermann U, Munzel T (2006) Endothelial nitric oxide synthase in vascular disease: from marvel to menace. Circulation 113:1708–1714

Gill PS, Wilcox CS (2006) NADPH oxidases in the kidney. Antioxid Redox Signal 8:1597–1607

Gill R, Tsung A, Billiar T (2010) Linking oxidative stress to inflammation: toll-like receptors. Free Radic Biol Med 48:1121–1132

Gorin Y, Cavaglieri RC, Khazim K, Lee DY, Bruno F, Thakur S et al (2015) Targeting NADPH oxidase with a novel dual Nox1/Nox4 inhibitor attenuates renal pathology in type 1 diabetes. Am J Physiol Renal Physiol 308:F1276–F1287
Handy DE, Loscalzo J (2012) Redox regulation of mitochondrial function. Antioxid Redox Signal 16:1323–1367
He T, Guan X, Wang S, Xiao T, Yang K, Xu X et al (2015) Resveratrol prevents high glucose-induced epithelial-mesenchymal transition in renal tubular epithelial cells by inhibiting NADPH oxidase/ROS/ERK pathway. Mol Cell Endocrinol 402:13–20
He T, Xiong J, Nie L, Yu Y, Guan X, Xu X et al (2016) Resveratrol inhibits renal interstitial fibrosis in diabetic nephropathy by regulating AMPK/Nox4/ROS pathway. J Mol Med (Berl, Ger) 94:1359–1371
Hong YA, Lim JH, Kim MY, Kim EN, Koh ES, Shin SJ et al (2014) Delayed treatment with oleanolic acid attenuates tubulointerstitial fibrosis in chronic cyclosporine nephropathy through Nrf2/HO-1 signaling. J Transl Med 12:50
Jefferson JA, Shankland SJ (2014) The pathogenesis of focal segmental glomerulosclerosis. Adv Chronic Kidney Dis 21:408–416
Jha JC, Gray SP, Barit D, Okabe J, El-Osta A, Namikoshi T et al (2014) Genetic targeting or pharmacologic inhibition of NADPH oxidase Nox4 provides renoprotection in long-term diabetic nephropathy. J Am Soc Nephrol 25:1237–1254
Jimenez-Osorio AS, Garcia-Nino WR, Gonzalez-Reyes S, Alvarez-Mejia AE, Guerra-Leon S, Salazar-Segovia J et al (2016) The effect of dietary supplementation with curcumin on redox status and Nrf2 activation in patients with nondiabetic or diabetic proteinuric chronic kidney disease: a pilot study. J Ren Nutr: Off J Counc Ren Nutr Natl Kidney Found 26:237–244
Jones DP (2006) Redefining oxidative stress. Antioxid Redox Signal 8:1865–1879
Katagiri D, Hamasaki Y, Doi K, Negishi K, Sugaya T, Nangaku M et al (2016) Interstitial renal fibrosis due to multiple cisplatin treatments is ameliorated by semicarbazide-sensitive amine oxidase inhibition. Kidney Int 89:374–385
Kida Y, Tchao BN, Yamaguchi I (2014) Peritubular capillary rarefaction: a new therapeutic target in chronic kidney disease. Pediatr Nephrol 29:333–342
Kim JH, Kim BK, Moon KC, Hong HK, Lee HS (2003) Activation of the TGF-/Smad signaling pathway in focal segmental glomerulosclerosis. Kidney Int 64:1715–1721
Kim SM, Kim YG, Jeong KH, Lee SH, Lee TW, Ihm CG et al (2012) Angiotensin II-induced mitochondrial Nox4 is a major endogenous source of oxidative stress in kidney tubular cells. PLoS ONE 7:e39739
Kim S, Kim SJ, Yoon HE, Chung S, Choi BS, Park CW et al (2015) Fimasartan, a novel angiotensin-receptor blocker, protects against renal inflammation and fibrosis in mice with unilateral ureteral obstruction: the possible Role of Nrf2. Int J Med Sci 12:891–904
Krause KH (2007) Aging: a revisited theory based on free radicals generated by Nox family NADPH oxidases. Exp Gerontol 42:256–262
Lassègue B, Clempus RE (2003) Vascular NAD(P)H oxidases: specific features, expression, and regulation. Am J Physiol Regul Integr Comp Physiol 285:R277–297
Lee DY, Wauquier F, Eid AA, Roman LJ, Ghosh-Choudhury G, Khazim K et al (2013) Nox4 NADPH oxidase mediates peroxynitrite-dependent uncoupling of endothelial nitric-oxide synthase and fibronectin expression in response to angiotensin II: role of mitochondrial reactive oxygen species. J Biol Chem 288:28668–28686
Lee WJ, Liu SH, Chiang CK, Lin SY, Liang KW, Chen CH et al (2016) Aryl hydrocarbon receptor deficiency attenuates oxidative stress-related mesangial cell activation and macrophage infiltration and extracellular matrix accumulation in diabetic nephropathy. Antioxid Redox Signal 24:217–231
Leo CH, Jelinic M, Ng HH, Marshall SA (2017) Vascular actions of relaxin: nitric oxide and beyond. Br J Pharmacol 174:1002–1014
Lichtenberg D, Pinchuk I (2015) Oxidative stress, the term and the concept. Biochem Biophys Res Commun 461:441–444

Lin EY, Bayarsengee U, Wang CC, Chiang YH, Cheng CW (2018) The natural compound 2,3,5,4'-tetrahydroxystilbene-2-O-beta-d glucoside protects against adriamycin-induced nephropathy through activating the Nrf2-Keap1 antioxidant pathway. Environ Toxicol 33:72–82
Liu G, Shi Y, Peng X, Liu H, Peng Y, He L (2015) Astaxanthin attenuates adriamycin-induced focal segmental glomerulosclerosis. Pharmacology 95:193–200
Lorenz G, Darisipudi MN, Anders HJ (2014) Canonical and non-canonical effects of the NLRP3 inflammasome in kidney inflammation and fibrosis. Nephrol Dial Transplant 29:41–48
Lu Q, Zhou Y, Hao M, Li C, Wang J, Shu F et al (2017) The mTOR promotes oxidative stress-induced apoptosis of mesangial cells in diabetic nephropathy. Mol Cell Endocrinol 473:31–43
Lushchak VI (2014) Free radicals, reactive oxygen species, oxidative stress and its classification. Chem Biol Interact 224:164–175
Lv W, Booz GW, Fan F, Wang Y, Roman RJ (2018) Oxidative stress and renal fibrosis: recent insights for the development of novel therapeutic strategies. Front Physiol 9:105
Madero M, Sarnak MJ, Wang X, Greene T, Beck GJ, Kusek JW et al (2009) Uric acid and long-term outcomes in CKD. Am J Kidney Dis 53:796–803
Martinon F, Pétrilli V, Mayor A, Tardivel A, Tschopp J (2006) Gout-associated uric acid crystals activate the NALP3 inflammasome. Nature 440:237–241
Mason RM (2003) Extracellular matrix metabolism in diabetic nephropathy. J Am Soc Nephrol 14:1358–1373
Massy ZA, Ceballos I, Chadefaux-Vekemens B, Nguyen-Khoa T, Descamps-Latscha B, Drueke TB et al (2001) Homocyst(e)ine, oxidative stress, and endothelium function in uremic patients. Kidney Int Suppl 78:S243–S245
Meotti FC, Jameson GN, Turner R, Harwood DT, Stockwell S, Rees MD et al (2011) Urate as a physiological substrate for myeloperoxidase: implications for hyperuricemia and inflammation. J Biol Chem 286:12901–12911
Montorfano I, Becerra A, Cerro R, Echeverria C, Saez E, Morales MG et al (2014) Oxidative stress mediates the conversion of endothelial cells into myofibroblasts via a TGF-beta1 and TGF-beta2-dependent pathway. Lab Invest 94:1068–1082
Nishikawa T, Edelstein D, Du XL, Yamagishi S, Matsumura T, Kaneda Y et al (2000) Normalizing mitochondrial superoxide production blocks three pathways of hyperglycaemic damage. Nature 404:787–790
Nishikimi T, Matsuoka H (2006) Molecular mechanisms and therapeutic strategies of chronic renal injury: renoprotective effect of rho-kinase inhibitor in hypertensive glomerulosclerosis. J Pharmacol Sci 100:22–28
Noh H, Kim JS, Han KH, Lee GT, Song JS, Chung SH et al (2006) Oxidative stress during peritoneal dialysis: implications in functional and structural changes in the membrane. Kidney Int 69:2022–2028
Oberg BP, McMenamin E, Lucas FL, McMonagle E, Morrow J, Ikizler TA et al (2004) Increased prevalence of oxidant stress and inflammation in patients with moderate to severe chronic kidney disease. Kidney Int 65:1009–1016
Okamura DM, Pennathur S (2015) The balance of powers: redox regulation of fibrogenic pathways in kidney injury. Redox biology 6:495–504
Okamura DM, Bahrami NM, Ren S, Pasichnyk K, Williams JM, Gangoiti JA et al (2014) Cysteamine modulates oxidative stress and blocks myofibroblast activity in CKD. J Am Soc Nephrol 25:43–54
Pacher P, Nivorozhkin A, Szabo C (2006) Therapeutic effects of xanthine oxidase inhibitors: renaissance half a century after the discovery of allopurinol. Pharmacol Rev 58:87–114
Qiao M, Zhao Q, Lee CF, Tannock LR, Smart EJ, LeBaron RG et al (2009) Thiol oxidative stress induced by metabolic disorders amplifies macrophage chemotactic responses and accelerates atherogenesis and kidney injury in LDL receptor-deficient mice. Arterioscler Thromb Vasc Biol 29:1779–1786
Qiao Y, Xu L, Tao X, Yin L, Qi Y, Xu Y et al (2018) Protective effects of dioscin against fructose-induced renal damage via adjusting Sirt3-mediated oxidative stress, fibrosis, lipid metabolism and inflammation. Toxicol Lett 284:37–45

Qin J, Xie YY, Huang L, Yuan QJ, Mei WJ, Yuan XN et al (2013) Fluorofenidone inhibits nicotinamide adenine dinucleotide phosphate oxidase via PI3 K/Akt pathway in the pathogenesis of renal interstitial fibrosis. Nephrology (Carlton, Vic) 18:690–699

Radi R (2013) Peroxynitrite, a stealthy biological oxidant. J Biol Chem 288:26464–26472

Ralto KM, Parikh SM (2016) Mitochondria in acute kidney injury. Semin Nephrol 36:8–16

Ratliff BB, Abdulmahdi W, Pawar R, Wolin MS (2016) Oxidant mechanisms in renal injury and disease. Antioxid Redox Signal 25:119–146

Sakai N, Wada T, Yokoyama H, Lipp M, Ueha S, Matsushima K et al (2006) Secondary lymphoid tissue chemokine (SLC/CCL21)/CCR72 signaling regulates fibrocytes in renal fibrosis. Proc Natl Acad Sci 103:14098–14103

Sánchez-Lozada LG, Lanaspa MA, Cristóbal-García M, García-Arroyo F, Soto V, Cruz-Robles D et al (2012) Uric acid-induced endothelial dysfunction is associated with mitochondrial alterations and decreased intracellular ATP concentrations. Nephron Exp Nephrol 121:e71–78

Sasser JM (2013) The emerging role of relaxin as a novel therapeutic pathway in the treatment of chronic kidney disease. Am J Physiol Regul Integr Comp Physiol 305:R559–R565

Sedeek M, Nasrallah R, Touyz RM, Hebert RL (2013) NADPH oxidases, reactive oxygen species, and the kidney: friend and foe. J Am Soc Nephrol 24:1512–1518

Seifert EL, Estey C, Xuan JY, Harper ME (2010) Electron transport chain-dependent and -independent mechanisms of mitochondrial H2O2 emission during long-chain fatty acid oxidation. J Biol Chem 285:5748–5758

Shahzad K, Bock F, Dong W, Wang H, Kopf S, Kohli S et al (2015) Nlrp3- inflammasome activation in non-myeloid-derived cells aggravates diabetic nephropathy. Kidney Int 87:74–84

Shankland SJ, Anders HJ, Romagnani P (2013) Glomerular parietal epithelial cells in kidney physiology, pathology, and repair. Curr Opin Nephrol Hypertens 22:302–309

Sies H (1985) Oxidative stress: introductory remarks. Oxidative stress. Academic press, London, pp 1–8

Sitar ME, Aydin S, Cakatay U (2013) Human serum albumin and its relation with oxidative stress. Clinical laboratory 59:945–952

Smeets B, Kuppe C, Sicking EM, Fuss A, Jirak P, van Kuppevelt TH et al (2011) Parietal epithelial cells participate in the formation of sclerotic lesions in focal segmental glomerulosclerosis. J Am Soc Nephrol 22:1262–1274

Soetikno V, Sari FR, Lakshmanan AP, Arumugam S, Harima M, Suzuki K et al (2013) Curcumin alleviates oxidative stress, inflammation, and renal fibrosis in remnant kidney through the Nrf2-keap1 pathway. Mol Nutr Food Res 57:1649–1659

Su H, Chen S, He FF, Wang YM, Bondzie P, Zhang C (2015) New insights into glomerular parietal epithelial cell activation and its signaling pathways in glomerular diseases. Biomed Res Int 2015:318935

Su H, Wan C, Lei CT, Zhang CY, Ye C, Tang H et al (2017) Lipid deposition in kidney disease: interplay among redox, lipid mediators, and renal impairment. Antioxid Redox Signal 28:1027–1043

Tang DQ, Wei YQ, Yin XX, Lu Q, Hao HH, Zhai YP et al (2011) In vitro suppression of quercetin on hypertrophy and extracellular matrix accumulation in rat glomerular mesangial cells cultured by high glucose. Fitoterapia 82:920–926

Tapia E, Garcia-Arroyo F, Silverio O, Rodriguez-Alcocer AN, Jimenez-Flores AB, Cristobal M et al (2016) Mycophenolate mofetil and curcumin provide comparable therapeutic benefit in experimental chronic kidney disease: role of Nrf2-Keap1 and renal dopamine pathways. Free Radical Res 50:781–792

Tbahriti HF, Kaddous A, Bouchenak M, Mekki K (2013) Effect of different stages of chronic kidney disease and renal replacement therapies on oxidant-antioxidant balance in uremic patients. Biochem Res Int 2013:358985

Tripathy D, Mohanty P, Dhindsa S, Syed T, Ghanim H, Aljada A et al (2003) Elevation of free fatty acids induces inflammation and impairs vascular reactivity in healthy subjects. Diabetes 52:2882–2887

Tsuda H, Kawada N, Kaimori J, Kitamura H, Moriyama T, Rakugi H et al (2012) Febuxostat suppressed renal ischemia-reperfusion injury via reduced oxidative stress. Biochem Biophys Res Commun 427:266–272
Tugcu V, Bas M, Ozbek E, Kemahli E, Arinci YV, Tuhri M et al (2008) Pyrolidium dithiocarbamate prevents shockwave lithotripsy-induced renal injury through inhibition of nuclear factor-kappa B and inducible nitric oxide synthase activity in rats. J Endourol 22:559–566
Tumur Z, Shimizu H, Enomoto A, Miyazaki H, Niwa T (2010) Indoxyl sulfate upregulates expression of ICAM-1 and MCP-1 by oxidative stress-induced NF-kappaB activation. Am J Nephrol 31:435–441
Tuttle KR, Bakris GL, Bilous RW, Chiang JL, de Boer IH, Goldstein-Fuchs J et al (2014) Diabetic kidney disease: a report from an ADA Consensus Conference. Diabetes Care 37:2864–2883
Wagner B, Tan C, Barnes JL, Ahuja S, Davis TL, Gorin Y et al (2012) Nephrogenic systemic fibrosis: evidence for oxidative stress and bone marrow-derived fibrocytes in skin, liver, and heart lesions using a 5/6 nephrectomy rodent model. Am J Pathol 181:1941–1952
Wan C, Su H, Zhang C (2016) Role of NADPH oxidase in metabolic disease-related renal injury: an update. Oxid Med Cell Longev 2016:7813072
Wang JY, Yin XX, Wu YM, Tang DQ, Gao YY, Wan MR et al (2006) Ginkgo biloba extract suppresses hypertrophy and extracellular matrix accumulation in rat mesangial cells. Acta Pharmacol Sin 27:1222–1230
Ward RA, McLeish KR (2003) Oxidant stress in hemodialysis patients: what are the determining factors? Artif Organs 27:230–236
Ward RA, Ouseph R, McLeish KR (2003) Effects of high-flux hemodialysis on oxidant stress. Kidney Int 63:353–359
Yang Z-Z, Zou A-P (2003) Homocysteine enhances TIMP-1 expression and cell proliferation associated with NADH oxidase in rat mesangial cells. Kidney Int 63:1012–1020
Yang S, Zhao L, Han Y, Liu Y, Chen C, Zhan M et al (2017) Probucol ameliorates renal injury in diabetic nephropathy by inhibiting the expression of the redox enzyme p66Shc. Redox biology 13:482–497
Zhao H, Liu YJ, Liu ZR, Tang DD, Chen XW, Chen YH et al (2017) Role of mitochondrial dysfunction in renal fibrosis promoted by hypochlorite-modified albumin in a remnant kidney model and protective effects of antioxidant peptide SS-31. Eur J Pharmacol 804:57–67
Zschiedrich S, Bork T, Liang W, Wanner N, Eulenbruch K, Munder S et al (2017) Targeting mTOR signaling can prevent the progression of FSGS. J Am Soc Nephrol 28:2144–2157

Part IV
Biomarkers and Treatment of Renal Fibrosis

Chapter 30
Urinary Biomarkers of Renal Fibrosis

Le-Ting Zhou, Lin-Li Lv and Bi-Cheng Liu

Abstract Renal fibrosis is the common pathological pathway of progressive CKD. The commonly used biomarkers in clinical practice are not optimal to detect injury or predict prognosis. Therefore, it is crucial to develop novel biomarkers to allow prompt intervention. Urine serves as a valuable resource of biomarker discovery for kidney diseases. Owing to the rapid development of omics platforms and bioinformatics, research on novel urinary biomarkers for renal fibrosis has proliferated in recent years. In this chapter, we discuss the current status and provide basic knowledge in this field. We present novel promising biomarkers including tubular injury markers, proteins related to activated inflammation/fibrosis pathways, CKD273, transcriptomic biomarkers, as well as metabolomic biomarkers. Furthermore, considering the complex nature of the pathogenesis of renal fibrosis, we also highlight the combination of biomarkers to further improve the diagnostic and prognostic performance.

Keywords Urinary biomarker · Renal fibrosis · Omics · Combined biomarker

30.1 Introduction

Renal fibrosis, characterized as a relentless deposition of extracellular matrix (ECM) with concomitant loss of the parenchyma, is the common pathological pathway of progressive CKD.

Generally speaking, there are two basic goals of developing biomarkers for renal fibrosis: (1) to diagnose renal fibrosis (diagnostic biomarkers); and (2) to predict the progression of renal fibrosis (prognostic biomarkers). Renal functional parameters such as serum creatinine (SCr) and SCr-based estimated glomerular-filtration rate (eGFR) measurements are most widely used biomarkers to evaluate renal fibrosis. However, these parameters cannot suggest the underlying molecular mechanism and usually change little at the onset of fibrosis (Puzantian and Townsend 2013). In

L.-T. Zhou · L.-L. Lv · B.-C. Liu (✉)
Institute of Nephrology, Zhong Da Hospital, Southeast University School of Medicine, DingJiaQiao Road, Nanjing, China
e-mail: liubc64@163.com

B.-C. Liu et al. (eds.), *Renal Fibrosis: Mechanisms and Therapies*,
Advances in Experimental Medicine and Biology 1165,
https://doi.org/10.1007/978-981-13-8871-2_30

addition, it is still difficult to identify subjects at high risk of developing advanced renal fibrosis accurately. These challenges all highlight the importance of finding novel biomarkers.

Urine is regarded as a "fountain" of valuable information of the kidney with easy access and therefore provides an ideal source of biomarkers for kidney diseases. With the advent of the omics era, biomarker research is increasingly facilitated by burgeoning high-throughput omics technologies, which permits providing urinary molecular profiles in unprecedented speed and details. Besides, the massive datasets generated by omics platforms also represent typical sources for developing combined biomarkers, which are expected to yield better performance than standalone molecules. This field has, therefore, witnessed enormous expansion in new findings.

In this Chapter, we will discuss the current status, as well as provide basic knowledge of urinary biomarkers for renal fibrosis, predominantly in the context of CKD. First, we will summarize some basic issues on this topic. In the second part, we will focus on the novel biomarkers, especially those identified by omics technologies in recent years. Lastly, we will talk about the opportunities and challenges of generating combined biomarkers using high-throughput technologies.

30.2 Basic Issues

30.2.1 An Overview of Developing Biomarkers: From Identification to Validation

We have mentioned in the introduction part that we aim to develop biomarkers that can diagnose renal fibrosis and predict disease progression. Generally speaking, a comprehensive pipeline for developing these biomarkers usually includes, but is not limited to, identification of candidate biomarkers, verification of differential expressions and validation of diagnostic/prognostic power (Fig. 30.1) (Latterich and Schnitzer 2011). Omics approaches are usually used for screening candidates in identification phase (Table 30.1). However, these high-throughput strategies may generate spurious candidates owing to the enormous amount of molecules and relatively small sample size. The field addressing this issue, namely feature selection, has become a hotspot in bioinformatics research. Several algorithms such as principal component analysis, clustering, or multidimensional scaling can help researchers to grasp several thousand parameters generated by high-throughput platforms (Wang et al. 2016, 2017a). Another feasible strategy to strengthen the effectiveness of selection is to testify the differential expressions of candidates using public databases. An excellent introduction of databases in nephrology can be found in a recent review (Papadopoulos et al. 2016).

In subsequent verification phase, more convenient and accurate assay methods (e.g., immunoassay for protein measurement) are applied to verify the association of putative biomarkers with renal fibrosis and those with best associations will move on

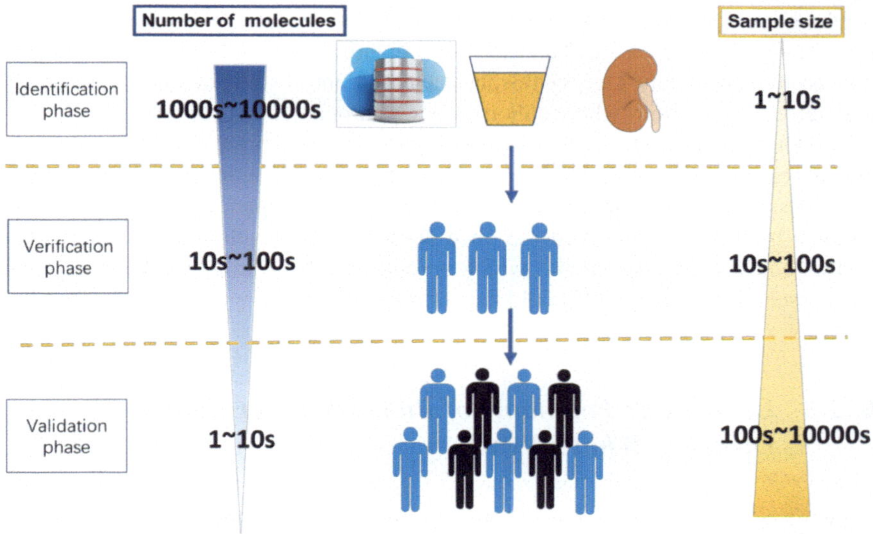

Fig. 30.1 A comprehensive pipeline for developing prognostic biomarkers using omics technologies

Table 30.1 An overview of urine omics

Category of Omics	Target	Major research platforms	Databases
Proteomics	Proteins and peptides degraded from urinary proteins	MS Microarray	Human kidney and urine proteome project (http://www.hkupp.org/) Urinary peptidomics and peak-maps (http://www.padb.org/updb) peptiCKDdb (www.peptickddb.com)
Transcriptomics	All coding and non-coding RNAs (including miRNA, lncRNA and circRNA)	RNA sequencing Microarray	GEO (http://www.ncbi.nlm.nih.gov/gds) SRA (https://www.ncbi.nlm.nih.gov/sra)
Metablomics	Products of cellular metabolism with molecular masses of 80–1200 Da	MS NMR	Human metabolome database (http://www.hmdb.ca/)

to subsequent validation. Case-control studies based on renal biopsies and prospective cohorts generate the ideal datasets for validating the diagnostic and prognostic biomarker, respectively. Despite this, nested case-control study or case-cohort study is preferred in situations where cohorts are too burdensome to carry out.

Validation is regarded as the bottleneck for translating biomarker discovery into clinical reality, owing to the difficulty of launching studies with large samples, ineffectiveness after adjusting known risk factors and discrepancy among the results obtained by different centers. Rifai once described the development of biomarker as "the long and uncertain path," which is somewhat frustrating but the reality that every researcher in this field should face (Rifai et al. 2006).

30.2.2 *Urine as a "Fountain" of Biomarkers for Kidney Diseases: What's New?*

Urine is a complex mixture of water, numerous dissolved compounds, cells and cell-derived particles derived from the kidney, urinary tract and circulation, as well as crystals, bacteria and sperms (Fig. 30.2). Owing to the development of high-throughput technologies, the urinary molecule profiles can be detected with high resolution in a

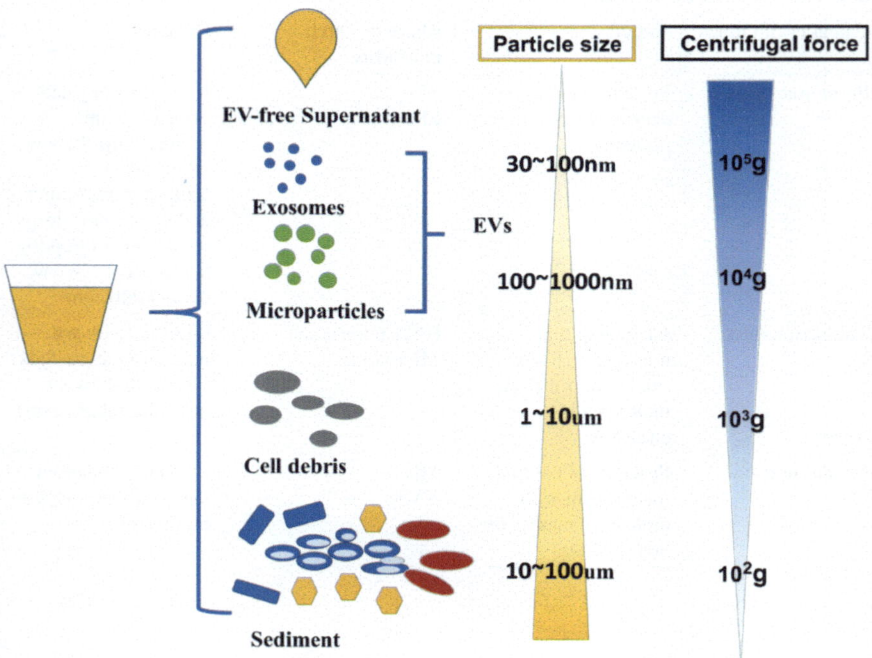

Fig. 30.2 Components of the urine isolated by differential centrifugation

relatively short time. In general, the cell-free supernatants are preferred for proteomic and metabolomic analysis, while cells or particles with membrane structures provide the ideal source for transcriptomic analysis (Papadopoulos et al. 2015). EVs are a heterogeneous group of vesicles with a pivotal role in cell-to-cell communication, which can be divided into microparticles and exosomes according to the diameter (Fig. 30.2). Recently, EVs are increasingly recognized as a novel abundant source for discovering biomarkers in addition to urinary sediment and supernatant (Erdbrugger and Le 2016). Interestingly, novel molecular types such as circular RNAs (circRNAs) and long non-coding RNAs (lncRNAs) are also enriched in EVs (Kim et al. 2017). Differential centrifugation is the most commonly used method to isolate EVs. Other isolation techniques include density gradient centrifugation, size exclusion chromatography, and immunoaffinity capture. (Mateescu et al. 2017). Currently, several technical issues remain to be addressed with respect to urinary EVs for biomarker discovery, including lack of optimized isolation method and normalization method of the molecular changes (Dear et al. 2013).

30.3 Novel Urinary Biomarkers for Renal Fibrosis

30.3.1 Proteomic Biomarkers

Proteins are both direct effectors of biological processes and targets of nearly all current drugs and therefore proteomics represents the classical realm for biomarker discovery in kidney diseases (Mischak et al. 2015). Sustained tubular injury, activated inflammation/fibrosis pathways and deposition of ECM are recognized as promoters of renal fibrosis. A growing number of putative protein biomarkers have been discovered based on these key pathological events. Lately, urinary peptidomics displays a good development momentum, casting new light on the clinical application of biomarkers in renal fibrosis.

30.3.2 Tubular Injury Markers

Neutrophil gelatinase-associated lipocalin (NGAL), kidney injury molecule 1 (KIM-1), N-acetyl-b-D-glucosamininidase (NAD) are representative tubular injury markers that have been demonstrated to be extremely valuable in the context of AKI. Considering the fact that tubular injury is not only a consequence but also a driving force of renal fibrosis progression, these biomarkers are expected to be potentially useful as predictors (Liu et al. 2018). Increased urinary levels of these tubular injury markers have been found in patients with various types of CKD and are associated with declined eGFR. However, conflicting results on their performance in predicting progression were obtained. A matched case-control study involving 286 participants from ARIC study showed that uNGAL but not KIM-1 was a predictor of incident CKD stage 3 in the general population (Bhavsar et al. 2012). In contrast, another nested case-control study from MESA study reported that KIM-1 but not NGAL was associated with incident CKD stage 3 and/or rapid kidney function decrease (Peralta et al. 2012). In terms of predicting ESKD, a recent meta-analysis showed that the concentration of uNGAL can be used as an independent predictor of ESKD among patients with CKD, whereas NAD and KIM-1 failed (Zhou et al. 2016). However, in a large prospective study published later, the CRIC study investigators reported that none of these biomarkers are capable of predicting CKD progression (the incidence of ESKD or halving of eGFR) (Hsu et al. 2017).

Other tubular injury markers are also faced with the same dilemma. In a small prospective study, urinary L-FABP levels might be a useful predictor of renal function deterioration in patients with glomerulonephritis (Mou et al. 2012). In a cross-sectional study, urinary TFF-3 was significantly elevated in stage 4 and 5 CKD (Lebherz-Eichinger et al. 2015). Moreover, a case-control study from ARIC study indicated urinary TFF-3 was associated with incident CKD stage 3 (with restriction of decline in eGFR $\geq$25%) (Astor et al. 2011). However, both L-FABP and TFF3 were not associated with incident CKD stage 3 in a large cohort from FHS (O'Seaghdha et al. 2013).

Overall, the prognostic power of urinary tubular injury markers, as revealed by large cohorts, is not as definitive as previously thought. On the other hand, novel tubular injury markers still hold the promise of predicting CKD progression efficiently. Recent study reveals that (TIMP-2*IGFBP-7), a tubular cell cycle arrest marker, is informative in predicting the occurrence of AKI (Aregger et al. 2014; Tan et al. 2016). Its utility as a biomarker for renal fibrosis remains to be examined.

30.3.3 Biomarkers Related to Activated Inflammation/Fibrosis Pathways

30.3.3.1 Chemokines

Chemokines play a pivotal role in amplifying inflammation during the early stage of renal fibrosis. CKD patients have increased urine excretion of chemokines such as CCL17, CCL20, CCL22, and CXCL11(Lebherz-Eichinger et al. 2014). Urinary chemokine (C-X-C motif) ligand CXCL1 concentration was suggested to be an independent risk predictor of renal function deterioration in IgAN (Zhao et al. 2015). CCL2, or monocyte chemoattractant protein-1 (MCP-1), is a chemokine that recruits monocytes to the sites of inflammation. Several earlier studies suggested that urinary CCL2 are significantly elevated in patients with DKD and might serve as an independent prognostic marker for renal function decline in this population (Shoukry et al. 2015; Titan et al. 2012) (Verhave et al. 2013). More recently, a nested case-control study from the ACCORD Trial reported that the addition of urinary MCP-1 to traditional predictors could significantly improve the power in discriminating patients with sustained renal decline from those with preserved renal function in the setting of T2DM (Nadkarni et al. 2016).

30.3.3.2 Angiotensinogen (AGT)

The activation of intrarenal renin–angiotensin system (RAS) is a hallmark of the development of renal fibrosis. Despite the kidney contains all components of the RAS, AGT is preferred for its good correlation with intrarenal RAS activation and stability in the urine (Yamamoto et al. 2007). Elevated urinary excretion of AGT has been reported in patients with DKD, IgAN, and MN. The use of RAS inhibitors can attenuate the increase of urinary AGT in patients with hypertension (Burns and Hiremath 2012). In a cohort of 80 CKD patients with a median follow-up of 23 months, elevated urinary AGT was associated with a higher risk of renal function deterioration (Yamamoto et al. 2007). Moreover, elevated urinary AGT was also associated with eGFR decline, chronic hemodialysis and cardiovascular events in T2DM patients (Sawaguchi et al. 2012). In a case-control study of 402 participants, higher levels of uAGT were independently associated with incidence of CKD after multiple adjustments (Mills et al. 2012). Interestingly, in another study including 141 T2DM patients, the middle tertile of uAGT/creatinine was shown to be associated with the lowest incidence of ESKD (Aflarian et al. 2015; Mills et al. 2012). Therefore, pending further validation, urinary uAGT may serve as not only a useful prognostic biomarker but also a therapeutic target for renal fibrosis.

30.3.3.3 Connective Tissue Growth Factor (CTGF)

CTGF is a secreted cysteine-rich protein implicated in a variety of fibrotic disorders including renal fibrosis. The utility of CTGF as a potential biomarker is mostly investigated in DKD. Higher urinary levels of CTGF were observed in diabetic rats and patients, especially in those with nephropathy (Riser et al. 2003). Notably, in a cross-sectional study involving 347 patients with type 1 diabetes, urinary CTGF levels correlated with urinary albumin excretion and inversely with eGFR, suggesting its potential role as a predictor of disease progression (Nguyen et al. 2006). Intuitively, higher urinary CTGF concentrations might be associated with increased risk of incident CKD. However, urinary CTGF concentrations showed a significant inverse association with incident CKD stage 3 in the general population (O'Seaghdha et al. 2011). The same result was also obtained by a large cohort involving 2948 participants from FHS (O'Seaghdha et al. 2013). The exact mechanism for this inverse association is unclear. One possible explanation is that lower CTGF is associated with attenuated recovery of mesangial injury, which contributes to the progression of renal fibrosis.

30.3.3.4 Periostin and MMP7

With the deepening of the research, several profibrotic molecules, such as periostin and MMP7, have been identified as novel mediators of renal fibrosis in recent years. Periostin is a 90 kDa protein with a role in regulating the formation of ECM (Prakoura and Chatziantoniou 2017). Expression of periostin is upregulated in the fibrotic areas of a variety of kidney diseases. Urinary levels of periostin were found elevated in patients with T2DM and correlated with impaired renal function in T2DM patients (Satirapoj et al. 2012, 2015). In a cohort involving 399 patients with IgAN, urinary periostin concentration is positively correlated with the severity of renal fibrosis and can predict the worsening of renal function (Hwang et al. 2016).

Similar to periostin, MMP-7 is also implicated in the regulation of ECM turnover. Wnt/beta-catenin signaling is responsible for enhancing the transcription of MMP-7, and urinary MMP-7 levels can serve as an indicator of renal Wnt/beta-catenin activity (He et al. 2012; Zhou et al. 2017a). In a cohort of 141 T2DM patients with a median follow-up of 3 years, uMMP7 was found to be associated with a higher risk of ESKD (Aflarian et al. 2015). However, the diagnostic role of these biomarkers remains to be established.

30.3.3.5 Other Novel Biomarkers

Most biomarkers introduced in the previous sections are discovered using a literature-based approach. The following two studies, on the other hand, not only provide novel candidate biomarkers but also exemplify the biomarker discovery approach for renal fibrosis in the omics era. In the first study, hundreds of dysregulated peptides were first identified by comparing the urinary peptidomic profiles in rat models. Next, urinary epidermal growth factor, one of the parent proteins, was confirmed to be decreased in rats by ELISA. The researchers further conducted a cohort involving over 600 patients with DKD with normal albuminuria at baseline to validate the prognostic power of urinary EGF. As a result, lower urinary EGF was independently associated with incident CKD stage 3 (Betz et al. 2016). Another study by Craciu et al. applied RNA sequencing instead to identify candidate genes from the kidneys of folic acid-induced nephropathy mice. Subsequently, by verification in three other distinct CKD mouse models, ten genes were selected as candidate biomarkers reflecting the course of renal fibrosis. Among these ten candidates, urinary concentration of CDH11, MRC1, and PLTP proteins were determined to significantly increase in CKD patients (Craciun et al. 2015).

Recently, urinary exosomes have emerged as a novel repository for proteomic biomarkers (Pisitkun et al. 2004). One of the first studies in nephrology performed proteomic analysis on urinary exosomes from cisplatin injected rats and found that urinary exosomal fetuin-A might be a prognostic biomarker of structural renal injury by subsequent verification (Zhou et al. 2006). A recent study identified a panel of urinary exosomal protein signature in response to hypertension and/or albuminuria including SERPINA5, S100A8 that might serve as prognostic marker for CKD progression (Gonzalez-Calero et al. 2017). However, the diagnostic value of exosomal proteins remains to be established.

30.3.4 Urinary Peptidomics

As a "subset" of the proteome, urinary peptidomics has gained increasing research interest for its several advantages over traditional proteomics, e.g., early change, better reproducibility, and stability (Klein et al. 2016). Good et al. compared the urinary peptidomics of 230 CKD patients with that of 379 healthy participants by CE-MS and identified a panel of 273 peptides with both statistical significance and known sequence (Good et al. 2010). The classifier generated by these peptides, named CKD273, was demonstrated to have a perfect discriminative power of CKD in a validation set of 128 participants (sensitivity of 85.5%, specificity of 100%) (Good et al. 2010). Subsequent studies indicated that CKD273 also correlated with eGFR and the extent of renal fibrosis (Argiles et al. 2013; Ovrehus et al. 2015; Magalhaes et al. 2017). These peptides are degradation fragments of both in situ and circulatory proteins, of which collagen-derived peptides account for the largest proportion. Other components of CKD273 are fragments of circulatory proteins, kidney-specific proteins and secreted proteins (Good et al. 2010). It has been shown that urinary levels of collagen-derived and circulatory peptides indicate renal fibrosis and nephron damage, respectively.

In terms of prognostic power, in a large multicenter cohort of 1990 participants with a mean follow-up of 54 months, CKD273 outperformed urinary albumin in predicting rapid progression (defined as a decrease in eGFR slope >5% per year) (Schanstra et al. 2015). Another large cohort including 2672 individuals at various CKD stages also found that CKD273 were superior to urinary albumin in discriminating progressors from those with stable renal function (Pontillo et al. 2016). The utility of CKD 273 as a prognostic marker of CKD progression was further supported by an independent assessment study with a high evidence level (1b) according to the Oxford Evidence-Based Medicine guideline (Critselis and Heerspink 2016).

30.3.5 Transcriptomic Biomarkers

30.3.5.1 mRNAs

A growing number of studies, mostly cross-sectional, have suggested that urinary mRNAs provide pathological information of renal fibrosis that is not captured by conventional risk factors and may serve as potential prognostic biomarkers (Lyu et al. 2017).

The utility of urinary mRNA as biomarker was first established by Li et al. in 2001 (Li et al. 2001). Later, Szeto et al. found that the level of TGF-β mRNA in urinary sediment correlated with intrarenal level, degree of tubulointerstitial fibrosis (TIF), and eGFR in CKD patients (Szeto et al. 2005). Our previous work showed that urinary podocyte-associated mRNAs including synaptopodin and podocalyxin increased with DN progression (Zheng et al. 2011). In a small cohort, CXCL9 mRNA levels were found to be independently associated with eGFR decline (Wang et al. 2015). Recently, targeted PCR array provides a high-throughput screening platform for discovering potential mRNA biomarkers. By analyzing urinary mRNA profiles of 39 CKD patients and 11 controls with a targeted PCR array of renal fibrosis, we identified 21 mRNAs that significantly altered in CKD, among which vimentin mRNA showed strong correlation with the score of TIF and could better discriminate none-to-mild renal fibrosis from moderate-to-severe fibrosis than eGFR (Zheng et al. 2012; Cao et al. 2015). Furthermore, we also developed a random forest classifier composed of four urinary mRNAs that can detect early renal fibrosis with excellent sensitivity (Zhou et al. 2017b).

To standardize the expression of mRNAs among different samples, housekeeping genes that ubiquitously expressed in nucleated cells are measured as the denominator. This common process, however, may cause the problem that the obtained mRNA expression is actually the weighted average of all cells, including those from lower urinary tract. To deal with the dilemma, Sato et al. introduced aquaporin 2 (AQP2) as the kidney reference gene and found mRNA: AQP2 ratios might be a candidate indicator of podocyte injury (Sato et al. 2009). However, AQP2 is predominantly expressed by renal collecting tubules and may not represent the overall contributions of the kidney. Reference genes that are extensively and specifically expressed by renal cells may yield better performance. In spite of this, the adjusted ratio is still a rough estimation of the actual expression. To analyze mRNA expression based on renal intrinsic cells in urine may unclog the current bottlenecks.

Besides, EVs also serve as a repository for urinary mRNAs. Our recent study revealed that CD2AP mRNA in exosome correlated negatively with 24-h urine protein and severity of tubulointerstitial fibrosis (Lv et al. 2014). Urinary EV-derived UMOD mRNA is significantly elevated in diabetic patients who subsequently develop renal injury (Yamamoto et al. 2018). But whether exosomal mRNAs yield better performance than those from urinary sediment still requires further investigation.

30.3.5.2 Non-coding RNAs

MicroRNAs (miRNAs) are a class of non-coding RNAs with an approximate sequence length of 22 nt. A growing list of miRNAs is demonstrated to participate in the pathogenesis of renal fibrosis, including miR-21, miR-29 (Van der Hauwaert et al. 2015; Lorenzen et al. 2011). Urinary sediment serves as a common resource for miRNA measurement. One of the first studies reported urinary miR-29b and miR-29c levels correlated with proteinuria and eGFR, and urinary miR-93 levels correlated with the extent of glomerulosclerosis in patients with IgAN (Wang et al. 2012). In a small cohort of patients with CKD, urinary miR-21 and miR-216a levels were associated with renal function decline and risk of progression to ESKD (Szeto et al. 2012). Using miRNA chip assay, a recent study further identified a urinary miRNA signature including miR-25-3p, miR-144-3p and miR-486-5p which were differentially expressed in patients with IgAN (Duan et al. 2016).

In addition to deciduous cell pellets, urinary miRNAs also exist in EVs and protein-bind form. Nevertheless, free miRNAs without membrane protection are likely to be gradually degraded by elevated urinary RNases during renal fibrosis (Cheng et al. 2014). Recent studies suggest that exosomes are rich sources of urinary miRNAs (Cheng et al. 2014). Our pilot study optimized the measurement of urinary exosomal microRNAs and found miR-29c could discriminate mild from moderate to severe tubulointerstitial fibrosis effectively (Lv et al. 2013a, b). Moreover, using RNA sequencing, Khurana et al. screened urinary exosomal miRNA profiles in patients with various CKD stages and healthy controls. As a result, miRNA-181a was found to be decreased by about 200-fold in exosomes of patients with CKD. Another study identified 384 differentially expressed mRNAs in urinary exosomes between patients with stage 4–5 CKD and those with earlier stages, which were mainly related to TGF-β signaling pathway (Muralidharan et al. 2017). The original data of both studies have been uploaded to sequence read archive (SRA, https://www.ncbi.nlm.nih.gov/sra) and a reanalysis of the urinary exosomal miRNA profiles might provide additional clues on developing biomarkers for renal fibrosis.

Long non-coding RNAs (lncRNAs) and circRNAs are novel classes of non-coding RNAs attracting increasing research interest for their emerging roles in mediating renal fibrosis (Bachmayr-Heyda et al. 2015; Zhou et al. 2014; Sun et al. 2017). In an experimental MN mouse model, urinary Xist (lncRNA) showed a strong correlation with disease severity (Huang et al. 2014). CircRNAs, in particular, are abundant and stable in mammalian cells and thus are regarded as promising resource for biomarkers discovery (Jeck and Sharpless 2014). CircRNAs can be detected in human blood with high expression and reproducibility. In the plasm of patients with chronic glomerulonephritis, over 700 circRNAs including hsa_circ_0006602 and hsa_circ_0091104 were dysregulated (Wang et al. 2017b). Though there is no publication on urinary circRNAs as biomarkers of renal fibrosis yet, it is expected that related research will spring up in the near future.

30.3.6 Metabolomic Biomarkers

Metabolomics, or metabolic profiling, is the quantitative measurement of all low-molecular-weight products of cellular metabolism, which is regarded as the link between genotypes and phenotypes (Zhao 2013; Fiehn 2002). A wide range of metabolic pathways are dysregulated in the context of CKD (Zhou et al. 2018). Although initial interest lay in examining metabolome of plasma or dialysate from ESKD patients, analysis of urinary metabolomics has become an emerging arena for biomarker discovery in renal fibrosis.

In adenine-induced CKD model, 12 urinary metabolites including phytosphingosine, adrenosterone, tryptophan, and creatinine were significantly altered as revealed by MS (Zhao et al. 2012). But this may not represent the real changes in patients. In fact, another seven urinary metabolites (5-oxoproline, glutamate guanidoacetate, α-phenylacetylglutamine, taurine, citrate, and trimethylamine N-oxide) were found significantly dysregulated in patients with CKD (Posada-Ayala et al. 2014). Besides, urinary excretions of proline and citrulline significantly increase with the progression of CKD, whereas urinary excretion of asymmetric dimethylarginine (ADMA) decreases (Duranton et al. 2014). Notably, in a case-control study of 386 participants from the Framingham Offspring cohort, urinary glycine and histidine were associated with a lower risk of incident CKD (McMahon et al. 2017). Despite their small sample sizes, these pilot studies still depict the profiles of urinary metabolites during renal fibrosis progression and provide potential biomarkers.

30.3.7 Toward Combined Biomarkers

Here, we highlight the notion that a biomarker can be a single measurement, such as the urinary protein level, or a combined one computed from many different variables, such as CKD273, which is a support vector machine (SVM) classifier calculated from the levels of 273 urinary peptides (Argiles et al. 2013). Given the inherent complexity of molecular mechanisms involved in renal fibrosis, it is not surprising that combined biomarkers yield better performance than standalone molecules. From a statistics standpoint, diagnosis is a classification problem in nature. With the rapid development of statistics, artificial intelligence, and computer science, a lot of machine learning methods have been developed and are expected to have a tremendous influence on the way biomarker research is done today. Most popular machine learning algorithms include k-nearest neighbor, naive bays, decision trees, random forests, support vector machines, and neural networks. Despite the fact that machine learning methods are more powerful in classification tasks, it should be kept in mind that no algorithm can yield best performance under any situation and the classification error rate can be quite different for the same model with different parameters. Therefore, understanding the strengths of the algorithms and setting the right parameters are important for generating an accurate model. We should also notice that there is widespread optimism in studies where novel combined biomarkers are discovered, which is largely resulted from selection bias and small sample size. This challenge is particularly prominent for the imbalanced datasets generated by high-throughput technologies, which contain small sample sizes and numerous measured quantities. In this context, good separation of the classes can be achieved even for sets of classifiers chosen randomly, which is known as "over-fitting". Cross validation is an easy way to test potential over-fitting. Several algorithms such as principal components analysis, clustering, or multidimensional scaling can also help researchers to grasp several thousand parameters generated by high-throughput platforms. Other strategies include using models with strong resistance to "over-fitting" such as random forest.

Traditionally, the prognostic power of a biomarker is determined by its independent relationship with the given outcome. This equals to testing the superiority of the new combined biomarker to the one with only established risk factors (in this case, the computing model is usually the cox model). Thereby, we are indeed seeking for a better combined biomarker when we are adjusting risk factors in a multivariable model. In this setting, cox model remains the ideal model for risk prediction. Tangri et al. provided a good example by proposing and validating a cox regression model containing routine clinical data, which showed good performance of predicting progression to ESKD among CKD patients (Tangri et al. 2011). For detailed procedures of developing a predictive model, the reader is directed to relevant reviews (McGeechan et al. 2008; Moons et al. 2012a, b).

Acknowledgements This work was financially supported by the National Key Research and Development Program of China (2018YFC1314004), the National Natural Science Foundation of China (No.81720108007, 81670696, 81470922 and 31671194), the Clinic Research Center of Jiangsu Province (No. BL2014080).

References

Aflarian M, Zelnick LR, Ruzinski J, Kestenbaum B, Himmelfarb J, de Boer IH et al (2015) Urine matrix metalloproteinase-7 and risk of kidney disease progression and mortality in type 2 diabetes. J Diabetes Complicat 29:1024–1031

Aregger F, Uehlinger DE, Witowski J, Brunisholz RA, Hunziker P, Frey FJ et al (2014) Identification of IGFBP-7 by urinary proteomics as a novel prognostic marker in early acute kidney injury. Kidney Int 85:909–919

Argiles A, Siwy J, Duranton F, Gayrard N, Dakna M, Lundin U et al (2013) CKD273, a new proteomics classifier assessing CKD and its prognosis. PLoS ONE 8:e62837

Astor BC, Koettgen A, Hwang S-J, Bhavsar NA, Fox CS, Coresh J (2011) Trefoil factor 3 predicts incident chronic kidney disease: a case-control study nested within the atherosclerosis risk in communities (ARIC) study. Am J Nephrol 34:291–297

Bachmayr-Heyda A, Reiner AT, Auer K, Sukhbaatar N, Aust S, Bachleitner-Hofmann T et al (2015) Correlation of circular RNA abundance with proliferation—exemplified with colorectal and ovarian cancer, idiopathic lung fibrosis, and normal human tissues. Sci Rep 5:8057

Betz BB, Jenks SJ, Cronshaw AD, Lamont DJ, Cairns C, Manning JR et al (2016) Urinary peptidomics in a rodent model of diabetic nephropathy highlights epidermal growth factor as a biomarker for renal deterioration in patients with type 2 diabetes. Kidney Int 89:1125–1135

Bhavsar NA, Koettgen A, Coresh J, Astor BC (2012) Neutrophil gelatinase-associated lipocalin (NGAL) and kidney injury molecule 1 (KIM-1) as predictors of incident CKD Stage 3: the atherosclerosis risk in communities (ARIC) Study. Am J Kidney Dis 60:233–240

Burns KD, Hiremath S (2012) Urinary angiotensinogen as a biomarker of chronic kidney disease: ready for prime time? Nephrol Dial Transplant 27:3010–3013

Cao YH, Lv LL, Zhang X, Hu H, Ding LH, Yin D et al (2015) Urinary vimentin mRNA as a potential novel biomarker of renal fibrosis. Am J Physiol Renal Physiol 309:F514–F522

Cheng L, Sun X, Scicluna BJ, Coleman BM, Hill AF (2014) Characterization and deep sequencing analysis of exosomal and non-exosomal miRNA in human urine. Kidney Int 86:433–444

Craciun FL, Bijol V, Ajay AK, Rao P, Kumar RK, Hutchinson J et al (2015) RNA sequencing identifies novel translational biomarkers of kidney fibrosis. J Am Soc Nephrol 27:1702–1713
Critselis E, Heerspink HL (2016) Utility of the CKD273 peptide classifier in predicting chronic kidney disease progression. Nephrol Dial Transplant 31:249–254
Dear JW, Street JM, Bailey MA (2013) Urinary exosomes: a reservoir for biomarker discovery and potential mediators of intrarenal signalling. Proteomics 13:1572–1580
Duan ZY, Cai GY, Bu R, Lu Y, Hou K, Chen XM (2016) Selection of urinary sediment miRNAs as specific biomarkers of IgA nephropathy. Sci Rep 6:23498
Duranton F, Lundin U, Gayrard N, Mischak H, Aparicio M, Mourad G et al (2014) Plasma and urinary amino acid metabolomic profiling in patients with different levels of kidney function. Clin J Am Soc Nephrol 9:37–45
Erdbrugger U, Le TH (2016) Extracellular vesicles in renal diseases: more than novel biomarkers? J Am Soc Nephrol 27:12–26
Fiehn O (2002) Metabolomics- the link between genotypes and phenotypes. Plant Mol Biol 48:155–171
Gonzalez-Calero L, Martinez PJ, Martin-Lorenzo M, Baldan-Martin M, Ruiz-Hurtado G, de la Cuesta F et al (2017) Urinary exosomes reveal protein signatures in hypertensive patients with albuminuria. Oncotarget 8:44217–44231
Good DM, Zuerbig P, Argiles A, Bauer HW, Behrens G, Coon JJ et al (2010) Naturally occurring human urinary peptides for use in diagnosis of chronic kidney disease. Mol Cell Proteomics 9:2424–2437
He W, Tan RJ, Li Y, Wang D, Nie J, Hou FF et al (2012) Matrix metalloproteinase-7 as a surrogate marker predicts renal Wnt/beta-Catenin activity in CKD. J Am Soc Nephrol 23:294–304
Hsu CY, Xie DW, Waikar SS, Bonventre JV, Zhang XM, Sabbisetti V et al (2017) Urine biomarkers of tubular injury do not improve on the clinical model predicting chronic kidney disease progression. Kidney Int 91:196–203
Huang YS, Hsieh HY, Shih HM, Sytwu HK, Wu CC (2014) Urinary Xist is a potential biomarker for membranous nephropathy. Biochem Biophys Res Commun 452:415–421
Hwang JH, Lee JP, Kim CT, Yang SH, Kim JH, An JN et al (2016) Urinary periostin excretion predicts renal outcome in IgA nephropathy. Am J Nephrol 44:481–492
Jeck WR, Sharpless NE (2014) Detecting and characterizing circular RNAs. Nat Biotechnol 32:453–461
Kim KM, Abdelmohsen K, Mustapic M, Kapogiannis D, Gorospe M (2017) RNA in extracellular vesicles. Wiley Interdisciplinary Reviews-Rna 8
Klein J, Bascands J-L, Mischak H, Schanstra JP (2016) The role of urinary peptidomics in kidney disease research. Kidney Int 89:539–545
Latterich M, Schnitzer JE (2011) Streamlining biomarker discovery. Nat Biotechnol 29:600–602
Lebherz-Eichinger D, Klaus DA, Reiter T, Hoerl WH, Haas M, Ankersmit HJ et al (2014) Increased chemokine excretion in patients suffering from chronic kidney disease. Transl Res 164:433–443
Lebherz-Eichinger D, Tudor B, Ankersmit HJ, Reiter T, Haas M, Roth-Walter F et al (2015) Trefoil Factor 1 excretion is increased in early stages of chronic kidney disease. PLoS ONE 10:e0138312
Li B, Hartono C, Ding R, Sharma VK, Ramaswamy R, Qian B et al (2001) Noninvasive diagnosis of renal-allograft rejection by measurement of messenger RNA for perforin and granzyme B in urine. N Engl J Med 344:947–954
Liu BC, Tang TT, Lv LL, Lan HY (2018) Renal tubule injury: a driving force toward chronic kidney disease. Kidney Int 93:568–579
Lorenzen JM, Haller H, Thum T (2011) MicroRNAs as mediators and therapeutic targets in chronic kidney disease. Nat Rev Nephrol 7:286–294
Lv LL, Cao YH, Ni HF, Xu M, Liu D, Liu H et al (2013a) MicroRNA-29c in urinary exosome/microvesicle as a biomarker of renal fibrosis. Am J Physiol-Ren Physiol 305:F1220–F1227
Lv LL, Cao Y, Liu D, Xu M, Liu H, Tang RN et al (2013b) Isolation and quantification of MicroRNAs from urinary exosomes/microvesicles for biomarker discovery. Int J Biol Sci 9:1021–1031

Lv LL, Cao YH, Pan MM, Liu H, Tang RN, Ma KL et al (2014) CD2AP mRNA in urinary exosome as biomarker of kidney disease. Clin Chim Acta 428:26–31

Lyu LL, Feng Y, Liu BC (2017) Urinary biomarkers for chronic kidney disease with a focus on gene transcript. Chin Med J 130:2251–2256

Magalhaes P, Pejchinovski M, Markoska K, Banasik M, Klinger M, Svec-Billa D et al (2017) Association of kidney fibrosis with urinary peptides: a path towards non-invasive liquid biopsies? Sci Rep 7:16915

Mateescu B, Kowal EJK, van Balkom BWM, Bartel S, Bhattacharyya SN, Buzas EI et al (2017) Obstacles and opportunities in the functional analysis of extracellular vesicle RNA- an ISEV position paper. J Extracell Vesicles 6:1286095

McGeechan K, Macaskill P, Irwig L, Liew G, Wong TY (2008) Assessing new biomarkers and predictive models for use in clinical practice a clinician's guide. Arch Intern Med 168:2304–2310

McMahon GM, Hwang SJ, Clish CB, Tin A, Yang Q, Larson MG et al (2017) Urinary metabolites along with common and rare genetic variations are associated with incident chronic kidney disease. Kidney Int 91:1426–1435

Mills KT, Kobori H, Hamm LL, Alper AB, Khan IE, Rahman M et al (2012) Increased urinary excretion of angiotensinogen is associated with risk of chronic kidney disease. Nephrol Dial Transplant 27:3176–3181

Mischak H, Delles C, Vlahou A, Vanholder R (2015) Proteomic biomarkers in kidney disease: issues in development and implementation. Nat Rev Nephrol 11:221–232

Moons KGM, Kengne AP, Grobbee DE, Royston P, Vergouwe Y, Altman DG et al (2012a) Risk prediction models: II. External validation, model updating, and impact assessment. Heart 98:691–698

Moons KGM, Kengne AP, Woodward M, Royston P, Vergouwe Y, Altman DG et al (2012b) Risk prediction models: I. Development, internal validation, and assessing the incremental value of a new (bio)marker. Heart 98:683–690

Mou S, Wang Q, Li J, Shi B, Ni Z (2012) Urinary excretion of liver-type fatty acid-binding protein as a marker of progressive kidney function deterioration in patients with chronic glomerulonephritis. Clin Chim Acta 413:187–191

Muralidharan J, Ramezani A, Hubal M, Knoblach S, Shrivastav S, Karandish S et al (2017) Extracellular microRNA signature in chronic kidney disease. Am J Physiol-Ren Physiol 312:F982–F991

Nadkarni GN, Rao V, Ismail-Beigi F, Fonseca VA, Shah SV, Simonson MS et al (2016) Association of urinary biomarkers of inflammation, injury, and fibrosis with renal function decline: the ACCORD trial. Clin J Am Soc Nephrol 11:1343–1352

Nguyen TQ, Tarnow L, Andersen S, Hovind P, Parving HH, Goldschmeding R et al (2006) Urinary connective tissue growth factor excretion correlates with clinical markers of renal disease in a large population of type 1 diabetic patients with diabetic nephropathy. Diabetes Care 29:83–88

O'Seaghdha CM, Hwang SJ, Bhavsar NA, Koettgen A, Coresh J, Astor BC et al (2011) Lower urinary connective tissue growth factor levels and incident CKD Stage 3 in the general population. Am J Kidney Dis 57:841–849

O'Seaghdha CM, Hwang SJ, Larson MG, Meigs JB, Vasan RS, Fox CS (2013) Analysis of a urinary biomarker panel for incident kidney disease and clinical outcomes. J Am Soc Nephrol 24:1880–1888

Ovrehus MA, Zuerbig P, Vikse BE, Hallan SI (2015) Urinary proteomics in chronic kidney disease: diagnosis and risk of progression beyond albuminuria. Clin Proteomics 12:21

Papadopoulos T, Belliere J, Bascands JL, Neau E, Klein J, Schanstra JP (2015) miRNAs in urine: a mirror image of kidney disease? Expert Rev Mol Diagn 15:361–374

Papadopoulos T, Krochmal M, Cisek K, Fernandes M, Husi H, Stevens R et al (2016) Omics databases on kidney disease: where they can be found and how to benefit from them. Clin Kidney J 9:343–352

Peralta CA, Katz R, Bonventre JV, Sabbisetti V, Siscovick D, Sarnak M et al (2012) Associations of urinary levels of kidney injury molecule 1 (KIM-1) and neutrophil gelatinase-associated lipocalin (NGAL) with kidney function decline in the multi-ethnic study of atherosclerosis (MESA). Am J Kidney Dis 60:904–911

Pisitkun T, Shen RF, Knepper MA (2004) Identification and proteomic profiling of exosomes in human urine. Proc Natl Acad Sci U S A 101:13368–13373

Pontillo C, Jacobs L, Staessen JA, Schanstra JP, Rossing P, Heerspink HJ et al (2016) A urinary proteome-based classifier for the early detection of decline in glomerular filtration. Nephrol Dial Transpl 32:1510–1516

Posada-Ayala M, Zubiri I, Martin-Lorenzo M, Sanz-Maroto A, Molero D, Gonzalez-Calero L et al (2014) Identification of a urine metabolomic signature in patients with advanced-stage chronic kidney disease. Kidney Int 85:103–111

Prakoura N, Chatziantoniou C (2017) Periostin in kidney diseases. Cell Mol Life Sci 74:4315–4320

Puzantian HV, Townsend RR (2013) Understanding kidney function assessment: the basics and advances. J Am Assoc Nurse Pract 25:334–341

Rifai N, Gillette MA, Carr SA (2006) Protein biomarker discovery and validation: the long and uncertain path to clinical utility. Nat Biotechnol 24:971–983

Riser BL, Cortes P, Denichilo M, Deshmukh PV, Chahal PS, Mohammed AK et al (2003) Urinary CCN2 (CTGF) as a possible predictor of diabetic nephropathy: preliminary report. Kidney Int 64:451–458

Satirapoj B, Wang Y, Chamberlin MP, Dai T, LaPage J, Phillips L et al (2012) Periostin: novel tissue and urinary biomarker of progressive renal injury induces a coordinated mesenchymal phenotype in tubular cells. Nephrol Dial Transpl 27:2702–2711

Satirapoj B, Tassanasorn S, Charoenpitakchai M, Supasyndh O (2015) Periostin as a tissue and urinary biomarker of renal injury in type 2 diabetes mellitus. PLoS ONE 10:e0124055

Sato Y, Wharram BL, Lee SK, Wickman L, Goyal M, Venkatareddy M et al (2009) Urine podocyte mRNAs mark progression of renal disease. J Am Soc Nephrol 20:1041–1052

Sawaguchi M, Araki S-i, Kobori H, Urushihara M, Haneda M, Koya D et al (2012) Association between urinary angiotensinogen levels and renal and cardiovascular prognoses in patients with type 2 diabetes mellitus. J Diabetes Investig 3:318–324

Schanstra JP, Zurbig P, Alkhalaf A, Argiles A, Bakker SJ, Beige J et al (2015) Diagnosis and prediction of CKD progression by assessment of urinary peptides. J Am Soc Nephrol 26:1999–2010

Shoukry A, Bdeer SE-A, El-Sokkary RH (2015) Urinary monocyte chemoattractant protein-1 and vitamin D-binding protein as biomarkers for early detection of diabetic nephropathy in type 2 diabetes mellitus. Mol Cell Biochem 408:25–35

Sun JZ, Zhang S, Shi BH, Zheng DX, Shi J (2017) Transcriptome identified lncRNAs associated with renal fibrosisin UUO rat model. Front Physiol 8:658

Szeto CC, Chan RW, Lai KB, Szeto CY, Chow KM, Li PK et al (2005) Messenger RNA expression of target genes in the urinary sediment of patients with chronic kidney diseases. Nephrol Dial Transplant 20:105–113

Szeto CC, Ching-Ha KB, Ka-Bik L, Mac-Moune LF, Cheung-Lung CP, Gang W et al (2012) MicroRNA expression in the urinary sediment of patients with chronic kidney diseases. Dis Markers 33:137–144

Tan HL, Yap JQ, Qian Q (2016) Acute kidney injury: tubular markers and risk for chronic kidney disease and end-stage kidney failure. Blood Purif 41:144–150

Tangri N, Stevens LA, Griffith J, Tighiouart H, Djurdjev O, Naimark D et al (2011) A predictive model for progression of chronic kidney disease to kidney failure. Jama-J Am Med Assoc 305:1553–1559

Titan SM, Vieira JM Jr, Dominguez WV, Moreira SRS, Pereira AB, Barros RT et al (2012) Urinary MCP-1 and RBP: independent predictors of renal outcome in macroalbuminuric diabetic nephropathy. J Diabetes Complicat 26:546–553

Van der Hauwaert C, Savary G, Hennino MF, Pottier N, Glowacki F, Cauffiez C (2015) MicroRNAs in kidney fibrosis. Nephrol Ther 11:474–482

Verhave JC, Bouchard J, Goupil R, Pichette V, Brachemi S, Madore F et al (2013) Clinical value of inflammatory urinary biomarkers in overt diabetic nephropathy: a prospective study. Diabetes Res Clin Pract 101:333–340

Wang G, Kwan BCH, Lai FMM, Chow KM, Li PKT, Szeto CC (2012) Urinary miR-21, miR-29, and miR-93: novel biomarkers of fibrosis. Am J Nephrol 36:412–418
Wang G, Lai FMM, Chow KM, Kwan BCH, Pang WF, Luk CCW et al (2015) Urinary mRNA levels of ELR-negative CXC chemokine ligand and extracellular matrix in diabetic nephropathy. Diabetes-Metablsm Res Rev 31:699–706
Wang L, Wang Y, Chang Q (2016) Feature selection methods for big data bioinformatics: a survey from the search perspective. Methods 111:21–31
Wang W, Sue ACH, Goh WWB (2017a) Feature selection in clinical proteomics: with great power comes great reproducibility. Drug Discovery Today 22(6):912–918
Wang X, Dai Y, Zhang WF, Sun DL, Zhang XZ (2017b) Microarray based circRNA expression profiles in uremic plasma and PBMCs due to chronic glomerulonephritis. Arch Biol Sci 69:523–534
Yamamoto T, Nakagawa T, Suzuki H, Ohashi N, Fukasawa H, Fujigaki Y et al (2007) Urinary angiotensinogen as a marker of intrarenal angiotensin II activity associated with deterioration of renal function in patients with chronic kidney disease. J Am Soc Nephrol 18:1558–1565
Yamamoto CM, Murakami T, Oakes ML, Mitsuhashi M, Kelly C, Henry RR et al (2018) Uromodulin mRNA from urinary extracellular vesicles correlate to kidney function decline in Type 2 diabetes mellitus. Am J Nephrol 47:283–291
Zhao YY (2013) Metabolomics in chronic kidney disease. Clin Chim Acta 422:59–69
Zhao YY, Liu J, Cheng XL, Bai X, Lin RC (2012) Urinary metabonomics study on biochemical changes in an experimental model of chronic renal failure by adenine based on UPLC Q-TOF/MS. Clin Chim Acta 413:642–649
Zhao Y, Zhu L, Zhou T, Zhang Q, Shi S, Liu L et al (2015) Urinary CXCL1: a novel predictor of IgA nephropathy progression. PLoS ONE 10:e0119033
Zheng M, Lv LL, Ni J, Ni HF, Li Q, Ma KL et al (2011) Urinary podocyte-associated mRNA profile in various stages of diabetic nephropathy. PLoS ONE 6:e20431
Zheng M, Lv LL, Cao YH, Liu H, Ni J, Dai HY et al (2012) A pilot trial assessing urinary gene expression profiling with an mRNA array for diabetic nephropathy. PLoS ONE 7:e34824
Zhou H, Pisitkun T, Aponte A, Yuen PST, Hoffert JD, Yasuda H et al (2006) Exosomal fetuin-a identified by proteomics: a novel urinary biomarker for detecting acute kidney injury. Kidney Int 70:1847–1857
Zhou Q, Chung ACK, Huang XR, Dong Y, Yu X, Lan HY (2014) Identification of novel long noncoding RNAs associated with TGF-β/Smad3-mediated renal inflammation and fibrosis by RNA sequencing. Am J Pathol 184:409–417
Zhou LT, Lv LL, Pan MM, Cao YH, Liu H, Feng Y et al (2016) Are urinary tubular injury markers useful in chronic kidney disease? a systematic review and meta analysis. PLoS ONE 11:e0167334
Zhou D, Tian Y, Sun L, Zhou LL, Xiao LX, Tan RJ et al (2017a) Matrix metalloproteinase-7 is a urinary biomarker and pathogenic mediator of kidney fibrosis. J Am Soc Nephrol 28:598–611
Zhou LT, Cao YH, Lv LL, Ma KL, Chen PS, Ni HF et al (2017b) Feature selection and classification of urinary mRNA microarray data by iterative random forest to diagnose renal fibrosis: a two-stage study. Sci Rep 7:39832
Zhou LT, Qiu S, Lv LL, Li ZL, Liu H, Tang RN et al (2018) Integrative bioinformatics analysis provides insight into the molecular mechanisms of chronic kidney disease. Kidney Blood Press Res 43:568–581

Chapter 31
New Therapies for the Treatment of Renal Fibrosis

Feng Liu and Shougang Zhuang

Abstract Renal fibrosis is the common pathway for progression of chronic kidney disease (CKD) to end stage of renal disease. It is now widely accepted that the degree of renal fibrosis correlates with kidney function and CKD stages. The key cellular basis of renal fibrosis includes activation of myofibroblasts, excessive production of extracellular matrix components, and infiltration of inflammatory cells. Many cellular mechanisms responsible for renal fibrosis have been identified, and some antifibrotic agents show a greater promise in slowing down and even reversing fibrosis in animal models; however, translating basic findings into effective antifibrotic therapies in human has been limited. In this chapter, we will discuss the effects and mechanisms of some novel antifibrotic agents in both preclinical studies and clinical trials.

Keywords Renal fibrosis · Mechanism · Anti-fibrosis treatment · Clinical trial

31.1 Introduction

Renal fibrogenesis manifests itself in the kidney both after acute kidney injury and in the progression of chronic kidney disease (CKD). Interstitial fibrosis accompanied by glomerulosclerosis, tubular atrophy and dilation, and vascular hyalinosis represents the final common pathway as CKD progresses to ESRD. It is now widely accepted that the degree of renal fibrosis correlates well with kidney function and CKD stages in native kidneys as well as in kidney allografts (Francois and Chatziantoniou 2018). It is standard teaching that fibrosis in the kidney is permanent and does not regress. The key cellular basis of renal fibrosis is the activation of myofibroblasts, resulting in the excessive and continual production of extracellular matrix (ECM) components, and

F. Liu · S. Zhuang (✉)
Department of Nephrology, Shanghai East Hospital, Tongji University School of Medicine, Shanghai, China
e-mail: shougang_zhuang@brown.edu

S. Zhuang
Department of Medicine, Rhode Island Hospital, Alpert Medical School of Brown University, Providence, RI, USA

B.-C. Liu et al. (eds.), *Renal Fibrosis: Mechanisms and Therapies*,
Advances in Experimental Medicine and Biology 1165,
https://doi.org/10.1007/978-981-13-8871-2_31

inflammatory cell infiltration (McVicker and Bennett 2017). Many cellular mechanisms responsible for renal fibrosis have been identified, and some antifibrotic agents show a greater promise in slowing down and even reversing fibrosis, but translating that knowledge into effective antifibrotic therapies in humans has been limited due to lack of sufficient effects, off-target effects, or both (Klinkhammer et al. 2017). In this chapter, we will discuss mechanisms and antifibrotic effects of those agents in both preclinical studies and clinical trials.

31.2 Potential Therapeutic Targets and Specific Antifibrotic Agents Tested in Clinical Trials

31.2.1 TGF-β and Its Inhibitors

TGF-β is a profibrotic mediator in all organs (Humphreys 2017). It stimulates tissue fibrosis mostly by inducing activation and proliferation of fibroblasts and subsequent overproduction of ECM proteins. Inhibition of TGF-β is highly effective in stopping the progression of experimental CKD in mice (Humphreys 2017). Because of the importance of TGF-β1 signaling in mediating renal fibrogenesis, numerous approaches have been designed and tested to block TGF-β signaling in preclinical models (Meng et al. 2015). Such approaches include genetic disruption, blocking antibodies, antisense oligodeoxynucleotides, soluble receptor inhibitors, and inhibitors of its kinases (Meng et al. 2015).

Several TGF-β-neutralizing antibodies and small-molecule inhibitors have been developed for clinical trials (Klinkhammer et al. 2017). Among them, metelimumab can neutralize the bioactivity of both TGF-β1 and TGF-β2, and fresolimumab can neutralize the biological activity of TGF-β1 (Klinkhammer et al. 2017). In phase II clinical trial of Fresolimumab in patients with focal segmental glomerulosclerosis (FSGS), proteinuria was not significantly reduced although estimated glomerular filtration rate (eGFR) was stable and not worse compared to the placebo group (Vincenti et al. 2017).

The effectiveness of LY2382770, an antibody that neutralizes the bioactivity of TGF-β1, was assessed in phase II clinical trial in patients with type 1 or type 2 diabetes mellitus (Klinkhammer et al. 2017). In this study, patients received LY2382770, subcutaneously (2, 10 or 50 mg monthly) for 12 months with the primary outcome being the change in serum creatinine from baseline to 12 months (Klinkhammer et al. 2017). The study was terminated prematurely in 2015, and no outcomes have been published yet.

Pirfenidone, a small synthetic molecule, was shown to reduce renal fibrosis in preclinical models. It is thought to act by blocking the TGF-β promotor (Cho and Kopp 2010). Although effective in animal models (RamachandraRao et al. 2009; Sharma et al. 2011), pirfenidone had mixed results in humans CKD. In Phase II trials in diabetic nephropathy (DN), eGFR increased only in patients treated with the lowest pirfenidone dose, but decreased at a similar rate in the group treated at a higher dose and in the placebo group (Sharma et al. 2011). In a phase II study in patients with FSGS, treatment with pirfenidone had no effect on blood pressure or proteinuria, but did demonstrate an effect in slowing eGFR decline by 25% (Cho et al. 2007). An ongoing phase III trial in diabetic patients has shown a slight improvement in eGFR (Klinkhammer et al. 2017).

The major barrier of the clinical application of TGF-β inhibitors is its dual role in cancer. TGF-β inhibition might lead to tumorigenesis, although it had a significant effect on blunting progression of advanced-stage tumors and metastases (Garber 2009). Biogen Idec (USA) is developing STX-100, a humanized monoclonal antibody that selectively targets integrin αvβ6, as a potential antifibrotic drug. STX-100 binding to integrin αvβ6 prevents integrin αvβ6 from binding and activating latent or inactive TGF-β complexes (Tampe and Zeisberg 2014b). In 2009, a phase II study of STX-100, in 48 renal transplant patients with interstitial fibrosis and tubular atrophy, was conducted to test the safety of subcutaneously administered STX-100 (Lo et al. 2013). The study was halted, however, before enrollment for reasons that were not announced. Additional strategies to inhibit TGF-β activity include the administration of antisense oligonucleotides like trabedersen by Isarna Therapeutics, Germany. But their ability to inhibit fibrogenesis has not been tested clinically (Tampe and Zeisberg 2014b).

31.2.2 CTGF and Its Inhibitors

Connective tissue growth factor (CTGF) is a profibrotic factor acting downstream of TGF-β (Kok et al. 2014). It is secreted by fibroblasts activated by TGF β and stimulates fibroblast growth with ECM production, but unlike TGF-β, it is not involved in anti-inflammatory effects (Kok et al. 2014). CTGF plasma levels seem to be an independent predictor of overall mortality in patients with DN and albuminuria (Tampe and Zeisberg 2014b). Inhibition of CTGF with antisense oligonucleotides, small interfering RNA (siRNA) or neutralizing antibodies can prevent expansion of ECM in experimental models of kidney disease (Kok et al. 2014). In one clinical study, treatment with the anti-CTGF monoclonal antibody FG 3019 (Fibrogen; USA) significantly decreased albuminuria in patients with diabetic kidney disease and was well tolerated (Adler et al. 2010).

31.2.3 BMP-7 and Its Agonists

BMP-7 is a member of the TGF-β superfamily, but it directly counteracts TGF-β1 signaling (Zeisberg et al. 2003). BMP-7 is a natural antagonist for TGF-β and has a potent renoprotective function, including reversing renal fibrosis in various animal models (Morrissey et al. 2002; Zeisberg et al. 2003; Manson et al. 2011).

Administration of exogenous BMP-7 not only ameliorates fibrosis, but also promotes renal regeneration (Zeisberg et al. 2003). Translational efforts to introduce BMP-7 to clinical use as antifibrotic therapy stalled due to the technical challenges in manufacturing bioactive BMP-7 in large enough quantities (Swencki-Underwood et al. 2008). A small-molecule AA123 that mimics BMP-7 activity through activating ALK3 signaling (Sugimoto et al. 2012) is sufficient to recapitulate the antifibrotic activity of recombinant BMP 7 in murine models of renal fibrosis (Sugimoto et al. 2012).

31.2.4 Galectin-3 (Gal-3) and Its Inhibitors

Galectin-3 (Gal-3), a 35kD-lectin ubiquitously expressed, including in the kidney, forms oligodimers upon ligand binding. These dimers consist of cross-linked oligosaccharides in the interstitial space that serve to strengthen the ECM during fibrogenesis (Drechsler et al. 2015). Gal-3 is also expressed in the distal tubules and the intercalated cells of the collecting ducts (Henderson et al. 2008; Desmedt et al. 2016). Its expression is upregulated in renal fibrosis; its serum concentration correlates with the decline in kidney function (Drechsler et al. 2015). In diseased kidneys, Gal-3 is produced by immune cells, tubular cells, endothelium, and myofibroblasts, but the galectin secreted by infiltrating macrophages appears to mediate myofibroblast activation (Desmedt et al. 2016). In experimental nephropathies, increases in Gal-3 concentration, as well as its absolute levels, correlated with interstitial fibrosis (Henderson et al. 2008). Many studies have demonstrated that either its genetic disruption or pharmacological inhibition prevents the development of fibrosis in experimental models of injury. Those models include ischemia-reperfusion injury (Fernandes Bertocchi et al. 2008), the lipid-induced kidney injury (Martinez-Martinez et al. 2016), the unilateral ureteral obstruction (UUO) (Henderson et al. 2008), and hypertensive nephropathy (Frenay et al. 2015). Depending on the model and the stage of the disease, other studies have shown that Gal-3 may prevent renal fibrosis (Tsuchiyama et al. 2000; Okamura et al. 2011). Thus, the role of Gal-3 in renal fibrosis remains controversial, as recently reviewed (Saccon et al. 2017).

Contradictory results not withstanding, clinical trials of Gal-3 inhibitors have been conducted. A phase 1 study investigated the safety of GCS-100, a modified citrus pectin and galectin-3 antagonist, in CKD patients. Inclusion criteria included a detectable level of Gal-3 in plasma before study entry (Klinkhammer et al. 2017). Although Gal-3 was not assessed in the kidneys, this study remains the only example of an effort to select patients based on the target expression. GCS-100 was administered once weekly by intravenous injection. The subsequent phase 2a study of weekly doses of GCS-100 in patients with CKD (primary outcome: eGFR) was completed in 2015 (Klinkhammer et al. 2017). Three phase 2 trials of GCS-100 in CKD sponsored by the same company started between 2014 and 2015 resulted in the promising outcomes (Klinkhammer et al. 2017). But one of these studies, aimed at determining the safety and tolerability of extended dosing with a fixed dose of 3 mg GCS-100 intravenously, was withdrawn before enrollment began (Klinkhammer et al. 2017). Gal-3 pharmacological inhibition with modified citrus pectin was studied in phase II clinical trials in DN (Francois and Chatziantoniou 2018). The results of these trials remain unpublished, and no phase III trials have been registered so far.

31.2.5 CC Motif Chemokine 2 (CCL2) and Its Inhibitors

Chemokines play an important role in the initiation of the inflammatory response and recruitment of fibrocytes (Charo and Ransohoff 2006). Despite the presence of 47 chemokines and 20 chemokine receptors, the CC motif chemokine 2 (CCL2)-CC chemokine receptor type 2 (CCR2) ligand-receptor axis is considered a therapeutic target for renal fibrosis (Anders et al. 2003). In the diseased kidney, CCL2 is primarily released by tubular epithelial cells, prompting an influx of CCR2-positive monocytes, T cells and fibrocytes (Reich et al. 2013). In preclinical studies using transgenic human CCR2 knock-in mice, treatment with CCX140 B, a CCR2 antagonist, ameliorated glomerular hypertrophy, podocyte loss and loss of renal function as well as improved blood glucose levels (Sullivan et al. 2013). However, the blood glucose reduction was also observed in a clinical trial with this inhibitor in DN (Tampe and Zeisberg 2014b). This may be related to the inhibitory effect of CCX140 B on monocyte infiltration in fat and the pancreas.

31.2.6 ACE-I/ARB

Angiotensin-converting enzyme inhibitors (ACE-I) or angiotensin receptor blockers (ARBs) have been the mainstays of therapy for CKD for more than 20 years (Breyer and Susztak 2016). The initial rationale for testing ACE-I/ARB therapy was based on preclinical and clinical studies that showed a reduction in intraglomerular hemodynamic pressure accompanied with resulting in a reduction of proteinuria (Hostetter et al. 2001). They remain the backbone of treatment in CKD patients although, at best, they can only slow the development of renal fibrosis. Notably, the therapeutic benefit of ACE-Is and ARBs is closely associated with an early reduction in eGFR coupled with a reduction in albuminuria (Brenner et al. 2001; Lewis et al. 2001; Parving et al. 2001), which is consistent with the capacity of these agents to reduce intraglomerular filtration pressure. While ACE-Is and ARBs slow the progression of renal disease, they do not halt this progress (Breyer and Susztak 2016). Nevertheless, ACE-Is and ARBs are now considered the standard of care for chronic proteinuric kidney disease; therefore, any novel therapy must prove added benefit on the background of ACE-I or ARB therapy.

Several trials investigating the combination of an ACE-I and an ARB have been conducted (Marquez et al. 2015). The ONTARGET trial compared the benefit of the ACE-I Ramipril, the ARB telmisartan and their combination in 25,920 patients with vascular disease and/or high-risk diabetes (Yusuf et al. 2008). Similarly, in patients with type 2 diabetes mellitus, the VA NEPHRON-D trial was conducted to identify whether renal outcomes were further improved when the combination of an ACE-I and an ARB (lisinopril and losartan, respectively) (Fried et al. 2013). In the year following randomization, a further decline in albuminuria was observed, which was significantly greater in the combination therapy group compared with patients receiving monotherapy (Fried et al. 2013). Although the combination of ACEI–ARB had a trend toward reducing renal failure events, there was no statistically significant difference. However, when compared to monotherapy, the combination may be related to doubling of the risk for hyperkaliemia and acute kidney injury (Fried et al. 2013) Therefore, the combination of ACE-I-ARB is not widely recommended in clinical practice although this combination might provide some benefit for selected patients with residual proteinuria by fastidious management of serum potassium (Yusuf et al. 2008).

Since renin-mediated cleavage of angiotensinogen to angiotensin I is the first step in the renin–angiotensin–aldosterone cascade, and its inhibition may be an alternative approach to RAS blockade. As a renin inhibitor, the effect of Aliskiren on renal outcomes was identified in patients with DN (the ALTITUDE trial) (Parving et al. 2012). Aliskiren administration resulted in greater albuminuria reduction than placebo, not along with improvement of renal function, but with the risk of hyperkalemia and stroke increasement (Parving et al. 2012). The ALTITUDE trial was prematurely terminated due to increasing adverse events and no benefit on renal function decline. In the conclusion, these results suggest the increasing considerations regarding the safety of complete RAS inhibition as to treat progressive renal disease.

31.2.7 Anti-inflammation Therapies

In most cases, fibrosis is associated with a robust mononuclear infiltration. However, in the absence of infectious agents or specific immunogens, sterile inflammation is considered to have a prominent role in the initiation of fibrotic responses (Kurts et al. 2013). This hypothesis is based on the observation that the inhibition of inflammation protects against fibrosis in experimental models of CKD (Anders et al. 2004). Macrophages, B cells, and T cells are considered the primary sources of profibrotic growth factors (Huaux et al. 2003; Ricardo et al. 2008). Although the immune-modulatory drugs targeting tumor necrosis factor $-\alpha$(TNF-α), IL 6 or $CD20^+$ B cells are available in the clinic, they have not been used to prevent fibrosis yet.

Membrane-associated TNFR1 and TNFR2 have important roles in activating chemokine and cytokine release (Breyer and Susztak 2016). Levels of both circulating TNFRs are increased and predictive of progression (Al-Lamki et al. 2005). TNFR1 and TNFR2 have been associated with different functions. A renal protective role for TNFR1 has also been postulated in studies showing that deletion of this receptor is associated with higher systolic pressure and urinary albumin excretion, as well as an altered GFR (Al-Lamki and Mayadas 2015). Currently, it remains uncertain whether

TNF or its receptors exerts a pathogenic role in CKD or diabetic nephropathy. There are only few studies to test the clinical efficacy of anti-TNF agents in lupus nephritis (LN) and FSGS (Al-Lamki et al. 2005; Al-Lamki and Mayadas 2015).

31.2.8 Endothelin-1 (ET-1) Blockers

Endothelins are a family of three peptides known to have powerful vasoconstrictor and vasopressor properties (Francois and Chatziantoniou 2018). One of these peptides, endothelin-1 (ET-1), is the only one shown to be expressed in the kidney (Dhaun et al. 2006). Systemic overexpression of ET-1 leads to renal inflammation and fibrosis (Dhaun et al. 2006). Biological activity of ET-1 is mediated via two receptors (endothelin-1 receptor [ETA] and endothelin B receptor [ETB]), both of which display ubiquitous expression in the kidney (Saito et al. 1990; Kuc and Davenport 2004). ET-1 promotes renal fibrosis mainly through its binding to ETR-A. Activation of ETR-A induces renal vasoconstriction and therefore increases glomerular pressure and renal ischemia, then in turn induces renal fibrosis (Dhaun et al. 2006). ETR-A can also directly promote ECM protein synthesis through increased inflammation, reactive oxygen species, and activation of the renin–angiotensin system (Dhaun et al. 2006; Barton 2008). In preclinical studies, treatment with a selective ETA antagonist reduced proteinuria and glomerulosclerosis (Benigni et al. 1993; Opocensky et al. 2006), as well as attenuated vascular fibrosis and collagen deposition in hypertensive mice (Boffa et al. 2001). Similar results were observed in various models of glomerulonephritis (Fukuda et al. 1996; Gomez-Garre et al. 1996; Benigni et al. 1998).

A clinical study in 286 patients with hypertension and CKD with or without diabetes mellitus (type 1 or type 2) showed that ETA antagonists could substantially reduce blood pressure and proteinuria (Wenzel et al. 2009; Saleh et al. 2011). However, a randomized controlled trial in patients type 2 diabetes mellitus was terminated prematurely (after 4 months), due to a high incidence of serious adverse effects, including pulmonary edema, congestive heart failure, and more death events in the experimental group than in the placebo group (Mann et al. 2010). A large phase III trial that sought to investigate the role of atrasentan in diabetic nephropathy was also stopped prematurely because of serious adverse cardiovascular events in the experimental arms (Egido et al. 2017). Thus, it seems that ETA antagonists cannot be safely used in patients with renal disease.

31.2.9 Aldosterone Blockers

Aldosterone is a mineralocorticoid hormone-promoting sodium retention and renal fibrosis by inducing renal vasoconstriction, oxidative stress, and inflammation (Francois and Chatziantoniou 2018). Aldosterone or mineralocorticoid (MR) blockers are still mainly used as diuretics in the treatment of hypertension (Francois and Chatziantoniou 2018). In experimental models of CKD, MR blockers have shown a significant decrease in renal fibrosis and proteinuria in subtotal nephrectomy, diabetic nephropathy, and various glomerulopathies as previously reviewed (Bertocchio et al. 2011). However, in CKD patients treated with an ACE-I or ARBs, increased aldosterone levels are observed (Francois and Chatziantoniou 2018); the application of MR blockers in addition to ACE-I and ARBs may thus reduce proteinuria and slow GFR decline (Bolignano et al. 2014; Hou et al. 2015) in this population of patients. A major drawback in this therapeutic strategy is hyperkalemia. A recent randomized controlled trial suggests that the addition of a non-absorbed potassium-binding agent, patiromer, and reduces hyperkalemia in patients treated by ACE-I, ARBs, or MR blockers (Bakris et al. 2015; Weir et al. 2016). Therefore, patiromer administration may help patients to remain on treatment of RAAS blockade and even to increase dosage of these drugs that achieve a better nephroprotection, by decreasing potassium levels and reducing aldosterone levels in patients on RAAS blockade.

31.2.10 Pirfenidone

Pirfenidone is an antifibrotic, anti-inflammatory, and antioxidant compound that was recently approved by the FDA for patients with idiopathic pulmonary fibrosis (IPF) (RamachandraRao et al. 2009). It inhibits TGF-α and free radical oxygen species (ROSs), as well as reduces IL-1, IL-6, IL-8, IL-12, and TNF-α levels (Taniguchi et al. 2010). Several preclinical studies and clinical trials have indicated that pirfenidone provides some promise in treating IPF and renal fibrosis (RamachandraRao et al. 2009; Li et al. 2017; Robalo-Cordeiro et al. 2017).

The effect of pirfenidone on renal fibrosis has also been investigated. Rao et al. showed that pirfenidone-inhibited mesangial matrix expansion and reduced levels of type I and IV collagen, and fibronectin gene expansion in kidneys of mice with DN (RamachandraRao et al. 2009). In 5/6 nephrectomy of the rats, Sharma et al. found that pirfenidone inhibits M1 and M2 macrophage infiltration (Sharma et al. 2011). In UUO rats and cultured HK-2 cells, pirfenidone significantly attenuated TGF-β1-induced EMT and ECM synthesis, as indicated by decreased expression of α-SMA, type I and III collagen, S100A4, fibronectin, and increased expression of E-cadherin (Li et al. 2017). All the preclinical studies indicated an antifibrotic effect of pirfenidone in the kidney. In a clinical trial that includes 77 patients with DN, (eGFR ranging from 20 mL/min/1.73 m^2 to 75 mL/min/1.73 m^2), pirfenidone treatment resulted in an average increase in GFR of +8.5 mL/min/1.73 m^2 compared with a mean eGFR decrease of 2.2 mL/min/1.73 m^2 in the placebo group after 1 year (Sharma et al. 2011). More experimental and clinical trials are needed to evaluate whether pirfenidone could be used for clinical treatment of fibrotic renal diseases.

31.2.11 PPARs Agonist

Peroxisome proliferator-activated receptors (PPARs), including three subtypes known as PPARα, PPARβ/δ, and PPARγ, are nuclear transcription factors that form obligate heterodimers with retinoid-X receptors to modulate transcription of target genes (Liu et al. 2014). Several preclinical studies and clinical trials have implicated all three PPARs as potential targets for antifibrotic therapy.

PPARγ agonists, known as thiazolidinedione (TZD) drugs, are targeted primarily as a treatment for diabetes, disorders of lipid metabolism, inflammation, and fibrosis (Wang et al. 2007). Many preclinical experiments have indicated that PPARγ agonists prevent and inhibit tissue fibrosis in liver, kidneys, heart, and lungs (McVicker and Bennett 2017). However, it is important to note the side effects of the TZDs can be significant and include weight gain, edema and bone density loss, increased risk of cardiovascular disease and bladder cancer (Tahrani et al. 2016). Clinical studies of non-TZD PPARγ agonists on fibrotic diseases are lacking. Fenofibrate, a non-TZD PPARα agonist, prevented pulmonary, hepatic, and renal fibrosis in preclinical studies (McVicker and Bennett 2017). In addition, fenofibrate, gemfibrozil, and BAY PP1 prevented interstitial renal fibrosis in a number of preclinical rodent models (McVicker and Bennett 2017). Few clinical studies of PPARα agonists in renal fibrotic diseases have been conducted. To date, there are only preclinical animal studies on the treatment of fibrotic diseases with PPARβ/δ agonists (McVicker and Bennett 2017). The PPARβ/δ agonist HPP593 may effectively abrogate renal fibrosis induced by chronic ischemia through a reduction in oxidative stress and preservation of mitochondrial function (Fedorova et al. 2013). Meanwhile, recent studies indicated that dual-, pan-, or mixed PPAR agonists show potential for the treatment of fibrotic disease (McVicker and Bennett 2017).

31.2.12 Pentoxifylline

Pentoxifylline is a methylxanthine derivative that has been recently approved for the treatment of vascular diseases with three main properties: improving the rheological properties of blood, anti-inflammatory, and antioxidative (Wen et al. 2017). Preclinical research has shown an inhibitory effect of pentoxifylline on fibrotic diseases, including radiation-induced fibrosis, schistosomiasis-induced hepatic granulomas, chronic pulmonary paracoccidioidomycosis, miscellaneous fibrotic conditions, peritoneal fibrosis, and tubulointerstitial fibrosis (Wen et al. 2017). The mechanism of action is through downregulation of TGF-β1 expression and reduction of proinflammatory and inflammatory factors (Wen et al. 2017).

Small clinical studies have demonstrated that pentoxifylline may be a potential therapeutic agent in CKD due to its anti-inflammatory and antiproteinuric effects. It has documented that pentoxifylline slows CKD progression in patients versus placebo as showed by an eGFR decline of -1.2 mL/min/1.73 m^2 during one-year period compared with −7.2 mL/min/1.73 m^2 in the placebo group (Perkins et al. 2009). In a prospective RCT in which 91 patients with eGFR <60 mL/min/1.73 m^2 were randomized to receive pentoxifylline or placebo, Goicoechea et al. found that serum CRP, fibrinogen, and TNF-α decreased significantly in the pentoxifylline groups (Goicoechea et al. 2012). A single-center retrospective analysis conducted by Chen examined the combination of pentoxifylline with ACE-Is or ARBs in patients with advanced CKD found no change in overall mortality or risk of incident cardiovascular events (Chen et al. 2014). However, the patients with the addition of pentoxifylline had better renal outcomes, demonstrating a 40% lower risk of developing ESRD than treatment with an ACE-I or ARB alone, especially in patients with large amounts of proteinuria (Chen et al. 2014).

31.2.13 Bardoxolone Methyl

Bardoxolone methyl is a promising antioxidant and anti-inflammatory transcription factor with antifibrotic effects in the kidney (Campbell and Weir 2015). A phase 2 and double-blinded RCT conducted by Pergola et al. showed that bardoxolone methyl was associated with improvement in eGFR in patients with advanced CKD and type 2 diabetes at 24 weeks of follow-up (Pergola et al. 2011). However, in a larger RCT including more than 2000 patients with stage 4 CKD and diabetes treated by bardoxolone methyl or placebo, bardoxolone methyl failed to reduce progression to ESRD or death, along with a higher rate of cardiovascular events (de Zeeuw et al. 2013). A 2018 post hoc analysis of bardoxolone methyl in patients with stage 4 CKD and type 2 diabetes study (BEACON) examined data from a multinational, randomized, double-blind, and placebo-controlled phase 3 trial from seven studies enrolling approximately 2600 patients (Chin et al. 2018). The use of bardoxolone methyl might sustain mean increasement in eGFR through study week 48, even though remain increasement in eGFR from baseline for 4 weeks after cessation of treatment (Chin et al. 2018). These patients were significantly less likely to experience the composite renal end point. Therefore, bardoxolone methyl may preserve kidney function and delay the onset of ESRD in patients with T2D and stage 4 CKD.

31.2.14 Lysophosphatidic Acid (LPA) Receptor Antagonists

Production of lysophosphatidic acid (LPA) and activation of its receptors (mainly LPA receptor 1 [LPAR1]) are associated with tissue fibrosis in a number of organs (Tager et al. 2008). The contribution of the LPAR1 subtype to the development of kidney, lung, vascular and dermal fibrosis has been demonstrated using genetic (knockout mice) and pharmacological approaches (Castelino et al. 2011). To date, LPAR1 antagonists have passed phase I and phase II clinical trials for IPF and systemic sclerosis (Llona-Minguez et al. 2015). The clinical trials of LPAR1 antagonists have not been performed in kidney disease.

31.2.15 Lysyl Oxidase (LOX) Inhibition

As an extracellular copper-dependent amine oxidase, lysyl oxidase (LOX) can catalyze the first step in the formation of cross-links in collagens and elastin. (Van Bergen et al. 2013). A new humanized monoclonal antibody binding to and inhibiting lysyl oxidase homologue 2 (LOXL2), simtuzumab, is currently developed in the clinical trials for lung and liver fibrosis (Van Bergen et al. 2013). Based on its mechanism of action, simtuzumab may also have beneficial effects for attenuation of renal fibrosis (Van Bergen et al. 2013). A clinical trial with the purpose to treat this disease is worthy of initiation.

31.2.16 Sodium-Glucose Co-transporter 2 Inhibition

The sodium-glucose co-transporter 2 (SGLT2) is responsible for the major renal glucose reuptake in the apical membrane of the renal proximal tubule (Ly et al. 2011). SGLT2 inhibitors have entered clinical practice as glucose control agents for patients with diabetes through their ability to inhibit renal glucose absorption and reduce hyperglycemia (Yale et al. 2014). The results from the studies in type 1 and type 2 diabetic mice showed that SGLT2 inhibition has beneficial effects on diabetic kidney disease in addition to reducing hyperglycemia effect, as indicated by reducing glomerular hyperfiltration, renal hypertrophy, albuminuria, and decreasing systolic blood pressure (Gembardt et al. 2014; Vallon et al. 2014). SGLT2 inhibition is thought to increase sodium delivery to the macula densa, activating constrictor signals to the afferent arteriole and thereby reducing glomerular filtration pressure (Cherney et al. 2014). While SGLT2 inhibition may target the afferent arteriole more selectively, the mechanisms of SGLT2 inhibitors with reducing glomerular filtration pressure overlap with ACE-I and ARB. Moreover, in a phase II trial of patients with type 2 diabetics and CKD, the SGLT2 inhibitor canagliflozin revealed an acute reduction in eGFR followed by eGFR stabilization, along with a significant and dose-dependent reduction in albuminuria, which is similar to ACE-I/ARB treatment (Yale et al. 2014). A phase III study to determine whether the observed decrease in albuminuria of canagliflozin improves the outcomes in patients with DN have been conducted by Janssen et al. 3700 patients at 531 sites globally were enrolled in the CREDENCE trial (Breyer and Susztak 2016). The primary composite end point of the study includes ESRD, doubling of serum creatinine, and renal or cardiovascular death. The results from this trial are expected to be issued in early 2020.

31.2.17 NOX1 and NOX4 Inhibitors

It has been documented that the overproduction of reactive oxygen species (ROS) is related to various pathologies, including hypertension, atherosclerosis, diabetes, and CKD (Wilcox 2002). Many molecules belong to the ROS family with divergent functions including regulation of cell growth and differentiation, modulation of ECM production, and stimulation of proinflammatory genes (Wilcox 2002). The NOX family includes seven members of NADP(H) oxidases, which is responsible for the generation of ROS. The subtypes of NOX family have tissue specific; e.g., NOX1 is found in colon and vascular cells, NOX2 is the catalytic subunit in respiratory phagocytes, NOX3 is expressed in fetal tissue, and NOX4 is expressed in the renal endothelial cells (Griendling 2006). The results from several models of experimental hypertension have demonstrated that ROS participates in the development and maintenance of hypertension (Griendling 2006). The expression of NOX1, NOX2, and NOX4 increased in rats and mice made hypertensive by angiotensin II infusion (Virdis et al. 2004). In humans, oxidative stress is thought to be increased in different types of hypertension (Higashi et al. 2002). The inhibition of the NOX enzymes could be a potential therapeutic strategy to improve endothelial function and lower blood pressure in human disease. Recently, a placebo-controlled, double-blind, randomized phase II clinical trial enrolling 120 patients with type 2 diabetes mellitus has been designed to determine the safety and efficacy of GKT137831 (Table 31.1).

31.3 New Fibrotic Targets and Antifibrotic Inhibitors in Preclinical Studies

31.3.1 RTKs and NRTK Inhibitors

Receptor tyrosine kinases (RTKs) and non-receptor tyrosine kinases (nRTKs), the two subtypes of tyrosine kinases, regulate many physiological processes at the cellular level, including metabolism, growth, differentiation, and apoptosis (Liu and Zhuang 2016). RTKs include a number of superfamilies, such as platelet-derived growth factor receptors (PDGFR), vascular endothelial growth factor receptors (VEGFR), fibroblast growth factor receptors (FGFR), epidermal growth factor receptors (EGFR), insulin-like growth factor receptors (IGFR), nRTKs which lack extracellular and transmembrane domains, include Src, c-Abl, and c-kit (Liu and Zhuang 2016). Increasing evidence indicates that RTKs and Src superfamily kinases are involved in the initiation and progression of tissue fibrosis (Liu and Zhuang 2016).

Tyrosine kinase inhibitors (TKIs) are FDA-approved first-line drugs for various malignancies. A number of preclinical studies have also confirmed the potential antifibrotic effects of TKIs in organ fibrosis, such as skin, lung, liver, and kidney. For example, imatinib, as a PDGFR inhibitor, can block fibroblast activation, reduce ECM synthesis, abrogate fibrogenesis, and improve existing fibrosis (Floege et al. 2008). In

Table 31.1 Antifibrotic drug tested in clinical trials

Antifibrotic drug	Disease	Targets	Efficacy	References
Fresolimumab (human monoclonal antibody against TGF-β)	FSGS (phase II)	TGF-β	– Well tolerated – Slight decline in eGFR	Trachtman et al. (2011)
Fresolimumab	Steroid-resistant primary FSGS (phase I)	TGF-β	– Partial or complete remission assessed by urinary protein/creatinine ratio – Trend to stabilized eGFR in the Fresolimumab group	Vincenti et al. (2017)
LY2382770 (an antibody that neutralizes the bioactivity of TGF-β1)	Type 1 or type 2 diabetes mellitus (phase II)	TGF-β	Change in serum creatinine from baseline to 12 months	Klinkhammer et al. (2017)
Pirfenidone	Diabetic nephropathy (phase I + II)	TGF-β and others	GFR increased only in the lowest pirfenidone dosage and decreased similarly than in the placebo group when a higher dose was used	Sharma et al. (2011)
Pirfenidone	FSGS (phase II)	TGF-β and others	– Significantly decreased GFR loss (25%) – No change in proteinuria and blood pressure	Cho et al. (2007)
Pirfenidone	Diabetic nephropathy (phase III)	TGF-β and others	– Decrease in albuminuria – A slight improvement in GFR (time frame: 12 months)	Klinkhammer et al. (2017), Sharma et al. (2011)

(continued)

Table 31.1 (continued)

Antifibrotic drug	Disease	Targets	Efficacy	References
STX-100 (a humanized monoclonal antibody that selectively targets integrin αvβ6, preventing its binding and activation of TGF-β complex	Renal transplant patients with interstitial fibrosis and tubular atrophy (phase II)	TGF-β	This study was halted before enrollment for unknown reasons	Lo et al. (2013)
FG-3019 (human monoclonal antibody to CTGF)	Incipient nephropathy Due to type 1 or type 2 Diabetes mellitus (phase I)	CTGF	– Treatment was well tolerated – Decrease in albuminuria (microalbuminuria)	Adler et al. (2010)
GCS-100 (a modified citrus pectin and galectin-3 antagonist)	CKD (phase I and II)	Gal-3	Not published or unknown	Klinkhammer et al. (2017)
MCP (a modified citrus pectin of pharmacological inhibitor for Gal-3)	Diabetic nephropathy (phase II)	Gal-3	Ongoing	Klinkhammer et al. (2017)
CCX140-B (chemokine receptor type 2 antagonist)	Diabetic nephropathy	CCR-2	Not published	Tampe and Zeisberg (2014b)
Lisinopril (ACEI) combination with losartan (ARB)	Diabetic nephropathy (phase II)	RAAS system	– A further decline in albuminuria – The effect of the combination therapy was better than monotherapy - The combination was associated with doubling of the risk for hyperkalemia and acute kidney injury compared to monotherapy	Fried et al. (2013)

(continued)

Table 31.1 (continued)

Antifibrotic drug	Disease	Targets	Efficacy	References
Aliskiren (the renin inhibitor)	Diabetic nephropathy	Renin	– Greater albuminuria reduction than placebo – Did not decrease renal events or improve the rate of eGFR loss – Increased hyperkalemia and stroke – Increased adverse events	Parving et al. (2012)
Avosentan (endothelin-1 receptor A antagonists)	Hypertension and CKD with or without diabetes mellitus (type 1 or type 2)	ET-1	– Reduce blood pressure – Decrease proteinuria	Saleh et al. (2011), Wenzel et al. (2009)
Avosentan	Type 2 diabetic nephropathy treated with RAAS blockade (phase III)	ET-1	– Reduced the albumin to creatinine ratio – Adverse cardiovascular events may be increased	Mann et al. (2010), Egido et al. (2017)
Spironolactone (aldosterone or mineralocorticoid (MR) blockers)	Diabetic nephropathy	Aldosterone or mineralocorticoid	– Add-on therapy to ACE or ARBs – Promotes a reduction in proteinuria and blunt GFR decline	Hou et al. (2015), Bolignano et al. (2014)
Pentoxifylline (methylxanthine derivative)	CKD	TGF-β1 and inflammatory factors	– eGFR decline of -1.2 mL/min/1.73 m^2 compared with -7.2 mL/min/1.73 m^2 in the control group	Perkins et al. (2009)

(continued)

Table 31.1 (continued)

Antifibrotic drug	Disease	Targets	Efficacy	References
Pentoxifylline	CKD patients with eGFR <60 mL/min/1.73 m^2	TGF-β1 and inflammatory factors	– Decreased serum CRP, fibrinogen and TNF-α	Goicoechea et al. (2012)
Pentoxifylline with ACEI or ARBs	Stage 3B-5 CKD patients who received ACEI or ARB treatment	TGF-β1 and inflammatory factors and RAAS system	– A better renal outcome: a 40% lower risk of developing ESRD than treatment with an ACE inhibitor or ARB alone, especially in patients with large amounts of proteinuria – No between-group difference regarding mortality and cardiovascular events	Chen et al. (2014)
Bardoxolone methyl	Advanced CKD and type 2 diabetes	Oxidant and inflammatory transcription factor	– Improvement in eGFR	Pergola et al. (2011)
Bardoxolone methyl	Stage 4 CKD and diabetes	Oxidant and inflammatory transcription factor	– Failed to reduce progression to ESRD or death – Increased rate of cardiovascular events	de Zeeuw et al. (2013)

(continued)

Table 31.1 (continued)

Antifibrotic drug	Disease	Targets	Efficacy	References
Bardoxolone methyl	Patients with stage 4 CKD and type 2 diabetes (phase III)	oxidant and inflammatory transcription factor	- Increased in eGFR that were sustained through study week 48 increased in eGFR from baseline were sustained 4 weeks after cessation of treatment	Chin et al. (2018)
Canagliflozin (sodium–glucose co-transporter 2 inhibitor)	patients with type 2 diabetics and CKD (phase II)	SGLT-2	– A significant and dose-dependent reduction in albuminuria – An acute reduction in eGFR followed by eGFR stabilization	Yale et al. (2014)
Canagliflozin	Patients with type 2 diabetics and CKD (phase III)	SGLT-2	– The primary composite end point of the study includes ESRD, doubling of serum creatinine, and renal or cardiovascular death – ongoing	Breyer and Susztak (2016)
GKT137831 (NOX1 and NOX4 inhibitors)	Type 2 diabetes mellitus	NOX1 and NOX4	Ongoing	Higashi et al. (2002)

the kidney, imatinib has also been shown to attenuate pathological changes in different experimental models of CKD, including glomerulonephritis (Gilbert et al. 2012), DN (Lassila et al. 2005), LN (Zoja et al. 2006), and chronic allograft nephropathy (Savikko et al. 2003). In addition, AG1296, a specific TKI for FGFR, also shows an inhibitory effect on FGF-2-induced renal fibroblast proliferation (Strutz et al. 2001). In addition, Erlotinib, a TKI for EGFR, can significantly decrease TGF-β-mediated fibrogenesis (Chen et al. 2012). In a rat model of hyperuricemic nephropathy, a highly selective EGFR inhibitor, gefitinib inhibits activation of renal interstitial fibroblasts, limits ECM overproduction and deposition, improves renal function, and reduces urine microalbumin (Liu et al. 2015). Src is activated in the kidney after chronic injury, and its inactivation with a chemical inhibitor (PP1) attenuates renal fibroblast activation and proliferation as well as renal fibrogenesis (Yan et al. 2015). The antifibrotic effects of Src inhibition involve suppressing activation of TGF-β1 and EGFR signaling pathways as well as blocking epithelial cell G2/M arrest and epithelial-to-mesenchymal cell transformation (Yan et al. 2015).

Nintedanib (also known as BIBF 1120) is a potent, indolinone-derived small molecule, and multiple-receptor TKI that simultaneously blocks the intracellular ATP-binding pocket of RTKs. As a result, it inhibits phosphorylation of several specific tyrosine kinases, including PDFGR-α and PDFGR-β, FGFR-1, 2, 3, VEGFR-1, 2, VEGFR-3, and Src family kinases (Src, Lyn, and Lck) (Liu and Zhuang 2016). In clinical trials, nintedanib was shown to be effective in improving the lung function, reducing the frequency of acute exacerbations, and improving the quality of life in patients with IPF (Woodcock et al. 2013; Richeldi et al. 2014). On this basis, nintedanib was designated a breakthrough therapy and approved by the FDA in October 2014 (McCormack 2015) and the European Commission Granted Marketing Authorization in January 2015 for the treatment of patients with IPF. Several in vitro and in vivo studies have also tested the efficacy of nintedanib in liver fibrosis, renal fibrosis, and dermal fibrosis and revealed its antifibrotic effect in those disorders. Notably, nintedanib attenuated liver fibroblasts and HSC activation, collagen deposition, HSC differentiation, contractility and migration, as well as suppressed intrahepatic inflammation and angiogenesis. Our study indicated that nintedanib also protects against renal fibrosis through the mechanisms associated with blockade of multiple RTKs, Src family kinases, TGF-β activation, and their downstream signaling pathways contributing to renal fibrogenesis (Liu et al. 2017). From the experience of clinical application of nintedanib on IPF and basic research on non-pulmonary fibrotic disease, it is suggested that nintedanib may have strong potential on the treatment of tissue fibrosis with higher effectiveness. A clinical trial is needed to evaluate the efficacy of nintedanib in human renal fibrotic diseases.

31.3.2 Epigenetic Targets DNA Methylation and Histone Modifications

The effect of acute metabolic changes may not be detected until later stages of disease as they can result in long-term epigenetic modifications. Several studies have reported differences in cytosine methylation levels between control and human kidney disease samples (Tampe and Zeisberg 2014a). Increased methylation of the RASAL1 gene, which encodes an inhibitor of the RAS oncoprotein, has been associated with fibroblast activation and fibrogenesis in the kidney (Tampe and Zeisberg 2014a). RASAL1 hypermethylation is mediated by the methyltransferase DNMT1. In mouse models of kidney fibrosis, genetic deletion of Dnmt1, or pharmacological inhibition of the enzyme protected animals from developing renal fibrosis (Tampe and Zeisberg 2014a). Studies from our laboratory have also shown that enhancer of zeste homolog 2 (EZH2), a histone methyltransferase that induces histone H3 lysine 27 trimethylation (H3K27me3), is highly expressed in the kidney of human diseases such as IgA nephropathy, FSGS, and in a murine model of UUO (Zhou et al. 2016). Pharmacologic inhibition of EZH2 with 3-deazaneplanocin A (3-DZNeP) abrogated deposition of ECM proteins and expression of α-SMA in the obstructed kidney (Zhou et al. 2016). In addition, the results from experiments by our group and others demonstrated that the administration of histone deacetylases (HDACs) inhibitors ameliorates fibrosis development in different models of renal fibrosis (Pang et al. 2009; Liu et al. 2013). Thus, histone modifications represent another important epigenetic regulator of renal fibrogenesis. Most of the currently available drugs targeting histone modifications, however, broadly interfere with transcription. Extended safety studies are needed to understand their side effect profiles.

31.3.3 The Cannabinoid Receptors 1 and 2 (CB1-2)

The cannabinoid system and its cannabinoid receptors (CB) 1 and -2 have recently emerged as potential targets in renal disease (Francois and Lecru 2018), with CB1 promoting fibrosis and CB2 inhibiting it. CB1 and CB2 are also involved in the regulation of appetite and metabolism and thus are promising targets in diabetes and obesity-induced metabolic syndrome (Nam et al. 2012). CB1 antagonists were recently shown to retain most of their efficacy in metabolic syndrome and diabetes (Tam 2016), including the reduction of albuminuria in early stages of diabetic renal disease (Francois and Lecru 2018). However, it has been demonstrated that CB1 promotes fibrosis independently from its function on metabolism. CB1 is mainly expressed in the endothelium where it regulates renal hemodynamics in the normal kidney tissue; however, the expression of CB1 is induced in most cells in animal models with kidney diseases, and in human diabetic, and IgA nephropathy (Lecru et al. 2015). CB1 inhibition (genetic or pharmacological) profoundly reduces renal fibrosis (Lecru et al. 2015). Interestingly, increased CB1 expression in podocytes was demonstrated to be directly involved in albuminuria and the development of DN, independent of its role in metabolism as well (Jourdan et al. 2014). CB1 inhibition may be more effective than CB2 stimulation in experimental models of nephropathies (Barutta et al. 2010, 2011, 2014; Lecru et al. 2015), although an additive effect of CB1 peripheral antagonism with a CB2 agonist in DN has also been reported (Barutta et al. 2017). The reason of these discrepancies may be due to variations in the potency of the CB1 pharmacological blockers used. Even though CB1 has demonstrated to play an important role in renal fibrosis, so far, there is still no definite evidence showing that CB1 inhibition can protect against kidney dysfunction and prevent CKD progression, especially in non-metabolic kidney diseases. A recent report shows that the specific deletion of CB1 in proximal tubules might inhibit renal fibrosis and preserve renal function in obesity-induced nephropathy in mice (Udi et al. 2017).

31.3.4 Discoidin Domain Receptor 1 (DDR1)

DDR1, a transmembrane tyrosine kinase receptor involved in collagen synthesis, is predominantly expressed in epithelial cells and is involved in cell migration, proliferation, and ECM protein synthesis, and degradation (Liu and Zhuang 2016). DDR2 expresses in apical membranes of select nephron segments, from the loop of Henle to the macula densa. DDR1 and DDR2 mRNA and protein expression are upregulated within the glomeruli in the remnant kidney (Liu and Zhuang 2016). Upon collagen binding, DDR1 has been shown to activate various downstream signaling pathways including the mitogen-activated protein kinase (MAPK) signaling pathway through ERK1/2 or JNK, NF-κB, PI3 Kinase (Liu and Zhuang 2016). There is now strong evidence of DDR1 involvement in renal fibrosis and CKD progression (Prakoura and Chatziantoniou 2017). DDR1 inhibition was found to prevent the development of proteinuria, glomerular and perivascular fibrosis, and renal inflammation in a hypertensive model of renal injury (Flamant et al. 2006). DDR1 was found to reduce fibrosis by impairing migration of inflammatory cells in the UUO model and in the anti-glomerular basement membrane (anti-GBM) model of glomerular nephritis, where inflammation is more prominent (Guerrot et al. 2011; Kerroch et al. 2012). DDR1 inhibition was also effective in reducing renal fibrosis and inflammation while protecting kidney function in an experimental model of Alport's syndrome (Gross et al. 2010) and in the remnant kidney model (Borza et al. 2017). Of note, DDR1 inhibition with antisense oligonucleotides after the establishment of the disease was protective in models of glomerulonephritis and ureteral obstruction (Kerroch et al. 2016), indicating that DDR1 inhibition may reverse fibrosis as well as prevent it.

31.3.5 Periostin

Periostin, a secreted ECM protein originally discovered in periosteum and periodontal ligaments (Horiuchi et al. 1999), is highly expressed during development, and in injury and repair tissues in adults, especially within the kidney. Periostin can bind to several ECM proteins, such as collagen, fibronectin, and several integrins, and plays an important role in many physiological processes, such as cell adhesion, migration, proliferation, and differentiation. Several important mediators of renal fibrosis and/or inflammation, such as TGF-β1, Ang II, PDGF-B, the interleukins IL-4 and IL-13, can induce periostin expression (Francois and Chatziantoniou 2018). Periostin, by its direct binding and interactions with collagens and fibronectin, affects ECM protein remodeling (Li et al. 2010; Francois and Chatziantoniou 2018) and is involved in renal fibrosis, inflammation, and development of CKD (Mael-Ainin et al. 2014; Prakoura and Chatziantoniou 2017). For instance, periostin was one of the most upregulated genes in three experimental models of renal fibrosis (Sen et al. 2011; Guerrot et al. 2012; Vethe et al. 2015). Moreover, periostin levels increase during diabetes, lupus nephritis, IgA nephropathy, and FSGS (Satirapoj et al. 2012; Hwang et al. 2016). Genetic disruption of periostin reduced renal inflammation and fibrosis in the UUO model and in a model of anti-GBM glomerulonephritis (Mael-Ainin et al. 2014; Prakoura and Chatziantoniou 2017). Periostin inhibition by antisense oligonucleotides also protected the kidney from fibrosis, even when given after the beginning of the disease (Prakoura and Chatziantoniou 2017). Similarly, in a mouse model of polycystic kidney diseases (PKD), genetic disruption of periostin reduces cysts number, reduces renal interstitial fibrosis, and improves renal function (Wallace et al. 2014).

31.3.6 MicroRNA (MiRNA)

As small noncoding mRNA molecules, micro (mi)RNAs can regulate gene expression at the posttranscriptional level and are involved in a wide range of pathophysiological processes (Francois and Chatziantoniou 2018). Some specific miRNAs have been shown to be involved in renal fibrosis, acting downstream in the TGF-β pathway (Gomez et al. 2016). miRNA 21 is the most abundantly expressed miRNA within the kidney and has been found to be upregulated in both acute and chronic kidney injury in animals (Zarjou et al. 2011) and in humans (Glowacki et al. 2013). It has pleiotropic functions associated with phosphatase and tensin homologue (PTEN), TGF-β, and tumor suppression. miRNA-29 and -200 have been shown to blunt the development of renal fibrosis, whereas miRNA-21, -214, -199, and -155 are directly involved in the fibrotic process (Gomez et al. 2016). Recently, these miRNAs have emerged as therapeutic targets in renal fibrosis (Gomez et al. 2016). An anti-miRNA-21 was able to prevent the development of renal fibrosis in an experimental model of Alport's syndrome (Zarjou et al. 2011; Gomez et al. 2015) and in the UUO model

Table 31.2 New inhibitors tested in animal models of CKD

New inhibitors	Diseases model	Targets	Efficacy	References
Imatinib	Rat acute anti-Thy 1.1 GN	PDGFR	Inhibits mesangial cells proliferation and matrix accumulation	Gilbert et al. (2012)
Imatinib	Murine streptozotocin-induced diabetes	PDGFR	Decreases albuminuria, glomerular and tubulointerstitial damage	Lassila et al. (2005)
Imatinib	Murine lupus	PDGFR	Improves survival, decreases proteinuria, glomerular and tubulointerstitial damage	Zoja et al. (2006)
Imatinib	Rat kidney transplantations	PDGFR	Improves histologic chronic changes, decreases creatinine values, and infiltration of inflammatory cells	Savikko et al. (2003)
AG1296	Human cortical fibroblasts and tubular epithelial cells	FGFR	Inhibits cell proliferation induced by TGF-β1	Strutz et al. (2001)
Erlotinib	Murine model with selective targeted deletion of EGFR in renal proximal tubules	EGFR	Decreases the proximal tubule dedifferentiation and tubulointerstitial fibrosis resulting from chronic Ang II administration	Chen et al. (2012)
Gefitinib	Rat model of hyperuricemic nephropathy	EGFR	Prevented renal dysfunction, reduced urine microalbumin, and inhibited activation of renal interstitial fibroblasts and expression of extracellular proteins	Liu et al. (2015)

(continued)

Table 31.2 (continued)

New inhibitors	Diseases model	Targets	Efficacy	References
PPI	Murine unilateral ureter obstruction	Src	Reduced renal fibroblast, activation and attenuated ECM protein deposition. Suppressed activation of TGF-β1 signaling, activation of EGFR and STAT3, reduced the number of renal epithelial cells arrested at the G2/M phase of the cell cycle	Yan et al. (2015)
Nintedanib	Murine unilateral ureter obstruction	PDGFR, FGFR, VEGFR, Src, Lck, Lyn	Attenuated renal fibrosis and inhibited activation of renal interstitial fibroblasts, decreased ECM protein overproduction and deposition, blocked activation of STAT3, NF-κB, and Smad-3, inhibited renal proinflammatory cytokine expression and macrophage infiltration.	Liu et al. (2017)
Nintedanib	Murine renal fibrosis induced by folic acid (FA) injection	PDGFR, FGFR, VEGFR, Src, Lck, Lyn	Attenuated renal fibrosis and inhibited activation of renal interstitial fibroblasts, decreased ECM protein overproduction and deposition	Liu et al. (2017)
5-azacytidine	Murine renal fibrosis induced by folic acid (FA) injection	RASAL1	Inhibited perpetuation of fibroblast activation and fibrogenesis in the kidney	Tampe and Zeisberg (2014a)
3-DZNeP	Murine unilateral ureter obstruction	EZH2	Abrogated deposition of ECM proteins and expression of α-SMA in the obstructed kidney	Zhou et al. (2016)

(continued)

Table 31.2 (continued)

New inhibitors	Diseases model	Targets	Efficacy	References
Trichostatin A (TSA)	Murine unilateral ureter obstruction	HDAC	Suppressed the expression ofα-SMA and fibronectin and attenuated the accumulation of renal interstitial fibroblasts; inhibited phosphorylation of STAT3; inhibited tubular cell apoptosis and caspase-3 activation	Pang et al. (2009)
MS-275	Murine unilateral ureter obstruction	HDAC	Suppressed the expression ofα-SMA, collagen type I and fibronectin inhibited TGF-β1, TGF-βR1, and phosphorylation of Smad-3; blocked phosphorylation and expression of EGFR and STAT3; reduced the number of renal tubular cells arrested in the G2/M phase of the cell cycle	Liu et al. (2013)
Sirtinol	Murine unilateral ureter obstruction	HDAC	Attenuated deposition of collagen fibrils, reduced expression of α-SMA, collagen I, and fibronectin; increased dephosphorylation of EGFR, PDGFRβ, and STAT3	Pang et al. (2009)
AM251	Streptozotocin-induced diabetic mice	Cannabinoid receptor 1 (CB1)	Decreased albuminuria; prevented diabetes-induced downregulation of nephrin, podocin, and ZO-1; inhibited overexpression of fibronectin, TGF-β1, and CTGF	Barutta et al. (2010)

(continued)

Table 31.2 (continued)

New inhibitors	Diseases model	Targets	Efficacy	References
AM1241	Streptozotocin-induced diabetic mice	CB2	Ameliorated albuminuria, podocyte protein downregulation, and glomerular monocyte infiltration	Barutta et al. (2011)
AM6545 combination with AM1241	Streptozotocin-induced diabetic mice	CB1R and CB2R	Reduced diabetes-induced albuminuria and prevented nephrin loss; Dual therapy performed better than monotherapies, abolished albuminuria, inflammation, tubular injury and markedly reduced renal fibrosis. Abolished diabetes-induced renal monocyte infiltration and M1/M2 macrophage imbalance	Barutta et al. (2017)
Antisense oligodeoxynucleotides	Murine unilateral ureter obstruction	Discoidin domain receptor 1 (DDR1)	Stopped the increase of proteinuria and protected animals against the progression of glomerulonephritis, as evidenced by functional, structural and cellular indexes	Kerroch et al. (2016)
Antisense oligonucleotides	Murine NTS-induced glomerulonephritis	Periostin	Reversed already established proteinuria, diminished tissue inflammation, and improved renal structure	Prakoura and Chatziantoniou (2017)

(continued)

Table 31.2 (continued)

New inhibitors	Diseases model	Targets	Efficacy	References
Anti-RNA oligonucleotide	Alport syndrome animal model	microRNA 21	Retarded the progression of disease, normalized tubular functions and led to an increase in life expectancy of nearly 50%; significant protection of mitochondrial function and marked activation of PPARα and PGC1α signaling pathways, which include FAO	Gomez et al. (2015)
Anti-RNA oligonucleotide	db/db mice	microRNA 21	Had beneficial effects on glomerular function and retarded the progression of diabetic kidney disease	Gomez et al. (2016)
Anti-RNA oligonucleotide	Murine unilateral ureter obstruction	microRNA 21	Attenuated UUO-induced renal fibrosis, presumably through diminishing the expression of profibrotic proteins and reducing infiltration of inflammatory macrophages in UUO kidneys	Zarjou et al. (2011)
Anti-RNA oligonucleotide	Streptozotocin-induced diabetic mice	microRNA 124	Reduced urinary podocytic nephrin, podocin and albumin excretion and up-regulate integrin $\alpha 3$ expression	Li et al. (2013)

(Zarjou et al. 2011; Denby et al. 2014). In addition, antagonism of miRNAs was also beneficial in streptozotocin-induced DN (with a miRNA-124 antisense) (Li et al. 2013) and in murine kidney of chronic allograft dysfunction (with a miRNA-21a-5p silencing probe) (Schauerte et al. 2017). Similarly, agomir or miRNA mimics (i.e., miRNA23b agomir) have been shown to exhibit beneficial effects in renal fibrosis induced by DN. However, there remains a major challenge to deliver the specific miRNA to the kidney without any side effects (Table 31.2).

31.4 Challenges and Prospects

Although great progress has been made in many preclinical studies on the mechanisms and therapeutic targets of renal fibrosis, there are still no agents beyond ACE-Is and ARBs approved to treat renal fibrosis in clinical practice. This is due to many challenges in translating basic research into clinical application in this field. For example, most animal models are conducted by a single hit in young and healthy male animals, whereas in human, fibrogenesis is a continuous process or comes in flares, striking kidney patients of both genders who are largely middle-aged or older and often have comorbid conditions. Moreover, despite advances in the effectiveness, safety and tolerability of novel antifibrotic agents are encouraging; clinical trials in renal disease are limited by variability in patient cohorts and specific patient selection, trial design and financially feasible longitudinal studies with appropriate end points. In addition, specific end point measures, novel imaging techniques, and biomarkers of renal fibrosis are also needed for clinical trials.

References

Adler SG, Schwartz S, Williams ME, Arauz-Pacheco C, Bolton WK, Lee T, Li D, Neff TB, Urquilla PR, Sewell KL (2010) Phase 1 study of anti-CTGF monoclonal antibody in patients with diabetes and microalbuminuria. Clin J Am Soc Nephrol: CJASN 5:1420–1428

Al-Lamki RS, Mayadas TN (2015) TNF receptors: signaling pathways and contribution to renal dysfunction. Kidney Int 87:281–296

Al-Lamki RS, Wang J, Vandenabeele P, Bradley JA, Thiru S, Luo D, Min W, Pober JS, Bradley JR (2005) TNFR1- and TNFR2-mediated signaling pathways in human kidney are cell type-specific and differentially contribute to renal injury. FASEB J: Off Publ Fed Am Soc Exp Biol 19:1637–1645

Anders HJ, Vielhauer V, Schlondorff D (2003) Chemokines and chemokine receptors are involved in the resolution or progression of renal disease. Kidney Int 63:401–415

Anders HJ, Belemezova E, Eis V, Segerer S, Vielhauer V, Perez de Lema G, Kretzler M, Cohen CD, Frink M, Horuk R, Hudkins KL, Alpers CE, Mampaso F, Schlondorff D (2004) Late onset of treatment with a chemokine receptor CCR4 antagonist prevents progression of lupus nephritis in MRL-Fas(lpr) mice. J Am Soc Nephrol: JASN 15:1504–1513

Bakris GL, Pitt B, Weir MR, Freeman MW, Mayo MR, Garza D, Stasiv Y, Zawadzki R, Berman L, Bushinsky DA (2015) Effect of patiromer on serum potassium level in patients with hyperkalemia and diabetic kidney disease: the AMETHYST-DN randomized clinical trial. JAMA 314:151–161

Barton M (2008) Reversal of proteinuric renal disease and the emerging role of endothelin. Nat Clin Pract Nephrol 4:490–501

Barutta F, Corbelli A, Mastrocola R, Gambino R, Di Marzo V, Pinach S, Rastaldi MP, Perin PC, Gruden G (2010) Cannabinoid receptor 1 blockade ameliorates albuminuria in experimental diabetic nephropathy. Diabetes 59:1046–1054

Barutta F, Piscitelli F, Pinach S, Bruno G, Gambino R, Rastaldi MP, Salvidio G, Di Marzo V, Cavallo Perin P, Gruden G (2011) Protective role of cannabinoid receptor type 2 in a mouse model of diabetic nephropathy. Diabetes 60:2386–2396

Barutta F, Grimaldi S, Franco I, Bellini S, Gambino R, Pinach S, Corbelli A, Bruno G, Rastaldi MP, Aveta T, Hirsch E, Di Marzo V, Gruden G (2014) Deficiency of cannabinoid receptor of type 2 worsens renal functional and structural abnormalities in streptozotocin-induced diabetic mice. Kidney Int 86:979–990

Barutta F, Grimaldi S, Gambino R, Vemuri K, Makriyannis A, Annaratone L, di Marzo V, Bruno G, Gruden G (2017) Dual therapy targeting the endocannabinoid system prevents experimental diabetic nephropathy. Nephrol Dialysis Transplant: Off Publ Eur Dial Transpl Assoc Eur Renal Assoc 32:1655–1665

Benigni A, Zoja C, Corna D, Orisio S, Longaretti L, Bertani T, Remuzzi G (1993) A specific endothelin subtype A receptor antagonist protects against injury in renal disease progression. Kidney Int 44:440–444

Benigni A, Corna D, Maffi R, Benedetti G, Zoja C, Remuzzi G (1998) Renoprotective effect of contemporary blocking of angiotensin II and endothelin-1 in rats with membranous nephropathy. Kidney Int 54:353–359

Bertocchio JP, Warnock DG, Jaisser F (2011) Mineralocorticoid receptor activation and blockade: an emerging paradigm in chronic kidney disease. Kidney Int 79:1051–1060

Boffa JJ, Tharaux PL, Dussaule JC and Chatziantoniou C (2001) Regression of renal vascular fibrosis by endothelin receptor antagonism. Hypertension (Dallas, Tex: 1979) **37**:490–496

Bolignano D, Palmer SC, Navaneethan SD and Strippoli GF (2014) Aldosterone antagonists for preventing the progression of chronic kidney disease. Cochrane database Syst Rev Cd007004

Borza CM, Su Y, Tran TL, Yu L, Steyns N, Temple KJ, Skwark MJ, Meiler J, Lindsley CW, Hicks BR, Leitinger B, Zent R, Pozzi A (2017) Discoidin domain receptor 1 kinase activity is required for regulating collagen IV synthesis. Matrix Biol: J Int Soc Matrix Biol 57–58:258–271

Brenner BM, Cooper ME, de Zeeuw D, Keane WF, Mitch WE, Parving HH, Remuzzi G, Snapinn SM, Zhang Z, Shahinfar S (2001) Effects of losartan on renal and cardiovascular outcomes in patients with type 2 diabetes and nephropathy. New Engl J Med 345:861–869

Breyer MD, Susztak K (2016) The next generation of therapeutics for chronic kidney disease. Nat Rev Drug Discovery 15:568–588

Campbell D, Weir MR (2015) Defining, treating, and understanding chronic kidney disease—a complex disorder. J Clin Hypertens (Greenwich, Conn) 17:514–527

Castelino FV, Seiders J, Bain G, Brooks SF, King CD, Swaney JS, Lorrain DS, Chun J, Luster AD, Tager AM (2011) Amelioration of dermal fibrosis by genetic deletion or pharmacologic antagonism of lysophosphatidic acid receptor 1 in a mouse model of scleroderma. Arthritis Rheum 63:1405–1415

Charo IF, Ransohoff RM (2006) The many roles of chemokines and chemokine receptors in inflammation. New Engl J Med 354:610–621

Chen J, Chen JK, Nagai K, Plieth D, Tan M, Lee TC, Threadgill DW, Neilson EG, Harris RC (2012) EGFR signaling promotes TGFbeta-dependent renal fibrosis. J Am Soc Nephrol: JASN 23:215–224

Chen PM, Lai TS, Chen PY, Lai CF, Wu V, Chiang WC, Chen YM, Wu KD and Tsai TJ (2014) Renoprotective effect of combining pentoxifylline with angiotensin-converting enzyme inhibitor or angiotensin II receptor blocker in advanced chronic kidney disease. J Formos Med Assoc = Taiwan yi zhi 113:219–226

Cherney DZ, Perkins BA, Soleymanlou N, Maione M, Lai V, Lee A, Fagan NM, Woerle HJ, Johansen OE, Broedl UC, von Eynatten M (2014) Renal hemodynamic effect of sodium-glucose cotransporter 2 inhibition in patients with type 1 diabetes mellitus. Circulation 129:587–597

Chin MP, Bakris GL, Block GA, Chertow GM, Goldsberry A, Inker LA, Heerspink HJL, O'Grady M, Pergola PE, Wanner C, Warnock DG, Meyer CJ (2018) Bardoxolone methyl improves kidney function in patients with chronic kidney disease stage 4 and type 2 diabetes: post-hoc analyses from bardoxolone methyl evaluation in patients with chronic kidney disease and type 2 diabetes study. Am J Nephrol 47:40–47

Cho ME, Kopp JB (2010) Pirfenidone: an anti-fibrotic therapy for progressive kidney disease. Expert Opin Invest Drugs 19:275–283

Cho ME, Smith DC, Branton MH, Penzak SR, Kopp JB (2007) Pirfenidone slows renal function decline in patients with focal segmental glomerulosclerosis. Clin J Am Soc Nephrol: CJASN 2:906–913

de Zeeuw D, Akizawa T, Audhya P, Bakris GL, Chin M, Christ-Schmidt H, Goldsberry A, Houser M, Krauth M, Lambers Heerspink HJ, McMurray JJ, Meyer CJ, Parving HH, Remuzzi G, Toto RD, Vaziri ND, Wanner C, Wittes J, Wrolstad D, Chertow GM (2013) Bardoxolone methyl in type 2 diabetes and stage 4 chronic kidney disease. New Engl J Med 369:2492–2503

Denby L, Ramdas V, Lu R, Conway BR, Grant JS, Dickinson B, Aurora AB, McClure JD, Kipgen D, Delles C, van Rooij E, Baker AH (2014) MicroRNA-214 antagonism protects against renal fibrosis. J Am Soc Nephrol: JASN 25:65–80

Desmedt V, Desmedt S, Delanghe JR, Speeckaert R, Speeckaert MM (2016) Galectin-3 in renal pathology: more than just an innocent bystander. Am J Nephrol 43:305–317

Dhaun N, Goddard J, Webb DJ (2006) The endothelin system and its antagonism in chronic kidney disease. J Am Soc Nephrol: JASN 17:943–955

Drechsler C, Delgado G, Wanner C, Blouin K, Pilz S, Tomaschitz A, Kleber ME, Dressel A, Willmes C, Krane V, Kramer BK, Marz W, Ritz E, van Gilst WH, van der Harst P, de Boer RA (2015) Galectin-3, renal function, and clinical outcomes: results from the LURIC and 4D studies. J Am Soc Nephrol: JASN 26:2213–2221

Egido J, Rojas-Rivera J, Mas S, Ruiz-Ortega M, Sanz AB, Gonzalez Parra E, Gomez-Guerrero C (2017) Atrasentan for the treatment of diabetic nephropathy. Expert Opin Invest Drugs 26:741–750

Fedorova LV, Sodhi K, Gatto-Weis C, Puri N, Hinds TD Jr, Shapiro JI, Malhotra D (2013) Peroxisome proliferator-activated receptor delta agonist, HPP593, prevents renal necrosis under chronic ischemia. PLoS ONE 8:e64436

Fernandes Bertocchi AP, Campanhole G, Wang PH, Goncalves GM, Damiao MJ, Cenedeze MA, Beraldo FC, de Paula Antunes Teixeira V, Dos Reis MA, Mazzali M, Pacheco-Silva A, Camara NO (2008) A role for galectin-3 in renal tissue damage triggered by ischemia and reperfusion injury. Transplant Int: Off J Eur Soc Organ Transplant 21:999–1007

Flamant M, Placier S, Rodenas A, Curat CA, Vogel WF, Chatziantoniou C, Dussaule JC (2006) Discoidin domain receptor 1 null mice are protected against hypertension-induced renal disease. J Am Soc Nephrol: JASN 17:3374–3381

Floege J, Eitner F, Alpers CE (2008) A new look at platelet-derived growth factor in renal disease. J Am Soc Nephrol: JASN 19:12–23

Francois H, Chatziantoniou C (2018) Renal fibrosis: recent translational aspects. Matrix Biol: J Int Soc Matrix Biol 68–69:318–332

Francois H, Lecru L (2018) The Role of cannabinoid receptors in renal diseases. Curr Med Chem 25:793–801

Frenay AR, Yu L, van der Velde AR, Vreeswijk-Baudoin I, Lopez-Andres N, van Goor H, Sillje HH, Ruifrok WP, de Boer RA (2015) Pharmacological inhibition of galectin-3 protects against hypertensive nephropathy. Am J Physiol Renal Physiol 308:F500–509

Fried LF, Emanuele N, Zhang JH, Brophy M, Conner TA, Duckworth W, Leehey DJ, McCullough PA, O'Connor T, Palevsky PM, Reilly RF, Seliger SL, Warren SR, Watnick S, Peduzzi P, Guarino

P (2013) Combined angiotensin inhibition for the treatment of diabetic nephropathy. New Engl J Med 369:1892–1903

Fukuda K, Yanagida T, Okuda S, Tamaki K, Ando T, Fujishima M (1996) Role of endothelin as a mitogen in experimental glomerulonephritis in rats. Kidney Int 49:1320–1329

Garber K (2009) Companies waver in efforts to target transforming growth factor beta in cancer. J Natl Cancer Inst 101:1664–1667

Gembardt F, Bartaun C, Jarzebska N, Mayoux E, Todorov VT, Hohenstein B, Hugo C (2014) The SGLT2 inhibitor empagliflozin ameliorates early features of diabetic nephropathy in BTBR ob/ob type 2 diabetic mice with and without hypertension. Am J Physiol Renal Physiol 307:F317–325

Gilbert RE, Zhang Y, Williams SJ, Zammit SC, Stapleton DI, Cox AJ, Krum H, Langham R, Kelly DJ (2012) A purpose-synthesised anti-fibrotic agent attenuates experimental kidney diseases in the rat. PLoS ONE 7:e47160

Glowacki F, Savary G, Gnemmi V, Buob D, Van der Hauwaert C, Lo-Guidice JM, Bouye S, Hazzan M, Pottier N, Perrais M, Aubert S, Cauffiez C (2013) Increased circulating miR-21 levels are associated with kidney fibrosis. PLoS ONE 8:e58014

Goicoechea M, Garcia de Vinuesa S, Quiroga B, Verdalles U, Barraca D, Yuste C, Panizo N, Verde E, Munoz MA, Luno J (2012) Effects of pentoxifylline on inflammatory parameters in chronic kidney disease patients: a randomized trial. J Nephrol 25:969–975

Gomez IG, MacKenna DA, Johnson BG, Kaimal V, Roach AM, Ren S, Nakagawa N, Xin C, Newitt R, Pandya S, Xia TH, Liu X, Borza DB, Grafals M, Shankland SJ, Himmelfarb J, Portilla D, Liu S, Chau BN, Duffield JS (2015) Anti-microRNA-21 oligonucleotides prevent Alport nephropathy progression by stimulating metabolic pathways. J Clin Invest 125:141–156

Gomez IG, Nakagawa N, Duffield JS (2016) MicroRNAs as novel therapeutic targets to treat kidney injury and fibrosis. Am J Physiol Renal Physiol 310:F931–944

Gomez-Garre D, Ruiz-Ortega M, Ortego M, Largo R, Lopez-Armada MJ, Plaza JJ, Gonzalez E and Egido J (1996) Effects and interactions of endothelin-1 and angiotensin II on matrix protein expression and synthesis and mesangial cell growth. Hypertension (Dallas, Tex: 1979) 27:885–892

Griendling KK (2006) NADPH oxidases: new regulators of old functions. Antioxid Redox Signal 8:1443–1445

Gross O, Girgert R, Beirowski B, Kretzler M, Kang HG, Kruegel J, Miosge N, Busse AC, Segerer S, Vogel WF, Muller GA, Weber M (2010) Loss of collagen-receptor DDR1 delays renal fibrosis in hereditary type IV collagen disease. Matrix Biol: J Int Soc Matrix Biol 29:346–356

Guerrot D, Kerroch M, Placier S, Vandermeersch S, Trivin C, Mael-Ainin M, Chatziantoniou C, Dussaule JC (2011) Discoidin domain receptor 1 is a major mediator of inflammation and fibrosis in obstructive nephropathy. Am J Pathol 179:83–91

Guerrot D, Dussaule JC, Mael-Ainin M, Xu-Dubois YC, Rondeau E, Chatziantoniou C, Placier S (2012) Identification of periostin as a critical marker of progression/reversal of hypertensive nephropathy. PLoS ONE 7:e31974

Henderson NC, Mackinnon AC, Farnworth SL, Kipari T, Haslett C, Iredale JP, Liu FT, Hughes J, Sethi T (2008) Galectin-3 expression and secretion links macrophages to the promotion of renal fibrosis. Am J Pathol 172:288–298

Higashi Y, Sasaki S, Nakagawa K, Matsuura H, Oshima T, Chayama K (2002) Endothelial function and oxidative stress in renovascular hypertension. New Engl J Med 346:1954–1962

Horiuchi K, Amizuka N, Takeshita S, Takamatsu H, Katsuura M, Ozawa H, Toyama Y, Bonewald LF, Kudo A (1999) Identification and characterization of a novel protein, periostin, with restricted expression to periosteum and periodontal ligament and increased expression by transforming growth factor beta. J Bone Mineral Res: Off J Am Soc Bone Mineral Res 14:1239–1249

Hostetter TH, Olson JL, Rennke HG, Venkatachalam MA, Brenner BM (2001) Hyperfiltration in remnant nephrons: a potentially adverse response to renal ablation. J Am Soc Nephrol: JASN 12:1315–1325

Hou J, Xiong W, Cao L, Wen X, Li A (2015) Spironolactone add-on for preventing or slowing the progression of diabetic nephropathy: a meta-analysis. Clin Ther 37:2086–2103.e2010

Huaux F, Liu T, McGarry B, Ullenbruch M, Xing Z and Phan SH (2003) Eosinophils and T lymphocytes possess distinct roles in bleomycin-induced lung injury and fibrosis. J immunol (Baltimore, Md: 1950) 171:5470–5481

Humphreys BD (2017) Mechanisms of renal fibrosis. Annu Rev Physiol 80:309–326

Hwang JH, Lee JP, Kim CT, Yang SH, Kim JH, An JN, Moon KC, Lee H, Oh YK, Joo KW, Kim DK, Kim YS, Lim CS (2016) Urinary periostin excretion predicts renal outcome in IgA nephropathy. Am J Nephrol 44:481–492

Jourdan T, Szanda G, Rosenberg AZ, Tam J, Earley BJ, Godlewski G, Cinar R, Liu Z, Liu J, Ju C, Pacher P, Kunos G (2014) Overactive cannabinoid 1 receptor in podocytes drives type 2 diabetic nephropathy. Proc Natl Acad Sci USA 111:E5420–5428

Kerroch M, Guerrot D, Vandermeersch S, Placier S, Mesnard L, Jouanneau C, Rondeau E, Ronco P, Boffa JJ, Chatziantoniou C, Dussaule JC (2012) Genetic inhibition of discoidin domain receptor 1 protects mice against crescentic glomerulonephritis. FASEB J: Off Publ Fed Am Soc Exp Biol 26:4079–4091

Kerroch M, Alfieri C, Dorison A, Boffa JJ, Chatziantoniou C, Dussaule JC (2016) Protective effects of genetic inhibition of Discoidin Domain Receptor 1 in experimental renal disease. Scientific reports 6:21262

Klinkhammer BM, Goldschmeding R, Floege J, Boor P (2017) Treatment of renal fibrosis-turning challenges into opportunities. Adv Chronic Kidney Dis 24:117–129

Kok HM, Falke LL, Goldschmeding R, Nguyen TQ (2014) Targeting CTGF, EGF and PDGF pathways to prevent progression of kidney disease. Nat Rev Nephrol 10:700–711

Kuc R, Davenport AP (2004) Comparison of endothelin-A and endothelin-B receptor distribution visualized by radioligand binding versus immunocytochemical localization using subtype selective antisera. J Cardiovasc Pharmacol 44(Suppl 1):S224–226

Kurts C, Panzer U, Anders HJ, Rees AJ (2013) The immune system and kidney disease: basic concepts and clinical implications. Nat Rev Immunol 13:738–753

Lassila M, Jandeleit-Dahm K, Seah KK, Smith CM, Calkin AC, Allen TJ, Cooper ME (2005) Imatinib attenuates diabetic nephropathy in apolipoprotein E-knockout mice. J Am Soc Nephrol: JASN 16:363–373

Lecru L, Desterke C, Grassin-Delyle S, Chatziantoniou C, Vandermeersch S, Devocelle A, Vernochet A, Ivanovski N, Ledent C, Ferlicot S, Dalia M, Said M, Beaudreuil S, Charpentier B, Vazquez A, Giron-Michel J, Azzarone B, Durrbach A, Francois H (2015) Cannabinoid receptor 1 is a major mediator of renal fibrosis. Kidney Int 88:72–84

Lewis EJ, Hunsicker LG, Clarke WR, Berl T, Pohl MA, Lewis JB, Ritz E, Atkins RC, Rohde R, Raz I (2001) Renoprotective effect of the angiotensin-receptor antagonist irbesartan in patients with nephropathy due to type 2 diabetes. New Engl J Med 345:851–860

Li G, Jin R, Norris RA, Zhang L, Yu S, Wu F, Markwald RR, Nanda A, Conway SJ, Smyth SS, Granger DN (2010) Periostin mediates vascular smooth muscle cell migration through the integrins alphavbeta3 and alphavbeta5 and focal adhesion kinase (FAK) pathway. Atherosclerosis 208:358–365

Li D, Lu Z, Jia J, Zheng Z, Lin S (2013) MiR-124 is related to podocytic adhesive capacity damage in STZ-induced uninephrectomized diabetic rats. Kidney Blood Press Res 37:422–431

Li Z, Liu X, Wang B, Nie Y, Wen J, Wang Q, Gu C (2017) Pirfenidone suppresses MAPK signalling pathway to reverse epithelial-mesenchymal transition and renal fibrosis. Nephrology (Carlton, Vic) 22:589–597

Liu F, Zhuang S (2016) Role of receptor tyrosine kinase signaling in renal fibrosis. Int J Mol Sci 17:972

Liu N, He S, Ma L, Ponnusamy M, Tang J, Tolbert E, Bayliss G, Zhao TC, Yan H, Zhuang S (2013) Blocking the class I histone deacetylase ameliorates renal fibrosis and inhibits renal fibroblast activation via modulating TGF-beta and EGFR signaling. PLoS ONE 8:e54001

Liu F, Mei X, Zhang Y, Qi H, Wang J, Wang Y, Jiang W, Zhang X, Yan H, Zhuang S (2014) Association of peroxisome proliferator-activated receptorgamma gene Pro12Ala and C161T poly-

morphisms with cardiovascular risk factors in maintenance hemodialysis patients. Mol Biol Rep 41:7555–7565
Liu N, Wang L, Yang T, Xiong C, Xu L, Shi Y, Bao W, Chin YE, Cheng SB, Yan H, Qiu A, Zhuang S (2015) EGF receptor inhibition alleviates hyperuricemic nephropathy. J Am Soc Nephrol: JASN 26:2716–2729
Liu F, Wang L, Qi H, Wang J, Wang Y, Jiang W, Xu L, Liu N and Zhuang S (2017) Nintedanib, a triple tyrosine kinase inhibitor, attenuates renal fibrosis in chronic kidney disease. Clin Sci (London, England: 1979) 131:2125–2143
Llona-Minguez S, Ghassemian A, Helleday T (2015) Lysophosphatidic acid receptor (LPAR) modulators: the current pharmacological toolbox. Prog Lipid Res 58:51–75
Lo DJ, Farris AB, Song M, Leopardi F, Anderson DJ, Strobert EA, Ramakrishnan S, Turgeon NA, Mehta AK, Turnbull B, Maroni B, Violette SM, Kirk AD (2013) Inhibition of alphavbeta6 promotes acute renal allograft rejection in nonhuman primates. Am J Transplant: Off J Am Soc Transplant Am Soc Transplant Surg 13:3085–3093
Ly JP, Onay T, Sison K, Sivaskandarajah G, Sabbisetti V, Li L, Bonventre JV, Flenniken A, Paragas N, Barasch JM, Adamson SL, Osborne L, Rossant J, Schnermann J, Quaggin SE (2011) The Sweet Pee model for Sglt2 mutation. J Am Soc Nephrol: JASN 22:113–123
Mael-Ainin M, Abed A, Conway SJ, Dussaule JC, Chatziantoniou C (2014) Inhibition of periostin expression protects against the development of renal inflammation and fibrosis. J Am Soc Nephrol: JASN 25:1724–1736
Mann JF, Green D, Jamerson K, Ruilope LM, Kuranoff SJ, Littke T, Viberti G (2010) Avosentan for overt diabetic nephropathy. J Am Soc Nephrol: JASN 21:527–535
Manson SR, Niederhoff RA, Hruska KA, Austin PF (2011) Endogenous BMP-7 is a critical molecular determinant of the reversibility of obstruction-induced renal injuries. Am J Physiol Renal Physiol 301:F1293–1302
Marquez DF, Ruiz-Hurtado G, Ruilope LM, Segura J (2015) An update of the blockade of the renin angiotensin aldosterone system in clinical practice. Expert Opin Pharmacother 16:2283–2292
Martinez-Martinez E, Ibarrola J, Calvier L, Fernandez-Celis A, Leroy C, Cachofeiro V, Rossignol P, Lopez-Andres N (2016) Galectin-3 blockade reduces renal fibrosis in two normotensive experimental models of renal damage. PLoS ONE 11:e0166272
McCormack PL (2015) Nintedanib: first global approval. Drugs 75:129–139
McVicker BL, Bennett RG (2017) Novel anti-fibrotic therapies. Front Pharmacol 8:318
Meng XM, Tang PM, Li J, Lan HY (2015) TGF-beta/Smad signaling in renal fibrosis. Front Physiol 6:82
Morrissey J, Hruska K, Guo G, Wang S, Chen Q, Klahr S (2002) Bone morphogenetic protein-7 improves renal fibrosis and accelerates the return of renal function. J Am Soc Nephrol: JASN 13(Suppl 1):S14–21
Nam DH, Lee MH, Kim JE, Song HK, Kang YS, Lee JE, Kim HW, Cha JJ, Hyun YY, Kim SH, Han SY, Han KH, Han JY, Cha DR (2012) Blockade of cannabinoid receptor 1 improves insulin resistance, lipid metabolism, and diabetic nephropathy in db/db mice. Endocrinology 153:1387–1396
Okamura DM, Pasichnyk K, Lopez-Guisa JM, Collins S, Hsu DK, Liu FT, Eddy AA (2011) Galectin-3 preserves renal tubules and modulates extracellular matrix remodeling in progressive fibrosis. Am J Physiol Renal Physiol 300:F245–253
Opocensky M, Kramer HJ, Backer A, Vernerova Z, Eis V, Cervenka L, Certikova Chabova V, Tesar V and Vaneckova I (2006) Late-onset endothelin-A receptor blockade reduces podocyte injury in homozygous Ren-2 rats despite severe hypertension. Hypertension (Dallas, Tex: 1979) 48:965–971
Pang M, Kothapally J, Mao H, Tolbert E, Ponnusamy M, Chin YE, Zhuang S (2009) Inhibition of histone deacetylase activity attenuates renal fibroblast activation and interstitial fibrosis in obstructive nephropathy. Am J Physiol Renal Physiol 297:F996–f1005

Parving HH, Lehnert H, Brochner-Mortensen J, Gomis R, Andersen S, Arner P (2001) The effect of irbesartan on the development of diabetic nephropathy in patients with type 2 diabetes. New Engl J Med 345:870–878

Parving HH, Brenner BM, McMurray JJ, de Zeeuw D, Haffner SM, Solomon SD, Chaturvedi N, Persson F, Desai AS, Nicolaides M, Richard A, Xiang Z, Brunel P, Pfeffer MA (2012) Cardiorenal end points in a trial of aliskiren for type 2 diabetes. New Engl J Med 367:2204–2213

Pergola PE, Raskin P, Toto RD, Meyer CJ, Huff JW, Grossman EB, Krauth M, Ruiz S, Audhya P, Christ-Schmidt H, Wittes J, Warnock DG (2011) Bardoxolone methyl and kidney function in CKD with type 2 diabetes. New Engl J Med 365:327–336

Perkins RM, Aboudara MC, Uy AL, Olson SW, Cushner HM, Yuan CM (2009) Effect of pentoxifylline on GFR decline in CKD: a pilot, double-blind, randomized, placebo-controlled trial. Am J Kidney Dis 53:606–616

Prakoura N, Chatziantoniou C (2017) Periostin and discoidin domain receptor 1: new biomarkers or targets for therapy of renal disease. Frontiers in medicine 4:52

RamachandraRao SP, Zhu Y, Ravasi T, McGowan TA, Toh I, Dunn SR, Okada S, Shaw MA, Sharma K (2009) Pirfenidone is renoprotective in diabetic kidney disease. J Am Soc Nephrol: JASN 20:1765–1775

Reich B, Schmidbauer K, Rodriguez Gomez M, Johannes Hermann F, Gobel N, Bruhl H, Ketelsen I, Talke Y, Mack M (2013) Fibrocytes develop outside the kidney but contribute to renal fibrosis in a mouse model. Kidney Int 84:78–89

Ricardo SD, van Goor H, Eddy AA (2008) Macrophage diversity in renal injury and repair. J Clin Invest 118:3522–3530

Richeldi L, du Bois RM, Raghu G, Azuma A, Brown KK, Costabel U, Cottin V, Flaherty KR, Hansell DM, Inoue Y, Kim DS, Kolb M, Nicholson AG, Noble PW, Selman M, Taniguchi H, Brun M, Le Maulf F, Girard M, Stowasser S, Schlenker-Herceg R, Disse B, Collard HR (2014) Efficacy and safety of nintedanib in idiopathic pulmonary fibrosis. New Engl J Med 370:2071–2082

Robalo-Cordeiro C, Campos P, Carvalho L, Borba A, Clemente S, Freitas S, Furtado S, Jesus JM, Leal C, Marques A, Melo N, Souto-Moura C, Neves S, Sousa V, Santos A, Morais A (2017) Idiopathic pulmonary fibrosis in the era of antifibrotic therapy: Searching for new opportunities grounded in evidence. Revista portuguesa de pneumologia 23:287–293

Saccon F, Gatto M, Ghirardello A, Iaccarino L, Punzi L, Doria A (2017) Role of galectin-3 in autoimmune and non-autoimmune nephropathies. Autoimmun Rev 16:34–47

Saito Y, Nakao K, Mukoyama M, Shirakami G, Itoh H, Yamada T, Arai H, Hosoda K, Suga S, Jougasaki M et al (1990) Application of monoclonal antibodies for endothelin to hypertensive research. Hypertension (Dallas, Tex: 1979) 15:734–738

Saleh MA, Pollock JS, Pollock DM (2011) Distinct actions of endothelin A-selective versus combined endothelin A/B receptor antagonists in early diabetic kidney disease. J Pharmacol Exp Ther 338:263–270

Satirapoj B, Wang Y, Chamberlin MP, Dai T, LaPage J, Phillips L, Nast CC, Adler SG (2012) Periostin: novel tissue and urinary biomarker of progressive renal injury induces a coordinated mesenchymal phenotype in tubular cells. Nephrol Dialysis Transplant: Off Publ Eur Dial Transpl Asso Eur Renal Assoc 27:2702–2711

Savikko J, Taskinen E, Von Willebrand E (2003) Chronic allograft nephropathy is prevented by inhibition of platelet-derived growth factor receptor: tyrosine kinase inhibitors as a potential therapy. Transplantation 75:1147–1153

Schauerte C, Hubner A, Rong S, Wang S, Shushakova N, Mengel M, Dettling A, Bang C, Scherf K, Koelling M, Melk A, Haller H, Thum T, Lorenzen JM (2017) Antagonism of profibrotic microRNA-21 improves outcome of murine chronic renal allograft dysfunction. Kidney Int 92:646–656

Sen K, Lindenmeyer MT, Gaspert A, Eichinger F, Neusser MA, Kretzler M, Segerer S, Cohen CD (2011) Periostin is induced in glomerular injury and expressed de novo in interstitial renal fibrosis. Am J Pathol 179:1756–1767

Sharma K, Ix JH, Mathew AV, Cho M, Pflueger A, Dunn SR, Francos B, Sharma S, Falkner B, McGowan TA, Donohue M, Ramachandrarao S, Xu R, Fervenza FC, Kopp JB (2011) Pirfenidone for diabetic nephropathy. J Am Soc Nephrol: JASN 22:1144–1151

Strutz F, Zeisberg M, Renziehausen A, Raschke B, Becker V, van Kooten C, Muller G (2001) TGF-beta 1 induces proliferation in human renal fibroblasts via induction of basic fibroblast growth factor (FGF-2). Kidney Int 59:579–592

Sugimoto H, LeBleu VS, Bosukonda D, Keck P, Taduri G, Bechtel W, Okada H, Carlson W Jr, Bey P, Rusckowski M, Tampe B, Tampe D, Kanasaki K, Zeisberg M, Kalluri R (2012) Activin-like kinase 3 is important for kidney regeneration and reversal of fibrosis. Nat Med 18:396–404

Sullivan T, Miao Z, Dairaghi DJ, Krasinski A, Wang Y, Zhao BN, Baumgart T, Ertl LS, Pennell A, Seitz L, Powers J, Zhao R, Ungashe S, Wei Z, Boring L, Tsou CL, Charo I, Berahovich RD, Schall TJ, Jaen JC (2013) CCR118 antagonist CCX140-B provides renal and glycemic benefits in diabetic transgenic human CCR118 knockin mice. Am J Physiol Renal Physiol 305:F1288–1297

Swencki-Underwood B, Mills JK, Vennarini J, Boakye K, Luo J, Pomerantz S, Cunningham MR, Farrell FX, Naso MF, Amegadzie B (2008) Expression and characterization of a human BMP-7 variant with improved biochemical properties. Protein Expr Purif 57:312–319

Tager AM, LaCamera P, Shea BS, Campanella GS, Selman M, Zhao Z, Polosukhin V, Wain J, Karimi-Shah BA, Kim ND, Hart WK, Pardo A, Blackwell TS, Xu Y, Chun J, Luster AD (2008) The lysophosphatidic acid receptor LPA1 links pulmonary fibrosis to lung injury by mediating fibroblast recruitment and vascular leak. Nat Med 14:45–54

Tahrani AA, Barnett AH, Bailey CJ (2016) Pharmacology and therapeutic implications of current drugs for type 2 diabetes mellitus. Nat Rev Endocrinol 12:566–592

Tam J (2016) The emerging role of the endocannabinoid system in the pathogenesis and treatment of kidney diseases. J Basic Clin Physiol Pharmacol 27:267–276

Tampe B, Zeisberg M (2014a) Evidence for the involvement of epigenetics in the progression of renal fibrogenesis. Nephrol Dialysis Transplant: Off Publ Eur Dial Transpl Asso Eur Renal Assoc 29(Suppl 1):i1–i8

Tampe D, Zeisberg M (2014b) Potential approaches to reverse or repair renal fibrosis. Nat Rev Nephrol 10:226–237

Taniguchi H, Ebina M, Kondoh Y, Ogura T, Azuma A, Suga M, Taguchi Y, Takahashi H, Nakata K, Sato A, Takeuchi M, Raghu G, Kudoh S, Nukiwa T (2010) Pirfenidone in idiopathic pulmonary fibrosis. Eur Respir J 35:821–829

Trachtman H, Fervenza FC, Gipson DS, Heering P, Jayne DR, Peters H, Rota S, Remuzzi G, Rump LC, Sellin LK, Heaton JP, Streisand JB, Hard ML, Ledbetter SR, Vincenti F (2011) A phase 1, single-dose study of fresolimumab, an anti-TGF-beta antibody, in treatment-resistant primary focal segmental glomerulosclerosis. Kidney Int 79:1236–1243

Tsuchiyama Y, Wada J, Zhang H, Morita Y, Hiragushi K, Hida K, Shikata K, Yamamura M, Kanwar YS, Makino H (2000) Efficacy of galectins in the amelioration of nephrotoxic serum nephritis in Wistar Kyoto rats. Kidney Int 58:1941–1952

Udi S, Hinden L, Earley B, Drori A, Reuveni N, Hadar R, Cinar R, Nemirovski A, Tam J (2017) Proximal tubular cannabinoid-1 receptor regulates obesity-induced CKD. J Am Soc Nephrol: JASN 28:3518–3532

Vallon V, Gerasimova M, Rose MA, Masuda T, Satriano J, Mayoux E, Koepsell H, Thomson SC, Rieg T (2014) SGLT2 inhibitor empagliflozin reduces renal growth and albuminuria in proportion to hyperglycemia and prevents glomerular hyperfiltration in diabetic Akita mice. Am J Physiol Renal Physiol 306:F194–204

Van Bergen T, Marshall D, Van de Veire S, Vandewalle E, Moons L, Herman J, Smith V, Stalmans I (2013) The role of LOX and LOXL2 in scar formation after glaucoma surgery. Invest Ophthalmol Vis Sci 54:5788–5796

Vethe H, Finne K, Skogstrand T, Vaudel M, Vikse BE, Hultstrom M, Placier S, Scherer A, Tenstad O, Marti HP (2015) Distinct protein signature of hypertension-induced damage in the renal proteome of the two-kidney, one-clip rat model. J Hypertens 33:126–135

Vincenti F, Fervenza FC, Campbell KN, Diaz M, Gesualdo L, Nelson P, Praga M, Radhakrishnan J, Sellin L, Singh A, Thornley-Brown D, Veronese FV, Accomando B, Engstrand S, Ledbetter S, Lin J, Neylan J, Tumlin J (2017) A phase 2, double-blind, placebo-controlled, randomized study of fresolimumab in patients with steroid-resistant primary focal segmental glomerulosclerosis. Kidney Int Rep 2:800–810

Virdis A, Neves MF, Amiri F, Touyz RM, Schiffrin EL (2004) Role of NAD(P)H oxidase on vascular alterations in angiotensin II-infused mice. J Hypertens 22:535–542

Wallace DP, White C, Savinkova L, Nivens E, Reif GA, Pinto CS, Raman A, Parnell SC, Conway SJ, Fields TA (2014) Periostin promotes renal cyst growth and interstitial fibrosis in polycystic kidney disease. Kidney Int 85:845–854

Wang W, Liu F, Chen N (2007) Peroxisome proliferator-activated receptor-gamma (PPAR-gamma) agonists attenuate the profibrotic response induced by TGF-beta1 in renal interstitial fibroblasts. Mediat Inflamm 2007:62641

Weir MR, Bakris GL, Gross C, Mayo MR, Garza D, Stasiv Y, Yuan J, Berman L, Williams GH (2016) Treatment with patiromer decreases aldosterone in patients with chronic kidney disease and hyperkalemia on renin-angiotensin system inhibitors. Kidney Int 90:696–704

Wen WX, Lee SY, Siang R, Koh RY (2017) Repurposing pentoxifylline for the treatment of fibrosis: an overview. Adv Therapy 34:1245–1269

Wenzel RR, Littke T, Kuranoff S, Jurgens C, Bruck H, Ritz E, Philipp T, Mitchell A (2009) Avosentan reduces albumin excretion in diabetics with macroalbuminuria. J Am Soc Nephrol: JASN 20:655–664

Wilcox CS (2002) Reactive oxygen species: roles in blood pressure and kidney function. Curr Hypertens Rep 4:160–166

Woodcock HV, Molyneaux PL, Maher TM (2013) Reducing lung function decline in patients with idiopathic pulmonary fibrosis: potential of nintedanib. Drug Des Dev Therapy 7:503–510

Yale JF, Bakris G, Cariou B, Nieto J, David-Neto E, Yue D, Wajs E, Figueroa K, Jiang J, Law G, Usiskin K, Meininger G (2014) Efficacy and safety of canagliflozin over 52 weeks in patients with type 2 diabetes mellitus and chronic kidney disease. Diabetes Obes Metab 16:1016–1027

Yan Y, Ma L, Zhou X, Ponnusamy M, Tang J, Zhuang MA, Tolbert E, Bayliss G, Bai J and Zhuang S (2015) Src inhibition blocks renal interstitial fibroblast activation and ameliorates renal fibrosis. Kidney Int

Yusuf S, Teo KK, Pogue J, Dyal L, Copland I, Schumacher H, Dagenais G, Sleight P, Anderson C (2008) Telmisartan, ramipril, or both in patients at high risk for vascular events. New Engl J Med 358:1547–1559

Zarjou A, Yang S, Abraham E, Agarwal A, Liu G (2011) Identification of a microRNA signature in renal fibrosis: role of miR-21. Am J Physiol Renal Physiol 301:F793–801

Zeisberg M, Hanai J, Sugimoto H, Mammoto T, Charytan D, Strutz F, Kalluri R (2003) BMP-7 counteracts TGF-beta1-induced epithelial-to-mesenchymal transition and reverses chronic renal injury. Nat Med 9:964–968

Zhou X, Zang X, Ponnusamy M, Masucci MV, Tolbert E, Gong R, Zhao TC, Liu N, Bayliss G, Dworkin LD, Zhuang S (2016) Enhancer of Zeste Homolog 2 Inhibition Attenuates Renal Fibrosis by Maintaining Smad7 and Phosphatase and Tensin Homolog Expression. J Am Soc Nephrol: JASN 27:2092–2108

Zoja C, Corna D, Rottoli D, Zanchi C, Abbate M, Remuzzi G (2006) Imatinib ameliorates renal disease and survival in murine lupus autoimmune disease. Kidney Int 70:97–103

Chapter 32
Renal Injury Repair: How About the Role of Stem Cells

Jian-Si Li and Bing Li

Abstract Renal failure is one of the most important causes of mortality and morbidity all over the world. Acute kidney injury (AKI) is a major clinical problem that affects up to 5% of all hospitalized patients. Although the kidney has a remarkable capacity for regeneration after acute injury, the mortality among patients with severe AKI remains dismally high, and in clinical practice, most patients cannot be cured completely and suffer from chronic kidney disease (CKD). Recently, the incidence and prevalence of CKD have increased, largely as a result of the enhanced prevalence of diabetes and obesity. The progressive nature of CKD and the ensuing end-stage renal disease (ESRD) place a substantial burden on global healthcare resources. Currently, dialysis and transplantation remain the only treatment options. Finding new therapeutic methods to fight AKI and CKD remains an ongoing quest. Although the human renal histological structure is complex, stem cell therapies have been applied to repair injured kidneys. The curative effects of mesenchymal stem cells (MSCs), hematopoietic stem cells (HSCs), induced pluripotent stem cells (iPSCs), and nephron progenitor cells (NPCs) on renal repair have also been reported by researchers. This review focuses on stem cell therapy and mechanisms for renal injury repair.

Keywords Stem cell · Renal repair · Acute kidney injury · Chronic kidney disease

32.1 Introduction

Tissue repair and regeneration is limited in the mammalian kidney, and it is necessary for kidney to reacquire functionality after ischemic, toxic, or inflammatory insults. Mesenchymal stem cells (MSCs) are multi-potent cells with self-renewal, proliferative, regenerative, and the potential of multi-lineage differentiation (Charbord 2010). MSCs which characterized by the expression of MSC markers can differentiate into adipocytes, osteocytes, and chondrocytes (Dominici et al. 2006). New evidence sup-

J.-S. Li · B. Li (✉)
Department of Nephrology, 2nd Affiliated Hospital, Harbin Medical University, Harbin, China
e-mail: icecreamlee@hotmail.com

B.-C. Liu et al. (eds.), *Renal Fibrosis: Mechanisms and Therapies*,
Advances in Experimental Medicine and Biology 1165,
https://doi.org/10.1007/978-981-13-8871-2_32

ports that the kidney-resident MSCs are originated from renal pericytes and form an extensive network around the microvasculature (Bruno et al. 2014). Kidney-resident MSCs play key roles in regulating renal blood flow, endothelial survival, capillary permeability, and immunologic surveillance (Kramann and Humphreys 2014). Hematopoietic stem cells (HSCs) are undifferentiated cells with the capability of self-renewal and stepwise differentiation into erythrocytes, thrombocytes, and other specialized blood cells.

The Nobel Prize in physiology or medicine in 2012 was awarded jointly to Sir John B. Gurdon and Shinya Yamanaka for the discovery that mature cells can be reprogrammed to become pluripotent. Their groundbreaking discoveries have completely changed our view of the development and specialization of cells, and their findings have shown that a mature cell can be returned to an immature, pluripotent state, becoming a pluripotent stem cell under certain conditions. In 2006, Yamanaka and his coworkers developed a reprogramming method for inducing mouse skin cells to become pluripotent stem cells and called them induced pluripotent stem cells (iPSCs). Comparing the ability of differentiation into kidney cells between iPSCs and embryonic stem cells (ESCs), the results showed that both of them can differentiate into mature kidney cells. However, compared with ESCs, iPSCs tended to remain undifferentiated and less sensitive for kidney differentiation (Morizane et al. 2009). Recent advances in stem cell have observed key signals that are necessary to maintain stemness of human nephron progenitor cells (NPCs) in vitro and have established the protocols to generate NPCs and nephron epithelial cells from human fetal kidneys and human pluripotent stem cells (hPSCs). hPSCs, on account of their unlimited self-renewal and capability to generate all kind of cells (Takahashi et al. 2007a, b; Thomson et al. 1998), are suited for the derivation of NPCs to generate functional human kidney cells and tissues. This article will focus on stem cells above in renal injury repair.

32.2 Mesenchymal Stem Cells on Renal Injury Repair

MSCs are fibroblast-like cells that are capable of self-renewal and multi-lineage differentiation (Uccelli et al. 2008). Preclinical studies have demonstrated that administration of MSCs can reduce renal ischemia-reperfusion injury (IRI) and improve renal function, owing to their anti-inflammatory and immunoregulatory properties (Humphreys and Bonventre 2008; Morigi et al. 2004; Semedo et al. 2007). MSCs have the ability to accelerate tissue repair by direct migration to the injured sites and paracrine factors (Togel et al. 2007; Zhao et al. 2014). Recently, MSC-based therapy in renal IRI has attracted a lot of attention (Jang et al. 2014; Shih et al. 2013; Wise et al. 2014).

According to clinical trials, MSC therapy has been used for AKI and CKD, including focal segmental glomerulosclerosis, diabetic nephropathy, systemic lupus erythematous, and kidney transplantation (Hickson et al. 2016; Peired et al. 2016; Westenfelder and Togel 2011). Patients at high risk of postoperative AKI after cardiac surgery were safely treated with allogeneic MSCs (Kaushal and Shah 2014; Togel and Westenfelder 2012). Many studies found that BMSCs have the potential to differentiate into glomerular mesangial cells (Imasawa et al. 2001; Ito et al. 2001; Wong et al. 2014). Other experiments showed that after adding colony-stimulating factor (CSF), the conversion rate of BMSCs to renal stem cells was increased (Jia et al. 2012). Other than paracrine factors, microvesicles, which are an important mechanism of cell–cell association (Collino et al. 2010), derived from stem cells play an important role in renal repair (Bruno et al. 2012; Gatti et al. 2011; He et al. 2012). BMSCs have strong effects on restraining inflammation and fibrosis and consequently improve kidney function (Qi and Wu 2013; Reinders et al. 2014; Yuen et al. 2013).

The repair effect of BMSCs in kidney tissue was confirmed, and the homing mechanism of BMSCs requires further study. After BMSCs are transplanted into blood, they anchor to the kidney through a combination of their own adhesion molecules and chemokine receptors and those expressed on the damaged kidney. The SDF-1/CXCR4 signaling pathway plays an important role in BMSC homing and tissue repair (Liu et al. 2012, 2013a, b; Marquez-Curtis and Janowska-Wieczorek 2013; Ponte et al. 2007; Si et al. 2014; Togel and Westenfelder 2011). Liu et al. added BMSCs and vitamin E separately to the culture medium of gentamycin and renal tubular epithelial cells. The results showed that the proliferative capacity of renal tubular epithelial cells increased, significantly when BMSCs and vitamin E were added at the same time. In in vivo experiments, similar results also proved that a better effect occurred when BMSCs and vitamin E were added together than when they were separately added (Liu et al. 2013c). Recently, some researchers found that CXCR7 was a new receptor of SDF-1. Currently, CXCR7 is known to improve tumor growth in mice. Mazzinghi et al. observed that CXCR7 might play a role in the process of progenitor cell migration to the kidney, and penetration to the kidney endothelium may be induced by SDF-1 (Mazzinghi et al. 2008). Some experiments showed that after hypoxic preconditioning, the expression of CXCR7 increased on the surface of BMSCs. Further study showed that CXCR4 mainly contributed to cell migration and that CXCR7 played an important role in cell survival (Dai et al. 2011; Liu et al. 2012).

32.3 Hematopoietic Stem Cells on Renal Injury Repair

Enhancements in the mobilization, propagation, and delivery of BMSCs to the kidney present the potential for BMSCs to serve as an entirely new approach for the treatment of AKI. Stem cell factor and granulocyte CSF (G-CSF) can induce HSC homing to the injured kidney, leading to a significant improvement of the kidney functional recovery (Iwasaki et al. 2005; Stokman et al. 2005). By comparison, some reports showed data against the use of granulocytosis-inducing HSC mobilization protocols for the treatment of ischemic injury. Unlike the research above, boosting the peripheral stem cell count was associated with increased severity of renal failure and mortality. Large quantities of activated granulocytes appear to obscure the potential renoprotective effects of HSCs (Togel et al. 2004). Data against the potential of BMSCs to trans-differentiate into tubular cells after injury are reported (Dekel et al. 2006). Based on the study from transgenic mice which express green fluorescent protein (GFP) in BMSCs (Duffield et al. 2005), in mature renal tubular epithelial cells (Lin et al. 2005) or in all mesenchyme-derived renal epithelial cells, it was observed that when BMSC recruitment occurs, kidney repair is significantly triggered via proliferation of endogenous renal cells. BMSCs might produce protective and regenerative factors, rather than replace damaged cells directly by differentiation to improve the regenerative process (Humphreys et al. 2008).

Due to limited methods of tracking BMSCs, especially in injured tissues, the results of localization of BMSCs and the contribution degree of BMSCs to kidney regeneration may be inconsistent. The protocols used in these studies are different. The differences included the species, transplantation methods, types of cells used for transplantation, the detection methods for bone marrow cell origin and the specificity and sensitivity of those methods. In addition, cell fusion may also account for this contradiction (Fang et al. 2005; Held et al. 2006). Data of bone marrow cell recruitment to damaged kidneys have not establish the lineage of the recruited BMSCs. Moreover, the beneficial effects of the recruitment of BMSCs on chronic renal damage remain to be determined.

32.4 Induced Pluripotent Stem Cells on Renal Injury Repair

iPSCs (Takahashi and Yamanaka 2006) have the potential for multi-lineage differentiation and provide stem-cell-based treatment. They are transformed from somatic cells via expression of selected transcription factors. Recently, unique methods have been developed for stimulating the differentiation of human iPSCs into cells with kidney lineages (Araoka et al. 2014; Lam et al. 2014; Mae et al. 2013), and three-dimensional structures of the kidney (Taguchi et al. 2014) have been developed. iPSCs have been established from normal human mesangial cells (Song et al. 2011), renal tubular cells present in urine (Zhou et al. 2011, 2012), and fibroblasts of patients with autosomal dominant polycystic kidney disease (Freedman et al. 2013). With the further study of reprogrammed kidney iPSCs, the research progress of genetic kidney diseases may improve and the development of novel therapies may occur soon. The therapeutic effect of iPSCs on renal ischemia was also reported. The expression of oxidative substances, proinflammatory cytokines, and apoptotic factors was reduced by transplantation of iPSCs. Eventually, iPSCs contributed to improve the survival rate of rats with ischemic AKI (Lee et al. 2012).

Recent study provided a new disease-modeling platform for functional genomics in the directed differentiation of human iPSCs (hiPSCs) toward renal lineages (Takasato et al. 2015). Now, the generation of iPSCs from human skin or blood samples is convenient (Takahashi et al. 2007a). In addition, the study of CRISPR/Cas9 corrected a candidate variant of unknown significance (VUS). It enables the generation of isogenic control iPSC lines. Meanwhile, within an experiment, the isolation of the effect of the VUS was corrected and created by a non-isogenic control the removal of the confounding influences of genomic variation was amended. Academically, with the isolation of the phenotypic effect of the proband's specific allele, the comparison of organoids generated from the proband and matched gene-corrected iPSC lines should be considered. A disease phenotype within a proband-derived organoid has not been proved previously. However, kidney organoids represent a model of the developing nephron currently. Previously, iPSC-based kidney disease model relied on the generation of mutations by CRISPR/Cas9-induced nonhomologous end joining in wild-type human cell lines has been reported (Cruz et al. 2017; Freedman et al. 2015). The proband iPSC-derived renal organoids model reveals the proband-specific alleles for any candidate variant. At the same time, the genomic background of that individual is presented. The finding allowed for contributions to phenotype from gene modifiers or allowed the dissection of multigene disorders. Importantly, if a modeled candidate VUS ultimately proves nondisease association, data based on the study of the proband-derived organoids will keep on expressing the VUS, potentially permitting a retrospective diagnosis.

32.5 Nephron Progenitor Cells on Renal Injury Repair

The discovery of endogenous stem/progenitor cell systems in other organs has forced the identification of innovative therapeutic strategies for regenerative medicine and tissue engineering (Radtke and Clevers 2005; Solanas and Benitah 2013). Recently, the identification of endogenous stem/progenitor systems in kidney was still a challenge. Patients with renal disease can readily generate hiPSCs which enable the development of immunocompatible tissues as well as patient-specific models of renal disease (Takahashi et al. 2007b). New findings of directed differentiation protocols toward kidney lineage cells have been reported, which lead to the generation of NPCs and kidney organoids from human ESCs (husks) and hiPSCs (Lam et al. 2014; Morizane et al. 2015; Taguchi et al. 2014; Takasato et al. 2014, 2015).

In humans, CD133+CD24+ renal epithelial cells represent a hierarchical population of NPCs, including parietal epithelial cells and a scattered population of tubular epithelial cells, represent approximately 2–4% of the total number of renal cells (Romagnani et al. 2013; Sagrinati et al. 2006). These cells possess the potential for self-renewal, the ability to resist senescence, the capacity to grow in culture as spheres, and the capability to differentiate into different types of renal epithelial cells in vitro, such as podocytes and tubular epithelial cells. They can also differentiate into adipocytes, endothelial cells, osteoblasts, and neuronal cells (Ronconi et al. 2009; Sagrinati et al. 2006). Recent advances in angiomyolipomas in tuberous sclerosis are derived from a multi-potent cancer stem cell that originates from the renal epithelium. The hypothesis that the renal epithelial cells have the potential of differentiation ability more than the epithelial phenotype has been confirmed. Formerly based on lineage-tracing experiments performed in mouse models of AKI, it was considered to be its only possible lineage (Cho et al. 2017; Goncalves et al. 2017). A few independent studies have certificated that adult renal progenitors had treatment effects in immunodeficient mice with rhabdomyolysis-induced AKI (Bongso et al. 1994; Bussolati et al. 2005; Grange et al. 2014; Mazzinghi et al. 2008; Sagrinati et al. 2006). The administration of CD133+CD24+ cells isolated from Bowman's capsule elevated the renal outcomes in a podocyte injury model of focal segmental glomerulosclerosis (FSGS) (Ronconi et al. 2009). In Adriamycin nephropathy mice model, these cells isolated from urine also differentiated into podocytes and reduced proteinuria (Lazzeri et al. 2015).

Molecules which have been demonstrated can regulate NPC-mediated kidney regeneration included the blockers of chemokine SDF-1 (Darisipudi et al. 2011), Notch signaling inhibitors (Peired et al. 2013), retinoic acid (Peired et al. 2013), and several drugs such as interferon (Migliorini et al. 2013), steroids (Zhang et al. 2013), renin–angiotensin–aldosterone system inhibitors (Rizzo et al. 2013), and leptin (Pichaiwong et al. 2013). The pharmacological agents above can improve renal progenitor differentiation into podocytes, support glomerular regeneration and block NPC hyperactivation.

32.6 Conclusion

Although the results are controversial, and many questions remain unanswered, stem-cell-based therapy is a potential strategy for acute and chronic kidney diseases (Asanuma et al. 2010). There are difficulties in the process of studying stem cells, but these studies may provide many new views on diagnosis and therapy.

Acknowledgements This study was supported by the National Natural Science Foundation of China (No. 81670616 and No. 81873598), the innovation fund for the clinical research of the Harbin Medical University (2017LCZX44).

References

Araoka T, Mae S, Kurose Y, Uesugi M, Ohta A, Yamanaka S et al (2014) Efficient and rapid induction of human iPSCs/ESCs into nephrogenic intermediate mesoderm using small molecule-based differentiation methods. PLoS ONE 9:e84881

Asanuma H, Meldrum DR, Meldrum KK (2010) Therapeutic applications of mesenchymal stem cells to repair kidney injury. J Urol 184:26–33

Bongso A, Fong CY, Ng SC, Ratnam S (1994) Isolation and culture of inner cell mass cells from human blastocysts. Hum Reprod 9:2110–2117

Bruno S, Chiabotto G, Camussi G (2014) Concise review: different mesenchymal stromal/stem cell populations reside in the adult kidney. Stem Cells Transl Med 3:1451–1455

Bruno S, Grange C, Collino F, Deregibus MC, Cantaluppi V, Biancone L et al (2012) Microvesicles derived from mesenchymal stem cells enhance survival in a lethal model of acute kidney injury. PLoS ONE 7:e33115

Bussolati B, Bruno S, Grange C, Buttiglieri S, Deregibus MC, Cantino D et al (2005) Isolation of renal progenitor cells from adult human kidney. Am J Pathol 166:545–555

Charbord P (2010) Bone marrow mesenchymal stem cells: historical overview and concepts. Hum Gene Ther 21:1045–1056

Cho JH, Patel B, Bonala S, Manne S, Zhou Y, Vadrevu SK et al (2017) Notch transactivates Rheb to maintain the multipotency of TSC-null cells. Nat Commun 8:1848

Collino F, Deregibus MC, Bruno S, Sterpone L, Aghemo G, Viltono L et al (2010) Microvesicles derived from adult human bone marrow and tissue specific mesenchymal stem cells shuttle selected pattern of miRNAs. PLoS ONE 5:e11803

Cruz NM, Song X, Czerniecki SM, Gulieva RE, Churchill AJ, Kim YK et al (2017) Organoid cystogenesis reveals a critical role of microenvironment in human polycystic kidney disease. Nat Mater 16:1112–1119

Dai X, Tan Y, Cai S, Xiong X, Wang L, Ye Q et al (2011) The role of CXCR11 on the adhesion, proliferation and angiogenesis of endothelial progenitor cells. J Cell Mol Med 15:1299–1309

Darisipudi MN, Kulkarni OP, Sayyed SG, Ryu M, Migliorini A, Sagrinati C et al (2011) Dual blockade of the homeostatic chemokine CXCL12 and the proinflammatory chemokine CCL2 has additive protective effects on diabetic kidney disease. Am J Pathol 179:116–124

Dekel B, Shezen E, Even-Tov-Friedman S, Katchman H, Margalit R, Nagler A et al (2006) Transplantation of human hematopoietic stem cells into ischemic and growing kidneys suggests a role in vasculogenesis but not tubulogenesis. Stem Cells 24:1185–1193

Dominici M, Le Blanc K, Mueller I, Slaper-Cortenbach I, Marini F, Krause D et al (2006) Minimal criteria for defining multipotent mesenchymal stromal cells. The International Society for Cellular Therapy position statement. Cytotherapy 8:315–317

Duffield JS, Park KM, Hsiao LL, Kelley VR, Scadden DT, Ichimura T et al (2005) Restoration of tubular epithelial cells during repair of the postischemic kidney occurs independently of bone marrow-derived stem cells. J Clin Invest 115:1743–1755

Fang TC, Alison MR, Cook HT, Jeffery R, Wright NA, Poulsom R (2005) Proliferation of bone marrow-derived cells contributes to regeneration after folic acid-induced acute tubular injury. J Am Soc Nephrol 16:1723–1732

Freedman BS, Brooks CR, Lam AQ, Fu H, Morizane R, Agrawal V et al (2015) Modelling kidney disease with CRISPR-mutant kidney organoids derived from human pluripotent epiblast spheroids. Nat Commun 6:8715

Freedman BS, Lam AQ, Sundsbak JL, Iatrino R, Su X, Koon SJ et al (2013) Reduced ciliary polycystin-2 in induced pluripotent stem cells from polycystic kidney disease patients with PKD1 mutations. J Am Soc Nephrol 24:1571–1586

Gatti S, Bruno S, Deregibus MC, Sordi A, Cantaluppi V, Tetta C et al (2011) Microvesicles derived from human adult mesenchymal stem cells protect against ischaemia-reperfusion-induced acute and chronic kidney injury. Nephrol Dial Transplant 26:1474–1483

Goncalves AF, Adlesic M, Brandt S, Hejhal T, Harlander S, Sommer L et al (2017) Evidence of renal angiomyolipoma neoplastic stem cells arising from renal epithelial cells. Nat Commun 8:1466

Grange C, Moggio A, Tapparo M, Porta S, Camussi G, Bussolati B (2014) Protective effect and localization by optical imaging of human renal CD133 + progenitor cells in an acute kidney injury model. Physiol Rep 2:e12009

He J, Wang Y, Sun S, Yu M, Wang C, Pei X et al (2012) Bone marrow stem cells-derived microvesicles protect against renal injury in the mouse remnant kidney model. Nephrology (Carlton) 17:493–500

Held PK, Al-Dhalimy M, Willenbring H, Akkari Y, Jiang S, Torimaru Y et al (2006) In vivo genetic selection of renal proximal tubules. Mol Ther 13:49–58

Hickson LJ, Eirin A, Lerman LO (2016) Challenges and opportunities for stem cell therapy in patients with chronic kidney disease. Kidney Int 89:767–778

Humphreys BD, Bonventre JV (2008) Mesenchymal stem cells in acute kidney injury. Annu Rev Med 59:311–325
Humphreys BD, Valerius MT, Kobayashi A, Mugford JW, Soeung S, Duffield JS et al (2008) Intrinsic epithelial cells repair the kidney after injury. Cell Stem Cell 2:284–291
Imasawa T, Utsunomiya Y, Kawamura T, Zhong Y, Nagasawa R, Okabe M et al (2001) The potential of bone marrow-derived cells to differentiate to glomerular mesangial cells. J Am Soc Nephrol 12:1401–1409
Ito T, Suzuki A, Imai E, Okabe M, Hori M (2001) Bone marrow is a reservoir of repopulating mesangial cells during glomerular remodeling. J Am Soc Nephrol 12:2625–2635
Iwasaki M, Adachi Y, Minamino K, Suzuki Y, Zhang Y, Okigaki M et al (2005) Mobilization of bone marrow cells by G-CSF rescues mice from cisplatin-induced renal failure, and M-CSF enhances the effects of G-CSF. J Am Soc Nephrol 16:658–666
Jang HR, Park JH, Kwon GY, Lee JE, Huh W, Jin HJ et al (2014) Effect of preemptive treatment with human umbilical cord blood-derived mesenchymal stem cells on the development of renal ischemia-reperfusion injury in mice. Am J Physiol Renal Physiol 307:F1149–F1161
Jia X, Xie X, Feng G, Lu H, Zhao Q, Che Y et al (2012) Bone marrow-derived cells can acquire renal stem cells properties and ameliorate ischemia-reperfusion induced acute renal injury. BMC Nephrol 13:105
Kaushal GP, Shah SV (2014) Challenges and advances in the treatment of AKI. J Am Soc Nephrol 25:877–883
Kramann R, Humphreys BD (2014) Kidney pericytes: roles in regeneration and fibrosis. Semin Nephrol 34:374–383
Lam AQ, Freedman BS, Morizane R, Lerou PH, Valerius MT, Bonventre JV (2014) Rapid and efficient differentiation of human pluripotent stem cells into intermediate mesoderm that forms tubules expressing kidney proximal tubular markers. J Am Soc Nephrol 25:1211–1225
Lazzeri E, Ronconi E, Angelotti ML, Peired A, Mazzinghi B, Becherucci F et al (2015) Human urine-derived renal progenitors for personalized modeling of genetic kidney disorders. J Am Soc Nephrol 26:1961–1974
Lee PY, Chien Y, Chiou GY, Lin CH, Chiou CH, Tarng DC (2012) Induced pluripotent stem cells without c-Myc attenuate acute kidney injury via downregulating the signaling of oxidative stress and inflammation in ischemia-reperfusion rats. Cell Transplant 21:2569–2585
Lin F, Moran A, Igarashi P (2005) Intrarenal cells, not bone marrow-derived cells, are the major source for regeneration in postischemic kidney. J Clin Invest 115:1756–1764
Liu H, Liu S, Li Y, Wang X, Xue W, Ge G et al (2012) The role of SDF-1-CXCR38/CXCR38 axis in the therapeutic effects of hypoxia-preconditioned mesenchymal stem cells for renal ischemia/reperfusion injury. PLoS ONE 7:e34608
Liu N, Han G, Cheng J, Huang J, Tian J (2013a) Erythropoietin promotes the repair effect of acute kidney injury by bone-marrow mesenchymal stem cells transplantation. Exp Biol Med (Maywood) 238:678–686
Liu N, Patzak A, Zhang J (2013b) CXCR40-overexpressing bone marrow-derived mesenchymal stem cells improve repair of acute kidney injury. Am J Physiol Renal Physiol 305:F1064–F1073
Liu P, Feng Y, Dong C, Liu D, Wu X, Wu H et al (2013c) Study on therapeutic action of bone marrow derived mesenchymal stem cell combined with vitamin E against acute kidney injury in rats. Life Sci 92:829–837
Mae SI, Shono A, Shiota F, Yasuno T, Kajiwara M, Gotoda-Nishimura N et al (2013) Monitoring and robust induction of nephrogenic intermediate mesoderm from human pluripotent stem cells. Nat Commun 4:1367
Marquez-Curtis LA, Janowska-Wieczorek A (2013) Enhancing the migration ability of mesenchymal stromal cells by targeting the SDF-1/CXCR43 axis. Biomed Res Int 2013:561098
Mazzinghi B, Ronconi E, Lazzeri E, Sagrinati C, Ballerini L, Angelotti ML et al (2008) Essential but differential role for CXCR44 and CXCR44 in the therapeutic homing of human renal progenitor cells. J Exp Med 205:479–490

Migliorini A, Angelotti ML, Mulay SR, Kulkarni OO, Demleitner J, Dietrich A et al (2013) The antiviral cytokines IFN-alpha and IFN-beta modulate parietal epithelial cells and promote podocyte loss: implications for IFN toxicity, viral glomerulonephritis, and glomerular regeneration. Am J Pathol 183:431–440

Morigi M, Imberti B, Zoja C, Corna D, Tomasoni S, Abbate M et al (2004) Mesenchymal stem cells are renotropic, helping to repair the kidney and improve function in acute renal failure. J Am Soc Nephrol 15:1794–1804

Morizane R, Lam AQ, Freedman BS, Kishi S, Valerius MT, Bonventre JV (2015) Nephron organoids derived from human pluripotent stem cells model kidney development and injury. Nat Biotechnol 33:1193–1200

Morizane R, Monkawa T, Itoh H (2009) Differentiation of murine embryonic stem and induced pluripotent stem cells to renal lineage in vitro. Biochem Biophys Res Commun 390:1334–1339

Peired A, Angelotti ML, Ronconi E, la Marca G, Mazzinghi B, Sisti A et al (2013) Proteinuria impairs podocyte regeneration by sequestering retinoic acid. J Am Soc Nephrol 24:1756–1768

Peired AJ, Sisti A, Romagnani P (2016) Mesenchymal stem cell-based therapy for kidney disease: a review of clinical evidence. Stem Cells Int 2016:4798639

Pichaiwong W, Hudkins KL, Wietecha T, Nguyen TQ, Tachaudomdach C, Li W et al (2013) Reversibility of structural and functional damage in a model of advanced diabetic nephropathy. J Am Soc Nephrol 24:1088–1102

Ponte AL, Marais E, Gallay N, Langonne A, Delorme B, Herault O et al (2007) The in vitro migration capacity of human bone marrow mesenchymal stem cells: comparison of chemokine and growth factor chemotactic activities. Stem Cells 25:1737–1745

Qi S, Wu D (2013) Bone marrow-derived mesenchymal stem cells protect against cisplatin-induced acute kidney injury in rats by inhibiting cell apoptosis. Int J Mol Med 32:1262–1272

Radtke F, Clevers H (2005) Self-renewal and cancer of the gut: two sides of a coin. Science 307:1904–1909

Reinders ME, de Fijter JW, Rabelink TJ (2014) Mesenchymal stromal cells to prevent fibrosis in kidney transplantation. Curr Opin Organ Transplant 19:54–59

Rizzo P, Perico N, Gagliardini E, Novelli R, Alison MR, Remuzzi G et al (2013) Nature and mediators of parietal epithelial cell activation in glomerulonephritides of human and rat. Am J Pathol 183:1769–1778

Romagnani P, Lasagni L, Remuzzi G (2013) Renal progenitors: an evolutionary conserved strategy for kidney regeneration. Nat Rev Nephrol 9:137–146

Ronconi E, Sagrinati C, Angelotti ML, Lazzeri E, Mazzinghi B, Ballerini L et al (2009) Regeneration of glomerular podocytes by human renal progenitors. J Am Soc Nephrol 20:322–332

Sagrinati C, Netti GS, Mazzinghi B, Lazzeri E, Liotta F, Frosali F et al (2006) Isolation and characterization of multipotent progenitor cells from the Bowman's capsule of adult human kidneys. J Am Soc Nephrol 17:2443–2456

Semedo P, Wang PM, Andreucci TH, Cenedeze MA, Teixeira VP, Reis MA et al (2007) Mesenchymal stem cells ameliorate tissue damages triggered by renal ischemia and reperfusion injury. Transplant Proc 39:421–423

Shih YC, Lee PY, Cheng H, Tsai CH, Ma H, Tarng DC (2013) Adipose-derived stem cells exhibit antioxidative and antiapoptotic properties to rescue ischemic acute kidney injury in rats. Plast Reconstr Surg 132:940e–951e

Si XY, Li JJ, Yao T, Wu XY (2014) Transforming growth factor-beta1 in the microenvironment of ischemia reperfusion-injured kidney enhances the chemotaxis of mesenchymal stem cells to stromal cell-derived factor-1 through upregulation of surface chemokine (C-X-C motif) receptor 4. Mol Med Rep 9:1794–1798

Solanas G, Benitah SA (2013) Regenerating the skin: a task for the heterogeneous stem cell pool and surrounding niche. Nat Rev Mol Cell Biol 14:737–748

Song B, Niclis JC, Alikhan MA, Sakkal S, Sylvain A, Kerr PG et al (2011) Generation of induced pluripotent stem cells from human kidney mesangial cells. J Am Soc Nephrol 22:1213–1220

Stokman G, Leemans JC, Claessen N, Weening JJ, Florquin S (2005) Hematopoietic stem cell mobilization therapy accelerates recovery of renal function independent of stem cell contribution. J Am Soc Nephrol 16:1684–1692
Taguchi A, Kaku Y, Ohmori T, Sharmin S, Ogawa M, Sasaki H et al (2014) Redefining the in vivo origin of metanephric nephron progenitors enables generation of complex kidney structures from pluripotent stem cells. Cell Stem Cell 14:53–67
Takahashi K, Okita K, Nakagawa M, Yamanaka S (2007a) Induction of pluripotent stem cells from fibroblast cultures. Nat Protoc 2:3081–3089
Takahashi K, Tanabe K, Ohnuki M, Narita M, Ichisaka T, Tomoda K et al (2007b) Induction of pluripotent stem cells from adult human fibroblasts by defined factors. Cell 131:861–872
Takahashi K, Yamanaka S (2006) Induction of pluripotent stem cells from mouse embryonic and adult fibroblast cultures by defined factors. Cell 126:663–676
Takasato M, Er PX, Becroft M, Vanslambrouck JM, Stanley EG, Elefanty AG et al (2014) Directing human embryonic stem cell differentiation towards a renal lineage generates a self-organizing kidney. Nat Cell Biol 16:118–126
Takasato M, Er PX, Chiu HS, Maier B, Baillie GJ, Ferguson C et al (2015) Kidney organoids from human iPS cells contain multiple lineages and model human nephrogenesis. Nature 526:564–568
Thomson JA, Itskovitz-Eldor J, Shapiro SS, Waknitz MA, Swiergiel JJ, Marshall VS et al (1998) Embryonic stem cell lines derived from human blastocysts. Science 282:1145–1147
Togel F, Isaac J, Westenfelder C (2004) Hematopoietic stem cell mobilization-associated granulocytosis severely worsens acute renal failure. J Am Soc Nephrol 15:1261–1267
Togel F, Weiss K, Yang Y, Hu Z, Zhang P, Westenfelder C (2007) Vasculotropic, paracrine actions of infused mesenchymal stem cells are important to the recovery from acute kidney injury. Am J Physiol Renal Physiol 292:F1626–F1635
Togel FE, Westenfelder C (2011) Role of SDF-1 as a regulatory chemokine in renal regeneration after acute kidney injury. Kidney Int 1(Suppl):87–89
Togel FE, Westenfelder C (2012) Kidney protection and regeneration following acute injury: progress through stem cell therapy. Am J Kidney Dis 60:1012–1022
Uccelli A, Moretta L, Pistoia V (2008) Mesenchymal stem cells in health and disease. Nat Rev Immunol 8:726–736
Westenfelder C, Togel FE (2011) Protective actions of administered mesenchymal stem cells in acute kidney injury: relevance to clinical trials. Kidney Int 1(Suppl):103–106
Wise AF, Williams TM, Kiewiet MB, Payne NL, Siatskas C, Samuel CS et al (2014) Human mesenchymal stem cells alter macrophage phenotype and promote regeneration via homing to the kidney following ischemia-reperfusion injury. Am J Physiol Renal Physiol 306:F1222–F1235
Wong CY, Tan EL, Cheong SK (2014) In vitro differentiation of mesenchymal stem cells into mesangial cells when co-cultured with injured mesangial cells. Cell Biol Int 38:497–501
Yuen DA, Connelly KA, Zhang Y, Advani SL, Thai K, Kabir G et al (2013) Early outgrowth cells release soluble endocrine antifibrotic factors that reduce progressive organ fibrosis. Stem Cells 31:2408–2419
Zhang J, Pippin JW, Krofft RD, Naito S, Liu ZH, Shankland SJ (2013) Podocyte repopulation by renal progenitor cells following glucocorticoids treatment in experimental FSGS. Am J Physiol Renal Physiol 304:F1375–F1389
Zhao JJ, Liu JL, Liu L, Jia HY (2014) Protection of mesenchymal stem cells on acute kidney injury. Mol Med Rep 9:91–96
Zhou T, Benda C, Dunzinger S, Huang Y, Ho JC, Yang J et al (2012) Generation of human induced pluripotent stem cells from urine samples. Nat Protoc 7:2080–2089
Zhou T, Benda C, Duzinger S, Huang Y, Li X, Li Y et al (2011) Generation of induced pluripotent stem cells from urine. J Am Soc Nephrol 22:1221–1228

Chapter 33
Antifibrotic Roles of RAAS Blockers: Update

Ying-Ying Zhang, Ying Yu and Chen Yu

Abstract The rennin–angiotensin–aldosterone system (RAAS) has been well documented in regulating blood pressure, fluid volume, and sodium balance. Overactivity of RAAS promotes both systemic and regional glomerular capillary hypertension, which could induce hemodynamic injury to the glomerulus, leading to kidney damage and renal fibrosis via profibrotic and proinflammatory pathway. Therefore, the use of RAAS inhibitors (i.e., ACEIs, ARBs, and MRAs) as the optional therapy has been demonstrated to prevent proteinuria, and kidney fibrosis and slow the decline of renal function effectively in the process of kidney disease during the last few decades. Recently, several new components of the RAAS have been discovered, including ACE2 and the corresponding ACE2/Ang (1-7)/Mas axis, which are also present in the kidney. Besides the classic RAAS inhibitors target the angiotensin-AT1-aldosterone axis, with the expanding knowledge about RAAS, a number of potential therapeutic targets in this system is emerging. Newer agents that are more specific are being developed. The present chapter outlines the insights of the RAAS agents (classic RAAS antagonists/the new RAAS drugs), and discusses its clinical application in the combat of renal fibrosis.

Keywords Renin–angiotensin–aldosterone system (RAAS) · Fibrosis · Antagonists

33.1 Introduction

Renal fibrosis is a common step in the progression of a variety of chronic kidney diseases to end-stage renal disease. It is characterized by excessive accumulation of extracellular matrix, representing the final target to treat chronic kidney disease (CKD). It is widely accepted that the degree of renal fibrosis correlates well with kidney function and CKD stage (Schainuck et al. 1970).

Y.-Y. Zhang · Y. Yu · C. Yu (✉)
Department of Nephrology, Tongji Hospital, Tongji University School of Medicine, Shanghai, China
e-mail: yuchen@tongji.edu.cn

B.-C. Liu et al. (eds.), *Renal Fibrosis: Mechanisms and Therapies*,
Advances in Experimental Medicine and Biology 1165,
https://doi.org/10.1007/978-981-13-8871-2_33

The rennin–angiotensin–aldosterone system (RAAS) plays a key role in regulating blood pressure, fluid volume, and sodium balance. Overactivity of the RAAS is involved in the pathology progression of a variety of diseases, such as hypertension, atherosclerosis, left ventricular hypertrophy, myocardial infarction, and heart failure. Researchers have demonstrated that the overactivity of RAAS contributed to the progression of renal fibrosis and that RAAS antagonists prevented renal fibrosis and slowed the decline in renal function in patients with kidney disease.

In 1971, Oparil, S et al. described the main cascade of the RAAS system (Oparil and Haber 1971). Plasma angiotensinogen is cleaved by renal renin, generating angiotensin I (AngI), which is then converted to angiotensin II (AngII) by endothelial angiotensin-converting enzyme (ACE). AngII is considered the most important RAAS peptide and is associated with vasoconstriction and high blood pressure. AngII binds to the type–1 AngII receptor (AT1) in a variety of tissues. Then, aldosterone is stimulated via the AT1 receptor in the adrenal gland, facilitating sodium retention by the kidney when aldosterone binds to the mineralocorticoid receptor. More recently, several new components of the RAAS have been discovered, including ACE2 and the corresponding ACE2/Ang (1-7)/Mas axis, which are also present in the kidney.

The classic RAAS inhibitors target the angiotensin-AT1-aldosterone axis. However, with the expanding knowledge about RAAS, the number of potential therapeutic targets in this system is increasing. In this secession, we discuss novel agonists and antagonists of the RAAS that might combat renal fibrosis (Fig. 33.1).

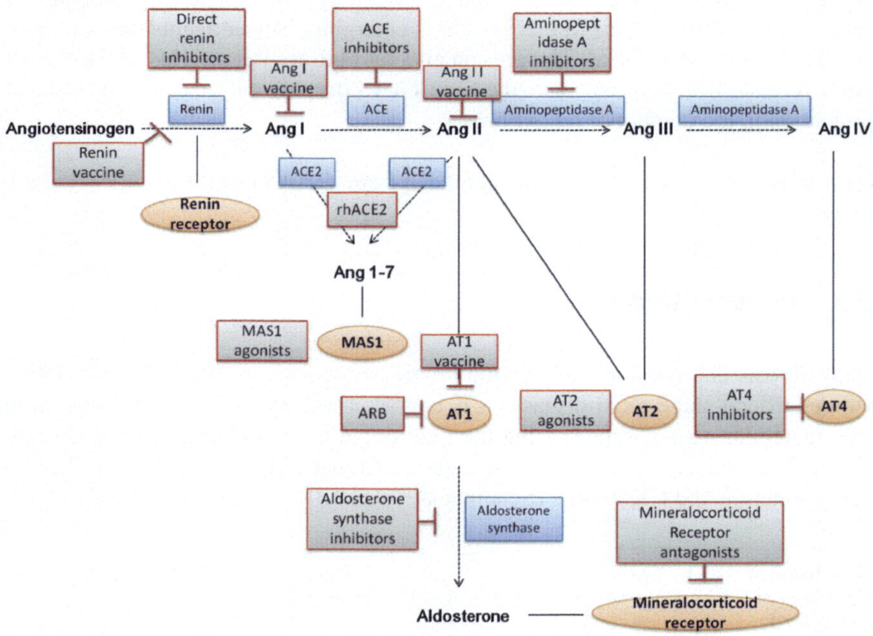

Fig. 33.1 Antifibrotic role of RAAS blockers in renal fibrosis

Multiple drugs have been well established to interfere with RAAS at different levels, such as renin inhibitors, ACE inhibitors, ARBs, and mineralocorticoid receptor antagonists, which directly inhibits renin, ACE, AT1R, and the mineralocorticoid receptor, respectively. Novel blockers are developed to target Aminopeptidase A, the enzyme that catalyzes the conversion of Ang II to Ang III, and Ang III to Ang IV. On the other hand, replenishment of RhACE2 are used to activate ACE2, the enzyme that catalyzes the conversion of Ang I and Ang II to Ang (1-7). Moreover, novel agonists have been designed to target AT2, AT4 and MAS1 receptors. In addition to inhibitors and agonists, alternative strategies such as vaccines specifically target rennin, AngI, AngII, and AT1 receptor have also been developed. ACE: angiotensin-converting enzyme; Ang: angiotensin; ARB: angiotensin receptor blocker; AT1: type-1 Ang II receptor; AT2: type-2 Ang II receptor; AT4: type-1 Ang II receptor; MAS1: proto-oncogene Mas; rh: recombinant human.

33.2 Classic RAAS Antagonists

33.2.1 Angiotensin-Converting Enzyme Inhibitors (ACEIs)

Stopping the activation of AT1 is an attractive antifibrosis target in RAAS. ACE inhibitors (ACEIs) block the synthesis of AngII, which is catalyzed by ACE, preventing the conversion of AngI to AngII, limiting the effect of AngII and further decreasing the secretion of aldosterone and vasopressin.

The effectiveness of ACEIs in preventing or attenuating kidney disease in the clinic may be partly due to hemodynamics and non-hemodynamic factors. ACEIs can reduce the intraglomerular pressure by reducing the afferent arterial pressure and slow the breakdown of bradykinin, decreasing the size and charge selectivity of the glomerular cell wall. In addition, ACEIs can reduce cytokine production, such as by transforming growth factor-beta (TGF-β), which induces glomerulosclerosis and renal fibrosis (Zhang et al. 2017).

Enalapril, an ACEI, significantly attenuated BSA-induced rat renal tubule-interstitial inflammation and fibrosis by suppressing NLRP3 inflammasome expression (Ding et al. 2014). Furthermore, in a UUO mouse model, the amelioration of Enalapril on renal fibrosis was mast cell-dependent, as there was no effect of Enalapril on mast cell-deficient mice developing renal fibrosis (Sun et al. 2016).

In 1993, the CAPTOPRIL trial studied the effect of ACEI captopril on people with type 1 diabetes with proteinuria and showed that, compared with the placebo, treatment with captopril led to a 30% reduction in proteinuria, a 43% reduction in the risk of the primary end point of doubling of serum creatinine and a 50% reduction in the combined end point of death and the need for dialysis or renal transplantation, depending on BP levels (Lewis et al. 1993). In the Bergamo Nephrology Diabetic Complications trial, 1204 type 2 diabetic patients with hypertension and normal urine albumin excretion were studied and followed for a median of 3.6 years. The results showed that trandolapril decreased the incidence of persistent micro-albuminuria compared to placebo and verapamil, indicating that ACEIs not only could delay the progression of diabetic nephropathy but also were able to prevent the onset of nephropathy (Ruggenenti et al. 2004).

Although ACEIs result in lower levels of AngII in the blood, AngII produced by alternative conversion pathways can still be combined with AT1. When AngII levels decline, AngII accumulates by non-ACE pathways, which is called "ACE escape." The phenomenon may be due to the increase in renin release reactively because of AngII loss. ACEIs could also upregulate Ang1-7 and bradykinin, which may be related to target organ protection by ACEIs (Sureshkumar 2008).

33.2.2 Angiotensin Receptor Blockers (ARBs)

Angiotensin receptor blockers (ARBs), which are highly selective for the AT1 receptor, increase AngII levels in circulation and retard AngII binding to the AT1R. ARBs have similar effects to those of ACEIs, including regulating blood pressure and maintaining endothelial function; however, they do not increase bradykinin production. In addition, there are no such ACE escape problems In addition, because the blockade of AT1R AngII binds with AT2R alternatively or is diverted to Ang1-7, ARBs exert beneficial effects.

Obata et al. concluded that in hypertensive glomerulosclerosis animal models, RAAS activation and sensitivity to AngII increase in glomeruli (Obata et al. 1997). In rats treated with candesartan for 12 weeks, UAE decreased, as did TGF-β, fibronectin (FN), and RAAS components.

Wang et al. (2013) investigated the therapeutic role of ARB in an IgA nephropathy rat model and found that administration of losartan decreased urinary protein levels and reduced serum BUN, Scr, TGF-β1, FN, α-SMA, and FGF-1, finally delaying the progression of advanced IgA nephropathy with impaired renal function. In the IDTN and RENAAL trials, ARB treatment reduced the relative risk of reaching the primary composite end point—death, doubling of serum creatinine, or end-stage renal disease—in people with type 2 diabetes with kidney involvement by 20 and 16%, respectively (Lewis et al. 2001; Brenner et al. 2001).

33.2.3 Mineralocorticoid Receptor Antagonists (MRAs)

Aldosterone is a mineralocorticoid hormone that promotes sodium retention and renal fibrosis by inducing renal vasoconstriction, oxidative stress, and inflammation (Bertocchio et al. 2011). MRAs are still mainly used as diuretics in the treatment of hypertension but also heart failure, with a significant reduction in mortality (Pitt et al. 1999). In CKD patients treated with an ACEI or ARBs, increased aldosterone levels were observed, called the "aldosterone escape phenomenon," and activated the mineralocorticoid receptor (MR) (Lijnen et al. 1982). This supported the use of mineralocorticoid receptor antagonists (MRAs) in addition to ACEI or ARBs on renal fibrosis.

MRAs are classified as conventional steroidal and nonsteroidal compounds. Steroidal compounds (spironolactone and eplerenone) inhibit the effects of aldosterone by competing for its ligand-binding domain on MR and prevent MR from adopting the active conformation.

Researchers demonstrated that administration of spironolactone to the STZ-induced diabetic rat reduced proteinuria and decreased the expression of collagen I/IV, TGF-β, and attenuated glomerular and tubulo-interstitial fibrosis, despite the absence of BP or blood glucose reduction (Kolkhof et al. 2015). In addition, spironolactone significantly alleviated cisplatin-induced nephrotoxicity and renal fibrosis both in vivo and in vitro (Elseweidy et al. 2018).

It was reported that inhibition of local aldosterone by eplerenone prevented the upregulation of transforming growth factor-β1, connective tissue growth factor, plasminogen activator inhibitor type 1, and collagen I in adrenalectomized rats (Sun et al. 2015).

33.2.4 Direct Renin Inhibitors (DRIs)

Many patients have elevated AngII levels due to AngII reactivation under ACE inhibition and ALDO escape during long-term treatment with an ACEI or an ARB. A similar phenomenon has been discovered for MRAs, called "aldosterone escape." The patients who take MRAs may have higher aldosterone compared to pretreatment. Direct renin inhibitors block the RAAS at an earlier stage in the cascade than do ACEIs and ARBs and prevent the formation of both AngI and AngII by both ACE and non-ACE pathways (Nadeem and Batisky 2014).

The development of renin inhibitors has provided an opportunity to evaluate the effects of direct renin inhibition (DRIs) as another means of RAAS blockade. Aliskiren is the first in a new class of orally effective DRIs approved for the treatment of hypertension (Jensen et al. 2008).

The safety and efficacy of aliskiren have been well defined through preclinical pharmacological safety studies. These were also assessed in adult patients with varying degrees of renal or hepatic impairment (Kelly et al. 2007). The results indicate that there is no need to adjust the dose of aliskiren in patients with renal or hepatic impairment; however, according to the package insert, it is recommended to exercise caution when prescribing aliskiren to patients who have moderate to severe renal dysfunction, a history of dialysis therapy, nephrotic syndrome, or renovascular hypertension due to the scarcity of data in these patients and the potential for other agents affecting the RAAS to increase serum creatinine and blood urea nitrogen levels (Miyata et al. 2014).

Several clinical studies have demonstrated that aliskiren is effective on CKD (Li et al. 2012; Miyata et al. 2014). Persson et al. reported that aliskiren (300 mg daily) treatment reduced UACR, with a 17% reduction by days 2–4, a 31% reduction by days 8–10, and a maximum reduction of 44% at the end of the treatment (day 28) (Persson et al. 2008). Wu et al. found that aliskiren at a dosage of 150 mg/day for six months in 103 Chinese CKD patients (both with and without diabetes) had a favorable effect on reducing residual proteinuria and inadequately controlled blood pressure (Wu et al. 2014). Mechanically, aliskiren prevents renal disease progression by suppressing both angiotensin I and II in RAAS-activated pathology; moreover, Renke revealed that aliskiren also attenuated oxidative stress and improved the functional status of tubules in nondiabetic nephropathy (Renke et al. 2014). Collectedly, these results strongly suggest that aliskiren has beneficial effects for renoprotection with or without BP-lowering effects in CKD patients. In addition, aliskiren can reduce sympathetic hyperactivity, which is often exhibited and contributes to the pathogenesis of hypertension and CVD in patients with CKD (Miyata et al. 2014).

DRIs, in theory, seem to have the same adverse effects as do other RAAS blockers. A pooled analysis showed that the most common adverse events thought to be related to aliskiren treatment were headache, diarrhea, nasopharyngitis, back pain, and dizziness. Hyperkalemia, which is a frequent concern in CKD patients, is the primary danger of RAAS-blocking medications. The blockade of the RAAS leads to a decrease in aldosterone levels. Because of the urinary potassium excretion caused by aldosterone, RAAS blockers can cause potassium retention. Several clinical trials targeting aliskiren in CKD patients with tracked potassium levels showed that there was no significant trend for increased hyperkalemia by aliskiren in patients with CKD in these trials (Vaidyanathan et al. 2007).

Currently, many controversies regarding the use of aliskiren in diabetic kidney disease are under discussion. Numerous animal models of DKD have shown significantly improved BP control and proteinuria with the use of aliskiren. At the level of the glomerulus, the use of aliskiren is associated with decreased podocytopathy and glomerular pressure. Tubulo-interstitial fibrosis and levels of profibrotic and oxidative markers were found to be reduced with aliskiren. The AVOID trial showed that aliskiren might have additional renoprotective effects that are independent of its blood pressure-lowering effects in patients with hypertension, type 2 diabetes and nephropathy when added to ARB. Unfortunately, another clinical trial, the ALTITUDE trial, due to a lack of apparent benefit and a higher risk of side effects, was prematurely stopped for futility in patients with type 2 diabetes and microalbuminuria, macro-albuminuria, or cardiovascular disease. Therefore, the FDA has given a formal recommendation about aliskiren: it should not be used in association with ARBs or ACEIs as dual therapy in patients with diabetes mellitus or renal disease (Parving et al., 2012).

A new drug, DRI-ACT-077825, was proved to be safe in the human body when administered once daily in 2013. ACT-077825 inhibited plasma renin activity, while the immunoreactive renin level increased. However, the decrease in plasma renin activity is short lived and is not affected by the dose within 7 days. The effect of DRI on blood pressure is controversial. More research is needed to identify inconsistencies (Nicolas et al. 2013).

Thus, aliskiren has beneficial effects for renoprotection, the control of BP, and the prevention of CVD in patients with CKD. However, the possible adverse effects and potential interactions with other drugs being used together have to be carefully considered. In DKD patients, it clearly has an antihypertensive effect and a proteinuria-reducing effect that are greater than placebo. These benefits are not superior to those offered by currently used ACEIs or ARBs. Thus, DRI is a reasonable alternative to the use of ACEI or ARB in DKD, especially in settings where ACEI or ARB may not be used. Another limitation is that there is still a lack of high-quality studies to confirm the effects of aliskiren in patients with CKD.

33.3 New RAAS Drugs of the AngII-AT1/2R Axis

33.3.1 The Angiotensin II Subtype-2 Receptor (AT2R) Agonist

AngII primarily activates two receptor subtypes, AT1R and AT2R. It is well established that activation of AT1R by AngII mediates pathophysiological effects such as vasoconstriction, proliferation, fibrosis, oxidative stress, and inflammation, which occur in multiple organs. On the other hand, activation of AT2R is thought to counter-regulate the pathophysiological effects induced by AT1R and exert vasodilator, antifibrotic, antiproliferative, and anti-inflammatory effects, as well as natriuretic and antihypertensive effects, in renal disease (Wang et al. 2017; Hallberg et al. 2018).

In contrast to AT1R, AT2R is very sparsely expressed and is abundant only in certain tissues, such as the vascular endothelium, kidney, and brain. In the kidney, AT2R is mainly localized to renal vessels, glomeruli, and tubules. AT2R activation also opposes the vasoconstrictor actions of AT1R by promoting dilation of the afferent and efferent arterioles. A previous study revealed that the appropriate balance between AT1R and AT2R activation might therefore play a key role in regulating the physiological functions of the renal systems (Ozono et al. 1997). Mechanistically, the following phenomena might contribute to the renoprotection of AT2R. First, AT2R stimulates mitogen-activated protein kinase phosphatase-1 (MKP-1) and inhibits ERK activity, resulting in the reduction in mitogen-activated protein kinase (MAPK) activity and growth inhibition. Second, AT2R regulates the production of renal nitric oxide (NO), guanosine cyclic 3′,5′-monophosphate (cGMP), and bradykinin (Siragy and Carey 1997; Abadir et al. 2003; Matavelli and Siragy 2015), indicating that it could be involved in the induction of vasodilation. In addition, Rompe et al. (2010) showed that targeting AT2R exerted a direct anti-inflammatory effect by inhibiting NF-κB activation, leading to reduced TNF-α-mediated IL-6 release from fibroblasts. Together, these data demonstrate that AT2R plays a substantial role in organ inflammation and tissue fibrosis and that its cellular signaling pathways go in opposite directions to those of AT1R.

In 2004, a novel AT2R agonist named Compound 21 (C21) was developed. C21 is a nonpeptidic compound that is orally and systemically active with an oral bioavailability of 20–30%. C21 is a highly selective AT2R agonist that is in the final stage of a preclinical trial (Steckelings et al. 2012). Many basic experiments have revealed that C21 can inhibit inflammatory responses and renal fibrosis via activation of AT2R. It lacks AT1R affinity and was demonstrated in human embryonic kidney cells to have 4000-fold selectivity to AT2R (Bosnyak et al. 2011). If clinical studies confirm its efficacy, this compound could be useful for the management of diverse cardiovascular and kidney diseases, including diabetic kidney disease and glomerulonephritis (Pandey and Gaikwad 2017).

More recently, additional AT2R ligands have been investigated in the clinic. Namely, MP-157 is an AT2R agonist in phase I clinical trials in Europe, aimed at the cardiovascular system, according to information from Mitsubishi Tanabe Pharma "State of New Product Development (as of August 2, 2016)." Unfortunately, its structural formula is still not disclosed. Another AT2R antagonist, EMA401, from Spinifex Pharmaceuticals Pty Ltd, Australia, and now acquired by Novartis, has been successfully tested in a phase II trial in patients with neuropathic pain (Rice et al. 2014). Thus, it is conceivable that AT2R agonists will play an important role in the future.

33.3.2 Vaccines Against the RAAS

The concept of immunization as a treatment for hypertension is not new, and several studies have yielded insights into an angiotensin vaccine as a strategy for inhibiting RAAS. As a specific target antigen, the antihypertension vaccine is complicated by the multifactorial etiology of hypertension compared to the vaccines generated against bacteria and viruses (Do et al. 2010). RAAS is potentially the most important regulator of systemic blood pressure and is believed to be a major factor in hypertension onset. Antibodies against renin, angiotensinogen, AngI, AngII, ACE, and AT1/2 have been studied in experiments in an attempt to create a vaccine that would chronically suppress RAAS activity.

The renin vaccine is the earliest vaccine that effectively reduces blood pressure in animal models; its binding to renin inhibits the interaction between renin and angiotensinogen and suppresses renin's enzymatic activity. Because the structure of renin is highly species specific, vaccination against heterologous renins led to the production of antibodies that produced only incomplete suppression of renin's enzymatic activity. Subsequent advances in human renin purification made it possible to actively immunize marmosets against human renin. Unfortunately, these experiments also revealed that the renin vaccine led to the onset of various autoimmune diseases that emerged as a major concern for any immunotherapy directed at endogenous RAAS components (Gradman and Pinto 2008).

Unlike renin, AngI and AngII are very small peptide molecules composed of ten and eight amino acids, respectively. It is well documented that smaller molecules will be less likely to stimulate autoimmune diseases. In animal models, AngI vaccines between two carriers, tetanus toxoid (TT) (PMD-2850) and keyhole limpet hemocyanin (KLH) (PMD3117), induced equivalent immune responses and inhibition of the pressor effects of exogenous AngI. In a phase IIa clinical trial, three or four injections of the angiotensin I vaccine PMD3117 resulted in antibody production with titers peaking 64 days after protocol initiation and no evidence of autoimmune disease. The results revealed that PMD3117 could block the RAAS system. Therefore, a new formulation of PMD3117 has been developed. Confirmation of safety is the priority, and renoprotection should be addressed by further clinical assessment in phase IIb and phase III trials (Gardiner et al. 2000; Downham et al. 2003; Do et al. 2010).

The AngII vaccine is a logical target for immunotherapy. In 2007, Cytos Biotechnology developed an AngII-specific vaccine (CYT006-AngQb) composed of an AngII peptide with an N-terminal Cys-Gly-Gly extension that is covalently coupled to virus-like particles (VLP) derived from the coat protein of the bacteriophage Qb. Ambühl PM showed that VLP (AngQb) reduces blood pressure in SHR to levels obtained with an ACE inhibitor and is well tolerated in humans. Therefore, vaccination against angiotensin II has the potential to become a useful antihypertensive treatment that provides long-lasting effects and improves patient compliance (Ambuhl et al. 2007). Some studies reported that SHR exhibited an SBP decrease of 17 mmHg after immunization with a peptide-based vaccine made of a seven-amino acid sequence (AFHYESR) from the second extracellular loop of rat AT-1A receptor (ATR12181), which alleviated cardiac hypertrophy and attenuation of kidney injuries. There were no signs of autoimmune diseases in the biopsied sections of kidney. Because Freund's adjuvant is not safe for use in humans, the next phase of the study will focus on the use of ATR12181 in combination with VLP in human subjects (Zhu et al. 2006).

In conclusion, vaccination to achieve chronic RAAS suppression represents a novel approach to treating hypertension. However, the effect of vaccines on renal fibrosis after controlling hypertension and blocking the RAAS system is still unknown.

33.4 New Drugs of the Ang1-7-Mas Axis

33.4.1 Human Recombinant ACE2 (hrACE2)

In the kidneys, ACE2 is expressed in the proximal tubules and less strongly in the glomeruli. The synthesis of inactive Ang 1-9 from AngI and the catabolism of AngII to produce Ang 1-7 are the main functions of ACE2 (Nishiyama et al. 2002; Tikellis et al. 2004). There may be two mechanisms to explain the beneficial effect of ACE2 in vascular diseases. First, ACE2 reduced the AngII interaction with AT1R by inducing AngII degradation; second, ACE2 reduced vasoconstriction, water retention, and reactive oxygen stress by increasing Ang1-7 synthesis (Clarke and Turner 2012). Zhong et al. found that daily treatment with recombinant human ACE2 (hrACE2) reduced the AngII-induced pressor response and normalized renal AngII levels and oxidative stress, which indicated that human ACE2 prevented AngII-mediated renal oxidative stress, inflammation, and tubulo-interstitial fibrosis in animal models (Zhong et al. 2011). Oudit GY found that treatment with hrACE2 attenuated diabetic kidney injury in the Akita mouse in association with a reduction in blood pressure and a decrease in NADPH oxidase activity (Oudit et al. 2010). Furthermore, accumulating evidence indicates that the ACE/ACE2 ratio regulates the production and accumulation of AngII and that ACE2 deficiency leads to increased AngII concentrations (Ye et al. 2004; Wakahara et al. 2007).

There is emerging evidence of the upregulation of ACE2 in urine from diabetic patients; this upregulation may reflect the pathological shedding of renal ACE2. Studies in experimental models have investigated the feasibility of pharmacological induction of ACE2 for improvement of renal function, inflammation, and fibrosis (Williams and Scholey 2018). The small-molecule ACE2 activator 1-[[2-(dimethylamino)ethyl] amino]-4-(hydroxymethyl)-7-[[(4-methylphenyl)sulfonyl]oxy]-9H-xantona-9 (XNT) prevented renal and myocardial hydroxyproline accumulation (Paulis et al. 2015). Other ACE2 activators, such as diminazene aceturate (DIZE) and recombinant human ACE2, have been tested in preclinical trials (Treml et al. 2010; Ferreira et al. 2011).

33.4.2 MAS1 Agonists

Similar to ACE2 administration, Ang1-7 binding to the MAS1 receptor has shown beneficial effects in animal models. Further, Ang1-7 ameliorates cardiac remodeling by decreasing hypertrophy and fibrosis, while genetic depletion of Mas results in dyslipidemia, insulin resistance and marked fibrotic and hypertrophic changes in rat myocardia (Wiemer et al. 2002; Grobe et al. 2007).

The major limitation of exogenous administration of Ang1-7 is that it is a peptide with a very short biological half-life, low oral bioavailability, and very low stability (Yamada et al. 1998). Because of these limitations, Ang1-7 is generally administered subcutaneously by osmotic mini pumps, which are quite expensive and are not readily available.

More than 10 years ago, the effects of nonpeptide compound AVE 0991 were similar to those of Ang1-7 in endothelial cells (Wiemer et al. 2002). Subsequent data demonstrated that AVE 0991 and Ang1-7 competitively bound the same MAS1 receptor in the kidneys, and AVE 0991 reduced blood pressure, alone or in combination with renin inhibitors (Singh et al. 2013). In 2014, AVE 0991 was shown to ameliorate kidney inflammation and improve cell infiltration, cytokine release, and histology in two rodent models of arthritis. In diabetic animal models, AVE-0991 produced cardio-renal protection, possibly by improving glucose and lipid metabolism in diabetic rats, independent of its blood pressure-lowering action (Singh et al. 2012). The first phase of clinical trials is being conducted to test the safety of combination therapy in humans (R. A. Santos, personal communication). Further clinical trials should be performed to test this promising therapeutic approach in humans (Barroso et al. 2012).

33.5 New Drugs of the AngIV-AT4R Axis

33.5.1 Aminopeptidase A

AngII is increased in the plasma and tissues either by increased ACE expression and activity or by decreased expression and activity of key enzymes that metabolize AngII. A balance between the formation and degradation of the effector peptide AngII is critical. The roles of AngII degradation in blood pressure and renal regulation have not been well studied. AngII is metabolized to form des aspartyl1-AngII, also called AngIII, primarily by aminopeptidase A (APA), which is involved in the degradation of AngII.

APA is expressed in glomerular podocytes and tubular epithelia and metabolizes AngII, a peptide known to promote glomerulosclerosis. APA activity is involved in the deterioration of salt-induced hypertension and renal injury (Nomura et al. 2005). Since APA is the key enzyme for the degradation of AngII, its implications for renal diseases have been studied in APA-knockout mice. AngII-treated APA-null mice developed a significant rise in albuminuria along with increased segmental and global sclerosis and/or collapse of juxtamedullary glomeruli, micro-cystic tubular dilation, and tubulo-interstitial fibrosis, which are blocked in AngII-treated APA-wild-type mice. The augmented AngII-mediated kidney injury observed in association with increased intrarenal AngII accumulation in the absence of APA suggests a protective, metabolizing role of APA in AngII-mediated glomerular diseases (Velez et al. 2014, 2017).

However, the clinical benefits of APA as a therapeutic target are most likely be limited, since it acts to decrease circulating and tissue AngII levels primarily by degrading AngII rather than by inhibiting AngII formation. ACE inhibitors have been widely used to block the RAAS to treat hypertensive and kidney diseases with proven clinically beneficial outcomes. If ACE inhibitors are inadequate to treat hypertension or kidney diseases associated with the formation of AngII, a potential pharmacological strategy to upregulate APA expression or enhance its activity may be alternative therapy (Gao et al. 2014).

QGC001 (originally named RB150) is the first drug candidate of a new class of antihypertensive agents targeting the brain RAAS, particularly brain APA, the enzyme generating brain AngIII. The phase I study of QGC001 showed that QGC001 offered a potential alternative therapeutic strategy for improving the BP control of hypertensive patients. However, the renoprotection of QGC001 warrants further investigation (Balavoine et al. 2014).

33.5.2 AT4 Receptor Inhibitors

The AT4 receptor (AT4R) was originally defined as the specific, high-affinity binding site for the hexapeptide angiotensin IV (AngIV). AT4R has a broad distribution and is found in a range of tissues, including the adrenal gland, kidney, lung, and heart. AT4R was generally found to be more abundant than AT1R in mammalian kidneys, whereas the Ang1-7 receptor was not detected in mammalian kidneys. Rats subjected to various chronic treatments were found to preferentially present decreased kidney AT4R density (furosemide, puromycin aminonucleoside, and nitro-L-arginine methylester), decreased kidney AT1R density (bilateral ureteral obstruction), or increased kidney AT1R distribution in the inner medulla (water diuresis). These results suggest that AT4R can be expressed in several renal cells within the normal kidney. Furthermore, many animal models of renal dysfunction and injury have been identified that selectively alter kidney AT4R density and may potentially aid in elucidating the role of this novel angiotensin receptor system in renal function (Handa et al. 2001).

IRAP, an AT4R that is inhibited by AngIV, was identified in 2001. A previous study demonstrated that IRAP is an AT4R and proposed that AT4R ligands may exert their effects by inhibiting the catalytic activity of IRAP, thereby extending the half-life of its neuropeptide substrates. HFI-419 is an IRAP-selective pyridinyl compound that enhances memory in rats. Studies have demonstrated that HFI-419 prevented cardiac and endothelial damage induced by AngII, independent of blood pressure. HFI-419 also exceeded the anti-inflammatory effect. Regarding the renoprotection of HFI-419, more studies are necessary (Albiston et al. 2008).

33.6 New Drugs Targeting Aldosterone

33.6.1 Nonsteroidal MRAs

Finerenone (BAY 94-8662) is a potent and highly selective nonsteroidal MRA (>500-fold more selective than for other steroid receptors) (Taira et al. 2008). Lattenist et al. (2017) evaluated the efficacy of finerenone to prevent the acute and chronic consequences of ischemic acute kidney injury. After 4 months of ischemia-reperfusion, finerenone fully prevented the transition from acute kidney injury to chronic kidney disease (kidney dysfunction, increased proteinuria and tubular dilation, extensive tubule-interstitial fibrosis, and an increase in kidney TGF-β and collagen I mRNA). Moreover, Barrera-Chimal J's lab (Barrera-Chimal et al. 2018) also reported that finerenone efficiently prevented IR-induced increases in plasma creatinine, urea, and proteinuria levels, as well as the expression of TGF-β as an indicator of kidney fibrosis, which means that finerenone protected against the transition from acute kidney injury to chronic kidney disease. Furthermore, they found that MR deficiency in myeloid cells protected against chronic dysfunction and fibrosis induced by ischemia-reperfusion. Finally, they found that the protection afforded by MR antagonism (finerenone) or myeloid MR deficiency was due to the promotion of macrophage polarization to a wound-healing phenotype after kidney IR, preventing the development of chronic kidney fibrosis and dysfunction.

Arai K and his colleagues (Arai et al. 2016) observed, in a hypertensive rat model, that esaxerenone (CS-3150), a novel nonsteroidal mineralocorticoid receptor antagonist, not only prevented but also ameliorated ongoing DOCA/salt loading-induced mRNA expression of fibrosis, inflammation, and oxidative stress markers. Moreover, they found that compared with spironolactone and eplerenone, CS-3150 had an equivalent antihypertensive effect but a superior ability to ameliorate glomerulosclerosis, tubular injury, and tubulo-interstitial fibrosis (Arai et al. 2015).

33.6.2 Aldosterone Synthase Inhibitor

Aldosterone is synthesized from cholesterol in the outermost layer of the adrenal cortex via a series of steroid hydroxylase and deoxygenase enzymes. Aldosterone synthase (also named CYP11B2) catalyzes the last and rate-limiting steps in aldosterone synthesis. The major glucocorticoid, cortisol, is synthesized in the zona fasciculata of the adrenal cortex, with CYP11B1 (11β-hydroxylase (cytochrome P450 type I)) as the rate-limiting enzyme. It is well known that aldosterone and cortisol biosynthesis share many common steps (Azizi et al. 2013; Tamargo et al. 2014).

Aldosterone stimulates the production of ROS, inflammation and fibrosis of the heart, vasculature, and kidney through both mineralocorticoid receptor (MR)-dependent and MR-independent mechanisms. Among them, the MR-independent effects occur via the angiotensin II receptor and via the G-protein-coupled receptor. Studies in rodents genetically deficient in aldosterone synthase or treated with a pharmacological aldosterone synthase inhibitor have provided insight into the relative contribution of aldosterone compared with the contribution of mineralocorticoid receptor activation in inflammation, fibrosis, and injury (Brown 2013).

LCI699 is an orally active and nonselective aldosterone synthase inhibitor that has been evaluated in humans (Andersen et al. 2012). With aldosterone synthase inhibition by LCI699, cortisol levels remain normal, and 11-deoxycorticosterone increases, which means that the adrenocorticotropic hormone–cortisol axis is activated by CYP11B1 gene inhibition. However, the effects of aldosterone synthase inhibition on renal injury have not been reported in humans to date (Brown 2013).

33.7 Combination Strategy

Despite the broad range of new possible therapeutic targets for hypertension described above, it is seemingly difficult to devise a new molecule that can be advanced to later phases of clinical investigation and that could successfully compete with the existing therapeutics. Thus, there is much room to take advantage of the already broad choice of molecules and to optimize their usage. The combination of current drugs is an alternative strategy. The use of combination therapy for the treatment of renal fibrosis is already established in practice and in the current guidelines. This positive effect has been amply confirmed for ACEI or ARB used as monotherapy, and two or more separate RAAS drugs have been combined in the hope that they would have an antifibrotic effect and offset each other's negative side effects.

33.7.1 ACEI + ARB Combination

In theory, the combination of RAAS inhibitors may have a greater inhibitory effect of renal fibrosis and a better clinical outcome. However, the combination use of ACEI and ARB may have synergistic or addictive effects. A meta-analysis found that a combination of ACEI and ARB resulted in a clinically significant reduction in proteinuria in patients with chronic kidney disease and diabetic nephropathy regardless of BP changes (Doulton and Macgregor 2005). Another systematic review and meta-analysis of randomized trials evaluating the combination of an ACEI and an ARB in patients with chronic proteinuric renal disease. This finding indicated that the combination of ACEI and ARB therapy in patients with chronic proteinuric renal disease was safe, without clinically meaningful changes in serum potassium levels or glomerular filtration rates. Combination therapy was also associated with a significant decrease in proteinuria, at least in the short term (MacKinnon et al. 2006). However, Schmerbach et al. (2012) observed the effect of ACEI/ARB combination therapy in spontaneously hypertensive stroke-prone rats on a salt-rich diet. Renal glomerulosclerosis and interstitial fibrosis were decreased by monotherapy (telmisartan or ramipril), whereas combination treatment (telmisartan and ramipril) failed to have a significant effect. In the ONTARGET trial, the combination therapy of ramipril plus telmisartan was compared to monotherapy in patients with DM and high cardiovascular risk. There was no difference in the composite primary outcome of death or hospitalization for cardiovascular disease, but combination therapy was associated with significantly more hypotension, syncopes, and a faster rate of decline in renal function compared to monotherapy. The ONTARGET trial also showed that dual RAAS blockade led to worsening of kidney function (Mann et al. 2008). As with ONTARGET, the VA NEPHRON D Trial, targeting patients with T2DM, compared combination therapy of losartan and lisinopril to monotherapy with losartan. Due to excess adverse events in the combination arm of ACEI and ARB, mostly acute kidney injury and hyperkalemia, the trial was prematurely stopped (Fried et al. 2013).

33.7.2 ACEI/ARB + MRA Combination

MRAs are used in combination with ACEI or ARB, particularly in patients with low renin or resistant hypertension, which is effective in reducing BP because of the inability of these drugs to reliably decrease aldosterone after chronic use as a consequence of the aldosterone breakthrough that follows the escape of the effects of ACEI or ARB. Eplerenone is also effective in resistant hypertension in association with ACEI or ARB. The demonstrated better tolerability of eplerenone over spironolactone could make the use of eplerenone more desirable, but the indication for hypertension is not recognized except in the presence of intolerance to spironolactone (Marquez et al. 2015).

Mehdi et al. (2009) published a double-blind, placebo-controlled trial that included 81 patients with diabetes, hypertension, and albuminuria (UAE > 300 mg/g). Patients who were treated with lisinopril 80 mg were randomized to placebo, losartan 100 mg or spironolactone 25 mg daily for 48 weeks. The results revealed that compared with placebo, the urinary albumin-to-creatinine ratio decreased by 34.0% in the spironolactone group and by 16.8% in the losartan group. However, the serum potassium level was significantly higher with the addition of either spironolactone or losartan, which indicated that the addition of spironolactone, but not losartan, to a regimen including a maximal dose of an ACEI generated better renoprotection in patients with diabetic nephropathy. Epstein et al. published similar RCT results in 2006 (Epstein et al. 2006). Additional large-scale trials with hard end points are needed in this field to prove that the addition of an aldosterone receptor blocker to standard therapy with an ACEI or ARB could improve renal outcomes, and the effect of the new MRA on renoprotection should be analyzed.

33.7.3 Dual Blockade of ARB-Neprilysin Inhibitor

Natriuretic peptides (NP) (especially ANP and BNP), playing an important role in sodium and water homeostasis, have antioxidant, anti-inflammatory, and antifibrotic properties, which may significantly contribute to their renoprotective properties. Neprilysin is a vasopeptidase that metabolizes NP and other peptides, such as bradykinins. Inhibition of neprilysin is accompanied by the increase in NP levels but has mild effects on BP (MacKinnon et al. 2006). The combination of a neprilysin inhibitor with a RAAS blocker represents a powerful tool. LCZ696 is a dual-acting ARB and neprilysin inhibitor (ARNI) that decreases BP in animals and healthy humans with low rates of side effects. Therefore, it is not surprising that in CKD animals treated with LCZ, there was a significant improvement in indices of oxidative stress, inflammation, fibrosis, and the Nrf2 system beyond that observed with ARB therapy alone. In a clinical study, a meta-analysis of LCZ696 revealed that it was better than ARB alone at reducing most blood pressure parameters without resulting in more adverse events in treating hypertension (Jing et al. 2017). To determine the renoprotective effect of LCZ696, additional clinical trials should address its ability to prevent renal fibrosis and the safety of its long-term use.

33.7.4 ACEI/ARB + DRI Combination

With respect to preclinical studies, the result of the ALTITUDE study has been unexpected; in particular, a renoprotective effect was expected from the aliskiren + ACEI/ARB combination. This combination therapy might reduce proteinuria to a great extent. However, further subgroup analysis revealed that dual blockade of the renin–angiotensin–aldosterone system with aliskiren and ARB did not improve the hard renal endpoints, so it should not be recommended in CKD patients (Parving et al. 2012; Rasche et al. 2018).

33.8 Conclusions

For years, a large number of experimental and clinical trials have shown that the traditional RAAS blocker (ACEI/ARB/MAR) could delay the development of renal fibrosis and chronic kidney disease, but the effect was limited. Improvements in RAAS blockade beyond the effects of ACEIs and ARBs will be challenging, but studies in this area are abundant. Some new nonpeptide drugs that modify RAAS activity to combat renal fibrosis have been developed, including several novel agonists (such as AT2R agonists and MAS1 agonists) and antagonists (such as AT4R inhibitors) of the RAAS, some of which have entered clinical applications and some of which are still in the stage of clinical trials. Furthermore, with the development of new technology, new blockers of the RAAS have been developed, such as human recombinant ACE2 and vaccines against renin, AngI or AngII, which may be more specific. However, large multicenter randomized controlled studies are needed to evaluate the beneficial effects and to explore the efficacy and safety of these new blockers of the RAAS.

References

Abadir PM, Carey RM, Siragy HM (2003) Angiotensin AT2 receptors directly stimulate renal nitric oxide in bradykinin B2-receptor-null mice. Hypertension 42:600–604

Albiston AL, Morton CJ, Ng HL, Pham V, Yeatman HR, Ye S et al (2008) Identification and characterization of a new cognitive enhancer based on inhibition of insulin-regulated aminopeptidase. FASEB J 22:4209–4217

Ambuhl PM, Tissot AC, Fulurija A, Maurer P, Nussberger J, Sabat R et al (2007) A vaccine for hypertension based on virus-like particles: preclinical efficacy and phase I safety and immunogenicity. J Hypertens 25:63–72

Andersen K, Hartman D, Peppard T, Hermann D, Van Ess P, Lefkowitz M et al (2012) The effects of aldosterone synthase inhibition on aldosterone and cortisol in patients with hypertension: a phase II, randomized, double-blind, placebo-controlled, multicenter study. J Clin Hypertens 14:580–587

Arai K, Morikawa Y, Ubukata N, Tsuruoka H, Homma T (2016) CS-3150, a novel nonsteroidal mineralocorticoid receptor antagonist, shows preventive and therapeutic effects on renal injury in deoxycorticosterone acetate/salt-induced hypertensive rats. J Pharmacol Exp Ther 358:548–557

Arai K, Tsuruoka H, Homma T (2015) CS-3150, a novel non-steroidal mineralocorticoid receptor antagonist, prevents hypertension and cardiorenal injury in Dahl salt-sensitive hypertensive rats. Eur J Pharmacol 769:266–273

Azizi M, Amar L, Menard J (2013) Aldosterone synthase inhibition in humans. Nephrol Dial Transplant 28:36–43

Balavoine F, Azizi M, Bergerot D, De Mota N, Patouret R, Roques BP et al (2014) Randomised, double-blind, placebo-controlled, dose-escalating phase I study of QGC001, a centrally acting aminopeptidase a inhibitor prodrug. Clin Pharmacokinet 53:385–395

Barrera-Chimal J, Estrela GR, Lechner SM, Giraud S, El Moghrabi S, Kaaki S et al (2018) The myeloid mineralocorticoid receptor controls inflammatory and fibrotic responses after renal injury via macrophage interleukin-4 receptor signaling. Kidney Int 93:1344–1355

Barroso LC, Silveira KD, Lima CX, Borges V, Bader M, Rachid M et al (2012) Renoprotective effects of AVE0991, a nonpeptide mas receptor agonist, in experimental acute renal injury. Int J Hypertens 2012:808726

Bertocchio JP, Warnock DG, Jaisser F (2011) Mineralocorticoid receptor activation and blockade: an emerging paradigm in chronic kidney disease. Kidney Int 79:1051–1060

Bosnyak S, Jones ES, Christopoulos A, Aguilar MI, Thomas WG, Widdop RE (2011) Relative affinity of angiotensin peptides and novel ligands at AT1 and AT2 receptors. Clin Sci (Lond) 121:297–303

Brenner BM, Cooper ME, de Zeeuw D, Keane WF, Mitch WE, Parving HH et al (2001) Effects of losartan on renal and cardiovascular outcomes in patients with type 2 diabetes and nephropathy. N Engl J Med 345:861–869

Brown NJ (2013) Contribution of aldosterone to cardiovascular and renal inflammation and fibrosis. Nat Rev Nephrol 9:459–469

Clarke NE, Turner AJ (2012) Angiotensin-converting enzyme 2: the first decade. Int J Hypertens 2012:307315

Do TH, Chen Y, Nguyen VT, Phisitkul S (2010) Vaccines in the management of hypertension. Expert Opin Biol Ther 10:1077–1087

Doulton TW, Macgregor GA (2005) Combination renin-angiotensin system blockade in hypertension. Kidney Int 68:1898

Downham MR, Auton TR, Rosul A, Sharp HL, Sjostrom L, Rushton A et al (2003) Evaluation of two carrier protein-angiotensin I conjugate vaccines to assess their future potential to control high blood pressure (hypertension) in man. Br J Clin Pharmacol 56:505–512

Elseweidy MM, Askar ME, Elswefy SE, Shawky M (2018) Nephrotoxicity induced by cisplatin intake in experimental rats and therapeutic approach of using mesenchymal stem cells and spironolactone. Appl Biochem Biotechnol 184:1390–1403

Epstein M, Williams GH, Weinberger M, Lewin A, Krause S, Mukherjee R et al (2006) Selective aldosterone blockade with eplerenone reduces albuminuria in patients with type 2 diabetes. Clin J Am Soc Nephrol 1:940–951

Ferreira AJ, Shenoy V, Qi Y, Fraga-Silva RA, Santos RA, Katovich MJ et al (2011) Angiotensin-converting enzyme 2 activation protects against hypertension-induced cardiac fibrosis involving extracellular signal-regulated kinases. Exp Physiol 96:287–294

Fried LF, Emanuele N, Zhang JH, Brophy M, Conner TA, Duckworth W et al (2013) Combined angiotensin inhibition for the treatment of diabetic nephropathy. N Engl J Med 369:1892–1903

Gao J, Marc Y, Iturrioz X, Leroux V, Balavoine F, Llorens-Cortes C (2014) A new strategy for treating hypertension by blocking the activity of the brain renin-angiotensin system with aminopeptidase A inhibitors. Clin Sci 127:135–148

Gardiner SM, Auton TR, Downham MR, Sharp HL, Kemp PA, March JE et al (2000) Active immunization with angiotensin I peptide analogue vaccines selectively reduces the presser effects of exogenous angiotensin I in conscious rats. Brit J Pharmacol 129:1178–1182

Gradman AH, Pinto R (2008) Vaccination: a novel strategy for inhibiting the renin-angiotensin-aldosterone system. Curr Hypertens Rep 10:473–479

Grobe JL, Der Sarkissian S, Stewart JM, Meszaros JG, Raizada MK, Katovich MJ (2007) ACE2 overexpression inhibits hypoxia-induced collagen production by cardiac fibroblasts. Clin Sci (Lond) 113:357–364

Hallberg M, Sumners C, Steckelings UM, Hallberg A (2018) Small-molecule AT2 receptor agonists. Med Res Rev 38:602–624

Handa RK, Handa SE, Elgemark MK (2001) Autoradiographic analysis and regulation of angiotensin receptor subtypes AT(4), AT(1), and AT((1-7)) in the kidney. Am J Physiol Renal Physiol 281:F936–F947

Jensen C, Herold P, Brunner HR (2008) Aliskiren: the first renin inhibitor for clinical treatment. Nat Rev Drug Discov 7:399–410

Jing W, Vaziri ND, Nunes A, Suematsu Y, Farzaneh T, Khazaeli M et al (2017) LCZ696 (Sacubitril/valsartan) ameliorates oxidative stress, inflammation, fibrosis and improves renal function beyond angiotensin receptor blockade in CKD. Am J Transl Res 9:5473–5484

Kelly DJ, Zhang Y, Moe G, Naik G, Gilbert RE (2007) Aliskiren, a novel renin inhibitor, is renoprotective in a model of advanced diabetic nephropathy in rats. Diabetologia 50:2398–2404

Kolkhof P, Nowack C, Eitner F (2015) Nonsteroidal antagonists of the mineralocorticoid receptor. Curr Opin Nephrol Hypertens 24:417–424

Lattenist L, Lechner SM, Messaoudi S, Le Mercier A, El Moghrabi S, Prince S et al (2017) Nonsteroidal mineralocorticoid receptor antagonist finerenone protects against acute kidney injury-mediated chronic kidney disease: role of oxidative stress. Hypertension 69:870–878

Lewis EJ, Hunsicker LG, Bain RP, Rohde RD (1993) The effect of angiotensin-converting-enzyme inhibition on diabetic nephropathy. The collaborative study group. N Engl J Med 329:1456–1462

Lewis EJ, Hunsicker LG, Clarke WR, Berl T, Pohl MA, Lewis JB et al (2001) Renoprotective effect of the angiotensin-receptor antagonist irbesartan in patients with nephropathy due to type 2 diabetes. N Engl J Med 345:851–860

Li SY, Chen YT, Yang WC, Tarng DC, Lin CC, Yang CY et al (2012) Effect of add-on direct renin inhibitor aliskiren in patients with non-diabetes related chronic kidney disease. BMC Nephrol 13:89

Lijnen P, Staessen J, Fagard R, Amery A (1982) Increase in plasma aldosterone during prolonged captopril treatment. Am J Cardiol 49:1561–1563

MacKinnon M, Shurraw S, Akbari A, Knoll GA, Jaffey J, Clark HD (2006) Combination therapy with an angiotensin receptor blocker and an ACE inhibitor in proteinuric renal disease: a systematic review of the efficacy and safety data. Am J Kidney Dis 48:8–20

Mann JF, Schmieder RE, McQueen M, Dyal L, Schumacher H, Pogue J et al (2008) Renal outcomes with telmisartan, ramipril, or both, in people at high vascular risk (the ONTARGET study): a multicentre, randomised, double-blind, controlled trial. Lancet 372:547–553

Marquez DF, Ruiz-Hurtado G, Ruilope LM, Segura J (2015) An update of the blockade of the renin angiotensin aldosterone system in clinical practice. Expert Opin Pharmacother 16:2283–2292

Matavelli LC, Siragy HM (2015) AT2 receptor activities and pathophysiological implications. J Cardiovasc Pharmacol 65:226–232

Mehdi UF, Adams-Huet B, Raskin P, Vega GL, Toto RD (2009) Addition of angiotensin receptor blockade or mineralocorticoid antagonism to maximal angiotensin-converting enzyme inhibition in diabetic nephropathy. J Am Soc Nephrol 20:2641–2650

Miyata K, Satou R, Inui D, Katsurada A, Seth D, Davis A et al (2014) Renoprotective effects of direct renin inhibition in glomerulonephritis. Am J Med Sci 348:306–314

Nadeem S, Batisky DL (2014) Aliskiren, the first direct renin inhibitor: assessing a role in pediatric hypertension and kidney diseases. Pediatr Nephrol 29:2105–2111

Nicolas LB, Gutierrez M, Binkert C, Dingemanse J (2013) Pharmacokinetics, pharmacodynamics, and tolerability of ACT-077825, a new direct renin inhibitor after multiple-ascending doses in healthy subjects. J Cardiovasc Pharm 61:42–50

Nishiyama A, Seth DM, Navar LG (2002) Renal interstitial fluid concentrations of angiotensins I and II in anesthetized rats. Hypertension 39:129–134

Nomura M, Nomura S, Mitsui T, Suzuki M, Kobayashi H, Ito T et al (2005) Possible involvement of aminopeptidase A in hypertension and renal damage in Dahl salt-sensitive rats. Am J Hypertens 18:538–543

Obata J, Nakamura T, Kuroyanagi R, Yoshida Y, Guo DF, Inagami T (1997) Candesartan prevents the progression of glomerulosclerosis in genetic hypertensive rats. Kidney Int 63(Suppl):S229–S231

Oparil S, Haber E (1971) Renin in differential diagnosis of hypertension. Am Heart J 82:568–570

Oudit GY, Liu GC, Zhong J, Basu R, Chow FL, Zhou J et al (2010) Human recombinant ACE2 reduces the progression of diabetic nephropathy. Diabetes 59:529–538

Ozono R, Wang ZQ, Moore AF, Inagami T, Siragy HM, Carey RM (1997) Expression of the subtype 2 angiotensin (AT2) receptor protein in rat kidney. Hypertension 30:1238–1246

Pandey A, Gaikwad AB (2017) AT2 receptor agonist Compound 21: a silver lining for diabetic nephropathy. Eur J Pharmacol 815:251–257

Parving HH, Brenner BM, McMurray JJV, de Zeeuw D, Haffner SM, Solomon SD et al (2012) Cardiorenal end points in a trial of aliskiren for type 2 diabetes. New Engl J Med 367:2204–2213

Paulis L, Rajkovicova R, Simko F (2015) New developments in the pharmacological treatment of hypertension: dead-end or a glimmer at the horizon? Curr Hypertens Rep 17:557

Persson F, Rossing P, Schjoedt KJ, Juhl T, Tarnow L, Stehouwer CD et al (2008) Time course of the antiproteinuric and antihypertensive effects of direct renin inhibition in type 2 diabetes. Kidney Int 73:1419–1425

Pitt B, Zannad F, Remme WJ, Cody R, Castaigne A, Perez A et al (1999) The effect of spironolactone on morbidity and mortality in patients with severe heart failure. Randomized aldactone evaluation study investigators. N Engl J Med 341:709–717

Rasche FM, Joel C, Ebert T, Frese T, Barinka F, Busch V et al (2018) Dual RAAS blockade with aliskiren in patients with severely impaired chronic kidney disease. Exp Clin Endocrinol Diab 126:39–52

Renke M, Lizakowski S, Tylicki L, Rutkowski P, Knap N, Heleniak Z et al (2014) Aliskiren attenuates oxidative stress and improves tubular status in non-diabetic patients with chronic kidney disease-Placebo controlled, randomized, cross-over study. Adv Med Sci 59:256–260

Rice ASC, Dworkin RH, McCarthy TD, Anand P, Bountra C, McCloud PI et al (2014) EMA401, an orally administered highly selective angiotensin II type 2 receptor antagonist, as a novel treatment for postherpetic neuralgia: a randomised, double-blind, placebo-controlled phase 2 clinical trial. Lancet 383:1637–1647

Rompe F, Artuc M, Hallberg A, Alterman M, Stroder K, Thone-Reineke C et al (2010) Direct angiotensin II type 2 receptor stimulation acts anti-inflammatory through epoxyeicosatrienoic acid and inhibition of nuclear factor kappaB. Hypertension 55:924–931

Ruggenenti P, Fassi A, Ilieva AP, Bruno S, Iliev IP, Brusegan V et al (2004) Preventing microalbuminuria in type 2 diabetes. N Engl J Med 351:1941–1951

Schainuck LI, Striker GE, Cutler RE, Benditt EP (1970) Structural-functional correlations in renal disease. II. The correlations. Hum Pathol 1:631–641

Schmerbach K, Kalk P, Wengenmayer C, Lucht K, Unger T, Hocher B et al (2012) Renal outcome in equipotent antihypertensive treatment with telmisartan, ramipril and in combination in SHR-SP rats. Clin Lab 58:625–633

Singh K, Sharma K, Singh M, Sharma PL (2012) Possible mechanism of the cardio-renal protective effects of AVE-0991, a non-peptide Mas-receptor agonist, in diabetic rats. J Renin-Angio-Aldo S 13:334–340

Singh Y, Singh K, Sharma PL (2013) Effect of combination of renin inhibitor and Mas-receptor agonist in DOCA-salt-induced hypertension in rats. Mol Cell Biochem 373:189–194

Siragy HM, Carey RM (1997) The subtype 2 (AT2) angiotensin receptor mediates renal production of nitric oxide in conscious rats. J Clin Invest 100:264–269

Steckelings UM, Paulis L, Namsolleck P, Unger T (2012) AT2 receptor agonists: hypertension and beyond. Curr Opin Nephrol Hypertens 21:142–146

Sun N, Zhai L, Li H, Shi LH, Yao Z, Zhang B (2016) Angiotensin-converting enzyme inhibitor (ACEI)-mediated amelioration in renal fibrosis involves suppression of mast cell degranulation. Kidney Blood Press Res 41:108–118

Sun QL, Li M, Rui HL, Chen YP (2015) Inhibition of local aldosterone by eplerenone reduces renal structural damage in a novel model of chronic cyclosporine A nephrotoxicity. J Renin Angiotensin Aldosterone Syst 16:301–310

Sureshkumar KK (2008) Renin inhibition with aliskiren in hypertension: focus on aliskiren/hydrochlorothiazide combination therapy. Vasc Health Risk Manag 4:1205–1220

Taira M, Toba H, Murakami M, Iga I, Serizawa R, Murata S et al (2008) Spironolactone exhibits direct renoprotective effects and inhibits renal renin-angiotensin-aldosterone system in diabetic rats. Eur J Pharmacol 589:264–271

Tamargo J, Solini A, Ruilope LM (2014) Comparison of agents that affect aldosterone action. Semin Nephrol 34:285–306

Tikellis C, Johnston CI, Forbes JM, Burns W, Thomas MC, Lew RA et al (2004) Identification of angiotensin converting enzyme 2 in the rodent retina. Curr Eye Res 29:419–427

Treml B, Neu N, Kleinsasser A, Gritsch C, Finsterwalder T, Geiger R et al (2010) Recombinant angiotensin-converting enzyme 2 improves pulmonary blood flow and oxygenation in lipopolysaccharide-induced lung injury in piglets. Crit Care Med 38:596–601

Vaidyanathan S, Bigler H, Yeh CM, Bizot MN, Dieterich HA, Howard D et al (2007) Pharmacokinetics of the oral direct renin inhibitor aliskiren alone and in combination with irbesartan in renal impairment. Clin Pharmacokinet 46:661–675

Velez JC, Janech MG, Hicks MP, Morinelli TA, Rodgers J, Self SE et al (2014) Lack of renoprotective effect of chronic intravenous angiotensin-(1-7) or angiotensin-(2-10) in a rat model of focal segmental glomerulosclerosis. PLoS ONE 9:e110083

Velez JCQ, Arif E, Rodgers J, Hicks MP, Arthur JM, Nihalani D et al (2017) Deficiency of the angiotensinase aminopeptidase A increases susceptibility to glomerular injury. J Am Soc Nephrol 28:2119–2132

Wakahara S, Konoshita T, Mizuno S, Motomura M, Aoyama C, Makino Y et al (2007) Synergistic expression of angiotensin-converting enzyme (ACE) and ACE2 in human renal tissue and confounding effects of hypertension on the ACE to ACE2 ratio. Endocrinology 148:2453–2457

Wang H, Fu W, Jin Z, Wang Y, Yao W, Yin P et al (2013) Advanced IgA nephropathy with impaired renal function benefits from losartan treatment in rats. Ren Fail 35:812–818

Wang Y, Del Borgo M, Lee HW, Baraldi D, Hirmiz B, Gaspari TA et al (2017) Anti-fibrotic potential of AT2 receptor agonists. Front Pharmacol 8:564

Wiemer G, Dobrucki LW, Louka FR, Malinski T, Heitsch H (2002) AVE 0991, a nonpeptide mimic of the effects of angiotensin-(1-7) on the endothelium. Hypertension 40:847–852

Williams VR, Scholey JW (2018) Angiotensin-converting enzyme 2 and renal disease. Curr Opin Nephrol Hy 27:35–41

Wu MT, Tung SC, Hsu KT, Lee CT (2014) Aliskiren add-on therapy effectively reduces proteinuria in chronic kidney disease: an open-label prospective trial. J Renin Angiotensin Aldosterone Syst 15:271–277

Yamada K, Iyer SN, Chappell MC, Ganten D, Ferrario CM (1998) Converting enzyme determines plasma clearance of angiotensin-(1-7). Hypertension 32:496–502

Ye M, Wysocki J, Naaz P, Salabat MR, LaPointe MS, Batlle D (2004) Increased ACE 2 and decreased ACE protein in renal tubules from diabetic mice: a renoprotective combination? Hypertension 43:1120–1125

Zhang F, Liu H, Liu D, Liu Y, Li H, Tan X, Liu F, Peng Y, Zhang H (2017) Effects of RAAS inhibitors in patients with kidney disease. Curr Hypertens Rep 19:72
Zhong J, Guo D, Chen CB, Wang W, Schuster M, Loibner H et al (2011) Prevention of angiotensin II-mediated renal oxidative stress, inflammation, and fibrosis by angiotensin-converting enzyme 2. Hypertension 57:314–322
Zhu F, Liao YH, Li LD, Cheng M, Wei F, Wei YM et al (2006) Target organ protection from a novel angiotensin II receptor (AT1) vaccine ATR12181 in spontaneously hypertensive rats. Cell Mol Immunol 3:107–114

Chapter 34
Extracellular Vesicles: Opportunities and Challenges for the Treatment of Renal Fibrosis

Tao-Tao Tang and Bi-Cheng Liu

Abstract Extracellular vesicles (EVs) are small lipid-based membrane-bound vesicles secreted by most cells under both physiological and pathological conditions. A key function of EVs is to mediate cell–cell communication via transferring mRNAs, miRNAs and proteins from parent cells to recipient cells. These unique features of EVs have spurred a renewed interest in their utility for therapeutics. Given the growing evidence for EV-mediated renal diseases, strategies that could block the release or uptake of pathogenic EVs will be discussed in this review. Then, the therapeutic potential of EVs predominantly from stem cells in renal diseases will be outlined. Finally, we will focus on the specific application of EVs as a novel drug delivery system and highlight the challenges of EVs-based therapies for renal diseases.

Keywords Extracellular vesicles · Treatment · Renal fibrosis · Drug delivery

34.1 Introduction

Extracellular vesicles (EVs) are nanoscale vesicles released by cells in physiological and pathological conditions. Depending on their size and biogenesis, EVs are classified into three major categories: exosomes, microvesicles and apoptotic bodies (van der Pol et al. 2012; Raposo and Stoorvogel 2013). Here, we focus on the first two classes of EVs. Exosomes, ranging from 30 to 150 nm in diameter, are formed by the fusion of intracellular multivesicular bodies with the plasma membrane (Colombo et al. 2014), whereas microvesicles, 50–1000 nm in size, are shed directly from the plasma membrane (Morel et al. 2011) (Fig. 34.1).

EVs were initially regarded as cell dust with no biological significance (Wolf 1967), but there is increasingly evidence for their important role in cell signalling and communication in normal and disease states (Karpman et al. 2017; Erdbrügger

T.-T. Tang · B.-C. Liu (✉)
Institute of Nephrology, Zhong Da Hospital, Southeast University School of Medicine, Nanjing, China
e-mail: liubc64@163.com

B.-C. Liu et al. (eds.), *Renal Fibrosis: Mechanisms and Therapies*, Advances in Experimental Medicine and Biology 1165,
https://doi.org/10.1007/978-981-13-8871-2_34

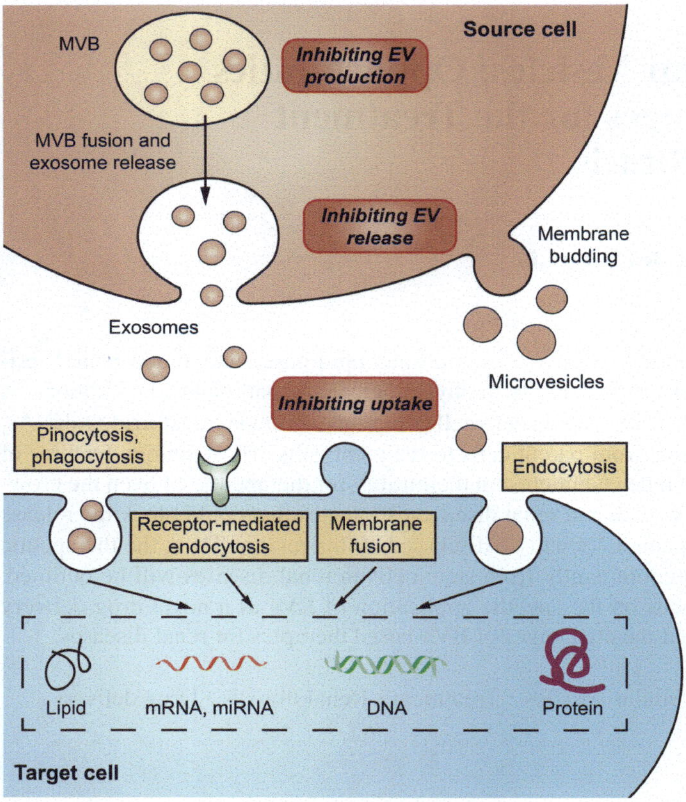

Type	Formation	Size	Markers	Contents
Exosomes	Endolysosomal pathway; internal budding and fusion of multivesicular body with cell membrane	30-150 nm	Tetraspanins (CD63, CD9, CD81), Alix, PDCD6IP, TSG101, flotillin, MFGE8	Proteins, lipids, mRNA, miRNA, DNA and cytosol
Microvesicles	Cell surface; outward budding of cell membrane	100-1000 nm	Integrins, selectins, CD40 ligand	Proteins, lipids, mRNA, miRNA, DNA and cytosol

Fig. 34.1 Biogenesis and characteristics of major classes of EVs. EVs can be classed as exosomes, microvesicles and apoptotic bodies based on their biogenesis and size. Exosomes are formed by the fusion of intracellular multivesicular bodies (MVBs) with the plasma membrane, whereas microvesicles are shed directly from the plasma membrane. EVs are taken up by cells by endocytosis, phagocytosis, pinocytosis or membrane fusion, and subsequently transfer cell membrane receptors or deliver effectors including mRNA, miRNA, DNA, lipid or protein into recipient cells. In addition, EVs could serve as a therapeutic target by inhibition of their production, release or cellular uptake

and Le 2016; Morrison et al. 2016; Zhang et al. 2016b; Camussi et al. 2010). In kidneys, EVs have been tightly linked to inflammation, fibrosis, thrombosis, adhesion, immune suppression, or growth and regeneration (Karpman et al. 2017; Erdbrügger and Le 2016; Morrison et al. 2016; Zhang et al. 2016b; Camussi et al. 2010). Therefore, EVs and their components could serve as the therapeutic targets, which can be inhibited to alleviate disease progression. Moreover, as EVs are suggested to participate in the tissue repair and immune modulation, they could be utilized directly as therapeutic agents in regenerative medicine and the treatment of autoimmune diseases. For example, EVs from mesenchymal stem cells protected against acute tubular injury and attenuated kidney inflammation (Bruno et al. 2009; Eirin et al. 2017; Rani et al. 2015).

Finally, given the natural role in transporting bioactive entities of EVs, they also have potential as drug carrier like a "Trojan horse" (van Dommelen et al. 2012; Fuhrmann et al. 2015). Recent studies indicate that EVs can function as efficient carriers of chemotherapeutic drugs (Tang et al. 2012; Yang et al. 2015), RNA drugs (Kamerkar et al. 2017; Alvarez-Erviti et al. 2011) and anti-inflammatory drugs (Sun et al. 2010; Zhuang et al. 2011). In this review, we will focus on recent developments in EV-based therapy as potential targets and as novel therapeutic agents, especially in the use of EVs as smart drug carriers.

34.2 Extracellular Vesicles as Potential Therapeutic Targets

Within the kidney, EVs can originate from blood cells, endothelial cells, podocytes or tubular epithelial cells (TECs), which have been strongly implicated in the pathogenesis of both acute kidney injury (AKI) and chronic kidney disease (CKD). Our group demonstrated that in the setting of proteinuric kidney disease, albumin triggered TECs to release exosomes packaged with CCL2 mRNA, which was delivered to macrophages and leads to interstitial inflammation (Lv et al. 2018a). Borges et al. identified that injured TECs released exosomes containing TGF-β mRNA to activate fibroblasts, contributing to the development of renal fibrosis in post-AKI kidneys (Borges et al. 2013). Moreover, microvesicle-mediated delivery of miR-21 among TECs could also drive the progressive renal fibrosis (Zhou et al. 2013a). Recent data found that transglutaminase-2, a matrix crosslinking enzyme for fibrotic remodelling, was secreted from TECs via exosomes (Furini et al. 2018). Thus, specifically inhibiting the biogenesis or uptake of these pathogenic EVs could be a potential therapeutic approach to alleviate disease progression (Fig. 34.1).

Various cellular components are known to be crucial for the biogenesis and release of EVs, and a number of possible therapeutic targets have been identified. For exosomes, ceramide is an important component in endosomal sorting and exosome biogenesis and its inhibition by GW4869 (neutral sphingomyelinase inhibitor) or amiloride (an antihypertensive agent) decreases exosome production (Trajkovic et al. 2008; Chalmin et al. 2010). GTPases Rab27b can regulate exosome release in some tumour cells, and this was demonstrated to be a therapeutic target (using RNAi) for reducing tumour progression (Peinado et al. 2012; Ostrowski et al. 2010; Bobrie et al. 2012). For microvesicles, the calpain inhibitor calpeptin or calpastatin can reduce the shedding of microvesicles (Zafrani et al. 2012; Yano et al. 1993), as well as blocking P2X receptors (Arvidsson et al. 2015). Furthermore, C1 inhibitor lessens the release of endothelial microvesicles, alleviating inflammatory diseases such as vasculitis (Mossberg et al. 2017). However, there are many limitations to target EV biogenesis and release because the precise mechanism remains elusive and is likely to vary among different cells.

In addition to reducing the level of EVs, inhibition of EV uptake into cells is also possible by certain substances and antibodies (Mulcahy et al. 2014). Blocking surface phosphatidylserine (which is important for cell adhesion) using Diannexin decreases the uptake of EVs derived from tumour cells (Lima et al. 2009; Al-Nedawi et al. 2009). Besides, an antibody to DEL1, annexin V, abciximab, chlorpromazine, cytochalasin D or cytochalasin B also have been demonstrated to block the uptake of EVs (Mulcahy et al. 2014; Dasgupta et al. 2012; Faille et al. 2012; Barrès et al. 2010), but it is difficult to translate these into therapeutic intervention due to the lack of specific mechanism regarding the key steps in EV trafficking and target definition.

34.3 Extracellular Vesicles as Therapeutic Agents

An increasing number of studies have demonstrated EVs, especially those derived from stem cells, and have innate therapeutic potential by virtue of their intrinsic cargoes, such as growth factors, soluble proteins and nucleic acids (Andaloussi et al. 2013). In kidney, mesenchymal stem cell-derived EVs of different origin also exhibit encouraging renoprotective efficacy, as shown in models of AKI, diabetic nephropathy, CKD and fibrosis. The application of these EVs in kidney diseases has been summarized in Table 34.1. For instance, Wang et al. showed that exosomes derived from bone marrow MSCs were able to transfer miR-let7c to damaged kidney cells and attenuate renal fibrosis in UUO mice (Wang et al. 2016). Kholia et al. reported that EVs derived from liver stem cells exhibited a regenerative, anti-inflammatory and anti-fibrotic role in aristolochic acid-induced kidney fibrosis (Kholia et al. 2018). In addition, EVs obtained from umbilical cord MSCs (Zhou et al. 2013b; Ju et al. 2015), Wharton's jelly MSCs (Zou et al. 2014; Gu et al. 2016; Zhang et al. 2016a), adipose-derived MSCs (Eirin et al. 2017; Lin et al. 2016), kidney MSCs (Choi et al. 2014; Ranghino et al. 2017; Choi et al. 2015), as well as urine-derived MSCs (Jiang et al. 2016) also showed potential therapeutic benefits on kidney diseases.

Mechanistically, the protective effect of MSC-EVs on kidney diseases depends on their transfer of genetic material including mRNA and miRNA (Rani et al. 2015; Grange et al. 2017; Nargesi et al. 2017). This was confirmed in many studies when degradation of the RNAs in MSC-EVs using RNase could abolish aforementioned therapeutic benefits (Bruno et al. 2009; Choi et al. 2015; Zou et al. 2016), suggesting RNA-dependent biological effect. EVs derived from the Drosha-knockdown MSCs also showed global downregulation of miRNAs, resulting in ineffective renal repair of

Table 34.1 Therapeutic application of extracellular vesicles in kidney disease

EV origin	Kidney injury model	EVs doses	Injection method	Effective molecules
BM-MSCs	Glycerol-induced AKI	15 μg	Intravenous injection	mRNA
		2.2×10^8 EVs	Intravenous injection	miRNA
	IRI-induced AKI	200 μg	Renal capsule injection	CCR2 protein
		30 μg	Intravenous injection	mRNA
	Cisplatin-induced AKI	100 μg	Intravenous injection	Not studied
	Diabetic nephropathy	5.3×10^7 EVs	Renal subcapsular	Not studied
	Unilateral ureteral obstruction	1×10^6 EVs	Intravenous injection	miR-let7c
		30 μg	Intravenous injection	miRNA
		30 mg	Intravenous injection	miRNA
UC-MSCs	Cisplatin-induced AKI	200 μg	Renal capsule injection	Not studied
	IRI-induced AKI	30 μg	Intravenous injection	HGF mRNA
WJ-MSCs	IRI-induced AKI	100 μg	Intravenous injection	Not studied
		100 μg	Intravenous injection	miR-30
		100 μg	Intravenous injection	Not studied
A-MSCs	IRI-induced AKI	100 μg	Intravenous injection	Not studied
	Metabolic syndrome + renal artery stenosis	1×10^{10} EVs	Stenotic renal artery injection	IL-10 protein
L-MSCs	Glycerol-induced AKI	$1.88 \pm 0.6 \times 10^9$ $5.53 \pm 2.15 \times 10^9$	Intravenous injection	Not studied

(continued)

Table 34.1 (continued)

EV origin	Kidney injury model	EVs doses	Injection method	Effective molecules
	Aristolochic acid-induced kidney fibrosis	1×10^{10} EVs	Intravenous injection	Not studied
K-MSCs	IRI-induced AKI	2×10^{7} EVs	Intravenous injection	VEGF, IGF, FGF mRNA
	IRI-induced AKI	4×10^{8} EVs	Intravenous injection	miRNA
	Unilateral ureteral obstruction	2×10^{7} EVs	Intravenous injection	mRNA
U-MSCs	Type I diabetes	100 μg	Intravenous injection	VEGF, TGF-β1, angiogenin and BMP7 protein
ECFCs	IRI-induced AKI	15 μg	Intravenous injection	Not studied
	IRI-induced AKI	20 μg	Intravenous injection	miR-486-5p
EPCs	IRI-induced AKI	30 μg	Intravenous injection	miR-126 miR-296
	Anti-Thy1.1 glomeru-lonephritis	30 μg	Intravenous injection	Factor H, CD55, CD59 mRNA
Hypoxic TECs	IRI-induced AKI	100 μg	Intravenous injection	mRNA
Scattered TECs	Renal artery stenosis	30 μg	Intravenous injection	Mitochondria

BM bone marrow; *UC* umbilical cord; *WJ* Wharton's jelly; *A* adipose tissue; *L* liver; *K* kidney; *U* urine; *ECFC* endothelial colony-forming cells; *EPC* endothelial progenitor cell

glycerol-induced AKI (Collino et al. 2015). Gene ontology analysis further showed that those genes shuttled by MSC-EVs were involved in healing pathways associated with renal regeneration (Collino et al. 2015). Moreover, EVs can also deliver proteins from MSCs to injured kidney cells. Proteins related to cell proliferation, adhesion, migration and morphogenesis have been identified in the vesicles by extensive proteomic analysis (Eirin et al. 2017; Shen et al. 2016; Jiang et al. 2016; Kim et al. 2012). In this regard, an elegant study showed that adipose-derived MSC-EVs attenuated renal inflammation in a porcine model of coexisting metabolic syndrome and renal artery stenosis by their cargo of IL-10 (Eirin et al. 2017).

In addition to MSC-EVs, other sources of cell-derived EVs, such as endothelial colony-forming cells (ECFCs), endothelial progenitor cells (EPCs) and hypoxic TECs, have shown significant beneficial effects as well (Table 34.1). In models of ischemic AKI, both ECFC-derived exosomes and EPC-derived EVs ameliorated renal injury via transfer of miRNAs (Viñas et al. 2016; Cantaluppi et al. 2012). In anti-Thy1.1-induced model of glomerulonephritis, EPC-derived EVs alleviated mesangial cell activation, leukocyte infiltration and apoptosis, which was related to its content of mRNAs coding for anti-apoptotic factors and the complement inhibitors (Cantaluppi et al. 2014). Interestingly, Dominguez et al. found that EVs derived from hypoxic TECs significantly improved renal tubular damage, fibrosis and microvascular pruning in established renal IRI (Dominguez et al. 2017). However, paradoxically, EVs from injured TECs also contribute to the progression of interstitial inflammation and fibrosis (Lv et al. 2018a; Borges et al. 2013; Zhou et al. 2013a; Furini et al. 2018), and the dual role of TEC-derived EVs needs to be further clarified.

34.4 Extracellular Vesicles as Smart Drug Carriers

Currently, the most preferred drug delivery systems are nanoparticle platforms based on liposomes, albumin, polymeric micelles and nanosized polymer-drug conjugates, which effectively improve the pharmacokinetics and biodistribution of drugs (Kamaly et al. 2016). However, their immunogenicity, stability and toxicity still remain elusive. In this case, EV-based drug delivery—with many of advantages, such as high permeability, less immunogenicity and non-cytotoxicity—appears to be a superior choice, overcoming the limitations observed with nanoparticles (Ha et al. 2016; Lv et al. 2018b). So far, EVs have been eloquently demonstrated to be as therapeutic nanocarriers for delivering a variety of cargos, including siRNAs, miRNAs, proteins and drugs (van Dommelen et al. 2012; Fuhrmann et al. 2015). But the application of EVs in kidney diseases has just begun its journey.

34.4.1 Cargo-Loading Techniques

In order to employ EV-based drug delivery, it is essential to consider the methods of cargo loading and their suitability under different circumstances. In brief, cargo encapsulation can be performed exogenously or endogenously (van Dommelen et al. 2012; Fuhrmann et al. 2015; Batrakova and Kim 2015) (Fig. 34.2). For exogenously loading, the cargos were packaged into pre-assembled EVs ex vitro. A number of methods, including electroporation, sonication, direct transfection and simple incubation, are valid strategies for drug incorporation in this regard (Syn et al. 2017). For example, simple incubation is a versatile and feasible approach employed in many cases, through which several small lipophilic molecules, such as curcumin (antioxidant agents) (Sun et al. 2010; Zhuang et al. 2011), doxorubicin (Rani et al. 2015; Tian et al. 2014) and paclitaxel (Yang et al. 2015) (anti-cancer agents), are passively loaded into exosomes, but the loading capacity is low. Besides, potential limitations of electroporation may include size-dependent loading efficiency, denaturation and

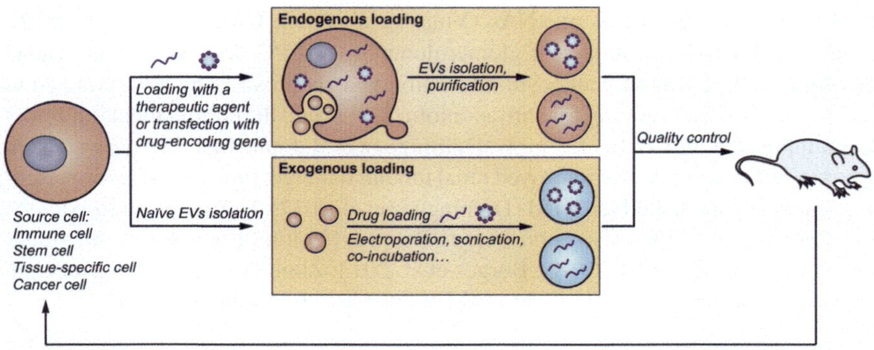

Fig. 34.2 Flow of the production of EV-based drug formulations. EV-based drug delivery requires the correct choice of source cell type for the specific application and should ideally be patient-derived to avoid triggering immune response. The therapeutic cargo can include different types of siRNA, miRNA, proteins or small-molecule compound such as curcumin or chemotherapeutics. Drug loading can be carried out either endogenously or exogenously. Endogenous loading is achieved by loading source cell with a therapeutic agent or transfecting source cell with drug-encoding gene which is then released in EVs upon collection. Exogenous loading allows the isolation of EVs before their loading with therapeutic cargo with the help of electroporation or by simple co-incubation. Importantly, the generation process should meet the quality requirements

degradation of organic molecules and colloidal stability of the exosomal preparation (Syn et al. 2017).

For endogenously loading, the drug-loaded EVs are isolated from the modified parent cells through genetic engineering or medication with cytotoxic drugs. This method is convenient and requires very few manipulation steps. It is reported that paclitaxel is incorporated by MSCs and released in exosomes (Pascucci et al. 2014), as well as other anti-cancer agents: etoposide, carboplatin, irinotecan, epirubicin and mitoxantrone (Lv et al. 2012), which are loaded in exosomes with strong anti-proliferative activity. Moreover, recent studies demonstrated that the therapeutic protein and its genetic material could be loaded into EVs when parental cells were transfected with drug-encoding gene (Zeelenberg et al. 2008; Lee et al. 2015; Yim et al. 2016), but that might confer risks of genotoxicity and adverse host immune response. Of note, each cargo-loading strategy has its advantages and limitations depending on the type of therapeutic cargo and site of the disease, and thus further nuanced understanding is needed to select the optimal approach for mass production.

34.4.2 EVs as Delivery Vehicles for Nucleic Acids

It is known that EVs naturally carry nucleic acids, making them stable in the circulation and protecting from degradation. Given this, EVs may offer unique advantages for genetic therapy, and key studies using EVs as carriers for genetic materials are highlighted below. The first report on EV-mediated transfer of exogenous nucleic acids was published in 2010, when it was shown that THP-1 cells, which were transfected with a miR-150 mimic, secreted miR-150-enriched EVs and that could be functionally delivered to recipient cells (Zhang et al. 2010). Subsequent study conducted by Akao et al. found that THP-1 monocytes transfected with miR-143 mimic ex vivo secreted miR-143-containing EVs in nude mice after intravenous injection (Akao et al. 2011). Furthermore, when injected intravenously into UUO mice, engineered MSCs that overexpressed miR-let7c attenuated renal fibrosis via secreting miR-let7c-loaded exosomes (Wang et al. 2016). All these studies have eloquently corroborated such modes of miRNA transfer.

Small interference RNA (siRNA) is used to disrupt genes of interest and has great potential for the treatment of a range of diseases. Several studies have been conducted to test the usefulness of EVs as delivery vehicle for siRNA, and the first study conducted by Alvarez-Erviti et al. found that by expressing a neuron-targeting protein on the surface of exosomes, they could specifically deliver siRNA to the brain and resulted in a specific gene knockdown (Alvarez-Erviti et al. 2011). Importantly, the treatment displayed minimal toxicity and immune stimulation, even following repeated administration, suggesting EVs are suitable to deliver vectors in RNA interference therapy. This notion has been further confirmed by Wahlgren et al. that the gene MAPK1 was selectively silenced in monocytes and lymphocytes by using siRNA-loaded exosomes derived from human plasma (Wahlgren et al. 2012). More recently, an elegant research employed fibroblast-like mesenchymal cell-derived exosomes to deliver siRNA or short hairpin RNA specific to oncogenic KRAS, achieving enhanced therapeutic efficacy in suppressing tumour growth and improving the overall survival (Kamerkar et al. 2017). Notably, the therapeutic effects of engineered exosomes were greater than siRNA-loaded liposomes (Kamerkar et al. 2017). Beyond miRNA and siRNA delivery, EVs were also exploited to encapsulate adeno-associated viruses (AAVs), which were substantially more efficient than free AAVs for the delivery of genetic cargo into recipient cells (Maguire et al. 2012). Collectively, these studies emphasize the potential of using EVs for the therapeutic delivery of nucleic acids.

34.4.3 EVs as Delivery Vehicles for Proteins

In addition to delivering nucleic acids, EVs are also used to deliver large molecules such as proteins. Haney and colleagues found that exosomes loaded with the antioxidant protein catalase were successfully delivered across the blood–brain barrier (BBB) and provided significant neuroprotective effects in a model of Parkinson's disease (Haney et al. 2015). In this study, catalase was incorporated into pre-assembled exosomes ex vivo using different methods, and identified sonication and extrusion approaches achieved better loading efficiency, sustained release and protein preservation (Haney et al. 2015). Similar results were reported by Yuan et al., showing that macrophage-derived exosomes efficiently crossed the BBB and delivered a cargo protein to the brain, further indicating the potency of EVs as nanocarriers for brain delivery of therapeutic proteins (Yuan et al. 2017). The cargo protein in the study was also loaded in an exogenous way by mixing with exosomes; in addition, the therapeutic protein can be packaged into EVs by transfecting parental cells as well. For example, HEK-293T cells transfected with suicide gene secreted EVs enriched in suicide mRNA and protein, which were subsequently used to treat schwannoma tumours in an orthotopic mouse model, leading to reduced tumour growth (Mizrak et al. 2013). Overall, these studies suggest that EVs can serve as novel nanocarriers to effectively deliver therapeutic proteins.

34.4.4 EVs as Delivery Vehicles for Drugs

EVs have been utilized as delivery vehicles for therapeutic drugs in extensive research (Tang et al. 2012; Yang et al. 2015; Sun et al. 2010; Zhuang et al. 2011). Early studies from the Zhang group (Sun et al. 2010; Zhuang et al. 2011) demonstrated an anti-inflammatory small-molecule compound curcumin could be incorporated into exosomes by mixing curcumin with murine tumour cell line (EL-4) or microglia cell (JSI124)-derived exosomes, and found that exosomal curcumin exhibited enhanced anti-inflammatory activity in LPS-induced septic shock mouse model. Interestingly, exosomal packaging leads to an increase in the solubility, stability and bioavailability of curcumin (Sun et al. 2010), suggesting EVs are capable to modify the bioavailability of the native drug. For another natural phytochemical compound celastrol, exosome-mediated delivery also improved drug biodistribution and subsequently enhanced its anti-tumour efficacy (Aqil et al. 2016). This study further highlighted the benefits of EVs in enhancing the properties of drugs, such as solubility, stability and bioavailability.

Besides, the deployment of EVs encapsulating chemotherapeutics such as paclitaxel and doxorubicin has yielded promising results, representing encouraging anticancer efficacy with minimal cytotoxicity towards non-cancerous cells (Tang et al. 2012; Yang et al. 2015; Syn et al. 2017; Tian et al. 2014; Pascucci et al. 2014; Jang et al. 2013; Saari et al. 2015; Toffoli et al. 2015; Srivastava et al. 2016; Martins-Marques et al. 2016). For example, anti-cancer drug-loaded exosomes or exosome-like vesicles were shown to traffic to tumour tissue and reduce tumour growth in mice without overt adverse effects (Tian et al. 2014; Jang et al. 2013). Importantly, exosomes had superior therapeutic effects when compared to liposomes (Jang et al. 2013). Moreover, the administration of doxorubicin loaded in exosomes resulted in significantly less drug accumulation in non-target organs and prevented the onset of off-target cardiotoxicity compared with mice treated with unmodified doxorubicin (Saari et al. 2015; Toffoli et al. 2015; Srivastava et al. 2016; Martins-Marques et al. 2016). Thus, the advantages of exosomes packaging may improve the safety profile of cytotoxic agents and present further opportunities to address cancer therapy.

Table 34.2 Advantages and limitations of extracellular vesicle-based therapy

Advantages	Limitations
☺ Nanoscale ☺ Natural lipid and surface protein composition ☺ Stable in biological fluids ☺ Low immunogenicity ☺ Cell-to-cell communicators ☺ Unidirectional targeting or active targeting by modification ☺ Suitable for multi-drug delivery ☺ Various drug encapsulation method ☺ Translocation through physical barriers	☹ Biochemical composition of EVs unclear ☹ Production or uptake mechanism yet poorly described ☹ Good manufacturing practice standards lacking ☹ High scale and efficient production difficulty ☹ Difficult to package through renal barriers ☹ (Pre)clinical evaluation lacking

34.5 Benefits and Challenges of Extracellular Vesicle Therapy

Unarguably, the field of EV-based therapeutics holds significant promise to enable targeted drug delivery with superior efficiency (Table 34.2). Compared with existing liposomes or polymeric nanoparticles, the outstanding advantage of EV-based therapy is their natural lipid and surface protein composition, which enable them to evade phagocytosis, extend blood half-life and reduce long-term safety issues. Moreover, the small size of EVs facilitates their extravasation, translocation through physical barriers and passage through extracellular matrix (van Dommelen et al. 2012; van den Boorn et al. 2011). Several studies have demonstrated that EVs successfully cross the BBB and deliver cargos into the brain, but whether EVs are able to pass through the glomerular filtration barrier remains unclear. In addition, EVs encapsulation also makes the new drug candidates such as proteins and nucleic acids more stable and targetable to treatment site (Zhu et al. 2012; Bruno et al. 2013).

However, before EV-based therapy can be translated to the clinic, several hurdles need to be overcome (Table 34.2). First, many properties and mechanisms about EV biology such as the biochemical composition of EV currently remain elusive, and the production or uptake mechanism yet poorly described. Even though from the same cell types, EVs may have contradictory effects as a consequence of differences in cell culture conditions and differences in the purification protocols used or due to a lack of robust extracellular vesicle characterization (Andaloussi et al. 2013; Zhu et al. 2012; Bruno et al. 2013). In addition, a major bottleneck in the translation of EV-based therapy into clinic is the lack of good manufacturing practice (GMP) standards. To develop clinical-grade EVs, sterile generation, high scale and efficient production of sufficient amounts of EVs with therapeutic payloads for clinical testing are required. Very recently, Mendt and colleagues have illustrated the process and feasibility of generating GMP-grade exosomes (Mendt et al. 2018). Finally, regarding the particularity of kidney, the glomerular filtration barrier is the primary obstacle that excludes EVs from accessing podocytes or tubular cells. The level of EVs accumulation in the kidney is highly restricted based on the injury degree of the glomerulus; thus, effective engineering of the size, shape and surface charge will conduce to EVs passing through renal barriers and their advancement to the clinic.

34.6 Conclusions

EVs are important conveyers of information between cells and have been strongly implicated in numerous biological and pathological processes. Targeting EVs directly to inhibit their pathogenic effects or exploiting their innate potential for renal regenerative medicine is promising therapeutic strategy. Moreover, although EV-based therapy has just begun its journey, they provide an enormous promise and a fresh therapeutic area for delivery of different drugs such as small-molecule compounds, particularly therapeutic nucleic acid delivery.

Acknowledgements This chapter was modified from a paper reported by our group in Front Physiol (Tang et al. 2019).
This study was supported by grants from the National Key Research and Development Program of China (2018YFC1314004), the National Natural Science Foundation of China (No.81720108007, 81670696, 81470922 and 31671194), the Clinical Research Center of Jiangsu Province (No. BL2014080) and the Postgraduate Research and Practice Innovation Program of Jiangsu Province (No. KYCX18_0171).

References

Akao Y, Iio A, Itoh T, Noguchi S, Itoh Y, Ohtsuki Y et al (2011) Microvesicle-mediated RNA molecule delivery system using monocytes/macrophages. Mol Ther 19:395–399
Al-Nedawi K, Meehan B, Kerbel RS, Allison AC, Rak J (2009) Endothelial expression of autocrine VEGF upon the uptake of tumor-derived microvesicles containing oncogenic EGFR. Proc Natl Acad Sci U S A 106:3794–3799
Alvarez-Erviti L, Seow Y, Yin H, Betts C, Lakhal S, Wood MJ (2011) Delivery of siRNA to the mouse brain by systemic injection of targeted exosomes. Nat Biotechnol 29:341–345
Andaloussi SEL, Mäger I, Breakefield XO, Wood MJ (2013) Extracellular vesicles: biology and emerging therapeutic opportunities. Nat Rev Drug Discov 12:347–357
Aqil F, Kausar H, Agrawal AK, Jeyabalan J, Kyakulaga AH, Munagala R, Munagala R et al (2016) Exosomal formulation enhances therapeutic response of celastrol against lung cancer. Exp Mol Pathol 101:12–21
Arvidsson I, Ståhl AL, Hedström MM, Kristoffersson AC, Rylander C, Westman JS et al (2015) Shiga toxin-induced complement-mediated hemolysis and release of complement-coated red blood cell-derived microvesicles in hemolytic uremic syndrome. J Immunol 194:2309–2318
Barrès C, Blanc L, Bette-Bobillo P, André S, Mamoun R, Gabius HJ et al (2010) Galectin-5 is bound onto the surface of rat reticulocyte exosomes and modulates vesicle uptake by macrophages. Blood 115:696–705
Batrakova EV, Kim MS (2015) Using exosomes, naturally-equipped nanocarriers, for drug delivery. J Control Release 219:396–405

Bobrie A, Krumeich S, Reyal F, Recchi C, Moita LF, Seabra MC et al (2012) Rab27a supports exosome-dependent and -independent mechanisms that modify the tumor microenvironment and can promote tumor progression. Cancer Res 72:4920–4930
Borges FT, Melo SA, Özdemir BC, Kato N, Revuelta I, Miller CA et al (2013) TGF-β1-containing exosomes from injured epithelial cells activate fibroblasts to initiate tissue regenerative responses and fibrosis. J Am Soc Nephrol 24:385–392
Bruno S, Grange C, Deregibus MC, Calogero RA, Saviozzi S, Collino F et al (2009) Mesenchymal stem cell-derived microvesicles protect against acute tubular injury. J Am Soc Nephrol 20:1053–1067
Bruno S, Collino F, Deregibus MC, Grange C, Tetta C, Camussi G (2013) Microvesicles derived from human bone marrow mesenchymal stem cells inhibit tumor growth. Stem Cells Dev 22:758–771
Camussi G, Deregibus MC, Bruno S, Cantaluppi V, Biancone L (2010) Exosomes/microvesicles as a mechanism of cell-to-cell communication. Kidney Int 78:838–848
Cantaluppi V, Gatti S, Medica D, Figliolini F, Bruno S, Deregibus MC et al (2012) Microvesicles derived from endothelial progenitor cells protect the kidney from ischemia-reperfusion injury by microRNA-dependent reprogramming of resident renal cells. Kidney Int 82:412–427
Cantaluppi V, Medica D, Mannari C, Stiaccini G, Figliolini F, Dellepiane S et al (2014) Endothelial progenitor cell-derived extracellular vesicles protect from complement-mediated mesangial injury in experimental anti-Thy1.1 glomerulonephritis. Nephrol Dial Transplant 30:410–422
Chalmin F, Ladoire S, Mignot G, Vincent J, Bruchard M, Remy-Martin JP et al (2010) Membrane-associated Hsp72 from tumor-derived exosomes mediates STAT3-dependent immunosuppressive function of mouse and human myeloid-derived suppressor cells. J Clin Invest 120:457–471
Choi HY, Moon SJ, Ratliff BB, Ahn SH, Jung A, Lee M et al (2014) Microparticles from kidney-derived mesenchymal stem cells act as carriers of proangiogenic signals and contribute to recovery from acute kidney injury. PLoS ONE 9:e87853
Choi HY, Lee HG, Kim BS, Ahn SH, Jung A, Lee M et al (2015) Mesenchymal stem cell-derived microparticles ameliorate peritubular capillary rarefaction via inhibition of endothelial-mesenchymal transition and decrease tubulointerstitial fibrosis in unilateral ureteral obstruction. Stem Cell Res Ther 6:18
Collino F, Bruno S, Incarnato D, Dettori D, Neri F, Provero P et al (2015) AKI recovery induced by mesenchymal stromal cell-derived extracellular vesicles carrying MicroRNAs. J Am Soc Nephrol 26:2349–2360
Colombo M, Raposo G, Théry C (2014) Biogenesis, secretion, and intercellular interactions of exosomes and other extracellular vesicles. Annu Rev Cell Dev Biol 30:255–289
Dasgupta SK, Le A, Chavakis T, Rumbaut RE, Thiagarajan P (2012) Developmental endothelial locus-1 (Del-1) mediates clearance of platelet microparticles by the endothelium. Circulation 125:1664–1672
Dominguez JH, Liu Y, Gao H, Dominguez JM 2nd, Xie D, Kelly KJ (2017) Renal tubular cell-derived extracellular vesicles accelerate the recovery of established renal ischemia reperfusion injury. J Am Soc Nephrol 28:3533–3544
Eirin A, Zhu XY, Puranik AS, Tang H, McGurren KA, van Wijnen AJ et al (2017) Mesenchymal stem cell-derived extracellular vesicles attenuate kidney inflammation. Kidney Int 92:114–124
Erdbrügger U, Le TH (2016) Extracellular vesicles in renal diseases: more than novel biomarkers? J Am Soc Nephrol 27:12–26
Faille D, El-Assaad F, Mitchell AJ, Alessi MC, Chimini G, Fusai T et al (2012) Endocytosis and intracellular processing of platelet microparticles by brain endothelial cells. J Cell Mol Med 16:1731–1738
Fuhrmann G, Herrmann IK, Stevens MM (2015) Cell-derived vesicles for drug therapy and diagnostics: opportunities and challenges. Nano Today 10:397–409
Furini G, Schroeder N, Huang L, Boocock D, Scarpellini A, Coveney C et al (2018) Proteomic profiling reveals the transglutaminase-2 externalization pathway in kidneys after unilateral ureteric obstruction. J Am Soc Nephrol 29:880–905

Grange C, Iampietro C, Bussolati B (2017) Stem cell extracellular vesicles and kidney injury. Stem Cell Investig 4:90

Gu D, Zou X, Ju G, Zhang G, Bao E, Zhu Y (2016) Mesenchymal stromal cells derived extracellular vesicles ameliorate acute renal ischemia reperfusion injury by inhibition of mitochondrial fission through miR-30. Stem Cells Int 2016:2093940

Ha D, Yang N, Nadithe V (2016) Exosomes as therapeutic drug carriers and delivery vehicles across biological membranes: current perspectives and future challenges. Acta Pharm Sin B 6:287–296

Haney MJ, Klyachko NL, Zhao Y, Gupta R, Plotnikova EG, He Z et al (2015) Exosomes as drug delivery vehicles for Parkinson's disease therapy. J Control Release 207:18–30

Jang SC, Kim OY, Yoon CM, Choi DS, Roh TY, Park J et al (2013) Bioinspired exosome-mimetic nanovesicles for targeted delivery of chemotherapeutics to malignant tumors. ACS Nano 7:7698–7710

Jiang ZZ, Liu YM, Niu X, Yin JY, Hu B, Guo SC et al (2016) Exosomes secreted by human urine-derived stem cells could prevent kidney complications from type I diabetes in rats. Stem Cell Res Ther 7:24

Ju GQ, Cheng J, Zhong L, Wu S, Zou XY, Zhang GY et al (2015) Microvesicles derived from human umbilical cord mesenchymal stem cells facilitate tubular epithelial cell dedifferentiation and growth via hepatocyte growth factor induction. PLoS ONE 10:e0121534

Kamaly N, He JC, Ausiello DA, Farokhzad OC (2016) Nanomedicines for renal disease: current status and future applications. Nat Rev Nephrol 12:738–753

Kamerkar S, LeBleu VS, Sugimoto H, Yang S, Ruivo CF, Melo SA et al (2017) Exosomes facilitate therapeutic targeting of oncogenic KRAS in pancreatic cancer. Nature 546:498–503

Karpman D, Ståhl AL, Arvidsson I (2017) Extracellular vesicles in renal disease. Nat Rev Nephrol 13:545–562

Kholia S, Herrera Sanchez MB, Cedrino M, Papadimitriou E, Tapparo M, Deregibus MC et al (2018) Human liver stem cell-derived extracellular vesicles prevent aristolochic acid-induced kidney fibrosis. Front Immunol 9:1639

Kim HS, Choi DY, Yun SJ, Choi SM, Kang JW, Jung JW et al (2012) Proteomic analysis of microvesicles derived from human mesenchymal stem cells. J Proteome Res 11:839–849

Lee J, Kim J, Jeong M, Lee H, Goh U, Kim H et al (2015) Liposome-based engineering of cells to package hydrophobic compounds in membrane vesicles for tumor penetration. Nano Lett 15:2938–2944

Lima LG, Chammas R, Monteiro RQ, Moreira ME, Barcinski MA (2009) Tumor-derived microvesicles modulate the establishment of metastatic melanoma in a phosphatidylserine-dependent manner. Cancer Lett 283:168–175

Lin KC, Yip HK, Shao PL, Wu SC, Chen KH, Chen YT et al (2016) Combination of adipose-derived mesenchymal stem cells (ADMSC) and ADMSC-derived exosomes for protecting kidney from acute ischemia-reperfusion injury. Int J Cardiol 216:173–185

Lv LH, Wan YL, Lin Y, Zhang W, Yang M, Li GL et al (2012) Anticancer drugs cause release of exosomes with heat shock proteins from human hepatocellular carcinoma cells that elicit effective natural killer cell antitumor responses in vitro. J Biol Chem 287:15874–15885

Lv LL, Feng Y, Wen Y, Wu WJ, Ni HF, Li ZL et al (2018a) Exosomal CCL2 from tubular epithelial cells is critical for albumin-induced tubulointerstitial inflammation. J Am Soc Nephrol 29:919–935

Lv LL, Wu WJ, Feng Y, Li ZL, Tang TT, Liu BC (2018b) Therapeutic application of extracellular vesicles in kidney disease: promises and challenges. J Cell Mol Med 22:728–737

Maguire CA, Balaj L, Sivaraman S, Crommentuijn MH, Ericsson M, Mincheva-Nilsson L et al (2012) Microvesicle-associated AAV vector as a novel gene delivery system. Mol Ther 20:960–971

Martins-Marques T, Pinho MJ, Zuzarte M, Oliveira C, Pereira P, Sluijter JP et al (2016) Presence of Cx43 in extracellular vesicles reduces the cardiotoxicity of the anti-tumour therapeutic approach with doxorubicin. J Extracell Vesicles 5:32538

Mendt M, Kamerkar S, Sugimoto H, McAndrews KM, Wu CC, Gagea M et al (2018) Generation and testing of clinical-grade exosomes for pancreatic cancer. JCI Insight 3:99263
Mizrak A, Bolukbasi MF, Ozdener GB, Brenner GJ, Madlener S, Erkan EP et al (2013) Genetically engineered microvesicles carrying suicide mRNA/protein inhibit schwannoma tumor growth. Mol Ther 21:101–108
Morel O, Jesel L, Freyssinet JM, Toti F (2011) Cellular mechanisms underlying the formation of circulating microparticles. Arterioscler Thromb Vasc Biol 31:15–26
Morrison EE, Bailey MA, Dear JW (2016) Renal extracellular vesicles: from physiology to clinical application. J Physiol 594:5735–5748
Mossberg M, Ståhl AL, Kahn R, Kristoffersson AC, Tati R, Heijl C et al (2017) C1-inhibitor decreases the release of vasculitis-like chemotactic endothelial microvesicles. J Am Soc Nephrol 28:2472–2481
Mulcahy LA, Pink RC, Carter DR (2014) Routes and mechanisms of extracellular vesicle uptake. J Extracell Vesicles 3:24641
Nargesi AA, Lerman LO, Eirin A (2017) Mesenchymal stem cell-derived extracellular vesicles for renal repair. Curr Gene Ther 17:29–42
Ostrowski M, Carmo NB, Krumeich S, Fanget I, Raposo G, Savina A et al (2010) Rab27a and Rab27b control different steps of the exosome secretion pathway. Nat Cell Biol 12:19–30
Pascucci L, Coccè V, Bonomi A, Ami D, Ceccarelli P, Ciusani E et al (2014) Paclitaxel is incorporated by mesenchymal stromal cells and released in exosomes that inhibit in vitro tumor growth: a new approach for drug delivery. J Control Release 192:262–270
Peinado H, Alečković M, Lavotshkin S, Matei I, Costa-Silva B, Moreno-Bueno G et al (2012) Melanoma exosomes educate bone marrow progenitor cells toward a pro-metastatic phenotype through MET. Nat Med 18:883–891
Ranghino A, Bruno S, Bussolati B, Moggio A, Dimuccio V, Tapparo M et al (2017) The effects of glomerular and tubular renal progenitors and derived extracellular vesicles on recovery from acute kidney injury. Stem Cell Res Ther 8:24
Rani S, Ryan AE, Griffin MD, Ritter T (2015) Mesenchymal stem cell-derived extracellular vesicles: toward cell-free therapeutic applications. Mol Ther 23:812–823
Raposo G, Stoorvogel W (2013) Extracellular vesicles: exosomes, microvesicles, and friends. J Cell Biol 200:373–383
Saari H, Lázaro-Ibáñez E, Viitala T, Vuorimaa-Laukkanen E, Siljander P, Yliperttula M (2015) Microvesicle- and exosome-mediated drug delivery enhances the cytotoxicity of paclitaxel in autologous prostate cancer cells. J Control Release 220:727–737
Shen B, Liu J, Zhang F, Wang Y, Qin Y, Zhou Z et al (2016) CCR68 positive exosome released by mesenchymal stem cells suppresses macrophage functions and alleviates ischemia/reperfusion-induced renal injury. Stem Cells Int 2016:1240301
Srivastava A, Amreddy N, Babu A, Panneerselvam J, Mehta M, Muralidharan R et al (2016) Nanosomes carrying doxorubicin exhibit potent anticancer activity against human lung cancer cells. Sci Rep 6:38541
Sun D, Zhuang X, Xiang X, Liu Y, Zhang S, Liu C et al (2010) A novel nanoparticle drug delivery system: the anti-inflammatory activity of curcumin is enhanced when encapsulated in exosomes. Mol Ther 18:1606–1614
Syn NL, Wang L, Chow EK, Lim CT, Goh BC (2017) Exosomes in cancer nanomedicine and immunotherapy: prospects and challenges. Trends Biotechnol 35:665–676
Tang K, Zhang Y, Zhang H, Xu P, Liu J, Ma J et al (2012) Delivery of chemotherapeutic drugs in tumour cell-derived microparticles. Nat Commun 3:1282
Tang TT, Lv LL, Lan HY, Liu BC (2019) Extracellular vesicles: opportunities and challenges for the treatment of renal diseases. Front Physiol 10:226
Tian Y, Li S, Song J, Ji T, Zhu M, Anderson GJ et al (2014) A doxorubicin delivery platform using engineered natural membrane vesicle exosomes for targeted tumor therapy. Biomaterials 35:2383–2390

Toffoli G, Hadla M, Corona G, Caligiuri I, Palazzolo S, Semeraro S et al (2015) Exosomal doxorubicin reduces the cardiac toxicity of doxorubicin. Nanomedicine 10:2963–2971

Trajkovic K, Hsu C, Chiantia S, Rajendran L, Wenzel D, Wieland F et al (2008) Ceramide triggers budding of exosome vesicles into multivesicular endosomes. Science 319:1244–1247

van den Boorn JG, Schlee M, Coch C, Hartmann G (2011) SiRNA delivery with exosome nanoparticles. Nat Biotechnol 29:325–326

van der Pol E, Böing AN, Harrison P, Sturk A, Nieuwland R (2012) Classification, functions, and clinical relevance of extracellular vesicles. Pharmacol Rev 64:676–705

van Dommelen SM, Vader P, Lakhal S, Kooijmans SA, van Solinge WW, Wood MJ et al (2012) Microvesicles and exosomes: opportunities for cell-derived membrane vesicles in drug delivery. J Control Release 161:635–644

Viñas JL, Burger D, Zimpelmann J, Haneef R, Knoll W, Campbell P et al (2016) Transfer of microRNA-486-5p from human endothelial colony forming cell-derived exosomes reduces ischemic kidney injury. Kidney Int 90:1238–1250

Wahlgren J, Karlson TDL, Brisslert M, Vaziri Sani F, Telemo E, Sunnerhagen P et al (2012) Plasma exosomes can deliver exogenous short interfering RNA to monocytes and lymphocytes. Nucleic Acids Res 40:e130

Wang B, Yao K, Huuskes BM, Shen HH, Zhuang J, Godson C et al (2016) Mesenchymal stem cells deliver exogenous MicroRNA-let7c via exosomes to attenuate renal fibrosis. Mol Ther 24:1290–1301

Wolf P (1967) The nature and significance of platelet products in human plasma. Br J Haematol 13:269–288

Yang T, Martin P, Fogarty B, Brown A, Schurman K, Phipps R et al (2015) Exosome delivered anticancer drugs across the blood-brain barrier for brain cancer therapy in Danio rerio. Pharm Res 32:2003–2014

Yano Y, Shiba E, Kambayashi J, Sakon M, Kawasaki T, Fujitani K et al (1993) The effects of calpeptin (a calpain specific inhibitor) on agonist induced microparticle formation from the platelet plasma membrane. Thromb Res 71:385–396

Yim N, Ryu SW, Choi K, Lee KR, Lee S, Choi H et al (2016) Exosome engineering for efficient intracellular delivery of soluble proteins using optically reversible protein-protein interaction module. Nat Commun 7:12277

Yuan D, Zhao Y, Banks WA, Bullock KM, Haney M, Batrakova E et al (2017) Macrophage exosomes as natural nanocarriers for protein delivery to inflamed brain. Biomaterials 142:1–12

Zafrani L, Gerotziafas G, Byrnes C, Hu X, Perez J, Lévi C et al (2012) Calpastatin controls polymicrobial sepsis by limiting procoagulant microparticle release. Am J Respir Crit Care Med 185:744–755

Zeelenberg IS, Ostrowski M, Krumeich S, Bobrie A, Jancic C, Boissonnas A et al (2008) Targeting tumor antigens to secreted membrane vesicles in vivo induces efficient antitumor immune responses. Cancer Res 68:1228–1235

Zhang Y, Liu D, Chen X, Li J, Li L, Bian Z et al (2010) Secreted monocytic miR-150 enhances targeted endothelial cell migration. Mol Cell 39:133–144

Zhang G, Zou X, Huang Y, Wang F, Miao S, Liu G et al (2016a) Mesenchymal stromal cell-derived extracellular vesicles protect against acute kidney injury through anti-oxidation by enhancing Nrf2/ARE activation in rats. Kidney Blood Press Res 41:119–128

Zhang W, Zhou X, Zhang H, Yao Q, Liu Y, Dong Z (2016b) Extracellular vesicles in diagnosis and therapy of kidney diseases. Am J Physiol Renal Physiol 311:F844–F851

Zhou Y, Xiong M, Fang L, Jiang L, Wen P, Dai C et al (2013a) miR-21-containing microvesicles from injured tubular epithelial cells promote tubular phenotype transition by targeting PTEN protein. Am J Pathol 183:1183–1196

Zhou Y, Xu H, Xu W, Wang B, Wu H, Tao Y et al (2013b) Exosomes released by human umbilical cord mesenchymal stem cells protect against cisplatin-induced renal oxidative stress and apoptosis in vivo and in vitro. Stem Cell Res Ther 4:34

Zhu W, Huang L, Li Y, Zhang X, Gu J, Yan Y et al (2012) Exosomes derived from human bone marrow mesenchymal stem cells promote tumor growth in vivo. Cancer Lett 315:28–37
Zhuang X, Xiang X, Grizzle W, Sun D, Zhang S, Axtell RC et al (2011) Treatment of brain inflammatory diseases by delivering exosome encapsulated anti-inflammatory drugs from the nasal region to the brain. Mol Ther 19:1769–1779
Zou X, Zhang G, Cheng Z, Yin D, Du T, Ju G et al (2014) Microvesicles derived from human Wharton's Jelly mesenchymal stromal cells ameliorate renal ischemia-reperfusion injury in rats by suppressing CX3CL1. Stem Cell Res Ther 5:40
Zou X, Gu D, Xing X, Cheng Z, Gong D, Zhang G et al (2016) Human mesenchymal stromal cell-derived extracellular vesicles alleviate renal ischemic reperfusion injury and enhance angiogenesis in rats. Am J Transl Res 8:4289–4299